D1256299

Withdrawn
University of Waterloo

Allen's Astrophysical Quantities

Fourth Edition

Allen's Astrophysical Quantities

Fourth Edition

Arthur N. Cox
Editor

Withdrawn
University of Waterloo

  Springer

Arthur N. Cox
Theoretical Division
Los Alamos National Laboratory MS B288
P.O. Box 1663
Los Alamos, NM 87545
USA
anc@lanl.gov

Cover illustration: An international team of astronomers, led by Dr. Wendy Freedman of the Observatories of the Carnegie Institution of Washington, Robert Kennicutt of the University of Arizona, and Jeremy Mould of the Australian National University observed this spiral galaxy NGC 4414 on 13 different occasions over the course of two months. (AURA/STScI/NASA)

In 1995, the majestic spiral galaxy NGC 4414 was imaged by the Hubble Space Telescope as part of the HST Key Project on the Extragalactic Distance Scale. Images were obtained with Hubble's Wide Field Planetary Camera 2 (WFPC2) through three different color filters. Based on their discovery and careful brightness measurements of variable stars in NGC 4414, the Key Project astronomers were able to make an accurate determination of the distance to the galaxy.

The resulting distance to NGC 4414, 19.1 megaparsecs or about 60 million light-years, along with similarly determined distances to other nearby galaxies, contributes to astronomers' overall knowledge of the rate of expansion of the universe. The Hubble constant (H_0) is the ratio of how fast galaxies are moving away from us to their distance from us. This astronomical value is used to determine distances, sizes, and the intrinsic luminosities for many objects in our universe, and the age of the universe itself.

Library of Congress Cataloging-in-Publication Data
Cox, Arthur
Allen's astrophysical quantities/editor, Arthur Cox.
    p.   cm.
   Includes bibliographical references.
   ISBN 0-387-98746-0 (hardcover: alk. paper)
   1. Astrophysical—Tables.   I. Cox, Arthur N.
   QB461.A7685. 1999
   523.01'021—dc21   98-53154

Printed on acid-free paper.

© 1976 C.W. Allen. Published by The Athlone Press, Ltd. London.

© 2000 The Estate of C.W. Allen.
AIP is an imprint of Springer-Verlag New York, Inc.
All rights reserved. This work may not be translated or copied in whole or in part without the written permission of the publisher (Springer-Verlag New York, Inc., 175 Fifth Avenue, New York, NY 10010, USA), except for brief excerpts in connection with reviews or scholarly analysis. Use in connection with any form of information storage and retrieval, electronic adaptation, computer software, or by similar or dissimilar methodology now known or hereafter developed is forbidden.
The use of general descriptive names, trade names, trademarks, etc., in this publication, even if the former are not especially identified, is not to be taken as a sign that such names, as understood by the Trade Marks and Merchandise Marks Act, may accordingly be used freely by anyone.
This publication is based on sources and information believed to be reliable, but the publisher and the authors disclaim any warranty or liability based on or relating to the contents of this publication.
The publisher does not endorse any product, manufacturers, or suppliers. Nothing in this book should be interpreted as implying such endorsement.
While the advice and information in this book are believed to be true and accurate at the date of going to press, neither the authors nor the editors nor the publisher can accept any legal responsibility for any errors or omissions that may be made. The publisher makes no warranty, express or implied, with respect to the material contained herein.

Production managed by Frank McGuckin; manufacturing supervised by Jeffrey Taub.
Photocomposed copy prepared from the editor's TeX files.
Printed and bound by Hamilton Printing Co., Rensselaer, NY.
Printed in the United States of America.

9 8 7 6 5 4 3 2 1

ISBN 0-387-98746-0 Springer-Verlag New York Berlin Heidelberg   SPIN 10709826

Preface

This handbook is the result of compilations and writing of ninety authors who have worked over a period of nine years to revise the famous Allen's *Astrophysical Quantities*. The need for such a revision had been known since shortly after the last edition edited by C.W. Allen in 1972. Even though his 1973 edition remained in print through the late 1980s, Allen himself called for help in revising the book in that third edition Preface. His death unfortunately prevented any revision, and only a few attempts known to me were made by interested astronomers. By 1990, with the third edition completely outdated, Arlo Landolt convinced the American Institute of Physics that they should undertake extensive revisions of the Allen book. How my name came up, in late 1990, I do not know, but once friends discovered I had been solicited by the AIP, they all encouraged me to find the various astrophysics experts to prepare this new edition, published jointly by the AIP and Springer-Verlag.

The task of finding suitable authors and anonymous referees for the chapters was made easier by the help of Peter Boyce at the American Astronomical Society and its publications board. Chairpersons Caty Pilachowski, Hugh Van Horn, Jim Liebert, and Bob Hanisch suggested and helped recruit many contributors. Numerous AAS officials, especially Roger Bell, helped me and the authors interface with AIP and Springer.

The basic structure of the earlier Allen editions has been followed, but many changes were necessary. For example, radio astronomy was represented by Allen with a page-long table of sources and a few supplementary ones plus some data about solar radio emission. Today a complete chapter is necessary, and even that does not seem to be as much as the author and I would have liked to include. Other advances in astrophysics have required us to include new chapters for infrared, ultraviolet, X-ray, and gamma-ray and neutrino astronomy. The explosion in observations of our solar system has resulted in a great expansion in information about these nearby bodies, as well as for our Sun itself. Later in the development of this book we found that we needed to add a chapter about stellar evolution because the level of understanding essentially the entire lives of stars had matured enormously. Most dramatically, modern large telescopes have revealed huge quantities of data about galaxies, galaxy clusters, and their exotic emissions. Three separate chapters cover different aspects of this material. A much expanded Cosmology chapter was needed to include our current understanding of the structure of the Universe. Finally, we have added many supplemental tables including an attempt to list the world's largest optical telescopes, with the help of Kari Parker, that surely will be out of date soon.

While writing the chapters, many authors found that they needed some specialists to supply and even write sections that were beyond their current knowledge. These section authors are not given in the table of contents, but only at the start of the sections where they contributed. Thanks are due to these scientists who have supplied important information that we found relevant, often rather late in the book development. Their submissions could easily merit a mention in the table of contents, but the complicated process of assembling this greatly revised handbook and keeping its structure in control has resulted in this special format.

Readers must realize that a project that involves ninety otherwise very busy astrophysicists is bound to be uneven. Some authors were able to get their material to me as early as mid-1992, while others were not even solicited by me for last-minute data until mid-1998. Our plan to include updates to a uniform date for all chapters could not be carried out because of its complexity, but some data as recent as the summer of 1999 are included. Readers are invited to contact individual authors directly for details. Our hope is that we have adequately pointed the way to the extensive literature for each subject.

Some astrophysicists have already decided to adopt our carefully compiled data as standard for their own special lists. This is reasonable, since this new Allen edition has been prepared by the world's experts in the various areas of astrophysics. One thing we have learned is that definitive data depend on interpretations for those last little details, and the best source for the most current and accurate data is always the experts. We hope our authors are these.

The contents of this new edition of Allen will be available in electronic form with many tables and graphs "live" for interactive searching, correlating, interpolating, and so forth. The electronic version will be available by subscription and kept up-to-date on the publisher's web site (www.springer-ny.com) and will also be available as a CD-ROM for use on a Windows PC. At the minimum, these electronic data will greatly assist in future editions.

Every publishing undertaking ends with regrets that some things could not be included. Thus all should realize that our book is a good reference book, but it still misses, for example, the newly published definitive NIST physical constants, the recent discovery of a satellite around the asteroid (45) Eugenia, the growing list of brown dwarf candidates, a new and unexpected class of intrinsic variable (Gamma Doradus) stars, and the latest gamma burst explosions now optically detected from the far reaches of our Universe. The organization of these new astrophysical quantities into an additional concise revised-again edition awaits future generations of authors, I hope as skilled and dedicated as ours.

Los Alamos, New Mexico
October 1999

Arthur N. Cox
anc@lanl.gov

Contents

Contributors

This list of the contributors gives the institution where the author did the bulk of the writing of the chapter or section for this handbook. For those authors who participated only in a topic section, that subject is briefly indicated.

Chapter 1

Arthur N. Cox
Los Alamos National Laboratory

Sarah Stevens-Rayburn
Space Telescope Science Institute
Astrophysical Journals

Chapter 2

Arthur N. Cox
Los Alamos National Laboratory

Alan D. Fiala
United States Naval Observatory
Astronomical Constants, General and Time

Chapter 3

Werner Dappen
University of Southern California

Chapter 4

Charles R. Cowley
University of Michigan

Wolfgang L. Wiese
National Institute of Standards and Technology

Jeffrey Fuhr
National Institute of Standards and Technology

Ludmila A. Kuznetsova
Moscow State University

Chapter 5

John J. Keady
Los Alamos National Laboratory

David P. Kilcrease
Los Alamos National Laboratory

Chapter 6

Robert M. Hjellming
National Radio Astronomy Observatory

Chapter 7

Alan T. Tokunaga
University of Hawaii

Chapter 8

Terry J. Teays
Goddard Space Fight Center

Chapter 9

Fredrick D. Seward
Smithsonian Astrophysical Observatory

Chapter 10

Richard E. Lingenfelter
University of California–San Diego

Richard E. Rothschild
University of California–San Diego

Thomas J. Bowles
Los Alamos National Laboratory
Neutrino Observatories

Wick C. Haxton
University of Washington
Neutrino Observations

Chapter 11

Gerald Schubert
University of California–Los Angeles

Richard L. Walterscheid
The Aerospace Corporation

David Crisp
Jet Propulsion Laboratory/California Institute
of Technology
*Earth Atmosphere Scattering, Absorption,
Emission*

Chapter 12

David J. Tholen
University of Hawaii

Victor G. Tejfel
Fessenkov Astrophysical Institute

Arthur N. Cox
Los Alamos National Laboratory

Glenn S. Orton
Jet Propulsion Laboratory/California Institute
of Technology
Physical Conditions on Planets

Dan Pascu
United States Naval Observatory
*Names, Designations, and Discoveries of
Satellites, Satellite Orbits and Physical
Elements*

Chapter 13

Richard P. Binzel
Massachusetts Institute of Technology

Martha S. Hanner
Jet Propulsion Laboratory/California Institute
of Technology

Duncan I. Steel
University of Salford, UK

Chapter 14

William C. Livingston
National Solar Observatory

Pierre Demarque
Yale University
Solar Model

David B. Guenther
St. Mary's University
Solar Model

Frank Hill
National Solar Observatory
Solar Oscillations

Eugene Avrett
Harvard–Smithsonian Center for Astrophysics
Photospheric Model

Oran R. White
High Altitude Observatory
Solar Spectral Lines

Heinz Neckel
Hamburger Sternwarte
Solar Spectral Energy Distribution

A. Keith Pierce
National Solar Observatory
Solar Limb Darkening

Serge Koutchmy
Institut d'Astrophysique
Solar Corona

Robert F. Howard
National Solar Observatory
Solar Rotation

Richard Muller
Observatoire Pic du Midi
Granulation

Peter V. Foukal
Cambridge Research and Instrumentation, Inc.
Surface Magnetism and Tracers

Sami Solanki
Institute of Astronomy, ETH-Zentrum
Surface Magnetism and Tracers, Sunspots

Jack B. Zirker
National Solar Observatory
Surface Magnetism and Tracers

Karen L. Harvey
Solar Physics Research Corporation
Sunspot Statistics

Peter Wilson
University of Sydney
Sunspot Statistics

Stephen W. Kahler
United States Air Force Research Laboratory
Flares, CMEs

Timothy Bastian
National Radio Astronomy Observatory
Solar Radio Emission

Chapter 15

John S. Drilling
Louisiana State University

Arlo U. Landolt
Louisiana State University
Normal Stars

James W. Liebert
University of Arizona
White Dwarf Spectral Classification

Edward M. Sion
Villanova University
White Dwarf Spectral Classification

Chapter 16

J. Donald Fernie
David Dunlap Observatory

Douglas S. Hall
Vanderbilt University
Variable Stars, Rotating Variables, Flare Stars

Paul A. Bradley
Los Alamos National Laboratory
Variable White Dwarf Tables

Gibor S. Basri
University of California–Berkeley
T Tauri Stars

Kenneth R. Brownsberger
University of Colorado
Wolf-Rayet and Luminous Blue Variable Stars

Peter S. Conti
University of Colorado
Wolf-Rayet and Luminous Blue Variable Stars

Arne Slettebak
Ohio State University
Be Stars

Myron Smith
Computer Science Corporation
Be Stars

Cecilia S. Barnbaum
University of California–Berkeley
Characteristics of Carbon-rich Stars

William Dean Pesnell
Nomad Research, Inc.
Barium, CH, and Subgiant CH Stars

Warrick Lawson
Australian Defence Force Academy
Hydrogen Deficient Carbon Stars

Peter J.T. Leonard
Goddard Space Flight Center
Blue Stragglers

Kaiyou Chen
Los Alamos National Laboratory
Pulsars

John Middleditch
Los Alamos National Laboratory
Pulsars

Jonathan E. Grindlay
Harvard College Observatory
Galactic Black Hole Candidate X-Ray Binaries

Chapter 17

Warren M. Sparks
Los Alamos National Laboratory

Sumner G. Starrfield
Arizona State University

Edward M. Sion
Villanova University

Steven N. Shore
Indiana University South Bend

Ganesh Chanmugam
Louisiana State University

Ronald F. Webbink
University of Illinois

Chapter 18
J. Craig Wheeler
University of Texas

Stefano Benetti
European Southern Observatory

Chapter 19
Gerard F. Gilmore
Cambridge University

Michael Zeilik
University of New Mexico

Chapter 20
Arthur N. Cox
Los Alamos National Laboratory

Stephen A. Becker
Los Alamos National Laboratory

William Dean Pesnell
Nomad Research, Inc.
Barium, CH, and Subgiant CH Stars

Chapter 21
John S. Mathis
University of Wisconsin

Donald P. Cox
University of Wisconsin
Cosmic Rays Excluding Photons and Neutrinos

Jonathan F. Ormes
Goddard Space Flight Center
Cosmic Rays Excluding Photons and Neutrinos

Chapter 22
Hugh C. Harris
United States Naval Observatory

William E. Harris
McMaster University

Chapter 23
Virginia Trimble
University of California–Irvine and University
of Maryland

Chapter 24
Belinda J. Wilkes
Smithsonian Astrophysical Observatory

Chapter 25
Neta A. Bahcall
Princeton University

Chapter 26
Douglas Scott
University of British Columbia

Joseph Silk
University of California–Berkeley

Edward W. Kolb
Fermi National Accelerator Laboratory

Michael S. Turner
The University of Chicago

Chapter 27
Alan D. Fiala
United States Naval Observatory

William F. van Altena
Yale University

Stephen T. Ridgway
National Optical Astronomy Observatory

Roger W. Sinnott
Sky and Telescope

Kari Parker
Sky and Telescope
Largest Optical Telescopes

Chapter 1

Introduction

Arthur N. Cox

1.1 BACKGROUND

This handbook is a revision of the third edition of Allen's *Astrophysical Quantities* [1], published in 1973, with further printings in 1976, 1981, and 1983. An attempt has been made to follow the original format, but the great advances in astronomical and astrophysical subfields have made this very difficult. More modern styles have been adopted, and many more subjects have been included. However, the original concept to present a rather concise, but still extensive, listing of astrophysical quantities has been retained. It is expected that scientists can use this book for quick information and also for key references to more detailed data sources.

One concept was that this handbook be a companion to *A Physicist's Desk Reference*, the second edition of *Physics Vade Mecum*, edited by Herbert L. Anderson. That book also has a long author list, and its currently planned revision will undoubtedly include even more.

To allow for space to present information for newly developed astronomical and astrophysical subfields, some more classical material has been deleted. Reference to the older Allen editions might be necessary.

The current fields are so extensive that we needed 90 authors. They are indicated either as chapter authors or authors of individual sections. All were asked to present their information using the electronic editing language LATEX 2$_\varepsilon$, so that the submissions to the publisher could be almost camera ready. Not all authors could completely comply with this policy, and the editor, with help from the publisher, occasionally needed to reformat the material. With so many involved, including the copyeditors, it is easy to see why it is important to keep the presentation style as consistent as possible.

One hope is that the electronic files now available for this book can be revised in the coming years, and an updated version can be published more easily. With the extensive use of the World Wide Web by so many scientists, it is conceivable that individual chapters can be updated after some years and made available there. At least future revisions should be easier to produce based on the very great efforts over most of eight years by all the authors.

The editor requested authors to update their information to about the end of 1997 or later. Obviously this has not been completely successful. For questions and corrections, readers should consult individual authors, mostly since no one single individual can know more than a fraction of the knowledge of the entire area of astronomy and astrophysics.

Just a small part of the Allen introduction chapter has been retained here.

1.2 ASTRONOMICAL SYMBOLS

The standard symbols for astronomical objects and zodiacal areas are given in Table 1.1.

Table 1.1. *Sun, Moon, planetary, zodiacal, and orbit symbols* [1, 2].

Symbol	Name	Symbol	Name	Symbol	Name
*	Star	☉	Sun	☾	Moon
☿	Mercury	♀	Venus	♁	Earth
♂	Mars	♃	Jupiter	♄	Saturn
♄	Saturn 2	♅	Uranus	♆	Neptune
♇	Neptune 2	♇	Pluto	♇	Pluto 2
♈	Aries (0°)	♉	Taurus (30°)	♊	Gemini (60°)
♋	Cancer (90°)	♌	Leo (120°)	♍	Virgo (150°)
♎	Libra (180°)	♎	Libra 2 (180°)	♏	Scorpio (210°)
♐	Sagittarius (240°)	♑	Capricornus (270°)	♒	Aquarius (300°)
≈	Aquarius 2 (300°)	≋	Aquarius 3 (300°)	♓	Pisces (330°)
☊	Ascending Node	☋	Descending Node		
♎	Autumnal Equinox	♈	Vernal Equinox		

References
1. Rahtz, S. & Rose, K. ftp to sunsite.unc.edu in directory pub/packages/TeX/cmastro
2. Schmitt, P. 1992, ftp to sunsite.unc.edu in directory pub/packages/TeX/astro

1.3 ASTRONOMICAL AND ASTROPHYSICAL JOURNALS
by Sarah Stevens-Rayburn

The names of 51 journals thought to be of interest to readers with an astronomy and astrophysics background is given together with their first publication dates and the current publisher. Many of these journals are now available electronically on the World Wide Web. Note that the older European journals, *Annales d'Astrophysique, Bulletin of the Astronomical Institutes of the Netherlands, Bulletin Astronomique, Journal des Observateurs, Zeitschrift für Astrophysik,* and a few other smaller ones have been discontinued and have been replaced by *Astronomy and Astrophysics.*

Acta Astronomica	1925	Warsaw: Copernicus Foundation for Polish Astronomy
Acta Astronomica Sinica = Tien wen hsüeh pao	1953	Beijing, China: Science Press Beijing: K'o hsüeh ch'u pan she
Acta Astrophysica Sinica = Tien t i wu li hsüeh pao	1981	Beijing, China: Science Press Beijing: K'o hsüeh ch'u pan she
Acta Cosmologica	1973	Krakow: Uniwersytet Jagiellonski
Annual Review of Astronomy and Astrophysics (ARA&A)	1963	Palo Alto, Calif.: Annual Reviews, Inc.
Annual Review of Earth and Planetary Sciences	1973	Palo Alto, Calif.: Annual Reviews, Inc.
The Astronomical Journal (AJ)	1894	Chicago: University of Chicago Press for the American Astronomical Society
Astronomische Nachrichten	1823	Berlin: Wiley-VCH
Astronomy and Astrophysics (A&A)	1869	Berlin: Springer, on behalf of the Board of Directors
The Astronomy and Astrophysics Review (A&A Rev)	1989	Berlin: Springer on behalf of the Board of Directors
Astronomy & Astrophysics Supplement series (A&AS)	1970	Les Ulis, France: EDP Sciences on behalf of the Board of Directors
Astronomy & Geophysics The Journal of the Royal Astronomical Society (continues QJRAS, 1960–1996)	1997	Bristol: Institute of Physics Publishing Ltd. for the Royal Astronomical Society
Astronomy Letters (continues *Soviet Astronomy Letters*, 1975–1992)	1993	Moscow: Maik Nauka/Interperiodica Publishing, distributed by the AIP (translation of *Pisma v astronomicheskii zhurnal*)
Astronomy Reports (continues *Soviet Astronomy*, 1974–1992, which continues *Soviet Astronomy AJ*, 1957–1973)	1993	Moscow: Maik Nauka/Interperiodica Publishing, distributed by the AIP (translation of *Astronomicheskii zhurnal*)
The Astrophysical Journal (ApJ)	1895	Chicago: University of Chicago Press for the American Astronomical Society
The Astrophysical Journal Supplement series (ApJS)	1954	Chicago: University of Chicago Press for the American Astronomical Society
Astrophysical Letters & Communications (Astrophys. Lett) (continues *Astrophysical Letters*, 1967–1987)	1987	New York: Gordon and Breach
Astrophysics	1965	New York: Consultants Bureau (translation of *Astrofizika*)
Astrophysics and Space Science (Ap&SS)	1968	Dordrecht: Kluwer Academic
Astrophysics Reports: Publications of the Beijing Astronomical Observatory (continues *Publications of the Beijing Astronomical Observatory*, 1987–1994)	1994	Beijing, China: Beijing Astronomical Observatory, Chinese Academy of Sciences
Astrophysics Reports: Publications of the Beijing Astronomical Observatory (Supplement series)	1997	Beijing, China: Beijing Astronomical Observatory, Chinese Academy of Sciences
Baltic Astronomy: An International Journal	1992	Vilnius: Institute of Theoretical Physics and Astronomy
Boletin / Asociacion Argentina de Astronomia	1958	La Plata: La Asociacion

Bulletin of the Astronomical Society of India	1973	Hyderabad: Astronomical Society of India
Celestial Mechanics and Dynamical Astronomy (continues *Celestial Mechanics*, 1969–1988)	1989	Dordrecht: Kluwer Academic
Chinese Astronomy and Astrophysics (continues *Chinese Astronomy*, 1977–1980)	1981	Kidlington, Oxford; Elsevier Science
Earth, Moon, and Planets (continues *The Moon and Planets*, 1978–1983, which continues *The Moon*, 1969–1977)	1984	Dordrecht: Kluwer Academic
Experimental Astronomy	1989	Dordrecht: Kluwer Academic
Icarus	1962	Orlando: Academic Press
International Journal of Modern Physics, D Gravitation, Astrophysics, Cosmology	1992	Singapore: World Scientific
The Irish Astronomical Journal	1950	Sheffield: IAJ Editorial Board
Journal for the History of Astronomy	1970	Cambridge: Science History Publications
Journal of Astrophysics and Astronomy	1980	Bangalore: Indian Academy of Sciences
Journal of Geophysical Research. A, Space Physics (J. Geophys. Res. A) (continues in part *Journal of Geophysical Research*, – 1949)	1949	Washington, DC: American Geophysical Union
Journal of Geophysical Research. E, Planets (J. Geophys. Res. E) (continues in part *Journal of Geophysical Research*, –1990)	1991	Washington, DC: American Geophysical Union
The Journal of the Royal Astronomical Society of Canada (JRASC)	1907	Toronto: Royal Astronomical Society of Canada
Memorie della Societa Astronomica Italiana (Mem. Soc. Astron. Italiana)	1872	Firenze: Societa Astronomica Italiana
Monthly Notes of the Astronomical Society of Southern Africa	1940	Observatory, S.A.: The Society
Monthly Notices of the Royal Astronomical Society (MNRAS) (continues *Monthly Notices of the Astronomical Society of London*, 1927–1931)	1931	Edinburgh: Blackwell Science for the Royal Astronomical Society
Nature	1869	London: Macmillan Magazines Ltd.
The Observatory	1877	Chilton, Didcot, Oxon: Editors of *The Observatory*
Planetary and Space Science (Planet. Space Sci.)	1959	Kidlington, Oxford: Elsevier Science Ltd.
Publications / Astronomical Society of Australia (continues *Proceedings of the Astronomical Society of Australia*, 1967–1994)	1995	Australia: Published for the Astronomical Society of Australia by CSIRO, Australia
Publications of the Astronomical Society of Japan (PASJ)	1949	Tokyo: Astronomical Society of Japan
Publications of the Astronomical Society of the Pacific (PASP)	1889	Chicago: University of Chicago Press for the Astronomical Society of the Pacific
Revista Mexicana de Astronomia y Astrofisica	1974	Mexico D.F.: Instituto de Astronomia, Universidad Nacional Autonoma de Mexico
Science	1883	Washington: American Association for the Advancement of Science

Solar Physics (Sol. Phys.)	1967	Dordrecht: Kluwer Academic
Southern Stars	1934	Wellington, N.Z.: Royal Astronomical Society of New Zealand
Space Science Reviews (Space Sci. Rev)	1962	Dordrecht: Kluwer Academic
Zvaigznota debess	1958	Riga: Zinatne

REFERENCES

1. Allen, C.W. 1973, *Astrophysical Quantities* (Athlone Press, London)

Chapter 2

General Constants and Units

Arthur N. Cox

2.1 MATHEMATICAL CONSTANTS [1–3]

Constant	Number	Log
π	3.141 592 653 6	0.497 149 872 7
2π	6.283 185 307 2	0.798 179 868 4
4π	12.566 370 614 4	1.099 209 864 0
π^2	9.869 604 401 1	0.994 299 745 4
$\sqrt{\pi}$	1.772 453 850 9	0.248 574 936 3
e or e	2.718 281 828 5	0.434 294 481 9
$\mathrm{mod} = M = \log e$	0.434 294 481 9	0.637 784 311 3 − 1
$1/M = \ln 10$	2.302 585 093 0	0.362 215 688 7
2	2.000 000 000 0	0.301 029 995 7
$\sqrt{2}$	1.414 213 562 4	0.150 514 997 8
$\sqrt{3}$	1.732 050 807 6	0.238 560 627 4
$\sqrt{10}$	3.162 277 660 2	0.500 000 000 0

Constant	Number	Log
$\ln \pi$	1.144 729 885 8	0.058 703 021 2
e^{π}	23.140 692 632 8	1.364 376 353 8
Euler constant γ	0.577 215 664 9	0.761 338 108 8 − 1
1 radian	rad = $57°.295 779 513 1$	1.758 122 632 4
	= $3437'.746 770 78$	3.536 273 882 8
	= $206 264''.806 25$	5.314 425 133 2
	$1° = 0.^{rad}017 453 292 5$	0.241 877 367 6 − 2
	$1' = 0.^{rad}000 290 888 2$	0.463 726 117 2 − 4
	$1'' = 0.^{rad}000 004 848 1$	0.685 574 866 8 − 6

Square degrees on a sphere = $129 600/\pi$ = 41 252.961 25.
Square degrees in a steradian = $32 400/\pi^2$ = 32 82.806 35.

for Gaussian distribution $\dfrac{1}{\sigma \sqrt{2\pi}} \exp \left(-\dfrac{x^2}{2\sigma^2} \right)$.

Probable error/Standard error = r/σ = 0.674 489 750 2.
Probable error/Average error = r/η = 0.845 347 539 4.
σ/η = 1.253 314 137.
$\rho = (r/\sigma)/\sqrt{2}$ = 0.476 936 276 2.

2.2   PHYSICAL CONSTANTS [4, 5]

These fundamental physical constants, mostly in SI units from [5], are the latest available. A revision by Cohen and Taylor is expected by the end of 1998. For many values, the standard error of the last digits follows in parentheses. In the formulations the electron charge e is in esu and e in emu = e/c.

Fundamental constants

Speed of light (exact)

$c = 2.997 924 58 \times 10^8$ m s^{-1}
$c^2 = 8.987 551 79 \times 10^{16}$ m^2 s^{-2}

Gravitation constant

$G = 6.672 59(85) \times 10^{-11}$ m^3 kg^{-1} s^{-2}

Standard acceleration of gravity (exact)

$g_n = 9.806 65$ m s^{-1}

Planck constant

$2\pi \hbar = h = 6.626 075 5(40) \times 10^{-34}$ J s
$\hbar = 1.054 572 66(63) \times 10^{-34}$ J s

Planck mass

$(\hbar c/G)^{1/2} = 2.176 71(14) \times 10^{-8}$ kg

Planck length

$(\hbar G/c^3)^{1/2} = 1.616 05(10) \times 10^{-35}$ m

Planck time

$(\hbar G/c^5)^{1/2} = 5.390 56(34) \times 10^{-44}$ s

Elementary charge

$e = 4.803 206 8(15) \times 10^{-19}$ C
e = $1.602 177 33(49) \times 10^{-20}$ emu
$e^2 = 23.070 796 \times 10^{-20}$ in esu
$e^4 = 5.322 616 1 \times 10^{-38}$ in esu

Mass of electron

$m_e = 9.109 389 7(54) \times 10^{-31}$ kg
= $5.485 799 03(13) \times 10^{-4}$ u

Mass of unit atomic weight (^{12}C = 12 scale)	$u = 1.660\,540\,2(10) \times 10^{-27}$ kg
Boltzmann constant	$k = 1.380\,658(12) \times 10^{-23}$ J K^{-1}
	$= 8.617\,385(73) \times 10^{-5}$ eV K^{-1}
	$k^{1/2} = 1.175\,014 \times 10^{-8}$ erg$^{1/2}$ K$^{-1/2}$
Gas constant (^{12}C scale)	$R = 8.314\,510(70)$ J K^{-1} mol^{-1}
	$= 1.987\,216$ cal K^{-1} mol^{-1}
	$= 82.057\,83(70)$
	cm^3 atm K^{-1} mol^{-1}
Joule equivalent (chemical, exact) [4]	$= 4.184$ J cal^{-1}
Avogadro constant	$N_A = 6.022\,136\,7(36) \times 10^{23}$ mol^{-1}
Loschmidt constant	$n_0 = 2.686\,763(23) \times 10^{25}$ m^{-3}
Volume of gram-molecule at STP ($T = 273.15$ K, $P = 101\,325$ Pa)	$N_A/n_0 = V_0 = 22.414\,10(19) \times 10^{-3}$ m^3 mol^{-1}
Standard atmosphere pressure (exact)	$P_0 = 1\,013\,250$ dyn cm^{-2}
	$= 760$ mmHg
Ice point	$0°$ C $= 273.150$ K
Triple point (H$_2$O)	$= 273.160$ K
Faraday	$N_A e/c = 96\,485.309(29)$ C mol^{-1}

Atomic constants

Rydberg constant for ^{1}H	$R_H = 10\,967\,758.306(13)$ m^{-1}
	$1/R_H = 911.763\,345\,0$ Å
Rydberg constant for infinite nuclear mass $2\pi^2 m_e e^4/ch^3$	$R_\infty = 10\,973\,731.534(13)$ m^{-1}
	$1/R_\infty = 911.267\,053\,4$ Å
	$cR_\infty = 3.289\,841\,950 \times 10^{15}$ s^{-1}
Fine structure constant $2\pi e^2/hc$	$\alpha = 7.297\,353\,08(33) \times 10^{-3}$
	$1/\alpha = 137.035\,989\,5(61)$
	$\alpha^2 = 5.325\,136\,20 \times 10^{-5}$
Radius for first Bohr orbit (infinite nuclear mass) $h^2/4\pi^2 m_e e^2$	$a_0 = 0.529\,177\,249(24) \times 10^{-10}$ m
Time for $(2\pi)^{-1}$ revolutions in first Bohr orbit $m_e^{1/2} a^{3/2} e^{-1} = h^3/8\pi^3 m_e e^4$	$\tau_0 = 2.418\,884\,4 \times 10^{-17}$ s
Frequency of first Bohr orbit	$= 6.579\,683\,7 \times 10^{15}$ s^{-1}
Area of first Bohr orbit	$\pi a_0^2 = 8.797\,356\,70 \times 10^{-21}$ m^2
Electron speed in first Bohr orbit	$a_0 \tau_0^{-1} = 2.187\,691\,4 \times 10^6$ m s^{-1}
Atomic unit of energy (Hartree = 2 Rydbergs) $e^2/a_0 = 2chR_\infty$	$= 4.359\,748\,2(26) \times 10^{-18}$ J
	$= 27.211\,396\,1(81)$ eV
Energy of Rydberg (often adopted as atomic unit)	ryd $= 2.179\,874\,1(13) \times 10^{-18}$ J
	$= 13.605\,698\,1(40)$ eV
Atomic unit of angular momentum $h/2\pi$	$\hbar = 1.054\,572\,66(63) \times 10^{-34}$ kg m^{2}s^{-1}
Classical electron radius $e^2/m_0 c^2$	$l = 2.817\,940\,92(38) \times 10^{-15}$ m
Schrödinger constant for fixed nucleus	$8\pi^2 m_e h^{-2} = 1.638\,197\,48 \times 10^{27}$ erg^{-1} cm^{-2}
Schrödinger constant for ^{1}H atom	$= 1.637\,305\,78 \times 10^{27}$ erg^{-1} cm^{-2}

Hyperfine structure splitting of ^{1}H ground state	$\nu_H = 1\,420.405\,751\,768 \times 10^6$ s^{-1}
Doublet separation in ^{1}H atom	$= 0.365\,866\,231$ cm^{-1}
$(1/16)R_H\alpha^2[1 + \alpha/\pi + (5/8 - 5.946/\pi^2)\alpha^2]$	$= 1.096\,839\,36 \times 10^{10}$ s^{-1}
Reduced mass of electron in ^{1}H atom	$m_e(m_p/m_H) = 9.104\,431\,3 \times 10^{-31}$ kg
Mass of ^{1}H atom	$= 1.673\,534\,4 \times 10^{-27}$ kg
	$= 1.007\,825\,050(12)$ u
Mass of proton	$= 1.672\,623\,1(10) \times 10^{-27}$ kg
	$= 1.007\,276\,470(12)$ u
Mass of neutron	$= 1.674\,928\,6(10) \times 10^{-27}$ kg
	$= 1.008\,664\,904(14)$ u
Mass of deuteron	$= 3.343\,586\,0(20) \times 10^{-27}$ kg
	$= 2.013\,553\,214(24)$ u
Mass energy of unit atomic mass	$uc^2 = 1.492\,419\,1 \times 10^{-10}$ J
	$= 931.494\,2(28)$ MeV
Rest mass energy of electron	$m_ec^2 = 8.187\,111\,1 \times 10^{-14}$ J
	$= 0.510\,999\,06(15)$ MeV
Mass ratio proton/electron	$= 1\,836.152\,701(37)$
Specific electron charge	$e/m_e = 1.758\,819\,62 \times 10^7$ emu g^{-1}
	$e/m_e = 5.272\,808\,6 \times 10^{17}$ esu g^{-1}
Quantum of magnetic flux	$h/e = 1.379\,510\,77 \times 10^{-17}$ erg s esu^{-1}
	$hc/e = 4.135\,669\,2 \times 10^{-7}$ gauss cm^2
Quantum of circulation	$h/m_e = 7.273\,896\,2$ erg s g^{-1}
Compton wavelength	$h/m_ec = 2.426\,310\,58(22) \times 10^{-12}$ m
	$h/2\pi m_ec = 3.861\,593\,23(35) \times 10^{-13}$ m
Band spectrum constant (moment of inertia/wave number)	$h/8\pi^2 c = 27.992\,774 \times 10^{-40}$ g cm
Atomic specific heat constant $c_2/c = h/k$	$= 4.799\,216 \times 10^{-11}$ s K
Magnetic moment of 1 Bohr magneton $\mu_B = \frac{1}{2}\alpha m_e^{1/2} a_0^{5/2} \tau_0^{-1} = he/4\pi m_ec$	$\mu_B = 9.274\,015\,4(31) \times 10^{-21}$ erg gauss^{-1}
Electron magnetic moment	$\mu_e = 1.001\,159\,652\,193(10)\mu_B$
Proton magnetic moment	$\mu_p = 1.521\,032\,202(15) \times 10^{-3}\mu_B$
Gyromagnetic ratio of proton corrected for diamagnetism of H_2O	$\gamma_p = 2.675\,221\,28(81) \times 10^4$ rad s^{-1} gauss^{-1}
Magnetic moment of 1 nuclear magneton $he/4\pi m_pc$	$\mu_n = 5.050\,786\,6(17) \times 10^{-24}$ erg gauss^{-1}
Atomic unit of magnetic moment $2\mu_B/\alpha$	$= 2.541\,747\,8 \times 10^{-18}$ erg gauss^{-1}
Magnetic moment per mole of 1 Bohr magneton per molecule	$= 5\,584.938\,8$ erg gauss^{-1} mol^{-1}
Zeeman displacement $3/4\pi m_ec$ (e in emu)	$= 4.668\,643\,7(14) \times 10^{-5}$ cm^{-1} gauss^{-1}
in frequency	$= 1.399\,624\,18(42) \times 10^6$ s^{-1} gauss^{-1}

The electron–volt and photons [5]

Wavelength associated with 1 eV $\qquad\qquad\qquad \lambda_0 = 12\,398.428\,2 \times 10^{-10}$ m

Wave number associated with 1 eV $\qquad\qquad s_0 = 8\,065.538\,51$ cm^{-1}

$\qquad\qquad\qquad\qquad\qquad\qquad\qquad\qquad\quad = 8.065\,538\,51$ kilo–kayser

Frequency associated with 1 eV $\qquad\qquad\quad \nu_0 = 2.417\,988\,36(72) \times 10^{14}$ s^{-1}

Energy of 1 eV $\qquad\qquad\qquad\qquad\qquad\quad E_0 = 1.602\,177\,33(49) \times 10^{-19}$ J

$\qquad\qquad\qquad\qquad\qquad\qquad\qquad\qquad\quad = 0.073\,498\,617\,6$ ryd

Photon energy associated with unit $\qquad\quad hc = 1.986\,448\,0 \times 10^{-23}$ J

$\quad$ wave number

Photon energy associated with $\qquad\qquad\qquad = 1.986\,448\,0 \times 10^{-8}/\lambda$ erg (λ in Å)

$\quad$ wavelength λ

Speed of 1 eV electron $\qquad\qquad\qquad\qquad\quad = 5.930\,968\,92 \times 10^5$ m s^{-1}

$\quad (2 \times 10^8 (e/m_e c))^{1/2}$

Speed^2 $\qquad\qquad\qquad\qquad\qquad\qquad\qquad = 3.517\,639\,23 \times 10^{11}$ m^2 s^{-2}

Wavelength of electron of energy V in eV $\quad = V^{-1/2}(12.264\,263 \times 10^{-8})$ cm

$\quad h(2m_e E_0)^{-1/2} V^{-1/2}$

Temperature associated with 1 eV $\qquad\qquad = 11\,604.45$ K

$\quad E_0/k$

Temperature associated with 1 eV $\qquad\qquad = 5039.75$ K

$\quad$ in common logs $= (E_0/k) \log e$

Temperature associated with 1 kilo-kayser $\quad = 624.849\,3$ K

$\quad$ in common logs $= 10^3(hc/k) \log e$

Energy of 1 eV per molecule $\qquad\qquad\qquad = 23\,060.054\,2$ cal mol^{-1}

Radiation constants

Radiation density constant $\qquad\qquad\qquad\quad a = 7.565\,91(25) \times 10^{-15}$

$\quad 8\pi^5 k^4/15c^3 h^3 \qquad\qquad\qquad\qquad\qquad\qquad$ erg cm^{-3} K^{-4}

Stefan–Boltzmann constant $= ac/4 \qquad\qquad \sigma = 5.670\,51(19) \times 10^{-5}$

$\qquad\qquad\qquad\qquad\qquad\qquad\qquad\qquad\quad$ erg cm^{-2} K^{-4} s^{-1}

First radiation constant $\qquad\qquad\qquad\qquad c_1 = 3.741\,774\,9(22) \times 10^{-5}$

$\quad$ (emittance) $= 2\pi hc^2 \qquad\qquad\qquad\qquad\qquad$ erg cm^2 s^{-1}

First radiation constant (radiation density) $\quad 8\pi hc = 4.992\,487\,0 \times 10^{-15}$ erg cm

Second radiation constant $= hc/k \qquad\qquad c_2 = 1.438\,769(12)$ cm K

Wien displacement law constant $\qquad\qquad\qquad = 0.289\,775\,5$

$\quad c_2/4.965\,114\,23$

Some general constants [1, 5]

Density of mercury (0° C, 760 mmHg) $\qquad\quad = 13.395\,080$ g cm^{-3}

Ratio, grating to Siegbahn scale

$\quad$ of X-ray wavelengths [5] $\qquad\qquad\quad \lambda_g/\lambda_s = 1.002\,077\,89(70)$

$\qquad\qquad\qquad\qquad\qquad\qquad\qquad\qquad\quad [\lambda_s$ (Cu Kα_1) $= 1.537\,400$ kXu]

Lattice spacing of Si (in vacuum, 22.5° C) $\qquad = 0.543\,101\,96(11) \times 10^{-9}$ m

Molar volume of Si $\qquad\qquad\qquad\qquad\qquad = 12.058\,817\,9(89)$ cm^3 mol^{-1}

Maximum density of water $\qquad\qquad\qquad\qquad = 0.999\,972$ g cm^{-3}

Cesium resonance frequency $\qquad\qquad\qquad\quad = 9\,192\,631\,770$ Hz

$\quad$ (defining the SI second) [6]

2.3   GENERAL ASTRONOMICAL CONSTANTS
by Alan D. Fiala

Astronomical unit of distance	= mean Sun–Earth distance
	= semimajor axis of Earth orbit [2, 6].
	AU = $1.495\,978\,706\,6 \times 10^{11}$ m.
Parsec (= 206 264.806 AU)	pc = $3.085\,677\,6 \times 10^{16}$ m.
	= 3.261 563 8 light (Julian) year.
Light (Julian) year	= $9.460\,730\,472 \times 10^{15}$ m.
Light time for 1 AU [6]	= 499.004 783 70 s = 0.005 775 518 33 d.
Solar mass	$\mathcal{M}_\odot = 1.9891 \times 10^{30}$ kg.
Solar radius	$\mathcal{R}_\odot = 6.955\,08 \pm 0.00026 \times 10^8$ m.
Solar radiation	$\mathcal{L}_\odot = 3.845(8) \times 10^{33}$ erg s^{-1}.
Earth mass	$M_\oplus = 5.974\,2 \times 10^{24}$ kg.
Earth mean density	$\bar{\rho}_\oplus = 5.515$ g cm^{-3}.
Earth equatorial radius [6]	= 6378.136 km.

Galactic pole (J2000.0)	$\alpha_3 = 192°859\,481\,23$	$\delta_3 = +27°128\,251\,20$
	$12^h51^m26^s2755$	$+27°7'41''704$
Direction of galactic center (J2000.0)	$\alpha_1 = 266.404\,996\,25$	$\delta_1 = -28°936\,172\,42$
	$17^h45^m37^s1991$	$-28°56'10''221$
Solar motion toward galactic center [7]	U = 10.00 ± 0.36 km s^{-1}	
toward direction of galactic rotation	V = 5.23 ± 0.62 km s^{-1}	
vertically up in north direction	W = 7.17 ± 0.38 km s^{-1}	
Galactic rotation [8]	$R_0 = 7.66 \pm 0.32$ kpc	
	$V_{circ} = 237 \pm 12$ km s^{-1}	
Sun's equatorial horizontal parallax [6]	= $8''794\,144(3)$	
	= $4.263\,521 \times 10^{-5}$ rad	
Moon's equatorial horizontal parallax	= $3422''608$	
at mean distance		
Constant of nutation [6]	= $9''2025$	
Constant of aberration [6]	= $20''495\,52$	

$\dfrac{2\pi \times 206\,265 \times \text{AU}}{ct(1 - e^2)^{1/2}}$

t = sidereal year, e = Earth orbital
  eccentricity
Gaussian gravitational constant k
  in $n^2a^3 = k^2(1 + m)$, where m = mass
  of planet in solar units, n = mean daily in AU $k = 0.0172\,020\,989\,5$ rad
  motion, and a = semimajor axis
  (a defining constant)   = $3548''187\,607$
        = $0°985\,607\,668\,6$

$k/86400 = 2\pi/$(sidereal year in sec) $k' = 1.990\,983\,675 \times 10^{-7}$ rad,
        for use with seconds of time

Heliocentric gravitational constant = AU$^3(k')^2$ = $1.327\,124\,40 \times 10^{26}$ cm^2 s^{-1}
Semimajor axis of Earth orbit in terms of AU   = $1.000\,001\,057\,266\,65$ AU

Mass ratios [6, 9]

$\mathcal{M}_\oplus/\mathcal{M}_\mathbb{C}$	$= 81.300\,59$
$\mathcal{M}_\odot/\mathcal{M}_\oplus$	$= 332\,946.05$
$\mathcal{M}_\odot/(\mathcal{M}_\oplus + \mathcal{M}_\mathbb{C})$	$= 328\,900\,56(2)$
Obliquity of ecliptic (fixed ecliptic of J2000.0)	$\epsilon = 23°26'\,21.''4119$

2.4 ASTRONOMICAL CONSTANTS INVOLVING TIME [6]
by Alan D. Fiala

The basic unit of time is the Système International (SI) second which is defined to be the duration of 9 192 631 770 cycles of one of the hyperfine transitions of the ground state of ^{133}Cs. Based on this defined unit, International Atomic Time (TAI) is formed from statistical analysis of individual frequency standards and time scales based on atomic clocks in many countries. It was introduced in January 1972, and is a coordinate time scale.

Universal Time (UT) is the measure of time used for all civil time keeping, and conforms closely to the mean diurnal motion of the Sun. It is directly related to sidereal time by means of an adopted numerical formula. It does not refer to the motion of the Earth and is not precisely related to the hour angle of the Sun.

UT0 is the uncorrected observed rotational time scale derived from observation of sidereal time at a particular station. When this time scale is corrected for the shift in longitudes caused by polar motion, it is designated UT1. This still contains the variable rotation of the Earth and is generally implied when the symbol "UT" is used without qualification.

Coordinated Universal Time (UTC) is the time scale distributed by radio signals, satellites, communication media, as the basis for civil time keeping around the world. UTC is maintained within 0.9 second of UT1 by the introduction of leap seconds. UTC differs from TAI by an integer number of seconds, which difference changes when leap seconds are introduced.

Dynamical time represents the independent variable of the equations of motion of the bodies in the Solar System. It depends on the theory of relativity being used, as does the transformation between barycentric and geocentric time scales. In the transformation, the constants can be chosen so that the timescales have only periodic variations with respect to each other. The dynamical time scale for apparent geocentric ephemerides was chosen to be unique and independent of the theories; the barycentric timescales are theory dependent.

Terrestrial Dynamical Time (TDT), or Terrestrial Time (TT), is the idealized time on the geoid of the Earth and is approximated as being equal to TAI + 32.184 seconds. Terrestrial Time is a continuation of Ephemeris Time (ET), beginning 1977 Jan. 1.0 TAI. The relationship between UT and TT changes according to the variations in the rotation of the Earth.

Barycentric Dynamical Time (TDB) is the relativistically transformed time for referring equations of motion to the barycenter of the Solar System. It is defined to contain only periodic variations with respect to TDT.

The time scales Geocentric Coordinate Time (TCG) and Barycentric Coordinate Time (TCB) are the time-like arguments appropriate for coordinate systems defined with respect to the geocenter of the Earth and the barycenter of the Solar System, respectively, including all relativistic transformations from terrestrial time.

Up to 1984, the tropical year was used as the basis of time for reference systems and the Besselian year was used as the epoch for such reference frames, thus designated, for example, as B1950.0. Since 1984, the Julian Century has been used as the time unit for reference frames and the standard epoch is then designated as, for example, J2000.0.

Sidereal time is defined by the hour angle of the equinox. The relationship between Greenwich Mean Sidereal Time (GMST) and UT1 is specified by an adopted equation, which is often considered to be the definition of UT1. At 0 hours UT1:

$$\text{GMST} = 24110.54841 + 8640184{.}^{\!s}812866 T_u + 0{.}^{\!s}093104 T_u^2 - 6.2 \times 10^{-6} T_u^3$$

seconds of time, where $T_u = d_u/36525$, d_u is the number of days of Universal Time elapsed since JD 2451545.0 UT1 (2000 January 1, 12 hrs UT1), taking on values ± 0.5, ± 1.5, etc.

The ratio of mean sidereal time to UT1 is

$$r' = 1.002\,737\,909\,350\,795 + 5.900\,6 \times 10^{-11} T - 5.9 \times 10^{-15} T^2,$$

where T is the number of Julian centuries elapsed since JD 2 451 545.0.

The ratio of UT1 to mean sidereal time is

$$1/r' = 0.997\,269\,566\,329\,084 - 5.868\,4 \times 10^{-11} T + 5.9 \times 10^{-15} T^2.$$

The relationships between time scales in seconds of time are:

$$\text{TT} = \text{TDT} = \text{ET} = \text{TAI} + 32.184,$$
$$\Delta T = \text{ET} - \text{UT} = \text{TDT} - \text{UT} = \text{TT} - \text{UT},$$
$$\Delta AT = \text{TAI} - \text{UTC},$$
$$\text{DUT} \sim \Delta UT = \text{UT1} - \text{UTC},$$
$$\text{TDB} = \text{TDT} + P,$$
$$\text{TCG} - \text{TT} = 6.969\,290\,4 \times 10^{-10} (JD - 2\,443\,144.5) \times 86\,400,$$
$$\text{TCB} - \text{TCG} = 1.480\,813 \times 10^{-8} (JD - 2\,443\,144.5) \times 86\,400 + \mathbf{v_e} \cdot (\mathbf{x} - \mathbf{x_e}) c^{-2} + P,$$
$$\text{TCB} - \text{TDB} = 1.550\,506 \times 10^{-8} (JD - 2\,443\,144.5) \times 86\,400,$$
$$P = 0.001\,656\,8 \sin(35\,999.37T + 357.5) + 0.000\,022\,4 \sin(32\,964.5T + 246)$$
$$+ 0.000\,013\,8 \sin(71\,998.7T + 355) + 0.000\,004\,8 \sin(3\,034.9T + 25)$$
$$+ 0.000\,004\,7 \sin(34\,777.3T + 230),$$

where T is the elapsed time from J2000.0 measured in Julian centuries and the coefficients are rounded at their last digits [6, 10, 11]. Arguments are in degrees. Here $\mathbf{x_e}$ and $\mathbf{v_e}$ denote the barycentric position and velocity of the Earth's center of mass, the difference $(\mathbf{x} - \mathbf{x_e})$ is the vector distance of the observer from this center of mass, and c is the speed of light.

2.4.1   Reduction of Time Scales

The variations in the Earth's rotation rate have resulted in differences between time based on it and that based on planetary orbits. The differences between the ephemeris and the (generally slower) universal time are given in Tables 2.1 and 2.2 for the last 130 years. For dates back to 1620, see the *Astronomical Almanacs* [12]. Even earlier to the year 1500 B.C., one can find a table in the *Canon of Lunar Eclipses* [13]. Before 1884, $\Delta T = \text{ET} - \text{UT}$, after 1984, $\Delta T = \text{TDT} - \text{UT}$, and after 1989, the differences are for exactly 1 Jan. 0^h UTC.

Table 2.1. *Reduction of time scales from 1870 to 1974.*

Year	ΔT	Year	ΔT	Year	ΔT	Year	ΔT	Year	ΔT
1870	+1.61	1895	−6.47	1920	+21.16	1945	+26.77	1970	+40.18
1875	−3.24	1900	−2.72	1925	+23.62	1950	+29.15	1971	+41.17
1880	−5.40	1905	+3.86	1930	+24.02	1955	+31.07	1972	+42.23
1885	−5.79	1910	+10.46	1935	+23.93	1960	+33.15	1973	+43.37
1890	−5.87	1915	+17.20	1940	+24.33	1965	+35.73	1974	+44.49

Table 2.2. *UTC leap seconds since 1971 and starting at the given date.*

Year	ΔT	Year	ΔT	Year	ΔT	Year	ΔT
1972, Jan. 1	10	1977, Jan. 1	16	1983, July 1	22	1992, July 1	27
1972, July 1	11	1978, Jan. 1	17	1985, July 1	23	1993, July 1	28
1973, Jan. 1	12	1979, Jan. 1	18	1988, Jan. 1	24	1994, July 1	29
1974, Jan. 1	13	1980, Jan. 1	19	1990, Jan. 1	25	1996, Jan. 1	30
1975, Jan. 1	14	1981, July 1	20	1991, Jan. 1	26	1997, July 1	31
1976, Jan. 1	15	1982, July 1	21				

Day

1 day = 24 hours = 1440 minutes = 86400 SI seconds

Period of rotation of Earth (referred to fixed stars)

In mean sidereal time	= 86 164.090 54 SI seconds.
In mean solar time	= $23^h 56^m 04^s_{.}090 549$.
1 day of mean sidereal time	= 0.997 269 566 33 of mean solar time.
1 day of mean solar time	= 1.002 737 909 35 of mean sidereal time.
Rate of rotation	= $15''041 067 178 669 10 \text{ s}^{-1}$.
	= $7.292 115 10 \times 10^{-5} \text{ rad s}^{-1}$.

Year

1 Julian year = 365.25 days = 8766 hours = 525 960 minutes = 31 557 600 SI seconds

	d	d	h	m	s
Tropical (equinox to equinox)	365.242 189 7	365	05	48	45.19
Sidereal (fixed star to fixed star)	365.256 36	365	06	09	10
Anomalistic (perihelion to perihelion)	365.259 64	365	06	13	53
Eclipse (Moon's node to Moon's node)	346.620 05	346	14	52	52
Gaussian (Kepler's law for $a = 1$)	365.256 90	365	06	09	56
Julian (based on Julian calendar)	365.25	365	06		
Gregorian (based on Gregorian calendar)	365.242 5	365	05	49	12

Calendar Julian Dates (*see Chapter* 27 *also*)
1900   January 0.5 = JD 2 415 020.0,
1925   January 0.5 = JD 2 424 151.0,
1950   January 0.5 = JD 2 433 282.0,
2000   January 0.5 = JD 2 451 544.0,
2050   January 0.5 = JD 2 469 807.0,
2100   January 0.5 = JD 2 488 069.0.

Length of the month

	d	d	h	m	s
Synodic (new Moon to new Moon)	29.530 59	29	12	44	03
Tropical (equinox to equinox)	27.321 58	27	07	43	05
Sidereal (fixed star to fixed star)	27.321 66	27	07	43	12
Anomalistic (perigee to perigee)	27.554 55	27	13	18	33
Draconic (node to node)	27.212 22	27	05	05	36

Orbit of the Moon about the Earth

Sidereal mean motion of Moon	$2.661\,699\,489 \times 10^{-6}$ rad s^{-1}
Mean distance of Moon from Earth	$3.844 \times 10 * 5$ km
	60.27 Earth radii
	0.002 570 AU
Equatorial horizontal parallax	$57'\,02.''608$
at mean distance	$3422.''608$
Mean distance of center of Earth	
from Earth–Moon barycenter	4.671×10^3 km
Mean eccentricity	0.054 90
Mean inclination to ecliptic	$5.°145\,396$
Mean inclination to lunar equator	$6°\,41'$
Limits of geocentric declination	$\pm 29°$
Saros = 223 lunations = 19 passages of Sun through node = $6585\frac{1}{3}$ days	
Period of revolution of node	6798 days
Period of revolution of perigee	3232 days
Mean orbital speed	1023 m s^{-1} = 0.000 591 AU day^{-1}
Mean centripetal acceleration	0.002 72 m s^{-2} = 0.0003 g.

Precession

Annual rates of precession (T in centuries from J2000.0)
   general precession in longitude $50.''290\,966 + 0.''022\,222\,6T,$
   lunisolar precession in longitude $50.''387\,784 + 0.''004\,926\,3T,$
   planetary precession $-0.''018\,862\,3 - 0.''047\,612\,8T,$
   geodesic precession (relativistic nonperiodic Coriolis effect) $1.''92T.$

2.5 UNITS

The seven SI base units are: meter (m), kilogram (kg), second (s), ampere (A), Kelvin (K), mole (mol), and candela (cd) [14]. All other units are derived from these. Units used with SI are: the time units of minute (min), hour (h), and day (d); the plane angle units of radian (rad), degree (°), minute ('), and (arc)second ("); the solid angle unit, steradian (sr); the volume unit liter; (L); the mass unit metric ton (t); and the land area hectare (ha). Other experimentally determined units used with SI are: the special energy unit (eV), and the atomic mass unit (u).

Units used in astronomy and astrophysics are often not standard but unique to the special subfield. This procedure is followed for many chapters in this book. They are frequently defined at the beginning of each chapter.

Table 2.3 gives the SI unit prefixes.

Table 2.3. *The SI prefixes.*

Factor	Prefix	Symbol	Factor	Prefix	Symbol
10^{24}	yotta	Y	10^{-1}	deci	d
10^{21}	zetta	Z	10^{-2}	centi	c
10^{18}	exa	E	10^{-3}	milli	m
10^{15}	peta	P	10^{-6}	micro	μ
10^{12}	tera	T	10^{-9}	nano	n
10^{9}	giga	G	10^{-12}	pico	p
10^{6}	mega	M	10^{-15}	femto	f
10^{3}	kilo	k	10^{-18}	atto	a
10^{2}	hecto	h	10^{-21}	zepto	z
10^{1}	deka	da	10^{-24}	yocto	y

Unconventional (nonstandard) units sometimes used in astronomy and astrophysics are listed below.

Length

Angstrom unit	$\text{Å} = 10^{-8} \text{ cm} = 10^{-10} \text{ m}$
Micron	$\mu = \mu\text{m} = 10^{-4} \text{ cm} = 10^{-6} \text{ m}$
Foot	$\text{ft} = 30.4800 \text{ cm} = 12 \text{ in}$
Inch	$\text{in} = 2.540\,000 \text{ cm}$
Mile	$= 1.609\,344 \text{ km} = 5280 \text{ ft}$
Nautical mile [2]	$= 1.853 \text{ km} = 6080 \text{ ft}$

Area

Square foot	$\text{ft}^2 = 929.03 \text{ cm}$
Acre	$= 4046.85 \text{ m}^2 = 43560 \text{ ft}^2$
Barn	$= 10^{-24} \text{ cm}^2$

Volume
 Cubic foot

$$ft^3 = 28\,316.8\ cm^3$$
$$= 6.229\ British\ gallons$$
$$= 7.481\ US\ gallons$$

 Fluid ounce

$$= 480\ minims\ (British\ and\ US)$$
$$= 28.413\ cm^3\ (British)$$
$$= 29.574\ cm^3\ (US)$$

Mass
 Kilogram (SI unit)
 Pound avoirdupois

$$kg = 1000\ g$$
$$British\ lb = 453.592\,37\ g = 7000\ grains$$
$$US\ lb = 453.592\,43\ g = 7000\ grains$$

 Pound troy and apothecary
 Grain (all systems)
 Carat
 Slug
 Ton = tonne

$$= 373.242\ g = 5760\ grains$$
$$= 0.064\,798\,9\ g$$
$$= 0.2000\ g$$
$$= 14.594\ kg$$
$$= 2240\ lb$$
$$= 1.016\,047 \times 10^6\ g$$

 Metric ton

$$= 10^6\ g$$

Energy
 Joule (SI unit)
 Calorie [4] (exact)
 Kilowatt-hour

$$J = 10^7\ erg$$
$$cal = 4.184\ J = 4.184 \times 10^7\ erg$$
$$= 3600 \times 10^3\ J$$
$$= 8.6042 \times 10^5\ cal$$

 British thermal unit
 Therm
 Foot-pound
 Kiloton of TNT

$$BTU = 1055\ J = 252.0\ cal$$
$$= 100\,000\ BTU$$
$$= 1.355\,82 \times 10^7\ erg$$
$$= 4.184 \times 10^{19}\ erg$$

Power
 Watt (SI unit)
 British horse-power
 Force de cheval

$$= 10^7\ erg/s = J\,s^{-1}$$
$$= 745.7\ W$$
$$= 735.5\ W$$

Force
 Newton (SI unit)
 Poundal
 Pound weight
 Slug
 Gram weight

$$N = 10^5\ dyn$$
$$= 1.3825 \times 10^4\ dyn$$
$$= 4.4482 \times 10^5\ dyn$$
$$= 14.594\ kg$$
$$= 980.665\ dyn$$

Acceleration
 Standard gravity
 Gravity (equator)
 Gravity (pole)

$$1\ gal = 1\ cm\ s^{-2}$$
$$g = 978.031\ cm\ s^{-2} = 32.09\ ft\ s^{-2}$$
$$g = 983.217\ cm\ s^{-2} = 32.\,26\ ft\ s^{-2}$$

Speed
 Mile per hour
 Knot

$$= 44.704\ cm\,s^{-1} = 1.4667\ ft\,s^{-1}$$
$$= 51.47\ cm\,s^{-1}$$

Pressure

Pascal (SI unit) $= 10 \text{ dyn cm}^{-2} = 10 \ \mu\text{b}$

Barye (occasionally called Bar) $\mu\text{b} = 1.000 \text{ dyn cm}^{-2}$

Bar $\text{bar} = 1.000 \times 10^6 \text{ dyn cm}^{-2}$
$= 0.986\,923 \text{ atm}$
$= 1.0197 \times 10^3 \text{ g-weight cm}^{-2}$

Millibar $\text{mb} = 10^{-3} \text{ bar} = 10^3 \ \mu\text{b}$
$= 10^3 \text{ dyn cm}^{-2}$

Atmosphere (standard) $\text{atm} = 1.013\,250 \times 10^6 \text{ dyn cm}^{-2}$
$= 760 \text{ mmHg} = 1013.25 \text{ mb}$

Millimeter of mercury ($= 1$ Torr) $\text{mmHg} = 1333.22 \text{ dyn cm}^{-2}$
$= 0.001\,315\,8 \text{ atm}$

Inch of mercury $= 3.386\,38 \times 10^4 \text{ dyn cm}^{-2}$
$= 0.033\,421 \text{ atm}$

Pound per square inch $= 6.8947 \times 10^4 \text{ dyn cm}^{-2}$
$= 0.068\,046 \text{ atm}$

Density

Kilogram/cubic meter (SI unit) $= 1.000 \times 10^{-3} \text{ g cm}^{-3}$

Density of water (4° C) $= 0.999\,972 \text{ g cm}^{-3}$

Density of mercury (0° C) $= 13.5951 \text{ g cm}^{-3}$

Solar mass/cubic parsec $= 6.770 \times 10^{-23} \text{ g cm}^{-3}$

STP gas density $= 4.4616 \times 10^{-5} \mu_0 \text{ g cm}^{-3}$
    where μ_0 is molecular weight

Temperature

Degree scales (Kelvin K, $\qquad$ $\text{K} = \text{deg C} = 1.8 \text{ deg F}$
    Celsius (centigrade) C, Fahrenheit F)

Temperature comparisons $\qquad$ $0° \text{ C} = 273.150 \text{ K} = 32° \text{ F}$
$100° \text{ C} = 373.150 \text{ K} = 212° \text{ F}$

Triple point of natural water $= 273.160 \text{ K} = 0.010° \text{ C}$

Viscosity (dynamic)

Poise $\qquad$ $\text{P} = 1 \text{ g cm}^{-1} \text{ s}^{-1} = 0.1 \text{ Pa s}$

SI unit $\qquad$ $\text{N s m}^{-2} = 10 \text{ g cm}^{-1} \text{ s}^{-1}$

Viscosity (kinematic)

Stokes $= 1 \text{ cm}^2 \text{ s}^{-1}$

SI unit $\qquad$ $\text{m}^2 \text{ s}^{-1} = 10000 \text{ cm}^2 \text{ s}^{-1}$

Frequency

Hertz $\qquad$ $\text{Hz} = \text{cycle s}^{-1}$

Kayser (a wave number unit) $\qquad$ $\text{cm}^{-1} = c \text{ Hz} \simeq 3 \times 10^{10} \text{ Hz}$

Rydberg frequency $\qquad$ $c R_\infty = 3.289\,84 \times 10^{15} \text{ Hz}$

Frequency in first Bohr orbit $\qquad$ $2c R_\infty = 6.5797 \times 10^{15} \text{ Hz}$

Frequency of free electron $= 2.7993 \times 10^6 \ \mathcal{H} \text{ Hz}$
    in magnetic field $\mathcal{H}$ (gauss)

Plasma frequency associated $= 8.979 \times 10^3 N_e^{1/2} \text{ Hz}$
    with electron density N_e in cm^{-3})

Angular velocity ($= 2\pi$ frequency)
  Unit of angular velocity $= 1 \text{ rad s}^{-1} = 2\pi \text{ Hz}$
  $1''$ of arc per tropical year $= 1.536\,314\,7 \times 10^{-13} \text{ rad s}^{-1}$
  $1''$ of arc per day $= 5.611\,269\,5 \times 10^{-11} \text{ rad s}^{-1}$
  Angular velocity of Earth on its axis $= 7.292\,115\,2 \times 10^{-5} \text{ rad s}^{-1}$
  Mean angular velocity of Earth in its orbit $= 1.990\,986\,7 \times 10^{-7} \text{ rad s}^{-1}$

Momentum
  Linear momentum, SI unit $= 10^5 \text{ g cm s}^{-1} = 1 \text{ kg m s}^{-1}$
  mc $= 2.730\,93 \times 10^{-17} \text{ g cm s}^{-1}$

Angular momentum
  SI unit $= 10^7 \text{ g cm}^2 \text{ s}^{-1} = 1 \text{ kg m}^2 \text{ s}^{-1}$
  Electron momentum in first Bohr orbit $= 1.993 \times 10^{-19} \text{ g cm s}^{-1}$
  Quantum unit $\hbar = 1.0546 \times 10^{-27} \text{ erg s}$
  Homogeneous sphere angular momentum $= (2/5) R^2 \mathcal{M} \omega$
    ($R = $ radius, $\mathcal{M} = $ mass,
    $\omega = $ angular velocity)
  Angular momentum of solar system $= 3.148 \times 10^{50} \text{ g cm}^{-2} \text{ s}^{-1}$

Luminous intensity
  Luminous intensity is defined
    as the luminous emission per sterad
  Candela (SI unit) $\text{cd} = (1/60)$ luminous intensity
                             of 1 projected cm^2
                             black body at the
                             temperature of melting
                             platinum (2044 K)
  Star, $m_v = 0$, outside Earth atmosphere $= 2.45 \times 10^{29} \text{ cd}$

Luminous flux
  Lumen (both SI and CGS unit) $= $ flux from 1 cd into 1 sr
                             $= $ flux from $(1/60\pi) \text{ cm}^2$ of
                             black body at 2044 K
  Lumen of maximum visibility radiation $= 1.470 \times 10^{-3} \text{ W}$
    at 5550 Å
    therefore 1 W at 5550 Å $= 680$ lumens
  Jansky $= 10^{-23} \text{ ergs cm}^{-2} \text{ s}^{-1} \text{ Hz}^{-1}$
                             $= 10^{-26} \text{ W m}^{-2} \text{ Hz}^{-1}$

Luminous energy
  Talbot (SI unit) $= 1$ lumerg (CGS unit)
                             $= 1$ lumen second

Surface brightness
  Stilb $\text{sb} = 1 \text{ cd cm}^{-2} = \pi \text{ lambert}$
                             $= 1 \text{ lumen cm}^{-2} \text{ sr}^{-1}$
  Lambert $= (1/\pi) \text{ cd cm}^{-2}$
                             $= 1000$ millilambert
                             $\equiv 1 \text{ lumen cm}^{-2}$ for a perfectly
                             diffusing surface

Apostilb

$= 1$ lumen m^{-2} for a perfectly
diffusing surface
$= 10^{-4}$ lambert

Nit (SI unit)

$= 10^{-4}$ sb $=$ cd m^{-2}

Candle per square inch

$= 0.487$ lambert $= 0.155$ stilb

Foot-lambert

$= 1.076 \times 10^{-3}$ lambert
$= 343 \times 10^{-4}$ stilb

$1 m_v = 0$ star per square degree outside
atmosphere

$= 0.84 \times 10^{-6}$ stilb
$= 0.84 \times 10^{-2}$ nit
$= 2.63 \times 10^{-6}$ lambert

$1 m_v = 0$ star per square degree inside clear
unit airmass

$= 0.69 \times 10^{-6}$ stilb

Luminous emittance (of a surface)
Lumen per square meter (SI unit)

$= 10^{-4}$ lumen cm^{-2}

Illuminance (light received per unit surface)
Phot (CGS unit)

$= 1$ lumen cm^{-2}

Lux (SI unit)

$1 x = 1$ lumen m$^{-2} = 10^{-4}$ phot
$= 1$ m-candle

Foot-candle

$= 10.76$ lux $= 1.076 \times 10^{-3}$ phot
$= 1$ lumen ft^{-2}

Star, $m_v = 0$, outside Earth atmosphere

$= 2.54 \times 10^{-10}$ phot

Electrical units
The general inter-relations between electric
and magnetic units are given in Sec. 2.6

Electrical charge
Coulomb (SI unit)

$C = 2.997\,925 \times 10^9$ esu $= 0.10$ emu
$= -6.241\,51 \times 10^{18}$ electrons

Electron charge

$e = -4.803\,25 \times 10^{-10}$ esu
$e = -1.602\,18 \times 10^{-19}$ C

Electrical potential
Volt (SI unit)

$V = 3.335\,64 \times 10^{-3}$ esu $= 10^8$ emu

Potential of electron at first Bohr orbit distance

$= 27.211$ volt $= 0.090\,767$ esu

Ionization potential from first Bohr orbit

$= 13.606$ volt $= 0.045\,384$ esu

Electric field
Volt per meter (SI unit)

$= 3.335\,64 \times 10^{-5}$ esu $= 10^6$ emu

Nuclear field at first Bohr orbit

$= 5.1402 \times 10^{11}$ volt m^{-1}
$= 1.715\,2 \times 10^7$ esu

Resistance
Ohm (SI unit)

$\Omega = 1.112\,65 \times 10^{-12}$ esu $= 10^9$ emu

Electric current
Ampere (SI unit)

$A = 2.997\,925 \times 10^9$ esu $= 0.10$ emu
$= -6.241\,51 \times 10^{18}$ electrons s^{-1}

Current in first Bohr orbit

$= 1.054 \times 10^{-3}$ A
$= 3.160 \times 10^6$ esu

Electric dipole moment

Coulomb-meter (SI unit)

Dipole moment of nucleus and electron in
first Bohr orbit

$C\,m = 2.9979 \times 10^{11}$ esu $= 10$ emu
$= 0.8478 \times 10^{-29}$ C m
$= 2.5417 \times 10^{-18}$ esu

Magnetic field

Ampere-turn per meter (SI unit)

$= 4\pi \times 10^{-3}$ oersted [emu]
$= 3.767 \times 10^{8}$ esu

Gauss (in free space)

$= 1$ oersted $= 79.58$ amp-turn m^{-1}

Gamma

$\gamma = 10^{-5}$ oersted

Atomic unit ($m_e^{1/2} a_0^{-1/2} \tau_0^{-1}$)

$= 1.715 \times 10^{7}$ gauss

Field at nucleus due to electron in first Bohr orbit
$\alpha m_e^{1/2} a_0^{-1/2} \tau_0^{-1}$, $\tau_0 = \omega_0^{-1} = 2.4189 \times 10^{-17}$ s

$= 1.252 \times 10^{5}$ oersted

Magnetic flux density, Magnetic induction

Tesla (SI unit)

$= 10^{4}$ gauss
$= 1$ weber m^{-2}

Magnetic moment

Weber-meter (SI unit)

Atomic unit ($m_e^{1/2} a_0^{5/2} \tau_0^{-1}$)

Bohr magneton, magnetic moment of
electron in first Bohr orbit
$\frac{1}{2}\alpha m_e^{1/2} a_0^{5/2} \tau_0^{-1}$

$= (1/4\pi)\, 10^{10}$ emu $= 0.02654$ esu
$= 2.541 \times 10^{-18}$ erg gauss^{-1}
$\mu_B = 0.9274 \times 10^{-20}$ erg gauss^{-1}

Nuclear magneton
$\mu_B (m_e/m_p)$

$\mu_K = 5.051 \times 10^{-24}$ erg gauss^{-1}

Earth magnetic moment

$= 7.98 \times 10^{25}$ emu

Radioactivity

Curie [4]

Roentgen

Rad

$= 3.700 \times 10^{10}$ disintegrations s^{-1}
$=$ exposure to radiation producing
2.082×10^{9} ion pairs in
0.001293 g of air $= 1$ esu cm^{-3}
$= 2.58 \times 10^{-4}$ C kg^{-1}
$= 10^{-2}$ J kg^{-1}

2.6   ELECTRIC AND MAGNETIC UNIT RELATIONS

Table 2.4 on pages 24 and 25 is adapted from the previous *Astrophysical Quantities* edition. Many of these quantities are now superseded by the SI units, but the older esu and emu units are still frequently used for special cases in astrophysics.

For the SI units, the permittivity (ϵ_0) and permeability (μ_0) of free space are defined to be exact as $(1/4\pi c^2) \times 10^{11}$ F m^{-1} = $8.854\,187\,817\ldots$ F m^{-1} and $4\pi \times 10^{-7}$ N A^{-2}, respectively. Here F is the farad capitance unit, N is the newton force unit, and A is the ampere current unit.

REFERENCES

1. *Astrophysical Quantities.* **1**, §7
2. *Astrophysical Quantities*, **3**, §10
3. Abramowitz, M. & Stegun, I.A. 1965, *Handbook of Mathematical Functions*, (Dover, New York), p. 2
4. Anderson, H.L. 1989, *A Physicist's Desk Reference*, the second edition of *Physics Vade Mecum* (American Institute of Physics, New York)
5. Cohen, E.R., & Taylor, B.N. 1998, http://physics.nist.gov/cuu/Reference/versioncon.html
6. Seidelmann, P.K. 1992, *Explanatory Supplement to the Astronomical Almanac* (University Science Books, Mill Valley, CA)
7. Dehnen, W., & Binney, J.J. 1998, MNRAS, 298, 387
8. Metzger, M.R., Caldwell, J.A., & Schechter, P.L. 1998, AJ, 115, 635
9. Standish M. 1995, Report of the IAU WGAS Sub-Group on Numerical Standards, in *Highlights of Astronomy* edited by Appenzeller (Kluwer Academic, Dordrecht)
10. Fairhead L., Bretagnon, P. & Lestrade, J.F. 1998, IAU Symposium 128 (Kluwer, Dordrecht) p. 419
11. Hirayama, Th. et al. 1987, Proc. IAG Symposia I., IUGG XIX General Assembly, Vancouver
12. *Astronomical Almanacs* (USNO, GPO)
13. Liu, Bao-Lin, & Fiala, A.D. 1992, Canon of Lunar Eclipses 1500 B.C.–A.D. 3000 (Willmann-Bell, Richmond)
14. Nelson, R.A. 1998, Physics today, BG11

Table 2.4. *Electric and*

Quantity	SI symbol and unit		in esu
Charge	Q	coulomb C	$= c \times 10^{-1}$ esu
Current	I	ampere A	$= c \times 10^{-1}$ esu
Potential, EMF	V	volt V	$= (1/c) \times 10^{-12}$ esu
Electric field	$\mathcal{E}$	volt/m	$= (1/c) \times 10^{-14}$ esu
Resistance	R	ohm Ω	$= (1/c^2) \times 10^9$ esu
Resistivity	ρ	ohm m	$= (1/c^2) \times 10^{-11}$ esu
Conductance	G	siemens, mho	$= c^2 \times 10^{-9}$ esu
Conductivity	σ	mho/m	$= c^2 \times 10^{-11}$ esu
Capitance	C	farad F	$= c^2 \times 10^{-9}$ cm
Electric flux	Ψ	coulomb C	$= 4\pi c \times 10^{-1}$ esu
Electric flux density, displacement	D	coulomb/m^2	$= 4\pi c \times 10^{-5}$ esu
Polarization	P	coulomb/m^2	$= c \times 10^{-5}$ esu
Electric dipole moment		coulomb/m	$= c \times 10^1$ esu
Permittivity, dielectric constant	ϵ	farad/m	$= 4\pi c^2 \times 10^{-11}$ esu
Permittivity of free space	ϵ_0	$(1/4\pi c^2) \times 10^{11}$ F/m	$= 1$ esu
Inductance	L	henry H	$= (1/c^2) \times 10^9$ esu
Magnetic pole strength	m	weber Wb	$= (1/4\pi c) \times 10^8$ esu
Magnetic flux	Φ	weber Wb	$= (1/c) \times 10^8$ esu
Magnetic field	$\mathcal{H}$	ampere turn/m	$= 4\pi c \times 10^{-3}$ esu
Magnetomotive force, magnetic potential	$\mathcal{F}$	ampere turn AT	$= 4\pi c \times 10^{-1}$ esu
Magnetic dipole moment	M	weber m	$= (1/4\pi c) \times 10^{10}$ esu
Electromagnetic moment	m	ampere m^2	
Mag. flux density, induction	B	tesla T	$= (1/c) \times 10^4$ esu
Intensity of magnetization	J	weber/m^2 T	$= (1/4\pi c) \times 10^{16}$ esu
Magnetic energy density	$B \times \mathcal{H}$	joule/m^3	
Permeance	Λ	henry H	$= (1/4\pi c^2) \times 10^9$ esu
Reluctance		1/henry	$= 4\pi c^2 \times 10^{-9}$ esu
Permeability	μ	henry/m	$= (1/4\pi c^2) \times 10^7$ esu
Permeability of free space	μ_0	$4\pi \times 10^{-7}$ H/m	$= (1/c^2)$ esu

magnetic units.

in emu, etc.	ESU				EMU				esu/emu	SI			
	L	M	T	κ	L	M	T	μ		L	M	T	I
$= 10^{-1}$ emu	3/2	1/2	−1	1/2	1/2	1/2	0	−1/2	$1/c$	0	0	1	1
$= 10^{-1}$ emu	3/2	1/2	−2	1/2	1/2	1/2	−1	−1/2	$1/c$	0	0	0	1
$= 10^{8}$ emu	1/2	1/2	−1	−1/2	3/2	1/2	−2	1/2	c	2	1	−3	−1
$= 10^{6}$ emu	−1/2	1/2	−1	−1/2	1/2	1/2	−2	1/2	c	1	1	−3	−1
$= 10^{9}$ emu	−1	0	1	−1	1	0	−1	1	c^2	2	1	−3	−2
$= 10^{11}$	0	0	1	−1	2	0	−1	1	c^2	3	1	−3	−2
$= 10^{-9}$ emu	1	0	−1	1	−1	0	1	−1	$1/c^2$	−2	−1	3	2
$= 10^{-11}$ emu	0	0	−1	1	−2	0	1	−1	$1/c^2$	−3	−1	3	2
$= 10^{-9}$ emu	1	0	0	1	−1	0	2	−1	$1/c^2$	−2	−1	4	2
$= 4\pi \times 10^{-1}$ emu	3/2	1/2	−1	1/2	1/2	1/2	0	−1/2	$1/c$	0	0	1	1
$= 4\pi \times 10^{-5}$ emu	−1/2	1/2	−1	1/2	−3/2	1/2	0	−1/2	$1/c$	−2	0	1	1
$= 10^{-5}$ emu	−1/2	1/2	−1	1/2	−3/2	1/2	0	−1/2	$1/c$	−2	0	1	1
$= 10$ emu	5/2	1/2	−1	1/2	3/2	1/2	0	−1/2	$1/c$	1	0	1	1
$= 4\pi \times 10^{-11}$ emu	0	0	0	1	−2	0	2	−1	$1/c^2$	−3	−1	4	2
$= (1/c^2)$ emu									$1/c^2$				
$= 10^{9}$ cm	−1	0	2	−1	1	0	0	1	c^2	2	1	−2	−2
$= (1/4\pi) \times 10^{8}$ emu	1/2	1/2	0	−1/2	3/2	1/2	−1	1/2	c	2	1	−2	−1
$= 10^{8}$ maxwell (Mx)	1/2	1/2	0	−1/2	3/2	1/2	−1	1/2	c	2	1	−2	−1
$= 4\pi \times 10^{-3}$ oersted (Oe)	1/2	1/2	−2	1/2	−1/2	1/2	−1	−1/2	$1/c$	−1	0	0	1
$= 4\pi \times 10^{-1}$ gilbert (Gb)	3/2	1/2	−2	1/2	1/2	1/2	−1	−1/2	$1/c$	0	0	0	1
$= (1/4\pi) \times 10^{10}$ emu	3/2	1/2	0	−1/2	5/2	1/2	−1	1/2	c	3	1	−2	−1
$= 10^{3}$ emu	7/2	1/2	−2	1/2	5/2	1/2	−1	−1/2	$1/c$	2	0	0	1
$= 10^{4}$ gauss (Gs)	−3/2	1/2	0	−1/2	−1/2	1/2	−1	1/2	c	0	1	−2	−1
$= (1/4\pi) \times 10^{4}$ emu	−3/2	1/2	0	−1/2	−1/2	1/2	−1	1/2	c	0	1	−2	−1
$= 40\pi$ Gs Oe	−1	1	−2	0	−1	1	−2	0	1	−1	1	−2	0
$= (1/4\pi) \times 10^{9}$ Mx/Gb	−1	0	2	−1	1	0	0	1	c^2	2	1	−2	−2
$= 4\pi \times 10^{-9}$ Gb/Mx	1	0	−2	1	−1	0	0	−1	$1/c^2$	−2	−1	2	2
$= (1/4\pi) \times 10^{7}$ emu	−2	0	2	−1	0	0	0	1	c^2	1	1	−2	−2
$= 1$ emu									c^2				

Chapter 3

Atoms and Molecules

Werner Däppen

3.1 ONLINE DATABASES AND OTHER SOURCES

The National Institute of Standards and Technology (NIST) gives access to extensive physical and atomic data (http://physics.nist.gov). The Plasma Laboratory of the Weizmann Institute (http://plasma-gate.weizmann.ac.il) and the Southwest Research Institute (http://espsun.space.swri.edu/spacephysics/www.atomic.html) provide, besides their own data, many useful links to other databases. For astrophysical applications, among the most extensive databases are those of the Harvard–Smithsonian Center for Astrophysics (http://cfa-www.harvard.edu/amp/data) (giving, e.g., searchable access to the data by R.L. Kurucz and R.L. Kelly) and of the Opacity Project (http://astro.u-strasbg.fr/OP) (with monochromatic opacities, collision strengths, and other atomic data). A further source of important data is the Iron Project (http://www.am.qub.ac.uk/projects/iron). Gary Ferland's Web Page (http://www.pa.uky.edu/gary/cloudy) has references to CLOUDY ("Photoionization Simulations for the Discriminating Astrophysicist"), which contains pointers to the atomic databases they use and maintain (e.g., http://www.pa.uky.edu/verner/atom.html, "Atomic Data for Astrophysics"). The CHIANTI group (http://www.solar.nrl.navy.mil/chianti) has installed a database with information suitable for extreme-UV applications. The Particle Data Group (http://pdg.lbl.com) makes available periodically its newest releases of particle properties. Other sources of information are the recent *Atomic, Molecular, and Optical Physics Handbook* [1], the results of the work of the Collaborative Computational Project No. 7 (United Kingdom) [2], and the *Handbook of Chemistry and Physics* [3].

3.2   ELEMENTS, ATOMIC MASS, AND SOLAR-SYSTEM ABUNDANCE

Atomic masses (weighted by the fractional abundances of the stable isotopes in normal terrestrial composition [4]) are scaled to ^{12}C $= 12.00$. Standard values abridged to five significant digits are given (from the International Union of Pure and Applied Chemistry (IUPAC); see [5]). For some elements, atomic masses can be accurately measured to seven or more significant figures. IUPAC regularly publishes these values irrespective of interest to any user. For many users, however, it is often desirable that the published data remain valid over an extended period, which is helpful for textbooks and numerical tables derived from atomic-mass data. IUPAC has recognized this need and approved the use of the designation *standard* to its abridged atomic-mass table, with the hope that the quoted values may survive for at least a decade.

The *solar-system abundances* (formerly denoted as *cosmic abundances*) are expressed logarithmically on a scale for which H is 12.00 dex. The intention is that they express cosmic abundance [6]. Thus, abundances are taken mainly from meteorites and the Sun's photosphere. In both cases, values by number are quoted. The agreement between meteoritic and solar data has improved remarkably since the 1970s. Discrepancies have mostly gone away as the solar values—thanks especially to improved transition probabilities and other atomic data—have become more accurate [6]. The two principal exceptions are the solar photospheric Li and Be abundances that are smaller than the meteoritic ones by 2.15 and 0.27 dex, respectively. The reason is that these elements are destroyed by nuclear reactions at the bottom of the solar convection zone. For most other elements the agreement is better than ± 0.04 dex (for this, and exceptions, see [6]). In the case of iron, a previous controversy has been solved: the solar and meteoritic values agree now [7]. For details on isotopic abundances, see [1] and [4].

The group abundance ratios given in Table 3.1 are derived from Table 3.2. The H ratio is set to 100.

Table 3.1. *Group abundance ratios.*

Element group	Number	Mass	Stripped electrons
H	100	100	100
He	9.8	39	20
C, N, O, Ne	0.145	2.19	1.1
Other	0.013	0.44	0.21
Total	109.96	141.63	121.3

The composition by mass [2] is as follows:

fraction of H, X	$0.707 \pm 2.5\%$
fraction of He, Y	$0.274 \pm 6\%$
fraction of other atoms, Z	$0.0189 \pm 8.5\%$

Mean atomic mass of cosmic material	1.30
Mean atomic mass per H atom	1.41
Mean atomic mass for fully ionized cosmic plasma	0.62

Table 3.2. *Atomic masses and solar-system abundances.*

Element	Symbol [1]	Atomic number	Atomic mass	Log abundance [2]	
				Meteoritic	Solar
Hydrogen	H	1	1.0079	12.00^a	12.00
Helium [3]	He	2	4.0026	10.99^a	10.99^b
Lithium	Li	3	6.941	3.31	1.16
Beryllium	Be	4	9.0122	1.42	1.15
Boron	B	5	10.811	2.8	2.6^c
Carbon	C	6	12.011	8.56^a	8.56
Nitrogen	N	7	14.007	8.05^a	8.05
Oxygen	O	8	15.999	8.93^a	8.93
Fluorine	F	9	18.998	4.48	4.56
Neon	Ne	10	20.180	8.09	8.09^d
Sodium	Na	11	22.990	6.31	6.33
Magnesium	Mg	12	24.305	7.58	7.58
Aluminum	Al	13	26.982	6.48	6.47
Silicon	Si	14	28.086	7.55	7.55
Phosphorus	P	15	30.974	5.57	5.45
Sulphur	S	16	32.066	7.27	7.21
Chlorine	Cl	17	35.453	5.27	5.5
Argon	Ar	18	39.948	6.56^d	6.56^d
Potassium	K	19	39.098	5.13	5.12
Calcium	Ca	20	40.078	6.34	6.36
Scandium	Sc	21	44.956	3.09	3.10
Titanium	Ti	22	47.88	4.93	4.99
Vanadium	V	23	50.942	4.02	4.00
Chromium	Cr	24	51.996	5.68	5.67
Manganese	Mn	25	54.938	5.53	5.39
Iron [2]	Fe	26	55.847	7.51	7.54
Cobalt	Co	27	58.933	4.91	4.92
Nickel	Ni	28	58.693	6.25	6.25
Copper	Cu	29	63.546	4.27	4.21
Zinc	Zn	30	65.39	4.65	4.60
Gallium	Ga	31	69.723	3.13	2.88
Germanium	Ge	32	72.61	3.63	3.41
Arsenic	As	33	74.922	2.37	
Selenium	Se	34	78.96	3.35	
Bromine	Br	35	79.904	2.63	
Krypton	Kr	36	83.80	3.23	
Rubidium	Rb	37	85.468	2.40	2.60
Strontium	Sr	38	87.62	2.93	2.90
Yttrium	Y	39	88.906	2.22	2.24
Zirconium	Zr	40	91.224	2.61	2.60
Niobium	Nb	41	92.906	1.40	1.42
Molybdenum	Mo	42	95.94	1.96	1.92
Technetium	Tc	43	98.906		
Ruthenium	Ru	44	101.07	1.82	1.84
Rhodium	Rh	45	102.91	1.09	1.12

Table 3.2. *(Continued.)*

Element	Symbol [1]	Atomic number	Atomic mass	Log abundance [2]	
				Meteoritic	Solar
Palladium	Pd	46	106.42	1.70	1.69
Silver	Ag	47	107.87	1.24	0.94[c]
Cadmium	Cd	48	112.41	1.76	1.86
Indium	In	49	114.82	0.82	1.66[c]
Tin	Sn	50	118.71	2.14	2.0
Antimony	Sb	51	121.76	1.04	1.0
Tellurium	Te	52	127.60	2.24	
Iodine	I	53	126.90	1.51	
Xenon	Xe	54	131.29	2.23	
Cesium	Cs	55	132.91	1.12	
Barium	Ba	56	137.33	2.21	2.13
Lanthanum	La	57	138.91	1.20	1.22
Cerium	Ce	58	140.12	1.61	1.55
Praseodymium	Pr	59	140.91	0.78	0.71
Neodymium	Nd	60	144.24	1.47	1.50
Promethium	Pm	61	146.92		
Samarium	Sm	62	150.36	0.97	1.00
Europium	Eu	63	151.96	0.54	0.51
Gadolinium	Gd	64	157.25	1.07	1.12
Terbium	Tb	65	158.93	0.33	−0.1
Dysprosium	Dy	66	162.50	1.15	1.1
Holmium	Ho	67	164.93	0.50	0.26[c]
Erbium	Er	68	167.26	0.95	0.93
Thulium	Tm	69	168.93	0.13	0.00[c]
Ytterbium	Yb	70	170.04	0.95	1.08
Lutetium	Lu	71	174.97	0.12	0.76[c]
Hafnium	Hf	72	178.49	0.73	0.88
Tantalum	Ta	73	180.95	0.13	
Tungsten	W	74	183.85	0.68	1.11[c]
Rhenium	Re	75	186.21	0.27	
Osmium	Os	76	190.2	1.38	1.45
Iridium	Ir	77	192.22	1.37	1.35
Platinum	Pt	78	195.08	1.68	1.8
Gold	Au	79	196.97	0.83	1.01[c]
Mercury	Hg	80	200.59	1.09	
Thallium	Tl	81	204.38	0.82	0.9[c]
Lead	Pb	82	207.2	2.05	1.85
Bismuth	Bi	83	208.98	0.71	
Polonium	Po	84	209.98		
Astatine	At	85	209.99		
Radon	Rn	86	222.02		
Francium	Fr	87	223.02		
Radium	Ra	88	226.03		
Actinium	Ac	89	227.03		
Thorium	Th	90	232.04	0.08	0.12
Protactinium	Pa	91	231.04		
Uranium	U	92	238.03	−0.49	< −0.45[c]
Neptunium	Np	93	237.05		
Plutonium	Pu	94	239.05		
Americium	Am	95	241.06		

Table 3.2. *(Continued.)*

Element	Symbol [1]	Atomic number	Atomic mass	Log abundance [2]	
				Meteoritic	Solar
Curium	Cm	96	244.06		
Berkelium	Bk	97	249.08		
Californium	Cf	98	252.08		
Einsteinium	Es	99	252.08		
Fermium	Fm	100	257.10		
Mendelevium	Md	101	258.10		
Nobelium	No	102	259.10		
Lawrencium	Lr	103	262.11		

Notes

[a] Based on solar data.

[b] Based on stellar observations and solar models [1, 3, 4].

[c] Uncertain.

[d] Based on other astronomical data.

References

1. IUPAC 1969, *Comptes Rendus XXV Conference*, p. 95
2. Anders, E., & Grevesse, N. 1989, *Geochim. Cosmochim. Acta*, **53**, 197; Grevesse, N., & Noels, A. 1993, in *Origin and Evolution of the Elements*, edited by N. Prantzos, E. Vangioni, & M. Cassé (Cambridge University Press, Cambridge), p. 15
3. Christensen-Dalsgaard, J., Däppen, W., & the GONG Team 1996, *Science*, **272**, 1286
4. Biémont, E., Baudoux, M., Kurucz, R.L., Ansbacher, W., & Pinnington, E.H. 1991, *A&A*, **249**, 539
5. Kosovichev, A.G., Christensen-Dalsgaard, J., Däppen, W., Dziembowski, W.A., Gough, D.O., & Thompson, M.J. 1992, *MNRAS*, **259**, 536

3.3 EXCITATION, IONIZATION, AND PARTITION FUNCTIONS

3.3.1 Introduction

Finding the occupation of individual levels of atoms and ions and the fractions of ions of any given chemical element in a plasma is a complex task. The difficulty arises from the interaction of the plasma with the atoms. Therefore, in principle, the problems of modified atoms and of their statistical occupation should be solved simultaneously and self-consistently. The typical task of quantum-statistical mechanics consists of the calculation of a density operator (ensemble) for the system of all particles. The partition function, i.e., the trace over the density operator, not only gives the occupation of all states, but it also leads to a thermodynamical potential.

It is evident that various approximations are necessary before this procedure can be carried out. One such approximation consists of treating the motion of the heavy particles (nuclei, atoms, ions) according to classical mechanics. Once the heavy particles are separated out, quantum-mechanical electrons remain. In the treatment of electrons, we find a bifurcation into two distinct classes of approach, the "chemical picture" and the "physical picture." While in the more conventional chemical picture, bound configurations (atoms, ions, and molecules) are introduced and treated as new and independent species, only *fundamental* particles (electrons and nuclei) appear in the physical picture. In the chemical picture, *reactions* between the various species occur. Thus the thermodynamical equilibrium must be sought among the stoichiometrically allowed set of concentration variables by means of a maximum entropy (or minimum free-energy) principle. In contrast, the physical picture has the aesthetic advantage that there is no need for a minimax principle. The question of bound states is dealt with implicitly through the Hamiltonian describing the interaction between the fundamental particles.

It is obvious that these self-consistent approaches require extensive analytical and numerical work. For a recent realization of the chemical-picture approach, see, e.g., [8–10]. For the physical picture, the most detailed work so far was done as part of the OPAL opacity project [11–14]. In the OPAL project, the physical picture was not only used to model excitation and ionization processes, but for the first time also to yield the highly accurate thermodynamic quantities needed in computations of stellar models [15]. The book by Ebeling et al. [16] contains further information and references on the physical picture. The most recent addition to the set of stellar equations of state is based on the formalism of the path integral in the framework of the Feynman–Kac representation. This formalism leads to a virial expansion of the thermodynamic functions in the power of the total density of a Coulomb plasma ([17, 18], and references therein).

For many lower-density applications, especially stellar spectroscopy, adequate qualitative and quantitative information can be extracted from simpler considerations, in which atoms are assumed to have an unperturbed structure. In this case, excitation fractions are given by the Boltzmann factor, and the ionization degree follows from the Saha equation, which is the mass-action law for the ionization reaction. The Saha equation contains the internal partition functions for bound systems. A fundamental theoretical flaw of this approximate approach is that isolated atoms would have infinite partition functions because of their infinite number of excited states. Many heuristic recipes to truncate partition functions exist. However, only the physical picture comes to a satisfactory solution, which then can often be used to justify the intuitive concepts [19]. In many cases, neglecting all excited states, that is, assuming ground-state-only internal partition functions, is a reasonable approximation. The following simple treatment of excitation, ionization, and partition functions is, with reasonable care, still very useful for many qualitative and semiquantitative astrophysical applications.

3.3.2   Approximate Methods and Results

For practical applications, a useful introduction to the statistical mechanics of plasmas is the book by Eliezer et al. [20]. The number of atoms existing in various atomic levels $0, 1, 2, \ldots$ when in thermal equilibrium at temperature T is approximately described by the Boltzmann distribution

$$N_2/N_1 = (g_2/g_1)\exp(-\chi_{1,2}/kT),$$
$$N_2/N = (g_2/U)\exp(-\chi_{0,2}/kT).$$

Numerically

$$\log(N_2/N_1) = \log(g_2/g_1) - \chi_{12}(5040/T) \quad (\chi_{12} \text{ in eV}),$$

where N is the total number of atoms per cm^3, N_0, N_1, and N_2 are the numbers of atoms per cm^3 in the zero and higher levels, g_0, g_1, and g_2 are the corresponding statistical weights, $\chi_{1,2}$ is the potential difference between levels 1 and 2, and U is the partition function.

The degree of ionization in conditions of thermal equilibrium is given by the Saha equation

$$\frac{N_{Y+1}}{N_Y}P_e = \frac{U_{Y+1}}{U_Y}2\frac{(2\pi m)^{3/2}(kT)^{5/2}}{h^3}\exp(-\chi_{Y,Y+1}/kT).$$

Numerically

$$\log\left(\frac{N_{Y+1}}{N_Y}P_e\right) = -\chi_{Y,Y+1}\frac{5040}{T} + \tfrac{5}{2}\log T - 0.4772 + \log\left(\frac{2U_{Y+1}}{U_Y}\right)$$

or

$$\log\left(\frac{N_{Y+1}}{N_Y}N_e\right) = -\chi_{Y,Y+1}\Theta - \tfrac{3}{2}\log\Theta + 20.9366 + \log\left(\frac{2U_{Y+1}}{U_Y}\right),$$

where N_Y and N_{Y+1} are the numbers of atoms per cm^3 in the Y and $Y+1$ stages of ionization ($Y = 1$, neutral; $Y = 2$nd, 1st ion; etc.), N_e is the number of electrons per cm^3, P_e is the electron pressure in $dyn\,cm^{-2}$, $\chi_{Y,Y+1}$ the ionization potential in eV from the Y to the $Y+1$ stage of ionization, $\Theta = 5040$ K/T, U_Y and U_{Y+1} are the partition functions, and the factor 2 represents the statistical weight of an electron.

The degree of ionization, when ionizations are caused by electron collisions and recombinations are radiative, can be approximately given by

$$N_{Y+1}/N_Y = S/\alpha,$$

where the effect of both collisional ionizations from state of ionization $Y+1$ and of recombinations of $Y+2$ in the abundance of ions in $Y+1$ is neglected, and the possibility of multiple-ionization events is excluded. In the formula, S is the collision ionization coefficient (such that SN_eN_Y = rate of collisional ionization, see Sec. 3.6), and α is the recombination coefficient (such that αN_eN_{Y+1} = rate of recombination, see Chap. 5).

Detailed calculations of partition functions are given by Irwin [21] (atoms and molecules) and Sauval and Tatum [22] (molecules). However, for the approximate purposes of this section, the partition function may simply be regarded as the effective statistical weight of the atom or ion under existing conditions of excitation. Except in extreme conditions it is approximately equal to the weight of the lowest ground term. The ground term weight g_0 is therefore given and this can normally be extrapolated along the isoelectronic sequences to give the approximate partition function for any ion. The partition functions, given in Table 3.3 in the form $\log U$ for $\Theta = 1.0$ and 0.5, are *not* intended to include the concentration of terms close to each series limit. The part of the partition function associated with these high-n terms is dependent on both T and P_e. This part is usually negligible unless the atom concerned is mainly ionized in which case the high-n terms may be counted statistically with the ion.

Lowering of $\chi_{Y,Y+1}$ in the Saha equation to allow for the merging of high-level spectrum lines gives [23]

$$\Delta\chi_{Y,Y+1} = 7.0 \times 10^{-7}N_e^{1/3}Y^{2/3},$$

with $\Delta\chi$ in eV and N_e in cm^{-3}, and where Y is the charge on the $Y+1$ ion.

Table 3.3. *Partition function* [1–3].

| | | | $Y = \mathrm{I}$ | | | $Y = \mathrm{II}$ | | $Y = \mathrm{III}$ |
| | | | $\log U$ | | | $\log U$ | | |
	Element	g_0	$\Theta = 1.0$	$\Theta = 0.5$	g_0	$\Theta = 1.0$	$\Theta = 0.5$	g_0
1	H	2	0.30	0.30	1	0.00	0.00	
2	He	1	0.00	0.00	2	0.30	0.30	1
3	Li	2	0.32	0.49	1	0.00	0.00	2
4	Be	1	0.01	0.13	2	0.30	0.30	1
5	B	6	0.78	0.78	1	0.00	0.00	2
6	C	9	0.97	1.00	6	0.78	0.78	1
7	N	4	0.61	0.66	9	0.95	0.97	6
8	O	9	0.94	0.97	4	0.60	0.61	9
9	F	6	0.75	0.77	9	0.92	0.94	4
10	Ne	1	0.00	0.00	6	0.73	0.75	9

Table 3.3. *(Continued.)*

	Element	g_0	Y = I log U $\Theta = 1.0$	Y = I log U $\Theta = 0.5$	g_0	Y = II log U $\Theta = 1.0$	Y = II log U $\Theta = 0.5$	Y = III g_0
11	Na	2	0.31	0.60	1	0.00	0.00	6
12	Mg	1	0.01	0.15	2	0.31	0.31	1
13	Al	6	0.77	0.81	1	0.00	0.01	2
14	Si	9	0.98	1.04	6	0.76	0.77	1
15	P	4	0.65	0.79	9	0.91	0.94	6
16	S	9	0.91	0.94	4	0.62	0.72	9
17	Cl	6	0.72	0.75	9	0.89	0.92	4
18	Ar	1	0.00	0.00	6	0.69	0.71	9
19	K	2	0.34	0.60	1	0.00	0.00	6
20	Ca	1	0.07	0.55	2	0.34	0.54	1
21	Sc	10	1.08	1.49	15	1.36	1.52	10
22	Ti	21	1.48	1.88	28	1.75	1.92	21
23	V	28	1.62	2.03	25	1.64	1.89	28
24	Cr	7	1.02	1.51	6	0.86	1.22	25
25	Mn	6	0.81	1.16	7	0.89	1.13	6
26	Fe	25	1.43	1.74	30	1.63	1.80	25
27	Co	28	1.52	1.76	21	1.46	1.66	28
28	Ni	21	1.47	1.60	10	1.02	1.28	21
29	Cu	2	0.36	0.58	1	0.01	0.18	10
30	Zn	1	0.00	0.03	2	0.30	0.30	1
31	Ga	6	0.73	0.77	1	0.00	0.00	2
32	Ge	9	0.91	1.01	6	0.64	0.70	1
34	Se	9	0.83	0.89	4			9
36	Kr	1	0.00	0.00	6	0.62	0.66	9
37	Rb	2	0.36	0.7	1	0.00	0.00	6
38	Sr	1	0.10	0.70	2	0.34	0.53	1
39	Y	10	1.08	1.50	1 + 15	1.18	1.41	10
40	Zr	21	1.53	1.99	28	1.66	1.91	21
48	Cd	1	0.00	0.02	2	0.30	0.30	1
50	Sn	9	0.73	0.88	6	0.52	0.61	1
56	Ba	1	0.36	0.92	2	0.62	0.85	1
57	La	10	1.41	1.85	21	1.47	1.71	10
70	Yb	1	0.02	0.21	2	0.30	0.31	
82	Pb	9	0.26	0.54	6	0.32	0.40	1

References

1. *Astrophysical Quantities*, **1**, §15; **2**, §15
2. Cayrel, R., & Jugaku, J. 1963, *Ann. d'Astrophys.*, **26**, 495
3. Bolton, C.T. 1970, *ApJ*, **161**, 1187

The degree of ionization in the material of stellar atmospheres is given in Table 3.4, relating gas pressure P_g, electron pressure P_e, and temperature T. The data are averaged from [24] (rather high heavy-element abundance) and [25] (rather low heavy-element abundance).

Table 3.4. log P_g.

		Θ and T							
	Θ	0.1	0.2	0.4	0.6	0.8	1.0	1.2	1.4
log P_e	T (K)	50 400	25 200	12 600	8 400	6 300	5 040	4 200	3 600
-2		-1.9	-1.8	-1.70	-1.67	-1.54	$+0.78$	$+2.0$	$+2.4$
-1		-0.8	-0.74	-0.70	-0.66	-0.01	2.57	3.1	3.9
0		$+0.27$	$+0.29$	$+0.31$	$+0.35$	$+1.90$	3.9	4.5	5.3
1		1.27	1.30	1.33	1.47	3.87	5.2	6.0	6.7
2		2.27	2.30	2.34	2.98	5.65	6.7	7.7	8.5
3		3.28	3.30	3.35	4.87	7.0	8.3	9.4	10.4
4		4.28	4.31	4.43	6.84	8.7	10.0	11.2	12.4
5		5.59	5.30	5.87	8.66	10.4	11.8	13.2	14.4

3.4 IONIZATION POTENTIALS

Table 3.5 gives the energy in eV required to ionize each element to the next stage of ionization. I ($Y = 1$) denotes the neutral atom, II the first ion, etc. Dividing lines between shells and subshells are added to assist interpolation. Part of the data is based on an especially accurate compilation for selected ions [6–20], made available by the National Institute of Standards and Technology (NIST, see Sec. 3.1 for online access). If the data are given in wave numbers, the currently recommended conversion factor to energy is 1 eV $= 8\,065.541$ cm^{-1} [26].

3.5 ELECTRON AFFINITIES

Electron affinity is the energy difference between the lowest state of the atom (or molecule or ion) and the lowest state of the corresponding negative ion (see Table 3.6). It is positive for those atoms or molecules that form stable negative ions. Regarding the astrophysically important H$^-$, it was thought earlier that a second stable state exists [27]. Later, however, it was proven rigorously that there is only one stable state [28, 29].

3.6 ATOMIC CROSS SECTIONS FOR ELECTRONIC COLLISIONS

Definitions of symbols are presented below:

Q	Atomic cross section $[= Q(v)]$.
v	Precollision electron velocity.
πa_0^2	Atomic unit cross section $= 8.797 \times 10^{-17}$ cm^2.
N_e, N_a, N_i	Electron, atom, ion densities (per cm^3).
$L = \overline{vQ}$	Collision rate for each atom per unit N_e.
$N_e L$	Collision rate per atom (or ion).
$N_e N_a L$	Collision rate per cm^3.
P_c	Collisions encountered by an electron per cm at 0° C and 1 mm Hg. pressure, then $Q = 2.828 \times 10^{-17} P_c = 0.321\,5\,\pi a_0^2 P_c$.

Table 3.5. *Ionization potentials (electron volts)* [1–20].

	Atom	Stage of ionization

	Atom	I	II	III	IV	V	VI	VII	VIII	IX	X	XI	XII	XIII	XIV
1	H	13.59844													
2	He	24.58741	54.41778												
3	Li	5.39172	75.64018	122.454											
4	Be	9.32263	18.21116	153.897	217.713										
5	B	8.29803	25.15484	37.931	259.366	340.22									
6	C	11.26030	24.38332	47.888	64.492	392.08	489.98								
7	N	14.53414	29.6013	47.449	77.472	97.89	552.06	667.03							
8	O	13.61806	35.11730	54.936	77.413	113.90	138.12	739.29	871.41						
9	F	17.42282	34.97082	62.708	87.140	114.24	157.17	185.19	953.91	1103.1					
10	Ne	21.56454	40.96328	63.45	97.12	126.21	157.93	207.28	239.10	1195.8	1362.2				
11	Na	5.13908	47.2864	71.620	98.91	138.40	172.18	208.50	264.25	299.9	1465.1	1648.7			
12	Mg	7.64624	15.03528	80.144	109.265	141.27	186.76	225.02	265.96	328.1	367.5	1761.8	1963		
13	Al	5.98577	18.82856	28.448	119.99	153.83	190.49	241.76	284.66	330.1	398.8	442.0	2086	2304	
14	Si	8.15169	16.34585	33.493	45.142	166.77	205.27	246.49	303.54	351.1	401.4	476.4	523	2438	2673
15	P	10.48669	19.7694	30.203	51.444	65.03	220.42	263.57	309.60	372.1	424.4	479.5	561	612	2817
16	S	10.36001	23.3379	34.79	47.222	72.59	88.05	280.95	328.75	379.6	447.5	504.8	564	652	707
17	Cl	12.96764	23.814	39.61	53.465	67.8	97.03	114.20	348.28	400.1	455.6	529.3	592	657	750
18	Ar	15.75962	27.62967	40.74	59.81	75.02	91.01	124.32	143.46	422.5	478.7	539.0	618	686	756
19	K	4.34066	31.63	45.806	60.91	82.66	99.4	117.56	154.88	175.8	503.8	564.7	629	715	787
20	Ca	6.11316	11.87172	50.913	67.27	84.50	108.78	127.2	147.24	188.5	211.3	591.9	657	727	818
21	Sc	6.56144	12.79967	24.757	73.489	91.65	111.68	138.0	158.1	180.0	225.2	249.8	688	757	831
22	Ti	6.8282	13.5755	27.492	43.267	99.30	119.53	140.8	170.4	192.1	215.9	265.1	292	788	863
23	V	6.7463	14.66	29.311	46.71	65.28	128.1	150.6	173.4	205.8	230.5	255.1	308	336	896
24	Cr	6.76664	16.4857	30.96	49.16	69.46	90.64	161.18	184.7	209.3	244.4	270.7	298	355	384
25	Mn	7.43402	15.63999	33.668	51.2	72.4	95.6	119.20	194.5	221.8	248.3	286.0	314	344	404
26	Fe	7.9024	16.1878	30.652	54.8	75.0	99.1	124.98	151.06	233.6	262.1	290.2	331	361	392
27	Co	7.8810	17.083	33.50	51.3	79.5	103	131	160	186.2	276.2	305	336	379	411
28	Ni	7.6398	18.16884	35.19	54.9	75.5	108	134	164	193	224.6	321	352	384	430
29	Cu	7.72638	20.29240	36.841	55.2	79.9	103	139	167	199	232	266	369	401	435
30	Zn	9.39405	17.96440	39.723	59.4	82.6	108	136	175	203	238	274	311	412	454

Table 3.5. (*Continued.*)

Atom		XV	XVI	XVII	XVIII	XIX	XX	XXI	XXII	XXIII	XXIV	XXV	XXVI	XXVII	XXVIII
1	H														
2	He														
3	Li														
4	Be														
5	B														
6	C														
7	N														
8	O														
9	F														
10	Ne														
11	Na														
12	Mg														
13	Al														
14	Si														
15	P	3 070													
16	S	3 224	3 494												
17	Cl	809	3 658	3 946											
18	Ar	855	918	4 121	4 426										
19	K	862	968	1 034	4 611	4 934									
20	Ca	895	974	1 087	1 157	5 129	5 470								
21	Sc	927	1 009	1 094	1 213	1 288	5 675	6 034							
22	Ti	941	1 044	1 131	1 221	1 346	1 425	6 249	6 626						
23	V	975	1 060	1 168	1 260	1 355	1 486	1 569	6 851	7 246					
24	Cr	1 011	1 097	1 185	1 299	1 396	1 496	1 634	1 721	7 482	7 895				
25	Mn	435	1 136	1 224	1 317	1 437	1 539	1 644	1 788	1 879	8 141	8 572			
26	Fe	457	489	1 266	1 358	1 456	1 582	1 689	1 799	1 950	2 045	8 828	9 278		
27	Co	444	512	547	1 402	1 500	1 602	1 734	1 846	1 962	2 119	2 218	9 544	10 030	
28	Ni	464	499	571	607	1 546	1 648	1 756	1 894	2 010	2 131	2 295	2 398	10 280	10 790
29	Cu	484	520	557	633	671	1 698	1 804	1 919	2 060	2 182	2 310	2 478	2 560	11 050
30	Zn	490	542	579	619	698	738	1 856	1 970	2 088	2 234	2 363	2 495	2 660	2 730

Stage of ionization

Table 3.5. *(Continued.)*

Atom		I	II	III	IV	V	VI	VII	VIII	IX	X
		stage									

Atom		I	II	III	IV	V	VI	VII	VIII	IX	X
31	Ga	5.999 30	20.514	30.71	64	87	116	140	170	212	243
32	Ge	7.900	15.935	34.224	45.71	93.5	112	144	174	207	250
33	As	9.815 2	18.633	28.351	50.13	62.63	127.6	147	179	212	242
34	Se	9.752 38	21.19	30.820	42.944	68.3	81.7	155.4	184	218	250
35	Br	11.813 81	21.8	36	47.3	59.7	88.6	103.0	192.8	224	257
36	Kr	13.999 61	24.360	36.95	52.5	64.7	78.5	111.0	126	230.9	263
37	Rb	4.177 13	27.285	40	52.6	71.0	84.4	9.2	136	150	277.1
38	Sr	5.694 84	11.030	42.89	57	71.6	90.8	106	122.3	162	177
39	Y	6.217	12.24	20.52	61.8	77.0	93	116	129	146.2	191
40	Zr	6.633 90	13.13	22.99	34.34	81.5	99	117	140	155	
41	Nb	6.758 85	14.32	25.04	38.3	50.55	102.6	125	142	161	
42	Mo	7.092 43	16.16	27.13	46.4	61.2	68	126.8	153	163	
43	Tc	7.28	15.26	29.54	46	55	80			187	
44	Ru	7.360 50	16.76	28.47	50	60	92				
45	Rh	7.458 90	18.08	31.06	48	65	97				
46	Pd	8.336 9	19.43	32.93	53	62	90	110	130	155	180
47	Ag	7.576 24	21.49	34.83	56	68	89	115	140	160	185
48	Cd	8.993 67	16.908	37.48	59	72	94	115	145	170	195
49	In	5.786 36	18.870	28.03	54.4	77	98	120	145	180	205
50	Sn	7.343 81	14.632	30.503	40.734	72.28	103	125	150	175	210
51	Sb	8.64	16.531	25.3	44.2	56	108	130	155	185	210
52	Te	9.009 6	18.6	27.96	37.41	58.75	70.7	137	165	190	220
53	I	10.451 26	19.131	33	42	66	81	100	170	200	230
54	Xe	12.129 87	21.21	32.123	46	57	82	100	120	210	240
55	Cs	3.893 9	23.157	35	46	62	74	100	120	145	250
56	Ba	5.211 70	10.004		49	62	80	95	120	145	160
57	La	5.577 0	11.06	19.177	52	66	80	100	115	145	165
58	Ce	.538 7	10.85	20.198	36.72	70	85	100	120	140	165
59	Pr	5.464	10.55	21.624	38.95	57.45	89	105	120	145	160
60	Nd	5.525 0	10.73					110	130	150	170
61	Pm	5.55	10.90						135	155	175
62	Sm	5.643 7	11.07							160	180
63	Eu	5.670 4	11.241								190
64	Gd	6.150 0	12.09								
65	Tb	5.863 9	11.52								
66	Dy	5.938 9	11.67								
67	Ho	6.021 6	11.80								
68	Er	6.107 8	11.93								
69	Tm	6.184 31	12.05	23.68							
70	Yb	6.254 16	12.176	25.05							
71	Lu	5.425 85	13.9	20.959							
72	Hf	6.825 07	14.9	23.3	33.3						
73	Ta	7.89	16	22	33	45					
74	W	7.98	18	24	35	48	61				
75	Re	7.88	17	26	38	51	64	79			
76	Os	8.7	17	25	40	54	68	83	100		
77	Ir	9.1	17	27	39	57	72	88	105	120	
78	Pt	9.0	18.563	28	41	55	75	92	110	125	145
79	Au	9.225 67	20.5	30	44	58	73	96	115	135	155
80	Hg	10.437 50	18.756	34.2	46	61	77	94	120	140	160
81	Tl	6.108 29	20.428	29.83	50.7	64	81	98	115	145	165
82	Pb	7.416 66	15.032	31.937	42.32	68.8	84	103	120	140	175
83	Bi	7.289	16.69	25.56	45.3	56.0	88.3	107	125	150	170
84	Po	8.416 71	19	27	38	61	73	112	130	155	175
85	At	9.3	20	29	41	51	78	91	140	160	185

Stage of ionization

Table 3.5. *(Continued.)*

Atom		Stage of ionization									
		I	II	III	IV	V	VI	VII	VIII	IX	X
86	Rn	10.748 50	21	29	44	55	67	97	110	165	190
87	Fr	4	22	33	43	59	71	84	115	135	195
88	Ra	5.278 92	10.147	34	46	58	76	89	105	140	155
89	Ac	5.17	12.1	20	49	62	76	95	110	125	165
90	Th	6.08	11.5	20.0	28.8	65	80	94	115	130	145
91	Pa	5.89					84	100	115	140	155
92	U	6.194 05						104	120	140	160
93	Np	6.265 7									
94	Pu	6.06									
95	Am	5.993									

References

1. *Astrophysical Quantities*, **1**, §16; **2**, §16; **3**, §16
2. Lotz, W. 1966, *Ionisierungsenergien von Ionen H bis Ni* (Inst. Plasmaphys, München)
3. Moore, C.E. 1970, *Ionization Potentials*, NSRDS-NBS 34, Washington
4. Finkelnberg, W., & Humbach, W. 1955, *Naturwiss.*, **42**, 35
5. *Handbook of Chemistry and Physics, 77th ed.* (CRC, Boca Raton, FL, 1996)
6. Martin, W.C. 1987, *Phys. Rev.* A, **36**, 3575 (He I)
7. Martin, W.C., Kaufman, V., & Musgrove, A. 1993, *J. Phys. Chem. Ref. Data*, **22**, 1179 (O II)
8. Martin, W.C., & Zalubas, R. 1981, *J. Phys. Chem. Ref. Data*, **10**, 153 (Na I–XI)
9. Martin, W.C., & Zalubas, R. 1980, *J. Phys. Chem. Ref. Data*, **9**, 1 (Mg I–XII)
10. Martin, W.C., & Zalubas, R. 1979, *J. Phys. Chem. Ref. Data*, **8**, 817 (Al I–XIII)
11. Martin, W.C., & Zalubas, R. 1983, *J. Phys. Chem. Ref. Data*, **12**, 323 (Si I–XIV)
12. Martin, W.C., Zalubas, R., & Musgrove, A. 1985, *J. Phys. Chem. Ref. Data*, **14**, 751 (P I–XV)
13. Martin, W.C. Zalubas, R., & Musgrove, A. 1990, *J. Phys. Chem. Ref. Data*, **19**, 821 (S I–XVI)
14. Sugar, J., & Corliss, C. 1985, *J. Phys. Chem. Ref. Data*, **14**, Suppl. No. 2 (K, Ca, Sc, Ti, V, Cr, Mn, Fe, Co, Ni)
15. Sugar, J., & Musgrove, A. 1990, *J. Phys. Chem. Ref. Data*, **19**, 527 (I–XXIX)
16. Sugar, J., & Musgrove, A. 1995, *J. Phys. Chem. Ref. Data*, **24**, 1803 (Zn I–XXX)
17. Sugar, J., & Musgrove, A. 1993, *J. Phys. Chem. Ref. Data*, **22**, 1213 (Ge I–XXXII)
18. Sugar, J., & Musgrove, A. 1991, *J. Phys. Chem. Ref. Data*, **20**, 859 (Kr I–XXXVI)
19. Sugar, J., & Musgrove, A. 1988, *J. Phys. Chem. Ref. Data*, **17**, 155 (Mo I–XLII)
20. Martin, W.C., Zalubas, R., & Hagan, L. 1978, Natl. Stand. Ref. Data Ser. (Natl. Bur. Stand., U.S.) 60 (Rare-Earth Elements)
21. Cohen, E.R., & Taylor, B.N. 1988, J. Phys. Chem. Ref. Data, **17**, 1795

Table 3.6. *Electron affinities* [1–2].

Atom	Electron affinity (eV)	Atom	Electron affinity (eV)	Molecule	Electron affinity (eV)
H	+0.754	Na	+0.479	O_2	+0.451
He	−0.3	Mg	−0.4	O_3	+2.102 8
Li	+0.618	Al	+0.441	OH	+1.827 67
Be	−0.4	Si	+1.385	SH	+2.314
B	+0.277	P	+0.747	C_2	+3.269
				C_3	+1.981
C	+1.263	S	+2.077		
N	−0.2	Cl	+3.612	CN	+3.862
O	+1.461	Br	+3.48	NH_2	+0.771
O^-	−6.7	I	+3.17	NO	+0.026
F	+3.401	K	+0.501	NO_2	+2.273
Ne	−0.7	Ca	+0.018	NO_3	+3.951
				CH	+1.238

References

1. *Astrophysical Quantities*, **1**, §17; **2**, §17; **3**, §17
2. *Handbook of Chemistry and Physics, 77th ed.* (CRC, Boca Raton, FL, 1996)

3.6.1 Ionization Cross Section

The classical cross section of atoms for ionization by electrons [30] is

$$Q_1 = 4n\pi a_0^2 \frac{1}{\chi \epsilon}\left(1 - \frac{\chi}{\epsilon}\right),$$

where χ is the ionization energy in rydbergs (Ry), ϵ the electronic energy before collision in Ry, and n the number of optical electrons.

The general approximation for cross sections of atoms for ionization by electrons (see, [30–33]) is

$$Q_1 = n\pi a_0^2 \frac{1}{\chi \epsilon} F(Y, \epsilon/\chi) = \frac{n\pi a_0^2}{\chi^2}q$$

$$= 1.63 \times 10^{-14} n(1/\chi_{eV}^2)(\chi/\epsilon)F(Y, \epsilon/\chi),$$

where Y is the charge on the ionized atom (or next ion stage) and χ_{eV} is the ionization energy in eV. The function $F(Y, \epsilon/\chi)$ is given and also $q = (\chi/\epsilon)F(Y, \epsilon/\chi)$, which is sometimes called the reduced cross section in Table 3.7. The $Y = 1$ and $Y = 2$ values are from experiment and $Y = \infty$ from calculation. About $\pm 10\%$ accuracy may be expected for hydrogenic ions. In other cases ± 0.3 dex may be expected. Other empirical forms have been suggested (see, e.g., [34–36]).

Table 3.7. *Numerical functions $F(Y, \epsilon/\chi)$ and $q(Y, \epsilon/\chi)$.*

ϵ/χ	1.0	1.2	1.5	2.0	3	5	10
F(classical) $= 4(1 - \chi/\epsilon)$	0.00	0.67	1.33	2.00	2.67	3.20	3.60
$F(1, \epsilon/\chi)$	0.0	0.31	0.78	1.60	2.9	4.6	6.4
$F(2, \epsilon/\chi)$	0.00	0.53	1.17	2.02	3.3	4.7	6.4
$F(\infty, \epsilon/\chi)$	0.00	0.74	1.54	2.56	3.8	5.0	6.4
q(classical) $= 4(\chi/\epsilon)(1 - \chi/\epsilon)$	0.00	0.56	0.89	1.00	0.89	0.64	0.36
$q(l, \epsilon/\chi)$	0.00	0.26	0.52	0.80	0.97	0.92	0.64
$q(2, \epsilon/\chi)$	0.00	0.44	0.78	1.01	1.09	0.94	0.64
$q(\infty, \epsilon/\chi)$	0.00	0.62	1.03	1.28	1.28	1.00	0.64

The maximum ionization cross section for the classical case is

$$Q_{max} = n\pi a_0^2 \chi^{-2} \quad \text{at } \epsilon = 2\chi.$$

The value of Q_{max} is approximately the same in actual cases but the maximum occurs near $\epsilon = 4\chi$.

The rate of ionization by electrons (see [30–32]) is

$$L_1 = \overline{vQ}_1.$$

The neutral atom approximation (with $kT <$ ionization energy) gives

$$L_1 = 1.1 \times 10^{-8} n T^{1/2} \chi_{eV}^{-2} 10^{-5040\chi_{eV}/T} \text{ cm}^3 \text{ s}^{-1}.$$

The coronal ion approximation (with $kT <$ ionization energy) gives

$$L_1 = 2.1 \times 10^{-8} n T^{1/2} \chi_{eV}^{-2} 10^{-5040\chi_{eV}/T} \text{ cm}^3 \text{ s}^{-1}.$$

3.6.2 Excitation Cross Section (Permitted Transitions)

An approximation for Q_{ex}, the excitation cross section of an atom (see [30, 37]), is given. The approximation applies fairly well when $\Delta n \geq 1$ (notation of Chap. 5). For $\Delta n = 0$ the approximation tends to be small:

$$Q_{ex} = \frac{8\pi}{\sqrt{3}}\pi a_0^2 \frac{f}{\epsilon W}b$$
$$= 1740\pi a_0^2 \lambda^2 (W/\epsilon)fb$$
$$= 1.28 \times 10^{-15}(f/\epsilon W)b \text{ cm}^2,$$

where f is the oscillator strength, W is the excitation energy in Ry ($= 0.0912/\lambda$ with λ in μm), and ϵ is the electron energy before collision, also in Ry. See Table 3.8.

Table 3.8. *Numerical factors b and bW/ϵ.*

ϵ/W	1.0	1.2	1.5	2.0	3	5	10	30	100
b, neutral atoms	0.00	0.03	0.06	0.11	0.21	0.33	0.56	0.98	1.33
b, ions	0.20	0.20	0.20	0.20	0.24	0.33	0.56	0.98	1.33
bW/ϵ, neutral atoms	0.00	0.03	0.04	0.06	0.07	0.07	0.06	0.03	0.01
bW/ϵ, ions	0.20	0.17	0.13	0.10	0.08	0.07	0.06	0.03	0.01

The maximum excitation cross section is as follows:

- The neutral atom approximation gives

$$Q_{max} = 125\pi a_0^2 \lambda^2 f \quad \text{near } \epsilon = 3W.$$

- The ion approximation gives

$$Q_{max} = 350\pi a_0^2 \lambda^2 f \quad \text{near } \epsilon = W \quad (\lambda \text{ in } \mu\text{m}).$$

The rate of excitation (see [34, 35, 37]) is

$$L = \overline{vQ_{ex}} = 17.0 \times 10^{-4} \frac{f}{T^{1/2} W_{eV}} 10^{-5040 W_{eV}/T} P(W/kT),$$

where W_{eV} and W are the excitation energy in eV and in ergs (with $11\,600 W_{eV}/kT = W/kT$) and $P(W/kT)$ is tabulated from [37] (see Table 3.9).

Table 3.9. *Numerical factors $P(W/kT)$ and W/kT.*

	$P(W/kT)$	
W/kT	Neutral atoms	Ions
< 0.01	$0.29E_1(W/kT)^a$	
0.01	1.16	1.16
0.02	0.96	0.98
0.05	0.70	0.74

Table 3.9. *(Continued.)*

	$P(W/kT)$	
W/kT	Neutral atoms	Ions
0.1	0.49	0.55
0.2	0.33	0.40
0.5	0.17	0.26
1	0.10	0.22
2	0.063	0.21
5	0.035	0.20
10	0.023	0.20
> 10	$0.066/(W/kT)^{1/2}$	0.20

Note
$^{a}E_1(\ )$ is the first exponential integral.

The tabulated $P(W/kT)$ are too small when the total quantum number of Chap. 5 is unchanged.

The approximations quoted should be replaced by quantum calculations when available (see [30, 38–40]). A Coulomb approximation for ions [41] gives $b = g_{\text{eff}}(2L + 1)/g_1$ (L in Chap. 5). The tabulations of g_{eff}, the effective Gaunt factor, range from 0.5 to 0.9.

3.6.3   Deexcitation Cross Sections

Deexcitation cross sections Q_{21} are related to excitation cross sections Q_{12} (2 being the upper level) through

$$g_2\epsilon_2 Q_{21} = g_1\epsilon_1 Q_{12},$$

where $\epsilon_2 = \epsilon_1 + W$, and g_2 and g_1 are statistical weights.

The deexcitation rate L_{21} and excitation rate L_{12} are related by

$$g_2 L_{21} = g_1 L_{12} \exp(W/kT).$$

3.6.4   Excitation Cross Sections (Forbidden Transitions)

The collision strength Ω for each line is defined by (see [33, 42])

$$Q_f = \pi\Omega/g_1 k_v^2 = \pi a_0^2 \Omega/g_1\epsilon$$
$$= \frac{h^2}{4\pi m^2}\frac{\Omega}{g_1 v^2} = 4.21\Omega/g_1 v^2,$$

where $k_v/2\pi$ is the wave number of the incident electron (then k_v^2 in atomic units $= \epsilon$ in Ry), v is the electron velocity, g_1 is the statistical weight of the initial (lower) level, and Q_f is the forbidden line cross section for atoms in this level. Then Ω_{12}(excitation) $= \Omega_{21}$(deexcitation).

3.6.5   Collision Strengths: Extensive Databases

Crude recipes to estimate the order of magnitude of collision strengths (for allowed and forbidden transitions) can be found in older references [43]. In recent years, however, a wealth of accurate collision strengths have been obtained for a very large number of transitions. They are based on extensive UV and IR emission-line observations and on theoretical calculations. Data are available, e.g., from the Opacity Project, the Iron Project, and the Harvard–Smithsonian Center for Astrophysics (see Sec. 3.1 for information about online access of these sources).

3.6.6 Total Atomic Cross Section (Elastic and Inelastic)

An approximation for the total cross section is (see [31, 32, 44])

$$Q \simeq 180\pi a_0^2 \lambda / \epsilon^{1/2} \quad (\lambda \text{ in } \mu\text{m}, \epsilon \text{ in Ry}),$$

where λ is the wavelength of the strongest low-level lines.

3.6.7 Ionic Collision Cross Section

Cross section for collision deflection of at least a right angle (see [45])

$$Q_\perp = \pi(Y-1)^2(e^2/mv^2)^2 = \pi(Y-1)^2(e^2/2\epsilon hcR)^2$$
$$= \pi a_0^2 (Y-1)^2/\epsilon^2 \quad (\epsilon \text{ in Ry}),$$

where $Y - 1$ is the ionic charge.

The effective ionic collision cross section is usually concerned with the more distant collision involving deflections much less than a right angle. These increase the effective Q by a factor depending logarithmically on the most distant collisions that enter the integration and also on the circumstances. The factor is usually between 10 and 50 (see Sec. 3.10). We may write a general approximation:

$$Q(\text{effective}) \simeq 20\pi a_0^2 (Y-1)^2/\epsilon^2.$$

3.7 ATOMIC RADII

Atomic radii are defined through the closeness of approach of atoms in the formation of molecules and crystals. The radius r so derived is approximately that of maximum radial density in the charge distribution of neutral atoms (see Table 3.10). For ions the appropriate radius measures to the point where the radial density falls to 10% of its maximum value. The atomic mass divided by the atomic volume $(4/3)\pi r^3$ gives the density of the more compact solids. $2r$ is approximately the gas-kinetic diameter of monoatomic molecules.

Table 3.10. *Atomic radii* [1–5].

Atom	r (Å)	Ion [3]	r (Å)	Atom	r (Å)	Ion [3]	r (Å)	Atom	r (Å)	Ion [3]	r (Å)
H	0.7	H$^-$	1.8	S	1.05	S^{2-}	1.70	Br	1.2	Br$^-$	1.82
He	1.2			Cl	1.02	Cl$^-$	1.67	Kr	1.82		
Li	1.58	Li$^+$	0.68	Ar	1.6			Rb	2.54	Rb$^+$	1.66
Be	1.06	Be^{2+}	0.39	K	2.37	K$^+$	1.52	Sr	2.3	Sr^{2+}	1.32
B	0.83	B^{3+}	0.28	Ca	1.97	Ca^{2+}	1.14	Ag	1.44	Ag$^+$	1.29

Table 3.10. *(Continued.)*

Atom	r (Å)	Ion [3]	r (Å)	Atom	r (Å)	Ion [3]	r (Å)	Atom	r (Å)	Ion [3]	r (Å)
C	0.77	C^{4+}	0.22	Sc	1.64	Sc^{3+}	0.89	Cd	1.6	Cd^{2+}	1.09
N	0.70	N^{3-}	1.92	Ti	1.46	Ti^{4+}	0.75	Sn	1.62	Sn^{4+}	0.76
O	0.66	O^{2-}	1.26	V	1.39	V^{4+}	0.61	I	1.4	I^-	2.06
F	0.62	F^-	1.19	Cr	1.28			Xe	2.00		
Ne	1.3			Mn	1.26	Mn^{2+}	0.81	Cs	2.73	Cs^+	1.81
Na	1.91	Na^+	1.16	Fe	1.27	Fe^{2+}	0.75	Ba	2.24	Ba^{2+}	1.49
Mg	1.62	Mg^{2+}	0.86	Co	1.25	Co^{2+}	0.79	Pt	1.38		
Al	1.43	Al^{3+}	0.67	Ni	1.29	Ni^{2+}	0.83	Au	1.44	Au^+	1.51
Si	1.09	Si^{4+}	0.47	Cu	1.28	Cu^+	0.91	Hg	1.57	Hg^{2+}	1.16
P	1.08	P^{3-}	2.3	Zn	1.39	Zn^{2+}	0.77				

References

1. *Astrophysical Quantities*, **1**, §19; **2**, §19; **3**, §19
2. Teatum, E., Gschneidner, K., & Waber, J. 1960, Los Alamos Scientific Laboratory, Report No. LA-2345
3. Shannon, R.D. 1976, *Acta Cryst.*, **A32**, 751
4. Allen, F.H., Kennard, O., Watson, D.G., Brammer, L., Orpen, A.G., & Taylor, R. 1987, *J. Chem. Soc. Perkin II*, S1
5. Alcock, N.W. 1990, *Bonding and Structure: Structural Principles in Inorganic and Organic Chemistry*, (Ellis Horwood, New York)

3.8 PARTICLES OF MODERN PHYSICS

A representative selection of particles is given in Table 3.11. Hadrons include mesons, nucleons, and baryons. Possible proton decay is not included. I denotes the isotopic spin, J the spin, and P the parity. The lifetime is that in free space. In the column labeled "Decay" are given the main decay products. The mean life τ for W and Z bosons is given as the linewidth Γ ($\tau \Gamma \approx h$).

Table 3.11. *Selected particles of modern physics* [1–3].

Name	Symbol	Charge	Mass (amu)	I	J^P	Mean life (s)	Decay
			Bosons				
Gauge bosons							
Photon	γ	0	0.000	0, 1	1^-	∞	
W	W	+1, −1	86.24		1	$\Gamma = 2.1$ GeV	$e\nu$, etc.
Z	Z	0	97.90		1	$\Gamma = 2.5$ GeV	e^+e^-, etc.
Mesons							
π-mesons (pion)	π^+, π^-	+1, −1	0.149 84	1	0^-	2.603×10^{-8}	$\mu\nu$
	π^0	0	0.144 90	1	0^-	0.83×10^{-16}	$\gamma\gamma$
K meson (kayon)	K_0^+, K^-	+1, −1	0.530 15	½	0^-	1.237×10^{-8}	$\mu\nu, \pi\pi^0$
	K_S^0	0	0.534 38	½	0^-	0.892×10^{-10}	$\pi^+\pi^-, \pi^0\pi^0$
	K_L^0	0	0.534 38	½	0^-	5.38×10^{-8}	$\pi e\nu, \pi\mu\nu, 3\pi^0$
			Fermions				
Leptons							
e Neutrino	ν_e	0	$< 5 \times 10^{-8}$		½	∞	
μ Neutrino	ν_μ	0	$< 5 \times 10^{-4}$		½	∞	
τ Neutrino	ν_τ	0	< 0.2		½	∞	
Electron, Positron	e	−1, +1	0.000 548 6		½	∞	
μ meson (muon)	μ	−1, +1	0.113 4		½	2.197×10^{-6}	$e\nu\bar{\nu}$
τ meson (tauon)	τ	−1, +1	1.915		½	$(3.4 \pm 0.5) \times 10^{-13}$	$e\nu\bar{\nu}$

Table 3.11. *(Continued.)*

Name	Symbol	Charge	Mass (amu)	I	J^P	Mean life (s)	Decay
Nonstrange baryons							
Proton	p	+1, −1	1.007 275	½	½⁺	∞	
Neutron	n	0	1.008 664	½	½⁺	0.932×10^3	$pe^- \nu$
Strangeness-1 baryons							
Λ	Λ	0	1.197 6	0	½⁺	2.632×10^{-10}	$p\pi^-$, $n\pi^0$, etc.
Σ^+	Σ^+	+1, −1	1.276 8	1	½⁺	0.800×10^{-10}	$p\pi^0$, $n\pi^+$, etc.
Σ^0	Σ^0	0	1.280 2	1	½⁺	$< 10^{-19}$	$\Lambda\gamma$, etc.
Σ^-	Σ^-	−1, +1	1.285 4	1	½⁺	1.482×10^{-10}	$n\pi^-$, etc.
Strangeness-2 baryons							
Ξ^0	Ξ^0	0	1.411 6	½	½⁺	2.90×10^{-10}	$\Lambda\pi^0$, etc.
Ξ^-	Ξ^-	−1, +1	1.418 5	½	½⁺	1.641×10^{-10}	$\Lambda\pi^-$, etc.
Strangeness-3 baryons							
Ω^-	Ω^-	−1, +1	1.795	0	3/2⁺	0.819×10^{-10}	ΛK^-, etc.
Nonstrange charmed baryons							
Λ_c	Λ_c	−1, +1	2.450	0	½⁺	2.3×10^{-13}	ΛK^-, etc.
			Composite particles				
Hydrogen ($^2S_{1/2}$)	^{1}H	0	1.007 82			∞	
Deuterium ($^2S_{1/2}$)	^{2}H	0	2.014 10			∞	
Deuteron	D	+1	2.013 55			∞	
α particle	α	+2	4.001 40			∞	

References

1. *Astrophysical Quantities*, **1**, §20; **2**, §20; **3**, §20
2. Barnett, R.M. et al. 1996, *Rev. Mod. Phys.*, **68**, 611
3. Barnett, R.M. et al. 1996, *Phys. Rev.*, **D54**, 1

3.9 MOLECULES

Some definitions follow:

N_A, N_B, N_{AB}	Number of atoms A, B, and molecules AB per cm^3.
m_{AB}	Reduced mass $= m_A m_B / (m_A + m_B)$.
r_0	Internuclear distance (lowest state).
D_0	Dissociation energy (lowest state).
g_0	Electronic statistical weight (lowest state), or Multiplicity, $= 2S + 1$ for Σ states, $= 2(2S + 1)$ for other states.
σ	$= 1$ for heteronuclear molecules, $= 2$ for homonuclear molecules.
v	Vibrational quantum number.
B_e, α_e	Rotational constants [46, 47].
ΔE	Energy change $= hcB = h^2/8\pi^2 I = h^2/8\pi^2 m_{AB} r_e^2$.
$\omega_e, \omega_e x_e$	Vibrational constants.
IP	Ionizational potential.
U_A, U_B	Atomic partition functions (Sec. 3.3).
Q_{AB}	Molecular partition function, $= Q_{rot} Q_{vib} Q_{el}$, each term dimensionless.
I	Moment of inertia, $= m_{AB} r_e^2$.

Molecular diameters (diatomic) are

$$\simeq 3r_0 \simeq 3.4 \text{ Å}.$$

Molecular dissociation is represented by

$$N_A N_B / N_{AB} = (2\pi m_{AB} kT / h^2)^{3/2} e^{-D/kT} U_A U_B / Q_{AB}.$$

Numerically,

$$\log(N_A N_B / N_{AB}) = 20.2735 + \tfrac{3}{2} \log m_{AB} + \tfrac{3}{2} \log T - 5040D/T + \log(U_A U_B / Q_{AB})$$

with m in amu, D in eV, N in cm^{-3},

$$Q_{rot} = kT / \sigma h c B_v = (T / 1.439 \text{ K}) \sigma B_v,$$
$$B_v = B_e - \alpha_e(v + \tfrac{1}{2}),$$
$$Q_{vib} = \sum_v \exp\left(-\frac{1.439 \text{ K}}{T} [\omega_e v - \omega_e x_e(v^2 + v)] \right),$$
$$Q_{el} = \sum_{el} g_{el} \exp\left(-\frac{1.439 \text{ K}}{T} T_{el} \right),$$

with B_v, ω_e, T_{el} (= electronic excitation energy) in cm^{-1}.

The main ground-level constants are given in Tables 3.12 and 3.13, but upper level constants [46, 47] are required for dissociation calculations.

Table 3.12. *Diatomic molecules* [1–3].[a]

Molecule	g_0	σ	D_0 (eV)	m_{AB} (amu)	B_e (cm^{-1})	α_e (cm^{-1})	ω_e (cm^{-1})	$\omega_e x_e$ (cm^{-1})	r_0 (Å)	IP (eV)
H$_2$	1	2	4.4781	0.504	60.85	3.06	4401	121	0.741	15.426
H$_2^+$	4	2	2.6507	0.504	30.2	1.68	2321	66.2	1.052	
He$_2$	1	2	0[b]	2.002				22.22		
BH	1	1	3.42	0.923	12.02	0.412	2367	49.4	1.232	9.77
BO	2	1	8.28	6.452	1.78	0.017	1886	11.8	1.205	7.0
C$_2$	1	2	6.296	6.003	1.82	0.018	1855	13.3	1.243	12.15
CH	4	1	3.465	0.930	14.46	0.53	2859	63.0	1.120	10.64
CH$^+$	1	1	4.085	0.930	14.18	0.49	2740		1.131	
CO	1	1	11.092	6.856	1.93	0.018	2170	13.29	1.128	14.01
CO$^+$	2	1	8.338	6.859	1.977	0.019	2214	15.16	1.115	26.8
CN	2	1	7.76	6.462	1.90	0.017	2068	13.09	1.172	14.17
N$_2$	1	2	9.759	7.002	1.998	0.017	2359	14.32	1.098	15.58
N$_2^+$	2	2	8.713	7.001	1.932	0.019	2207	16.10	1.116	27.1
NH	3	1	3.47	0.940	16.699	0.649	3282	78.35	1.036	13.63
NO	4	1	6.497	7.467	1.672	0.017	1904	14.08	1.151	9.26
O$_2$	3	2	5.116	7.997	1.445	0.016	1580	11.98	1.208	12.07
O$_2^+$	4	2	6.663	7.997	1.691	0.020	1905	16.26	1.116	24.2
OH	4	1	4.392	0.948	18.91	0.724	3738	84.88	0.970	12.90
OH$^+$	3	1	5.09	0.948	16.79	0.749	3113	78.52	1.029	
MgH	2	1	1.34	0.967	5.826	0.185	1495	31.89	1.730	
AlH	1	1	3.06	0.972	6.391	0.186	1683	29.09	1.648	
AlO	2	1	5.27	10.042	0.641	0.006	979	6.97	1.618	9.53
SiH	4	1	3.06	0.973	7.500	0.219	2042		1.520	8.04

Table 3.12. *(Continued.)*

Molecule	g_0	σ	D_0 (eV)	m_{AB} (amu)	B_e (cm^{-1})	α_e (cm^{-1})	ω_e (cm^{-1})	$\omega_e x_e$ (cm^{-1})	r_0 (eV)	IP
SiO	1	1	8.26	10.177	0.727	0.005	1242	5.97	1.510	11.43
SiN	2	1	4.5	9.332	0.731	0.006	1151	6.47	1.572	
SO	3	1	5.359	10.661	0.721	0.006	1149	5.63	1.481	10.29
CaH	2	1	1.70	0.983	4.276	0.097	1298	19.10	2.003	5.86
CaO	1	1	4.8	11.423	0.445	0.003	732	4.81	1.822	
ScO	2	1	6.96	11.797	0.513	0.003	965	4.20		
TiO	6	1	6.87	11.994	0.535	0.003	1009	4.50	1.620	6.4
VO	4	1	6.4	12.173	0.548	0.004	1011	4.86	1.589	
CrO		1	4.4	12.229	0.541	0.005	898	6.75	1.615	8.2
FeO		1	4.20	12.438	0.513	0.004	965	8.71		
YO	2	1	7.29	13.556	0.388	0.002	861	2.93	1.790	
ZrO	6	1	7.85	13.579	0.423	0.002	970	4.90	1.712	
LaO	2	1	8.23	14.343	0.353	0.001	812	2.22	1.825	4.95

Notes

[a] See Sec. 4.11 for further molecular data and references.

[b] The lowest electronic state supports no bound state. However, the ground-state energy (as a function of nuclear separation) has a potential well. Its depth is $D_e = 0.0009$ eV.

References

1. *Astrophysical Quantities*, **1**, §21; **2**, §22; **3**, §21
2. Herzberg, G. 1950, *Spectra of Diatomic Molecules* (Van Nostrand, New York)
3. Huber, K.P., & Herzberg, G. 1979, *Molecular Spectra and Molecular Structure IV. Constants of Diatomic Molecules* (Van Nostrand, New York)

Table 3.13. *Selected polyatomic molecules* [1–2].

Molecule	IP (eV)	D (eV)	Diameter (Å)
H_2O	12.61	5.11	3.5
N_2O	12.89	1.68	4.0
CO_2	13.77	5.45	3.8
NH_3	10.15	4.3	3.0
CH_4	13.0	4.4	3.5
HCN	13.91	5.6	

References

1. *Astrophysical Quantities*, **1**, §21; **2**, §22; **3**, §21
2. Herzberg, G. 1966, *Electronic Spectra of Polyatomic Molecules* (Van Nostrand, New York)

3.10 PLASMAS

Some definitions follow:

N_e, N_i, N_p, N	Electron, ion, proton, total heavy-particle densities.
Z_i	Charge on i ion (denoted $Y_i - 1$ in other sections).
L	Characteristic size (e.g., diameter) of plasma.
T, B, ρ	Temperature, magnetic field, density.
A	Mass in amu.

The Debye length, electron screening, the distance from an ion over which N_e can differ appreciably from $\sum_i N_i Z_i$ is

$$D = (kT/4\pi e^2 N_e)^{1/2} = 6.92(T/N_e)^{1/2} \text{ cm},$$

with T in K and N_e in cm^{-3}.

The *plasma oscillation frequency* is

$$\nu_{pl} = (Ne^2/\pi m_e)^{1/2} = 8.978 \times 10^3 N_e^{1/2} s^{-2} \quad \text{(in cgs)}.$$

The *gyrofrequency* for electrons is

$$\begin{aligned} \nu_{gy} &= (e/2\pi m_e c)B \\ &= 2.7994 \times 10^6 B \ s^{-1}, \end{aligned}$$

and for ions is

$$\begin{aligned} \nu_{gy} &= (Ze/2\pi m_i c)B \\ &= 1.535 \times 10^3 Z_i B/A \ s^{-1}, \end{aligned}$$

with B in G.

The *gyroradius* for electrons is

$$\begin{aligned} a_e &= m_e v_\perp c/eB \\ &= 5.69 \times 10^{-8} v_\perp B \text{ cm} \\ &\simeq 2.21 \times 10^{-2} T^{1/2}/B \text{ cm}, \end{aligned}$$

and for ions is

$$\begin{aligned} a_1 &= m_i v_\perp c/Z_i eB \\ &= 1.036 \times 10^{-4} v_\perp A/Z_i B \text{ cm} \\ &\simeq 0.945 T^{1/2} A^{1/2}/Z_i B \text{ cm}, \end{aligned}$$

where $v_\perp$ is the velocity normal to B.

The most probable *thermal velocity* for electrons is

$$\begin{aligned} v &= (2kT/m_e)^{1/2} \\ &= 5.506 \times 10^5 T^{1/2} \text{ cm/s}, \end{aligned}$$

and for atoms and ions is

$$\begin{aligned} v &= (2kT/m)^{1/2} \\ &= 1.290 \times 10^4 (T/A)^{1/2} \text{ cm/s}. \end{aligned}$$

For rms velocities increase v by the factor $\sqrt{3/2} = 1.225$.

The velocity of *sound* is

$$v_s = (\gamma kT/m)^{1/2}[(N + N_e)/N]^{1/2},$$

comparable with thermal velocity.

The Alfvén speed (magnetohydrodynamic or hydromagnetic wave) is

$$v_A = B/(4\pi\rho)^{1/2} = 0.282 B/\rho^{1/2}.$$

The phase velocity is $c(1 + 4\pi\rho c^2/B^2)^{1/2}$.

The electron drift velocity in crossed magnetic and electric fields is $10^8 E_\perp/B$ cm/s, with $E_\perp$ in V/cm and B in G.

The electron drift velocity in magnetic and gravitational fields is

$$m_e gc/eB = 5.686 \times 10^{-8} g/B \text{ cm/s,}$$

with g in cm/s^2 and B in G.

The collision radius p for right-angle deflection of electrons by an ion is

$$p_0 = Z_i e^2/m_e v_e^2 \simeq \tfrac{1}{2} Z_1 e^2/kT$$
$$= 8.3 \times 10^{-4} Z_1/T \text{ cm.}$$

The corresponding collision cross section is

$$\pi p_0^2 = 2.16 \times 10^{-6} Z_1^2 T^{-2} \text{ cm}^2.$$

The cross section for all electron collisions with an ion is

$$\pi p_0^2 \ln \Lambda,$$

with

$$\ln \Lambda = \ln(d/c) = \int_c^d p^{-1} \, dp$$

and where c is the minimum of p in circumstances and d is the maximum of p in circumstances.

c is the largest of

$$c_1 = 8.3 \times 10^{-4} Z_1/T \text{ cm} \quad \text{from the right-angle definition}$$

or

$$c_2 = 1.06 \times 10^{-6} T^{-1/2} \text{ cm} \quad \text{from electron size.}$$

d is the smallest of

$$d_1 = N^{-1/3} \text{ cm} \quad \text{from ion spacing}$$

or

$$d_2 = D = 6.9 T^{1/2} N^{-1/2} \quad \text{(the Debye length)}$$

or

$$d_3 = 1.8 \times 10^5 T^{1/2}/\nu \quad \text{for collisions giving free–free absorption of frequency } \nu \text{ radiation.}$$

The most general approximation for Λ is

$$\ln \Lambda = 9.00 + 3.45 \log T - 1.15 \log N_e.$$

The collision cross section for neutral atoms and molecules is $\simeq 10^{-15}$ cm^2.

The collision frequency for electrons is $N_1 v_e \times$(cross section) $= 2.5(\ln \Lambda) N_e T^{-3/2} Z_i$ s^{-1}.

The collision frequency for ions with ions is $8 \times 10^{-2}(\ln \Lambda) N_e A^{-1/2} T^{-3/2} Z_1^2$ s^{-1}.

The mean free path of electrons among charged particles is $4.7 \times 10^5 T^2 N_1^{-1} N_1^{-2}$ cm.

The mean free path of electrons among neutral particles is $10^{15} N^{-1}$ cm.

The electrical resistivity [48] is

$$\eta = 8 \times 10^{12}(\ln \Lambda)T^{-3/2} \quad \text{(emu)}$$
$$= 9 \times 10^{-9}(\ln \Lambda)T^{-3/2} \quad \text{(esu)},$$

applying when the energy gains during free path $< kT$.

The thermal conductivity [48–50] is $1.0 \times 10^{-6}T^{5/2} \, \text{erg cm}^{-1}\,\text{s}^{-1}\,\text{K}^{-1}$.

The life of a magnetic field in a plasma is

$$\tau = 4\pi L^2/\eta \quad (\eta \text{ in emu})$$
$$= 1.5 \times 10^{-12}L^2(\ln \Lambda)^{-1}T^{3/2} \, \text{s}.$$

For approximate parameters for some plasmas, see Table 3.14.

Table 3.14. *Approximate parameters for some plasmas.[a] Values are logarithmic.*

Definition	Quantity	Unit	Ion.[b]	Intpl.[c]	⊙ Cor.[d]	⊙ Rev.[e]	Interstellar.[f] H I[g]	H II[h]
	$\log L$	cm	7.0	13.0	10.0	7.0	19.5	19.5
	$\log N_e$	cm^{-3}	5.5	0.5	8.0	12.5	−3.0	0.0
	$\log N$	cm^{-3}	11.0	0.5	8.0	16.5	0.0	0.0
	$\log T$	K	3.0	5.0	6.0	3.7	2.0	4.0
	$\log B$	G	−1.0	−5.0	0.0	0.0	−5.0	−5.0
Plasma freq.	$4.0 + \frac{1}{2}\log N_e$	s^{-1}	6.8	4.2	8.0	10.2	2.5	4.0
Debye length	$0.7 + \frac{1}{2}\log T - \frac{1}{2}\log N_e$	cm	−0.6	3.0	−0.3	−3.6	3.2	2.7
Gyro freq.								
Electron	$6.4 + \log B$	s^{-1}	5.4	1.4	6.4	6.4	1.4	1.4
Ion	$3.2 + \log B$	s^{-1}	2.2	−1.8	0.7	3.2	−1.8	−1.8
Collision freq.								
Electron	$1.7 + \log N_e - \frac{3}{2}\log T$	s^{-1}	2.2	−5.9	3.2	8.7	−1.8	−4.3
Ion	$0.2 + \log N_e - \frac{3}{2}\log T$	s^{-1}	1.2	−7.4	−0.8	7.2	−5.8	−5.8
Electrical conductivity	$6.3 + \frac{3}{2}\log T$	esu	10.8	13.8	15.3	11.9	9.3	12.3
	$-14.6 + \frac{3}{2}\log T$	emu	−10.1	−7.1	−5.6	−9.0	−11.6	−8.6
Mean free path								
Ion	$5.7 + 2\log T - \log N_e$	cm	6.2	15.2	9.7	0.6	12.7	13.7
Neutron	$15.0 - \log N$	cm	4.0	14.5	7.0	−1.5	15.0	15.0
Gyroradius								
Electron	$-1.7 + \frac{1}{2}\log T - \log B$	cm	0.8	5.8	1.3	0.1	4.3	5.3
Proton	$0.0 + \frac{1}{2}\log T - \log B$	cm	2.5	7.5	3.0	1.8	6.0	7.0
Alfvén v	$11.3 - \frac{1}{2}\log N + \log B$	cm/s	7.5	6.1	7.3	5.1	7.8	6.3
Sound v	$4.2 + \frac{1}{2}\log T$	cm/s	5.7	6.7	7.2	6.0	5.2	6.2
B decay	$-13.1 + 2\log L + \frac{3}{2}\log T$	s	5.4	19.4	15.9	6.5	29.9	31.9
		yr	−2.1	11.9	8.4	−1.0	22.4	24.4

Notes

[a]For spectral emission from high-temperature plasmas, see Chap. 14.

[b] Ion. denotes ionosphere.
[c] Intpl. denotes interplanetary space.
[d] ⊙ Cor. denotes solar corona.
[e] ⊙ Rev. denotes solar reversing layer.
[f] Interstellar denotes interstellar space.
[g] H I denotes the H I region.
[h] H II denotes the H II region.

ACKNOWLEDGMENT

The author was supported in part by Grant No. AST-9315112 of the National Science Foundation.

REFERENCES

1. Drake, G., editor, 1996, *Atomic, Molecular, and Optical Physics Handbook* (AIP, New York)
2. Jeffery, C.S. 1996, *QJRAS*, **37**, 39, and references therein
3. Lide, D.R., editor-in-chief 1996, *Handbook of Chemistry and Physics, 77th ed.* (CRC, Boca Raton, FL)
4. DeLaeter, J.R., Heumann, K.G., & Rosman, K.J.R. 1991, *Pure Appl. Chem.*, **63**, 991, reprinted in *J. Phys. Chem. Ref. Data*, **20**, No. 6, 1327
5. DeLaeter, J.R., & Heumann, K.G. 1991, *Pure Appl. Chem.*, **63**, 975, reprinted in *J. Phys. Chem. Ref. Data*, **20**, No. 6, 1313
6. Anders, E., & Grevesse, N. 1989, *Geochim. Cosmochim. Acta* **53**, 197; Grevesse, N., & Noels, A. 1993, in *Origin and Evolution of the Elements*, edited by N. Prantzos, E. Vangioni, and M . Cassé (Cambridge University Press, Cambridge), p. 15
7. Biémont, E., Baudoux, M., Kurucz, R.L., Ansbacher, W., & Pinnington, E.H. 1991, *A&A*, **249**, 539
8. Hummer, D.G., & Mihalas, D. 1988, *ApJ*, **331**, 794
9. Mihalas, D., Däppen W., & Hummer, D.G. 1988, *ApJ*, **331**, 815
10. Däppen, W., Anderson, L.S., & Mihalas, D. 1987, *ApJ*, **319**, 195
11. Rogers, F.J. 1981, *Phys. Rev. A*, **24**, 1531
12. Rogers, F.J. 1986, *ApJ*, **310**, 723
13. Iglesias, C.A., & Rogers, F.J. 1991, *ApJ*, **371**, L73
14. Rogers, F.J., & Iglesias, C.A. 1992, *ApJS*, **401**, 361
15. Rogers, F.J., Swenson, F.J., & C.A. Iglesias 1996, *ApJ*, **456**, 902
16. Ebeling, W., Förster, A., Fortov, V.E., Gryaznov, V.K., & Polishchuk, A.Ya. 1991, *Thermodynamic Properties of Hot Dense Plasmas* (Teubner, Stuttgart)
17. Alastuey, A., & Perez, A. 1992, *Europhys. Lett.*, **20**, 19
18. Alastuey, A., & Perez, A. 1996, *Phys. Rev. E* **53**, 5714
19. Iglesias, C.A., & Rogers, F.J. 1992, in *Astrophysical Opacities*, edited by C. Mendoza and C. Zeippen (Revista Mexicana de Astronomía y Astrofísica, Mexico City), Vol. 23, p. 161
20. Eliezer, S., Ghatak, A., & Hora, H. 1986, *An Introduction to Equations of State: Theory and Applications* (Cambridge University Press, Cambridge)
21. Irwin, A.W. 1981, *ApJS*, **45**, 621
22. Sauval, A.J., & Tatum, J.B. 1984, *ApJS*, **56**, 193
23. Lochte-Holtgreven, W. 1958, *Rep. Prog. Phys.*, **21**, 312
24. Rosa, A. 1948, *Z. Ap.*, **25**, 1
25. Aller, L.H. 1961, *Stellar Atmospheres*, edited by J. Greenstein (University of Chicago Press, Chicago.), p. 232
26. Cohen, E.R., & Taylor, B.N. 1988, *J. Phys. Chem. Ref. Data*, **17**, 1795
27. Hyleraas, E. 1950, *ApJ*, **111**, 209
28. Thirring, W. 1979, *Lehrbuch der mathematischen Physik, Vol. 3: Quantenmechanik von Atomen und Molekülen* (Springer, Berlin), p. 207
29. Hill, R.N. 1977, *J. Math. Phys.*, **18**, 2316
30. Bely, O., & Van Regemorter, H. 1970, *ARA&A*, **8**, 329
31. *Astrophysical Quantities*, **1**, §18
32. *Astrophysical Quantities*, **2**, §18
33. Seaton, M.J. 1962, *Atomic and Molecular Processes*, edited by D. R. Bates (Academic Press, New York), p. 375
34. Lotz, W. 1967, *ApJS*, **14**, 207
35. Sampson, D.H. 1969, *ApJ*, **155**, 575
36. Bely, O., & Faucher, P. 1972, *A&A*, **18**, 487
37. Van Regemorter, H. 1962, *ApJ*, **136**, 906
38. Moiseiwitsch, B.L., & Smith, S.L. 1968, *Rev. Mod. Phys*, **40**, 238
39. Seaton, M.J. 1970, *Trans. IAU*, **XIVA**, 128 and references therein
40. Kieffer, L.J. 1969, *JILA* Rep. 7
41. Blaha, M. 1969, *ApJ*, **157**, 473
42. Hebb, M.H., & Menzel, D.H. 1940, *ApJ*, **92**, 408
43. *Astrophysical Quantities*, **3**, §18, and references therein
44. Bederson, B., & Kieffer, L.J. 1971, *Rev. Mod. Phys.*, **43**, 601
45. Spitzer, L. 1956, *Physics of Fully Ionized Gases* (Interscience, New York)
46. Herzberg, G. 1950, *Spectra of Diatomic Molecules* (Van Nostrand, New York)
47. Tatum, J.B. 1966, *Pub. Dom. Ap. Obs.*, **13**, 1
48. Spitzer, L. 1962, *Physics of Fully Ionized Gases* (Interscience, New York)
49. Astrophysical Quantities, **1**; **2**; **3**, §22
50. Delcroix, A., & Lemaire, A. 1969, *ApJ*, **156**, 787

Chapter 4

Spectra

Charles Cowley, Wolfgang L. Wiese, Jeffrey Fuhr, and Ludmila A. Kuznetsova

4.1 ONLINE DATABASE

Extensive data and references are available online through the Internet [1]. A comprehensive, critically evaluated database, whose address is given below, is maintained by the National Institute of Standards and Technology (NIST). Files of special relevance to atomic spectroscopy are the *Atomic Spectroscopic*

Database by J.R. Fuhr, W.C. Martin, A. Musgrove, J. Sugar, and W.L. Wiese, *Bibliographic Database on Atomic Transition Probabilities* by J.R. Fuhr and H.R. Felrice, and *Program of the NIST Atomic Data Centers* by W.L. Wiese and W.C. Martin. The uniform resource locator, or URL, is currently http://physics.nist.gov/PhysRefData/contents.html. File names and locations are subject to change. The above files might be found by first "opening" http://physics.nist.gov/ and following the appropriate links, or simply by doing a network search for the keywords "NIST atomic data."

4.2   TERMINOLOGY FOR ATOMIC STATES, LEVELS, TERMS, ETC.

The angular momenta of atoms are *vector* quantities describing the orbital angular momenta $(\mathbf{l}, \mathbf{L})$, the spin $(\mathbf{s}, \mathbf{S})$, and the sum of the two $(\mathbf{j}, \mathbf{J})$. Lowercase letters are used for individual electrons, and uppercase letters refer to corresponding sums (e.g., $\mathbf{L} = \sum \mathbf{l}$). The magnitudes of these vectors are specified by quantum numbers usually written with lightface italic symbols. For example, $|\mathbf{l}| = \sqrt{l(l+1)}\hbar$. Spectroscopists often interchange the meaning of the vector quantities and the associated quantum numbers, and say, for example, that L is the sum of the l's, although the relation is only valid for the vector quantities. This loose usage is convenient and is followed here. Spectroscopic levels are typically described by quantum numbers based on LS (Russell–Saunders) coupling. For other coupling schemes, see [2] and [3]. Often, levels are expressed as mixtures of LS terms, where the *leading component* is the single LS term that best describes the level.

Orbital angular momentum (or azimuthal quantum number), L = vector sum of orbital angular momenta l of individual electrons. The unit is $h/2\pi \equiv \hbar$, and the designations are

L (or l)	0	1	2	3	4	5	6	7	8	9	10	11	12	13	14	15
Designation (L)	*S*	*P*	*D*	*F*	*G*	*H*	*I*	*K*	*L*	*M*	*N*	*O*	*Q*	*R*	*T*	*U*
Designation (l)	*s*	*p*	*d*	*f*	*g*	*h*	*i*	*k*	*l*	*m*	*n*	*o*	*q*	*r*	*t*	*u*

Spin angular momentum, S = vector sum of s for individual electrons. The *multiplicity* of terms $= 2S + 1$.

The effects of the atomic nucleus on atomic structure, including nuclear spin I, are treated in a separate section below.

Total angular momentum quantum number, J = vector sum $L + S$ (in LS coupling). In jj coupling, j = vector sum $l + s$ for each electron, and $J = \sum j$. The total angular momentum J is said to be a "good quantum number," independent of the coupling scheme.

Electron shells are described by the principal quantum number n as follows:

n	1	2	3	4	5	6	7
Shell designation	*K*	*L*	*M*	*N*	*O*	*P*	*Q*

Only the magnitude of an angular momentum (e.g., $|\mathbf{L}|$) and one of its components (e.g., L_z) are observables. The z component is chosen arbitrarily. Quantum numbers corresponding to these z components are designated, for example, by m_l, M_L, or $M_J = M$. If the atom is in a magnetic field, it is convenient to choose the z direction along the field, so the m's have been called magnetic quantum numbers.

Maximum values of various quantum numbers are limited as follows:

$$l \le n - 1; \qquad s = \tfrac{1}{2}; \qquad J \le S + L; \qquad M_L \le L; \qquad M_S \le S; \qquad M \le J;$$

$$S \le \tfrac{1}{2} n_a; \qquad L \le l_1 + l_2 + \cdots + l_{n_a},$$

where there are n_a electrons in open shells.

Interpretation of a typical symbol for an atomic level, e.g., $2p^3 \quad {}^2D^0_{1\frac{1}{2}}$:

2	Principal quantum number of outer electrons $= 2$; i.e., L shell.
p^3	Three outer electrons with $l = 1$.
2	Multiplicity $= 2$, whence $S = \tfrac{1}{2}$.
D	Orbital momentum $L = 2$.
$1\frac{1}{2}$	$J = 1\frac{1}{2}$, whence statistical weight $g = 2J + 1 = 4$.
0	The level is odd (omitted when level is even).

Possible J values for given L and S:

Singlets	${}^1S_0, {}^1P_1, {}^1D_2, {}^1F_3, {}^1G_4, {}^1H_5, \ldots.$
Doublets	${}^2S_{\frac{1}{2}}, {}^2P_{\frac{1}{2},1\frac{1}{2}}, {}^2D_{1\frac{1}{2},2\frac{1}{2}}, {}^2F_{2\frac{1}{2},3\frac{1}{2}}, {}^2G_{3\frac{1}{2},4\frac{1}{2}}, \ldots.$
Triplets	${}^3S_1, {}^3P_{0,1,2}, {}^3D_{1,2,3}, {}^3F_{2,3,4}, {}^3G_{3,4,5}, \ldots.$
Quartets	${}^4S_{1\frac{1}{2}}, {}^4P_{\frac{1}{2},1\frac{1}{2},2\frac{1}{2}}, {}^4D_{\frac{1}{2},1\frac{1}{2},2\frac{1}{2},3\frac{1}{2}}, {}^4F_{1\frac{1}{2},2\frac{1}{2},3\frac{1}{2},4\frac{1}{2}}, \ldots.$
Quintets	${}^5S_2, {}^5P_{1,2,3}, {}^5D_{0,1,2,3,4}, {}^5F_{1,2,3,4,5}, {}^5G_{2,3,4,5,6}, \ldots.$
Sextets	${}^6S_{2\frac{1}{2}}, {}^6P_{1\frac{1}{2},2\frac{1}{2},3\frac{1}{2}}, {}^6D_{\frac{1}{2},1\frac{1}{2},2\frac{1}{2},3\frac{1}{2},4\frac{1}{2}}, {}^6F_{\frac{1}{2},1\frac{1}{2},2\frac{1}{2},3\frac{1}{2},4\frac{1}{2},5\frac{1}{2}},$
	${}^6G_{1\frac{1}{2},2\frac{1}{2},3\frac{1}{2},4\frac{1}{2},5\frac{1}{2},6\frac{1}{2}}, \ldots.$
Septets	${}^7S_3, {}^7P_{2,3,4}, {}^7D_{1,2,3,4,5}, {}^7F_{0,1,2,3,4,5,6}, {}^7G_{1,2,3,4,5,6,7}, \ldots.$

The magnetic quantum numbers are usually not indicated unless the level is split by a magnetic field. In the absence of such a perturbation, the energies of all levels with a given J are the same, and are therefore $(2J + 1)$-*fold degenerate*.

Classical atomic spectroscopists have used the following hierarchial scheme to describe energy states, combinations thereof, and transitions among such states, as given in Table 4.1.

Table 4.1. *Hierarchy of designations.*

Atomic division	Specification	Statistical weight g	Transition
State	Specified by L, S, J, M, or L, S, M_L, M_S	1	Component (of line)
Level	Specified by L, S, J, e.g., ${}^4S_{1\frac{1}{2}}$	$2J + 1$	Line
Term	Group of levels specified by L, S	$(2S + 1)(2L + 1)$	Multiplet

Table 4.1. *(Continued.)*

Atomic division	Specification	Statistical weight g	Transition
Polyad	Group of terms from one parent term, and with same multiplicity or S		Supermultiplet
Configuration	Specified by n and l of all electrons	See text	Transition array

Nowadays, spectroscopists rarely use the term *polyad*. Very complicated level structures arise with the filling of the $3d$ (iron group), $4d$ (palladium group), $5d$ (platinum group), $4f$ (lanthanides), and $5f$ (actinides) subshells. Johansson refers to a *subconfiguration* for all of the levels that result from the addition of an electron (nl) to a *parent* term. For example, if we use S_p and L_p to designate the spin and orbital angular momentum of the parent, $3d^2(^{S_p}L_p)nl$ has five subconfigurations corresponding to the five allowed terms from d^2. Similarly, $3d^4(^{S_p}L_p)nl$ would have 16 subconfigurations. He uses the term *supermultiplet* to mean all transitions between levels belonging to subconfigurations of opposite parity [4].

4.2.1 Terms from Various Configurations

Table 4.2 gives the multiplicities and orbital angular momenta of the various terms arising in LS coupling from the configurations listed [2, 3]. When a term can appear more than once, the number of possible terms is written below the symbol. Complete shells s^2, p^6, d^{10}, f^{14}, etc., give rise to only ^1S terms. They need not be considered for possible terms due to outer electrons.

Electrons with the same n and l are said to be equivalent. Terms arising from complementary numbers of equivalent electrons are the same; e.g., terms from p^2 and p^4 are the same, since six electrons complete the p shell.

The total weight of an electron configuration may be written [2]

$$g(l_1^{w_1} l_2^{w_2} \cdots l_n^{w_n}) = \prod_{i=1}^{n} \binom{4l_i + 2}{w_i}.$$

Here, w_i is the number of (equivalent) electrons with angular momentum l_i. A number of examples are given below. If a single electron with angular momentum l is added to a parent with L_p and S_p, the total weight of the resulting terms of both resulting multiplicities is

$$g = (4S_p + 2)[(b+1)^2 - a^2],$$

where $b = |L_p + l|$ and $a = |L_p - l|$.

Often, atomic energy levels are not well described by a single electronic configuration. In such cases, *configuration interaction* or *configuration mixing* is said to occur.

Table 4.2. *Allowed terms for equivalent electrons.*

Configuration	Terms	Total weight
	Equivalent s electrons	
s	^2S	2
s^2	^1S	1

Table 4.2. *(Continued.)*

Configuration	Terms	Total weight
Equivalent p electrons		
p p^5	$^2P^0$	6
p^2 p^4	^1SD ^3P	15
p^3	$^2PD^0$ $^4S^0$	20
Equivalent d electrons		
d d^9	^2D	10
d^2 d^8	^1SDG ^3PF	45
d^3 d^7	^2PDFGH ^4PF	120
	2	
d^4 d^6	^1SDFGI ^3PDFGH ^5D	210
	2 2 2 2 2	
d^5	$^2SPDFGHI$ ^4PDFG ^6S	252
	3 2 2	
f f^{13}	$^2F^0$	14
f^2 f^{12}	^1SDGI ^3PFH	91
f^3 f^{11}	$^2PDFGHIKL^0$ $^4SDFGI^0$	364
	2 2 2 2	
f^4 f^{10}	$^1SDFGHIKLN$ $^3PDFGHIKLM$ ^5SDFGI	1 001
	2 4 423 2 3 2 4 3 4 2 2	
f^5 f^9	$^2PDFGHIKLMNO^0$ $^4SPDFGHIKLM^0$ $^6PFH^0$	2 002
	4 5 7 6 7 5 5 3 2 2 3 4 4 3 3 2	
f^6 f^8	$^1SPDFGHIKLMNQ$ $^3PDFGHIKLMNO$ $^5SPDFGHIKL$ ^7F	3 003
	4 6 4 8 4 7 3 4 2 2 6 5 9 7 9 6 6 6 3 3 3 2 3 2 2	
f^7	$^2SPDFGHIKLMNOQ^0$ $^4SPDFGHIKLMN^0$ $^6PDFGHI^0$ $^8S^0$	3 432
	2 5 710 10 9 9 7 5 4 2 2 2 6 5 7 5 5 3 3	

4.3   ELECTRONIC CONFIGURATIONS

Tables 4.3 and 4.4 give the electronic configurations for ground-level atoms [5]. The inner core of electrons is not explicitly shown for heavier elements. Extensive tabulations of energy levels are available [6, 7].

Table 4.3. *Ground-level configurations.*

Atom	K 1s	L 2s	2p	M 3s	3p	3d	N 4s	4p	4d	O 5s	Ground level
H 1	1										$^2S_{1/2}$
He 2	2										$^1S_0$
Li 3	2	1									$^2S_{1/2}$
Be 4	2	2									$^1S_0$
B 5	2	2	1								$^2P^0_{1/2}$
C 6	2	2	2								$^3P_0$
N 7	2	2	3								$^4S^0_{1\,1/2}$
O 8	2	2	4								$^3P_1$
F 9	2	2	5								$^2P^0_{1\,1/2}$

Atom	N 4f	O 5s	5p	5d	5f	P 6s	6p	6d	Q 7s	Ground level
Ag 47		1								$^2S_{1/2}$
Cd 48		2								$^1S_0$
In 49		2	1							$^2P^0_{1/2}$
Sn 50		2	2							$^3P_0$
Sb 51		2	3							$^4S^0_{1\,1/2}$
Te 52		2	4							$^3P_2$
I 53		2	5							$^2P^0_{1\,1/2}$
Xe 54		2	6							$^1S_0$
Cs 55		2	6			1				$^2S_{1/2}$

Table 4.3. *(Continued.)*

	K	L		M			N			O	Ground
Atom	1s	2s	2p	3s	3p	3d	4s	4p	4d	5s	level
Ne 10	2	2	6								$^1S_0$
Na 11	2	2	6	1							$^2S_{1/2}$
Mg 12				2							$^1S_0$
Al 13				2	1						$^2P^0_{1/2}$
Si 14			10	2	2						$^3P_0$
P 15				2	3						$^4S^0_{1\frac12}$
S 16			Ne core	2	4						$^3P_2$
Cl 17				2	5						$^2P^0_{1\frac12}$
Ar 18				2	6						$^1S_0$
K 19	2	2	6	2	6		1				$^2S_{1/2}$
Ca 20							2				$^1S_0$
Sc 21						1	2				$^2D_{1\frac12}$
Ti 22						2	2				$^3F_2$
V 23					18	3	2				$^4F_{1\frac12}$
Cr 24						5	1				$^7S_3$
Mn 25					Ar core	5	2				$^6S_{2\frac12}$
Fe 26						6	2				$^5D_4$
Co 27						7	2				$^4F_{4\frac12}$
Ni 28						8	2				$^3F_4$
Cu 29	2	2	6	2	6	10	1				$^2S_{1/2}$
Zn 30							2				$^1S_0$
Ga 31							2	1			$^2P^0_{1/2}$
Ge 32						28	2	2			$^3P_0$
As 33							2	3			$^4S^0_{1\frac12}$
Se 34							2	4			$^3P_2$
Br 35							2	5			$^2P^0_{1\frac12}$
Kr 36							2	6			$^1S_0$
Rb 37	2	2	6	2	6	10	2	6		1	$^2S_{1/2}$
Sr 38										2	$^1S_0$
Y 39									1	2	$^2D_{1\frac12}$
Zr 40									2	2	$^3F_2$
Nb 41						36			4	1	$^6D_{1/2}$
Mo 42									5	1	$^7S_3$
Tc 43						Kr core			5	2	$^6S_{2\frac12}$
Ru 44									7	1	$^5F_5$
Rh 45									8	1	$^4F_{4\frac12}$
Pd 46									10		$^1S_0$

	N	O				P			Q	Ground
Atom	4f	5s	5p	5d	5f	6s	6p	6d	7s	level
Ba 56			8			2				$^1S_0$
La 57				1		2				$^2D_{1\frac12}$
Ce 58	1	2	6	1		2				$^1G^0_4$
Pr 59	3					2				$^4I^0_{4\frac12}$
Nd 60	4					2				$^5I_4$
Pm 61	5					2				$^6H^0_{2\frac12}$
Sm 62	6					2				$^7F_0$
Eu 63	7					2				$^8S^0_{3\frac12}$
Gd 64	7			1		2				$^9D_2$
Tb 65	9					2				$^6H^0_{7\frac12}$
Dy 66	10					2				$^5I_8$
Ho 67	11					2				$^4I^0_{7\frac12}$
Er 68	12					2				$^3H_6$
Tm 69	13					2				$^2F^0_{3\frac12}$
Yb 70	14					2				$^1S_0$
Lu 71	14			1		2				$^2D_{1\frac12}$
Hf 72	14	2	6	2		2				$^3F_2$
Ta 73				3		2				$^4F_{1\frac12}$
W 74				4		2				$^5D_0$
Re 75	46 + 22			5		2				$^6S_{2\frac12}$
Os 76				6		2				$^5D_4$
Ir 77				7		2				$^4F_{1\frac12}$
Pt 78				9		1				$^3D_3$
Au 79	14	2	6	10		1				$^2S_{1/2}$
Hg 80						2				$^1S_0$
Tl 81						2	1			$^2P^0_{1/2}$
Pb 82	46 + 32					2	2			$^3P_0$
Bi 83						2	3			$^4S^0_{1\frac12}$
Po 84						2	4			$^3P_2$
At 85						2	5			$^2P^0_{1\frac12}$
Rn 86						2	6			$^1S_0$
Fr 87	14	2	6	10		2	6		1	$^2S_{1/2}$
Ra 88									2	$^1S_0$
Ac 89	46 + 32							1	2	$^2D_{1\frac12}$
Th 90								2	2	$^3F_2$
Pa 91					2			1	2	$^4K_{5\frac12}$
U 92					3			1	2	$^5L^0_6$

Table 4.4. *Transuranic elements.*

Atom		O			P	Q	Ground level
		$5f$	$6s$	$6p$	$6d$	$7s$	
Np	93	4	2	6	1	2	$^6L_{5\,1/2}$
Pu	94	6	2	6		2	$^7F_0$
Am	95	7	2	6		2	$^8S^0_{3\,1/2}$
Cm	96	7	2	6	1	2	$^9D^0_2$
Bk	97	9	2	6		2	$^6H^0_{7\,1/2}$
Cf	98	10	2	6		2	$^5I_8$
Es	99	11	2	6		2	$^4I^0_{7\,1/2}$
Fm	100	12	2	6		2	$^3H_6$
Md	101	13	2	6		2	$^2F^0_{3\,1/2}$
No	102	14	2	6		2	$^1S_0$
Lr	103	14	2	6	1	2	$^2D_{3\,1/2}$

Table 4.5 of first ions (Sc II, etc.), is restricted to those ions whose ground levels differ from those of the preceding atom. Table 4.5 gives outer and incomplete shells only.

Table 4.5. *First ions.*

Element	Configuration	Ground level	Element	Configuration	Ground level	Element	Configuration	Ground level
Sc	$3d4s$	$^3D_1$	La	$5d^2$	$^3F_2$	Ta	$5d^36s$	$^5F_1$
Ti	$3d^24s$	$^4F_{1\,1/2}$	Ce	$4f5d^2$	$^4H^0_{3\,1/2}$	W	$5d^46s$	$^6D_{1/2}$
V	$3d^4$	$^5D_0$	Pr	$4f^36s$	$^5I^0_4$	Re	$5d^56s$	$^7S_3$
Cr	$3d^5$	$^6S_{2\,1/2}$	Nd	$4f^46s$	$^6I_{3\,1/2}$	Os	$5d^66s$	$^6D_{4\,1/2}$
Mn	$3d^54s$	$^7S_3$	Pm	$4f^56s$	$^7H^0_2$	Ir	$5d^76s$	$^5F_5$
Fe	$3d^64s$	$^6D_{4\,1/2}$	Sm	$4f^66s$	$^8F_{1\,1/2}$	Pt	$5d^9$	$^2D_{2\,1/2}$
Co	$3d^8$	$^3F_4$	Eu	$4f^76s$	$^9S^0_4$	Au	$5d^{10}$	$^1S_0$
Ni	$3d^9$	$^2D_{2\,1/2}$	Gd	$4f^75d6s$	$^{10}D^0_{2\,1/2}$			
Cu	$3d^{10}$	$^1S_0$	Tb	$4f^96s$	$^7H^0_8$	Th	$6d^27s$	$^4F_{1\,1/2}$
			Dy	$4f^{10}6s$	$^6I_{8\,1/2}$	Pa	$5f^27s^2$	$^3H_4$
Zr	$4d^25s$	$^4F_{1\,1/2}$	Ho	$4f^{11}6s$	$^5I_8$	U	$5f^37s^2$	$^4I^0_{4\,1/2}$
Nb	$4d^4$	$^5D_0$	Er	$4f^{12}6s$	$^4H_{6\,1/2}$	Np	$5f^57s$	$^7H^0_2$
Mo	$4d^5$	$^6S_{2\,1/2}$	Tm	$4f^{13}6s$	$^3F^0_4$	Pu	$5f^67s$	$^8F_{1\,1/2}$
Tc	$4d^55s$	$^7S_3$	Yb	$4f^{14}6s$	$^2S_{1/2}$	Am	$5f^77s$	$^9S^0_4$
Ru	$4d^7$	$^4F_{4\,1/2}$						
Rh	$4d^8$	$^3F_4$						
Pd	$4d^9$	$^2D_{2\,1/2}$						

4.4   SPECTRUM LINE INTENSITIES

4.4.1   Definitions

We use the symbol Ð to mean "dimensionally equal to" or "has dimensions of"; Ð 0 means "dimensionless."

g = (dimensionless) statistical weight for a level = $2J + 1$. Subscripts denote levels.

f = (dimensionless) oscillator strength, or simply f value. Unless otherwise stated, this is the absorption oscillator strength f_{abs}, related to the emission oscillator strength f_{em} (which is often taken to be negative) by $g_1 f_{abs} = -g_2 f_{em}$. Here, g_1 and g_2 are the statistical weights of the lower level and upper level, respectively.

gf = weighted oscillator strength = $g_1 f_{12} = -g_2 f_{21}$. gf is symmetrical between emission and absorption.

A = Einstein's A Ðs^{-1}; spontaneous transition probability (for a downward transition).

B_{12}, B_{21} = Einstein's B; induced transition probability upward and downward. Bu_ν = probability of transition where u_ν is the radiation energy density at the frequency ν of the transition. Then Bu_ν Ðs^{-1}. The B coefficients are sometimes defined with specific intensity I_ν, whence BI_ν Ðs^{-1}.

S = line strength. Sum of the matrix elements of the electric dipole operator Ð$e^2 |x|^2$. Also used for higher-order radiation (see below).

γ_{cl} = classical damping constant (Ðs^{-1}). γ_{cl} is the full width at half maximum (FWHM) in units of circular frequency ($\omega = 2\pi\nu$) of an absorption line due to a *classical oscillator*.

γ_2 = reciprocal mean life of level 2
= $\sum_1 A_{21} + \sum_1 B_{21} u(\nu_{21}) + \sum_3 B_{23} u(\nu_{23})+$ collision terms, where level 1 is below and level 3 is above 2.

γ = damping constant = $\gamma_1 + \gamma_2$ for transition $1 \rightarrow 2$. It is convenient to define damping constants γ_ν and γ_λ, for use when profiles are expressed in frequency or wavelength units. Then $\gamma_\nu/\nu = \gamma_\lambda/\lambda = \gamma/\omega$.

σ_ν = atomic cross section (Ðcm^2) near an absorption line. Note: $\sigma_\lambda = \sigma_\nu$. Traditionally, σ_ν is written in terms of γ, not γ_ν. Often a_ν or α_ν is used for atomic cross sections.

N_1 = number of atoms per unit volume in level 1 (the lower level).

κ = $N_1 \sigma_\nu$, the line absorption coefficient, which must be corrected for stimulated emission: $\kappa_{corr} = \kappa[1 - \exp(-h\nu/kT)]$.

ν_0 = frequency at line center.

R_i, R_f = Initial and final radial wave functions of the active electron. For bound levels, $R_{i,f}$ Ð cm$^{-3/2}$. Commonly, $r R_{i,f} \equiv P_{i,f}$, where r = radius.

σ = proportional to radial transition moment (see below), not related to σ_ν or σ_λ.

$\mathcal{S}$ = multiplet strength, scale as given in Table 4.9.

E = energy emitted due to spontaneous, bound–bound transitions in all directions, per unit volume and time.

ε_{Ry} = photon energy in rydbergs ($e^2/2a_0 = 2\pi^2 m_e e^4 / h^2$).

n^* = effective principal quantum number; describes the energy of an atomic level.

4.4.2 Formulas

For a spectral line that arises from a transition between levels αLSJ and $\alpha'L'S'J'$, the *line strength* for a dipole transition is defined as

$$S = \sum_{MM'q} |\langle \alpha LSJM|ex_q|\alpha'L'S'J'M'\rangle|^2 \tag{4.1}$$

$$= \sum_{MM'q} |\langle \alpha LSJM|erC_q^{(1)}|\alpha'L'S'J'M'\rangle|^2. \tag{4.2}$$

Here, α and α' stand for unspecified quantum numbers. q runs from 1 to 3 for the three components of the position vector of the active electron, or equivalently, the three components of the *spherical tensor* of rank 1, $rC_q^{(1)}$. The C's are proportional to the spherical harmonics of corresponding order:

$$C_m^{(l)} \equiv \sqrt{\frac{4\pi}{2m+1}} Y_l^m; \qquad x^2 + y^2 + z^2 = |rC_{-1}^{(1)}|^2 + |rC_0^{(1)}|^2 + |rC_{+1}^{(1)}|^2.$$

Consider a simple electronic transition, where there is a single active optical electron ($L_p l \rightarrow L_p l'$), where L_p stands for the orbital angular momentum of the parent. The greater of l and l' is usually written $l_>$. In a nonrelativistic, single-configuration approximation, the line strength can be written with the help of two Wigner $6-j$ symbols [2]:

$$S = (2J+1)(2J'+1)(2L+1)(2L'+1)$$

$$\times \begin{Bmatrix} L & S & J \\ J' & 1 & L' \end{Bmatrix}^2 \begin{Bmatrix} L_p & l & L \\ 1 & L' & l' \end{Bmatrix}^2 l_> \left(\int_0^\infty R_i r R_f r^2\, dr \right)^2.$$

The line strength S is often taken to be in *atomic units* ($e = a_0 = m_e = 1$), but that is not the case in the following relations (the B's used here are defined with energy density; we use $m \equiv m_e$):

$$g_2 A_{21} = g_2 \frac{8\pi h\nu^3}{c^3} B_{21} = g_1 \frac{8\pi h\nu^3}{c^3} B_{12} = \frac{64\pi^4}{3h\lambda^3} S_{12\ or\ 21}$$

$$= 3\gamma_{cl} g_1 f_{12} = -3\gamma_{cl} g_2 f_{21} = \frac{8\pi^2 e^2 \nu^2}{mc^3} g_1 f_{12},$$

$$\gamma_{cl} = \frac{8\pi^2 e^2 \nu^2}{3mc^3} = \frac{8\pi^2 e^2}{3mc\lambda^2},$$

$$gf = g_1 f_{12} = -g_2 f_{21} = \frac{mh\nu}{\pi e^2} g_1 B_{12} = \frac{8\pi^2 m\nu}{3he^2} S_{12},$$

$$g_1 B_{12} = g_2 B_{21} = \frac{8\pi^3}{3h^2} S_{12},$$

$$E = N_2 A_{21} h\nu = \frac{N_2}{g_2} \frac{8\pi^2 e^2 h\nu^3}{mc^3} g_1 f_{12} = N_2 \frac{8\pi^2 e^2 h\nu^3}{mc^3}(-f_{21})$$

$$= N_2 \frac{8\pi^2 e^2 h}{m\lambda^3}(-f_{21}),$$

$$\int \sigma_\nu\, d\nu = \frac{\pi e^2}{mc} f_{abs},$$

$$\kappa_\nu = \frac{\pi e^2}{mc} f_{abs} \frac{\gamma}{4\pi^2} \frac{N_1}{(\nu - \nu_0)^2 + (\gamma/4\pi)^2} \text{\DH cm}^{-1},$$

$$\kappa_\lambda = \frac{\pi e^2}{mc^2} f_{abs} \frac{\gamma_\lambda}{2\pi} \frac{\lambda_0^2 N_1}{(\lambda - \lambda_0)^2 + (\gamma_\lambda/2)^2} \ \text{cm}^{-1},$$

$$\kappa_{\nu_0} = \frac{4}{\gamma} \frac{\pi e^2}{mc} f_{abs} N_1, \qquad \kappa_{\lambda_0} = \frac{2e^2}{mc^2} \frac{\lambda_0^2}{\gamma_\lambda} N_1 f_{abs},$$

$$\int \kappa_\nu \, d\nu = \frac{\pi e^2}{mc} f_{abs} N_1, \qquad \int \kappa_\lambda \, d\lambda = \frac{\pi e^2}{mc^2} \lambda_0^2 f_{abs} N_1,$$

$$(R_{n'l'}^{nl})^2 = \left(\int_0^\infty R_i R_f r^3 \, dr \right)^2 = \left(\int_0^\infty P_i r P_f \, dr \right)^2,$$

$$(R_{n'l'}^{nl})^2 = (l_> P_{ll'}^{(1)})^2 \qquad \text{(notation of [2])},$$

$$\sigma^2 = \frac{1}{4l_>^2 - 1} (R_{n'l'}^{nl})^2,$$

$$n_{nl}^* = Z \sqrt{\frac{\chi_H}{\chi_{ion} + \chi_p - \chi_{nl}}}.$$

The *effective principal quantum number* n^* may be defined for each level with excitation χ_{nl}. The core, or parent excitation, χ_p, if present, must be added to the ionization energy χ_{ion}. For example, the $2s^2 2p^2(^1D)3s$ level of N I at 12.36 eV (99 663 cm^{-1}) is built on an excited parent in N II. Therefore, in calculating n^*, one must add $\chi_p = 1.90$ eV to the (first) ionization energy 14.53 eV of N I.

4.4.3   Numerical Relations

The following relations are based on the above formulas, which are derived from an approximate, nonrelativistic radiation theory. The numerical factors are given only to four figures. Physical constants are from [8]. Note that the line strength S is in atomic units in the following:

$$gf = 303.8S/\lambda = 1.499 \times 10^{-16} g_2 A\lambda^2 \quad (\lambda \text{ in Å, } A \text{ in s}^{-1}),$$
$$S = 0.003\,292gf\lambda = 4.936 \times 10^{-19} g_2 A\lambda^3 \quad (\lambda \text{ in Å, } A \text{ in s}^{-1}),$$
$$g_2 A = 2.026 \times 10^{18} S/\lambda^3 = 2.678 \times 10^9 \varepsilon_{Ry}^3 S$$
$$= 0.6670 \times 10^{16} gf/\lambda^2 \quad (\lambda \text{ in Å, } A \text{ in s}^{-1}),$$
$$\pi e^2/mc = 0.026\,54 \text{ cm}^2 \text{ s}^{-1}, \qquad \pi e^2/mc^2 = 8.853 \times 10^{-13} \text{ cm},$$
$$\gamma_{cl} = 2.223 \times 10^{15}/\lambda^2 \text{ s}^{-1} \quad (\lambda \text{ in Å}),$$
$$8\pi h\nu^3/c^3 = 8\pi h/\lambda^3 = 0.1665/\lambda^3 \quad (\lambda \text{ in Å}),$$
$$a_0^2 e^2 = 6.460 \times 10^{-36} \text{ cm}^2 \text{ esu}^2.$$

4.4.4   Forbidden Transitions: Electric Quadrupole (*E*2) and Magnetic Dipole (*M*1)

In astronomical usage, a line is called "forbidden" when it violates the rules for an electric dipole (*E*1)-induced transition. The lines are designated with a bracket notation, e.g., [O III] for transitions among the low-level, even-parity states of doubly ionized oxygen. *E*1 transitions with $\Delta S \neq 0$ occur frequently and in the astronomical literature are often written with a single bracket. For example, the transition $2s^2\ {}^1S_0$–$2s2p\ {}^3P_1^0$ at $\lambda 1909$ is written C III]. Such a transition is sometimes called *semiforbidden*, or *spin forbidden*. In complex spectra the rule against *intercombination of multiplicities* is violated so frequently that this notation is not particularly useful, and it is rarely employed. In the formulas below, α is the *fine structure constant* ($\oplus 0$), and σ is the *wave number* of the photon ($\oplus \mathrm{cm}^{-1}$). The gyromagnetic ratio of the electron spin has been assumed to be 2.000 [2]. For magnetic dipole radiation,

$$g_2 A_{21} = \frac{4\pi^2 h e^2 \sigma^3}{3m^2 c^2} \sum_{qMM'} |\langle \gamma JM | J_q^{(1)} + S_q^{(1)} | \gamma' J'M' \rangle|^2$$

$$= 2.697\,4 \times 10^{-11} \sigma^3 \sum_{qMM'} |\langle \gamma JM | J_q^{(1)} + S_q^{(1)} | \gamma' J'M' \rangle|^2.$$

For electric quadrupole radiation, we show the explicit sum over i electrons. These sums are implicit in the symbols $J_q^{(1)}$ and $S_q^{(1)}$ above. In practice, only one electron is important. We have

$$g_2 A_{21} = \frac{64\pi^4 e^2 a_0^4 \sigma^5}{15h} \sum_{qMM'i} |\langle \gamma JM | r_i^2 C_q^{(2)}(i) | \gamma' J'M' \rangle|^2$$

$$= 1.120\,0 \times 10^{-22} \sigma^5 \sum_{qMM'i} |\langle \gamma JM | r_i^2 C_q^{(2)}(i) | \gamma' J'M' \rangle|^2.$$

4.4.5 Selection Rules

Selection rules for atomic transitions are summarized in Table 4.6, including rules for LS coupling. When levels are not accurately described by single values of L and S, rules involving these quantum numbers are no longer valid. However, even in complex atoms it is often the case that transitions that violate the LS selection rules are weak. Configuration interaction can cause the selection rule on Δl to be violated. An example is found in Si I, $\lambda 5621.61$ of multiplet 17.01 [9]. This appears to be a jump from $3p4s$ to $3p4f$ ($\Delta l = 3$). The transition occurs because the $3p4s$ configuration is mixed with $3p3d$.

Table 4.6. *Selection rules for atomic transitions.*[a]

Electric Dipole ($E1$)
$\Delta l = \pm 1$, parity change
Δn arbitrary
$\Delta J = 0 \pm 1$, $J = 0 \nleftrightarrow J = 0$
$\Delta L = 0, \pm 1$, $L = 0 \nleftrightarrow L = 0$
$\Delta S = 0$

Magnetic Dipole ($M1$)
$\Delta J = 0, \pm 1$, $J = 0 \nleftrightarrow J = 0$
$\Delta M = 0, \pm 1$
$\Delta l = 0$, $\Delta n = 0$, for all electrons
ΔS, $\Delta L = 0$

Table 4.6. *(Continued.)*

Electric Quadrupole ($E2$)

$\Delta J = 0, \pm 1, \pm 2, \quad 0 \not\leftrightarrow 0, \frac{1}{2} \not\leftrightarrow \frac{1}{2}, 0 \not\leftrightarrow 1$

$\Delta l = 0, \pm 2, \quad$ no parity change

Δn arbitrary

$\Delta M = 0, \pm 1, \pm 2$

$\Delta S = 0, \quad \Delta L = 0, \pm 1, \pm 2, \quad L = 0 \not\leftrightarrow L = 0, 1$

Note

[a]Rules for L and S hold for LS coupling, while those for J are independent of the coupling conditions.

4.4.6   Radial Integrals and Related Calculations

The *Coulomb approximation* [10, 11] to the radial integral for a single electron is still of heuristic interest. It uses the *effective principal quantum numbers* n^*. Let Z be the charge seen by the active electron at large distances from the nucleus. $Z = 1$ for a neutral species, 2 for a first ion, etc. Set $a = (Z/n^*)$. The normalizations of the wave functions in the Coulomb approximation are

$$N = \frac{1}{n^*}\sqrt{\frac{Z}{\Gamma(n+l+1)\Gamma(n-l)}},$$

where Γ is the gamma function. We shall use numbers 1 and 2 to distinguish upper and lower levels in the relations below (as above, $l_>$ means the greater of the two values, l_1 and l_2):

$$\sigma = \frac{1}{\sqrt{4l_>^2 - 1}} N_1(n_1^*, l_1) N_2(n_2^*, l_2) \frac{(2a_1)^{n_1^*}(2a_2)^{n_2^*}}{(a_1+a_2)^{n_1^*+n_2^*+2}} \sum_{p=0}^{\max} G_p \sum_{q=0}^{p} C_{p-q}(1) C_q(2).$$

The coefficients G and C are easily obtained from recurrence relations:

$$C_0 = 1, \qquad C_k(i) = -\frac{(-n_i^* - l_i + k - 1)(l_i - n_i^* + k)}{k} \frac{a_1 + a_2}{2a_i} C_{k-1}(i), \qquad i = 1, 2,$$

$$G_0 = \Gamma(n_1^* + n_2^* + 2), \qquad G_k = \frac{G_{k-1}}{n_1^* + n_2^* + 2 - k}.$$

For integral n^*'s the coefficients C are 0 for k above $n - l - 1$. Then, if the sum includes all nonvanishing terms, the results are identical with those well known for hydrogen and hydrogenlike ions. The Coulomb approximation usually gives good results when $n_{nl}^* > l + 1$ with max $< n_1^* + n_2^* - 1$. Useful tables are given in [11].

Kurucz and Bell [12] have made extensive calculations of radial integrals for complex atoms using scaled Thomas–Fermi–Dirac potentials. Results from the international "Opacity Project" are becoming available [13].

4.4.7   Sum Rules

The Kuhn–Thomas–Reiche f-sum rule states

$$\sum_1 f_{21} + \sum_3 f_{23} = z,$$

where the summations are for level 1 below the selected level 2, and 3 above that level (including an integral over continuum). z is the number of atomic or ionic electrons. f_{21} is negative and hence for upward transitions $\sum_3 f_{23} \geq z$. The rule is rigorous for nonrelativistic quantum mechanics, but the sum includes physically unrealistic states. Restricted and approximate forms of the sum rule are of more practical importance, as for more complex spectra where the lines concerned are mainly the lowest members of their series and contain most of the total oscillaator strength.

The Wigner–Kirkwood rule for a one-electron jump [2] is

$$\sum f = -\frac{1}{3} \frac{l(2l-1)}{2l+1} \quad \text{for} \quad l \to l-1,$$

$$\sum f = \frac{1}{3} \frac{(l+1)(2l+3)}{2l+1} \quad \text{for} \quad l \to l+1$$

(l is the orbital quantum number); for example,

$$\sum f = -\frac{1}{9}, \qquad p \to ns,$$

$$\sum f = 1, \qquad s \to np,$$

$$\sum f = -\frac{2}{5}, \qquad d \to np,$$

$$\sum f = \frac{10}{9}, \qquad p \to nd.$$

The above rule may sometimes be used for complex spectra, but it applies precisely for hydrogen.

The J file and J group sum rules refer to a transition array, e.g., $sp \leftrightarrow pp$. A J file refers to all transitions that begin or end on a specified level. Let all line strengths $S(\gamma' L' S' J', \gamma L S J)$ within a transition array be entered in an $i \times f$ matrix, with i being the number of initial levels and f the number of final levels. A J file is any single row or column in this matrix. The J file sum rule states that

$$\sum_J S(\gamma' L' S' J', \gamma L S J) \propto 2J' + 1$$

and

$$\sum_{J'} S(\gamma' L' S' J', \gamma L S J) \propto 2J + 1.$$

These two rules are independent of the coupling conditions, but apply only to simple transition arrays, where either the moving electron has no equivalent congeners or the electron configuration with the *summed* J or J' values does not contain equivalent electrons.

A J group consists of all lines in a transition array connecting a level with a given J (e.g., initial) with one with a given J' (e.g., final). The J group sum rule states that the sum of the strengths of the lines in a J group are independent of the coupling conditions.

4.5 RELATIVE STRENGTHS WITHIN MULTIPLETS

Table 4.9 gives the relative strengths of lines in multiplets. The notation used here is for LS-coupling multiplets for a transition $LSJ \to L'S'J'$. It is important to note that the relative strengths apply much more generally to any case where two angular momenta, say j_1, and j_2, couple to a third j_3, where j_2 commutes with the dipole operator $e\mathbf{r}$ [2]. As a result of this generality, these same relative intensity tables may be used for lines in a *hyperfine* "multiplet" by the following substitution of quantum numbers: $J \to F, L \to J$, and $S \to I$, where F is the total angular momentum including the nuclear spin I. Similarly, the relative intensities of what were once called "supermultiplets" may be computed by making the following exchange: $J \rightarrow L$, $L \rightarrow l$, and $S \rightarrow L_p$. Here, we assume a single optical electron with angular momentum l that couples to a parent core with angular momentum L_p. It therefore turns out that tables [14] giving relative multiplet strengths are unnecessary.

The entries are all proportional to

$$(2J' + 1)(2J + 1)W^2(LL'JJ'; 1S) = (2J' + 1)(2J + 1) \begin{Bmatrix} L & S & J \\ J' & 1 & L' \end{Bmatrix}^2 ,$$

Table 4.7. *Normal multiplets SP, PD, DF, etc.*

	J_m	$J_m - 1$	$J_m - 2$	$J_m - 3$	$J_m - 4$
$J_m - 1$	x_1	y_1	z_1		
$J_m - 2$		x_2	y_2	z_2	
$J_m - 3$			x_3	y_3	z_3
$J_m - 4$				x_4	y_4

where the W is a Racah coefficient, and the symbol on the right is a Wigner $6 - j$ symbol [2]. We normalize so that the sum of the entries for a given multiplet is $S = (2S + 1)(2L + 1)(2L' + 1)$. The entries are therefore proportional to the *line strengths* as defined above and do not contain wavelength-dependent factors. Therefore, they are only approximately proportional to relative line intensities in real (LS-coupling) multiplets.

The following qualitative rules describing the intensities in LS multiplets are of practical value. The most intense lines are those where L and J change in the same sense, for example, $J \rightarrow J + 1$ while $L \rightarrow L + 1$. These strong lines are called the *principal*, or sometimes diagonal, lines of the multiplet. In Tables 4.7 and 4.8, their intensities are called $x_1, x_2, x_3, \ldots$. The intensity of the *strongest* line on the (principal) diagonal is called x_1, and it belongs to the line involving the *largest* J value, called J_m below. With a few exceptions that may be seen in Table 4.9, the intensities diminish down the diagonal. Lines that fall off the main diagonal are called satellite, or off-diagonal, lines. There are two kinds of multiplets to consider, the symmetrical ones ($P \rightarrow P$, $D \rightarrow D$, etc.), and "normal multiplets" ($L \rightarrow L + 1$, such as $S \rightarrow P$ or $P \rightarrow D$). Since the line strength factors are independent of which level is upper and which is lower, we are free to choose J_m to belong to the largest L. For the symmetrical multiplets, we call the intensities of the lines for which $J_m \rightarrow J_m - 1$, y_1, those for which $J_m - 1 \rightarrow J_m - 2$, y_2, etc. Lines with identical intensities fall on the complementary side of the diagonal, as shown below. In normal multiplets, there are a second series of satellites with $J_m - 1 \rightarrow J_m - 2$, etc., which are designated z_1, etc. We remark that in a breakdown of LS coupling, the weaker lines typically deviate more strongly from the LS intensities, so that a calculation in LS coupling may yield reasonable results for a line on the main diagonal, but could be badly off for a satellite.

Table 4.8. *Symmetrical multiplets PP, DD, etc.*

	J_m	$J_m - 1$	$J_m - 2$	$J_m - 3$
J_m	x_1	y_1		
$J_m - 1$	y_1	x_2	y_2	
$J_m - 2$		y_2	x_3	y_3
$J_m - 3$			y_3	x_4

Table 4.9. *Intensities in LS-coupling multiplets.*

	\multicolumn{11}{c}{Multiplicity}									
	1	2	3	4	5	6	7	8	9	10	11
						SP					
$\mathcal{S} =$	3	6	9	12	15	18	21	24	27	30	33
x_1	3.00	4.00	5.00	6.00	7.00	8.00	9.00	10.00	11.00	12.00	13.00
y_1		2.00	3.00	4.00	5.00	6.00	7.00	8.00	9.00	10.00	11.00
z_1			1.00	2.00	3.00	4.00	5.00	6.00	7.00	8.00	9.00
						PP					
$\mathcal{S} =$	9	18	27	36	45	54	63	72	81	90	99
x_1	9.00	10.00	11.25	12.60	14.00	15.43	16.88	18.33	19.80	21.27	22.75
x_2		4.00	2.25	1.60	1.25	1.03	0.88	0.76	0.67	0.61	0.55
x_3				1.00	2.25	3.60	5.00	6.43	7.87	9.33	10.80
y_1		2.00	3.75	5.40	7.00	8.57	10.13	11.67	13.20	14.73	16.25
y_2			3.00	5.00	6.75	8.40	10.00	11.57	13.13	14.67	16.20
						PD					
$\mathcal{S} =$	15	30	45	60	75	90	105	120	135	150	165
x_1	15.00	18.00	21.00	24.00	27.00	30.00	33.00	36.00	39.00	42.00	45.00
x_2		10.00	11.25	12.60	14.00	15.43	16.88	18.33	19.80	21.27	22.75
x_3			5.00	5.00	5.25	5.60	6.00	6.43	6.88	7.33	7.80
y_1		2.00	3.75	5.40	7.00	8.57	10.13	11.67	13.20	14.73	16.25
y_2			3.75	6.40	8.75	10.97	13.13	15.24	17.33	19.39	21.45
y_3				5.00	6.75	8.40	10.00	11.57	13.13	14.67	16.20
z_1			0.25	0.60	1.00	1.43	1.88	2.33	2.80	3.27	3.75
z_2				1.00	2.25	3.60	5.00	6.43	7.88	9.33	10.80
z_3				3.00	6.00	9.00	12.00	15.00	18.00	21.00
						DD					
$\mathcal{S} =$	25	50	75	100	125	150	175	200	225	250	275
x_1	25.00	28.00	31.11	34.29	37.50	40.74	44.00	47.27	50.56	53.85	57.14
x_2		18.00	17.36	17.29	17.50	17.88	18.38	18.94	19.56	20.21	20.89
x_3			11.25	8.00	6.25	5.14	4.37	3.81	3.38	3.03	2.75
x_4				5.00	1.25	0.22		0.14	0.49	0.95	1.50
x_5						2.22	5.00	8.00	11.11	14.29	17.50
y_1		2.00	3.89	5.71	7.50	9.26	11.00	12.73	14.44	16.15	17.86
y_2			3.75	7.00	10.00	12.86	15.63	18.33	21.00	23.64	26.25
y_3				5.00	8.75	12.00	15.00	17.86	20.63	23.33	26.00
y_4					5.00	7.78	10.00	12.00	13.89	15.71	17.50
$\mathcal{S} =$	35	70	105	140	175	210	245	280	315	350	385
x_1	35.00	40.00	45.00	50.00	55.00	60.00	65.00	70.00	75.00	80.00	85.00
x_2		28.00	31.11	34.29	37.50	40.74	44.00	47.27	50.56	53.85	57.14
x_3			21.00	22.40	24.00	25.71	27.50	29.33	31.20	33.09	35.00
x_4				14.00	14.00	14.40	15.00	15.71	16.50	17.33	18.20
x_5					7.00	6.22	6.00	6.00	6.11	6.29	6.50
y_1		2.00	3.89	5.71	7.50	9.26	11.00	12.73	14.44	16.15	17.86
y_2			3.89	7.31	10.50	13.54	16.50	19.39	22.24	25.06	27.86
y_3				5.60	10.00	13.89	17.50	20.95	24.30	27.58	30.80
y_4					7.00	11.38	15.00	18.29	21.39	24.38	27.30
y_5						7.78	10.00	12.00	13.89	15.71	17.50

Table 4.9. *(Continued.)*

	1	2	3	4	5	6	7	8	9	10	11
z_1			0.11	0.29	0.50	0.74	1.00	1.27	1.56	1.85	2.14
z_2				0.40	1.00	1.71	2.50	3.33	4.20	5.09	6.00
z_3					1.00	2.40	4.00	5.71	7.50	9.33	11.20
z_4						2.22	5.00	8.00	11.11	14.29	17.50
z_5							5.00	10.00	15.00	20.00	25.00
						FF					
$S =$	49	98	147	196	245	294	343	392	441	490	539
x_1	49.00	54.00	59.06	64.17	69.30	74.45	79.62	84.81	90.00	95.20	100.41
x_2		40.00	41.17	42.67	44.36	46.20	48.12	50.13	52.18	54.28	56.41
x_3			31.11	28.90	27.56	26.74	26.25	25.98	25.88	25.90	26.00
x_4				22.40	17.50	14.40	12.25	10.67	9.45	8.48	7.70
x_5					14.00	7.62	4.37	2.50	1.36	0.67	0.26
x_6						6.22	0.87		0.49	1.60	3.06
x_7								3.50	7.88	12.60	17.50
y_1		2.00	3.94	5.83	7.70	9.55	11.38	13.19	15.00	16.80	18.59
y_2			3.89	7.50	10.94	14.26	17.50	20.68	23.82	26.92	30.00
y_3				5.60	10.50	15.00	19.25	23.33	27.30	31.18	35.00
y_4					7.00	12.60	17.50	22.00	26.25	30.33	34.30
y_5						7.78	13.13	17.50	21.39	25.00	28.44
y_6							7.00	10.50	13.13	15.40	17.50
						FG					
$S =$	63	126	189	252	315	378	441	504	567	630	693
x_1	63.00	70.00	77.00	84.00	91.00	98.00	105.00	112.00	119.00	126.00	133.00
x_2		54.00	59.06	64.17	69.30	74.45	79.62	84.81	90.00	95.20	100.41
x_3			45.00	48.21	51.56	55.00	58.50	62.05	65.63	69.23	72.86
x_4				36.00	37.50	39.29	41.25	43.33	45.50	47.73	50.00
x_5					27.00	27.00	27.50	28.29	29.25	30.33	31.50
x_6						18.00	16.88	16.50	16.50	16.71	17.06
x_7							9.00	7.50	6.88	6.60	6.50
y_1		2.00	3.94	5.83	7.70	9.55	11.38	13.19	15.00	16.80	18.59
y_2			3.94	7.62	11.14	14.55	17.88	21.15	24.38	27.57	30.74
y_3				5.79	10.94	15.71	20.25	24.62	28.88	33.04	37.14
y_4					7.50	13.71	19.25	24.38	29.25	33.94	38.50
y_5						9.00	15.63	21.21	26.25	30.95	35.44
y_6							10.13	16.00	20.63	24.69	28.44
y_7								10.50	13.13	15.40	17.50
z_1			0.06	0.17	0.30	0.45	0.62	0.81	1.00	1.20	1.41
z_2				0.21	0.56	1.00	1.50	2.05	2.63	3.23	3.86
z_3					0.50	1.29	2.25	3.33	4.50	5.73	7.00
z_4						1.00	2.50	4.29	6.25	8.33	10.50
z_5							1.88	4.50	7.50	10.71	14.06
z_6								3.50	7.87	12.60	17.50
z_7									7.00	14.00	21.00

4.6   WAVELENGTHS AND WAVE NUMBERS

Angstrom units (Å) and microns (micrometers, μm) are used for wavelengths in the tables presented in the following sections. Astronomers often indicate wavelengths in angstrom units by the λ symbol. Wavelengths may be truncated after the last unit of an angstrom or they may be rounded off. We have tried to follow the latter procedure here, but there is no uniformity in the literature. Thus a line of Ca I at 4 226.73 Å might be called either λ4226 or λ4227. Wave numbers are almost always given in units of cm^{-1}, although reciprocal microns are occasionally used. Common symbols for wave numbers are ν, $\tilde{\nu}$, and σ.

Many workers use the SI unit nanometer (nm) for wavelengths, 1 nm = 10 Å. Wavelengths here are given "in air" for (air) wavelengths greater than 2 000 Å. Air and vacuum wavelengths are related by the index of refraction of air, n: $\lambda_{\text{vacuum}} = n\lambda_{\text{air}}$. An extensive tabulation [15] is based on Edlen's formula for n,

$$n = 1 + 6432.8 \times 10^{-8} + \frac{2\,949\,810}{146 \times 10^8 - \sigma^2} + \frac{25\,540}{41 \times 10^8 - \sigma^2},$$

where σ is the wave number in cm^{-1}. This formula suffices for conversions from air to vacuo when no more than eight-figure accuracy is desired [16]. For shorter wavelengths reciprocal wave numbers, or "vacuum wavelengths" are used. With the advent of space astronomy, some workers have suggested the exclusive use of vacuum wavelengths, but this has not been adopted here.

Reader and Corliss [17] give a modern table of wavelengths of the chemical elements. They include lines that are suitable for use in calibration of most spectrographs. Extensive references to wavelength standards are given by Wiese and Martin [18].

4.7 ATOMIC OSCILLATOR STRENGTHS FOR ALLOWED LINES

Atomic hydrogen is considered separately here, in the nonrelativistic approximation in Table 4.10. Exact numerical values have been available from the early days of quantum mechanics [19]; they remain of heuristic as well as of practical value. For most cases of astrophysical interest, it is permissible to ignore the electron spin in hydrogen. Each level with a given l is then $(2l + 1)$- rather than $2(2l + 1)$-fold degenerate, and the weight of all states belonging to a principal quantum number n is n^2 rather than $2n^2$. The corresponding *partition function* at low temperatures is then 1 and not 2; this must be used in the Boltzmann and Saha formulas when absorption or emission coefficients are calculated. Authors sometimes use a notation that is valid if spin is ignored with statistical weights that take spin into account [19, 20]. Transitions may be designated $nl \rightarrow n'l'$, e.g., $1s \rightarrow 2p$ for Lyman α in absorption. If spin is ignored, then the line strength $S = l_>(R_{nl}^{n'l'})^2$. The values for S in [20] and [21] allow for the spin degeneracy and are twice this value. The Einstein coefficients, line strengths, and f values must be defined in such a way that the intensities or equivalent widths of lines do not depend on whether spin is included in the level-counting scheme. For example, consider the equivalent width W_λ' of a *weak* hydrogen absorption line when light passes through a uniform slab of thickness H. If we use the Boltzmann formula to express the population of the lower level N_i as a function of the total population of neutrals N_T, we have

$$W_\lambda' = \frac{\pi e^2}{mc^2}\lambda_0^2 N_T \frac{g_i f_{ik}}{u(T)}[1 - \exp(h\nu/kT)]e^{-\chi_i/kT} H.$$

Table 4.10. *Radial integrals and absorption oscillator strengths for hydrogen.*

Line	Transition	Wavelength (Å)	$(R_{nl}^{n'l'})^2$	f_{abs}
L_α	$1s$–$2p$	1 215.67	1.664 79	0.416 2
L_β	$1s$–$3p$	1 025.72	0.266 968	0.079 10
L_γ	$1s$–$4p$	972.54	0.092 771	0.028 99

Table 4.10. *(Continued.)*

Line	Transition	Wavelength (Å)	$(R_{nl}^{n'l'})^2$	f_{abs}
H_α	$2s–3p$	6562.74	9.3931	0.4349
H_α	$2p–3s$	6562.86	0.8806	0.01359
H_α	$2p–3d$	6562.81	22.5434	0.6958
H_β	$2s–4p$	4861.29	1.6444	0.1028
H_β	$2p–4s$	4861.35	0.1462	0.003045
H_β	$2p–4d$	4861.33	2.9231	0.1218

The factor $[1 - \exp(h\nu/kT)]$ allows for stimulated emission.

Spin doubles the value of all statistical weights *and* the partition function $u(T)$. Therefore, the sum of the gf's for transitions including spin must be double the corresponding sum of the gf's with spin ignored ($g_l = 2l + 1$), in order to keep W_λ' the same.

We use the convention that when a double subscript is written for an f or A value, the first subscript belongs to the *initial* level. A few authors follow a convention from atomic spectroscopy that the *lower level is written first*. Spin is ignored in calculating the absorption f values in Table 4.10 condensed from [21].

It is also possible to ignore the l degeneracy of hydrogen, so that only transitions of the form $n \leftrightarrow n'$ are considered. Let n and l be the initial levels and let n' and l' be the final ones. Then one defines average values of A as follows:

$$A_{nn'} = (1/n^2) \sum_{ll'} (2l + 1)A_{nl \to n'l'}.$$

For example,

$$A_{32} = \tfrac{1}{9}(1A_{3s \to 2p} + 3A_{3p \to 2s} + 5A_{3d \to 2p}).$$

A similar definition holds for the absorption $f_{nn'}$, but with the weights for the initial, *lower* level. Thus:

$$f_{23} = \tfrac{1}{4}(1f_{2s \to 3p} + 3f_{2p \to 3s} + 3f_{2p \to 3d}).$$

Data for the major series in hydrogen from [21] are given in Table 4.11.

Table 4.11. *Average Einstein A's and absorption f's.*

Line	Transition	Wavelength (Å)	A (s^{-1})	f_{abs}
L_α	1–2	1215.67	4.699×10^8	0.4162
L_β	1–3	1025.72	5.575×10^7	7.910×10^{-2}
L_γ	1–4	792.54	1.278×10^7	2.899×10^{-2}
L_{limit}	1–∞	911.8		
H_α	2–3	6562.80	4.410×10^7	0.6407
H_β	2–4	4861.32	8.419×10^6	0.1193
H_γ	2–5	4340.46	2.530×10^6	4.467×10^{-2}
H_δ	2–6	4101.73	9.732×10^5	2.209×10^{-2}
H_ϵ	2–7	3970.07	4.389×10^5	1.270×10^{-2}
H_8	2–8	3889.05	2.215×10^5	8.036×10^{-3}
H_{limit}	2–∞	3646		
P_α	3–4	18751.0	8.986×10^6	0.8421
P_β	3–5	12818.1	2.201×10^6	0.1506
P_γ	3–6	10938.1	7.783×10^5	5.584×10^{-2}
P_{limit}	3–∞	8204		

Table 4.11. *(Continued.)*

Line	Transition	Wavelength (Å)	A (s^{-1})	f_{abs}
B_α	4–5	40 512.0	2.699×10^6	1.038
B_β	4–6	26 252.0	7.711×10^5	0.179 3
B_γ	4–7	21 655.0	3.041×10^5	6.549×10^{-2}
B_{limit}	4–∞	14 584		

In Table 4.12 for L_α and H_α, the additional states and lines due to electron spin are shown explicitly. The levels are designated with quantum numbers n, l, s, L, S, and J (appropriate to LS coupling).

Wavelengths and levels are from [22]. The f and A values were generated from the expression for the line strength given above, using $L_p = 0$ in the appropriate $6 - j$ symbol, and the numerical constants from [8]. Use of these constants accounts for small differences with other tabulated values. For example, our sum of the gf values for the two L_α lines is 0.8321, while twice the value for $1s$–$2p$ given above is 0.8324.

In Table 4.13, multiplet numbers are mostly from [23]. They are labeled with u when the ultraviolet table [24] is used. The values of $\log(gf)$ given without explicit references were derived from [25]. Asterisks preceding the wavelengths indicate blends, in which case the gf is for the blend as a whole. Accuracy assessments are indicated by letters [21]. References for the table are collected separately at its end. Uncertainties in the range of 25%–50% are indicated by the letter D, those from 10%–25% by a C, 3%–10% by B, 1%–3% by A, and within 1% by AA. The letter E is used for accuracies below 50%. The same scheme is followed for other sources when accuracy estimates are available.

Table 4.12. *L_α and H_α transitions with doublet structure.*

Wavelength (Å)	Lower	Upper	χ_1 (cm^{-1})	χ_2 (cm^{-1})	$g_1 f_{12}$	$g_2 A_{21} \times 10^{-8}$ (s^{-1})
1 215.673 7	$1s\,^2S_{1/2}$	$2p\,^2P_{1/2}$	0	82 258.913	0.277 4	12.51
1 215.668 3	$1s\,^2S_{1/2}$	$2p\,^2P_{3/2}$	0	82 259.279	0.554 7	25.03
6 562.272 0	$2s\,^2S_{1/2}$	$3p\,^2P_{1/2}$	82 258.949	97 492.205	0.289 8	0.448 5
6 562.725 6	$2s\,^2S_{1/2}$	$3p\,^2P_{3/2}$	82 258.949	97 492.313	0.579 6	0.897 0
6 562.752 0	$2p\,^2P_{1/2}$	$3s\,^2S_{1/2}$	82 258.913	97 492.215	0.027 17	0.042 04
6 562.909 9	$2p\,^2P_{3/2}$	$3s\,^2S_{1/2}$	82 259.279	97 492.215	0.054 34	0.084 08
6 562.710 1	$2p\,^2P_{1/2}$	$3d\,^2D_{3/2}$	82 258.913	97 492.313	1.391	2.153
6 562.867 5	$2p\,^2P_{3/2}$	$3d\,^2D_{3/2}$	82 259.279	97 492.313	0.278 2	0.430 5
6 562.852 0	$2p\,^2P_{3/2}$	$3d\,^2D_{5/2}$	82 259.279	97 492.349	2.504	3.875

Table 4.13. *Atomic oscillator strengths for allowed lines.*

Atom	Transition	Multiplet			Line			
		No.	Designation	$J_i - J_k$	λ (Å)	$\log(gf)$	Accuracy	Reference
He II,	Hydrogen-like ions have nearly the same f values as those for hydrogen.							
Li III,	See discussion in [1] and [2] for Sc XXI–Ni XXVIII for higher-order effects.							
Be IV,								
B V, etc.								
He I	$1s^2$–$1s2p$	2u	^1S–$^1P^0$	0–1	584.33	$-0.558\,8$	AA	[3]
	$1s^2$–$1s3p$	3u	^1S–$^1P^0$	0–1	537.03	$-1.134\,1$	AA	[3]
	$1s^2$–$1s4p$	4u	^1S–$^1P^0$	0–1	522.21	$-1.524\,9$	AA	[3]

Table 4.13. *(Continued.)*

Atom	Transition	Multiplet			Line			Reference
		No.	Designation	$J_i - J_k$	λ (Å)	$\log(gf)$	Accuracy	
He I	$1s2s-1s2p$	1	$^3S-^3P^0$	1–0, 1, 2	*10 830	0.208 8	AA	[3]
(*Cont.*)			$^1S-^1P^0$	0–1	20 581	−0.424 3	AA	[3]
	$1s2s-1s3p$	2	$^3S-^3P^0$	1–0, 1, 2	*3 889	−0.713 5	AA	[3]
		4	$^1S-^1P^0$	0–1	5 016	−0.820 0	AA	[3]
	$1s2s-1s4p$	3	$^3S-^3P^0$	1–0, 1, 2	*3 188	−1.111 9	AA	[3]
		5	$^1S-^1P^0$	0–1	3 965	−1.308 5	AA	[3]
	$1s2p-1s3s$	10	$^3P^0-^3S$	0, 1, 2–1	*7 065	−0.203 7	AA	[3]
		45	$^1P^0-^1S$	1–0	7 281	−0.837 3	AA	[3]
	$1s2p-1s4s$	12	$^3P^0-^3S$	0, 1, 2–1	*4 713	−1.021 6	AA	[3]
		47	$^1P^0-^1S$	1–0	5 048	−1.587 3	AA	[3]
	$1s2p-1s3d$	11	$^3P^0-^3D$	0, 1, 2–1, 2, 3	*5 876	0.739 7	AA	[3]
		46	$^1P^0-^1D$	1–2	6 678	0.328 5	AA	[3]
	$1s2p-1s4d$	14	$^3P^0-^3D$	0, 1, 2–1, 2, 3	*4 472	0.043 6	AA	[3]
		48	$^1P^0-^1D$	1–2	4 922	−0.442 7	AA	[3]
	$1s2p-1s5d$	18	$^3P^0-^3D$	0, 1, 2–1, 2, 3	*4 026	−0.373 7	AA	[3]
		51	$^1P^0-^1D$	1–2	4 388	−0.886 6	AA	[3]
	$1s3s-1s3p$		$^3S-^3P^0$	1–0, 1, 2	*42 947	0.427 0	AA	[3]
			$^1S-^1P^0$	0–1	74 355	−0.203 2	AA	[3]
	$1s3s-1s4p$		$^3S-^3P^0$	1–0, 1, 2	*12 527	−0.823 4	AA	[3]
			$^1S-^1P^0$	0–1	15 084	−0.841 9	AA	[3]
Li I	$2s-2p$	1	$^2S-^2P^0$	$^1/_2$–1$^1/_2$	6 708	−0.001 2	AA	[3]
	$2s-2p$	1	$^2S-^2P^0$	$^1/_2$–$^1/_2$	6 708	−0.302 3	AA	[3]
Be II	$2s-2p$	1	$^2S-^2P^0$	$^1/_2$–1$^1/_2$	3 130	−0.177 2	AA	[3]
	$2s-2p$	1	$^2S-^2P^0$	$^1/_2$–$^1/_2$	3 131	−0.478 3	AA	[3]
C I	$2p3s-2p3p$	1	$^3P^0-^3D$	2–3	10 691	0.345	B	[3]
	$2p^2-2s2p^3$	31u	$^3P-^3S^0$	2–1	945.6	−0.118	C$^+$	[3]
		3u	$^3P-^3D^0$	2–3	1 561	−0.521	A	[3]
	$2p^2-2p3s$	2u	$^3P-^3P^0$	2–2	1 657	−0.285	A	[3]
	$2p^2-2p3d$	u7	$^3P-^3D^0$	2–3	1 278	−0.403	B$^-$	[3]
	$2p3s-2p3p$	10	$^1P^0-^1S$	1–0	8 335	−0.437	B$^+$	[3]
	$2p3s-2p4p$	6	$^3P^0-^3P$	2–2	4 772	−1.866	C	[3]
	$2p3s-2p4p$	11	$^1P^0-^1P$	1–1	5 380	−1.615	B	[3]
		12	$^1P^0-^1D$	1–2	5 052	−1.304	B	[3]
		13	$^1P^0-^1S$	1–0	4 932	−1.658	B	[3]
C II	$2s^22p-2s2p^2$	1u	$^2P^0-^2P$	1$^1/_2$–1$^1/_2$	904.1	0.224	B	[3]
	$2s^22p-2s2p^2$	2u	$^2P^0-^2S$	1$^1/_2$–$^1/_2$	1 037	−0.310	B	[3]
	$2s^22p-2s2p^2$	3u	$^2P^0-^2D$	1$^1/_2$–2$^1/_2$	1 336	−0.341	B	[3]
	$2p-3s$	4u	$^2P^0-^2S$	1$^1/_2$–$^1/_2$	858.6	−1.284	B	[3]
	$2p-3d$	5u	$^2P^0-^2D$	1$^1/_2$–2$^1/_2$	687.3	0.082	B	[3]
	$3s-3p$	2	$^2S-^2P^0$	$^1/_2$–1$^1/_2$	6 578	−0.026	B	[3]
	$3p-4s$	4	$^2P^0-^2S$	1$^1/_2$–$^1/_2$	3 921	−0.232	B	[3]
	$3p-3d$	3	$^2P^0-^2D$	1$^1/_2$–2$^1/_2$	7 236	0.298	B	[3]
	$3d-4f$	6	$^2D-^2F^0$	2$^1/_2$–3$^1/_2$	4 267	0.717	C$^+$	[3]
C III	$2s^2-2s2p$	1u	$^1S-^1P^0$	0–1	977.0	−0.120 0	A$^+$	[3]
	$2s^2-2s3p$	2u	$^1S-^1P^0$	0–1	386.2	−0.634	B	[3]
	$2s3s-2s3p$	1	$^3S-^3P^0$	1–2	4 647	0.070	B$^+$	[3]

Table 4.13. *(Continued.)*

Atom	Transition	No.	Designation	$J_i - J_k$	λ (Å)	log(gf)	Accuracy	Reference
C IV	$2s-2p$	1u	$^2S-^2P^0$	$1/2-11/2$	1 548	−0.419	A	[3]
	$2s-2p$	1u	$^2S-^2P^0$	$1/2-1/2$	1 551	−0.721	A	[3]
	$2s-3p$	2u	$^2S-^2P^0$	$1/2-1/2, 11/2$	*312.4	−0.391	A⁻	[3]
	$3s-3p$	1	$^2S-^2P^0$	$1/2-11/2$	5 801	−0.194	A	[3]
C V	$1s^2-1s2p$		$^1S-^1P^0$	$0-1$	40.27	−0.189 1	AA	[3]
N I	$2p^3-2p^23s$	2u	$^4S^0-^4P$	$11/2-21/2$	1 199.6	−0.285	B⁺	[3]
	$2p^23s-2p^23p$	1	$^4P-^4D^0$	$21/2-31/2$	8 680	0.346	B⁺	[3]
		8	$^2P-^2P^0$	$11/2-11/2$	8 629	0.075	B	[3]
	$2p^23s-2p^24p$	6	$^4P-^4S^0$	$21/2-11/2$	4 152	−1.981	C⁺	[3]
N II	$2s^22p^2-2s2p^3$	2u	$^3P-^3P^0$	$0-1$	915.6	−0.782	B⁺	[3]
	$2s^22p^2-2s2p^3$	1u	$^3P-^3D^0$	$1-1$	1 085	−1.071	B⁺	[3]
	$2p3s-2p3p$	3	$^3P^0-^3D$	$2-3$	5 679.6	0.250	A	[3]
		12	$^1P^0-^1D$	$1-2$	3 995	0.215	B⁺	[3]
	$2p3p-2p3d$	19	$^3D-^3F^0$	$3-4$	500.2	0.592	C⁺	[3]
N III	$2s^22p-2s2p^2$	1u	$^2P^0-^2D$	$11/2-21/2$	991.6	−0.357	B	[3]
	$2s^22p-2s2p^2$	1u	$^2P^0-^2D$	$11/2-11/2$	991.5	−1.317	B	[3]
	$3s-3p$	1	$^2S-^2P^0$	$1/2-11/2$	4 097	−0.057	B	[3]
	$2s2p3s-2s2p3p$	3	$^4P^0-^4D$	$21/2-31/2$	4 515	0.221	B	[3]
N IV	$2s^2-2s2p$	1u	$^1S-^1P^0$	$0-1$	765.1	−0.214 0	A⁺	[3]
	$2s^2-2s3p$	2u	$^1S-^1P^0$	$0-1$	247.2	−0.486	B	[3]
	$2s3s-2s3p$	1	$^3S-^3P^0$	$1-0, 1, 2$	*3 481	0.238	B	[3]
	$2s3p-2s3d$	3	$^1P^0-^1D$	$1-2$	4 058	−0.088	B	[3]
N V	$2s-2p$	1u	$^2S-^2P^0$	$1/2-11/2$	1 239	−0.505	A	[3]
	$2s-2p$	1u	$^2S-^2P^0$	$1/2-1/2$	1 243	−0.807	A	[3]
	$2s-3p$	2u	$^2S-^2P^0$	$1/2-1/2, 11/2$	*209.3	−0.321	A	[3]
	$3s-3p$	1	$^2S-^2P^0$	$1/2-11/2$	4 604	−0.278	A	[3]
N VI	$1s^2-1s2p$		$^1S-^1P^0$	$0-1$	28.79	−0.171 2	AA	[3]
O I	$2p^4-2p^33s$	2u	$^3P-^3S^0$	$2-1$	1 302	−0.585	A	[3]
		5u	$^3P-^3D^0$	$2-3$	988.8	−0.634	B	[3]
	$2p^33s-2p^33p$	1	$^5S^0-^5P$	$2-3$	7 772	0.369	A	[3]
		4	$^3S^0-^3P$	$1-2$	8 446	0.236	B	[3]
	$2p^33s-2p^34p$	5	$^3S^0-^3P$	$1-2$	4 368	−1.983	C	[3]
	$2p^33p-2p^34d$	10	$^5P-^5D^0$	$3-4$	6 158	−0.409	B⁺	[3]
O II	$2p^3-2p^23d$	3u	$^4S^0-^4P$	$11/2-21/2$	430.2	−0.139	B⁺	[3]
	$2s^22p^3-2s2p^4$	1u	$^4S^0-^4P$	$11/2-21/2$	834.5	−0.268	B⁺	[3]
	$2p^23s-2p^23p$	1	$^4P-^4D^0$	$21/2-31/2$	4 649	0.307	B⁺	[3]
		3	$^4P-^4S^0$	$21/2-11/2$	3 749	−0.105	B⁺	[3]
	$2p^23p-2p^23d$	20	$^4P^0-^4D$	$21/2-31/2$	4 119	0.433	B⁺	[3]
O III	$2s^22p^2-2s2p^3$	1u	$^3P-^3D^0$	$2-3$	835.3	−0.358	A	[3]
		2u	$^3P-^3P^0$	$2-2$	703.9	−0.293	A	[3]
	$2p3s-2p3p$	2	$^3P^0-^3D$	$2-3$	3 760	0.162	C⁺	[3]
	$2p3p-2p3d$	14	$^3P-^3D^0$	$2-3$	3 715	0.149	C⁺	[3]
O IV	$2p-3d$	5u	$^2P^0-^2D$	$11/2-21/2$	238.6	0.258	B	[3]
	$2s^22p-2s2p^2$	1u	$^2P^0-^2D$	$11/2-21/2$	790.2	−0.401	B	[3]
	$2s2p3s-2s2p3p$	3	$^4P^0-^4D$	$21/2-31/2$	3 386	0.148	B	[3]

Table 4.13. (*Continued.*)

Atom	Transition	No.	Designation	$J_i - J_k$	λ (Å)	log(gf)	Accuracy	Reference
		\multicolumn{2}{Multiplet}	Line					
O v	$2s^2$–$2s2p$	1u	^1S–$^1P^0$	0–1	629.7	−0.2905	A$^+$	[3]
	$2s^2$–$2s3p$	2u	^1S–$^1P^0$	0–1	172.2	−0.407	B	[3]
	$2p3s$–$2p3p$	4	$^3P^0$–^3D	2–3	4124	−0.066	B	[3]
	$2p3p$–$2p3d$	11	^3S–$^3P^0$	1–2	4159	−0.356	B	[3]
O vi	$2s$–$2p$	1u	^2S–$^2P^0$	$\frac{1}{2}$–1$\frac{1}{2}$	1032	−0.576	A	[3]
	$2s$–$2p$	1u	^2S–$^2P^0$	$\frac{1}{2}$–$\frac{1}{2}$	1038	−0.879	A	[3]
	$2s$–$3p$	2u	^2S–$^2P^0$	$\frac{1}{2}$–1$\frac{1}{2}$	150.1	−0.451	A$^-$	[3]
	$3s$–$3p$	1	^2S–$^2P^0$	$\frac{1}{2}$–1$\frac{1}{2}$	3811	−0.349	A	[3]
O vii	$1s^2$–$1s2p$		^1S–$^1P^0$	0–1	21.60	−0.1584	AA	[3]
Ne i	$2p^53s$–$2p^53p$	1		2–3	6402	0.345	B	
Ne ii	$2p^43s$–$2p^43p$	1	^4P–$^4P^0$	2$\frac{1}{2}$–2$\frac{1}{2}$	3694	0.09	D	
Ne vi	$2p$–$3d$		$^2P^0$–^2D	1$\frac{1}{2}$–2$\frac{1}{2}$	122.7	0.313	D	[3]
Ne vii	$2s^2$–$2s2p$		^1S–$^1P^0$	0–1	465.2	−0.410	C	[3]
Ne viii	$2s$–$2p$		^2S–$^2P^0$	$\frac{1}{2}$–1$\frac{1}{2}$	770.4	−0.689	B$^+$	[3]
Ne ix	$1s^2$–$1s2p$		^1S–$^1P^0$	0–1	13.45	−0.141	A	[3]
Na i	$3s$–$3p$	1	^2S–$^2P^0$	$\frac{1}{2}$–1$\frac{1}{2}$	5890	0.104	A	[3]
	$3s$–$3p$	1	^2S–$^2P^0$	$\frac{1}{2}$–$\frac{1}{2}$	5896	−0.197	A	[3]
	$3s$–$4p$	2	^2S–$^2P^0$	$\frac{1}{2}$–1$\frac{1}{2}$	3302	−1.736	C	
	$3p$–$4s$	3	$^2P^0$–^2S	1$\frac{1}{2}$–$\frac{1}{2}$	11404	−0.163	C	
	$3p$–$5s$	5	$^2P^0$–^2S	1$\frac{1}{2}$–$\frac{1}{2}$	6161	−1.23	C	
	$3p$–$6s$	8	$^2P^0$–^2S	1$\frac{1}{2}$–$\frac{1}{2}$	5153	−1.732	C	
	$3p$–$3d$	4	$^2P^0$–^2D	1$\frac{1}{2}$–2$\frac{1}{2}$	8195	0.51	C	
	$3p$–$4d$	6	$^2P^0$–^2D	1$\frac{1}{2}$–2$\frac{1}{2}$	5688	−0.46	C	
	$3p$–$5d$	9	$^2P^0$–^2D	1$\frac{1}{2}$–2$\frac{1}{2}$	4983	−0.962	C	
Mg i	$3s^2$–$3s3p$	1u	^1S–$^1P^0$	0–1	2852	0.29	D	
	$3s3p$–$3s4s$	2	$^3P^0$–^3S	2–1	5184	−0.158	B	
		6	$^1P^0$–^1S	1–0	11828	−0.27	D	[4]
	$3s3p$–$3s5s$	4	$^3P^0$–^3S	2–1	3337	−1.10	C	
	$3s3p$–$3s3d$	3	$^3P^0$–^3D	2–3	3838	0.414	B	[4]
		7	$^1P^0$–^1D	1–2	8807	−0.08	D	[4]
	$3s3p$–$3p^2$	6u	$^3P^0$–^3P	0, 1, 2–0, 1, 2	*2780	0.73	C	
Mg ii	$3s$–$3p$	1u	^2S–$^2P^0$	$\frac{1}{2}$–1$\frac{1}{2}$	2796	0.09	C	
	$3p$–$4s$	2u	$^2P^0$–^2S	1$\frac{1}{2}$–$\frac{1}{2}$	2937	−0.23	C	
	$3p$–$3d$	3u	$^2P^0$–^2D	1$\frac{1}{2}$–1$\frac{1}{2}$	2798	−0.43	D	
	$3d$–$4f$	4	^2D–$^2F^0$	1$\frac{1}{2}$, 2$\frac{1}{2}$–2$\frac{1}{2}$, 3$\frac{1}{2}$	*4481	0.973	C	
	$4s$–$4p$	1	^2S–$^2P^0$	$\frac{1}{2}$–1$\frac{1}{2}$	9218	0.26	C	
Mg ix	$2s^2$–$2s2p$		^1S–$^1P^0$	0–1	368.1	−0.493	B	
Mg x	$2s$–$2p$		^2S–$^2P^0$	$\frac{1}{2}$–1$\frac{1}{2}$	609.8	−0.775	B	
	$2s$–$3p$		^2S–$^2P^0$	$\frac{1}{2}$–1$\frac{1}{2}$	57.88	−0.377	B	
Mg xi	$1s^2$–$1s2p$		^1S–$^1P^0$	0–1	9.169	−0.128	B	
Al i	$3p$–$4s$	1	$^2P^0$–^2S	1$\frac{1}{2}$–$\frac{1}{2}$	3962	−0.34	C	
	$4s$–$5p$	5	^2S–$^2P^0$	$\frac{1}{2}$–1$\frac{1}{2}$	6696	−1.343	C	
	$3p$–$3d$	3	$^2P^0$–^2D	1$\frac{1}{2}$–2$\frac{1}{2}$	3093	0.263	C	

Table 4.13. *(Continued.)*

Atom	Transition	No.	Designation	$J_i - J_k$	λ (Å)	log(gf)	Accuracy	Reference
Al II	$3s^2$–$3s3p$	2u	^1S–$^1P^0$	0–1	1671	0.263	B	
	$3s3p$–$3s4s$	4u	$^3P^0$–^3S	2–1	1862	−0.192	B	
Al III	$3s$–$3p$	1u	^2S–$^2P^0$	½–1½	1855	0.047	B	
	$4s$–$4p$	2	^2S–$^2P^0$	½–1½	5696	0.235	B	[4]
Al X	$2s^2$–$2s2p$		^1S–$^1P^0$	0–1	332.8	−0.55	C	
Si I	$3p^2$–$3p4s$	1u	^3P–$^3P^0$	2–2	2516	−0.241	C	
		43u	^1D–$^1P^0$	2–1	2882	−0.151	C	
		3	^1S–$^1P^0$	0–1	3906	−1.092	C	
	$3s^23p^2$–$3s3p^3$	3u	^3P–$^3D^0$	2–3	2217	−0.55	C	
	$3p4s$–$3p4p$	4	$^3P^0$–^3D	2–3	12031	0.41	D	[4]
		5	$^3P^0$–^3P	2–2	10827	0.16	D	[4]
		6	$^3P^0$–^3S	2–1	10585	−0.19	D	[4]
Si II	$4s$–$4p$	2	^2S–$^2P^0$	½–1½	6347	0.23	C	
	$3d$–$4f$	3	^2D–$^2F^0$	2½–3½	4131	0.463	C	
	$3s^23p$–$3s3p^2$	1u	$^2P^0$–^2D	½–1½	1808	−2.14	D	
	$3s^23p$–$3s3p^2$	5u	$^2P^0$–^2P	1½–1½	1195	0.49	D	
	$3p$–$3d$	4u	$^2P^0$–^2D	1½–2½	1265	0.52	D	
	$3p$–$4s$	2u	$^2P^0$–^2S	1½–½	1534	−0.28	C	
	$3p$–$4d$	6u	$^2P^0$–^2D	1½–2½	992.7	−0.15	D	
Si III	$3s^2$–$3s3p$	2u	^1S–$^1P^0$	0–1	1207	0.22	B	[5]
	$3s4s$–$3s4p$	2	^3S–$^3P^0$	1–2	4553	0.292	C	
		4	^1S–$^1P^0$	0–1	5740	−0.16	D	
Si IV	$3s$–$3p$	1u	^2S–$^2P^0$	½–1½	1394	0.01	B	[5]
	$3s$–$4p$	2u	^2S–$^2P^0$	½–1½	457.8	−1.34	D	
	$4s$–$4p$	1	^2S–$^2P^0$	½–1½	4089	0.195	B	[4]
Si XI	$2s^2$–$2s2p$		^1S–$^1P^0$	0–1	303.3	−0.576	C	
Si XII	$2s$–$2p$		^2S–$^2P^0$	½–1½	499.4	−0.845	B	[6]
S I	$3p^34s$–$3p^34p$	1	$^5S^0$–^5P	2–3	9213	0.42	D	[4]
S II	$3s^23p^3$–$3s3p^4$	1u	$^4S^0$–^4P	1½–2½	1260	−1.31	C	
S IV	$3p$–$4s$	5u	$^2P^0$–^2S	1½–½	554.1	−0.425	C	
S V	$3s^2$–$3s3p$	1u	^1S–$^1P^0$	0–1	786.5	0.165	B	
	$3s3p$–$3s3d$	3u	$^3P^0$–^3D	0, 1, 2–1, 2, 3	*661.5	0.802	B	
K I	$4s$–$4p$	1	^2S–$^2P^0$	½–1½	7665	0.135	B	
	$4s$–$5p$	3	^2S–$^2P^0$	½–1½	4044	−1.915	C	
Ca I	$4s^2$–$4s4p$	2	^1S–$^1P^0$	0–1	4227	0.243	B	
	$4s4p$–$4s5s$	3	$^3P^0$–^3S	2–1	6162	−0.089	C	
	$4s4p$–$4s6s$	6	$^3P^0$–^3S	2–1	3974	−0.906	C	
	$4s4p$–$4s4d$	4	$^3P^0$–^3D	2–3	4455	0.26	C	
	$4s4p$–$4s5d$	9	$^3P^0$–^3D	2–3	3644	−0.306	C	
	$4s4p$–$4s6d$	11	$^3P^0$–^3D	2–3	3362	−0.578	C	
	$4s4p$–$4p^2$	5	$^3P^0$–^3P	2–2	4303	0.276	C	
	$3d4s$–$3d4p$	21	^3D–$^3D^0$	3–3	5589	0.21	D	

Table 4.13. *(Continued.)*

Atom	Transition	Multiplet No.	Designation	$J_i - J_k$	Line λ (Å)	$\log(gf)$	Accuracy	Reference
Ca II	$4s$–$4p$	1	^2S–$^2P^0$	$^1/_2$–$1^1/_2$	3 934	0.135	C	
	$3d$–$4p$	2	^2D–$^2P^0$	$2^1/_2$–$1^1/_2$	8 542	−0.365	C	[4]
	$4p$–$5s$	3	$^2P^0$–^2S	$1^1/_2$–$^1/_2$	3 737	−0.15	C	
	$4p$–$4d$	4	$^2P^0$–^2D	$1^1/_2$–$2^1/_2$	3 179	0.51	C	
Sc I	$3d^24s$–$3d^24p$	12	^4F–$^4G^0$	$4^1/_2$–$5^1/_2$	5 672	0.49	D	
		14	^4F–$^4D^0$	$4^1/_2$–$3^1/_2$	4 744	0.42	D	
		15	^2F–$^2G^0$	$3^1/_2$–$4^1/_2$	5 521	0.29	D	
		16	^2F–$^2F^0$	$3^1/_2$–$3^1/_2$	5 482	0.27	D	
		6	^2D–$^2P^0$	$2^1/_2$–$1^1/_2$	4 082	−0.57	C	
Ti I	$3d^34s$–$3d^34p$	38	^5F–$^5G^0$	5–6	4 982	0.504	C+	[1]
		42	^5F–$^5F^0$	5–5	4 533	0.476	C+	[1]
		145	^5P–$^5D^0$	3–4	4 617	0.389	C+	[1]
		12	^3F–$^3F^0$	4–4	3 999	−0.056	B	[1]
		24	^3F–$^3G^0$	4–5	3 371	0.13	C	[1]
	$3d^34s$–$3d^24s4p$	110	^3F–$^3G^0$	3–4	5 036	0.130	C+	[1]
Ti II	$3d^24s$–$3d^24p$	1	^4F–$^4G^0$	$3^1/_2$–$4^1/_2$	3 361	0.28	C	[1]
		2	^4F–$^4F^0$	$4^1/_2$–$4^1/_2$	3 235	0.336	C+	[1]
	$3d^3$–$3d^24p$	7	^4F–$^4F^0$	$4^1/_2$–$4^1/_2$	3 323	−0.183	C+	[1]
V I	$3d^44s$–$3d^44p$	21	^6D–$^6P^0$	$4^1/_2$–$3^1/_2$	4 460	−0.15	C−	[1]
		22	^6D–$^6F^0$	$4^1/_2$–$5^1/_2$	4 379	0.58	C	[1]
		27	^6D–$^6D^0$	$4^1/_2$–$4^1/_2$	4 112	0.408	B	[1]
		88	^4H–$^4H^0$	$6^1/_2$–$6^1/_2$	4 269	0.65	C−	[1]
		109	^4F–$^4G^0$	$4^1/_2$–$5^1/_2$	4 545	0.45	C−	[1]
		14	^4F–$^4G^0$	$4^1/_2$–$5^1/_2$	3 185	0.69	C	[1]
	$3d^44s$–$3d^34s4p$	29	^6D–$^6P^0$	$4^1/_2$–$3^1/_2$	3 704	0.18	C−	[1]
		41	^4D–$^4F^0$	$3^1/_2$–$4^1/_2$	4 091	0.33	C	[1]
	$3d^34s4p$–$3d^34s5s$	125	$^6F^0$–^6F	$5^1/_2$–$5^1/_2$	5 193	0.29	C−	[1]
	$3d^34s4p$–$3d^34s4d$	114	$^6G^0$–^6H	$6^1/_2$–$7^1/_2$	3 695	0.97	C−	[1]
V II	$3d^34s$–$3d^34p$	11	^3P–$^5D^0$	2–3	3 903	−0.89	B	[1]
		5	^3F–$^3D^0$	4–3	3 557	−0.17	B	[1]
		25	^5P–$^5D^0$	3–4	4 202	−1.75	D	[1]
Cr I	$3d^54s$–$3d^54p$	1	^7S–$^7P^0$	3–4	4 254	−0.114	B	[1]
		7	^5S–$^5P^0$	2–3	5 208	0.158	B	[1]
		38	^5G–$^5H^0$	6–7	3 964	0.67	D−	[1]
	$3d^44s^2$–$3d^44s4p$	22	^5D–$^5F^0$	4–5	4 351	−0.44	C	[1]
	$3d^54s$–$3d^44s4p$	4	^7S–$^7P^0$	3–4	3 579	0.409	B	[1]
		43	^5G–$^5G^0$	6–6	3 744	0.318	B	[1]
Mn I	$3d^64s$–$3d^64p$	5	^6D–$^6D^0$	$4^1/_2$–$4^1/_2$	4 041	0.285	C+	[1]
		6	^6D–$^6F^0$	$4^1/_2$–$5^1/_2$	3 807	0.19	B	[1]
	$3d^54s^2$–$3d^54s4p$	2	^6S–$^6P^0$	$2^1/_2$–$3^1/_2$	4 031	−0.47	C+	[1]
		1u	^6S–$^6P^0$	$2^1/_2$–$3^1/_2$	2 795	0.53	C	[1]
Fe I	$3d^74s$–$3d^74p$	20	^5F–$^5D^0$	5–4	3 820	0.119	B+	[2]
		23	^5F–$^5G^0$	5–6	3 581	0.406	B+	[2]
		41	^3F–$^5G^0$	4–5	4 384	0.200	B+	[2]
		42	^3F–$^3G^0$	4–5	4 272	−0.164	B+	[2]
		43	^3F–$^3F^0$	4–4	4 046	0.280	B+	[2]
		45	^3F–$^3D^0$	4–3	3 816	0.237	B	[7]

Table 4.13. *(Continued.)*

Atom	Transition	No.	Designation	$J_i - J_k$	λ (Å)	log(gf)	Accuracy	Reference
			Multiplet		**Line**			
Fe I	$3d^64s^2-3d^64s4p$	4	$^5D-^5D^0$	4–4	3 860	−0.710	B$^+$	[2]
(*Cont.*)		5	$^5D-^5F^0$	4–5	3 720	−0.431	B$^+$	[2]
	$3d^74s-3d^64s4p$	68	$^5P-^5D^0$	3–4	4 529	−0.822	B$^+$	[2]
	$3d^64s4p-3d^64s5s$	152	$^7D^0-^7D$	5–5	4 260	0.077	B	[7]
Fe II	$3d^64s-3d^64p$	27	$^4P-^4D^0$	2½–3½	4 233	−2.00	C	[2]
		38	$^4F-^4D^0$	4½–3½	4 584	−2.02	D	[2]
Co I	$3d^84s-3d^84p$	22	$^4F-^4G^0$	4½–5½	3 454	0.38	C$^+$	[2]
		23	$^4F-^4F^0$	4½–4½	3 405	0.25	C$^+$	[2]
		35	$^2F-^2F^0$	3½–3½	3 569	0.37	C	[2]
	$3d^74s^2-3d^74s4p$	5	$^4F-^4G^0$	4½–5½	3 466	−0.70	C	[2]
	$3d^84s-3d^74s4p$	28	$^2F-^2G^0$	3½–4½	4 121	−0.32	C	[2]
Ni I	$3d^94s-3d^94p$	19	$^3D-^3F^0$	3–4	3 415	−0.06	C	[2]
		35	$^1D-^1F^0$	2–3	3 619	−0.04	C	[2]
	$3d^84s^2-3d^84s4p$	7	$^3F-^3G^0$	4–5	3 233	−0.90	C	[2]
	$3d^94s-3d^84s4p$	25	$^3D-^3F^0$	3–4	3 051	−0.12	C$^+$	[2]
	$3d^84s4p-3d^84s5s$	111	$^5F^0-^5F$	5–5	5 018	−0.08	D	[2]
	$3d^84s4p-3d^84s4d$	17	$^3D-^5F^0$	3–3	3 374	−1.76	C	[2]
	$3d^94p-3d^94d$	130	$^3P^0-^3P$	2–2	4 855	0.00	D	[2]
		143	$^3F^0-^3G$	4–5	5 081	0.13	D	[2]
		162	$^3D^0-^3F$	3–4	5 084	0.03	D$^-$	[2]
		194	$^1F^0-^1G$	3–4	5 081	0.30	D$^-$	[2]
Cu I	$4s-4p$	1	$^2S-^2P^0$	½–1½	3 248	−0.056	C	
	$3d^94s^2-3d^{10}4p$	2	$^2D-^2P^0$	2½–1½	5 106	−1.50	D	
	$4p-4d$	7	$^2P^0-^2D$	1½–2½	5 218	0.26	D	
Zn I	$4s4p-4s4d$	4	$^3P^0-^3D$	2–3	3 345	0.30	B	
	$4s4p-4s4d$	6	$^1P^0-^1D$	1–2	6 362	0.158	C	
Sr I	$5s^2-5s5p$	2	$^1S-^1P^0$	0–1	4 607	0.283	C	
Sr II	$5s-5p$	1	$^2S-^2P^0$	½–1½	4 078	0.151	C	
	$5d-6s$	3	$^2P^0-^2S$	1½–½	4 306	−0.11	D	
Ba I	$6s^2-6s6p$	2	$^1S-^1P^0$	0–1	5 536	0.215	C	
Ba II	$6s-6p$	1	$^2S-^2P^0$	½–1½	4 554	0.163	C	
	$5d-6p$	2	$^2D-^2P^0$	2½–1½	6 142	−0.08	D	
	$6p-6d$	4	$^2P^0-^2D$	1½–2½	4 131	0.441	C	
Hg I	$6s6p-6s7s$	1	$^3P^0-^3S$	0–1	4 047	−0.81	D	
		1		1–2	4 348	−0.92	D	
		1		2–1	5 461	−0.185	C	
Pb I	$6p^2-6p7s$	1	$^3P-^3P^0$	2–1	4 058	−0.18	D	

References

1. Martin, G.A., Fuhr, J.R., & Wiese, W.L. 1988, *J. Phys. Chem. Ref. Data*, **17**, Suppl. 3
2. Fuhr, J.R., Martin, G.A., & Wiese, W.L. 1988, *J. Phys. Chem. Ref. Data*, **17**, Suppl. 4
3. Wiese, W.L., Fuhr, J.R., & Deters, T.M. 1996, *J. Phys. Chem. Ref. Data Monograph*, **7**; and other data to be published
4. Wiese, W.L., Smith, M.W., & Miles, B.M. 1969, *Atomic Transition Probabilities, Sodium Through Calcium, NSRDS-NBS*, **22**
5. Morton, D.C. 1991, *ApJS*, **77**, 119
6. Wiese, W.L., Smith, M.W., & Glennon, B.M. 1966, *Atomic Transition Probabilities*, H *Through* Ne, *NSRDS-NBS*, **4**
7. O'Brian, T.R., Wickliffe, M.E., Lawler, J.E., Whaling, W., & Brault, J.W. 1991, *J. Opt. Soc. Am.*, **B8**, 1185

4.8 NUCLEAR SPIN AND HYPERFINE STRUCTURE: LOW-LEVEL HYPERFINE TRANSITIONS

The angular momentum, or spin, of the ground levels of nuclei [26] can be of importance in atomic spectra and structure. Nonzero spins result from unpaired nucleons and occur for *some* isotopes of most elements. In elements with odd Z, the most abundant isotope will have a nonzero spin, so hyperfine structure is most important for these species. However, secondary (odd-N) isotopes of even-Z elements may make a significant contribution to the overall line shape.

If the spin $\mathbf{I}$ is taken into account, the total angular momentum of an atomic level is $\mathbf{F} = \mathbf{J} + \mathbf{I}$. The vectors $\mathbf{J}$ and $\mathbf{I}$ are added using the same rules as when $\mathbf{L}$ and $\mathbf{S}$ are added to form $\mathbf{J}$. The quantum numbers F and I play analogous roles to J and S. Thus, for a given J and I there are $2I + 1$ values of F if $J > I$, and $2J + 1$ if $I > J$. The number of elementary states belonging to a level with a given F is $2F + 1$, corresponding to the number of possible values of M_F, the projection of $\mathbf{F}$ on the z axis in units of $\hbar$. When there should be no ambiguity, M_F may be written without the subscript M.

Nuclear spin broadens spectral lines and adds $2I + 1$ additional states to an atomic system. The first factor, known as hyperfine splitting, may usually be ignored if the resultant width is much smaller than that due to other broadening mechanisms, such as pressure or Doppler broadening. The additional atomic states cancel in the Boltzmann and Saha formulas and usually are not accounted for explicitly.

The splitting of atomic levels due to nuclear spin (ΔE_M) may be augmented (ΔE_Q) if the nucleus has an electric quadrupole moment [27]:

$$\Delta E_M = \tfrac{1}{2}A[F(F+1) - J(J+1) - I(I+1)] \equiv \tfrac{1}{2}AC,$$
$$\Delta E_Q = B[C(C+1) - \tfrac{4}{3}J(J+1)I(I+1)].$$

Values of A and B are given by [28].

Nuclear mass effects may be treated as follows. Let $\mathbf{P}$ be the momentum of the nucleus with mass M, and let $\mathbf{p}_i$ be the momentum of the ith electron. The kinetic energy is then

$$E = \frac{\mathbf{P}^2}{2M} + \sum_i \frac{\mathbf{p}_i^2}{2m}.$$

We can eliminate $\mathbf{P}$ using $\mathbf{P} + \sum_i \mathbf{p}_i = \mathbf{0}$, whence

$$E = \left(\frac{1}{2M} + \frac{1}{2m}\right)\sum_i \mathbf{p}_i^2 + \frac{1}{M}\sum_{i,j>i} \mathbf{p}_i \cdot \mathbf{p}_j.$$

The first term is called the *normal mass shift* and gives rise to well-known displacements of lines in very light elements [29]. The second term, called the *specific mass shift*, is difficult to calculate [27] but may be measured in the laboratory. It can be significant even for heavy atoms [30].

Finally, nuclear volume, *field* effects, or isotope shifts occur because the potential at small r departs from a pure $1/r$ dependence due to the finite size of the nucleus. Astrophysically important consequences have been documented [31].

While the hyperfine width is difficult to calculate, the relative intensities of lines in a hyperfine multiplet follow readily from the quantum theory of angular momentum. The relative line strengths are written simply with a Wigner $6 - j$ symbol:

$$S(J'IF' \leftrightarrow JIF) \propto (2F + 1)(2F' + 1)\begin{Bmatrix} J & I & F \\ F' & 1 & J' \end{Bmatrix}^2 .$$

The relative intensities are identical to those discussed for LS coupling, and the tables of Sec. 4.5 may be used with the substitutions $J \to F$, $S \to I$, and $L \to J$.

The celebrated 21-cm line in atomic hydrogen is an example of a pure magnetic dipole transition. Similar transitions occur in ionized ^{3}He, as well as in deuterium. Results are summarized in Table 4.14, with 1986 constants and transition frequencies from [32]. The formula for magnetic dipole radiation simplifies in this case to

$$g_{F'}A(F', F) = \frac{4\pi^2 he^2 g_e^2 \nu^3}{3m^2 c^5} \sum_{MM'q} |\langle s_n' s_e' F' M' | S_q^{(1)} | s_n s_e F M \rangle|^2.$$

Here, $S_q^{(1)}$ is a spherical tensor, analogous to $C_m^{(1)}$, which operates in electron spin space. Quantum numbers s_n and s_e describe the spin states of the nucleus and electron. For ^{1}H I and ^{3}He II, $g_{F'} = 3$, while for ^{2}H I it is 4. The sums over M, M', and q are 3/4 for ^{1}H I and ^{3}He II and 4/3 for ^{2}H I. The numerical coefficient is $4.01367 \times 10^{-42} \nu^3$ (cgs). We have neglected the magnetic moment of the nucleus. The ground state *orbital* functions are not indicated, since they contribute only a trivial multiplicative factor of unity.

4.9 FORBIDDEN LINE TRANSITION PROBABILITIES

Most of the lines in Table 4.15 are forbidden in the sense that they involve no change in parity. A few *intersystem* lines are included. Both magnetic dipole ($M1$) and electric quadrupole ($E2$) lines are possible at the same wavelength in many cases. The dominant radiation is indicated, but the A value is for the *sum* over all mechanisms, including electric dipole radiation (for intersystem lines). When both magnetic dipole and electric quadrupole transitions are permitted by their selection rules, the Einstein A coefficient for the magnetic dipole will usually dominate for optical transitions. Generally, $A_m/A_q \approx 3 \times 10^{11}/\sigma^2$, where σ is the wave number of the transition. Typical A values for electric dipole transitions are 10^5 times larger than their magnetic dipole congeners. Accuracy estimates from [33–36] are indicated where available. The notation is the same as in Sec. 4.7.

Table 4.14. *Hyperfine transitions.*

	I	F'	F	ν (Hz)	A_{21} (s^{-1})
^{1}H I	½	1	0	$1.420\,405\,752 \times 10^9$	2.876×10^{-15}
^{2}H I	1	1½	½	$3.273\,843\,523 \times 10^8$	4.695×10^{-17}
^{3}He II	½	1	0	$8.665\,649\,867 \times 10^9$	6.530×10^{-13}

Table 4.15. *Forbidden and intercombination lines.*

Atom	Array	Designation lower–upper	J_i–J_k	λ (Å)	A (s^{-1})	Accuracy	$M1$ or $E2$	Reference
He I]	$1s^2$–$1s2p$	^1S–$^3P^0$	0–1	591.4	$1.76 \times 10^{+2}$			[1]
[C I]	$2p^2$	^1D–^1S	2–0	8727	0.634	B	$E2$	[2]
C II]	$2s^2 2p$–$2s2p^2$	$^2P^0$–^4P	1½–2½	2325.4	52.6	B		[2]
C III]	$2s^2$–$2s2p$	^1S–$^3P^0$	0–1	1908.7	114	B$^+$		[2]
[N I]	$2p^3$	$^4S^0$–$^2D^0$	1½–1½	5198	2.26×10^{-5}	C	$M1$	[2]
		$^4S^0$–$^2D^0$	1½–2½	5200	5.77×10^{-6}	B	$E2$	[2]

Table 4.15. *(Continued.)*

Atom	Array	Designation lower–upper	J_i–J_k	λ (Å)	A (s⁻¹)	Accuracy	$M1$ or $E2$	Reference
[N II]	$2p^2$	^1D–^1S	2–0	5755	1.17	B	$E2$	[2]
		^3P–^1S	2–0	3071	1.40×10^{-4}	B	$E2$	[2]
		^3P–^1S	1–0	3063	3.15×10^{-2}	B	$M1$	[2]
		^3P–^1D	1–2	6548	9.20×10^{-4}	B	$M1$	[2]
		^3P–^1D	2–2	6583	2.73×10^{-3}	B	$M1$	[2]
		^3P–^1D	0–2	6527	5.45×10^{-7}	B	$E2$	[2]
		^3P–^3P	1–2	121.8 μm	7.40×10^{-6}	B	$M1$	[2]
		^3P–^3P	0–2	76.45 μm	9.69×10^{-13}	C	$E2$	[2]
		^3P–^3P	0–1	205.3 μm	2.07×10^{-6}	B	$E2$	[2]
N II]	$2s^22p^2$–$2s2p^3$	^3P–$^5S^0$	2–2	2143	$1.27 \times 10^{+2}$	B⁻		[2]
	$2s^22p^2$–$2s2p^3$	^3P–$^5S^0$	1–2	2139	$5.49 \times 10^{+1}$	B⁻		[2]
N III]	$2s^22p$–$2s2p^2$	$^2P^0$–^4P	1½–2½	1749.7	3.08×10^2	C⁺		[2]
N IV]	$2s^2$–$2s2p$	^1S–$^3P^0$	0–1	1486	$1.02 \times 10^{+3}$	B		[2]
[O I]	$2p^4$	^1D–^1S	2–0	5577	1.26	B⁺	$E2$	[2]
		^3P–^1S	2–0	2958	2.42×10^{-4}	C⁺	$E2$	[2]
		^3P–^1S	1–0	2972	7.54×10^{-2}	B⁺	$M1$	[2]
		^3P–^1D	2–2	6300	5.65×10^{-3}	B⁺	$M1$	[2]
		^3P–^1D	1–2	6364	1.82×10^{-3}	B⁺	$M1$	[2]
		^3P–^1D	0–2	6392	8.60×10^{-7}	B⁺	$E2$	[2]
		^3P–^3P	1–0	145.5 μm	1.75×10^{-5}	B⁺	$M1$	[2]
		^3P–^3P	2–0	44.06 μm	1.34×10^{-10}	C⁺	$E2$	[2]
		^3P–^3P	2–1	63.19 μm	8.91×10^{-5}	B⁺	$M1$	[2]
[O II]	$2p^3$	$^2D^0$–$^2P^0$	2½–1½	7320	9.91×10^{-2}	B	$E2$	[2]
		$^2D^0$–$^2P^0$	1½–1½	7331	5.34×10^{-2}	B	$E2$	[2]
		$^2D^0$–$^2P^0$	2½–½	7319	5.19×10^{-2}	B	$E2$	[2]
		$^2D^0$–$^2P^0$	1½–½	7330	8.67×10^{-2}	B	$E2$	[2]
		$^4S^0$–$^2P^0$	1½–1½	2470	5.22×10^{-2}	C⁺	$M1$	[2]
		$^2S^0$–$^2P^0$	1½–½	2470	2.12×10^{-2}	C⁺	$M1$	[2]
		$^4S^0$–$^2D^0$	1½–2½	3729	3.06×10^{-5}	C	$E2$	[2]
		$^4S^0$–$^2D^0$	1½–1½	3726	1.78×10^{-4}	C	$M1$	[2]
[O III]	$2p^2$	^1D–^1S	2–0	4363	1.71	B	$E2$	[2]
		^3P–^1D	1–2	4959	6.21×10^{-3}	B	$M1$	[2]
		^3P–^1D	2–2	5007	1.81×10^{-2}	B	$M1$	[2]
		^3P–^1D	0–2	4931	2.41×10^{-6}	C⁺	$E2$	[2]
		^3P–^3P	1–2	51.81 μm	9.76×10^{-5}	B⁺	$M1$	[2]
		^3P–^3P	0–2	32.66 μm	3.17×10^{-11}	B⁺	$E2$	[2]
		^3P–^3P	0–1	88.18 μm	2.62×10^{-5}	B⁺	$M1$	[2]
O III]	$2s^22p^2$–$2s2p^3$	^3P–$^5S^0$	2–2	1666	$5.48 \times 10^{+2}$	B		[2]
	$2s^22p^2$–$2s2p^3$	^3P–$^5S^0$	1–2	1661	$2.20 \times 10^{+2}$	B		[2]
O IV]	$2s^22p$–$2s2p^2$	$^2P^0$–^4P	1½–2½	1401.2	$1.47 \times 10^{+3}$			[3]
O V]	$2s^2$–$2s2p$	^1S–$^3P^0$	0–1	1218.3	$3.68 \times 10^{+3}$	B		[2]
[F IV]	$2p^2$	^3P–^1D	2–2	4060	9.25×10^{-2}	B	$M1$	[2]
[Ne III]	$2p^4$	^1D–^1S	2–0	3342	2.72	B	$E2$	[2]
		^3P–^1S	2–0	1794	1.88	B	$E2$	[2]
		^3P–^1S	1–0	1815	2.02	A	$M1$	[2]
		^3P–^1D	2–2	3869	0.159	A	$M1$	[2]
		^3P–^1D	1–2	3967	4.92×10^{-2}	A	$M1$	[2]
		^3P–^1D	0–2	4012	9.60×10^{-6}	B⁺	$E2$	[2]
		^3P–^3P	1–0	36.02 μm	1.15×10^{-3}	B⁺	$M1$	[2]
		^3P–^3P	2–0	10.86 μm	2.19×10^{-8}	B	$E2$	[2]
		^3P–^3P	2–1	15.55 μm	5.97×10^{-3}	A	$M1$	[2]

Table 4.15. *(Continued.)*

Atom	Array	Designation lower–upper	J_i-J_k	λ (Å)	A (s⁻¹)	Accuracy	M1 or E2	Reference
[Ne IV]	$2p^3$	$^2D^0$–$^2P^0$	$2\frac{1}{2}$–$1\frac{1}{2}$	4 714	0.380	B	$M1$	[2]
		$^2D^0$–$^2P^0$	$1\frac{1}{2}$–$1\frac{1}{2}$	4 724	0.421	B	$M1$	[2]
		$^2D^0$–$^2P^0$	$2\frac{1}{2}$–$\frac{1}{2}$	4 716	0.105	B	$E2$	[2]
		$^2D^0$–$^2P^0$	$1\frac{1}{2}$–$\frac{1}{2}$	4 726	0.372	B	$M1$	[2]
		$^4S^0$–$^2P^0$	$1\frac{1}{2}$–$1\frac{1}{2}$	1 602	1.23	B	$M1$	[2]
		$^4S^0$–$^2P^0$	$1\frac{1}{2}$–$\frac{1}{2}$	1 602	0.499	B	$M1$	[2]
		$^4S^0$–$^2D^0$	$1\frac{1}{2}$–$2\frac{1}{2}$	2 424	4.12×10^{-4}	C⁺	$E2$	[2]
		$^4S^0$–$^2D^0$	$1\frac{1}{2}$–$1\frac{1}{2}$	2 422	5.76×10^{-3}	C	$M1$	[2]
		$^2D^0$–$^2P^0$	$1\frac{1}{2}$–$\frac{1}{2}$	4 726	0.372	B	$M1$	[1]
[Ne V]	$2p^2$	^1D–^1S	2–0	2 973	2.89	B	$E2$	[2]
		^3P–^1D	2–2	3 426	0.351	B	$M1$	[2]
		^3P–^1D	1–2	3 345	0.126	B	$M1$	[2]
		^3P–^1D	0–2	3 300	2.44×10^{-5}	B	$E2$	[2]
		^3P–^3P	1–2	14.33 μm	4.59×10^{-3}	A	$M1$	[2]
		^3P–^3P	0–2	9.008 μm	5.12×10^{-9}	B⁺	$E2$	[2]
		^3P–^3P	0–1	24.25 μm	1.28×10^{-3}	A	$M1$	[2]
Ne V]	$2s^22p^2$–$2s2p^3$	^3P–$^5S^0$	2–2	1 146	$6.06 \times 10^{+3}$			[1]
	$2s^22p^2$–$2s2p^3$	^3P–$^5S^0$	1–2	1 137	$2.37 \times 10^{+3}$			[1]
Si III]	$3s^2$–$3s3p$	^1S–$^3P^0$	0–1	1 892	$1.67 \times 10^{+4}$			[3]
[S I]	$3p^4$	^1D–^1S	2–0	7 725	1.53		$E2$	[1]
[S II]	$3p^3$	$^2D^0$–$^2P^0$	$2\frac{1}{2}$–$1\frac{1}{2}$	10 320	0.179		$E2$	[1]
		$^2D^0$–$^2P^0$	$1\frac{1}{2}$–$1\frac{1}{2}$	10 287	0.133		$M1$	[1]
		$^2D^0$–$^2P^0$	$2\frac{1}{2}$–$\frac{1}{2}$	10 370	7.79×10^{-2}		$E2$	[1]
		$^2D^0$–$^2P^0$	$1\frac{1}{2}$–$\frac{1}{2}$	10 336	0.163		$E2$	[1]
		$^4S^0$–$^2P^0$	$1\frac{1}{2}$–$1\frac{1}{2}$	4 069	0.225		$M1$	[1]
		$^4S^0$–$^2P^0$	$1\frac{1}{2}$–$\frac{1}{2}$	4 076	9.06×10^{-2}		$M1$	[1]
		$^4S^0$–$^2D^0$	$1\frac{1}{2}$–$2\frac{1}{2}$	6 716	2.60×10^{-4}		$E2$	[1]
		$^4S^0$–$^2D^0$	$1\frac{1}{2}$–$1\frac{1}{2}$	6 731	8.82×10^{-4}		$E2$	[1]
[S III]	$3p^2$	^1D–^1S	2–0	6 312	2.22		$E2$	[1]
		^3P–^1S	2–0	3 797	1.05×10^{-2}		$E2$	[1]
		^3P–^1S	1–0	3 722	0.796		$M1$	[1]
		^3P–^1D	2–2	9 531	5.76×10^{-2}		$M1$	[1]
		^3P–^1D	1–2	9 069	2.21×10^{-2}		$M1$	[1]
		^3P–^1D	0–2	8 830	5.82×10^{-6}		$E2$	[1]
		^3P–^3P	1–2	18.71 μm	2.07×10^{-3}		$M1$	[1]
		^3P–^3P	0–2	12.00 μm	4.61×10^{-8}		$E2$	[1]
		^3P–^3P	0–1	33.48 μm	4.72×10^{-4}		$M1$	[1]
S III]	$3s^3p^2$–$3s3p^3$	^3P–$^5S^0$	2–2	1 729	7.32×10^3			[3]
		^3P–$^5S^0$	1–2	1 713	2.66×10^3			[3]
[Cl III]	$3p^3$	$^4S^0$–$^2D^0$	$1\frac{1}{2}$–$1\frac{1}{2}$	5 538	4.83×10^{-3}		$M1$	[1]
[Cl IV]	$3p^2$	^1D–^1S	2–0	5 323	2.80		$E2$	[1]
		^3P–^1D	1–2	7 531	7.23×10^{-2}		$M1$	[1]
		^3P–^1D	2–2	8 046	0.179		$M1$	[1]
[Ar III]	$3p^4$	^1D–^1S	2–0	5 192	2.59		$E2$	[1]
		^3P–^1S	2–0	3 005	4.17×10^{-2}		$E2$	[1]
		^3P–^1S	1–0	3 109	3.91		$M1$	[1]
		^3P–^1D	2–2	7 136	0.314		$M1$	[1]
		^3P–^1D	1–2	7 751	8.23×10^{-2}		$M1$	[1]
		^3P–^1D	0–2	8 036	2.15×10^{-5}		$E2$	[1]
		^3P–^3P	1–0	21.83 μm	5.17×10^{-3}		$M1$	[1]
		^3P–^3P	2–0	6.369 μm	2.37×10^{-6}		$E2$	[1]
		^3P–^3P	2–1	8.992 μm	3.08×10^{-2}		$M1$	[1]

Table 4.15. (Continued.)

Atom	Array	Designation lower–upper	J_i–J_k	λ (Å)	A (s⁻¹)	Accuracy	M1 or E2	Reference
[Ar IV]	$3p^3$	$^2D^0$–$^2P^0$	$2\frac{1}{2}$–$1\frac{1}{2}$	7 237	0.598		M1	[1]
		$^2D^0$–$^2P^0$	$1\frac{1}{2}$–$1\frac{1}{2}$	7 171	0.789		M1	[1]
		$^2D^0$–$^2P^0$	$2\frac{1}{2}$–$\frac{1}{2}$	7 331	0.119		E2	[1]
		$^2D^0$–$^2P^0$	$1\frac{1}{2}$–$\frac{1}{2}$	7 263	0.603		M1	[1]
		$^4S^0$–$^2P^0$	$1\frac{1}{2}$–$1\frac{1}{2}$	2 854	2.11		M1	[1]
		$^4S^0$–$^2P^0$	$1\frac{1}{2}$–$\frac{1}{2}$	2 868	0.862		M1	[1]
		$^4S^0$–$^2D^0$	$1\frac{1}{2}$–$2\frac{1}{2}$	4 711	1.77×10^{-3}		E2	[1]
		$^4S^0$–$^2D^0$	$1\frac{1}{2}$–$1\frac{1}{2}$	4 740	2.23×10^{-2}		M1	[1]
[Ar V]	$3p^2$	^1D–^1S	2–0	4 626	3.29		E2	[1]
		^3P–^1S	2–0	2 786	5.69×10^{-2}		E2	[1]
		^3P–^1S	1–0	2 691	6.55		M1	[1]
		^3P–^1D	2–2	7 006	0.476		M1	[1]
		^3P–^1D	1–2	6 435	0.204		M1	[1]
		^3P–^1D	0–2	6 133	3.50×10^{-5}		E2	[1]
		^3P–^3P	1–2	7.903 μm	2.72×10^{-2}		M1	[1]
		^3P–^3P	0–2	4.928 μm	1.24×10^{-6}		E2	[1]
		^3P–^3P	0–1	13.09 μm	7.99×10^{-3}		M1	[1]
[Ar X]	$2p^5$	$^2P^0$–$^2P^0$	$1\frac{1}{2}$–$\frac{1}{2}$	5 533	$1.06 \times 10^{+2}$	B	M1	[4]
[Ar XIV]	$2p$	$^2P^0$–$^2P^0$	$\frac{1}{2}$–$1\frac{1}{2}$	4 412	$1.04 \times 10^{+2}$	A	M1	[4]
[K IV]	$3p^4$	^1D–^1S	2–0	4 511	3.18		E2	[1]
		^3P–^1D	1–2	6 795	0.203		M1	[4]
		^3P–^1D	2–2	6 102	0.838		M1	[4]
[Ca V]	$3p^4$	^3P–^1D	2–2	5 309	1.90		M1	[1]
[Ca XII]	$2p^5$	$^2P^0$–$^2P^0$	$1\frac{1}{2}$–$\frac{1}{2}$	3 328	$4.87 \times 10^{+2}$		M1	[4]
[Ca XIII]	$2p^4$	^3P–^3P	2–1	4 087	$3.19 \times 10^{+2}$		M1	[4]
[Ca XV]	$2p^2$	^3P–^3P	1–2	5 446	$7.9 \times 10^{+1}$		M1	[4]
		^3P–^3P	0–1	5 694	$9.4 \times 10^{+1}$		M1	[4]
[Fe II]	$3d^64s$–$3d^7$	^6D–^4P	$3\frac{1}{2}$–$2\frac{1}{2}$	4 890	0.36	E	M1	[5]
		^6D–^4F	$4\frac{1}{2}$–$4\frac{1}{2}$	4 416	0.46	E	M1	[5]
	$3d^64s$–$3d^54s^2$	^6D–^6S	$4\frac{1}{2}$–$2\frac{1}{2}$	4 287	1.5	E	E2	[5]
		^6D–^6S	$3\frac{1}{2}$–$2\frac{1}{2}$	4 359	1.1	E	E2	[5]
	$3d^7$–$3d^64s$	^4D–^2P	$\frac{1}{2}$–$\frac{1}{2}$	5 528	0.12	E	M1	[5]
		^4F–^4G	$4\frac{1}{2}$–$5\frac{1}{2}$	4 244	0.90	E	E2	[5]
[Fe III]	$3d^6$	^5D–^3F	4–4	4 658	0.44	D	M1	[5]
		^5D–^3P	3–2	5 270	0.40	D	M1	[5]
[Fe IV]	$3d^5$	^4G–^4F	$5\frac{1}{2}$–$4\frac{1}{2}$	4 907	0.32	E	E2	[5]
[Fe V]	$3d^4$	^5D–$^3P2$	3–2	3 895	0.71	D	M1	[5]
		^5D–$^3F2$	4–4	3 891	0.74	D	M1	[5]
[Fe VI]	$3d^3$	^4F–^4P	$4\frac{1}{2}$–$2\frac{1}{2}$	5 677	0.052	E	E2	[5]
		^4F–^2G	$4\frac{1}{2}$–$4\frac{1}{2}$	5 176	0.62	D	M1	[5]
[Fe VII]	$3d^2$	^3F–^3P	4–2	5 276	0.050	E	E2	[5]
		^3F–^1D	2–2	5 721	0.36	D	M1	[5]
			3–2	6 087	0.58	D	M1	[5]
[Fe X]	$3p^5$	$^2P^0$–$^2P^0$	$1\frac{1}{2}$–$\frac{1}{2}$	6 375	69.2	B	M1	[5]
[Fe XI]	$3p^4$	^3P–^1D	1–2	3 987	9.5	D⁻	M1	[5]
		^3P–^3P	2–1	7 892	$4.36 \times 10^{+1}$	C⁺	M1	[5]
[Fe XIII]	$3p^2$	^3P–^3P	0–1	10 747	$1.4 \times 10^{+1}$	C⁺	M1	[5]
		^3P–^3P	1–2	10 798	9.86	C⁺	M1	[5]
		^3P–^1D	2–2	3 389	$7.5 \times 10^{+1}$	E	M1	[5]
[Fe XIV]	$3p$	$^2p^0$–$^2p^0$	$\frac{1}{2}$–$1\frac{1}{2}$	5 303	$6.01 \times 10^{+1}$	C⁺	M1	[5]
[Fe XV]	$3s3p$	$^3P^0$–$^3P^0$	1–2	7 059	$3.80 \times 10^{+1}$	C⁺	M1	[5]

Table 4.15. *(Continued.)*

Atom	Array	Designation lower–upper	J_i–J_k	λ (Å)	A (s^{-1})	Accuracy	M1 or E2	Reference
[Ni II]	$3d^9$–$3d^8(^3F)4s$	^2D–^2F	2½–2½	6668	0.099	E	E2	[5]
	$3d^9$–$3d^8(^3P)4s$	^2D–^4P	2½–2½	4326	0.35	E	E2	[5]
[Ni III]	$3d^8$	^3F–^3P	4–2	6000	0.050	E	E2	[5]
[Ni XII]	$3p^5$	$^2P^0$–$^2P^0$	1½–½	4231	$2.37 \times 10^{+2}$	B	M1	[5]
[Ni XIII]	$3p^4$	^3P–^1D	1–2	3637	$1.8 \times 10^{+1}$	E	M1	[5]
		^3P–^3P	2–1	5116	$1.57 \times 10^{+2}$	C$^+$	M1	[5]
[Ni XV]	$3p^2$	^3P–^3P	0–1	6702	$5.65 \times 10^{+1}$	C$^+$	M1	[5]
		^3P–^3P	1–2	8024	$2.27 \times 10^{+1}$	C$^+$	M1	[5]
[Ni XVI]	$3p$	$^2P^0$–$^2P^0$	½–1½	3601	$1.92 \times 10^{+2}$	C$^+$	M1	[5]

References

1. Mendoza, C. 1983, in *Planetary Nebulae*, edited by D.R. Flower, IAU Symposium No. 103 (Reidel, Dordrecht), p. 143
2. Wiese, W.L., Fuhr, J.R., & Deters, T.M. 1996, *J. Phys. Chem. Ref. Data Monograph*, **7**; and other data to be published
3. Morton, D.C. 1991, *ApJS*, **77**, 119
4. Kaufman, V., & Sugar, J. 1986, *J. Phys. Chem. Ref. Data Ser.*, **15**, 321
5. Fuhr, J.R., Martin, G.A., & Wiese, W.L. 1988, *J. Phys. Chem. Ref. Data Ser.*, **17**, Suppl. 4

4.10 SPECTRA OF DIATOMIC MOLECULES

4.10.1 General Remarks

Realistic calculations of astronomical spectra today involve the use of extensive databases such as HITRAN [37], RADEN [38], or the material assembled by Kurucz [39]. The proceedings of IAU Commission 14 [40, 41] describe these sources and contain additional material, also covering polyatomic molecules. Recent texts [42, 43] treat diatomic molecules.

4.10.2 Approximate Wave Function

It is often assumed that the total wave function of a diatomic molecule may be written as a product containing electronic, vibrational, rotational, and nuclear spin components: $\psi = \psi_e \psi_v \psi_r \psi_n$. A more general situation is considered below. Traditionally, electronic spin is included in ψ_e, but the nuclear spin wave functions are written separately. In the simplest cases, ψ_v and ψ_r are the functions describing the quantum oscillator and rotator. The latter are spherical harmonics. Sophisticated treatments of ψ_v use realistic potential functions. In general, the rotational function ψ_r may include electronic angular momentum. In this case, ψ_r is described by symmetrical top wave functions [43, 44]. For the rotational functions to have the proper behavior with respect to parity operations, it is often necessary to use linear combinations of symmetrical top functions.

4.10.3 Quantum Numbers and Notation

Angular momentum vectors **L** and **S** have the same meanings as for atoms. These, and other (e.g., **J**) angular momenta, are often loosely referred to by the associated quantum numbers (L, S, J).

R or O = angular momentum of nuclear (end over end) rotation. $R = 0, 1, \ldots$.

N = total angular momentum apart from spin; formerly called K.

S = total electron spin; $(2S + 1)$ is given as a pre-superscript.

Σ = projection of S on internuclear axis (can be positive or negative).

J = total angular momentum exclusive of nuclear spin.

Λ = component of electron orbital angular momentum along one internuclear axis, symbolized by Σ ($\Lambda = 0$), Π ($\Lambda = 1$), Δ ($\Lambda = 2$), ...

$\Omega = |\Lambda + \Sigma|$. $\Lambda + \Sigma$ is used as a term subscript (e.g., $^4\Pi_{-1/2}$, $^4\Pi_{3/2}$).

I = total nuclear spin.

F = total angular momentum including nuclear spin [not $F(J)$; cf. below].

M = projection of vector $\mathbf{J}$ (M_J) or $\mathbf{F}$ (M_F) on the z axis of the laboratory coordinate system.

$F(J)$ = rotational energy in cm^{-1}, F_1, F_2, ...

A = spin-coupling constant; tabulated by [45] in footnotes.

$Y = A/B_v$ describes intermediate coupling; small $|Y| \Rightarrow$ case (b).

v = vibrational quantum number, $v = 0, 1, \dots$.

T_e = equilibrium electronic energy (or "term value") in cm^{-1}.

$G(v)$ = vibrational energy in cm^{-1}.

v_{00} = wave number of the 0–0 band of a band system.

+, −   describe the parity of *electronic* wave functions of Σ states, viz., Σ^+ and Σ^-, with respect to reflection in plane of nuclei.

g, u describe the parity of *electronic* wave functions in homonuclear diatomic molecules with respect to inversion of electronic coordinates.

+, −   describe the total parity of $\psi_e \psi_v \psi_r$ for rotational levels with respect to inversion of all coordinates in the laboratory frame.

s, a describe the parity of $\psi_e \psi_v \psi_r \psi_n$ of homonuclear molecules with respect to exchange of two nuclei.

4.10.4   Angular Momenta and Hund's Cases [42–44, 46, 47]

The quantum numbers Λ, Σ, and Ω all derive from the projection of vectors and are similar in nature to the numbers M_L, M_S, and M_J of atoms. In the nomenclature of molecular spectroscopy, only positive values of these projections are commonly used. However, just as in the atomic case, positive and negative projections occur, and it is often necessary to employ both signs in the theoretical description of a molecular state.

Case (a): $\mathbf{J} = \mathbf{L} + \mathbf{S} + \mathbf{R}$. The projection of $\mathbf{L}$, whose absolute magnitude is called Λ, is well defined, as is the projection of $\mathbf{S}$, called Σ. Unlike atoms, molecules have their full multiplicity, and $\Lambda \pm \Sigma$ is written as a subscript, e.g., $^4\Pi_{5/2}$, $^4\Pi_{3/2}$, $^4\Pi_{1/2}$, $^4\Pi_{-1/2}$.

Case (b): $\mathbf{L} + \mathbf{R} = \mathbf{N}$ (formerly called $\mathbf{K}$). $\mathbf{N} + \mathbf{S} = \mathbf{J}$. Rotational levels, which may be labeled by the quantum number N, are split into $2S + 1$ sublevels if $N > S$, and $2N + 1$ sublevels if $S > N$.

Case (c): $\mathbf{L} + \mathbf{S} = \mathbf{J}_a$. The quantum numbers Λ and Σ are not "good," but the projection of $\mathbf{J}_a$ on the internuclear axis, Ω, is well defined. $\mathbf{N} + \mathbf{J}_a = \mathbf{J}$, the total angular momentum. Case (c) is common for heavier molecules.

Case (d): $\mathbf{L} + \mathbf{R} = \mathbf{N}$ as in case (b), but the energy splitting due to spin and orbital angular momentum is very small. The vector $\mathbf{J} = \mathbf{S} + \mathbf{N}$ does not differ significantly from $\mathbf{N}$, and energy levels are proportional to $B_v R(R + 1)$.

4.11   ENERGY LEVELS

Approximate energy levels (in cm^{-1}) may be calculated from the following formulas:

$$T = T_e + G(v) + F(J),$$
$$G(v) = \omega_e(v + \tfrac{1}{2}) - \omega_e x_e(v + \tfrac{1}{2})^2 + \cdots,$$
$$F(J) = J(J + 1)B_v - J^2(J + 1)^2 D_v,$$
$$B_v = B_e - \alpha_e(v + \tfrac{1}{2}) + \cdots,$$
$$D_v \approx D_e.$$

For accurate work it is necessary to consult relations specialized for individual molecules (see [45]). Electron spin manifests itself on molecular energy levels in a variety of ways that are not easily described by general formulas (see [44]). The splitting of $^2\Pi$ levels due to spin, for example, may be approximately described by the formulas below. Here, F_1 and F_2 refer to the levels with $J = N + \tfrac{1}{2}$ and $J - \tfrac{1}{2}$, respectively. $Y = A/B$, as above.

$$F_1(J) = B_v\left[(J + \tfrac{1}{2})^2 - \Lambda^2 - \tfrac{1}{2}\sqrt{4(J + \tfrac{1}{2})^2 + Y(Y - 4)\Lambda^2}\right] + \cdots,$$
$$F_2(J) = B_v\left[(J + \tfrac{1}{2})^2 - \Lambda^2 + \tfrac{1}{2}\sqrt{4(J + \tfrac{1}{2})^2 + Y(Y - 4)\Lambda^2}\right] + \cdots.$$

Levels with $\Lambda > 0$ are twofold degenerate $(\pm|M_L|)$. Rotation can lift this degeneracy, giving rise to Λ-doubled pairs of levels with opposite parity. See [42, 48] for additional comments and notation (a, b, c, d, e, f) used to describe rotational levels.

4.11.1   Molecular Constants

Tables 4.16 and 4.17 give the more important constants for selected electronic states of some common diatomic molecules of astrophysical interest. These constants are sufficient for approximate and heuristic work. For example, one may use them to locate lower-order bands and define their character (red or violet degredation). Accurate work would require the use of more elaborate formulas than can be written with these constants alone. Higher-order constants may be found in the papers cited.

Table 4.16. *Selected constants for diatomic molecules.[a]*

State	Te	ω_e	$\omega_e x_e$	B_e	α_e	D_e	r_e (Å)
			$^1H_2$, $D_0^0 = 4.478\,075$ eV				
$C\,^1\Pi_u 2p\pi$	100 089.8	2 444.66	65.58	31.324	1.599	1.994(−2)	1.03
$B\,^1\Sigma_u^+ 2p\sigma$	91 700.0	1 357.19	20.15	19.984	1.115	1.656(−2)	1.29
$X\,^1\Sigma_g^+ 1s\sigma^2$	0.0	4 402.93	123.07	60.847	3.053	4.644(−2)	0.74

Table 4.16. *(Continued,)*

State	T_e	ω_e	$\omega_e x_e$	B_e	α_e	D_e	r_e (Å)
			$^{12}C_2$,	$D_0^0 = 6.29_6$ eV			
$d\,^3\Pi_g$	20 024.597	1 788.222 0	16.457 4	1.755 523	0.019 07	6.72(−6)	1.27
$B^1\,^1\Sigma_g^+$	15 409.139	1 424.119	2.571 1	1.481 0	0.117$_5$	6.86(−6)	1.38
$B\,^1\Delta_g$	12 082.336	1 407.465	11.479 4	1.463 685	0.016 816	6.319(−6)	1.39
$c\,^3\Sigma_u^+$	9 124.212	2 085.899	18.623	1.921	0.012 55	6.517(−6)	1.21
$A\,^1\Pi_u$	8 391.408	1 608.199	12.060	1.616 628	0.016 969	6.509(−6)	1.32
$b\,^3\Sigma_g^-$	6 435.736	1 470.415	11.155	1.498 64	0.016 312	6.196(−6)	1.37
$a\,^3\Pi_u$	718.318	1 641.329 59	11.651 95	1.632 365	0.016 662 5	6.463(−6)	1.31
$X\,^1\Sigma_g^+$	0.0	1 855.014	13.555	1.820 10	0.018 01	6.964(−6)	1.24
			$^{12}C\,^{14}N$,	$D_0^0 = 7.74$ eV			
$D\,^2\Pi_i$	54 486.3	1 004.7$_1$	8.7$_8$	1.162	0.013	7(−6)	1.50
$b\,^4\Pi_i$	44 317	1 148	18.1	1.170	0.016		1.49
$a\,^4\Sigma^+$	(36 400)	(1 400)	(20)				
$B\,^2\Sigma^+$	25 753.22	2 160.38	17.74	1.968 79	0.019 96	6.58(−6)	1.15
$A\,^2\Pi_i$	9 243.308	1 813.235	12.751	1.715 62	0.017 12	6.129(−6)	1.23
$X\,^2\Sigma^+$	0.0	2 068.648	13.097	1.899 783 2	0.017 372	6.406(−6)	1.17
			$^{12}C\,^{16}O$,	$D_0^0 = 11.108$ eV			
$a'\,^3\Sigma^+$	55 825.4$_9$	1 228.60	10.468	1.344 6	0.018 9$_2$	6.41(−6)	1.352 3
$a\,^3\Pi_r$	48 686.70	1 743.4$_1$	14.3$_6$	1.691 24	0.019 04	6.36(−6)	1.205 74
$X\,^1\Sigma^+$	0.0	2 169.814	13.288 3	1.931 3	0.017 5	6.121(−6)	1.128
			$^{16}O\,^1H$,	$D_0^0 = 4.392$ eV			
$X\,^2\Pi_i$	0.0	3 737.76$_1$	84.881$_3$	18.910$_8$	0.724 2	19.38(−4)	0.969 66

Note

[a] Units are cm^{-1} except as indicated. The power of ten to be applied to the entry for D_e is shown in parentheses.
References: H$_2$ [1–4]; C$_2$ [5–15]; CO [1,16]; CN [17–21].

References

1. Huber, K.P., & Herzberg, G. 1979, *Molecular Spectra and Molecular Structure IV. Constants of Diatomic Molecules* (Van Nostrand, New York)
2. Dabrowski, I. 1984, *Can. J. Phys.*, **62**, 1639
3. Abgrall, H., Roueff, E., Launay, F., Roucin, J.-Y., & Subtil, J.-L. 1993, *J. Mol. Spectrosc.*, **157**, 512
4. Balakrishnan, A., Smith, V., & Stoicheff, B.P. 1992, *Phys. Rev. Lett.*, **68**, 2149
5. Douay, M., Nietmann, R., & Bernath, P.F. 1988, *J. Mol. Spectrosc.*, **131**, 250, 261
6. Prasad, C.V.V., & Bernath, P.F. 1994, *ApJS*, **426**, 812
7. Davis, S.P., Abrams, M.C., Phillips, J.G., & Rao, M.L.P. 1988, *J. Opt. Soc. Am.*, **B5**, 2280
8. Galehouse, D.C., Brault, J.W., & Davis, S.P. 1980, *ApJ*, **42**, 241
9. Simard, B., & Hackett, P.A. 1991, *J. Mol. Spectrosc.*, **148**, 128
10. Phillips, J.G. 1973, *ApJS*, **26**, 313
11. Hocking, W.H., Gerry, M.C.L., & Merer, A.J. 1979, *Can. J. Phys.*, **57**, 54
12. Veseth, L. 1975, *Can. J. Phys.*, **53**, 299
13. Urdahl, R.S., Bao, Y., & Jackson, W.M. 1991, *J. Chem. Phys. Lett.*, **178**, 425
14. Amiot, C., Chauville, J., & Maillard, J.-P. 1979, *J. Mol. Spectrosc.*, **75**, 19
15. Davis, S.P., Abrams, M.C., Sandalphon, X.X., Brault, J.W., & Rao, M.L.P. 1988, *J. Opt. Soc. Am.*, **B5**, 1838
16. Eidelsberg, M., Roncin, J.-Y, LeFloch, A., Launay, F., Letzelter, C., & Rostas, J. 1987, *J. Mol. Spectrosc.*, **121**, 309
17. Ito, H., Ozaki, Y., Suzuki, K., Kondow, T., & Kuchitsu, K. 1992, *J. Chem. Phys.*, **96**, 4195
18. Huang, Y., Barts, S.A., & Halpern, J.B. 1992, *J. Phys. Chem.*, **96**, 425
19. Ito, H., Ozaki, Y., Nagata, T., Kondow, T., & Kuchitsu, K. 1984, *Can. J. Phys.*, **62**, 1586
20. Prasad, C.V.V., & Bernath, P.F. 1992, *J. Mol. Spectrosc.*, **156**, 327
21. Kotlar, A.J., Field, R.W., Steinfeld, J.I., & Coxon, J.A. 1980, *J. Mol. Spectrosc.*, **80**, 86

Table 4.17. *Selected constants continued:* TiO.[a]

State	T_0	ω_e	$\omega_e x_e$	B_e	α_e	D_e	r_e (Å)
			^{48}Ti ^{16}O, $D_0^0 = 6.87$ eV				
$C\,^3\Delta_3$	[19 536.63]	838.256 7	4.759 2	0.489 888	0.063 062	6.627(−7)	1.69
$C\,^3\Delta_2$	[19 441.47]						
$C\,^3\Delta_1$	[19 341.68]						
$B\,^3\Pi_2$	[16 266.797]	[863.563]		0.506 223	0.003 18	6.97(−7)	1.67
$B\,^3\Pi_1$	[16 247.951]						
$B\,^3\Pi_0$	[16 255.986]						
$b\,^1\Pi$	[14 721.14]	919.759 3	4.279 9				
$A\,^3\Phi_4$	[14 365.60]	867.779 9	3.942 2	0.507 390	0.003 145	6.918(−7)	1.66
$A\,^3\Phi_3$	[14 193.69]						
$A\,^3\Phi_2$	[14 019.43]						
$E\,^3\Pi_2$	[12 016.13]	924	5.1	[0.515 5]			1.65
$E\,^3\Pi_1$	[11 925.26]						
$E\,^3\Pi_0$	[11 840.15]						
$d\,^1\Sigma^+$	[5 667.10]	1 023.058 5	4.893 5	0.549 320	0.003 348	6.337(−7)	1.60
$a\,^1\Delta$	[3 448.32]	1 018.273	4.521	0.537 602	0.002 916	5.9(−7)	1.62
$X\,^3\Delta_3$	[202.617 7]	1 009.169 7	4.564 0	0.535 431	0.003 022	6.32(−7)	1.62
$X\,^3\Delta_2$	[97.817 7]						
$X\,^3\Delta_1$	0.0						

Note

[a]Units are cm^{-1} except as indicated. The power of ten to be applied to the entry for D_e is shown in parentheses. For TiO the square brackets indicate that T_0 is given rather than the usual T_e. These apply to the $v = 0$ vibrational level. The constants ω_e, etc., are the same for the levels split by spin–orbit interaction. References: TiO [1–4].

References

1. Gustavsson, T., Amiot, C., & Vergès, J. 1991, *J. Mol. Spectrosc.*, **145**, 56
2. Hildebrand, D.L. 1976, *Chem. Phys. Lett.*, **44**, 281
3. Merer, A.J. 1989, *Annu. Rev. Phys. Chem.*, **40**, 407
4. Brandes, G.R., & Galehouse, D.C. 1985, *J. Mol. Spectrosc.*, **109**, 345

4.12 TRANSITIONS

The upper level is written first, for both absorption and emission. Symbols describing the upper level have a single prime, while a double prime is used for the lower level.

4.12.1 Rotation and Vibration

Rotational transitions in emission or absorption are assigned to P, Q, and R *branches* designated as follows for dipole radiation:

$$P: \quad J'' \leftrightarrow J' = J'' - 1,$$
$$Q: \quad J'' \leftrightarrow J' = J'',$$
$$R: \quad J'' \leftrightarrow J' = J'' + 1.$$

Transitions forbidden for electric dipole radiation can give rise to lines in an O branch ($J'' \leftrightarrow J' = J'' - 2$) and an S branch ($J'' \leftrightarrow J' = J'' + 2$).

In case (b), when the spin splitting is small with respect to the rotational separation of the energy levels, one can have P-, Q-, and R-form branches whose nomenclature depends on N' and N''. For example, a line in a P-form Q branch would arise when $J' \leftrightarrow J''$, but $N'' \leftrightarrow N' = N'' - 1$. It would be labeled $^P Q$. The branch labels also contain subscripts. The symbol $^Q R_{12}$ would designate a transition in a Q-form R branch from a lower level labeled F_2 to an upper F_1. See [49] for additional notation.

A common designation of rotational lines uses the J value of the lower level. Thus $R(0)$ arises in transitions between $J' = 1$ and $J'' = 0$ (in absorption from $J'' = 0$ and in emission from $J' = 1$). Since $J' = 0 \leftrightarrow J'' = 0$ is forbidden for electric dipole radiation, $Q(0)$ does not occur. The corresponding wave number is, however, called the *band origin*, v_{00} or $\tilde{v}_{00}$.

Vibrational transitions are designated by the corresponding quantum numbers. For example, the (0–0) band means a transition from $v' = 0$ to $v'' = 0$. The quantum number for the upper vibrational state is written first.

4.12.2 Electronic Transitions

In the spectra of diatomic molecules *line strengths* are defined in the same way as for atomic transitions, by a sum over the degenerate elementary states of both the upper and lower levels, which are labeled by M' and M'':

$$S_{J'J''} = \sum_{M'M''} |\langle \psi_{M'} | \boldsymbol{\mu} | \psi_{M''} \rangle|^2 .$$

This "line strength" is symmetrical in the upper and lower levels. The electric dipole moment, here written as $\boldsymbol{\mu}$, is the sum of the electric moments (charge times displacement) of the electrons and nuclei. This vector must be in the fixed or laboratory frame. For convenience, it is transformed to the frame of the molecule with the help of Euler angles. In practice, for a given transition, only one electron is important.

The line strength for an electronic transition may be written as a product of three factors [50, 51]

$$S_{J'J''} = |R_e|^2 q_{v'v''} S_{J'J''} .$$

The quantity R_e is called the electronic transition moment. Its definition, consistent with the Hönl–London factor S (see below), is such that

$$|R_e| = |\langle \Lambda'S'\Sigma'|z|\Lambda''S''\Sigma''\rangle|, \qquad \Delta\Lambda = 0,$$
$$= \left|\langle \Lambda'S'\Sigma'| (x \pm iy)/\sqrt{2} |\Lambda''S''\Sigma\rangle\right|, \qquad \Delta\Lambda = \pm 1,$$
$$\sum S_{J'J''} = (2 - \delta_{0,\Lambda'}\delta_{0,\Lambda''})(2S + 1)(2J + 1).$$

This normalization [50] holds for absorption or emission. In the former case, the value of J on the right-hand side is J'', while in the latter it is J'. The Kronecker δ functions are zero if Λ' or Λ'' is not equal to zero, and are unity otherwise. Consider a given J (J' or J''). It is necessary to sum the rotational strengths S for all allowed transitions from the $(2 - \delta_{0,\Lambda})(2S + 1)$ levels with a given J for which transitions are allowed. Thus the sum extends over more than one energy level in general and includes lines with the same J that arise from Λ doubling. If Λ doubling is present in *both* upper and lower levels, the number of allowed lines is exactly twice that which would result if there were no degeneracy. However, if only *one* of the upper or lower levels is doubled, the resulting number of allowed lines is the same as if neither upper nor lower were doubled because of the selection rule on parity. The sum of these strengths may not equal the theoretical value for low levels where the full spin multiplicity (of levels) has not developed [43].

The recommended normalization follows naturally if the rotational strengths are written with $n - j$ symbols [47, 52]. Thus for Hund's case (a), we have

$$S = (2J' + 1)(2J'' + 1) \begin{pmatrix} J' & 1 & J'' \\ \Omega' & \Omega'' - \Omega' & -\Omega'' \end{pmatrix}^2.$$

The symbol in the large parentheses is a Wigner $3 - j$ symbol. This formula also holds for cases (c) and (d), in the latter instance with the replacement of Ω by Λ. These cases are of less importance for molecules of astrophysical interest. For case (b), it is necessary to decouple the electron spin, and this introduces a $6 - j$ (curly bracket) symbol:

$$S = (2J' + 1)(2N' + 1)(2J'' + 1)(2N'' + 1) \begin{Bmatrix} N' & 1 & N'' \\ J'' & S & J' \end{Bmatrix}^2 \begin{pmatrix} N' & 1 & N'' \\ \Lambda' & \Lambda'' - \Lambda' & -\Lambda'' \end{pmatrix}^2.$$

Pure Hund cases are only approximations to the more general description of molecular levels by *intermediate coupling*. Intensity formulas have been given by various authors, e.g., [53–55], and Whiting [56] has published a program for the S's consistent with the above summation rules. We recommend use of the Whiting code for all but Σ–Σ transitions, which are inherently case (b).

It is often useful to have guides to the rotational structure of electronic transitions. In addition to the basic reference [44] useful diagrams may be found in [49, 51, 57].

Oscillator strengths and Einstein coefficients are related to $S_{J'J''}$ by the same formulas as for atoms (Sec. 4.4).

4.13 SELECTION RULES: DIPOLE RADIATION

Many selection rules for diatomic molecules can be inferred from the properties of the $n - j$ symbols of Sec. 4.12; the relevant $3 - j$ or $6 - j$ symbol will vanish for the forbidden transition. For example, we can infer for electric dipole radiation $\Delta J = 0, \pm 1$ with $J = 0 \nleftrightarrow J = 0$. Similarly, we have $\Delta \Omega = 0$, ± 1 for case (a) and $\Delta \Lambda = 0, \pm 1$ for case (b). Case (b) also has $\Delta N = 0, \pm 1$ and $N = 0 \nleftrightarrow N = 0$. The $3 - j$ symbol vanishes if $N' = N''$ while $\Lambda' = \Lambda'' = 0$; consequently, $\Delta N = 0$ is forbidden for $\Sigma \leftrightarrow \Sigma$ transitions [case (b)]. Similarly $\Delta J = 0$ for $\Omega = 0 \leftrightarrow \Omega = 0$ [case (a)].

The total spin operator commutes with the dipole moment; consequently, ΔS is forbidden for electric dipole radiation. In case (a), where Σ is well defined, we also have $\Delta \Sigma = 0$.

Symmetry of the electronic wave functions prevent Σ^+ from combining with Σ^-, while symmetry of the overall wave functions prevent positive–positive and negative–negative transitions. For homonuclear molecules *gerade–gerade* and *ungerade–ungerade* transitions are prohibited, while symmetric–antisymmetric rotational transitions cannot occur.

4.13.1 Parameters for Selected Electronic Transitions

Table 4.18 gives parameters for line-strength calculations in a few diatomic band systems of astrophysical interest. The material is primarily for heuristic use. For detailed calculations it is necessary to consult the sources cited. Entries are primarily from the RADEN database [38]. The first three columns identify the systems and give wavelength ranges, following [58]. A very useful table of persistent band heads is given in [59]. The fourth column contains the band origin for the 0–0 band. The following columns provide information relevant to line-strength calculations. Entries are for r-centroid $[R_e(r_{v'v''})]$ and ab initio $[R_e(r)]$ calculations. The former are used with Franck–Condon factors ($q_{v'v''}$) while the latter involve an integration of $R_e(r)$ over the vibrational wave functions. The final column of the table gives square of the transition moment for the 0–0 band, by the two methods, with vibration included. A superscript a in this column indicates $R_{00}^2 = R_e^2(r_{00})q_{00}$, while b indicates $R_{00}^2 = |\langle v' = 0 | R_e(r) | v'' = 0 \rangle|^2$.

Table 4.18. *Parameters for molecular transition strengths.*

Molecule	System	Approx. range (Å)	v_{00} (cm⁻¹)	q_{00}	r_{00} (Å)	Recommended electronic transition moment (ea_0)	R_{00}^2 (ea_0)²
C_2	Swan $d\,^3\Pi_g - a\,^3\Pi_u$	λλ3400–7850	19 378.44	0.7244		$R_e(r_{v'v''}) = (2.380 \pm 0.28)(1 - 0.52\,r_{v'v''})$ for $r_{v'v''} = 1.12$–1.50 Å.	0.441[a]
	Deslandres–d'Azambuja $C\,^1\Pi_g - A\,^1\Pi_u$	λλ3390–4110	25 969.19	0.5445	1.2938	See [1]. Figure of $R_e(r)$ for $r = (2.0$–$3.5)a_0$.	0.54[b]
CH	$A\,^2\Delta - X\,^2\Pi$	λλ4314–4890	23 217.5	0.9907	1.2913	$(R_e) = 0.280 \pm 0.008$. See [2] for $R_e(r)$, $r = (1.3$–$4.0)a_0$.	0.078[a] 0.084[a]
CN	Red system $A\,^2\Pi - X\,^2\Sigma^+$	λλ4370–15050	9 117.38	0.4957	1.1313	$R_e(r_{v'v''}) = (0.19 \pm 0.03)(1 + 0.571\,r_{v'v''})$ for $r_{v'v''} = 1.05$–1.27 Å. See [3] for $R_e(r)$, $r = (1.4$–$4.0)a_0$.	0.0511[a]
CN	Violet $B\,^2\Sigma^+ - X\,^2\Sigma^+$	λλ3440–4600	25 797.84	0.9180	1.2062	$R_e(r_{v'v''}) = (0.72 \pm 0.02)(1 - 0.03\,r_{v'v''})$ for $r_{v'v''} = 0.95$–1.35 Å. See [3] for $R_e(r)$, $r = (1.6$–$4.0)a_0$.	0.0426[b] 0.442[a]
CO	4th positive $A\,^1\Pi - X\,^1\Sigma^+$	λλ1115–1544 (abs.) λλ2006–2785 (emiss.)	64 748.48	0.1150	1.1658	$R_e(r_{v'v''}) = (2.94 \pm 0.15)(1 - 0.68\,r_{v'v''})$ for $r_{v'v''} = 1.0$–1.3 Å. See [4] for $R_e(r)$, $r = (1.8$–$8.0)a_0$.	0.427[b] 0.0386[a]
CO^+	Comet tail $A\,^2\Pi - X\,^2\Sigma^+$	λλ3080–8500	20 407.6	0.0422	1.1806	See [5] for $R_e(r)$, $r = (1.4$–$3.1)a_0$.	0.0371[b] 0.00158[b]
H_2	Lyman $B\,^1\Sigma_u^+ - X\,^1\Sigma_g^+$	λλ955–1674	90 203.55	0.0037	1.1783	See [2] for $R_e(r)$, $r = (1.0$–$12)a_0$.	0.00549[b]
H_2	Werner $C\,^1\Pi_u - X\,^1\Sigma_g^+$	λλ1028–1239	99 120.17	0.1217	0.9624	See [6] for $R_e(r)$, $r = (1.0$–$10.0)a_0$.	0.0785[b]
LaO	$B\,^2\Sigma^+ - X\,^2\Sigma^+$	λλ5015–6450	17 837.8	0.8604	0.8937	$R_e(r_{v'v''}) = 348(-1 + 1.74275\,r_{v'v''} - 0.99636\,r_{v'v''}^2 + 1.8803\,r_{v'v''}^3)$ for $r_{v'v''} = 1.6$–2.1 Å.	2.23[a]
MgH	$A\,^2\Pi - X\,^2\Sigma^+$	λλ4700–6100	19 278.4	0.944	1.8424	$(R_e) = 1.22$.	1.42[a]
N_2	Second positive $C\,^3\Pi_u - B\,^3\Pi_g$	λλ2680–5450	29 671.0	0.4545	1.7270 1.1843	$R_e(r_{v'v''}) = (1.86 \pm 0.17)(1 - 0.51\,r_{v'v''})$ for $r_{v'v''} = 1.05$–1.35 Å. $R_e(r) = 0.887 \exp[-3.30(r - 0.95)]$ for $r = 0.95$–1.40 Å.	0.24[a] 0.25[b]
N_2^+	First negative $B\,^2\Sigma_u^+ - X\,^2\Sigma_g^+$	λλ2800–5900	25 566.04	0.6636	1.0995	$R_e(r_{v'v''}) = (12.12 \pm 0.01)(1 - 1.63\,r_{v'v''} + 0.70\,r_{v'v''})$ for $r_{v'v''} = 0.97$–1.16 Å. $R_e(r) = 1.051 + 0.2033\,r - 0.4646\,r^2$ for $r = 0.85$–2.65 Å.	0.325[a] 0.326[b]
NH	$A\,^3\Pi - X\,^3\Sigma^-$	λλ3020–3680	29 776.76	0.9999	1.0559	$(R_e) = 0.210 \pm 0.006$. See [7] for $R_e(r)$, $r = (1.25$–$20.0)a_0$.	0.0441[a] 0.0463[b]
NO	γ system $A\,^2\Sigma^+ - X\,^2\Pi$	λλ1950–3400	44 080.5 44 200.2	0.1639	1.1091	$R_e(r_{v'v''}) = (26.8 \pm 1.5)(1 - 2.898\,r_{v'v''} + 2.7499\,r_{v'v''}^2 - 0.8597\,r_{v'v''}^3)$ for $r_{v'v''} = (1.0$–$1.20)a_0$. See [8] for $R_e(r)$, $r = (1.6$–$2.4)a_0$.	0.0031[a] 0.0026[b]

Table 4.18. (*Continued.*)

Molecule	System	Approx. range (Å)	ν_{00} (cm^{-1})	q_{00} r_{00} (Å)	Recommended electronic transition moment (ea_0)	R_{00}^2 (ea_0)^2
O_2	Schumann–Runge $B\,^3\Sigma_u^- - X\,^3\Sigma_g^-$	λλ1750–5350 (bands) λλ1300–1750 (continuum)	49358.15	For band (15,0) 0.000272 1.3107	$R_e(r_{v'v''}) = 1.86 - 0.8069\,r_{v'v''}$ for $r_{v'v''} = 1.30$–2.16 Å.	For (15,0) band 0.000175[a]
OH	$A\,^2\Sigma^+ - X\,^2\Pi$	λλ2608–4107	32402.39	0.9067	$R_e(r_{v'v''}) = (0.42 \pm 0.01)(1.0 - 0.75\,r_{v'v''})$ for $r_{v'v''} = 0.8$–1.2 Å.	0.009 53[a]
SiH	$A\,^2\Delta - X\,^2\Pi$	λλ3863–4278	24193.04	1.0080 0.9932 1.5459	See [9] for $R_e(r)$, $r = (1.3$–4.4$)a_0$. $\langle R_e \rangle = 0.25 \pm 0.03$.	0.0109[b] 0.0621[a]
TiO	α system; $C\,^3\Delta - X\,^3\Delta$	λλ4050–6300	19334.03 19343.66 19341.68	0.4092 1.6590	$R_e(r_{v'v''}) = (102 \pm 25)\exp(-2.57\,r_{v'v''})$ for $r_{v'v''} = 1.58$–1.72 Å.	0.084[a]
	β system; $c\,^1\Phi - a\,^1\Delta$	λλ4900–5800	17840.6	0.9152	$\langle R_e \rangle = 2.25$.	4.63[a]
	γ system; $A\,^3\Phi - X\,^3\Delta$	λλ5700–8650	14163.00 14095.88 14019.43	1.6313 0.7191 1.6448	$\langle R_e \rangle = 2.7$.	5.24[a]
YO	$A\,^2\Pi - X\,^2\Sigma^+$	λλ5700–6800	16722.75 16294.72	0.9950 1.7940	$\langle R_e \rangle = 1.83$.	3.33[a]
VO	$C\,^4\Sigma^- - X\,^4\Sigma^-$	λλ4400–7000	17420.2	0.3130 1.6331	$\langle R_e \rangle = 0.52$.	0.085[a]
ZrO	α system; $d\,^3\Delta - a\,^3\Delta$	λλ4200–5600	21631.48 21548.46 21536.36	0.672	See [10], for $R_e(r)$, $r = (2.8$–3.8$)a_0$.	0.97[b]
	γ system; $b\,^3\Phi - a\,^3\Delta$	λλ5110–7600	16033.81 15741.31 15426.78	0.973 1.7564 1.7448	See [10], for $R_e(r)$ for $r = (2.8$–3.8$)a_0$.	2.63[b]

References

1. Chabalowski, C.F., Peyerimhof, S.D., & Buenker, R.J. 1983, *J. Chem. Phys.*, **81**, 57
2. van Dishoeck, E. 1987, *J. Chem. Phys.*, **86**, 196
3. Bauschlicher, C.W., Langhoff, S.R., & Taylor, P.R. 1988, *ApJ*, **332**, 531
4. Kirby, K., & Cooper, D.L. 1989, *J. Chem. Phys.*, **90**, 4895
5. Marian, C.M., Larsson, M., Olsson, B.J., & Sigray, P. 1989, *J. Chem. Phys.*, **130**, 361 *ApJ*, **332**, 531
6. Dressler, H., Wolniewicz, L. 1985, *J. Chem. Phys.*, **82**, 4720
7. Kirby, K.P., & Goldfield, E.M. 1991, *J. Chem. Phys.*, **94**, 1271
8. Langhoff, S.R., Bauschlicher, C.W., & Partridge, H. 1988, *J. Chem. Phys.*, **89**, 4909
9. Bauschlicher, C.W., & Langhoff, S.R. 1987, *J. Chem. Phys.*, **87**, 4665
10. Langhoff, S.R., & Bauschlicher, C.W. 1990, *ApJ*, **349**, 369

ACKNOWLEDGMENTS

One of our authors (C.R.C.) thanks the following for advice and help of various kinds: P.F. Bernath, T.M. Dunn, K.T. Hecht, Sveneric Johansson, R.L. Kurucz, W.C. Martin, D.C. Morton, R.W. Nicholls, L.S. Rothman, R.L. Sears, R.H. Tipping, and E.E. Whiting.

REFERENCES

1. Clark, D. 1996, *Student's Guide to the Internet*, 2nd ed. (QUE, Indianapolis). (There are also many other books that deal with the Internet and its applications.)
2. Cowan, R.D. 1981, *The Theory of Atomic Structure and Spectra* (University of California Press, Berkeley)
3. Martin, W.C., Zalubas, R., & Hagan, L. 1978, *Atomic Energy Levels—The Rare Earths*, *NSRDS-NBS* No. 60
4. Johansson, S., & Cowley, C.R. 1988, *J. Opt. Soc. Am.*, **5B**, 2664
5. Wiese, W.L., & Martin, G.A. 1989, in *A Physicist's Desk Reference*, edited by H.L. Anderson (AIP, New York)
6. Moore, C.E. 1993, in *Tables of Spectra of Hydrogen, Carbon, Nitrogen, and Oxygen Atoms and Ions*, edited by J.W. Gallagher (CRC, Boca Raton, FL)
7. Martin, W.C. 1992, in *Atomic and Molecular Data for Space Astronomy*, edited by P.L. Smith and W.L. Wiese (Springer, Berlin)
8. Cohen, E.R., & Taylor, B.N. 1987, *Rev. Mod. Phys.*, **59**, 1121
9. Moore, C.E. 1967, *Selected Tables of Atomic Spectra*, *NSRDS-NBS* No. 3, Sec. 2
10. Bates, D.R., & Damgaard, A. 1949, *PTRSL A*, **242**, 101
11. Oertel, G.K., & Shomo, L.P. 1968, *ApJS*, **16**, 175
12. Kurucz, R.L., & Bell, B. 1995, "*Atomic Line List*," *Kurucz* CD-ROM No. 23 (Smithsonian Ap. Obs., Cambridge, MA)
13. Seaton, M.J., Yan, Y., Mihalas, D., & Pradhan, A.K. 1994, *MNRAS*, **266**, 805
14. Allen, C.W. 1976, *Astrophysical Quantities*, 3rd ed. (Athlone, London), Sec. 27
15. Coleman, C.D., Bozman, W.R., & Meggers, W.F. 1960, *Table of Wavenumbers*, NBS Monograph No. 3 (two volumes)
16. Peck, E.R., & Reeder, K. 1972, *J. Opt. Soc. Am.*, **62**, 958
17. Reader, J., & Corliss, C.H. 1991, in *Handbook of Chemistry and Physics*, 72nd ed., edited by R. Lide (CRC, Boca Raton, FL), p. 10-1
18. Wiese, W.L., & Martin, G.A. 1989, in *A Physicist's Desk Reference*, edited by H.L. Anderson (AIP, New York), p. 97
19. Bethe, H., & Salpeter, E.E. 1957, *Quantum Mechanics of One- and Two-Electron Atoms* (Academic Press, New York)
20. Green, L.C., Rush, P.R., & Chandler, C.D. 1957, *ApJS*, **3**, 37
21. Wiese, W.L., Smith, M.W., & Glennon, B.M. 1966, *Atomic Transition Probabilities*, H Through Ne, *NSRDS-NBS* 4
22. Moore, C.E. 1993, in *Tables of Spectra of Hydrogen, Carbon, Nitrogen, and Oxygen Atoms and Ions*, edited by J.W. Gallagher (CRC, Boca Raton, FL)
23. Moore, C.E. 1959, A Multiplet Table of Astrophysical Interest, revised ed., NBS Tech. Note 36
24. Moore, C.E. 1952, An Ultraviolet Multiplet Table, NBS Circ. No. 488, 1950
25. Fuhr, J.R., & Wiese, W.L. 1991, *Handbook of Chemistry and Physics*, 72nd ed., edited by D.R. Lide (CRC, Boca Raton, FL), pp. 10–128
26. Holden, N.E. 1991, in *Handbook of Chemistry and Physics*, 72nd ed., edited by D.R. Lide (CRC, Boca Raton, FL), Sec. 11–128
27. Cowan, R.D. 1981, *The Theory of Atomic Structure and Spectra* (University of California Press, Berkeley), Chaps. 15 and 17
28. Landolt-Börnstein, 1952, *Zahlenwerte und Funktionen*, 6th ed. (Springer, Berlin), Vol. I, Part 5
29. Cowley, C.R. 1995, *A Textbook of Cosmochemistry* (Cambridge University Press, Cambridge), Sec. 11.8
30. Rosberg, M., Litzén, U., & Johansson, S. 1993, *MNRAS*, **262**, L1
31. Leckrone, D.S., Wahlgren, G.M., & Johansson, S. 1991, *ApJ*, **377**, L37
32. Hinds, E.A. 1988, in *The Spectrum of Atomic Hydrogen: Advances*, edited by G.W. Series (World Scientific, Singapore)
33. Fuhr, J.R., Martin, G.A., & Wiese, W.L. 1988, *J. Phys. Chem. Ref. Data. Ser.*, **17**, Suppl. 4
34. Kaufman, V., & Sugar, J. 1986, *J. Phys. Chem. Ref. Data. Ser.*, **15**, 321
35. Morton, D.C. 1991, *ApJS* **77**, 119
36. Wiese, W.L., Fuhr, J.R., & Deters, T.M., 1996, *J. Phys. Chem. Ref. Data Monograph*, **7**
37. Rothman, L.S., Gamache, R.R., Tipping, R.H., Rinsland, C.P., Smith, M.A.H., Benner, D.C., Malathy Devi, V., Flaud, J.-M., Brown, L.R., & Toth, R.A. 1992, *J. Quant. Spectrosc. Rad. Transf.*, **48**, 469
38. Kuznetsova, L.A. et al. 1993 (Russian) *J. Phys. Chem.*, **67**, 11
39. Kurucz, R.L. 1994, in *Molecules in the Stellar Environment*, Lecture Notes in Physics, edited by U.G. Jørgensen (Springer, Berlin), Vol. 428, p. 282; see also Kurucz CD-ROM No. 15 (Smithsonian Ap. Obs., Cambridge, MA), 1993
40. Parkinson, W.H. 1992, in *Atomic and Molecular Data for Space Astronomy*, Lecture Notes in Physics, edited by P.L. Smith and W.L. Wiese (Springer, Berlin), Vol. 407, p. 149
41. Jørgensen, U.G. 1994, in *Molecules in the Stellar Environment*, Lecture Notes in Physics, edited by U.G. Jørgensen (Springer, Berlin), Vol. 428, p. 29

42. Bernath, P. 1995, *Spectra of Atoms and Molecules* (Oxford University Press, Oxford)
43. Zare, R.N. 1988, *Angular Momentum* (Wiley, New York)
44. Herzberg, G. 1950, *Spectra of Diatomic Molecules*, 2nd edited by (Van Nostrand, New York)
45. Huber, K.P., & Herzberg, G. 1979, *Molecular Spectra and Molecular Structure IV. Constants of Diatomic Molecules* (Van Nostrand-Reinhold, New York)
46. Gordy, W., & Cook, R.L. 1984, *Microwave Molecular Spectroscopy* (Wiley, New York)
47. Judd, B. 1975, *Angular Momentum Theory for Diatomic Molecules* (Academic Press, New York), see pp. 184–186
48. Brown, J.M., Hougen, J.T., Huber, K.-P., Johns, J.W.C., Kopp, I., Lefebvre-Brion, H., Merer, A.J., Ramsay, D.A., Rostas, J., & Zare, R.N. 1975, *J. Mol. Spectrosc.*, **55**, 500
49. Tatum, J.B. 1967, *ApJS*, **24**, 3
50. Whiting, E.E., Schadee, A., Tatum, J.B., Hougen, J.T., & Nicholls, R.W. 1980, *J. Mol. Spectrosc.*, **80**, 249
51. Morton, D.C. 1994, *ApJS*, **95**, 301
52. Edmonds, A.R. 1960, *Angular Momentum in Quantum Mechanics* (Princeton University Press, Princeton)
53. Schadee, A. 1964, *Bull. Astron. Netherlands*, **17**, 311
54. Kovács, I. 1969, *Rotational Structure in the Spectra of Diatomic Molecules* (Elsevier, Amsterdam)
55. Whiting, E.E., Paterson, J.A., Kovács, I., & Nicholls, R.W. 1973, *J. Mol. Spectrosc.*, **47**, 84
56. Whiting, E.E. 1973, NASA Tech. Note D-7268
57. Herzberg, G. 1971, *The Spectra and Structure of Simple Free Radicals* (Cornell University Press, Ithaca)
58. Rosen, B. 1970, *Spectroscopic Data Relative to Diatomic Molecules* (Pergamon, New York)
59. Pearse, R.W.B., & Gaydon, A.G. 1976, *The Identification of Molecular Spectra*, 4th ed. (Chapman and Hall, London)

Chapter 5

Radiation

J.J. Keady and D.P. Kilcrease

5.1 RADIATION QUANTITIES AND INTERRELATIONS

The quantitative concepts of radiation are defined [1] in terms of I, the flux of radiation at a given point in a given direction across a unit surface normal to that direction per unit time and per unit solid angle. This is called *specific intensity*, or simply *intensity*.

The flux of radiation through a unit surface is the *surface flux*, or *flux density*,

$$\mathcal{F} = \int_{4\pi} I \cos \theta \, d\omega,$$

where θ is the angle between the ray and the outward normal and integration is in all directions.

The emittance is the flux of radiation emitted from a unit surface,

$$\mathcal{F} = \int_{2\pi} I \cos \theta \, d\omega$$

for isotropic radiation πI, where in this case the integration is over the outward hemisphere.

The radiation density is

$$u = (1/c) \int_{4\pi} I \, d\omega = (4\pi/c)\bar{I}.$$

The radiation quantities per unit frequency and wavelength ranges are written I_ν, I_λ, $\mathcal{F}_\nu$, etc.:

$$I = \int I_\nu \, d\nu = \int I_\lambda \, d\lambda,$$

$$I_\lambda = \frac{c}{\lambda^2} I_\nu = \frac{\nu^2}{c} I_\nu, \qquad \lambda I_\lambda = \nu I_\nu,$$

$$d\lambda = -\frac{\lambda^2}{c} d\nu = -\frac{c}{\nu^2} d\nu, \qquad c = \lambda \nu.$$

The linear absorption coefficient is κ_s:

$$dI/ds = -\kappa_s I.$$

The scattering coefficient σ_s, is similar to the absorption coefficient but applies to the radiation scattered. It is used in the sense that $\kappa_s - \sigma_s$ represents absorption and transference into heat.

The mass absorption coefficient is κ_m (the subscript is usually omitted):

$$dI/ds = -\rho \kappa_m I,$$

where ρ is the density.

The atomic or particle absorption coefficient or cross section is a:

$$dI/ds = -Na I,$$

where there are N atoms or particles per unit volume and a represents the effective area over which the incident radiation if fully absorbed.

The emission coefficient j is the radiant flux emitted per unit volume and unit solid angle. For uniform scattering,

$$j = (\sigma/4\pi) \int_{4\pi} I \, d\omega,$$

where the first term represents scattering and the integral represents incident radiation.

For scattering by electrons, atoms, molecules,

$$j = (\sigma/4\pi) \int_{4\pi} \tfrac{3}{4}(1 + \cos^2 \theta) I \, d\omega,$$

where θ is the angle between incident and scattered light. I is assumed to be unpolarized, and the scattered radiation when viewed at the angle θ is polarized with intensity proportional to $\cos^2 \theta$ in the plane of scattering and proportional to I in the direction perpendicular to the plane of scattering.

The optical thickness or depth is

$$\tau = \int \kappa_s \, ds = \int \rho \kappa_m \, ds.$$

The source function is

$$S = j/\kappa_s.$$

The intensity emitted from an absorbing medium is

$$I = \int j \exp(-\tau) \, ds = \int S \exp(-\tau) \, d\tau.$$

We show two forms of the Kirchoff law:

(a) In a volume element,
$$j_\nu = \kappa_{s,\nu} B_\nu(T),$$
where $B_\nu(T)$ is blackbody intensity at temperature T.

(b) At a surface element,
$$I_\nu = A_\nu B_\nu(T),$$
where A_ν is the fraction of incident radiation absorbed, i.e., $1 - A_\nu$ is the reflection coefficient and analogous to albedo.

The atomic polarizability α is the induced dipole moment per unit electric field ($\bar{\alpha}$ for a steady or low-frequency field):

$$\bar{\alpha} = 4a_0^3 \sum_n f_n/(\nu_n/cR_\infty)^2$$
$$= 5.927 \times 10^{-25} \sum_n f_n/(\nu_n/cR_\infty)^2 \text{ cm}^3$$
$$= 7.138 \times 10^{-23} \sum_n f_n \lambda_n^2 \text{ cm}^3 \quad (\lambda \text{ in } \mu m),$$

where ν_n/cR_∞ is the frequency in rydbergs of lines connecting the ground level and f_n is the corresponding oscillator strength.

For scattering,

$$\alpha_s = (128\pi^5/3)N(\nu/c)^4\alpha^2$$
$$= (128\pi^5/3\lambda^4)N\alpha^2$$
$$= 1.3057 \times 10^{20} N\alpha^2/\lambda^4 \quad (\lambda \text{ in } \mu m).$$

The index of refraction is n:

$$n - 1 = 2\pi N\alpha$$
$$= 1.688 \times 10^{20}\alpha \quad \text{at STP}.$$

The molecular refraction is

$$R = \frac{n^2 - 1}{n^2 + 2} \frac{M}{\rho} = \frac{4\pi}{3} N_0 \alpha,$$

where M is the molecular weight, ρ is the density, and N_0 is the Avogadro number.

The radiation constants are

$$c_1 = 2\pi h c^2 = 3.741\,77 \times 10^{-5} \text{ erg cm}^2 \text{ s}^{-1},$$
$$c_2 = hc/k = 1.438\,77 \text{ cm K}.$$

The Stefan–Boltzmann constant is

$$\sigma = 2\pi^5 k^4/(15c^2 h^3) = \pi^4 c_1/(15 c_2^4)$$
$$= 5.6705 \times 10^{-5} \text{ erg cm}^{-2} \text{ s}^{-1} \text{ K}^{-4}.$$

The blackbody emittance is

$$\mathcal{F} = \sigma T^4.$$

The blackbody intensity is

$$B = (\sigma/\pi)T^4 = 1.804\,98 \times 10^{-5} T^4 \text{ erg cm}^{-2} \text{ s}^{-1} \text{ sr}^{-1} \text{ K}^{-4}.$$

The radiation density u in a cavity at temperature T is

$$u = aT^4 = (4\sigma/c)T^4 = 7.565\,91 \times 10^{-15} T^4 \text{ erg cm}^{-3} \text{ K}^{-4}.$$

In a medium of refractive index n,

$$B = n^2(\sigma/\pi)T^4,$$
$$u = n^3(4\sigma/c)T^4,$$

with similar factors applying for the Planck law with n_ν and n_λ.

The photon emission constant is

$$p = 4\pi\zeta(3)c/c_2^3$$
$$= 1.520\,486 \times 10^{11} \text{ photons cm}^{-2} \text{ s}^{-1} \text{ K}^{-3},$$

where $\zeta(n)$ is the Riemann zeta function.

The photon flux from a unit blackbody surface is

$$N = pT^3.$$

Blackbody radiation is unpolarized, hence the intensity of radiation linearly polarized in a specific direction will be half the value quoted in the formulas.

The Planck function in wavelength units is

$$(c/4)u_\lambda = \pi B_\lambda = \mathcal{F}_\lambda = 2\pi hc^2\lambda^{-5}/(e^{hc/k\lambda T} - 1)$$
$$= c_1\lambda^{-5}/(e^{c_2/\lambda T} - 1) \quad (\lambda \text{ in cm}),$$

where u_λ, B_λ, and $\mathcal{F}_\lambda$ are the radiation density, intensity, and emittance for unit wavelength ranges.

The Planck function in frequency units is

$$(c/4)u_\nu = \pi B_\nu = \mathcal{F}_\nu = 2\pi h\nu^3 c^{-2}/(e^{h\nu/kT} - 1).$$

The photon distribution law is

$$N_\lambda = 2\pi c\lambda^{-4}/(e^{c_2/\lambda T} - 1),$$
$$N_\nu = 2\pi c^{-2}\nu^2/(e^{h\nu/kT} - 1),$$

where N_λ and N_ν are the emittance of photons per squared centimeter per second and per unit wavelength and frequency ranges, respectively.

The Rayleigh–Jeans distribution (for the red end of the spectrum) is

$$\mathcal{F}_\lambda = 2\pi ckT\lambda^{-4} = (c_1/c_2)T\lambda^{-4},$$
$$\mathcal{F}_\nu = 2\pi c^{-2}kT\nu^2 = 2\pi kT\lambda^{-2}.$$

The Wien distribution (for the violet end of the spectrum) is

$$\mathcal{F}_\lambda = 2\pi hc^2\lambda^{-5}e^{-c_2/\lambda T} = c_1\lambda^{-5}e^{-c_2/\lambda T},$$
$$\mathcal{F}_\nu = 2\pi hc^{-2}\nu^3 e^{-h\nu/kT}.$$

Wien law: The wavelength of maximum $\mathcal{F}_\lambda$ or B_λ is $\lambda_{\max}$:

$$T\lambda_{\max} = 0.201\,405\,2c_2 = 0.289\,78 \text{ cm K}.$$

The wavelength of maximum photon emission is λ_m:

$$T\lambda_m = 0.255\,057\,1c_2 = 0.366\,97 \text{ cm K}.$$

The frequency of maximum $\mathcal{F}_\nu$ or B_ν is ν_m:

$$Tc/\nu_m = 0.354\,429\,0c_2 = 0.509\,94 \text{ cm K}.$$

The three numerical constants above are $1/y$ in $y = 5(1 - e^{-y})$, $y = 4(1 - e^{-y})$, and $y = 3(1 - e^{-y})$, respectively.

5.2 REFRACTIVE INDEX AND AVERAGE POLARIZABILITY

The refractive index and polarizability of atomic and molecular gases are given in Tables 5.1 and 5.2, where n is the refractive index at STP,

$$n - 1 = A(1 + B/\lambda^2) \quad (\lambda \text{ in } \mu m),$$

and $\bar{\alpha}$ is the polarizability at low frequency.

Table 5.1. *Refractive index and polarizability of atomic gases.*

Atom	$\bar{\alpha}$ [1] (10^{-25} cm^3)	n (D lines)	A (units of 10^{-5})	B (units of 10^{-3})	Atom	$\bar{\alpha}$ [1] (10^{-25} cm^3)	n (D lines)	A (units of 10^{-5})	B (units of 10^{-3})
H	6.67				Cl	21.8			
He	2.05	1.000 035 0	3.48	2.3	Ar	16.4	1.000 283 7	27.92	5.6
Li	243				K	434			
Be	56				Ca	250			
C	17.6				Sc	169			
N	11.0				Ti	136			
O	8.03				V	114			
Ne	3.95	1.000 067 1	6.6	2.4	Cr	68			
Na	236				Mn	86			
Mg	106				Fe	75			
Al	83.4				Co	68			
Si	53.8				Ni	65			
P	36.3				Cu	61			
S	29.0				Zn	71			

Reference
1. Miller, T.M., & Bederson, B. 1977, *Adv. Atom. Mol. Phys.* **13**, 1

Table 5.2. *Refractive index of molecular gases.*

Molecule	n (D lines)	A (units of 10^{-5})	B (units of 10^{-3})	Molecule	n (D lines)	A (units of 10^{-5})	B (units of 10^{-3})
Air	1.000 291 8	28.71	5.67	CO$_2$	1.000 449 8	43.9	6.4
H$_2$	1.000 138 4	13.58	7.52	CO	1.000 334	32.7	8.1
O$_2$	1.000 272	26.63	5.07	NH$_3$	1.000 375	37.0	12.0
N$_2$	1.000 297	29.06	7.7	NO	1.000 297	28.9	7.4
H$_2$O	1.000 254	516 (radio freq.)		CH$_4$	1.000 441		

The refractive indices quoted in Table 5.3 are relative to air at 15° C. The temperatures of the media are about 18° C and the temperature coefficients quoted are the change of D-line refractive index for a 1° C temperature rise. Manufacturers' reports must be consulted for indices that are accurate enough for optical design. The table also gives the spectral limits (λ in μm) within which the absorption is less than 2.72 cm^{-1} (i.e., 1 cm transmission > 37%).

Table 5.3. *Refractive indices of optical media* [1, 2].

λ (μm)	Calcite Ord. ray	Calcite Extr. ray	Glass BSC crown	Glass DF flint	Fluorite CaF$_2$	Quartz Ord. ray	Quartz Extr. ray	Fused silica	Rock salt	Water
0.2	1.91	1.58			1.495	1.651	1.663	1.550	1.792	1.423
0.3	1.722	1.515	1.557		1.455	1.579	1.589	1.489	1.602	1.358
0.4	1.683	1.499	1.531	1.650	1.442	1.558	1.567	1.471	1.568	1.343
0.5	1.666	1.491	1.522	1.627	1.437	1.549	1.558	1.463	1.552	1.336
0.6	1.657	1.486	1.517	1.616	1.434	1.544	1.553	1.458	1.543	1.332
0.7	1.652	1.483	1.513	1.610	1.432	1.541	1.550	1.455	1.538	1.330
0.8	1.648	1.481	1.511	1.605	1.430	1.539	1.548		1.535	1.328
1.0	1.643	1.479	1.507	1.600	1.429	1.536	1.544		1.532	1.325
2	1.626	1.476	1.496		1.424	1.520	1.528		1.526	1.315
5					1.398	1.42			1.519	
10					1.303				1.494	
Temp. coef.	+0.000005	+0.000014	−0.000001	+0.000003	−0.00001	−0.000005	−0.000006	−0.000003	−0.00004	−0.00008
Limits [2] Low λ	0.23	0.23	0.32	0.37	0.13	0.17	0.17	0.16	0.20	< 0.2
High λ	2.2	4	2.2	2.8	9.0	3.6	3.6	21	17	1.14

Note

For information on atmospheric refraction, see Table 5.2.

References

1. Allen, C.W. 1973, *Astrophysical Quantities*, 3rd ed. 1, Sec. 34; 2, Sec. 35; 3, Sec. 36
2. Garton, W.R.S. 1966, *Adv. Atom. Mol. Phys.* **2**, 93

5.3   ABSORPTION AND SCATTERING BY PARTICLES

For scattering of free electrons, σ_e (Thomson scattering) is [2]

$$\sigma_e = \frac{8\pi}{3}\left(\frac{e^2}{mc^2}\right)^2\left(1 - 2\frac{h\nu}{mc^2}\right) = 0.665\,24 \times 10^{-24}\left(1 - 2\frac{h\nu}{mc^2}\right) cm^2,$$

where σ_e is the (exponential) scattering coefficient per electron (Sec. 5.1) with the relativistic term $2h\nu/mc^2$. At the high densities and temperatures found in stellar interiors, further corrections due to correlations and thermal motion may be required [3, 4].

For Rayleigh scattering of atoms or molecules,

$$\sigma_s = \frac{32\pi^3}{3N}\frac{(n - 1)^2}{\lambda^4}\frac{6 + 3\Delta}{6 - 7\Delta}$$
$$= 3.307 \times 10^{18}(n - 1)^2\delta/\lambda^4 N \ cm^{-1} \quad (\lambda \ in \ \mu m),$$

where N is the number of atoms or molecules per unit volume, n is the refractive index of the medium, σ_s is the linear scattering coefficient, and $\delta = (6 + 3\Delta)/(6 - 7\Delta) =$ depolarizing factor [5, 6]. $\Delta = 0.030$ for N_2 and 0.054 for O_2 [7].

The Rayleigh scattering cross section of an atom or a molecule is

$$\sigma_a = \frac{32\pi^3\delta}{3\lambda^4}\left(\frac{n - 1}{N}\right)^2 = \frac{128\pi^5}{3\lambda^4}\delta\alpha^2$$
$$= 1.306 \times 10^{20}\delta\alpha^2/\lambda^4 \ cm^2 \quad (\lambda \ in \ \mu m),$$

where the polarizability $\alpha = (n - 1)/(2\pi N)$.

For atomic scattering at some distance from any absorption line,

$$\sigma_a = \frac{8\pi}{3}\left(\frac{e^2}{mc^2}\right)^2\left(\sum_2 \frac{f_{12}\nu^2}{\nu_{12}^2 - \nu^2}\right)^2,$$

where f_{12} is the oscillator strength (1 is the ground level when excitation is low).

For the absorption of small particles (spherical) of radius a in terms of πa^2 [5], the efficiency factors ($Q = \sigma/\pi a^2$) for extinction, scattering, absorption, and radiation pressure are Q_{ext}, Q_{sca}, Q_{abs}, and Q_{pr}, respectively, with

$$Q_{ext} = Q_{sca} + Q_{abs},$$
$$Q_{pr} = Q_{ext} - \langle\cos\theta\rangle Q_{sca},$$

where $\langle\cos\theta\rangle$ is the forward asymmetry of scattering [8]. For large objects $Q_{ext} = 2.0$ of which 1.0 is intercepted and 1.0 scattered with $\langle\cos\theta\rangle = 1.0$. The extinction coefficient k is related to the complex dielectric constant ϵ through

$$\epsilon = \epsilon_1 \pm i\epsilon_2 = (n \pm ik)^2.$$

Here n is the refractive index and ϵ_1, ϵ_2, n, and k are real. k and n are therefore given by

$$\left.\begin{matrix} k \\ n \end{matrix}\right\} = \frac{1}{\sqrt{2}}[(\epsilon_1^2 + \epsilon_2^2)^{1/2} \mp \epsilon_1]^{1/2}.$$

For measurements performed in vacuum, with $2\pi a/\lambda \ll 1$ everywhere, the measured extinction coefficient is related to the indices of refraction via

$$\frac{Q_{abs}}{a} \approx \frac{Q_{ext}}{a} = \frac{2\pi}{\lambda} \frac{24nk}{(n^2 - k^2 + 2) + 4n^2k^2}.$$

See Tables 5.4–5.10, extinction efficiency factors for various compounds. The mean particle radius for the spheroids in the amorphous carbon sample was 40 Å.

Table 5.4. *Extinction efficiency factor Q_{ext} for water droplets as a function of particle radius and wavelength* [1].

λ (μm)	$a = 0.3\mu$m	1.0 μm	3.0 μm	10.0 μm	λ (μm)	$a = 0.3\mu$m	1.0 μm	3.0 μm	10.0 μm
0.70	1.39	2.79	2.47	2.07	4.50	0.0182	0.342	3.02	2.46
0.80	1.07	3.37	2.13	1.99	4.66	0.0197	0.319	2.88	2.26
0.90	0.80	3.76	1.95	2.05	4.80	0.0180	0.291	2.76	2.21
1.00	0.58	3.90	2.37	2.14	5.00	0.0131	0.242	2.56	2.02
1.10	0.44	3.89	2.68	2.18	5.26	0.0098	0.189	2.22	1.85
1.20	0.35	3.60	2.80	2.05	5.50	0.0111	0.162	1.93	2.02
1.30	0.28	3.41	2.28	2.25	5.80	0.026	0.177	1.50	2.65
1.40	0.23	3.08	2.04	2.25	6.00	0.076	0.39	1.95	2.40
1.50	0.18	2.83	1.86	2.01	6.05	0.083	0.419	2.02	2.38
1.60	0.15	2.56	1.83	2.27	6.40	0.030	0.215	1.95	2.30
1.80	0.099	2.07	2.09	2.11	7.00	0.022	0.145	1.47	2.73
2.00	0.067	1.65	2.87	2.13	7.50	0.021	0.120	1.19	3.09
2.10	0.054	1.45	3.16	2.42	8.00	0.020	0.106	1.02	3.24
2.20	0.043	1.25	3.51	2.46	8.50	0.020	0.097	0.886	3.30
2.30	0.035	1.09	3.67	2.28	9.00	0.020	0.091	0.734	3.21
2.50	0.022	0.69	3.66	2.34	9.50	0.021	0.087	0.596	2.95
2.60	0.0127	0.33	2.61	1.96	10.00	0.025	0.094	0.509	2.50
2.70	0.0136	0.189	1.58	2.26	10.50	0.031	0.112	0.487	2.08
2.75	0.0477	0.297	1.61	2.34	11.00	0.044	0.155	0.547	1.78
2.80	0.1754	0.940	2.58	2.23	11.50	0.063	0.220	0.706	1.83
2.90	0.4178	1.65	2.45	2.21	12.00	0.083	0.286	0.884	1.97
2.95	0.458	1.74	2.44	2.21	12.50	0.091	0.317	0.992	2.09
3.00	0.423	1.76	2.49	2.22	13.00	0.103	0.360	1.13	2.20
3.10	0.293	1.79	2.67	2.24	13.50	0.107	0.376	1.20	2.26
3.20	0.159	1.93	2.74	2.26	14.00	0.114	0.400	1.29	2.32
3.30	0.0806	1.52	3.14	2.33	15.00	0.123	0.434	1.44	2.42
3.40	0.0429	1.21	3.53	2.42	16.00	0.104	0.371	1.38	2.51
3.50	0.0264	0.944	3.97	2.17	17.50	0.086	0.309	1.30	2.64
3.60	0.0189	0.763	3.99	1.97	18.00	0.080	0.289	1.29	2.71
3.75	0.0130	0.569	3.84	2.38	20.00	0.061	0.223	1.23	2.92
3.83	0.0114	0.497	3.76	2.66	30.00	0.037	0.129	0.528	2.75
4.00	0.0114	0.428	3.58	2.80	100.00	0.014	0.048	0.149	0.758

Reference
1. Irvine, W.R., & Pollack, J.B. 1968, *Icarus*, **8**, 324

Table 5.5. *Extinction efficiency factor Q_{ext} for ice particles as a function of particle radius and wavelength* [1].

λ (μm)	$a = 0.3\mu$m	1.0 μm	3.0 μm	10.0 μm	λ (μm)	$a = 0.3\mu$m	1.0 μm	3.0 μm	10.0 μm
0.95	0.572	2.92	2.41	2.02	3.60	0.025	0.76	3.91	2.04
1.00	0.49	3.80	2.66	2.08	3.80	0.017	0.52	3.70	2.57
1.20	0.295	3.35	2.28	2.15	3.90	0.018	0.46	3.51	2.63
1.50	0.158	2.52	1.82	2.24	4.00	0.019	0.41	3.28	2.64
2.00	0.063	1.52	3.10	2.35	4.10	0.021	0.38	3.04	2.43
2.35	0.029	0.94	3.77	1.96	4.20	0.023	0.35	2.84	2.30

Table 5.5. *Continued.*

λ (μm)	a = 0.3μm	1.0 μm	3.0 μm	10.0 μm	λ (μm)	a = 0.3 μm	1.0 μm	3.0 μm	10.0 μm
2.40	0.025	0.823	3.81	1.93	4.30	0.027	0.34	2.67	2.16
2.45	0.022	0.724	3.73	2.12	4.40	0.032	0.33	2.46	2.11
2.50	0.019	0.626	3.58	2.46	4.50	0.036	0.32	2.31	2.10
2.55	0.016	0.522	3.35	2.71	4.60	0.031	0.28	2.17	2.10
2.60	0.013	0.433	3.09	2.58	4.70	0.023	0.23	2.02	2.06
2.625	0.012	0.397	2.96	2.36	4.80	0.018	0.20	1.86	2.16
2.65	0.012	0.364	2.79	2.18	4.90	0.015	0.17	1.73	2.27
2.800	0.027	0.261	1.84	2.07	5.00	0.014	0.16	1.61	2.39
2.85	0.066	0.360	1.68	2.31	5.70	0.026	0.16	1.20	2.95
2.90	0.177	0.675	1.78	2.19	6.00	0.048	0.23	1.31	2.73
2.95	0.306	0.986	1.92	2.14	6.40	0.040	0.19	1.11	2.86
3.00	0.377	1.151	2.00	2.14	6.70	0.039	0.18	1.02	2.89
3.05	0.514	1.516	2.22	2.18	7.00	0.033	0.15	0.92	2.96
3.075	0.547	1.63	2.28	2.19	8.00	0.021	0.10	0.67	3.01
3.10	0.511	1.69	2.36	2.20	9.00	0.019	0.08	0.51	2.74
3.15	0.379	2.30	2.54	2.25	10.00	0.020	0.07	0.33	1.76
3.20	0.244	2.23	2.57	2.26	11.00	0.038	0.15	0.77	2.79
3.25	0.151	2.10	2.62	2.26	12.00	0.039	0.16	1.25	3.00
3.30	0.109 6	1.91	2.78	2.25	15.00	0.019	0.076	0.80	3.58
3.35	0.081 7	1.74	2.92	2.25	20.00	0.005	0.019	0.21	3.06
3.40	0.060 8	1.50	3.14	2.36	40.00	0.017	0.058	0.20	1.37
3.45	0.046	1.16	3.62	2.37	62.00	0.016	0.055	0.18	1.02
3.50	0.035	0.97	3.87	2.19	100.00	0.002 7	0.009	0.028	0.17
3.55	0.029	0.85	3.94	2.06	150.00	0.000 57	0.002	0.007	0.04

Reference
1. Irvine, W.R., & Pollack, J.B. 1968, *Icarus*, **8**, 324

Table 5.6. *Extinction efficiencies for amorphous carbon* [1, 2].[a]

λ (μm)	Q_{ext}/a (cm^{-1})	λ (μm)	Q_{ext}/a (cm^{-1})	λ (μm)	Q_{ext}/a (cm^{-1})
0.12	5.12[5]	0.27	1.49[5]	7.14	4.49[3]
0.13	3.21[5]	0.30	1.28[5]	10.00	3.18[3]
0.14	2.02[5]	0.50	7.07[4]	15.40	1.74[3]
0.15	2.18[5]	0.70	4.81[4]	20.00	1.35[3]
0.16	1.84[5]	1.00	3.52[4]	30.50	8.80[2]
0.18	1.47[5]	1.60	2.14[4]	50.80	5.38[2]
0.20	1.43[5]	2.00	1.69[4]	70.50	4.04[2]
0.22	1.56[5]	3.13	1.05[4]	101.00	3.17[2]
0.23	1.63[5]	4.00	7.96[3]	205.00	1.66[2]
0.25	1.60[5]	5.00	6.06[3]	289.00	1.19[2]

Note
[a]Numbers in square brackets denote powers of 10.

References
1. Bussoletti, E. et al. 1987, *A&AS*, **70**, 257
2. Maron, M. 1990, *ApS&S*, **172**, 21

Table 5.7. *Extinction efficiency factor Q_{ext} for graphite as a function of particle radius and wavelength* [1].[a]

λ (μm)	$a = 0.01\mu$m	0.1 μm	λ (μm)	$a = 0.01\mu$m	0.1 μm
0.12	0.425	2.58	2.00	1.00[−2]	0.19
0.15	0.244	2.32	3.00	5.37[−3]	0.084
0.18	0.512	2.49	4.00	3.44[−3]	0.048
0.20	0.995	2.83	5.00	2.43[−3]	0.032
0.21	1.51	2.92	6.00	1.83[−3]	0.024
0.215	1.25	3.03	9.00	4.65[−6]	1.23[−2]
0.2175	1.22	3.06	10.00	9.00[−4]	1.07[−2]
0.22	1.17	3.08	11.00	8.21[−4]	9.61[−3]
0.225	1.01	3.11	11.52	2.98[−3]	3.13[−2]
0.23	0.88	3.13	11.54	8.82[−4]	1.01[−2]
0.24	0.63	3.12	12.00	7.60[−4]	8.78[−3]
0.26	0.40	3.02	20.00	6.91[−4]	7.45[−3]
0.28	0.31	2.98	40.00	6.51[−4]	6.87[−3]
0.30	0.25	3.01	60.00	4.28[−4]	4.41[−3]
0.33	0.20	3.09	80.00	2.77[−4]	2.80[−3]
0.365	0.160	3.11	100.0	1.89[−4]	1.90[−3]
0.4861	0.093	3.34	200.0	5.24[−5]	5.13[−4]
0.6562	0.058	2.65	400.0	1.40[−5]	1.29[−4]
0.80	0.043	1.88	700.0	4.71[−6]	4.24[−5]
1.00	0.030	1.09	1000.0	2.33[−6]	2.08[−5]
1.40	0.017	0.45	2000.0	5.87[−7]	5.22[−5]

Note
[a] Numbers in square brackets denote powers of 10.

Reference
1. Draine, B.L. 1985, *ApJS*, **57**, 587

Table 5.8. *Extinction efficiencies for silicon carbide* [1].[a]

λ (μm)	Q_{ext}/a (cm^{-1})	λ (μm)	Q_{ext}/a (cm^{-1})	λ (μm)	Q_{ext}/a (cm^{-1})
0.10	3.46[6]	9.12	4.24[2]	12.74	2.53[3]
0.20	7.65[4]	9.90	9.65[2]	13.04	1.94[3]
0.40	1.29[4]	10.14	1.66[3]	13.34	1.50[3]
0.78	8.85[4]	10.37	3.06[3]	13.65	1.18[3]
0.99	8.13[3]	10.61	5.76[3]	13.96	9.69[2]
1.96	7.01[3]	10.86	9.35[3]	14.29	7.96[2]
3.09	4.15[3]	11.11	1.37[4]	15.30	5.76[2]
3.97	2.55[3]	11.37	1.45[4]	22.00	3.23[2]
4.69	1.66[3]	11.63	1.04[4]	36.90	1.57[2]
6.02	6.97[2]	11.90	7.30[3]	52.00	9.74[1]
7.10	4.83[2]	12.17	4.84[3]	103.3	3.79[1]
8.39	4.03[2]	12.45	3.31[3]	205.3	1.23[1]

Note
[a] Numbers in square brackets denote powers of 10.

Reference
1. Pégourié, B. 1988, *A& A*, **194**, 335

Table 5.9. *Extinction efficiency factors* Q_{ext} *for silicate as a function of particle radius and wavelength* [1].[a]

λ (μm)	$a = 0.01 \mu m$	0.1 μm	λ (μm)	$a = 0.01 \mu m$	0.1 μm
0.12	0.943	2.56	1.00	3.22[−3]	0.11
0.14	0.51	2.62	1.65	4.19[−4]	3.18[−2]
0.16	0.154	2.67	2.00	2.25[−4]	2.31[−2]
0.20	0.032	3.13	2.60	9.40[−5]	1.67[−2]
0.23	0.0146	3.98	3.00	5.75[−5]	1.46[−2]
0.30	1.09[−2]	3.41	4.00	2.06[−5]	1.14[−2]
0.40	7.99[−3]	2.26	5.00	8.64[−4]	1.01[−2]
0.55	5.76[−3]	0.78	6.00	9.95[−4]	1.00[−2]

Note

[a]Numbers in square brackets denote powers of 10.

Reference

1. Draine, B.L. 1985, *ApJS*, **57**, 587

Table 5.10. *Extinction efficiencies for silicate for* λ > 6 μm [1].[a]

λ (μm)	Q_{ext}/a (cm^{-1})	λ (μm)	Q_{ext}/a (cm^{-1})	λ (μm)	Q_{ext}/a (cm^{-1})
7.0	1.04[3]	13.0	3.74[3]	27.5	2.10[3]
8.0	3.26[3]	14.0	2.79[3]	30.0	1.75[3]
8.5	6.75[3]	15.0	3.08[3]	40.0	9.80[2]
9.0	1.20[4]	16.0	3.70[3]	50.0	6.15[2]
9.5	1.32[4]	17.0	4.34[3]	80.0	2.26[2]
10.0	1.20[4]	18.0	4.70[3]	100.0	1.41[2]
10.5	1.05[4]	19.0	4.62[3]	200.0	3.40[1]
11.0	8.38[3]	20.0	4.19[3]	500.0	5.38
11.5	6.80[3]	23.0	3.06[3]	1 000.0	1.34
12.0	5.58[3]	25.0	2.57[3]	2 000.0	3.36[−1]

Note

[a]Numbers in square brackets denote powers of 10.

Reference

1. Draine, B.L. 1985, *ApJS*, **57**, 587

5.4 PHOTOIONIZATION AND RECOMBINATION

5.4.1 Photoionization Fit Parameters for Ground States

The following parameters are taken from [9] and are used in the formula for the photoionization cross section:

$$a_E = a_T \times 10^{-18}[R(E_T/E)^s + (1 - R)(E_T/E)^{s+1}] \text{ cm}^2, \qquad E > E_T,$$

where E_T is the threshold energy in eV and a_T is the threshold cross section, divided by 1.0×10^{-18} cm^2. The fitting coefficients R and s are found in Table 5.11.

Table 5.11. *Photoionization cross-section fits* [1].

Parent	Resulting ion	E_T	a_T	R	s
$H(^2S)$	$H^+(^1S)$	13.6	6.30	1.34	2.99
$He(^1S)$	$He^+(^2S)$	24.6	7.83	1.66	2.05
$He^+(^2S)$	$He^{2+}(^1S)$	54.4	1.58	1.34	2.99
$C(^3P)$	$C^+(^2P)$	11.3	12.2	3.32	2.00
$C^+(^2P)$	$C^{2+}(^1S)$	24.4	4.60	1.95	3.00
$C^{2+}(^1S)$	$C^{3+}(^2S)$	47.9	1.60	2.60	3.00
$C^{3+}(^2S)$	$C^{4+}(^1S)$	64.5	0.68	1.00	2.00
$N(^4S)$	$N^+(^3P)$	14.5	11.4	4.29	2.00
$N^+(^3P)$	$N^{2+}(^2P)$	29.6	6.65	2.86	3.00
$N^{2+}(^2P)$	$N^{3+}(^1S)$	47.5	2.06	1.63	3.00
$N^{3+}(^1S)$	$N^{4+}(^2S)$	77.5	1.08	2.60	3.00
$N^{4+}(^2S)$	$N^{5+}(^1S)$	97.9	0.48	1.00	2.00
$O(^3P)$	$O^+(^4S)$	13.6	2.94	2.66	1.00
$O(^3P)$	$O^+(^2D)$	16.9	3.85	4.38	1.50
$O(^3P)$	$O^+(^2P)$	18.6	2.26	4.31	1.50
$O^+(^4S)$	$O^{2+}(^3P)$	35.2	7.32	3.84	2.50
$O^{2+}(^3P)$	$O^{3+}(^2P)$	54.95	3.65	2.01	3.00
$O^{3+}(^2P)$	$O^{4+}(^1S)$	77.4	1.27	0.83	3.00
$O^{4+}(^1S)$	$O^{5+}(^2S)$	113.9	0.78	2.60	2.00
$O^{5+}(^2S)$	$O^{6+}(^1S)$	138.1	0.36	1.00	2.10
$Ne(^1S)$	$Ne^+(^2P)$	21.6	5.35	3.77	1.00
$Ne^+(^2P)$	$Ne^{2+}(^3P)$	41.1	4.16	2.72	1.50
$Ne^+(^2P)$	$Ne^{2+}(^1D)$	42.3	2.71	2.15	1.50
$Ne^+(^2P)$	$Ne^{2+}(^1S)$	47.99	0.52	2.13	1.50
$Ne^{2+}(^3P)$	$Ne^{3+}(^4S)$	63.74	1.80	2.28	2.00
$Ne^{2+}(^3P)$	$Ne^{3+}(^2D)$	68.8	2.50	2.35	2.50
$Ne^{2+}(^3P)$	$Ne^{3+}(^2P)$	71.5	1.48	2.23	2.50
$Ne^{3+}(^4S)$	$Ne^{4+}(^3P)$	97.2	3.11	1.96	3.00
$Ne^{4+}(^3P)$	$Ne^{5+}(^2P)$	126.5	1.40	1.47	3.00
$Ne^{5+}(^2S)$	$Ne^{6+}(^1S)$	138.1	0.36	1.00	2.10
$Ne^{5+}(^2P)$	$Ne^{6+}(^1S)$	157.96	0.49	1.15	3.00

Reference
1. Osterbrock, D.E. 1974, *Astrophysics of Gaseous Nebulae* (Freeman, San Francisco)

5.4.2 Photoionization of Light Hydrogenic Ions

A semiempirical expression [10] for the photoionization cross section per K-shell electron for hydrogenic light elements is given by

$$a_v = \frac{2^9 \pi^2 a_0^2}{3 \alpha^2 Z^5} \left(\frac{-E_1}{h\nu}\right)^4 (\eta \beta \gamma)^3 \left[1 + \frac{3}{4} \frac{\gamma(\gamma-2)}{\gamma+1} \left(1 - \frac{1}{2\beta\gamma^2} \ln \frac{1+\beta}{1-\beta}\right)\right] \frac{\exp(-4\eta \operatorname{arccot} \eta)}{1 - \exp(-2\pi\eta)},$$

where $\eta = [-E_1/(h\nu + E_1)]^{1/2}$. E_1 is the negative binding energy of the $1s$ electron, v is the electron velocity, and

$$\beta = \frac{v}{c} = \frac{[(h\nu + E_1)^2 + 2(h\nu + E_1)mc^2]^{1/2}}{h\nu + E_1 + mc^2} \sim [2(h\nu + E_1)/mc^2]^{1/2} \sim \alpha Z/\eta,$$

$$\gamma = (1 - \beta^2)^{-1/2} = 1 + (h\nu + E_1)/mc^2, \qquad \alpha \simeq 1/137.036.$$

5.4.3   Radiative Recombination

Given an absorption coefficient a_ν for a particular level, microscopic reversibility demands that the recombination cross section into that same level be given by

$$\sigma(v) = \frac{g_i}{g_{i+1}} \frac{(h\nu)^2}{(mcv)^2} a_\nu.$$

This is the Milne relation. Here g_i is the statistical weight for the particular level or term i in the recombined ion, g_{i+1} is that of the original ion.

Using the Milne relation and the above analytic form for the photoionization cross section, the recombination rate coefficient can be expressed as [9]

$$\alpha_i(T) = \frac{4}{\sqrt{\pi}} \frac{g_i}{g_{i+1}} \left(\frac{m}{2kT}\right)^{3/2} e^{E_T/kT} \frac{E_T^3}{m^3c^2} a_T \left[R\mathcal{E}_{s-2}(E_T/kT) + (1-R)\mathcal{E}_{s-3}(E_T/kT)\right],$$

where $\mathcal{E}_n(x)$ is the exponential integral function. If s is noninteger, the relation $\mathcal{E}_n(x) \to x^{n-1}\Gamma(1 - n, x)$ can be used, where $\Gamma(a, x)$ is the incomplete gamma function.

Recombination into excited states is generally at least as, if not more, important than recombination into the ground state. If the excited state can be approximated as hydrogenic (often a good approximation), then the cumulative recombination coefficient $\alpha(n)$ for principal quantum number n and higher is $\alpha(n) = \sum_{n'=n}^{\infty} \alpha_{n'}$ and is given in Table 5.12 as a function of temperature for the first four values of n for $Z = 1$ [9]. For an arbitrary ionic charge Z, $\alpha(n; Z, T) = Z\alpha(n; 1, T/Z^2)$.

Table 5.12. *Recombination coefficients $\alpha(n)$ in cm^3 s^{-1} for hydrogen* [1].

n	1250 K	2500 K	5000 K	10000 K	20000 K
1	1.74[−12]	1.10[−12]	6.82[−13]	4.18[−13]	2.51[−13]
2	1.28[−12]	7.72[−13]	4.54[−13]	2.60[−13]	1.43[−13]
3	1.03[−12]	5.99[−13]	3.37[−13]	1.83[−13]	9.50[−14]
4	8.65[−13]	4.86[−13]	2.64[−13]	1.37[−13]	6.83[−14]

Reference
1. Osterbrock, D.E. 1974, *Astrophysics of Gaseous Nebulae* (Freeman, San Francisco)

Recombination into the nth hydrogenic level can be written as (taking $s = 3$ and $R = 1$)

$$\alpha_n = \frac{4}{\sqrt{\pi}} \frac{g_i}{g_{i+1}} \left(\frac{m}{2kT}\right)^{3/2} \frac{E_T^3}{m^3c^2} a_T \, e^{E_T/kT} \mathcal{E}_1(E_T/kT).$$

Note that $\exp(x)\mathcal{E}_1(x) \approx 1/x$ for $x > 5$.

The threshold photoionization cross section a_T can be obtained from the table of photoionization cross-section fitting parameters or from the Kramers–Gaunt formula:

$$a_T(\text{Kramers \& Gaunt}) = \frac{8h^3 gn}{3\sqrt{3}\pi^2 m^2 ce^2 Z^2} = 7.907 \times 10^{-18} \frac{ng}{Z^2} \, cm^2,$$

where g is the Gaunt factor [11] given in Table 5.13.

Table 5.13. *Bound–free Gaunt factors for the hydrogen atom* [1].

Configuration	g at absorption edge	g level average
$1s$	0.80	0.80
$2s$	0.96	0.89
$2p$	0.88	
$3s$	1.14	0.92
$3p$	1.14	
$3d$	0.73	
$4s$	1.3	0.94
$4p$	1.3	
$4d$		
$4f$		
5		0.95
6		0.96
7		0.97

Reference
1. Gaunt, J.A. 1930, *Philos. Trans.* **229**, 163

5.4.4 Dielectronic Recombination

For dielectric recombination into ion X^+, with excited state X^{+*} and charge Z, we have the Burgess formula [12] for the recombination coefficient α_d,

$$\alpha_d = 3.0 \times 10^{-3}\, T^{-3/2} f\, A(x) B(Z) \exp[-\chi C(Z)/T]\ \text{cm}^3\,\text{s}^{-1},$$

where f is the oscillator strength for the transition $X^+ \to X^{+*}$ with T in K and

$$\begin{aligned}
x &= 2[E(X^{+*}) - E(x^+)]/[(Z+1)\mathcal{E}_0], & \mathcal{E}_0 &\equiv 27.2\ \text{eV}, \\
A(x) &= x^{1/2}/(1 + 0.105x + 0.015x^2), & x &> 0.05, \\
B(Z) &= \left[Z(Z+1)^5/(Z^2 + 13.4)\right]^{1/2}, & Z &\leq 20, \\
C(Z) &= 1.58 \times 10^5\,(Z+1)/\left[1 + 0.015Z^3/(Z+1)^2\right], & xC(Z)/T &\lesssim 5.0.
\end{aligned}$$

5.5 X-RAY ATTENUATION

The smoothed fits in Table 5.14 provide an approximate representation ($\sim 10\%$ or better) to both the Henke experimental data [13] and to relativistic calculations [14] for the photoionization cross section σ (1 barn = 10^{-24} cm^2). The photon energy E is measured in keV, and

$$\log_{10}[\sigma\,(\text{barn/atom})] = \sum_{m=0}^{n} a_m\,(\log_{10} E)^m.$$

Table 5.14. *Cold material X-ray attenuation (total cross section) fits.*[a]

	E range (keV)	n	a_0	a_1	a_2	a_3	a_4
H	0.1089–8.0470	5	1.000 266	−2.874 876	1.337 216	0.690 5931	−0.273 4547
	8.0470–44.77	2	−0.105 7762	−8.314 851[−2]			
He	0.1086–8.0470	5	2.596 995	−3.278 818	−5.823 671[−2]	0.637 2270	0.344 4271
C	0.0305–0.2885	3	3.113 512	−3.093 789	−0.505 4382		
	0.2888–30.000	4	4.641 890	−2.929 378	−0.391 9827	0.364 0990	
N	0.0105–0.4027	4	3.653 780	−2.036 035	0.592 8674	0.429 1417	
	0.4031–32.20	4	4.880 570	−2.726 891	−0.317 1900	6.076 411[−2]	
O	0.0305–0.5374	4	3.851 303	−2.158 767	0.745 1610	0.548 5890	
	0.5380–30.320	5	5.087 381	−2.624 299	−0.341 0862	−0.197 2559	0.244 2756
Na	0.0415–0.0708	5	2.522 877[2]	7.578 575[2]	8.701 549[2]	4.394 846[2]	82.195 58
	0.0724–1.079	5	4.372 625	−2.803 046	−0.783 3634	−1.137 645	−0.794 4890
	1.0800–9.886	3	5.551 365	−2.508 414	−0.295 8825		
Mg	0.0305–0.0574	3	1.819 299	−3.725 268	−0.968 6183		
	0.0576–0.0824	4	0.759 7808	2.040 594[2]	2.000 915[2]	65.192 85	
	0.0824–1.3113	5	4.556 145	−2.709 325	−0.624 9299	−1.125 062	−0.961 1755
	1.3130–30.921	4	5.664 605	−2.340 560	−0.642 0745	0.235 7794	
Al	0.0305–0.0813	3	1.516 313	−5.356 061	−1.787 405		
	0.0815–0.1144	4	4.047 249[2]	1.227 899[3]	1.261 952[3]	4.319 780[2]	
	0.1144–1.5680	5	4.713 295	−2.629 794	−0.421 7032	−1.136 888	−1.261 846
	1.5690–30.000	4	5.732 064	−2.167 852	−0.759 3566	0.254 8878	
Si	0.0305–0.1087	4	4.653 385	1.585 146	4.064 835	1.700 534	
	0.1089–0.1300	4	1.704 974[3]	5.668 094[3]	6.305 044[3]	2.337 491[3]	
	0.1303–1.8470	4	4.842 629	−2.691 588	0.127 3064	0.821 4728	
	1.849–30.800	4	5.818 893	−2.084 606	−0.807 0467	0.261 1724	
S	0.0394–0.1740	4	4.862 989	1.407 408	4.580 753	2.192 223	
	0.1742–0.1932	3	−3.287 565[2]	−9.238 824[2]	−6.359 921[2]		
	0.1996–2.4772	4	5.105 629	−2.701 358	−0.226 9985	0.697 1264	
	2.479–31.970	4	5.957 286	−1.944 676	−0.839 6531	0.246 5677	
Ar	0.0504–0.1085	4	0.103 8804	0.158 0149	0.180 2517	6.544 871	
	0.1085–0.2497	3	3.766 572	−3.816 305	−1.506 623		
	0.2653–3.2002	4	5.325 805	−2.604 915	−0.411 9090	0.483 1536	
	3.2033–32.200	3	6.236 304	−2.427 534	−0.185 3513		

Note

[a] Numbers in square brackets denote powers of 10.

See Sec. 5.4 for a semiempirical equation for hydrogen and hydrogenic ion cross sections.

5.6   ABSORPTION OF MATERIAL OF STELLAR INTERIORS

The opacity of stellar interiors is usually expressed as the Rosseland mean of the mass absorption coefficient $\bar{\kappa}$. Tabulations are available [15, 16] for a wide range of compositions expressed by X, Y, and Z. Tables 5.15–5.17 give $\log_{10} \bar{\kappa}$ in $cm^2 \, g^{-1}$ as a function of $\log_{10} \rho$ where density ρ is in $g \, cm^{-3}$ and temperature T in units of 10^{-6} K. These tables are based on interpolations of data in [15].

Table 5.15. *Hydrogen Rosseland mean opacity* $\log_{10} \bar{\kappa}$ [1].

T^a	$\log \rho = -10.0$	-9.0	-8.0	-7.0	-6.0	-5.0	-4.0	-3.0	-2.0	-1.0	0.0
0.006	-1.67	-1.41	-1.00	-0.61	-0.13						
0.007	-0.66	-0.49	-0.19	0.16	0.57						
0.008	0.18	0.33	0.54	0.82	1.15						
0.009	0.71	1.01	1.20	1.40	1.65						
0.010	0.85	1.42	1.71	1.89	2.09	2.36					
0.011	0.76	1.56	2.05	2.27	2.46	2.68					
0.012	0.60	1.50	2.21	2.57	2.77	2.98					
0.014	0.31	1.19	2.16	2.86	3.22	3.47					
0.016	0.12	0.89	1.91	2.86	3.48	3.82					
0.018	-0.01	0.68	1.68	2.73	3.57	4.07					
0.020	-0.08	0.52	1.48	2.56	3.55	4.23					
0.025	-0.15	0.30	1.17	2.25	3.37	4.35	4.96				
0.030	-0.17	0.21	1.00	2.06	3.19	4.29	5.10				
0.035	-0.21	0.15	0.89	1.92	3.04	4.15	5.07				
0.040	-0.27	0.07	0.78	1.78	2.88	3.96	4.90				
0.045	-0.32	-0.02	0.65	1.62	2.67	3.72					
0.050	-0.35	-0.11	0.50	1.43	2.44	3.46					
0.055	-0.37	-0.18	0.35	1.21	2.20	3.20					
0.060	-0.38	-0.24	0.21	1.01	1.97	2.96					
0.070	-0.40	-0.31	0.00	0.66	1.56	2.53					
0.080	-0.40	-0.35	-0.14	0.39	1.22	2.17					
0.090	-0.40	-0.37	-0.22	0.19	0.94	1.86					
0.100	-0.40	-0.38	-0.28	0.05	0.71	1.60	2.56				
0.120		-0.39	-0.34	-0.13	0.37	1.17	2.11				
0.150		-0.40	-0.37	-0.26	0.07	0.72	1.60				
0.200		-0.40	-0.39	-0.35	-0.17	0.26	1.01				
0.250			-0.40	-0.38	-0.28	0.02	0.61	1.45			
0.300			-0.40	-0.39	-0.33	-0.13	0.34	1.10			
0.400			-0.40	-0.40	-0.37	-0.27	0.03	0.62			
0.500				-0.40	-0.39	-0.33	-0.14	0.32	1.07		
0.600				-0.40	-0.39	-0.36	-0.22	0.13	0.79		
0.800				-0.40	-0.40	-0.39	-0.32	-0.10	0.40		
1.000				-0.40	-0.40	-0.39	-0.35	-0.21	0.17	0.85	
1.200					-0.40	-0.40	-0.37	-0.27	0.02	0.60	
1.500					-0.40	-0.40	-0.39	-0.33	-0.13	0.34	1.06
2.000					-0.40	-0.40	-0.40	-0.37	-0.25	0.07	0.68
2.500						-0.40	-0.40	-0.38	-0.31	-0.08	0.41
3.000						-0.40	-0.40	-0.39	-0.34	-0.17	0.23
4.000	-0.40	-0.41	-0.40	-0.38	-0.28	0.00	0.60				
5.000		-0.41	-0.40	-0.39	-0.33	-0.14	0.34				
6.000		-0.41	-0.41	-0.40	-0.36	-0.22	0.16				
8.000		-0.41	-0.41	-0.41	-0.38	-0.31	-0.05	0.49			
10.00		-0.41	-0.41	-0.41	-0.40	-0.35	-0.17	0.25			
15.00			-0.42	-0.42	-0.41	-0.39	-0.31	-0.07	0.43		
20.00			-0.42	-0.42	-0.42	-0.41	-0.37	-0.22	0.14		
30.00				-0.43	-0.43	-0.43	-0.41	-0.35	-0.16		
40.00				-0.44	-0.44	-0.44	-0.43	-0.40	-0.29	-0.01	
60.00					-0.46	-0.46	-0.46	-0.44	-0.41	-0.29	
80.00					-0.47	-0.47	-0.47	-0.47	-0.45	-0.42	-0.17
100.0					-0.49	-0.49	-0.49	-0.49	-0.48	-0.48	-0.35

Note

^{a}Units of 10^6 K.

Reference

1. Iglesias, C.A., Rogers, F.J., & Wilson, B.G. 1992, *ApJ*, **397**, 717

Table 5.16. *Helium Rosseland mean opacity* $\log_{10} \bar{\kappa}$ [1].

T^a	$\log \rho = -10.0$	-9.0	-8.0	-7.0	-6.0	-5.0	-4.0	-3.0	-2.0	-1.0	0.0
0.006	-5.79	-5.81	-5.71								
0.007	-4.95	-5.14	-5.13								
0.008	-4.13	-4.43	-4.52								
0.009	-3.36	-3.73	-3.94								
0.010	-2.71	-2.99	-3.24	-3.15							
0.011	-2.04	-2.31	-2.43	-2.41							
0.012	-1.49	-1.67	-1.76	-1.73							
0.014	-0.82	-0.72	-0.68	-0.65							
0.016	-0.74	-0.33	-0.01	0.13							
0.018	-0.83	-0.37	0.24	0.62							
0.020	-0.88	-0.51	0.19	0.86							
0.025	-0.93	-0.72	-0.17	0.72	1.49						
0.030	-0.73	-0.64	-0.27	0.46	1.36						
0.035	-0.53	-0.28	0.02	0.52	1.24						
0.040	-0.60	-0.20	0.39	0.85	1.34						
0.045	-0.64	-0.34	0.43	1.17	1.63						
0.050	-0.66	-0.45	0.17	1.19	1.90	2.39					
0.055	-0.68	-0.52	-0.03	0.95	1.98	2.57					
0.060	-0.67	-0.57	-0.16	0.69	1.83	2.69					
0.070	-0.67	-0.61	-0.33	0.35	1.40	2.57					
0.080	-0.67	-0.62	-0.42	0.15	1.11	2.26					
0.090	-0.67	-0.62	-0.46	0.02	0.93	2.03					
0.100	-0.68	-0.62	-0.47	-0.06	0.80	1.89	3.04				
0.120		-0.65	-0.49	-0.15	0.63	1.68	2.83				
0.150		-0.68	-0.56	-0.23	0.47	1.48	2.58				
0.200		-0.69	-0.66	-0.45	0.12	1.03	2.04				
0.250			-0.69	-0.59	-0.21	0.52	1.45	2.41			
0.300			-0.69	-0.65	-0.42	0.13	0.97	1.91			
0.400			-0.70	-0.69	-0.60	-0.30	0.32	1.18			
0.500				-0.69	-0.66	-0.49	-0.06	0.68	1.58		
0.600				-0.70	-0.68	-0.58	-0.28	0.32	1.17		
0.800				-0.70	-0.69	-0.66	-0.50	-0.10	0.60		
1.000				-0.70	-0.70	-0.68	-0.60	-0.33	0.22	1.02	
1.200					-0.70	-0.69	-0.64	-0.46	-0.03	0.68	
1.500					-0.70	-0.70	-0.67	-0.57	-0.27	0.32	
2.000					-0.70	-0.70	-0.69	-0.64	-0.47	-0.05	0.65
2.500						-0.70	-0.70	-0.67	-0.56	-0.26	0.34
3.000						-0.70	-0.70	-0.69	-0.62	-0.40	0.09
4.000	-0.70	-0.70	-0.70	-0.66	-0.53	-0.19	0.43				
5.000		-0.70	-0.70	-0.68	-0.60	-0.36	0.15				
6.000		-0.70	-0.70	-0.69	-0.64	-0.46	-0.05				
8.000		-0.71	-0.71	-0.70	-0.68	-0.57	-0.29	0.29			
10.00		-0.71	-0.71	-0.71	-0.69	-0.63	-0.43	0.04			
15.00			-0.71	-0.71	-0.71	-0.69	-0.59	-0.31	0.24		
20.00			-0.72	-0.72	-0.72	-0.71	-0.66	-0.48	-0.07		
30.00				-0.73	-0.73	-0.72	-0.71	-0.63	-0.39		
40.00				-0.74	-0.74	-0.74	-0.73	-0.69	-0.54	-0.22	
60.00					-0.76	-0.75	-0.75	-0.74	-0.68	-0.51	
80.00					-0.77	-0.77	-0.77	-0.77	-0.74	-0.64	-0.40
100.0					-0.79	-0.79	-0.79	-0.79	-0.77	-0.72	-0.57

Note

^{a}Units of 10^6 K.

Reference

1. Iglesias, C.A., Rogers, F.J., & Wilson, B.G. 1992, *ApJ*, **397**, 717

Table 5.17. *Solar composition* ($X = 0.73$, $Z = 0.018$) *Rosseland mean opacity* $\log_{10} \bar{\kappa}$ [1].

T^a	$\log \rho = -10.0$	-9.0	-8.0	-7.0	-6.0	-5.0	-4.0	-3.0	-2.0	-1.0	0.0
0.006	-1.77	-1.53	-1.10	-0.62	0.00						
0.007	-0.78	-0.62	-0.32	0.07	0.52						
0.008	0.05	0.20	0.42	0.72	1.08						
0.009	0.55	0.88	1.08	1.30	1.59						
0.010	0.67	1.27	1.60	1.80	2.03	2.33					
0.011	0.61	1.41	1.94	2.20	2.41	2.65					
0.012	0.47	1.35	2.08	2.48	2.71	2.95					
0.014	0.24	1.06	2.00	2.74	3.14	3.40					
0.016	0.13	0.82	1.76	2.71	3.36	3.71					
0.018	0.04	0.66	1.56	2.56	3.42	3.94					
0.020	-0.04	0.54	1.39	2.40	3.37	4.08					
0.025	-0.13	0.33	1.14	2.15	3.20	4.16	4.80				
0.030	-0.12	0.25	0.97	1.97	3.06	4.11	4.93				
0.035	-0.14	0.25	0.92	1.85	2.94	4.03	4.95				
0.040	-0.21	0.20	0.90	1.80	2.83	3.92	4.88				
0.045	-0.26	0.09	0.81	1.74	2.73	3.78					
0.050	-0.31	0.00	0.66	1.62	2.62	3.63					
0.055	-0.34	-0.09	0.50	1.43	2.47	3.47					
0.060	-0.35	-0.15	0.38	1.25	2.29	3.32					
0.070	-0.36	-0.22	0.18	0.95	1.96	3.02	3.93				
0.080	-0.37	-0.25	0.06	0.74	1.70	2.76	3.75				
0.090	-0.36	-0.27	0.00	0.60	1.51	2.56	3.58				
0.100	-0.33	-0.26	-0.04	0.50	1.36	2.39	3.44				
0.120		-0.19	-0.05	0.38	1.18	2.15	3.17				
0.150		-0.04	0.09	0.38	0.99	1.90	2.85	3.66			
0.200		-0.20	-0.01	0.33	0.84	1.56	2.42	3.12			
0.250			-0.24	0.02	0.53	1.25	2.05	2.74			
0.300			-0.38	-0.19	0.24	0.97	1.77	2.45			
0.400			-0.46	-0.40	-0.18	0.48	1.34	2.03	2.52		
0.500				-0.44	-0.37	0.05	0.94	1.76	2.26		
0.600				-0.45	-0.40	-0.20	0.53	1.49	2.06		
0.800				-0.44	-0.44	-0.34	0.05	0.93	1.74	2.23	
1.000				-0.45	-0.44	-0.38	-0.15	0.52	1.43	2.03	
1.200					-0.44	-0.39	-0.25	0.27	1.15	1.87	
1.500					-0.44	-0.40	-0.29	0.05	0.83	1.65	2.10
2.000					-0.45	-0.45	-0.35	-0.11	0.48	1.29	1.78
2.500						-0.46	-0.42	-0.25	0.26	0.98	1.50
3.000						-0.47	-0.44	-0.36	0.05	0.73	1.24
4.000	-0.46	-0.46	-0.43	-0.26	0.29	0.82	1.19				
5.000		-0.47	-0.45	-0.38	-0.03	0.50	0.88				
6.000		-0.47	-0.46	-0.43	-0.22	0.23	0.64				
8.000		-0.47	-0.47	-0.46	-0.38	-0.11	0.29	0.72			
10.00		-0.47	-0.47	-0.46	-0.43	-0.27	0.06	0.48			
15.00			-0.48	-0.47	-0.46	-0.39	-0.21	0.11	0.53		
20.00			-0.48	-0.48	-0.48	-0.44	-0.32	-0.09	0.22		
30.00				-0.49	-0.49	-0.49	-0.44	-0.32	-0.11		
40.00				-0.50	-0.50	-0.50	-0.49	-0.42	-0.28		
60.00					-0.52	-0.52	-0.52	-0.50	-0.44	-0.30	
80.00					-0.54	-0.54	-0.54	-0.53	-0.50	-0.44	-0.18
100.0					-0.56	-0.56	-0.56	-0.55	-0.54	-0.51	-0.37

Note

^{a}Units of 10^6 K.

Reference

1. Iglesias, C.A., Rogers, F.J., & Wilson, B.G. 1992, *ApJ*, **397**, 717

The opacity due to electron scattering alone for a completely ionized plasma, with hydrogen mass fraction X, is given by $\bar{\kappa}_e = 0.200(1 + X)$ [17].

5.7   ABSORPTION OF MATERIAL OF THE SOLAR PHOTOSPHERE

The Rosseland mean opacity for the solar photosphere including diatomic species is given by Table 5.18, as $\log_{10} \bar{\kappa}$ with $\bar{\kappa}$ in $\mathrm{cm^2\,g^{-1}}$. The assumed microturbulent velocity is 2 km/s.

Table 5.18. *Solar photospheric Rosseland mean opacity* $\log_{10} \bar{\kappa}$ [1].

$T\,(10^{-6}\,\mathrm{K})$	$\log \rho = -10.0$	-9.0	-8.0	-7.0	-6.0	-5.0	-4.0
0.0021	−5.31	−5.27	−5.16	−4.95	−4.61	−4.03	−3.12
0.0030	−4.12	−3.66	−3.04	−2.49	−2.14	−1.94	−1.72
0.0040	−3.42	−2.72	−2.00	−1.38	−0.84	−0.38	−0.11
0.0050	−2.83	−2.42	−1.74	−0.92	−0.14	0.47	0.86
0.0062	−1.60	−1.34	−0.92	−0.43	0.17	0.85	1.43
0.0071	−0.70	−0.53	−0.21	0.19	0.65	1.18	1.72
0.0081	0.15	0.33	0.54	0.83	1.20	1.63	2.10
0.0093	0.61	1.05	1.29	1.48	1.76	2.11	
0.0100	0.66	1.27	1.61	1.79	2.03	2.33	
0.0126	0.35	1.24	2.05	2.56	2.81	3.04	
0.0158	0.07	0.78	1.73	2.64	3.28	3.63	
0.0200	−0.08	0.49	1.34	2.35	3.29	4.01	
0.0316	−0.17	0.20	0.87	1.83	2.92	3.98	
0.0398	−0.23	0.16	0.83	1.71	2.71	3.80	
0.0501	−0.32	−0.03	0.62	1.54	2.50		
0.0631	−0.36	−0.21	0.26	1.11	2.09		
0.0708	−0.35	−0.26	0.11	0.87	1.84		
0.0794	−0.36	−0.28	0.01	0.66	1.61		
0.0891	−0.37	−0.30	−0.06	0.50	1.39		
0.1000	−0.37	−0.31	−0.11	0.38	1.21		

Reference
1. Kurucz, R.L. 1992, *Rev. Mexicana Astron. Af.*, **23**, 181

5.8   SOLAR PHOTOIONIZATION RATES

The solar photoionization rates are calculated as [18]

$$R = \int_{\nu_0}^{\infty} \sigma_\nu(\nu) F_\nu \, d\nu,$$

where $\sigma_\nu(\nu)$ is the photoionization cross section having threshold ν_0 and F_ν is the solar flux.

The unattenuated photoionization rate for the quiet and active Sun for some monatomic species, at heliocentric distance 1 AU, is shown in Table 5.19.

Table 5.19. *Solar photoionization rate R in* $\mathrm{s^{-1}}$ [1].[a]

Species	Quiet Sun	Active Sun	Species	Quiet Sun	Active Sun
H$^-$	14.	14.	O (^1S)	2.0[−7]	7.5[−7]
H	7.3[−8]	1.9[−7]	F	2.1[−7]	9.5[−7]
He	5.2[−8]	2.2[−7]	Na (expt.)	1.6[−5]	1.7[−5]

Table 5.19. *(Continued.)*

Species	Quiet Sun	Active Sun	Species	Quiet Sun	Active Sun
C (^3P)	4.1[−7]	1.0[−6]	Na (theor.)	5.9[−6]	6.6[−6]
C (^1D)	3.6[−6]	1.0[−5]	S(^3P)	1.1[−6]	2.6[−6]
C (^1S)	4.3[−6]	1.2[−5]	S(^1D)	1.1[−6]	2.6[−6]
N	1.9[−7]	6.3[−7]	S(^1S)	1.0[−6]	2.5[−6]
O (^3P)	2.1[−7]	8.5[−7]	Cl	5.7[−7]	1.5[−6]
O (^1D)	1.8[−7]	7.4[−7]	K	2.2[−5]	2.3[−5]

Note
[a]Numbers in square brackets denote powers of 10.

Reference
1. Huebner, W.F., Keady, J.J., & Lyon, S.P. 1992, *AP&SS*, **195**, 1

5.9 FREE–FREE ABSORPTION AND EMISSION

The free–free linear absorption coefficient [19, 20] is

$$\kappa_s = \frac{4\pi}{3\sqrt{3}} \frac{Z^2 e^6}{hcm^2 v} \cdot \frac{g}{v^3} N_e N_i \quad (\kappa \text{ in } \exp \text{cm}^{-1})$$

$$= 1.802 \times 10^{14} (Z^2 g / v^3 \upsilon) N_e N_i \quad (\upsilon \text{ in cm/s})$$

$$= 6.686 \times 10^{-18} Z^2 g \lambda^3 N_e N_i / \upsilon \quad (\lambda \text{ in cm}),$$

where υ is the electron velocity, g is the Gaunt factor representing the departure from Kramers's theory, Z is the ionic charge, and N_e and N_i are the electronic and ionic densities in cm^{-3}. The mean $1/\upsilon$ is $(2m/\pi kT)^{1/2}$, whence

$$\kappa_s = 3.692 \times 10^8 Z^2 g T^{-1/2} v^{-3} N_e N_i$$

$$= 1.370 \times 10^{-23} Z^2 \lambda^3 g N_e N_i / T^{1/2} \quad (\lambda \text{ in cm}).$$

The effective linear absorption coefficient κ' after allowance for stimulated emission is

$$\kappa' = 3.692 \times 10^8 [1 - \exp(-h\nu/kT)] Z^2 g T^{-1/2} v^{-3} N_e N_i.$$

For small $h\nu/kT$ ($= 1.438/\lambda T$), e.g., for radio waves,

$$\kappa' = \frac{8}{3} \left(\frac{\pi}{6}\right)^{1/2} \frac{e^6}{c(mkT)^{3/2}} \frac{Z^2 g}{v^2} N_e N_i \quad (\kappa' \text{ in } \exp \text{cm}^{-1})$$

$$= 0.017\,8 Z^2 g v^{-2} T^{-3/2} N_e N_i$$

$$= 1.98 \times 10^{-23} Z^2 g \lambda^2 N_e N_i T^{-3/2} \quad (\lambda \text{ in cm}).$$

The Gaunt factor for radio waves is [19, 21]

$$g = 10.6 + 1.90 \log_{10} T - 1.26 \log_{10} Z\nu.$$

Gaunt-factor calculations incorporating relativistic effects and electron degeneracy can be found in [22, 23]. The temperature parameter γ^2 is defined by

$$\gamma^2 = Z_j^2 \, \text{Ry}/kT, \qquad Z_j = 1 \quad \text{(for H)}, \qquad Z_j = 2 \quad \text{(for He)},$$

$$= Z_j^2 \frac{1.579 \times 10^5 \, \text{K}}{T} \quad (T \text{ in K}).$$

For $-3 \leq \log_{10} \gamma^2 \leq 2.0$ we list in Table 5.20 approximate Gaunt factors for both hydrogen and helium, where η (defined as the chemical potential divided by kT) is the degeneracy parameter and $u \equiv h\nu/kT$.

Table 5.20. *Relativistic thermally averaged free–free Gaunt factors.*

$\log u$	$\eta = -6.0$	-2.0	0.0	1.0	2.0	3.0
-4.0	5.5	5.13	3.77	2.77	1.94	1.36
-3.5	4.89	4.55	3.35	2.48	1.74	1.22
-3.0	4.26	3.97	2.95	2.18	1.54	1.09
-2.0	3.03	2.84	2.14	1.61	1.15	0.82
-1.4	2.37	2.22	1.69	1.28	0.93	0.67
-1.0	1.95	1.85	1.42	1.10	0.80	0.58
-0.7	1.69	1.59	1.25	0.96	0.72	0.52
-0.2	1.30	1.25	1.03	0.83	0.63	0.47
0.0	1.17	1.13	0.96	0.80	0.62	0.46

These smoothed numbers are least accurate (~ 10–20%) for large u and small $\log_{10} \gamma^2$ (≈ -3), improving to a few percent for smaller u and larger $\log_{10} \gamma^2$. For larger u, Gaunt factors for hydrogen and helium can differ significantly [22, 23].

When degeneracy is not important ($\eta \leq -4$, say),

$$g \cong \begin{cases} \dfrac{\sqrt{3}}{\pi} \ln\left(\dfrac{4}{\Gamma u}\right), & \Gamma = 1.781 \quad \text{for } u \ll 1, \\ u^{-0.4} & \text{for } u \approx 1. \end{cases}$$

The free–free emission (bremsstrahlung) per unit solid angle, volume, time, and frequency range is

$$j_\nu = \kappa'_\nu B_\nu \quad \text{(black body)}$$

$$= \frac{16}{3}\left(\frac{\pi}{6}\right)^{1/2} \frac{e^6 Z^2}{c^3 m^2}\left(\frac{m}{kT}\right)^{1/2} g \exp\left(-\frac{h\nu}{kT}\right) N_e N_i$$

$$= 5.444 \times 10^{-39} Z^2 g \exp\left(-\frac{c_2}{\lambda T}\right) T^{-1/2} N_e N_i \ \text{erg cm}^{-3}\,\text{s}^{-1}\,\text{sr}^{-1}\,\text{Hz}^{-1} \quad (T \text{ in K}, \ N \text{ in cm}^{-3}).$$

The free–free emission from a cosmic plasma is

$$j_\nu = 6.2 \times 10^{-39} \, g \exp\left(-\frac{c_2}{\lambda T}\right) T^{-1/2} \int N_e^2 \, dV \ \text{erg s}^{-1}\,\text{sr}^{-1}\,\text{Hz}^{-1} \quad (T \text{ in K}, \ N_e \text{ in cm}^{-3}),$$

where $\int N_e^2 \, dV$ (integrated over volume) is called the emission measure.

The integrated free–free emission is

$$4\pi \int j_\nu \, d\nu = \frac{64\pi}{3}\left(\frac{\pi}{6}\right)^{1/2} \frac{e^6 Z^2}{hc^3 m}\left(\frac{kT}{m}\right)^{1/2} g N_e N_i$$

$$= 1.426 \times 10^{-27} Z^2 T^{1/2} g N_e N_i \ \text{erg cm}^{-3}\,\text{s}^{-1}.$$

For a completely ionized plasma with solar abundance [24], $\sum_i N_i N_e Z_i^2 \cong 1.4 N_e^2$, and thus

$$4\pi \int j_\nu \, d\nu = 2.0 \times 10^{-27} g T^{1/2} \int N_e^2 \, dV \ \text{erg s}^{-1}.$$

Free–free absorption from neutral atoms: For highly polarizable target atoms, an approximation for the thermally averaged free–free absorption coefficient valid for long wavelength and low temperature is [25]

$$k = 1.62 \times 10^{-19} N_e N_a \alpha^{1/2} \lambda^3 T (1 - e^{-hc/\lambda kT}) \text{ cm}^{-1},$$

where $N_e (N_a)$ are the number of free electrons (neutral atoms) per cm^3, α is the polarizability (cm^3), λ is the wavelength (cm), and T is the temperature (K).

The validity criterion is $0.633/\alpha < \varepsilon < 2/\alpha^{1/3}$, where the free-electron kinetic energy ε is measured in Rydbergs and α is in units of a_0^3.

5.10 REFLECTION FROM METALLIC MIRRORS

In Table 5.21, no attempt has been made to differentiate between different methods of deposition [26].

Table 5.21. *Reflection from metallic mirrors.*

λ (μm)	Ag (%)	Al (%)	Speculum (%)	Hg (%)	Ni (%)	Cu (%)	Au (%)	Si (%)	Pt (%)	Steel (%)	W (%)
0.20	20	72			35	34	18	68	20	24	15
0.22	25	78			40	34	27	68	29	27	16
0.24	27	81	26		42	31	32	68	35	30	18
0.26	27	82	33	58	40	29	34	68	37	33	20
0.28	23	82	38	61	39	28	34	67	38	36	21
0.30	12	82	44	64	39	29	35	65	39	39	23
0.32	7	82	48	67	41	30	33	61	40	41	25
0.34	63	83	51	69	43	32	33	56	42	44	27
0.36	77	83	54	71	45	34	33	50	43	46	30
0.38	82	84	56	73	47	36	34	41	45	49	34
0.40	85	85	58	74	50	38	34	35	48	51	38
0.45	90	86	61	74	57	42	37	30	56	55	45
0.50	91	87	63	73	61	47	51	30	59	57	49
0.55	92	88	65	73	63	60	77	30	60	57	52
0.60	93	89	66	74	65	74	84	30	61	56	51
0.65	94	88	67	74	67	82	89	30	63	55	52
0.70	95	87	68	75	69	85	93	30	66	56	53
0.80	97	85	70	70	70	89	95	29	70	59	56
1.00	98	93	72	73	73	92	97	28	74	63	60
2.0	98	96	82	82	84	96	98	28	81	77	87
5.0	99	97	89	89	92	98	99	28	91	90	95
10.0	99	98	92	92	96	99	99	28	95	93	98

Reflections in the EUV [27, 28] are strongly dependent on the details of deposition, the age of the surface, and the reflection angle. No summary can be made.

5.11 VISUAL PHOTOMETRY

Units of visual photometry are given in Chapters 2 and 15. For values of the relative visibility factor K_λ for normal brightness (about 5×10^{-4} stilb or greater), the photopic curve (international) (cone vision at fovea) is given in Table 5.22 [29]. This and the following table are actually one-dimensional tables so that the last column and first row entry of the first table applies to a wavelength of 3 900 Å.

Table 5.22. *Relative visibility factors.*

λ (Å)	0	100	200	300	400	500	600	700	800	900
3000									0.000 04	0.000 12
4000	0.000 4	0.001 2	0.004 0	0.011 6	0.023	0.038	0.060	0.091	0.139	0.208
5000	0.323	0.503	0.710	0.862	0.954	0.995	0.995	0.952	0.870	0.757
6000	0.631	0.503	0.381	0.265	0.175	0.107	0.061	0.032	0.017	0.008 2
7000	0.004 1	0.002 1	0.001 05	0.000 52	0.000 25	0.000 12	0.000 06	0.000 03		

The equivalent width of the K_λ curve is $\int K_\lambda \, d\lambda = 1\,068$ Å.

The mechanical equivalent of light (experimental) [29] is

$$K_\lambda \text{ lumens} \equiv 0.001\,47 \text{ W}.$$

The luminous energy (in lumergs) is $680 \int K_\lambda e_\lambda \, d\lambda$, where $e_\lambda \, d\lambda$ is the element of energy in joules.

$$1 \text{ lumen } (5\,550 \text{ Å radiation}) = 4.11 \times 10^{15} \text{ photons s}^{-1},$$
$$1 \text{ nanolambert } (5\,550 \text{ Å radiation}) = 1.31 \times 10^6 \text{ photons s}^{-1} \text{ cm}^{-2} \text{ sr}^{-1}.$$

The number of lumens L entering a telescope of diameter D in cm for a star of visual apparent magnitude V near the zenith (clear conditions) is

$$\log_{10} L = 2 \log_{10} D - 0.4V - 9.86.$$

For relative visibility for dark-adapted eyes (about 10^{-7} stilb or less), the scotopic curve (rod vision) is given in Table 5.23.

Table 5.23. *Relative visibility factors.*

λ (Å)	0	100	200	300	400	500	600	700	800	900
4000	0.018 5	0.040	0.076	0.132	0.212	0.302	0.406	0.520	0.650	0.770
5000	0.900	0.985	0.960	0.840	0.680	0.500	0.350	0.228	0.140	0.083
6000	0.049 0	0.030 0	0.017 5	0.010 0	0.005 8	0.003 2	0.001 7	0.000 87	0.000 44	0.000 21
7000	0.000 11									

For dark-adapted eyes,

$$1 \text{ lumen at } 5\,100 \text{ Å (scotopic)} \equiv 0.000\,58 \text{ W}.$$

The quantum thresholds for a single scintillation with most favorable conditions for human eye are 4 quanta in 0.15 s (absorbed) and 60 quanta in 0.15 s (incident).

The threshold intensity for a large steady source [30] is 1.4×10^{-10} stilb.

The size of the retinal image for $1'$ arc is 4.9 μm.

The eye resolving power $\simeq 1' \simeq 5$ μm at fovea.

Density of rods and cones in the retina [29]:

$$\text{Rods} \quad 30 \times 10^6 \text{ rods/sr} = 2.7 \text{ rods/(minutes of arc)}^2,$$
$$\text{Cones} \quad 1.2 \times 10^6 \text{ cones/sr} = 0.1 \text{ cones/(minutes of arc)}^2.$$

The density of cones in the fovea $\simeq 50 \times 10^6$ cones/sr.
The equivalent diameter of the fovea region containing no rods [31] is $1° \, 40'$.
The diameters of individual cones are 2 μm $\equiv 25''$ (variable).
The diameters of individual rods are 1 μm $\equiv 12''$.
The approximate brightness of common objects is given in Table 5.24 [32].

Table 5.24. *Object and brightness (stilb).*

Candle	0.6
Acetylene (Kodak burner)	10.8
Welsbach (high-pressure) mantle	25
Tungsten lamp filament	800
Sodium vapor lamp	70
Mercury vapor lamp (high pressure)	150
Arc crater (plain carbon)	16 000
Clear blue sky	0.2–0.6
Overcast sky	0.3–0.7
Zenith Sun	165 000

Table 5.25 gives approximate albedos for common objects.

Table 5.25. *Approximate albedos* [1, 2].

White cartridge paper	0.80
Magnesium oxide (or carbonate)	0.98
Black cloth	0.012
Black velvet	0.004

References
1. Walsh, J.W.T. *Photometry*, 3rd ed. (Dover, New York), p. 529
2. Houston, R.A. 1924, *Treatise on Light* (Longmans, London)

REFERENCES

1. Allen, C.W. 1973, *Astrophysical Quantities*, 3rd ed., Sec. 35 (Athlone Press, London)
2. Allen, C.W. 1973, *Astrophysical Quantities*, 3rd ed., Sec. 37 (Athlone Press, London)
3. Sampson, D.H. 1959, *ApJ*, **129**, 734
4. Boercker, D.B. 1987, *ApJ*, **316**, L95
5. van de Hulst, H.C. 1957, *Light Scattering by Small Particles* (Wiley, Chapman and Hall, New York)
6. Stergis, C.G. 1966, *J. Atm. Terr. Phys.*, **28**, 273
7. Penndorf, R. 1957, *J. Opt. Soc. Am.*, **47**, 176
8. Irvine, W.R. 1965, *J. Opt. Soc. Am.*, **55**, 16
9. Osterbrock, D.E. 1974, *Astrophysics of Gaseous Nebulae* (Freeman, San Francisco)
10. Huebner, W.F. 1986, *Physics of the Sun*, Vol. I, edited by P. Sturrock (Reidel, Dordrecht)
11. Gaunt, J.A. 1930, *Philos. Trans.*, **229**, 163
12. Burgess, A. 1965, *ApJ*, **141**, 1588
13. Henke, B.L. et al. 1982, *At. Data Nucl. Data Tables*, **27**, 1
14. Saloman, E.B., & Hubbell, J.H. 1986, *X-Ray Attenuation Coefficients (Total Cross Sections): Comparison of the Experimental Data Base with the Recommended*

Values of Henke and the Theoretical Values of Scofield for Energies between 0.1–100 keV, U.S. Dept. of Commerce Report No. NBSIR 86-3431

15. Iglesias, C.A., Rogers, F.J., & Wilson, B.G. 1992, *ApJ*, **397**, 717

16. Rogers, F.J., & Iglesias, C.A. 1992, *ApJ*, **7**, 507

17. Cox, A.N. 1965, in *Stellar Structure*, 3rd ed., edited by L.H. Aller & D.B. McLaughlin (University of Chicago Press, Chicago), 195

18. Huebner, W.F., Keady, J.J., & Lyon, S.P. 1992, *AP&SS*, **195**, 1

19. Allen, C.W. 1973, *Astrophysical Quantities*, 3rd ed., Sec. 43 (Athlone Press, London)

20. Spitzer, L. 1962, *Physics of Fully Ionized Gases*, 2nd ed. (Interscience, New York), p. 148

21. Chambe, G., & Lantos, P. 1971, *Sol. Phys.*, **17**, 97

22. Itoh, N., Nakagawa, M., & Kohyama, Y. 1985, *ApJ*, **294**, 17

23. Nakagawa, M., Kohyama, Y., & Itoh, N. 1987, *ApJ Supp.*, **63**, 661

24. Grevese, N., & Anders. E. 1991, in *Solar Interior and Atmosphere*, edited by A.N. Cox, W.C. Livingston, M.S. Matthews (University of Arizona Press, Tucson), p. 1227

25. Hyman, H.A., Kivel, B., & Bethe, H.A. 1973, *Inverse Neutral Bremsstrahlung for Highly Polarizable Atoms*, AVCO–Everett Research Laboratory Report No. SAMSO-TR-73-98

26. Allen, C.W. 1973, *Astrophysical Quantities*, 3rd ed., Sec. 45 (Athlone Press, London)

27. Hass, G., & Tousey, R., 1958, *J. Opt. Soc. Am.*, **49**, 593

28. Garton, W.R.S. 1966, *Adv. Atom. Mol. Phys.*, **2**, 93

29. Allen, C.W. 1973, *Astrophysical Quantities*, 3rd ed., Sec. 46 (Athlone Press, London)

30. Pirenne, M.H. 1961, *Endeavour*, **20**, 197

31. Martin, L.C. 1948, *Technical Optics* (Pitman, London), pp. 1, 144

32. Walsh, J.W.T. *Photometry*, 3rd ed. (Dover, New York), p. 529

Chapter 6

Radio Astronomy

Robert M. Hjellming

6.1 INTRODUCTION

Radio astronomy is defined by three things. First is the range of frequencies that constitute the radio windows of the Earth's atmosphere, ranging roughly from 20 MHz to 1000 GHz. Second are the astronomical objects that emit radio waves with sufficient strength to be detectable at the Earth; some emit by thermal processes, but most are seen because of the emission from relativistic electrons (cosmic rays) interacting with local magnetic fields. Third, radio radiation behaves more like waves than particles (photons), allowing the measurement of both amplitude and phase of the radiation field. The capability to measure phase gives radio interferometry the capability to do the highest resolution imaging of astronomical objects currently possible in astronomy.

The variety and number of radio sources is so large that in Table 6.1 we merely summarize a number of source catalogs devoted to these topics.

Table 6.1. *List of radio source catalogs.*

Type of Source	Contents
4C Radio Survey [1, 2]	4844 56 cm sources > 2 Jy, $-7° \leq \delta \leq 80°$
Bologna B2 Sky Survey [3–5]	408 MHz, $21.7° \leq \delta \leq 40°$
Bologna B3 Sky Survey [6, 7]	408 MHz, $37° \leq \delta \leq 47°$
20 cm N. Sky Catalog [8]	365, 1400, 4850 MHz, $0 \leq \delta \leq 75°$
6 cm Gal. Plane Survey [9]	Parkes 64 m, $\|b\| \leq 2°$, $l = 190° - 360° - 40°$
6 cm South & Tropical Surveys [10, 11]	Parkes 64 m, $-78° \leq \delta \leq -10°$
11 cm All Sky Catalog [12]	Bright sources
11 cm N. Sky Catalog [13]	6483 sources, $\|b\| \leq 5°$, $240° \leq l \leq 357°$
Bright Galaxies [14]	All radio data pre-1975, normal galaxies
20 cm Gal. Plane Survey [15]	VLA B Conf. Survey
21 cm Gal. Plane Survey [16]	Eff. 100m, $\|b\| \leq 4°$, $95° \leq l \leq 357°$
Galaxy HI [17, 18]	Galaxies, $v_{radial} \leq 3000$ km/s
Atlas of Galactic HI [19]	21 cm H emission line profiles
Molecular lines [20, 21]	Frequencies, other molecular data
Pulsars [22]	Catalog of known pulsars
Opt. Pos. Radio Stars [23]	221 radio stars
20 cm Radio Sources $\delta > -5°$ [24, 25]	1400 MHz sky atlas

References

1. Pilkington, J.D.H., & Scott, P.F. 1965, *Mem. R.A.S.*, **69**, 183
2. Gower, J.F.R., Scott, P.F., & Willis D. 1967, *Mem. R.A.S.*, **71**, 144
3. Colla, G. et al. 1970, *A&AS*, **1**, 281
4. Colla, G. et al. 1972, *A&AS*, **7**, 1
5. Colla, G. et al. 1973, *A&AS*, **11**, 291
6. Fanti, C. et al. 1974, *A&AS*, **18**, 147
7. Ficcara, A., Grueff, G., & Tomesetti, G. 1985, *A&AS*, **59**, 255
8. White, R.L., & Becker, R.H. 1992, *ApJS*, **79**, 331
9. Haynes, R.F., Caswell, J.L., & Simons, L.W.J. 1978, *Aust. J. Phys. Suppl.*, **48**, 1
10. Griffith, M.R. et al. 1994, *ApJS*, **90**, 179
11. Wright, A.E. et al. 1994, *ApJS*, **91**, 111
12. Wall, J.V., & Peacock, J.A. 1985, *MNRAS*, **216**, 173
13. Fürst, E. et al. 1990, *A&AS*, **85**, 805
14. Haynes, R.F., Caswell, J.L., & Simons, L.W.J. 1978, *Aust. J. Phys. Suppl.*, **45**, 1
15. Zoonematkermani, S. et al. 1990, *ApJS*, **74**, 181
16. Reich, W., Reich, P., & Fürst, E. 1990, *A&AS*, **83**, 539
17. Tully, R.B. 1988, *Nearby Galaxies Catalog* (Cambridge University Press, Cambridge)
18. Huchtmeier, W.K., & Richter, O.-G. 1989, *A General Catalog of* HI *Observations of Galaxies: The Reference Catalog* (Springer-Verlag, New York)
19. Hartman, D., & Burton, W.D. 1995, *Atlas of Galactic* HI (Cambridge University Press, Cambridge)
20. Lovas, F.J. 1992, *J. Phys. Chem. Ref. Data*, **21**, 181
21. Lovas, F.J., Snyder, L.E., & Johnson, D.R. 1979, *ApJS*, **41**, 451
22. Taylor, J.H., Manchester, R.N., & Lyne, A.G. 1993, *ApJS*, **66**, 529
23. Réquième, Y. & Mazurier, J.M. 1991, *A&AS*, **89**, 311
24. Condon, J.J., Broderick, J.J., & Seielstad, G.A. 1989, *AJ*, **97**, 1064
25. Condon, J.J., Broderick, J.J., & Seielstad, G.A. 1991, *AJ*, **102**, 2041

Other information and images from surveys can be obtained from Internet pages maintained by the National Center for Super Computing Applications (NCSA) Astronomy Digital Image Library (ADIL, http://imagelib.ncsa.uiuc.edu/imagelib.html) and the National Radio Astronomy Observatory (NRAO, http://info.aoc.nrao.edu). More extensive radio (and other) catalogs can be found from Internet pages for the NASA Astrophysics Data System (http://adswww.harvard.edu/index.html).

In Figure 6.1 schematic radio spectra representative of a variety of continuum radio sources are plotted in the form of flux density (in units of Jy = Jansky = 10^{-23} ergs cm^{-2} s^{-1} Hz^{-1} = 10^{-26} W m^{-2} Hz^{-1}) as a function of frequency. The sources and spectra in Figure 6.1 are schematic

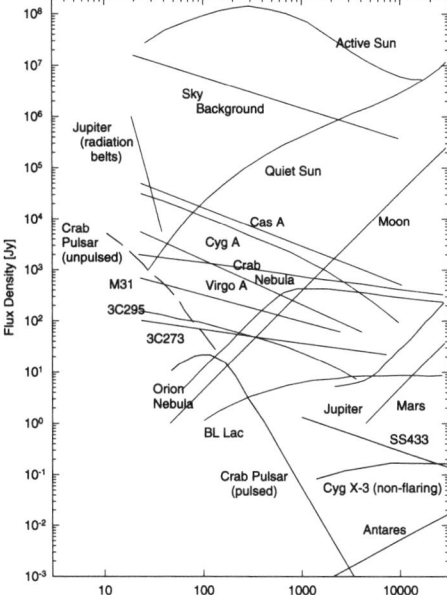

Figure 6.1. Radio spectra of various types of sources in the form of flux density as a function of frequency v.

indicators of typical behavior for a wide variety of objects. Major solar-system radio sources are the active Sun, the quiet Sun, the Earth's Moon, Mars, the surface of Jupiter, and Jupiter's radiation belts. Stellar system sources are: pulsars, with the pulsed and unpulsed (dashed line) emission from the Crab pulsar; the partially ionized coronal emission of the red supergiant Antares; and the X-ray binaries SS433 and Cyg X-3 (quiescent). The optically thick and thin portions of the radio spectrum of the ionized gas from an HII region is indicated for the Orion Nebula. Cas A is a remnant of a supernova explosion in our Galaxy; and the Crab nebula spectrum is representative of the relatively rare plerion, i.e., nebulosities energized by pulsars. The radio spectrum for M31, the Andromeda nebula, is representative of spiral galaxy behavior. Strong radio galaxies are represented by Cyg A and Virgo A; BL Lac is a blazar, while 3C273 and 3C295 are strong quasars.

Defining the convention that the spectral index α is the power law index for the flux density, $S_v \propto v^{\alpha}$, then positive spectral indices indicate optically thick emission while negative spectral indices indicate optically thin emission. Spectral indices near zero indicate optically thin emission for thermal sources, while similarly flat spectra for synchrotron radiation sources indicate optically thick emission from a wide range of optical depths.

6.2 ATMOSPHERIC WINDOW AND SKY BRIGHTNESS

6.2.1 Atmospheric Window

The low-frequency end of the radio window is set by ionospheric absorption. Since the ionosphere has diurnal variations, solar cycle variations, variations depending upon the effect of particle storms impacting the atmosphere, and variations with Earth latitude and longitude, the lowest frequency where the atmosphere is transparent varies considerably over the range 10 to 30 MHz. From that low-frequency cut-off, up to about 22 GHz, the Earth's atmosphere does not absorb radio waves, although variation in the index of refraction affects the phase of incoming radio waves at the higher frequencies.

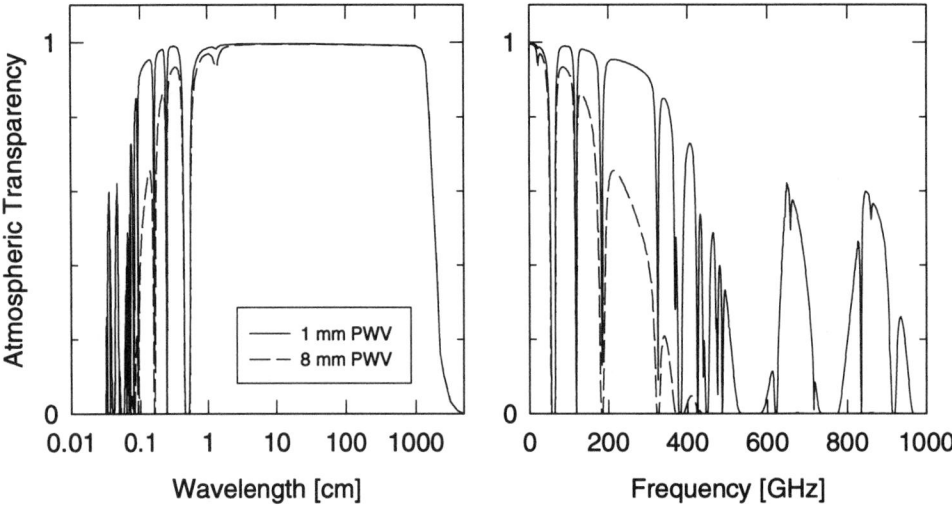

Figure 6.2. The radio window for the Earth's atmosphere shown in terms of plots of atmospheric transparency as a function of both wavelength and frequency. The solid line is for excellent conditions with 1 mm PWV and the dashed line is for normal conditions with 8 mm PWV.

At the lowest frequencies plasma effects in the ionosphere induce variable Faraday rotation in incoming radio waves.

In Figure 6.2 we plot models for the transparency of the Earth's atmosphere for two values of precipitable water vapor (PWV), 1 and 8 mm, representing extremely low and normal levels, respectively.

6.2.2 Surface Brightness and Brightness Temperature

All astronomy is based upon measurements of surface brightness on the celestial sphere: $I_\nu(\alpha, \delta, t)$, where t is time, (α, δ) is position specified by right ascension and declination, ν is frequency, and I is the specific intensity which can be any of the four Stokes parameters. The surface brightness of a black body of temperature T is $B_\nu = (2h\nu^3/c^2)(1/[e^{h\nu/kT} - 1])$, where h and k are the Planck and Boltzmann constants. For small values of $h\nu/kT$, the Rayleigh–Jeans approximation is valid, and $B_\nu(T) \cong 2kT/\lambda^2$; this is valid at longer radio wavelengths. For reasons related to the antenna measurement equation, radio astronomy often uses brightness temperature rather than surface brightness to describe measurements, where the brightness temperature is

$$T_b \equiv \frac{\lambda^2}{2k} B_\nu. \tag{6.1}$$

Equation (6.1) is often used even when black body or long wavelength approximations are not applicable. Because of this, the concept of radiation temperature ($T_r \equiv [h\nu/k]/[e^{h\nu/kT} - 1]$) is commonly used; where T is a "true" physical temperature.

6.2.3 Sources of Background Radiation

At each wavelength there are sources of background radiation detectable by radio telescopes, principally the Earth's atmosphere, the cosmic radiation background, and diffuse emission from our

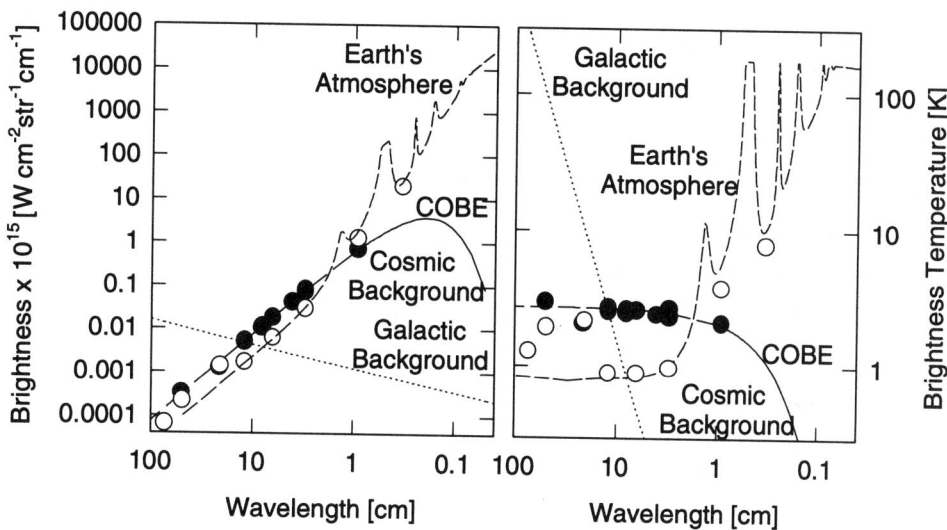

Figure 6.3. Apparent brightness (left) and brightness temperature (right) plotted as a function of wavelength for: the Earth's atmosphere (dashed line, model; open circle, measurements); cosmic background radiation from COBE (solid line) and other instruments (filled circles); and galactic background radiation (dotted line).

Galaxy. In Figure 6.3 we plot data and models for these sources of radiation. At wavelengths shorter than ~ 1 cm the Earth's atmosphere dominates the background, while above 13 cm the galactic background dominates. Between 1 and ~ 13 cm the cosmic background dominates. However, in many cases man-made interference, or solar radio emission, can be more important "background emission," particularly at wavelengths longer than 10 cm.

6.3 RADIO WAVE PROPAGATION

6.3.1 Radiators-Absorbers, Fields, and Coupling Equations

Radiation measurement at radio wavelengths allows direct measurement of the amplitude and phase of electromagnetic fields; this is the reason radiation field analysis in radio astronomy is usually done in terms of fields.

If we identify a distant point in a radio source with the vector $\mathbf{R}$, and the location of the oscillating current element with the vector, the fields between the two points are proportional to the propagator [1]

$$P_\nu(\mathbf{r}) = \frac{e^{2\pi i \nu |\mathbf{R}-\mathbf{r}|/c}}{|\mathbf{R} - \mathbf{r}|}, \tag{6.2}$$

so by Huygens' Principle, and the fact that all radiation appears as if it originates from the celestial sphere, the field on the celestial sphere at $\mathbf{r}$ is the superposition

$$E_\nu(\mathbf{r}) = \int E(\mathbf{R}) \frac{e^{2\pi i \nu |\mathbf{R}-\mathbf{r}|/c}}{|\mathbf{R} - \mathbf{r}|} dS, \tag{6.3}$$

where dS is a surface element on the celestial sphere.

The radiation flux for propagating fields is determined by the real part of the Poynting flux averaged over time, so $\langle S_r \rangle = \langle \mathbf{E} \times \mathbf{H} \rangle = \mathrm{Re}(E_\theta H\phi - E_\phi H_\theta)/2 = E_\theta H_\phi$. The power dP at distance r

passing through solid angle $d\Omega = dA/r^2$ is the Poynting flux through dA at distance r. From the field distribution over the surface of an antenna we compute $dP/d\Omega$ which allows us to calculate the important properties of antennas. The normalized antenna pattern is defined by

$$P_{n}(\theta, \phi) = \frac{dP/d\Omega}{(dP/d\Omega)_{max}} \tag{6.4}$$

so an infinitesimal current element has $P_{n}(\theta, \phi) = \sin^2 \theta$. The *beam solid angle* of an antenna is $\Omega_a = \int_{4\pi} P_{n}(\theta, \phi) \, d\Omega$, and the *directivity* is $D = 4\pi / \Omega_a$ [2].

For real antennas we integrate over the collecting surface and from that compute $dP/d\Omega$ from which antenna parameters may be computed.

6.3.2   Radio Noise and Detection Limits

The measured quantities for radio telescopes are total power or correlated total power at some point in the signal path of the electronics. This is one of the sources of the predominance of using temperature as a measurement variable, because of Nyquist's law which relates the total power P_a to an antenna temperature T_a by

$$P_a = 4kT_a \, \Delta\nu, \tag{6.5}$$

where $k \, \Delta\nu$ is the frequency range (or bandwidth) over which the total power measurement is made. P_a is the measure of the sum of the power due to radiation sources external to the antenna and the power generated in the antenna and its electronics system. In modern systems the latter is dominated by the receiver temperature, but is generally called system temperature (T_{sys}) which constitutes an irreducible minimum level of noise in the measured signal. The power being measured is that of a fluctuating voltage, originally induced by the external electric fields in the form of oscillating currents in the antenna's collecting surface, which radiate with a focus on either a feed at the prime focus of the antenna or on one or more subreflectors directed at a feed. The feed absorbs radiation and produces an output signal which is transferred to an electronic receiver followed (usually) by a complicated series of amplifiers, frequency converters, and other electronics elements. Then the fluctuating total power is then measured over a specific frequency range with a bandwidth ($\Delta\nu$), and over a specific time interval Δt. For a system temperature T_{sys} the noise component of the measured signal is

$$\Delta T_{noise} = \frac{T_{sys}}{\sqrt{\Delta\nu \, \Delta t}}, \tag{6.6}$$

which indicates how noise changes with bandwidth and integration time. The same principle applies in the case of interferometry where signals are either added or (almost all the time) correlated (or multiplied), so it is the correlated power from two antennas that is being measured.

The noise component of the system temperature is usually the limiting factor in the measurements being made. The fundamental limit to electronic noise temperatures is the quantum limit: $h\nu/k = 0.048\nu_{GHz}$ K, where ν_{GHz} is the frequency in units of GHz. Real system temperatures must be above this limit. In the 1990s the state of the art of modern electronics is roughly capable of producing noise temperatures of $4h\nu/k$. It is expected that at the beginning of the twenty-first century noise temperatures will be $\sim 2h\nu/k$, but it is unlikely that they will ever get very close to the quantum limit.

6.3.3   Effects of Ionosphere, ISM, and Source Environment

Because radio signals are best dealt with by an analysis based upon electric fields, one should think of radio sources as three-dimensional regions radiating electric fields. Once radiation that will eventually reach a radio telescope on, or near, the Earth is produced, one then must think of propagation phenomena which affect these fields. Absorption and re-radiation by free electrons, ions, and atomic or molecular species can change the original radiation field into one modified by many effects. One can decompose this process into propagation effects: inside radio sources; in the intergalactic and interstellar medium (IGM and ISM); in the interplanetary medium of the solar system; and in the Earth's atmosphere and ionosphere.

The ionized component of the intervening medium between a radio source and an observing telescope affects propagation by introducing time delays and Faraday rotation. The time of arrival of a pulse of radiation is $t_{\text{pulse}} = \int_0^L ds/v_{\text{group}}$ where L is the propagation path length, ds is a segment of the line of sight, and v_{group} is the group velocity. The delay per unit frequency introduced by the electron concentration N_e is approximated by

$$\frac{dt_{\text{pulse}}}{d\nu} \cong \frac{e^2}{4\pi\epsilon_0 mc\nu^3} \int_0^L N_e \, ds. \tag{6.7}$$

The pulsed emission of radio pulsars allows the measurement of the changing time of arrival of a pulse with frequency, so the *dispersion measure*, $D_m = \int_0^L N_e \, ds$, is routinely measured for the lines of sight to pulsars in our galaxy. If coexistent with magnetic fields, the same electrons that cause propagation delays induce Faraday rotation of the field vectors by an angle $\phi = \lambda^2 R_m$ where λ is the wavelength. R_m is the rotation measure, which depends on the product of the local electron density, N_e, and the component of the magnetic field parallel to the line of sight, $B_{||}$, and is given by

$$R_m = 8.1 \times 10^5 \int N_e B_{||} \, ds \tag{6.8}$$

in units of radians/m^2, where $B_{||}$ is in Gauss, N_e is in cm^{-3}, ds is in parsecs, and the integral is over the path length along the line of sight. An estimate for the magnitude of these effects in can be made using mean values for the Galactic ISM such as $\langle N_e \rangle \cong 0.003$ cm^{-3}, $B_{||} \cong 2$ μG, and typically $R_m \sim -18|\cot b|\cos(l - 94°)$ radians/m^2, where l and b are the galactic longitude and latitude.

The scattering of radiation from electrons in the ISM, which produces interstellar scintillation, also has the effect of increasing the apparent size of point sources of radio emission. This effect produces an angular size due to scintillation which is roughly $7.5\lambda^{11/5}$ milliarcsec for $|b| \leq 0.°6$, $0.5(|\sin b|)^{-3/5}\lambda^{11/5}$ milliarcsec for $0.°6 < |b| < 4°$, $13(|\sin b|)^{-3/5}\lambda^{11/5}$ milliarcsec for $15° > |b| > 4°$, and $15\lambda^2/\sqrt{|\sin b|}$ milliarcsec for $|b| > 15°$, where b is the galactic latitude and λ is the wavelength in meters.

At long wavelengths, and for source line of sights close to the Sun, the solar corona and solar wind contribute very strongly to propagation delay, Faraday rotation, and scintillation. This can be estimated from the above formulas by $N_e \cong (1.55R^{-6} + 2.99R^{-16}) \times 10^{14}$ m^{-3} for $R < 4$, and $N_e \cong 5 \times 10^{11} R^{-2}$ m^{-3} for $4 < R < 20$, where R is the radial distance from the center of the Sun. The scattering size for an unresolved radio source due to the interplanetary medium is approximately $50(\lambda/R)^2$ arcmin where λ is in meters.

The ionosphere of the Earth is a major, and highly variable, contributor of electrons along the line of sight that affects the propagation of radio waves at long wavelengths. Figure 6.4 shows typical, but idealized, distributions of the ionospheric electron density for night and day, during sunspot maximum, and at temperate latitudes.

The troposphere of the Earth's atmosphere also has the effect of absorbing the radiation from distant sources. If $I_{0,\nu}$ is the surface brightness of a source seen from outside the Earth's atmosphere, and $\tau_{0,\nu}$ is the optical depth at the zenith, then at a zenith angle (z)

$$I(z, \nu) = I_{0,\nu} e^{-\tau_{0,\nu} X(z)}, \tag{6.9}$$

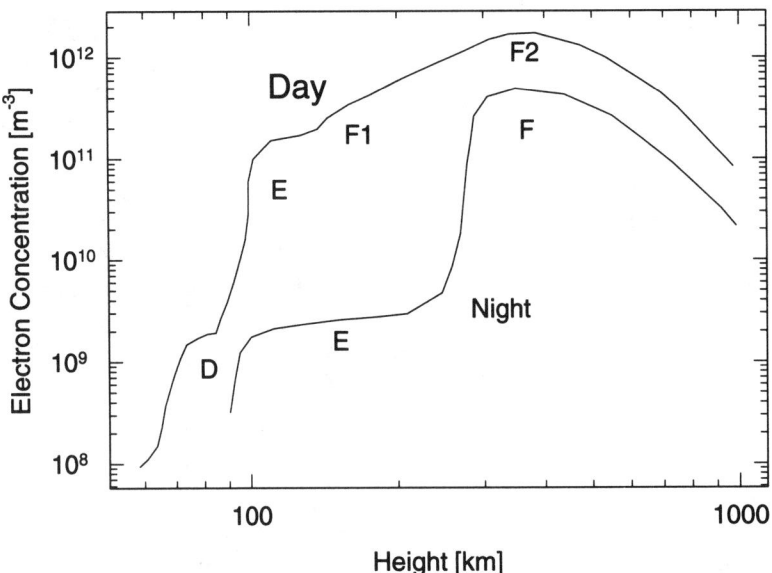

Figure 6.4. The electron concentration profile of the Earth's ionosphere for day and night plotted against height above the Earth's surface, at solar maximum, and at mid-latitudes.

where $X(z)$ is the relative air mass in units of the air mass at the zenith. To first order, $X(z) \cong \sec(z) = 1/\cos z$. For $X < 5$ the formula $X(z) = -0.0045 + 1.00672 \sec z - 0.002234 \sec^2 z - 0.000\,624\,7 \sec^3 z$ has an error less than 6×10^{-4}. Also, $\tau_{0,\nu} \cong 0.12$, 0.05, and 0.04 at 20, 6, and 2 cm, but can be very large and variable at mm wavelengths.

Even more important than air mass in affecting radio observations are the delay and scattering effects in the Earth's troposphere that produce the radio equivalent of seeing disks. However, since the 1970s radio astronomers have devised algorithms that allow so-called self-calibration of phase variations produced by the troposphere for radio interferometry measurements with four or more antennas in an array. Phase self-calibration by interferometric arrays are methods by which one can, for any observed field with a strong source, solve for the differences in atmospheric phase variations over each antenna, and then remove these phase variations from interferometric data.

6.4 RADIO TELESCOPES AND ARRAYS

6.4.1 Properties of Antennas

The normalized antenna pattern $P_n(\theta, \phi)$ (Equation (6.4)) is used to define many of the important antenna properties. Although real antenna patterns cannot be exactly described by a mathematical function, many are close to that for a uniformly illuminated circular aperture of diameter D for which

$$P_n(\theta) = \left\{ \frac{2J_1\left(\dfrac{\pi D}{\lambda} \sin \theta\right)}{\dfrac{\pi D}{\lambda} \sin \theta} \right\}^2, \tag{6.10}$$

where J_1 is a first-order Bessel function. In Table 6.2 we list various definitions of antenna and source properties that depend upon the antenna pattern, and give the approximate values for uniform circular apertures. Since some of the directly measured quantities for radio sources are dependent upon the antenna pattern, some of these are also listed in Table 6.2 [2].

Table 6.2. *Antenna properties.*

Quantity	Definition	Uniform Circular Ap.
Source solid angle	$\Omega_s = \int_{\text{source}} d\Omega$	
Effective source solid angle	$\Omega_s = \int_{\text{source}} P_n(\theta, \phi)\, d\Omega$	
Surface brightness	$\langle B_\nu \rangle = \dfrac{\int_{\text{main lobe}} B_\nu(\theta, \phi) P_n\, d\Omega}{\int_{\text{main lobe}} P_n\, d\Omega}$	
Half power beam width	$P_n(\Theta_{\text{HPBW}}) = 1/2$	$\Theta_{\text{HPBW}} = 1.02(\lambda/D) = 58°(\lambda/D)$
Beam width at first nulls	$P_n(\Theta_{\text{BWFN}}) = 0$	$\Theta_{\text{BWFN}} = 2.44(\lambda/D) = 140°(\lambda/D)$
Beam solid angle	$\Omega_A = \int_{4\pi} P_n(\theta, \phi)\, d\Omega$	
Main beam solid angle	$\Omega_M = \int_{\text{main lobe}} P_n(\theta, \phi)\, d\Omega$	
Directivity	$4\pi/\Omega_A$	
Effective area	$A_e = \lambda^2/\Omega_A$	
Aperture efficiency	$\epsilon_A = A_e/(\pi D^2/4)$	
Beam efficiency	$\epsilon_M = \Omega_M/\Omega_A$	

6.4.2 Major Radio Telescopes and Arrays

There are many tens of radio telescope systems in the world that are, have been, or are about to be, important in radio astronomy. Many are listed in Table 6.3, where the size of the telescope or array, its main operational wavelengths, and its type or specialized role may be listed in abbreviated form. For a complete list of most of these telescopes and arrays, see [3].

Table 6.3. *Radio observatories.*

Name of Observatory/Inst.	Location	Description
Australia Tel. Nat. Facility	Culgoora, Australia	6 km EW Arr. 1 + (6) 22 m; 0.3–24 cm
	Parkes, Australia	64 m; 0.7–75 cm
Basovizza-Solar Radio Stn.	Trieste, Italy	10 m; 38–130 cm solar
Bleien Radio Ast. Obs.	Zurich, Switzerland	5 m; 30–300 cm solar
		7 m; 10–300 cm solar
Caltech Submillimeter Obs.	Mauna Kea, Hawaii	10.4 m; 1.3 mm–350 μm
Crawford Hill Obs.	Holmdel, New Jersey	7 m Horn Tunable; 21 cm
		7 m Tunable; 1.3–3 mm
Decameter Wave Radio Obs.	Gauribidanur, India	1.5 km T Arr.; 200–900 cm
Deep Space Network Sta.	Goldstone, California	34 m; 3.6–13 cm
		70 m; 3.6–13 cm
Deep Space Network Sta.	Robledo, Spain	34 m; 3.6–13 cm
		70 m; 3.6–13 cm
Deep Space Network Sta.	Tidbinbilla, Australia	34 m; 3.6–13 cm
		70 m; 3.6–13 cm

Table 6.3. *(Continued.)*

Name of Observatory/Inst.	Location	Description
Dominion Radio Astro. Obs.	Penticton, BC, Canada	600 m EW Arr. 4 9 m; 21,90 cm
		26 m; 18–21 cm
Dwingeloo Radio Obs.	Dwingeloo, Netherlands	25 m; 6–29 cm
Eur. Incoh. Scatt. Fac.	Kiruna, Sweden	32 m; 32 cm
		4 40 × 120 m; 130 cm
Fleurs Radio Tel.	Fleurs, NSW, Aus.	EW Arr. 32 5.8 m + 6 14 m; 21 cm
Five College Radio Ast. Obs.	Quabbin Res., Mass.	13.7 m; 1–7 mm
Giant Metrewave Radio Tel.	Pune District, India	25 km Irr. Arr. 36 45 m; 21–790 cm
Hat Creek Radio Ast. Obs.	Cassel, California	3 + (3) T Arr. 6 m; 1–7 mm
Haystack Obs.	Westford, Massachusetts	36 m; 0.7–18 cm
Hiraiso Solar-Terr. Res. Ctr.	Nakaminato, Ibaraki, Japan	2 10 m + 6 m + 1 m; 300–950 cm
Humain Radio Ast. Stn.	Humain, Belgium	7.5 m; 50 cm solar
		T Arr. 48 4 m; 74 cm solar
High Time Resolving Tel.	Nanjing, China	2 m; 3.2 cm solar
Interplanetary Scint. Obs.	Toyakawa, Japan	100 × 20 m; 92 cm
Instituto Argentino de Rad.	Parque Pereya Iraola, Arg.	2 30 m; 17–21 cm
Inst. de Radio Astron. Mill.	Plateau de Bure, France	0.4 km T Arr. (3) 15 m; 1.3–4 mm
	Pico Veleta, Spain	30 m; 0.8–4 mm
IPS Fleurs Solar Obs.	Fleurs, NSW, Aus.	2.5 m + 12-Yagi; 6,11,20, 120 cm
Itapetinga Radio Obs.	Atibaia, Brazil	13.7 m; 0.3–28 cm
		1.5 m; 4.3 cm solar
James Clerk Maxwell Tel.	Mauna Kea, Hawaii	15 m; 1.3 mm-300 μm
Kashima Space Res. Ctr.	Kashima, Japan	26 m; 3–150 cm
Kisaruzu College Obs.	Kisaruza chiba, Japan	1.5 m; CO 2.6 mm Line
Maryland Point Obs.	Riverside, Maryland	26 m; 0.7–90 cm
Staz. Rad. di Medicina	Medicina, Italy	32 m; 1.3–20 cm VLBI
		T Parab. Cyl.; 74 cm
Metsahovi Obs. Radio Sta.	Metsähovi, Finland	14 m; 0.3–1.3 cm
Molonglo Obs. Synth. Tel.	Hoskinstown, NSW, Aus.	1570 × 12 m EW Cyl. Par.; 36 cm
Mount Pleasant Radio Obs.	Cambridge, Tasmania, Aus.	14 m; 30–50 cm pulars
		26 m; 2.5–50 cm
Mullard Radio Ast. Obs.	Cambridge, England	5 km EW Arr. (4) 13 m; cm λ
National Astro. and Ion. Obs.	Arecibo, Puerto Rico	305 m; λ > 6 cm
National Obs. Ast. Center	Yebes, Spain	14 m; mm λ
National Radio Ast. Obs.	Green Bank, West Va.	43 m; 1.3–600 cm
		100 m; 0.7–90 cm
NRAO mm Telescope	Kitt Peak, Arizona	12 m; 0.9–7 mm
Very Large Array	Socorro, New Mexico	1–35 km Y Arr. (27) 25 m; 0.7–90 cm
Very Long Baseline Array	Socorro, New Mexico	5000 km Arr. 10 25 m; 0.7–90 cm VLBI
Netherlands Found. Res. Ast.	Dwingeloo, Netherlands	22 m Ant.
Nobeyama Solar Radio Obs.	Nobeyama, Japan	0.5 km T Arr. 16 1.2 m, 1.8 cm solar
		T Arr 17 6 m, 190 cm
Nobeyama Radio Obs.	Nobeyama, Japan	45 m, 0.26–3 cm
	Nobeyama, Japan	0.7 km T Arr. (5) 10 m, 0.25–1.3 cm
Noto Radio Ast. Sta.	Noto, Sicily	32 m; 0.3–1.3 cm VLBI
Nuffield Radio Ast. Labs.	England	134 km Irr. MERLIN Arr. 7 Ant.:
	Jodrell Bank, Cheshire	76 m; 5–200 cm
	Jodrell Bank, Cheshire	38 × 26 m Par.; 1.3–200 cm
	Wardle, Cheshire	38 × 26 m; 17–370 cm
	Defford, Worchestershire	25 m; 6–200 cm
	Darnhall, Cheshire	25 m; 1.3–200 cm
	Knockin, Shropshire	25 m; 1.3–200 cm
	Pickmere, Cheshire	25 m; 1.3–200 cm
	Jodrell Bank, Cheshire	13 m; 20–50 cm pulsars
Obs. de Bordeaux	Floirac, France	2.5 m; 2.5–4 mm
Obs. de Grenoble	Plateau de Bure, France	2.5 m; 1.3 mm

Table 6.3. *(Continued.)*

Name of Observatory/Inst.	Location	Description
Obs. für Solare Astr.	Tremsdorf, Germany	1.5 + 4 + 10.5 m + Yagi; 3.2–750 cm
Obs. Radioastron. de Maipú	Maipú, Chile	160 × 73 m; 670 cm
Ohio State Radio Obs.	Delaware, Ohio	100 × 30 m; 1.9 cm
Ondrejov Ast. Obs.	Ondrejov, Czech.	3 m; 1.5–3 cm solar
		7.5 m; 37–120 cm solar
		7.5 m; 25–300 cm solar
Onsala Space Obs.	Onsala, Sweden	20 m; 3.7–270 cm
		25 m; 30–260 cm
Ooty Radiotelescope	Ootacamund, India	530 × 30 m Par. Cyl.; 90 cm
Ooty Synthesis Radiotelescope		(ORT) + 8 23 × 9 m; 90 cm
Owens Valley Radio Obs.	Big Pine, California	0.4 km T Arr. (4) 10 m; 1.3–3 mm
		40 m; 0.7–90 cm
		Arr. of 2 27 m; 4–30 cm
Purple Mountain Obs.	Delingha, China	13.7 m; 0.3,1.3 cm
Purple Mt. Obs. Solar Fac.	Nanjing, China	1.5 m; 3.2–11 cm solar
		2 m; 6 cm solar
Puschino Radio Ast. Sta.	Puschino, Russia	22 m Tel.
		Cross-type 1 km decimetric array
		18 acre decimetric phased array
Radioobs. Effelsberg	Effelsberg, Germany	100 m; 0.6–49 cm
RATAN-600	Zelenchukskaya, SU	0.6 cm circle of 895 elem.; 0.8–30 cm
Solar Radiospec. Obs.	Ravensburg, Germany	7 m + 8 dipoles; 30–1000 cm
Sta. de Rad. de Nancay	Nancay, France	16 1 m Parab.; 3.2 cm
		24 Parab.; 67–200 cm solar
		16 + 2 Parab.; 67–200 cm solar
		299 × 40 m; 9–21 cm
		144 Con. Log. Per.; 380–2000 cm
Swedish-ESO Submm. Tel.	La Silla, Chile	15 m; 0.8–3 mm
Tonantzintla Solar Radio Int.	Puebla, Mexico	2 1.1 m Parab.; 4 cm solar
Toyokaya Obs.	Toyokawa, Japan	0.85 m; 3.2 cm solar
		1.5 m; 8 cm solar
		2 m; 14 cm solar
		3 m; 2.8 cm solar
		3 arrays; 3.2,7.8 cm solar
U. of Mich. Radio Obs.	Dexter, Michigan	25 m; cm λ
URAN-1 Interferometer	Kharkov, SU	Linear Arr, Xed dipoles; 1200–3000 cm
UTR-2 Array	Kharkov, SU	T Arr 1800 × 54 m + 900 × 54 m; 1200–3000 cm
Westerbork Synth. Rad. Tel.	Westerbork, Netherlands	4 km EW Arr. 10 + (4) 25 m; 6–90 cm
Yunnan Obs.	Kunming, China	2.5, 3, 3.2, & 10 m; 8.1–130 cm

6.5 RADIO EMISSION AND ABSORPTION PROCESSES

6.5.1 Source Models and Prediction of Observables

The relationship between the emission and absorption coefficients, which are used to describe the microphysics of radiation processes, and the theoretical quantities corresponding to the direct observables, is important in discussing the principle physical processes in radio astronomy. There are three principal observables: surface brightness B_ν (and the related brightness temperature $T_{b,\nu}$), the integrated flux density S_ν for a source with a closed boundary of solid angle Ω_{source}, and V_ν the coherence (or visibility) function measured by radio interferometers. All are computed as integrals over the true sky brightness, or specific intensity, I_ν, and are given by

$$B_\nu = \frac{\int_{\text{source}} I_\nu \, d\Omega_{\text{source}}}{\int_{\text{source}} d\Omega_{\text{source}}}, \tag{6.11}$$

$$S_\nu = \int_{\text{source}} I_\nu \, d\Omega_{\text{source}} \cong \frac{2k}{\lambda^2} \int_{\text{source}} T_b \, d\Omega_{\text{source}}, \tag{6.12}$$

(where the approximation is valid for the Rayleigh–Jeans limit and most radio wavelengths) and

$$V_\nu(u, v) = \iint I_\nu(\alpha, \delta) P_n(\alpha - \alpha_0, \delta - \delta_0) e^{-2\pi i/\lambda[\mathbf{L}_j - \mathbf{L}_k]\cdot[\mathbf{s}(\alpha,\delta)-\mathbf{s}_0(\alpha_0,\delta_0)]} \, d\Omega, \tag{6.13}$$

where $\mathbf{L}_j$ and $\mathbf{L}_k$ are vector locations of antennas j and k, and $\mathbf{s}$ is a unit vector pointing to locations on the celestial sphere such as (α, δ) or the reference position (α_0, δ_0). For a spherically symmetric brightness distribution the latter equation simplifies to a Hankel transform:

$$V_\nu(u, v) = \int I_\nu(\theta) P_n(\theta) J_0(2\pi q\theta) 2\pi \theta \, d\theta, \tag{6.14}$$

where $(\mathbf{L}_j - \mathbf{L}_k) \cdot (\mathbf{s} - \mathbf{s}_0)/\lambda = ux + vy + wz$, $q = (u^2 + v^2 + w^2)^{1/2} \cong (u^2 + v^2)^{1/2}$, and $\theta = (x^2 + y^2 + z^2)^{1/2} \cong (x^2 + y^2)^{1/2}$.

In Table 6.4 a number of models and the associated observables are listed.

Table 6.4. *Observables for simple source models.*

Model	$T_b(\theta)/T_{b,\text{max}}$	$S_\nu(Jy)$	$V_\nu(q)/S_\nu$
Gaussian	$e^{-4\ln 2(\theta/\theta_H)^2}$	$\dfrac{T_{b,\text{max}}(K)\theta_H^2 \text{ (arcsec)}}{1360\lambda^2(\text{cm})}$	$e^{-(\pi^2 4\ln 2)(q\theta_H)^2}$
Uniform disk	$1, \theta \le \theta_H/2$ $0, \theta > \theta_H/2$	$\dfrac{T_{b,\text{max}}(K)\theta_H^2 \text{ (arcsec)}}{1961\lambda^2(\text{cm})}$	$2J_1(\pi q\theta_H)/(\pi q\theta_H)^a$
Limb-brightened (shot glass)	$2\theta_H/(\theta_H^2 - \theta^2),$ $\theta \le \theta_H/2$ $0, \theta > \theta_H/2$	$\dfrac{T_{b,\text{max}}(K)\theta_H^2 \text{ (arcsec)}}{981\lambda^2(\text{cm})}$	$\sin(\pi q\theta_H)/(\pi q\theta_H)$
"Thin" ring	$\delta(\theta - \theta_H/2)$	$\dfrac{420 T_{b,\text{max}}(K)\theta_H^2 \text{ (arcsec)}}{\lambda^2(\text{cm})}$	$J_0(\pi q\theta_H)$

Note

^{a} J_n is a Bessel function of order n, and assuming that source is at the center of the field, then $\theta = r/d$ (d is the distance to the object) and θ_H is the half-intensity point in these spherically symmetric models.

The resolution of an instrument of size D is set by diffraction theory to be λ/D, and for radio astronomy the best resolution of an antenna is determined by its diameter (D_m in units of meters) and the shortest wavelength it can observe ($\sim$ [surface rms]/16). For arrays the size is the maximum separation of antennas (D_{km} in units of km). Thus

$$\Theta_{\text{resolution}} = \frac{\lambda}{D} = 34' \frac{\lambda_{\text{cm}}}{D_m} = 2'' \frac{\lambda_{\text{cm}}}{D_{km}}. \tag{6.15}$$

The resolution of the instrument and the characteristics of the radio source, which we will discuss in terms of the models in Table 6.4, together with the other parameters determining the sensitivity of the antenna or array, determine the minimum surface brightness that can be detected. For modern paraboloid-shape antennas the 5σ detection level for flux density during a time interval Δt_{sec} is

$$\sigma_{\text{detection}} = \frac{20 T_{\text{sys}}}{F_{\text{K/Jy}} D_{\text{m}}^2} \left(\frac{1}{\sqrt{\Delta \nu_{\text{MHz}} \, \Delta t_{\text{sec}}}} \right) \text{Jy}, \tag{6.16}$$

where $F_{\text{K/Jy}}$ is a fixed, empirically determined constant for each antenna–receiver combination, while for an array of N antennas of this size,

$$\sigma_{\text{detection}} = \frac{2.5 T_{\text{sys}}}{\eta_c \epsilon_a D_{\text{m}}^2} \left(\frac{1}{\sqrt{\Delta \nu_{\text{MHz}} \, \Delta t_{\text{sec}} N(N - 1)/2}} \right) \text{Jy}, \tag{6.17}$$

where η_c is the correlator efficiency (0.82 for 3-level correlations), ϵ_a is the aperture efficiency (~ 0.5 for simple paraboloids, but 0.6–0.7 for specially designed antennas), $\Delta \nu_{\text{MHz}}$ is the bandwidth in MHz, and Δt_{sec} is the integration time in seconds.

6.5.2 Thermal Free–Free and Free–Bound Transitions

For thermal bremsstrahlung radiation, the emission and absorption coefficients $j_\nu \rho$ and $\kappa_\nu \rho$ are proportional to ρ^2, where ρ is the mass density. The source function j_ν / κ_ν, at radio wavelengths, is the Planck function for an electron temperature T_e. Thermal bremsstrahlung or free–free radiation processes are caused by interactions between free electrons and positive ions in a partially or fully ionized plasma. The emission coefficient for free–free emission is given by

$$j_{\text{ff},\nu} \rho = 5.4 \times 10^{-39} N_e^2 T_e^{1/2} g_{\text{ff}} e^{h\nu/kT_e} \cong 7.45 \times 10^{-39} N_e^2 T_e^{0.34} \nu^{-0.11}, \tag{6.18}$$

where we have used

$$g_{\text{ff}}(\nu, T_e) = \frac{3^{1/2}}{\pi} \left[17.7 + \ln \left(\frac{T_e^{3/2}}{\nu} \right) \right] \cong 1.38 \times T_e^{-0.34} \nu_{\text{GHz}}^{-0.11} \tag{6.19}$$

for the free–free Gaunt factor, with an approximation valid at radio wavelengths [4,5]. In (6.18)–(6.19), N_e is the electron concentration in units of cm^{-3}, and T_e is the electron temperature. The Planck function relates the emission and absorption coefficients for black body radiation, i.e.,

$$S_\nu = \frac{j_\nu}{\kappa_\nu} = \frac{2h\nu^3}{c^2} \frac{1}{e^{h\nu/kT_e} - 1} \equiv 2kT_r \left(\frac{\nu}{c} \right)^2 \cong 2kT_e \left(\frac{\nu}{c} \right)^2, \tag{6.20}$$

where the approximation that the radiation temperature $T_r \cong T_e$ is valid for longer radio wavelengths, but becomes invalid at mm wavelengths.

For bound–free transitions the emission coefficient and Gaunt factor are given by

$$j_{\text{fb},\nu} \rho = 1.72 \times 10^{-33} N_e^2 T_e^{-3/2} \sum_{n=m}^{\infty} n^{-3} g_{\text{fb}}(\nu, T_e, n) e^{157890/n^2 T_e}, \tag{6.21}$$

where n is the principal quantum number and m is the integer portion of $(3.789\,89 \times 10^{15} \text{ Hz}/\nu)$ and

$$g_{\text{bf}}(\nu, T_e, n) = 1 + \frac{0.1728(u_n - 1)}{n(u_n + 1)^{2/3}} - \frac{0.0496(u_n^2 + \frac{4}{3}u_n + 1)}{n(u_n + 1)^{4/3}}, \tag{6.22}$$

where $u_n = n^2(\nu/3.289\,89 \times 10^{15} \text{ Hz})$ [6].

6.5.3   Spectral Lines—Thermal Bound–Bound Transitions

For bound–bound transitions between any two levels with energies $E_n = E_1, ..., E_{\text{ionization}}$ and statistical weights g_n, radiation absorption and emission involves photons of energy $h\nu_{nm} = E_m - E_n$. The properties of a line transition are determined by the Einstein coefficients A_{mn}, B_{mn}, and B_{nm}, which are related by

$$B_{mn} = \frac{c^2}{2h\nu^3} A_{mn} \quad \text{and} \quad B_{nm} = \frac{g_m}{g_n} B_{mn}. \tag{6.23}$$

The emission and absorption coefficients for line transitions are

$$j_{\text{bb}, \nu_{nm}} = \frac{h\nu_{nm}}{4\pi} N_m A_{mn} \phi_{mn}(\nu_{nm}) \tag{6.24}$$

and

$$\kappa_{\text{bb}, \nu_{nm}} = \frac{h\nu_{nm}}{4\pi} N_m B_{nm} \phi_{mn}(\nu_{nm}) \left[1 - \frac{N_n g_n \phi_{mn}(\nu_{mn})}{N_m g_m \phi_{mn}(\nu_{nm})} \right], \tag{6.25}$$

where $\phi_{nm}(\nu_{mn})$ and $\phi_{mn}(\nu_{mn})$ are absorption and emission "line" profile functions, which are usually equal to each other. Note that in general the source function can be expressed as

$$S_{\nu_{nm}} = \frac{j_{\text{bb}, \nu_{nm}}}{\kappa_{\text{bb}, \nu_{nm}}} = \frac{2h\nu^3}{c^2} \left(\frac{1}{\dfrac{N_n g_n \phi_{mn}(\nu_{mn})}{N_m g_m \phi_{mn}(\nu_{nm})} - 1} \right). \tag{6.26}$$

If the line is in Local Thermodynamic Equilibrium (LTE) with temperature T, then $N_n/N_m = (g_n/g_m) \exp(h\nu_{nm}/kT)$ and $S = B_\nu$ so

$$\kappa_{\text{bb,LTE}, \nu_{nm}} = \frac{h\nu_{nm}}{4\pi} N_m B_{nm} \phi_{mn}(\nu_{nm}) \left[1 - e^{-h\nu_{nm}/kT} \right]. \tag{6.27}$$

Most of the known spectral lines from molecules in the interstellar medium have been detected at radio wavelengths. Many are in Table 6.5. For current lists and information about the parameters of various molecular species see the Internet URL http://spec.jpl.nasa.gov.

Table 6.5. *Known interstellar molecules.*

Name of molecule	Chemical symbol	Wavelength region	Date and telescope of discovery
methyladyne	CH	4 300 Å	1937 – Mt. Wilson 2.5 m
cyanogen radical	CN	3 875 Å	1940 – Mt. Wilson 2.5 m
methyladyne ion	CH^+	4 232 Å	1941 – Mt. Wilson 2.5 m
hydroxyl radical	OH	18 cm	1963 – Lincoln Lab. 26 m
ammonia	NH_3	1.3 cm	1968 – Hat Creek 6 m
water	H_2O	1.4 cm	1968 – Hat Creek 6 m
formaldehyde	H_2CO	6.2 cm	1969 – NRAO 43 m
carbon monoxide	CO	2.6 mm	1970 – NRAO 11 m
hydrogen cyanide	HCN	3.4 mm	1970 – NRAO 11 m
cyanoacetylene	HC_3N	3.3 cm	1970 – NRAO 43 m
hydrogen	H_2	1013–1108 Å	1970 – NRL rocket
methanol	CH_3OH	36 cm	1970 – NRAO 43 m
formic acid	HCOOH	18 cm	1970 – NRAO 43 m
formyl radical ion	HCO^+	3.4 mm	1970 – NRAO 11 m

Table 6.5. (*Continued.*)

Name of molecule	Chemical symbol	Wavelength region	Telescope of discovery
formamide	NH_2CHO	6.5 cm	1971 – NRAO 43 m
carbon monosulfide	CS	2.0 mm	1971 – NRAO 11 m
silicon monoxide	SiO	2.3 mm	1971 – NRAO 11 m
carbonyl sulfide	OCS	2.7 mm	1971 – NRAO 11 m
methyl cyanide	CH_3CN	2.7 mm	1971 – NRAO 11 m
isocyanic acid	HNCO	3.4 mm	1971 – NRAO 11 m
methyl acetylene	CH_3CCH	3.5 mm	1971 – NRAO 11 m
acetaldehyde	CH_3CHO	28 cm	1971 – NRAO 43 m
thioformaldehyde	H_2CS	9.5 cm	1971 – Parkes 64 m
hydrogen isocyanide	HNC	3.3 mm	1971 – NRAO 11 m
hydrogen sulfide	H_2S	1.8 mm	1972 – NRAO 11 m
methanimine	CH_2NH	5.7 cm	1972 – Parkes 64 m
sulfur monoxide	SO	3.0 mm	1973 – NRAO 11 m
diazenylium	N_2H^+	3.2 mm	1974 – NRAO 11 m
ethynyl radical	C_2H	3.4 mm	1974 – NRAO 11 m
methylamine	CH_3NH_2	3.5 mm	1974 – NRAO 11 m
		4.1 mm	1974 – Tokyo 6 m
dimethyl ether	$(CH_3)_2O$	9.6 mm	1974 – NRAO 11 m
ethanol	CH_3CH_2OH	2.9–3.5 mm	1974 – NRAO 11 m
sulfur dioxide	SO_2	3.6 mm	1975 – NRAO 11 m
silicon sulfide	SiS	2.8, 3.3 mm	1975 – NRAO 11 m
vinyl cyanide	CH_2CHCN	22 cm	1975 – Parkes 64 m
methyl formate	CH_3OCHO	18 cm	1975 – Parkes 64 m
nitrogen sulfide	NS	2.6 mm	1975 – Texas 5 m
cyanamide	NH_2CN	3.7 mm	1975 – NRAO 11 m
cyanodiacetylene	HC_5N	3.0 cm	1976 – Algonquin 46 m
formyl radical	HCO	3.5 mm	1976 – NRAO 11 m
acetylene	C_2H_2	infrared	1976 – KPNO 4 m
cyanoethynyl radical	C_3N	3.4 mm	1976 – NRAO 11 m
ketene	H_2CCO	2.9 mm	1976 – NRAO 11 m
cyanotriacetylene	HC_7N	2.9 cm	1977 – Algonquin 46 m
nitrosyl radical	HNO	3.7 mm	1977 – NRAO 11 m
confirmed		1.9 mm	1990 – FCRAO 14 m
ethyl cyanide	CH_3CH_2CN	2.7–4.0 mm	1977 – NRAO 11 m
cyano-octatetra-yne	HC_9N	2.9 cm	1977 – Algonquin 46 m
methane	CH_4	infrared	1977 – KPNO 4 m
nitric oxide	NO	2.0 mm	1978 – NRAO 11 m
butadiynyl radical	C_4H	2.6–3.5 mm	1978 – NRAO 11 m
methyl mercaptan	CH_3SH	3 mm	1979 – BTL 7 m
isothiocyanic acid	HNCS	3 mm	1979 – BTL 7 m
thioformyl radical ion	HCS^+	3 mm	1980 – BTL 7 m
protonated carbon dioxide	$HOCO^+$	3 mm	1980 – BTL 7 m
ethylene	C_2H_4	infrared	1980 – KPNO 4 m
cyanotetraacetylene	$HC_{11}N$	many (cm)	1981 – Algonquin 46 m
silicon dicarbide	SiC_2	many (mm)	1984 – BTL 7 m
propynylidyne	C_3H	many (3 mm)	1984 – BTL 7 m
methyl diacetylene	CH_3C_4H	several (cm)	1984 – Haystack 37 m
methyl cyanoacetylene	CH_3C_3N	several (cm)	1984 – Haystack 37 m
tricarbon monoxide	C_3O	2 cm	1984 – Haystack 37 m
silane	SiH_4	infrared	1984 – KPNO 4 m
protonated HCN	$NCNH^+$	3, 2, 1 mm	1984 – NRAO 12 m
			1984 – Texas 5 m
cyclopropynylidene	C_3H_2	3, 2 mm	1985 – many
—	H_2D^+ ?	0.62 mm	1985 – KAO
hydrogen chloride	HCl ?	0.48 mm	1985 – KAO
protonated water	H_3O^+	1.0 mm	1986 – NRAO 12 m
confirmed		0.8 mm	1990 – CSO 10 m

Table 6.5. *(Continued.)*

Name of molecule	Chemical symbol	Wavelength region	Telescope of discovery
pentynylidyne radical	C_5H	radio	1986 – IRAM 30 m
hexatriynyl radical	C_6H	radio	1986 – IRAM 30 m
phosphorus nitride	PN	3, 2, 1 mm	1986 – NRAO 12 m
			1986 – FCRAO 14 m
—	C_2S	3, 7 mm	1986 – Nobeyama 45 m
—	C_3S	3, 7 mm	1986 – Nobeyama 45 m
acetone	$(CH_3)_2CO$?	3.4 mm	1987 – IRAM 30 m
sodium chloride	NaCl	2, 3 mm	1987 – IRAM 30 m
aluminum chloride	AlCl	2, 3 mm	1987 – IRAM 30 m
potassium chloride	KCl	2, 3 mm	1987 – IRAM 30 m
aluminum fluoride	AlF	2, 3 mm	1987 – IRAM 30 m
methyl isocyanide	CH_3NC	3 mm	1987 – IRAM 30 m
cyanomethyl radical	CH_2CN	6.5 mm	1988 – Nobeyama 45 m
		1.3 cm	1988 – NRAO 43 m
silicon carbide	SiC	2 mm	1989 – IRAM 30 m
propynal	HCCCHO	8 mm	1989 – Nobeyama 45 m
		1.6 cm	1989 – NRAO 43 m
—	SiC_4	3.5, 7.5 mm	1989 – Nobeyama 45 m
phosphorus carbide	CP	1.2, 3.0 mm	1989 – IRAM 30 m
propadienylidene	H_2CCC	3 mm	1990 – IRAM 30 m
		1.4 cm	1990 – NRAO 43 m
butatrienylidene	H_2CCCC	many (3 mm)	1990 – IRAM 30 m
silicon nitride	SiN	1.1, 1.4 mm	1990 – NRAO 12 m
silylene (pend. conf.)	SiH_2 ?	0.8, 1.0 mm	1990 – CSO 10 m
—	HCCN	3 mm	1991 – IRAM 30 m
carbon suboxide	C_2O	7 mm	1991 – Nobeyama 45 m
		1.3 cm	1991 – NRAO 43 m
isocyanoacetylene	HCCNC	7 mm	1992 – Nobeyama 45 m
sulfur oxide ion	SO^+	1.3, 3 mm	1992 – NRAO 12 m
ethinylisocyanide	HNCCC	1.1, 0.8, 0.64 cm	1992 – Nobeyama 45 m
magnesium isocyanide	MgNC	radio	1992 – Nobeyama 45 m
protonated HC_3NH^+	NC_3H^+	radio	1993 – Nobeyama 45 m
—	NH_2	0.645 mm	1993 – CSO
carbon monoxide ion	CO^+	1.3, 0.8 mm	1993 – NRAO 12 m, CSO
sodium cyanide	NaCN	3, 2.4 mm	1993 – NRAO 12 m
—	CH_2D^+	0.8–4.0 mm	1993 – NRAO 12 m, CSO
nitrous oxide	N_2O	2.0–4.0 mm	1994 – NRAO 12 m
magnesium cyanide	MgCN	radio	1995 – NRAO 12 m
prot. formaldehyde	H_2COH^+	radio	1995 – Nobeyama 45 m
octatetraynyl radical	C_8H	radio	1996 – IRAM 30 m
octatetraynyl radical	C_7H	radio	1996 – IRAM 30 m
protonated hydrogen	H_3^+	IR-3.67 μm	1996 – UKIRT
hexapentaenylidene	H_2C_6	K-band	1997 – NASA 34 m
ethylene oxide	$c-C_2H_4O$	K, Q, 3 mm,1 mm	1997 – NRO, SEST, Haystack
hydrogen flouride	HF	121.7 μm	1997 – ISO

6.5.4 Magneto-Bremsstrahlung Emission and Absorption

The other radiation process of dominant importance in radio astronomy is magneto-bremsstrahlung emission and absorption which results from the interaction between fast-moving electrons and the ambient magnetic field. There are three major varieties of magneto-bremsstrahlung emission. When the moving electrons are very relativistic, the emission process is the very efficient, highly beamed synchrotron emission which dominates the emission from many galactic and most extragalactic radio sources. However, in some stars and certain solar system environments, two other magneto-bremsstrahlung processes occur when the moving electrons are nonrelativistic or only mildly relativistic. When mildly relativistic ($\gamma_{\text{Lorentz}} \equiv (1 - (v/c)^2)^{-1/2} \cong 2\text{–}3$) electrons undergo interactions with magnetic fields, the emission process is called gyro synchrotron, which is much less beamed and much less efficient at producing radio emission than synchrotron processes; however, it tends to be highly circularly polarized, a characteristic commonly found in stellar radio emission. Even less efficient is the magneto-bremsstrahlung resulting from less relativistic particles, with $\gamma_{\text{Lorentz}} \leq 1$, which is called cyclotron or gyro resonance emission. The details of nonrelativistic (cyclotron) and mildly relativistic (gyro synchrotron) emission and absorption processes are more complicated than the highly relativistic case. They are summarized by [7], and extensively discussed by [8] and [9].

Much of the analysis of radio source data uses very simple models for the behavior of synchrotron radiating sources. This section summarizes formulas used in such analysis. We assume that the density of relativistic electrons can be described by a power law distribution in energy, $N(E) = KE^{-\gamma}$ in the energy range E to $E + dE$, where γ is a constant. If these electrons are mixed in with a uniform distribution of magnetic fields of strength H, which can be described as having uniformly random directions on a large scale, then the emission and absorption coefficients are given by

$$j_\nu \rho = 1.35 \times 10^{-22} a(\gamma) K H^{(\gamma+1)/2} \left(\frac{6.36 \times 10^{18}}{\nu} \right)^{(\gamma-1)/2} \tag{6.28}$$

and

$$\kappa_\nu \rho = 0.019 g(\gamma)(3.5 \times 10^9)^\gamma K H^{(\gamma+2)/2} \nu^{-(\gamma+4)/2}, \tag{6.29}$$

where H is in gauss and all other variables are in cgs units, so that the source function is

$$S_\nu = \frac{j_\nu}{\kappa_\nu} = 2.84 \times 10^{-30} \frac{a(\gamma)}{g(\gamma)} 2^{\gamma/2} H^{-1/2} \nu^{5/2} \text{ erg s}^{-1} \text{ cm}^{-3} \tag{6.30}$$

[10]. Table 6.6 gives some of the values of the functions $a(\gamma)$, $g(\gamma)$, the brightness temperature source function T_s, and the angular size (Θ_0) of a uniform source with flux density S_0 and magnetic field H_{mG} in units of milligauss.

Table 6.6. *Incoherent synchrotron emission constants.*

γ	$a(\gamma)$	$g(\gamma)$	α	$T_s \nu_{\text{GHz}}^{-1/2} H_{\text{mG}}^{1/2}$	$\Theta_0 S_0^{-1/2} H_{\text{mG}}^{-1/4} \nu_{\text{GHz}}^{5/4}$
1.0	0.283	0.96	0.0	1.9×10^{12} K	0.015 mas
1 5	0.147	0.79	0.25	1.0×10^{12} K	0.021 mas
2.0	0.103	0.70	0.50	6.6×10^{11} K	0.026 mas
2.5	0.0852	0.66	0.75	4.9×10^{11} K	0.030 mas
3.0	0.0742	0.65	1.0	3.6×10^{11} K	0.035 mas
4.0	0.0725	0.69	1.5	2.3×10^{11} K	0.044 mas
5.0	0.0922	0.83	2.0	1.7×10^{11} K	0.051 mas

The specific intensity is given by

$$I_\nu = \int_0^L S_\nu (1 - \exp(-\tau_\nu)) \, d\tau_\nu, \tag{6.31}$$

where $d\tau_\nu = \kappa_\nu \rho \, ds$ in the above integral, which is along the line of sight of length L.

We see from (6.30) that the important case where j_ν/κ_ν is constant occurs if the magnetic field is uniform in strength and geometry. Equations (6.28)–(6.30) also tell us that for the optically thick and thin limits, I_ν is proportional to $H^{-1/2}\nu^{5/2}$ and $KH^{(\gamma+1)/2}\nu^{-(\gamma-1)/2}L$, respectively, where L is a line-of-sight path length. Defining the spectral index α by $S_\nu \propto \nu^\alpha$, then $\alpha = 2.5$ and $-(\gamma-1)/2$ in the optically thick and thin limits, respectively. The values of α in the latter case for various values of γ are listed in Table 6.6.

The evolution of the relativistic electron energy distribution $N(E,t)$ is the primary factor in the evolution or synchrotron radiation sources. In its most general form this evolution is described by

$$\frac{\partial N(E,t)}{\partial t} + \frac{\partial}{\partial E}\left[N(e,t)\frac{dE}{dt}\right] = \Gamma(E,T) - \Lambda(E,t), \tag{6.32}$$

where $\Gamma(E,T)$ and $\Lambda(E,t)$ are functions describing particle "injection" and "escape," with escape time scales T, respectively. Usually Γ and Λ are zero, as they will be for all models discussed below, or applies to only small portions of the radio sources. The equation for evolution of $N(E,t)$ is supplemented by the energy loss equation:

$$\frac{dE}{dt} = \phi(E) = -\zeta - \eta E - \xi E^2, \tag{6.33}$$

where the first two terms are due to losses by ionization in the ambient, medium, and free–free interaction with this medium. The last term, which is usually the most important, includes both energy loss due to synchrotron radiation and energy loss due to inverse Compton scattering off photons in the local radiation field. The coefficients of the first two terms in (6.33) are:

$$\zeta \approx 3.33 \times 10^{-20} n_H \left(6.27 + \ln\left(\frac{E}{mc^2}\right)\right) + 1.22 \times 10^{-20} n_e \left(73.4\ln\left(\frac{E}{mc^2}\right) - \ln n_e\right) \tag{6.34}$$

and

$$\eta \approx 8 \times 10^{-16} n_H n_e \left(0.36 + \ln\left(\frac{E}{mc^2}\right)\right), \tag{6.35}$$

where n_H is the concentration of neutral hydrogen atoms and n_e is the concentration of thermal electrons.

For synchrotron radiation losses $\xi_{\text{synch}} \cong 2.37 \times 10^{-3} H_\perp^2$, where $H_\perp$ is the magnetic field perpendicular to the velocity vector of the radiating relativistic electron. The time scale for synchrotron losses is $t_{\text{synch}} \equiv E/(-dE/dt) = 1/(\zeta_{\text{synch}}E) \sim (5.1 \times 10^8/H_\perp^2)(mc^2/E)$ s. Synchrotron radiation losses dominate in regions with large magnetic fields and low radiation fields.

For inverse Compton losses $\xi_{\text{IC}} \approx 3.97 \times 10^{-4} u_{\text{rad}}$, where u_{rad} is the radiation energy density. Inverse Compton losses dominate when brightness temperature approaches $T_b \leq 10^{12}$ K. Above 10^{12} K all relativistic electrons loss their energy rapidly due to this process.

For steady-state synchrotron radiation sources where one assumes $\Lambda(E,t) = \Lambda(E)$, and $N(E,t) = N(E)$, and $\partial N/\partial t = 0$, the relativistic particle energy equation has the solution $N(E) = \phi^{-1}(E)\int \Lambda(E)\,dE$. So if one can assume a power law for the particle injection, $\Lambda(E) = \Lambda_0 E^{-\gamma}$, then

$$N(E) = \Lambda_0(\gamma-1)^{-1}E^{-\gamma}(\zeta E^{-1} + \eta + \zeta E)^{-1}. \tag{6.36}$$

The relativistic particle spectra and observable radio spectrum for such steady-state radio sources can be described by three segments of power laws as shown in Figure 6.5. The principle effect of more complex models is replacement of the sharp spectral breaks with smooth transitions.

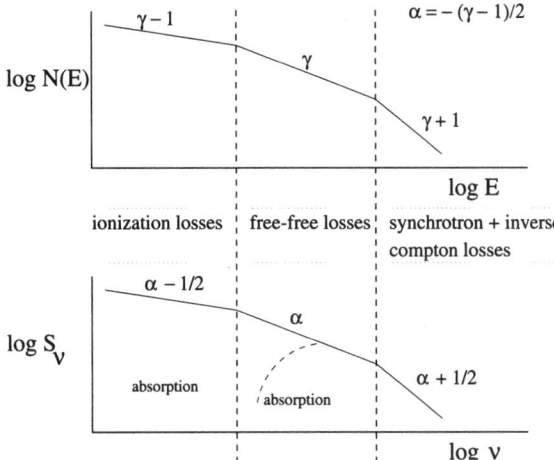

Figure 6.5. Relativistic particle (top), and observable flux density (bottom), spectrum for steady-state synchrotron radiation sources. The changes in slope indicate regions where ionization, free–free, and synchrotron (+ inverse Compton) losses dominate. By each curve segment is the power law index for the energy and flux density spectra. External or self-absorption can occur at any point in the flux density spectrum; two possibilities are indicated in the figure.

A synchrotron radio source with losses only due to synchrotron radiation has an energy loss equation $d(E^{-1})/dt = \zeta_{\text{synch}}$ which leads to a spectrum described by $E = E_0/[1 + \zeta_{\text{synch}}(t - t_0)E_0]$. Then if at $t = t_0$, $N(E) = N_0 E^{-\gamma}$, then at later times $N(E) = N_0 E^{-\gamma}(1 - \zeta_{\text{synch}}Et)^{\gamma-1}$ for $E < E_T(t)$, and $N(E, T) = 0$ for $E > E_T(t)$, where $E_T(t) = 1/(\zeta_{\text{synch}}t)$. The frequency above which synchrotron losses dominate is $\nu_b \approx B_{\text{Gauss}}^{-3} t_{\text{year}}^{-2}$ GHz.

Time-dependent synchrotron radiation source models with complex geometry and complex energy loss mechanisms require extensive computation. However in the case of simple geometries and only adiabatic losses the evolution of the flux density of a radio source can be expressed in analytic or near-analytic equations. For a spherical region which "suddenly" has a distribution of relativistic particles given by $N_0 = K_0 E^{-\gamma}$ and which expands with an outer radius $r_2(t)$ the flux density of the source is described by the quantities in Table 6.7 [11–13]. An analogous model can be expressed for a conical jet with ejection of adiabatically expanding plasma [13]. The jet ejection parameters are such that at an initial distance z_0 the radius of the cross section of the jet is r_0 (jet Mach number $M_0 = z_0/r_0$) and the relativistic particle density is N_0. The quantities that determine the flux density for an observer for whom the jet is oriented at an inclination angle Θ are in Table 6.7; for a continuous jet $z_2 = \infty$ and $z_1 = z_0$; however, a time-dependent jet which starts ejection with velocity v_2 at time t_{start} and ends ejection with velocity v_1 at time t_{stop} is described by the same equations with the necessary time dependence in the limits of integral for the flux density.

Table 6.7. *Radio source models—only adiabatic energy losses.*

Expanding sphere	Conical jet
$E = E_0(r_2/r_0)^{-1}$	$E = E_0[r_2(z_2)/r_0]^{-2/3}$
$H = H_0(r_2/r_0)^{-2}$	$H = H_0[r_2(z_2)/r_0]^{-1}$
$K = K_0(r_2/r_0)^{-(\gamma+2)}$	$K = K_0[r(z_2)/r_0]^{-2(\gamma+2)/3}$
$\tau_0 = 0.038g(\gamma)(3.5 \times 10^9)^{\gamma} K_0 H_0^{(\gamma+2)/2} r_0$	same

Table 6.7. *(Continued.)*

Expanding sphere	Conical jet
$\xi(\tau; \tau > 20) = 1$	$\xi(\tau; \tau > 20) = 1$
$\xi(\tau; \tau < 20) = 0.66584 + 0.09089\tau$ $\quad - 0.009989\tau^2 + 0.0005208\tau^3$ $\quad - 0.00001268\tau^4 + 0.000000115\tau^5$	$\xi(\tau; \tau < 20) = 0.78517 + 0.06273\tau$ $\quad - 0.007242\tau^2 + 0.0003905\tau^3$ $\quad - 0.00000973\tau^4 + 0.00000009\tau^5$
$\tau_\nu' = \tau_0 \left(\dfrac{r}{r_0}\right)^{-(2\gamma+3)} \left(\dfrac{\nu}{\nu_0}\right)^{-(\gamma+4)/2}$	$\tau_\nu' = \left(\dfrac{\tau_0}{\sin(\Theta)}\right)\left(\dfrac{\nu}{\nu_0}\right)^{-(\gamma+4)/2} \left[\dfrac{r_2(z_2)}{r_0}\right]^{-(7\gamma+8)/6}$
$S_\nu = 1.42 e^{-0.65\gamma} H_{0,\mathrm{mG}}^{-1/2} \nu_{\mathrm{GHz}}^{5/2} \Theta_{\mathrm{mas}}^3$ $\quad \times \left[1 - \exp(-\tau_\nu') \times \xi(\tau_\nu')\right]$	$S_\nu = S_0 M_0 \left(\dfrac{\nu}{\nu_0}\right)^{5/2} \sin(\Theta)$ $\quad \times \int_{z_1(t)}^{z_2(t)} \zeta^{3/2} \left[1 - \exp(-\tau_\nu')\right]\xi(\tau_\nu')\,d\zeta$
$r_2(t; \text{free exp.}) = r_0 \left(\dfrac{t}{t_0}\right)$	$r_2(t; \text{free exp.}) = r_0 \left(\dfrac{z_2}{z_0}\right)$
$r_2(t; \text{E cons.}) = r_0 \left(\dfrac{t}{t_0}\right)^{2/5}$	$r_2(t; \text{E cons.}) = r_0 \left(\dfrac{z_2}{z_0}\right)^{1/2}$
$r_2(t; \text{Mom. cons.}) = r_0 \left(\dfrac{t}{t_0}\right)^{1/4}$	$r_2(t; \text{Mom. cons.}) = r_0 \left(\dfrac{z_2}{z_0}\right)^{1/3}$
	$z_2 = v_2(t - t_{\mathrm{start}})$ $z_1 = v_1(t - t_{\mathrm{stop}})$

6.6   RADIO ASTRONOMY REFERENCES

Table 6.8 gives further references for radio astronomy.

Table 6.8. *List of radio astronomy references.*

Topic	Reference
Radio astronomy research field reviews	[1]
Properties of radio telescopes	[2]
Interferometry and aperture synthesis	[3]
Aperture synthesis	[4, 5]
Observation-oriented radio astronomy textbooks	[6, 7]
Synchrotron radiation physics	[8]
Radio astrophysical theory	[9]
General astrophysical theory	[10]

References

1. Verschuur, V.L., & Kellermann, K.I. 1988, *Galactic and Extragalactic Radio Astronomy* (Springer-Verlag, New York)
2. Christiansen, W.N., & Högbom, J.A. 1985, *Radio Telescopes* (Cambridge University Press, Cambridge)
3. Thompson, A.R., Moran, J.M., & Swenson, G.W. 1986, *Interferometry and Synthesis in Radio Astronomy* (Wiley, New York)
4. Perley, R.A., Schwab, F.R., & Bridle, A.H. 1989, *Synthesis Imaging in Radio Astronomy*, A.S.P. Conference Series, 6
5. Cornwell, T.J., & Perley, R.A. 1991, *Radio Interferometry: Theory, Techniques, and Applications*, A.S.P. Conference Series, 19
6. Rohlfs, K. 1986, *Tools of Radio Astronomy* (Springer-Verlag, Berlin)
7. Krause, J.D. 1986, *Radio Astronomy*, 2nd ed. (Cygnus-Quasar Books, Powell)
8. Ginzburg, V.L., & Syrovatskii, S.I. 1965, *Ann. Rev. Astron. Astrophys.*, **3**, 297
9. Pacholczyk, A.G. 1970, *Radio Astrophysics* (W.H. Freeman, San Francisco)
10. Longair, M.S. 1981, *High Energy Astrophysics: an informal introduction for students of physics and astronomy* (Cambridge University Press, Cambridge)

REFERENCES

1. Clark, B.G. 1989, *Synthesis Imaging in Radio Astronomy*, A.S.P. Conference Series, 6, p. 1
2. Christiansen, W.N., & Högbom, J.A. 1985, *Radio Telescopes* (Cambridge University Press, Cambridge)
3. Price, R.M. et al. 1989, *Radio Astronomy Observatories* (National Academy Press, Washington, DC)
4. Karzas W.J., & Latter, R. 1961, *ApJS*, **6**, 167
5. Hjellming, R.M., Wade, C.M., Vandenberg, N.R., & Newell, R.T. 1979, *AJ*, **84**, 1619
6. Aller, L.H. & Liller, W. 1968, *Nebulae and Interstellar Matter*, edited by B. Middlehurst & L.H. Aller (University of Chicago Press, Chicago), Chapter 9
7. Dulk, G.A. 1985, *ARA&A*, **23**, 169
8. Kundu, M.R. 1965, *Solar Radio Astronomy* (Interscience, New York)
9. Zheleznyakov, V.V. 1970, *Radio Emission of the Sun and Planets* (Pergamon Press, Oxford)
10. Ginzburg, V.L., & Syrovatskii, S.I. 1965, *ARA&A*, **3**, 297
11. van der Laan, H. 1966, *Nature*, **211**, 1131
12. Kellermann, K.I. 1966, *ApJ*, **146**, 621
13. Hjellming, R.M., & Johnston, K.J. 1988, *ApJ*, **328**, 600

Chapter 7

Infrared Astronomy

A.T. Tokunaga

7.1 USEFUL EQUATIONS; UNITS

The Planck function in wavelength units ($\lambda_{\mu m}$ in μm; T in K) is

$$B_\lambda = 2hc^2\lambda^{-5}/(e^{hc/k\lambda T} - 1)$$
$$= 1.1910 \times 10^8 \lambda_{\mu m}^{-5}/(e^{14\,387.7/\lambda_{\mu m}T} - 1)\ \mathrm{W\,m^{-2}\,\mu m^{-1}\,sr^{-1}}.$$

The Planck function in frequency units (ν in Hz) is

$$B_\nu = 2h\nu^3 c^{-2}/(e^{h\nu/kT} - 1)$$
$$= 1.4745 \times 10^{-50} \nu^3/(e^{4.799\,22 \times 10^{-11}\,\nu/T} - 1)\ \mathrm{W\,m^{-2}\,Hz^{-1}\,sr^{-1}}.$$

The Rayleigh–Jeans approximation (for $h\nu \ll kT$) is

$$B_\lambda = 2ckT\lambda^{-4} = 8.2782 \times 10^3\, T/\lambda_{\mu m}^4 \; \text{W m}^{-2}\, \mu m^{-1}\, \text{sr}^{-1},$$
$$B_\nu = 2c^{-2}kT\nu^2 = 3.0724 \times 10^{-40}\, T\nu^2 \; \text{W m}^{-2}\, \text{Hz}^{-1}\, \text{sr}^{-1}.$$

The Stefan–Boltzmann law is

$$B = \int B_\lambda\, d\lambda = (\sigma/\pi)T^4 = 1.8050 \times 10^{-8}\, T^4 \; \text{W m}^{-2}\, \text{sr}^{-1}.$$

The wavelength of maximum B_λ (Wien law) is

$$\lambda_{\max} = 2898/T, \qquad \lambda_{\max} \text{ in } \mu m.$$

The frequency of maximum B_ν is

$$\nu_{\max} = 5.878 \times 10^{10}\, T, \qquad \nu_{\max} \text{ in Hz}.$$

The conversion equations (Ω in sr) are $F_\lambda = \Omega B_\lambda$, $F_\nu = \Omega B_\nu$, $F_\lambda = 3.0 \times 10^{14} F_\nu/\lambda_{\mu m}^2$. Other units are 1 Jansky (Jy) $= 10^{-26}$ W m^{-2} Hz^{-1}. Units details are given in Table 7.1.

Table 7.1. *Units* [1–4].

Units	Radiometric name	Astronomical name
W	Flux	Luminosity
W m^{-2}	Irradiance; radiant exitance	Flux
W sr^{-1}	Intensity	$\ldots$
W m^{-2} sr^{-1}	Radiance	Intensity
W m^{-2} μm^{-1}; W m^{-2} Hz^{-1}	Spectral irradiance	Flux density
W m^{-2} μm^{-1} sr^{-1}; W m^{-2} Hz^{-1} sr^{-1}	Spectral radiance	Surface brightness; specific intensity

References
1. Boyd, R.W. 1983, *Radiometry and the Detection of Optical Radiation* (Wiley, New York)
2. Dereniak, E.L., & Crowe, D.G. 1984, *Optical Radiation Detectors* (Wiley, New York)
3. Wolfe, W.L., & Zissis, G.J. 1985, *The Infrared Handbook*, rev. ed. (Office of Naval Research, Washington, DC)
4. Rieke, G.H. 1994, *Detection of Light; From the Ultraviolet to the Submillimeter* (Cambridge University Press, Cambridge)

7.2 ATMOSPHERIC TRANSMISSION

The major atmospheric absorbers and central wavelengths of absorption bands are H_2O (0.94, 1.13, 1.37, 1.87, 2.7, 3.2, 6.3, $\lambda > 16$ μm); CO_2 (2.0, 4.3, 15 μm); N_2O (4.5, 17 μm); CH_4 (3.3, 7.7 μm); O_3 (9.6 μm). See Figures 7.1 and 7.2.

For atmospheric transmission at airborne and balloon altitudes, see [6, 11]. For water-vapor measurements at observatory sites, see [12–14]. For atmospheric extinction, see [2, 15–17].

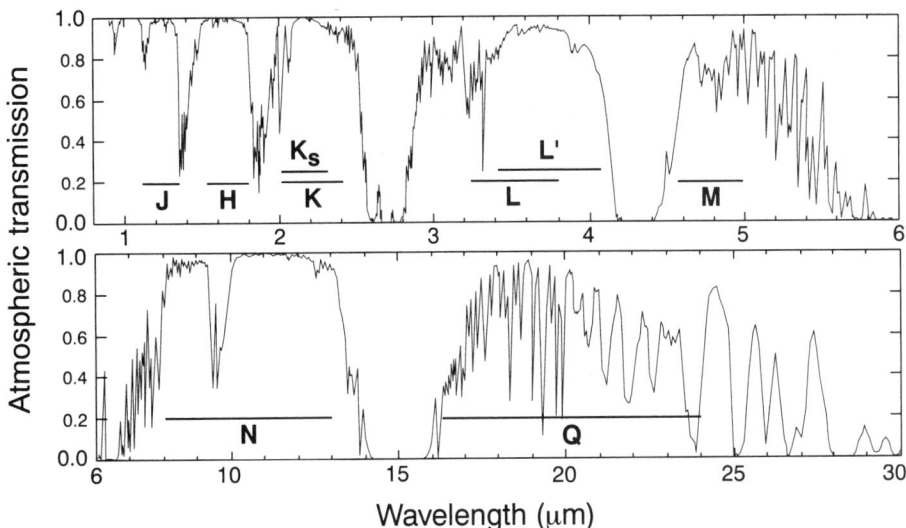

Figure 7.1. Atmospheric transmission from 0.9 to 30 μm under conditions appropriate for Mauna Kea, Hawaii. Altitude = 4.2 km, zenith angle = 30° (air mass = 1.15), precipitable water vapor overhead = 1 mm. $\lambda/\Delta\lambda$ = 300 for 1–6 μm and 150 for 6–30 μm. Spectra are calculated by Lord [1]. The infrared filter band passes are shown as horizontal lines; see Table 7.5 for definitions. Note that the filter transmission is modified by the atmospheric absorption. For the atmospheric transmission at Kitt Peak, see [2]. For ESO, see [3]. See also [4].

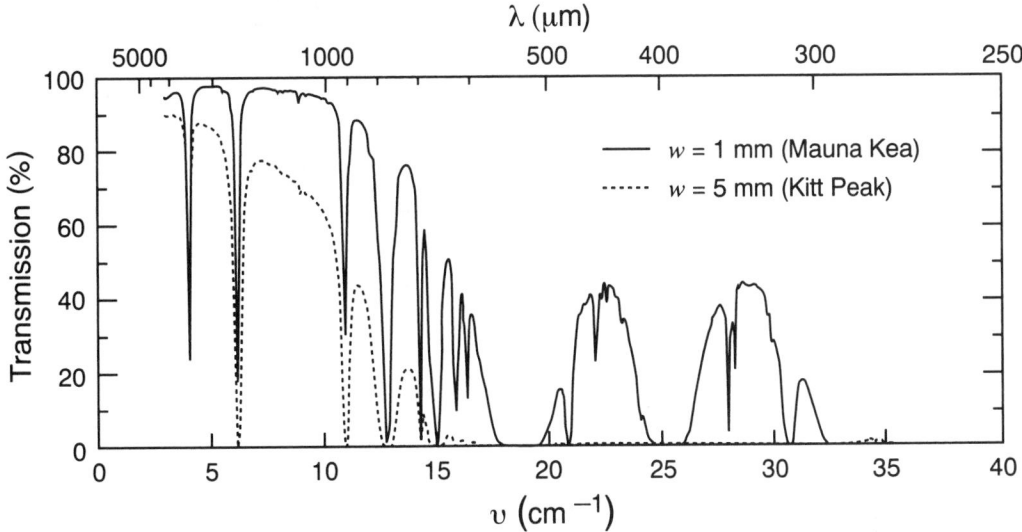

Figure 7.2. Atmospheric transmission from 0.25 to 3 mm, adapted from [5]. The precipitable water vapor is denoted by w. See also [6–9]. For the South Pole, see [10].

7.3 BACKGROUND EMISSION

7.3.1 Background Emission Sources

Table 7.2 gives the background emission from a ground-based telescope. The main background emission sources are shown in Figure 7.3. Where specified they are blackbody functions reduced by a multiplying factor ϵ. In most cases, only the minimum background levels are plotted.

OH   OH airglow. Average OH emission of 15.6 and 13.8 mag arcsec^{-2} at J and H, respectively [18–21].

GBT   Ground-based telescope thermal emission, optimized for the thermal infrared and approximated as a 273 K blackbody with $\epsilon = 0.02$. Emission from the Earth's atmosphere at 1.5–25 μm is shown [22].

ZSL   Zodiacal scattered light at the ecliptic pole, approximated as a 5 800 K blackbody with $\epsilon = 3 \times 10^{-14}$ (based on data from [23]).

ZE   Zodiacal emission from interplanetary dust at the ecliptic pole, approximated as a 275 K blackbody with $\epsilon = 7.1 \times 10^{-8}$. Based on observations from the Infrared Astronomical Satellite (IRAS) [24].

GBE   Galactic background emission from interstellar dust in the plane of the Galaxy. In the plane of the Galaxy away from the Galactic Center, it can be approximated by a 17 K blackbody and $\epsilon = 10^{-3}$ [25, 26].

SEP   South ecliptic pole emission as measured by the Cosmic Background Explorer (COBE) spacecraft [27].

CST   Cryogenic space telescope, cooled to 10 K with $\epsilon = 0.05$.

CBR   Cosmic background radiation, 2.73 K blackbody with $\epsilon = 1.0$ [28].

Table 7.2. *Combined sky, telescope, and instrument background emission at the 3.0 m IRTF [1].*[a]

Band	λ (μm)	$\Delta\lambda$	Surface brightness (mag arcsec^{-2})	Band	λ (μm)	$\Delta\lambda$	Surface brightness (mag arcsec^{-2})
J	1.26	0.31	15.9	L	3.50	0.61	4.9
H	1.62	0.28	13.4	L'	3.78	0.59	4.5
K_s	2.15	0.35	14.1	M'	4.78	0.22	0.3
K	2.21	0.39	13.7	M	4.85	0.62	−0.7

Note
[a]Telescope emissivity at the time of the observations was about 7%.

Reference
1. Shure, M. et al. 1994, *Proc. SPIE*, **2198**, 614

7.3.2 OH Emission Spectrum

The OH emission is often given in Rayleigh units. To convert to other units, use the following equations, with $\lambda_{\mu m}$ in μm [29]:

$$1 \text{ Rayleigh unit} = 10^{10}/4\pi \text{ photons s}^{-1} \text{ m}^{-2} \text{ sr}^{-1}$$
$$= 1.580\,8 \times 10^{-10}/\lambda_{\mu m} \text{ W m}^{-2} \text{ sr}^{-1},$$

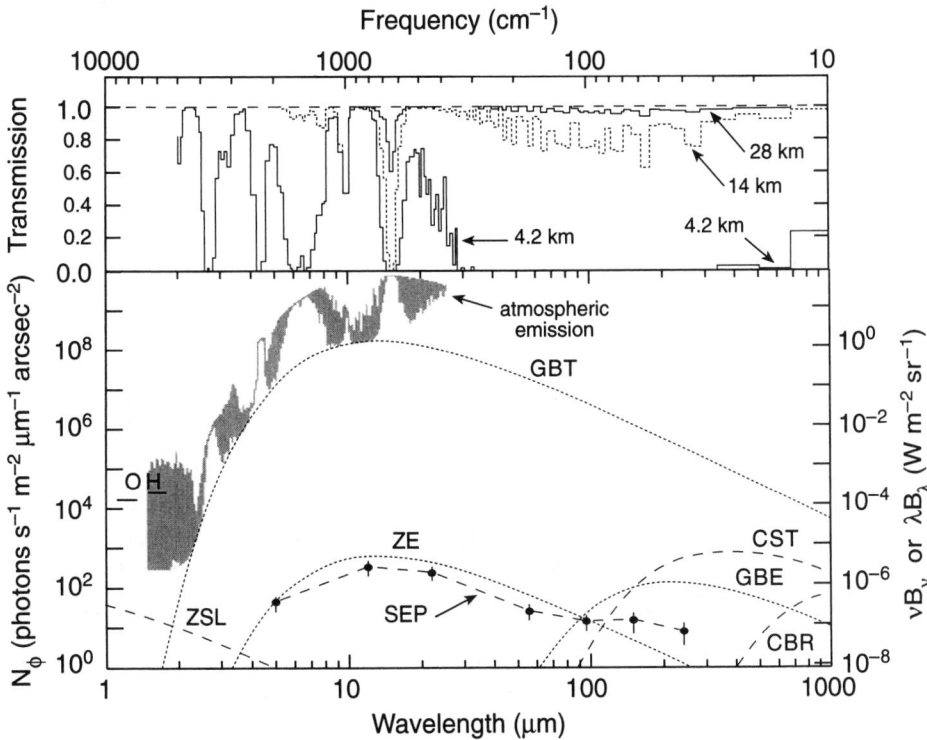

Figure 7.3. Top: Transmission of the Earth's atmosphere at Mauna Kea (4.2 km), airborne (14 km), and balloon altitudes (28 km), adapted from [6]. Bottom: Background emission sources. The surface brightness is calculated from $N_\phi = \epsilon \lambda_{\mu m} B_\lambda/(hc) = 1.41 \times 10^{16} \epsilon \lambda_{\mu m}^{-4}/(e^{14387.7/\lambda_{\mu m}T} - 1)$ ($\lambda_{\mu m}$ in μm, T in K). The intensity is derived from $\lambda_{\mu m} B_\lambda = 8.45 \times 10^{-9} N_\phi$.

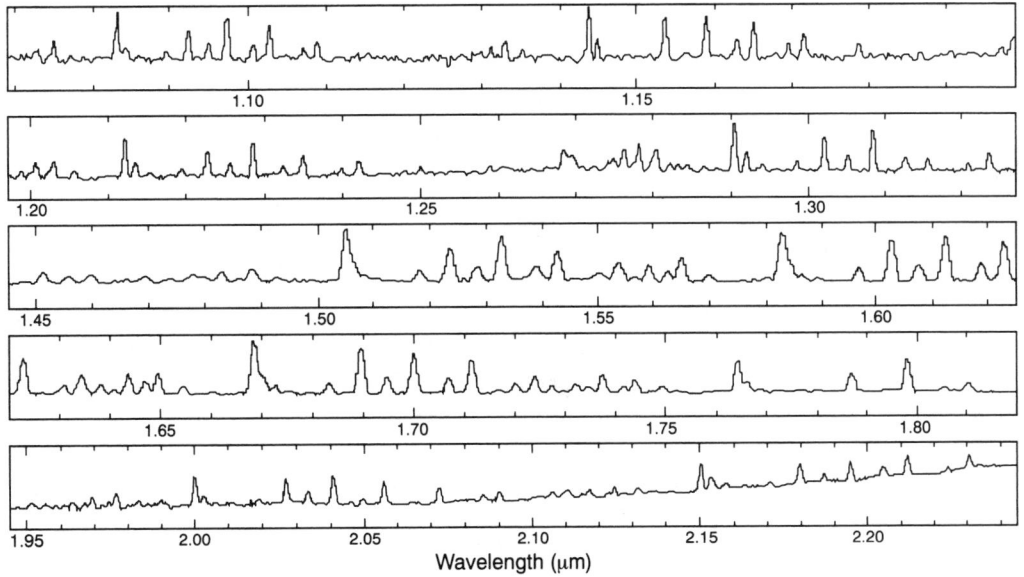

Figure 7.4. Observed OH airglow spectrum adapted from [30]. See also [19, 31–34, 29].

$$1 \text{ Rayleigh unit/Å} = 1.580\,8 \times 10^{-6}/\lambda_{\mu m} \text{ W m}^{-2} \, \mu m^{-1} \text{ sr}^{-1}$$
$$= 3.718\,4 \times 10^{-17}/\lambda_{\mu m} \text{ W m}^{-2} \, \mu m^{-1} \text{ arcsec}^{-2}.$$

The OH airglow spectrum is given in Figure 7.4.

7.4 DETECTORS AND SIGNAL-TO-NOISE RATIOS

Tables 7.3 and 7.4 list the basic detector types for infrared observations.

Table 7.3. *Basic detector types and typical useful wavelength ranges* [1–4].

Material	Type[a]	Wavelength range (μm)[b]	Material	Type[a]	Wavelength range (μm)[b]
Si	PD	< 1.1	Si:Sb	IBC	14–38
Ge	PD	< 1.8	Ge:Be	PC	30–50
HgCdTe	PD	1–2.5	Ge:Ga	PC	40–120
PtSi	SD	1–4	Ge:Ga	PC (stressed)	120–200
InSb	PD	1–5.6	Ge or Si	TD (bolometer)	200–1000
Si:As	IBC	6–27			

Notes

[a] PD = photodiode, PC = photoconductor, SD = Schottky diode, IBC = impurity band conduction photoconductor [also known as blocked impurity band (BIB) photoconductor]; TD = thermal detector.

[b] The HgCdTe long-wavelength cutoff is determined by the Hg/Cd ratio and can be extended to 25 μm.

References
1. Rieke, G.H. 1994, *Detection of Light: From the Ultraviolet to the Submillimeter* (Cambridge University Press, Cambridge)
2. Fazio, G.G. 1994, *Infrared Phys. Technol.*, **35**, 107
3. Herter, T. 1994, in *Infrared Astronomy with Arrays*, edited by I. McLean (Kluwer Academic, Dordrecht), p. 409
4. Haller, E.E. 1994, *Infrared Phys. Technol.*, **35**, 127

For an object that is distributed over n pixels, the signal photocurrent for photodiodes, photoconductors, and Schottky diodes is [35]

$$\sum_n i_s = A\tau\eta G\lambda F_\lambda \Delta\lambda/(hc) = A\tau\eta G(\Delta\lambda/\lambda)F_\nu/h \text{ electrons s}^{-1},$$

where i_s is the photocurrent from an individual pixel, A (m^2) is the telescope collecting area, τ is the transmission of instrument, telescope, and atmosphere, η is the detector quantum efficiency, G is the photoconductive gain (= 1 for a photodiode; $\leq$ 0.5 for a photoconductor), $\Delta\lambda/\lambda$ is the fractional spectral bandwidth, F_λ (W m^{-2} μm^{-1}) $= \Omega_{\text{source}} B_\lambda =$ source flux density, and F_ν (W m^{-2} Hz^{-1}) $= \Omega_{\text{source}} B_\nu =$ source flux density.

The background photocurrent per pixel is

$$i_{\text{bg}} = A\tau\eta G N_\phi \Delta\lambda \Omega_{\text{pix}} \text{ electrons s}^{-1},$$

where N_ϕ (photons s^{-1} m^{-2} μm^{-1} arcsec^{-2}) is the background surface brightness and Ω_{pix} (arcsec^2) is the solid angle on the sky viewed by one pixel.

The RMS noise per pixel is

$$\left[N_r^2 + xG(i_s + i_{bg} + i_{dc})\,t \right]^{1/2} \text{ electrons,}$$

where N_r (electrons) is the detector read noise, i_{dc} (electrons s^{-1}) is the detector dark current, t (s) is the integration time, and $x = 1$ for a photodiode or IBC photoconductor or $x = 2$ for a photoconductor. The signal-to-noise ratio before sky subtraction is

$$S/N = \left(\sum_n i_s \right) t \left(\sum_n [\,N_r^2 + xG(i_s + i_{bg} + i_{dc})t\,] \right)^{-1/2}.$$

An alternative signal-to-noise ratio equation including the noise introduced by sky subtraction is [36]:

$$S/N = N_{obj}[N_{obj} + n_{pix}(1 + n_{pix}/n_{bg})(N_r^2 + N_{bg} + N_{dc} + N_{dig}^2)]^{-1/2},$$

where N_{obj} is the total number of signal electrons from the object ($= \sum i_s\, t$), n_{pix} is the number of pixels summed for the object, n_{bg} is the number of pixels summed for the sky subtraction, N_r is the read noise in electrons per pixel, N_{bg} is the sky background in electrons per pixel ($= xGi_{bg}t$), N_{dc} is the dark current in electrons per pixel ($= xGi_{dc}t$), and N_{dig} is the digitization noise in electrons per pixel (usually negligible).

Table 7.4. *Far-infrared heterodyne detectors* [1,2].

Type	Wavelength range (μm)
Schottky diode	100–300
Superconducting–insulator–superconducting (SIS)	300–3000

References
1. Phillips, T.G. 1988, in *Millimetre and Submillimetre Astronomy*, edited by R.D. Wolstencroft and W.B. Burton (Kluwer Academic, Dordrecht), p. 1
2. White, G.J. 1988, in *Millimetre and Submillimetre Astronomy*, edited by R.D. Wolstencroft and W.B. Burton (Kluwer Academic, Dordrecht), p. 27

For a heterodyne receiver [37],

$$S/N = (T_S/T_N)\sqrt{t\,\Delta \nu},$$

where T_S is the source temperature (K), T_N is the equivalent Rayleigh–Jeans noise temperature (K) of the receiver, and $\Delta \nu$ (Hz) is the channel width of the radio integrator.

7.5 PHOTOMETRY ($\lambda < 30\ \mu$m)

There is no common infrared photometric (radiometric) system. As a result, filter central wavelengths, filter bandwidths, and instrumental responses are different at each observatory, as are the effects of the atmospheric transmission. The flux density of Vega established by Cohen et al. [38] is presented in Table 7.5. It is based upon an atmospheric model for Vega and the flux density calibration at $0.555\,6\ \mu$m given by Hayes [39]. It is consistent with ground-based absolute flux density measurements to within $\leq 2\sigma$ of the measurement errors.

Table 7.5. *Filter wavelengths, bandwidths, and flux densities for Vega.*[a]

Filter name	λ_{iso}[b] (μm)	$\Delta\lambda$[c] (μm)	F_λ (W m^{-2} μm^{-1})	F_ν (Jy)	N_ϕ (photons s^{-1} m^{-2} μm^{-1})
V	0.5556[d]	...	3.44×10^{-8}	3 540	9.60×10^{10}
J	1.215	0.26	3.31×10^{-9}	1 630	2.02×10^{10}
H	1.654	0.29	1.15×10^{-9}	1 050	9.56×10^{9}
K_s	2.157	0.32	4.30×10^{-10}	667	4.66×10^{9}
K	2.179	0.41	4.14×10^{-10}	655	4.53×10^{9}
L	3.547	0.57	6.59×10^{-11}	276	1.17×10^{9}
L'	3.761	0.65	5.26×10^{-11}	248	9.94×10^{8}
M	4.769	0.45	2.11×10^{-11}	160	5.06×10^{8}
8.7	8.756	1.2	1.96×10^{-12}	50.0	8.62×10^{7}
N	10.472	5.19	9.63×10^{-13}	35.2	5.07×10^{7}
11.7	11.653	1.2	6.31×10^{-13}	28.6	3.69×10^{7}
Q	20.130	7.8	7.18×10^{-14}	9.70	7.26×10^{6}

Notes

[a]Cohen et al. [1] recommend the use of Sirius rather than Vega as the photometric standard for $\lambda > 20$ μm because of the infrared excess of Vega at these wavelengths. The magnitude of Vega depends on the photometric system used, and it is either assumed to be 0.0 mag or assumed to be 0.02 or 0.03 mag for consistency with the visual magnitude.

[b]The infrared isophotal wavelengths and flux densities (except for K_s) are taken from Table 1 of [1], and they are based on the UKIRT filter set and the atmospheric absorption at Mauna Kea. See Table 2 of [1] for the case of the atmospheric absorption at Kitt Peak. The isophotal wavelength is defined by $F(\lambda_{iso}) = \int F(\lambda)S(\lambda)\,d\lambda / \int S(\lambda)\,d\lambda$, where $F(\lambda)$ is the flux density of Vega and $S(\lambda)$ is the (detector quantum efficiency) $\times$ (filter transmission) $\times$ (optical efficiency) $\times$ (atmospheric transmission) [2]. λ_{iso} depends on the spectral shape of the source and a correction must be applied for broadband photometry of sources that deviate from the spectral shape of the standard star [3]. The flux density and λ_{iso} for K_s were calculated here. For another filter, K', at 2.11 μm, see [4].

[c]The filter full width at half maximum.

[d]The wavelength at V is a monochromatic wavelength; see [5].

References

1. Cohen, M. et al. 1992, *AJ*, **104**, 1650
2. Golay, M. 1974, *Introduction to Astronomical Photometry* (Reidel, Dordrecht), p. 40
3. Hanner, M.S., et al. 1984, *AJ*, **89**, 162
4. Wainscoat, R.J., & Cowie, L.L. 1992, *AJ*, **103**, 332
5. Hayes, D.S. 1985, in *Calibration of Fundamental Stellar Quantities*, edited by D.S. Hayes, et al., Proc. IAU Symp. No. 111 (Reidel, Dordrecht), p. 225

Absolute calibration. (a) For 1.2–5 μm, see [40]. (b) For 10–20 μm, see [41].

Photometric systems and standard star observations. For AAO, 1.2–3.8 μm, see [42]; for CIT, 1.2–3.5 μm, see [43]; for ESO, 1.2–3.8 μm, see [44]; for ESO, 1.2–4.8 μm, see [45]; for IRTF, 10–20 μm, see [46]; for KPNO, 1.2–2.2 μm, see [47]; for MSO, 1.2–2.2 μm, see [48]; for OAN, 1.2–2.2 μm, see [49]; for SAAO, 1.2–3.4 μm, see [50]; for UA, IRTF, WIRO, 1.2–20 μm, see [51]; for UKIRT, 1–2.2 μm, faint standards, see [52]; for WIRO, 1.2–33 μm, see [53].

Color transformations. For *JHKLL'M*; SAAO–Johnson, SAAO-ESO, SAAO–AAO, AAO–MSO, AAO–CIT, see [17]; for *JHKLM*; ESO–SAAO; ESO–AAO; ESO–MSSSO; ESO–CTIO, see [44]; for *JHK*; OAN–CIT, OAN–AAO, OAN–ESO, OAN–Johnson, see [49]; for *JHKL*; SAAO–ESO, SAAO–AAO, SAAO–MSSSO, SAAO–CTIO, see [50]; for *JHKL*; CIT–AAO, CIT–SAAO, CIT–Johnson, see [54]; for *JHKLM*; Johnson–ESO, Johnson–SAAO, see [55]; for *JHK*; CIT–IRTF, CIT–UKIRT, CIT–CTIO, CIT–ESO, CIT–KPNO, CIT–HCO, CIT–AAO, CIT–Johnson/Glass, see [56].

Acronyms. AAO = Anglo-Australian Observatory; CIT = California Institute of Technology; CTIO = Cerro Tololo Inter-American Observatory; ESO = European Southern Observatory; HCO =

Harvard College Observatory (Mt. Hopkins); IRTF = NASA Infrared Telescope Facility; KPNO = Kitt Peak National Observatory; MSO = Mt. Stromlo Observatory; MSSSO = Mt. Stromlo/Siding Springs Observatory; OAN = San Pedro Mártir National Observatory; SAAO = South African Astronomical Observatory; UA = University of Arizona; UKIRT = United Kingdom Infrared Telescope; WIRO = Wyoming Infrared Observatory.

Tables 7.6–7.8 give intrinsic colors and effective temperatures for stars.

Table 7.6. *Intrinsic colors and effective temperatures for the main sequence (class* V*).[a]*

Spectral type	$V-K$	$J-H$	$H-K$	$K-L$	$K-L'$	$K-M$	T_{eff}[b]
O9	−0.87	−0.14	−0.04	−0.06			35 900
O9.5	−0.85	−0.13	−0.04	−0.06			34 600
B0	−0.83	−0.12	−0.04	−0.06			31 500
B1	−0.74	−0.10	−0.03	−0.05			25 600
B2	−0.66	−0.09	−0.03	−0.05			22 300
B3	−0.56	−0.08	−0.02	−0.05			19 000
B4	−0.49	−0.07	−0.02	−0.05			17 200
B5	−0.42	−0.06	−0.01	−0.04			15 400
B6	−0.36	−0.05	−0.01	−0.04			14 100
B7	−0.29	−0.03	−0.01	−0.04			13 000
B8	−0.24	−0.03	0.00	−0.04			11 800
B9	−0.13	−0.01	0.00	−0.03			10 700
A0	0.00	0.00	0.00	0.00	0.00	0.00	9 480
A2	0.14	0.02	0.01	0.01	0.01	0.01	8 810
A5	0.38	0.06	0.02	0.02	0.02	0.03	8 160
A7	0.50	0.09	0.03	0.03	0.03	0.03	7 930
F0	0.70	0.13	0.03	0.03	0.03	0.03	7 020
F2	0.82	0.17	0.04	0.03	0.03	0.03	6 750
F5	1.10	0.23	0.04	0.04	0.04	0.02	6 530
F7	1.32	0.29	0.05	0.04	0.04	0.02	6 240
G0	1.41	0.31	0.05	0.05	0.05	0.01	5 930
G2	1.46	0.32	0.05	0.05	0.05	0.01	5 830
G4	1.53	0.33	0.06	0.05	0.05	0.01	5 740
G6	1.64	0.37	0.06	0.05	0.05	0.00	5 620
K0	1.96	0.45	0.08	0.06	0.06	−0.01	5 240
K2	2.22	0.50	0.09	0.07	0.07	−0.02	5 010
K4	2.63	0.58	0.11	0.09	0.10	−0.04	4 560
K5	2.85	0.61	0.11	0.10	0.11		4 340
K7	3.16	0.66	0.15	0.11	0.13		4 040
M0	3.65	0.67	0.17	0.14	0.17		3 800
M1	3.87	0.66	0.18	0.15	0.21		3 680
M2	4.11	0.66	0.20	0.16	0.23		3 530
M3	4.65	0.64	0.23	0.20	0.32		3 380
M4	5.28	0.62	0.27	0.23	0.37		3 180
M5	6.17	0.62	0.33	0.29	0.42		3 030
M6	7.37	0.66	0.38	0.36	0.48		2 850

Notes

[a] Colors given in the Johnson–Glass system as established by Bessell and Brett [1]. References used: O, B, [2]; A, F, G, K, [1]; K, M, [3]. Did not use $K-M$ from [2] because there is a large offset compared to [1]. Approximate uncertainties (one standard deviation): ±0.02 (O−K); ±0.03 (M).

[b] T_{eff} is an average of values from the following sources: for O, B, [4]; for B, A, F, G, K, [5]; for B, G, K, [6]; for A, F, [7]; for A, F, G, K, [8]; for A, F, G, [9]; for G, K, [10]; for K, M, [3]; for M, [11], [7], [12]. Approximate uncertainties (one standard deviation): ±1000 K (O9−B2); ±250 K (B3−B9); ±100 K (A0−M6).

References

1. Bessell, M.S., & Brett, J.M. 1988, *PASP*, **100**, 1134

2. Koornneef, J. 1983, *A&A*, **128**, 84
3. Bessell, M.S. 1991, *AJ*, **101**, 662
4. Vacca, W.D. et al. 1996, *ApJ*, **460**, 914
5. Popper, D.M. 1980, *ARA&A*, **18**, 115
6. Böhm-Vitense, E. 1981, *ARA&A*, **19**, 295
7. Böhm-Vitense, E. 1982, *ApJ*, **255**, 191
8. Blackwell, D.E. et al. 1991, *A&A*, **245**, 567
9. Fernley, J.A. 1989, *MNRAS*, **239**, 905
10. Bell, R.A., & Gustafsson, B. 1989, *MNRAS*, **236**, 653
11. Jones, H.R.A. et al. 1995, *MNRAS*, **277**, 767
12. Leggett, S.K. et al. 1996, *ApJS*, **104**, 117

Table 7.7. *Intrinsic colors and effective temperatures for giant stars (class III).*[a]

Spectral type	$V - K$	$J - H$	$H - K$	$K - L$	$K - L'$	$K - M$	T_{eff}[b]
G0	1.75	0.37	0.07	0.04	0.05	0.00	5 910
G4	2.05	0.47	0.08	0.05	0.06	−0.01	5 190
G6	2.15	0.50	0.09	0.06	0.07	−0.02	5 050
G8	2.16	0.50	0.09	0.06	0.07	−0.02	4 960
K0	2.31	0.54	0.10	0.07	0.08	−0.03	4 810
K1	2.50	0.58	0.10	0.08	0.09	−0.04	4 610
K2	2.70	0.63	0.12	0.09	0.10	−0.05	4 500
K3	3.00	0.68	0.14	0.10	0.12	−0.06	4 320
K4	3.26	0.73	0.15	0.11	0.14	−0.07	4 080
K5	3.60	0.79	0.17	0.12	0.16	−0.08	3 980
M0	3.85	0.83	0.19	0.12	0.17	−0.09	3 820
M1	4.05	0.85	0.21	0.13	0.17	−0.10	3 780
M2	4.30	0.87	0.22	0.15	0.19	−0.12	3 710
M3	4.64	0.90	0.24	0.17	0.20	−0.13	3 630
M4	5.10	0.93	0.25	0.18	0.21	−0.14	3 560
M5	5.96	0.95	0.29	0.20	0.22	−0.15	3 420
M6	6.84	0.96	0.30				3 250
M7	7.80	0.96	0.31				

Notes

[a] Colors given in the Johnson–Glass system as established by Bessell and Brett in [1]. Approximate uncertainties (one standard deviation): ±0.02.

[b] T_{eff} is an average of values from the following sources: for G, K, M, [2]; for K, M, [3]; for G, K, [4]; for G, K, M, [5]. Approximate uncertainties (one standard deviation): ±50 K (G2−K5); ±70 K (M0−M6). For O and B stars, see [6].

References

1. Bessell, M.S., & Brett, J.M. 1988, *PASP*, **100**, 1134
2. Ridgway, S.T. et al. 1980, *ApJ*, **235**, 126
3. Di Benedetto, G.P., & Rabbia, Y. 1987, *A&A*, **188**, 114
4. Bell, R.A., & Gustafsson, B. 1989, *MNRAS*, **236**, 653
5. Blackwell, D.E. et al. 1991, *A&A*, **245**, 567
6. Vacca, W.D. et al. 1996, *ApJ*, **460**, 914

Table 7.8. *Intrinsic colors and effective temperatures for supergiant stars (class I).*[a]

Spectral type	$V - K$	$J - H$	$H - K$	$K - L$	T_{eff}[b]
O9	−0.82	−0.05	−0.13	−0.08	32 500
B0	−0.69	−0.04	−0.10	−0.07	26 000
B1	−0.55	−0.03	−0.06	−0.07	20 700
B2	−0.40	−0.04	0.00	−0.07	17 800
B3	−0.28	−0.03	0.03	−0.05	15 600
B4	−0.20	−0.01	0.01	−0.01	13 900

Table 7.8. *(Continued.)*

Spectral type	$V - K$	$J - H$	$H - K$	$K - L$	T_{eff}[b]
B5	−0.13	0.01	0.00	0.02	13 400
B6	−0.07	0.04	−0.02	0.03	12 700
B7	0.01	0.06	−0.02	0.04	12 000
B8	0.07	0.07	−0.02	0.05	11 200
B9	0.13	0.08	−0.02	0.06	10 500
A0	0.19	0.09	−0.02	0.07	9 730
A1	0.26	0.11	−0.01	0.07	9 230
A2	0.32	0.12	−0.01	0.08	9 080
A5	0.48	0.13	0.02	0.07	8 510
F0	0.64	0.15	0.04	0.06	7 700
F2	0.75	0.18	0.05	0.06	7 170
F5	0.93	0.22	0.06	0.07	6 640
F8	1.21	0.28	0.07	0.07	6 100
G0	1.44	0.33	0.08	0.08	5 510
G3	1.67	0.38	0.09	0.08	4 980
G8	1.99	0.43	0.11	0.09	4 590
K0	2.15	0.46	0.12	0.10	4 420
K1	2.28	0.49	0.13	0.11	4 330
K2	2.43	0.52	0.13	0.12	4 260
K3 Iab	2.90	0.59	0.13	0.15	4 130
K5 Iab	3.50	0.67	0.14	0.18	3 850
M0 Iab	3.80	0.73	0.18	0.20	3 650
M1 Iab	3.90	0.73	0.20	0.22	3 550
M2 Iab	4.10	0.73	0.22	0.24	3 450
M3 Iab	4.60	0.74	0.24	0.26	3 200
M4 Iab	5.20	0.78	0.26	0.28	2 980
M0 Ib	3.80	0.76	0.18	0.12	
M1 Ib	3.90	0.76	0.20	0.14	
M2 Ib	4.10	0.76	0.22	0.16	
M3 Ib	4.60	0.77	0.24	0.18	
M4 Ib	5.20	0.81	0.26	0.20	
M0 Ia	3.80	0.61	0.18	0.27	
M1 Ia	3.90	0.61	0.20	0.29	
M2 Ia	4.10	0.61	0.22	0.31	
M3 Ia	4.60	0.62	0.24	0.33	
M4 Ia	5.20	0.66	0.26	0.35	

Notes

[a] Colors given in the Johnson–Glass system as established by Bessell and Brett [1]. References used: For O, A, [2]; for A, F, G, K, [3]; for K, M, [4]. Approximate uncertainties (one standard deviation): ±0.03.

[b] T_{eff} is an average of values from the following references: For O–M, [5]; for O–K, [6]; for O, B, [7]. Approximate uncertainties (one standard deviation): ±1000 K (O9–B2); ±250 K (B3–B9); ±200 K (A–M).

References

1. Bessell, M.S., & Brett, J.M. 1988, *PASP*, **100**, 1134
2. Whittet, D.C.B., & van Breda, I.G. 1980, *MNRAS*, **192**, 467
3. Koornneef, J. 1983, *A&A*, **128**, 84
4. Elias, J. et al. 1985, *ApJS*, **57**, 91
5. Schmidt-Kaler, Th. 1982, in *Landolt-Börnstein*, New Series, edited by K. Schaifer & H.H. Voigt (Springer-Verlag, Berlin), Vol. VI/2b, p. 451
6. Böhm-Vitense, E. 1981, *ARA&A*, **19**, 295
7. Remie, H., & Lamers, H.J.G.L.M. 1982, *A&A*, **105**, 85

7.6 PHOTOMETRY ($\lambda > 30\ \mu$m)

The primary flux density calibrator for ground-based submillimeter and millimeter observations is Mars [57]. The main secondary calibrators are Uranus [58, 59] and Jupiter [5, 59, 60]. Other secondary calibrators consist of astronomical sources [59, 61].

Instrument details for the IRAS satellite are given in Table 7.9.

Table 7.9. *Infrared Astronomical Satellite (IRAS) summary information.*[a]

Effective wavelength (μm)	12	25	60	100
Bandwidth (FWHM) (μm)	7.0	11.15	32.5	31.5
Typical detector field of view, (in scan) × (cross scan) (arcmin)	0.76 × 4.55	0.76 × 4.65	1.51 × 4.75	3.03 × 5.05
Point Source Catalog, with 2 coverages, 90% completeness limits (Jy)[b]	0.45	...	0.64	...
Faint Source Catalog median 90% completeness limits (Jy)[b]	0.18	0.29	0.26	...

Notes

[a] IRAS observations are sensitive to dust with $T > 25$ K. For IRAS catalogs, see [1, 2].
[b] Completeness limits vary according to the amount of sky coverage obtained.

References

1. Infrared Astronomical Satellite (IRAS) Catalogs and Atlases, 1988, ed. Joint IRAS Science Working Group (U.S. Government Printing Office, Washington, DC), Vols. 1–7
2. The Infrared Processing & Analysis Center (IPAC) WWW Home Page (http://www.ipac.caltech.edu/) has numerous databases and information on IRAS catalogs

The following formulas give the IRAS four-band and two-band fluxes. For galactic sources [62]

$$F_{\mathrm{ir}}(7 - 135\ \mu\mathrm{m}) = 1.0 \times 10^{-14}(20.653\,f_{12} + 7.538\,f_{25} + 4.578\,f_{60} + 1.762\,f_{100})\ \mathrm{W\,m}^{-2}.$$

For extragalactic sources [63, 64]

$$F_{\mathrm{ir}}(8 - 1\,000\ \mu\mathrm{m}) = 1.8 \times 10^{-14}(13.48\,f_{12} + 5.16\,f_{25} + 2.58\,f_{60} + f_{100})\ \mathrm{W\,m}^{-2},$$
$$F_{\mathrm{fir}}(43 - 123\ \mu\mathrm{m}) = 1.26 \times 10^{-14}(2.58\,f_{60} + f_{100})\ \mathrm{W\,m}^{-2},$$

where f_{12}, f_{25}, f_{60}, and f_{100} are the IRAS flux densities in Jy at 12, 25, 60, and 100 μm. These formulas are approximations based on assumptions about the intrinsic source spectrum and dust emissivity. It is recommended that the original references be consulted for details.

The luminosity (in solar luminosities) is

$$L_{\mathrm{ir,fir}} = 3.127 \times 10^7 D^2 F_{\mathrm{ir,fir}} L_\odot,$$

where D is in pc and $F_{\mathrm{ir,fir}}$ is in W m^{-2}.

The far-infrared emission–radio emission correlation [65] is

$$q = \log\{[F_{\mathrm{fir}}/(3.75 \times 10^{12}\ \mathrm{Hz})]/f_{1.4\ \mathrm{GHz}}\} = 2.14,$$

where $f_{1.4\ \mathrm{GHz}}$ is the 1.4 GHz flux density in W m^{-2} Hz^{-1}.

7.7 INFRARED LINE LIST

Table 7.10 presents data for a sample of infrared lines.

Table 7.10. *Selected infrared lines.*

$\lambda\ (\mu m)^a$	$\nu\ (cm^{-1})^a$	Species	Transition[b]	Reference[c]
1.005 21	9 948.17	H I	$n = 7\text{--}3$ (Paδ)	[1, 2, 3]
1.012 64	9 875.18	He II	$n = 5\text{--}4$	[1, 3]
1.083 3	9 231.2	He I	$2p\,^3P^o\text{--}2s\,^3S$	[1, 3]
1.094 11	9 139.85	H I	$n = 6\text{--}3$ (Paγ)	[1, 2, 3]
1.112 86	8 985.84	Fe II	$b\,^4G_{5/2}\text{--}z\,^4F_{3/2}$	[4, 5, 6]
1.129 0	8 857.4	O I	$3d\,^3D^o\text{--}3p\,^3P$	[2, 3, 4]
1.162 96	8 598.75	He II	$n = 7\text{--}5$	[1, 3]
1.167 64	8 564.28	He II	$n = 11\text{--}6$	[1, 7]
1.252	7 987.0	[Si IX]	$^3P_1 - {}^3P_2$	[8, 9, 10, 11]
1.257 02	7 955.30	[Fe II]	$a\,^4D_{7/2}\text{--}a\,^6D_{9/2}$	[4, 5, 6]
1.282 16	7 799.34	H I	$n = 5\text{--}3$ (Paβ)	[1, 2, 3]
1.316 82	7 594.03	O I	$4s\,^3S^o\text{--}3p\,^3P$	[3, 6]
1.476 44	6 773.05	He II	$n = 9\text{--}6$	[1, 3]
1.526 47	6 551.08	H I	$n = 19\text{--}4$ (Br19)	[1, 2, 3]
1.588 48	6 295.29	H I	$n = 14\text{--}4$ (Br14)	[1, 2, 3]
1.611 37	6 205.92	H I	$n = 13\text{--}4$ (Br13)	[1, 2, 3]
1.618 9	6 177.0	CO	$v = 6\text{--}3$ band head	[12]
1.626 46	6 148.32	OH	$v = 2\text{--}0\ P_{1d}(15)$	[12]
1.641 17	6 093.21	H I	$n = 12\text{--}4$ (Br12)	[1, 2, 3]
1.644 00	6 082.73	[Fe II]	$a\,^4D_{7/2}\text{--}a\,^4F_{9/2}$	[4, 5, 6]
1.681 11	5 948.45	H I	$n = 11\text{--}4$ (Br11)	[1, 2, 3]
1.687 78	5 924.94	Fe II	$c\,^4F_{9/2}\text{--}z\,^4F_{9/2}$	[4, 5, 6]
1.692 30	5 909.12	He II	$n = 12\text{--}7$	[1, 7]
1.700 76	5 879.74	He I	$4d\,^3D - 3p\,^3P^o$	[3, 6]
1.736 69	5 758.08	H I	$n = 10\text{--}4$ (Br10)	[1, 2, 3]
1.741 88	5 740.94	Fe II	$c\,^4F_{7/2}\text{--}z\,^4D_{7/2}$	[3, 6]
1.817 91	5 500.82	H I	$n = 9\text{--}4$ (Br9)	[1, 2, 3]
1.875 61	5 331.60	H I	$n = 4\text{--}3$ (Paα)	[2, 3]
1.945 09	5 141.15	H I	$n = 8\text{--}4$ (Brδ)	[1, 2, 3]
1.957 56	5 108.40	H_2	$v = 1\text{--}0\ S(3)$	[13, 14]
1.963 4	5 093.2	[Si VI]	$^2P_{1/2}\text{--}^2P_{3/2}$	[9, 10, 15]
2.033 76	4 917.01	H_2	$v = 1\text{--}0\ S(2)$	[14, 16]
2.040	4 902.0	[Al IX]	$^2P^o_{3/2}\text{--}^2P^o_{1/2}$	[9, 10]
2.040 65	4 900.39	H_3^+	$v = 2\nu_2(2)\text{--}0; (4, 6, +2)\text{--}(3, 3)$	[17]
2.058 69	4 857.45	He I	$2p\,^1P^o\text{--}2s\,^1S$	[3, 18]
2.060 59	4 852.99	Fe II	$c\,^4F_{5/2}\text{--}z\,^4F_{3/2}$	[5, 6, 18]
2.089 38	4 786.11	Fe II	$c\,^4F_{3/2}\text{--}z\,^4F_{3/2}$	[5, 6, 16]
2.093 26	4 777.23	H_3^+	$v = 2\nu_2(2)\text{--}0; (7, 9, +2)\text{--}(6, 6)$	[17]
2.112 7	4 733.4	He I	$4s\,^3S\text{--}3p\,^3P^o$	[3, 6]
2.121 83	4 712.91	H_2	$v = 1\text{--}0\ S(1)$	[14, 16]
2.137 48	4 678.41	Mg II	$5p\,^2P^o_{3/2}\text{--}5s\,^2S_{1/2}$	[18]
2.143 80	4 664.61	Mg II	$5p\,^2P^o_{1/2}\text{--}5s\,^2S_{1/2}$	[18]
2.166 12	4 616.55	H I	$n = 7\text{--}4$ (Brγ)	[2, 3, 16]
2.189 11	4 568.07	He II	$n = 10\text{--}7$	[1, 7]
2.206 24	4 532.59	Na I	$4p\,^2P^o_{3/2}\text{--}4s\,^2S_{1/2}$	[16, 19, 20]
2.208 97	4 527.00	Na I	$4p\,^2P^o_{1/2}\text{--}4s\,^2S_{1/2}$	[16, 19, 20]
2.223 29	4 497.84	H_2	$v = 1\text{--}0\ S(0)$	[14, 16]
2.247 72	4 448.96	H_2	$v = 2\text{--}1\ S(1)$	[14, 16]
2.263 11	4 418.69	Ca I	$4f\,^3F^o_3\text{--}4d\,^3D_2$	[19]

Table 7.10. *(Continued.)*

λ (μm)[a]	ν (cm^{-1})[a]	Species	Transition[b]	Reference[c]
2.265 73	4 413.58	Ca I	$4f\,^3F_4^o$–$4d\,^3D_3$	[19]
2.293 53	4 360.09	CO	$v = 2$–0 band head	[16]
2.322 65	4 305.42	CO	$v = 3$–1 band head	[16]
2.343 27	4 267.54	CO	$v = 2$–0 R(1)	[21]
2.345 31	4 263.84	CO	$v = 2$–0 R(0)	[21]
2.349 50	4 256.22	CO	$v = 2$–0 P(1)	[21]
2.351 67	4 252.30	CO	$v = 2$–0 P(2)	[21]
2.352 46	4 250.87	CO	$v = 4$–2 band head	[16]
2.382 95	4 196.48	CO	$v = 5$–3 band head	[16]
2.406 59	4 155.25	H$_2$	$v = 1$–0 Q(1)	[13, 14]
2.413 44	4 143.47	H$_2$	$v = 1$–0 Q(2)	[13, 14]
2.423 73	4 125.87	H$_2$	$v = 1$–0 Q(3)	[13, 14]
2.483 3	4 026.9	[Si VII]	$^3P_1$–$^3P_2$	[9, 10, 15]
2.499 95	4 000.08	H$_2$	$v = 1$–0 Q(7)	[14, 22]
2.625 87	3 808.26	H I	$n = 6$–4 (Brβ)	[2, 22]
2.626 88	3 806.80	H$_2$	$v = 1$–0 O(2)	[14, 22]
3.027 9	3 302.6	[Mg VIII]	$^2P_{3/2}^o$–$^2P_{1/2}^o$	[8, 9, 10, 15]
3.039 20	3 290.34	H I	$n = 10$–5 (Pfϵ)	[2]
3.091 69	3 234.48	He II	$n = 7$–6	[3]
3.133	3 192.0	OH	$v = 1$–0, K=9 multiplet	[23]
3.296 99	3 033.07	H I	$n = 9$–5 (Pfδ)	[2, 24]
3.418 84	2 924.97	He II	$n = 25$–11	[3, 24]
3.484 01	2 870.26	He II	$n = 17$–10	[3, 24]
3.501 16	2 856.20	H I	$n = 24$–6 (Hu24)	[2, 24]
3.522 03	2 839.27	H I	$n = 23$–6 (Hu23)	[2, 24]
3.624 6	2 758.9	H$_2$	$v = 0$–0 S(15)	[25, 26]
3.645 92	2 742.79	H I	$n = 19$–6 (Hu19)	[2, 25]
3.661	2 731.0	[Al VI]	$^3P_1 - \,^3P_2$	[8–10]
3.692 63	2 708.10	H I	$n = 18$–6 (Hu18)	[2, 25]
3.724 0	2 685.3	H$_2$	$v = 0$–0 S(14)	[26]
3.740 56	2 673.40	H I	$n = 8$–5 (Pfγ)	[2]
3.807 41	2 626.46	H$_2$	$v = 1$–0 O(7)	[14, 27]
3.846 2	2 600.0	H$_2$	$v = 0$–0 S(13)	[26, 27]
3.935	2 541	[Si IX]	$^3P_1$–$^3P_0$	[9, 10, 15]
3.953 00	2 529.72	H$_3^+$	$v = v_2(1)$–0; (1, 0, −1)–(1, 0)	[28]
4.004 5	2 497.2	SiO	$v = 2 - 0$ band head	[29]
4.020 87	2 487.02	H I	$n = 14$–6 (Hu14)	[2, 16]
4.037 81	2 476.59	He I	$5f\,^3F^o$–$4d\,^3D$	[3, 30]
4.049 00	2 469.75	He I	$5g\,^1G$–$4f\,^1F^o$; $5g\,^3G$–$4f\,^3F^o$	[30]
4.052 26	2 467.76	H I	$n = 5$–4 (Brα)	[2, 16]
4.170 79	2 397.63	H I	$n = 13$–6	[2, 16]
4.649 31	2 150.86	CO	$v = 1$–0 R(1)	[31]
4.653 78	2 148.79	H I	$n = 7$–5 (Pfβ)	[2, 16]
4.657 48	2 147.08	CO	$v = 1$–0 R(0)	[31]
4.674 15	2 139.43	CO	$v = 1$–0 P(1)	[31]
4.682 62	2 135.55	CO	$v = 1$–0 P(2)	[31]
4.694 62	2 130.10	H$_2$	$v = 0$–0 S(9)	[14]
5.053 1	1 979.0	H$_2$	$v = 0$–0 S(8)	[14, 26]
6.634	1 507	[Ni II]	$a\,^2D_{3/2}$–$a\,^2D_{5/2}$	[32]
6.985	1 432	[Ar II]	$^2P_{1/2}^o$–$^2P_{3/2}^o$	[33]
7.642	1 309	[Ne VI]	$^2P_{3/2}$–$^2P_{1/2}$	[9, 10]
8.991 35	1 112.18	[Ar III]	$^3P_1$–$^3P_2$	[33, 34]
10.51	951.5	[S IV]	$^2P_{3/2}^o$–$^2P_{1/2}^o$	[33]
10.521	950.48	[Co II]	$a\,^3F_3$–$a\,^3F_4$	[11, 32]
12.278 6	814.425	H$_2$	$v = 0$–0 S(2)	[14]
12.372 0	808.283	H I	$n = 7$–6 (Huα)	[2]

Table 7.10. *(Continued.)*

$\lambda\ (\mu m)^a$	$\nu\ (cm^{-1})^a$	Species	Transition[b]	Reference[c]
12.8135	780.424	[Ne II]	$^2P^o_{1/2}$–$^2P^o_{3/2}$	[33, 34]
15.56	642.7	[Ne III]	$^3P_1$–$^3P_2$	[33]
17.0348	587.032	H_2	$v = 0$–0 S(1)	[14]
18.7130	534.387	[S III]	$^3P_2$–$^3P_1$	[34, 35]
24.3158	411.256	[Ne V]	$^3P_1$–$^3P_0$	[34, 35]
25.87	386.5	[O IV]	$^2P_{3/2}$–$^2P_{1/2}$	[33]
28.2188	354.374	H_2	$v = 0$–0 S(0)	[14]
33.482	298.67	[S III]	$^3P_1$–$^3P_0$	[35, 36]
51.816	192.99	[O III]	$^3P_2$–$^3P_1$	[33, 35]
57.317	174.47	[N III]	$^2P^o_{3/2}$–$^2P^o_{1/2}$	[33]
63.1837	158.269	[O I]	$^3P_1$–$^3P_2$	[37]
77.059	129.77	CO	$J = 34$–33	[37]
88.355	113.18	[O III]	$^3P_1$–$^3P_0$	[33, 35]
119.23	83.872	OH	$^2\Pi_{3/2}\ J = 5/2$–$3/2$	[37]
119.44	83.724	OH	$^2\Pi_{3/2}\ J = 5/2$–$3/2$	[37]
121.898	82.0358	[N II]	$^3P_2$–$^3P_1$	[33, 38]
124.65	80.225	NH_3	$K = 3, J = 4$–$3, a - s$	[37]
145.526	68.7162	[O I]	$^3P_0$–$^3P_1$	[33]
157.741	63.3951	[C II]	$^2P^o_{3/2}$–$^2P^o_{1/2}$	[33]
162.81	61.421	CO	$J = 16$–15	[37]
205.178	48.7382	[N II]	$^3P_1$–$^3P_0$	[38]
370.415	26.9967	[C I]	$^3P_2$–$^3P_1$	[33]
371.65	26.907	CO	$J = 7$–6	[39]
609.135	16.4167	[C I]	$^3P_1$–$^3P_0$	[33]

Notes

[a] Vacuum wavelengths and frequencies are given.

[b] Transition shown is (upper level)–(lower level).

[c] Because of space limitations, only a few transitions of each species are shown; see references for additional lines. Wavelength and frequencies were calculated or obtained from primary references where possible. For additional information, see [40–45].

References

1. Treffers, R.R. et al. 1976, *ApJ*, **209**, 793
2. Moore, C.E. 1993, in *Tables of Spectra of Hydrogen, Carbon, Nitrogen, and Oxygen Atoms and Ions*, edited by J.W. Gallagher (CRC, Boca Raton, FL); van Hoof, P.A.M., private communication
3. Bashkin, S., & Stoner, J.O. 1975, *Atomic Energy Levels and Grotian Diagrams* (North-Holland, Amsterdam); van Hoof, P.A.M., private communication
4. Allen, D.A. et al. 1985, *ApJ*, **291**, 280
5. Johansson, S. 1978, *Phys. Scr.*, **18**, 217
6. Hamann, F. et al. 1994, *ApJ*, **422**, 626
7. Moore, C.E. 1971, *Atomic Energy Levels*, NSRDS-NBS Publication No. 35; van Hoof, P.A.M., private communication
8. Woodward, C.E. et al. 1995, *ApJ*, **438**, 921
9. Greenhouse, M.A. et al. 1993, *ApJS*, **88**, 23
10. Gehrz, R.D. 1988, *ARA&A*, **26**, 377
11. Roche, P.F. et al. 1993, *MNRAS*, **261**, 522
12. Hinkle, K.H. 1978, *ApJ*, **220**, 210
13. Gautier III, T.H. et al. 1976, *ApJ*, **207**, L129
14. Black, J.H., & van Dishoeck, E.F. 1987, *ApJ*, **322**, 412
15. Reconditi, M., & Oliva, E. 1993, *A&A*, **274**, 662; Oliva, E. et al. 1994, *A&A*, **288**, 457
16. Scoville, N. et al. 1983, *ApJ*, **275**, 201
17. Drossart, P. et al. 1989, *Nature*, **340**, 539; see also Kao, L. et al. 1991, *ApJS*, **77**, 317
18. Simon, M., & Cassar, L. 1984, *ApJ*, **283**, 179
19. Kleinman, S.G., & Hall, D.N.B. 1986, *ApJS*, **62**, 501

20. Martin, W.C., & Zalubas, R. 1981, *J. Phys. Chem. Ref. Data Ser.*, **10**, 153
21. Black, J.H., & Willner, S.P. 1984, *ApJ*, **279**, 673
22. Davis, D.S. et al. 1982, *ApJ*, **259**, 166
23. Beer, R. et al. 1972, *ApJ*, **172**, 89
24. Lowe, R.P. et al. 1991, *ApJ*, **368**, 195
25. Sanford, S.A. 1991, *ApJ*, **376**, 599
26. Knacke, R.F., & Young, E.T. 1981, *ApJ*, **249**, L65
27. Brand, P.W.J.L. et al. 1989, *MNRAS*, **236**, 929
28. Oka, T., & Geballe, T.R. 1990, *ApJ*, **351**, L53
29. Hinkle, K.H. et al. 1976, *ApJ*, **210**, L141
30. Hamann, F., & Simon, M. 1986, *ApJ*, **311**, 909
31. Guelachvili, G. 1979, *J. Mol. Spectrosc.*, **75**, 251; Mitchell, G.F. et al. 1989, *ApJ*, **341**, 1020
32. Wooden, D.H. et al. 1993, *ApJS*, **88**, 477
33. Genzel, R. 1988, in *Millimetre and Submillimetre Astronomy*, edited by R.D. Wolstencroft and W.B. Burton (Kluwer Academic, Dordrecht), p. 223
34. Kelly, D.M., & Lacy, J.H. 1995, *ApJ*, **454**, L161
35. Emery, R.J., & Kessler, M.F. 1984, in *Galactic and Extragalactic Infrared Spectroscopy*, edited by M.F. Kessler and J.P. Phillips (Reidel, Dordrecht), p. 289
36. Stacy, G.J. et al. 1993, *Proc. SPIE*, **1946**, 238
37. Watson, D.M. 1984, in *Galactic and Extragalactic Infrared Spectroscopy*, edited by M.F. Kessler and J.P. Phillips (Reidel, Dordrecht), p. 195; Townes, C.H., & Melnick, G. 1990, *PASP*, **102**, 357
38. Colgan, S.W.J. et al. 1993, *ApJ*, **413**, 237
39. Howe, J.E. et al. 1993, *ApJ*, **410**, 179
40. H: Wynn-Williams, C.G. 1984, in *Galactic and Extragalactic Infrared Spectroscopy*, edited by M.F. Kessler and J.P. Phillips (Reidel, Dordrecht), p. 133
41. H_2: Schwartz, R.D. et al. 1987, *ApJ*, **322**, 403; Black, J.H., & van Dishoeck, E.F. 1987, *ApJ*, **322**, 412
42. CO: Goorvitch, D. 1994, *ApJS*, **95**, 535
43. Solar atlases: Livingston, W., & Wallace, L. 1991, *An Atlas of the Solar Spectrum in the Infrared from 1850 to 9000 cm^{-1} (1.1–5.4 μm)*, NSO Technical Report No. 91-001 (NOAO, Tucson); Wallace, L., & Livingston, W. 1992, *An Atlas of a Dark Sunspot Umbral Spectrum from 1970 to 8640 cm^{-1} (1.16–5.1 μm)*, NSO Technical Report No. 92-001 (NOAO, Tucson)
44. Infrared spectra: Jourdain de Muizon, M. et al. 1994, *Database of Astronomical Infrared Spectroscopic Observations* (University of Leiden, Leiden)
45. Infrared wavelength calibration: Outred, M. 1978, *J. Phys. Chem. Ref. Data Ser.*, **7**, 1; Rao, K.N. et al. 1966, *Wavelength Standards in the Infrared* (Academic Press, New York)

7.8 DUST

For the infrared interstellar reddening law, see [66–69].

The total to selective absorption ([66–68], for $R = A_V/\mathrm{E}(B-V) = 3.1$) is

$$A_V/\mathrm{E}(J-K) = 5.82 \pm 0.1, \qquad A_V/\mathrm{E}(H-K) = 15.3 \pm 0.6,$$
$$A_V/\mathrm{E}(V-K) = 1.13 \pm 0.03,$$
$$A_\lambda/\mathrm{E}(J-K) = 2.4(\lambda)^{-1.75} \quad \text{(for } 0.9 < \lambda < 6 \ \mu\text{m)}.$$

The color excess ratio [67] is

$$\mathrm{E}(J-H)/\mathrm{E}(H-K) = 1.70 \pm 0.05.$$

The ratio of visual extinction to silicate band optical depth (τ_{Si}) [68, 70, 71] is

$$A_V/\tau_{\mathrm{Si}} = 19 \pm 1 \quad \text{(local interstellar medium)},$$
$$A_V/\tau_{\mathrm{Si}} = 11 \pm 2 \quad \text{(Galactic Center region)}.$$

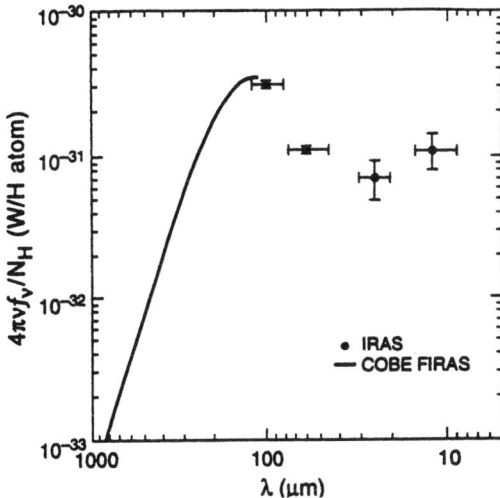

Figure 7.5. Emission spectrum of interstellar dust. Adapted from [78]. See also [26, 79, 80].

The average visual extinction to the Galactic Center region is 34 mag [72] and to individual sources it ranges from 23 to 35 mag [67].

The extinction cross section per H nucleus in the local interstellar medium [68] is

$$N_H/E(J - K) = 1.1 \times 10^{22} \text{ nuclei cm}^{-2} \text{ mag}^{-1}.$$

The interstellar linear polarization [73–75]:

$$P(\lambda)/P_{max} = \exp[-K \ln^2(\lambda_{max}/\lambda)] \quad \text{(for } \lambda < 2 \text{ } \mu\text{m)},$$
$$P(\lambda) \propto \lambda^{-\beta}, \quad \beta = 1.6 - 2.0 \quad \text{(for } 2 < \lambda < 5 \text{ } \mu\text{m)},$$

where $P(\lambda)$ is the percentage polarization, P_{max} is the maximum percentage polarization occurring at λ_{max}, and $K = 0.01 \pm 0.05 + (1.66 \pm 0.09)\lambda_{max}$.

Table 7.11 and Figure 7.5 present data on the interstellar dust emission. Table 7.12 presents far-infrared dust properties.

Table 7.11. *Average galactic diffuse emission* [1].[a]

λ (μm)	νI_ν (10^{-7} W m^{-2} sr^{-1})	λ (μm)	νI_ν (10^{-7} W m^{-2} sr^{-1})
3.5	0.21	60	0.88
4.9	0.13	100	2.0
12	0.80	140	3.8
25	0.41	240	2.5

Note

[a]For galactic latitudes $-6°$ to $-4°$ and $+4°$ to $+6°$. Emission is highly variable on small spatial scales [1, 2].

References

1. Bernard, J.P. et al. 1994, *A&A*, **291**, L5
2. Cutri, R.M., & Latter, W.B., editors, 1993, *The First Symposium on the Infrared Cirrus and Diffuse Interstellar Clouds*, ASP Conf. Ser. (ASP, San Francisco), Vol. 58

The dust mass estimate from the 100 μm flux density is

$$M_{\text{dust}} = 4.81 \times 10^{-12} f_{100} D^2 (e^{143.88/T_d} - 1) \, M_\odot,$$

where f_{100} is the 100 μm flux density in Jy, D is the distance in pc, and T_d is the dust temperature in K. The derivation follows from [76], using a mass absorption coefficient of 2.5 m^2 kg^{-1} at 100 μm. The dust mass absorption coefficient at submillimeter wavelengths is estimated in [68, 76, 77].

The equilibrium dust temperature of a particle with albedo A at a distance r (in pc) from a source of luminosity L (in $L_\odot$) is

$$T_e = 0.612(1 - A)^{0.25} L^{0.25} r^{-0.5} \text{ K}.$$

The nonequilibrium emission from extremely small particles is discussed in [81–83].

Table 7.12. *Galactic dust properties at 140–240 μm. Mean values in the galactic plane ($|b| < 1°$) [1].*[a]

Quantity	Inner galaxy ($270° < \ell < 350°$; $10° < \ell < 90°$)	Outer galaxy ($90° < \ell < 270°$)	Entire galaxy
Dust temperature (K)	20 ± 1	17 ± 1	19 ± 1
240 μm optical depth	$(5.0 \pm 2.0) \times 10^{-3}$	$(9.5 \pm 3.0) \times 10^{-4}$	$(3.0 \pm 1.0) \times 10^{-3}$
Total FIR radiance (W m^{-2} sr^{-1})	$(3.7 \pm 0.3) \times 10^{-5}$	$(2.4 \pm 0.2) \times 10^{-6}$	$(2.0 \pm 0.2) \times 10^{-5}$
Gas-to-dust ratio	140 ± 50	190 ± 60	160 ± 60
FIR luminosity per H mass ($L_\odot / \mathcal{M}_\odot$)	3.0 ± 0.3	0.9 ± 0.1	2.0 ± 0.2

Note
[a] Data from the Cosmic Background Explorer (COBE) satellite; for additional information, see the COBE WWW Home Page: http://www.gsfc.nasa.gov/astro/cobe/cobe_home.html

Reference
1. Sodroski, T.J. et al. 1994, *ApJ*, **428**, 638

Spectral features of dust and ice in the infrared are listed in Table 7.13.

Table 7.13. *Major dust and ice features* [1–7].

λ (μm)	Identification	Where observed
3.08	H$_2$O ice	Molecular clouds; OH–IR stars
3.29, 6.2, 7.7, 8.65, 11.25	Aromatic hydrocarbons[a]	H II regions, planetary nebulae, reflection nebulae, young and evolved stars, starburst galaxies
4.62	"X–CN"	Molecular clouds
4.67	CO ice	Molecular clouds
6.0	H$_2$O ice	Molecular clouds
6.85	CH$_3$OH + other	Molecular clouds
~ 9.7	Amorphous silicates	H II regions, molecular clouds
~ 11.2	SiC	Circumstellar shells; planetary nebulae
11.5	H$_2$O ice	OH–IR stars
~ 18	Amorphous silicates	H II regions; Galactic center
~ 34	MgS (?)	Planetary nebulae; carbon stars
43	H$_2$O ice	OH–IR stars

Note
[a] The nature of the "aromatic hydrocarbons" is not known precisely [7]; it is commonly assumed to be polycyclic aromatic hydrocarbons (PAHs).

References

1. Willner, S.P. 1984, in *Galactic and Extragalactic Infrared Spectroscopy*, edited by M.F. Kessler and J.P. Phillips (Reidel, Dordrecht), p. 37
2. Roche, P.F. 1989, in *Proc. 22nd ESLAB Symp. on Infrared Spectroscopy in Astronomy*, ESA SP-290, p. 79
3. Tokunaga, A.T., & Brooke, T.Y. 1990, *Icarus*, **86**, 208
4. Whittet, D.C.B. 1992, *Dust in the Galactic Environment* (Institute of Physics, Bristol), p. 147
5. Allamandola, L.J. et al. 1989, *ApJS*, **71**, 733
6. Léger, A., & d'Hendecourt, L. 1987, in *Polycyclic Aromatic Hydrocarbons and Astrophysics*, edited by A. Léger et al. (Reidel, Dordrecht), p. 223
7. Sellgren, K. 1994, in *The First Symposium on the Infrared Cirrus and Diffuse Interstellar Clouds*, edited by R.M. Cutri and W.B. Latter, ASP Conf. Ser. (ASP, San Francisco), Vol. 58, p. 243

7.9   SOLAR SYSTEM

The solar colors are [84]

$$J - H = 0.310, \quad H - K = 0.060, \quad K - L = 0.034, \quad L - M = -0.053, \quad V - K = 1.486.$$

Solar analogs [85] are 16 Cyg B, VB64, HD 105590, HR 2290.
The blackbody temperature of an object without an atmosphere in the solar system is

$$T_b = 278.8(1 - A)^{0.25} r^{-0.5} \text{ K},$$

where A is the albedo and r is the distance from the Sun in AU.

For thermal emission from asteroids, see [86–88].
For the infrared spectra of planetary atmospheres, see [89–92].
For the infrared spectra of comets, see [93, 94].
For near-infrared spectra of satellites, see [95, 96].
For near-infrared spectra of asteroids, see [97, 98].
The infrared magnitudes and colors of many solar system objects are given in Table 7.14.

Table 7.14. *Magnitudes of selected solar system bodies.*[a]

Object	Ref.	$V(1, 0)$[b]	ΔV[c]	$V - J$	$J - H$	$H - K$	$K - L$	$V - N$	$V - Q$	T (K)[d]
J1 Io	[1–4]	−1.68	0.15	1.3	0.35	0.08	0.00	4.70	9.29	137[e]
J2 Europa (L)	[1–5]	−1.37	0.3	1.2	−0.31	−0.35	−2.24	3.91	8.81	130[e]
J2 Europa (T)	[1–5]			1.4	−0.37	−0.53	−2.35			
J3 Ganymede (L)	[1–5]	−2.08	0.15	1.0	−0.10	−0.08	−1.90	5.69	10.26	142[e]
J3 Ganymede (T)	[1–5]				−0.07	−0.07	−1.44			
J4 Callisto	[1–5]	−0.95	0.13	1.5	−0.27	0.07	−1.01	7.26	11.72	152[e]
S2 Enceladus	[6–8]	1.9	0.5	1.06	−0.05	−0.24	< −0.5			
S3 Tethys	[4, 6, 7]	0.7	0.1	0.9	−0.20	−0.16				
S4 Dione	[4, 6, 7]	0.88	0.3	0.8	−0.20	−0.12				
S5 Rhea	[4, 5, 8, 9]	0.1	0.2	1.06	−0.05	−0.24	−1.6		8.5	
S6 Titan	[4, 10–13]	−1.3	0.0	0.2	−0.31	−0.38	−1.7	6.3		76[f]
S8 Iapetus (L)	[13–15]	2.4		1.60	0.4	0.05				
S8 Iapetus (T)	[13–15]	0.6		0.8	−0.11	−0.13				
U1 Ariel	[4, 16]	1.7		1.20	0.21	−0.04				
U2 Umbriel	[7, 9]	2.4		1.30	0.25	−0.09			10.4	
U3 Titania	[4, 7, 9]	1.3		1.30	0.20	−0.14			10.0	
U4 Oberon	[7, 9]	1.6		1.35	0.20	−0.14			10.4	

Table 7.14. *(Continued.)*

Object	Ref.	$V(1, 0)^b$	ΔV^c	$V - J$	$J - H$	$H - K$	$K - L$	$V - N$	$V - Q$	T (K)^d
N1 Triton	[5, 8, 17, 18]	−1.0		1.3	0.31	−0.24			> 8.2	38^d
Pluto, Charon	[17, 19–21]	−0.76	0.30	1.3	−0.01	−0.36			> 9.9	55^g
1 Ceres	[22–28]	3.72	0.04	1.2	0.31	0.05		10.0	12.8	245^h
2 Pallas	[22–28]	4.45	0.16	1.2	0.21	0.04		9.9	12.4	270^h
3 Juno	[22–28]	5.73	0.22			0.05		8.7	12.0	230^h
4 Vesta	[22–28]	3.55	0.12	1.4	0.17	0.01		8.4	11.2	250^h

Notes

^{a}Average magnitude given unless indicated otherwise; (L) = leading hemisphere, (T) = trailing hemisphere. Approximate filter wavelengths: V (0.55 μm), J (1.25 μm), H (1.65 μm), K (2.2 μm), L (3.45 μm), N (10 μm), Q (20 μm); see references for details.

$^b V(1, 0)$ = absolute visual magnitude at a distance of 1 AU from the Earth and 1 AU from the Sun at 0° phase angle. The apparent visual magnitude of an object is $V(r, \Delta, \alpha) = V(1, 0) + C\alpha + 5\log(r\Delta)$, where r is the heliocentric distance and Δ is the geocentric distance (both in AU), C is the phase coefficent in mag deg^{-1}, and α is the phase angle (deg). The opposition effect, occurring when $\alpha \approx 0°$, is not included in this table.

$^c \Delta V$ = visual light curve amplitude (peak to peak).

$^d T_B$ = brightness temperature; T_S = surface or subsolar temperature.

$^e T_B$ (10 μm).

$^f T_B$ (100 μm).

$^g T_B$ (60 μm).

$^h T_S$ (10 μm).

References

1. Morrison, D. et al. 1976, *ApJ*, **207**, L213
2. Morrison, D. 1977, in *Planetary Satellites*, edited by J.A. Burns (University of Arizona, Tuscon), p. 269
3. Morrison, D., & Morrison, N.D. 1977, in *Planetary Satellites*, edited by J.A. Burns (University of Arizona, Tuscon), p. 363
4. Morrison, D., & Cruikshank, D.P. 1974, *SSRv*, **15**, 641
5. Hartmann, W.K. et al. 1982, *Icarus*, **52**, 377
6. Franz, O.G., & Millis, R.L. 1975, *Icarus*, **24**, 433
7. Cruikshank, D.P. 1980, *Icarus*, **41**, 246, and private communication
8. Cruikshank, D.P. et al. 1977, *ApJ*, **217**, 1006
9. Brown, R.H. et al. 1982, *Nature*, **300**, 423
10. Andersson, L.E. 1977, in *Planetary Satellites*, edited by J.A. Burns (University of Arizona, Tucson), p. 451
11. Noll, K.S., & Knacke, R.F. 1993, *Icarus*, **101**, 272
12. Gillett, F.C. 1975, *ApJ*, **201**, L41
13. Loewenstein, R.F. et al. 1980, *Icarus*, **43**, 283
14. Cruikshank, D.P. et al. 1983, *Icarus*, **53**, 90
15. Cruikshank, D.P. et al. 1979, *Rev. Geophys. Space Phys.*, **17**, 165
16. Nicholson, P.D., & Jones, T.J. 1980, *Icarus*, **42**, 54
17. Morrison, D. et al. 1982, *Nature*, **300**, 425
18. Tryka, K.A. et al. 1993, *Science*, **261**, 751
19. Binzel, R.P., & Mulholland, J.D. 1984, *AJ*, **89**, 1759
20. Soifer, B.T. et al. 1980, *AJ*, **85**, 166
21. Jewitt, D.C. 1994, *AJ*, **107**, 372
22. Veeder, G.J. et al. 1978, *AJ*, **83**, 651, and private communication
23. Lagerkvist, C.-I. et al. 1989, in *Asteroids* II, edited by R.P. Binzel et al. (University of Arizona, Tucson), p. 1162
24. Tedesco, E. 1979, in *Asteroids*, edited by T. Gehrels (University of Arizona, Tuscon), p. 1098
25. Johnson, T.V. et al. 1975, *ApJ*, **197**, 527
26. Morrison, D. 1974, *ApJ*, **194**, 203
27. Lebofsky, L.A. et al. 1986, *Icarus*, **68**, 239
28. McCheyne, R.S. et al. 1985, *Icarus*, **61**, 443

7.10 STARS

Molecular features seen in cool stars are listed in Table 7.15.

Table 7.15. *Molecular bands in cool stars* [1, 2].

Molecule	Bands	Wavelength range (μm)	Selected references
CO	$\Delta v = 1, 2, 3$	1.5–4.7	[3, 4, 5, 6, 7, 8, 9]
H_2	$\Delta v = 1$ (quadrapole vib-rot)	1.7–2.5	[3]
H_2O	$v_3, 2v_2, v_2 + v_3 - v_2,$ $v_2 + v_3, v_1 + v_2$	1.3–3.6	[10, 11]
CN	$A\,^2\Pi - X\,^2\Sigma$	< 4	[3, 4, 6, 12, 13, 14, 15]
C_2	$b\,^1\Pi_u - x\,^1\Sigma_g^+$ (Phillips) $A'\,^3\Sigma_g^- - X'\,^3\Pi_u$ (Ballik–Ramsey)	< 2.5	[3, 6, 14, 16]
C_3, C_5	v_3	4–5	[12, 17, 18]
HCN	$v_2, v_3, 2v_2, 3v_2, 2v_1 + v_2$	2–5, 7.1, 14	[13, 15, 16, 19]
C_2H_2	$v_3, v_5, v_1 + v_5$	2.5–4, 14	[13, 16, 19]
SiO	$\Delta v = 1, 2$	4–4.2, 8.0–8.3	[9, 20, 21, 22, 23]
OH	$\Delta v = 1, 2$	1.6–2.0, 3.1–4.0	[8, 22, 24]
CH	$\Delta v = 1$	3.3–4.0	[3, 22]
CS	$\Delta v = 2$	3.8–4.0	[22, 23]

References

1. Merrill, K.M., & Ridgway, S.T. 1979, *ARA&A*, **17**, 9
2. Tsuji, T. 1986, *ARA&A*, **24**, 89
3. Lambert, D.L. et al. 1986, *ApJS*, **62**, 373
4. Thompson, R.I. et al. 1972, *PASP*, **84**, 779
5. Ridgway, S.T. et al. 1974, in *Highlights of Astronomy*, edited by G. Contopoulos (Reidel, Dordrecht), Vol. 3, p. 327
6. Querci, M., & Querci, F. 1975, *A&A*, **42**, 329
7. Geballe, T.R. et al. 1977, *PASP*, **89**, 840
8. Hinkle, K.H. 1978, *ApJ*, **220**, 210
9. Cohen, M. et al. 1992, *AJ*, **104**, 2045
10. Strecker, D.W. et al. 1978, *AJ*, **83**, 26
11. Hinkle, K.H., & Barnes, T.G. 1979, *ApJ*, **227**, 923
12. Goebel, J.H. et al. 1978, *ApJ*, **222**, L129
13. Goebel, J.H. et al. 1981, *ApJ*, **246**, 455
14. Goebel, J.H. et al. 1983, *ApJ*, **270**, 190
15. Wiedemann, G.R. et al. 1991, *ApJ*, **282**, 321
16. Goebel, J.H. et al. 1980, *ApJ*, **235**, 104
17. Hinkle, K.H. et al. 1988, *Science*, **241**, 1319
18. Bernath, P.F. et al. 1989, *Science*, **244**, 562
19. Ridgway, S.T. et al. 1978, *ApJ*, **255**, 138
20. Geballe, T.R. et al. 1979, *ApJ*, **230**, L47
21. Rinsland, C.P., & Wing, R.F. 1982, *ApJ*, **262**, 201
22. Ridgway, S.T. et al. 1984, *ApJS*, **54**, 177
23. Lambert, D.L. et al. 1990, *AJ*, **99**, 1612
24. Beer, R. et al. 1972, *ApJ*, **172**, 89

For the spectrophotometry of standard stars, see [99–102].

For the infrared star count models, see [103–105].

Useful catalogs are found in [106–109].

For near-infrared spectra of young stars, see [110–118].

For spectral energy distributions of young stellar objects and pre–main sequence stars, see [119–124].

Figure 7.6 shows the color–color diagram for stars.

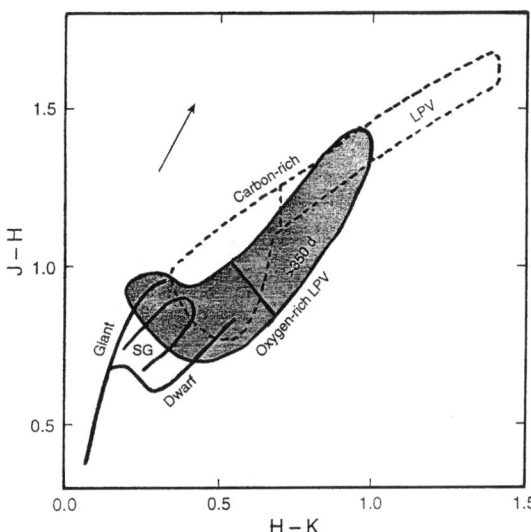

Figure 7.6. Color–color diagram for various classes of stars, adapted from [17]. The dark line indicates the location of G5 to M6 main sequence dwarf and giant stars. The dashed lines indicate the boundary for most carbon-rich stars; the carbon long-period variable (LPV) stars lie to the right. The oxygen-rich (M type) LPV stars fall within the boundary of the solid line, and the LPV stars with periods greater than 350 days are to the right and overlap the carbon-rich LPV stars. The supergiant M stars (SG) lie in a region below and to the right of the giant sequence. The arrow indicates the direction of the interstellar reddening.

7.11 EXTRAGALACTIC OBJECTS

7.11.1 Energy Distributions and Colors

Infrared energy distributions of galaxies vary widely. Representative examples may be found in [125, 126]. At least five different physical causes have been identified for the continuum infrared emission from galaxies:

(a) Photospheric emission from evolved stars (usually dominant in the 1–3 μm region) [127, 128]: Mean colors of elliptical galaxies (CIT photometric system): $V–K = 3.33$ mag; $J–H = 0.69$ mag; $H–K = 0.21$ mag. Molecular absorption bands in elliptical galaxies H_2O (1.95 μm) = 0.12 mag; CO (2.3 μm) = 0.16 mag. For additional near-infrared colors, see [129–132].

(b) Dust shells around evolved stars [133]: This is the main cause of 10–12 μm emission in elliptical galaxies, for which $f_\nu(12\ \mu m) = 0.13 f_\nu(2.2\ \mu m)$. Units of f_ν are Jy.

(c) Emission from interstellar dust [134, 135]: Transiently heated "small" grains dominate at about 10 μm; "large" grains in thermal equilibrium dominate at 50–100 μm. A typical energy distribution from dust emission in a starburst galaxy normalized to 60 μm is $f_\nu(12\ \mu m) = 0.035$; $f_\nu(25\ \mu m) = 0.18$; $f_\nu(60\ \mu m) = 1.0$; $f_\nu(100\ \mu m) = 1.41$ [136].

(d) Seyfert nucleus: Seyfert galaxies exhibit infrared emission from dust heated by the central source, as well as emission from starburst or nonthermal components. Seyfert galaxies tend to be most prominent at 60 μm, but energy distributions vary widely. The IRAS 25–60 μm spectral slope has been found useful for selecting Seyfert galaxies [137, 138].

(e) Blazar component: Nonthermal, approximately power-law emission ($f_\nu \propto \nu^\alpha$). Mean values are $\alpha(1\ \mu m) = -1.42 \pm 0.95$; $\alpha(10\ \mu m) = -1.12 \pm 0.47$; $\alpha(100\ \mu m) = -0.88 \pm 0.43$; $\alpha(1\ mm) = -0.18 \pm 0.42$ [139].

For far-infrared colors of extragalactic objects, see [125, 140–143].

7.11.2 Statistics of Galaxies at Infrared Wavelengths

Galaxy number counts at 2.2 μm. The number of galaxies per square degree per magnitude is [144]:

$$dN/dK = 4\,000 \times 10^{\alpha(K-17)},$$

where $\alpha = 0.67$ for $10 < K < 17$, $\alpha = 0.26$ for $17 < K < 23$, and $K = 2.2\ \mu$m mag.

Luminosity function at 60 μm [125, 145]. The density of galaxies per cubic megaparsec per magnitude interval at 60 μm is

$$\log(\rho) = -3.2 - \alpha \left\{\log[\nu L_\nu(60\,\mu\text{m})] - 10.2\right\},$$

where $\nu L_\nu(60\,\mu$m) is given in units of $L_\odot$, and $\alpha = 0.8$ for $\log[\nu L_\nu(60\,\mu\text{m})] < 10.2$ and $\alpha = 2.0$ for $\log[\nu L_\nu(60\,\mu\text{m})] > 10.2$. H_0 is assumed to be 75 km s^{-1} Mpc^{-1}.

The total infrared energy output of the local universe from 8 to 1000 μm is $1.24 \times 10^8 L_\odot$ Mpc^{-3} [146].

ACKNOWLEDGMENTS

Many people have helped with their comments and suggestions. I thank in particular the following persons for valuable comments and contributions to this chapter: E. Becklin, M. Cohen, D. Cruikshank, M. Hanner, T. Herter, J. Hora, E. Hu, T. Geballe, I. Glass, R. Knacke, S. Leggett, P. Léna, C. Lonsdale, S. Lord, J. Mazzarella, J. Pipher, S. Ridgway, K. Robertson, P. Roche, K. Sellgren, M. Simon, G. Veeder, M. Werner, G. Wynn-Williams, W. Vacca, and D. Van Buren.

REFERENCES

1. Lord, S. 1992, NASA Tech. Mem. 103957
2. Manduca, A., & Bell, R.A. 1979, *PASP*, **91**, 858
3. Bersanelli, M. et al. 1991, *A&A*, **252**, 854
4. Wolfe, W.L., & Zissis, G.J. 1985, *The Infrared Handbook*, rev. ed. (U.S. Government Printing Office, Washington, DC), p. 5-1
5. Griffin, M.J. et al. 1986, *Icarus*, **65**, 244
6. Traub, W.A., & Stier, M.T. 1976, *Appl. Opt.*, **15**, 364
7. Taylor, D.A. et al. 1991, *MNRAS*, **251**, 199
8. Taylor, D.A. et al. 1994, *Proc. SPIE*, **2198**, 703
9. Melnick, G.J. 1993, *Adv. Space Res.*, **13**, 535
10. Bally, J. 1989, in *Astrophysics in Antarctica*, AIP Conf. No. 198 (AIP, New York), p. 100; Townes, C.H., & Melnick, G. 1990, *PASP*, **102**, 357
11. Buhl, D. 1984, in *Galactic and Extragalactic Infrared Spectroscopy*, edited by M.F. Kessler and J.P. Phillips (Reidel, Dordrecht), p. 221
12. Warner, J.W. 1977, *PASP*, **89**, 724
13. Beckers, J.M. 1979, *PASP*, **91**, 857
14. Wallace, L. 1984, *PASP*, **96**, 184 & 836
15. Krisciunas, K. et al. 1987, *PASP*, **99**, 887
16. Volk, K. et al. 1989, in *Lecture Notes in Physics*, Vol. 341, edited by E.F. Milone (Springer-Verlag, Berlin), p. 15
17. Bessell, M.S., & Brett, J.M. 1988, *PASP*, **100**, 1134
18. Hodapp, K.-W. et al. 1992, *PASP*, **104**, 441
19. Maihara, T. et al. 1993, *PASP*, **105**, 940

20. Shure, M. et al. 1994, *Proc. SPIE*, **2198**, 614
21. Becklin, E., private communication
22. Roche, P.F., & Glasse, A.C.H. 1990, United Kingdom Large Telescope Report
23. Binzel, R.P. et al., this volume
24. Hauser, M.G. et al. 1984, *ApJ*, **278**, L15
25. Sodroski, T.J. et al. 1994, *ApJ*, **428**, 638
26. Cox, P., & Mezger, P.G. 1989, *A&A Rev.*, **1**, 49
27. Wright, E.L. 1993, *SPIE*, **2019**, 158
28. Mather, J.C. et al. 1994, *ApJ*, **420**, 439
29. Baker, D.J., & Romick, G.J. 1976, *Appl. Opt.*, **15**, 1966
30. Oliva, E., & Origlia, L. 1992, *A&A*, **254**, 466
31. Steed, A.J., & Baker, D.J. 1979, *Appl. Opt.*, **18**, 3386
32. McCaughrean, M.J. 1988, Ph.D. thesis, University of Edinburgh, Edinburgh
33. Ramsay, S.K. et al. 1992, *MNRAS*, **259**, 751
34. Osterbrock, D.E., & Martel, A. 1992, *PASP*, **104**, 76
35. Gillett, F.C. 1987, in *Infrared Astronomy with Arrays*, edited by C.G. Wynn-Williams and E.E. Becklin (University of Hawaii, Honolulu), p. 3
36. Merline, W.J., & Howell, S.B. 1995, *Exp. Astron.*, **6**, 163
37. Phillips, T.G. 1988, in *Millimeter and Submillimeter Astronomy*, edited by R. Wolstencroft and W.B. Burton (Kluwer Academic, Dordrecht), p. 1
38. Cohen, M. et al. 1992, *AJ*, **104**, 1650

39. Hayes, D.S. 1985, in *Calibration of Fundamental Stellar Quantities*, edited by D.S. Hayes et al., Proc. IAU Symp. No. 111 (Reidel, Dordrecht), p. 225

40. Blackwell, D.A. et al. 1990, *A&A*, **232**, 396; 1991, ibid., **245**, 567; Campins, H. et al. 1985, *AJ*, **90**, 896; Mountain, C.M. et al. 1985, *A&A*, **151**, 399

41. Rieke, G.H. et al. 1985, *AJ*, **90**, 900

42. Allen, D.A., & Cragg, T.A. 1983, *MNRAS*, **203**, 777

43. Elias, J.H. et al. 1982, *AJ*, **87**, 1029

44. Bersanelli, M. et al. 1991, *A&A*, **252**, 854

45. Bouchet, P. et al. 1991, *A&AS*, **91**, 409

46. Tokunaga, A.T. 1984, *AJ*, **89**, 172

47. Caillault, J.-P., & Patterson, J. 1990, *AJ*, **100**, 825

48. Jones, T.J., & Hyland, A.R. 1982, *MNRAS*, **200**, 509

49. Carrasco, L. et al. 1991, *PASP*, **103**, 987

50. Carter, B.S. 1990, *MNRAS*, **242**, 1

51. Hanner, M.S., & Tokunaga, A.T. 1991, in *Comets in the Post-Halley Era*, edited by R.L. Newburn et al. (Kluwer Academic, Dordrecht), p. 67

52. Casali, M., & Hawardsen, T. 1992, JCMT-UKIRT Newsletter, No. 4, p. 33

53. Gehrz, R.D. et al. 1987, in *Encyclopedia of Physical Science and Technology*, edited by R.A. Meyers (Academic Press, Orlando), Vol. 2, p. 53

54. Elias, J. et al. 1985, *ApJS*, **57**, 91

55. Koornneef, J. 1983, *A&AS*, **51**, 489

56. Leggett, S.K. et al. 1993, *IAU Colloq.*, **136**, 66

57. Wright, E.L. 1976, *ApJ*, **210**, 250

58. Orton, G.S. et al. 1986, *Icarus*, **67**, 289

59. Ulrich, B.L. 1981, *AJ*, **86**, 1619

60. Hildebrand, R.H. et al. 1985, *Icarus*, **64**, 64

61. Sandell, G. 1994, *MNRAS*, **271**, 75

62. Emerson, J.P. 1988, in *Formation and Evolution of Low Mass Stars*, edited by A.K. Dupree and M.T.V.T. Lago (Kluwer Academic, Dordrecht), Vol. 241, p. 193

63. Helou, G. et al. 1988, *ApJS*, **68**, 151

64. Pérault, M. 1987, Ph.D. thesis, University of Paris, Paris; Lonsdale, C., & Mazzarella, J., private communication

65. Helou, G. et al. 1985, *ApJ*, **298**, L7

66. Mathis, J.S., this volume

67. Rieke, G.H., & Lebofsky, M.J. 1985, *ApJ*, **288**, 618

68. Draine, B.T. 1989, in *Proceedings of the 22nd ESLAB Symposium on Infrared Spectroscopy in Astronomy*, edited by B.H. Kaldeich (ESA, Noordwijk), ESA SP-290, p. 93

69. He, L. et al. 1995, *ApJS*, **101**, 335

70. Roche, P.F., & Aitken, D.K. 1984, *MNRAS*, **208**, 481

71. Roche, P.F., & Aitken, D.K. 1985, *MNRAS*, **215**, 425

72. Henry, J.P. et al. 1984, *ApJ*, **285**, L27

73. Nagata, T. et al. 1994, *ApJ*, **423**, L113

74. Martin, P.G. et al. 1992, *ApJ*, **392**, 691

75. Whittet, D.C.B. et al. 1992, *ApJ*, **386**, 562

76. Hildebrand, R.H. et al. 1983, *QJRAS*, **24**, 267

77. Beckwith, S.V.W., & Sargent, A.I. 1991, *ApJ*, **381**, 250

78. Draine, B.T. 1994, in *First Symposium on the Infrared Cirrus and Diffuse Interstellar Clouds*, ASP Conf. Ser. Vol. 58, edited by R.M. Cutri and W.B. Latter (ASP, San Francisco), p. 227

79. Désert, F.X. et al. 1990, *A&A*, **237**, 215

80. Beichman, C.A. 1987, *ARA&A*, **25**, 521

81. Sellgren, K. 1984, *ApJ*, **277**, 623

82. Léger, A., & Puget, J.L. 1984, *A&A*, **137**, L5

83. Draine, B.T., & Anderson, N. 1985, *ApJ*, **292**, 494

84. Campins, H. et al. 1985, *AJ*, **90**, 896

85. Hardorp, J. 1982, *A&A*, **105**, 120

86. Lebofsky, L.A. et al. 1986, *Icarus*, **68**, 239

87. Morrison, D. 1974, *ApJ*, **194**, 203

88. Tedesco, E.F. et al. 1992, *The IRAS Minor Planet Survey*, Phillips Laboratory Report No. PL-TR-92-2049

89. Hanel, R.A. et al. 1992, *Exploration of the Solar System by Infrared Remote Sensing* (Cambridge University Press, Cambridge)

90. Encrenaz, Th., & Bibring, J.P. 1990, *The Solar System* (Springer-Verlag, Berlin)

91. Encrenaz, Th. 1992, in *Infrared Astronomy with ISO*, edited by Th. Encrenaz and M.F. Kessler (Nova Science, Commack), p. 173

92. Larson, H.P. 1980, *ARA&A*, **18**, 43

93. Crovisier, J. 1992, in *Infrared Astronomy with ISO*, edited by Th. Encrenaz and M.F. Kessler (Nova Science, Commack), p. 221

94. Mumma, M.J. et al. 1993, in *Protostars and Planets III*, edited by E.H. Levy and J.I. Lunine (University of Arizona, Tucson), p. 1177

95. Clark, R.N. et al. 1986, in *Satellites*, edited by J.A. Burns and M.S. Matthews (University of Arizona, Tucson), p. 437

96. Cruikshank, D.P., & Brown, R.H., in *Satellites*, edited by J.A. Burns and M.S. Matthews (University of Arizona, Tucson), p. 836

97. Larson, H.P., & Veeder, G.J. 1979, in *Asteroids* (University of Arizona, Tucson), p. 724

98. Gaffey, M.F. et al. 1989, in *Asteroids* II, edited by R.P. Binzel et al. (University of Arizona, Tucson), p. 98

99. Merrill, K.M., & Stein, W.A. 1976, *PASP*, **88**, 285

100. Strecker, D.W. et al. 1979, *ApJS*, **41**, 501

101. Cohen, M. et al. 1992, *AJ*, **104**, 2030

102. Walker, R.G., & Cohen, M. 1992, *An Atlas of Selected Calibrated Stellar Spectra*, NASA Contractor Report No. 177604

103. Wainscoat, R.J. et al. 1992, *ApJS*, **83**, 111

104. Cohen, M. 1993, *AJ*, **105**, 1860

105. Cohen, M. 1994, *AJ*, **107**, 582

106. Gezari, D.Y. et al. 1993, *Catalog of Infrared Observations*, 3rd ed. NASA Reference Publication No. 1294

107. Gezari, D.Y. et al. 1993, *Far Infrared Supplement, Catalog of Infrared Observations* ($\lambda > 4.6$ μm), 3rd ed., Rev. 1, NASA Reference Publication No. 1295

108. Jourdain de Muizon, M. et al. 1994, *Database of Astronomical Infrared Spectroscopic Observations* (Leiden University, Leiden)

109. Kleinmann, S.G., & Hall, D.N.B. 1986, *ApJS*, **62**, 501

110. Scoville, N.Z. et al. 1983, *ApJ*, **275**, 201

111. Simon, M., & Cassar, L. 1984, *ApJ*, **283**, 179

112. Persson, S.E. et al. 1984, *ApJ*, **286**, 289

113. Hamann, F., & Simon, M. 1986, *ApJ*, **311**, 909

114. Schwartz, R.D. et al. 1987, *ApJ*, **322**, 403

115. Carr, J.S. 1989, *ApJ*, **345**, 522

116. Mitchell, G.F. et al. 1990, *ApJ*, **363**, 554

117. Chandler, C.J. et al. 1993, *ApJ*, **412**, L71

118. Hamann, F. et al. 1994, *ApJ*, **422**, 626
119. Lada, C.J. 1988, in *Formation and Evolution of Low Mass Stars*, edited by A.K. Dupree and M.T.V.T. Lago (Kluwer Academic, Dordrecht), p. 93
120. Shu, F.H. et al. 1987, *ARA&A*, **25**, 23
121. Wilking, B.A. 1989, *PASP*, **101**, 229
122. Lada, E.A. et al. 1993, in *Protostars and Protoplanets* III, edited by E.H. Levy and J.I. Lunine (University of Arizona, Tucson), p. 245
123. Zinnecker, H. et al. 1993, in *Protostars and Protoplanets* III, edited by E.H. Levy and J.I. Lunine (University of Arizona, Tucson), p. 429
124. André, P., & Montmerle, T. 1994, *ApJ*, **420**, 837
125. Soifer, B.T. et al. 1987, *ARA&A*, **25**, 187
126. Telesco, C.M. 1988, *ARA&A*, **26**, 343
127. Frogel, J.A. et al. 1978, *ApJ*, **220**, 75
128. Aaronson, M. et al. 1978, *ApJ*, **220**, 442
129. Glass, I.S. 1984, *MNRAS*, **211**, 461
130. Glass, I.S., & Moorwood, A.F.M. 1985, *MNRAS*, **214**, 429

131. Sanders, D.B. et al. 1988, *ApJ*, **325**, 74
132. Carico, D.P. et al. 1988, *AJ*, **95**, 356
133. Knapp, G.R. et al. 1992, *ApJ*, **399**, 76
134. Désert, F.X. et al. 1990, *A&A*, **237**, 215
135. Knapp, G.R. 1995, in *Airborne Astronomy Symposium on the Galactic Ecosystem: From Gas to Stars to Dust*, edited by M.R. Haas et al., ASP Conf. Ser. No. 73, p. 121
136. Roche, P.F. et al. 1991, *MNRAS*, **248**, 606
137. de Grijp, M.H.K. et al. 1985, *Nature*, **314**, 240
138. Glass, I.S. 1985, *MNASSA*, **44**, 60
139. Impey, C.D., & Neugebauer, G. 1988, *AJ*, **95**, 307
140. Soifer, B.T. et al. 1989, *AJ*, **98**, 766
141. Mazzarella, J.M. et al. 1991, *AJ*, **101**, 2034
142. Cohen, M. 1992, *AJ*, **103**, 1734
143. Walker, H.J. et al. 1989, *AJ*, **98**, 2163
144. Gardner, J.P. et al. 1993, *ApJ*, **415**, L9
145. Soifer, B.T. et al. 1987, *ApJ*, **320**, 238
146. Soifer, B.T., & Neugebauer, G. 1991, *AJ*, **101**, 354

Chapter 8

Ultraviolet Astronomy

Terry J. Teays

8.1 ULTRAVIOLET WAVELENGTHS

The Earth's atmosphere is an efficient absorber of ultraviolet radiation, and so astronomical observations in this wavelength regime are pretty well limited to space-based instruments. As such, I adopt the nomenclature that "ultraviolet" refers to the wavelengths in the region from the atmospheric cutoff at $\approx 3\,200$ Å down to 100 Å. (The terms "far ultraviolet" and "extreme ultraviolet" are frequently used to refer to the shorter end of the ultraviolet wavelength range, but the usage has not been consistent in the literature. Generally one thinks of the far ultraviolet as referring to wavelengths shorter than that of the Lyman limit at 912 Å, and the extreme ultraviolet as being the region between 912 and 100 Å.) Note that wavelengths given in this chapter will always be *vacuum* ones. In the past ultraviolet wavelengths shorter than 2 000 Å were expressed as vacuum values, while those longward of this were given with regard to wavelengths in air. This convention has been continued in the *International Ultraviolet Explorer* (IUE) Project, but is currently being changed in their newest pipeline processing system, and eventually the entire archive will make use of only vacuum wavelengths. Newer missions such as the *Hubble Space Telescope* (HST) and *Extreme Ultraviolet Explorer* (EUVE) are using vacuum wavelengths exclusively. This practice conforms to Resolution C15 of the 21st General Assembly of the International Astronomical Union. Equation (8.1) is the algorithm for calculating the index of refraction (n) of standard air as a function of vacuum wavelength. This algorithm was derived by Edlén [1], and was the one officially adopted by the International Astronomical Union (IAU) [2]. The wavelength in air is the vacuum wavelength divided by the index of refraction:

$$n = 1 + 6.4328 \times 10^{-5} + \frac{2.949\,81 \times 10^{-2}}{146 \times 10^8 - \sigma^2} + \frac{2\,554.0 \times 10^{-4}}{41 \times 10^8 - \sigma^2},$$ (8.1)

where σ represents the wave number in vacuum, expressed in reciprocal Å.

8.2   ULTRAVIOLET ASTRONOMY SATELLITE MISSIONS

There have been numerous balloon and rocket flights devoted to ultraviolet astronomy, as well as various short-term studies, such as those conducted from manned space missions. The first ultraviolet spectrum of the Sun was obtained in 1946 using a captured V2 rocket, while the first stellar ultraviolet observations took place during 1955–1957. The first stellar ultraviolet spectrophotometry, by Stecher and Milligan [3], was accomplished by a rocket-borne instrument, while the first ultraviolet stellar spectroscopy (i.e., wavelength resolution sufficient to resolve individual spectral lines) was achieved in a 1965 rocket flight [4]. A balloon-borne stellar spectrograph first examined the very important Mg II resonance doublet in 1971 [5]. The principal long-term ultraviolet astronomy missions are summarized in Table 8.1. Note that the extensive number of missions that have been devoted to ultraviolet solar studies have not been included in the table. The first column in Table 8.1 gives the mission's name or acronym. OAO-2 stands for the second satellite in the *Orbiting Astronomical Observatory* series (the first having failed). It was the first instrument to carry out an extensive survey of the ultraviolet sky. The fourth satellite in this series was named *Copernicus*. It made substantial contributions to our understanding of the interstellar medium, hot stars, and stellar chromospheres. The TD-1 mission (named after the launch vehicle—a Thor Delta) was a European Space Agency (ESA) mission which had two ultraviolet experiments on board, including the S2/68 Ultraviolet Sky Survey Telescope. TD-1's primary legacy is the catalog of ultraviolet fluxes, which is cited in Table 8.2. ANS, the *Astronomy Netherlands Satellite*, had one ultraviolet experiment. Though well known for their spectacular success in planetary encounter missions, each of the two *Voyager* spacecraft have an ultraviolet spectrometer (UVS) that has been used for stellar spectroscopy, now that the primary mission objectives are completed. IUE, the *International Ultraviolet Explorer*, was a joint project of NASA, ESA, and the British SERC. It was originally intended for a three-year mission, but it continued to operate for over 18 years. One of the first major international satellites, IUE was operated in real-time from NASA's Goddard Space Flight Center for 16 hours per day, and from the ESA tracking station near Madrid for the remaining 8 hours. It is in an eccentric geosynchronous orbit. *Röntgensatellit* (ROSAT) is primarily an X-ray mission, but it has a wide field camera which operates in the ultraviolet wavelength range and has been used to produce an all-sky survey. The *Hubble Space Telescope* contains a battery of instruments, most with a number of configurations, which operate at ultraviolet wavelengths. For example, the *Goddard High Resolution Spectrograph* (GHRS) had a number of gratings and echelle cross-dispersers, which have not been detailed specifically in the table, rather representative ranges have been listed. These instruments, referred to by their acronyms in Table 8.1, are the GHRS, Faint Object Spectrograph (FOS), Wide Field/Planetary Camera (WF/PC), Faint Object Camera (FOC), High Speed Photometer (HSP), and the Space Telescope Imaging Spectrograph (STIS).

Table 8.1. *Major long-term ultraviolet astronomy missions.*

Mission	Operational dates	Tel. apert. (cm)	Instrument	λ (Å)	Spect. resol. (Å)	Reference
OAO-2	12/07/68–2/13/73	20	Photometer	1 430		[1]
		20	Photometer	1 550		
		20	Photometer	1 910		
		20	Photometer	2 460		
		20	Photometer	2 980		
		20	Photometer	3 320		
		40	Nebular photometer	1 200–4 000		
		30	Vidicon			
		30	Vidicon			
		30	Vidicon			
		30	Vidicon			
			Spectrometer	1 160–1 850	12	
			Spectrometer	1 850–3 600	22	
Copernicus	8/21/72–12/31/80	80	Spectrometer	912–1 500	0.05	[2]
			Spectrometer	912–1 645	0.2	
			Spectrometer	1 640–3 185	0.01	
			Spectrometer	1 480–3 275	0.04	
TD-1	3/12/72–1/9/80	27.5	Photometer	2 740		[3]
			Spectrophotometer	1 350–2 550		
ANS	8/30/74–6/14/77	22	Photometer	1 550		[4]
			Photometer	1 800		
			Photometer	2 200		
			Photometer	2 500		
			Photometer	3 300		
IUE	1/26/78–9/30/96	45	Echelle spectrograph	1 145–3 230	0.2	[5, 6]
			Spectrograph	1 150–3 300	6	
HST	4/24/90–	240	GHRS	1 110–3 200	0.01-3.5	[7]
			FOS	1 150–7 000	1.2-7	
			WF/PC	1 200–10 000		
			FOC	1 200–6 500		
			HSP	1 150–8 000		
			STIS	1 150–10 000		
ROSAT	6/1/90–	a[a]	Wide field camera	60–140		[8]
			Wide field camera	112–200		
			Wide field camera	150–220		
			Wide field camera	530–720		
EUVE	6/7/92–		Scanning photometer	44–360		[9]
		a[a]	Scanning photometer	44–360		
		a[a]	Scanning photometer	400–750		
		a[a]	Deep survey	40–385		
		a[a]	Spectrometer	70–190	0.5	
		a[a]	Spectrometer	140–380	1	
		a[a]	Spectrometer	280–760	2	

Note

[a] See text for aperture discussion.

References

1. Code, A.S., Houck, T.E., McNall, J.F., Bless, R.C., & Lillie, C.F. 1970, *ApJ*, **161**, 377
2. Rogerson, J.B., Spitzer, L., Drake, J.F., Dressler, K., Jenkins, E.B., Morton, D.C., & York, D.G. 1973, *ApJ*, **181**, 97
3. Jamar, C., Macau-Hercot, D., Monfils, A., Thompson, G.I., Houziaux, L., & Wilson, R. 1976, *Ultraviolet Bright-Star Spectrophotometric Catalogue* (ESA, Paris)

4. Wesselius, P.R., van Duinen, R.J., de Jonge, A.R.W., Aalders, J.W.G., Luinge, W., & Wildeman, K.J. 1982, *A&AS*, **49**, 427

5. Kondo, Y., editor, 1987, *Exploring the Universe with the IUE Satellite* (Reidel, Dordrecht).

6. Newmark, J.S., Holm, A.V., Imhoff, C.I., Oliversen, N.A., Pitts, R.E., & Sonneborn, G. 1992, *NASA IUE Newslett.*, **47**, 1

7. Bless, R.C. 1992, in *The Astronomy and Astrophysics Encyclopedia*, edited by S.P. Maran (Van Nostrand, New York), pp. 912–915

8. Pye, J.P., Watson, M.G., Pounds, K.A., & Wells, A. 1991, in *Extreme Ultraviolet Astronomy*, edited by R.F. Malina and S. Bowyer (Pergamon, New York), p. 409

9. EUVE Guest Observer Center 1992, EUVE Guest Observer Program Handbook (Appendix G of NASA NRA 92-OSS-5)

This configuration will change as a result of servicing missions for HST. The *Extreme Ultraviolet Explorer* (EUVE) is still in operation at the time of writing. The ROSAT and EUVE missions provided the first extensive and detailed look at this wavelength regime. HST and EUVE are in low-Earth orbits. Column 2 of Table 8.1 gives the mission's operational dates (the first date is the launch date, and so science operations will have begun somewhat later). Column 3 gives, when applicable, the size of the telescope objective (in cm) for the satellite or specific instrument. The notation "a" is used for the ROSAT and EUVE instruments to indicate that the matter of aperture is not as straightforward in the case of those instruments. They make use of various types of segmented filter masks which allow a given instrument to make use of a specific fraction of the aperture. Column 4 indicates the type of instrument, and column 5 gives the experiment's wavelength range (for spectrographic and spectrophotometric instruments) or the effective and/or central wavelength (for photometric instruments). Column 6 gives the approximate average spectral resolution (in Å) for spectrographic instruments. (This will, of course, vary with wavelength in each instrument, so the entries in column 6 are intended to be representative only.) Finally, column 7 lists a representative reference which gives information about the mission.

8.3   SIGNIFICANT ATLASES AND CATALOGS

Table 8.2 gives titles and references for some of the more important catalogs and atlases of ultraviolet astronomical data.

Table 8.2. *Important atlases and catalogs of ultraviolet data.*

The Variation of Galactic Interstellar Extinction in the Ultraviolet [1]
Atlas of the Wavelength Dependence of Ultraviolet Extinction in the Galaxy [2]
IUE-ULDA Access Guide No. 2: Comets [3]
ANS Ultraviolet Photometry, Catalogue of Point Sources [4]
An Atlas of Extreme Ultraviolet Explorer (EUVE) Sources [5]
IUE Low-Dispersion Spectra Reference Atlas. Part 1. Normal Stars [6]
IUE Ultraviolet Spectral Atlas of Selected Astronomical Objects [7]
Ultraviolet Bright-Star Spectrophotometric Catalogue [8]
Supplement to the Ultraviolet Bright-Star Spectrophotometric Catalogue [9]
Catalogue of Stellar Ultraviolet Fluxes [10]
Ultraviolet Photometry from the Orbiting Astronomical Observatory. XXXII. An Atlas of Ultraviolet
    Stellar Spectra [11]
IUE Ultraviolet Spectral Atlas [12]
IUE Ultraviolet Spectral Atlas [13]
The Extreme Ultraviolet Explorer Stellar Spectral Atlas [14]
Spectral Synthesis in the Ultraviolet. I. Far-Ultraviolet Stellar Library [15]
An Atlas of High Resolution IUE Ultraviolet Spectra of 14 Wolf–Rayet Stars [16]
The Hopkins Ultraviolet Telescope Far-Ultraviolet Spectral Atlas of Wolf–Rayet Stars [17]
International Ultraviolet Explorer Atlas of O Type Spectra from 1200 to 1900 Å [18]
Ultraviolet Spectral Morphology of the O Stars. II. The Main Sequence [19]
P Cygni and Related Profiles in the Ultraviolet Spectra of O-Stars [20]

Table 8.2. *(Continued.)*

An Atlas of Ultraviolet P Cygni Profiles [21]
Identification of Lines in the Satellite Ultraviolet: The Spectrum of Tau Scorpii [22]
Spectral Classification with the International Ultraviolet Explorer: An Atlas of B-Type Spectra [23]
The IUE Spectral Atlas of Two Normal B Stars: π Ceti and ν Capricorni (125–198 nm) [24]
Identification Lists of the Far UV Spectra of 7 Solar Chemical Composition Main Sequence Stars in the
    Spectral Range B2-B9.5 [25]
A Catalog of 0.2 Å Resolution Far-Ultraviolet Stellar Spectra Measured with Copernicus [26]
The Copernicus Ultraviolet Spectral Atlas of Vega [27]
The Copernicus Ultraviolet Spectral Atlas of Sirius [28]
Early Type Strong Emission-Line Supergiants of the Magellanic Clouds: A Spectroscopic Zoology [29]
Chromospheric Mg II Emission in A5 to K5 Main Sequence Stars from High Resolution IUE Spectra [30]
Atlas of High Resolution IUE Spectra of Late-Type Stars, 2500–3230 Å [31]
The Spectra of Late-Type Dwarfs and Sub-Dwarfs in the Near Ultraviolet. I. Line Identifications [32]
Outer Atmospheres of Cool Stars. VII. High Resolution Absolute Flux Profiles of the Mg II h and k Lines
    in Stars of Spectral Types F8 to M5 [33]
UV Fluxes of Pop II Stars [34]
IUE Low Dispersion Observations of Symbiotic Objects [35]
A Far-Ultraviolet Atlas of Symbiotic Stars Observed with IUE. I. The SWP Range [36]
A Spectrophotometric Atlas of White Dwarfs Compiled from the IUE Archives [37]
Ultraviolet Observations of Cataclysmic Variables: The IUE Archive [38]
A Catalogue of Low-Resolution IUE Spectra of Dwarf Novae and Nova-Like Stars [39]
An Atlas of UV Spectra of Supernovae [40]
UV Observations of SN 1987a [41]
International Ultraviolet Explorer Atlas of Planetary Nebulae, Central Stars, and Related Objects [42]
UV Spectra of the Central Stars of Large Planetary Nebulae [43]
A Survey of Ultraviolet Interstellar Absorption Lines [44]
Galactic Interstellar Abundance Surveys with IUE. II. The Equivalent Widths & Column Densities [45]
An Ultraviolet Spectral Atlas of Interstellar Lines toward SN 1987a [46]
IUE UV Spectra of Extra Galactic H II Regions. I. The Catalogue & the Atlas [47]
UV Observations by IUE of 31 Clusters of the LMC [48]
IUE-ULDA Access Guide No. 3: Normal Galaxies [49]
An Atlas of Hubble Space Telescope Ultraviolet Images of Nearby Galaxies [50]
An Atlas of Ultraviolet Spectra of Star-Forming Galaxies [51]
IUE-ULDA Access Guide No. 4: Active Galactic Nuclei [52]
The Ultraviolet Variability of Seyfert I Galaxies [53]
An Ultraviolet Atlas of Quasar and Blazar Spectra [54]

References

1. Witt, A.N., Bohlin, R.C., & Stecher, T.P. 1984, *ApJ*, **279**, 698
2. Aiello, S., Barsella, B., Chlewicki, G., Greenberg, J.M., Patriarchi, P., & Perinotto, M. 1988, *A&AS*, **73**, 195
3. Festou, M.C. 1990, IUE-ULDA Access Guide No. 2: Comets (ESA SP-1134)
4. Wesselius, P.R., van Duinen, R.J., de Jonge, A.R.W., Aalders, J.W.G., Luinge, W., & Wildeman, K.J. 1982, *A&AS*, **49**, 427
5. Shara, M.M., Bergeron, I.E., Christian, C.A., Craig, N., & Bowyer, S. 1997, *PASP*, **109**, 998
6. Heck, A. 1987, in *Exploring the Universe with the IUE* Satellite, edited by Y. Kondo (Reidel, Dordrecht) p. 121
7. Wu, C.-C. et al. 1992, IUE Ultraviolet Spectral Atlas of Selected Astronomical Objects, NASA Tech. Memo. No. 1285
8. Jamar, C., Macau-Hercot, D., Monfils, A., Thompson, G.I., Houziaux, L., & Wilson, R. 1976, *Ultraviolet Bright-Star Spectrophotometric Catalogue* (ESA, Paris)
9. Macau-Hercot, D., Jamar, C., Monfils, A., Thompson, G.I., Houziaux, L., & Wilson, R. 1978, *Supplement to the Ultraviolet Bright-Star Spectrophotometric Catalogue* (ESA, Paris)
10. Thompson, G.I., Nandy, K., Jamar, D., Monfils, A., Houziaux, L., Carnochan, D.J., & Wilson, R. 1978, *Catalogue of Stellar Ultraviolet Fluxes* (Science Research Council, London)
11. Code, A.D., & Meade, M.R. 1979, *ApJS*, **39**, 195
12. Wu, C.-C. et al. 1983, *NASA IUE Newslett.*, **22**, 1
13. Wu, C.-C. et al. 1991, *NASA IUE Newslett.*, **43**, 1
14. Craig, N., Abbott M., Finley, D., Jessop, H., Howell, S.B., Mathioudakis, M., Sommers, J., Vallerga, J.V., & Malina, R.F. 1997, *ApJS*, **113**, 131

15. Fanelli, M.N., O'Connell, R.W., & Thuan, T.X. 1987, *ApJ*, **221**, 768
16. Willis, A.J., van der Hucht, K.A., Conti, P.S., & Garmany, D. 1986, *A&AS*, **43**, 417
17. Schulte-Ladbeck, R.E., Hillier, D.J., & Herald, J.E. 1995, *ApJ*, **454**, L51
18. Walborn, N.R., Nichols-Bohlin, J., & Panek, R.J. 1985, IUE Atlas of O-Type Spectra from 1200 to 1900 Å, NASA RP-1155
19. Walborn, N.R., & Panek, R.J. 1984, *ApJ*, **286**, 718
20. Costero, R., & Stalio, R. 1984, *A&AS*, **58**, 95
21. Snow, T.P., Lamers, H.J.G.L.M., Lindholm, D.M., Odell, A.P. 1994, *ApJS*, **95**, 163
22. Cowley, C.R., & Merritt, D.R. 1987, *ApJ*, **321**, 553
23. Rountree, J. & Sonneborn, G. 1993, NASA Reference Publication No. 1312 (NASA, Washington)
24. Artru, M.-C., Borsenberger, J., & Lanz, T. 1989, *A&AS*, **80**, 17
25. Ramella, M., Castelli, F., Malagnini, M.L., Morossi, C., & Pasian, F. 1987, *A&AS*, **69**, 1
26. Snow, Jr., T.P., & Jenkins, E.B. 1977, *ApJS*, **33**, 269
27. Rogerson, J.B. 1989, *ApJS*, **71**, 1011
28. Rogerson, J.B. 1987, *ApJS*, **63**, 369
29. Shore, S.N., & Sanduleak, N. 1984, *ApJS*, **55**, 1
30. Blanco, C., Bruca, L., Catalano, S., & Marilli, E. 1982, *A&A*, **115**, 280
31. Wing, R.F., Carpenter, K.G., & Wahlgren, G.M. 1983, Perkins Obs. Special Pub. No. 1
32. Beckman, J.E., Crivellari, L., & Selvelli, P.L. 1982, *A&AS*, **47**, 295
33. Stencel, R.E., Mullan, D.J., Linsky, J.L., Basri, G.S., & Worden, S.P. 1980, *ApJS*, **44**, 383
34. Cacciari, C. 1985, *A&AS*, **61**, 407
35. Sahade, J., Brandi, E., & Fountenla, J.M. 1984, *A&AS*, **56**, 17
36. Meier, S.R., Kafatos, M., Fahey, R.P., Michalitsianos, A.G. 1994 *ApJS*, **94**, 183
37. Wegner, G., & Swanson, S.R. 1991, *ApJS*, **75**, 507
38. Verbunt, F. 1987, *A&AS*, **71**, 339
39. La Dous, C. 1990, *Space Sci. Rev.*, **52**, 203
40. Benvenuti, P., Sanz Fernandez de Cordoba, L., Wamsteker, W., Macchetto, F., Palumbo, G.C., & Panagia, N. 1982, ESA Special Pub. No. 1046
41. Kirschner, R.P., Sonneborn, G., Crenshaw, D.M., & Nassiopoulos, G.E. 1987, *ApJ*, **320**, 602
42. Feibelman, W.A., Oliversen, N.A., Nichols-Bohlin, J., & Garhart, M.P. 1988, NASA Ref. Pub. No. 1203
43. Kaler, J.B., & Feibelman, W.A. 1985, *ApJ*, **297**, 724
44. Bohlin, R.C., Hill, J.K., Jenkins, E.B., Savage, B.D., Snow, Jr., T.P., Spitzer, Jr., L.S., & York, D.G. 1983, *ApJS*, **51**, 277
45. Van Steenberg, M.E., & Shull, J.M. 1988, *ApJS*, **67**, 225
46. Blades, J.C., Wheatley, J.M., Panagia, N., Grewing, M., Pettini, M., & Wamstecker, W. 1988, *ApJ*, **334**, 308
47. Rosa, M., Joubert, M., & Benvenuti, P. 1984, *A&AS*, **57**, 361
48. Cassatella, A., Barbero, J., & Geyer, E.H. 1987, *ApJS*, **64**, 83
49. Longo, G., & Capaccioli, M. 1992, IUE-ULDA Access Guide No. 3: Normal Galaxies, ESA SP-1152
50. Maoz, D., Filippenko, A.V., Ho, L.C., Macchetto, F.D., Rix, H.-W. & Schneider, D.P. 1996, *ApJS*, **107**, 215
51. Kinney, A.L., Bohlin, R.C., Calzetti, D., Panagia, N., & Wyse, R.F.G. 1993, *ApJS*, **86**, 5
52. Courvoisier, T.J.-L., & Paltani, S. 1992, IUE-ULDA Access Guide No. 4: Active Galactic Nuclei, ESA SP-1153
53. Chapman, G.N.F., Geller, M.J., & Huchra, J.P. 1985, *ApJ*, **297**, 151
54. Kinney, A.L., Bohlin, R.C., Blades, J.C., & York, D.G. 1991, *ApJS*, **75**, 645

8.4   INTERSTELLAR EXTINCTION IN THE ULTRAVIOLET

Since interstellar extinction is significantly stronger in the ultraviolet than at visual wavelengths, correcting for its effects is very important. The most prominent feature in the ultraviolet extinction curves is a broad peak centered at $\approx 2\,175$ Å.

Equation (8.2) [6] gives some useful analytic functions which can be used to determine A_λ in the ultraviolet. Equation (8.2) is broken into three wavelength domains, and is parametrized in terms of σ, the wave number expressed in microns:

$$2.70 \le \sigma \le 3.65, \qquad A_\lambda/E_{B-V} = 1.56 + 1.048\sigma + \frac{1.01}{[(\sigma - 4.60)^2 + 0.280]}, \qquad (8.2a)$$

$$3.65 \leq \sigma \leq 7.14, \qquad A_\lambda/E_{B-V} = 2.29 + 0.848\sigma + \frac{1.01}{[(\sigma - 4.60)^2 + 0.280]}, \qquad (8.2b)$$

$$7.14 \leq \sigma \leq 10, \qquad A_\lambda/E_{B-V} = 16.17 - 3.20\sigma + 0.2975\sigma^2. \qquad (8.2c)$$

Savage and Mathis [7] adopt 3.1 for the value of $A_V/E_{(R-V)}$, while Seaton [6] uses 3.2. More detailed information is available in the review by Savage and Mathis [7], and additional references concerning ultraviolet extinction as a function of location in the sky are cited in Table 8.2.

8.5 COMMONLY OBSERVED ULTRAVIOLET EMISSION LINES

Table 8.3 (which is an expanded version of one given in Wu et al. [8]) gives a list of some of the more prominent ultraviolet emission lines observed in astronomical objects. The organization of Table 8.3 is as follows. Column 1 gives the wavelength (in Å) of the line, using the convention that a reasonably precise value (to 0.01 Å) is given for single lines, while an approximate value is given for lines formed of closely spaced individual lines of a given element. This value corresponds to the approximate location of the (blended) line which would be seen in low-resolution spectra, such as those taken in IUE's low-dispersion mode. In cases where there is a spectral region which contains a large number of lines due to a single element, then the range of wavelengths is given in column 1. In the cases of multiple lines, column 4 gives more accurate wavelengths for the individual components that may be present. Column 2 specifies the ion which is the source of the emission line, while column 3 lists the type of objects in which this emission line is generally observed. The abbreviations used in column 3 to specify object type are given at the bottom of Table 8.3.

Table 8.3. *Emission lines commonly found in ultraviolet spectra.*

λ (Å)^a	Ion	Type of object where observed ^{b}	Individual components in multiplets
538	O II	C	537.83, 538.26, 538.32, 539.13
584.33	He I	SSO	
834	O III	C	832.93, 833.74, 835.29
916	N II	C	915.61, 915.96, 916.02, 916.10, 916.35, 916.70, 916.71
933.4	S VI	SNR	
977.02	C III	SNR	
1 033	O IV	SNR	1 031.93, 1 033.82, 1 037.62
1 066.66	Ar I	SSO, C	
1 085	N II	C	1 083.99, 1 084.56, 1 084.58, 1 085.53, 1 085.55, 1 085.70, 1 085.12
1 175	C III	WR, PN, CS, SS	1 174.93, 1 175.26, 1 175.59, 1 175.71, 1 175.99, 1 176.37
1 199	S III	SSO	1 190.21, 1 194.06, 1 194.46, 1 197.56, 1 200.97, 1 201.73, 1 202.13
1 215.67	H I	(all sources)	
1 240	N V	PN, SS, WR, CV, XRB, SN, IB, N, SQ, SNR	1 238.82, 1 240.15, 1 242.80
1 247.38	C III	SS, WR	
1 256	S II	SSO	1 250.58, 1 253.81, 1 256.12, 1 259.52
1 279	C I	TT, LTS	1 276.48, 1 276.75, 1 277.19, 1 277.25, 1 277.28, 1 277.46, 1 277.51, 1 277.55, 1 277.72, 1 277.95, 1 279.06, 1 279.23, 1 279.50, 1 279.89, 1 280.14, 1 280.33, 1 280.36, 1 280.40, 1 280.60, 1 280.85
1 299	Si III	SS, IB, TT	1 298.89, 1 298.96
1 304	O I	RS, LTS, N, SQ, C	1 302.17, 1 303.49, 1 304.86, 1 306.03
1 309	Si II	PN	1 304.37, 1 307.64, 1 309.28

Table 8.3. *(Continued.)*

λ (Å)[a]	Ion	Type of object where observed [b]	Individual components in multiplets
1 335	C II	TT, PN, LTS, RS, WR, CV, N, SNR, C	1 334.53, 1 335.31, 1 335.66, 1 335.71
1 342	O IV	CS, SS, WR, XRB	1 342.99, 1 343.51
1 371.29	O V	PN, CS, SS, XRB, SNR	
1 394	Si IV	PN, LTS, RS, TT, XRB, CV, IV, N, SQ	1 393.76, 1 396.75, 1 398.13
1 397–1 407	O IV	PN, SS, N	1 397.23, 1 399.78, 1 401.16, 1 404.81, 1 407.38
1 402.77	Si IV	PN, LTS, RS, TT, XRB, CV, IV, N, SQ	
1 460	C I	TT	1 459.03, 1 463.34, 1 467.40, 1 467.88, 1 468.41
1 473	S I	RS, LTS	1 472.97, 1 473.02, 1 473.07, 1 473.99, 1 474.38, 1 474.57, 1 478.50
1 483.32	N IV	PN, SS, WR, N	
1 486	S I	RS, LTS	1 485.62, 1 487.15
1 487	N IV	PN, SS, WR, N, SNR	1 486.50, 1 487.89
1 550	C IV	TT, PN, LTS, SS, N, WR, CV, IB, XRB, SQ, SNR	1 548.20, 1 550.77
1 561	C I	C	1 560.31, 1 560.68, 1 560.71, 1 561.05, 1 561.34, 1 561.37, 1 561.44
1 574.77	Ne V	PN, N	
1 577	C III	SS	1 576.48, 1 577.30, 1 577.89
1 602	Ne IV	PN, N, SS	1 601.50, 1 601.68
1 640	He II	TT, PN, LTS, RS, WR, XRB, SQ, SNR	1 640.47, 1 640.49
1 641.31	O I	RS, LTS, SS, N	
1 657	C I	C, RS, LTS, TT	1 656.28, 1 656.93, 1 657.01, 1 657.38, 1 657.59, 1 657.91, 1 658.12
1 663	O III	PN, WR, SQ, N, LTS, HII, SS, SNR	1 660.81, 1 666.15
1 670.79	Al II	IB, LTS	
1 710	Si II	PN, WR	1 710.83, 1 711.30
1 718.55	N IV	PN, WR, XRB, CV, N	
1 728.94	S III	SSO	
1 750	N III	WR, TT, HII, N, SN, SNR	1 746.82, 1 748.65, 1 749.67, 1 752.16, 1 754.00
1 760	C II	PN	1 760.47, 1 760.82
1 815	Si II	TT, PN, RS, LTS	1 808.01, 1 816.93, 1 817.45
1 814.63	Ne III	PN, N	
1 860	Al III	LTS, IB	1 854.72, 1 862.79
1 882.71	Si III	PN, LTS, HII, SN, N, SQ, SNR	
1 892.03	Si III	TT, PN, LTS, HII, SN, N, SQ, SNR	
1 900.29	S I	RS, LTS, SN, HII	
1 908.73	C III	TT, PN, LTS, WR, HII, N, SN, SQ, ELG	1 906.68, 1 908.73, 1 909.60
1 914.70	S I	RS, LTS	
1 993.62	C I	RS, LTS	
2 321.67	O III	PN	
2 326	C II	RS, LTS, SQ	2 324.21, 2 325.40, 2 326.11, 2 327.65, 2 328.84

Table 8.3. *(Continued.)*

λ (Å)[a]	Ion	Type of object where observed [b]	Individual components in multiplets
2 328–2 414	Fe II	LTS, TT	2 328.11, 2 333.52, 2 338.73, 2 344.21, 2 344.70, 2 345.00, 2 349.02, 2 359.83, 2 365.55, 2 367.59, 2 374.46, 2 381.49, 2 382.77, 2 383.79, 2 389.36, 2 394.98, 2 396.15, 2 396.36, 2 399.97, 2 405.16, 2 405.62, 2 407.39, 2 411.25, 2 411.80, 2 414.05
2 329.23	Si II	RS, LTS	
2 335	Si II	PN	2 335.12, 2 335.32, 2 344.92, 2 350.89
2 381.13	He II	PN	
2 424	Ne IV	SQ	2 422.51, 2 425.15
2 471.04	O II	PN, SN, HII	
2 511.96	He II	PN	
2 586–2 632	Fe II	LTS, TT, N, MSG	2 586.65, 2 599.15, 2 600.17, 2 607.87, 2 611.41, 2 612.65, 2 614.61, 2 618.40, 2 621.19, 2 622.45, 2 626.45, 2 629.08, 2 631.83, 2 632.11
2 664.06	He I	PN	
2 696.92	He I	PN	
2 724.00	He I	PN	
2 734.14	He I	PN	
2 764.62	He I	PN, HII	
2 783.03	Mg V	PN	
2 786.81	Ar V	PN	
2 794	Mg II	PN	2 798.81, 2 791.59
2 800	Mg II	PN, LTS, RS, TT, IB, N, SQ, ELG	2 796.35, 2 803.53
2 829.91	He I	PN, HII	
2 838	C II	PN	2 837.54, 2 838.44
2 852.96	Mg I	PN, HII	
2 854.48	Ar IV	PN	
2 869.00	Ar IV	PN	
2 928.34	Mg V	PN	
2 933	Mg II	PN	2 929.49, 2 937.36
2 945.97	He I	PN	
2 950.07	Mn II	PN, TT	
2 973.15	O I	C	
2 978	N III	PN	2 973.43, 2 979.70
3 005.36	Ar III	PN	
3 024.33	O III	PN	
3 046	O III	PN	3 043.91, 3 048.02
3 068	N II	PN	3 063.72, 3 071.44
3 109	Ar III	PN	3 109.16, 3 110.06
3 133.77	O III	PN, TT, N, LTS	
3 188.67	He I	PN	
3 204.03	He II	PN	

Notes

[a] Wavelengths (in vacuum) are taken from: Aller, L.H. 1984, *Physics of Thermal Gaseous Nebula* (Reidel, Dordrecht); Kelly, R.L. 1979, *Atomic Emission Lines in the Near Ultraviolet; Hydrogen through Krypton*, NASA Tech. Memo. No. 80268; Kelly, R.L. 1987, *Atomic and Ionic Spectrum Lines Below 2000 Å: Hydrogen through Krypton* (American Chemical Society, New York); Kelly, R.L. & Palumbo, L.J. 1973, *Atomic and Ionic Emission Lines Below 2000 Angstroms* (Naval Research Lab., Washington, DC); Köppen, J., & Aller, L.H. 1987, in *Exploring the Universe with the IUE Satellite*, edited by Y. Kondo (Reidel, Dordrecht), p. 589; and Morton, D.C. 1991, *ApJS*, **77**, 119.

[b] The astronomical objects where these lines are frequently seen in emission are noted by the abbreviated code in column 3. They are: C, comets; CS, carbon stars; CV, cataclysmic variables (N.B. novae have a separate listing); ELG, emission line galaxies; HII, H II regions; IB, interacting binaries; LTS, late-type stars; MSG, massive supergiants; N, novae; PN, planetary nebulae; RS, RS CVn stars; SQ, Seyfert galaxies and QSOs; SN, supernovae; SS, symbiotic stars; SSO, solar system objects; TT, T Tau stars; WR, Wolf–Rayet stars; XRB, low-mass X-ray binaries.

8.6   ULTRAVIOLET SPECTRAL CLASSIFICATION

Studies of spectral classification of O and B stars based on ultraviolet spectra have been made using Copernicus data and the extensive IUE archive. Low-dispersion spectra were used by Heck et al. [9], Heck [10], and Jaschek and Jaschek [11]. High-dispersion studies have been conducted by Snow and Morton [12], Walborn and Panek [13], Walborn et al. [14], Walborn and Nichols-Bohlin [15], Massa [16], Bates and Gilheany [17], Prinja [18], and Rountree and Sonneborn [19]. For detailed quantitative comparisons, the papers by Massa and Prinja are convenient, because they give tables and/or figures which show the equivalent widths as a function of spectral type or temperature. Prinja [18] gives two useful formulas relating equivalent widths (W_a) in mÅ to effective temperature. The most sensitive diagnostic for O stars temperatures is Si III λ1299:

$$\log(W_a) = 17.89 - 3.43 \log T_{\text{eff}}. \tag{3}$$

For B stars, the Si II λ1265 is the most sensitive temperature indicator [16]:

$$\log(W_a) = 20.57 - 4.21 \log T_{\text{eff}}. \tag{4}$$

The information in Table 8.4 is taken from these studies. Table 8.4 gives the approximate wavelength and identification for classification lines in its first two columns, and summarizes their changing characteristics as a function of spectral type and luminosity in the final column. (More accurate wavelengths can be found in Table 8.3.)

Table 8.4. *Lines useful for spectral classification of O and B stars.*

λ (Å)	Ion	Comments
1 175	C III	In low dispersion this blend of six lines (λλ 1174.933–1176.370) is seen to increase from O4 to a maximum at B1, and disappears at B6 into the *Ly α* wing. In high dispersion one can see dramatic P Cygni profiles for all supergiants from O4 I–B0.5 Ia, for bright giants as late as O9.5, and for giants as late as O8.
1 216	H I	When not affected by interstellar or circumstellar components has a half-width at half-maximum which increases from 10 Å at O9 to 100 Å at B8.
1 240	N V	λλ 1239, 1243 show wind profiles in most O stars. Shows a dependence on luminosity at O9.5, since the stellar wind effects have declined by then.
1 247	C III	Blended with Fe II λλ 1246.8, 1247.8, and can be severely affected by emission component of NV λ1240 P Cygni lines in luminous stars. Generally increases in strength from early to late O. Strongest in early B (B0–B1), and then slowly declines. The ratio C III λ1247/O IV λ1339 depends on luminosity class, being higher for more luminous stars. This ratio can be as large as 4 between supergiants and main-sequence stars at a given temperature (Prinja, R.K. 1990, *MNRAS*, **246**, 392). The comparison of this line with Si II λ1265 shows a slight dependence on luminosity class (N.B.: can be affected by a reseaux mark in high-dispersion IUE spectra).
1 255	Fe V	Decreases from O3 to O7.
1 264	Si II	Becomes visible at B1; at B1.5 it is clearly present but weaker than λ1247; at B2 it is as strong as λ1247; and by B4 it is much stronger. Continues to increase through B9. Does not show any luminosity effect.
1 300	Si III	Probably the most sensitive diagnostic of O star temperatures. Increases sharply from O3 to B2, then levels out in strength from B2 to B5.
1 310	Si II	Useful diagnostic in B stars. It is weaker than λ1300 at B2, greater than or equal to 1300 at B3–B4, and dominates the spectrum at B5–B8.
1 336	C II	Doublet, which increases from B0 to a maximum at B8. The wind profiles achieve maximum strength at B1–B2 Ia. There is a very strong interstellar contribution to this line.
1 339	O IV	Shows a well-defined temperature sequence for luminosity classes I and V in O stars, decreasing as temperature declines. Generally only the λ1339 line is used in this doublet, since the λ1343 line is blended with a nearby Si III line (as well as lying in an awkward location in IUE echelle spectra).
1 371	O V	This line declines from O3 until it disappears at O7.

Table 8.4. *(Continued.)*

λ (Å)	Ion	Comments
1 400	Si IV	Blend of the λ1394 and λ1403 lines of Si IV. In low-dispersion spectra this blended pair is a useful luminosity indicator for late O, and a spectral type discriminator for B. First strongly visible in low-dispersion spectra at O7, and gets stronger as surface gravity decreases. In high dispersion, at O6.5 lines display stellar wind effects which increase with luminosity, from none at V to a full P Cyg profile at Ia. At O9.5 the doublet shows no stellar-wind effect in luminosity classes V–III, but it develops gradually as a function of luminosity from classes II through Ia. In the B stars, Si IV is strong in B0 and B1 and decreases in strength until it disappears at about B6. The intensity ratio Si IV λ1400/C IV λ1550 is very sensitive to the O star spectral type, being ≈ 1 at O6, and greater than 1 for O6.5–O9.7.
1 428	C III	(In low dispersion the λ1426 and λ1428 lines are blended, though they are never especially strong. They increase from O4 to a maximum at B1.) Especially fine discriminator in the O7–B1 region, where it can be compared to λ1430.
1 430	Fe V	The ratio λ1429/λ1430 = 1 for this Fe V doublet between O3 and O4, and declines at O5 and later.
1 453	Blend	In low dispersion, has a maximum at O4 and disappears at B0.
1 485	Si II	Blend of three lines. First present at B2 and becomes stronger through B9.
1 527	Si II	Absorption feature becomes prominent in late B.
1 533	Si II	Absorption feature becomes prominent in late B.
1 550	C IV	Resonance doublet is one of the most prominent UV lines. Strong in O stars, decreasing from O3 to B2 (in dwarfs) where it disappears. If seen in mid-B, indicates a supergiant. Saturated P Cyg profiles from O3–O6, declining at O7. Continues to show strong wind absorption through O9, becoming purely photospheric at B1. At the transition type O9.5 there is an increase in strength with luminosity class.
1 608	Fe II	A large collection of Fe II lines exist in the λλ 1600–1610 region. These blends increase in strength with increasing luminosity, while showing little temperature effect. In O stars there is a noticeable interstellar component.
1 640	He II	Present throughout the O star regime, is still strong at B0, still noticeable, but declining in B0.5–B1, weak at B1.5, and weak to absent at B2.
1 655	C I	Increases in strength as spectral type gets later. It is a prominent line in B5–B9.
1 670	Al II	Becomes prominent in late B (N.B.: there is frequently a strong interstellar line seen in O stars, due to this ion).
1 718	N IV	Unsaturated subordinate line which shows P Cyg profiles through O6, then declines in strength with increasingly later spectral type. It is still strong at B0, much less prominent at B0.5, and weak to absent at B1. At B0 it is stronger in giants than dwarfs.
1 723	Al II	Blend. The components are at λλ 1719.44, 1721.24, 1721.27, 1724.95, 1724.98. Line strength increase with luminosity in B stars.
1 750	N III	Doublet at λλ 1748, 1752. The strength of both lines increases between O3 and O4, and the ratio λ1748/λ1752 increases dramatically between O3 and O4. The pair remains distinct through B0, but starts to weaken at B0.5, and disappears as B1.
1 859	Al III	Doublet at λλ 1855, 1862. Purely interstellar in O stars. In B stars increases with increasing luminosity class. There is a strong wind maximum at B1-2 Ia.
1 862	Al II	Strong in O stars. Blended with λ1855 in low-resolution spectra. Shows an increased strength with increased luminosity class.
1 891	Fe III	Present in early B stars. Shows a positive luminosity effect. There are many Fe III lines in this wavelength region. The use of this line and others below is most generally useful in low-dispersion spectra.
1 926	Fe III	Similar to λ1891.
1 967	Fe III	Similar to λ1891.

The ultraviolet is particularly suitable for classifying O and B stars, due to the strong fluxes for these objects in that wavelength regime. Difficulties with classifying OB stars include the contamination of some lines by strong interstellar components, and the fact that ultraviolet resonance lines are frequently severely affected by stellar winds. Snow and Morton [12] found that all O and B supergiants exhibited mass loss, with P Cygni profiles being seen to as late as B1. For bright giants and giants, strong P Cygni profiles were noted as late as O9.5 and O9, respectively, and all main-noted sequence O stars showed evidence of mass loss. A further complication is that the wind profiles of some B supergiants have been found to be variable. Exactly how much of the dispersion in wind line strengths is due to variations in the intrinsic stellar properties, and how much is due to variability or abundance anomalies, is uncertain [17, 20].

8.7 ULTRAVIOLET SPECTROPHOTOMETRIC STANDARDS

Spectrophotometric calibration has always been a thorny problem for long-term ultraviolet satellite missions. Early efforts tended to focus on using hot subdwarfs as reasonably line-free continuum sources, which were not generally variable, and had very small or negligible interstellar reddening. The current IUE absolute calibration is based on comparison with the earlier measurements of some baseline standard stars made by OAO-2 and TD-1, and normalized to the flux values for the fundamental calibration star, η UMa. The stars used were HD 60753, BD + 75° 325, HD 93521, BD + 33° 2642, and BD + 28° 4211 for the low-dispersion data, while ζ Cas, λ Lep, and τ Sco were used for the high-dispersion data. It should be noted that both ζ Cas and η UMa have shown some indications of microvariability [21]. A more complete list of IUE standards can be found in [22], while the HST standards are cited in [23]. More recently a shift has been made to using hot DA (i.e., essentially pure helium) white dwarfs as fundamental calibrators. The reasoning behind this is that the models for these stars are very simple and well understood, as well as being unaffected by spectral lines. The IUE Project's Final Archive is making use of white dwarfs for their new absolute calibration. The EUVE used this approach from the very beginning. The fundamental calibrator that is being used is G191B2B. Table 8.5 lists some of the ultraviolet standard stars that have been used in common by many missions. Columns 1 and 2 give the star's catalog number and common name, while columns 3 and 4 list the star's coordinates. Columns 5 and 6 give the spectral type and visual magnitude, while column 7 indicates which missions have observed this star for calibration purposes.

Table 8.5. *Selected ultraviolet spectrophotometric standard stars.*

Catalog ID	Common name	α(2000)	δ(2000)	Sp. Type	V	Observed by[a]
HD 2151	β Hyi	00:25:45.4	−77:15:16	G2 IV	2.80	H
HD 3360	ζ Cas	00:36:58.3	+53:53:49	B2 IV	3.68	OTAVI
BPM 16274		00:50:03.2	−52:08:17	DA	14.2	H
Feige 11		01:04:21.6	+04:13:38	B0 VI	12.06	OIH
HD 10144	α Eri	01:37:42.9	−57:14:12	B3 Vpe	0.46	OCTI
HD 11636	β Ari	01:54:38.3	+20:48:29	A5 V	2.64	OTI
HD 15318	ξ^2 Cet	02:28:09.5	+08:27:36	B9 III	4.29	H
GD 50		03:48:50.1	−00:58:30	DA	14.06	H
HZ 4		03:55:21.7	+09:47:19	DA	14.52	H
LB 227		04:09:28.8	+17:07:54.4	DA	15.34	H
HZ 2		04:12:43.5	+11:51:50	DA	13.86	H
G191B2B		05:01:31.0	+52:45:48	DA	11.78	VIHE
HD 32630	η Aur	05:06:30.8	+41:14:04	B3 V	3.17	OTAI
HD 34816	λ Lep	05:19:34.4	−13:10:37	B0.5 IV	4.29	OTAI
HD 35468	γ Ori	05:25:07.8	+06:20:59	B2 III	1.64	OTI
HD 35580	κ Pic	05:22:22:2	−56:08:04	B8–9 V	6.11	TI
HD 38666	μ Col	05:44:08.4	−32:19:27	O9.5 IV	5.17	VIH
PG 0549 + 158	GD 71	05:52:27.5	+15:53:17	DA	13.04	VIE
HD 45557		06:24:13.7	−60:16:52	A0 V	5.80	TI
HD 49798		06:48:04.6	−44:18:59	sdO6	8.30	VIH
HD 60753		07:33:27.3	−50:35:04	B3 IV	6.69	TIH
CD −31° 4800		07:36:30.2	−32:12:45	O8 AI	10.50	AI
HD 61421	α CMi	07:39:18.1	+05:13:30	F5 IV–V	0.38	OCTAI
HD 66811	ζ Pup	08:03:35.1	−40:00:12	O5f	2.26	OCTVIH
BD +75° 325		08:10:49.3	+74:57:58	O5p	9.54	OTAVIH
HD 80007	β Car	09:13:12.1	−69:43:02	A2 IV	1.68	OTI
AGK +81° 266		09:21:19.1	+81:43:29	sdO	11.92	AIH
BD +48° 1777		09:30:46.6	+48:16:26	O VI	10.37	AI
HD 87901	α Leo	10:08:22.3	+11:58:02	B7 V	1.35	OCTAVIH
Feige 34		10:39:36.7	+43:06:10	DO	11.18	VIH

Table 8.5. *(Continued.)*

Catalog ID	Common name	α(2000)	δ(2000)	Sp. Type	V	Observed by[a]
HD 93521		10:48:23.5	+37:34:13	O9 Vp	7.04	TAVIH
HD 100889	θ Crt	11:36:40.8	−09:48:08	B9.5 Vn	4.70	IH
HD 103287	γ UMa	11.53:49.8	+53:41:41	A0 Ve	2.44	IH
HZ 21		12:13:56.4	+32:56:31	DO	14.68	H
PG 1254 + 223	GD 153	12:57:04.5	+22:12:45	DA	13.4	VIE
HZ 44		13:23:35.4	+36:08:00	sdO	11.66	VH
Grw +70° 5824		13:38:51.8	+70:17:09	DA	12.77	H
HD 120315	η UMa	13:47:32.4	+49:18:48	B3 V	1.86	OCTVIH
HD 121263	ζ Cen	13:55:32.3	−47:17:18	B2.5 IV	2.55	OCTAI
HD 122451	β Cen	14:03:49.5	−60:22:23	B1 III	0.61	H
HD 125924		14:22:43.0	−08:14:54	B2 IV	9.70	TAI
HD 128801		14:38:48.1	+07:54:44	B9	8.80	TAI
HD 137389		15:22:37.1	+62:02:50	A0pSi	5.98	TAI
HD 137744	ι Dra	15:24:55.7	+58:57:58	K2 III	3.29	H
BD +33° 2642		15:51:59.9	+32:56:55	B2 IV	10.81	OTAIH
HD 142669	ρ Sco	15:56:53.0	−29:12:50	B2 IV–V	3.88	OTAI
HD 145454		16:06:19.5	+67:48:36	A0 Vn	5.44	TI
G153−41		16:17:55.4	−15:35:49	DA	13.42	VIH
HD 149438	τ Sco	16:35:52.9	−28:12:58	B0 V	2.82	OCTAVI
HD 149757	ζ Oph	16:37:09.5	−10:34:02	O9.5 Vn	2.56	OCTVIH
HD 155763	ζ Dra	17:08:47.1	+65:42:53	B6 III	3.17	OCTAI
HD 164058	γ Dra	17:56:30.4	+51:29:20	K5 III	2.22	H
HD 172167	α Lyr	18:36:56.3	+38:47:01	A0 V	0.03	OCTAVIH
HD 172883		18:39:52.7	+52:11:46	A0pHg	6.00	TI
HD 177724	ζ Aql	19:05:24.5	+13:51:48	A0 Vn	2.99	OTAI
HD 186427	16 Cyg B	19:41:52.0	+50:31:03	G1.5 V	6.20	IH
HD 196519	υ Pav	20:41:57.1	−66:45:39	B9 III	5.15	TAI
HD 197637		20:36:00.6	+79:25:49	B3	6.78	TI
HD 201908		21:05:29.2	+78:07:35	B8 Vn	5.91	OTI
LDS 749B		21:32:15.8	+00:15:14	DB	14.67	H
BD +28° 4211		21:51:11.1	+28:51:52	sdOp	10.51	OTAVIH
G93−48		21:52:25.3	+02:23:24	DA	12.74	H
HD 209952	α Gru	22:08:13.9	−46:57:40	B7 IV	1.74	OCTI
NGC 7293		22:29:38.5	−20:50:13	PNN	13.51	VIH
HD 214680	10 Lac	22:39:15.6	+39:03:01	O9 V	4.88	OCTAI
HD 214923	ζ Peg	22:41:27.7	+10:49:53	B8 V	3.40	H
PG 2309 + 105	GD 246	23:12:35.3	+10:50:27	DA	13.10	IHE
Feige 110		23:19:58.4	−05:09:56	DOp	11.82	H

Note

[a] Observations were made of these standards by many of the ultraviolet astronomy missions, and they are listed in column 7, where the letters refer to O = OAO-2, C = Copernicus, T = TD-1, A = ANS, V = Voyager UVS, I = IUE, H = HST, E = EUVE.

REFERENCES

1. Edlén, B. 1953, *JOSA*, **43**, 339
2. Oosterhoff, P.T. 1957, *Trans. IAU*, **9**, 202
3. Stecher, T.P., & Milligan, J.E. 1962, *ApJ*, **136**, 1
4. Morton, D.C., & Spitzer, L. 1966, *ApJ*, **144**, 1
5. Kondo, Y., Giuli, T., Modisette, J.L., & Rydgren, A.E. 1972, *ApJ*, **176**, 153
6. Seaton, M.J. 1979, *MNRAS*, **187**, 73
7. Savage, B.D., & Mathis, J.S. 1979, *ARA&A*, **17**, 73
8. Wu, C.-C. et al. 1992, *IUE Ultraviolet Spectral Atlas of Selected Astronomical Objects*, NASA Tech. Memo. No. 1285
9. Heck, A., Egret, D., Jaschek, M., & Jaschek, C. 1984, *IUE Low-Resolution Spectra Reference Atlas: Part 1. Normal Stars* (ESA, Paris)
10. Heck, A. 1987, in *Exploring the Universe with the IUE Satellite*, edited by Y. Kondo (Reidel, Dordrecht), p. 121
11. Jaschek, C., & Jaschek, M. 1987, *The Classification of Stars* (Cambridge University Press, Cambridge)
12. Snow, Jr., T.P., & Morton, D.C. 1976, *ApJS*, **32**, 429
13. Walborn, N.R., & Panek, R.J. 1984, *ApJ*, **286**, 718

14. Walborn, N.R., Nichols-Bohlin, J., & Panek, R.J. 1985, *IUE Atlas of O-Type Spectra from* 1200 *to* 1900 Å, NASA RP-1155

15. Walborn, N.R., & Nichols-Bohlin, J. 1987, *PASP*, **99**, 40

16. Massa, D. 1989, *A&A*, **224**, 131

17. Bates, B., & Gilheany, S. 1990, *MNRAS*, **243**, 320

18. Prinja, R.K. 1990, *MNRAS*, **246**, 392

19. Rountree, J., & Sonneborn, G. 1991, *ApJ*, **369**, 515

20. Massa, D., Altner, B., Wynne, D., & Lamers, H.J.G.L.M. 1991, *A&A*, **242**, 188

21. Taylor, B.J. 1984, *ApJS*, **54**, 259

22. Pérez, M.R., Oliversen, N.A., Garhart, M.P., & Teays, T.J. 1990, in *Evolution in Astrophysics: IUE Astronomy in the Era of New Space Missions*, edited by E.J. Rolfe (ESA, Noordwijk), p. 349

23. Turnshek, D.A., Bohlin, R.C., Williamson, R.L., Lupie, O.L., & Koorneef, J. 1990, *ApJ*, **99**, 1243

Chapter 9

X-Ray Astronomy

Frederick D. Seward

9.1 USEFUL CONVERSIONS

$1 \ (\text{keV}) = 1.6021 \times 10^{-9} \ \text{erg} = 1.6021 \times 10^{-16} \ \text{J}$: the kilo-electron-volt.

$E \ (\text{keV}) = 12.398 \ [\lambda \ (\text{Å})]^{-1}$: the energy of a photon.

$E \ (\text{keV}) = 0.862T \ (10^7 \text{K})$: the characteristic energy, kT, of a thermal source.

$\nu \ (\text{Hz}) = 2.998 \times 10^{18} \ [\lambda(\text{Å})]^{-1} = 2.418 \times 10^{17} E \ (\text{keV})$.

$T \ (\text{K}) = 1.160 \times 10^7 \ [kT \ (\text{keV})]$.

$1 \ \mu\text{Jy} = 10^{-29} \ \text{erg cm}^{-2} \text{s}^{-1} \text{Hz}^{-1} = 10^{-32} \ \text{W m}^{-2} \text{Hz}^{-1}$: the micro-Jansky.

Spectra are usually presented as the dependence of spectral irradiance (spectral flux density) I, on wavelength λ (Å), frequency ν (Hz), or photon energy E (keV or erg). To convert from one to the other:

$$I_\lambda(\text{erg cm}^{-2}\text{s}^{-1}\text{Å}^{-1}) = 3.336 \times 10^{-19}\nu^2(\text{Hz}) \ I_\nu(\text{erg cm}^{-2}\text{s}^{-1}\text{Hz}^{-1})$$
$$= 5.034 \times 10^7 E^2(\text{erg}) \ I_E(\text{erg cm}^{-2}\text{s}^{-1}\text{erg}^{-1}),$$

$$I_\nu(\mathrm{erg\,cm^{-2}\,s^{-1}\,Hz^{-1}}) = 3.336 \times 10^{-19}\lambda^2(\text{Å})\,I_\lambda(\mathrm{erg\,cm^{-2}\,s^{-1}\,Å^{-1}})$$
$$= 6.626 \times 10^{-27} I_E\ (\mathrm{keV\,cm^{-2}\,s^{-1}\,keV^{-1}}),$$

$$I_E(\mathrm{keV\,cm^{-2}\,s^{-1}\,keV^{-1}}) = 1.509 \times 10^{26} I_\nu\ (\mathrm{erg\,cm^{-2}\,s^{-1}\,Hz^{-1}})$$
$$= 5.034 \times 10^7 \lambda^2(\text{Å})\,I_\lambda(\mathrm{erg\,cm^{-2}\,s^{-1}\,Å^{-1}}),$$

$$N_p(\mathrm{photon\,cm^{-2}\,s^{-1}\,keV^{-1}}) = I_E(\mathrm{keV\,cm^{-2}\,s^{-1}\,keV^{-1}})E^{-1}(\mathrm{keV}).$$

9.2   CHARACTERISTIC X-RAY TRANSITIONS

Energies of absorption edges and emission lines are given in Table 9.1. All energies are in keV.

9.3   EMISSION MECHANISMS AND SPECTRA

9.3.1   Continuum Models

X-ray spectra have historically been compared to three simple models that imply emission from: (a) high-energy electrons moving in a magnetic field; (b) thermal electrons in an optically thin plasma with temperature, $T > 3 \times 10^7$ K; and (c) thermal radiation from an optically thick object. These spectra are:

(a) Power law,
$$I(E) = A E^{-\alpha}, \quad \alpha = \text{spectral index.}$$

(b) Thermal bremsstrahlung,

$$I(E, T) = AG(E, T)Z^2 n_e n_i (kT)^{-1/2} e^{-E/kT}.$$

Densities of electrons and positive ions are n_e and n_i, respectively, and G is the Gaunt factor, a slowly varying function with increasing value as E decreases [1, 2].
When $E \ll kT$, $G \approx 0.55 \ln(2.25kT/E)$, and when $E \sim kT$, $G \approx (E/kT)^{-0.4}$ is an adequate approximation [3].
When electrons are relativistic, the Gaunt factor can be approximated as

$$G = [0.9 + 0.75(kT/mc^2)](E/kT)^{-1/4} + 1.9(kT/mc^2)(E/kT)^{-1/6} + 3.4(kT/mc^2)^2(E/kT),$$

an approximation better than 20% in the range $(kT/mc^2) \leq 1$, $(E/kT) \leq 6$ [4].
(c) Blackbody radiation,

$$I(E, T) = (2E^3/h^2c^2)(e^{E/kT} - 1)^{-1}.$$

Early observations were usually well fit using these simple models. Spectra of actual sources are, of course, more complex. There are emission lines, absorption edges, and, usually, scattering and absorption in material surrounding, or close to, the sources. Observations with high spectral resolution and good counting statistics, or those covering a broad spectral range, require more complex models for good fits [5].

Table 9.1. *Energies of characteristic lines and edges* [1].

Z	$K(ab)$	$K\beta_3$	$K\beta_1$	$K\beta_2$	$K\alpha_1$	$K\alpha_2$
			K series			
1 H	0.0136					
2 He	0.025					
3 Li	0.055				0.052	
4 Be	0.112				0.110	
5 B	0.192				0.185	
6 C	0.283				0.277	
7 N	0.399				0.392	
8 O	0.531				0.525	
9 F	0.687				0.677	
10 Ne	0.867				0.848	
11 Na	1.072		1.067		1.041	
12 Mg	1.305		1.295		1.253	
13 Al	1.559		1.553		1.486	1.486
14 Si	1.838		1.829		1.740	1.739
15 P	2.142		2.136		2.013	2.012
16 S	2.472		2.464		2.307	2.306
17 Cl	2.822				2.622	2.620
18 Ar	3.202	3.190			2.957	2.955
19 K	3.607	3.589			3.313	3.310
20 Ca	4.038	4.012			3.691	3.687
21 Sc	4.496	4.460			4.090	4.085
22 Ti	4.965	4.931			4.510	4.504
23 V	5.465	5.426			4.951	4.944
24 Cr	5.989	5.946			5.414	5.405
25 Mn	6.540	6.489			5.898	5.887
26 Fe	7.112	7.057			6.403	6.390
27 Co	7.709	7.648			6.929	6.914
28 Ni	8.333	8.263			7.477	7.460
29 Cu	8.979	8.904			8.046	8.026
30 Zn	9.659	9.570		9.656	8.637	8.614
31 Ga	10.368	10.259	10.263	10.365	9.250	9.223
32 Ge	11.104	10.976	10.980	11.099	9.885	9.854
33 As	11.868	11.718	11.724	11.862	10.542	10.506
34 Se	12.658	12.437	12.494	12.650	11.220	11.179
35 Br	13.474	13.282	13.289	13.467	11.922	11.876
36 Kr	14.322	14.102	14.110	14.312	12.648	12.596
37 Rb	15.201	14.949	14.959	15.183	13.393	13.333
38 Sr	16.105	15.822	15.833	16.082	14.163	14.095
39 Y	17.037	16.723	16.735	17.013	14.956	14.880
40 Zr	17.998	17.651	17.665	17.967	15.772	15.688
41 Nb	18.986	18.603	18.619	18.949	16.612	16.518
42 Mo	20.002	19.587	19.605	19.962	17.476	17.371
43 Tc	21.054	20.595	20.615	21.002	18.364	18.248
44 Ru	22.118	21.631	21.653	22.070	19.276	19.147
45 Rh	23.224	22.695	22.720	23.169	20.213	20.070
46 Pd	24.350	23.787	23.815	24.295	21.174	21.017
47 Ag	25.514	24.907	24.938	25.452	22.159	21.987
48 Cd	26.711	26.057	26.091	26.639	23.170	22.980
49 In	27.940	27.233	27.271	27.856	24.206	23.998

Table 9.1. *(Continued.)*

Z	K(ab)	Kβ₃	Kβ₁	Kβ₂	Kα₁	Kα₂
			K series			
50 Sn	29.200	28.439	28.481	29.104	25.267	25.040
51 Sb	30.491	29.674	29.721	30.388	26.355	26.106
52 Te	31.813	30.939	30.990	31.698	27.468	27.197
53 I	33.169	32.234	32.289	33.036	28.607	28.312
54 Xe	34.582	33.556	33.619	34.408	29.774	29.453
55 Cs	35.959	34.913	34.981	35.815	30.968	30.620
56 Ba	37.441	36.298	36.372	37.251	32.188	31.812
57 La	38.925	37.714	37.795	38.723	33.436	33.028
58 Ce	40.449	39.163	39.251	40.226	34.714	34.273
59 Pr	41.998	40.646	40.741	41.767	36.020	35.544
60 Nd	43.571	42.159	42.264	43.327	37.355	36.841
61 Pm	45.207	43.705	43.818	44.929	38.718	38.165
62 Sm	46.835	45.281	45.405	46.566	40.111	39.516
63 Eu	48.515	46.896	47.030	48.248	41.535	40.895
64 Gd	50.240	48.547	48.688	49.952	42.989	42.302
65 Tb	51.996	50.221	50.374	51.715	44.474	43.737
66 Dy	53.789	51.949	52.110	53.500	45.991	45.200
67 Ho	55.615	53.702	53.868	55.315	47.539	46.692
68 Er	57.483	55.485	55.672	57.204	49.119	48.213
69 Tm	59.390	57.293	57.506	59.085	50.733	49.764
70 Yb	61.332	59.141	59.356	60.974	52.380	51.345
71 Lu	63.304	61.037	61.272	62.956	54.061	52.956
72 Hf	65.351	62.969	63.222	64.969	55.781	54.602
73 Ta	67.414	64.938	65.212	67.001	57.523	56.267
74 W	69.524	66.940	67.233	69.089	59.308	57.972
75 Re	71.662	68.983	69.298	71.219	61.130	59.708
76 Os	73.860	71.065	71.401	73.390	62.990	61.476
77 Ir	76.112	73.190	73.548	75.606	64.885	63.276
78 Pt	78.395	75.355	75.735	77.864	66.821	65.112
79 Au	80.723	77.567	77.971	80.172	68.792	66.978
80 Hg	83.103	79.809	80.240	82.530	70.807	68.883
81 Tl	85.528	82.104	82.562	84.933	72.859	70.820
82 Pb	88.006	84.436	84.922	87.351	74.956	72.792
83 Bi	90.572	86.819	87.328	89.846	77.095	74.802
84 Po	93.112	89.231	89.781	92.383	79.279	76.851
85 At	95.740	91.707	92.287	94.974	81.499	78.930
86 Rn	98.418	94.230	94.850	97.622	83.768	81.051
87 Fr	101.147	96.791	97.460	100.307	86.089	83.217
88 Ra	103.927	99.415	100.113	103.051	88.454	85.419
89 Ac	106.759	102.084	102.829	105.849	90.868	87.660
90 Th	109.649	104.813	105.591	108.699	93.334	89.938
91 Pa	112.581	107.576	108.409	111.605	95.852	92.271
92 U	115.603	110.387	111.281	114.587	98.422	94.649
93 Np	118.619		113.725	118.057	100.781	96.844
94 Pu	121.760		116.943	120.350	103.300	99.168
95 Am	124.876		120.350	123.960	105.949	101.607
96 Cm	128.088		122.733	126.490	108.737	104.168
97 Bk	131.357		126.490	130.484	111.676	106.862
98 Cf	134.683		127.794	133.290	114.778	109.699

Table 9.1. (*Continued.*)

Z	L_I series				L_II series				L_III series				
	L_I(ab)	$L\gamma_3$	$L\beta_3$	$L\beta_4$	L_{II}(ab)	$L\gamma_1$	$L\beta_1$	$L\eta$	L_{III}(ab)	$L\beta_2$	$L\alpha_1$	$L\alpha_2$	Ll
19 K								0.262					0.260
20 Ca	0.400					0.350	0.345	0.306	0.346		0.341		0.303
21 Sc	0.463					0.407	0.400	0.353	0.403		0.395		0.348
22 Ti	0.530					0.460	0.458	0.401	0.454		0.452		0.395
23 V	0.604		0.585			0.520	0.519	0.453	0.513		0.511		0.446
24 Cr	0.682		0.654			0.583	0.583	0.510	0.574		0.573		0.500
25 Mn	0.754		0.721			0.652	0.649	0.567	0.641		0.637		0.556
26 Fe	0.842		0.792			0.721	0.718	0.628	0.709		0.705		0.615
27 Co	0.929		0.866			0.794	0.791	0.694	0.799		0.776		0.678
28 Ni	1.012		0.941			0.872	0.869	0.762	0.855		0.851		0.743
29 Cu	1.100		1.023			0.952	0.950	0.832	0.932		0.930		0.811
30 Zn	1.196		1.107			1.044	1.034	0.906	1.021		1.012		0.884
31 Ga	1.300		1.197			1.134	1.125	0.984	1.117		1.098		0.957
32 Ge	1.420		1.294	1.286		1.249	1.218	1.068	1.218		1.188		1.036
33 As	1.530		1.388			1.360	1.317	1.155	1.325		1.282		1.120
34 Se	1.653		1.490			1.477	1.419	1.244	1.436		1.379		1.204
35 Br	1.794		1.596		1.596		1.526	1.399	1.550		1.480		1.293
36 Kr	1.920		1.706	1.697	1.756		1.636		1.675		1.586		
37 Rb	2.067		1.826	1.817	1.866		1.752	1.542	1.806		1.694	1.692	1.482
38 Sr	2.216		1.947	1.936	2.007		1.871	1.649	1.940		1.806	1.804	1.582
39 Y	2.369		2.072	2.060	2.145		1.995	1.761	2.079		1.922	1.920	1.685
40 Zr	2.547		2.201	2.187	2.307	2.302	2.124	1.876	2.223	2.219	2.042	2.040	1.792
41 Nb	2.698		2.334	2.319	2.465	2.461	2.257	1.996	2.371	2.367	2.166	2.163	1.902
42 Mo	2.866		2.473	2.455	2.625	2.623	2.394	2.120	2.520	2.518	2.293	2.289	2.015
43 Tc	3.054				2.795		2.536		2.677		2.424		
44 Ru	3.236		2.763	2.741	2.996	2.964	2.683	2.382	2.837	2.835	2.558	2.554	2.252
45 Rh	3.419		2.915	2.890	3.146	3.143	2.834	2.519	3.003	3.001	2.696	2.692	2.376

Table 9.1. (*Continued.*)

Z	L_I series				L series — L_II series				L_III series				
	L_I(ab)	$L\gamma_3$	$L\beta_3$	$L\beta_4$	L_{II}(ab)	$L\gamma_1$	$L\beta_1$	$L\eta$	L_{III}(ab)	$L\beta_2$	$L\alpha_1$	$L\alpha_2$	Ll
46 Pd	3.617		3.072	3.045	3.330	3.328	2.990	2.660	3.173	3.171	2.838	2.833	2.503
47 Ag	3.806	3.749	3.234	3.203	3.524	3.519	3.150	2.806	3.351	3.347	2.984	2.978	2.633
48 Cd	4.019		3.401	3.367	3.727	3.716	3.316	2.956	3.537	3.528	3.133	3.126	2.767
49 In	4.237		3.572	3.535	3.938	3.920	3.487	3.112	3.730	3.713	3.286	3.279	2.904
50 Sn	4.465		3.750	3.708	4.156	4.130	3.662	3.272	3.929	3.904	3.443	3.435	3.044
51 Sb	4.698		3.932	3.886	4.381	4.347	3.843	3.436	4.132	4.100	3.604	3.595	3.188
52 Te	4.939		4.120	4.069	4.612	4.570	4.029	3.605	4.341	4.301	3.769	3.758	3.335
53 I	5.188		4.313	4.257	4.852	4.800	4.220	3.780	4.557	4.507	3.937	3.925	3.484
54 Xe	5.452				5.100				4.781		4.109		
55 Cs	5.720	5.552	4.716	4.649	5.358	5.279	4.619	4.141	5.011	4.935	4.286	4.272	3.794
56 Ba	5.955	5.808	4.926	4.851	5.624	5.530	4.827	4.330	5.247	5.156	4.465	4.450	3.953
57 La	6.267	6.073	5.143	5.061	5.891	5.788	5.041	4.524	5.483	5.383	4.650	4.633	4.124
58 Ce	6.549	6.340	5.364	5.276	6.165	6.051	5.261	4.731	5.724	5.612	4.839	4.822	4.287
59 Pr	6.846	6.615	5.591	5.497	6.443	6.321	5.488	4.935	5.968	5.849	5.033	5.013	4.452
60 Nd	7.126	6.900	5.828	5.721	6.722	6.601	5.721	5.145	6.208	6.088	5.229	5.207	4.632
61 Pm	7.448		6.070		7.018	6.891	5.960		6.466	6.338	5.432	5.407	
62 Sm	7.737	7.485	6.317	6.195	7.312	7.177	6.205	5.588	6.717	6.586	5.635	5.607	4.994
63 Eu	8.069	7.795	6.570	6.438	7.624	7.479	6.455	5.816	6.983	6.842	5.845	5.816	5.176
64 Gd	8.376	8.104	6.830	6.686	7.931	7.784	6.712	6.049	7.243	7.102	6.056	6.024	5.361
65 Tb	8.708	8.422	7.095	6.939	8.252	8.100	6.977	6.283	7.515	7.365	6.272	6.237	5.546
66 Dy	9.083	8.752	7.369	7.203	8.621	8.417	7.246	6.533	7.850	7.634	6.494	6.457	5.742
67 Ho	9.395	9.086	7.650	7.470	8.919	8.746	7.524	6.787	8.071	7.910	6.719	6.679	5.942
68 Er	9.776	9.429	7.938	7.744	9.263	9.087	7.809	7.057	8.364	8.188	6.947	6.904	6.152
69 Tm	10.116	9.778	8.229	8.024	9.618	9.424	8.100	7.308	8.648	8.467	7.179	7.132	6.341
70 Yb	10.486	10.141	8.535	8.312	9.978	9.778	8.400	7.579	8.943	8.757	7.414	7.366	6.544
71 Lu	10.867	10.509	8.845	8.605	10.345	10.142	8.708	7.856	9.241	9.047	7.654	7.604	6.752
72 Hf	11.264	10.889	9.162	8.904	10.739	10.514	9.021	8.138	9.561	9.346	7.898	7.843	6.958

Table 9.1. (*Continued.*)

	L_I series				L_{II} series				L_{III} series				
												L series	
Z	L_I(ab)	$L\gamma_3$	$L\beta_3$	$L\beta_4$	L_{II}(ab)	$L\gamma_1$	$L\beta_1$	$L\eta$	L_{III}(ab)	$L\beta_2$	$L\alpha_1$	$L\alpha_2$	Ll
73 Ta	11.680	11.276	9.486	9.211	11.139	10.893	9.342	8.427	9.881	9.650	8.145	8.086	7.172
74 W	12.098	11.672	9.817	9.524	11.542	11.284	9.671	8.723	10.204	9.960	8.396	8.334	7.386
75 Re	12.522	12.080	10.158	9.845	11.955	11.683	10.008	9.026	10.531	10.274	8.651	8.585	7.602
76 Os	12.965	12.498	10.509	10.174	12.383	12.093	10.354	9.335	10.869	10.597	8.910	8.840	7.821
77 Ir	13.424	12.922	10.866	10.509	12.824	12.510	10.706	9.649	11.215	10.919	9.174	9.098	8.040
78 Pt	13.892	13.359	11.233	10.852	13.273	12.940	11.069	9.973	11.564	11.249	9.441	9.360	8.267
79 Au	14.353	13.807	11.608	11.203	13.733	13.379	11.440	10.307	11.918	11.583	9.712	9.626	8.493
80 Hg	14.846	14.262	11.993	11.561	14.209	13.828	11.821	10.649	12.284	11.922	9.987	9.896	8.720
81 Tl	15.344	14.734	12.388	11.929	14.698	14.289	12.211	10.992	12.657	12.270	10.267	10.171	8.952
82 Pb	15.860	15.215	12.791	12.304	15.198	14.762	12.612	11.347	13.035	12.621	10.550	10.448	9.183
83 Bi	16.385	15.708	13.208	12.689	15.708	15.245	13.021	11.710	13.418	12.978	10.837	10.729	9.419
84 Po	16.935		13.635	13.083	16.244	15.741	13.445		13.817	13.338	11.129	11.014	9.662
85 At	17.490		14.065		16.784	16.249	13.874		14.215		11.425	11.303	
86 Rn	18.058		14.509		17.337	16.768	14.313		14.618		11.725	11.596	
87 Fr	18.638		14.973		17.904	17.300	14.768		15.028	14.448	12.029	11.893	
88 Ra	19.233	18.354	15.442	14.745	18.480	17.845	15.233	13.661	15.442	14.839	12.338	12.194	10.620
89 Ac	19.842		15.929		19.078	18.405	15.710		15.865		12.650	12.499	
90 Th	20.470	19.503	16.423	15.640	19.692	18.979	16.199	14.507	16.300	15.621	12.987	12.807	11.117
91 Pa	21.102	20.094	16.927	16.101	20.311	19.565	16.699	14.944	16.731	16.022	13.288	13.120	11.364
92 U	21.756	20.709	17.452	16.573	20.947	20.164	17.217	15.397	17.167	16.425	13.612	13.437	11.616
93 Np	22.417	21.336	17.986	17.058	21.596	20.781	17.747	15.874	17.614	16.837	13.942	13.757	11.887
94 Pu	23.095	21.979	18.537	17.553	22.263	21.414	18.291	16.330	18.053	17.252	14.276	14.082	12.122
95 Am	23.793		19.103	18.060	22.944	22.061	18.849		18.526	17.673	14.615	14.409	12.381
96 Cm	24.503				23.640	22.703	19.399		18.990	18.096	14.953	14.740	
97 Bk	25.230				24.352	23.389	19.961		19.461	18.529	15.304	15.080	
98 Cf	25.971				25.080	24.070	20.557		19.938	18.983	15.652	15.418	

Table 9.1. *(Continued.)*

		M series			
	M_{IV} series		M_V series		
Z	M_{IV}(ab)	$M\beta$	M_V(ab)	$M\alpha_1$	$M\alpha_2$
57 La	0.851	0.854			
58 Ce	0.902	0.902			
59 Pr		0.949			
60 Nd	1.004	0.996			
61 Pm					
62 Sm	1.108	1.100			
63 Eu		1.153			
64 Gd	1.221	1.209			
65 Tb	1.280	1.266			
66 Dy		1.325			
67 Ho	1.390	1.383			
68 Er		1.443			
69 Tm	1.515	1.503			
70 Yb	1.578	1.567			
71 Lu		1.631			
72 Hf	1.718	1.697			
73 Ta	1.793	1.765			
74 W	1.871	1.835	1.809	1.775	1.773
75 Re		1.906			
76 Os		1.978			
77 Ir	2.116	2.053	2.041	1.980	1.975
78 Pt	2.202	2.127	2.122	2.050	2.046
79 Au	2.291	2.204	2.206	2.123	2.118
80 Hg	2.385	2.282	2.295		
81 Ti	2.485	2.362	2.389	2.270	2.265
82 Pb	2.586	2.442	2.484	2.345	2.339
83 Bi	2.687	2.525	2.579	2.422	2.416
84 Po					
85 At					
86 Rn					
87 Fr					
88 Ra					
89 Ac					
90 Th	3.491	3.145	3.332	2.996	2.986
91 Pa		3.239		3.082	3.072
92 U	3.728	3.336	3.552	3.170	3.159

Reference
1. Woldseth, R. 1973, *X-Ray Energy Spectroscopy* (Kevex Corp., Burlingame, CA)

9.3.2 Line Emission

When the source is "thin" and $T < 3 \times 10^7$ K, the dominant radiation mechanism is line emission. Elemental abundances [6] and temperature determine which lines are strongest. In the X-ray band, lines from H-like and He-like ions and from ions of O and Fe are usually prominent. Figures 9.1 through 9.3 show the relative numbers of selected ions as a function of temperature [7]. Collisional equilibrium is assumed. Some sources, however, have ages smaller than the time required to achieve equilibrium. Young supernova remnants are the most obvious examples. In these cases, the ion populations are quite different from those expected at the temperature indicated by the continuum radiation. Emission from such nonequilibrium models has been calculated [8].

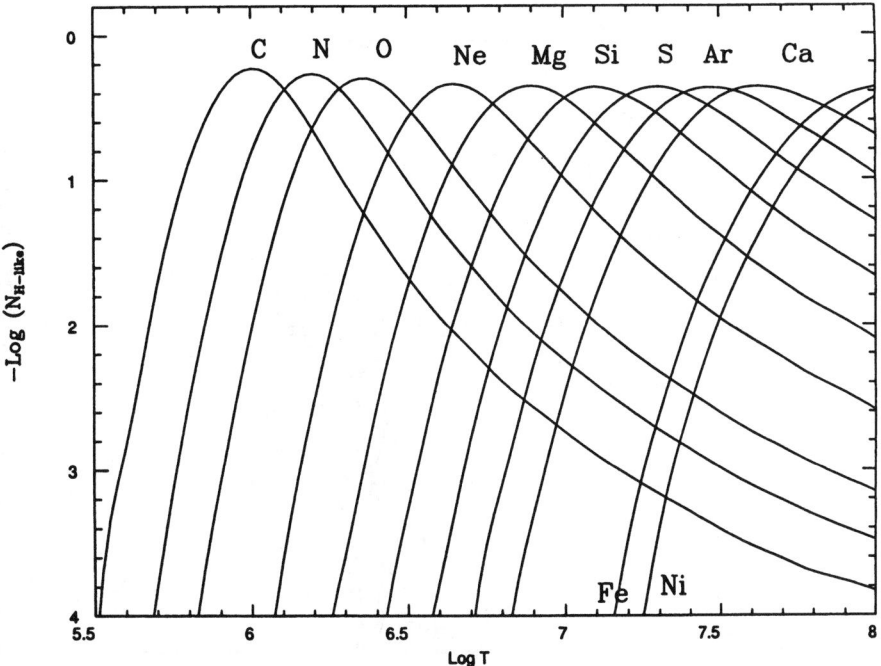

Figure 9.1. Concentration of hydrogen-like ions versus temperature.

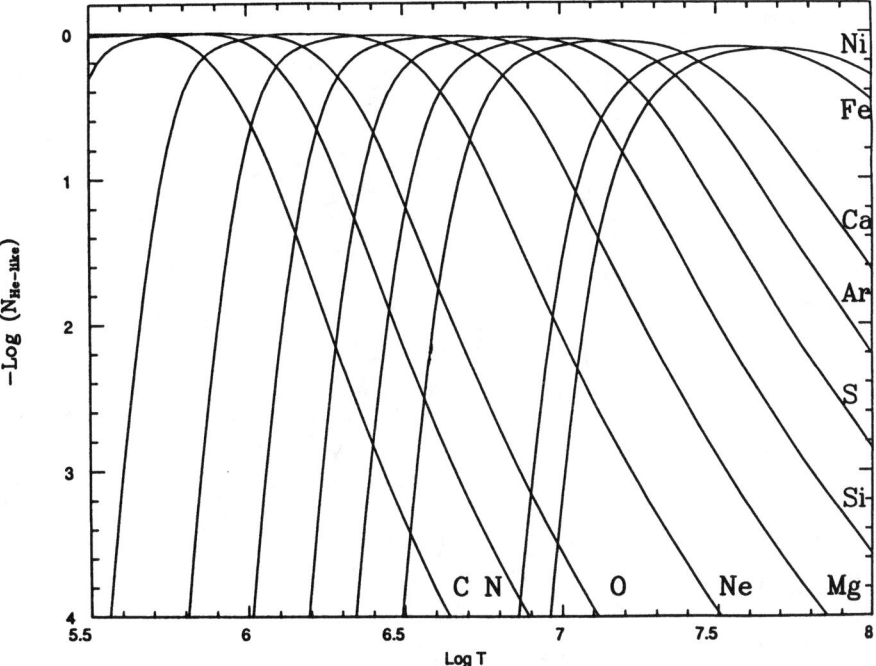

Figure 9.2. Concentration of helium-like ions versus temperature.

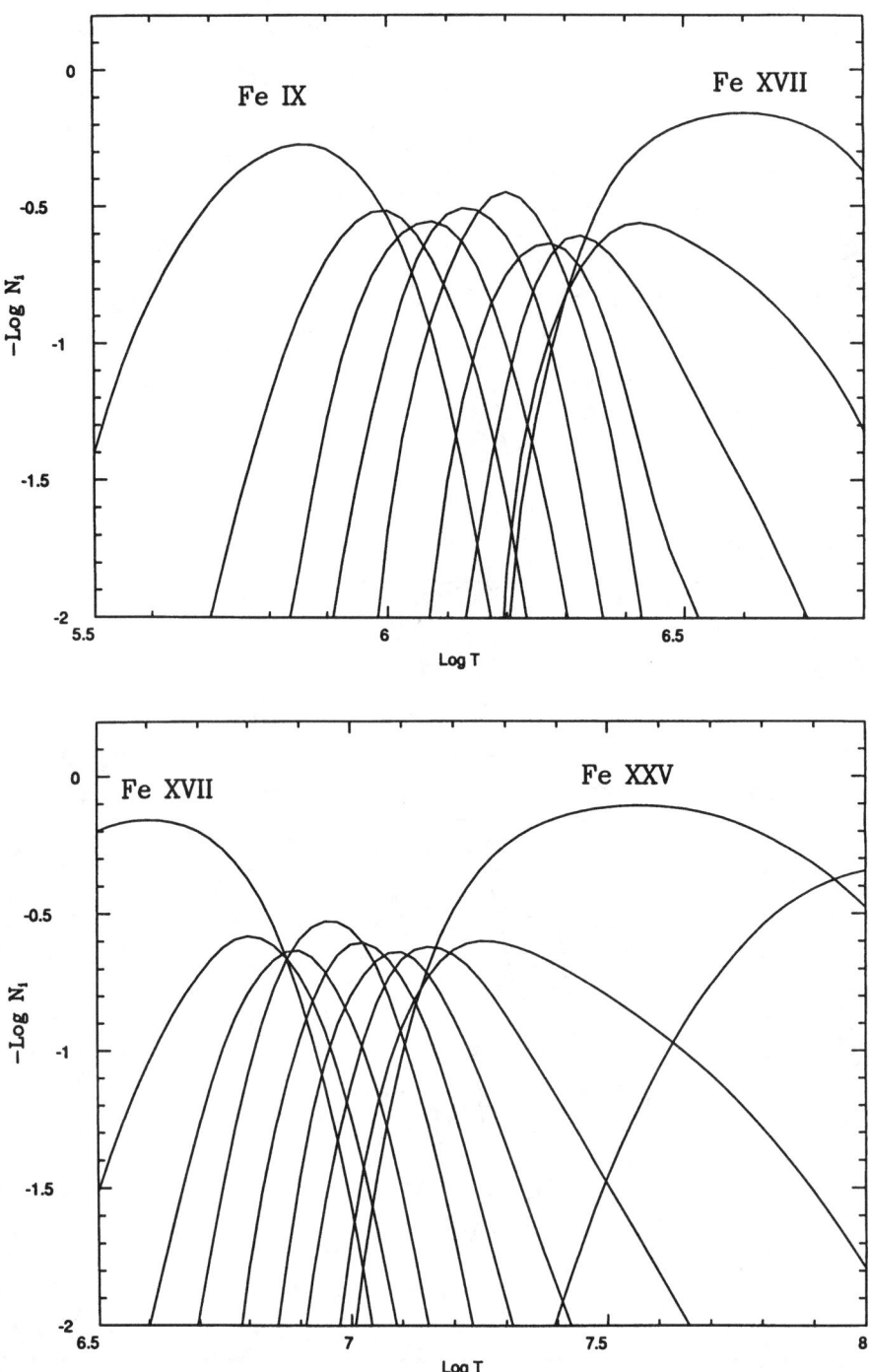

Figure 9.3. Concentration of iron ions versus temperature. Above: low temperatures; below: high temperatures.

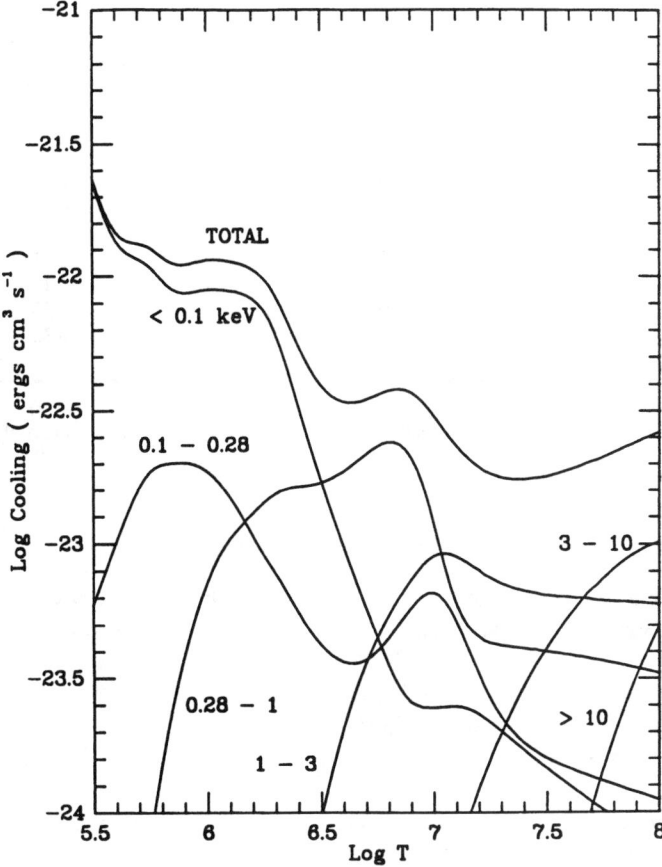

Figure 9.4. Power radiated from a low-density plasma.

Table 9.2 gives the power from a low-density plasma, with solar abundances, in units of 10^{-23} ergs cm^3 s^{-1}. Figure 9.4 plots these data.

Table 9.2. *Power radiated from a low-density plasma* [1].

log T	< 0.10 keV	0.10–0.28	0.28–1.00	1.00–3.00	3.0–10.0	> 10.0	Total
5.500	22.75	0.49	0.00	0.00	0.00	0.00	23.24
5.600	13.71	0.92	0.02	0.00	0.00	0.00	14.66
5.700	11.66	1.52	0.12	0.00	0.00	0.00	13.30
5.800	9.84	1.90	0.37	0.00	0.00	0.00	12.11
5.900	8.69	1.96	0.79	0.00	0.00	0.00	11.44
6.000	8.93	1.85	1.29	0.00	0.00	0.00	12.07
6.100	8.82	1.49	1.67	0.00	0.00	0.00	11.98
6.200	8.25	1.08	1.82	0.00	0.00	0.00	11.15
6.300	6.15	0.79	1.89	0.01	0.00	0.00	8.84
6.400	3.35	0.56	1.87	0.03	0.00	0.00	5.81
6.500	1.75	0.42	1.85	0.09	0.00	0.00	4.12
6.600	1.00	0.36	1.99	0.20	0.00	0.00	3.55
6.700	0.60	0.37	2.28	0.35	0.00	0.00	3.59
6.800	0.37	0.45	2.51	0.52	0.01	0.00	3.85
6.900	0.29	0.55	2.20	0.72	0.01	0.00	3.78

Table 9.2. *(Continued.)*

log T	< 0.10 keV	0.10–0.28	0.28–1.00	1.00–3.00	3.0–10.0	> 10.0	Total
7.000	0.38	0.53	1.28	0.89	0.03	0.00	3.11
7.100	0.43	0.35	0.67	0.89	0.07	0.00	2.41
7.200	0.34	0.23	0.52	0.80	0.12	0.00	2.01
7.300	0.26	0.18	0.48	0.71	0.20	0.00	1.83
7.400	0.20	0.16	0.47	0.67	0.30	0.01	1.81
7.500	0.16	0.15	0.46	0.65	0.41	0.02	1.85
7.600	0.14	0.14	0.44	0.63	0.55	0.05	1.96
7.700	0.12	0.13	0.43	0.62	0.68	0.11	2.10
7.800	0.10	0.12	0.41	0.62	0.81	0.20	2.26
7.900	0.09	0.12	0.39	0.61	0.93	0.32	2.46
8.000	0.08	0.11	0.37	0.60	1.02	0.48	2.66

Reference
1. Raymond, J.C. 1992, current version of code described by Raymond, J.C., & Smith, B.W. 1977, *ApJ*, **35**, 419

9.3.3   X-Ray Sources for In-Flight Calibration

The best-known X-ray spectrum is that of the Crab Nebula. It has small angular extent, and for our purpose, is not time variable (the pulsar contribution is only ∼ 5% in the soft X-ray band and the diffuse nebula is probably only slowly decreasing in flux). It has a simple power-law spectral continuum. The Crab spectrum between 3 and 30 keV is $I(E) = (9.7 \pm 1)E^{-1.10 \pm .03}$ keV cm^{-2} s^{-1} keV^{-1} [9]. Below 3 keV this spectrum is decreased by interstellar absorption, which is not well measured. Most observations fall within the range $N_H = (2 - 3.5) \times 10^{21}$ atoms cm^{-2} (Morrison and McCammon abundances).

Other sources which, because of small angular extent, simple spectra, and constant flux, are suitable for calibration purposes are: the supernova remnant G21.5-0.9, the clusters A1795 and A401, and, probably, the white dwarf HZ43. The supernova remnant N132D in the Large Magellanic Cloud is in a favorable location (close to the ecliptic pole) and often used, but has a thermal spectrum with detailed spectral structure.

9.4   TRANSMISSION OF X-RAYS THROUGH THE INTERSTELLAR MEDIUM

The transmission of X-rays through interstellar gas (which is cold and atomic) depends on column density, usually expressed as the number of hydrogen atoms cm^{-2}, N_H, and elemental composition. Two models with different elemental composition and absorption cross sections have been used extensively in the literature: that of Brown and Gould (BG) [10], and that of Morrison and McCammon (MM) [11]. The BG gas contains H, He, C, N, O, Ne, Mg, Si, S, and Ar. The MM gas contains these same elements with updated cross sections [12] plus Ca, Fe, and Ni. Figure 9.5 shows the relative importance of these elements. At 0.7 keV, for example, half the absorption is in O. At 10 keV, half the absorption is in Fe. Since very little of the absorption occurs in H, the use of N_H is somewhat misleading. This is sometimes referred to as the number of "equivalent H atoms" cm^{-2}.

9.4.1   Transmission of the MM Gas

Figure 9.6 shows graphs of transmission through the MM gas over a large range of column densities. Only photoelectric absorption has been considered. Compton scattering will become important at energies above 5 keV, so 10^{24} atoms cm^{-2} is the highest column density shown. The two prominent absorption edges are O K at 0.53 keV and Fe K at 7.1 keV.

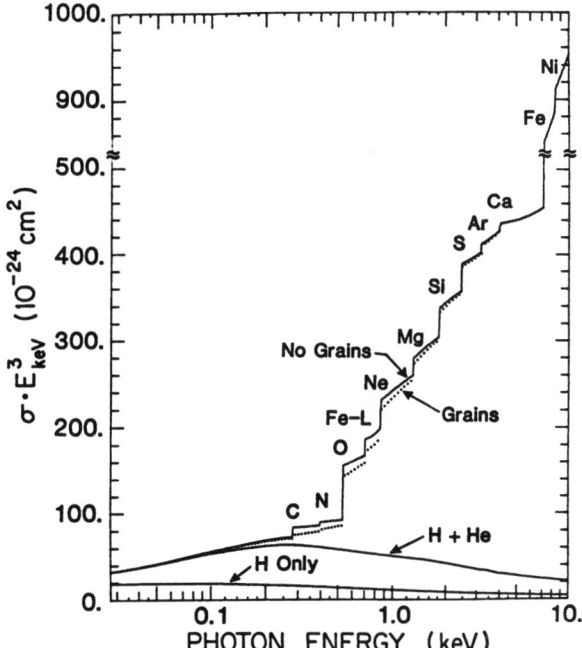

Figure 9.5. X-ray absorption coefficient versus energy for the ISM. The structure is caused by atomic absorption edges [11].

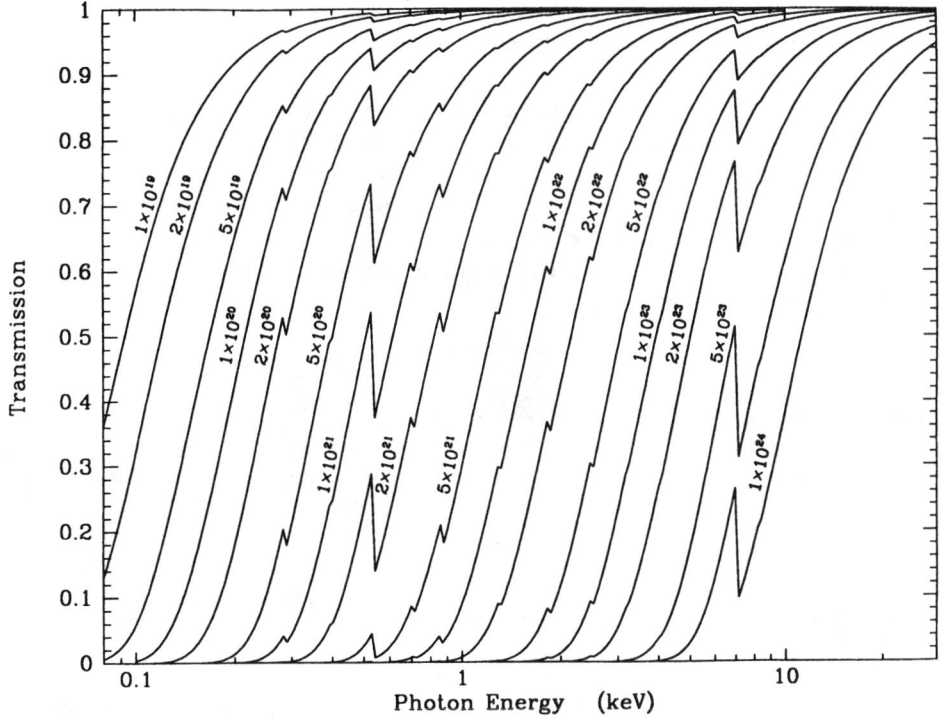

Figure 9.6. Transmission versus energy for various interstellar medium column densities.

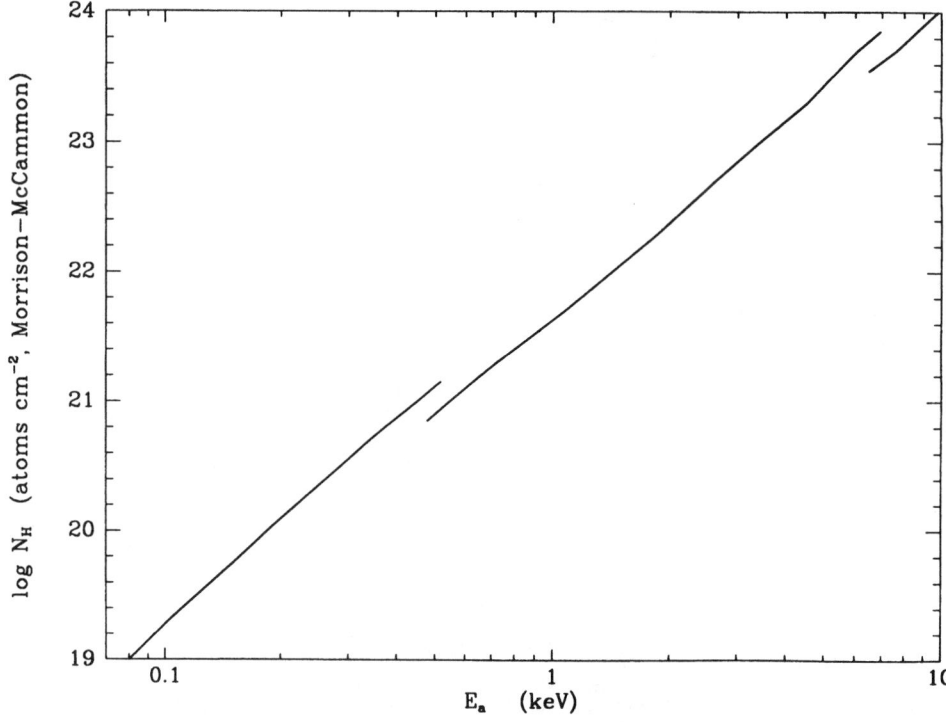

Figure 9.7. Conversion of the parameter E_a to column density.

9.4.2   Comparison of Different Models

In order to compare published analyses, a conversion between different absorption-measure models is needed. Table 9.3 gives a comparison between the measured BG and MM column densities. The conversion is not precise because the BG and MM transmission curves do not have the same shape. Transmission through the column densities in Table 9.3 cross at ~ 0.5, but the BG transmission curve is steeper at low N_H, and the MM curve is steeper at high N_H.

In early work, the parameter E_a was used to quantify observed absorption. The transmission T_r through interstellar gas was expressed as

$$T_r = e^{-(Ea/E)^{8/3}}.$$

Figure 9.7 gives the conversion of E_a to MM column density. It is assumed that the measurement is made over that energy range where T_r drops from 0.8 to 0.2, and there is some ambiguity when the O or Fe absorption edge falls in this range. The value of the conversion factor at these energies will depend on the characteristics of the X-ray detector used to make the observation.

Table 9.3. *Equivalent transmission column densities for MM and BG gas.*

Model	Column					
MM gas	1×10^{19}	1×10^{20}	1×10^{21}	1×10^{22}	1×10^{23}	1×10^{24}
BG gas	0.90×10^{19}	1.18×10^{20}	1.15×10^{21}	1.25×10^{22}	0.95×10^{23}	0.80×10^{24}
E_a	0.080	0.187	~ 0.50	1.41	3.4	9.8

9.4.3 Relation to Optical Extinction

An empirical relation between X-ray absorption and optical extinction has been noted. Using diffuse supernova remnants as calibrators, the relationship to extinction, A_V, and to color excess, E_{B-V}, were found to be:

$$N_H/A_V = 1.9 \times 10^{21} \text{ atoms cm}^{-2} \text{ mag}^{-1} \text{ [13]},$$
$$N_H/E_{B-V} = (5.9 \pm 1.6) \times 10^{21} \text{ atoms cm}^{-2} \text{ mag}^{-1} \text{ [14]}.$$

Agreement with *Copernicus* observations of absorption in atomic (H_1) and molecular (H_2) hydrogen is quite good. The average measured column density toward 100 stars (90 closer than 2 kpc) is

$$\langle N(H_I + H_2)/E_{B-V} \rangle = 5.8 \times 10^{21} \text{ atoms cm}^{-2} \text{ mag}^{-1} \text{ [15]}.$$

ROSAT (Röntgensatellit) observations of bright, strongly absorbed sources have given a measure of both soft X-ray absorption and of the brightness and extent of dust-scattering halos, a direct relation between dust and gas. The result is

$$N_H/A_V = 1.79 \pm 0.03 \times 10^{21} \text{ atoms cm}^{-2} \text{ mag}^{-1} \text{ [16]}.$$

Table 9.4. *Conversion of count rate to millicrab, μJy, and energy flux.*

Satellite	Instrument	Crab rate (count s^{-1})	1 count s^{-1} =	erg cm^{-2} count^{-1}	Energy range (keV)
Vela 5B		40 [1]	27 μJy @ 5 keV	4.5×10^{-10}	3–12
Uhuru	survey	947 [2]	1.15 μJy @ 5 keV	1.6×10^{-11}	2–6
Ariel V	survey	403 [3, 4]	2.70 μJy @ 5 keV	5.3×10^{-11}	2–10
HEAO-1	A1	13 600^a [5]	0.080 μJy @ 5 keV	2.7×10^{-12}	1–20
	A2 (1 keV band)	575 [6]	4.8 μJy @ 2 keV	2.7×10^{-11}	0.5–3
	A4 (channel A)	11.2 [7]	21 μJy @ 20 keV	6.8×10^{-10}	13–25
Einstein	HRI	120 [8]	23 μJy @ 2 keV	1.6×10^{-10}	0.5–4
	IPC	684 [8]	4.0 μJy @ 2 keV	2.8×10^{-11}	0.5–4
	MPC	1 383 [8]	0.79 μJy @ 5 keV	1.6×10^{-11}	1–6
EXOSAT	ME	1 740 [9]	0.63 μJy @ 5 keV	2.1×10^{-11}	1–20
Tenma	GSPC(A + B)	1 640 [10]	0.66 μJy @ 5 keV	2.2×10^{-11}	1–20
Ginga	LAC	10 500 [11]	0.104 μJy @ 5 keV	2.8×10^{-12}	2–20
ROSAT	HRI	362 [12]	7.6 μJy @ 2 keV	3.7×10^{-11}	0.5–2.4
	PSPC	964 [12]	2.8 μJy @ 2 keV	1.4×10^{-11}	0.5–2.4
ASCA	SIS	1020 [12]	1.07 μJy @ 5 keV	3.3×10^{-11}	0.4–12
	GIS	890 [12]	1.22 μJy @ 5 keV	3.8×10^{-11}	0.4–12
XTE	PCA	11 800 [12]	0.092 μJy @ 5 keV	3.1×10^{-12}	1–20
	HEXTE	200 [12]	1.2 μJy @ 20 keV	9.0×10^{-11}	12–60
BeppoSAX	MECS	328 [12]	3.3 μJy @ 5 keV	8.1×10^{-11}	1.3–10
	HPGSPC	334 [12]	0.71 μJy @ 20 keV	7.8×10^{-11}	4–34
	PDS	210 [12]	1.13 μJy @ 20 keV	9.5×10^{-11}	13–80
Chandra	HRC-I	1 200 [12]	2.3 μJy @ 2 keV	2.8×10^{-11}	0.4–10
	ACIS-I	3 200 [12]	0.34 μJy @ 5 keV	1.1×10^{-11}	0.4–10

Note

^{a}HEAO-1 A1 rates are cataloged in counts cm^{-2} s^{-1}; to convert, use area = 3 300 cm^2.

References

1. Priedhorsky, W.C., Terrell, J. & Holt, S.S. 1983, *ApJ*, **270**, 233
2. Forman, W. et al. 1978, *ApJS*, **38**, 357

3. Warwick, R.W. et al. 1981, *MNRAS*, **197**, 865
4. McHardy, I.M. et al. 1981, *MNRAS*, **197**, 893
5. Wood, K.S. et al. 1984, *ApJS*, **56**, 507
6. Nugent, J.J. et al. 1983, *ApJS*, **51**, 1
7. Levine, A.M. et al. 1984, *ApJS*, **54**, 581
8. Seward, F.D. 1990, *ApJS*, **73**, 781
9. White, N. 1992, the ME catalog accessed through the HEASARC online database
10. Koyama, K. et al. 1984, *PASJ*, **36**, 659
11. Turner, M.J.L. et al. 1989, *PASJ*, **41**, 345
12. PIMMS program, 1997, NASA/GSFC HEASARC online software

9.5 COSMIC X-RAY SOURCES

9.5.1 All-Sky Surveys

Sensitive surveys have been completed by Uhuru (2–6 keV) [17]; Ariel V (2–18 keV) [18, 19]; HEAO-1 (1–20 keV) [20]; *Einstein* in the slew mode (0.2–4 keV) [21]; and ROSAT (0.2–2.4 keV) [22]. Table 9.4 gives sensitivities of the survey instruments. Figure 9.8 shows the 842 sources found by HEAO-1. HEAO-1 also covered lower [23] and higher [24] energy bands and, because of its 10-year coverage, the Vela 5B data are unique [25].

Count rates from these surveys and from other missions can be converted to spectral irradiance assuming a Crab-like spectrum [a well-known power law (see Sec. 9.3.3) modified by absorption below 3 keV and with no spectral flux below 0.4 keV]. The Crab spectral irradiance at 2, 5, 10, and 20 keV is 2750, 1090, 510 and 238 μJy. The energy flux per count listed in Table 9.4 was calculated using a Crab-like spectrum, including interstellar absorption. This conversion is reasonable for most sources with absorbing column $> 5 \times 10^{20}$ atoms cm^{-2}. For soft sources and the *Einstein* and ROSAT detectors, which have appreciable sensitivity down to 0.1 or 0.2 keV, the energy conversion factors in Table 9.4 can be an order of magnitude too large.

9.5.2 Types of Sources

Tables 9.5–9.10 give characteristics of sources in several broad categories, and are by no means complete. They include some of the brightest sources and others chosen to illustrate the variety of objects in each category [26].

Table 9.5. *Brightest X-ray emitting supernova remnants* [1].

Galactic coordinates	Name	Flux density 0.2–4 keV (millicrab)	Flux density 1.5–10 keV (millicrab)	Distance (LY[a])	X-ray diameter (arcmin)	L_X 0.2–4 keV (erg s^{-1})
184.6−5.8	Crab Nebula	1 000	1 000	6 500	2	3×10^{37}
74.3−8.5	Cygnus Loop	965	10	2 500	160	2×10^{36}
263.5−2.7	Vela XYZ	760	39	1 500	420	3×10^{35}
260.4−3.4	Puppis A	365	26	6 500	50	6×10^{36}
111.7−2.1	Cassiopeia A	89	80	8 000	4	1×10^{37}
120.1+1.4	Tycho SNR	32	20	8 000	8	3×10^{36}

Note

[a]LY = light year = 0.31 parsec $\approx 1 \times 10^{16}$ m.

Reference

1. Seward, F.D. 1990, *ApJS*, **73**, 781

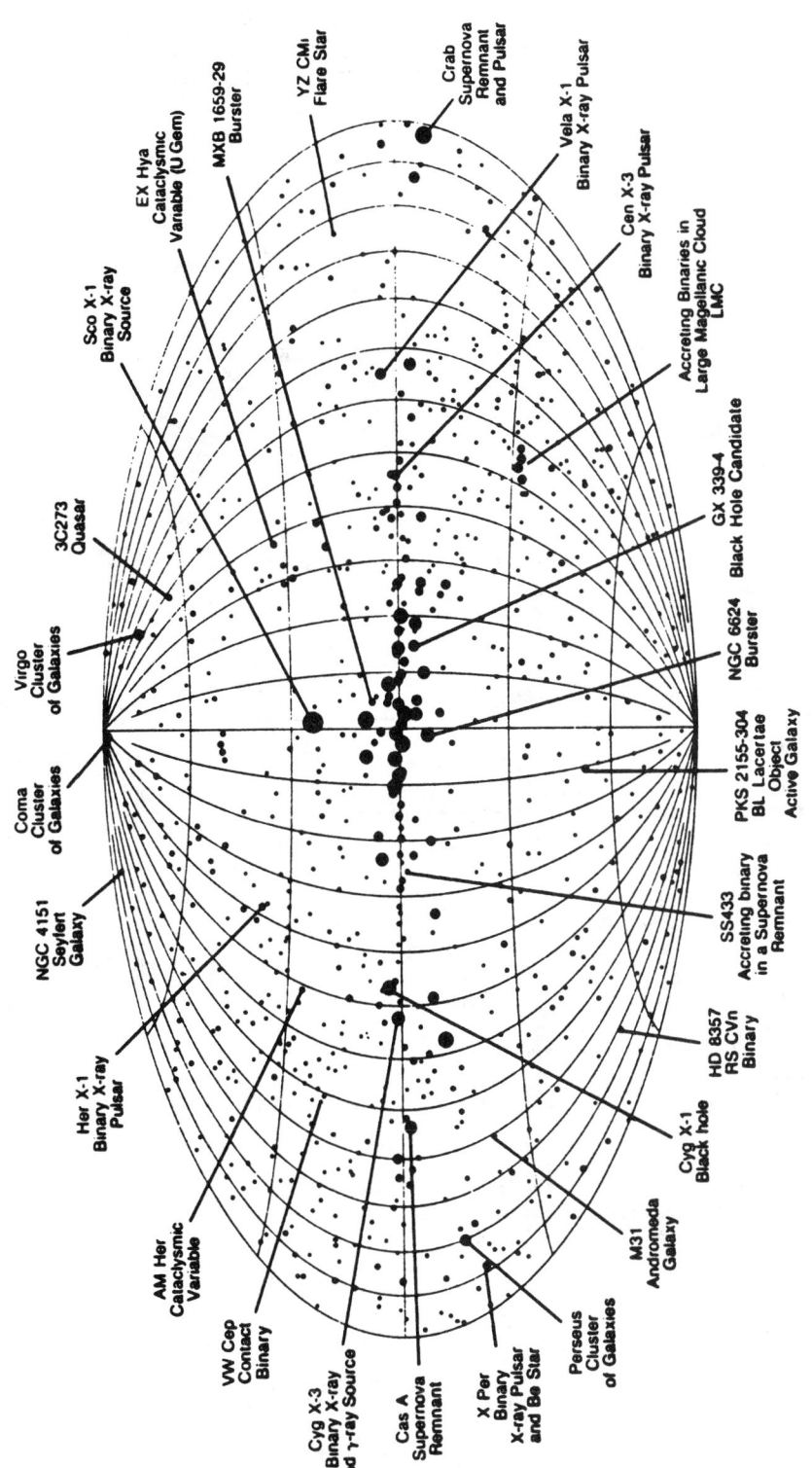

Figure 9.8. The HEAO-1 X-ray sky [20].

Table 9.6. *Selected X-ray emitting stars.*

Name	Flux density 0.2–4 keV [1] (millicrab)	Stellar type	Distance (LY[a])	L_X [2] 0.2–4 keV (erg s^{-1})
Capella (α Aur)	6.1	G8V+FV	44	2.0×10^{30}
HZ 43	2.2	WD	210	4×10^{31}
Algol (β Per)	1.8	B8V	100	5×10^{30}
Wolf 630	1.1	M4V+M5V	20	2.0×10^{29}
24 UMa	0.70	G1V	80	1.0×10^{30}
Prox Cen	0.70	M5V	4.3	2.5×10^{27}
YY Gem	0.67	MV	50	4×10^{29}
ϵ Eri	0.63	K2V	10.8	2.0×10^{28}
EQ Peg	0.60	MV	20	6×10^{28}
ζ Ori	0.58	O9.5I	1600	3.5×10^{32}
α Cen	0.58	K5V+G2V	4.4	3.2×10^{27}
ϵ Ori	0.44	B0I	1500	2.0×10^{32}
α Tri	0.35	F2V	60	3.2×10^{29}
Sirius (α CMa B)	0.34	WD	8.6	6×10^{28}

Note

[a] LY = light year = 0.31 parsec $\approx 1 \times 10^{16}$ m.

References

1. Harris, D.E. et al. 1990, *The Einstein Observatory Catalog of IPC X-Ray Sources*, NASA TM 108401, Vol. 1, 24
2. Vaiana, G.S. et al. 1981, *ApJ*, **245**, 163

Table 9.7. *Selected X-ray emitting normal galaxies* [1, 2].

Galaxy	Type	Flux density 0.2–4 keV [1] (millicrab)	Distance[a] (MLY[b])	L_X 0.2–4 keV (erg s^{-1})
NGC 507	S0	0.46	320	1.1×10^{43}
NGC 720	E	0.10	105	2.2×10^{41}
NGC 4382	S0	0.032	90	5.2×10^{40}
NGC 4472	E/S0	0.73	90	1.1×10^{42}
M31	Sb	2.3	2.2	3.6×10^{39}
NGC 253	Sc	0.31	10	7.4×10^{39}
M 81	Sb	0.35	11	1.3×10^{40}
M 82	Irr	0.88	11	3.5×10^{40}

Notes

[a] $H_0 = 50$ km s^{-1} Mpc^{-1}, $q_0 = 0.5$. [b] MLY = one million light years = 0.31×10^6 pc $\approx 1 \times 10^{22}$ m.

References

1. Fabbiano, G., Kim, D.-W., & Trinchieri, G. 1992, *ApJS*, **80**, 531
2. Harris, D.E. et al. 1990, *The Einstein Observatory Catalog of IPC X-Ray Sources*, NASA TM 108401, Vol. 1, p. 24

Table 9.8. *Selected X-ray emitting accretion-powered binaries* [1, 2].[a]

Source	Name	Optical counterpart	Galactic coordinates l	b	Max Fl. Den. 2–10 keV (millicrab)	Max. V (mag)	P_{orb}	P_{spin} (s)	Type [3]	Comment
3A0535+26		V725 Tau	181.4	−2.6	2 700 [4]	8.9	111 d	104	HMXB	Tr; NS+O9.7IIe
4U0538−64	LMC X−3		273.6	−32.1	25 [5]	16.7	1.70 d		HMXB	In LMC; BH+B3ve
3A0620−00		V616 Mon	210.0	−6.5	125 000 [4]	11.2	7.75 h		LMXB	Tr; BH+K5V
EXO0748−676		UY Vol	280.0	−19.3	30 [1]	16.9	3.82 h		LMXB	Ecl; ADC; Bur
H0752+22		U Gem	199.2	23.4		9.0	4.22 h	25	CV	Bright 0.25 keV; DN; WD+M5V
4U0900−40	Vela X−1	GP Vel	263.1	3.9	250 [5]	6.9	8.96 d	283	HMXB	Ecl; NS+B0.5Ib
4U1118−60	Cen X−3	V779 Cen	292.1	0.3	300 [4]	13.3	2.09 d	4.84	HMXB	NS+O6.5II–III; Ecl
1455−31	Cen X−4	V822 Cen	332.2	23.9	20 000 [6]	12.8	15.1 h		LMXB	Tr; Bur; NS+K5V
4U1516−56	Cir X−1	BR Cir	322.1	0.0	850 [4]		16.6 h		LMXB	Tr; Bur
4U1617−15	Sco X−1	V818 Sco	359.1	23.8	17 000 [5]	12.2	18.9 h		LMXB	
4U1626−67		KZ TrA	321.8	−13.1	20 [7]	18.5	0.69 h	7.7	LMXB	Bur
4U1636−53		V801 Ara	332.9	−4.8	310 [4]	17.5	3.80 h		LMXB	
GROJ1655−40			345.0	2.5	1 200 [1]	14.4	2.61 d		LMXB	Tr; BH+F1V; Ecl; jets
4U1656+35	Her X−1	HZ Her	58.2	37.5	160 [7]	13.0	1.70 d	1.24	LMXB	Ecl; NS+(B0V to F0V)
4U1658−48	GX339−4	V821 Ara	338.9	−4.3	350 [5]	15.5	14.8 h		LMXB	BH ?
1730−335	Rapid Bur		354.8	−0.2	250 [5]				LMXB	In GC Liller 1, type II Bur
4U1758−25	GX 5−1		5.1	−.0	1 300 [4]				LMXB	
4U1814+49		AM Her	77.9	25.9	3 [8]	11.3	3.09 h		CV	Bright at 0.25 keV; WD+M5V
4U1820−30	Sgr X−4		2.8	−7.9	320 [5]		0.19 h		LMXB	In GC NGC 6624; Bur
3A1909+048	SS433		39.7	−2.2	10 [4]	14.2	13.1 d		HMXB	BH ?; jets
4U1956+35	Cyg X−1	V1357 Cyg	71.3	3.1	1 175 [5]	8.9	5.60 d		HMXB	BH+O9.7Iab
GS2023+338		V404 Cyg	73.2	−2.2	20 000 [9]	12.7	6.47 d		LMXB	Tr; BH+(K0III—V)
4U2030+40	Cyg X−3	V1521 Cyg	79.9	0.7	380 [5]		4.79 h		HMXB	NS+Helium star
4U2129+12		AC211	65.0	−27.3	6 [1]	15.8	17.1 h		LMXB	In GC M15;Bur
4U2129+47		V1727 Cyg	91.6	−3.0	20 [5]	16.4	5.24 h		LMXB	Tr; ADC; Bur; Triple?
1H2140+433	SS Cyg		90.6	−7.1	4 [8]	8.2	6.50 h	9	CV	Bright 0.25 keV; DN; WD+K5V
4U2142+38	Cyg X−2	V1341 Cyg	87.3	−11.3	750 [7]	14.7	9.84 d		LMXB	NS+(A5III to F2III)

Note

[a]HMXB = high-mass X-ray binary; LMXB = low-mass X-ray binary; CV = cataclysmic variable; DN = dwarf nova; Bur = Burster; GC = globular cluster; Tr = transient; NS = neutron star; LMC = Large Magellanic Cloud; BH = black hole; Ecl = eclipse; ADC = accretion disk corona; WD = white dwarf.

References

1. van Paradijs, J. 1995, in *X-Ray Binaries*, edited by W. Lewin, J. van Paradijs, and E.P.J. van den Heuvel, (Cambridge University Press, Cambridge) p. 536
2. Ritter, H. 1987, *A&AS*, **70**, 335
3. Bradt, H., & McClintock, J. 1983, *ARA&A*, **21**, 3
4. Warwick, R.W. et al. 1981, *MNRAS*, **197**, 865
5. Forman, W. et al. 1978, *ApJ*, **38**, 357
6. Conner, J.P., Evans, W.D., & Belian, R.D. 1969, *ApJ*, **157**, L157
7. McHardy, I.M. et al. 1981, *MNRAS*, **197**, 893
8. Wood, K.S. et al. 1984, *ApJS*, **56**, 507
9. Makino, F. 1989, IAU Circular No. 4786

Table 9.9. *The brightest X-ray emitting clusters of galaxies* [1].

Cluster	Source	Flux density 2–10 keV (millicrab)	Redshift z	Distance^a (MLY^b)	L_X 2–10 keV (erg s^{-1})
A426 (Perseus)	4U 0316+41	47 ^{c}	0.0183	360	1.4×10^{45}
Ophiuchus Cluster	4U 1708−23	30	0.028	550	2.5×10^{45}
M87 (Virgo)	4U 1228+12	22 ^{d}	0.0037	73	3×10^{43}
A1656 (Coma)	4U 1257+28	15	0.0235	460	9×10^{44}
Centaurus Cluster	4U 1246−41	5	0.0107	210	6×10^{43}
A2199	4U 1627+39	4	0.0305	600	3×10^{44}
A496	4U 0431−12	3	0.0316	620	3×10^{44}
A85	4U 0037−10	3	0.0518	1000	8×10^{44}

Notes

^{a} $H_0 = 50 \text{km s}^{-1} \text{Mpc}^{-1}$, $q_0 = 0.5$.

^{b} MLY = one million light years = 0.31×10^6 pc $\approx 1 \times 10^{22}$ m.

^{c} Includes the nucleus of the galaxy NGC1275 and diffuse emission from the Cluster.

^{d} Includes the active nucleus and diffuse emission from M87, and emission from the surrounding Virgo Cluster.

Reference

1. Forman, W. et al. 1978, *ApJS*, **38**, 357

Table 9.10. *Selected X-ray emitting active galaxies.*

Name		Flux den. 2–10 keV (millicrab)	Red-shift z	Dist.^{a} (MLY^b)	L_X 2–10 keV (erg s^{-1})	Type
2E 189	MRK 348	1.3 [1]	0.014	270	1×10^{43}	Seyfert 2, highly obscured
1H 0244+001	NGC 1068	0.8 [2]	0.0037	73	9×10^{41}	Seyfert 2, Compton thick
2E 1007	Q0420−388	0.02 [3]	3.12	80 000	2×10^{46}	High redshift quasar
4U 0432+05	3C 120	2.3 [3]	0.033	650	2×10^{44}	Superlum VLBI radio gal^c
2E 2195	M81	0.16 [1]	0.0006	11.7	5×10^{39}	Low luminosity AGN^d
4U 1206+39	NGC 4151	4.3 [3]	0.0033	65	4×10^{42}	Seyfert 1.5
1H 1226+022	3C 273	3.1 [2]	0.158	3200	6×10^{45}	Radio loud quasar
1H 1226+128	M87	22.4^e [3]	0.0037	73	3×10^{43}	Radio galaxy
2E 2900	3C 279	0.23 [1]	0.538	11 000	6×10^{45}	OVV (Blazar),^{f} γ-rays
4U 1322-42	Cen A	8.4 [3]	0.0008	15.6	5×10^{41}	Radio galaxy
4U 1414+25	NGC 5548	1.7 [3]	0.0017	33	4×10^{41}	Seyfert 1
2E 4066	E1821+643	0.80 [1]	0.297	6200	7×10^{45}	Radio quiet quasar
1H 1937-106	NGC 6814	1.9 [2]	0.005	98	4×10^{42}	Seyfert 1, highly variable
1H 2156-304	PKS 2155−304	8.4 [2]	0.17	3500	2×10^{46}	BL Lac

Notes

^{a} $H_0 = 50 \text{ km s}^{-1} \text{Mpc}^{-1}$, $q_0 = 0.5$.

^{b} MLY = one million light years = 0.31×10^6 pc $\approx 1 \times 10^{22}$ m.

^{c} VLBI = very long baseline interferometry.

^{d} AGN = active galactic nucleus.

^{e} Includes diffuse emission from M87 and from Virgo cluster.

^{f} OVV = optically violent variable.

References

1. Harris, D.E. et al. 1990, *The Einstein Obs. Catalog of IPC X-Ray Sources*, NASA TM 108401, Vol. 1, p. 24

2. Wood, K.S. et al. 1984, *ApJS*, **56**, 507

3. Forman, W. et al. 1978, *ApJS*, **38**, 357

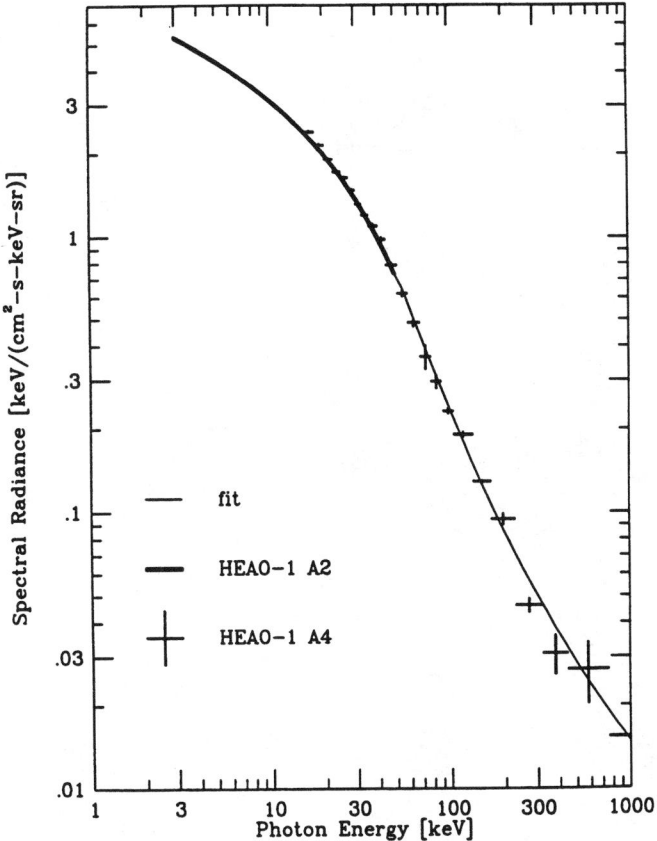

Figure 9.9. The diffuse background spectrum above 3 keV [28].

9.6 DIFFUSE BACKGROUND

Above 3 keV, surface brightness fluctuations of the high-latitude extragalactic diffuse background, measured in $5° \times 5°$ blocks, are less than 2% above those expected from an extrapolation of the observed distribution of sources [27]. The spectrum from the HEAO-1A2 detectors is one of the best measured in astrophysics and, in the range 3–50 keV, is given by

$$I(E) = 7.8E^{-0.29}e^{-E/40} \text{ keV cm}^{-2}\text{ s}^{-1}\text{ st}^{-1}\text{ keV}^{-1} \text{ [28]}.$$

Figure 9.9 shows the HEAO-1 A2 and A4 data. Statistical error for the A2 instrument is less than the width of the heavy solid line. The lighter curve is an empirical fit to all the data and is

$$I(E) = \begin{cases} 7.877E^{-0.29}e^{-E/41.13}, & E < 60 \text{ keV}, \\ 1652E^{-2} + 1.75E^{-0.7}, & E > 60 \text{ keV [29]}. \end{cases}$$

In the range 2–60 keV there is an additional component of 2%–10% associated with the galactic plane. Intensity is concentrated in the plane and is strongest in the direction of the center [30]. Below 2 keV there is structure in the background at all latitudes, which changes appreciably with energy. Both absorption and local emission contribute to this structure. Figure 9.10 shows the background at 0.25 keV. Similar maps are available from 0.13 to 2.2 keV [31, 32].

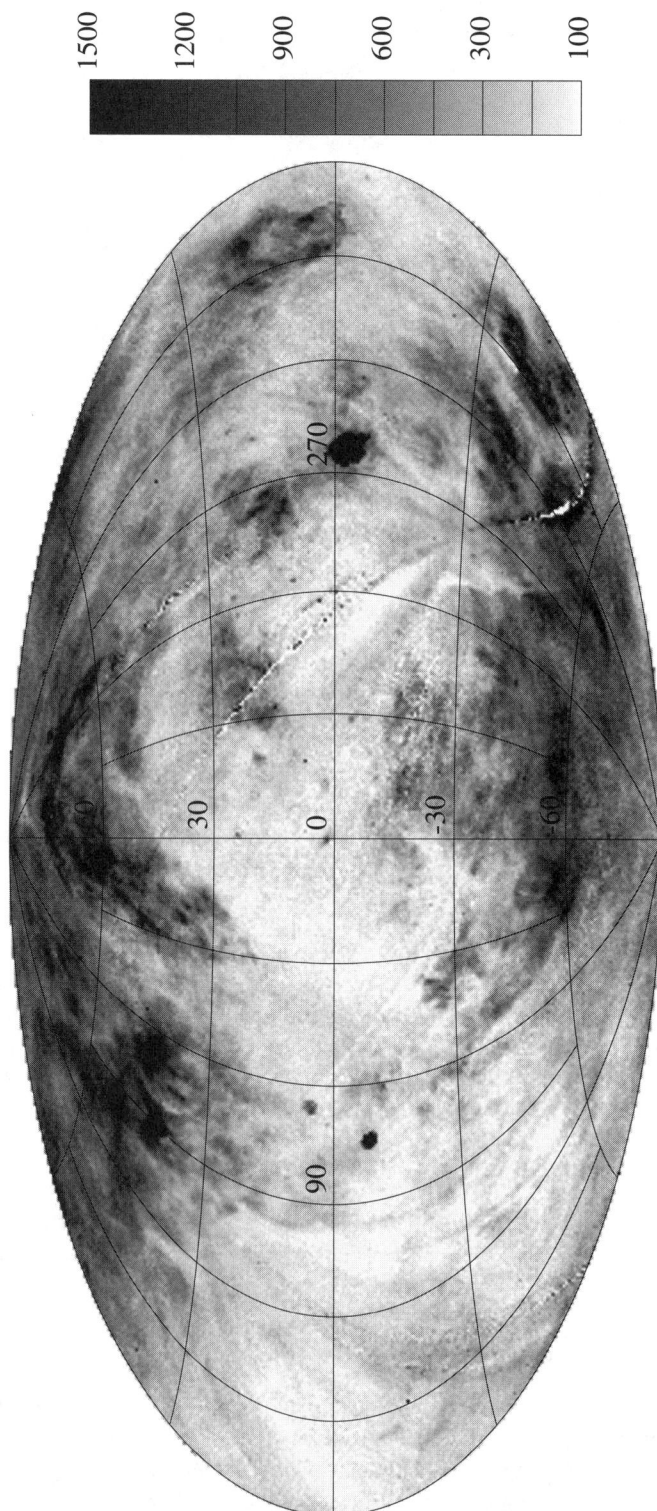

Figure 9.10. The diffuse background at $\frac{1}{4}$ keV measured by ROSAT [32]. Two bright sources, the Cygnus Loop (74, −8) and the Vela supernova remnant (263, −3) are also clearly visible. Color bar values are 10^{-6} counts s^{-1} arcmin^{-2}.

9.7 X-RAY ASTRONOMY MISSIONS

Characteristics of some prominent X-ray satellites are given in Tables 9.11 and 9.12.

Table 9.11. *Some X-ray astronomy satellites* [1, 2].

Satellite	Country	Launch	Last data	Type[a]	Energy range[b] (keV)
Vela 5A,B	USA	May 69	June 79	Scanning, small SC	3–12
Uhuru	USA	Dec. 70	Jan. 75	Scanning, large PC	2–10
OSO-7	USA	Sep. 71	May 73	Scanning, PC	1–40
Copernicus	USA/UK	Aug. 72	Dec. 80	Pointed, small concentrator	0.2–10, 0.2–0.6
ANS	Netherlands	Aug. 74	July 76	Pointed, PC, Bragg crystal	2–40
Ariel-V	UK	Oct. 74	Mar. 80	Scanning, pointed, ASM, large PC	2–10
SAS-3	USA	May 75	Apr. 80	Scanning, RMC, large PC	1.5–10
HEAO-1	USA	Aug. 77	Jan. 79	Scanning, very large PC, SC	1–20
Einstein	USA	Nov. 78	Apr. 81	Pointed, telescope, PC, HRI, Si detector	0.2–4
Hakucho	Japan	Feb. 79	Apr. 85	Scanning, RMC, large PC	0.1–2, 2–20
Tenma	Japan	Feb. 83	Nov. 85	Scanning, GSPC, ASM	2–10
EXOSAT	ESA	May 83	Apr. 86	Pointed, small telescope, large PC	0.05–2.0, 1.5–10
Spartan 101	USA	June 85	June 85	PC	2–10
Ginga	Japan	Feb. 87	Oct. 91	Pointed, large PC	1.5–30
KVANT	USSR	June 87		Pointed, GSPC, SC, coded mask	2–30
Granat	USSR	Dec. 89		Pointed, Coded masks, ASM	40–100, 0–1300
ROSAT	Germany	June 90		Scanning, pointed, telescope, PC, HRI	0.1–2.5
Astro-1	USA	Dec. 90	Dec. 90	Pointed, BBXRT collector, Si detector,	0.3–12
DXS	USA	Jan. 93	Jan. 93	Spectrometer to study background	0.15–0.28
ASCA	Japan	Feb. 93		Pointed, telescope, CCD, GSPC	0.4–12
Alexis	USA	Apr. 93		Scanning, small multilayer telescopes	0.06–0.10
RXTE	USA	Dec. 95		Pointed, large PC, large SC, ASM	2–50, 15–200
IXAE	India	Mar. 96		Pointed, large PC, ASM	2–18
BeppoSAX	Italy	Apr. 96		Pointed, GSPC, SC, WFC [3]	0.1–300
Chandra	USA	Jul. 99		Pointed, telescope, CCD, HRI, gratings	

Notes

[a] ASM = all-sky monitor; BBXRT = broad-band X-ray telescope; CCD = charged coupled detector; GSPC = gas scintillation proportional counter; HRI = high resolution, channel plate detector; PC = proportional counter; RMC = rotating modulation collimator; SC = scintillation counter; WFC = wide field camera.

[b] These data apply to the main survey instrument(s). Since most satellites carried several detectors, data were usually collected over a broader energy range than that specified here. The range for a given detector also depends on source strength. Since the background obscures the higher-energy photons from weaker sources, this range is larger for strong sources than for weaker sources.

References

1. Bradt, H., Ohashi, T., & Pounds, K.A. 1992, *ARA&A*, **30**, 391
2. Charles, P.A., & Seward, F.D. 1995, *Exploring the X-Ray Universe* (Cambridge University Press, Cambridge)
3. Boella, G. et al. 1997, *A&AS*, **122**, 299

Table 9.12. *Characteristics of X-ray telescope mirrors.*

Mission	Aperture diameter (cm)	Mirrors	Geometric area (cm^2)	Field (arcmin diam.)	Reflection angles (arcmin)	Focal length (m)	Mirror coating	Highest energy (keV)	On-axis resolution (arcsec)
Einstein	58	4 nested	350	60	40–70	3.45	Ni	4.5	4
EXOSAT	28	2 nested	80	120	90–110	1.09	Au	2	18
ROSAT	83	4 nested	1140	120	83–135	2.4	Au	2.4	3
ACSA	34	120 nested	1300	40	14–42	3.5	Au	12	180
(4 mod.)	(1 mod.)	(1 mod.)	(4 mod.)						
Chandra	123	4 nested	1145	30	27–52	10.0	Ir	10	0.5

ACKNOWLEDGMENTS

Information and advice for this section was kindly supplied by M. Elvis, G. Fabbiano, F.R. Harnden Jr., J.P. Hughes (Fig. 9.6), C. Jones, J. McClintock, J. McDowell, J. Raymond (Figs. 9.1–9.4), W. Tucker, D. Worrall, and M. Zombeck at the Harvard-Smithsonian Center for Astrophysics, E. Boldt and S. Snowden (Fig. 9.10) at the Goddard Space Flight Center, D. Gruber of the University of California, San Diego (Fig. 9.9), D. Cox and D. McCammon of the University of Wisconsin (Figs. 9.5 and 9.10), H. Bradt at MIT, and K. Wood of the Naval Research Laboratory (Fig. 9.8).

REFERENCES

1. Karzas, W., & Latter, R. 1961, *ApJS*, **6**, 157
2. Chodil, G. et al. 1968, *ApJ*, **154**, 645
3. Blumenthal, G., & Tucker, W., 1974, in *X-Ray Astronomy*, edited by R. Giacconi and H. Gursky (Reidel, Dordrecht), p. 99
4. Schwartz, D.A., & Tucker, W.H. 1988, *ApJ*, **332**, 157
5. Holt, S.S., & McCray, R. 1982, *ARA&A*, **20**, 323
6. Allen, C.W. 1973, *Astrophysical Quantities* (Athlone Press, London), p. 30
7. Raymond, J.C. 1992, current version of code described by Raymond, J.C., & Smith, B.W. 1977, *ApJS*, **35**, 419
8. Hamilton, A.J.S., Sarazin, C.L., & Chevalier, R.A. 1983, *ApJS*, **51**, 115
9. Toor, A., & Seward, F.D. 1974, *AJ*, **79**, 995
10. Brown, R.L., & Gould, R. 1970, *Phys. Rev.*, **D1**, 2252
11. Morrison, R., & McCammon, D. 1983, *ApJ*, **270**, 119
12. Henke, B.L. et al. 1982, *Atomic Data Nucl. Data*, **27**, 1
13. Gorenstein, P. 1975, *ApJ*, **198**, 95 (corrected to MM absorption)
14. Ryter, C., Cezarsky, C., & Andouze, J. 1975, *ApJ*, **198**, 103 (corrected to MM absorption)
15. Bohlin, R., Savage, B., & Drake, J. 1978, *ApJ*, **224**, 132
16. Predehl, P., & Schmitt, J.H.M.M. 1995, *A&A*, **293**, 889
17. Forman, W. et al. 1978, *ApJS*, **38**, 357
18. Warwick, R.W. et al. 1981, *MNRAS*, **197**, 865
19. McHardy, I.M. et al. 1981, *MNRAS*, **197**, 893
20. Wood, K.S. et al. 1984, *ApJS*, **56**, 507
21. Elvis, M.S., et al. 1992, *ApJS*, **80**, 257
22. Voges, W. 1992, *Proceedings of Satellite Symposium 3: Space Sciences with Particular Emphasis on High Energy Astrophysics*, from the "International Space Year" conference, Munich, Germany, March 1992 (ESA, ISY-3), p. 9
23. Nugent, J.J. et al. 1983, *ApJS*, **51**, 1
24. Levine, A.M. et al. 1984, *ApJS*, **54**, 581
25. Priedhorsky, W.C., Terrell, J., and Holt, S.S, 1983, *ApJ*, **270**, 233
26. Charles, P.A. & Seward, F.D. 1995, *Exploring the X-Ray Universe* (Cambridge University Press, Cambridge)
27. Fabian, A.C., & Shafer, R.A. 1983, in *Early Evolution of the Universe*, edited by G. Abell and G. Chincarini (Reidel, Dordrecht), p. 333
28. Boldt, E. 1992, in *Proceedings of the International Workshop on the X-Ray Background*, edited by X. Barcons and A. Fabian (Cambridge University Press, Cambridge), p. 115
29. Gruber, D. 1992, in *Proceedings of the International Workshop on the X-Ray Background*, edited by X. Barcons and A. Fabian (Cambridge University Press, Cambridge), p. 44
30. Iwan, D. et al. 1982, *ApJ*, **260**, 111
31. McCammon, D. et al. 1983, *ApJ*, **269**, 107
32. Snowden, S. L. et al. 1997, *ApJ*, **485**, 125

γ-Ray and Neutrino Astronomy

R.E. Lingenfelter and R.E. Rothschild

10.1 CONTINUUM EMISSION PROCESSES

Important processes for continuum emission at γ-ray energies are bremsstrahlung, magneto-bremsstrahlung, and Compton scattering of blackbody radiation by energetic electrons and positrons [1–6].

10.1.1 Bremsstrahlung

The bremsstrahlung luminosity spectrum of an optically thin thermal plasma of temperature T in a volume V is [3]

$$L(v)_{\text{brem}} = \frac{32\pi e^6}{3m^2c^4} \left(\frac{2\pi mc^2}{3kT} \right)^{1/2} Z^2 n_e n_i V g(v, T) \exp(-hv/kT),$$

where the index of refraction is assumed to be unity, m is the electron mass, Z is the mean atomic charge, n_e and n_i are the electron and ion densities, and the Gaunt factor $g(v, T) \approx (3kT/\pi hv)^{1/2}$ for $hv > kT$ and $T > 3.6 \times 10^5 Z^2$ K, or

$$L(v)_{\text{brem}} \approx 6.8 \times 10^{-38} Z^2 n_e n_i V g(v, T) T^{-1/2} \exp(-hv/kT) \text{ erg s}^{-1} \text{ Hz}^{-1}.$$

10.1.2 Magnetobremsstrahlung

The synchrotron luminosity spectrum of an isotropic, optically thin nonthermal distribution of relativistic electrons with a power-law spectrum, $N(\gamma) = N_0 \gamma^{-S}$, interacting with a homogeneous magnetic field of strength, H, is [5]

$$L(\nu)_{\text{synch}} \approx \frac{0.8e^3}{3mc^2} \left(\frac{3e}{4\pi mc} \right)^{(S-1)/2} V N_0 H^{(S+1)/2} \nu^{(1-S)/2}$$

or

$$L(\nu)_{\text{synch}} \approx 3.60 \times 10^{-23} V N_0 H^{(S+1)/2} (4.2 \times 10^6 / \nu)^{(S-1)/2} \text{ erg s}^{-1} \text{ Hz}^{-1}.$$

10.1.3 Compton-Scattered Blackbody Radiation

The Compton-scattering (cs) luminosity spectrum of an optically thin, isotropic nonthermal distribution of relativistic electrons with a power-law spectrum, $N(\gamma) = N_0 \gamma^{-S}$, interacting with blackbody photons having a temperature T is [5]

$$L(\nu)_{\text{cs}} \approx \frac{4e^4}{3m^2c^3} \left(\frac{h}{3.6k} \right)^{(3-S)/2} V N_0 w_{\text{bb}} T^{(S-3)/2} \nu^{(1-S)/2}$$

or

$$L(\nu)_{\text{cs}} \approx 4.22 \times 10^{-26} V N_0 w_{\text{bb}} T^{(S-3)/2} (7.5 \times 10^{10} / \nu)^{(S-1)/2} \text{ erg s}^{-1} \text{ Hz}^{-1},$$

where w_{bb} is the energy density of the blackbody radiation.

10.2 LINE EMISSION PROCESSES

Important processes for line emission at γ-ray energies are electron–positron annihilation, nuclear deexcitation, decay of radio nuclei, and radiative capture (see Tables 10.1–10.3).

10.2.1 Electron–Positron Annihilation Radiation

Positron annihilation can occur either via a direct interaction with a free electron or via positronium formed by charge exchange with a bound electron or by radiative combination with a free electron (e.g., [7–12]). See Figure 10.1.

Direct annihilation (da) leads to line emission, $e^+ e^- \rightarrow 2\gamma$, at a mean energy,

$$h\nu_{\text{da}} = m_e c^2 \begin{cases} +kT_e/2, & T_e \ll 10^7 \text{ K,} \\ +3kT_e/4, & 10^7 < T_e < 10^{10} \text{ K,} \\ +kT_e, & T_e > 10^{10} \text{ K,} \end{cases}$$

where $m_e c^2 = 510.9991$ keV and T_e is the temperature of the annihilating electrons and positrons.

The direct-annihilation line spectrum can be approximated by a Gaussian with a linewidth [12] $\Gamma_{\text{da}} \approx 0.87(T_e/10^4 \text{ K})^{0.50}$ keV, for $T_e \ll 10^9$ K, and at higher temperatures the width [10] $\Gamma_{\text{da}} \approx kT_e$, for $T_e \gg 10^9$ K.

The cross section for direct annihilation of a positron of energy $\gamma m_e c^2$ with an electron at rest [1] is

$$\sigma(\gamma)_{\text{da}} = \frac{3\sigma_T}{8(\gamma + 1)} \left(\frac{\gamma^2 + 4\gamma + 1}{\gamma^2 - 1} \ln(\gamma + \sqrt{\gamma^2 - 1}) - \frac{\gamma + 3}{\sqrt{\gamma^2 - 1}} \right),$$

where the Thomson cross section, $\sigma_T = 8\pi e^4/(3m^2c^4) = 0.6652$ barn (b).

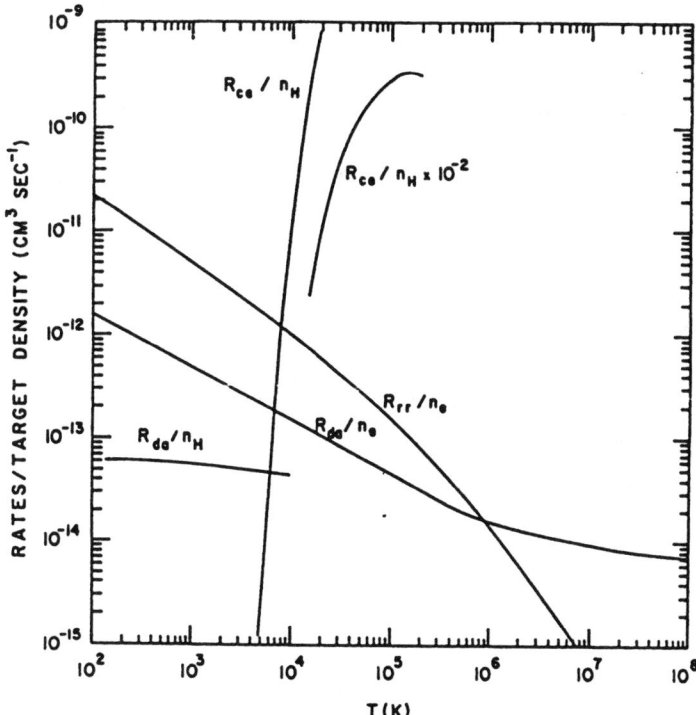

Figure 10.1. Positron-annihilation rates in a thermal medium per unit density as a function of temperature, for annihilation directly with free electrons (R_{da}/n_e) or with bound electrons (R_{da}/n_H), and via positronium formation by radiative combination with free electrons (R_{rc}/n_e) or by charge exchange with neutral hydrogen (R_{ce}/n_H), from [8].

Annihilation via positronium formation leads to line emission only from the singlet parapositronium, $para\text{-}Ps \rightarrow 2\gamma$, which forms 25% of the time. The mean energy of the positronium line,

$$h\nu_{ps} = m_e c^2 - (\mathcal{R}/4n^2),$$

where the Rydberg $\mathcal{R} = 0.0136$ keV, and n is 1 for the ground state.

The parapositronium annihilation line spectrum can be approximated by a Gaussian with a linewidth $\Gamma_{rc} \approx 0.80(T/10^4 \text{ K})^{0.44}$ keV for radiative combination (rc), valid at least from 8 000 to 10^6 K, and a Gaussian linewidth $\Gamma_{ce} \approx 6.4$ keV for charge exchange (ce), since the parapositronium mean life of $\sim 10^{-10}$ s is much less than the energy loss time [12].

The total number of 511 keV line photons emitted per positron annihilation,

$$\gamma_{511}/e^+ = 2 - 1.5 f_{ps},$$

where f_{ps} is the fraction of positrons that annihilate via positronium.

Annihilation via positronium formation leads to three-photon continuum emission from the triplet orthopositronium, $ortho\text{-}Ps \rightarrow 3\gamma$, which forms 75% of the time. The spectrum [7] of this emission is

$$P(h\nu)_{3\gamma} = \frac{2}{(\pi^2 - 9)m_e c^2}\left(\frac{\eta(1-\eta)}{(2-\eta)^2} + \frac{2(1-\eta)}{\eta^2}\ln(1-\eta) - \frac{2(1-\eta)^2}{(2-\eta)^3}\ln(1-\eta) + \frac{2-\eta}{\eta}\right),$$

where $\eta = h\nu/mc^2$ is the photon energy, and the spectrum is normalized to unity.

Table 10.1. *Nuclear deexcitation γ-ray lines.*[a,b]

Energy (MeV)	Emission mechanism	Excitation processes	Mean life (s)
0.429 1	^{7}Be$^{*0.429} \to$ g.s.	^{4}He$(\alpha, n)^7$Be*	1.9×10^{-13}
0.477 6	^{7}Li$^{*0.478} \to$ g.s.	^{4}He$(\alpha, p)^7$Li*	1.1×10^{-13}
		^{4}He$(\alpha, n)^7$Be$(\epsilon)^7$Li*(10%)	6.6×10^6
0.718 3	^{10}B$^{*0.718} \to$ g.s.	^{12}C$(p, x)^{10}$B*	1.0×10^{-9}
		^{16}O$(p, x)^{10}$B*	1.0×10^{-9}
		^{12}C$(p, x)^{10}$C$(e^+)^{10}$B*	27.78
		^{16}O$(p, x)^{10}$C$(e^+)^{10}$B*	27.78
0.846 8	^{56}Fe$^{*0.847} \to$ g.s.	^{56}Fe$(p, p')^{56}$Fe*	9.1×10^{-12}
		^{56}Fe$(p, n)^{56}$Co$(e^+; \epsilon)^{56}$Fe*	9.6×10^6
1.238 3	^{56}Fe$^{*2.085} \to {}^{56}$Fe$^{*0.847}$	^{56}Fe$(p, p')^{56}$Fe*	1.0×10^{-12}
		^{56}Fe$(p, n)^{56}$Co$(e^+; \epsilon)^{56}$Fe*(67%)	9.6×10^6
1.274 5	^{22}Ne$^{*1.275} \to$ g.s.	^{22}Ne$(p, p')^{22}$Ne*	5.2×10^{-12}
		^{22}Ne$(\alpha, \alpha')^{22}$Ne*	5.2×10^{-12}
		^{22}Ne$(p, n)^{22}$Na$(e^+; \epsilon)^{22}$Ne*	1.2×10^8
		^{24}Mg$(p, x)^{22}$Na$(e^+; \epsilon)^{22}$Ne*	1.2×10^8
		^{25}Mg$(p, x)^{22}$Na$(e^+; \epsilon)^{22}$Ne*	1.2×10^8
		^{28}Si$(p, x)^{22}$Na$(e^+; \epsilon)^{22}$Ne*	1.2×10^8
1.368 5	^{24}Mg$^{*1.369} \to$ g.s.	^{24}Mg$(p, p')^{24}$Mg*	1.9×10^{-12}
		^{24}Mg$(\alpha, \alpha')^{24}$Mg*	1.9×10^{-12}
		^{28}Si$(p, x)^{24}$Mg*	1.9×10^{-12}
1.408 3	^{55}Fe$^{*1.408} \to$ g.s.	^{56}Fe$(p, pn)^{55}$Fe*	5.5×10^{-11}
		^{56}Fe$(p, 2n)^{55}$Co$(e^+; \epsilon)^{55}$Fe*(18%)	9.1×10^4
1.408 4	^{54}Fe$^{*1.408} \to$ g.s.	^{56}Fe$(p, x)^{54}$Fe*	1.2×10^{-12}
1.434 1	^{52}Cr$^{*1.434} \to$ g.s.	^{56}Fe$(p, x)^{52}$Cr*	9.8×10^{-13}
		^{56}Fe$(p, x)^{52}$Mn$^*(e^+; \epsilon)^{52}$Cr*	1.8×10^3
		^{56}Fe$(p, x)^{52}$Mn$(e^+; \epsilon)^{52}$Cr*	7.0×10^5
1.633 6	^{20}Ne$^{*1.634} \to$ g.s.	^{20}Ne$(p, p')^{20}$Ne*	1.0×10^{-12}
		^{20}Ne$(\alpha, \alpha')^{20}$Ne*	1.0×10^{-12}
		^{20}Ne$(p, n)^{20}$Na$(e^+)^{20}$Ne*(80%)	6.4×10^{-1}
		^{24}Mg$(p, x)^{20}$Ne*	1.0×10^{-12}
		^{28}Si$(p, x)^{20}$Ne*	1.0×10^{-12}
1.635 2	^{14}N$^{*3.948} \to {}^{14}$N$^{*2.313}$	^{14}N$(p, p')^{14}$N*	6.9×10^{-15}
		^{14}N$(\alpha, \alpha')^{14}$N*	6.9×10^{-15}
		^{16}O$(p, x)^{14}$N*	6.9×10^{-15}
1.779 0	^{28}Si$^{*1.779} \to$ g.s.	^{28}Si$(p, p')^{28}$Si*	6.8×10^{-13}
		^{28}Si$(\alpha, \alpha')^{28}$Si*	6.8×10^{-13}
		^{32}S$(p, x)^{28}$Si*	6.8×10^{-13}
1.808 6	^{26}Mg$^{*1.809} \to$ g.s.	^{26}Mg$(p, p')^{26}$Mg*	6.9×10^{-13}
		^{26}Mg$(\alpha, \alpha')^{26}$Mg*	6.9×10^{-13}
		^{26}Mg$(p, n)^{26}$Al$(e^+; \epsilon)^{26}$Mg*	3.2×10^{13}
		^{27}Al$(p, pn)^{26}$Al$(e^+; \epsilon)^{26}$Mg*	3.2×10^{13}
		^{28}Si$(p, x)^{26}$Al$(e^+; \epsilon)^{26}$Mg*	3.2×10^{13}
2.230 2	^{32}S$^{*2.230} \to$ g.s.	^{32}S$(p, p')^{32}$S*	2.4×10^{-13}
		^{32}S$(\alpha, \alpha')^{32}$S*	2.4×10^{-13}
2.312 6	^{14}N$^{*2.313} \to$ g.s.	^{14}N$(p, p')^{14}$N*	9.8×10^{-14}
		^{14}N$(\alpha, \alpha')^{14}$N*	9.8×10^{-14}
		^{14}N$(p, n)^{14}$O$(e^+)^{14}$N*	101.9
		^{16}O$(p, x)^{14}$N*	8.7×10^{-14}
		^{16}O$(p, x)^{14}$O$(e^+)^{14}$N*	101.9

Table 10.1. *(Continued.)*

Energy (MeV)	Emission mechanism	Excitation processes	Mean life (s)
2.613 8	$^{20}\text{Ne}^{*4.248} \rightarrow {}^{20}\text{Ne}^{*1.634}$	$^{20}\text{Ne}(p, p')^{20}\text{Ne}^*$	9.2×10^{-14}
		$^{20}\text{Ne}(\alpha, \alpha')^{20}\text{Ne}^*$	9.2×10^{-14}
		$^{24}\text{Mg}(p, x)^{20}\text{Ne}^*$	9.2×10^{-14}
		$^{28}\text{Si}(p, x)^{20}\text{Ne}^*$	9.2×10^{-14}
2.741 2	$^{16}\text{O}^{*8.872} \rightarrow {}^{16}\text{O}^{*6.130}$	$^{16}\text{O}(p, p')^{16}\text{O}^*$	1.8×10^{-13}
2.754 0	$^{24}\text{Mg}^{*4.123} \rightarrow {}^{24}\text{Mg}^{*1.369}$	$^{24}\text{Mg}(p, p')^{24}\text{Mg}^*$	3.5×10^{-14}
		$^{24}\text{Mg}(\alpha, \alpha')^{24}\text{Mg}^*$	3.5×10^{-14}
3.736 5	$^{40}\text{Ca}^{*3.737} \rightarrow \text{g.s.}$	$^{40}\text{Ca}(p, p')^{40}\text{Ca}^*$	6.8×10^{-11}
		$^{40}\text{Ca}(\alpha, \alpha')^{40}\text{Ca}^*$	6.8×10^{-11}
4.438 0	$^{12}\text{C}^{*4.439} \rightarrow \text{g.s.}$	$^{12}\text{C}(p, p')^{12}\text{C}^*$	6.1×10^{-14}
		$^{12}\text{C}(\alpha, \alpha')^{12}\text{C}^*$	6.1×10^{-14}
		$^{14}\text{N}(p, x)^{12}\text{C}^*$	6.1×10^{-14}
		$^{14}\text{N}(\alpha, x)^{12}\text{C}^*$	6.1×10^{-14}
		$^{16}\text{O}(p, x)^{12}\text{C}^*$	6.1×10^{-14}
		$^{16}\text{O}(\alpha, x)^{12}\text{C}^*$	6.1×10^{-14}
4.443 9	$^{11}\text{B}^{*4.445} \rightarrow \text{g.s.}$	$^{12}\text{C}(p, 2p)^{11}\text{B}^*$	1.1×10^{-15}
		$^{12}\text{C}(\alpha, x)^{11}\text{B}^*$	1.1×10^{-15}
5.104 9	$^{14}\text{N}^{*5.106} \rightarrow \text{g.s.}$	$^{14}\text{N}(p, p')^{14}\text{N}^*$	6.3×10^{-12}
		$^{14}\text{N}(\alpha, \alpha')^{14}\text{N}^*$	6.3×10^{-12}
		$^{16}\text{O}(p, x)^{14}\text{N}^*$	6.3×10^{-12}
		$^{16}\text{O}(\alpha, x)^{14}\text{N}^*$	6.3×10^{-12}
6.129 1	$^{16}\text{O}^{*6.130} \rightarrow \text{g.s.}$	$^{16}\text{O}(p, p')^{16}\text{O}^*$	2.7×10^{-11}
		$^{16}\text{O}(\alpha, \alpha')^{16}\text{O}^*$	2.7×10^{-11}
		$^{20}\text{Ne}(p, x)^{16}\text{O}^*$	2.7×10^{-11}
6.877 8	$^{28}\text{Si}^{*6.879} \rightarrow \text{g.s.}$	$^{28}\text{Si}(p, p')^{28}\text{Si}^*$	2.6×10^{-12}
		$^{28}\text{Si}(\alpha, \alpha')^{28}\text{Si}^*$	2.6×10^{-12}
6.917 4	$^{16}\text{O}^{*6.919} \rightarrow \text{g.s.}$	$^{16}\text{O}(p, p')^{16}\text{O}^*$	6.8×10^{-15}
		$^{16}\text{O}(\alpha, \alpha')^{16}\text{O}^*$	6.8×10^{-15}
7.115 2	$^{16}\text{O}^{*7.117} \rightarrow \text{g.s.}$	$^{16}\text{O}(p, p')^{16}\text{O}^*$	1.2×10^{-14}
		$^{16}\text{O}(\alpha, \alpha')^{16}\text{O}^*$	1.2×10^{-14}

Notes

[a]Updated from Ramaty, R., Kozlovsky, B., & Lingenfelter, R.E. 1979, *ApJS*, **40**, 487, with data from Firestone, R.B. et al. 1996, *Table of Isotopes* (Wiley, New York).

[b]Because of recoil the observed γ-ray energy $h\nu' = h\nu(1 - h\nu/2Mc^2)$, where $h\nu$ is the transition energy and M is nuclear mass.

Table 10.2. *Nucleosynthetic radioactive decay lines.*[a]

Radioactive decay	Dominant decay mean life	Line energy (MeV)	Branching ratio (%)
$^{56}\text{Ni}(\epsilon)^{56}\text{Co}$	8.80 days	0.1584	98.8
		0.8119	86.0
		0.7500	49.5
		0.2695	36.5
		0.4805	36.5
		1.5618	14.0

Table 10.2. *(Continued.)*

Radioactive decay	Dominant decay mean life	Line energy (MeV)	Branching ratio (%)
^{48}V$(e^+; \epsilon)^{48}$Ti	23.0 days	0.9835	100.
		1.3121	96.6
		0.5110	50.0n^c
^{56}Co$(e^+; \epsilon)^{56}$Fe	111.3 days	0.8468	99.9
		1.2383	68.4
		$\langle 0.0064 \rangle^b$	21.7
		0.5110	19.0n^c
		2.5986	17.4
		1.7715	15.5
		1.0379	14.1
		$\langle 3.244 \rangle$	12.4
		$\langle 2.029 \rangle$	11.3
^{65}Zn$(e^+; \epsilon)^{65}$Cu	352.4 days	1.1155	50.6
		0.0080	34.2
^{57}Co$(\epsilon)^{57}$Fe	392.1 days	0.1221	85.5
		$\langle 0.0064 \rangle$	48.9
		0.1365	10.3
		0.0144	9.5
^{22}Na$(e^+; \epsilon)^{22}$Ne	3.754 yr	1.2745	99.9
		0.5110	89.4n^c
^{125}Sb$(e^-)^{125}$Te	3.979 yr	$\langle 0.0274 \rangle$	62.1
		0.4279	29.4
		0.6006	17.8
		0.6360	11.3
		0.4634	10.5
^{44}Ti$(\epsilon)^{44}$Sc	91 ± 4 yr	0.0679	100
		0.0783	99.3
		$\langle 0.0041 \rangle$	16.7
^{44}Sc$(e^+; \epsilon)^{44}$Ca	(0.236 day)	1.1570	99.9
		0.5110	94.0n^c
^{60}Fe$(e^-)^{60}$Co	2.2×10^6 yr	0.0586	2.0
^{60}Co$(e^-)^{60}$Ni	(7.60 yr)	1.3325	100
		1.1732	99.9
^{26}Al$(e^+; \epsilon)^{26}$Mg	1.03×10^6 yr	1.8086	99.7
		0.5110	82.1n^c
^{40}K$(\epsilon)^{40}$Ar	1.84×10^9 yr	1.4608	10.7

Notes

[a]Based on data from Browne E., & Firestone, R.B. 1986, *Table of Radioactive Isotopes* (Wiley, New York), Firestone, R.B. 1996, *Table of Isotopes* (Wiley, New York), and Norman, E.B. et al. 1997, *Nuc. Phys.*, **A621**, 92 for the ^{44}Ti mean-life.

[b]Bracketed $\langle \rangle$ line energies are the mean of two or more close lines.

[c]The number of 0.5110 MeV photons per positron annihilation, $n = 2 - 1.5 f_{ps}$, where f_{ps} is the fraction of annihilation occurring via positronium formation.

Table 10.3. *Radiative capture γ-ray lines.*[a]

Radiative capture	Thermal cross section (b)	Line energy (MeV)	Branching ratio (%)
$^1H(n, \gamma)^2H$	0.332	2.2233	100
$^{56}Fe(n, \gamma)^{57}Fe$	2.6	0.0144	64
		7.6316	30
		7.6456	24
		0.3525	12
		5.9205	9
		6.0185	9
		1.7252	9

Note

[a]Based on data Nuclear Data Group, 1973, Nuclear Level Schemes $A = 45$ through $A = 257$ from Nuclear Data Sheets (Academic Press, New York).

10.3 SCATTERING AND ABSORPTION PROCESSES

γ-Ray emission spectra can be modified by several processes: photoelectric absorption, electron–positron pair production, Compton scattering, and Landau-level electron scattering in intense magnetic fields [1–4, 13–21]. See Figure 10.2.

10.3.1 Photoelectric Absorption

The cross section for photoelectric absorption of a photon by the ejection of a K-shell electron from an atom of nuclear charge Z is [1]

$$\sigma(h\nu)_K = \frac{3\sigma_T Z^5 \alpha^4}{2} \left(\frac{mc^2}{h\nu}\right)^5 (\gamma^2 - 1)^{3/2}$$

$$\times \left\{\frac{4}{3} + \frac{\gamma(\gamma - 2)}{\gamma + 1}\left[1 - \frac{1}{2\gamma\sqrt{\gamma^2 - 1}}\ln\left(\frac{\gamma + \sqrt{\gamma^2 - 1}}{\gamma - \sqrt{\gamma^2 - 1}}\right)\right]\right\},$$

where the Thomson cross section, $\sigma_T = 8\pi e^4/(3m^2 e^4) = 0.665\,2$ b, and the Lorenz factor of the ejected electron $\gamma = 1 + h\nu/mc^2$.

10.3.2 Pair Production

The cross section for electron–positron pair production (pp) by a photon in the presence of a nucleus of charge Z is [14]

$$\sigma(h\nu)_{pp} = \frac{3\alpha Z^2 \sigma_T}{2\pi}\left[\frac{7}{9}\ln\left(\frac{2h\nu}{mc^2}\right) - \frac{109}{54}\right]$$

for no screening when $1 \ll h\nu/mc^2 \ll 1/\alpha Z^{1/3}$, and

$$\sigma(h\nu)_{pp} = \frac{3\alpha Z^2 \sigma_T}{2\pi}\left[\frac{7}{9}\ln\left(\frac{183}{Z^{1/3}}\right) - \frac{1}{54}\right]$$

for complete screening when $h\nu/mc^2 \gg 1/\alpha Z^{1/3}$.

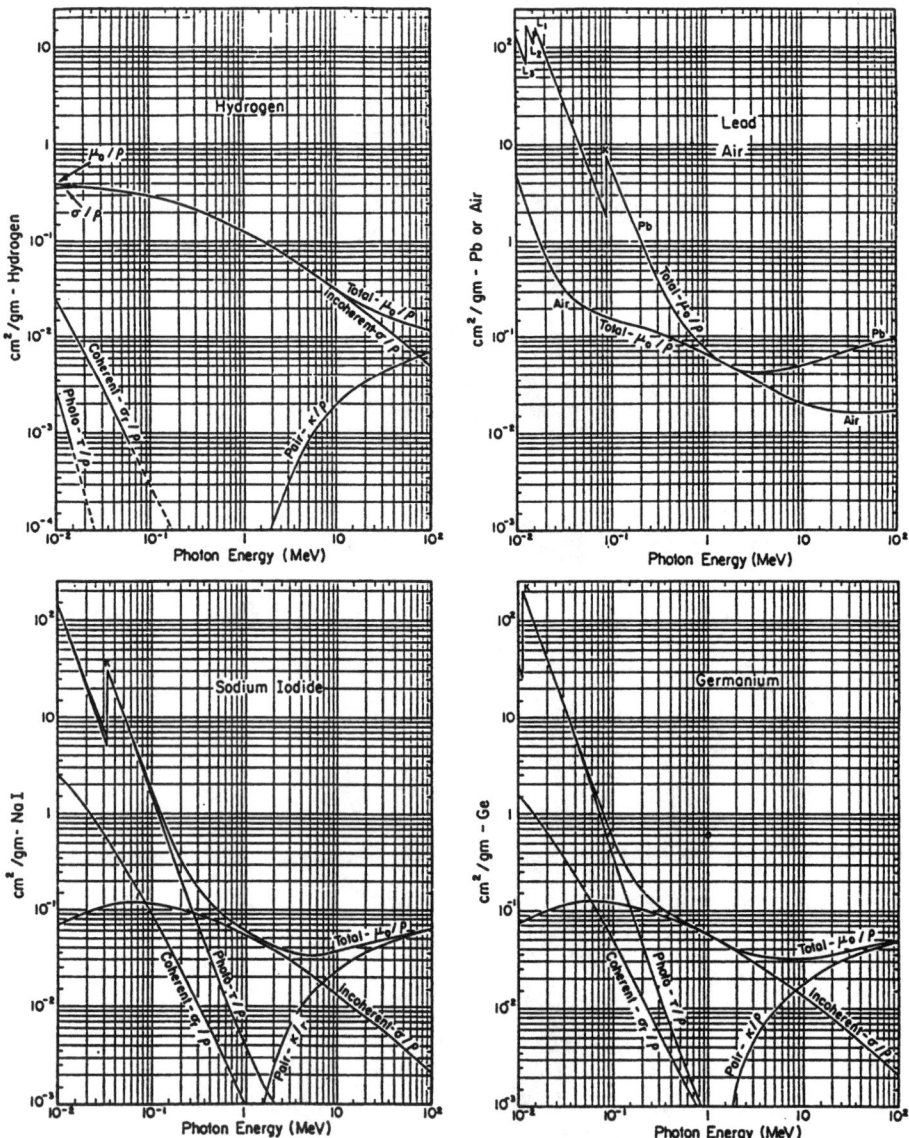

Figure 10.2. Macroscopic cross sections for γ-ray attenuation by photoelectric absorption, Compton scattering and pair production in hydrogen, air, NaI, and Ge, as a function of photon energy from [21].

The cross section for electron–positron pair production by the interaction of two photons of energy $h\nu$ and $h\nu'$ when $h\nu h\nu' > m^2 c^4$ is [1]

$$\sigma(h\nu, h\nu')_{\mathrm{pp}} = \frac{3\sigma_{\mathrm{T}}}{16}(1 - \beta^2)\left[2\beta(\beta^2 - 2) + (3 - \beta^4)\ln\left(\frac{1+\beta}{1-\beta}\right)\right],$$

where $\beta = (1 - m^2 c^4 / h\nu h\nu')^{1/2}$.

The attenuation coefficient for electron–positron pair production by a photon in a strong magnetic field, in the limit $h^2 v^2/2m^2 c^4 B_* \gg 1$ with $B_* = B/4.414 \times 10^{13}$ G, is [15]

$$R_{1\gamma} = \frac{\alpha mc}{2h} B_* \sin\theta = \begin{cases} 0.377 \exp\left(-\dfrac{4}{3\chi}\right), & \chi \ll 1, \\ 0.6\chi^{-1/3}, & \chi \gg 1, \end{cases}$$

where $\chi \equiv (h\nu/2mc^2)B_* \sin\theta$ and the threshold energy is $2mc^2/\sin\theta$.

10.3.3 Compton Scattering

The cross section for Compton scattering (cs) of photons by electrons is [13]

$$\sigma(h\nu)_{cs} = \frac{3\sigma_T}{8\eta}\left[\left(1 - \frac{2\eta + 2}{\eta^2}\right)\ln(2\eta + 1) + \frac{1}{2} + \frac{4}{\eta} - \frac{1}{2(2\eta + 1)^2}\right],$$

where $\eta = h\nu/mc^2$ is the initial photon energy.

The angular distribution of the scattered photons, in terms of the scattering angle ϕ, is

$$f(\cos\phi) = \frac{3\sigma_T}{8\sigma(h\nu)_{cs}}\left[\frac{(1 + \eta + \eta^2 - \eta\cos\phi)(1 + \cos^2\phi) - 2\eta^2\cos\phi}{(1 + \eta - \eta\cos\phi)^3}\right].$$

The energy of the Compton-scattered photon $h\nu'$ relative to the initial photon energy $h\nu$ is

$$r = h\nu'h\nu = 1/(1 + \eta - \eta\cos\phi),$$

and the energy distribution of the Compton-scattered photons is

$$f(r) = \frac{3\sigma_T}{8\eta\sigma(h\nu)_{cs}}\left[r - 1 + \frac{1}{r} + \frac{(\eta r + r - 1)^2}{\eta^2 r^2}\right],$$

for $1/(2\eta + 1) \le r \le 1$, corresponding to scattering angles $0° \le \phi \le 180°$, and $f(r) = 0$ for other values of r.

In a magnetic field, where the electron energies are quantized in Landau states, the total scattering cross section for unpolarized photons in the Thomson limit is [16]

$$\sigma = \frac{\sigma_T}{2}\left[\sin^2\theta + \frac{1}{2}(1 + \cos^2\theta)\left(\frac{h^2 v^2}{(h\nu + h\nu_B)^2} + \frac{h^2 v^2}{(h\nu - h\nu_B)^2}\right)\right],$$

where θ and $h\nu$ are the angle and energy of the incident photon with respect to the magnetic field in the electron rest frame, and $h\nu_B = eB/mc$ is the cyclotron frequency. When $(h\nu/h\nu_B)B > 10^{12}$ G, relativistic effects modify the cross section [17, 18].

10.3.4 Cyclotron Absorption

In a magnetic field, the cross section for absorption of photons by electron scattering from ground state to higher Landau levels is [19]

$$\sigma_{abs}^n(\theta) = \frac{\alpha\pi^2\hbar^2 c^2}{E_n}\delta(h\nu - h\nu_n)\frac{e^{-Z}Z^{n-1}}{(n-1)!}\left((1 + \cos^2\theta) + \frac{Z}{n}\sin^2\theta\right),$$

where $Z = h^2\nu^2\sin^2\theta/2mc^2B_*$, $E_n = (m^2c^4 + h^2\nu^2\cos^2\theta + 2nB_*m^2c^4)^{1/2}$, and $B_* = B/4.414 \times 10^{13}$ G. The photons are absorbed at the resonant energies

$$h\nu_n = mc^2[(1 + 2nB_*\sin^2\theta)^{1/2} - 1]/\sin^2\theta.$$

In the nonrelativistic limit, $nB_* = nB/4.414 \times 10^{13}$ G $\ll 1$, the absorption cross section is [20]

$$\sigma_{\rm abs}^n(\theta) \approx \frac{\alpha\pi^2\hbar^2c^2}{m}\left(\frac{n^2}{2}B_*\sin^2\theta\right)^{n-1}\frac{1+\cos^2\theta}{(n-1)!},$$

where photons are absorbed at harmonics $h\nu_n = neB/mc$.

10.4 ASTROPHYSICAL γ-RAY OBSERVATIONS

The γ-ray sky is extremely variable. Unlike the sources seen at longer wavelengths, most of the astrophysical γ-ray sources have been seen only in their transient emission. Out of roughly a thousand γ-ray sources less than 10% are relatively steady, persistent sources. The latter include a wide variety of sources such as the Sun, supernova remnants, the interstellar medium, and the cosmic background emission, but they are mostly compact objects: radio pulsars, accreting neutron stars, and blackhole candidates, ranging from stellar mass objects in our own galaxy to supermassive, active galactic nuclei.

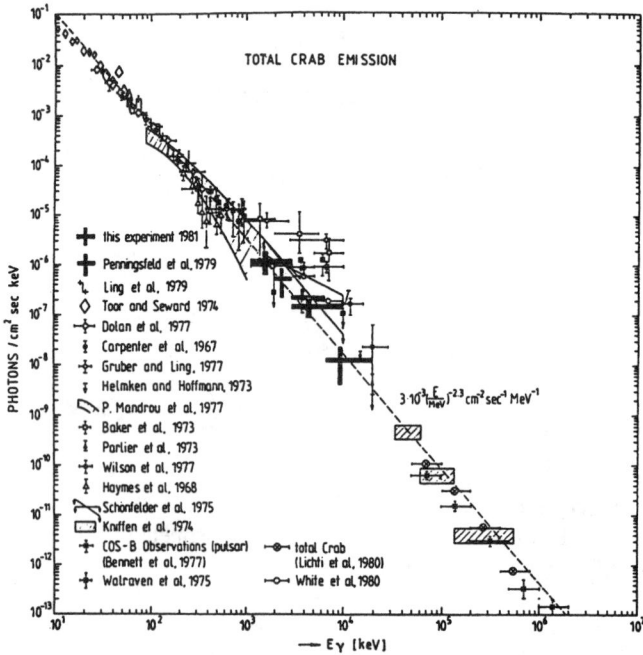

Figure 10.3. Total Crab Nebula and pulsar emission from 10 keV to 2 GeV. The Crab flux is the de facto standard for the expression of source fluxes, e.g., 10 milliCrabs. This figure is provided to relate Crab fluxes at various energies to the more useful photons cm^{-2} s^{-1} $^{-1}$keV. The plot is from Graser, U., & Schönfelder, V. 1982, *ApJ*, **263**, 677, and references to observations contained within the plot can be found in that paper.

The vast majority of γ-ray sources, however, have been seen only briefly for times as short as a few milliseconds to as much as 1000 s. These are collectively known as γ-ray bursts, but because of their diverse properties, they may arise from a variety of sources and processes.

General reviews of astrophysical γ-ray sources are given in [22–25]. Figure 10.3 displays the famous Crab Nebula spectrum.

Many γ-ray bursts are reviewed in [26–31].

The locations and properties of selected galactic and extragalactic γ-ray sources are listed in Tables 10.4–10.9 and basic data on the major hard X-ray and γ-ray instruments are included in Tables 10.10–10.12.

Table 10.4. *Selected galactic sources* > 100 keV.

Source name	Type[a] period[b]	α^c δ	$l^{II\,d}$ b^{II}	Dist.[e]	Flux[f]	Energy	Lum.[g]	Refs.
X Persei	XRBe	58.06	163.08	0.35	4×10^{-5}	30 keV	8×10^{32}	[1]
0352+308	835 s	+30.90	−17.14		1×10^{-5}	100 keV	2×10^{33}	[1]
0422+328	BHC	64.63	165.89	2	2×10^{-2}	30 keV	1×10^{37}	[2]
		+32.79	−11.91		2×10^{-3}	100 keV	1×10^{37}	[2]
					1×10^{-4}	300 keV	6×10^{36}	[2]
Crab (total)	SNR and	82.88	184.56	2	8×10^{-3}	30 keV	5×10^{36}	[3]
0531+219	Pulsar	+21.98	−5.79		6×10^{-4}	100 keV	5×10^{36}	[3]
					5×10^{-5}	300 keV	3×10^{36}	[3]
					5×10^{-6}	1 MeV	4×10^{36}	[3]
					2×10^{-8}	10 MeV	2×10^{36}	[3]
					6×10^{-11}	100 MeV	5×10^{35}	[3]
					2×10^{-13}	1 GeV	2×10^{35}	[3]
Crab (pulsar)	Pulsar	82.88	184.56	2	1×10^{-3}	30 keV	7×10^{35}	[4]
0531+219	0.0332 s	+21.98	−5.79		1×10^{-4}	100 keV	8×10^{35}	[3]
					2×10^{-5}	300 keV	1×10^{36}	[3]
					6×10^{-7}	1 MeV	5×10^{35}	[3]
					5×10^{-9}	10 MeV	4×10^{35}	[3]
					2×10^{-11}	100 MeV	2×10^{35}	[3]
					2×10^{-13}	1 GeV	2×10^{35}	[3]
					1×10^{-15}	10 GeV	8×10^{34}	[5]
					3×10^{-21}	1 TeV	2×10^{33}	[6]
0535+262	XRBe	84.06	181.47	1.8	1×10^{-2}	30 keV	6×10^{36}	[7]
	104 s	+26.32	−2.54		5×10^{-5}	100 keV	3×10^{35}	[7]
SN1987A	SN[h]	83.96	279.71	50	2×10^{-4}	30 keV	1×10^{38}	[8]
0536−693	(in LMC)	−69.30	−31.94		4×10^{-5}	100 keV	2×10^{38}	[8]
					7×10^{-6}	300 keV	3×10^{38}	[8]
0620−00	BHC	95.05	209.96	0.87	3×10^{-3}	30 keV	4×10^{35}	[9]
		−0.32	−6.54		2×10^{-4}	100 keV	3×10^{35}	[9]
Geminga	Pulsar	97.75	195.14	< 0.4†	3×10^{-11}	100 MeV	$< 6 \times 10^{33}$	[10]
0630+178	0.2371 s	+17.81	+4.27		6×10^{-13}	1 GeV	$< 2 \times 10^{34}$	[10]

Table 10.4. (*Continued.*)

Source name	Type[a] period[b]	α^c δ	$l^{II\,d}$ b^{II}	Dist.[e]	Flux[f]	Energy	Lum.[g]	Refs.
Vela (pulsar) 0833−45	Pulsar 0.0892 s	128.40 −45.05	263.58 −2.82	0.5	4×10^{-7}	100 keV	2×10^{32}	[11]
					2×10^{-7}	300 keV	1×10^{33}	[11]
					1×10^{-8}	3 MeV	5×10^{33}	[11]
					3×10^{-9}	10 MeV	2×10^{34}	[12]
					6×10^{-10}	30 MeV	3×10^{34}	[12]
					1×10^{-10}	100 MeV	5×10^{34}	[12]
					2×10^{-12}	1 GeV	8×10^{34}	[12]
1009−45	BHC	153.37 −45.06	275.85 +9.35	3†	1×10^{-4}	100 keV	2×10^{36}	[13]
					7×10^{-6}	300 keV	1×10^{36}	[13]
1055−52	Pulsar	164.50 −52.45	286.00 6.65	1.53	2×10^{-12}	100 MeV	1×10^{34}	[14]
Nova Muscae 1124−684	BHC	171.08 −68.40	295.31 −7.07	1†	4×10^{-3}	30 keV	7×10^{35}	[15]
					2×10^{-4}	100 keV	4×10^{35}	[15]
					1×10^{-5}	300 keV	2×10^{35}	[15]
1509−58	Pulsar 0.1502 s	227.50 −58.95	320.33 −1.16	1†	3×10^{-5}	30 keV	5×10^{33}	[16]
					4×10^{-6}	100 keV	8×10^{33}	[16]
					8×10^{-7}	300 keV	1×10^{34}	[16]
1543−47	BHC	235.96 −47.56	330.92 +5.43	4	8×10^{-3}	30 keV	2×10^{37}	[17]
					2×10^{-4}	100 keV	6×10^{36}	[17]
1655−40	BHC	253.50 −39.85	344.98 +2.46	3.2	2×10^{-4}	100 keV	4×10^{36}	[13]
					1×10^{-5}	300 keV	2×10^{36}	[13]
Her X−1 1656+354	LMXB 1.24 s	254.01 +35.42	58.15 +37.52	5	1×10^{-3}	30 keV	4×10^{36}	[18]
					1×10^{-5}	100 keV	5×10^{35}	[18]
					3×10^{-20}	1 TeV	1×10^{35}	[6]
GX 339−4 1659−487	BHC	254.76 −48.72	338.94 −4.33	10†	2×10^{-3}	30 keV	4×10^{37}	[17]
					2×10^{-4}	100 keV	4×10^{37}	[17]
					1×10^{-5}	300 keV	2×10^{37}	[17]
1700−37	HMXB	255.14 −37.78	347.76 +2.17	1.7	1×10^{-3}	30 keV	5×10^{35}	[19]
					3×10^{-5}	100 keV	2×10^{35}	[19]
Nova Oph '77 1705−250	BHC	256.29 −25.03	358.59 +9.06	10†	2×10^{-3}	30 keV	3×10^{37}	[20]
					1×10^{-4}	100 keV	2×10^{37}	[20]
1706−44	Pulsar 0.1024 s	256.52 −44.42	343.10 −2.68	1.82	7×10^{-12}	100 MeV	3×10^{34}	[5]
					1×10^{-13}	1 GeV	6×10^{34}	[5]
1716−249	BHC	259.94 −24.97	0.20 +6.99	2.4	4×10^{-4}	100 keV	5×10^{36}	[13]
					2×10^{-5}	300 keV	2×10^{36}	[13]
Terzian 2 1724−308	LMXB	261.08 −30.76	356.32 +2.30	14†	2×10^{-4}	40 keV	1×10^{37}	[21]
					3×10^{-5}	100 keV	1×10^{37}	[21]
GX 1+4 1728−247	LMXB 114 s	262.15 −24.70	1.90 +4.87	10†	2×10^{-3}	30 keV	3×10^{37}	[19]
					4×10^{-5}	100 keV	8×10^{36}	[19]
1737.9−2952	BHC	264.48 −29.52	358.97 +0.52	10†	1×10^{-3}	30 keV	2×10^{37}	[22]
					4×10^{-4}	100 keV	8×10^{37}	[22]

Table 10.4. (*Continued.*)

Source name	Type[a] period[b]	α^c δ	$l^{II\,d}$ b^{II}	Dist.[e]	Flux[f]	Energy	Lum.[g]	Refs.
1740.7−2942	BHC	265.18 −29.71	359.12 −0.10	10†	5×10^{-4} 7×10^{-5} 1×10^{-5}	40 keV 100 keV 300 keV	2×10^{37} 1×10^{37} 2×10^{37}	[23] [23] [23]
"Galactic Center" 1742−294	BHC	266.24 −29.38	359.89 −0.71	10†	3×10^{-3} 1×10^{-4} 2×10^{-5}	30 keV 100 keV 300 keV	5×10^{37} 2×10^{37} 3×10^{37}	[24] [24] [25]
1743−322	XRT	265.44 −32.21	357.13 −1.61	10†	6×10^{-4} 4×10^{-5}	30 keV 100 keV	1×10^{37} 8×10^{36}	[19] [19]
GX 5−1 1758−250	LMXB	269.51 −25.08	5.08 −1.02	10†	8×10^{-4} 4×10^{-5}	30 keV 100 keV	1×10^{37} 8×10^{36}	[19] [19]
1758−258	BHC	269.53 −25.74	4.52 −1.36	10†	4×10^{-4} 6×10^{-5} 3×10^{-6}	30 keV 100 keV 300 keV	7×10^{36} 1×10^{37} 5×10^{36}	[25] [25] [25]
1915+105	BHC	288.80 +10.95	45.37 −0.22	12.5	8×10^{-5} 2×10^{-6}	100 keV 300 keV	3×10^{37} 6×10^{36}	[13] [13]
Cyg X−1 1956+350	BHC	299.04 +35.05	71.29 +3.12	2.5	9×10^{-3} 1×10^{-3} 4×10^{-5} 1×10^{-5}	30 keV 100 keV 300 keV 1 MeV	1×10^{37} 1×10^{37} 4×10^{36} 1×10^{37}	[26] [26] [27] [27]
2000+25	BHC	300.18 +25.10	63.38 −3.00	2†	2×10^{-3} 2×10^{-4} 2×10^{-5}	30 keV 100 keV 300 keV	1×10^{36} 2×10^{36} 1×10^{36}	[28] [28] [28]
2023+338	BHC	305.53 +33.71	73.13 −2.09	2†	1×10^{-2} 1×10^{-3} 1×10^{-4}	30 keV 100 keV 300 keV	7×10^{36} 8×10^{36} 7×10^{36}	[13] [13] [13]
Cyg X−3 2030+407	HMXB	307.52 +40.76	79.76 +0.77	10†	1×10^{-3} 2×10^{-5} 5×10^{-20} 2×10^{-26}	30 keV 100 keV 1 TeV 1 PeV	2×10^{37} 4×10^{36} 1×10^{36} 4×10^{35}	[18] [18] [6] [6]

Notes

[a]BHC, black hole candidate; HMXB, high mass X-ray binary; LMXB, low-mass X-ray binary system; SN, supernova; SNR, supernova remnant; XRBe denotes Be star plus collapsed object binary system; XRT, X-ray transient.

[b]Pulsar periods in seconds are from Taylor, J.H., Manchester, R.N., & Lyne, A.G. 1993, *ApJS*, **88**, 529, and an update to be found at pulsar.princeton.edu. Binary pulse periods are from Nagase, F. 1989, *PASJ*, **41**, 1.

[c]Celestial coordinates in degrees from Wood, K.S. et al. 1984, *ApJS*, **56**, 507, except for SN1987A (West, R. 1987, ESO Workshop on the SN1987A, 5); A0620−00 (Boley, F.I. et al. 1976, *ApJ*, **203**, L13); Geminga (Bignami, G.F. et al. 1983, *ApJ*, **272**, L9); Vela Pulsar (Forman, W.R. et al. 1978, *ApJS*, **38**, 357); Nova Muscae (West, R. 1991, IAU Circ. No. 5165); GRS1227+0229 (Jourdain, E. et al. 1991, *Int. Cosmic Ray Conf.*, **1**, 173); PSR1509−58 (Princeton Pulsar List, 1992); A1524−62 (Murdin, P. et al. 1977, *MNRAS*, **178**, 27); 4U1700−37 (Forman, W.R. et al. 1978, *ApJS*, **38**, 357); PSR1706−44 (Princeton Pulsar List, 1992); Terzian 2 (Hertz, P.L., & Grindlay, J.E. 1983, *ApJ*, **275**, 105); 1740.7−2942 (Hertz, P.L., & Grindlay, J.E. 1984, *ApJ*, **278**, 137); GRS1758−258 (Sunyaev, R. et al. 1991, *Sov. Astron. Lett.*, **17**, 50); Briggs Source (Briggs, M.S. et al. 1995, *ApJ*, **442**, 638); GS2000+25 (Tsunemi, H. et al. 1989, *ApJ*, **337**, L81); GS2023+338 (Wagner, R.M. et al. 1989, IAU Circ. No. 4783).

[d]Galactic coordinates in degrees.

[e]All distances in kiloparsecs. Those marked with a dagger (†) are assumptions, some of which are based on optical limitations and some of which are unknown in which case the value of 10 kpc is used. Known distance references are Crab (Trimble, V. 1968, *AJ*, **73**, 535); X Persei (Brucato, R.J., & Kristian, J. 1972, *ApJ*, **173**, L105); A0535+26 (Giangrande, A. et al. 1980, *A&AS*, **40**, 289); SN1987A (Arnett, W.D. et al. 1989, *ARA&A*, **27**, 629); A0620−00 (Oke,

J.B. 1977, *ApJ*, **217**, 181); Vela (Grenier, I.A. et al. 1988, *A&A*, **204**, 117); A1524−62 (Murdin, P. et al. 1977, *MNRAS*, **178**, 27); Her X-1 (Bahcall, N.A. 1973, *Sixth Texas Symp.*, **224**, 178); 4U1700−37 (Bradt, H.V., & McClintock, J.E. 1983, *ARA&A*, **21**, 13); Terzian 2 (Malkan, M.A. et al. 1980, *ApJ*, **237**, 432); GX 1+4 (Davidsen, A.F. et al. 1977, *ApJ*, **211**, 866); Cyg X-1 (Margon, B.H. et al. 1973, *ApJ*, **185**, L117); Cyg X-3 (Breas, L.L.E. et al. 1973, *NaturePS*, **242**, 66).

^{f} Observed flux in photons/cm^2 s keV.

^{g} Inferred luminosity per logarithmic interval assuming isotropic emission, $E^2 \times \text{(Flux)} = E^2$ (keV^2) $\times$ Distance^2 (kpc^2) $\times$ Flux (phot./cm^2 s keV) $\times 2 \times 10^{35}$ erg/s ln E.

^{h} Peak flux from supernova explosion in the Large Magellanic Cloud (LMC).

References

1. Worrall, D.M. et al. 1981, *ApJ*, **247**, L31
2. Paciesas, W.S. et al. 1992, IAU Circ. No. 5580; Harmon, B.A. et al. 1992, IAU Circ. No. 5584; McCrosky, R.E. 1992, IAU Circ. No. 5597
3. Graser, U., & Schonfelder, V. 1982, *ApJ*, **263**, 677
4. Knight, F.K. 1982, *ApJ*, **260**, 538
5. Kniffen, D.A. et al. 1992, *ApJ*, **383**, L49
6. Weekes, T.C. 1988, *Phys. Rep.*, **160**, 1; Weekes, T.C. 1992, *Space Sci. Rev.*, **59**, 315
7. Ricker, G.R. et al. 1976, *ApJ*, **204**, L73
8. Sunyaev, R. et al. 1988, *Sov. Astron. Lett.*, **14**, 247
9. Coe, M.J. et al. 1976, *Nature*, **259**, 544
10. Hermsen, W. 1980, Ph.D. thesis, Leiden University; Bertsch, D.L. et al. 1992, *Nature*, **357**, 306
11. Strickman, M.S. et al. 1996, *ApJ*, **460**, 735
12. Hermsen, W. et al. 1992, *AIP Conf. Proc.*, **280**, 204
13. Grove, J.E. et al. 1997, *AIP Conf. Proc.*, **410**, 122
14. Thompson, D.J. et al. 1995, *ApJS*, **101**, 259
15. Sunyaev, R. et al. 1992, *ApJ*, **389**, L75
16. Ulmer, M.P. et al. 1992, *ApJ*, **417**, 738; Matz, S.M. et al. 1994, *ApJ*, **434**, 288; Marsden, D.C. et al. 1996, *ApJ*, **491**, L39
17. Harmon, B.A. et al. 1992, *AIP Conf. Proc.*, **280**, 314, 350
18. Trumper, J. et al. 1978, *ApJ*, **219**, L105
19. Levine, A.M. et al. 1984, *ApJS*, **54**, 581
20. Wilson, C.S., & Rothschild, R.E. 1983, *ApJ*, **274**, 717
21. Barret, P.E. et al. 1991, *ApJ*, **379**, L21
22. Grindlay, J.E. et al. 1992, *A&AS*, **97**, 155
23. Cook, M.C. et al. 1991, *ApJ*, **372**, L75; Sunyaev, R. et al. 1992, *ApJ*, **383**, L49
24. Slassi, S. et al. 1991, *22nd International Cosmic Ray Conference*, **1**, OG3.2.8
25. Sunyaev, R. et al. 1991, *Sov. Astron. Lett.*, **17**, 50
26. Nolan, P.L., & Matteson, J.L. 1983, *ApJ*, **265**, 389
27. Ling, J.C. et al. 1987, *ApJ*, **321**, L117
28. Sunyaev, R. et al. 1988, *Sov. Astron. Lett.*, **14**, 327

Table 10.5. *Brightest annihilation and nuclear line sources.*

Process	Line E (MeV)	FWHM (keV)	Line source	Max. flux (ph./cm^2 s)	Lum. (erg/s)	Refs.
$e^{\pm}$ Annihilation Radiation						
	0.511	2	Interstellar gas	$1.5 \times 10^{-3\,a}$	7×10^{36}	[1]
	0.511	3	BH? near GC^b	2×10^{-3}	1×10^{37}	[2]
	0.511	$< 10^c$	Solar flares	2×10^{-2}	5×10^{19}	[3]
Redshifted	0.430	100	GBS 0526−66	100	2×10^{43}	[4]
Redshifted	0.480	240	1E 1740.7−2942	1×10^{-2}	6×10^{37}	[5]
Redshifted	0.481	60	Nova Muscae	6×10^{-3}	$6 \times 10^{35\,d}$	[6]
Redshifted	0.404	3	CrabPulsar transient	7×10^{-3}	2×10^{36}	[7]
Redshifted	0.413	15	10June74 transient	7×10^{-3}	$6 \times 10^{35\,d}$	[8]
Blueshifted	0.500–2.0		Cygnus X-1	2×10^{-2}	2×10^{37}	[9]
Backscattered	0.170	12	BH? near GC^b	7×10^{-4}	2×10^{34}	[10]
Backscattered	0.19	40	Nova Muscae	2×10^{-3}	$7 \times 10^{34\,d}$	[6]

Table 10.5. *(Continued.)*

Process	Line E (MeV)	FWHM (keV)	Line source	Max. flux (ph./cm^2 s)	Lum. (erg/s)	Refs.
Radioactive Decay						
^{56}Co($\epsilon\gamma$, $\beta^+\gamma$)^{56}Fe	0.847	~ 9	Supernova 1987A	1×10^{-3}	4×10^{38}	[11]
	1.238	~ 11	Supernova 1987A	1×10^{-3}	6×10^{38}	[11]
	2.598	$\sim 26^c$	Supernova 1987A	3×10^{-4}	4×10^{38}	[12]
	3.244	$\sim 32^c$	Supernova 1987A	2×10^{-4}	3×10^{38}	[12]
^{57}Co($\epsilon\gamma$)^{57}Fe	0.122	$\sim 1^c$	Supernova 1987A	4×10^{-5}	3×10^{36}	[13]
^{44}Ti($\epsilon\gamma$)^{44}Sc	0.068	$\sim 2^c$	SN Remnant CasA	4×10^{-5}	4×10^{33}	[14]
	0.078	$\sim 2^c$	SN Remnant CasA	4×10^{-5}	5×10^{33}	[14]
^{44}Sc($\epsilon\gamma$, $\beta^+\gamma$)^{44}Ca	1.157	$\sim 30^c$	SN Remnant CasA	4×10^{-5}	8×10^{34}	[15]
^{26}Al($\beta^+\gamma$)^{26}Mg	1.809	5.4	Interstellar medium	$4.5 \times 10^{-4\,a}$	8×10^{36}	[16]
Nuclear Excitation						
^{4}He(α, n)^{7}Be*	0.429	25^c	Solar flares	3×10^{-2}	6×10^{19}	[17]
^{4}He(α, p)^{7}Li*	0.478	30^c	Solar flares	3×10^{-2}	7×10^{19}	[17]
^{56}Fe(p, $p'\gamma$)	0.847	5^c	Solar flares	1×10^{-2}	4×10^{19}	[2]
^{12}C, ^{16}O(p, x)^{10}B*	1.023	30^c	Solar flares	5×10^{-3}	2×10^{19}	[2]
^{56}Fe(p, $p'\gamma$)	1.238	7^c	Solar flares	1×10^{-2}	6×10^{19}	[2]
^{24}Mg(p, $p'\gamma$)	1.369	15^c	Solar flares	2×10^{-2}	1×10^{20}	[2]
^{20}Ne(p, $p'\gamma$)	1.634	22^c	Solar flares	4×10^{-2}	3×10^{20}	[2]
^{28}Si(p, $p'\gamma$)	1.779	20^c	Solar flares	5×10^{-2}	4×10^{20}	[2]
^{12}C(p, $p'\gamma$)	4.438	97^c	Solar flares	5×10^{-2}	1×10^{21}	[2]
^{16}O(p, $p'\gamma$)	6.129	114^c	Solar flares	4×10^{-2}	1×10^{21}	[2]
Neutron Capture						
^{1}H(n, γ)^{2}H	2.223	$< 0.1^c$	Solar flares	~ 1	1×10^{22}	[2, 18]
	2.223	70	10June74 transient	1.5×10^{-2}	$6 \times 10^{36\,d}$	[8]
Redshifted	1.790	95	10June74 transient	3×10^{-2}	$1 \times 10^{37\,d}$	[8]
^{56}Fe(n, γ)^{57}Fe Redshifted	5.947	25	10June74 transient	1.5×10^{-2}	$2 \times 10^{37\,d}$	[8]

Notes

^{a} Per radian of longitude in the Galactic Plane.

^{b} Black hole? Near Galactic Center.

^{c} Theoretical widths for unresolved lines.

^{d} For a nominal distance of 1 kpc.

References

1. Haymes, R.C. et al. 1975, *ApJ*, **201**, 593; Leventhal, M. et al. 1978, *ApJ*, **225**, L11; Riegler, G.R. et al. 1981, *ApJ*, **248**, L13; Share, G.H. et al. 1988, *ApJ*, **326**, 717; Wallyn, P. et al. 1993, *ApJ*, **403**, 621; Leventhal, M. et al. 1993, *ApJ*, **405**, L25; Purcell, W.R. et al. 1997, *ApJ*, **491**, 725; Harris, M.J. et al. 1998, *ApJ*, **501**, L55

2. Riegler, G.R. et al. 1981, *ApJ*, **248**, L13; Leventhal, M. et al. 1982, *ApJ*, **260**, L1; Leventhal, M. et al. 1986, *ApJ*, **302**, 459

3. Chupp, E.L. et al. 1973, *Nature*, **241**, 333; Chupp, E.L. 1984, *ARA&A*, **22**, 359; Murphy, R. et al. 1990, *ApJ*, **358**, 298

4. Mazets, E.P. et al. 1982, *Ap&SS*, **84**, 173

5. Bouchet, F.R. et al. 1991, *ApJ*, **383**, L45

6. Goldwurm, A. et al. 1992, *ApJ*, **389**, L79; Sunyaev, R. et al. 1992, *ApJ*, **389**, L75

7. Leventhal, M. et al. 1977, *ApJ*, **216**, 491; Ayre, C.A. et al. 1983, *MNRAS*, **205**, 285

8. Ling, J.C. et al. 1982, *AIP Conf. Proc.*, **77**, 143

9. Nolan, P.L., & Matteson, J.L. 1983, *ApJ*, **265**, 389; Ling, J.C. et al. 1987, *ApJ*, **321**, L117; Ling, J.C., & Wheaton, W.A. 1989, *ApJ*, **343**, L57

10. Leventhal, M., & MacCallum, C.J. 1980, *Ann. N.Y. Acad. Sci.*, **336**, 248; Matteson, J. et al. 1991, *AIP Conf. Proc.*, **232**, 45; Lingenfelter, R.E., & Hua, X.M. 1991, *ApJ*, **381**, 426; Smith, D. et al. 1993, *ApJ*, **414**, 165

11. Mahoney, W.A. et al. 1988, *ApJ*, **334**, L81; Matz, S.M. et al. 1988, *Nature*, **331**, 416; Sandie, W.G. et al. 1988, *ApJ*, **334**, L91; Rester, A.C. et al. 1989, *ApJ*, **342**, L71; Tueller, J. et al. 1990, *ApJ*, **351**, L41

12. Tueller, J. et al. 1990, *ApJ*, **351**, L41; Leising, M.D., & Share, G.H. 1990, *ApJ*, **357**, 638
13. Kurfess, J.D. et al. 1992, *ApJ*, **399**, L137
14. Rothschild, R.E. et al. 1998, *NucPhys B Proc. Suppl.* 69, 68
15. Iyudin, A.F. et al. 1994, *A&A*, **284**, L1
16. Mahoney, W.A. et al. 1984, *ApJ*, **286**, 578; Harris, M.J. et al. 1990, *ApJ*, **362**, 135; Diehl, R. et al. 1995, A&A. **298**, 445; Naya, J. et al. 1991, *Nature*, **384**, 44
17. Murphy, R. et al. 1990, *ApJ*, **351**, 299
18. Hudson, H.S. et al. 1980, *ApJ*, **236**, L91; Prince, T.A. et al. 1982, *ApJ*, **255**, L81

Table 10.6. *Cyclotron line sources.*

Source name	Object type	α (deg) δ (deg)	l (deg) b (deg)	Centroid (keV)	FWHM (keV)	Field (10^{12} G)	Refs.
0115+634	X ray Binary	19.82 +63.82	126.00 +1.11	12.1 ± 0.2 22.6 ± 0.4	3.1 ± 0.6 4.3 ± 0.9	1.0	[1]
0332+530	X ray Binary	53.75 +53.18	146.05 −2.19	28.5 ± 0.5 52.6 ± 1.4	11.0 ± 0.9 10 ± 3	2.5	[2]
NP0531 0531+219	Pulsar	83.63 +22.01	184.56 −5.79	73.3 ± 1.0[a,b]	< 4.9	6.7	[3]
0535+262	X ray Pulsar	83.95 +26.29	181.09 −3.24	∼ 55 ∼ 110		4.3	[4]
Vel X-1 0900−403	X ray Binary	135.53 −40.56	263.06 3.93	25.6 ± 0.9 57.9 ± 1.0	7.2 ± 2.6 24.0 ± 1	2.2	[5]
Cen X-3 1119−603	X ray Binary	170.31 −60.62	292.09 0.34	28.5 ± 0.5	6.3 ± 2.0	2.5	[6]
1538−522	X ray Binary	235.60 −52.39	327.42 +2.16	20.9 ± 0.2[c]	5.1 ± 0.3[c]	1.7	[7]
4U1626−67 1627−673	X ray Binary	248.07 −67.46	321.79 −13.09	∼ 7 ± 1[b] ∼ 18 ± 1[b] 36.5 ± 1.0	... ∼ 15 7 ± 2.8	∼ 3	[8]
Her X-1 1656+354	X ray Binary	254.46 +35.34	58.15 +37.52	34.7 ± 0.9[c]	12.0 ± 2.0[c]	2.9	[9]
1907+097	X ray Binary	287.41 +9.83	43.74 0.48	20.0 ± 1.0	4.1 ± 2.6	1.7	[10]
Cep X-4 2137+579	X ray Binary	324.88 +57.99	99.68 +4.06	30.5 ± 0.4	15.0 ± 1.4	2.6	[11]
GRB870303	γ burst	...	...	20.4 ± 0.7 40.6 ± 2.6	3.5 ± 2.7 12.3 ± 6.3	∼ 1.7	[12]
GRB880205	γ burst	...	...	19.3 ± 0.7 38.6 ± 1.6	4.1 ± 2.2 14.4 ± 4.6	∼ 1.7	[12]
GRB890929	γ burst	...	...	26.3 ± 1.5 46.6 ± 1.7	$7.5^{+4.5}_{-4.1}$ $12.7^{+5.8}_{-5.1}$	∼ 2.1	[13]

Notes

[a] Transient line seen between 73 and 79 keV.

[b] Emission line.

[c] Line centroid and width are observed to vary with pulse phase.

References

1. Nagase, F. et al. 1991, *ApJ*, **375**, L49
2. Makishima, K. et al. 1990, *ApJ*, **365**, L59
3. Ling, J.C. et al. 1979, *ApJ*, **231**, 896; Ayre, C.A. et al. 1983, *MNRAS*, **205**, 285
4. Grove, J.E. et al. 1995, *ApJ*, **438**, L25; Maisack, M. et al. 1997, *A&A*, **325**, 212
5. Makishima, K., & Mihara, T. 1992, *Frontiers of X-Ray Astronomy* (University Academy Press, Tokyo) p. 23; Mihara, T. 1995, Thesis, University of Tokyo; Kretschmar, P. et al. 1997, *A&A*, **325**, 623; Dal Fiume, D. et al. 1998, *Nuc Phys B Proc. Suppl.*, **69**, 145
6. Dal Fiume, D. et al. 1998, *Nuc Phys B Proc. Suppl.*, **69**, 145
7. Clark, G.W. et al. 1990, *ApJ*, **353**, 274
8. Pravdo, S.H. et al. 1979, *ApJ*, **231**, 912; Orlandini, M. 1998, *ApJ*, **500**, L163
9. Mihara, T. et al. 1990, *Nature*, **346**, 250
10. Makishima, K., & Mihara, T. 1992, *Frontiers of X-Ray Astronomy* (University Academy Press, Tokyo) p. 23; Mihara, T. 1995, Thesis, University of Tokyo
11. Mihara, T. et al. 1991, *ApJ*, **379**, L61
12. Murakami, T. et al. 1988, *Nature*, **335**, 234
13. Yoshida, A. et al. 1991, *PASJ*, **43**, L69

Table 10.7. γ-*Ray burst source positions* < 100 arcmin^2.[a,b]

Burst source	Date (yr mo day)	Time (s)	$F > 30$ keV (erg/cm^2)	α (deg)	δ (deg)	l (deg)	b (deg)	Error box (arcmin^2)
GBS0010−160	79 11 16	51 400	2×10^{-4}	3.20	−15.69	82.85	−75.46	4
GBS0026−630	98 01 09[e]	4 341	4×10^{-6}	6.48	−63.02	307.50	−53.86	50
GBS0117−289	78 11 19	34 021	3×10^{-4}	19.72	−28.64	228.50	−83.75	8
GBS0502+118	97 02 28[d]	10 681	1×10^{-5}	75.43	+11.78	188.91	−17.95	2
GBS0526−661	79 03 05b[c]	57 125	1×10^{-3}	81.51	−66.08	276.09	−33.24	0.05
GBS0615−461	79 03 13	62 636	6×10^{-5}	94.1	−46.1	253.8	−25.0	24
GBS0625−346	79 10 14	40 412	1×10^{-5}	96.7	−34.6	242.6	−19.7	82
GBS0653+793	97 05 08[d]	78 106	4×10^{-6}	103.37	+79.29	134.94	+26.71	28
GBS0702+388	98 03 29[d]	13 478	5×10^{-5}	105.65	+38.84	178.12	+18.65	3
GBS0723−271	91 11 09	12 458	7×10^{-6}	110.8	−27.1	240.6	−5.6	6
GBS0813−326	92 05 01	76 695	4×10^{-5}	123.34	−32.59	250.80	+0.96	4
GBS0836−189	98 03 26[d]	76 733	1×10^{-6}	129.14	−18.86	242.37	+13.03	80
GBS0847−361	92 03 11	08 423	1×10^{-4}	131.8	−36.1	257.8	+4.5	4
GBS0912−510	91 05 22	44 036	3×10^{-5}	137.9	−51.0	271.9	−1.9	4
GBS1028+459	79 03 29	80 512	7×10^{-5}	157.8	+45.6	169.9	+56.6	41
GBS1104−229	91 11 18	68 252	5×10^{-5}	166.0	−22.9	272.9	+33.6	20
GBS1156+652	97 12 14[d]	84 041	1×10^{-5}	179.13	+65.20	132.02	+50.95	48
GBS1205+239	78 11 24	14 130	4×10^{-5}	181.94	+23.65	229.93	+79.54	48
GBS1257+592	97 12 27[d]	30 187	7×10^{-7}	194.31	+59.40	121.55	+57.71	7
GBS1327+375	92 07 20	11 524	2×10^{-5}	201.8	+37.5	89.2	+77.2	6
GBS1330−164	92 05 17	11 875	4×10^{-5}	202.6	−16.4	316.3	+45.5	12
GBS1400−468	79 03 07	80 330	2×10^{-4}	210.69	−46.99	315.37	+14.15	10
GBS1407+353	91 11 04	54 282	1×10^{-5}	211.8	+35.3	64.4	+71.9	16
GBS1412+789	79 06 13	50 755	4×10^{-7}	213.1	+78.9	118.0	+37.7	0.8
GBS1450−693	97 04 02[e]	80 352	$\sim 10^{-5}$	222.53	−69.33	313.11	−8.84	2
GBS1528+196	97 01 11[e]	35 040	$\sim 10^{-5}$	232.06	+19.60	29.63	+53.39	28
GBS1625−583	91 07 17	16 378	7×10^{-6}	246.3	−58.3	328.1	−6.3	10
GBS1630−765	79 01 13	27 360	1×10^{-4}	249.2	−76.6	314.7	−19.2	78
GBS1703+006	78 11 21a	05 736	9×10^{-5}	256.4	+0.5	20.7	+23.6	~ 100
GBS1730+491	96 07 20	41 813	3×10^{-6}	262.65	+49.10	75.76	+33.09	28
GBS1756−261	91 04 21	33 246	4×10^{-6}	268.9	26.1	51.5	+23.2	~ 100
GBS1806−207	79 01 07[c]	20 155	1×10^{-6}	272.17	−20.41	10.0	−0.24	6

Table 10.7. *(Continued.)*

Burst source	Date (yr mo day)	Time (s)	$F > 30$ keV (erg/cm^2)	α (deg)	δ (deg)	l (deg)	b (deg)	Error box (arcmin^2)
GBS1808+593	97 08 28[e]	63 877	$\sim 10^{-5}$	272.13	+59.13	87.95	+28.45	0.8
GBS1810+314	79 03 25b	49 500	5×10^{-5}	273.0	+31.4	58.2	+21.6	2
GBS1847+728	92 07 11	58 166	8×10^{-6}	281.8	+72.8	103.7	+26.1	100
GBS1900+145	79 03 24[c]	58 010	1×10^{-6}	286.83	+9.45	43.08	+0.81	7
GBS1912−577	92 04 06	09 915	1×10^{-4}	288.0	−57.7	339.0	−25.3	4
GBS1926+036	79 03 31	76 172	8×10^{-5}	292.0	+3.7	40.4	−6.4	20
GBS2000−427	92 05 25	12 427	1×10^{-4}	300.0	−42.7	357.2	−30.1	6
GBS2006−216	78 11 04b	58 667	3×10^{-4}	302.2	−21.5	21.1	−26.2	14
GBS2142−414	79 06 22	02 665	7×10^{-5}	326.4	−41.2	0.3	−49.6	~ 100
GBS2252−025	79 11 05b	48 862	1×10^{-5}	343.55	−2.26	69.45	−52.51	35
GBS2311+319	79 05 04	31 464	6×10^{-6}	348.4	+32.1	99.9	−26.3	58
GBS2311−499	79 04 06	42 447	1×10^{-6}	348.51	−49.66	336.03	−60.74	0.3
GBS2320+128	92 03 25	62 261	3×10^{-5}	349.9	+12.8	90.8	−44.3	7

Notes

[a]Quiescent X-ray counterparts have been suggested for the three repeater burst sources GBS0526−661, GBS1806−207 and GBS1900+145, which are associated with supernova remnants N49, G10.0−0.3, and G42.8+0.6 (see note c below and Rothschild, R.E., & Lingenfelter, R.E. 1996, *High Velocity Neutron Stars and Gamma-Ray Bursts* (AIP, New York)). No quiescent counterparts have been identified for the "classical" bursts, but fading afterglow sources have been seen following several bursts (see note d) and underlying "host" galaxies have been reported.

[b]Locations (2000 coordinates) for bursts prior to 1990 are based on catalog of Atteia, J.L. et al. 1987, *ApJS*, **64**, 305, and fluences from Mazets, E.P. et al. 1981, *Ap&SS*, **80**, 1, except as follows: GBS1550+762 data from Hueter, G.J. 1987, Ph.D. Dissertation, University of California, San Diego; GBS1806−207 position from Atteia, J.L. et al. 1987, *ApJ*, **320**, L110, and private communication; GBS1900+145 position also from Mazets, E.P. et al. 1981; GBS0746−672 data from Katoh, T. et al. 1984, in *AIP Conf. Proc.* **115**, 390; locations of bursts after 1990 are from Hurley, K., private communication on behalf of the 3rd Interplanetary Network; and from BeppoSAX burst detections listed in notes d and e. Fluences are from Third BATSE Catalog (Meegan, C.A. et al. 1996, *ApJS*, **106**, 65, and the online update of that catalog.

[c]Repeaters: 17 bursts have been observed from the source GBS0526−661 (Golenetskii, S.V. et al. 1979, *Sov. Astron. Lett.*, **13**, 166) associated with supernova remnant N49 in LMC and possibly an X-ray source at α 05^{h}26^{m}0.55^s, δ −66°4'35.56" (Rothschild, R.E., Kulkarni, S.R., & Lingenfelter, R.E. 1994, *Nature*, **368**, 432); > 100 bursts from GBS1806−204 (Atteia, J.L. et al. 1987, *ApJ*, **320**, L105; Laros, J.G. et al. 1987, *ApJ*, **320**, L111) associated with Galactic supernova remnant G10.0−0.3 and an X-ray source at α 18^{h}8^{m}40.34^s, δ −20°24'41.67" (Murakami, T. et al. 1994, *Nature*, **368**, 127), and six bursts from GBS1900+145 (Mazets, E.P. et al. 1979, *Sov. Astron. Lett.*, **5**, 343; Kouveliotou, C. et al. 1993, *Nature*, **362**, 728; Hurley, K. et al. 1994, *ApJ*, **431**, L31) associated with Galactic supernova remnant G42.8+0.6 and possibly an X-ray source at α 19^{h}7^{m}17^s, δ +9°19'18" (Vasisht, G. et al. 1994, *ApJ*, **431**, L35).

[d]Fading optical sources have been observed for GRB0502+118 (Costa, E. et al. 1997, IAU Circ. No. 6572) at $V = 21.3$ discovered 0.9 days after burst at α 05^{h}01^{m}46.61^s, δ +11°46'53.4" (van Paradijs, J. et al. 1997, *Nature*, **386**, 686); GRB0653+793 (Heise, J. et al. 1997, IAU Circ. No. 6654) at $V = 20.5$ discovered 1.28 days after burst at α 06^{h}53^{m}49.43^s, δ +79°16'19.6" (Bond 1997, IAU Circ. No. 6654) and red-shifted absorption lines observed with $z = 0.835$ (Metzger, M.R. et al. 1997, *Nature*, **387**, 878); GBS0702+388 (in't Zand, J. et al. 1998, IAU Circ. No. 6854) at 250 μJy at 8.4 GHz discovered 2.9 days after burst at α 07^{h}02^{m}38.02170^s, δ +38°50'44.0170" (Taylor, G.B. et al. 1998, GCN. No. 40) and at $K = 21.4$ after 4 days (Metzger, M.R. et al. 1998, IAU Circ. No. 6874) GBS0836−189 (Celidonio, G. et al. 1998, IAU Circ. No. 6851) at $R = 21.7$ discovered 0.5 days after burst at α 8^{h}36^{m}34.28^s, δ −18°51'23.9" (Groot, P.J. et al. 1998, IAU Circ. No. 6852) GRB1156+652 (Heise, J. et al. 1997, IAU Circ. No. 6787) at $I = 21.2$ discovered 0.5 days after burst at α 11^{h}56^{m}26.4^s, δ +65°12'00.5" (Halpern, J. et al. 1997, IAU Circ. No. 6788) and red-shifted emission lines observed with $z = 3.4$ (Kulkarni, S. et al. 1998, *Nature*, **393**, 35) GBS1257+592 (Piro. L. et al. 1997, IAU Circ. No. 6797) at $R = 19.5$ discovered 0.6 days after burst at α 12^{h}57^{m}10.6^s, δ +59°24'43" (Castro-Tirado, A.J. et al. 1997, IAU Circ. No. 6800)

[e]No fading optical sources were observed for GBS0026−630 (in't Zand, J. et al. 1998, IAU Circ. No. 6805) with $I < 21$ (Sahu, K.C., & Sterken, C. 1998, IAU Circ. No. 6808) GBS1450−693 (Piro. L. et al. 1997, IAU Circ. No. 6617) with $V < 22.5$ (Pedersen, H. et al. 1997, IAU Circ. No. 6628) GBS1528+196 (in't Zand, J. et al. 1997, IAU Circ. No. 6569) with $R < 22.6$ (Castro-Tirado, A.J. et al. 1997, IAU Circ. No. 6598) GBS1808+593 (Murakami, T. et al. 1997, IAU Circ. No. 6732) with $R < 24.5$ (Odewahn, S.C. et al. 1997, IAU Circ. No. 6735)

Table 10.8. γ-*Ray burst properties.*[a]

Property	Observed values	Comments	References
	"Soft" Repeating Bursts		
Energy range	~ 1 keV–1 MeV	γ–γ opacity constraints	[1]
Energy spectra	$\phi(h\nu) \propto \exp(-h\nu/\epsilon)$	with $\epsilon \geq 25$ keV	[1]
Emission features	~ 430 keV	Redshifted $e^- e^+$ Annihilation radiation	[2]
Rise times	As short as 0.2 ms	Size < 60 km	[3]
Durations	$\sim 10^{-2}$–$\sim 10^2$ s		[1]
Periodicity	8.0 s	Burst GB790305b	[3, 4]
	~ 23 ms	Burst GB790305b	[5]
Source	Off-center in Supernova remnants	high-velocity neutron stars?	[6]
	"Classical" Bursts		
Energy range	~ 1 keV–20 GeV	γ–γ opacity constraints	[7]
Energy spectra	$\phi(h\nu) \sim (h\nu)^s$ $\phi(h\nu) \sim (h\nu)^s$ $E_o \sim 50$–1000 keV	with $s \leq -1$ for $(h\nu)^s < E_o$ with $s \leq -2$ for $(h\nu)^s > E_o$	[8]
Absorption features	20–50 keV	Cyclotron absorption $\sim$ few 10^{12} G fields	[9]
Rise times	As short as 0.2 ms	Size < 60 km	[10]
Durations	$\sim 10^{-2}$–$\sim 10^4$ s		[7]
$\langle V/V_{max} \rangle$	0.33 ± 0.01	Spatially nonuniform	[11]
$\langle \cos\theta \rangle$ Galactocentric angle θ	-0.01 ± 0.02	Isotropic $= 0$	[11]
Source	Optical transient and host galaxies? for several bursts	at $z \sim 0.8$–3.4	[12] [12]

Note

[a]For general reviews, see also Higdon, J.C., & Lingenfelter, R.E. 1990, *ARA&A*, **28**, 401; Harding, A.K. 1991, *Phys. Rep.*, **206**, 327; Fishman, G.J., & Meegan, C.A. 1995, *ARA&A*, **33**, 415; Rothschild, R.E., & Lingenfelter, R.E. 1995, *High Velocity Neutron Star and Gamma-Ray Bursts* (American Institute of Physics, New York) 282 pp.; Kouveliotou, C., Briggs, M.F., & Fishman, G.J. 1996, *Gamma-Ray Bursts* (American Institute of Physics, New York) 1008 pp.

References

1. Mazets, E.P. et al. 1981, *Ap&SS*, **80**, 1; Mazets, E.P., & Golenetski, S.V. 1981, *Ap&SpPhysRev*, **1**, 205; Mazets, E.P. et al. 1982, *Ap&SS*, **82**, 261; Atteia, J.L. et al. 1987, *ApJ*, **320**, L105; Laros, J.G. et al. 1987, *ApJ*, **320**, L111; Murakami, T. et al. 1994, *Nature*, **368**, 127
2. Mazets, E.P. et al. 1982, *Ap&SS*, **84**, 173
3. Cline, T.L. et al. 1980, *ApJ*, **237**, L1
4. Mazets, E.P. et al. 1979, *Nature*, **282**, 587; Barat, C. et al. 1979, *A&A*, **79**, L24
5. Barat, C. et al. 1983, *A&A*, **126**, 400
6. Rothschild, R.E., & Lingenfelter, R.E. 1995, *High Velocity Neutron Star and Gamma-Ray Bursts* (American Institute of Physics, New York) 282 pp.; and previous Table 10.7
7. Mazets, E.P. et al. 1981, *Ap&SS*, **80**, 1; Mazets, E.P., & Golenetski, S.V. 1981, *Ap&SpPhysRev*, **1**, 205; Meegan, C.A. et al. 1996, *ApJS*, **106**, 65; Hurley, K. et al. 1979, *Nature*, **372**, 652
8. Mazets, E.P. et al. 1981, *Ap&SS*, **80**, 1; Band, D. et al. 1993, *ApJ*, **413**, 281; Higdon, J.C., & Lingenfelter, R.E. 1986, *ApJ*, **307**, 197
9. Murakami, T. et al. 1988, *Nature*, **335**, 234; Mazets, E.P. et al. 1982, *Ap&SS*, **82**, 261; Hueter, G.J. 1987, Ph.D. thesis, University of California, San Diego
10. Walker, K.C., & Schaefer, B.E. 1998, "*Gamma Ray Bursts*," *AIP Conf. Proc.*, **428**, edited by C.

Meegan, R. Preece, and T. Koshut (AIP, New York) p. 34

11. Meegan, C.A. et al. 1996, *ApJS*, **106**, 65
12. See previous Table 10.7

Table 10.9. *Extragalactic hard X-ray or γ-ray sources.[a]*

Source name	Object type	α[b] δ	z d^c	Flux[d]	Energy	Lum.[e]	Refs.
NGC 253 0045−255	Starburst galaxy	11.27 −25.56	0.6 0.0036	2×10^{-3}	100 keV	5×10^{46}	[1]
4C+15.05 0202+149	QSO blazar	30.53 +15.00	0.833 3.25	3×10^{-9e}	100 MeV	1×10^{47}	[2]
0208−512	QSO blazar	32.24 −51.25	1.003 6.0	5×10^{-8} 7×10^{-9} 1×10^{-9} 1×10^{-10} 2×10^{-11}	30 MeV 100 MeV 300 MeV 1 GeV 3 GeV	3×10^{47} 5×10^{47} 6×10^{47} 7×10^{47} 1×10^{48}	[3] [3] [3] [3] [3]
3C 66A 0219+428	BL Lac	34.88 +42.81	0.833 3.25	1×10^{-9f}	100 MeV	1×10^{46}	[4]
4C+28.07 0234+285	BL Lac	38.73 +28.59	1.213 3.97	3×10^{-9f}	100 MeV	3×10^{47}	[4]
0235+164	BL Lac	38.97 +16.40	0.94 5.6	2×10^{-8} 6×10^{-9} 8×10^{-10} 8×10^{-11} 1×10^{-11}	50 MeV 100 MeV 300 MeV 1 GeV 3 GeV	3×10^{47} 4×10^{47} 4×10^{47} 5×10^{47} 6×10^{47}	[5] [5] [5] [5] [5]
NGC 1275 0316+413	Seyfert-2	49.12 +41.33	0.0172 0.10	2×10^{-1} 3×10^{-2} 5×10^{-3}	30 keV 100 keV 300 keV	3×10^{44} 6×10^{44} 1×10^{45}	[6] [6] [6]
CTA 26 0336−019	QSO blazar	54.25 −1.94	0.852 3.29	1×10^{-8f}	100 MeV	5×10^{48}	[4]
3C 111 0415+379	Seyfert-1	63.75 +37.90	0.0485 0.283	3×10^{-3f}	100 keV	5×10^{44}	[7]
OA 129 0420−014	QSO blazar	65.18 −1.46	0.915 5.5	4×10^{-9f}	100 MeV	2×10^{47}	[2]
3C 120 0433+052	Seyfert-1	67.63 +5.25	0.0330 0.194	3×10^{-3f}	100 keV	2×10^{44}	[7]
NRAO 190 0440−003	QSO blazar	70.02 −0.39	0.844 3.27	9×10^{-9f}	100 MeV	4×10^{47}	[4]
0454−463	QSO	73.60 −46.34	0.86 5.2	3×10^{-9f}	100 MeV	1×10^{47}	[2]
4C−02.19 0458−020	QSO blazar	74.67 −2.06	2.286 4.98	3×10^{-9f}	100 MeV	1×10^{48}	[4]
0521−365	BL Lac	81.00 −36.49	0.055 0.32	2×10^{-9f}	100 MeV	4×10^{44}	[4]

Table 10.9. *(Continued.)*

Source name	Object type	α^b δ	z d^c	Flux^d	Energy	Lum.^{e}	Refs.
0528+134	QSO	82.03	2.06	3×10^{-7}	30 MeV	8×10^{48}	[8]
	blazar	+31.50	12.4	2×10^{-8}	100 MeV	6×10^{48}	[8]
				1×10^{-9}	300 MeV	3×10^{48}	[8]
				4×10^{-11}	1 GeV	1×10^{48}	[8]
				3×10^{-12}	3 GeV	8×10^{47}	[8]
0537−441	BL Lac	84.34	0.894	2×10^{-9}	100 MeV	1×10^{47}	[9]
		−44.11	5.4	2×10^{-10}	300 MeV	1×10^{47}	[9]
				2×10^{-11}	1 GeV	1×10^{47}	[9]
MCG 8-11-11	Seyfert-1	87.79	0.0205	2×10^{-1}	30 keV	5×10^{44}	[10]
0551+464		+46.43	0.12	6×10^{-2}	100 keV	2×10^{45}	[10]
				2×10^{-2}	300 keV	5×10^{45}	[10]
				6×10^{-3}	1 MeV	2×10^{46}	[10]
				3×10^{-5}	10 MeV	1×10^{46}	[10]
0716+714	BL Lac	109.05	$\cdots$	2×10^{-9}	100 MeV	$\cdots$	[2]
		+71.44					
OI 158	BL Lac	113.81	0.424	3×10^{-9f}	100 MeV	4×10^{46}	[4]
0735+178		+17.82	2.04				
0827+243	QSO	127.80	2.046	7×10^{-9f}	100 MeV	2×10^{48}	[4]
	blazar	+24.05	4.83				
OJ 49	BL Lac	127.30	0.18	2×10^{-9f}	100 MeV	5×10^{45}	[4]
0829+046		+4.66	0.98				
4C+71.07	QSO	129.09	2.172	3×10^{-9f}	100 MeV	1×10^{48}	[2]
0836+710	blazar	+71.07	4.92				
0917+449	QSO	139.43	2.18	3×10^{-9f}	100 MeV	1×10^{48}	[4]
	blazar	+44.91	4.51				
MCG -5-23-16	Seyfert-2	146.37	0.0485	4×10^{-3f}	100 keV	2×10^{43}	[7]
0945−307		−30.72	0.283				
4C+55.17	QSO	148.56	0.909	5×10^{-9f}	100 MeV	3×10^{47}	[4]
0954+556	blazar	+55.62	3.42				
0954+658	BL Lac	148.74	0.368	2×10^{-9f}	100 MeV	1×10^{46}	[4]
		+65.80	1.82				
MRK 421	BL Lac	165.42	0.0308	1×10^{-1}	30 keV	6×10^{44}	[11]
1101+384		+38.48	0.18	4×10^{-2}	100 keV	3×10^{45}	[11]
				7×10^{-9}	50 MeV	1×10^{44}	[12]
				2×10^{-9}	100 MeV	1×10^{44}	[12]
				2×10^{-10}	300 MeV	1×10^{44}	[12]
				2×10^{-11}	1 GeV	1×10^{44}	[12]
				2×10^{-12}	3 GeV	1×10^{44}	[12]
				3×10^{-17f}	500 GeV	5×10^{43}	[13]
4C+29.45	QSO	179.24	0.729	2×10^{-8f}	100 MeV	7×10^{47}	[4]
1156+295	blazar	+29.52	2.99				

Table 10.9. *(Continued.)*

Source name	Object type	α^b δ	z d^c	Flux^d	Energy	Lum.^{e}	Refs.
NGC 4151 1208+396	Seyfert-1	182.00 +39.68	0.003 0.018	2×10^{-1} 5×10^{-2} 1×10^{-2} 8×10^{-3}	30 keV 100 keV 300 keV 1 MeV	1×10^{43} 3×10^{43} 6×10^{43} 5×10^{44}	[14] [14] [14] [14]
W Comae 1219+285	BL Lac	184.76 +28.51	0.102 0.58	$5 \times 10^{-9\,f}$	100 MeV	4×10^{45}	[4]
4C+21.35 1222+216	QSO blazar	185.60 +21.66	0.435 2.08	$5 \times 10^{-9\,f}$	100 MeV	7×10^{46}	[4]
NGC 4388 1223+126	Seyfert-2	185.81 +12.94	0.00842 0.051	$6 \times 10^{-3\,f}$	100 keV	3×10^{43}	[7]
3C 273 1226+023	QSO	186.64 +2.33	0.158 0.95	1×10^{-1} 1×10^{-2} 5×10^{-3} 2×10^{-4} 2×10^{-5} 2×10^{-6} 2×10^{-7} 1×10^{-8} 1×10^{-9} 3×10^{-11}	30 keV 100 keV 300 keV 1 MeV 3 MeV 10 MeV 30 MeV 100 MeV 300 MeV 1 GeV	2×10^{46} 2×10^{46} 8×10^{46} 3×10^{46} 3×10^{46} 3×10^{46} 3×10^{46} 2×10^{46} 2×10^{46} 5×10^{45}	[15] [15] [16] [17] [17] [17] [17] [17] [17] [17]
1227+023	QSO	186.83 +2.41	0.57 3.4	3×10^{-1} 2×10^{-2}	40 keV 100 keV	1×10^{48} 5×10^{47}	[18, 19] [18, 19]
4C−02.55 1229−021	QSO blazar	187.36 −2.13	1.045 3.68	$2 \times 10^{-9\,f}$	100 MeV	1×10^{47}	[4]
M87 1228+124	NELG	187.08 +12.67	(0.0042) 0.025	1×10^{-1} 6×10^{-3}	30 keV 100 keV	1×10^{43} 7×10^{42}	[20] [20]
3C 279 1253−055	QSO	193.40 −5.52	0.538 3.2	2×10^{-5} 3×10^{-6} 2×10^{-7} 2×10^{-8} 3×10^{-9} 3×10^{-10} 4×10^{-11} 4×10^{-12}	3 MeV 10 MeV 30 MeV 100 MeV 300 MeV 1 GeV 3 GeV 10 GeV	4×10^{47} 6×10^{47} 4×10^{47} 4×10^{47} 5×10^{47} 6×10^{47} 7×10^{47} 8×10^{47}	[17] [17] [21] [21] [21] [21] [21] [21]
X Comae 1257+286	Seyfert-1	194.49 +28.67	0.092 0.55	2×10^{-1} 3×10^{-2}	30 keV 100 keV	1×10^{46} 2×10^{46}	[22] [22]
1313−333	QSO blazar	198.33 −33.39	1.21 3.96	$2 \times 10^{-9\,f}$	100 MeV	3×10^{47}	[4]
Cen A 1322−427	Radio galaxy	200.74 −42.71	(0.001825) 0.0073	1×10^{0} 1×10^{-1} 2×10^{-2} 2×10^{-3} 7×10^{-5}	30 keV 100 keV 300 keV 1 MeV 10 MeV	1×10^{43} 1×10^{43} 2×10^{43} 2×10^{43} 7×10^{43}	[23] [23] [23] [24] [24]
OP 151 1331+170	QSO blazar	202.79 +17.07	2.084 4.86	$1 \times 10^{-9\,f}$	100 MeV	3×10^{47}	[4]

Table 10.9. *(Continued.)*

Source name	Object type	α[b] δ	z d[c]	Flux[d]	Energy	Lum.[e]	Refs.
MCG -6-30-15 1314−340	Seyfert-1	203.26 −34.04	0.00775 0.048	5×10^{-3}[f]	100 keV	2×10^{43}	[7]
IC 4329A 1346−300	Seyfert-1	206.62 −30.06	0.01605 0.094	7×10^{-3}[f]	100 keV	1×10^{44}	[7]
MRK 279 1348+700	Seyfert-1	207.97 +69.55	0.0294 0.175	3×10^{-3}[f]	100 keV	2×10^{44}	[7]
OQ−010 1406−076	QSO blazar	211.58 −7.64	1.494 4.34	1×10^{-8}[f]	100 MeV	2×10^{48}	[4]
NGC 5548 1415+255	Seyfert-1	214.50 +25.14	0.0168 0.100	4×10^{-3}[f]	100 keV	8×10^{43}	[7]
1424−418	QSO blazar	216.00 +41.80	1.522 4.37	6×10^{-9}[f]	100 MeV	1×10^{48}	[4]
OR−017 1510-089	QSO blazar	227.54 −8.91	0.361 1.79	5×10^{-9}[f]	100 MeV	5×10^{46}	[4]
4C+15.54 1604+159	BL Lac	241.21 +15.99	0.357 1.78	4×10^{-9}[f]	100 MeV	4×10^{46}	[4]
OS 319 1611+343	QSO blazar	242.95 +34.34	1.401 4.23	7×10^{-9}[f]	100 MeV	1×10^{48}	[4]
1622−253	QSO blazar	245.68 −25.35	0.786 3.14	7×10^{-9}[f]	100 MeV	3×10^{47}	[4]
1622−297	QSO blazar	246.36 −29.92	0.815 3.21	3×10^{-8}[f]	100 MeV	1×10^{48}	[4]
4C 38.41 1633+382	QSO	248.38 +38.24	1.814 10.9	2×10^{-8}	50 MeV	1×10^{48}	[25]
				6×10^{-9}	100 MeV	1×10^{48}	[25]
				7×10^{-10}	300 MeV	1×10^{48}	[25]
				8×10^{-11}	1 GeV	2×10^{48}	[25]
				1×10^{-11}	3 GeV	2×10^{48}	[25]
				1×10^{-12}	10 GeV	2×10^{48}	[25]
NRAO 530 1730−130	QSO blazar	262.56 −13.05	0.902 3.40	1×10^{-8}[e]	100 MeV	6×10^{47}	[4]
4C+51.37 1739+522	QSO blazar	264.87 +52.22	1.375 4.19	4×10^{-9}[f]	100 MeV	5×10^{47}	[4]
OT−68 1741−038	QSO blazar	265.34 −3.81	1.054 3.70	4×10^{-9}[f]	100 MeV	3×10^{47}	[4]
3C 390.3 1845+797	Seyfert-1	281.41 +79.75	0.0561 0.326	3×10^{-3}[f]	100 keV	7×10^{44}	[7]
1933−400	QSO blazar	293.46 −40.08	0.966 3.53	1×10^{-8}[f]	100 MeV	7×10^{47}	[4]
NGC 6814 1942−102	Seyfert-1	295.67 −10.32	0.00521 0.030	3×10^{-3}[f]	100 keV	6×10^{42}	[7]
NRAO 629 2022−077	QSO blazar	305.75 −7.76	1.388 4.21	7×10^{-9}[f]	100 MeV	9×10^{47}	[4]

Table 10.9. *(Continued.)*

Source name	Object type	α^b δ	z d^c	Flux^d	Energy	Lum.^{e}	Refs.
MRK 509 2041-107	Seyfert-1	310.36 −10.91	0.0344 0.203	$4 \times 10^{-3 f}$	100 keV	3×10^{44}	[7]
2052-474	QSO blazar	314.52 −46.96	1.489 4.33	$3 \times 10^{-9 f}$	100 MeV	5×10^{47}	[4]
2155−304	BL Lac	328.99 −30.47	0.116 0.655	$3 \times 10^{-9 f}$	100 MeV	3×10^{45}	[4]
BL Lacertae 2200+420	BL Lac	330.16 +42.04	0.0686 0.398	$4 \times 10^{-9 f}$	100 MeV	1×10^{45}	[4]
2209+236	QSO blazar	332.51 +23.97	1.489 4.33	$1 \times 10^{-9 f}$	100 MeV	2×10^{47}	[4]
CTA 102 2230+114	QSO	337.53 +11.47	1.037 6.2	4×10^{-9}	100 MeV	3×10^{47}	[2]
3CR 454.3 2251+158	QSO	342.87 +15.88	0.859 5.2	8×10^{-9}	100 MeV	4×10^{47}	[2]
NGC 7582 2318−422	Seyfert-2	344.18 −43.23	0.00525 0.033	$3 \times 10^{-3 f}$	100 keV	6×10^{42}	[7]
OZ 193 2356+196	QSO blazar	359.05 +19.64	1.066 3.72	$3 \times 10^{-9 f}$	100 MeV	2×10^{47}	[4]
Diffuse background				5×10^{1}/sr	30 keV		[26]
				2×10^{0}/sr	100 keV		[26]
				1×10^{-1}/sr	300 keV		[26]
				1×10^{-2}/sr	1 MeV		[26]
				1×10^{-4}/sr	10 MeV		[26]
				2×10^{-7}/sr	100 MeV		[26]

Notes

^{a} Source type, position, and redshift are from Hewitt, A., & Burbidge, G. 1987, *ApJS*, **63**, 1; 1989, *ApJS*, **69**, 1; and 1991, *ApJS*, **75**, 297, except for M87 and Cen A from Tully, R. 1988, *Nearby Galaxies Catalog* (Cambridge University Press, Cambridge) for which the redshifts are corrected for local motion, and for GRS1227+0229 from Grindlay, J.E. 1993, *A&AS*, **97**, 113.

^{b} Positions in degrees.

^{c} Distances in Gpc assume cosmological redshifts with $H_0 = 50$ km/s Mpc. d (Gpc) $= 6 \times \frac{(1+z)^2-1}{(1+z)^2+1}$

^{d} Flux in photons/cm^2 s MeV at the energy denoted.

^{e} Assuming isotropic emission, $E^2 \times$ (flux) $= E^2$ (keV^2) $\times z^2 \times$ [flux (phot./cm)^2 s MeV] $\times 7 \times 10^{45}$ ergs/s ln E.

^{f} Differential flux determined from integral flux assuming a differential spectrum of the form E^{-2}.

References

1. Bhattacharya, D. et al. 1992, *AIP Conf. Proc.*, **280**, 498
2. Fichtel, C.E. et al. 1992, *AIP Conf. Proc.*, **280**, 461
3. Bertsch, D.L. et al. 1993, *ApJ*, **405**, L21
4. Hartman, R.C. et al. 1997, *AIP Conf. Proc.*, **410**, 307
5. Hunter, S.D. et al. 1992, *A&A*, **272**, 59
6. Rothschild, R.E. et al. 1981, *ApJ*, **243**, L9
7. Kurfess, J.D. et al. 1995, *NATO ASI Series C*, **461**, 233
8. Hunter, S.D. et al. 1993, *ApJ*, **409**, 134
9. Thompson, D.L. et al. 1992, *ApJ*, **410**, 87
10. Perotti, F. et al. 1981, *Nature*, **292**, 133
11. Ubertini, P. et al. 1984, *ApJ*, **284**, 54
12. Lin, Y.C. et al. 1993, *ApJ*, **401**, L61

13. Punch, M. et al. 1992, *Nature*, **358**, 477
14. Perotti, F. et al. 1981, *ApJ*, **247**, L63
15. Primini, F.A. et al. 1979, *Nature*, **278**, 234
16. Bassani, L. et al. 1991, *22nd Int. Cosmic Ray Conf.*, **1**, 173
17. Hermsen, W. et al. 1993, *A&AS*, **97**, 97
18. Bassani, L. et al. 1991, *22nd Int. Cosmic Ray Conf.*, **1**, 173
19. Grindlay, J.E. 1993, *A&AS*, **97**, 113
20. Lea, S. et al. 1981, *ApJ*, **246**, 369
21. Kniffen, D.A. et al. 1993, *ApJ*, **411**, 133
22. Bazzano, A. et al. 1990, *ApJ*, **362**, L51
23. Baity, W.A. et al. 1981, *ApJ*, **244**, 429
24. von Ballmoos, P. et al. 1987, *ApJ*, **312**, 134
25. Mattox, J.R. et al. 1993, *ApJ*, **410**, 609
26. Rothschild, R.E. et al. 1983, *ApJ*, **269**, 423

Table 10.10. *Hard X-ray and γ-ray instruments in space since* 1970.

Instrument	Mission	Energy range $\Delta E/E$	Field of view resolution	Area (cm^2)	Date	PI institution	Refs.
Cosmic X-ray telescope	OSO-7	6–500 keV 33% @ 60 keV	6.5°	64	1971–73	Peterson UCSD	[1]
Solar X-ray telescope	OSO-7	10–350 keV 18% @ 60 keV	90° × 20°	9.6	1971–73	Peterson UCSD	[2]
γ-ray monitor	OSO-7	0.3–10 MeV < 8% @ 662 keV	120° × 70°	45	1971–73	Chupp UNH	[3]
γ-ray telescope	SAS-2	30–200 MeV ∼ 50%	30° ∼ 2°	115	1972–73	Fichtel GSFC	[4]
Scintillator telescope	Ariel-V	26 keV–1.2 MeV 30% @ 662 keV	8°	8	1974–80	Imperial College	[5]
Celestial X-ray detector	OSO-8	15 keV–3 MeV 50% @ 60 keV	5°	28	1975–78	Frost GSFC	[6]
γ-ray detector	COS-B	50 MeV–2 GeV 40% @ 100 MeV	∼ 30° ∼ 1°	75	1975–82	Caravane Collaboration	[7]
A-4 LED	HEAO-1	15–180 keV 25% @ 60 keV	1.2° × 20°	206	1977–79	Peterson–Lewin UCSD–MIT	[8]
A-4 MED	HEAO-1	0.1–2 MeV 10% @ 1 MeV	16.5°	160	1977–79	Peterson UCSD	
A-4 HED	HEAO-1	0.2–10 MeV 10% @ 1 MeV	40°	120	1977–79	Peterson UCSD	
C-1 germanium spectrometer	HEAO-3	50 keV–10 MeV 0.2% @ 1.8 MeV	30°	64	1979–80	Jacobson JPL	[9]
GRS	SMM	0.3–9 MeV 7% @ 662 keV	180°	310	1979–89	Chupp UNH	[10]
HXRBS	SMM	20–260 keV 30% @ 122 keV	40°	71	1979–89	Frost GSFC	[11]
HEXE	MIR KVANT	15–200 keV 30% @ 60 keV	1.6° × 1.6°	800	1987–	Trumper MPI	[12]
Pulsar X-1	KVANT	50–800 keV	3° × 3°	1256	1987–	Sunyaev IKI	[13]
GSPC	KVANT	3–100 keV 3%@60 keV	2.3°	∼ 150	1987–	Schnopper SRL	[14]

Table 10.10. *(Continued.)*

Instrument	Mission	Energy range $\Delta E/E$	Field of view resolution	Area (cm^2)	Date	PI institution	Refs.
SIGMA	GRANAT	30 keV–1.3 MeV 8% @ 511 keV	4.7° × 4.3° 0.2°	797	1989–	Paul–Mandrou CESR–Saclay	[15]
WATCH	GRANAT	6–180 KeV	4 sr	30	1989–	Lund DSRI	[16]
ART-P	GRANAT	4–100 keV 14% @ 60 keV	1.8° × 1.8° 0.1°	2520	1989–	Sunyaev IKI	[17]
ART-S	GRANAT	3–100 keV 11% @ 60 keV	2.1° × 2.1°	800	1989–	Sunyaev IKI	[17]
BATSE occultation	CGRO	20 keV–1.8 MeV 30% @ 88 keV	2π sr 1°	1800	1991–	Fishman MSFC	[18]
OSSE	CGRO	50 keV–10 MeV 8% @ 511 keV	3.8° × 11.4°	2620	1991–	Kurfess NRL	[19]
COMPTEL	CGRO	0.8–30 MeV 9% @ 1.3 MeV	∼ 1 sr ∼ 1.5°	45	1991–	Schonfelder MPI	[20]
EGRET	CGRO	20 MeV–30 GeV ∼ 20% 0.1–5 GeV	∼ 40° 0.1°–0.4°	1600	1991–	Fichtel GSFC	[21]
HEXTE	RXTE	15 KeV–250 KeV 15% @ 60 keV	∼ 1°	1600	1995–	Rothschild UCSD	[22]
PDS	BeppoSAX	15 KeV–300 KeV ∼ 15% 60 keV	∼ 1.4°	800	1996–	TeSRE/IAS	[22]

References
1. Peterson, L.E. 1972, *IAU Symp. No. 55*, 51
2. Harrington, T. et al. 1972, *IEEE Trans. Nucl. Sci.*, **NS-19**, 596
3. Higbie, P.R. et al. 1972, *IEEE Trans. Nucl. Sci.*, **NS-19**, 606
4. Derdeyn, S. et al. 1972, *Nucl. Instrum. Methods*, **98**, 557
5. Engel, A.R., & Coe, M.J. 1977, *Space Sci. Instrum.*, **3**, 407
6. Dennis, B.R. et al. 1977, *Space Sci. Instrum.*, **3**, 325
7. Bignami, G.F. et al. 1975, *Space Sci. Instrum.*, **1**, 245
8. Jung, G.V. 1989, *ApJ*, **338**, 972; Knight, F.K. 1982, *ApJ*, **260**, 538
9. Mahoney, W.A. et al. 1980, *Nucl. Instrum. Methods*, **178**, 363
10. Forrest, D.J. et al. 1980, *Solar Phys.*, **65**, 15
11. Orwig, L. et al. 1980, *Solar Phys.*, **65**, 25
12. Reppin, C. et al. 1985, in *Nonthermal and Very High Temperature Phenomena in X-ray Astronomy*, edited by G.C. Perola and M. Salvati (Instituto Astronomico, Roma) p. 279
13. Sunyaev, R. et al. 1990, *Adv. Space Sci.*, **10**, 41
14. Smith, A. 1985, in *Nonthermal and Very High Temperature Phenomena in X-ray Astronomy*, edited by G.C. Perola and M. Salvati (Instituto Astronomico, Roma) p. 271
15. Paul, J.A. et al. 1991, *Adv. Space Res.*, **11**, (8) 289
16. Lund, N. 1991, *Adv. Space Res.*, **11**, (8) 17
17. Sunyaev, R. et al. 1990, *Adv. Space Res.*, **10**, (2) 233
18. Fishman, G.J. et al. 1992, *NASA Conf. Publ. 3137*, 26
19. Kurfess, J.D. et al. 1991, *Adv. Space Res.*, **11**, (8) 323
20. Schonfelder, V. 1991, *Adv. Space Sci.*, **11**, (8) 313
21. Kanbach, G. et al. 1988, *Space Sci. Instrum.*, **49**, 69
22. Rothschild, R.E. et al. 1998, *ApJ*, **496**, 538
23. Frontera, F. et al. 1997, *A&AS*, **122**, 357

Table 10.11. γ-Ray burst instruments.

Satellite	Dates	Orbit[a]	Detectors	Energy range (MeV)	Time resolution (s)	Trigger Time (s)	Trigger Energy (MeV)	Refs.
Vela 5A/B Vela 6A/B	5/69–3/84	GC	6–10 cm^3 CsI	0.2–1	≥ 0.016	0.25, 1.5	0.03–0.1.5	[1]
Helios-2	1/76–12/79	H	21.5 cm^3 CsI	> 0.1	≥ 0.004	0.004 0.032 0.250	> 0.1	[2]
Solrad-11A/B	4/76–6/77	GC	2–43 cm^3 CsI	0.2–2	≥ 0.0003	0.625	0.2–2	[3]
Signe-3	6/77–3/78	GC	950 cm^2 CsI[b]	> 0.06	0.008			[4]
HEAO-1	8/77–2/79	GC	2000 cm^2 CsI[b] 280 cm^2 NaI 3300 cm^2 PC	0.1–1.6 0.03–6 0.0005-0.02	≥ 0.05 0.32 0.1	~ 0.3	0.13–1.7	[5] [5] [6]
Prognoz-6	9/77–3/78	G	63 cm^2 NaI 750 cm^2 CsI[b] 16 cm^3 NaI	0.08–1 > 0.3 0.02–> 0.3	≥ 0.002 4 0.25	0.02	0.08–0.4	[7] [7] [7]
ICE	8/78–3/87	H	22 cm^2 NaI	0.02–1.25	≥ 0.004	0.00025–0.008	0.132–1.25	[8]
(ISEE–3)			35 cm^3 Ge	0.2–3	0.001	0.00013–0.001	0.2–3	[9]
PVO	5/78–9/92	V	2–36 cm^3 NaI	0.1–2	≥ 0.012	0.25, 1, 4	0.1–2	[10]
Venera 11/12 (Konus)	9/78–1/80	H	2–63 cm^3 NaI 6–50 cm^2 NaI	0.1–2.5 0.03–2	> 0.002 ≥ 0.016	0.02 0.25, 1.5	0.08–0.4 0.05–0.15	[11] [12]
Prognoz-7	11/78–6/79	G	63 cm^2 NaI 750 cm^2 CsI[b]	0.1–2.5 > 0.1	≥ 0.002 0.002	0.25	0.08–0.4	[7] [7]
Venera 13/14 (Konus)	11/81–4/83	H	2–63 cm^3 NaI 6–50 cm^2 NaI	0.05–1 0.03–2	≥ 0.002 ≥ 0.004	0.25 0.25, 1.5	0.08–0.4 0.05–0.15	[11] [13]
Prognoz-9	7/83–2/84	G	2–178 cm^2 NaI	0.04–8	≥ 0.016	0.5, 2	0.073–0.966	[14]
Ginga	2/87–11/91	GC	60 cm^2 NaI 63 cm^2 PC	0.014–0.40 0.002–0.030	0.031 0.031	0.25, 1, 4 1, 4	0.014–0.4 0.002–0.03	[15] [15]
GRANAT –SIGMA –SIGMA –WATCH –Konus-B –Phebus	12/89– 12/89–2/90 	G	 800 cm^2 NaI 8–2400 cm^2 CsI 4–30 cm^2 NaI/CsI 6–314 cm^2 NaI 6–573 cm^3 BGO	 0.03–2 0.1–1 0.006–0.18 0.01–8 0.1–100	 $\cdots$ ≥ 0.000008 ≥ 0.0001 0.002 ≥ 0.00003	 0.25, 2 0.25, 2 0.004–32 0.25, 1.5 0.008	 0.03–2 0.1–1 0.006–0.18 0.05–0.2 0.075–1.6	 [16] [16] [17] [18] [19]
Ulysses	11/90–	H	41 cm^2 CsI	0.015–0.150	≥ 0.008	0.125-4.0	0.015–0.150	[20]
Compton GRO BATSE–LAD BATSE–SD	4/91–	GC	 8–2025 cm^2 NaI 8–127 cm^2 NaI	 0.03–1.9 0.015–110	 ≥ 0.000002 0.000128	 0.06, 0.25, 1	 0.06–0.3	 [21] [21]
BeppoSAX WFC	4/96–	GC	 2-250 cm^2 Xe	 0.002–0.028	 0.0005	 —	 0.002–0.028	 [22]

Notes

[a] G, geocentric; GC, geocentric circular; H, heliocentric; V: venuscentric.

[b] Anticoincidence shield used as burst detector.

References

1. Klebesadel, R.W. et al. 1973, *ApJ*, **182**, L85
2. Cline, T.L. et al. 1979, *ApJ*, **229**, L47
3. Laros, J.G. et al. 1977, *Nature*, **267**, 131

4. Chambon, G. et al. 1979, *X-Ray Astronomy* (Pergamon, Oxford), p. 509
5. Hueter, G.J. 1987, Ph.D. thesis, University of California, San Diego
6. Wood, K.S. et al. 1984, *ApJS*, **56**, 507
7. Chambon, G. et al. 1979, *Space Sci. Instrum.*, **5**, 73
8. Anderson, R.D. et al. 1978, *IEEE Trans.*, **GE-16**, 157
9. Teegarden, B., & Cline, T.L. 1980, *ApJ*, **236**, L67
10. Klebesadel, R.W. et al. 1980, *IEEE Trans.*, **GE-18**, 76
11. Barat, C. et al. 1981, *Space Sci. Instrum.*, **5**, 229
12. Mazets, E.P. et al. 1981, *Ap&SS*, **80**, 3
13. Mazets, E.P. et al. 1983, *AIP Conf. Proc. No. 101*, 36
14. Boer, M. et al. 1986, *Adv. Space Sci.*, **6**, 97
15. Murakami, T. et al. 1989, *PASJ*, **41**, 405
16. Guerry, H. et al. 1986, *Adv. Space Sci.*, **6**, 103
17. Brandt, S. et al. 1990, *Adv. Space Sci.*, **10**, 239
18. Golenetskii, S.V. et al. 1991, *Adv. Space Sci.*, **11**, 125
19. Terekhov, O. et al. 1991, *Adv. Space Sci.*, **11**, 129
20. Hurley, K. et al. 1992, *A&ASS*, **92**, 401
21. Fishman, G.J. et al. 1989, *Proc. Gamma Ray Observatory Sci. Workshop*, 2–39
22. Jager, R. et al. 1997, *A&AS*, **125**, 557

Table 10.12. *Very-high-energy and ultrahigh-energy γ-ray experiments: Atmospheric Cherenkov and particle arrays.*[a]

Array	Country	Lat. (deg)	Long. (deg)	Elev. (km)	Area (10^4 m^2)	Threshold (TeV)	$\Delta\Theta$ (deg)	Began
Themis	France	43N	1W	1.5		0.1		1986
Albuquerque	USA	35N	107W	1.5		0.2		1986
Mt. Hopkins	USA	32N	111W	2.3		0.3	0.1	1983
Narrabri	Australia	31S	145E	0.21		0.3		1986
Haleakala	USA	21N	156W	3.0		0.5		1985
Pachmarchi	India	23N	78E	1.1		0.5		1987
Gulmarg	India	35N	77E	2.7		1		1985
Potchefstroom	South Africa	27S	27E	1.4		1		1985
White Cliffs	Australia	32S	143E	0.16		1		1986
Crimea	Ukraine	45N	34E	0.6	3.5	1	1.4	1986
Beijing	China	40N	117E	1.0		1		1987
Plateau Rosa	Italy	46N	8E	3.5	1	10	5.5	1981
Gran Sasso	Italy	42N	14E	2.0	10	10	1	1988
Tibet	China	30N	90E	4.2	2.0	10	0.8	1990
Tien Shan	Kirghiz	42N	75E	3.3	0.5	100	3	1974
Ooty	India	11N	77E	2.2	0.5	100	3	1984
Mt. Hopkins	USA	32N	111W	2.3	~ 0.5	100	1	1985
La Palma	Spain	29N	18W	2.2	4	100	1	1986
Mt. Aragats	Armenia	40N	44E	3.2		100	1	1987
South Pole	Antarctica	90S	0W	2.5	~ 1	100	1	1988
Mt. Norikura	Japan	36N	137E	2.8	≤ 1	100	1	1988
Dugway	USA	40N	112W	1.5	$\sim 2/25$	100	0.5–1	1989
Mt. Chacaltaya	Bolivia	16S	68W	5.2	> 0.5	200	1–3	1986
Cygnus	USA	36N	106W	2.1	> 8	200	1	1986
Baksan	Kab-Balkar	43N	43E	1.7	0.5	300	1.5	1984
Kolar	India	13N	78E	0.9	1.66	500	1.5	1984
Haverah Park	UK	54N	1W	0	> 1	500	1	1986
Akeno Ranch	Japan	35N	138E	0.9	~ 1	1000	3	1981
Moscow	Russia	56N	37E	0		1000	3	1982
Buckland Park	Australia	35S	138W	0	1.0	1000	2.5	1984
Janzos	New Zealand	41N	172E	0.9	> 0.23	1000	2	1988

Note

[a]Based on Weekes, T.C. 1988, *Phys. Rep.*, **160**, 1; Yodh, G. 1992, private communication; and Stepanian, A.A. 1992, private communication.

10.5 NEUTRINOS IN ASTROPHYSICS
by Wick C. Haxton

Perhaps the original motivation for studying astrophysical neutrinos was the prospect of directly probing the interior of our Sun: neutrinos produced as a byproduct of nuclear fusion pass undistorted through the outer layers of the Sun, carrying in their flux and spectrum a detailed memory of the nuclear reactions that produced them. As the competition between the three cycles comprising the pp chain (the process that dominates solar burning of four protons into ^{4}He) depends sensitively on the solar core temperature T_c, one can deduce T_c by measuring the various components of the solar neutrino flux.

Results from the ^{37}Cl detector, which has operated for nearly 30 years, and from three more recent experiments, SAGE and GALLEX (radiochemical detectors containing ^{71}Ga) and Kamioka II/III (an active water Cerenkov detector sensitive to higher energy solar neutrinos), have revealed some surprises. The results are consistent with a flux of high-energy ^{8}B neutrinos reduced to about 50% of the standard solar model value and a greatly suppressed flux of neutrinos produced from electron capture on ^{7}Be. This is a surprising pattern because a reduction in T_c tends to suppress the ^{8}B solar neutrino flux more than the ^{7}Be flux, not less. In fact, detailed fits seem to show that the ^{7}Be neutrinos must be completely absent to account the experimental results.

One popular explanation for this puzzle is the phenomenon of neutrino oscillations: if neutrinos have nonzero masses and mix (so that the electron, muon, and tauon neutrinos are not identical to the mass eigenstates, but linear combinations of these), solar electron neutrinos can oscillate into muon neutrinos and escape detection. While once it was thought that neutrino oscillations would most likely produce only a small reduction in the solar electron neutrino flux, it was discovered about a decade ago that oscillation effects can be greatly enhanced within the Sun. This phenomenon, known as the Mikheyev–Smirnov–Wolfenstein or MSW mechanism, arises because the effective masses of neutrinos change when the neutrinos pass through matter. The MSW solution that best reproduces the results of the ^{37}Cl, SAGE/GALLEX, and KamiokaII/III experiments is consistent with oscillations of a very light electron neutrino into a muon neutrino with a mass of about 0.003 electron volts (eV).

Two new detectors, SuperKamiokande and the *Sudbury Neutrino Observatory* (*SNO*), should be able to confirm or rule out neutrino oscillations as a solution to the solar neutrino problem. SuperKamiokande is an enormous (22.5 kiloton fiducial volume) ultrapure water Cerenkov detector located in a Japanese mine. It began operations in the Spring of 1996. By making a precision measurement of the spectrum of recoil electrons following neutrino–electron scattering, the experimentalists hope to find subtle distortions characteristic of the MSW mechanism. SNO, which should be fully operational by the end of 1998, is a Canadian–US–UK detector located deep within a nickel mine in Sudbury, Ontario. The inner volume of this water Cerenkov detector contains heavy water. Reactions on the deuterium nuclei provide separate charged and neutral current signals. Thus, in addition to spectrum distortions, the experimentalists hope to measure directly the neutrinos of a different flavor that are generated by the MSW mechanism.

SuperKamiokande, SNO, and similar detectors are sensitive to another source of neutrinos, those produced in the atmosphere by the interactions of cosmic rays impinging on the Earth. For some years most such detectors have found a puzzling result, an unexpected ratio of muon neutrino to electron neutrino events given our understanding of cosmic ray neutrino production. Very recently the SuperKamiokande group, by comparing upward- to downward-going neutrinos, have claimed that this anomaly is definitive evidence for neutrino oscillations and thus of massive neutrinos.

Another source of neutrinos is associated with one of the most spectacular events in astrophysics, the sudden collapse of the core of a massive star. This collapse triggers the ejection of the star's mantle, producing the spectacular display known as a supernova. However 99% of the energy released in such a collapse, an enormous 3×10^{53} ergs, is invisible optically as it is carried by an intense three-second burst of neutrinos emitted by the cooling protoneutron star forming at the star's center.

We were extremely fortunate to have two large water Cerenkov detectors, Kamioka II and IMB, operating at the time of Supernova 1987A. The free protons in water absorb electron antineutrinos, emitting relativistic positrons that can be detected readily in such detectors. In each detector approximately 10 events were detected from a star that collapsed in the Large Magellanic Cloud 150 000 light years from earth. The characteristics of the detected neutrinos—the number of events, the spectrum, the duration of the neutrino pulse—were in good accord with supernova theory.

There were no detectors operating that had the necessary characteristics and sensitivities to record the electron neutrinos or the muon and tauon neutrinos and antineutrinos. This was unfortunate because supernova electron neutrinos may hold the key to one of the central problems in cosmology, the dark matter. Studies on a variety of astrophysical scales—galaxies, clusters of galaxies, etc.—indicate that at least 90% of the mass in the Universe is dark, not emitting or absorbing electromagnetic radiation. Most estimates of the dark matter lead to a minimum mean density in the Universe of 20% of the closure density, the density that would keep the Universe from expanding forever. As the standard theory of big bang nucleosynthesis argues that at least some of this dark matter is nonbaryonic, massive neutrinos seem a natural explanation for this component. In particular, a heavy tauon neutrino with a mass of about 5–10 eV could comprise an important fraction of the dark matter and would also help to explain how galaxies and other structures in the Universe formed.

Such a mass is quite consistent with a theoretical model for generating neutrino masses known as the seesaw mechanism. If the solar neutrino problem involves oscillations between the electron neutrino and a 0.003 eV muon neutrino, then the seesaw mechanism predicts that the tauon neutrino mass might be in the range required to explain large scale structure.

How can one test the hypothesis of a tauon neutrino mass of a few eV? Just as the densities available in the Sun enhance oscillations between electron and muon neutrinos, the much larger densities found near the core of a supernova can enhance oscillations between electron neutrinos and massive tauon neutrinos. Because the tauon neutrinos emitted by a supernova tend to be substantially more energetic than supernova electron neutrinos, such oscillations would produce an anomalously energetic electron neutrino spectrum. Thus the detection of these electron neutrinos could demonstrate that massive tauon neutrinos make up an important component of the dark matter. As the standard model of electroweak interactions cannot accommodate massive neutrinos, such a discovery would also have a profound impact on particle physics.

Neutrinos also play a crucial role in nuclear astrophysics. Arguments based on big-bang nucleosynthesis provided early evidence that there were only a few (three or four) light neutrino flavors, a result now beautifully confirmed by measurements of the width of the Z_0. Neutrinos govern much of the nucleosynthesis that occurs in a supernova. For example, the process of rapid neutron capture, by which about half of the heavy elements and all of the transuranics are synthesized, is now believed to depend on conditions in the hot bubble that resides just above the surface of the protoneutron star. The entropy and neutron/proton ratio in this bubble are largely determined by neutrino interactions. Neutrinos also directly synthesize nuclei like ^{19}F and ^{11}B by scattering off the neon and carbon in the mantle of the collapsing star. The subsequent supernova explosion is the mechanism by which these newly synthesized metals are ejected into the interstellar medium.

Finally, there is an enormous density of very low energy neutrinos—about 300/cm^3—throughout the Universe, a relic of the big bang similar to the background microwave photons. Recent precision measurements of the microwave background allow us to look backward to the time of recombination, when electrons condensed on nuclei to form neutral atoms, providing a snapshot of conditions in the early Universe, 100 000 years after the big bang. Were we ever to find a method to detect the relic neutrinos, this would provide a probe of the Universe at the time the neutrinos decoupled from matter, early in the first minute in the history of the Universe. Detection of these relic neutrinos is likely to remain a challenge for many decades.

10.6 CURRENT NEUTRINO OBSERVATORIES
by Thomas J. Bowles

Table 10.13 lists the existing neutrino observatories and a description of each one. Some of these are still under development.

Table 10.13. *Existing neutrino observatories.*

Detector	Main aims[a]	"Size" of target	Depth (mwe)[b]	Sensors[c] Detection techniques	Remarks
Antarctica AMANDA	Heν	9 000 m^2	1 800–2 400	Čerenkov	Under development
Baksan, Caucusus Russia	SN, HEν	330 tons 250 m^2	$\approx$ 1 000	LS	One of the oldest underground neutrino observatories
Homestake Mine S. Dakota	HEν, ND	140 ton	4 000	LS	Experiment no longer in operation
Artyomovsk Ukraine	SN	100 ton		LS	
Mt. Blanc, Italy NUSEX	ND, SN	150 ton	5 000	Plastic tubes in limited streamer mode	Experiment no longer in operation
Mt. Blanc, Italy LSD	ND, SN	90 ton	5 000	LS	Experiment no longer in operation
Frejus France	ND, SN	912 ton	4 850	Flash chambers, Geiger tubes	Experiment no longer in operation
Gran Sasso, Italy MACRO	SN, HEν	3 240 m^2	3 800	LS, streamer tubes	Full operation began in 1996
Gran Sasso, Italy LVD	SN, HEν	1 800 ton	3 800	LS, streamer tubes	
Greece NESTOR	HEν	1 × 10^4 m^2	3 700	Čerenkov	Under development
Hawaii DUMAND	HEν	2 × 10^4 m^2	4 700	Čerenkov	Under development
Lake Baikal, Siberia NT-200	HEν	500 m^2	1 000	Čerenkov	"NT" stands for neutrino telescope
Soudan, Minnesota SOUDAN II	ND, HEν	1 100 ton	7 200	Honeycomb drift chamber	Iron calorimeter
Soudan, Minnesota MINOS	ND, HEν, LB	10 000 ton	7 200	Honeycomb drift chamber	Iron calorimeter Under development
Kolar Gold Fields (2) India	ND, HEν	140 ton	7 200	Proportional counters, calorimeter	Experiment no longer in operation
Kamiokande Japan	ND, SN, HEν	4 500 ton	2 400	Čerenkov	Detected ν_e from SN 1 987a Detects ^{8}B neutrinos Experiment no longer in operation
SuperKamiokande Japan	ND, SN, NEν, LB	50 000 ton	2 400	Čerenkov	Detects ^{8}B neutrinos Operational in 1996
IMB, Ohio	ND, SN, HEν	3 300 ton	1 580	Čerenkov	Detected ν_e from SN 1 987a Experiment no longer in operation
Homestake Mine, S. Dakota	sol	615 ton	4 900 (perchlorethylene)	Radiochemical	^{37}Cl + $\nu_e \rightarrow ^{37}$Ar + e^- Detects ^{7}Be and ^{8}B neutrinos
Homestake mine S. Dakota	sol	100 tons	4 900 (NaI solution)	Radiochemical	^{127}I + $\nu_e \rightarrow ^{127}$Xe + e^- Detects ^{7}Be and ^{8}B neutrinos
Baksan, Russia SAGE	sol	60 tons Ga	4 815	Radiochemical	^{71}Ga + $\nu_e \rightarrow ^{37}$Ar + e^- Detects p–p neutrinos
Gran Sasso, Italy Borexino	sol	300 tons	3 800	LS	$\nu_x + e^- \rightarrow \nu_x + e^-$ Detects ^{7}Be neutrinos Operational in 2001
Gran Sasso, Italy GALLEX	sol	30 tons Ga	3 800	Radiochemical	Detects p–p neutrinos Experiment completed in 1997
Gran Sasso, Italy GNO	sol	30 tons Ga	3 800	Radiochemical	Detects p–p neutrinos Operation began in 1998
Gran Sasso, Italy ICARUS	sol, ND, LB	1 600 tons	3 800	Liquid argon	Time production chamber Under development

Table 10.13. *(Continued.)*

Detector	Main aims[a]	"Size" of target	Depth (mwe)[b]	Sensors[c] Detection techniques	Remarks
Sudbury, Canada SNO	sol, SN	100 ton D_2O 5 000 ton H_2O	5 900	Čerenkov	$\nu_e + d \rightarrow p + p + e^-$ $\nu_x + d \rightarrow n + p + \nu_x$ $\nu_{x+c} + e^- \rightarrow \nu_x + e^-$ $\nu_e + d \rightarrow n + n + e^+$ Operational in 1998

Notes

[a] SN, supernova bursts; ND, nucleon decay; HEν, high-energy neutrinos; sol, solar neutrinos; LB, long baseline experiment using an accelerator neutrino source.

[b] mwe, meters water equivalent.

[c] Sensors means detectors of neutrino secondaries, e.g., muons; LS, liquid scintillator; Čerenkov light from charged secondaries is observed by photomultipliers.

ACKNOWLEDGMENTS

We wish to thank Ed Chupp, Carl Fichtel, Gerry Fishman, Alice Harding, Wick Haxton, Jim Higdon, Kevin Hurley, John Laros, Chip Meegan, Larry Peterson, Reuven Ramaty, A. Stepanian, and Trevor Weekes for valuable comments and contributions.

REFERENCES

1. Heitler, W. 1954, *The Quantum Theory of Radiation* (Clarendon Press, Oxford)
2. Jauch, J.M., & Rohrlich, F. 1976, *The Theory of Photons and Electrons* (Springer-Verlag, Berlin)
3. Rybicki, G.B., & Lightman, A.P. 1979, *Radiative Processes in Astrophysics* (Wiley, New York)
4. Lang, K.R. 1980, *Astrophysical Formulae* (Springer-Verlag, Berlin).
5. Felten, J.E., & Morrison, P. 1966, *ApJ*, **146**, 686
6. Blumenthal, G.R., & Gould, R.J. 1970, *Rev. Mod. Phys.*, **42**, 237
7. Ore, A., & Powell, J.L. 1949, *Phys. Rev.* **75**, 1696
8. Bussard, R.W. et al. 1979, *ApJ*, **228**, 928
9. Zdziarski, A.A. 1980, *Acta Astron.*, **30**, 371
10. Ramaty, R., & Meszaros, P. 1981, *ApJ*, **250**, 384
11. Gould, R.J. 1989, *ApJ*, **344**, 232
12. Guessoum, N. et al. 1991, *ApJ*, **378**, 170
13. Klein, O., & Nishina, Y. 1929, *Z. Phys.* **52**, 853
14. Marmier, P., & Sheldon, E. 1969, *Physics of Nuclei and Particles* (Academic Press, New York)
15. Erber, T. 1966, *Rev. Mod. Phys.*, **38**, 626
16. Canuto, V. et al. 1971, *Phys. Rev. D*, **3**, 2303
17. Bussard, R.W. et al. 1986, *Phys. Rev. D*, **34**, 440
18. Daugherty, J.K., & Harding, A.K. 1986, *ApJ*, **309**, 362
19. Harding, A.K., & Daugherty, J.K. 1991, *ApJ*, **374**, 687
20. Canuto, V., & Ventura, J. 1977, *Fund. Cosmic Phys.*, **2**, 203
21. Chupp, E.L. 1972, *Gamma-Ray Astronomy* (Reidel, Dordrecht)
22. Ramaty, R., & Lingenfelter, R.E. 1982, *Ann. Rev. Nucl. Part. Phys.*, **32**, 235
23. Bignami, G.F., & Hemsen, W. 1983, *ARA&A*, **21**, 67
24. Chupp, E.L. 1984, *ARA&A*, **22**, 359
25. Ramaty, R., & Lingenfelter, R.E. 1994, Chapter 3 in *High Energy Astrophysics* (World Scientific, New York), p. 32
26. Harding, A.K. 1991, *Phys. Rep*, **206**, 327
27. Higdon, J.C., & Lingenfelter, R.E. 1991, *ARA&A*, **28**, 401
28. Fishman, G.J., & Meegan, C.A. 1995, *ARA&A*, **33**, 415
29. Rothschild, R.E., & Lingenfelter, R.E. 1995, *High Velocity Neutron Star and Gamma-Ray Bursts* (American Institute of Physics, New York), 282 pp.
30. Kouveliotou, C., Briggs, M.F., & Fishman, G.J. 1996, *Gamma-Ray Bursts* (American Institute of Physics, New York), 1008 pp.
31. Rees, M.J. 1998, *Proc. 18th Texas symposium Rel. Astrophys.*. edited by A.V. Olinto, J.A. Frieman, and D.N. Schramm (World Scientific, Singapore) p. 34; Rees, M.J. 1999, *NucPhys B, Proc. Suppl.*, **69** 681

Chapter 11

Earth

Gerald Schubert and Richard L. Walterscheid

11.1 OBLATE ELLIPSOIDAL REFERENCE FIGURE [1, 2]

Equatorial radius $a = 6.378\,136 \times 10^6$ m.
Polar radius $c = 6.356\,753 \times 10^6$ m.
Mean radius $R_\oplus = (a^2 c)^{1/3} = 6.371\,000 \times 10^6$ m.
Length of equatorial quadrant $= 1.001\,875 \times 10^7$ m.
Length of meridional quadrant $= 9.985\,164 \times 10^6$ m.
Ellipticity or Flattening $(a - c)/a = 1/298.257 = 0.003\,352\,8$.
Eccentricity $e = (a^2 - c^2)^{1/2}/a = 0.081\,818$.

$$\text{Surface Area} = 2\pi \left\{ a^2 + c^2 \left(1 - c^2/a^2\right)^{-1/2} \ln\left[a/c + \left(a^2/c^2 - 1\right)^{1/2}\right] \right\}$$

$$= 5.100\,657 \times 10^{14} \text{ m}^2.$$

Volume $= \frac{4}{3}\pi a^2 c = 1.083\,207 \times 10^{21}$ m^3.

11.2 MASS AND MOMENTS OF INERTIA [1–3]

Earth mass $M_\oplus = 5.973\,7 \times 10^{24}$ kg.
Moon–Earth mass ratio $M_{\text{Moon}}/M_\oplus = 0.012\,300\,034$.
Sun–Earth mass ratio $M_\odot/M_\oplus = 332\,946.038$.
Earth mass multiplied by the gravitational constant:

$\quad GM_\oplus = 3.986\,004\,41 \times 10^{14}$ m^3 s^{-2},
$\quad (GM_\oplus)^{1/2} = 1.996\,498 \times 10^7$ m$^{3/2}$ s^{-1}.

Earth mean density $\bar{\rho}_\oplus = 5514.8$ kg m^{-3}.
Moments of inertia (see below):

$\quad$ about rotation axis $C = 8.035\,8 \times 10^{37}$ kg m^2,
$\quad$ average about equatorial axis $(A + B)/2 = 8.009\,5 \times 10^{37}$ kg m^2,
$\quad$ dynamical ellipticity or flattening $\{C - (A + B)/2\}/C = 0.003\,272\,9$,

$J_2 = \{C - (A + B)/2\}/M_\oplus a^2 = 1.082\,626 \times 10^{-3}$,
$C/M_\oplus a^2 = 0.330\,78$,
$M_\oplus a^2 = 2.430\,14 \times 10^{38}$ kg m^2.

11.3   GRAVITATIONAL POTENTIAL AND RELATION TO PRODUCTS OF INERTIA [1–3]

The gravitational potential is

$$U = \frac{GM_\oplus}{r}\left\{1 + \sum_{l=2}^{\infty}\left(\frac{a}{r}\right)^l \sum_{m=0}^{l} \overline{P}_l^m(\sin\phi)\left[\overline{C}_l^m\cos m\lambda + \overline{S}_l^m\sin m\lambda\right]\right\},$$

r = radial distance from Earth center of mass,

$\overline{P}_l^m$ = fully normalized associated Legendre polynomials, i.e., the mean square value of $\overline{P}_l^m(\sin\phi)(\cos m\lambda, \sin m\lambda)$ over a spherical surface is unity,

$\overline{P}_l^m = \{(2-\delta_{m,0})(2l+1)[(l-m)!/(l+m)!]\}^{1/2}P_l^m$, where P_l^m is the ordinary associated Legendre polynomial,

l, m = degree and order of normalized spherical harmonic $\overline{P}_l^m(\sin\phi)(\cos m\lambda, \sin m\lambda)$,

ϕ = latitude,

λ = longitude,

$\overline{C}_l^m, \overline{S}_l^m$ = coefficients in spherical harmonic expansion of Earth's gravitational potential using fully normalized functions.

With coordinate system origin at the center of mass $\overline{C}_1^0 = \overline{C}_1^1 = \overline{S}_1^1 = 0$. Table 11.1 gives the values of the zonal coefficients $\overline{C}_l^0$ in a spherical harmonic expansion of the gravitational potential using fully normalized functions.

Table 11.1. *Zonal coefficients $\overline{C}_l^0$ in units of 10^{-6}.*

l	$\overline{C}_l^0$	l	$\overline{C}_l^0$
2.	−484.165	3.	0.957 20
4.	0.539 52	5.	0.068 343
6.	−0.149 51	7.	0.091 301
8.	0.048 883	9.	0.026 862
10.	0.054 065	11.	−0.049 464
12.	0.035 629	13.	0.040 112
14.	−0.021 555	15.	0.003 227 5
16.	−0.006 189 1	17.	0.017 427
18.	0.008 524 6	19.	−0.002 155 1
20.	0.019 924		

Table 11.2 gives values of the coefficients $\overline{C}_l^m, \overline{S}_l^m$ in a spherical harmonic expansion of the gravitational potential using fully normalized functions. Note that $\overline{C}_2^1 = 0$ and $\overline{S}_2^1 = 0$.

Table 11.2. *Coefficients* $\overline{C}_l^m$, $\overline{S}_l^m$ *in units of* 10^{-6}.

l, m	$\overline{C}_l^m$,	$\overline{S}_l^m$	l, m	$\overline{C}_l^m$,	$\overline{S}_l^m$	l, m	$\overline{C}_l^m$,	$\overline{S}_l^m$
2, 2	2.439,	−1.4001						
3, 1	2.0277,	0.2492	3, 2	0.9045,	−0.6194	3, 3	0.7203,	1.4139
4, 1	−0.5362,	−0.4734	4, 2	0.3502,	0.6630	4, 3	0.9909,	−0.2009
4, 4	−0.1888,	0.3094						
5, 1	−0.0583,	−0.0961	5, 2	0.6527,	−0.3239	5, 3	−0.4523,	−0.2153
5, 4	−0.2956,	0.0497	5, 5	0.1738,	−0.6689			
6, 1	−0.0769,	0.0270	6, 2	0.0487,	−0.3740	6, 3	0.0572,	0.0094
6, 4	−0.0868,	−0.4713	6, 5	−0.2673,	−0.5368	6, 6	0.0097,	−0.2371
7, 1	0.2749,	0.0975	7, 2	0.3278,	0.0932	7, 3	0.2512,	−0.2153
7, 4	−0.2756,	−0.1238	7, 5	0.0013,	0.0186	7, 6	−0.3588,	0.1517
7, 7	0.0010,	0.0241						
8, 1	0.0236,	0.0588	8, 2	0.0776,	0.0660	8, 3	−0.0178,	−0.0863
8, 4	−0.2463,	0.0702	8, 5	−0.0250,	0.0895	8, 6	−0.0649,	0.3091
8, 7	0.0675,	0.0751	8, 8	−0.1242,	0.1202			
9, 1	0.1461,	0.0200	9, 2	0.0225,	−0.0336	9, 3	−0.1613,	−0.0760
9, 4	−0.0101,	0.0190	9, 5	−0.0171,	−0.0538	9, 6	0.0639,	0.2226
9, 7	−0.1190,	−0.0970	9, 8	0.1871,	−0.0024	9, 9	−0.0481,	0.0987
10, 1	0.0815,	−0.1303	10, 2	−0.0913,	−0.0511	10, 3	−0.0086,	−0.1550
10, 4	−0.0853,	−0.0787	10, 5	−0.0510,	−0.0511	10, 6	−0.0371,	−0.0784
10, 7	0.0076,	−0.0034	10, 8	0.0401,	−0.0917	10, 9	0.1243,	−0.0380
10, 10	0.0998,	−0.0225						

A simplified expression for the gravitational potential is

$$U \approx \frac{GM_\oplus}{r}\left\{1 - \sum_{l=2}^{\infty}\left(\frac{a}{r}\right)^l J_l P_l(\sin\phi)\right\},$$

where P_l is the Legendre polynomial of degree l. Values of the zonal coefficients J_l, defined by

$$J_l \equiv -\overline{C}_l^0 (2l+1)^{1/2}, \qquad l \geq 2,$$

are given in Table 11.3.

Table 11.3. *Zonal coefficients* J_l, *in units of* 10^{-6}.

l	J_l	l	J_l
2.	1 082.626	3.	−2.533
4.	−1.618 6	5.	−0.226 7
6.	0.539 1	7.	−0.353 6
8.	−0.201 5	9.	−0.117 1
10.	−0.247 8	11.	0.237 2

Table 11.3. *(Continued.)*

l	J_l	l	J_l
12.	−0.178 1	13.	−0.208 4
14.	0.116 1	15.	−0.017 97
16.	0.003 555	17.	−0.103 10
18.	−0.005 185 3	19.	0.013 459
20.	−0.127 58		

The relation of the second degree coefficients in a spherical harmonic expansion of the gravitational potential to products of inertia I_{ij} is

$$-\sqrt{5}\,\overline{C}_2^0 = J_2 = \frac{1}{M_\oplus a^2}\left\{I_{33} - \frac{I_{11} + I_{22}}{2}\right\},$$

$$\sqrt{\tfrac{5}{3}}\,\overline{C}_2^1 = \frac{I_{13}}{M_\oplus a^2}, \qquad \sqrt{\tfrac{5}{3}}\,\overline{S}_2^1 = \frac{I_{23}}{M_\oplus a^2},$$

$$\sqrt{\tfrac{5}{12}}\,\overline{C}_2^2 = \frac{I_{22} - I_{11}}{4 M_\oplus a^2}, \qquad \sqrt{\tfrac{5}{12}}\,\overline{S}_2^2 = \frac{I_{12}}{2 M_\oplus a^2}.$$

The principal products of inertia I_{11}, I_{22}, I_{33} are often denoted A, B, C with $C > B > A$ or $I_{33} > I_{22} > I_{11}$,

$$I_{11} = A = 8.009\,4 \times 10^{37} \text{ kg m}^2,$$
$$I_{22} = B = 8.009\,6 \times 10^{37} \text{ kg m}^2,$$
$$I_{33} = C = 8.035\,8 \times 10^{37} \text{ kg m}^2.$$

11.4 TOPOGRAPHY [2, 4, 5]

The topography of solid Earth, T, is:

$$T \text{ (in } 10^3 \text{ m)} = \sum_{l=0}^{\infty} \sum_{m=0}^{l} \overline{P}_l^m(\sin\phi)\left[\overline{CT}_l^m \cos m\lambda + \overline{ST}_l^m \sin m\lambda\right].$$

$\overline{P}_l^m(\sin\phi), \phi, \lambda$ are defined in the expression for the gravitational potential in Section 11.3. The coefficients are given in Table 11.4.

Table 11.4. *Values of the coefficients $\overline{CT}_l^m$ and $\overline{ST}_l^m$ (in units of 10^3 m).*

l, m	$\overline{CT}_l^m$,	$\overline{ST}_l^m$	l, m	$\overline{CT}_l^m$,	$\overline{ST}_l^m$	l, m	$\overline{CT}_l^m$,	$\overline{ST}_l^m$
0, 0	−2.3890,	—						
1, 0	0.6605,	—	1, 1	0.6072,	0.4062			
2, 0	0.5644,	—	2, 1	0.3333,	0.3173	2, 2	0.4208,	0.0839
3, 0	−0.1683,	—	3, 1	−0.1518,	0.1244	3, 2	0.4477,	0.4589
3, 3	0.1299,	0.5733						

Table 11.4. (*Continued.*)

l, m	$\overline{CT_l^m}$,	$\overline{ST_l^m}$	l, m	$\overline{CT_l^m}$,	$\overline{ST_l^m}$	l, m	$\overline{CT_l^m}$,	$\overline{ST_l^m}$
4, 0	0.3162,	—	4, 1	−0.2241,	−0.2563	4, 2	−0.3928,	0.0716
4, 3	0.3761,	−0.1291	4, 4	−0.6387,	0.4703			
5, 0	−0.5514,	—	5, 1	−0.0406,	−0.0770	5, 2	−0.0216,	−0.1577
5, 3	0.1232,	0.0386	5, 4	0.5254,	−0.0654	5, 5	−0.0549,	0.2276
6, 0	0.2567,	—	6, 1	0.0013,	−0.0171	6, 2	0.0247,	−0.1323
6, 3	0.0601,	0.1865	6, 4	0.1960,	−0.1737	6, 5	−0.1076,	−0.2075
6, 6	0.0354,	0.0282						

Area $= 5.100\,657 \times 10^{14}$ m^2.
Land area $= 1.48 \times 10^{14}$ m^2.
Water area $= 3.62 \times 10^{14}$ m^2.
Continental area including margins $= 2.0 \times 10^{14}$ m^2.
Mean land elevation $= 825$ m.
Mean ocean depth $= 3770$ m.

11.5   ROTATION (SPIN) AND REVOLUTION ABOUT THE SUN [1, 2, 6, 7]

Rotational period with respect to fixed stars $= 24^{\text{h}}00^{\text{m}}00^{\text{s}}.008\,4$ mean sidereal time,
$\qquad\qquad\qquad\qquad\qquad\qquad\qquad = 23^{\text{h}}56^{\text{m}}04^{\text{s}}.098\,9$ mean solar time.
Mean angular velocity $= 7.292\,115 \times 10^{-5}$ rad s^{-1}, 15.041 067 arcsec s^{-1}.
Equatorial rotational velocity $= 465.10$ m s^{-1}.
Centrifugal acceleration at equator $= 3.391\,57 \times 10^{-2}$ m s^{-2}.
Angular momentum $= \omega C = 5.859\,8 \times 10^{33}$ m^2 kg s^{-1}.
Rotational energy $= \frac{1}{2}C\omega^2 = 2.136\,5 \times 10^{29}$ J.

The general precession in longitude per Julian century for J2000.0 is $p = 5\,029''.096\,6$, where p is the long period motion of the mean pole of the equator about the pole of the ecliptic with a period of about 26,000 years. The general precession is due to the gravitational torques of the Sun, Moon, and planets on the Earth's dynamical figure.

Nutations are the motions of the Earth's rotation axis with respect to inertially fixed axes. Nutation includes the general precession and shorter period motions. A nutation induced by the Moon has a period of 18.6 years and an amplitude of about 9 arcsec. The gravitation of the Sun causes the lunar orbit to precess with respect to the plane of the ecliptic with a period of 18.6 years. Smaller nutations have periods of a solar year and a lunar month and harmonics thereof.

Length of Day (LOD) variations comprise an overall linear increase from tidal dissipation (of about 1 to 2 ms per century). There are large irregular fluctuations with amplitudes of milliseconds and time scales of decades, and smaller oscillations with shorter time scales. LOD variations with periods of a year and less are generally attributable to exchange of angular momentum between the solid Earth and the atmosphere–ocean system and to effects of solid Earth and ocean tides. LOD fluctuations with decade time scales may be due to angular momentum exchange between the solid Earth and the liquid outer core.

Polar motion or wobble is the motion of the solid Earth with respect to the spin axis of the Earth. Polar motion is dominated by nearly circular oscillations at periods of one year, the annual wobble with an amplitude of about 100 milliarcseconds, and at about 434 days, the Chandler wobble with an amplitude of about 200 milliarcseconds. The Chandler wobble is a free oscillation of the Earth; its excitation mechanism is uncertain. Other components of polar motion occur over a wide range of time scales from weeks to thousands of years. Loading of the solid Earth by the redistribution of mass in the atmosphere, oceans, groundwater, and ice caps contributes to polar motion.

Mean orbital speed $= 2.978\,48 \times 10^4$ m s^{-1}.
Mean centripetal acceleration $= 5.930\,1 \times 10^{-3}$ m s^{-2}.
Mean distance from Sun $= 1.000\,001\,057$ AU $= 1.495\,980\,29 \times 10^{11}$ m.
Mean eccentricity of orbit about the Sun $= 0.016\,708\,617$.
Obliquity of the ecliptic at J2000.0 $= 23°26'21''.411\,9$.
1 AU $= 1.495\,978\,706\,6 \times 10^{11}$ m.
Light time for 1 AU $= 499.004\,783\,53$ s.

11.6 GRAVITY [5, 7]

Gravity includes the gravitational attraction of the Earth's mass and the centrifugal acceleration of the Earth's rotation.

Surface gravity on reference ellipsoid g(m s^{-2}) $= 9.806\,21 - 0.025\,93\cos 2\phi + 0.000\,03\cos 4\phi$
$$= 9.780\,31 + 0.051\,86\sin^2\phi - 0.000\,06\sin^2 2\phi.$$

ϕ is the geodetic latitude of point p, i.e., the angle between the equator of the reference ellipsoid and the normal from p to the ellipsoid. Gravity anomalies are actual values of g minus the reference g given above. A practical unit for the measurement of gravity anomalies is the mgal $= 10^{-5}$ m s^{-2}.

Reference equatorial gravity $= 9.780\,31$ m s^{-2}.
Reference polar gravity $= 9.832\,17$ m s^{-2}.
Reference gravity at $\phi = 45° = 9.806\,18$ m s^{-2}.
Gravitation at the equator $= GM_{\oplus}/a^2 = 9.798\,29$ m s^{-2}.
Centrifugal acceleration at equator/gravitation at equator $= 3.461\,39 \times 10^{-3}$.
Variation of g with altitude at the Earth's surface $= 0.308\,6 \times 10^{-5}$ s^{-2}
$$= 3.086 \text{ mm s}^{-2} \text{ km}^{-1}$$
$$= 0.308\,6 \text{ mgal m}^{-1}.$$
g decreases by 3.086 mm s^{-2} per kilometer of elevation at the Earth's surface.

Gravity anomalies corrected for altitude, i.e., evaluated on the reference ellipsoid, are known as free-air gravity anomalies.

11.7 GEOID [2, 5, 7]

The gravity potential is the sum of the gravitational potential U (see above) and the centrifugal potential $\frac{1}{2}\omega^2 r^2\cos^2\phi$, where ω is the mean angular velocity.

    The geoid is the equipotential of gravity that coincides with mean sea level in the oceans. The geoid lies generally below the topography.

    The height of the geoid N is given with respect to a reference ellipsoid with the observed flattening of the Earth 1/298.257 and with the Earth's equatorial radius 6 378.136 km.

    The equation of the reference ellipsoid is $r = a\{1 + [(2f - f^2)/(1 - f)^2]\sin^2\phi\}^{-1/2}$, where f is the flattening. With $f = 1/298.257$

$$r = a\left\{1 + 0.673\,95\sin^2\phi\right\}^{-1/2}$$
$$\approx a\left\{1 - 0.336\,98\sin^2\phi + 0.170\,33\sin^4\phi\right\}.$$

11.8 COORDINATES [7]

Geodetic latitude (ϕ) − geocentric latitude (ϕ') = $692''.74 \sin 2\phi - 1''.16 \sin 4\phi$.

Geocentric latitude of a point p is the angle between the equator of the reference ellipsoid and a line from p to the center of the ellipsoid. Geodetic latitude is defined above.

$$1° \text{ of latitude} = 110.575 + 1.110 \sin^2 \phi, \, 10^3 \text{ m}.$$
$$1° \text{ of longitude} = (111.320 + 0.373 \sin^2 \phi) \cos \phi, \, 10^3 \text{ m}.$$
$$\tan \phi' = \frac{(1 - e^2) N_\phi + h}{N_\phi + h} \tan \phi.$$

e is the eccentricity of the reference ellipsoid

$$e^2 = 2f - f^2.$$

f is the flattening of the ellipsoid.

N_ϕ is the ellipsoidal radius of curvature in the meridian

$$N_\phi = \frac{a}{\left(1 - e^2 \sin^2 \phi\right)^{1/2}}.$$

h is the height of a point p above the reference ellipsoid.

With $f = 1/298.257$, $e^2 = 6.694\,385 \times 10^{-3}$, $e^2 \ll 1$, $N_\phi \approx a$,

$$\tan \phi' \approx \tan \phi \left(1 - e^2 + e^2 \frac{h}{a}\right)$$
$$\approx \tan \phi \left(0.993\,306 + 1.049\,583 \times 10^{-9} h(m)\right).$$

11.9 SOLID BODY TIDES [7, 8]

The tidal potential due to the gravitation of the Sun and the Moon U_T is the gravitational potential of these bodies expressed in the coordinate system of the Earth's gravitational potential, but without the $l = 1$ spherical harmonic terms. These $l = 1$ terms determine the orbital motion of the Earth. The tidal potential is a differential gravitational potential. Each spherical harmonic component of the tidal potential has contributions with different periods and amplitudes. Table 11.5 lists contributions to the $l = 2$ tidal potential, the dominant tidal component.

Table 11.5. *Periods and amplitudes for the $l = 2$ tidal potential.*

m	Tidal contribution	Period	(Amplitude) g^{-1}, 10^{-2} m
Long Period $m = 0$	Lunar nodal tides	18.613 years	2.79
	S_a	365.26 d	0.49
	S_{sa}	182.62 d	3.10
	M_m	27.555 d	3.52
	M_f	13.661 d	6.66

11.9 SOLID BODY TIDES / 247

Table 11.5. *(Continued.)*

m	m	Tidal contribution	Period	(Amplitude) g^{-1}, 10^{-2} m
Diurnal		O_1	25.819 h	26.22
$m = 1$		P_1	24.066 h	12.20
		S_1	24. h	0.29
		K_1	23.934 h	36.88
		Ψ_1	23.869 h	0.29
		Φ_1	23.804 h	0.52
Semi-Diurnal		N_2	12.658 h	12.10
$m = 2$		M_2	12.421 h	63.19
		S_2	12. h	29.40
		K_2	11.967 h	8.00

The perturbation in the Earth's second degree gravitational potential at the surface of the Earth due to tidal deformation of the Earth's interior is the product of the second degree tidal potential evaluated at the Earth's surface with the second degree potential Love number k.

The product of the second degree body tide displacement Love number h with the second degree component of U_T/g evaluated at the Earth's surface gives the tidally induced radial displacement of the surface.

Southward and eastward displacements of the tidally deformed surface of the Earth are given in terms of the body tide displacement Love number l by

$$\frac{-l}{g} \frac{\partial U_T}{\partial \theta}$$

and

$$\frac{l}{g \sin \theta} \frac{\partial U_T}{\partial \lambda},$$

respectively, where θ is colatitude, λ is eastward longitude, and g, U_T and its derivatives are evaluated at the Earth's surface. Second degree contributions are understood here.

Second degree tidal effects on surface gravity and surface tilt are represented by the gravimetric factor

$$\delta = 1 - \tfrac{3}{2}k + h$$

and the tilt factor

$$\eta = 1 + k - h,$$

respectively, similar to the above. Table 11.6 gives these Love numbers for a model of the Earth.

Table 11.6. *Second degree Love numbers for a spherical, rotating, ellipsoidal, elastic, oceanless Earth.*

m	Tidal contributions	k	h	l	δ	η
0	Any long period tide	0.299	0.606	0.0840	1.155	0.689
1	O_1	0.298	0.603	0.0841	1.152	0.689
	P_1	0.287	0.581	0.0849	1.147	0.700
	S_1	0.280	0.568	0.0853	1.144	0.707
	K_1	0.256	0.520	0.0868	1.132	0.730

Table 11.6. *(Continued.)*

m	Tidal contributions	k	h	l	δ	η
	Ψ_1	0.466	0.937	0.0736	1.235	0.523
	Φ_1	0.328	0.662	0.0823	1.167	0.660
2	Any semi-diurnal tide	0.302	0.609	0.0852	1.160	0.692

Values of the Love numbers for the real Earth are strongly modified by ocean tides and slightly modified by anelasticity in the solid Earth.

11.10   GEOLOGICAL TIME SCALE [9]

Age of Earth $= 4.5 - 4.7$ Ga
Oldest Geological Dates:

Rocks at Isua in southern West Greenland have yielded dates of metamorphic events at about 3750 Ma.

Sand River gneisses in the Limpopo belt of Southern Africa have been dated at about 3800 Ma.

Detrital zircons from Western Australia have yielded dates of about 4200 Ma, indicative of pre-existing crust.

Table 11.7 gives dates of various geologic eras in the Phanerozoic eon, and Table 11.8 gives dates in the Precambrian eon. Table 11.9 lists the major geological and biological events in the Earth's history.

Table 11.7. *The Phanerozoic Eon (Present–570 Million Years Ago).*

Period	Duration
Cenozoic Era	Present–65 Ma
Quaternary Sub-Era	Present–1.64 Ma
Holocene Epoch	Present–0.01 Ma
Pleistocene Epoch	0.01–1.64 Ma
Tertiary Sub-Era	1.64–65 Ma
Neogene Period	1.64–23.3 Ma
Pliocene Epoch	1.64–5.2 Ma
Miocene Epoch	5.2–23.3 Ma
Paleogene Period	23.3–65 Ma
Oligocene Epoch	23.3–35.4 Ma
Eocene Epoch	35.4–56.5 Ma
Paleocene Epoch	56.5–65 Ma
Mesozoic Era	65–245 Ma
Cretaceous Period	65–145.6 Ma
Senonian Epoch	65–88.5 Ma
Gallic Epoch	88.5–131.8 Ma
Neocomian Epoch	131.8–145.6 Ma
K2 Gulf Epoch	65–97 Ma,
K1	97–145.6 Ma)
Jurassic Period	145.6–208 Ma
J3, Malm Epoch	145.6–157.1 Ma
J2, Dogger Epoch	157.1–178 Ma
J1, Lias Epoch	178–208 Ma
Triassic Period	208–245 Ma
Tr3 Epoch	208–235 Ma
Tr2 Epoch	235–241.1 Ma
Tr1, Scythian Epoch	241.1–245 Ma

Table 11.7. *(Continued.)*

Period	Duration
Paleozoic Era	245–570 Ma
Permian Period	245–290 Ma
Zechstein Epoch	245–256 Ma
Rotliegendes Epoch	256–290 Ma
Carboniferous Period	290–362.5 Ma
Pennsylvanian Subperiod	290–323 Ma
Gzelian, Kasimovian, Moscovian, Bashkirian Epochs	
Mississippian Subperiod	323–362.5 Ma
Serpukhovian, Visean, Tounaisian Epochs	
Devonian Period	362.5–408.5 Ma
D_3 Epoch	362.5–377.5 Ma
D_2 Epoch	377.5–386 Ma
D_1 Epoch	386–408.5 Ma
Silurian Period	408.5–439 Ma
Pridoli, Ludlow, Wenlock, Llandovery Epochs	
Ordovician Period	439–510 Ma
Bala Subperiod	439–464 Ma
Ashgill, Caradoc Epochs	
Dyfed Subperiod	464–476 Ma
Llandeilo, Llanvirn Epochs	
Canadian Subperiod	476–510 Ma
Arenig, Tremadoc Epochs	
Cambrian Period	510–570 Ma
Merioneth Epoch	510–517 Ma
St. David's Epoch	517–536 Ma
Caerfai Epoch	536–570 Ma

Table 11.8. *The Precambrian Eon (570–4550–4570 Ma)[a].*

Period	Duration
Sinian Era	570–800 Ma
Vendian Period	570–610 Ma
Sturtian Period	610–800 Ma
Riphean Era	800–1650 Ma
Karatau Period	800–1050 Ma
Yurmatin Period	1050–1350 Ma
Burzyan Period	1350–1650 Ma
Animikean Era	1650–2200 Ma
Gunflint Period	1650–2200 Ma
Huronian Era	2200–2400–2500 Ma
Cobalt, Qurke Lake, Hough Lake, Eliot Lake Periods	
Randian Era	2400–2500–2800 Ma
Ventersdorp, Central Rand, Dominion Periods	
Swazian Era	2800–3500 Ma
Pongola, Moodies, Figtree, Onverwacht Periods	
Isuan Era	3500–3800 Ma
Hadean Era	3800–4550–4570 Ma
Imbrian (pars) Period	3800–3850 Ma
Nectarian Period	3850–3950 Ma
Pre-Nectarian Period	3950–4150 Ma
Cryptic Division	4150–4550–4570 Ma

Note

[a]The Precambrian is also divided as follows: Proterozoic Eon (570–2500 Ma); Pt_3 (570–900 Ma), Pt_2 (900–1600 Ma), Pt_1 (1600–2500 Ma) Subeons; Archean Eon (2500–4000 Ma); Ar_3 (2500–3000 Ma), Ar_2 (3000–3500 Ma), Ar_1 (3500–4000 Ma) Subeons; Priscoan Eon (4000–4550–4570 Ma).

Table 11.9. *Major "events" in Earth history.*

Event	Approximate age (Ma, million years ago)
Homo sapiens, Neanderthal man, Homo erectus, Australopithecus africanus, worldwide glaciations	0–3 Ma
Gulf of California opens, Calabria collides Italy–Sicily	3–5 Ma
Mediterranean desiccation, Panama collides NW Columbia, Red Sea Opens	5–10 Ma
FA (First Appearance) Hipparion (horse), FA hominids, Sivapithecus, Kenyapithecus, Khabylies collides Africa	10–15 Ma
Andaman Sea opens, South China Sea spreading ceases, Calabria rifts SE from Sardinia, Corsica–Sardinia collide Apulia, Main Himalayan Orogeny	15–20 Ma
Okinawa trough opens, Japanese Sea opens, Corsica–Sardinia parts France, East African and Red Sea rifting begins, Balearics/Khabalirs rift from Iberia	20–25 Ma
Norwegian Sea opens east of Jan Mayen, Main Alpine Orogeny	25–30 Ma
South China Sea opens, Scotia Sea opens Drake Passage opens, Caribbean Plate moves east	30–35 Ma
Late Eocene extinction, FA proboscideans (mastodons, elephants), early anthropoids, Labrador Sea/Baffin Bay cease spreading, Jan Mayen Ridge rifts from Greenland	35–45 Ma
FA rodents, Cuba collides Bahama Bank, India Eurasia collision begins, Indian–Australian plates united, Eurasia Basin opens, Norwegian Sea opens, Tasman Sea opens	45–55 Ma
FA horses, FA grasses, mammals diversify, FA primates	55–60 Ma
North Atlantic lavas, Indian Ocean spreads northwest of Seychelles, Yucatan Basin opens as Cuba moves north, Laramide Orogeny	60–65 Ma
Terminal Cretaceous extinction, Deccan lavas	65–70 Ma
FA early grasses, LA (last appearance) pteridosperms (seed ferns)	70-75 Ma
Cretaceous anoxic event, Labrador Sea opens, India–Madagascar separate, Australia parts Antarctica	85–95 Ma
FA diatoms (one-cell marine organisms), equatorial Atlantic opens, Bay of Biscay opens, Iberia parts Grand Banks	105–120 Ma
FA angiosperms (flowering plants), South Atlantic opens, East Indian Ocean opens, India parts from Australia–Antarctica, FA placental mammals	125–135 Ma
FA birds, Paleo Tethys closed	145–155 Ma
India–Madagascar Antarctica separate, Gulf of Mexico opens, Neo-Tethys opens, central Atlantic opens, East Gondwana (India, Australia, Antarctica) parts West Gondwana (Africa, South America)	155–170 Ma
Karoo volcanism	185–195 Ma
Early mammals, terminal Triassic extinction, Rifting between Gondwana and Laurasia	205–215 Ma
Iran, Crete, Turkey part from Gondwana, FA dinosaurs, Siberian lavas	235–250 Ma
Gondwana Laurasia collide, Appalachian Ocean finally closed	265–280 Ma
Iran, Tibet rift from Gondwana, FA conifers	280–300 Ma
FA winged insects, FA pelycosaurs (early mammal-like reptile)	300–320 Ma
South China rifts from Gondwana, FA sharks	350–380 Ma
FA wingless insects, firns, Iapetus Ocean finally closed	380–400 Ma
FA lungfish, land plants, jawed fish, North China rifts from Gondwana	400–430 Ma

Table 11.9. *(Continued.)*

Event	Approximate age (Ma, million years ago)
Ediacaran metazoans (soft body multicell animals), Skilogalee microbiota, Grenvilian Orogeny	570–1000 Ma
Keweenawan, Mackenzie Volcanics, Duluth Muskox intrusives, Oldest megascopic algae (large-celled algae), algal coals	1100–1400 Ma
Hudsonian and Penokean Orogenies, FA common red beds, Sudbury intrusion, Banded iron formations, Oxygen buildup in atmosphere	1700–2000 Ma
Bushveld intrusion, Gunflint microbial structures in chert, Hammersley & Fortescue biota, Kenoran Orogeny	2000–2500 Ma
FA red beds, Ventersdorp biota, Stilwater volcanics and intrusives	2500–2800 Ma
Kaap Valley Granite, Fig Tree Group with bacteria and blue green algae, Barberton Gneisses	3200–3300 Ma
FA stromatolites (bacterial algal mats) in Onverwacht Group and Australia	≈ 3400 Ma
Amitsoq & Kaapvaal gneisses, evidence life well established (carbon isotopic ratios)	≈ 3800 Ma
Basin formation on the Moon	3800–4200 Ma
Zircons from early crust	4200–4300 Ma
Lunar melting and differentiation of anorthositic crust	≈ 4500 Ma
Accretion of Earth and Moon	4500–4600 Ma

11.11   GLACIATIONS [9–11]

The geological record contains evidence of major glaciations as listed in Table 11.10.

Table 11.10. *Ages and locations of major glaciations.*

Age (Ma)	Locations
0–15, Holocene, Pleistocene	Antarctica, North America, Eurasia
250–380, Permian, Carboniferous, Devonian	Gondwana
430–450, Silurian, Ordovician	Gondwana
600, Vendian	China, North Europe, North and South America
650, Sturtian	Eurasia, South Africa, Australia
800, Sturtian	Australia, North America, South Africa
900, Karatau	Africa
2300–2400, Huronian	North America, South Africa
2800, Randian, Swazian	South Africa

Some glaciations may be related to plate tectonics, e.g., Gondwana moved over the South Pole in the Paleozoic.

The Quaternary glaciations (most geologically recent glaciations) may be related to cyclical changes in the Earth's orbital motion about the Sun and in the motion of the Earth's rotation axis (Milankovitch or astronomical theory of ice ages). The tilt of the Earth's equator to the ecliptic varies from 21.5° to 24.5° with a period of about 41,000 years. The eccentricity of the Earth's orbit varies with periods of about 100,000 years and 400,000 years and the Earth's axis of rotation wobbles with a period of about 22,000 years. Pleistocene glaciations have occurred cyclically with a period of about

10^5 years. Typically there has been a relatively slow glaciation phase lasting about 9×10^4 years and a relatively fast deglaciation phase lasting about 10^4 years. The last deglaciation event of the current ice age began about 18,000 years ago and ended about 7000 years ago.

11.12 PLATE TECTONICS [5, 12]

Earth's outer shell is divided into units known as tectonic plates that behave essentially rigidly on geological time scales. Plates move with respect to each other and the underlying mantle which deforms like a very viscous fluid on geological time scales. Tectonic plates comprise the lithosphere or rheologically stiff outer shell of the Earth. Plates are separated by four types of boundaries: (1) midocean ridges or sites of seafloor spreading and generation of new oceanic crust; (2) subduction zones or sites of plate submergence into the mantle; (3) transform faults or sites of fault-parallel relative horizontal motion or sliding; and (4) collisional zones or sites of horizontal convergence characterized by strong deformation and mountain building. Nonrigid deformation of the lithosphere occurs mainly at plate boundaries.

Major tectonic plates include Eurasia, Pacific, Antarctic, North America, South America, Africa, Australia, Philippine, Arabia, Nazca, Cocos, Caribbean, and Juan de Fuca.

Plate motions are well described by rigid body rotations of the plates about axes through the center of the Earth and intersecting the surface at poles of rotation generally located remotely from the plates (Euler's theorem). The angular velocity vector of plate rotation is known as the Euler vector. Each plate rotates counterclockwise relative to the fixed Pacific plate (PA). These main plates are given in Table 11.11.

Table 11.11. *NUVEL–1 Euler vectors of plate rotation.*

Plate	Latitude of rotation pole °N	Longitude of rotation pole °E	Magnitude of rotation rate ω (deg. Myr^{-1})
Africa, AF	59.16	−73.174	0.9695
Antarctica, AN	64.315	−83.984	0.9093
Arabia, AR	59.658	−33.193	1.1616
Australia, AU	60.080	+1.742	1.1236
Caribbean, CA	54.195	−80.802	0.8534
Cocos, CO	36.823	−108.629	2.0890
Eurasia, EU	61.066	−85.819	0.8985
India, IN	60.494	−30.403	1.1539
Nazca, NZ	55.578	−90.096	1.4222
North America, NA	48.709	−78.167	0.7829
South America, SA	54.999	−85.752	0.6657
Juan de Fuca[a]	35.0	+26.0	0.53
Philippine[a]	0.	−47.	1.0

Note

[a] Listed Euler vectors are not part of the NUVEL-1 model.

11.13 EARTH CRUST [5, 11]

The crust is the outermost layer of the Earth. The rocks of the crust are chemically and physically distinct from underlying mantle rocks; the major distinction between crust and mantle is compositional. Crustal rocks are less dense than mantle rocks and contain greater concentrations of heat-producing radiogenic elements. The base of the crust is defined by a discontinuity in the depth profiles of seismic velocities known as the Mohorovičić discontinuity or Moho.

There are two major subdivisions of the crust—the oceanic crust and the continental crust. Both types of crust generally consist of a sediment layer, an upper layer, and a lower layer. The average properties of these crustal layers are given in Table 11.12.

Table 11.12. *Average properties of oceanic and continental crust.*

Property	Oceanic	Continental
Sediment layer thickness (km)	0–1	0–5
Upper layer thickness (km)	1.5(0.7–2)	17(10–20)
Lower layer thickness (km)	5(3–7)	21(15–25)
Total thickness (km)	7(5–15)	36(30–80)
Areal abundance (%)	59	41
Volume abundance (%)	21	79
Heat flow (mW m^{-2})	78	56.5
Bouguer anomaly (mgal)^a	250	−100
v_p, upper layer (km s^{-1})^b	5.1	6.1
v_p, lower layer (km s^{-1})^b	6.6	6.8

Notes

^{a}Bouguer anomaly = free air gravity anomaly (see above) $-2\pi G \rho_c h$ (a correction for the gravitational attraction of topography with elevation h and density ρ_c, G is the universal gravitational constant).

$^b v_p$ = velocity of seismic P or compressional waves; 1 mgal = 10^{-2} mm s^{-2}. Seismic shear velocities of crustal rocks v_s are about 3.7 km s^{-1}

The average composition of the oceanic crust is primarily that of a tholeiitic basalt (Table 11.13). Oceanic tholeiitic basalt is extruded and intruded at mid-ocean ridges as a consequence of pressure-release melting of upper mantle material that rises beneath the ridges. Oceanic basalts undergo varying degrees of alteration by reactions with seawater and hydrothermal fluids especially at and near mid-ocean ridges.

The average composition of the upper layer of the continental crust is similar to that of granodiorite. The lower layer of the continental crust may be largely similar to mafic granulites in composition though a more felsic composition is possible. Whereas the oceanic crust is produced in a one stage melting of the upper mantle, continental crustal rocks involve multiple melting events.

Table 11.13. *Estimated average composition of the oceanic and continental crust (excluding sediments).*

	Continental crust			Oceanic crust
	Upper	Lower, mafic	Lower, felsic	
	Oxides (in weight %)			
SiO_2	65.5	49.2	61.0	49.6
TiO_2	0.5	1.5	0.5	1.5
Al_2O_3	15.0	15.0	15.6	16.8
FeOT	4.3	13.0	5.3	8.8
MgO	2.2	7.8	3.4	7.2
CaO	4.2	10.4	5.6	11.8
Na_2O	3.6	2.2	4.4	2.7
K_2O	3.3	0.5	1.0	0.2
MnO	0.1	0.2	0.1	0.2
P_2O_5	0.2	0.2	0.2	0.2

Table 11.13. *(Continued.)*

	Continental crust			Oceanic crust
	Upper	Lower, mafic	Lower, felsic	
	Trace Elements (in ppm)			
Rb	110.	2.	10.	4.
Ba	800.	50.	780.	60.
Sr	325.	500.	570.	180.
La	30.	10.	20.	3.5
Yb	2.0	1.0	1.2	2.7
Zr	220.	30.	200.	100.
Nb	25.	3.	5.	5.
U	2.5	0.1	0.1	0.2
Th	11.	0.3	0.5	0.6
Cr	35.	200.	90.	230.
Ni	20.	150.	60.	80.

Properties of the main crustal rocks are given in Table 11.14.

Table 11.14. *Properties of crustal rocks.[a][b]*

Rock	Density $(kg\ m^{-3})$	Young's modulus $(10^{11}\ Pa)$	Shear modulus $(10^{11}\ Pa)$	Poisson's ratio	Thermal conductivity $W\ m^{-1}\ K^{-1}$	Thermal expansivity $10^{-5}\ K^{-1}$
			Sedimentary			
Shale	2100–2700	0.1–0.3	0.14	—	1.2–3	—
Sandstone	2200–2700	0.1–0.6	0.04–0.3	0.2–0.3	1.5–4.2	3.
Limestone	2200–2800	0.6–0.8	0.2–0.3	0.25–0.3	2–3.4	2.4
Dolomite	2200–2800	0.5–0.9	0.3–0.5	—	3.2–5	—
Marble	2200–2800	0.3–0.9	0.2–0.35	0.1–0.4	2.5–3	—
			Metamorphic			
Gneiss	2700	0.04–0.7	0.1–0.35	0.04–0.15	2.1–4.2	—
Amphibole	3000	—	0.5 - 1.0	0.4	2.5–3.8	—
			Igneous			
Basalt	2950	0.6–0.8	0.3	0.25	1.3–2.9	—
Granite	2650	0.4–0.7	0.2–0.3	0.1–0.25	2.4–3.8	2.4
Diabase	2900	0.8–1.1	0.3–0.45	0.25	1.7–2.5	—
Gabbro	2950	0.6–1.0	0.2–0.35	0.15–0.2	1.9–2.3	1.6
Diorite	2800	0.6–0.8	0.3–0.35	—	2.8–3.6	—
Anorthosite	2750	0.83	0.35	0.25	1.7–2.1	—
Granodiorite	2700	—	—	—	2.6–3.5	—

Notes
[a] The specific heats of crustal rocks are all approximately $1\ kJ\ kg^{-1}\ K^{-1}$.
[b] Mean density of the continental crust $= 2\,750\ kg\ m^{-3}$. Mean density of the oceanic crust $= 2\,900\ kg\ m^{-3}$.

The radioactive heat sources in the Earth's interior are listed in Table 11.15.

Table 11.15. *Radiogenic heat production rates per unit mass H and half-lives $\tau_{1/2}$ of the important radioactive isotopes in the Earth's interior.[a]*

Isotope or element	H (W kg^{-1})	$\tau_{1/2}$ (Gyr)	Mantle concentration (kg kg^{-1})
^{238}U	9.37×10^{-5}	4.47	25.5×10^{-9}
^{235}U	5.69×10^{-4}	0.704	1.85×10^{-10}
U	9.71×10^{-5}	—	25.7×10^{-9}
^{232}Th	2.69×10^{-5}	14.0	1.03×10^{-7}
^{40}K	2.79×10^{-5}	1.25	3.29×10^{-8}
K	3.58×10^{-9}	—	2.57×10^{-4}

Note

[a]U is 99.27% by weight ^{238}U and 0.72% ^{235}U. Th is 100% ^{232}Th. K is 0.0128% ^{40}K. Assumes kg K/kg U $= 10^4$, kg Th/kg U $= 4$, and $H = 6.18 \times 10^{-12}$ W kg^{-1} in present mantle. [1]

Reference

1. Turcotte D.L., & Schubert, G. 1982, *Geodynamics* (Wiley, New York)

The abundances of uranium, thorium and potassium in the Earth and meteorite rocks is given in Table 11.16.

Table 11.16. *Representative concentrations (by weight) of heat-producing elements in several rocks and chondritic meteorites.[a]*

Rock	Concentrations		
	U (ppm)	Th (ppm)	K (%)
Depleted Peridotites	0.012	0.035	0.004
Tholeiitic Basalt	0.1	0.35	0.2
Granite	4.	17.	3.2
Chondritic Meteorites	0.013	0.04	0.078

Note

[a]Radiogenic elements are highly concentrated in the continental crust.

11.14   EARTH INTERIOR [13]

The structure of the Earth's interior has been determined mainly from seismology. Table 11.17 summarizes the values of the physical properties of a spherically symmetric model of the Earth as a function of radius from the center of the Earth based on seismological data. The major divisions of the solid Earth model are the core (radius $r = 0$ to 3480 km), the mantle ($r = 3480$ to 6346.6 km), and the crust ($r = 6346.6$ to 6368 km). The model core is divided into a solid inner core ($r = 0$ to 1221.5 km) and a liquid outer core. The model mantle is divided into the lower mantle ($r = 3480$ to 5701 km) and upper mantle (5701 to 6368 km). Subregions of the model mantle are the D''-layer at the base of the mantle ($r = 3480$ to 3630 km), the transition zone in the mid-mantle ($r = 5701$ to 5971), the seismic low velocity zone ($r = 6151$ to 6291 km) and the lithosphere or lid ($r = 6291$ to 6346.6 km). Similar terms are used to describe regions of the real Earth whose radial thicknesses are not so readily defined. The real Earth is, of course, laterally heterogeneous.

Table 11.17. *Physical properties of the Earth's interior according to PREM (Preliminary Earth Reference Model).*[a]

Region	Radius (km)	v_p (m s^{-1})	v_s (m s^{-1})	ρ (kg m^{-3})	K_s (GPa)	μ (GPa)	ν	p (GPa)	g (m s^{-2})
Inner core	0.	11266.20	3667.80	13088.48	1425.3	176.1	0.4407	363.85	0
	200.	11255.93	3663.42	13079.77	1423.1	175.5	0.4408	362.90	0.7311
	400.	11237.12	3650.27	13053.64	1416.4	173.9	0.4410	360.03	1.4604
	600.	11205.76	3628.35	13010.09	1405.3	171.3	0.4414	355.28	2.1862
	800.	11161.86	3597.67	12949.12	1389.8	167.6	0.4420	348.67	2.9068
	1000.	11105.42	3558.23	12870.73	1370.1	163.0	0.4428	340.24	3.6203
	1200.	11036.43	3510.02	12774.93	1346.2	157.4	0.4437	330.05	4.3251
	1221.5	11028.27	3504.32	12763.60	1343.4	156.7	0.4438	328.85	4.4002
Outer core	1221.5	10355.68	0.	12166.34	1304.7	0.	0.5	328.85	4.4002
	1400.	10249.59	0.	12069.24	1267.9	0.	0.5	318.75	4.9413
	1600.	10122.91	0.	11946.82	1224.2	0.	0.5	306.15	5.5548
	1800.	9985.54	0.	11809.00	1177.5	0.	0.5	292.22	6.1669
	2000.	9834.96	0.	11654.78	1127.3	0.	0.5	277.04	6.7715
	2200.	9668.65	0.	11483.11	1073.5	0.	0.5	260.68	7.3645
	2400.	9484.09	0.	11292.98	1015.8	0.	0.5	243.25	7.9425
	2600.	9278.76	0.	11083.35	954.2	0.	0.5	224.85	8.5023
	2800.	9050.15	0.	10853.21	888.9	0.	0.5	205.60	9.0414
	3000.	8795.73	0.	10601.52	820.2	0.	0.5	185.64	9.5570
	3200.	8512.98	0.	10327.26	748.4	0.	0.5	165.12	10.0464
	3400.	8199.39	0.	10029.40	674.3	0.	0.5	144.19	10.5065
	3480.	8064.82	0.	9903.49	644.1	0.	0.5	135.75	10.6823
D''	3480.	13716.60	7264.66	5566.45	655.6	293.8	0.3051	135.75	10.6823
	3600.	13687.53	7265.75	5506.42	644.0	290.7	0.3038	128.71	10.5204
	3630.	13680.41	7265.97	5491.45	641.2	289.9	0.3035	126.97	10.4844
Lower mantle	3630.	13680.41	7265.97	5491.45	641.2	289.9	0.3035	126.97	10.4844
	3800.	13447.42	7188.92	5406.81	609.5	279.4	0.3012	117.35	10.3095
	4000.	13245.32	7099.74	5307.24	574.4	267.5	0.2984	106.39	10.1580
	4200.	13015.79	7010.53	5207.13	540.9	255.9	0.2957	95.76	10.0535
	4400.	12783.89	6919.57	5105.90	508.5	244.5	0.2928	85.43	9.9859
	4600.	12544.66	6825.12	5002.99	476.6	233.1	0.2898	75.36	9.9474
	4800.	12293.16	6725.48	4897.83	444.8	221.5	0.2864	65.52	9.9314
	5000.	12024.45	6618.91	4789.83	412.8	209.8	0.2826	55.90	9.9326
	5200.	11733.57	6563.70	4678.44	380.3	197.9	0.2783	46.49	9.9467
	5400.	11415.60	6378.13	4563.07	347.1	185.6	0.2731	37.29	9.9698
	5600.	11065.57	6240.46	4443.17	313.3	173.0	0.2668	28.29	9.9985
	5600.	11065.57	6240.46	4443.17	313.3	173.0	0.2668	28.29	9.9985
	5701.	10751.31	5945.08	4380.71	299.9	154.8	0.2798	23.83	10.0143
Transition zone	5701.	10266.22	5570.20	3992.14	255.6	123.9	0.2914	28.83	10.0143
	5771.	10157.82	5516.01	3975.84	248.9	121.0	0.2909	21.04	10.0038
	5771.	10157.82	5516.01	3975.84	248.9	121.0	0.2909	21.04	10.0038
	5871.	9645.88	5224.28	3849.80	218.1	105.1	0.2924	17.13	9.9883
	5971.	9133.97	4932.59	3723.78	189.9	90.6	0.2942	13.35	9.9686
	5971.	8905.22	4769.89	3543.25	173.5	80.6	0.2988	13.35	9.9686
	6061.	8732.09	4706.90	3489.51	163.0	77.3	0.2952	10.20	9.9361
	6151.	8558.96	4643.91	3435.78	152.9	74.1	0.2914	7.11	9.9048
Low-velocity zone	6151.	7989.70	4418.85	3359.50	127.0	65.6	0.2796	7.11	9.9048
	6221.	8033.70	4443.61	3367.10	128.7	66.5	0.2796	4.78	9.8783
	6291.	8076.88	4469.53	3374.71	130.3	67.4	0.2793	2.45	9.8553
Lid	6291.	8076.88	4469.53	3374.71	130.3	67.4	0.2793	2.45	9.8553

Table 11.17. *(Continued.)*

Region	Radius (km)	v_p (m s^{-1})	v_s (m s^{-1})	ρ (kg m^{-3})	K_s (GPa)	μ (GPa)	ν	p (GPa)	g (m s^{-2})
	6346.6	8110.61	4490.94	3380.76	131.5	68.2	0.2789	0.604	9.8394
Crust	6346.6	6800.00	3900.00	2900.00	75.3	44.1	0.2549	0.604	9.8394
	6356.	6800.00	3900.00	2900.00	75.3	44.1	0.2549	0.337	9.8332
	6356.	5800.00	3200.00	2600.00	52.0	26.6	0.2812	0.337	9.8332
	6368.	5800.00	3200.00	2600.00	52.0	26.6	0.2812	0.300	9.8222
Ocean	6368.	1450.00	0.	1020.00	2.1	0.	0.5	0.300	9.8222
	6371.	1450.00	0.	1020.00	2.1	0.	0.5	0.	9.8156

Note

^{a} K_s is the bulk modulus, μ is the shear modulus, and ν is Poisson's ratio.

11.15   EARTH ATMOSPHERE, DRY AIR AT STANDARD TEMPERATURE AND PRESSURE (STP) [14, 15]

Standard temperature $T_0 = 273.15$ K.
Standard pressure $p_0 = 1\,013.250 \times 10^2$ Pa $= 1\,013.25$ mbar.
Standard gravity $g_0 = 9.806\,65$ m s^{-2}.
Mass density of air $\rho_0 = 1.292\,8$ kg m^{-3}.
Molecular weight $M_0 = 28.964 \times 10^{-3}$ kg mole^{-1}.
Mean molecular mass $m_0 = 4.810 \times 10^{-26}$ kg.
Molecular root-mean-square velocity $(3RT_0/M_0)^{1/2} = 4.850 \times 10^2$ m s^{-1}.
Speed of sound $(\gamma p_0/\rho_0)^{1/2} = (\gamma RT_0/M_0)^{1/2} = 3.313 \times 10^2$ m s^{-1}.
Specific heat at constant pressure $c_p = 1005$ J kg^{-1} K^{-1}.
Specific heat at constant volume $c_v = 717.6$ J kg^{-1} K^{-1}.
Ratio of specific heats $\gamma = c_p/c_v = 1.400$.
Number density of air $N_0 = 2.688 \times 10^{25}$ m^{-3}.
Molecular diameter $\sigma = 3.65 \times 10^{-10}$ m.
Mean free path $L = 1/(2^{1/2}\pi N\sigma^2) = 6.285 \times 10^{-8}$ m.
Coefficient of viscosity $= 1.72 \times 10^{-5}$ Pa s.
Thermal conductivity $= 2.41 \times 10^{-2}$ W m^{-1} K^{-1}.
Refractive index n

$$(n-1) \times 10^6 = \frac{288.15}{273.15}\left(\frac{64.328 + 29\,498.1 \times 10^{-6}}{146 \times 10^{-6} - \sigma^2} + \frac{255.4 \times 10^{-6}}{41 \times 10^{-6} - \sigma^2}\right).$$

$$\sigma = 1/\lambda(m).$$

Rayleigh scattering (molecular) volume attenuation coefficient

$$k = 1.06\frac{32\pi^3}{3N\lambda^4}(n-1)^2.$$

11.16 COMPOSITION OF THE ATMOSPHERE [14, 16–21]

Table 11.18 gives the composition of the atmospheric gases.

Table 11.18. *Gases in the well-mixed atmosphere.*

Gas	Molecular weight	Fraction of dry air at surface		Column amount (atm-cm)^a
		volume percent	weight percent	
$N_2{}^b$	28.013	78.08	75.52	6.24×10^5
$O_2{}^c$	31.999	20.95	23.14	1.67×10^5
H_2O^{def}	18.015	$2 \times 10^{-6} - 3 \times 10^{-2}$	$3 \times 10^{-6} - 5 \times 10^{-2}$	1760
Ar^g	39.948	9.34×10^{-3}	12.9×10^{-3}	7470
$CO_2{}^c$	44.010	3.45×10^{-4}	5.24×10^{-4}	276
Ne^g	20.183	18.2×10^{-6}	12.7×10^{-6}	14.6
He^g	4.003	5.24×10^{-6}	0.724×10^{-6}	4.2
$CH_4{}^h$	16.043	1.72×10^{-6}	0.95×10^{-6}	1.3
Kr^g	83.80	1.14×10^{-6}	3.30×10^{-6}	0.91
CO^{de}	28.010	1.5×10^{-7}	1.5×10^{-7}	0.089
$SO_2{}^{de}$	64.06	3×10^{-10}	7×10^{-10}	1.1×10^{-4}
$H_2{}^i$	2.016	5.0×10^{-7}	0.35×10^{-7}	0.4
N_2O^j	44.012	3.1×10^{-7}	4.7×10^{-7}	0.25
$O_3{}^{dek}$	47.998	$3.0–6.5 \times 10^{-8}$	$5.0–11 \times 10^{-8}$	0.343
Xe^g	131.30	8.7×10^{-8}	39.4×10^{-8}	0.07
$NO_2{}^d$	46.006	2.3×10^{-11}	3.9×10^{-11}	2.0×10^{-4}
$HNO_3{}^d$	63.02	5×10^{-11}	11×10^{-11}	3.6×10^{-4}
NO^{de}	30.006	3×10^{-10}	3×10^{-10}	3.1×10^{-4}
$CFCl_3{}^l$	137.37	2.8×10^{-10}	13×10^{-10}	2.2×10^{-4}
$CF_2Cl_2{}^l$	120.91	4.8×10^{-10}	20×10^{-10}	3.8×10^{-4}

Notes

^{a}1 atm-cm = thickness of gas column when reduced to STP = 2.687×10^{23} molecules m^{-2}. Gases are well mixed (constant fractional amount with altitude) in the troposphere unless otherwise noted. Column amounts are nominal mid-latitude values [1, 2, 3]. Values for fractional amounts are from [1, 2, 3, 4, 5].

^{b}Photochemical dissociation in the thermosphere (see Table 11.20 for definition of thermosphere). Well mixed at lower levels [4].

^{c}Photochemical dissociation above 95 km. Well mixed at lower levels [4].

^{d}Considerable tropospheric vertical variation in the fractional amount. Very dry above the tropopause [1, 2]. See Table 11.20 for definitions of troposphere and tropopause.

^{e}Factor of 10^2 or more local variability related to local sources such as anthropogenic pollution and geothermal activity [1, 2, 4, 5, 6].

^{f}Fractional amounts are 1% extremes [1].

^{g}Well mixed up to ~ 110 km (turbopause). Diffusive separation at higher levels [4].

^{h}Dissociated in the mesosphere (see Table 11.20 for definition of mesosphere). Well mixed at lower levels [4, 7].

^{i}Increase with altitude in the mesosphere because of dissociation of H_2O. Minimum value in the stratosphere (see Table 11.20 for definition of stratosphere) [1, 7].

^{j}Dissociated in the stratosphere and mesosphere [4].

^{k}Range in fractional amount refers to monthly averages [5].

^{l}Dissociated in the stratosphere [4].

References

1. *COESA, U.S. Standard Atmosphere* 1976, (Government Printing Office, Washington DC)
2. Anderson, G.P. et al. 1986, AFGL-TR-86-0110, Atmospheric Constituent Profiles 10–120 km, *Air Force Geophysics Laboratory* (now Air Force Research Laboratory).
3. Allen, C.W. 1973, *Astrophysical Quantities*, 3rd ed. (Athlone Press, London)
4. Goody R.M., & Yung, Y.L. 1989, *Atmospheric Radiation: Theoretical Basis*, 2nd ed. (Oxford

University Press, New York)

5. Watson, R.F. et al. 1990, Greenhouse gases and aerosols, in *Climate Change: The IPCC Scientific Assessment* edited by J.T. Houghton, G.H. Henkins, and J.H. Ephraums (Cambridge University Press, New York)
6. Logan, J.A. et al. 1981, *J. Geophys. Res.*, **86**, 7210
7. Allen, M., Lunine, J.I., & Yung, Y.L. 1984, *J. Geophys. Res.* **89**, 4841

11.17 WATER VAPOR [22, 23]

The water vapor pressure in saturated air is given in Table 11.19.

Table 11.19. *Water vapor pressure e in saturated air.*

Over pure water							
T (°C)	−30	−20	−10	0	10	20	30	40
e (Pa)	50.88	125.4	286.3	610.8	1227	2337	4243	7378

Over ice			
T (°C)	−30	−20	−10	0
e (Pa)	37.98	103.2	259.7	610.7

Water vapor density (perfect gas law) $= (2.167 \times 10^{-3} e/T)$ kg m^{-3} with T in K and e the water vapor pressure in Pa.

1 cm precipitable water = 1245 cm STP water vapor.

Density of moist air (perfect gas law) $= 3.484 \times 10^{-3}(p - 0.378e)/T$ (kg m^{-3}) with P the total pressure, p the water vapor pressure, e in Pa, and T in K.

Mean change of water vapor pressure with height h

$$\log(e_h/e_0) = -h/2, \qquad h \leq 7.2 \text{ km}$$
$$= -(h - 2.16)/1.4, \qquad 7.2 \text{ km} \leq h \leq 13.6 \text{ km}.$$

h = height above surface (km).
e_h = water vapor pressure at height h.
e_0 = water vapor pressure at surface.

11.18 HOMOGENEOUS ATMOSPHERE, SCALE HEIGHTS AND GRADIENTS [17]

The scale height of the atmosphere (height for e-fold change of pressure in an isothermal atmosphere)

$$RT/g = R^*T/MWg = 2.93 \times 10^{-2}T \text{ (km)},$$

where R is the gas constant of dry air $= 287.05$ J kg^{-1} K^{-1}, R^* is the universal gas constant $= 8.314$ kJ K^{-1} kmole^{-1}, MW is the molecular weight of dry air $= 28.964$ kg kmole^{-1}, g is the acceleration of gravity $= 9.8$ m s^{-2}, and T is in K.

Height of homogeneous atmosphere. (An idealized atmosphere of finite height, constant temperature equal to the surface temperature, and constant density equal to the surface density.) $= H = R^*T/MWg$.

Surface Air T (°C)	−30	−15	0	15	30
H km	7.11	7.55	7.99	8.43	8.87

Mass of atmosphere per m^2 = 1.035×10^4 kg.
Total mass of Earth's atmosphere = 5.136×10^{18} kg.
Moment of inertia of the Earth's atmosphere = 1.413×10^{32} kg m^2.
Magnitude of the dry adiabatic temperature gradient g/c_p (c_p is the specific heat at constant pressure = $1\,005$ J kg^{-1} K^{-1} for dry air) = 9.75 K km^{-1}.
Mean temperature gradient in troposphere = -6.5 K km^{-1}.
Mass per unit area of 1 atm-cm of gas of molecular weight $MW = 4.462 \times 10^{-4} MW$ (kg m^{-2}) where MW is in kg kmole^{-1}.

11.19 REGIONS OF EARTH'S ATMOSPHERE AND DISTRIBUTION WITH HEIGHT [14, 17, 24]

The Earth's atmospheric layers are detailed in Table 11.20.

Table 11.20. *Atmospheric layers and transition levels.*

Layer	Height, h (km)	Characteristics
Troposphere	0–11	Weather, T decreases with h, radiative-convective equilibrium
Tropopause	11	Temperature minimum, limit of upward mixing of heat
Stratosphere	11–48	T increases with h due to absorption of solar UV by O_3, dry
Stratopause	48	Maximum heating due to absorption of solar UV by O_3
Mesosphere	48–85	T decreases with h
Mesopause	85	Coldest part of atmosphere, noctilucent clouds
Thermosphere	85–exobase	T increases with h, solar cycle and geomagnetic variations
Exobase	500–1000 km	
Exosphere	> exobase	Region of Rayleigh–Jeans escape
Ozonosphere	15–35 km	Ozone layer (full width at e^{-1} of maximum)
Ionosphere	> 70 km	Ionized layers
Homosphere	< 85 km	Major constituents well-mixed
Heterosphere	> 85 km	Constituents diffusively separate

Radiation belts	$r/R_\oplus$ at magnetic equator
Inner belt	~ 1.3–2.4
Outer belt	~ 3.5–11

Magnetosphere	$r/R_\oplus$ at magnetic equator
In direction of Sun	10
Bow shock in direction of Sun	12
In direction normal to ecliptic	18

Profiles of physical quantities in the atmosphere are given in Table 11.21.

Table 11.21. *Altitude profiles of mean physical conditions at latitude* $45°$ [1].

Altitude (km)	log P (Pa)	T (K)	log ρ (kg m^{-3})	log N (m^{-3})	H^a (km)	log l^b (m)
0	+5.006	288	+0.0881	25.41	8.4	−7.2
1	+4.95	282	+0.0460	25.36	8.3	−7.1
2	+4.90	275	+0.00286	25.32	8.1	−7.1
3	+4.85	269	−0.0413	25.28	7.9	−7.0
4	+4.79	262	−0.087	25.23	7.7	−7.0
5	+4.73	256	−0.133	25.19	7.5	−7.0
6	+4.67	249	−0.180	25.14	7.3	−6.9
8	+4.55	236	−0.279	25.04	6.9	−6.8
10	+4.42	223	−0.384	24.93	6.6	−6.7
15	+4.08	217	−0.71	24.61	6.4	−6.4
20	+3.74	217	−1.05	24.27	6.4	−6.0
30	+3.08	227	−1.73	23.58	6.7	−5.4
40	+2.46	250	−2.40	22.92	7.4	−4.7
50	+1.90	271	−2.99	22.33	8.0	−4.1
60	+1.34	247	−3.51	21.81	7.4	−3.6
70	+0.72	220	−4.08	21.24	6.6	−3.0
80	+0.022	199	−4.73	20.58	6.0	−2.4
90	−0.74	187	−5.47	19.85	5.6	−1.6
100	−1.49	195	−6.25	19.08	6.0	−0.85
110	−2.15	240	−7.01	18.33	7.7	−0.10
120	−2.60	360	−7.65	17.71	12.1	+0.52
150	−3.34	634	−8.68	16.71	23.	+1.52
220	−4.07	855	−9.59	15.86	36.	+2.38
250	−4.61	941	−10.22	15.28	45.	+2.95
300	−5.06	976	−10.72	14.81	51.	+3.41
400	−5.84	996	−11.55	14.02	60.	+3.80
500	−6.52	999	−12.28	13.34	69.	+4.89
700	−7.50	1000	−13.51	12.36	131.	+5.86
1000	−8.12	1000	−14.45	11.74	288.	+6.49

Notes

[a] H = pressure scale height (km).

[b] l = mean free path (m).

Reference

1. *COESA, U.S. Standard Atmosphere* 1976, (Government Printing Office, Washington DC)

Variations in physical quantities during the day and during the solar cycle are given in Table 11.22.

Table 11.22. *Diurnal and solar cycle variations from mean values* [1].[ab]

Altitude (km)	Diurnal ±$\delta\rho$ (%)	Solar	Diurnal ±δN (%)	Solar	Diurnal ±δp (%)	Solar	Diurnal ±δT (K)	Solar	Diurnal ±δMW (kg kmol)$^{-1}$	Solar
200	6.0	33	6.2	32	12.3	45	59	145	0.041	0.32
500	46.	84	44.	80	52.	87	121	207	0.49	1.62
1000	43.	71	25.	51	35.	64	122	207	0.99	1.40

Notes

[a] δ is the maximum departure in absolute value from mean values.

[b] Values obtained from the Mass Spectrometer Incoherent Scatter (MSIS) model for the following conditions. Diurnal: Solar Activity Index $F_{10.7} = 150$, geomagnetic activity index $A_p = 10$, day of year = 91, latitude = $45°$N ; Solar: Maximum $F_{10.7} = 200$, minimum $F_{10.7} = 75$, $A_p = 10$, day of year = 91, latitude = $45°$N, local time of day = 0900 h.

Reference

1. Hedin, A.E. 1983, *J. Geophys. Res. A*, **88**, 170

Composition and other atmosphere profile data are given in Table 11.23.

Table 11.23. *Mean molecular weight, composition and molecular collision frequency ν [1, 2].*[a]

Altitude (km)	MW (kg kmol^{-1})	Composition (% by volume)						$\log(\nu)$ ν in s^{-1}
		N$_2$	O$_2$	O	He	Ar	H	
100	28.44	77.	19.	3.4	< 0.05	0.8	< 0.05	3.42
150	24.18	61.	5.6	34.	< 0.05	0.1	< 0.05	1.36
200	21.55	42.	3.0	55.	.01	< 0.05	< 0.05	0.59
300	18.11	17.	0.8	81.	0.8	< 0.05	< 0.05	−0.377
400	16.42	6.0	0.2	91.	2.7	< 0.05	< 0.05	−1.143
500	15.23	1.9	< 0.05	90.	8.2	< 0.05	0.2	−1.796
700	10.63	0.1	< 0.05	55.	43.	< 0.05	1.6	−2.66
1000	4.48	< 0.05	< 0.05	5.7	88.	< 0.05	6.7	−3.12

Note

[a]Quantities obtained from the MSIS model for the following conditions: Solar activity index $F_{10.7} = 150$, geomagnetic index $A_p = 10$, day of year = 91, latitude = 45°N, and local time of day = 0900 h.

References

1. *COESA, U.S. Standard Atmosphere* 1976, (Government Printing Office, Washington DC)
2. Hedin, A.E. 1983, *J. Geophys. Res. A*, **88**, 170

11.20   ATMOSPHERIC REFRACTION AND AIR PATH

The refractive index n of dry air at pressure $p_s = 1\,013.25 \times 10^2$ Pa and temperature $T_s = 288.15$ K is given by

$$(n_s - 1) \times 10^6 = 64.328 + \frac{29498.1 \times 10^{-6}}{146 \times 10^{-6} - \sigma^2} + \frac{255.4 \times 10^{-6}}{41 \times 10^{-6} - \sigma^2},$$

where $\sigma = \lambda^{-1}$ and λ is the vacuum wavelength in nm [15]. For other temperatures and pressures the refractive index is found from

$$n - 1 = (pT_s/p_sT)(n_s - 1).$$

Water vapor reduces the refractive index by

$$43.49\left[1 - 7.956 \times 10^3\sigma^2\right]\frac{p_w}{p_s},$$

where p_w is the partial pressure of water vapor [15].

Refractive index of air for radio waves [25]

$$(n - 1) \times 10^6 = 0.776\frac{p}{T} - 0.056\frac{p_w}{T} + 3.75 \times 10^3\frac{p_w}{T^2}.$$

Atmospheric refraction R is defined by

$$R \equiv z_t - z_a,$$

where z_t is true zenith distance and z_a is apparent zenith distance.

The constant of refraction R_0 is

$$R_0 = \frac{n_0^2 - 1}{2n_0^2} = 0.000\,292\,6 = 60.35'',$$

where n_0 refers to n evaluated at $p_0 = 1\,013.25 \times 10^2$ Pa and $T_0 = 273.15$ K.

For $n = n_0$ refraction is [26]

$$R_{n_0} \cong R_0 \tan z_t, \qquad z_t \lesssim 80°,$$

$$R_{n_0} \cong R_0 \left(\frac{2.06}{0.058\,9 + (\pi/2 - z_t)} - 3.71 \right), \qquad z_t \gtrsim 80°.$$

For other temperature and pressure conditions

$$R = R_{n_0}(pT_0/p_0 T).$$

Table 11.24 presents refraction data for the atmosphere.

Table 11.24. *Refractive index n and refraction R versus wavelength λ.[a]*

λ(nm)	$(n_d - 1) \times 10^6$	$-(n_w - 1) \times 10^6$	$(n - 1) \times 10^6$	R (arcsec)
200	341.9	0.19	341.7	70.44
220	329.4	0.20	329.2	67.87
240	321.2	0.20	321.0	66.18
260	315.3	0.21	315.1	64.96
280	310.9	0.21	310.7	64.06
300	307.6	0.22	307.2	63.34
320	304.9	0.22	304.7	62.82
340	302.7	0.22	302.5	62.37
360	301.0	0.22	300.8	62.02
380	299.5	0.22	299.3	61.71
400	298.3	0.22	298.1	61.46
450	295.9	0.23	295.7	60.97
500	294.3	0.23	294.1	60.64
550	293.1	0.23	292.9	60.39
600	292.2	0.23	292.0	60.20
650	291.5	0.23	291.3	60.06
700	290.9	0.23	290.7	59.94
800	290.1	0.23	289.9	59.77
900	289.6	0.23	289.4	59.67
1000	289.2	0.23	289.0	59.58
1200	288.7	0.23	288.5	59.48
1400	288.4	0.24	288.2	59.42
1600	288.2	0.24	288.0	59.38
1800	288.1	0.24	287.9	59.36
2000	288.0	0.24	287.8	59.34
3000	287.7	0.24	287.5	59.28
4000	287.7	0.24	287.5	59.28
5000	287.6	0.24	287.4	59.26
7000	287.6	0.24	287.4	59.26
10000	287.6	0.24	287.4	59.26

Note

[a]Refractive index n_d is for dry air at $T_0 = 273.15$ K and $P_0 = 1\,013.25 \times 10^2$ Pa and the correction n_w for water vapor is for $P_w = 550$ Pa. For other temperatures and pressures multiply $n_d - 1$ by PT_0/P_0T and for other vapor pressures multiply $n_w - 1$ by p_w/p_0. Refraction $R = (n^2 - 1)/(2n^2) \cong n - 1$ in arc seconds.

For radio waves and dry air with $P_0 = 1\,013.25 \times 10^2$ Pa, $T_0 = 273.15$ K, n_d is $(n_d - 1) \times 10^6 = 288.0$. The correction n_w for water vapor with $P_w = 550$ Pa is $(n_w - 1) \times 10^6 = 30.2$. The refractive index n is $(n - 1) \times 10^6 = 318.2$.

Transmission data for atmosphere components are in Table 11.25.

Table 11.25. *Atmosphere transmission—absorber/scatterer* [1, 2, 3, 4, 5, 6, 7, 8, 9, 10].

$\lambda\ (\mu m)$	H_2O	CO_2	O_3	H_2O continuum	Molecular scattering	Aerosols[a]	Other	Total
10.00	0.971	0.995	0.851	0.946	1.000	0.977	0.999[b]	0.759
7.50	0.126	0.723	1.000	0.280	1.000	0.983	0.993[cd]	0.025
5.00	0.415	0.994	0.999	0.728	1.000	0.979	1.000	0.294
4.00	0.994	0.970	1.000	0.983	1.000	0.975	0.906[e]	0.837
3.00	0.462	0.980	1.000	0.859	1.000	0.966	1.000	0.376
2.00	0.828	0.565	1.000	0.982	1.000	0.961	1.000	0.441
1.00	0.990	1.000	1.000	1.000	0.991	0.862	1.000	0.846
0.90	0.790	1.000	1.000	0.990	0.987	0.836	1.000	0.645
0.80	0.967	1.000	1.000	1.000	0.979	0.811	1.000	0.767
0.70	0.943	1.000	0.993	0.999	0.964	0.787	1.000	0.709
0.65	0.981	1.000	0.978	1.000	0.952	0.765	1.000	0.699
0.60	0.990	1.000	0.959	1.000	0.934	0.744	1.000	0.659
0.55	1.000	1.000	0.972	1.000	0.908	0.723	1.000	0.637
0.50	1.000	1.000	0.990	1.000	0.867	0.689	1.000	0.591
0.45	1.000	1.000	0.999	1.000	0.802	0.657	1.000	0.527
0.40	1.000	1.000	1.000	1.000	0.698	0.627	1.000	0.438
0.38	1.000	1.000	1.000	1.000	0.641	0.615	1.000	0.394
0.36	1.000	1.000	1.000	1.000	0.572	0.604	1.000	0.345
0.34	1.000	1.000	0.986	1.000	0.492	0.592	1.000	0.287
0.32	1.000	1.000	0.765	1.000	0.399	0.578	1.000	0.177
0.30	1.000	1.000	0.037	1.000	0.298	0.564	1.000	0.0062
0.28	1.000	1.000	0.0000	1.000	0.196	0.551	1.000	0.0000
0.26	1.000	0.525	0.0000	1.000	0.105	0.538	1.000	0.0000
0.24	1.000	0.0014	0.0000	1.000	0.040	0.526	1.000	0.0000
0.22	1.000	0.0000	0.0000	1.000	0.0083	0.514	1.000	0.0000
0.20	1.000	0.0000	0.055	1.000	0.0005	0.502	1.000	0.0000

Notes

[a] Lowtran rural aerosol model.
[b] Trace gasses.
[c] Trace gasses (0.999).
[d] HNO_3 (0.934).
[e] N_2 continuum.

References

1. Anderson, G.P. et al. 1986, AFGL-TR-86-0110, Atmospheric Constituent Profiles 10–120 km, *Air Force Geophysics Laboratory* (now Air Force Research Laboratory).
2. Kneizys, F.X. et al. 1983, AFGL-TR-0187, Atmospheric Transmittance/Radiance: Computer Code LOWTRAN6, *Air Force Geophysics Laboratory* (now Air Force Research Laboratory)
3. McClatchey, R.A. et al. 1973, AFCRL-TR-73-0096, Atmospheric Absorption Line Parameters Compilation, Air Force Cambridge Research Laboratory (now Air Force Research Laboratory)
4. Rothman, L.S. & McClatchey, R.A. 1976, *Appl. Optics*, **15**, 2616
5. Rothman, L.S. 1978, *Appl. Optics*, **17**, 507
6. Rothman, L.S. 1978, *Appl. Optics*, **17**, 3517
7. Rothman, L.S. 1981, *Appl. Optics*, **20**, 791
8. Rothman, L.S. 1981, *Appl. Optics*, **20**, 1323
9. Rothman, L.S. 1983, *Appl. Optics*, **22**, 1616
10. Rothman, L.S. 1985, *Appl. Optics*, **22**, 2247

11.21 ATMOSPHERIC SCATTERING AND CONTINUUM ABSORPTION
by David Crisp

At wavelengths shorter than about 300 nm, scattering and continuum absorption by gases and airborne particles (aerosols) renders the Earth's atmosphere virtually opaque to incoming radiation. The depth of penetration of ultraviolet radiation is shown in Figure 11.1. For cloud-free conditions, Rayleigh scattering by the atmosphere's principal molecular constituents, N_2 and O_2, accounts for the majority of the scattering, while continuum absorption is produced primarily by O_2 and O_3.

The extinction (scattering and absorption) at these wavelengths obeys the Beer–Bougher–Lambert law, which states that the intensity I at wavelength λ and altitude z is given by

$$I(z, \Theta, \lambda) = I(\infty, \Theta, \lambda) \exp\{-M(\Theta)\tau(z)\},$$

$I(\infty, \Theta, \lambda)$ is the intensity at the top of the atmosphere at zenith angle Θ, $M(\Theta)$ is the air-mass factor ($M(\Theta) \sim \sec \Theta$ for Θ for $< 80°$), and $\tau(z)$ is the vertical extinction optical depth

$$\tau(z) = \sum_{i=1}^{m} \int_0^\infty N(i, z)\sigma(i, z)\, dz,$$

$N(i, z)$ is the altitude-dependent number density (particles m^{-3}) and $\sigma(i, z)$ is the effective extinction cross section of a particle (molecule or aerosol m^2).

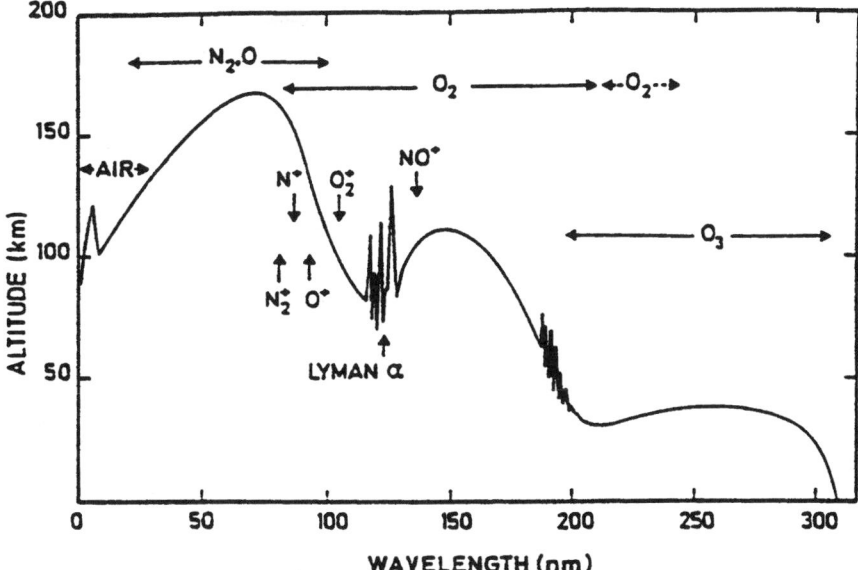

Figure 11.1. Depth of penetration of solar radiation as a function of wavelength. Altitudes correspond to an attenuation of l/e. The principal absorbers and ionization limits are indicated.

11.21.1 Rayleigh Scattering

The Rayleigh scattering cross section per molecule $\sigma_R(\lambda)$ is given by

$$\sigma_R(\lambda) = \frac{8\pi^3(n_g - 1)(6 + 3\delta)}{3\lambda^4 N^2(6 - 7\delta)},$$

δ is the depolarization factor and n_g is the wavelength-dependent refractive index of air. The Rayleigh scattering optical depth for air can be approximated by

$$\sigma_R(\lambda) \cong 0.008\,569\lambda^{-4}\left(1 + 0.001\,13\lambda^{-2} + 0.000\,13\lambda^{-4}\right)p/p_0,$$

p is the pressure (mbar) at altitude z, and $p_0 = 1\,013.25$ mbar is the sea-level pressure. The slight difference from the λ^{-4} dependence is introduced by the wavelength dependence of n_g [27].

11.21.2 Aerosol Extinction

The continuum absorption and scattering by aerosols cannot be specified uniquely because the aerosol abundance, composition, and size distribution can vary dramatically with location and time. However, representative global-annual-average values of the wavelength-dependent aerosol extinction optical depths have been derived for climate modeling studies. Tropospheric aerosols considered in these models include sea salt, sulfates, natural dust, hydrocarbons, and other more minor constituents. The stratospheric aerosols include sulfuric acid and silicates from volcanic eruptions, ammonium sulfates and persulfates and ammonium hydrates.

The integrated aerosol optical depths above sea-level (0 km), 3 km, and 12 km from one such modeling study [28] are shown in Figure 11.2. For hazy conditions, actual values of optical depth can be more than an order of magnitude larger. The wavelength dependence of the optical depths results from the particle size distribution (particles usually produce the most extinction at wavelengths

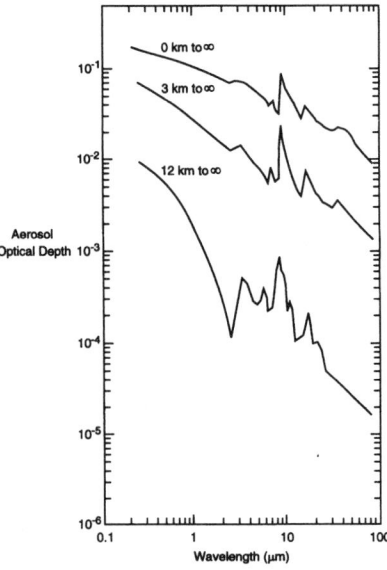

Figure 11.2. The calculated aerosol optical depth of the atmosphere.

comparable to their radius) as well as wavelength-dependent variations in the complex refractive indices of these materials.

11.21.3 Continuum Absorption by Gases at UV and Visible Wavelengths

Molecular oxygen O_2 and ozone O_3 are the principal continuum absorbers at ultraviolet and visible wavelengths. The principal O_2 features include the ionization continuum at $\lambda < 120$ nm, the Schumann–Runge continuum at $140 < \lambda < 180$ nm, the Schumann–Runge bands at $180 < \lambda < 200$ nm, and the Herzberg continuum at $\lambda > 200$ nm [29]. Several other gases, such as H_2O, CO_2, N_2O, and NO_2 also contribute absorption at these wavelengths.

The wavelength-dependent absorption optical depths for these gases can be derived from their cross sections once their number densities are known. If we neglect the temperature dependence of the gas continuum cross sections, the column-integrated optical depths can be simplified further and expressed as the product of the mean cross section, and the gas column abundance X which can be derived from the pressure-dependent gas mixing ratios, $r(p)$,

$$X = \int_0^\infty N(z)\,dz = \frac{A_0}{\mu_a g} \int_0^p r(p')\,dp',$$

A_0 is Avogadro's number (6.02×10^{23} molecules mol^{-1}), μ_a is the molecular weight of air (≈ 29 kg $kmol^{-1}$), g is the gravitational acceleration, and p is pressure. The wavelength-dependent, column-integrated optical depth is then given by

$$\tau(\lambda) = \sigma(\lambda)X.$$

Global-annual average gas mixing ratio profiles for the gases mentioned above are shown in Figure 11.3. Column abundances derived from these profiles are included in Table 11.26.

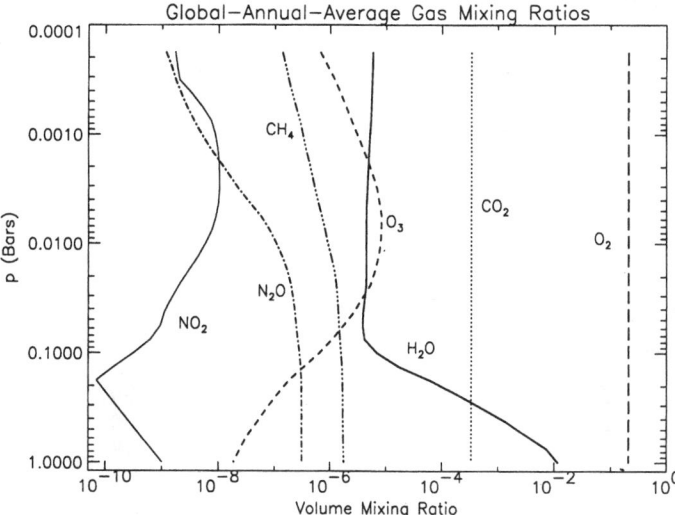

Figure 11.3. Global annual average gas mixing ratios.

Table 11.26. *Column abundances of atmospheric gases.*

Gas	X (molecules cm^{-2})
O_2	4.47×10^{24}
O_3	7.97×10^{18}
H_2O	8.12×10^{22}
CO_2	7.04×10^{21}
N_2O	6.36×10^{18}
NO_2	1.27×10^{16}

11.22   ABSORPTION BY ATMOSPHERIC GASES AT VISIBLE AND INFRARED WAVELENGTHS
by David Crisp

At wavelengths longer than 500 nm, the principal sources of atmospheric extinction are the vibration–rotation bands of gases. Unlike the slowly-varying ultraviolet gas absorption features described in the previous section, these bands consist of large numbers of narrow, overlapping absorption lines. Because the cores of these lines can become completely opaque while their wings remain much more transparent, the absorption within these bands does not strictly obey the Beer–Bougher–Lambert absorption law, except in spectral regions that are sufficiently narrow to completely resolve the individual line profiles (< 0.1 cm^{-1}). The absorption coefficients within vibration–rotation bands also vary much more strongly with pressure and temperature than those at ultraviolet wavelengths. The absorption by these gases has therefore been characterized by an effective vertical optical depth.

Figures 11.4–11.6 show the vertical optical depth above sea level (top, thick line) and above a high-altitude site, e.g., Mauna Kea Observatory in Hawaii ($z = 4$ km, $p = 600$ mbar, lower thin line). These synthetic spectra were generated with an atmospheric line-by-line model. This model employs a spectral resolution adequate to completely resolve the individual absorption lines (0.1 to 10^{-4} cm^{-1}), but the spectra shown here were then smoothed with a rectangular slit function with a full-width of 10 cm^{-1} (Figure 11.4) or 5 cm^{-1} (Figures 11.5 and 11.6). These figures therefore do not resolve individual absorption lines. Absorption line parameters for all gases are from the HITRAN database [30].

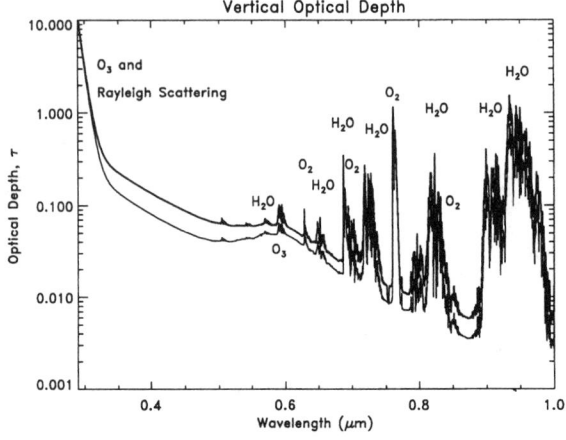

Figure 11.4. Vertical optical depth versus wavelength.

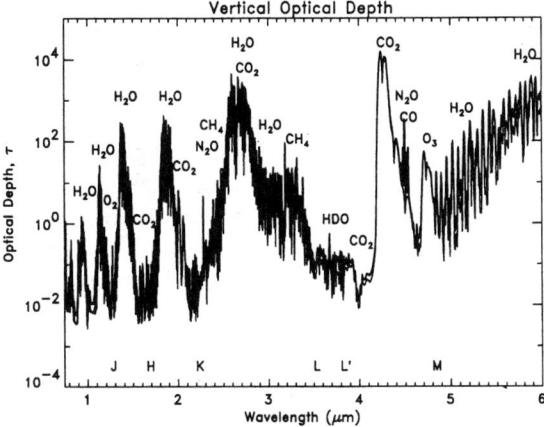

Figure 11.5. Vertical optical depth for near-infrared wavelengths.

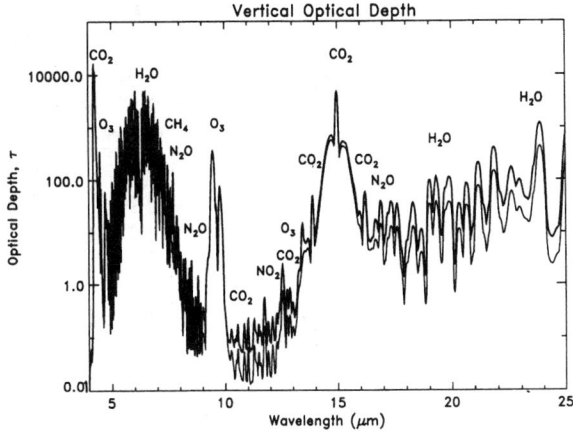

Figure 11.6. Vertical optical depth at long wavelengths.

Figure 11.4 confirms that Rayleigh scattering and O_3 continuum absorption dominate the extinction optical depth at wavelengths less than 0.5 μm. At longer wavelengths, water vapor is the principal absorber with its strongest features near 0.7, 0.82, and 0.94 μm. O_2 also has four significant bands between 0.65 and 1 μm. This figure also illustrates the advantage of working at a high-altitude site, where the atmospheric pressure and scattering optical depth are only 60% of their sea level values. Much less of an advantage is seen within the strong gas absorption bands, which are opaque even at the high-altitude site.

Figure 11.5 shows that water is also the principal absorber at near-infrared wavelengths between 1 and 6 μm, with very strong bands centered near 1.1, 1.38, 1.88, 2.7, and beyond 6 μm. CO_2 is the next most important absorber at these wavelengths, with strong bands near 2.0, 2.7, and 4.3 μm, and much weaker absorption near 1.22, 1.4, 1.6, 4.0, 4.8, and 5.2 μm. Other trace gases including CH_4 (2.4 and 3.3 μm), O_3 (3.3, 3.57, and 4.7 μm), and N_2O (2.1, 2.2, 2.47, 2.6, 2.9, and 4.7 μm) also produce some extinction at these wavelengths.

Water vapor absorption continues to dominate the spectrum at wavelengths beyond 5 μm (Figure 11.6). The most prominent water vapor bands at thermal wavelengths are the ν_2 fundamental centered near 6.3 μm and the rotation band beyond 20 μm, but this gas contributes significant absorption throughout this wavelength region. For example, the far wings of water vapor lines in the ν_2 and rotation bands provide much of the absorption in the atmospheric window regions near 8.5 and 12 μm. Within these windows, the high-altitude site (thin solid line) has up to a factor of 5 less absorption than the sea-level site (thick solid line), because the H_2O absorption coefficients at these wavelengths are very strong functions of pressure (proportional to density-squared), and the high-altitude site is above the majority of the water vapor. CO_2 and O_3 are the next most important absorbers at thermal wavelengths, with strong features near 15 and 9.6 μm, respectively. CH_4, N_2O, and NO_2 also have strong absorption bands at these wavelengths, but their bands are largely obscured by the stronger water vapor bands.

11.23   THERMAL EMISSION BY THE ATMOSPHERE
by David Crisp

The atmosphere emits as well as absorbs thermal radiation. This emission can enhance the sky brightness significantly at some wavelengths and reduce the detectability of faint astronomical sources. The intensities of the downwelling thermal radiance at a zenith angle of 21° are shown for a sea-level site (solid line), and a high-altitude site, e.g., Mauna Kea, Hawaii ($z = 4$ km, $p = 600$ mbar, dotted line) in Figures 11.7 and 11.8. At wavelengths within strong absorption bands, the atmosphere emits almost like a black body. Within the atmospheric window regions centered near 3.5 and 10 μm, the atmosphere emits much less radiation. The downward thermal radiation above a high-altitude site is substantially less than over the sea-level site because the overlying atmosphere is both cooler and less opaque.

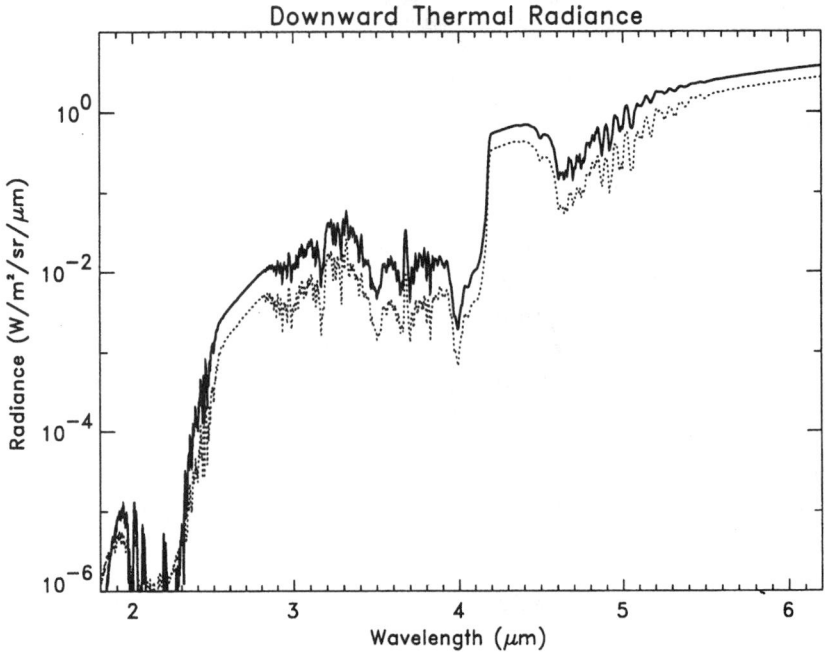

Figure 11.7. Downward thermal radiance in the near-infrared part of the spectrum.

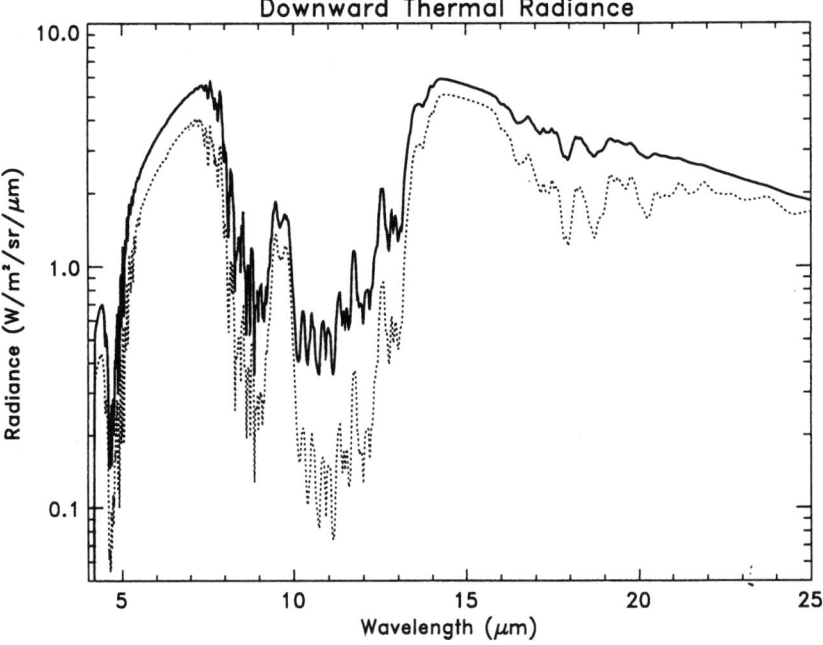

Figure 11.8. Downward thermal radiance at long wavelengths.

11.24 IONOSPHERE [17, 31]

The Earth's ionosphere is the partially ionized part of its atmosphere. It is divided into layers or regions, the main ones being the D, E, F1, and F2 regions, based principally on the altitude (z) profile of the electron density n_e (the number of electrons per unit volume). Ionospheric structure, $n_e(z)$, varies strongly with time of day and month, latitude and solar activity. At night, the D and F1 regions vanish, the E region weakens considerably, and the F2 region tends to persist at reduced intensity. Table 11.27 summarizes the characteristics of the ionospheric layers. The quantities are explained in the text below the table.

Table 11.27. *Properties of daytime ionospheric layers at middle and low latitudes.*

Quantity	R	D	E	F1	F2
Approx. altitude range (km)	⋯	60–95	105–160	160–180	> 180
Approx. height of max. n_e (km)		At top	105–110	170	200–400
Range of max. n_e (m^{-3})	⋯	10^8–10^{10}	10^{11}	10^{11}–10^{12}	10^{12}
f_0 (MHz), $\chi = 0$,	0	0.2	3.3	4.25	6.9
	100	0.28	3.82	5.34	11.95
Max. n_e (m^{-3}), $\chi = 0$,	0	5.0×10^8	1.35×10^{11}	2.24×10^{11}	5.91×10^{11}
	100	10^9	1.81×10^{11}	3.54×10^{11}	1.77×10^{12}
q (m^{-3} s^{-1})	0	2×10^5	5×10^8	7×10^8	10^8
	100	—	10^9	1.5×10^9	3×10^8
Layer thickness (km)	⋯	15	25	60	300
$\int q \, dz$ (m^{-2} s^{-1})	0	1.2×10^{13}	4×10^{13}	3×10^{13}	
	100	—	2.5×10^{13}	9×10^{13}	9×10^{13}
Ionizing emission at Sun Surface (photons m^{-2} s^{-1})	0	—	5×10^{17}	18×10^{17}	14×10^{13}
	100	—	12×10^{17}	40×10^{17}	40×10^{13}

Table 11.27. (*Continued.*)

Quantity	R	D	E	F1	F2
Neutral density at height of maximum n_e (m^{-3})	$\cdots$	4×10^{20}	10^{18}	2×10^{16}	10^{15}
T at height of max. n_e (K)		180	300	900	1100
Behavior		Regular	α-Chapman layer	Chapman layer	Anomalous, strongly variable
Recombination coefficient α $(m^3\ s^{-1})$		10^{-12}	10^{-14}–10^{-13}	4×10^{-14}	10^{-15}
Attachment β (s^{-1}), day		—	10^{-3}	10^{-3}	3×10^{-4}
ν_{ei} (s^{-1})		3	400	200	400
ν_{en} (s^{-1})		7×10^5	3×10^3	250	10

f_0 = critical frequency = maximum plasma frequency of an ionospheric layer = $(e^2(\text{maximum } n_e)/4\pi^2\epsilon_0 m_e)^{1/2}$, m_e = electron mass, e = electron charge, ϵ_0 = permittivity of free space, n_e = electron number density.

$(f_0\ (\text{Hz}))^2 = 80.5\ (\text{maximum } n_e\ (m^{-3}))$.

R = Wolf sunspot number = $k(f + 10\ g)$, f = total number of spots seen, g = number of disturbed regions (either single spots or groups of spots), k = a constant for a particular observatory.

q = ionization rate = rate of production of ion–electron pairs per unit volume (derived, e.g., from the Sun's spectrum and ionospheric absorption coefficients).

α = recombination coefficient, rate of electron loss by recombination = $\alpha n_i n_e$ (n_i = number density of ions) = αn_e^2 (normally, $n_i = n_e$). Electron loss rate αn_e^2 has units of number per unit volume per unit time.

α-Chapman layer = idealized model of an ionospheric layer, single species neutral atmosphere with constant scale height H, solar radiation absorption $\propto$ neutral gas number density, absorption coefficient is constant, $q = q_{mo}\exp(1 - z' - (\sec\chi)e^{-z'})$, $z' = (z - z_{mo})/H$, z is altitude, z_{mo} is the height of maximum production rate when the Sun is overhead ($\chi = 0$), q_{mo} is the production rate at z_{mo} (when $\chi = 0$), χ = solar zenith angle, production = loss, $q = \alpha n_e^2$, $n_e = n_e(z_{mo})\exp\frac{1}{2}(1 - z' - e^{-z'}\sec\chi)$, q_m is the maximum production rate = $q_{mo}\cos\chi$, z_m is the height of maximum production = $z_{mo} + H\ln(\sec\chi)$, $n_e(z_m) = n_e(z_{mo})\cos^{1/2}\chi$.

β = attachment coefficient, rate of electron loss by attachment to neutral particles to form negative ions = βn_e (neutral species number density $\gg n_e$). β has units of inverse time.

β-Chapman layer = similar to α- Chapman layer except for electron loss which occurs by attachment, $q = \beta n_e$, $n_e = n_e(z_{mo})\exp(1 - z' - e^{-z'}\sec\chi)$, $n_e(z_m) = n_e(z_{mo})\cos\chi$.

$dn_e/dt = q - \alpha n_e^2 - \beta n_e$, usually either α or β.

ν_{ei}, ν_{en} = collision frequency of mean electron with ions, and neutral particles.

ν_{en} (s^{-1}) = $(6.93 \times 10^5 n(N_2) + 4.37 \times 10^5 n(O_2)) u$, u is electron energy in J, $n(N_2)$ and $n(O_2)$ are number densities in m^{-3}.

w_B (rad s^{-1}) = gyrofrequency = QB/m, Q = charge on particle (C), B = magnetic flux density (T), m = charged particle mass (kg).

f_B (Hz) = w_B (rad s^{-1})/2π.

w_B (rad s^{-1}) for an electron = $1.759 \times 10^{11} B$ (T).

f_B (Hz) for an electron = $2.799 \times 10^{10} B$ (T).

w_N (rad s^{-1}) plasma frequency = $(n_e e^2 / \epsilon_0 m_e)^{1/2}$, e = charge on an electron.

$\omega_N^2 = 3182 n_e$ (ω_n in rad s^{-1} and n_e in m^{-3}).

ϕ = Faraday rotation = rotation of the polarization angle of a radio wave propagating through the ionosphere

$$= \frac{1}{2c\omega^2} \int \omega_N^2 \omega_B \hat{\underline{B}} \cdot \underline{dl} = \frac{e^3}{2c\epsilon_0 m_e^2} \frac{1}{\omega^2} \int n_e \underline{B} \cdot \underline{dl} = \frac{9.327 \times 10^5}{\omega^2} \int n_e \underline{B} \cdot \underline{dl}$$

$$= \frac{e^3 \mu_0}{2c\epsilon_0 m_e^2 4\pi^2 f^2} \int n_e \underline{H} \cdot \underline{dl} = \frac{2.969 \times 10^{-2}}{f^2} \int n_e \underline{H} \cdot \underline{dl},$$

c is the speed of light, $\hat{\underline{B}}$ is a unit vector in the direction of $\underline{B}$, $\underline{dl}$ is a path increment along the wave propagation direction, integration is along the path of the radio wave, ω is the circular frequency of the wave, f is the frequency of the wave in Hz, μ_0 is the permeability of free space, $\underline{H}$ is the magnetic field strength (A m^{-1}), all units are SI, it is assumed that $\omega \gg \omega_B$, the formula is approximate for cross-field propagation but accurate to within a few degrees of the normal to $\underline{B}$, the rotation follows the right-hand rule.

Photon efficiency of ionization η is the ratio of the rate of production of ion–electron pairs (number m^{-3} s^{-1}) to the total number of photons absorbed per unit volume and per unit time. Ionization of atomic species yields one ion–election pair for every 5.45×10^{-18} J absorbed. Accordingly,

$$\eta = \frac{1}{5.45 \times 10^{-18} \text{ J}/(hc/\lambda)} = 36.5/\lambda \text{ (nm)}, \qquad 2 < \lambda < 100 \text{ nm},$$

h is Planck's constant and λ is the wavelength of the radiation. For $\lambda < 2$ nm, η is approximately 20.

11.24.1 Ionosphere as a Whole

The total electron content of the ionosphere is

$$I \equiv \int_0^\infty n_e \, dz,$$

where z is altitude. More generally, I can be defined as a line integral along an arbitrary path. Typically, I is about 10^{17} electrons m^{-2}.

The equivalent thickness or slab thickness τ of the ionosphere is

$$\tau \equiv \frac{I}{\max n_e}.$$

This is the thickness of a hypothetical layer with uniform electron density equal to the maximum value of n_e and total electron content equal to I. Typically, τ is about 250 km.

11.24.2 Effects of Earth Curvature

The factor $\sec \chi$ in the formulas for ionization and absorption should be replaced by $Ch(x, \chi)$ to account for Earth curvature, where $x = (a + z)/H$, H = scale height, a = Earth radius, z = altitude, and χ is the zenith angle.

Curvature effects of the atmosphere are listed in Table 11.28.

Table 11.28. *The function* $Ch(x, \chi)$.[a]

		$Ch(x, \chi)$							
	$\chi =$	30°	45°	60°	75°	80°	85°	90°	95°
Q	$\sec \chi =$	1.155	1.414	2.000	3.864	5.76	11.47	∞	
50		1.148	1.389	1.901	3.228	4.19	5.82	8.93	16
100		1.151	1.401	1.946	3.473	4.70	7.07	12.58	30
200		1.153	1.407	1.972	3.646	5.10	8.28	17.76	68
400		1.154	1.411	1.985	3.742	5.38	9.33	25.09	220
800		1.154	1.412	1.993	3.800	5.55	10.15	35.46	1476
1000		1.155	1.413	1.994	3.812	5.59	10.35	39.65	

Note

[a] $Q \equiv (a + z_0)/H$, z_0 = altitude of maximum ionization rate.

11.24.3 International Reference Ionosphere (IRI)

IRI is an empirical reference model of ionospheric electron density, electron and ion temperatures, and ion composition recommended by COSPAR (Committee on Space Research) and URSI (International Union of Radio Science). It is updated bi-yearly; the 1990 model is used below. IRI is distributed by the National Space Science Data Center and World Data Center A for Rockets and Satellites (NSSDC/WDC-A-R&S) in Greenbelt, Md. IRI is available online on SPAN (Space Physics Analysis Network) now called NSI-DECNET (NASA Science Internet) and can be accessed interactively on NSSDC's Online Data Information Service (NODIS) account. Tables 11.29 to 11.36 give data about this IRI ionosphere model.

Table 11.29. *IRI-90 electron density.*[a]

z (km)	Noon		Midnight	
	n_e (m^{-3})	n_e/n_e F2(max)	n_e (m^{-3})	n_e/n_e F2(max)
65	8.3×10^7	3×10^{-4}	0	0
70	2.1×10^8	7×10^{-4}	0	0
75	3.7×10^8	1.3×10^{-3}	0	0
80	5.1×10^8	1.8×10^{-3}	0	0
85	1.1×10^9	3.7×10^{-3}	2.6×10^8	2.3×10^{-3}
90	1.2×10^{10}	4.0×10^{-2}	4.8×10^8	4.2×10^{-3}
95	5.2×10^{10}	1.8×10^{-1}	1.6×10^9	1.4×10^{-2}
100	1.1×10^{11}	3.7×10^{-1}	1.6×10^9	1.4×10^{-2}

Note

[a]Latitude = 45°, Longitude = 260°E, $R = 0$, Day = 6/22, F10.7 = 63.8, $\chi = 21.6°$ (Noon), 111.6° (Midnight).

Table 11.30. *IRI-90 electron density.*[a]

z (km)	Noon		Midnight	
	n_e (m^{-3})	n_e/n_e F2(max)	n_e (m^{-3})	n_e/n_e F2(max)
65	2.7×10^8	4×10^{-4}	0	0
70	6.7×10^8	1×10^{-3}	0	0
75	1.2×10^9	1.7×10^{-3}	0	0
80	1.6×10^9	2.4×10^{-3}	0	0
85	3.4×10^9	5.0×10^{-3}	2.6×10^8	6×10^{-4}
90	3.3×10^{10}	4.8×10^{-2}	4.8×10^8	1.1×10^{-3}
95	1.1×10^{11}	1.6×10^{-1}	2.8×10^9	6.2×10^{-3}
100	1.8×10^{11}	2.6×10^{-1}	3.9×10^9	8.8×10^{-3}

Note

[a]Latitude = 45°, Longitude = 260°E, $R = 150$, Day = 6/22, F10.7 = 193, $\chi = 21.6°$ (Noon), 111.6° (Midnight).

Table 11.31. *IRI-90 electron density.*[a]

	3/21	6/22	9/23	12/22
		$R = 0$		
n_e (70 km)	1.9×10^8	2.1×10^8	1.8×10^8	1.5×10^8
n_e (80 km)	4.5×10^8	5.1×10^8	4.4×10^8	3.8×10^8
n_e (90 km)	1.0×10^{10}	1.2×10^{10}	1.0×10^{10}	8.0×10^9
n_e (100 km)	9.5×10^{10}	1.1×10^{11}	9.5×10^{10}	7.0×10^{10}
		$R = 150$		
n_e (70 km)	5.9×10^8	6.7×10^8	5.8×10^8	1.6×10^8
n_e (80 km)	1.4×10^9	1.6×10^9	1.0×10^9	4.1×10^8
n_e (90 km)	2.9×10^{10}	3.3×10^{10}	2.9×10^{10}	9.0×10^9
n_e (100 km)	1.6×10^{11}	1.8×10^{11}	1.6×10^{11}	1.0×10^{11}

Note

[a]Time = Noon, Latitude = 45°, Longitude = 260°E, F10.7 = 63.8 ($R = 0$), 193 ($R = 150$), $\chi = 44.5°$(3/21), 21.6°(6/22), 45.4°(9/23), 68.5°(12/22), units of n_e are number m^{-3}.

Table 11.32. *IRI-90 electron density.*[a]

Latitude (°N)	45	60	75	90
		$R = 0$		
n_e (70 km)	1.5×10^8	1.2×10^8	0	—
n_e (80 km)	3.8×10^8	3.7×10^8	0	—
n_e (90 km)	8.0×10^9	6.0×10^9	4.7×10^8	—
n_e (100 km)	7.0×10^{10}	3.8×10^{10}	4.4×10^9	—
		$R = 150$		
n_e (70 km)	1.6×10^8	1.2×10^8	0	1.6×10^8
n_e (80 km)	4.1×10^8	3.7×10^8	0	3.8×10^8
n_e (90 km)	9.0×10^9	6.3×10^9	4.7×10^8	3.4×10^9
n_e (100 km)	1.0×10^{11}	5.4×10^{10}	6.3×10^9	4.0×10^9

Note

[a]Time = Noon, Longitude = 260°E, F10.7 = 63.8 ($R = 0$), 193 ($R = 150$), $\chi = 68.5°(45°N)$, $83.5°(60°N)$, $98.5°(75°N)$, $113.5°(90°N)$, Day = 12/22, units of n_e are number m^{-3}.

Table 11.33. *IRI-90 model ionosphere.*[a]

z (km)	n_e (m^{-3})	T_n (K)	T_i (K)	T_e (K)	O$^+$	H$^+$	He$^+$	O$_2^+$	NO$^+$
100	1.08×10^{11}	—	—	—	0	0	0	48	52
200	2.74×10^{11}	786	786	1419	23	0	0	21	56
300	2.40×10^{11}	821	1011	2689	99	0	0	0	1
400	1.11×10^{11}	822	1237	2831	100	0	0	0	0
500	4.63×10^{10}	822	1513	2835	96	4	0	0	0
600	2.41×10^{10}	822	1813	2846	88	10	1	0	0
700	1.61×10^{10}	822	2113	2936	80	18	2	0	0
800	1.28×10^{10}	822	2413	3042	69	28	3	0	0
900	1.14×10^{10}	822	2712	3148	59	37	4	0	0
1000	1.07×10^{10}	822	3012	3254	50	45	5	0	0

Note

[a]Latitude = 45°, Longitude = 260°E, $R = 0$, Day = 6/22, F10.7 = 63.8, $\chi = 21.6°$, Time = Noon, T_n = neutral temperature, T_i = ion temperature, T_e = electron temperature, ion composition is given in percent.

Table 11.34. *IRI-90 model ionosphere.*[a]

z (km)	n_e (m^{-3})	T_n (K)	T_i (K)	T_e (K)	O$^+$	H$^+$	He$^+$	O$_2^+$	NO$^+$
100	1.79×10^{11}	—	—	—	0	0	0	58	42
200	3.92×10^{11}	1187	1187	1421	59	0	0	6	35
300	6.85×10^{11}	1361	1361	2689	100	0	0	0	0
400	6.25×10^{11}	1385	1385	2831	100	0	0	0	0
500	4.63×10^{11}	1389	1513	2835	96	4	0	0	0
600	3.29×10^{11}	1390	1813	2846	88	10	1	0	0
700	2.51×10^{11}	1390	2113	2936	80	18	2	0	0
800	2.10×10^{11}	1390	2413	3042	69	28	3	0	0
900	1.89×10^{11}	1390	2712	3148	59	37	4	0	0
1000	1.77×10^{11}	1390	3012	3254	50	45	5	0	0

Note

[a]Latitude = 45°, Longitude = 260°E, $R = 150$, Day = 6/22, F10.7 = 193, $\chi = 21.6°$, Time = Noon, T_n = neutral temperature, T_i = ion temperature, T_e = electron temperature, ion composition is given in percent.

Table 11.35. *IRI-90 model ionosphere.*[a]

	Noon	Midnight	Noon	Midnight
	$R = 0$		$R = 150$	
n_e (m^{-3})	2.87×10^{11}	9.07×10^{10}	5.94×10^{11}	3.70×10^{10}
T_n (K)	815	692	1314	1090
T_i (K)	899	817	1314	1090
T_e (K)	2076	1010	2076	1090
%O$^+$	53[b]	42[c]	99	99
%H$^+$	0	0	0	0
%He$^+$	0	0	0	0

Note

[a]Latitude = 45°, Longitude = 260°E, Day = 6/22, F10.7 = 63.8 ($R = 0$), 193 ($R = 150$), T_n = neutral temperature, T_i = ion temperature, T_e = electron temperature, Altitude = 250 km, $\chi = 21.6°$ (Noon), 111.6° (Midnight).

[b]The other ions in this case are 46% NO$^+$ and 1% O$_2^+$.

[c]The other ions in this case are 57% NO$^+$ and 1% O$_2^+$.

Table 11.36. *IRI-90 model ionosphere.*[a]

Date	3/21	6/22	9/23	12/22
	$R = 0$			
n_e (m^{-3})	3.07×10^{11}	2.87×10^{11}	3.27×10^{11}	4.27×10^{11}
T_n (K)	777	815	773	687
T_i (K)	867	899	864	802
T_e (K)	1882	2076	1882	1780
%O$^+$	76	53	76	95
%H$^+$	0	0	0	0
%He$^+$	0	0	0	0
%O$_2^+$	0	1	0	0
%NO$^+$	23	46	23	5
	$R = 150$			
n_e (m^{-3})	1.15×10^{12}	5.94×10^{11}	8.60×10^{11}	1.63×10^{12}
T_n (K)	1221	1314	1219	1089
T_i (K)	1221	1314	1219	1089
T_e (K)	1882	2076	1882	1782
%O$^+$	95	99	95	95
%H$^+$	0	0	0	0
%He$^+$	0	0	0	0
%O$_2^+$	0	0	0	0
%NO$^+$	5	0	5	5

Note

[a]Latitude = 45°, Longitude = 260°, F10.7 = 63.8 ($R = 0$), 193 ($R = 150$), T_n = neutral temperature, T_i = ion temperature, T_e = electron temperature, Altitude = 250 km, $\chi = 44.5°$(3/21), 21.6°(6/22), 45.4°(9/23), 68.5°(12/22).

11.24.4   Irregularities of Ionospheric Behavior [31]

Storm	A severe departure from normal behavior lasting from one to several days.
Magnetic Storm	A magnetic storm consists of three phases: (1) an increase of magnetic field lasting a few hours; (2) a large decrease in the horizontal component of magnetic field building up to a maximum in about a day; (3) a recovery to normal over a few days. The initial phase (1) is caused by the compression of the magnet sphere by a burst of solar plasma. The main phase (2) is due to the ring current in the magnetosphere which flows around the Earth from east to west.
F-Region Ionospheric Storm	This storm is characterized by an initial positive phase of increasing electron density lasting a few hours followed by a main or negative phase of decreasing n_e. The ionosphere gradually returns to normal over one to several days during the recovery phase.

11.24.5 Sq Current System

The Sq current system is an ionospheric current system due to neutral winds blowing ions across magnetic field lines. The Sq winds and currents are driven by solar (S) tides under quiet (q) geomagnetic conditions. The winds have speeds of tens of meters per second and associated electric fields are a few millivolts per meter. The Sq currents produce daily magnetic field variations at the Earth's surface.

Node of EW currents is at latitude 38°.

Current between node and either pole or equator (at equinox and zero sunspots) $= 5.9 \times 10^4 A$.

11.24.6   Magnetic Indices [31]

K_p is based on the range of variation within 3 hour periods of the day observed in the records from about a dozen selected magnetic observatories. The K_p value for each 3 hour interval of the day is reported on a scale from 0 (very quiet) to 9 (very disturbed). Integer values are subdivided into thirds by use of the symbols $+$ and $-$. The K_p scale is quasi-logarithmic.

a_p—similar to K_p, but a linear scale of geomagnetic activity. The value of a_p is approximately half the range of variation of the most disturbed magnetic component measured in nT.

The relation between K_p and A_p is shown in Table 11.37.

Table 11.37. Relation between K_p and a_p.

K_p	0	1	2	3	4	5	6	7	8	9
a_p	0	3	7	15	27	48	80	140	240	400

A_p is a daily index, the average of a_p over a day.

AE is a geomagnetic index measuring the activity level of the auroral zone, particularly valuable as an indicator of magnetic substorms.

$\sum K_p$ is the sum of the eight K_p values over a UT day.

11.25 NIGHT SKY AND AURORA [17, 32–36]

The units for expressing the night sky brightness of spectroscopic features (lines or bands of restricted extent in wavelength) are:

$$1 \text{ Rayleigh} = R = 10^6 \text{ photons emitted in } 4\pi \text{ sr per cm}^2 \text{ vertical column per sec}$$
$$= 1.58 \times 10^{-7}\lambda^{-1} \text{ J m}^{-2} \text{ sr}^{-1} \text{ s}^{-1} \text{ at zenith } (\lambda \text{ in nm})$$
$$= 1.95 \times 10^{-7} \text{ nit for } \lambda = 555 \text{ nm.}$$
$$1 \text{ Photon} = 1.986 \times 10^{-16}\lambda^{-1} \text{ J } (\lambda \text{ in nm})$$
$$1 \, m_v = 10 \text{ star deg}^{-2} \text{ near 550 nm through clear atmosphere}$$
$$= 3.6 \times 10^{-2} R \text{ nm}^{-1}$$
$$= 7.1 \times 10^{-7} \text{ nit (for a bandwidth of 100 nm).}$$

Components of the night sky brightness are given in Table 11.38.

Table 11.38. *Night sky brightness.*

Source	Photographic 10th mag stars	Visual deg $^{-2}$	Photometry 10^{-5} nit
Airglow (near zenith)			
Atomic lines		40	3
Bands and continuum	30	50	4
Zodiacal light (away from zodiac)	60	100	6
Faint stars, $m > 6$ (galactic pole)	16	30	2
(mean sky)	48	95	7
(gal. equator)	140	320	23
Diffuse galactic light	10	20	1
Total brightness (zenith, mean sky)	145	290	21
(15° lat, mean sky)	190	380	28

Color index of night sky $C \cong 0.7$ ($C = B - V - 0.11$, where B is the apparent magnitude at 555 nm and V is the apparent magnitude at 435 nm).

Airglow variation with latitude: Generally brighter at middle and high latitudes than at low latitudes, a factor of ~ 2 increases with latitude for some emissions [35].

Airglow variation with solar cycle activity: Good correlation with sunspot activity for OI red line (630 nm), ambiguous evidence for variation in green line (557.7 nm) [35].

Van Rijin function: Off-zenith path length through a spherically symmetric airglow layer is increased relative to the zenith viewing by a factor

$$V = \left[1 - (r/(r + h))^2 \sin^2 z \right]^{-1/2},$$

where r is the Earth's radius, h is the height of the emitting layer above the Earth's surface, and z is the zenith angle.

The full moon brightness is 1100 tenth magnitude stars per square degree in the photographic spectral region and 100 in the visual band. For other phases of the Moon multiply by $\phi(\alpha)$, where α is the phase angle, the angle between the Sun and Earth seen from the Moon, and $\phi(\alpha)$ is the phase law or the change of the Moon's brightness with $\alpha(\phi(0) = 1)$ [17].

The sky brightness during twilight is given in Table 11.39.

Table 11.39. *Variation of sky brightness throughout twilight relative to* 0° *solar depression angle* [1].

Solar depression angle	0°	6°	12°	18°
Log relative brightness	0	−2.7	−4.7	−5.8

Reference
1. Allen, C.W. 1973, *Astrophysical Quantities*, 3rd ed. (Athlone Press, London)

Table 11.40 lists the night sky emissions from various components.

Table 11.40. *Spectral emissions in the night sky* [1, 2, 3, 4, 5].[a]

Emitter	λ, etc. nm	Intensity Night R	Twilight R	Aurora kR	Remarks
OI	557.7	250	180	100	< 100 R to > 500 R night to night variation
OI	630.0–636.4	100	1000	2–100	Sporadic enhancements in tropical nightglow
OI	297.2			6	ICB III Aurora
OI	130.4–135.6	150		30 ± 20	Observed from satellites, ICB III Aurora
OI	777.4			10	
OI	844.6		13	12	
NI	1040			6	ICB III Aurora
NI	346.6			1	ICB III Aurora
NI	519.9		10	0.1–2	ICB III Aurora
NII	VIS. and FUV			45	
NaI	589.0–589.6				NaD, Strong seasonal variation
	summer	30	1000	1	
	winter	200	5000	1	
HI	656.3	15		10	Hα
HI	121.6	2500		100	Lα
CaII	393.3–396.7		150		
LiI	670.8		30		
N₂	IR			880	1st positive, ICB III Aurora
N₂	UV			110	2nd positive, ICB III Aurora
N₂	FUV			200–400	LBH bands, ICB III Aurora
N₂	Blue	100		55	VK bands, ICB III Aurora
N₂	EUV			2000	BH, WK, ICB III Aurora, rough value deduced from photometer data
N₂⁺	NUV, VIS.		1000	150	1st negative, ICB III Aurora
N₂⁺	630–890			630	M, ICB III Aurora
O₂	300–400	1500			Hertzberg bands
O₂	864.5	500		60	Atm. (0–1), ICB III Aurora
O₂	1270	6000		1200	Atm. (0–0), not seen at ground, ICB III Aurora
O₂	1580		20 000	2500	IR Atm, ICB III Aurora
O₂⁺	VIS., IR			26	1st negative, ICB III Aurora
OH	1580	150 000			(4–2) Strongest bands are in NIR
OH	VIS.	130			(5–0) (7–1) (8–2) (9–3) bands
OH	8342	2000			(6, 2) band
OH	Total	4.5 × 10⁶			
NO₂	500–650	250			Nightglow continuum
NOγ	MUV			20–60	ICB III Aurora
HeI	1083		1000		

Note
[a]LBH = Lyman–Birge–Hopfield, M = Meinel, VK = Vegard–Kaplan, WK = Watson–Koontz, BH = Birge–Hopfield, ICB III = OI(557.7) = 100 kR [1, 2, 3, 4].

References
1. Allen, C.W. 1973, *Astrophysical Quantities*, 3rd ed. (Athlone Press, London)
2. Vallance Jones, A. 1974, *Aurora* (Reidel, Boston)
3. Roach, F.E., & Gordon, J.L. 1973, *The Light of the Night Sky*, (Reidel, Boston)
4. Krassovsky, V.I. et al. 1962, *Planet. Space Sci.*, **9**, 883
5. Chamberlain, J.W. 1961, *Physics of the Aurora and Airglow*, (Academic Press, New York)

Zone of maximum auroral activity $= 60$–$75°$ geomagnetic latitude [32].

Seasonal variation: Minima in auroral frequency at solstices, maxima at equinoxes (approximately a factor of 2 increase from minima to maxima as seen from Yerkes Observatory) [36].

Table 11.41 gives details of the types of aurorae.

Table 11.41. *Auroral heights* [1, 2].

	Aurora	Height
Type c (normal aurora)		
	Lower border strong aurora	95 km
	Lower border weak aurora	114 km
	Average value	105–108 km
	Average height of maximum emission	110 km
	Vertical extents	20–40 km
	Upper extremity	frequently > 200 km
Sunlit upper extremity		700 km (1000 km in extreme cases)
Type b: red lower border		80–100 km
Type d (red overall) lower border		250 km

References
1. Allen, C.W. 1973, *Astrophysical Quantities*, 3rd ed. (Athlone Press, London)
2. Meinel, A.B. et al. 1954, *J. Geophys. Res.*, **59**, 407

The proton input needed to produce auroral Hα is given in Table 11.42.

Table 11.42. *Flux of monoenergetic protons required to produce* 10 kR *of* Hα *in the zenith* [1].

Initial energy keV	Minimum penetration height km	Hα photons	Proton flux cm^{-2} s^{-1}	Total incident energy flux eV cm^{-2} s^{-1}
130	100	60	1.6×108	2.1×10^{13}
27	110	27	5×10^8	1.4×10^{13}
8.5	120	7	14×10^8	1.2×10^{13}

Reference
1. Chamberlain, J.W. 1961, *Physics of the Aurora and Airglow* (Academic Press, New York)

Auroral International Coefficients of Brightness:

I.C.B.	I	557.7 brightness $= 1$ kR $\approx 10^{-4}$ nit,
	II	557.7 brightness $= 10$ kR $\approx 10^{-3}$ nit,
	III	557.7 brightness $= 100$ kR $\approx 10^{-2}$ nit,
	IV	557.7 brightness $= 1000$ kR $\approx 10^{-1}$ nit.

11.26   GEOMAGNETISM [37–39]

The geomagnetic field arises from sources both interior and exterior to the solid Earth, including electric currents in the liquid outer core and the ionosphere and the magnetization of crustal rocks. Models of the global magnetic field are intended to describe the field originating in the core (the main field). The description of the main field is based on a spherical harmonic description of the potential $V(r, \theta, \phi, t)$ for magnetic induction $\underline{B}(r, \theta, \phi, t)$

$$\underline{B} = -\nabla V,$$

where r, θ, ϕ are spherical polar coordinates and t is time. The spherical harmonic expansion of V is

$$V(r, \theta, \phi, t) = a \sum_{l=1}^{L} \sum_{m=0}^{l} \left(\frac{a}{r}\right)^{l+1} \left\{ g_l^m(t) \cos m\phi + h_l^m(t) \sin m\phi \right\} \tilde{P}_l^m(\cos \theta),$$

where a is the mean radius of the Earth ($a = 6371.2$ km), L is the truncation level of the expansion, and the $\tilde{P}_l^m(\cos \theta)$ are Schmidt quasi-normalized associated Legendre functions, i.e., the integral of $\tilde{P}_l^m$ squared over all solid angles is $4\pi/(2l+1)$. The quantities $g_l^m(t)$ and $h_l^m(t)$ are known as Gauss geomagnetic coefficients; they vary with time over a broad range of time scales from less than a year to hundreds of millions of years. The core dynamo responsible for generating the main magnetic field is fundamentally time dependent in its behavior. If the small electrical conductivity of the mantle is neglected, then the above representation of the main geomagnetic field can be used to extrapolate the surface field down to the core–mantle boundary.

The components of the magnetic field are given by

$$B_r = -\frac{\partial V}{\partial r} = \sum_{l=1}^{L} \sum_{m=0}^{l} (l+1) \left(\frac{a}{r}\right)^{l+2} \left\{ g_l^m(t) \cos m\phi + h_l^m(t) \sin m\phi \right\} \tilde{P}_l^m(\cos \theta),$$

$$B_\theta = -\frac{1}{r}\frac{\partial V}{\partial \theta} = -\sum_{l=1}^{L} \sum_{m=0}^{l} \left(\frac{a}{r}\right)^{l+2} \left\{ g_l^m(t) \cos m\phi + h_l^m(t) \sin m\phi \right\} \frac{d\tilde{P}_l^m}{d\theta}(\cos \theta),$$

$$B_\phi = -\frac{1}{r \sin \theta}\frac{\partial V}{\partial \phi} = \sum_{l=1}^{L} \sum_{m=0}^{l} \left(\frac{a}{r}\right)^{l+2} \left\{ g_l^m(t) \sin m\phi - h_l^m(t) \cos m\phi \right\} \frac{m\tilde{P}_l^m(\cos \theta)}{\sin \theta}.$$

Magnetic field observations are generally described in terms of the quantities:

$X = -B_\theta =$ north magnetic field component,
$Y = B_\phi =$ east magnetic field component,
$Z = -B_r =$ vertically downward magnetic field component,
$H = (X^2 + Y^2)^{1/2} =$ horizontal magnetic field intensity,
$F = (X^2 + Y^2 + Z^2)^{1/2} =$ total magnetic field intensity,
$I = \arctan(Z/H) =$ magnetic inclination,
$D = \arctan(Y/X) =$ magnetic declination.

Historically, there has been much discussion of the westward drift of the main field or components thereof, particularly the nondipole part of the field (see below). While some features of the field may participate in a westward drift, the secular variation of the main field is more complex than a simple westward drift.

11.26.1 Geomagnetic Dipole

The contributions to V of the $l = 1$ terms in the spherical harmonic representation of V are from magnetic dipoles situated at $r = 0$ and oriented along the coordinate axes.

$\dfrac{a^3 \cos \theta}{r^2} g_1^0$ is the potential of a magnetic dipole in the $+z$-direction (along the Earth's rotation axis),

$\dfrac{a^3 \cos \phi \sin \theta}{r^2} g_1^1$ is the potential of a magnetic dipole in the $+x$-direction (along the Greenwich meridian),

$\dfrac{a^3 \sin \phi \sin \theta}{r^2} h_1^1$ is the potential of a magnetic dipole in the $+y$-direction.

The total dipole potential V^{dipole} is the sum of the above terms.

The total dipole magnetic field $\underline{B}^{\text{dipole}} = -\nabla V^{\text{dipole}}$ has components

$$B_r^{\text{dipole}} = \frac{2a^3}{r^3} \left(g_1^0 \cos \theta + \sin \theta (g_1^1 \cos \phi + h_1^1 \sin \phi) \right),$$

$$B_\theta^{\text{dipole}} = \frac{a^3}{r^3} \left(g_1^0 \sin \theta - \cos \theta (g_1^1 \cos \phi + h_1^1 \sin \phi) \right),$$

$$B_\phi^{\text{dipole}} = \frac{a^3}{r^3} (g_1^1 \sin \phi - h_1^1 \cos \phi).$$

The magnetic dipole moment $\underline{m}$ has magnitude

$$m = 4\pi a^3 \left((g_1^0)^2 + (g_1^1)^2 + (h_1^1)^2 \right)^{1/2}.$$

The magnetic dipole moment pierces the surface of the Earth at colatitude θ_m and longitude ϕ_m, given by

$$\theta_m = \arctan \left\{ \frac{((g_1^1)^2 + (h_1^1)^2)^{1/2}}{g_1^0} \right\}, \qquad \phi_m = \arctan \left(\frac{h_1^1}{g_1^1} \right).$$

Table 11.43 lists the values of $m/4\pi a^3$, θ_m, and ϕ_m for the years 1945–1990.

The orientation and magnitude of the centered, tilted, magnetic dipole from the $l = 1$ terms in the spherical harmonic representation of the main field is given in Table 11.43.

Table 11.43. *Orientation and magnitude of the centered, tilted, magnetic dipole.*

	1945	1950	1955	1960	1965	1970	1975	1980	1985	1990
$(m/4\pi a^3)$, $10^4 nT$	3.122	3.118	3.113	3.104	3.095	3.083	3.070	3.057	3.043	3.032
Colatitude of Geomagnetic Pole (θ_m, degrees)	11.53	11.53	11.54	11.49	11.47	11.41	11.31	11.19	11.03	10.87
Longitude of Geomagnetic Pole (ϕ_m, degrees)[a]	291.5	291.2	290.8	290.5	290.1	289.8	289.5	289.2	289.1	288.9

Note

[a] East from the Greenwich meridian.

The time rate of change of the magnetic dipole is obtained by differentiating the above expressions for m, θ_m, and ϕ_m with respect to time.

11.26.2 Eccentric Dipole

The Cartesian coordinates (x_0, y_0, z_0) of the eccentric dipole that best represents the main field are given by

$$x_0 = \frac{a(L_1 - g_1^1 T)}{3(m/4\pi a^3)^2},$$

$$y_0 = \frac{a(L_2 - h_1^1 T)}{3(m/4\pi a^3)^2},$$

$$z_0 = \frac{a(L_0 - g_1^0 T)}{3(m/4\pi a^3)^2},$$

where

$$L_0 = 2g_1^0 g_2^0 + \sqrt{3}(g_1^1 g_2^1 + h_1^1 h_2^1),$$
$$L_1 = g_1^1 g_2^0 + \sqrt{3}(g_1^0 g_2^1 + g_1^1 g_2^2 + h_1^1 h_2^2),$$
$$L_2 = -h_1^1 g_2^0 + \sqrt{3}(g_1^0 h_2^1 + g_1^1 h_2^2 - h_1^1 g_2^2),$$
$$T = \frac{L_0 g_1^0 + L_1 g_1^1 + L_2 h_1^1}{4(m/4\pi a^3)^2}.$$

11.26.3 Dipole Coordinate System

A coordinate system with its z-axis along the direction of the centered, tilted dipole is the dipole coordinate system or the geomagnetic coordinate system. The pole of this coordinate system is located at θ_m, ϕ_m, given above. This is the geomagnetic pole or dipole pole. If $\underline{x}_d$ is a vector in the dipole coordinate system and $\underline{x}$ is a vector in the standard coordinate system, then

$$\underline{x}_d = \underline{\underline{R}} \cdot \underline{x},$$

where $\underline{\underline{R}}$ is the rotation matrix with elements

$$\underline{\underline{R}} = \begin{bmatrix} \cos\theta_m \cos\phi_m & \cos\theta_m \sin\phi_m & -\sin\theta_m \\ -\sin\phi_m & \cos\phi_m & 0 \\ \sin\theta_m \cos\phi_m & \sin\theta_m \sin\phi_m & \cos\theta_m \end{bmatrix}.$$

11.26.4 Magnetic Dip-Poles

A magnetic dip-pole is a location at which the horizontal magnetic field is zero. At the north and south dip-poles the magnetic potential has its maximum and minimum values, respectively. Table 11.44 gives the coordinates of the dip-poles at different times.

Table 11.44. *Coordinates of the magnetic dip-poles.*

Year	Latitude (N)	Longitude (W)
	North Dip-Pole	
1831.4	70°05′	96°46′
1904.5	70°30′	95°30′
1948.0	73°00′	100°00′
1962.5	75°06′	100°48′
1973.5	76°00′	100°36′
	South Dip-Pole	
1841.0	75°05′	154°08′
1899.8	72°40′	152°30′
1909.0	72°55′	155°16′
1912.0	71°10′	150°45′
1931.0	70°20′	149°00′
1952.0	68°42′	143°00′
1962.1	67°30′	140°00′

11.26.5 Centered, Tilted Dipole Field [39]

Vertical magnetic field at geomagnetic poles, at $r = a$,

$$= 2\left(\frac{m}{4\pi a^3}\right) = 6.064 \times 10^4 nT.$$

Horizontal magnetic field at geomagnetic equator, at $r = a$,

$$= \frac{m}{4\pi a^3} = 3.032 \times 10^4 nT.$$

In the dipole coordinate system

$$V = \frac{m}{4\pi r^2}\cos\theta_{mag}$$

$$= \frac{a^3 \cos\theta_{mag}}{r^2}\left(\frac{m}{4\pi a^3}\right),$$

$$V(r = a) = \left(\frac{m}{4\pi a^3}\right)a\cos\theta_{mag},$$

where θ_{mag} is the magnetic colatitude. Numerical values are for the IGRF (1991 Revision).

11.27 METEORITES AND CRATERS [17, 40–44]

Classes of meteorites (natural objects of extraterrestrial origin that survive passage through the atmosphere) and statistics on falls and finds are given in Table 11.45. Falls refer to meteorites that were seen to fall; they are usually recovered soon after fall. Finds refer to meteorites that were not seen to fall but were found and recognized subsequently. Meteorites are broadly classified into stones, irons (pure metal, essentially nickel–iron alloy), and stony–irons. Additional classifications are required because of the great diversity of objects in these broad classes. Stony meteorites are divided into chondrites (meteorites containing distinctive features known as chondrules with compositions very similar to that of the solar photosphere for all but the most volatile elements) and achondrites (differentiated meteorites with compositions considerably different from the Sun).

Table 11.45. *Meteorite classes and statistics on falls and finds* [1].

Class	Falls	Fall frequency (%)[b]	Finds[a] Non-Antarctic	Finds[a] Antarctic[c]
Chondrites				
CI	5	0.60	0	0
CM	18	2.2	5	34
CO	5	0.60	2	6
CV	7	0.84	4	5
H	276	33.2	347	671
L	319	38.3	286	224
LL	66	7.9	21	42
EH	7	0.84	3	6
EL	6	0.72	4	1
Other	3	0.36	3	3
Anchondrites				
Eucrites	25	3.0	8	13
Howardites	18	2.2	3	4
Diogenites	9	1.1	0	9
Ureilites	4	0.48	6	9
Aubrites	9	1.1	1	17
Shergottites	2	0.24	0	2
Nakhlites	1	0.12	2	0
Chassignites	1	0.12	0	0
Anorthositic breccias	0	0	0	1
Stony-irons				
Mesosiderites	6	0.72	22	2
Pallasites	3	0.36	34	1
Irons				
IAB	6	0.73	97	4
IC	0	0.08	11	0
IIAB	5	0.45	60	6
IIC	0	0.05	7	0
IID	3	0.09	12	0
IIE	1	0.10	13	0
IIF	1	0.03	4	0
IIIAB	8	1.42	189	0
IIICD	2	0.14	19	0
IIIE	0	0.10	13	0
IIIF	0	0.05	6	0
IVA	3	0.39	52	1
IVB	0	0.09	12	0
Other irons	13	1.32	175	0

Notes

[a] Data for finds are given to provide an indication of available material. The unusual conditions in the Antarctic favor the recovery of large numbers of meteorites without the selection biases of non-Antarctic regions (e.g., in non-Antarctic regions, stony meteorites, especially anchondrites are more easily confused with terrestrial rocks than iron meteorites). The statistics for Antarctic finds, therefore, more closely resemble those of falls than non-Antarctic finds. In fact, several rarer classes are overrepresented in the Antarctic collections.

[b] Iron–meteorite fall statistics calculated from finds, scaled to percentage of total iron–meteorite falls.

[c] US finds in the Antarctic. In addition, > 6000 meteorites have been recovered from the Antarctic by Japanese teams.

Reference

1. Sears, D.W.G., & Dodd, R.T. 1988, in *Meteorites and the Early Solar System*, edited by J.F. Kerridge and M.S. Matthews (University of Arizona Press, Tucson, Arizona), pp. 3–31

11.27.1 Meteorite Infall Rates

Fall of meteorites large enough to be seen and found, ≈ 2 meteorites per day over the whole Earth.
The cumulative flux of meteoroids F in the vicinity of the Earth–Moon system is given by

$$F\left(\frac{\#}{10^6 \text{ km}^2 \text{ yr}}\right) = 7.9(m \text{ (kg)})^{-1.16}, \qquad 10^{-10} < m < 10^5 \text{ kg},$$

where F is the number of meteoroids with mass greater than m per 10^6 km^2 per year. Accordingly, meteoroids with masses greater than about 6 kg will arrive in the vicinity of the Earth–Moon system at a rate of about one per 10^6 km^2 per year.

11.27.2 Meteorite Masses

The most probable size of found meteorites for iron is 15 kg and for stones 3 kg. Meteoroid masses before entry to the Earth's atmosphere are ≈ 100 kg. The mass of the greatest known meteorite (Hoba, an iron meteorite) is 6×10^4 kg.

11.27.3 Cratering Efficiency

Mass displaced from crater/mass of impactor = cratering efficiency

$$= 0.2\left(\frac{1.61gL}{v_i^2}\right)^{-0.65},$$

$$g = \text{gravity}(\text{m s}^{-2}),$$
$$L = \text{projectile diameter (m)},$$
$$v_i = \text{impact velocity (m s}^{-1}).$$

11.27.4 Crater Diameter Scaling Relations for Terrestrial Craters [42]

$$D = 0.0133W^{1/3.4} + 1.51\rho_P^{1/2}\rho_T^{-1/2}L,$$
$$D = 1.8\rho_P^{0.11}\rho_T^{-1/3}g^{-0.2}L^{0.13}W^{0.22},$$
$$D = 0.2\rho_P^{1/6}\rho_T^{-1/2}W^{0.28}, \qquad D \gtrsim 1 \text{ km}.$$

All units in the above formulas are SI.

$$D = \text{diameter of a transient impact crater},$$
$$\rho_P = \text{impactor density},$$
$$\rho_T = \text{target density},$$
$$W = \text{impactor energy},$$
$$L = \text{impactor diameter}.$$

Formulas valid for vertical impacts.

$$\text{Energy of 1 kiloton of TNT} = 4.2 \times 10^{12} \text{ J}.$$

11.27.5 Crater Dimensions

Rim height h_R above original ground surface of many fresh (unrelaxed) lunar, terrestrial, explosion, and laboratory impact craters with diameter (rim to rim), $D \lesssim 15$ km,

$$h_R \text{ (m)} = 0.036(D \text{ (m)})^{1.014}.$$

For craters with $D > 15$ km on the Moon (collapsed craters)

$$h_R \text{ (m)} = 0.236(D \text{ (m)})^{0.399}.$$

Crater depth H (rim to floor) of fresh lunar craters with diameter $D \lesssim 11$ km

$$H \text{ (m)} = 0.196(D \text{ (m)})^{1.01}.$$

Crater depth of collapsed lunar craters

$$H \text{ (m)} = 1.044(D \text{ (m)})^{0.301} \quad 11 \text{ km} < D < 400 \text{ km}.$$

Crater depth of simple (relatively young) terrestrial impact craters (e.g., Meteor Crater, Arizona)

$$H \text{ (m)} = 0.14(D \text{ (m)})^{1.02}.$$

Crater depth of collapsed or complex terrestrial impact craters

$$H \text{ (m)} = 0.27(D \text{ (m)})^{0.16}.$$

Estimated cratering rate from relatively young (< 120 Myr) large craters on the North American and European cratons

$$(5.4 \pm 2.7) \times 10^{-15} \text{ km}^{-2} \text{ yr}^{-1} \quad \text{for} \quad D \geq 20 \text{ km}.$$

Estimated cratering rate from smaller craters on a nonglaciated area in the U.S.

$$(2.2 \pm 1.1) \times 10^{-14} \text{ km}^{-2} \text{ yr}^{-1} \quad \text{for} \quad D \geq 10 \text{ km}.$$

Important impact craters are listed in Table 11.46.

Table 11.46. *Terrestrial impact structures* [1].

Name	Latitude	Longitude	Diameter (km)	Age (Myr)
Amguid, Algeria	26°05′N	004°23′E	0.45	< 0.1
Aouelloul, Mauritania[a]	20°15′N	012°41′W	0.37	3.1 ± 0.3
Araguainha Dome, Brazil	16°46′S	052°59′W	40.	< 250
Azuara, Spain	41°01′N	000°55′W	30.	< 130
Barringer, Arizona, USA[a]	35°02′N	111°01′W	1.2	0.025
Bee Bluff, Texas, USA	29°02′N	099°51′W	2.4	< 40
Beyenchime-Salaatin, Russia	71°50′N	123°30′E	8.	< 65
Bigatch, Kazakhstan	48°30′N	082°00′E	7.	6 ± 3
Boltysh, Ukraine	48°45′N	032°10′E	25.	100 ± 5
Bosumtwi, Ghana	06°32′N	001°25′W	10.5	1.3 ± 0.2
Boxhole, Northern Territory, Australia[a]	22°37′S	135°12′E	0.18	—
B.P. Structure, Libya	25°19′N	024°20′E	2.8	< 120
Brent, Ontario, Canada[a]	46°05′N	078°29′W	3.8	450 ± 30

Table 11.46. *(Continued.)*

Name	Latitude	Longitude	Diameter (km)	Age (Myr)
Campo del Cielo, Argentina (20)[ab]	27°38′S	061°42′W	0.09	—
Carswell, Saskatchewan, Canada	58°27′N	109°30′W	37.	117 ± 8
Charlevoix, Quebec, Canada	47°32′N	070°18′W	46.	360 ± 25
Clearwater Lake East, Quebec, Canada	56°05′N	074°07′W	22.	290 ± 20
Clearwater Lake West, Quebec, Canada	56°13′N	074°30′W	32.	290 ± 20
Connolly Basin, Western Australia, Australia[a]	23°32′S	124°45′E	9.	< 60
Crooked Creek, Missouri, USA	37°50′N	091°23′W	5.6	320 ± 80
Dalgaranga, Western Australia, Australia[a]	27°43′S	117°05′E	0.21	—
Decaturville, Missouri, USA	37°54′N	092°43′W	6.	< 300
Deep Bay, Saskatchewan, Canada	56°24′N	102°59′W	12.	100 ± 50
Dellen, Sweden	61°55′N	016°32′E	15.	109.6 ± 1
Eagle Butte, Alberta, Canada	49°42′N	110°30′W	10.	< 65
El'gygytgyn, Russia	67°30′N	172°05′E	23.	3.5 ± 0.5
Flynn Creek, Tennessee, USA	36°17′N	085°40′W	3.8	360 ± 20
Glover Bluff, Wisconsin, USA	43°58′N	089°32′W	6.	< 500
Goat Paddock, Western Australia, Australia	18°20′S	126°40′E	5.	< 50
Gosses Bluff, Northern Territory, Australia	23°50′S	132°19′E	22.	142.5 ± 0.5
Gow Lake, Saskatchewan, Canada	56°27′N	104°29′W	5.	< 250
Gusev, Russia	≈ 54°N	≈ 22°E	3.	65
Haughton, Northwest Territories, Canada	75°22′N	089°40′W	20.	21.5 ± 1.2
Haviland, Kansas, USA[a]	37°35′N	099°10′W	0.011	—
Henbury, Northern Territory, Australia (14)[ab]	24°34′S	133°10′E	0.15	—
Holleford, Ontario, Canada	44°28′N	076°38′W	2.	550 ± 100
Ile Rouleau, Quebec, Canada	50°41′N	073°53′W	4.	< 300
Ilintsy, Ukraine	49°06′N	029°12′E	4.5	395 ± 5
Ilumetsy, Estonia	57°58′N	025°25′E	0.08	0.002
Janisjärvi, Russia	61°58′N	030°55′E	14.	698 ± 22
Kaalijärvi, Estonia (7)[ab]	58°24′N	022°40′E	0.11	0.004
Kaluga, Russia	54°30′N	036°15′E	15.	380 ± 10
Kamensk, Russia	48°20′N	040°15′E	25.	65
Kara, Russia[a]	69°10′N	065°00′E	60.	57 ± 9
Karla, Russia	57°54′N	048°00′E	10.	10
Kelly West, Northern Territory, Australia	19°30′S	132°50′E	2.5	< 550
Kentland, Indiana, USA	40°45′N	087°24′W	13.	< 300
Kjardla, Estonia	57°00′N	022°42′E	4.	510 ± 30
Kursk, Russia	51°40′N	036°00′E	5.	250 ± 80
Lac Couture, Quebec, Canada	60°08′N	075°20′W	8.	425 ± 25
Lac La Moinerie, Quebec, Canada	57°26′N	066°36′W	8.	400 ± 50
Lappajärvi, Finland[a]	63°09′N	023°42′E	14.	77 ± 4
Liverpool, Northern Territory, Australia	12°24′S	134°03′E	1.6	150 ± 70
Logancha, Russia	65°30′N	095°50′E	20.	50 ± 20
Logoisk, Byelorussia	54°12′N	027°48′E	17.	40 ± 5
Lonar, India	19°58′N	076°31′E	1.83	0.05

Table 11.46. (*Continued.*)

Name	Latitude	Longitude	Diameter (km)	Age (Myr)
Machi, Russia (5)[b]	57°30'N	116°00'E	0.3	< 1
Manicouagan, Quebec, Canada	51°23'N	068°42'W	100.	210 ± 4
Manson, Iowa, USA	42°35'N	094°31'W	32.	61 ± 9
Middlesboro, Kentucky, USA	36°37'N	083°44'W	6.	< 300
Mien, Sweden[a]	56°25'N	014°52'E	5.	118 ± 3
Misarai, Lithuania	54°00'N	023°54'E	5.	395 ± 145
Mishina Gora, Russia	58°40'N	028°00'E	2.5	< 360
Mistastin, Newfoundland, and Labrador, Canada	55°53'N	063°18'W	28.	38 ± 4
Monturaqui, Chile[a]	23°56'S	068°17'W	0.46	1
Morasko, Poland (7)[ab]	52°29'N	016°54'E	0.1	0.01
New Quebec, Quebec, Canada	61°17'N	073°40'W	3.2	< 5
Nicholson Lake, Northwest Territories, Canada[a]	62°40'N	102°41'W	12.5	< 400
Oasis, Libya	24°35'N	024°24'E	11.5	—
Obolon', Ukraine	49°30'N	032°55'E	15.	215 ± 25
Odessa, Texas, USA (3)[ab]	31°45'N	102°29'W	0.168	—
Ouarkziz, Algeria	29°00'N	007°33'W	3.5	< 70
Piccaninny, Western Australia, Australia	17°32'S	128°25'E	7.	< 360
Pilot Lake, Northwest Territories, Canada	60°17'N	111°01'W	6.	440 ± 2
Popigai, Russia	71°30'N	111°00'E	100.	39 ± 9
Puchezh-Katunki, Russia	57°06'N	043°35'E	80.	183 ± 5
Red Wing Creek, North Dakota, USA	47°36'N	103°33'W	9.	200
Riacho Ring, Brazil	07°43'S	046°39'W	4.	—
Ries, Germany[a]	48°53'N	010°37'E	24.	14.8 ± 0.7
Rochechouart, France[a]	45°30'N	000°56'E	23.	160 ± 5
Rogozinskaja, Russia	58°18'N	062°00'E	8.	55 ± 5
Rotmistrovka, Ukraine	49°00'N	032°00'E	2.5	140 ± 20
Sääksjärvi, Finland[a]	61°23'N	022°25'E	5.	< 330
Saint Martin, Manitoba, Canada	51°47'N	098°32'W	23.	225 ± 40
Serpent Mound, Ohio, USA	39°02'N	083°24'W	6.4	< 320
Serra da Canghala, Brazil	08°05'S	046°52'W	12.	< 300
Shunak, Kazakhstan	42°42'N	072°42'E	2.5	12
Sierra Madera, Texas, USA	30°36'N	102°55'W	13.	100
Sikhote Alin, Russia (122)[ab]	46°07'N	134°40'E	0.0265	—
Siljan, Sweden	61°02'N	014°52'E	52.	368 ± 1
Slate Island, Ontario, Canada	48°40'N	087°00'W	30.	< 350
Sobolev, Russia[a]	46°18'N	138°52'E	0.05	—
Söderfjärden, Finland	63°02'N	021°35'E	5.5	< 600
Spider, Western Australia, Australia	16°30'S	126°00'E	5.	—
Steen River, Alberta, Canada	59°31'N	117°38'W	25.	95 ± 7
Steinheim, Germany	48°41'N	010°04'E	3.4	14.8 ± 0.7
Strangways, Northern Territory, Australia[a]	15°12'S	133°35'E	24.	< 472
Sudbury, Ontario, Canada	46°36'N	081°11'W	140.	1850 ± 150
Tabun-Khara-Obo, Mongolia[a]	44°06'N	109°36'E	1.3	< 30
Talemzane, Algeria	33°19'N	004°02'E	1.75	< 3
Teague, Western Australia, Australia	25°50'S	120°55'E	28.	1685 ± 5
Tenoumer, Mauritania	22°55'N	010°24'W	1.9	2.5 ± 0.5
Ternovka, Ukraine	48°01'N	033°05'E	8.	330 ± 30
Tin Bider, Algeria	27°36'N	005°07'E	6.	< 70
Ust-Kara, Russia	69°18'N	065°18'E	25.	57 ± 9

Table 11.46. *(Continued.)*

Name	Latitude	Longitude	Diameter (km)	Age (Myr)
Upheaval Dome, Utah, USA	38°26′N	109°54′W	5.	—
Veevers, Western Australia, Australia[a]	22°58′S	125°22′E	0.08	< 450
Vepriaj, Lithuania	55°06′N	024°36′E	8.	160 ± 30
Vredefort, South Africa	27°00′S	027°30′E	140.	1970 ± 100
Wabar, Saudi Arabia (2)[ab]	21°30′N	050°28′E	0.097	—
Wanapitei Lake, Ontario, Canada[a]	46°44′N	080°44′W	8.5	37 ± 2
Wells Creek, Tennessee, USA	36°23′N	087°40′W	14.	200 ± 100
West Hawk Lake, Manitoba, Canada	49°46′N	095°11′W	2.7	100 ± 50
Wolf Creek, Western Australia, Australia[a]	19°10′S	127°47′E	0.85	—
Zeleny Gai, Ukraine	48°42′N	035°54′E	1.4	120 ± 20
Zhamanshin, Kazakhstan	48°24′N	060°48′E	10.	0.75 ± 0.06

Notes

[a] Structures with meteoritic fragments or geochemical anomalies considered to have a meteoritic source.

[b] Sites with multiple craters, with (n) indicating number of craters. Diameter given corresponds to largest crater.

Reference

1. Grieve, R.A.F. 1987, *Ann. Rev. Earth Planet. Sci.*, **15**, 245

There is increasing acceptance of the importance of impacts in the evolution of the Earth and planets. Examples of possible impact-related events in the Earth's history include the formation of the Moon by a Mars-sized impactor early in the Earth's evolution and the Cretaceous–Tertiary extinctions (about 65 Ma) by the effects of an impactor of mass about 10^{15} kg, energy about 10^{23} J, and diameter about 100 km.

REFERENCES

1. Lerch, F.J. et al. 1992, NASA Technical Memorandum **104555** Geopotential Models of Earth from Satellite Tracking and Altimeter and Surface Gravity Observations GEM-T3 and GEM-T3S

2. Cazenave, A. 1994, *American Geophysical Union Handbook of Physical Constants* (American Geophysical Union, Washington, DC)

3. International Earth Rotation Service Standards, Central Bureau of IERS (Observatoire de Paris)

4. ETOPO 5 Data Base (distributed by NOAA), 1986, (National Geophysical Data Center, Boulder, CO)

5. Turcotte D.L., & Schubert, G. 1982, *Geodynamics* (Wiley, New York)

6. Dickey, J.O. 1994, *American Geophysical Union Handbook of Physical Constants* (American Geophysical Union, Washington, DC)

7. *Explanatory Supplement to the Astronomical Almanac*, edited by P.K. Seidelmann 1992, (University Science Books, Mill Valley, CA)

8. Wahr, J. 1994, *American Geophysical Union Handbook of Physical Constants* (American Geophysical Union, Washington, DC)

9. Harland, W.B. et al. 1990, *A Geologic Timescale 1989* (Cambridge University Press, Cambridge)

10. Imbrie, J. 1985, *J. Geol. Lond.*, **142**, 417

11. Condie, K.C. 1989, *Plate Tectonics and Crustal Evolution* (Pergamon Press, Oxford)

12. DeMets, C. et al. 1990, *Geophys. J. Int.*, **101**, 425

13. Dziewonski, A.M. & Anderson, D.L. 1981, *Phys. Earth Planet. Int.*, **25**, 297

14. *COESA, U.S. Standard Atmosphere* 1976, (Government Printing Office, Washington, DC)

15. Edlén, B. 1953, *J. Opt. Soc. Amer.*, **43**, 339

16. Anderson, G.P. et al. 1986, AFGL-TR-86-0110, Atmospheric Constituent Profiles 10–120 km, *Air Force Geophysics Laboratory* (now Air Force Research Laboratory).

17. Allen, C.W. 1973, *Astrophysical Quantities*, 3rd ed. (Athlone Press, London)

18. Goody R.M., & Yung, Y.L. 1989, *Atmospheric Radi-*

ation: *Theoretical Basis*, 2nd ed. (Oxford University Press, New York)

19. Watson, R.F. et al. 1990, Greenhouse gases and aerosols, in *Climate Change: The IPCC Scientific Assessment*, edited by J.T. Houghton, G.H. Henkins, and J.H. Ephraums (Cambridge University Press, New York)

20. Logan, J.A. et al. 1981, *J. Geophys. Res.*, **86**, 7210

21. Allen, M., Lunine, J.I., & Yung, Y.L. 1984, *J. Geophys. Res.*, **89**, 4841

22. *Smithsonian Meteorological Tables*, 5th rev. 1949, (Smithsonian Institution Press, Washington, DC)

23. Walterscheid, R.L., DeVore J.G., & Venkateswaran, S.V. 1980, *J. Atmos. Sci.*, **37**, 455

24. Champion, K.S.W., Cole, A.E., & Kantor, A.J. 1985, Standard and reference atmospheres, in *Handbook of Geophysics and the Space Environment*, edited by A.S. Jursa (Air Force Geophysics Laboratory, Available from National Technical Information Service, **ADA 167000** 1985).

25. Smith E.K. Jr., & Wientraub, S. 1953, *Institute of Radio Engineers Proc.*, **41**, No. 8, 1035

26. Berman, A.L., & Rockwell, S.T. 1992, New Optical and Radio Frequency Angular Troposphere Refraction Models for Deep Space Applications, NASA TR-32-1601

27. Hansen, J.E., & Travis, L.D. 1974, *Space Sci. Rev.* **16**

28. Toon, O.B., & Pollack, J.B. 1976, *J. Appl. Met.* **15**

29. Brasseur, G., & Solomon, S. 1986, *Aeronomy of the Middle Atmosphere: Chemistry and Physics of the Stratosphere and Mesosphere*, 2nd ed. (Kluwer Academic, Amsterdam)

30. Rothman, L.S. et al. 1992, *JQSRT*, **48**, 469

31. Hargreaves, J.K. 1992, *The Solar–Terrestrial Environment* (Cambridge University Press, Cambridge)

32. Vallance Jones, A. 1974, *Aurora* (Reidel, Boston).

33. Roach, F.E., & Gordon, J.L. 1973, *The Light of the Night Sky* (Reidel, Boston)

34. Krassovsky, V.I. et al. 1962, *Planet. Space Sci.*, **9**, 883

35. Chamberlain, J.W. 1961, *Physics of the Aurora and Airglow* (Academic Press, New York)

36. Meinel, A.B. et al. 1954, *J. Geophys. Res.*, **59**, 407

37. Bloxham, J. 1995, *Global Earth Physics: A Handbook of Physical Constants*. T.J. Ahrens, Editor, AGU Reference Shelf 1. American Geophysical Union, Washington, DC)

38. Langel, R.A. 1987, in *Geomagnetism*, edited by J.A. Jacobs (Academic Press, Orlando), Vol. 1, p. 249

39. 1992 IAGA Division V, Working Group 8, IGRF, 1991 Revision, *EOS Trans. American Geophys. Union*, **73**, 182

40. Sears, D.W.G., & Dodd, R.T. 1988, in *Meteorites and the Early Solar System*, edited by J.F. Kerridge and M.S. Matthews, (University of Arizona Press, Tucson, AZ), pp. 3–31

41. Wasson, J.T. 1985, *Meteorites—Their Record of Early Solar System History* (W.H. Freeman, New York)

42. Melosh, H.J. 1989, *Impact Cratering—A Geologic Process*, (Oxford University Press, New York)

43. Pesonen, L.J., Terho, M, & Kukkonen, T.T. 1993, Physical properties of 368 meteorites: Implications for meteorite magnetism and planetary geophysics, *Proc. NIPR Symp. Antarct. Meteorites*, **6**, 401

44. Grieve, R.A.F. 1987, *Ann. Rev. Earth Planet. Sci.*, **15**, 245

Chapter 12

Planets and Satellites

David J. Tholen, Victor G. Tejfel, and Arthur N. Cox

12.1 PLANETARY SYSTEM

Total mass of planets [1]	$446.6 \mathcal{M}_\oplus (\mathcal{M}_\oplus = 5.9742 \times 10^{27}$ g)
Total mass of satellites [2]	6.2×10^{26} g $= 0.104 \mathcal{M}_\oplus$
Total mass of asteroids	1.8×10^{24} g $= 0.000301 \mathcal{M}_\oplus$
Total mass of meteoric and cometary matter	$10^{-9} \mathcal{M}_\oplus$
Total mass of entire planetary system	2.669×10^{30} g $= 446.7 \mathcal{M}_\oplus = 0.00134 \mathcal{M}_\odot$
Total angular momentum of planetary system	3.148×10^{50} g cm^2 s^{-1}
Total translational kinetic energy of planetary system	1.99×10^{42} erg
Total rotational energy of planets	0.7×10^{42} erg

Invariable (Laplacian) plane of the solar system [1, 3]
    with respect to the ecliptic and equinox of J2000.0

Longitude of ascending node	$107°34'\ 57\rlap{.}''7$
Inclination	$1°34'\ 43\rlap{.}''3$
North pole longitude	$17°34'\ 57\rlap{.}''7$
North pole latitude	$88°25'\ 16\rlap{.}''7$

    with respect to the equator and equinox of J2000.0
    (International Celestial Reference Frame)

Longitude of ascending node	$3°51'\ 09\rlap{.}''4$
Inclination	$23°00'\ 32\rlap{.}''0$
North pole right ascension	$273°51'\ 09\rlap{.}''4$
North pole declination	$66°59'\ 28\rlap{.}''0$

Gaussian period of comet or asteroid
    where a is the semimajor axis of orbit in AU $1.000\,040\,27a^{3/2}$ tropical years

12.2   ORBITS AND PHYSICAL CHARACTERISTICS OF PLANETS

The orbital elements in Tables 12.1 and 12.2 are given in [3]. They are given with respect to the mean ecliptic and equinox of J2000.0 at the epoch J2000 (JD 2451545.0). The longitudes of the ascending node, Ω, and perihelion, $\tilde{\omega}$, are measured from a mean γ, the intersection of the ecliptic and equator. Therefore, $\tilde{\omega} = \Omega + \omega$, where ω is the argument of perihelion measured from the ascending node along the orbit. L, the planet longitude for noon January 1, 2000, is also measured the same way from the mean γ. For the Earth, data are given for the Earth–Moon barycenter.

Table 12.1. *Planetary orbit data* [1].

Planet	Semimajor axis of orbit (AU)	Semimajor axis of orbit $(10^6$ km$)^a$	Period Sidereal (Julian years)	Period Synodic (days)	Mean daily motion (deg.)	Mean orbit vel. (km s^{-1})
Mercury ☿	0.387 098 93	57.909 175	0.240 844 45	115.877 5	4.092 377 06	47.8725
Venus ♀	0.723 331 99	108.208 93	0.615 182 57	583.921 4	1.602 168 74	35.0214
Earth ⊕	1.000 000 11	149.597 89	0.999 978 62		0.985 647 36	29.7859
Mars ♂	1.523 662 31	227.936 64	1.880 711 05	779.936 1	0.524 071 09	24.1309
Jupiter ♃	5.203 363 01	778.412 02	11.856 525 02	398.884 0	0.083 129 44	13.0697
Saturn ♄	9.537 070 32	1426.725 4	29.423 519 35	378.091 9	0.033 497 91	9.6724
Uranus ♅	19.191 263 93	2870.972 2	83.747 406 82	369.656 0	0.011 769 04	6.8352
Neptune ♆	30.068 963 48	4498.252 9	163.723 204 5	367.486 7	0.006 020 076	5.4778
Pluto ♇	39.481 686 77	5906.376 2	248.020 8	366.720 7	0.003 973 966	4.7490

Note
^{a}Calculated using 1 AU = $1.495\,978\,706\,6 \times 10^{11}$ m.

Reference
1. *Explanatory Supplement to the Astronomical Almanac 1992*, edited by P.K. Seidelmann (University Science, Mill Valley, CA), pp. 316, 704

Table 12.2. *Additional planetary orbit data.*

Planet	Eccentricity e (2000.0) [1,2]	Inclination to ecliptic i (2000.0) [1,2] (deg.)	Mean longitude of ascending node Ω [1,2] (deg.)	$''T^a$	perihelion $\tilde{\omega}$ [1,2] (deg.)	$''T^a$	Planet L 2000.0 Jan. 1.5 [1,2] (deg.)	Perihelion latest date before 1999 [3]
Mercury	0.205 630 69	7.004 87	48.331 67	−446	77.456 45	+574	252.250 84	1998 Dec. 2
Venus	0.006 773 23	3.394 71	76.680 69	−997	131.532 98	−109	181.979 73	1998 Sep. 7
Earth	0.016 710 22	0.000 05	−11.260 64	−18228	102.947 19	+1198	100.464 35	1998 Jan. 4
Mars	0.093 412 33	1.850 61	49.578 54	−1020	336.040 84	+1560	355.453 32	1998 Jan. 7
Jupiter	0.048 392 66	1.305 30	100.556 15	+1217	14.753 85	+840	34.404 38	1987 Jul. 10
Saturn	0.054 150 60	2.484 46	113.715 04	−1591	92.431 94	−1949	49.944 32	1974 Jan. 8
Uranus	0.047 167 71	0.769 86	74.229 68	+1681	170.964 24	+1312	313.232 18	1966 May 20
Neptune	0.008 585 87	1.769 17	131.721 69	−151	44.971 35	−844	304.880 03	1876 Sep. 2
Pluto	0.248 807 66	17.141 75	110.303 47	−37	224.066 76	−132	238.928 81	1989 Sep. 5

Note

[a] T is in centuries.

References

1. *Explanatory Supplement to the Astronomical Almanac 1992*, edited by P.K. Seidelmann, (University Science, Mill Valley, CA), p. 316
2. Lang, K.R. 1991, *Astrophysical Data: Planets and Satellites* (Springer-Verlag, New York), p. 937
3. *Astronomical Almanac* (USNO, Government Printing Office)

In Table 12.3, the closest approach is at inferior conjunction for Mercury and Venus, at opposition for the other planets. Note that the total mass of the Earth and Moon is 1.012300034 that of the Earth alone. For Venus, Uranus, and Pluto the rotation is retrograde with respect to the orbit. Table 12.4 gives additional physical information for the planets.

Table 12.3. *Physical characteristics of planets* [1].

Planet	Semi-diameter equator at 1 AU ($''$)	at closest approach ($''$)	Radius equator R_e (km)	$\oplus = 1$	Oblateness $\dfrac{R_e - R_p}{R_p}$	Volume $\oplus = 1$ [2]	Reciprocal mass (including satellites) [3] $1/\odot = 1$	Mass 10^{27} g (excluding satellites) [1]	Mass (excluding satellites) $\oplus = 1$
Mercury	3.37	5.5	2 439.7	0.3825	0.0	0.054	6 023 600	0.330 22	0.055 274
Venus	8.34	30.1	6 051.8	0.9488	0.0	0.88	408 523.71	4.869 0	0.815 00
Earth	8.794		6 378.14	1.000	0.003 353 64	1.00	328 900.56	5.974 2	1.000 000
Mars	4.69	8.95	3 397	0.5326	0.006 476 30	0.149	3 098 708	0.641 91	0.107 44
Jupiter	98.48	23.43	71 492	11.209	0.064 874 4	1 316.	1 047.348 6	1 898.7	317.82
Saturn	83.3	9.76	60 268	9.449	0.097 962 4	755.	3 497.898	568.51	95.161
Uranus	32.7	1.95	25 559	4.007	0.022 927 3	52.	22 902.98	86.849	14.371
Neptune	33.4	1.15	24 764	3.883	0.017 1	44.	19 412.24	102.44	17.147
Pluto	1.5	0.04	1 195	0.180	0	0.005	135 000 000	0.013	0.002 200

References

1. *Explanatory Supplement to the Astronomical Almanac 1992*, edited by P.K. Seidelmann (University Science, Mill Valley, CA)
2. Allen, C.W. 1973, *Astrophysical Quantities* (Athlone Press, London)
3. Standish, E.M. 1995, Report of the IAU WGAS sub-group on numerical standards. In *Highlights of Astronomy*, edited by I. Appenzeller (Kluwer Academic, Dordrecht)

Table 12.4. *Additional physical characteristics of planets* [1].

Planet	Density $(g\,cm^{-3})$	Surface gravity [2] $(cm\,s^{-2})$ attractive	centrifugal	Equatorial escape velocity [2] $(km\,s^{-1})$	Sidereal[ab] rotation period (equatorial) [3–8] (day)	Inclination of equator to orbit (deg.)	Moment of inertia [2] $(\mathcal{M}_\oplus R_\oplus^2)$
Mercury	5.43	370	−0.0	4.25	58.646 2	0.0	0.4
Venus	5.24	887	−0.0	10.36	−243.018 7	177.3	0.34
Earth	5.515	980	−3.391	11.18	0.997 269 68	23.45	0.3335
Mars	3.94	371	−1.706	5.02	1.025 956 75	25.19	0.377
Jupiter	1.33	2312	−224.841	59.54	0.413 54	3.12	0.25
Saturn	0.70	896	−175.310	35.49	0.444 01	26.73	0.22
Uranus	1.30	869	−26.195	21.29	−0.718 33	97.86	0.23
Neptune	1.76	1100	−29.065	23.71	0.671 25	29.58	0.29
Pluto [9]	1.1	81	−0.014	1.27	−6.387 18	119.61	

Notes

[a]Equatorial Jupiter I: $9^h50^m30^s$. High latitudes Jupiter II: $9^h55^m40\overset{s}{.}63$. Deep interior Jupiter III: $9^h55^m29\overset{s}{.}37$. This deep interior rotation rate is given in the table.

[b]Equatorial Saturn: I 10^h14^m. High latitudes Saturn: II 10^h38^m. Deep interior Saturn: III $10^h39^m24^s$.

References

1. *Explanatory Supplement to the Astronomical Almanac 1992*, edited by P.K. Seidelmann (University Science, Mill Valley, CA), p. 369
2. Allen, C.W. 1973, *Astrophysical Quantities* (Athlone Press, London)
3. Davies, M.E. et al. 1994, *Cel. Mech.*, **63**, 127
4. Klassen, K.P. 1976, Mercury's rotation axis and period. *Icarus*, **28**, 469
5. Shapiro, I.I., Campbell, D.B., & De Campli, W.M. 1979, Nonresonance rotation of Venus? *ApJL*, **230**, L123
6. Lindal, G.F. et al. 1987, The atmosphere of Uranus—results of radio occultation measurements with Voyager 2. *J. Geophys. Res.*, **92**, 14937
7. Warwick et al. 1986, Voyager 2 radio observations of Uranus. *Science*, **233**, 102
8. Warwick et al. 1989, Voyager planetary radio astronomy at Neptune. *Science*, **246**, 498
9. Tholen, D.J. & Buie, M.W. 1997, The orbit of Charon. I. New Hubble space telescope observations. *Icarus*, **125**, 245

12.2.1 Rotation Axes

Table 12.5 lists the right ascension and declination of the angular momentum vectors at epoch J2000.0 with the equinox J2000.0 for the Sun and the nine planets. Also listed are even more recent and accurate sidereal rotation periods. Those given in Table 12.4 were used for recent Astronomical Almanacs. The reference frame for the coordinates is the mean equator and equinox of J2000.0. The rotation periods for Jupiter, Saturn, Uranus, and Neptune with no visible markings on a hard surface refer to their magnetic fields. For information about prime meridians, see [3, 4].

Table 12.5. *Solar system cartographic data* [1].

Object	α (deg.)	δ (deg.)	Sidereal Period (days)
Sun [1,2]	286°.13	+63°.87	25.38
Mercury	281°.01	+61°.45	58.646 225
Venus	272°.76	+67°.16	−243.019 99

Table 12.5. *(Continued.)*

Object	α (deg.)	δ (deg.)	Sidereal Period (days)
Earth	0°.00	+90°.00	0.997 269 632 3
Mars	317°.681	+52°.886	1.025 956 754 3
Jupiter	268°.05	+64°.49	0.413 538 325 8
Saturn	40°.589	+83°.537	0.444 009 259 2
Uranus	257°.311	−15°.175	−0.718 333 333 3
Neptune	299°.36	+43°.46	0.671 250 000 0
Pluto	313°.02	−9°.09	−6.387 246 00

References

1. Davies, M.E. et al. 1994, *Cel. Mech.*, **63**, 127
2. Carrington, R.C. 1863, *Observations of Spots on the Sun* (Williams and Norgate, London)

12.2.2 Gravity Fields

Table 12.6 gives the spherical harmonic terms in the gravitational potential for the Earth and the outer planets. See Chapter 11 for more details for the Earth.

Table 12.6. *Coefficients of potential.*[a]

Planet	J2	J3	J4	J6	Reference
Earth	+0.001 082 63	−0.000 002 54	−0.000 001 61		[1]
Mars	+0.001 964	+0.000 036			
Jupiter	+0.014 736(1)	+0.000 000 14(50)	−0.000 587(5)	+0.000 031(20)	[2]
Saturn	+0.016 298(10)		−0.000 915(40)	+0.000 103(50)	[3]
Uranus	+0.012				[4]
Neptune	+0.003 411(10)		−0.000 026(+12/−20)		[5]

Note

[a] Numbers in parentheses are uncertainties in the last digits as given.

References

1. *Explanatory Supplement to the Astronomical Almanac, 1992*, edited by P.K. Seidelmann, (University Science, Mill Valley, CA), p. 697
2. Campbell, J.K., & S.P. Synnott 1985, Gravity field of the Jovian system from Pioneer and Voyager tracking data. *Astron. J.*, **90**, 364
3. Campbell, J.K., & J.D. Anderson 1989, Gravity field of the Saturnian system from Pioneer and Voyager tracking data. *Astron. J.*, **97**, 1485
4. Anderson, J.D., J.K. Campbell, R.A. Jacobson, D.N. Sweetnam, & A.H. Taylor 1990, Radio science with Voyager 2 at Uranus: Results on masses and densities of the planet and five principal satellites. *J. Geophys. Res.*, **92**, 14877
5. Tyler, G.L., D.N. Sweetnam, J.D. Anderson, S.E. Borutzki, J.K. Campbell, V.R. Eshleman, D.L. Gresh, E.M. Gurrola, D.P. Hinson, N. Kawashima, E.R. Kurinski, G.S. Levy, G.F. Lindal, J.R. Lyons, E.A. Marouf, P.A. Rosen, R.A. Simpson, & G.E. Wood 1989, Voyager radio science observations of Neptune and Triton. *Science*, **246**, 1466

12.2.3 Planetary Magnetic Fields

The dipole field strength at the planet surface in units of tesla-meter^3 are given in Table 12.7. Quadrupole and octapole strengths are the Schmidt normalized coefficients relative to the dipole moment.

Table 12.7. *Planet magnetic fields and angles.*

Planet	Dipole	Quadrapole (%)	Octapole (%)	Angle (deg.)	Reference
Mercury	2–6×10^{12}				[1]
Venus	$< 10^{11}$				[2]
Earth	7.84×10^{15}	13	10	10.8	[3]
Mars	$< 10^{12}$				[4]
Jupiter	1.55×10^{20}	23	20	10	[5]
Saturn	4.6×10^{18}	14	7	0	[6]
Uranus	3.9×10^{17}			59	[7]
Neptune	2.2×10^{17}			47	[8]
Pluto	?				

References

1. Connerney, J.E.P., & N.F. Ness 1988, Mercury's magnetic field and interior. In *Mercury* edited by Vilas, C.R. Chapman, and M.S. Matthews, (University of Arizona Press, Tucson) pp. 494–513
2. Phillips, J.L., & C.T. Russell 1987, Upper limit on the intrinsic magnetic field of Venus. *J. Geophys. Res.*, **92**, 2253
3. Barton, C.E. 1989, Geomagnetic secular variation: Direction and intensity. In *The Encyclopedia of Solid Earth Geophysics*, (Van Nostrand Reinhold, New York) pp. 561–577
4. Russell, C.T. 1978, The magnetic field of Mars: Mars 5 evidence re-examined. *Geophys. Res. Lett.*, **5**, 85
5. Connerney, J.E.P. 1981, The magnetic field of Jupiter: A generalized inverse approach. *J. Geophys. Res.*, **86**, 7679
6. Connerney, J.E.P., M.H. Acuna, & N.F. Ness 1984, The Z3 model of Saturn's magnetic field and the Pioneer 11 vector helium observations. *J. Geophys. Res.*, **89**, 7541
7. Connerney, J.E.P., M.H. Acuna, & N.F. Ness 1987, The magnetic field of Uranus. *J. Geophys. Res.*, **92**, 15379
8. Connerney, J.E.P., M.H. Acuna, & N.F. Ness 1991, The magnetic field of Neptune. *J. Geophys. Res.*, **96**, 19023

12.3 PHOTOMETRY OF PLANETS AND ASTEROIDS

Table 12.8 gives the photometry data for the planets and some asteroids.

Table 12.8. *Photometry of the planets and five asteroids* [1,2].

Planet or asteroid	Visual geometric albedo	Opposition V	B − V [1,3,4]	U − V	V(1, 0) or H (mag.)[a]	Variation of V with phase (α, L in deg.)[a,b]
Mercury	0.106	···	0.93	0.41	−0.42	$+0.038\alpha - 2.73 \times 10^{-4}\alpha^2$ $+2.00 \times 10^{-6}\alpha^3$
Venus	0.65	···	0.82	0.50	−4.40	$+0.0009\alpha + 2.39 \times 10^{-4}\alpha^2$ $-0.65 \times 10^{-6}\alpha^3$
Earth	0.367	···	0.2	···	−3.86	
Mars	0.150	−2.01	1.36	0.58	−1.52	$+0.016\alpha$
Jupiter	0.52	−2.70	0.83	0.48	−9.40	$+0.005\alpha$
Saturn	0.47	+0.67[c]	1.04	0.58	−8.88	$+0.044l - 2.6\sin b$ $+1.25\sin^2 b$
Uranus	0.51	+5.52	0.56	0.28	−7.19	$+0.0028\alpha$
Neptune	0.41	+7.84	0.41	0.21	−6.87	···
Pluto [5,6]	variable[d]	+15.12	0.842[d]	0.31	−0.81	0.037α
(1) Ceres [7,8,11,12]	0.113	+6.78	0.72	0.48	+3.34	0.12
(2) Pallas [7,8,11,12]	0.159	+7.60	0.67	0.28	+4.13	0.11
(3) Juno [7,8,11,12]	0.238	+8.57	0.79	0.52	+5.53	0.32
(4) Vesta [7,9,11,12]	0.423	+5.73	0.81	0.66	+3.20	0.32
(10) Hygiea [10–12]	0.072	+9.56	0.69	0.39	+5.43	
(243) Ida [10–12]	0.238	+13.57	0.81	0.61	+9.94	
(253) Mathilde [10–12]	0.044	+13.39			+10.2	
(433) Eros [7,10–12]		+10.28	0.92	0.80	+11.16	0.46
(951) Gaspra [10–12]		+13.59	0.87	0.77	+11.46	

Notes

[a] For the asteroids the V(1.0) is designated as H, and the phase function is given by the slope parameter G [13].

[b] α is the phase angle between the Sun and Earth as seen from the planet. l is the Saturnicentric longitude difference of the Sun and Earth and lies between −6° and 6°. b is the Saturnicentric ring latitude of the Earth that lies between −27° and 27°.

[c] V refers to the Saturn disk only.

[d] The Pluto visual geometric albedo is variable by 30%. The Pluto color is the combination of the planet and its satellite Charon.

References

1. *Astronomical Almanac*, 1998 (USNO, Government Printing Office)
2. Allen, C.W. 1973, *Astrophysical Quantities*, 3rd ed. Athlone Press, London
3. Irvine, W.M. et al. 1968, *AJ*, **73**, 251, 807
4. Harris, D.L. 1961, in *Planets and Satellites*, edited by G. Kuiper and B. Middlehurst (University of Chicago, Chicago), p. 272
5. Tholen D.J. & Tedesco, E.F. 1994, *Icarus*, **108**, p. 200
6. Buie, M.W., Tholen, D.J., & Wasserman L.H. 1997, Separate lightcurves of Pluto and Charon, *Icarus*, **125**, 233
7. Haupt, H. 1951, *Mitt. U.S. Wien*, **5**, 31
8. Watson, F.G. 1956, *Between the Planets*, rev. ed. (Harvard University Press, Boston)
9. Gehrels, T. 1967, *AJ*, **72**, 929
10. Tedesco, E.F. & Veeder, G.J. 1992, in *The IRAS Minor Planet Survey* edited by E.F. Tedesco, G.J. Veeder, J.W. Fowler, and J.R. Chillemi (Phillips Laboratory, Hanscom Air force Base)
11. Tedesco, E.F., Marsden, B.G., & Williams, G.V. 1990, *Minor Planet Circulars 17256–17273* (Smithsonian Astrophysical Observatory, Cambridge)
12. Bowell, E. Hapke, B, Domingue, D. Lumme, K., Peltoniemi, J., & Harris, A.W. 1989, in *Asteroids II* edited by R.P. Binzel, T. Gehrels, and M.S. Matthews (University of Arizona Press, Tucson)
13. Zellner, B., Tholen, D.J., & Tedesco, E.F. 1985, *Icarus*, **61**, 355

12.4 PHYSICAL CONDITIONS ON PLANETS
by Glenn S. Orton

Planetary atmosphere and surface conditions are given in Table 12.9:

T_e = Effective temperature of the planet.

T_a = Atmospheric temperature at the level with pressure 1 bar.

T_s = Mean temperature at the solid surface.

P_s = Atmospheric pressure at the solid surface for the terrestrial planets and satellites or at visible cloud surface for major planets.

H = Scale height.

So, Cl = Solid, cloud; for lowest visible surface.

Table 12.9. *Planetary and selected satellite atmosphere and surface conditions.*

Planet	Visible surface	T_e (K)	T_a (1bar) (K)	T_s (K)	P_s (bar)	H (km)
Mercury	So			440		
Venus	Cl	~230		730	90	15
Earth	So, Cl	~255	288	288–293	1	8
Mars	So	~212		183–268	0.007–0.010	11
Jupiter	Cl	124.4±0.3	165		~0.3	19–25
Saturn	Cl	95.0±0.4	134		~0.4	35–50
Uranus	Cl	59.1±0.3	76			22–29
Neptune	Cl	59.3±0.8	73			18–22
Pluto	So	50–70		57.8 [1]	8×10^{-5} [1]	
Moon	So			120–380		
Io	So	99		100–140 (300 local)		
Europe	So	97				
Ganymede	So	107			10^{-6}?	
Callisto	So	117				
Titan [2]	Cl	86	83	94	1.496±0.020	20–22
Triton [3]	So			38	$\sim1.5 \times 10^{-5}$	

References
1. Trafton L. & Stern S.A. 1983, *Ap. J.*, **267**, 872
2. *Saturn*, 1984, edited by T. Gehrels (University of Arizona Press, Tucson)
3. *Science*, 1990, **250**, 4979

Compositions of planetary atmospheres are given in Table 12.10.

Table 12.10. *Selected gas components of planetary atmospheres [1].*

Gas	Venus [2]	Earth [2]	Mars [2]	Jupiter [2–4]	Saturn [2,5,6]	Uranus [2,4,5]	Neptune [2,5,7]	Titan [8]
N_2	0.035	0.78084	0.027					0.73–0.99
O_2		0.20948	1.3×10^{-3}					
CO_2	0.965	3.33×10^{-4}	0.953		$\leq 3 \times 10^{-10}$	$< 3 \times 10^{-10}$	$\leq 5 \times 10^{-10}$	1.5×10^{-9}
CO	3×10^{-7}	2×10^{-7}	2.7×10^{-3}	1×10^{-10}	$(1-3) \times 10^{-9}$	$< 5 \times 10^{-7}$	$\sim 6 \times 10^{-7}$	$(0.6\text{–}1.5) \times 10^{-4}$
CH_4		2.0×10^{-6}		$(2.4 \pm 0.5) \times 10^{-3}$	$(1-4) \times 10^{-3}$	≤ 0.02	$\leq 0.02\text{–}0.04$	$\leq 0.02\text{–}0.10$
NH_3		4×10^{-9}		$\leq 7 \times 10^{-4}$	$\leq 1 \times 10^{-4}$	$< 10^{-4}$		
H_2O	2×10^{-5}	$\sim 10^{-6}$	3×10^{-4}	$\leq 6.5 \pm 2.9 \times 10^{-5}$	$(0.2\text{–}2) \times 10^{-7}$	$\leq (0.5\text{–}1.2) \times 10^{-8}$	$< (1.5\text{–}3.5) \times 10^{-9}$	$\leq (0.4\text{–}1.4) \times 10^{-9}$
H_2		5×10^{-7}		0.863 ± 0.007	0.94	0.85	0.85	2×10^{-3}
He		5.24×10^{-6}		0.156 ± 0.006	0.06	0.15	0.15	(0–0.25)
Ar	1.2×10^{-5}	9.34×10^{-3}	0.016	$(1.0 \pm 0.4) \times 10^{-5}$				
Ne	7×10^{-5}	1.818×10^{-5}	2.5×10^{-6}	$(2.3 \pm 0.25) \times 10^{-5}$				
Kr	1×10^{-5}	1.14×10^{-6}	3×10^{-7}	$< (8.5 \pm 4) \times 10^{-9}$				
Xe	3×10^{-9}	8.7×10^{-8}	8×10^{-8}	$< (5 \pm 2.5) \times 10^{-9}$	$< 10^{-6}$			
H_2S	3×10^{-6}	2×10^{-8}		$(1-8) \times 10^{-5}$				
SO_2	1.5×10^{-4}	1×10^{-9}						
HD				$\sim 1.1 \times 10^{-5}$	$(0.7\text{–}1.7) \times 10^{-8}$	$(6\text{–}10) \times 10^{-5}$	$\lesssim 1 \times 10^{-8}$	
PH_3				$\leq 6 \times 10^{-6}$	$\leq 4.5 \times 10^{-6}$			
CH_3D				$\sim 2 \times 10^{-8}$	3.4×10^{-7}	$\sim 2 \times 10^{-6}$	$\leq 1 \times 10^{-6}$	
GeH_4				6×10^{-10}	2×10^{-9}			
C_2H_2				$\leq 3 \times 10^{-8}$	$\leq 1 \times 10^{-7}$	$\leq 4 \times 10^{-7}$		$\leq 2 \times 10^{-6}$
C_2H_6				$\leq 3 \times 10^{-6}$	$\leq 5 \times 10^{-6}$			$\leq 2 \times 10^{-5}$
HCN				$< 1 \times 10^{-9}$	$< 10^{-9}$			$\leq 2 \times 10^{-7}$
C_3H_8				$\leq 10^{-7}$				4×10^{-6}
C_2H_4					10^{-9}			4×10^{-7}
HC_3N								$\leq (0.1\text{–}1) \times 10^{-7}$
C_2N_2				$< 10^{-9}$				$\leq (0.1\text{–}1) \times 10^{-7}$
CH_3NH_2					$< 10^{-6}$			$\leq (0.1\text{–}1) \times 10^{-7}$

References

1. Lewis, J.S. 1995, *Physics and Chemistry of the Solar System* (Academic Press, San Diego)
2. Encrenaz, T. & Bibring J.-P. 1990, *The Solar System* (Springer-Verlag, Berlin), 330 pp.
3. Niemann H.B. et al. 1996, *Science*, **272**, 846; Niemann H.B. 1998, *JGR*, **103**, 22831; Folkner, W.M., Woo, R., & Nandi, S. 1998, *JGR*, **103**, 22847; Encrenaz, T. et al. 1996, *A&A*, **315**, L347; Bèzard, B. et al. 1998, *A&A*, **334**, L41
4. Encrenaz, T. et al. 1998, *A&A*, **338**, L48
5. Feuchtgruber, H. et al. 1997, *Nature* **232**, 139
6. Griffin, M. et al. 1996, *A&A*, **315**, L389; Davis, G. et al. 1996, *A&A*, **313**, L393; de Graauw, T. et al. 1997, *A&A*, **321**, L43
7. Orton, G.S. 1992, *Icarus*, **100**, 541
8. *Science*, 1989, **246**, N4936; *Saturn*, 1984, edited by T. Gehrels (University of Arizona Press, Tucson); *Science*, 1990, **250**, N 4979

12.5 NAMES, DESIGNATIONS, AND DISCOVERIES OF SATELLITES
by Dan Pascu

Table 12.11 lists the names, designations, and discoveries of the planetary satellites.

Table 12.11. *Names, designations, and discoveries* [1, 2, 3].

Satellite	Name	Discovery date	Discoverer	Satellite	Name	Discovery date	Discoverer
Earth				XV	Atlas	1980	R. Terrile/Voyager 1
	Moon			XVI	Prometheus	1980	S.A. Collins,
Mars							D. Carlson/Voyager 1
I	Phobos	1877	A. Hall	XVII	Pandora	1980	S.A. Collins,
II	Deimos	1877	A. Hall				D. Carlson/Voyager 1
Jupiter				XVIII	Pan	1990	M. Showalter/Voyager 2
I	Io	1610	Galileo	*Uranus*			
II	Europa	1610	Galileo	I	Ariel	1851	W. Lassell
III	Ganymede	1610	Galileo	II	Umbriel	1851	W. Lassell
IV	Callisto	1610	Galileo	III	Titania	1787	W. Herschel
V	Amalthea	1892	E.E. Barnard	IV	Oberon	1787	W. Herschel
VI	Himalia	1904	C. Perrine	V	Miranda	1948	G. Kuiper
VII	Elara	1905	C. Perrine	VI	Cordelia	1986	R. Terrile/Voyager 2
VIII	Pasiphae	1908	P. Melotte	VII	Ophelia	1986	R. Terrile/Voyager 2
IX	Sinope	1914	S. Nicholson	VIII	Bianca	1986	Voyager 2
X	Lysithea	1938	S. Nicholson	IX	Cressida	1986	S. Synnott/Voyager 2
XI	Carme	1938	S. Nicholson	X	Desdemona	1986	S. Synnott/Voyager 2
XII	Ananke	1951	S. Nicholson	XI	Juliet	1986	S. Synnott/Voyager 2
XIII	Leda	1974	C. Kowal	XII	Portia	1986	S. Synnott/Voyager 2
XIV	Thebe	1980	S. Synnott/Voyager 1	XIII	Rosalind	1986	S. Synnott/Voyager 2
XV	Adrastea	1979	D. Jewitt,	XIV	Belinda	1986	S. Synnott/Voyager 2
			E. Danielson/Voyager 2	XV	Puck	1986	S. Synnott/Voyager 2
XVI	Metis	1980	S. Synnott/Voyager 2	XVI	Caliban[a]	1997	B.J. Gladman, P.D. Nicholson,
Saturn							J.A. Burns, J.J. Kavelaars
I	Mimas	1789	W. Herschel	XVII	Sycorax[a]	1997	P.D. Nicholson, B.J. Gladman,
II	Enceladus	1789	W. Herschel				J.A. Burns, J.J. Kavelaars
III	Tethys	1684	G.D. Cassini	*Neptune*			
IV	Dione	1684	G.D. Cassini	I	Triton	1846	W. Lassell
V	Rhea	1672	G.D. Cassini	II	Nereid	1949	G. Kuiper
VI	Titan	1655	C. Huygens	III	Naiad	1989	Voyager 2
VII	Hyperion	1848	W. & G. Bond/W. Lassell	IV	Thalassa	1989	R. Terrile/Voyager 2
VIII	Iapetus	1671	G.D. Cassini	V	Despina	1989	S. Synnott/Voyager 2
IX	Phoebe	1898	W. Pickering	VI	Galatea	1989	S. Synnott/Voyager 2
X	Janus	1966	A. Dollfus	VII	Larissa	1982	H. Reitsema, W. Hubbard,
XI	Epimetheus	1966/1978	R. Walker/J. Fountain, S. Larson				L. Lebofsky, D. Tholen
XII	Helene	1980	P. Laques, J. Lecacheux	VIII	Proteus	1989	S. Synnott/Voyager 2
XIII	Telesto	1980	B. Smith, H. Reitsema,	*Pluto*			
			S. Larson, J. Fountain	I	Charon	1978	J. Christy
XIV	Calypso	1980	D. Pascu, P.K. Seidelmann,				
			W. Baum, D. Currie				

Note

[a]The two distant satellites of Uranus, Caliban and Sycorax still have provisional names. They will be accepted or changed at the IAU General Assembly in 2000.

References

1. Burns, J.A. 1986, *Satellites*, edited by J.A. Burns and M.S. Matthews (University of Arizona Press, Tucson)
2. Pasachoff, J. 1998, *From the Earth to the Universe*, 5th ed. (Saunders College Pub., Fort Worth)
3. Veverka, J.M. 1998, *Observers Handbook*, edited by R.L. Bishop (University of Toronto Press, Toronto)

12.6 SATELLITE ORBITS AND PHYSICAL ELEMENTS
by Dan Pascu

The main orbital and physical elements for the planetary satellites are given in Tables 12.12 and 12.13. For comparison with observations, several factors are related to terrestrial opposition, labeled Op. Synodic periods are relative to the main planet.

The inclinations of satellite orbits are complicated by precession around the "proper plane," which is normally close to the planet's equator. Inclinations are measured from the planet's equator and values greater than 90° indicate that the motion is retrograde. The inclination of the Moon to the ecliptic is only $5°.145396$. Reciprocal mass of satellite totals:

$$
\begin{array}{ll}
\text{Jupiter} & 4831 \ (\text{Jupiter})^{-1}, \\
\text{Saturn} & 4050 \ (\text{Saturn})^{-1}, \\
\text{Uranus} & 9571 \ (\text{Uranus})^{-1}, \\
\text{Neptune} & 4780 \ (\text{Neptune}^{-1}, \\
\text{Pluto} & 8.3 \ (\text{Pluto})^{-1},
\end{array}
$$

Total mass of all satellites 7.34×10^{26} g.

The following commensurabilities exist among the mean motions n_i of planetary satellites [5–7]:

$$
\begin{array}{ll}
\text{Jupiter} & n_1 - 3n_2 + 2n_3 = 0, \\
\text{Saturn} & 5n_1 - 10n_2 + n_3 + 4n_4 = 0, \\
\text{Uranus} & n_5 - 3n_1 + 2n_2 = 0, \\
& n_1 - n_2 - 2n_3 + n_4 = 0.
\end{array}
$$

Table 12.12. *Planetary satellite orbits* [1, 2, 3].

Satellite		Semimajor axis (10^3 km)	Semimajor axis (10^{-3}) AU	Max elong. at mean opposition ′ ″	Sidereal period[a,b] (days)
Earth					
	Moon	384.400	2.5696		27.321661
Mars					
I	Phobos	9.378	0.0627	0 25	0.31891023
II	Deimos	23.459	0.1568	1 02	1.2624407
Jupiter					
I	Io	422.	2.8209	2 18	1.769137786
II	Europa	671.	4.4854	3 40	3.551181041
III	Ganymede	1070.	7.1525	5 51	7.15455296
IV	Callisto	1883.	12.5871	10 18	16.6890184
V	Amalthea	181.	1.2099	0 59	0.49817905
VI	Himalia	11480.	76.7391	62 46	250.5662
VII	Elara	11737.	78.4570	64 10	259.6528
VIII	Pasiphae	23500.	157.0878	128 26	R735.
IX	Sinope	23700.	158.4247	129 31	R758.
X	Lysithea	11720.	78.3434	64 04	259.22

Table 12.12. *(Continued.)*

Satellite		Semimajor axis		Max elong. at mean opposition	Sidereal period[a,b]
		$(10^3$ km)	(10^{-3}) AU	′ ″	(days)
Jupiter (Cont.)					
XI	Carme	22 600.	151.071 7	123 31	R692.
XII	Ananke	21 200.	141.713 2	115 52	R631.
XIII	Leda	11 094.	74.158 8	60 39	238.72
XIV	Thebe	222.	1.484 0	1 13	0.674 5
XV	Adrastea	129.	0.862 3	0 42	0.298 26
XVI	Metis	128.	0.855 6	0 42	0.294 780
Saturn					
I	Mimas	185.52	1.240 1	0 30	0.942 421 813
II	Enceladus	238.02	1.591 0	0 38	1.370 217 855
III	Tethys	294.66	1.969 7	0 48	1.887 802 160
IV	Dione	377.40	2.522 8	1 01	2.736 914 742
V	Rhea	527.04	3.523 0	1 25	4.517 500 436
VI	Titan	1 221.83	8.167 4	3 17	15.945 420 68
VII	Hyperion	1 481.1	9.900 5	3 59	21.276 608 8
VIII	Iapetus	3 561.3	23.805 3	9 35	79.330 182 5
IX	Phoebe	12 952.	86.578 8	34 51	R550.48
X	Janus	151.472	1.012 5	0 24	0.694 5
XI	Epimetheus	151.422	1.012 2	0 24	0.694 2
XII	Helene	377.40	2.522 8	1 01	2.736 9
XIII	Telesto	294.66	1.969 7	0 48	1.887 8
XIV	Calypso	294.66	1.969 7	0 48	1.887 8
XV	Atlas	137.670	0.920 3	0 22	0.601 9
XVI	Prometheus	139.353	0.931 5	0 23	0.613 0
XVII	Pandora	141.700	0.947 2	0 23	0.628 5
XVIII	Pan	133.583	0.892 9	0 21	0.575 0
Uranus					
I	Ariel	191.02	1.276 9	0 14	2.520 379 35
II	Umbriel	266.30	1.780 1	0 20	4.144 177 2
III	Titania	435.91	2.913 9	0 33	8.705 871 7
IV	Oberon	583.52	3.900 6	0 44	13.463 238 9
V	Miranda	129.39	0.864 9	0 10	1.413 479 25
VI	Cordelia	49.77	0.332 7	0 04	0.335 033 8
VII	Ophelia	53.79	0.359 6	0 04	0.376 400
VIII	Bianca	59.17	0.395 5	0 04	0.434 578 99
IX	Cressida	61.78	0.413 0	0 05	0.463 569 60
X	Desdemona	62.68	0.419 0	0 05	0.473 649 60
XI	Juliet	64.35	0.430 2	0 05	0.493 065 49
XII	Portia	66.09	0.441 8	0 05	0.513 195 92
XIII	Rosalind	69.94	0.467 5	0 05	0.558 459 53
XIV	Belinda	75.26	0.503 1	0 06	0.623 527 47
XV	Puck	86.01	0.574 9	0 07	0.761 832 87
XVI	Caliban[c]	7169.	47.29	8 56	R579.
XVII	Sycorax[c]	12214.	81.64	15 26	R1289.
Neptune					
I	Triton	354.76	2.371 4	0 17	R5.876 854 1
II	Nereid	5 513.4	36.854 8	4 21	360.136 19
III	Naiad	48.23	0.322 4	0 02	0.294 396
IV	Thalassa	50.07	0.334 7	0 02	0.311 485
V	Despina	52.53	0.351 1	0 02	0.334 655
VI	Galatea	61.95	0.414 1	0 03	0.428 745
VII	Larissa	73.55	0.491 7	0 03	0.554 654
VIII	Proteus	117.65	0.786 4	0 06	1.122 315

Table 12.12. *(Continued.)*

Satellite		Semimajor axis (10^3 km)	(10^{-3}) AU	Max elong. at mean opposition $'$ $''$	Sidereal period[a,b] (days)
Pluto					
I	Charon	19.6	0.131 0	<0 01	6.387 25

Notes

[a] R before the period indicates a retrograde orbit.
[b] Tropical periods are given for the Saturn satellites I to VIII.
[c] Provisional names.

References

1. *1998 Astronomical Almanac* (USNO, Government Printing Office)
2. Jacobson, R.A. 1998, *AJ*, **115**, 1195
3. Owen, W.M., Vaughn, R.M., & Synnott, S. 1991, *AJ*, **101**, 1511

Table 12.13. *Additional satellite data* [1, 2, 3, 4].

Satellite		Orbit Inclination (deg)	Eccentricity	Radius (km)	Mass (1/Planet)	Mass (g)
Earth						
	Moon	18.28–28.58	0.054 90 049	1 737.4	0.012 300 034	7.3483×10^{25}
Mars						
I	Phobos	1.0	0.015	$13.4 \times 11.2 \times 9.2$	1.654×10^{-8}	1.063×10^{19}
II	Deimos	0.9–2.7	0.000 5	$7.5 \times 6.1 \times 5.2$	3.71×10^{-9}	2.38×10^{18}
Jupiter						
I	Io	0.04	0.004	$1830.0 \times 1818.7 \times 1815.3$	4.7041×10^{-5}	8.9316×10^{25}
II	Europa	0.47	0.009	1 565	$2.528 0 \times 10^{-5}$	$4.799 82 \times 10^{25}$
III	Ganymede	0.21	0.002	2 634	$7.804 6 \times 10^{-5}$	$1.481 86 \times 10^{26}$
IV	Callisto	0.51	0.007	2 403	$5.666 7 \times 10^{-5}$	$1.075 93 \times 10^{26}$
V	Amalthea	0.40	0.003	$131.0 \times 73.0 \times 67.0$	3.8×10^{-9}	7.2×10^{21}
VI	Himalia	27.63	0.157 98	85	5.0×10^{-9}	9.5×10^{21}
VII	Elara	24.77	0.207 19	40	4×10^{-10}	$8. \times 10^{20}$
VIII	Pasiphae	145	0.378	18	1×10^{-10}	$2. \times 10^{20}$
IX	Sinope	153	0.275	14	4×10^{-11}	$8. \times 10^{19}$
X	Lysithea	29.02	0.107	12	4×10^{-11}	$8. \times 10^{19}$
XI	Carme	164	0.206 78	15	5×10^{-11}	$9. \times 10^{19}$
XII	Ananke	147	0.168 70	10	2×10^{-11}	$4. \times 10^{19}$
XIII	Leda	26.07	0.147 62	5	3×10^{-12}	$6. \times 10^{18}$
XIV	Thebe	0.8	0.015	55×45	4×10^{-10}	$8. \times 10^{20}$
XV	Adrastea			$13 \times 10 \times 8$	1×10^{-11}	$2. \times 10^{19}$
XVI	Metis			20×20	5×10^{-11}	$9. \times 10^{19}$
Saturn						
I	Mimas	1.53	0.020 2	$209.1 \times 196.2 \times 191.4$	6.60×10^{-8}	3.75×10^{22}
II	Enceladus	0.00	0.004 52	$256.3 \times 247.3 \times 244.6$	$1. \times 10^{-7}$	$7. \times 10^{22}$
III	Tethys	1.86	0.000 00	$535.6 \times 528.2 \times 525.8$	1.10×10^{-6}	6.27×10^{23}
IV	Dione	0.02	0.002 230	560	1.95×10^{-6}	1.10×10^{24}
V	Rhea	0.35	0.001 00	764	4.06×10^{-6}	2.31×10^{24}
VI	Titan	0.33	0.029 192	2 575	$2.366 7 \times 10^{-4}$	$1.345 5 \times 10^{26}$
VII	Hyperion	0.43	0.104	$180 \times 140 \times 112.5$	4×10^{-8}	$2. \times 10^{22}$
VIII	Iapetus	14.72	0.028 28	718	2.8×10^{-6}	1.6×10^{24}

Table 12.13. *(Continued.)*

Satellite		Orbit Inclination (deg)	Eccentricity	Radius (km)	Mass (1/Planet)	Mass (g)
Saturn (cont.)						
IX	Phoebe	177^a	0.163 26	110	7×10^{-10}	$4. \times 10^{20}$
X	Janus	0.14	0.007	$97.0 \times 95.0 \times 77.0$	3.38×10^{-9}	1.92×10^{21}
XI	Epimetheus	0.34	0.009	$69 \times 55 \times 55$	9.5×10^{-10}	5.4×10^{20}
XII	Helene	0.0	0.005	$18 \times 16 \times 15$		
XIII	Telesto			$15 \times 12.5 \times 7.5$		
XIV	Calypso			$15.0 \times 8.0 \times 8.0$		
XV	Atlas	0.3	0.000	$18.5 \times 17.2 \times 13.5$		
XVI	Prometheus	0.0	0.003	$74.0 \times 50.0 \times 34.0$		
XVII	Pandora	0.0	0.004	$55.0 \times 44.0 \times 31.0$		
XVIII	Pan			10		
Uranus						
I	Ariel	0.3	0.0034	$581.1 \times 577.9 \times 577.7$	1.55×10^{-5}	1.35×10^{24}
II	Umbriel	0.36	0.0050	584.7	1.35×10^{-5}	1.17×10^{24}
III	Titania	0.14	0.0022	788.9	4.06×10^{-5}	3.53×10^{24}
IV	Oberon	0.10	0.0008	761.4	3.47×10^{-5}	3.01×10^{24}
V	Miranda	4.2	0.0027	$240.4 \times 234.2 \times 232.9$	7.6×10^{-7}	6.6×10^{22}
VI	Cordelia	0.08	0.00026	13		
VII	Ophelia	0.10	0.0099	15		
VIII	Bianca	0.19	0.0009	21		
IX	Cressida	0.01	0.0004	31		
X	Desdemona	0.11	0.00013	27		
XI	Juliet	0.07	0.00066	42		
XII	Portia	0.06	0.0000	54		
XIII	Rosalind	0.28	0.0001	27		
XIV	Belinda	0.03	0.00007	33		
XV	Puck	0.32	0.00012	77		
XVI	Caliban^b	139.2^a	0.082	30		
XVII	Sycorax^b	152.7^a	0.509	60		
Neptune						
I	Triton	157.345	0.000016	1352.6	2.089×10^{-4}	2.140×10^{25}
II	Nereid	27.6^c	0.7512	170	2×10^{-7}	$2. \times 10^{22}$
III	Naiad	4.74	0.000	29:		
IV	Thalassa	0.21	0.000	40:		
V	Despina	0.07	0.000	74		
VI	Galatea	0.05	0.000	79		
VII	Larissa	0.20	0.00139	104×89		
VIII	Proteus	0.55	0.0004	$218 \times 208 \times 201$		
Pluto						
I	Charon	96.16^c		593	0.125	1.62×10^{24}

Notes

^{a}Relative to the ecliptic plane.

^{b}Provisional names.

^{c}Referred to the Earth equator of 1950.0 (Nereid) and of J2000 (Charon).

References

1. *1998 Astronomical Almanac* (USNO, Government Printing Office)
2. Davies, M.E. et al. 1995, *Cel. Mech.*, **63**, 127
3. Jacobson, R.A. 1998, *AJ*, **115**, 1195
4. Owen, W.M., Vaughn, R.M., & Synnott, S. 1991, *AJ*, **101**, 1511

Rotation and photometric data for many of the planetary satellites are given in Table 12.14.

Table 12.14. *Satellite rotation and photometric data* [1, 2, 3].

Satellite		Sidereal period of rotation (d)[a]	Geometric albedo $(V)^b$	$V(1,0)^c$	V_0^d	$B - V$	$U - B$
Earth							
	Moon	S	0.12	0.21	−12.74	0.92	0.46
Mars							
I	Phobos	S	0.07	11.8	11.3	0.6	
II	Deimos	S	0.08	12.89	12.40	0.65	0.18
Jupiter							
I	Io	S	0.63	−1.68	5.02	1.17	1.30
II	Europa	S	0.67	−1.41	5.29	0.87	0.52
III	Ganymede	S	0.44	−2.09	4.61	0.83	0.50
IV	Callisto	S	0.20	−1.05	5.65	0.86	0.55
V	Amalthea	S	0.07	7.4	14.1	1.50	
VI	Himalia	0.4	0.03	8.14	14.84	0.67	0.30
VII	Elara	0.5	0.03	10.07	16.77	0.69	0.28
VIII	Pasiphae		0.10	10.33	17.03	0.63	0.34
IX	Sinope		0.05	11.6	18.3	0.7	
X	Lysithea		0.06	11.7	18.4	0.7	
XI	Carme		0.06	11.3	18.0	0.7	
XII	Ananke		0.06	12.2	18.9	0.7	
XIII	Leda		0.07	13.5	20.2	0.7	
XIV	Thebe	S	0.04	9.0	15.7	1.3	
XV	Adrastea		0.05	12.4	19.1		
XVI	Metis		0.05	10.8	17.5		
Saturn							
I	Mimas	S	0.5	3.3	12.9		
II	Enceladus	S	1.0	2.1	11.7	0.70	0.28
III	Tethys	S	0.9	0.6	10.2	0.73	0.30
IV	Dione	S	0.7	0.8	10.4	0.71	0.31
V	Rhea	S	0.7	0.1	9.7	0.78	0.38
VI	Titan	S	0.22	−1.28	8.28	1.28	0.75
VII	Hyperion		0.3	4.63	14.19	0.78	0.33
VIII	Iapetus	S	0.2^e	1.5	11.1	0.72	0.30
IX	Phoebe	0.4	0.06	6.89	16.45	0.70	0.34
X	Janus	S	0.9:	4.4:	14.:		
XI	Epimetheus	S	0.8:	5.4:	15.:		
XII	Helene		0.7:	8.4:	18.:		
XIII	Telesto		1.0:	8.9:	18.5:		
XIV	Calypso		1.0:	9.1:	18.7:		
XV	Atlas		0.8:	8.4:	18.:		
XVI	Prometheus		0.5:	6.4:	16.:		
XVII	Pandora		0.7:	6.4:	16.:		
XVIII	Pan		0.5:				
Uranus							
I	Ariel	S	0.35	1.45	14.16	0.65	
II	Umbriel	S	0.19	2.10	14.81	0.68	
III	Titania	S	0.28	1.02	13.73	0.70	0.28
IV	Oberon	S	0.25	1.23	13.94	0.68	0.20
V	Miranda	S	0.27	3.6	16.3		
VI	Cordelia		0.07:	11.4	24.1		
VII	Ophelia		0.07:	11.1	23.8		

Table 12.14. *(Continued.)*

Satellite		Sidereal period of rotation (d)[a]	Geometric albedo $(V)^b$	$V(1,0)^c$	V_0^d	$B - V$	$U - B$
Uranus (cont.)							
VIII	Bianca		0.07:	10.3	23.0		
IX	Cressida		0.07:	9.5	22.2		
X	Desdemona		0.07:	9.8	22.5		
XI	Juliet		0.07:	8.8	21.5		
XII	Portia		0.07:	8.3	21.0		
XIII	Rosalind		0.07:	9.8	22.5		
XIV	Belinda		0.07:	9.4	22.1		
XV	Puck		0.075	7.5	20.2		
XVI	Caliban^f		0.07:		22.4		
XVII	Sycorax^f		0.07:		20.9		
Neptune							
I	Triton	S	0.77	−1.24	13.47	0.72	0.29
II	Nereid		0.4	4.0	18.7	0.65	
III	Naiad		0.06:	10.0:	24.7		
IV	Thalassa		0.06:	9.1:	23.8		
V	Despina		0.06:	7.9	22.6		
VI	Galatea		0.06:	7.6:	22.3		
VII	Larissa		0.06	7.3	22.0		
VIII	Proteus		0.06	5.6	20.3		
Pluto							
I	Charon	S	0.5	0.9	16.8		

Notes

^{a}S means the rotation is synchronous with the orbit period.

^{b}The solar V used is −26.75.

^{c}The apparent V magnitude with the planet 1 AU from both Sun and Earth at zero phase angle.

^{d}The apparent mean opposition V magnitude.

^{e}Bright side 0.5, faint side 0.05.

^{f}Provisional names.

References

1. *1998 Astronomical Almanac* (USNO, Government Printing Office)
2. Burns, J.A. 1986, *Satellites*, edited by J.A. Burns and M.S. Matthews (University of Arizona Press, Tucson)
3. Veverka, J.M. 1998, *Observers Handbook*, edited by R.L. Bishop (University of Toronto Press, Toronto)

12.7 MOON

Mean distance from Earth [8]	384 401 ± 1 km
Extreme range	356 400–406 700 km
Mean equatorial horizontal parallax	3 422.″608
Eccentricity of orbit	0.054 90
Inclination of orbit to ecliptic oscillating ±9′ with period of 173 d	5° 8′ 43.″42
Sidereal period (fixed stars)	27.321 661
Mean orbital speed [9]	1.023 km s^{-1}
Synodical month (new moon to new moon)	29.530 588

Tropical month
   (equinox to equinox)   27.321 582
Anomalistic month
   (perigee to perigee)   27.554 550
Nodical month (node to node)   27.212 220 days
Period of Moon's node (nutation period,
   retrograde)   18.61 Julian years
Period of rotation Moon's perigee (direct) [3]   8.849 Julian years
Moon's sidereal mean daily motion $47\,434\rlap{.}{''}889$
                                        $13\rlap{.}{°}176\,358$

Mean transit interval $24^{\mathrm{h}}50\rlap{.}{^{\mathrm{m}}}47$
Main periodic terms in the motion [10]
   Principle elliptic term in longitude $22\,639'' \sin g$
   Principle elliptic term in latitude $18\,461'' \sin u$
   Evection $4\,586'' \sin(2D - g)$
   Variation $2\,370'' \sin 2D$
   Annual inequality $-669'' \sin g'$
   Parallactic inequality $-125'' \sin D$
   where g = Moon's mean anomaly
      g' = Sun's mean anomaly
      D = Moon's age
      u = distance of mean Moon
      from ascending node
Physical libration [11]   longitude   latitude
   Displacement (selenocentric) $\pm 66''$ $\pm 105''$ [9]
   Period   1 yr   6 yr
   Optical libration [11]
   Displacement (selenocentric) $\pm 7° \, 53'$ $\pm 6° \, 51'$
   Period   approximately sidereal lunar
Surface area of Moon at some time
   visible from earth   59%
Inclination of lunar equator [11]
   To ecliptic $1° \, 32' \, 32\rlap{.}{''}7$
   To orbit $6° \, 41'$
Moon radii: a toward Earth,
   b along orbit, c toward pole.
Mean Moon radius $(b + c)/2$ [8]   1 738.2 km
                                         0.272 52 Earth equatorial radius
                                         $a - c = 1.09$ km
                                         $a - b = 0.31$ km
                                         $b - c = 0.78$ km
Moon mass [12] $\mathcal{M}_{\oplus}/81.301 = 7.353 \times 10^{25}$ g
Moon semi-diameter at mean distance
   geocentric $15' \, 32\rlap{.}{''}6$
   topocentric, zenith $15' \, 48\rlap{.}{''}3$
Mean volume 2.200×10^{25} cm^3
Moon mean density   3.341 g cm^{-3}

Surface gravity	162.2 cm s^{-2}
Surface escape velocity	2.38 km s^{-1}
Moment of inertia (about rotation axis) [13]	$0.396 \mathcal{M}_\oplus b^2$
Moment of inertia differences [13–15]	
$(\alpha + \gamma = \beta)$	$\alpha = (C - B)/A = 0.000\,400$
	$\beta = (C - A)/B = 0.000\,628$
	$\gamma = (B - A)/C = 0.000\,228$

A-axis toward Earth, B along orbit, C toward pole.

Gravitational potential term [13]	$J_2 = 2.05 \times 10^{-4}$
Mascons [16]	
Number of strong mascons on the near side of the Moon	4 exceeding 80 milligals
Mean surface temperature [12]	$+107$ C (day), 153 C (night)
Temperature extremes [12]	-233 C(?), $+123$ C
Flow of heat through Moon's surface [12]	29mW m^{-2}
Moon's atmospheric density [12]	$\sim 10^4$ molecules cm^{-3} (day)
	2×10^5 molecules cm^{-3} (night)

Number of maria and craters on lunar surface
with diameters greater than d [8, 17–19] $5 \times 10^{10} d^{-2.0}$ per 10^6 km^2 (d in m)

This rule extends from the largest maria ($d \simeq 1000$ km) to the smallest holes ($d \simeq 1$ cm).

Lunar surface and photometric data are given in Tables 12.9 and 12.14.

Table 12.15 gives the integral phase function for the Moon.

Table 12.15. *Lunar integral phase function* [1].

Phase angle (deg.)	Before full Moon	After full Moon	Phase angle (deg.)	Before full Moon	After full Moon
0	1.000	1.000	80	0.120	0.111
10	0.787	0.759	90	0.0824	0.0780
20	0.603	0.586	100	0.0560	0.0581
30	0.466	0.453	110	0.0377	0.0405
40	0.356	0.350	120	0.0249	0.0261
50	0.275	0.273	130	0.0151	0.0158
60	0.211	0.211	140		0.0093
70	0.161	0.156	150		0.0046

Reference
1. Hapke B. 1974, Optical properties of lunar surface. In *Physics and Astronomy of the Moon*, edited by Zd. Kopal (Academic Press, New York)

12.8 PLANETARY RINGS

12.8.1 Rings of Jupiter

The four rings of Jupiter are described in Table 12.16.

Table 12.16. *Rings of Jupiter* [1].

Ring	Distance (km)	(R_j)	Optical depth	Albedo
Halo ring	100 000–122 000	1.40–1.71	3×10^{-6}	
Main ring	122 000–129 000	1.71–1.81	5×10^{-6}	~0.015
Gossamer ring (inner)	129 200–182 000	1.81–2.55	1.0×10^{-7}	
Gossamer ring (outer)	182 000–224 900	2.55–3.15		

Reference

1. http://nssdc.gsfc.nasa.gov/planetary/factsheet/jupringfact.html, and private communication from G.S. Orton

12.8.2 Rings of Saturn

Table 12.17 lists the details of the rings of Saturn.

Table 12.17. *Rings of Saturn*[a] [1, 2].

Zone	Radius (km)	Distance R/R_{Saturn}	Optical depth	Albedo
D-ring inner edge	>66 900	1.11		
C-ring inner edge[b]	74 658	1.239	0.05–0.35	0.12–0.30
Titan ringlet	77 871	1.292		
Maxwell ringlet	87 491	1.452		
B-ring inner edge	91 975	1.526	0.4–2.5	0.4–0.6
B-ring outer edge	117 507	1.950		
Cassini division			0.05–.015	0.2–0.4
A-ring inner edge[c]	122 340	2.030	0.4–1.0	0.4–0.6
Encke gap center	133 589	2.216		
A-ring outer edge	136 775	2.269		
F-ring center	140 374	2.329	0.1	0.6
G-ring center	170 000	2.82	1.0×10^{-6}	
E-ring inner edge	~180 000	3	1.5×10^{-5}	
E-ring outer edge	~480 000	8		

Notes

[a]Total mass of rings $6 \times 10^{-8} M_{Saturn} = 3.4 \times 10^{22}$ g.

[b]Thickness of C-ring no more than 10 m.

[c]Thickness of A-ring about 50 m.

References

1. Zebker, H.A. & Tyler, G.L. 1984, *Science*, **223**, 396
2. http://nssdc.gsfc.nasa.gov/planetary/factsheet/satringfact.html

12.8.3 Rings of Uranus

Table 12.18 lists the details of the rings of Uranus.

Table 12.18. *Rings of Uranus* [1, 2, 3].

Zone	Radius (km)	Distance R/R_{Uranus}	Optical depth	Albedo
6	41 837	1.637	∼0.3	∼15 × 10⁻³
5	42 235	1.652	∼0.5	∼15 × 10⁻³
4	42 571	1.666	∼0.3	∼15 × 10⁻³
α	44 718	1.750	∼0.4	∼15 × 10⁻³
β	45 661	1.786	∼0.3	∼15 × 10⁻³
η	47 176	1.834	∼0.4−	∼15 × 10⁻³
γ	47 626	1.863	∼1.5+	∼15 × 10⁻³
δ	48 303	1.900	∼0.5	∼15 × 10⁻³
λ	50 024	1.957	∼0.1	∼15 × 10⁻³
ϵ	51 149	2.006	0.5–2.3	∼18 × 10⁻³

References

1. Stone, E.C. & Miner, E.D. 1986, *Science*, **233**, 39
2. *Astronomical Almanac*, 1996 (USNO, Government Printing Office)
3. http://nssdc.gsfc.nasa.gov/planetary/factsheet/uranringfact.html

12.8.4 Rings of Neptune

Table 12.19 lists the details of the rings of Neptune.

Table 12.19. *Rings of Neptune* [1, 2].

Zone	Radius (km)	Distance $R/R_{Neptune}$	Optical depth	Albedo
Galle (1989 N3R)	∼41 900	1.692	∼0.000 08	∼15 × 10⁻³
LeVerrier (1989 N2R)	∼53 200	2.148	∼0.002	∼15 × 10⁻³
Lassell (1989 N4R)[a]	∼53 200	2.148	∼0.000 15	∼15 × 10⁻³
Arago (1989 N4R)[a]	∼57 200	2.310		
Unnamed (indistinct)	61 950	2.501		
Adams (1989 N1R)[b]	62 933	2.541	∼0.004 5	∼15 × 10⁻³

Notes

[a] LeVerrier and Lassell were originally identified as one ring, designated 1989N4R.
[b] Arcs in the Adams Ring with optical depths of 0.12 and albedos of about 0.04 are: Courage, Liberté, Egalité 1, Egalité 2, and Fraternité.

References

1. Lang, K.R. 1991, *Astrophysical Data: Planets and Satellites*, (Springer-Verlag, New York), p. 937
2. http://nssdc.gsfc.nasa.gov/planetary/factsheet/nepringfact.html

REFERENCES

1. Myles Standish, DE 405, private communication
2. Robert Jacobson, private communication
3. *Explanatory Supplement to the Astronomical Almanac* 1992, edited by P.K. Seidelmann (University Science, Mill Valley, CA)
4. Davies, M.E. et al. 1994, *Cel. Mech.*, **63**, 127
5. *Handbook of the British Astronomy Association* (Annual)
6. Roy, A.E. & Ovenden, M.W. 1954, *MNRAS*, **114**, 232
7. Roy, A.E. & Ovenden, M.W. 1955, *MNRAS*, **115**, 296
8. *Astrophysical Quantities*, **1**, Sec. 86; **2**, Sec. 69
9. Lang, K.R. 1991, *Astrophysical Data: Planets and Satellites*, (Springer-Verlag, New York) p. 937
10. *Landolt–Börnstein Tables*, 1962 (Springer-Verlag, New York), pp. 3, 83
11. *Astronomical Almanacs* (USNO, Government Printing Office)
12. *Lunar Source Book*, 1991, edited by G. Heiken, D. Vaniman, and B.M. French (Cambridge University Press, Cambridge)
13. Cook, A.H. 1970, *MNRAS*, **150**, 187
14. Goudas, C.L. 1967, *AJ*, **72**, 955
15. Koziel, K. 1967, *Proc. R. Soc. London.*, **296**, 248
16. Mutch, T.A. 1970, *Geology of the Moon* (Princeton University Press, Princeton, NJ), pp. 80, 217, 265
17. Jaffe, L.D. 1969, *SSRv*, **9**, 508
18. Cross, C.A. 1966, *MNRAS*, **134**, 245
19. Marcus, A. 1966, *MNRAS*, **134**, 269

Chapter 13

Solar System Small Bodies

Richard P. Binzel, Martha S. Hanner, and Duncan I. Steel

13.1 ASTEROIDS OR MINOR PLANETS

13.1.1 Populations and Locations [1–3]

Number of minor planets having well-determined orbits, cataloged by permanent designations (numbers) as of 1998, January 1: 8125.

Number of known minor planets having less well-determined orbits, cataloged by provisional designations: $> 25,000$.

Most are located in the *Main-belt*, between Mars and Jupiter.

Semimajor axis, range 2.06 to 3.28 AU, mean $a = 2.68$.
Mean orbital eccentricity: $e = 0.142$.
Mean orbital inclination: $i = 7.92$ deg.
Mean orbital period: 4.40 yr.

Number of main-belt asteroids larger than 100 km in diameter: 188, 50 km: 475.

Estimated population of main-belt asteroids larger than diameter D (in km):

$$N(> D) = 9.1 \times 10^6 D^{-2.52}.$$

Near-Earth Asteroids (NEAs) are those approaching within 0.3 AU of the Earth's orbit.

Atens: $a < 1.0$ AU, aphelion $Q > 0.983$ AU.
Number known as of 1998, January $1 = 27$.
Apollos: $a \geq 1.0$ AU, perihelion $q \leq 1.017$ AU.
Number known as of 1998, January $1 = 213$.
Amors: $a > 1.0$ AU, $1.017 < q \leq 1.3$ AU.
Number known as of 1998, January $1 = 207$.
Aten and Apollo asteroids have orbits which cross the Earth's orbit.
Orbits of many Amor asteroids can evolve to become Earth-crossing.

Estimated population of Earth-crossing asteroids having diameter:

> 1 km: 2100.
> 100 m: 320,000. (A size likely to survive passage through the terrestrial atmosphere.)

Typical collisional frequency (per object) with Earth, for an NEA having an Earth-crossing orbit:
$P_i = 2.2$ per 10^9 yr.
Mean collision velocity with Earth: $V_c = 22.5$ km/s.

Trojan asteroids are located in the vicinities of the L_4 and L_5 Lagrange points of Jupiter.

Mean semimajor axis: $a = 5.20$ AU.
Mean eccentricity: $e = 0.080$.
Mean inclination: $i = 15.9$ deg.
Number known as of 1998, January 1: 413.

13.1.2 Magnitudes [4]

An asteroid's absolute magnitude, H, is defined as its mean V magnitude (neglecting rotational and aspect variations), if it were observed at a distance $r = 1$ AU from the Sun, $\Delta = 1$ AU from the Earth, and a phase angle (Earth–object–Sun angle) $\alpha = 0$. For other locations, an asteroid's mean apparent V magnitude can be expressed by

$$V = H(\alpha) + 5 \log r \Delta,$$

where

$$H(\alpha) = H - 2.5 \log[(1 - G)\Phi_1(\alpha) + G\Phi_2(\alpha)].$$

G is called the slope parameter which accounts for an asteroid's nonlinear change in brightness as a function of phase angle only. Φ_1 and Φ_2 are described by

$$\Phi_i = \exp\{-A_i[\tan(\alpha/2)]^{B_i}\}; \qquad i = 1, 2,$$
$$A_1 = 3.33, \qquad A_2 = 1.87,$$
$$B_1 = 0.63, \qquad B_2 = 1.22.$$

An asteroid's diameter D (in km) can be estimated by

$$\log D = 3.129 - 0.5 \log p - 0.2H,$$

where p is its geometric albedo in the V passband.

An asteroid's Bond albedo, A, is related to the geometric albedo by the phase integral, q, where

$$A = pq,$$
$$q = 0.290 + 0.684G; \qquad 0 \leq G \leq 1.$$

13.1.3 Physical Properties [5]

Estimated total mass of the asteroids $= 1.8 \times 10^{24}$ g.
Estimated densities for most asteroids, $1.0 - 3.5$ g cm^{-3}.
Possible compositions, typical albedos, slope parameters, and color indices for selected taxonomic types of asteroids.

C-types: Carbonaceous chondrite, $p = 0.05$, $G = 0.15$, $B - V = 0.70$, $U - B = 0.35$.
S-types: Stony-Iron? Ordinary chondrite?, $p = 0.19$, $G = 0.25$, $B - V = 0.85$, $U - B = 0.44$.
M-types: Metal-rich?, $p = 0.10$, $G = 0.20$, $B - V = 0.70$, $U - B = 0.25$.

Typical rotation period, $P \sim 9$ h. Observed range: 2 to > 1000 h.
Typical rotation light curve amplitude variation, $\Delta M \sim 0.2$ mag. Observed range: 0 to > 1 mag.
Typical shape, modeled by a triaxial ellipsoid with axes a, b, c, where $a > b > c$:

$$a : b : c = 2 : \sqrt{2} : 1.$$

Lowest energy rotation state occurs about the c-axis.

13.1.4 Data Tables

Tables 13.1 and 13.2 give the 100 largest and 147 of the nearest asteroids.

Table 13.1. *The* 100 *largest asteroids* [1].

No.	Name	Year of Discovery	D (km)	a	e	i	P (h)	ΔM (mag)	Type	p	H	G	$U - B$	$B - V$
1	Ceres	1801	913	2.77	0.078	10.6	9.075	0.04	G	0.10	3.32	0.11	0.43	0.72
2	Pallas	1802	523	2.77	0.234	34.8	7.811	0.03–0.16	B	0.14	4.13	0.15	0.29	0.66
4	Vesta	1807	501	2.36	0.091	7.1	5.342	0.12	V	0.38	3.16	0.34	0.50	0.80
10	Hygiea	1849	429	3.14	0.120	3.8	18.4	0.09–0.18	C	0.07	5.27	−0.04	0.35	0.69
511	Davida	1903	337	3.18	0.178	15.9	5.13	0.06–0.25	C	0.05	6.17	0.02	0.36	0.72
704	Interamnia	1910	333	3.06	0.148	17.3	8.727	0.03–0.11	F	0.06	6.00	0.02	0.26	0.64
52	Europa	1858	312	3.10	0.100	7.4	5.631	0.09–0.10	CF	0.05	6.25	0.00	0.33	0.66
15	Eunomia	1851	272	2.64	0.185	11.8	6.083	0.4–0.56	S	0.19	5.22	0.20	0.46	0.84
87	Sylvia	1866	271	3.49	0.083	10.9	5.183	0.30–0.62	P	0.04	6.95	0.28	0.25	0.70
16	Psyche	1852	264	2.92	0.134	3.1	4.196	0.03–0.42	M	0.10	5.99	0.22	0.25	0.70
24	Themis	1853	249	3.13	0.134	0.8	8.38	0.10–0.14	C		7.07	0.10	0.35	0.68
31	Euphrosyne	1854	248	3.14	0.228	26.3	5.531	0.09–0.13	C	0.07	6.53	0.15	0.32	0.67
65	Cybele	1861	245	3.44	0.104	3.5	6.07	0.04–0.12	P	0.05	6.79	0.15	0.27	0.67
3	Juno	1804	244	2.67	0.258	13.0	7.21	0.14–0.22	S	0.22	5.31	0.30	0.41	0.81
324	Bamberga	1892	242	2.68	0.341	11.1	29.43	0.07	CP	0.05	6.82	0.10	0.30	0.70
107	Camilla	1868	237	3.48	0.084	9.9	4.84	0.32–0.52	C	0.06	6.80	−0.17	0.30	0.70
624	Hektor	1907	233	5.18	0.024	18.2	6.921	0.1–1.1	D		7.47	0.15	0.24	0.79
532	Herculina	1904	231	2.77	0.176	16.4	9.405	0.08–0.18	S	0.16	5.78	0.25	0.41	0.85
451	Patientia	1899	230	3.06	0.071	15.2	9.727	0.05–0.10	CU	0.07	6.65	0.20	0.33	0.65
48	Doris	1857	225	3.11	0.069	6.5	11.89	0.35	CG	0.06	6.83	−0.05	0.43	0.72
19	Fortuna	1852	221	2.44	0.158	1.6	7.445	0.22–0.35	G		7.09	0.10	0.39	0.75
29	Amphitrite	1854	219	2.55	0.072	6.1	5.39	0.01–0.15	S	0.16	5.84	0.21	0.42	0.83
121	Hermione	1872	217	3.44	0.143	7.6	6.1	0.03	C	0.04	7.39	0.15	0.39	0.72
423	Diotima	1896	217	3.08	0.034	11.2	4.622	0.06–0.18	C	0.03	7.48	0.68	0.30	0.67
13	Egeria	1850	215	2.58	0.086	16.5	7.045	0.12	G	0.09	6.47	−0.02	0.46	0.75
45	Eugenia	1857	214	2.72	0.083	6.6	5.699	0.08–0.41	FC	0.04	7.27	0.15	0.27	0.66
94	Aurora	1867	212	3.16	0.082	8.0	7.22	0.12	CP	0.03	7.55	0.09	0.30	0.66
88	Thisbe	1866	210	2.77	0.164	5.2	6.042	0.08–0.21	CF		7.05	0.17	0.29	0.66
7	Iris	1847	203	2.39	0.230	5.5	7.139	0.04–0.29	S	0.21	5.76	0.51	0.48	0.85
702	Alauda	1910	202	3.19	0.029	20.6	8.36	0.07–0.10	C	0.05	7.23	0.13	0.32	0.66

Table 13.1. *(Continued).*

No.	Name	Year of Discovery	D (km)	a	e	i	P (h)	ΔM (mag)	Type	p	H	G	U − B	B − V
375	Ursula	1893	200	3.13	0.102	15.9	16.83	0.05–0.17	C		7.43	0.23	0.34	0.68
372	Palma	1893	195	3.14	0.264	23.9	6.58	0.12	BFC	0.05	7.33	0.25		
128	Nemesis	1872	194	2.75	0.126	6.2	39	0.10	C	0.04	7.55	0.15	0.36	0.68
6	Hebe	1847	192	2.43	0.202	14.8	7.274	0.05–0.20	S	0.25	5.70	0.24	0.38	0.83
154	Bertha	1875	192	3.18	0.095	21.1				0.07	7.09	0.15		
76	Freia	1862	190	3.42	0.169	2.1	9.98	0.15–0.2	P	0.02	8.08	0.44	0.29	0.70
130	Elektra	1873	189	3.11	0.219	22.9	5.225	0.19–0.58	G	0.08	6.86	−0.04	0.47	0.75
22	Kalliope	1852	187	2.91	0.098	13.7	4.147	0.04–0.30	M	0.12	6.50	0.22	0.25	0.69
259	Aletheia	1886	185	3.15	0.112	10.7			CP	0.03	7.86	0.15	0.28	0.67
776	Berbericia	1914	183	2.93	0.166	18.2	7.672	0.13–0.23	C		7.68	0.34	0.39	0.70
41	Daphne	1856	182	2.76	0.273	15.8	5.988	0.16–0.38	C	0.07	7.16	−0.06	0.37	0.73
2060	Chiron[a]	1977	180	13.68	0.380	6.9			B		6.62	0.25	0.28	0.70
9	Metis	1848	179	2.39	0.122	5.6	5.078	0.04–0.36	S		6.32	0.29	0.51	0.86
120	Lachesis	1872	178	3.12	0.064	7.0			C	0.04	7.73	0.17	0.38	0.70
747	Winchester	1913	178	3.00	0.343	18.2	9.4	0.13	PC	0.04	7.68	0.15	0.32	0.71
790	Pretoria	1912	176	3.41	0.154	20.6	10.37	0.16	P	0.03	8.05	0.15	0.30	0.70
566	Stereoskopia	1905	175	3.39	0.093	4.9			C	0.03	8.15	0.43	0.30	0.70
911	Agamemnon	1919	175	5.21	0.068	21.8	7	0.2–0.4	D	0.04	7.88	0.15	0.22	0.77
96	Aegle	1868	174	3.05	0.140	16.0			T	0.03	7.97	0.15	0.34	0.77
194	Prokne	1879	174	2.62	0.238	18.5	15.67	0.27	C	0.05	7.66	0.15	0.35	0.73
59	Elpis	1860	173	2.71	0.117	8.6	13.69	0.1	CP	0.04	7.72	0.01	0.29	0.67
386	Siegena	1894	173	2.90	0.169	20.3	9.763	0.11	C	0.06	7.42	0.23	0.40	0.74
54	Alexandra	1858	171	2.71	0.196	11.8	7.04	0.12	C	0.05	7.70	0.15	0.36	0.70
1437	Diomedes	1937	171	5.11	0.046	20.6	18	0.35–0.42	DP	0.02	8.30	0.15	0.24	0.70
334	Chicago	1892	170	3.87	0.041	4.7			C	0.06	7.46	−0.06	0.36	0.72
444	Gyptis	1899	170	2.77	0.173	10.3	6.214	0.15	C	0.04	7.85	0.23	0.30	0.68
241	Germania	1884	169	3.05	0.103	5.5			CP	0.06	7.50	0.04	0.29	0.69
409	Aspasia	1895	168	2.58	0.070	11.2	9.03	0.10–0.14	CX	0.05	7.60	0.28	0.34	0.72
185	Eunike	1878	165	2.74	0.127	23.2	10.83		C	0.05	7.73	0.27	0.33	0.68
11	Parthenope	1850	162	2.45	0.100	4.6	7.83	0.07–0.12	S	0.15	6.62	0.27	0.42	0.85
139	Juewa	1874	162	2.78	0.177	10.9	41.8	0.18	CP	0.05	7.79	0.15	0.29	0.70
354	Eleonora	1893	162	2.80	0.116	18.4	4.277	0.12–0.30	S	0.19	6.32	0.32	0.54	0.95
804	Hispania	1915	161	2.84	0.138	15.3	7.42	0.19	PC	0.04	7.87	0.22	0.38	0.71
165	Loreley	1876	160	3.13	0.070	11.2	7.6	0.12	CD	0.06	7.49	0.15	0.31	0.74
39	Laetitia	1856	159	2.77	0.115	10.4	5.138	0.08–0.53	S	0.29	5.94	−0.03	0.50	0.89
89	Julia	1866	159	2.55	0.181	16.1	11.39	0.10–0.25	S	0.16	6.57	0.14	0.48	0.88
173	Ino	1877	159	2.74	0.209	14.2	5.93	0.04–0.11	C	0.05	7.79	0.12	0.32	0.70
488	Kreusa	1902	158	3.14	0.179	11.5			C	0.05	7.83	0.15	0.36	0.70
536	Merapi	1904	158	3.50	0.090	19.4			X	0.04	8.08	0.15	0.28	0.69
85	Io	1865	157	2.65	0.194	12.0	6.875	0.15	FC	0.06	7.56	0.05	0.28	0.66
150	Nuwa	1875	157	2.98	0.125	2.2	8.14	0.09	CX	0.03	8.32	0.15	0.27	0.71
238	Hypatia	1884	156	2.91	0.089	12.4	8.9	0.12	C	0.03	8.38	0.51	0.38	0.73
145	Adeona	1875	155	2.67	0.146	12.6	8.1	0.08	C	0.04	8.05	0.01	0.36	0.69
49	Pales	1857	154	3.08	0.236	3.2	10.42	0.15–0.20	CG	0.05	7.91	0.39	0.39	0.75
117	Lomia	1871	154	2.99	0.023	14.9			XC	0.04	8.18	0.48	0.30	0.68
168	Sibylla	1876	154	3.38	0.049	4.6			C	0.05	7.93	0.16	0.38	0.75
14	Irene	1851	153	2.59	0.166	9.1	9.35	0.04–0.1	S		6.27	0.09	0.39	0.84
51	Nemausa	1858	153	2.37	0.065	10.0	7.785	0.14–0.25	CU	0.08	7.36	0.06	0.47	0.77
106	Dione	1868	152	3.16	0.182	4.6			G	0.08	7.42	0.17	0.47	0.74
20	Massalia	1852	151	2.41	0.144	0.7	8.098	0.17–0.27	S	0.19	6.52	0.26	0.42	0.81
1172	Aneas	1930	151	5.16	0.104	16.7			D	0.03	8.26	0.15	0.26	0.73
137	Meliboea	1874	150	3.11	0.224	13.4			C	0.04	8.04	0.10	0.33	0.70
283	Emma	1889	150	3.04	0.151	8.0	6.888	0.31	X	0.02	8.73	0.15	0.30	0.71
209	Dido	1879	149	3.14	0.067	7.2	8	0.20	C	0.04	8.15	−0.09	0.29	0.69
361	Bononia	1893	149	3.95	0.216	12.7			DP	0.03	8.27	0.15	0.19	0.75
617	Patroclus	1906	149	5.23	0.139	22.0			P	0.04	8.17	0.15	0.21	0.70
18	Melpomene	1852	148	2.30	0.218	10.1	11.57	0.22–0.35	S	0.22	6.41	0.18	0.39	0.85
211	Isolda	1879	148	3.05	0.155	3.9			C	0.05	7.84	0.03	0.36	0.72
308	Polyxo	1891	148	2.75	0.038	4.4	12.03	0.20	T	0.04	8.18	0.28	0.37	0.79
508	Princetonia	1903	147	3.16	0.023	13.3			C	0.03	8.30	0.15	0.33	0.73
895	Helio	1918	147	3.20	0.149	26.1			FCB	0.02	8.64	0.15		
93	Minerva	1867	146	2.75	0.142	8.6	5.97	0.10	CU	0.08	7.47	−0.11	0.25	0.73
144	Vibilia	1875	146	2.66	0.233	4.8	13.81	0.13	C	0.05	7.87	0.08	0.39	0.72
196	Philomela	1879	146	3.11	0.027	7.3	8.333	0.07–0.33	S	0.18	6.64	0.48	0.46	0.86
420	Bertholda	1896	146	3.41	0.047	6.7			P	0.03	8.35	0.04	0.23	0.69

Table 13.1. *(Continued).*

No.	Name	Year of Discovery	D (km)	a	e	i	P (h)	ΔM (mag)	Type	p	H	G	U − B	B − V
95	Arethusa	1867	145	3.07	0.144	12.9	8.688	0.24	C	0.06	7.84	0.08	0.37	0.71
489	Comacina	1902	144	3.16	0.032	12.9			C	0.03	8.36	0.15	0.36	0.69
69	Hesperia	1861	143	2.98	0.169	8.6	5.655	0.20	M	0.12	7.10	0.15	0.23	0.70
349	Dembowska	1892	143	2.92	0.091	8.3	4.701	0.08–0.47	R	0.34	5.98	0.33	0.54	0.93
762	Pulcova	1913	142	3.16	0.092	13.0			F	0.03	8.58	0.50	0.31	0.65

Note

[a] Object 2060 Chiron is known to exhibit cometary activity, e.g., IAUC 4770, and is catalogued as comet 95p.

Reference

1. Binzel, R.P., Gehrels, T., & Matthews, M.S., editors, 1989, Asteroids II Database, in *Asteroids II* (University of Arizona Press, Tucson), pp. 997–1190

Table 13.2. *Near-earth asteroids having permanent designations* [1–3].[a]

No.	Name	Provisional designation	q (AU)	a	e	i	H	Type	D (km)	P_i (10⁹ yr)	V_c (km/s)	Category
433	Eros	1898 DQ	1.133	1.458	0.223	10.8	11.2	S	17			Amor
719	Albert	1911 MT	1.189	2.584	0.540	10.8	16.0		2			Amor
887	Alinda	1918 DB	1.087	2.486	0.563	9.3	13.8	S	5			Amor
1036	Ganymed	1924 TD	1.226	2.658	0.539	26.6	9.5	S	41			Amor
1221	Amor	1932 EA1	1.083	1.919	0.436	11.9	17.7		1	1.5	15.4	Amor
1566	Icarus	1949 MA	0.187	1.078	0.827	22.9	16.9	S	2	1.8	30.6	Apollo
1580	Betulia	1950 KA	1.119	2.195	0.490	52.1	14.5	C	8	0.5	30.6	Amor
1620	Geographos	1951 RA	0.828	1.245	0.335	13.3	15.6	S	2	3.8	16.7	Apollo
1627	Ivar	1929 SH	1.124	1.863	0.397	8.4	13.2	S	7	1.8	14.0	Amor
1685	Toro	1948 OA	0.771	1.367	0.436	9.4	14.2	S	12	4.0	17.2	Apollo
1862	Apollo	1932 HA	0.647	1.471	0.560	6.4	16.3	Q	1	2.8	20.3	Apollo
1863	Antinous	1948 EA	0.891	2.260	0.606	18.4	15.5	S	2	1.3	19.9	Apollo
1864	Daedalus	1971 FA	0.563	1.461	0.615	22.2	14.9	SQ	3	1.0	26.0	Apollo
1865	Cerberus	1971 UA	0.576	1.080	0.467	16.1	16.8	S	1	2.5	20.9	Apollo
1866	Sisyphus	1972 XA	0.873	1.893	0.539	41.2	13.0	S	10			Apollo
1915	Quetzalcoatl	1953 EA	1.081	2.537	0.574	20.4	19.0	S	0.5			Amor
1916	Boreas	1953 RA	1.250	2.272	0.450	12.8	14.9	S	3			Amor
1917	Cuyo	1968 AA	1.066	2.150	0.504	23.9	13.9		6	1.3	17.8	Amor
1943	Anteros	1973 EC	1.064	1.430	0.256	8.7	15.8	S	2	3.5	13.4	Amor
1980	Tezcatlipoca	1950 LA	1.085	1.710	0.365	26.9	13.9	S	13			Amor
1981	Midas	1973 EA	0.622	1.776	0.650	39.8	15.5	S	3	3.8	30.7	Apollo
2059	Baboquivari	1963 UA	1.256	2.651	0.526	11.0	15.8		3			Amor
2061	Anza	1960 UA	1.048	2.265	0.537	3.8	16.6	TCG	3	1.3	14.2	Amor
2062	Aten	1976 AA	0.790	0.967	0.183	18.9	16.8	S	1	7.1	16.0	Aten
2063	Bacchus	1977 HB	0.701	1.078	0.349	9.4	17.1		1	6.5	15.8	Apollo
2100	Ra-Shalom	1978 RA	0.469	0.832	0.437	15.8	16.1	C	4	6.3	17.9	Aten
2101	Adonis	1936 CA	0.441	1.874	0.765	1.4	18.7		1	2.8	25.4	Apollo
2102	Tantalus	1975 YA	0.905	1.290	0.299	64.0	16.2		2	2.5	34.8	Apollo
2135	Aristaeus	1977 HA	0.794	1.599	0.503	23.0	17.9		1	2.0	21.0	Apollo
2201	Oljato	1947 XC	0.626	2.174	0.712	2.5	15.3	S?	2	2.3	26.4	Apollo
2202	Pele	1972 RA	1.119	2.292	0.512	8.8	17.6		1	0.1	14.8	Amor
2212	Hephaistos	1978 SB	0.362	2.168	0.833	11.8	13.9	SG	5	0.4	34.6	Apollo
2329	Orthos	1976 WA	0.820	2.402	0.659	24.4	14.9		4	1.8	23.3	Apollo
2340	Hathor	1976 UA	0.464	0.844	0.450	5.8	19.2	CSU	1	14.0	16.3	Aten
2368	Beltrovata	1977 RA	1.234	2.105	0.414	5.3	15.2	SQ	3			Amor
2608	Seneca	1978 DA	1.044	2.491	0.581	15.3	17.5	S	1			Amor
3102	Krok	1981 QA	1.188	2.152	0.448	8.4	15.6	QRS	2			Amor
3103	Eger	1982 BB	0.907	1.406	0.355	20.9	15.4	E	3	3.9	17.3	Apollo
3122	Florence	1981 ET3	1.021	1.769	0.423	22.2	14.2		6	2.1	17.0	Amor
3199	Nefertiti	1982 RA	1.128	1.574	0.284	33.0	14.8	S	3			Amor
3200	Phaethon	1983 TB	0.140	1.271	0.890	22.1	14.6	F	5	1.4	35.0	Apollo
3271	Ul	1982 RB	1.271	2.102	0.395	25.0	16.7		2			Amor
3288	Seleucus	1982 DV	1.102	2.032	0.458	5.9	15.3	S	3	1.4	21.0	Amor
3352	McAuliffe	1981 CW	1.186	1.879	0.369	4.8	15.8		3			Amor
3360		1981 VA	0.633	2.465	0.743	21.7	16.3		2	0.7	26.6	Apollo

Table 13.2. (Continued.)

No.	Name	Provisional designation	q (AU)	a	e	i	H	Type	D (km)	P_i (10^9 yr)	V_c (km/s)	Category
3361	Orpheus	1982 HR	0.819	1.209	0.323	2.7	19.0	V	1	21.0	14.0	Apollo
3362	Khufu	1984 QA	0.526	0.989	0.469	9.9	18.3		1	5.3	19.8	Aten
3551	Verenia	1983 RD	1.073	2.092	0.487	9.5	16.8	V	1			Amor
3552	Don Quixote	1983 SA	1.209	4.233	0.714	30.8	13.0	D	18			Amor
3553	Mera	1985 JA	1.117	1.645	0.321	36.8	16.5		2			Amor
3554	Amun	1986 EB	0.701	0.974	0.280	23.4	15.8	M	3	5.4	17.4	Aten
3671	Dionysus	1984 KD	1.003	2.196	0.543	13.6	16.3		2	1.5	16.0	Amor
3691		1982 FT	1.270	1.774	0.284	20.4	14.9		4			Amor
3752	Camillo	1985 PA	0.986	1.414	0.302	55.6	15.5		3	1.0	27.0	Apollo
3753		1986 TO	0.484	0.998	0.515	19.8	15.1		4	3.0	22.0	Aten
3757		1982 XB	1.017	1.835	0.446	3.9	19.0	S	0.5	5.1	13.4	Amor
3838	Epona	1986 WA	0.449	1.505	0.702	29.3	15.5		3	1.0	29.0	Apollo
3908		1980 PA	1.043	1.926	0.459	2.2	17.4	V	1	4.0	14.7	Amor
3988		1986 LA	1.055	1.545	0.317	10.8	18.2		1			Amor
4015[b]	Wilson-Harrington	1979 VA	1.000	2.644	0.622	2.8	16.0	CF	4	0.8	15.5	Amor
4034		1986 PA	0.589	1.060	0.444	11.2	18.1		1			Apollo
4055	Magellan	1985 DO2	1.226	1.820	0.326	23.2	14.8	V	3			Amor
4179	Toutatis	1989 AC	0.920	2.512	0.634	0.5	15.3		3			Apollo
4183	Cuno	1959 LM	0.718	1.980	0.638	6.8	14.4		5	1.3	21.3	Apollo
4197		1982 TA	0.523	2.298	0.773	12.2	14.6		5	1.0	27.0	Apollo
4257	Ubasti	1987 QA	0.876	1.647	0.468	40.7	16.2		2			Apollo
4341	Poseidon	1987 KF	0.588	1.835	0.679	11.9	15.5		3	1.5	23.3	Apollo
4401	Aditi	1985 TB	1.117	2.578	0.567	26.7	15.8		3			Amor
4450	Pan	1987 SY	0.596	1.442	0.586	5.5	17.2		1	6.4	22.2	Apollo
4486	Mithra	1987 SB	0.743	2.200	0.663	3.0	15.6		3	5.5	18.5	Apollo
4487	Pocahontas	1987 UA	1.217	1.731	0.297	16.4	17.1		1			Amor
4503	Cleobulus	1989 WM	1.279	2.703	0.527	2.5	15.6		3			Amor
4544	Xanthus	1989 FB	0.781	1.042	0.250	14.1	17.1		1	6.6	15.5	Apollo
4581	Asclepius	1989 FC	0.657	1.022	0.357	4.9	20.4		0.5	13.3	15.4	Apollo
4596		1981 QB	1.077	2.239	0.519	37.1	16.0		2	0.8	21.8	Amor
4660	Nereus	1982 DB	0.953	1.490	0.360	1.4	18.2		1	22.5	13.1	Apollo
4688		1980 WF	1.081	2.232	0.516	6.4	19.0	SQ	0.5	1.1	20.9	Amor
4769	Castalia	1989 PB	0.549	1.063	0.483	8.9	16.9		2	4.2	18.9	Apollo
4947	Ninkasi	1988 TJ1	1.139	1.370	0.168	15.6	18.7		1			Amor
4953		1990 MU	0.555	1.621	0.658	24.4	14.1		6	0.6	26.5	Apollo
4954	Eric	1990 SQ	1.104	2.001	0.448	17.5	12.6	S	12			Amor
4957	Brucemurray	1990 XJ	1.223	1.565	0.219	35.0	15.1		4			Amor
5011	Ptah	6743 P-L	0.818	1.635	0.500	7.4	17.1		1	3.8	16.7	Apollo
5131		1990 BG	0.639	1.486	0.570	36.4	14.1		6	0.6	26.3	Apollo
5143	Heracles	1991 VL	0.419	1.835	0.771	9.2	14.0		6			Apollo
5189		1990 UQ	0.810	1.551	0.478	3.6	17.3		1	5.8	17.2	Apollo
5324	Lyapunov	1987 SL	1.136	2.958	0.616	19.5	15.2		3			Amor
5332		1990 DA	1.176	2.163	0.456	25.4	13.9	S	6			Amor
5370	Taranis	1986 RA	1.228	3.344	0.633	19.0	15.7	C	5			Amor
5381	Sekhmet	1991 JY	0.667	0.947	0.296	49.0	16.5		2	2.9	26.7	Aten
5496		1973 NA	0.881	2.433	0.638	68.0	15.3		3	0.5	40.5	Apollo
5587		1990 SB	1.080	2.392	0.548	18.1	13.6		7			Amor
5590		1990 VA	0.710	0.985	0.279	14.2	19.7		0.5	5.4	16.5	Aten
5604		1992 FE	0.551	0.927	0.405	4.8	16.4		2			Aten
5620		1990 OA	1.247	2.159	0.422	7.8	17.0		2			Amor
5626		1991 FE	1.201	2.196	0.453	3.9	14.7		4	1.6	16.4	Amor
5645		1990 SP	0.830	1.355	0.387	13.5	17.0		2	4.0	17.0	Apollo
5646		1990 TR	1.205	2.143	0.438	7.9	14.3	S	5			Amor
5653		1992 WD5	1.248	1.794	0.304	6.9	15.4		3			Amor
5660		1974 MA	0.424	1.786	0.763	38.0	15.7		3	0.9	32.0	Apollo
5693		1993 EA	0.527	1.272	0.585	5.1	17.0		2			Apollo
5731	Zeus	1988 VP4	0.786	2.267	0.653	11.5	15.8		3	1.0	20.1	Apollo
5751	Zao	1992 AC	1.215	2.104	0.423	16.1	14.8		4			Amor
5786	Talos	1991 RC	0.187	1.081	0.827	23.3	17.0		2			Apollo
5797	Bivoj	1980 AA	1.053	1.893	0.444	4.2	19.1	S	1	0.3	13.2	Amor
5828		1991 AM	0.517	1.698	0.696	30.0	16.3		2	0.5	28.0	Apollo
5836		1993 MF	1.143	2.443	0.532	8.0	13.9		6			Amor
5863	Tara	1983 RB	1.097	2.222	0.506	19.4	15.5		3			Amor
5869	Tanith	1988 VN4	1.231	1.812	0.321	17.9	17.0		2			Amor
5879		1992 CH1	1.154	1.625	0.289	21.6	17.9		1			Amor
6037		1988 EG	0.636	1.269	0.499	3.5	18.7		1	8.8	18.3	Apollo
6047		1991 TB1	0.942	1.454	0.352	23.5	17.0		2			Apollo

Table 13.2. *(Continued.)*

No.	Name	Provisional designation	q (AU)	a	e	i	H	Type	D (km)	P_i (10^9 yr)	V_c (km/s)	Category
6050		1992 AE	1.240	2.202	0.437	6.4	15.4		3			Amor
6053		1993 BW3	1.010	2.146	0.529	21.6	15.1		4			Amor
6063	Jason	1984 KB	0.522	2.216	0.764	4.8	15.3	S	3	1.1	28.8	Apollo
6178		1986 DA	1.174	2.817	0.583	4.3	15.1	M	4			Amor
6239	Minos	1989 QF	0.676	1.151	0.413	3.9	17.9		1	6.4	17.9	Apollo
6455		1992 HE	0.959	2.241	0.572	37.4	13.8		7			Apollo
6456	Golombek	1992 OM	1.298	2.194	0.409	8.2	15.9		3			Amor
6489	Golevka	1991 JX	1.011	2.517	0.598	2.3	19.2		1			Amor
6491		1991 OA	1.036	2.508	0.587	5.5	18.5		1			Amor
6569		1993 MO	1.267	1.626	0.221	22.6	16.5		2			Amor
6611		1993 VW	0.873	1.695	0.485	8.7	16.5		2			Apollo
7025		1993 QA	1.011	1.476	0.315	12.6	18.3		1			Amor
7088	Ishtar	1992 AA	1.208	1.981	0.390	8.3	16.7		2			Amor
7092	Cadmus	1992 LC	0.744	2.522	0.705	17.8	15.4		3			Apollo
7236		1987 PA	1.185	2.717	0.564	16.4	18.4		1			Amor
7335		1989 JA	0.913	1.771	0.484	15.2	17.0		2	2.8	17.5	Apollo
7336		1989 RS1	1.195	2.305	0.481	7.2	18.7		1			Amor
7341		1991 VK	0.909	1.842	0.507	5.4	16.7		2			Apollo
7350		1993 VA	0.825	1.356	0.391	7.3	17.3		1			Apollo
7358		1995 YA3	1.095	2.198	0.502	4.7	14.4		5			Amor
7474		1992 TC	1.108	1.566	0.292	7.1	18.0		1			Amor
7480	Norwan	1994 PC	1.071	1.568	0.317	9.5	17.2		1			Amor
7482		1994 PC1	0.905	1.346	0.328	33.5	16.8		2			Apollo
7753		1988 XB	0.761	1.468	0.482	3.1	18.6		1	6.2	17.0	Apollo
7822		1991 CS	0.938	1.123	0.165	37.1	17.4		1	5.0	22.0	Apollo
7839		1994 ND	1.047	2.166	0.517	27.2	17.9		1			Amor
7888		1993 UC	0.819	2.436	0.664	26.0	15.3		3			Apollo
7889		1994 LX	0.825	1.262	0.346	36.9	15.3		3			Apollo
7977		1977 QQ5	1.189	2.226	0.466	25.2	15.4		3			Amor
8013		1990 KA	1.246	2.198	0.433	7.6	16.6		1			Amor
8014		1990 MF	0.950	1.746	0.456	1.9	18.7		0.5	16.6	14.1	Apollo
8034		1992 LR	1.082	1.830	0.409	2.0	17.9		1			Amor
8035		1992 TB	0.721	1.342	0.462	28.3	17.3		1			Apollo
8037		1993 HO1	1.159	1.987	0.417	5.9	16.6		1			Amor
	Hermes[c]	1937 UB	0.618	1.644	0.624	6.1	18.0		1	2.2	21.7	Apollo

Notes

[a]Collision probabilities are available only for objects discovered prior to mid-1991. These values are presented only for objects which can evolve into an Earth-intersecting orbit.

[b]Object 4015 Wilson–Harrington is also catalogued as comet 107P.

[c]Object Hermes received a permanent name upon discovery, but is currently lost.

References

1. IAU Minor Planet Center web page as of 1998 January 1. http://cfa.www.harvard.edu/cfa/ps/mpc.html
2. The Spaceguard Survey, Report of the NASA International Near-Earth Object Detection Workshop (1992)
3. T. Gehrels, editor, 1994, *Hazards Due to Comets and Asteroids*, (University of Arizona Press, Tucson)

13.2 COMETS

13.2.1 Locations and Populations [6, 7, 2]

The source region for long-period and high-inclination, short-period comets is the Oort cloud.

Estimated distance: 10^3 to 10^5 AU.
Estimated number of comets: 10^{11}–10^{13}.
Estimated total mass: 10^{25}–10^{27} kg.

The primary source for low-inclination, short-period comets is the Kuiper belt.

Estimated distance: 30 to 1000 AU.
Estimated number of comets: 10^8–10^{12}.

Estimated total mass: 10^{22}–10^{26} kg.
Total number known as of 1998, January 1: 60.

Short-period comets, defined as orbital period $P < 200$ yr.

Total number known as of 1998, January 1: 193.
Average number of apparitions per year: 17.
Typical discovery rate per year for new comets: 6.
Mean semimajor axis: $a = 5.8$ AU.
Mean orbital eccentricity: $e = 0.6$.
Mean inclination: $i = 19$ deg.

Long-period comets, defined as orbital period $P > 200$ yr.

Total number known as of 1998, January 1: 756.
Typical discovery rate per year for new comets: 6.
Estimated semimajor axes: 10^2–10^5 AU.
Typical orbital eccentricity: $e \sim 1$.
Inclinations are isotropic.

13.2.2 Magnitudes [6]

A comet's absolute magnitude, H_o, is defined as its integrated V magnitude if it were observed at a distance $r = 1$ AU from the Sun, $\Delta = 1$ AU from the Earth, and zero phase angle. At other distances, a comet's integrated V magnitude can be estimated by

$$V = H_o + 2.5n \log r + 5 \log \Delta.$$

Typical range for n: 2 to 8.
Average value: $n \sim 4$.
For a body with no coma, tail, or emission: $n = 2$.

13.2.3 Physical Properties [6–8]

Nucleus:

Diameter range: 1.0–40 km (Halley = $16 \times 8 \times 7$ km).
Mass range: 10^{14}–10^{19} g (Halley = 10^{17}–10^{18} g).
Density range: 0.1–1.1 g cm^{-3} (Halley estimates: 0.2 to 1.1 g cm^{-3}).
Estimated albedo range: 0.01–0.05 (Halley = 0.035).
Typical rotation period: 12 h (Halley = 2.2 and 7.4 days).
Typical dust production rate at 1 AU: 10^4–10^6 g/s.
Typical gas production rate at 1 AU: 10^{28}–10^{30} molecules/s.
Gas/dust expansion rate at 1 AU: 0.5 to 1.0 km/s.
Typical dust/gas ratio (by mass): 1.0 to 2.0.
Typical mass loss per apparition: 0.05 to 1.0 percent of total mass.
Estimated composition of ices: H_2O (80%), CO (3–7%), CO_2 (3%), CH_3OH (1–6%), plus CH_3CN, $(H_2CO)_n$, HCN.
Estimated composition of grains: Mg-rich silicates, refractory organics.

Coma:

Typical radius: 10^4–10^5 km.

Typical composition: H_2O, CO, CO_2, OH, H_2CO, CH_3OH, CH_3CN, CN, C_2, C_3.

Hydrogen cloud:

Typical radius: 10^7 km.

Typical production rate at 1 AU: 10^{28}–10^{30} H atoms/s.

Ion Tail (Type I):

Typical length: 10^6–10^8 km.

Direction: antisolar.

Principal species: CO^+, H_2O^+, CO_2^+, OH^+, H_3O^+.

Dust Tail (Type II):

Typical length: 10^6–10^7 km.

Direction: Initially antisolar, becoming curved as dust particles follow independent orbits.

Particle size range: 0.1 to 100 microns.

Typical particle composition: silicates and refractory organics.

13.2.4 Comet Data Tables

Table 13.3 lists short-period comets with more than one apparition, while Table 13.4 lists those with only one appearance. Table 13.5 gives selected long-period comets. Table 13.6 lists probable cometary nature objects.

Table 13.3. *Short period comets having more than one known apparition* [1].

Comet	Name	Perihelion date (Year)	Orbital period (yrs)	Perihelion (AU)	Orbital eccentricity	Longitude of perihelion	Longitude of asc. node	Orbital inclination	Apehelion (AU)
2P	Encke	1994.1	3.28	0.33	0.850	186.3	334.7	11.9	4.09
107P[a]	Wilson–Harrington	1992.6	4.29	1.00	0.623	90.9	271.1	2.8	4.29
26P	Grigg–Skjellerup	1992.6	5.10	1.00	0.664	359.3	213.3	21.1	4.93
79P	du Toit–Hartley	1987.4	5.21	1.20	0.601	251.6	309.3	2.9	4.81
96P	Machholz	1991.6	5.24	0.13	0.958	14.5	94.5	60.1	5.91
45P	Honda–Mrkos–Pajdusakova	1990.7	5.30	0.54	0.822	325.8	89.3	4.2	5.54
73P	Schwassmann–Wachmann 3	1990.4	5.35	0.94	0.694	198.8	69.9	11.4	5.18
25D	Neujmin 2	1927.0	5.43	1.34	0.567	193.7	328.7	10.6	4.84
5D	Brorsen	1879.2	5.46	0.59	0.810	14.9	103.0	29.4	5.61
41P	Tuttle–Giacobini–Kresak	1990.1	5.46	1.07	0.656	61.6	141.6	9.2	5.14
10P	Tempel 2	1994.2	5.48	1.48	0.522	194.9	118.2	12.0	4.73
9P	Tempel 1	1994.5	5.50	1.49	0.520	178.9	69.0	10.6	4.73
46P	Wirtanen	1991.7	5.50	1.08	0.652	356.2	82.3	11.7	5.15
71P	Clark	1989.9	5.51	1.56	0.501	209.0	59.7	9.5	4.68
88P	Howell	1993.2	5.58	1.41	0.552	234.8	57.7	4.4	4.88
11D	Tempel–Swift	1908.8	5.68	1.15	0.638	113.4	291.8	5.4	5.22
100P	Hartley 1	1991.4	6.02	1.82	0.451	178.8	38.9	25.7	4.80
83P	Russell 1	1985.5	6.10	1.61	0.517	0.4	230.8	22.7	5.06
37P	Forbes	1993.2	6.13	1.45	0.568	310.5	334.5	7.2	5.25
116P	Wild 4	1996.7	6.16	1.99	0.408	170.8	22.1	3.7	4.73
103P	Hartley 2	1991.7	6.26	0.95	0.719	174.9	226.8	9.3	5.84
54P	de Vico–Swift	1965.3	6.31	1.62	0.524	325.4	25.1	3.6	5.21
81P	Wild 2	1991.0	6.37	1.58	0.541	41.6	136.2	3.2	5.30
7P	Pons–Winnecke	1989.6	6.38	1.26	0.634	172.3	93.4	22.3	5.62
6P	d'Arrest	1989.1	6.39	1.29	0.625	177.1	139.5	19.4	5.59
57P	du Toit–Neujmin–Delporte	1989.8	6.39	1.72	0.502	115.3	189.1	2.8	5.17
104P	Kowal 2	1991.8	6.39	1.50	0.564	189.5	247.8	15.8	5.38
31P	Schwassmann–Wachmann 2	1994.1	6.39	2.07	0.399	358.2	126.2	3.8	4.82
76P	West–Kohoutek–Ikemura	1994.0	6.41	1.58	0.543	360.0	84.2	30.5	5.33

Table 13.3. *(Continued.)*

Comet	Name	Perihelion date (Year)	Orbital period (yrs)	Perihelion (AU)	Orbital eccentricity	Longitude of perihelion	Longitude of asc. node	Orbital inclination	Apehelion (AU)
105P	Singer Brewster	1992.8	6.43	2.03	0.414	46.6	192.6	9.2	4.89
22P	Kopff	1990.1	6.46	1.59	0.543	162.9	120.9	4.7	5.35
43P	Wolf–Harrington	1991.3	6.51	1.61	0.539	187.0	254.9	18.5	5.37
87P	Bus	1994.5	6.52	2.18	0.375	24.4	182.2	2.6	4.80
114P	Wiseman–Skiff	1993.4	6.53	1.51	0.568	171.9	271.7	18.2	5.47
94P	Russell 4	1990.5	6.57	2.22	0.366	93.0	71.0	6.2	4.79
67P	Churyumov–Gerasimenko	1989.5	6.59	1.30	0.630	11.4	51.0	7.1	5.73
21P	Giacobini–Zinner	1992.3	6.61	1.03	0.706	172.5	195.4	31.8	6.01
3D	Biela	1852.7	6.62	0.86	0.756	223.2	248.0	12.5	6.19
44P	Reinmuth 2	1994.5	6.64	1.89	0.464	45.9	296.2	7.0	5.17
112P	Urata–Niijima	1993.5	6.64	1.46	0.588	21.5	31.9	24.2	5.61
75P	Kohoutek	1987.8	6.65	1.78	0.498	175.7	269.7	5.9	5.30
62P	Tsuchinshan 1	1991.7	6.65	1.50	0.576	22.8	96.8	10.5	5.57
18P	Perrine–Mrkos	1968.8	6.72	1.27	0.643	166.0	240.9	17.8	5.85
51P	Harrington	1994.6	6.78	1.57	0.561	233.5	119.3	8.7	5.59
49P	Arend–Rigaux	1991.8	6.82	1.44	0.600	329.1	122.1	17.9	5.75
60P	Tsuchinshan 2	1992.4	6.82	1.78	0.504	203.1	288.3	6.7	5.41
65P	Gunn	1989.7	6.84	2.47	0.314	197.0	68.5	10.4	4.74
110P	Hartley 3	1994.4	6.84	2.46	0.317	168.4	287.9	11.7	4.75
19P	Borrelly	1988.0	6.86	1.36	0.624	353.3	75.4	30.3	5.86
16P	Brooks 2	1994.7	6.89	1.84	0.491	198.0	176.9	5.5	5.40
86P	Wild 3	1994.6	6.91	2.30	0.366	179.3	72.6	15.5	4.96
15P	Finlay	1988.4	6.95	1.09	0.699	322.3	42.4	3.7	6.19
84P	Giclas	1992.7	6.96	1.85	0.493	276.5	112.5	7.3	5.44
48P	Johnson	1990.9	6.97	2.31	0.366	208.3	117.3	13.7	4.98
69P	Taylor	1991.0	6.97	1.95	0.466	355.6	108.9	20.6	5.35
77P	Longmore	1988.8	7.00	2.41	0.341	195.7	15.7	24.4	4.91
33P	Daniel	1992.7	7.06	1.65	0.552	11.0	69.1	20.1	5.71
17P	Holmes	1993.3	7.09	2.18	0.410	23.2	328.0	19.2	5.21
113P	Spitaler	1994.1	7.10	2.13	0.422	50.2	14.5	5.8	5.25
98P	Takamizawa	1991.6	7.22	1.59	0.575	147.7	124.9	9.5	5.88
108P	Ciffreo	1993.1	7.23	1.71	0.543	358.0	53.7	13.1	5.77
106P	Schuster	1992.7	7.26	1.54	0.590	355.7	50.6	20.1	5.96
102P	Shoemaker 1	1992.0	7.26	1.99	0.470	18.8	340.0	26.2	5.51
30P	Reinmuth 1	1988.4	7.29	1.87	0.503	13.1	119.8	8.1	5.65
4P	Faye	1991.9	7.34	1.59	0.578	203.9	199.6	9.1	5.96
89P	Russell 2	1994.8	7.38	2.28	0.400	249.2	42.5	12.0	5.31
47P	Ashbrook–Jackson	1993.5	7.49	2.32	0.395	348.7	2.7	12.5	5.34
61P	Shajn–Schaldach	1993.9	7.49	2.35	0.388	216.6	166.9	6.1	5.31
91P	Russell 3	1990.4	7.50	2.52	0.343	353.2	248.7	14.1	5.15
52P	Harrington–Abell	1991.5	7.59	1.77	0.540	138.7	337.3	10.2	5.95
97P	Metcalf–Brewington	1991.0	7.76	1.59	0.594	208.0	187.8	13.0	6.25
70P	Kojima	1994.1	7.85	2.40	0.393	348.5	154.8	0.9	5.50
39P	Oterma	1958.4	7.88	3.39	0.144	354.9	155.8	4.0	4.53
78P	Gehrels 2	1989.8	7.94	2.35	0.410	183.5	216.3	6.7	5.61
50P	Arend	1991.4	7.99	1.85	0.537	47.1	356.2	19.9	6.14
83P	Gehrels 3	1993.6	8.11	3.43	0.151	231.6	243.3	1.1	4.64
80P	Peters–Hartley	1990.5	8.13	1.63	0.598	338.3	260.1	29.8	6.46
111P	Helin–Roman–Crockett	1996.8	8.16	3.49	0.139	10.2	92.0	4.2	4.61
24P	Schaumasse	1993.2	8.22	1.20	0.705	57.5	81.1	11.8	6.94
14P	Wolf	1992.7	8.25	2.43	0.406	162.3	204.1	27.5	5.74
58P	Jackson–Neujmin	1987.4	8.42	1.44	0.653	196.6	163.8	14.1	6.84
36P	Whipple	1995.0	8.53	3.09	0.259	201.9	182.5	9.9	5.25
74P	Smirnova–Chernykh	1992.6	8.57	3.57	0.147	89.0	77.5	6.6	4.81
115P	Maury	1994.2	8.74	2.03	0.522	119.8	176.8	11.7	6.46
32P	Comas Sola	1987.6	8.78	1.83	0.570	45.5	61.1	13.0	6.68
59P	Kearns–Kwee	1990.9	8.96	2.22	0.487	131.8	315.8	9.0	6.42
72P	Denning–Fujikawa	1978.8	9.01	0.78	0.820	334.1	41.6	8.7	7.88
93P	Lovas 1	1989.8	9.09	1.68	0.614	73.6	342.4	12.2	7.03
64P	Swift–Gehrels	1991.2	9.21	1.36	0.692	84.8	314.4	9.3	7.43
42P	Neujmin 3	1993.9	10.6	2.00	0.586	147.0	150.4	4.0	7.67
40P	Vaisala 1	1993.3	10.8	1.78	0.635	47.4	135.1	11.6	7.98
68P	Klemola	1987.6	10.9	1.77	0.640	154.5	176.5	10.9	8.09
34P	Gale	1938.5	11.0	1.18	0.761	209.2	67.9	11.7	8.70
85P	Boethin	1986.0	11.2	1.11	0.778	11.7	26.5	5.8	8.91
56P	Slaughter–Burnham	1993.5	11.6	2.54	0.504	44.1	346.4	8.2	7.70
53P	Van Biesbroeck	1991.3	12.4	2.40	0.553	134.2	149.1	6.6	8.33

Table 13.3. *(Continued.)*

Comet	Name	Perihelion date (Year)	Orbital period (yrs)	Perihelion (AU)	Orbital eccentricity	Longitude of perihelion	Longitude of asc. node	Orbital inclination	Apehelion (AU)
92P	Sanguin	1990.2	12.5	1.81	0.663	162.8	182.5	18.7	8.96
63P	Wild 1	1973.5	13.3	1.98	0.647	167.9	358.9	19.9	9.24
8P	Tuttle	1994.5	13.5	1.00	0.824	206.7	270.5	54.7	10.3
101P	Chernykh	1992.1	14.0	2.36	0.594	263.2	130.4	5.1	9.24
29P	Schwassmann–Wachmann 1	1989.8	14.9	5.77	0.045	49.9	312.8	9.4	6.31
66P	du Toit	1974.2	15.0	1.29	0.787	257.2	22.8	18.7	10.9
99P	Kowal 1	1992.2	15.0	4.67	0.233	174.5	28.8	4.4	7.50
90P	Gehrels 1	1987.6	15.1	2.99	0.510	28.5	13.6	9.6	9.21
28P	Neujmin 1	1984.8	18.2	1.55	0.776	346.8	347.0	14.2	12.3
27P	Crommelin	1984.1	27.4	0.74	0.919	195.8	250.9	29.1	17.4
55P	Tempel–Tuttle	1965.3	32.9	0.98	0.904	172.6	235.1	162.7	19.6
38P	Stephan–Oterma	1980.9	37.7	1.57	0.860	358.2	79.2	18.0	20.9
95P[b]	Chiron	1996.1	50.7	8.45	0.383	339.6	209.4	6.9	19.0
20D	Westphal	1913.9	61.9	1.25	0.920	57.1	348.0	40.9	30.0
13P	Olbers	1956.5	69.6	1.18	0.930	64.6	86.1	44.6	32.6
23P	Brorsen–Metcalf	1989.7	70.5	0.48	0.972	129.6	311.6	19.3	33.7
121P	Pons–Brooks	1954.4	70.9	0.77	0.955	199.0	255.9	74.2	33.5
1P	Halley	1986.1	76.0	0.59	0.967	111.9	58.9	162.2	35.3
109P	Swift–Tuttle	1993.0	135.	0.96	0.964	153.0	139.4	113.4	51.7
35P	Herschel–Rigollet	1939.6	155.	0.75	0.974	29.3	356.0	64.2	56.9

Notes

[a] Object 107P, Wilson–Harrington is also catalogued as minor planet 4015.

[b] Object 95P, Chiron is also catalogued as minor planet 2060.

Reference

1. Marsden, B.G., & Williams, G.V. 1995, *Catalogue of Cometary Orbits*, 10th ed., IAU Central Bureau for Astronomical Telegrams and Minor Planet Center

Table 13.4. *Short-period comets having one known apparition* [1].

Comet	Name	Perihelion date (Year)	Orbital period (yrs)	Perihelion (AU)	Orbital eccentricity	Longitude of perihelion	Longitude of asc. node	Orbital inclination	Apehelion (AU)
D/1766 G1	Helfenzrieder	1766.3	4.35	0.41	0.848	178.7	76.3	7.9	4.92
D/1819 W1	Blanpain	1819.9	5.10	0.89	0.699	350.3	79.8	9.1	5.03
P/1994 P1	Machholz 2	1994.7	5.23	0.75	0.750	149.3	246.2	12.8	5.27
D/1884 O1	Barnard 1	1884.6	5.38	1.28	0.583	301.1	6.8	5.5	4.86
D/1886 K1	Brooks 1	1886.4	5.44	1.33	0.571	176.9	55.1	12.7	4.86
P/1991 R2	Spacewatch	1991.0	5.59	1.54	0.511	87.1	153.4	10.0	4.76
D/1770 L1	Lexell	1770.6	5.60	0.67	0.786	225.0	134.5	1.6	5.63
P/1991 F1	Mrkos	1991.2	5.64	1.41	0.555	180.4	1.7	31.5	4.92
D/1783 W1	Pigott	1783.9	5.89	1.46	0.552	354.7	58.7	45.1	5.06
D/1978 R1	Haneda–Campos	1978.8	5.97	1.10	0.665	240.5	132.2	5.9	5.48
P/1990 R2	Holt–Olmstead	1990.8	6.16	2.04	0.392	2.6	15.3	14.9	4.68
D/1978 C2	Tritton	1977.8	6.35	1.44	0.580	147.7	300.8	7.0	5.42
D/1952 B1	Harrington–Wilson	1951.8	6.36	1.66	0.515	343.0	128.5	16.3	5.20
P/1991 C2	Shoemaker–Levy 4	1990.5	6.51	2.02	0.421	302.2	152.1	8.5	4.95
D/1892 T1	Barnard 3	1892.7	6.52	1.43	0.590	170.0	208.0	31.3	5.55
P/1990 R1	Mueller 2	1990.9	6.56	2.08	0.406	171.0	218.9	7.1	4.93
D/1896 R2	Giacobini	1896.8	6.65	1.46	0.588	140.5	194.9	11.4	5.62
D/1918 W1	Schorr	1918.8	6.67	1.88	0.469	279.3	119.0	5.6	5.21
P/1991 S1	McNaught–Hughes	1991.4	6.70	2.12	0.404	223.2	90.2	7.3	4.99
P/1991 V2	Shoemaker–Levy 7	1991.8	6.72	1.63	0.542	91.7	313.0	10.3	5.49
P/1986 W1	Lovas 2	1986.7	6.75	1.46	0.592	71.3	283.8	1.5	5.69
D/1895 Q1	Swift	1895.6	7.20	1.30	0.652	167.8	171.8	3.0	6.16
P/1991 C1	Shoemaker–Levy 3	1990.9	7.25	2.81	0.250	181.7	303.8	5.0	4.68
D/1984 H1	Kowal–Mrkos	1984.4	7.32	1.95	0.483	338.0	249.3	3.0	5.59
P/1994 A1	Kushida	1994.0	7.36	1.37	0.639	214.5	245.9	4.2	6.20
P/1993 X1	Kushida–Muramatsu	1993.9	7.40	2.75	0.277	348.3	93.7	2.4	4.85
D/1894 F1	Denning	1894.1	7.42	1.15	0.698	46.4	85.7	5.5	6.46
P/1992 G2	Shoemaker–Levy 8	1992.5	7.47	2.71	0.291	22.4	213.4	6.1	4.93
D/1977 C1	Skiff–Kosai	1976.6	7.54	2.85	0.259	26.6	80.8	3.2	4.84
P/1991 V1	Shoemaker–Levy 6	1991.8	7.57	1.13	0.706	333.1	37.9	16.9	6.58

Table 13.4. *(Continued.)*

Comet	Name	Perihelion date (Year)	Orbital period (yrs)	Perihelion (AU)	Orbital eccentricity	Longitude of perihelion	Longitude of asc. node	Orbital inclination	Aphelion (AU)
P/1989 E3	West–Hartley	1988.8	7.59	2.13	0.449	102.7	46.8	15.4	5.59
D/1984 W1	Shoemaker 2	1984.7	7.84	1.32	0.666	317.6	55.5	21.6	6.57
P/1989 E2	Shoemaker–Holt 2	1988.6	8.01	2.65	0.339	5.9	99.8	17.7	5.36
P/1989 U1	Helin–Roman–Alu 2	1989.8	8.19	1.93	0.525	200.7	203.0	7.4	6.19
P/1987 U2	Mueller 1	1987.9	8.45	2.75	0.338	30.3	4.6	8.8	5.55
P/1990 S1	Mueller 3	1990.6	8.65	3.00	0.288	226.0	138.0	9.4	5.43
P/1991 T1	Shoemaker–Levy 5	1991.9	8.66	1.98	0.529	6.0	29.7	11.8	6.45
P/1989 E1	Parker–Hartley	1987.6	8.85	3.03	0.292	181.3	244.3	5.2	5.53
P/1992 G3	Mueller 4	1992.1	8.97	2.64	0.389	43.6	145.4	29.8	6.00
P/1990 UL3	Shoemaker–Levy 2	1990.7	9.28	1.84	0.582	140.1	236.0	4.6	6.99
P/1993 K2	Helin–Lawrence	1993.5	9.45	3.09	0.309	163.7	92.0	9.9	5.85
P/1989 T2	Helin–Roman–Alu 1	1987.8	9.50	3.71	0.174	216.3	73.5	9.8	5.27
P/1987 U1	Shoemaker–Holt 1	1988.4	9.55	3.05	0.322	210.4	214.6	4.4	5.95
P/1992 Q1	Brewington	1992.4	10.7	1.60	0.671	47.8	343.7	18.1	8.12
P/1988 V1	Ge–Wang	1988.4	11.3	2.52	0.501	176.1	180.5	11.7	7.58
P/1983 M1	IRAS	1983.6	13.2	1.70	0.696	356.9	357.9	46.2	9.45
P/1993 W1	Mueller 5	1994.7	13.8	4.25	0.261	30.0	100.7	16.5	7.24
P/1987 G3	Helin	1987.6	14.5	2.57	0.567	216.3	143.7	4.7	9.30
P/1994 J3	Shoemaker 4	1994.8	14.6	2.94	0.507	192.2	92.9	24.8	8.99
D/1960 S1	van Houten	1961.3	15.6	3.96	0.367	14.4	23.6	6.7	8.54
P/1983 C1	Bowell–Skiff	1983.2	15.7	1.95	0.689	169.0	346.3	3.8	10.6
P/1983 J3	Kowal–Vavrova	1983.2	15.9	2.61	0.588	19.5	202.6	4.3	10.1
P/1986 A1	Shoemaker 3	1986.0	16.9	1.79	0.728	14.9	97.3	6.4	11.4
P/1990 V1	Shoemaker–Levy 1	1990.7	17.3	1.52	0.772	310.6	52.0	24.3	11.8
D/1993 F2	Shoemaker–Levy 9	1994.2	17.7	5.38	0.207	355.0	220.9	5.8	8.20
P/1994 X1	McNaught–Russell	1994.7	18.4	1.28	0.817	171.1	218.0	29.1	12.70
P/1994 N2	McNaught–Hartley	1994.9	20.8	2.49	0.671	312.2	36.0	17.6	12.60
P/1983 V1	Hartley–IRAS	1984.0	21.5	1.28	0.834	47.1	1.5	95.7	14.2
P/1991 L3	Levy	1991.5	51.3	0.98	0.929	41.5	329.4	19.2	26.6
D/1827 M1	Pons–Gambart	1827.4	57.5	0.81	0.946	19.2	320.0	136.5	29.0
D/1921 H1	Dubiago	1921.3	62.3	1.12	0.929	97.4	67.2	22.3	30.3
D/1846 D1	de Vico	1846.2	76.3	0.66	0.963	12.9	79.7	85.1	35.3
D/1989 A3	Bradfield 2	1988.9	81.9	0.42	0.978	194.7	28.4	83.1	37.3
D/1942 EA	Vaisala 2	1942.1	85.4	1.29	0.934	335.2	172.3	38.0	37.5
D/1889 M1	Barnard 2	1889.5	145.	1.11	0.960	60.2	272.6	31.2	54.2
D/1917 F1	Mellish	1917.3	145.	0.19	0.993	121.3	88.7	32.7	55.1
D/1984 A1	Bradfield 1	1984.0	151.	1.36	0.952	219.2	356.9	51.8	55.5
D/1937 D1	Wilk	1937.1	187.	0.62	0.981	31.5	58.3	26.0	64.9

Reference

1. Marsden, B.G., & Williams, G.V. 1995, *Catalogue of Cometary Orbits*, 10th ed., IAU Central Bureau for Astronomical Telegrams and Minor Planet Center

Table 13.5. *Selected long-period comets [1, 2].*

Comet	Name	Designation	Discovery date (Year)	Perihelion (AU)	Orbital eccentricity	Orbital inclination
C/1843 D1	Great March Comet of 1843	1843 I	1843	0.005	1.000	144.3
C/1858 L1	Donati	1858 VI	1858	0.58	0.996	117.0
C/1882 R1	Great September Comet of 1882	1882 II	1882	0.008	1.000	142.0
C/1908 R1	Morehouse	1908 III	1908	0.95	1.001	140.2
C/1956 R1	Arend–Roland	1957 III	1956	0.32	1.000	119.9
C/1965 S1	Ikeya–Seki	1965 VIII	1965	0.008	1.000	141.9
C/1969 Y1	Bennett	1970 II	1970	0.54	0.996	90.0
C/1973 E1	Kohoutek	1973 XII	1973	0.14	1.000	14.3
C/1975 V1	West	1976 VI	1976	0.20	1.000	43.1
C/1980 E1	Bowell	1982 I	1980	3.36	1.057	1.7
C/1983 H1	IRAS–Araki–Alcock	1983 VII	1983	0.99	0.990	73.3
C/1988 F1	Levy	1987 XXX	1988	1.17	0.998	62.8

Table 13.5. *(Continued.)*

Comet	Name	Designation	Discovery date (Year)	Perihelion (AU)	Orbital eccentricity	Orbital inclination
C/1988 J1	Shoemaker–Holt	1988 III	1988	1.17	0.998	62.8
C/1990 K1	Levy	1990 XX	1990	0.94	1.000	131.6
C/1991 C3	McNaught–Russell	1990 XIX	1991	4.78	1.002	113.4
C/1992 J2	Bradfield	1992 XIII	1992	0.59	1.000	158.6
C/1995 O1	Hale–Bopp	—	1995	0.91	0.996	89.4
C/1996 B2	Hyakutake	—	1996	0.23	1.0	124.9

References

1. Marsden, B.G., & Williams, G.V. 1995, *Catalogue of Cometary Orbits*, 10th ed., IAU Central Bureau for Astronomical Telegrams and Minor Planets
2. Beatty, J.K., & Chaikin, A., editors. 1990, in *The New Solar System* (Sky Publishing, Cambridge), p. 292

Table 13.6. *Outer solar system objects of probable cometary nature.*[a,b,c]

Number	Name	Provisional designation	Perihelion (AU)	Aphelion (AU)	a	e	i	H	D (km)
Centaurs									
2060	Chiron	1977 UB	8.45	18.8	13.648	0.381	6.9	6.5	180
5145	Pholus	1992 AD	8.67	31.8	20.226	0.571	24.7	7.0	150
7066	Nessus	1993 HA2	11.8	37.4	24.594	0.519	15.7	9.6	75
		1994 TA	11.7	22.0	16.843	0.304	5.4	11.5	25
		1995 DW2	18.9	31.0	24.916	0.243	4.2	9.0	100
		1995 GO	6.84	29.3	18.069	0.622	17.6	9.0	100
		1997 CU26	13.1	18.4	15.712	0.169	23.4	6.0	300
Trans-Neptunian Objects									
		1992 QB1	40.9	47.7	44.298	0.077	2.2	7.0	250
		1993 FW	41.5	45.5	43.522	0.045	7.8	7.0	250
		1993 RO	31.5	47.7	39.608	0.205	3.7	8.0	150
		1993 RP	34.9	43.8	39.329	0.114	2.6	9.0	100
		1993 SB	26.9	52.4	39.633	0.321	1.9	8.0	150
		1993 SC	32.3	47.5	39.880	0.191	5.1	7.0	250
		1994 ES2	40.3	50.8	45.530	0.115	1.1	7.5	200
		1994 EV3	40.8	44.7	42.763	0.046	1.7	7.0	250
		1994 GV9	41.0	46.0	43.495	0.058	0.6	7.0	250
		1994 JQ1	41.8	46.1	43.959	0.049	3.8	7.0	250
		1994 JR1	34.8	44.1	39.434	0.119	3.8	7.5	200
		1994 JS	33.0	51.6	42.289	0.219	14.1	8.0	150
		1994 JV	35.3	35.3	35.251	0	18.1	7.0	250
		1994 TB	27.1	52.6	39.845	0.321	12.1	7.0	250
		1994 TG	42.3	42.3	42.254	0	6.8	7.0	250
		1994 TG2	42.4	42.4	42.448	0	2.2	7.0	250
		1994 TH	40.9	40.9	40.940	0	16.1	7.0	250
		1994 VK8	41.7	44.0	42.830	0.027	1.5	6.5	300
		1995 DA2	33.7	38.7	36.181	0.069	6.6	8.0	150
		1995 DB2	40.1	52.5	46.290	0.134	4.1	7.5	200
		1995 DC2	40.8	46.9	43.850	0.070	2.3	7.0	250
		1995 FB21	42.4	42.4	42.426	0	0.7	7.5	200
		1995 GA7	34.8	44.2	39.455	0.119	3.5	7.5	200
		1995 GJ	39.0	46.8	42.907	0.091	22.9	7.0	250
		1995 GY7	41.3	41.3	41.347	0	0.9	7.5	200
		1995 HM5	29.5	49.3	39.369	0.251	4.8	8.0	150
		1995 KJ1	43.5	43.5	43.468	0	2.7	6.5	300
		1995 KK1	32.0	47.0	39.475	0.190	9.3	8.5	125

Table 13.6. *(Continued.)*

Number	Name	Provisional designation	Perihelion (AU)	Aphelion (AU)	a	e	i	H	D (km)
		1995 QY9	29.2	51.0	40.115	0.271	4.8	7.5	200
		1995 QZ9	33.7	45.8	39.769	0.153	19.5	7.5	200
		1995 WY2	40.6	52.3	46.432	0.126	1.7	7.0	250
		1995 YY3	30.7	48.1	39.389	0.221	0.4	8.5	125
		1996 KV1	41.2	44.7	42.966	0.041	8.4	7.0	250
		1996 KW1	46.6	46.6	46.602	0	5.5	7.0	250
		1996 KX1	35.7	43.4	39.543	0.097	1.5	8.5	125
		1996 KY1	35.7	43.3	39.517	0.096	30.9	8.0	150
		1996 RQ20	39.2	49.4	44.291	0.115	31.6	7.0	250
		1996 RR20	32.8	47.1	39.936	0.180	5.3	7.0	250
		1996 SZ4	29.6	50.1	39.817	0.257	4.7	8.0	150
		1996 TK66	42.9	43.2	43.035	0.004	3.3	7.0	250
		1996 TL66	35.1	134.0	84.457	0.585	24.0	5.0	600
		1996 TO66	38.1	49.3	43.700	0.128	27.3	4.5	750
		1996 TP66	26.4	53.0	39.703	0.335	5.7	6.5	300
		1996 TQ66	34.6	44.7	39.667	0.127	14.6	6.5	300
		1996 TR66	33.2	52.1	42.636	0.222	12.3	7.5	200
		1996 TS66	38.5	49.7	44.100	0.126	7.4	6.0	400
		1997 CQ29	41.2	47.7	44.412	0.073	2.9	6.5	300
		1997 CR29	42.0	42.0	41.996	0	20.2	6.5	300
		1997 CS29	43.4	44.0	43.703	0.006	2.3	5.0	600
		1997 CT29	42.3	44.9	43.580	0.030	1.0	5.0	600
		1997 CU29	41.9	44.8	43.331	0.034	1.5	6.5	300
		1997 CV29	40.0	48.5	44.227	0.096	7.8	7.0	250
		1997 CW29	36.3	42.5	39.375	0.079	19.0	6.5	300
		1997 QH4	41.3	47.4	44.359	0.070	12.8	7.0	250
		1997 QJ4	34.8	44.3	39.568	0.121	16.0	7.5	200
		1997 RT5	42.2	42.2	42.239	0	12.6	7.0	250
		1997 RX9	42.1	42.1	42.135	0	29.8	8.0	150
		1997 RY6	41.4	41.4	41.360	0	12.4	7.5	200
		1997 SZ10	31.6	47.6	39.584	0.201	12.7	8.5	125
		1997 TX8	32.0	46.6	39.312	0.186	9.0	8.5	125

Notes

[a] IAU Minor Planet Center web page as of 1998, January 1. URL http://cfa.www.harvard.edu/cfa/ps/mpc.html.

[b] For explanation of symbols, see section on Minor Planets.

[c] Object 2060 Chiron is known to exhibit cometary activity, e.g., IAUC 4770 and is catalogued as comet 95P.

13.3 ZODIACAL LIGHT

The zodiacal light is due to sunlight scattered by the interplanetary dust cloud. Zodiacal light brightness is a function of viewing direction, wavelength, heliocentric distance (r) and position of the observer relative to the dust symmetry plane. The brightness does not vary with the solar cycle [9, 10]. A comprehensive review is given in [11].

Table 13.7 presents the surface brightness (radiance) and degree of linear polarization of the zodiacal light at $\lambda 5000$ Å for an observer at $r = 1$ AU in the dust symmetry plane as a function of helioecliptic longitude ($\lambda - \lambda_\odot$) and latitude (β) [11–15].

Table 13.7. *Zodiacal light brightness and polarization.*

$\lambda - \lambda_\odot(°)$ \ $\beta(°)$	0	5	10	15	20	25	30	45	60	75
0				2450	1260	770	500	215	117	78
				.08	.10	.11	.12	.16	.19	.20
5				2300	1200	740	490	212	117	78
				.09	.10	.11	.12	.16	.19	.20
10			3700	1930	1070	675	460	206	116	78
			.11	.11	.12	.13	.14	.17	.19	.20
15	9000	5300	2690	1450	870	590	410	196	114	78
	.13	.13	.13	.13	.13	.14	.15	.17	.19	.20
20	5000	3500	1880	1100	710	495	355	185	110	77
	.14	.14	.14	.15	.15	.15	.15	.17	.19	.20
25	3000	2210	1350	860	585	425	320	174	106	76
	.15	.15	.16	.16	.16	.16	.16	.18	.19	.20
30	1940	1460	955	660	480	365	285	162	102	74
	.16	.16	.16	.16	.16	.17	.17	.18	.19	.20
35	1290	990	710	530	400	310	250	151	98	73
	.17	.17	.17	.17	.17	.17	.17	.18	.20	.20
40	925	735	545	415	325	264	220	140	94	72
	.17	.17	.17	.17	.18	.18	.18	.19	.20	.20
45	710	570	435	345	278	228	195	130	91	70
	.18	.18	.18	.18	.18	.18	.18	.19	.20	.20
60	395	345	275	228	190	163	143	105	81	67
	.19	.19	.19	.19	.19	.20	.20	.20	.20	.20
75	264	248	210	177	153	134	118	91	73	64
	.18	.18	.18	.18	.18	.19	.19	.19	.19	.19
90	202	196	176	151	130	115	103	81	67	62
	.16	.16	.16	.16	.16	.16	.17	.18	.18	.19
105	166	164	154	133	117	104	93	75	64	60
	.12	.12	.12	.12	.13	.13	.14	.15	.17	.19
120	147	145	138	120	108	98	88	70	60	58
	.08	.08	.09	.09	.09	.10	.11	.13	.15	.18
135	140	139	130	115	105	95	86	70	60	57
	.05	.05	.05	.06	.06	.07	.08	.11	.14	.17
150	140	139	129	116	107	99	91	75	62	56
	.02	.02	.02	.03	.03	.04	.05	.08	.12	.16
165	153	150	140	129	118	110	102	81	64	56
	−.02	−.02	−.01	−.01	0	.02	.03	.07	.11	.16
180	180	166	152	139	127	116	105	82	65	56
	0	−.02	−.03	−.02	−.01	0	.02	.06	.11	.16

The brightness is given in S_{10} (V), the equivalent number of tenth visual magnitude solar-type stars per square degree. One S_{10} (V) $= 1.26 \times 10^{-8}$ W m^{-2} sr^{-1} μm^{-1} at 5000 Å. The uncertainty in brightness and polarization is 10% in the bright regions, to 20% in the faint regions. Negative values mean that the direction of polarization lies in the scattering plane. The brightness at the ecliptic pole ($\beta = 90°$) is 60 S_{10} (V) and the degree of linear polarization is 0.19 [11, 12].

The component of the solar corona due to scattering by interplanetary dust is known as the F corona. The brightness of the solar F corona in S_{10} (V) is given in Table 13.8 as a function of elongation (ϵ) [16, 17], for the line of sight in the ecliptic plane ($i = 0°$) and line of sight in a plane perpendicular to the ecliptic plane ($i = 90°$).

Table 13.8. *Brightness of the solar F corona.*

ϵ	$i = 0°$	$i = 90°$
1°	3.9×10^6	2.6×10^6
2	8.6×10^5	4.3×10^5
5	1.2×10^5	4.8×10^4
10	2.4×10^4	8300

UBV colors of the zodiacal light are given by [15]

$$\frac{I_V}{I_B} = 1.14 - 5.5 \times 10^{-4}\epsilon, \qquad \frac{I_B}{I_U} = 1.11 - 5.0 \times 10^{-4}\epsilon,$$

where $\epsilon =$ solar elongation in degrees. An intensity ratio of 1.0 corresponds to solar color.

The dependence of intensity on heliocentric distance for an observer at r AU ($0.3 \leq r \leq 1.0$) as measured from the Helios probe is [15]

$$\frac{I(r)}{I(1 \text{ AU})} = r^{-2.3}.$$

The dependence of polarization on heliocentric distance [15] can be approximated by

$$\frac{P(r)}{P(1 \text{ AU})} = r^{+0.3}.$$

For $1 < r < 3.3$ AU, the $I(r)$ is given by

$$\frac{I(r)}{I(1 \text{ AU})} = r^{-2.5\pm0.5},$$

as measured from Pioneer 10 [18].

The plane of symmetry of the zodiacal light deviates from the ecliptic by a few degrees, causing annual variations of 10%–20% (peak to peak) in the zodiacal light brightness as viewed from Earth. The symmetry plane differs in the inner and outer solar system; at $r > 1$ AU it is close to the invariant plane

$$\text{for } r < 1 \text{ AU}, \quad i = 3°.0 \pm 0°.3, \quad \Omega = 87° \pm 4°, \quad [19],$$

$$\text{for } r \geq 1 \text{ AU}, \quad i = 1°.5 \pm 0°.4, \quad \Omega = 96° \pm 15°, \quad [9],$$

where $i =$ inclination to the ecliptic and $\Omega =$ ecliptic longitude of the ascending node.

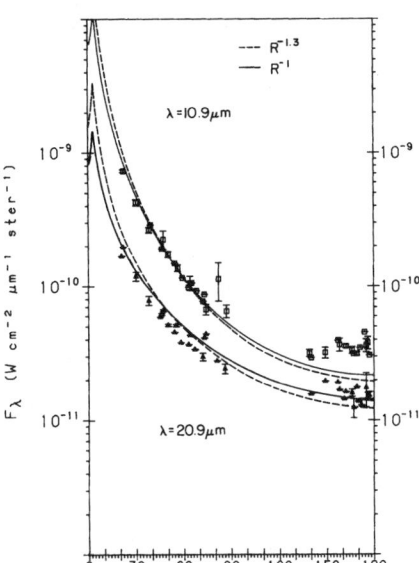

Figure 13.1. Zodiacal emission (radiance) as a function of solar elongation in the ecliptic plane [20]. □: 10.9 μm; Δ: 20.9 μm.

13.4 INFRARED ZODIACAL EMISSION

At $\lambda \gtrsim 3$ μm, thermal emission from the interplanetary dust (zodiacal emission, or ZE) dominates over scattered light. The zodiacal emission at 1 AU has been measured from rockets [20], from the *Infrared Astronomical Satellite* (IRAS) [21, 22], and from the Diffuse Infrared Background Experiment (DIRBE) on the *Cosmic Background Explorer* (COBE) satellite [23].

The observed variation in the 10.9 μm and 20.9 μm radiance along the ecliptic plane is presented in Figure 13.1 [20]. Absolute calibration accuracy is approximately 20%. Model fits for assumed radial dust distribution $\propto r^{-1.3}$ and $r^{-1.0}$ are shown by the dashed and solid lines.

Figure 13.2 shows the variation of zodiacal emission with ecliptic latitude at or near $\epsilon = 90°$ (i.e., in a plane perpendicular to the Earth–Sun line) as determined from survey observations of the IRAS satellite between February and November 1983 [21, 24]. Only the smooth component of the ZE is shown, represented by the following slowly-varying empirical function [25]. To remove zodiacal dust bands [22], point sources, and the diffuse emission of the Galaxy, the function was fitted in a lower envelope sense to IRAS scans that extended nearly from one ecliptic pole to the other:

$$I(\beta) = I_0 - \delta I\{1 - \delta\beta \mid \text{cosec}(\beta) \mid [1 - \exp(-\beta/\delta\beta - (\beta/\delta\beta)^2/3)]\},$$

where

        $I(\beta)$ = brightness at ecliptic latitude β,

        β = geocentric ecliptic latitude,

        I_0 = peak brightness,

        δI = parameter with units of brightness, and

        $\delta\beta$ = angle parameter characterizing the width of the brightness distribution.

The parameter values shown in Table 13.9 represent an annual average of the ZE at $\epsilon = 90°$. The position of peak emission deviates sinusoidally from the ecliptic plane by about two degrees on

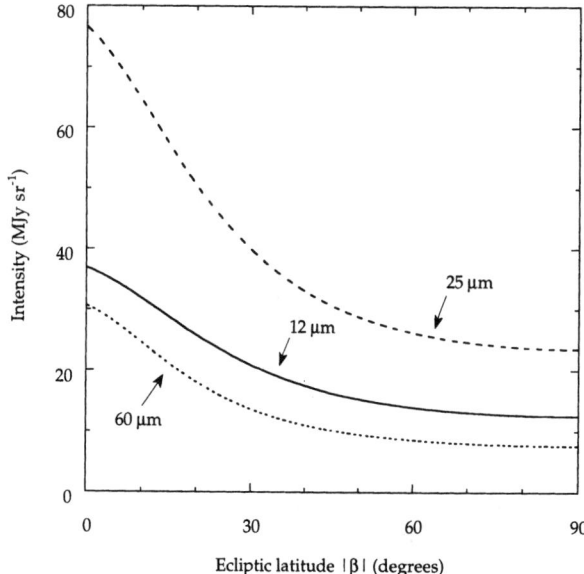

Figure 13.2. Intensity of the smooth component of the zodiacal emission as a function of the ecliptic latitude at solar elongation 90°: annual average from IRAS data [25].

a yearly cycle owing to the Earth's orbital motion in a plane inclined with respect to the approximate symmetry plane of the interplanetary dust; the peak brightness of the ZE near the ecliptic plane, I_0, and the ecliptic pole brightness, given by $I_0 - \delta I (1 - \delta\beta)$, similarly vary modestly on an annual cycle [25].

Table 13.9. *Empirical function parameters for the ZE at $\epsilon = 90°$.*

	Wavelength		
Fitted parameter	12 μm	25 μm	60 μm
I_0 (MJy sr^{-1})	37	77	31
δI (MJy sr^{-1})	34	70	29
$\delta\beta$ (degrees)	15.6	14.0	12.0

At 12 and 25 μm the diffuse infrared emission of the sky is dominated by zodiacal emission; at 60 μm, the ZE becomes less prominent, and by 100 μm emission from the galactic plane dominates the appearance of the sky, and the ZE is too weak compared with emission from the Galaxy to permit reliable separation by this method.

A linear transformation converts the IRAS values in Table 13.9 and Figure 13.2 to the somewhat different DIRBE calibration to an rms accuracy of several percent [26]. (Unlike IRAS, DIRBE has an instrumentally established zero point, an ability to measure electrical and radiative offsets, and superior stray light rejection.) The transformation is given as

$$(\text{DIRBE value}) = \text{Gain} \times (\text{IRAS value}) + \text{Offset},$$

where, at 12, 25, and 60 μm, respectively, Gain = 1.06, 1.01, and 0.87, and Offset = −0.48, −1.32, and 0.13 MJy sr^{-1}.

13.5 METEOROIDS AND INTERPLANETARY DUST

This section deals with the characteristics of meteoroids and interplanetary dust as determined from studies of their ablation or collection in the Earth's atmosphere, and from detections of impacts on spacecraft. The remote sensing of the space dust population through observations of the zodiacal light, or infrared studies such as from IRAS, COBE, ISO, etc., are covered in the preceding section.

Solid particles in space smaller than about 10 m in size are termed *meteoroids*, larger bodies being *asteroids*. Meteoroids produce *meteors* (synonym *shooting star*) when they enter the atmosphere. The term "meteor" encompasses the atmospheric phenomena resulting (optical emission, train of ionization, etc.). Dependent upon composition, entry angle, speed, and density, particles smaller than about 100 μm in size do not ablate, but remain intact and gradually settle to the Earth's surface. These particles are termed *interplanetary dust*. Such a size limit is also convenient because the majority of the zodiacal light is the result of scattering by particles in the 10–100 μm range.

The absolute visual magnitude of a meteor (M) is the observed magnitude corrected to a standard height of 100 km at the observer's zenith. Meteor activity (i.e., detection rate) is normally expressed in terms of the zenithal hourly rate. *Sporadic* (nonshower) activity is of the order of 5–10 per hour to $M = 6.5$, although there is a seasonal variation which depends upon the solar longitude and the observer's latitude. Meteor *shower* activity may be detectable at rates as low as a few per hour, although most well-known showers have zenithal rates of order 20–50 per hour. The prominent meteor showers, occurring when the Earth passes through a meteoroid *stream*, are listed in Tables 13.10 and 13.11. Every so often an exceptional shower will occur, with rates up to many thousands per hour being seen. At the time of writing the next such events, termed meteor *storms*, are anticipated in 1998 and/or 1999 November when the Leonid *storm* is due. For a more extensive discussion of all of the above, see [27].

Table 13.10. *Principal meteor showers.[a]*

Shower name[b]	Activity period	Solar long.[c]	Radiant RA	Radiant Dec	Diurnal drift RA	Diurnal drift Dec	Local time of transit	V_g^d	r^e	Peak ZHR	Number density[f]
Quadrantids	Jan 01–05	283.3	230	+49	+0.4	−0.2	08.5	39	2.2	120	80
Lyrids[g]	Apr 16–25	32.1	271	+34	+1.1	0.0	04.0	48	2.9	20	8–10
η Aquarids	Apr 19–May 28	43.1	336	−02	+0.9	+0.4	07.6	65	2.7	50	4–5
Arietids[h]	May 29–Jun 19	77	44	+23	+0.7	+0.6	09.9	35	—	—	—
ζ Perseids[h]	Jun 01–17	77	62	+23	+1.1	+0.4	11.0	25	—	—	—
β Taurids[h]	Jun 07–Jul 07	97	86	+19	+0.8	+0.4	11.2	28	—	—	—
α Capricornids	Jul 03–Aug 19	127	307	−10	+0.9	+0.3	00.0	20	2.5	10	150
S δ Aquarids	Jul 15–Aug 28	126	339	−16	+0.7	+0.2	02.2	39	3.2	20	20–25
Perseids[i]	Jul 17–Aug 24	139.9	46	+58	+1.3	+0.1	05.7	58	2.6	100	10–20
κ Cygnids	Aug 03–31	146	286	+59	+0.3	+0.1	21.3	22	3.0	5	125
S Taurids	Sep 15–Nov 25	221	50	+14	+0.8	+0.2	00.5	25	2.3	10	50
N Taurids	Sep 15–Nov 25	231	60	+23	+0.9	+0.2	00.5	27	2.3	8	30
Orionids	Oct 02–Nov 07	208	95	+16	+0.7	+0.1	04.3	65	2.9	25	2
Draconids[j]	Oct 06–10	197.0	262	+54	+0.4	0	16.1	17	2.6	—	—
Leonids[k]	Nov 14–21	235.2	152	+22	+0.7	−0.4	06.4	70	2.5	25	1–2
Geminids	Dec 07–17	262.0	112	+33	+1.0	−0.1	01.9	33	2.6	110	290
Ursids[l]	Dec 17–26	270.9	217	+75	0	0	08.4	31	3.0	20	80

Notes

[a]Courtesy J. Rendtel, M. Gyssens, P. Roggemans, and P. Brown, International Meteor Organization. All angles are in degrees, and referred to the 1950.0 equinox.

[b]See Table 13.11 for parent comet identifications.

[c]The solar longitude is that at the time of peak shower activity.

[d]V_g is the geocentric velocity of the meteoroid; the velocity at the top of the atmosphere after acceleration by the Earth is given by $V^2 = V_g^2 + 125$ (in km/s).

[e]The mass index s is related to the population index r by $s = 1 + 2.3 \log_{10} r$ (see [1] for details).

[f]The number density gives the number of particles of $m > 10^{-3}$ g per 10^9 km^3 [2].

[g]ZHR to 90.
[h]Daytime showers.
[i]ZHR > 200 in 1992–94 near parent comet return.
[j]Also known as the Giacobinids; periodic shower with ZHR > 200 occurring near alternate parent comet returns.
[k]Meteor *storms* anticipated in 1998 and 1999 near parent comet return with ZHR > 1000.
[l]ZHR to 50.

References
1. Hughes, D.W. 1978, in *Cosmic Dust*, edited by J.A.M. McDonnell (Wiley, New York), p. 123
2. Hughes, D.W. 1987, *A&A*, **187**, 879

Table 13.11. *Orbits of meteoroid streams* [1].

Shower name	a^a (AU)	e^b	q^c (AU)	ω^d (°)	Ω^e (°)	i^f (°)	Parent objects
Quadrantids	3.08	0.683	0.977	170	283.3	72.5	96P/Machholz 1 & 1491 I?
Lyrids	28	0.968	0.919	214	32.1	79	C/Thatcher (1861 G1)
η Aquarids	13	0.958	0.560	95	43.1	163.5	1P/Halley
Arietids	1.6	0.94	0.09	29	77	21	96P/Machholz 1 & 1491 I?
ζ Perseids	1.6	0.79	0.34	59	77	0	2P/Encke & various asteroids
β Taurids	2.2	0.85	0.34	246	277	6	—
α Capricornids	2.53	0.77	0.59	269	127	7	—
S δ Aquarids	2.86	0.976	0.069	153	306	27.2	96P/Machholz 1 & 1491 I?
Perseids	28	0.965	0.953	152	139.9	113.8	109P/Swift–Tuttle
κ Cygnids	3.09	0.68	0.99	194	146	38	—
S Taurids	1.93	0.806	0.375	113	41	5.2	2P/Encke & various asteroids
N Taurids	2.59	0.861	0.359	292	231	2.4	
Orionids	15	0.962	0.571	83	28	163.9	1P/Halley
Draconids	3.51	0.717	0.996	172	197.0	30.7	21P/Giacobini–Zinner
Leonids	11.5	0.915	0.985	173	235.2	162.6	55P/Tempel–Tuttle
Geminids	1.36	0.896	0.142	324	262.0	23.6	(3200) Phaethon
Ursids	5.70	0.85	0.939	206	270.9	53.6	8P/Tuttle

Notes
[a]a is the semimajor axis.
[b]e is the orbital eccentricity.
[c]q is the perihelion distance, $q = a(1 - e)$.
[d]ω is the argument of perihelion.
[e]Ω is the longitude of the ascending node (equinox 1950.0).
[f]i is the inclination to the ecliptic.

Reference
1. Cook, A.F. 1973, in *Evolutionary and Physical Properties of Meteoroids*, NASA SP-319, edited by C.L. Hemenway and A.F. Cook (NASA, Washington, DC)

The above discussion pertains to visual meteors, mostly produced by meteoroids larger than ~ 1 cm in size. Fainter meteors may be detected through HF/VHF radio wave scattering from their trains of ionization [27, 28]. Such meteors are due to smaller meteoroids, typically 100 μm–1 mm in size. The limiting magnitude is about +15 (corresponding to the micrometeor limit at ~ 100 μm); radars sensitive to such magnitudes may detect meteors at rates of one per few seconds, and especially powerful radars covering large areas at rates exceeding one per second [29]. It was thought for some years (see [27]) that the deficit of meteors detected in the radar regime (masses 10^{-6}–10^{-2} g) was due to the reduced ionizing efficiency of low-speed meteoroids (that efficiency varies as $\sim V^{3.5-4.0}$, V being the top-of-the-atmosphere velocity), but it is now known that the finite "echo ceiling" [28] of HF/VHF radars has led to only those ablating lower than ~ 105 km being detected, meaning that the majority ablating higher have been missed, but are detectable using MF radars [29, 30].

The magnitude of a meteor is given in [27, 28]:

$$M = 40 - 2.5 \log_{10} \alpha_z,$$

where α_z is the zenithal electron line density (per meter) in the train.

There have been many determinations of the relationship between α_z, V, and the initial meteoroid mass m [27, 31], both from theory and from observations. The form of the expression is generally given as

$$\alpha_z = C_1 m^x V^y \quad (m^{-1}),$$

where the normalizing constant C_1 has values typically in the range $2-8 \times 10^{-10}$, $x = 0.9-1.1$, and $y = 3.2-4.0$. Dependent upon the velocity, one finds [27]:

$$\log_{10} \alpha_z = C_2 + \log_{10} m,$$

where $C_2 = 16-17$. This implies that a meteor of zenithal magnitude zero ($M = 0$) has a mass of $\sim 0.1-1$ g.

The above assumed that the mean sporadic meteoroid speed is $\sim 30-40$ km/s; in fact the initial analysis of the Harvard Radar Meteor Project results [32] implied that the mean speed, at least for faint radar meteors, is < 20 km/s, but apparently an error was made such that the real mean speed is somewhat higher than 20 km s^{-1}[33]. Particles arriving from heliocentric elliptical orbits may impact the Earth at speeds between 11 and 73 km/s.

The *composition* of meteoroids and dust is still a matter of uncertainty. Spectroscopic observations of meteors indicate highly differentiated material similar to various meteorite classes, whilst dust collection in the stratosphere also indicates compositions similar to meteorite classes although volatile components may have been lost through heating in atmospheric entry; hypervelocity spacecraft impacts are unlikely to leave traces of any but the most refractory components. A variety of recent papers on these topics, and other features of meteoroids and interplanetary dust, may be found in [34–39]. The present state of knowledge indicates that the particles under consideration are largely comprised of meteoritic-type materials (silicates, nickel–iron) but with a significant fraction of heavy organics (kerogens) that are thermodynamically stable over periods of $\sim 10^4$ yr after release from their parent bodies, but which are destroyed on atmospheric entry.

The *origin* of at least some meteoroids is indicated by the association of various meteor showers with specific comets through orbit similarity [40]. The orbits of meteoroids determined in various surveys are reviewed and cataloged in [41], where evidence linking showers with various Earth-crossing asteroids (see Table 13.11) is also discussed. Larger meteoroids in the 5–10 m size range may also be cometary fragments [42]. While many meteoroids appear to be of low *density* ($\rho < 1$ g/cm^3), there is also a high-density component with $\rho = 3-8$ g/cm^3 [43], [44]. The evolution of meteoroid streams is reviewed in [45]. The origin of sporadic meteors appears to be gravitational stirring of streams, in particular by Jupiter; small meteoroids and dust are also subject to orbital circularization/inspiralling toward the Sun under the influence of the Poynting–Robertson drag force, with various other effects also being significant.

Meteoroids tend to end their lives through impacts upon smaller dust particles, their comminution maintaining the interplanetary dust supply (although it is not clear whether the present complex is in balance [46]), which in turn is depleted through collisions, inspiralling, and eventual ejection from the solar system by radiation/solar wind pressure.

The terrestrial mass accretion rate of small meteoroids and dust has been established from impact data collected with the Long Duration Exposure Facility [47] and other satellites [48], the small particle influx being $40 \pm 20 \times 10^6$ kg per year (see Figure 13.3), in accord with the influx determined by radar

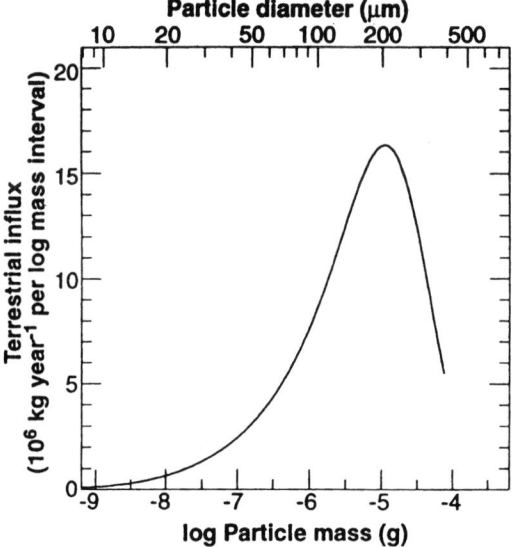

Figure 13.3. The logarithmic incremental mass influx to the Earth, in units of 10^6 kg per year per logarithmic mass interval. For these small particles (meteoroids and dust) the peak influx is at $\sim 10^{-5}$ g, and the integral under this curve is $\sim 40,000$ tonnes/year [47], although larger particles (asteroids and comets) dominate the long-term averaged mass influx [50]. From [47], Figure 4.

meteor techniques [29]. The small particle influx can also be measured from ice cores [49]. The influx over the whole mass spectrum (from dust through to large asteroids and comets) is reviewed in [50]. Whilst the interplanetary complex of meteoroids and dust is significant in a number of ways (such as its effect upon atmospheric chemistry and the light it scatters producing a diffuse background), its total mass is only equivalent to an asteroid or comet a few tens of kilometers in diameter.

REFERENCES

1. Through Minor Planet Circular 31044, 1997, December 14
2. Marsden, B.G., & Williams, G.V., and http://cfa.www.harvard.edu/cfa/ps/mpc.html
3. *The Spaceguard Survey*, Report of the NASA International Near-Earth Object Detection Workshop
4. Bowell, E., Hapke, B., Domingue, D., Lumme, K., Peltoniemi, J., & Harris, A. 1989, in *Asteroids II*, edited by R.P. Binzel, T. Gehrels, and M.S. Matthews (University of Arizona Press, Tucson), p. 524
5. Binzel, R.P., Gehrels, T., & Matthews, M.S., editors, 1989, Asteroids II Database, in *Asteroids II* (University of Arizona Press, Tucson), p. 997
6. Wilkening, L.L., editor, 1982, *Comets* (University of Arizona Press, Tucson)
7. Newburn, R.L., Neugebauer, M., & Rahe, J., editors, 1991, *Comets in the Post Halley Era* (Kluwer Academic, Dordrecht)
8. Brandt, J.C. 1990, in *The New Solar System*, edited by J.K. Beatty and A. Chaikin (Sky Publishing, Cambridge), p. 217
9. Dumont, R., & Levasseur-Regourd, A.C. 1978, *A&A*, **64**, 9
10. Leinert, Ch., Richter, I., & Planck, B. 1982, *A&A*, **110**, 111
11. Leinert, Ch. et al. 1998, *A&AS*, **127**, 1
12. Levasseur-Regourd, A.C., & Dumont, R. 1980, *A&A*, **84**, 277
13. Dumont, R., & Sánchez, F. 1976, *A&A*, **51**, 393
14. Frey, A., Hofmann, W., Lemke, D., & Thum, C. 1974, *A&A*, **36**, 447
15. Leinert, Ch., Richter, E., Pitz, E., & Hanner, M. 1982, *A&A*, **110**, 355
16. Fechtig, H., Leinert, Ch., & Grün, E. 1981, in *Landolt-Börnstein Numerical Data and Functional Relationships in Science and Technology*, Vol. 2, edited by K. Schaifers and H.H. Voigt (Springer-Verlag, Berlin), p. 235
17. Blackwell, D.E., Dewhirst, D.W., & Ingham, M.F. 1967, *Adv. Astron. Astrophys.*, **5**, 1
18. Hanner, M.S., Sparrow, J.G., Weinberg, J.L., & Beeson, D.E. 1976, in *Interplanetary Dust and Zodiacal Light*.

Lecture Notes Phys. 48, edited by H. Elsässer and H. Fechtig (Springer-Verlag, Heidelberg), p. 29

19. Leinert, Ch., Hanner, M., Richter, I., & Pitz, E. 1980, *A&A*, **82**, 328

20. Murdock, T.L., & Price, S.D. 1985, *AJ*, **90**, 375

21. Hauser, M.G. et al. 1984, *ApJ*, **278**, L15

22. Low, F.J. et al. 1984, *ApJ*, **278**, L19

23. Hauser, M.G. 1996, "COBE Observations of Zodiacal Emission," *IAU Colloquium 150 "Physics, Chemistry, and Dynamics of Interplanetary Dust."* ASP Conf. Ser. 104, edited by A.S. Gustafson & M.S. Hanner (Astronomical Society of the Pacific, San Francisco), p. 309

24. Beichman, C.A., Neugebauer, G., Habing, H.J., Clegg, P.E., & Chester, T.J., editors, *IRAS Catalogs and Atlases, Explanatory Supplement*, 1988, NASA RP-1190

25. Vrtilek, J.M., & Hauser, M.G. 1995, *ApJ*, **455**, 677

26. *Cosmic Background Explorer Diffuse Infrared Background Experiment Explanatory Supplement: The Galactic Plane Maps*, 1993, National Space Science Data Center (NASA/Goddard Space Flight Center)

27. Hughes, D.W., 1978, in *Cosmic Dust*, edited by J.A.M. McDonnell (Wiley, New York), p. 123

28. McKinley, D.W.R., 1961, *Meteor Science and Engineering* (McGraw-Hill, New York)

29. Thomas, R.M., Whitham, P.S., & Elford, W.G. 1988, *J. Atmos. Terr. Phys.*, **50**, 703

30. Steel, D., & Elford, W.G. 1991, *J. Atmos. Terr. Phys.*, **53**, 409

31. Bronshten, V.A. 1983, *Physics of Meteoric Phenomena* (Reidel, Dordrecht)

32. Cook, A.F., Flannery, M.R., Levy, H., McCrosky, R.E., Sekanina, Z., Shao, C.-Y., Southworth, R.B., & Williams, J.T. 1972, *Meteor Research Program*, NASA CR-2109 (Smithsonian Astrophysical Observatory, Cambridge)

33. Taylor, A.D. 1995, *Icarus*, **116**, 154

34. Levasseur-Regourd, A.C., & Hasegawa, H., editors, 1991, *Origin and Evolution of Interplanetary Dust*, IAU colloquium 126 (Kluwer Academic, Dordrecht)

35. Štohl, J., & Williams, I.P., editors, 1993, *Meteoroids and Their Parent Bodies* (Slovak Academy of Sciences, Bratislava)

36. Milani, A., Di-Martino, M., & Cellini, A., editors, 1994, *Asteroids, Comets, Meteors 1993*, IAU Symposium 160 (Kluwer Academic, Dordrecht)

37. Porubčan, V., Williams, I.P., & Vanýsek, V., editors, 1995, *Meteoroids: Earth, Moon & Planets*, Vol. 68, Nos. 1–3

38. Levasseur-Regourd, A.C., editor, 1997, *Asteroids, Comets, Meteors*, COSPAR Conference Proceedings, Vol. 10

39. Gustafson, B.Å.S., & Hanner, M.S., editors, 1996, *Physics, Chemistry, and Dynamics of Interplanetary Dust*, ASP Conference Series Vol. 104

40. Cook, A.F. 1973, in *Evolutionary and Physical Properties of Meteoroids*, NASA SP-319, edited by C.L. Hemenway, P.M. Millman, and A.F. Cook (NASA, Washington, DC), p. 183

41. Steel D.I., 1996, *Space Sci. Rev.*, **78**, 507

42. Rabinowitz, D.L. 1993, *ApJ*, **407**, 412

43. Babadzhanov, P.B. 1993, in *Meteoroids and Their Parent Bodies*, edited by Štohl, J. and Williams, I.P. (Slovak Academy of Sciences, Bratislava), p. 295

44. Babazdhanov, P.B. 1994, in *Asteroids, Comets, Meteors 1993*, edited by Milani, A., Di Martino, M, and Cellini, A., IAU Symposium 160 (Kluwer Academic, Dordrecht), p. 45

45. Williams, I.P. 1996 in *Physics, Chemistry, and Dynamics of Interplanetary Dust*, edited by Gustafson, B.Å.S. and Hanner, M.S., ASP Conference Series Vol. 104, p. 89

46. Grün, E., Zook, H.A., Fechtig, H., & Giese, R.H. 1985, *Icarus*, **62**, 244

47. Love, S.G., & Brownlee, D.E. 1993, *Science*, **262**, 550

48. Laurence, M.R., & Brownlee, D.E. 1986, *Nature*, **323**, 136

49. Maurette, M., Jehanno, C., Robin, E., & Hammer, C. 1987, *Nature*, **328**, 699

50. Ceplecha, Z. 1992, *A&A*, **263**, 361

Chapter 14

Sun

William C. Livingston

14.1 BASIC DATA

14.1.1 Global

Solar radius	$R_\odot = 6.955\,08 \pm 0.000\,26 \times 10^{10}$ cm [1]
Volume	$V_\odot = 1.4122 \times 10^{33}$ cm^3
Surface area	6.087×10^{22} cm^2
Solar mass	$\mathcal{M}_\odot = 1.989 \times 10^{33}$ g
Mean density	$\rho_\odot = 1.409$ g cm^{-3}
Gravity at surface	2.740×10^4 cm s^{-2}
Moment of inertia	5.7×10^{53} g cm^2
Angular rotation velocity at equator	2.85×10^{-6} rad s^{-1}
Angular momentum (based on surface rotation)	1.63×10^{48} g cm^2 s^{-1}
Work required to dissipate solar matter to infinity	6.6×10^{48} erg
Sun's total internal radiant energy	2.8×10^{47} erg
Escape velocity at solar surface	6.177×10^7 cm s^{-1}

14.1.2 Viewed from Earth

Mean equatorial horizontal parallax [2]	$8''794\,18$
	$= 4.263\,54 \times 10^{-5}$ rad
Mean distance from Earth	1 AU $= 1.495\,979 \times 10^{13}$ cm
(A = AU = astronomical unit)	
Distance at	
Perihelion	1.4710×10^{13} cm
Aphelion	1.5210×10^{13} cm
Semidiameter of Sun	$959''63$
At mean Earth distance	$0.004\,652\,4$ rad
Oblateness: Semidiameter equator–pole difference [3, 4]	$0''0086$
Solid angle of Sun, mean distance	6.8000×10^{-5} sr
	$A/R_\odot = 214.94$
	$(A/R_\odot)^2 = 46\,200$
	$(A/R_\odot)^{1/2} = 14.661$
Surface area of sphere of unit radius	$4\pi A^2 = 2.8123 \times 10^{27}$ cm^2
In heliographic coordinates	$1° = 12\,147$ km
At mean distance A	$1'$ of arc $= 4.352 \times 10^4$ km
	$1''$ of arc $= 725.3$ km

14.1.3 Total Solar Radiation

Solar constant S (total solar irradiance) = flux of total radiation received outside the Earth's atmosphere per unit area at the mean Sun–Earth distance [5–9]:

$$S = 1.365\text{--}1.369 \text{ W m}^{-2} = 1.365\text{--}1.369 \times 10^6 \text{ erg cm}^{-2} \text{s}^{-1},$$

Radiation from whole Sun	$L_\odot = 3.845 \times 10^{26}$ W $= 3.845 \times 10^{33}$ erg s^{-1}.
Radiation per unit mass	$L_\odot/\mathcal{M}_\odot = 1.933 \times 10^{-4}$ W kg^{-1} $= 1.933$ erg s^{-1} g^{-1}.

| Radiation emittance at Sun's surface | $\mathcal{F} = 6.312 \times 10^7 \, \mathrm{W \, m^{-2}} = 6.317 \times 10^{10} \, \mathrm{erg \, cm^{-2} \, s^{-1}}$. |

Mean radiation intensity of Sun's disk
$$F = \mathcal{F}/\pi = 2.009 \times 10^7 \, \mathrm{W \, m^{-2} \, sr^{-1}}.$$
$$= 2.009 \times 10^{10} \, \mathrm{erg \, cm^{-2} \, s^{-1}}.$$

14.1.4 Sun as a Star

Magnitudes of the Sun in three wavelength bands and the bolometric magnitude are given in Table 14.1 [10–13].

Table 14.1. *Solar magnitudes.*

	Apparent	Modulus	Absolute
Visual (m_v)	$V = -26.75$	31.57	$M_V = +4.82$
Blue	$B = -26.10$		$M_B = +5.47$
Ultraviolet	$U = -25.91$		$M_U = +5.66$
Bolometric	$m_{\mathrm{bol}} = -26.83$		$m_{\mathrm{bol}} = +4.74$

Color indices [10–14]:

$$B - V = +0.650,$$
$$U - B = +0.195,$$
$$U - V = +0.845,$$
$$V - R = +0.54,$$
$$V - I = +0.88,$$
$$V - K = +1.49.$$

Bolometric correction	$\mathrm{BC} = -0.08$.	
Spectral type	G2 V.	
Effective temperature	5777 K.	
Velocity relative to near stars	$19.7 \, \mathrm{km \, s^{-1}}$	
Solar apex	$A = 271°,$	$D = 30°(1900),$
	$L^{\Pi} = 57°,$	$B^{\Pi} = 22°.$
Age of Sun [15, 16]	$(4.5–4.7) \times 10^9$ yr.	
Mean magnetic field [17]		
Average	0 G,	
Peak	± 1 G.	

14.2 INTERIOR MODEL
by Pierre Demarque and David Guenther

The tabulated data in Table 14.2 are for a standard model of the Sun (no rotation, no diffusion), from Table 3B in [18]. This model was constructed using opacities from [19] and the solar mixture from [20]. Other similar recent models can be found in [21] and [22].

Central values

Temperature	$T_c = 15.7 \times 10^6$ K.
Density	$\rho_c = 151$ g cm^{-3}.
Pressure	$P_c = 2.33 \times 10^{17}$ dyn cm^{-2}.
Central hydrogen content by mass	$X_c = 0.355.$

Surface composition parameters

$$X = 0.6937,$$
$$Z = 0.0188.$$

The fraction of the radius at the base of the surface convection (SCZ or surface convection zone) can be determined by helioseismology [23, 24], which is within 1% of model [18]:

$$r_{SCZ}/R_\odot = 0.71.$$

Table 14.2. *Model of solar interior.*

r ($R_\odot$)	r (cm)	T (10^6 K)	ρ (g cm^{-3})	$\mathcal{M}r$ ($\mathcal{M}_\odot$)	Lr (erg s^{-1})	Lr ($L_\odot$)	P (dyn cm^{-2})	log P (dyn cm^{-2})
0.007	4.87×10^8	15.7	150	0.00003	1.01×10^{30}	0.0002	2.33×10^{17}	17.369
0.02	1.39×10^9	15.6	146	0.001	3.97×10^{31}	0.010	2.27×10^{17}	17.355
0.09	6.24×10^9	13.6	95.73	0.057	1.39×10^{33}	0.361	1.50×10^{17}	17.177
0.22	1.53×10^{10}	8.77	28.72	0.399	3.72×10^{33}	0.966	3.35×10^{16}	16.525
0.32	2.23×10^{10}	6.42	9.77	0.656	3.85×10^{33}	1.000	5.29×10^{15}	15.724
0.42	2.92×10^{10}	4.89	3.22	0.817	3.85×10^{33}	1.000	2.10×10^{15}	15.324
0.52	3.62×10^{10}	3.77	1.05	0.908	3.85×10^{33}	1.000	5.28×10^{14}	14.722
0.60	4.18×10^{10}	3.15	0.500	0.945	3.85×10^{33}	1.000	2.10×10^{14}	14.322
0.71	4.94×10^{10}	2.23	0.177	0.977	3.85×10^{33}	1.000	5.26×10^{13}	13.721
0.81	5.64×10^{10}	1.29	0.0766	0.992	3.85×10^{33}	1.000	1.32×10^{13}	13.119
0.91	6.33×10^{10}	0.514	0.0194	0.999	3.85×10^{33}	1.000	1.32×10^{12}	12.119
0.96	6.68×10^{10}	0.208	4.85×10^{-3}	0.9999	3.85×10^{33}	1.000	1.31×10^{11}	11.118
0.99	6.89×10^{10}	0.00441	2.56×10^{-4}	1.0000	3.85×10^{33}	1.000	1.31×10^9	9.118
0.995	6.93×10^{10}	0.00266	4.83×10^{-5}	1.0000	3.85×10^{33}	1.000	1.31×10^8	8.118
0.999	6.95×10^{10}	0.00135	1.29×10^{-6}	1.0000	3.85×10^{33}	1.000	1.31×10^6	6.118
1.000	6.96×10^{10}	0.00060	2.18×10^{-7}	1.0000	3.85×10^{33}	1.000	8.27×10^4	4.918

14.3 SOLAR OSCILLATIONS
by Frank Hill

$\mathcal{R}_\odot$ = solar radius.

g = gravitational acceleration at solar surface.

ℓ = spherical harmonic degree of mode of oscillation.

m = spherical harmonic azimuthal degree of mode.

n = radial order of mode.

ν = frequency of mode.

ω = angular frequency of mode, $\omega = 2\pi\nu$.

k_h = horizontal wave number of mode, $k_h = \sqrt{\ell(\ell+1)}/\mathcal{R}_\odot$.

P_i = Legendre polynomial of degree i.

$A(\nu, \ell)$ = amplitude of mode.

$\Gamma(\nu, \ell)$ = full width at half maximum of mode.

Characteristic period of p (pressure) modes	5 min.
Characteristic photospheric amplitude of p modes	10 cm s^{-1}.
Characteristic lifetime of p modes	7 days.
Estimated number of excited p modes	10^7.

14.3.1 Approximations for Frequencies $\nu_{n,\ell}$ of Zonal ($m = 0$) p Modes

(a) Tassoul first-order asymptotic approximation for low-degree modes with $\ell \leq 3$ and $11 \leq n \leq 33$ [25]:

$$\nu(n, \ell) = \nu_0 \left(n + \frac{\ell}{2} + \delta \right)$$

with measured coefficients in Table 14.3 [26] and accuracy of 2.8–4.1 μHz.

Table 14.3. *Fit values.*

ℓ	ν_0 (μHz)	δ
0	135.4	1.43
1	135.7	1.36
2	135.4	1.36
3	135.7	1.24

(b) Tassoul second-order asymptotic approximation for low-degree modes with $\ell \leq 3$ and $11 \leq n \leq 33$ [25]:

$$\nu(n, \ell) = \nu_0 \left(n + \frac{\ell}{2} + \delta - \frac{\ell(\ell+1)\alpha - \beta}{n + \ell/2 + \delta} \right),$$

with measured coefficients in Table 14.4 [26] and accuracy of 1.4–2.0 μHz.

Table 14.4. *Second-order fit values.*

ℓ	ν_0 (μHz)	α	β	δ
0	137.0		5.6	0.90
1	137.9	0.20	7.8	0.62
2	137.4	0.15	7.8	0.70
3	137.0	0.20	7.4	0.80

(c) Polynomial approximation for low-degree modes with $\ell \leq 3$ and $11 \leq n \leq 33$ [27]:

$$v(n, \ell) = \Delta v_\ell + \bar{v}_\ell \left(n + \frac{\ell}{2} - n_0 \right) + \gamma_\ell \left(n + \frac{\ell}{2} - n_0 \right)^2,$$

using $n_0 = 22$ as a reference order, measured coefficients listed in Table 14.5 in μHz [26] and accuracy of 1.0–1.2 μHz.

Table 14.5. *Polynomial fit values.*

ℓ	Δv_ℓ	$\bar{v}_\ell$	γ_ℓ
0	3169.4	135.31	0.090
1	3166.2	135.52	0.105
2	3160.5	135.35	0.085
3	3150.8	135.52	0.070

The quantities Δv_ℓ and $\bar{v}_\ell$ are linear functions of $\ell(\ell + 1)$:

$$\Delta v_\ell = \Delta v_0 - \ell(\ell + 1) D_0,$$
$$\bar{v}_\ell = v_0 + \ell(\ell + 1) d_0,$$

with fitted values [26]

$$\Delta v_0 = 3169.4 \ \mu\text{Hz},$$
$$v_0 = 135.35 \ \mu\text{Hz},$$
$$D_0 = 1.54 \ \mu\text{Hz},$$
$$d_0 = 0.012 \ \mu\text{Hz}.$$

(d) Parabolic fit for intermediate-degree modes with $4 \leq \ell \leq 100$, $3 \leq n \leq 24$, and accuracy of 1–10 μHz [26]:

$$v(n, \ell) = a_0(n) + a_1(n)\ell + a_2(n)\ell^2,$$

where coefficients a_i are fitted to second-order polynomials in n expressed in matrix form as

$$\begin{pmatrix} a_0 \\ a_1 \\ a_2 \end{pmatrix} = \begin{pmatrix} 643 & 101.3 & 0.71 \\ 8.6 & 2.9 & -0.047 \\ -0.025 & -0.008 & -0.0002 \end{pmatrix} \begin{pmatrix} 1 \\ n \\ n^2 \end{pmatrix}.$$

(e) Empirical fit for low- and intermediate-degree modes with $1 \leq \ell \leq 200$, 1.7 mHz $\leq v \leq$ 5.0 mHz, $\mathcal{R}_\odot$ in km, and accuracy of 10 μHz [28]:

$$v(n, \ell) = 2354.2(n + 1.57) e^{0.2053[(\ln x - 14.523)^2 + 4.1175]^{1/2} - \ln x} \ \mu\text{Hz},$$
$$x = (n + 1.57)\pi \mathcal{R}_\odot [\ell(\ell + 1)]^{-1/2}.$$

The Duvall dispersion law [29] collapses all p-mode ridges in an k_h–ω diagram to a single ridge via a transformation of coordinates. This transformation is

$$\frac{(n + \alpha)\pi}{\omega} = f\left(\frac{\omega}{k_h} \right)$$

with fitted value

$$\alpha = 1.67.$$

The dependence of v on m for p modes is [30]

$$v(\ell, m, n) = v(\ell, n) + \sqrt{\ell(\ell+1)} \sum_{i=1}^{i=5} a_i(v, \ell) P_i \left(\frac{-m}{\sqrt{\ell(\ell+1)}} \right),$$

where the splitting coefficients a_i depend on $v(n, \ell)$, in mHz [31]:

$$a_i(v, \ell) = a_i^*(\ell) + b_i^*(\ell)[v(n, \ell) - 2.5].$$

Some of these coefficients are given in Table 14.6.

Table 14.6. *Selected splitting coefficients* [1]. *All coefficients in* nHz.

ℓ	a_1^*	b_1^*	a_2^*	b_2^*	a_3^*	b_3^*	a_4^*	b_4^*	a_5^*	b_5^*
11	436.7	−1.0	−3.5	−1.7	12.0	−2.1	1.3	6.8	1.3	3.2
20	438.4	−0.7	−0.9	0.2	16.9	0.3	0.8	1.3	−3.1	2.2
29	439.5	−0.5	−0.5	−0.5	19.9	−5.1	1.2	−1.1	−3.4	1.0
38	440.7	0.3	−0.1	−0.8	21.3	−2.0	0.9	1.4	−2.5	−0.1
47	441.4	0.1	−0.3	−1.0	21.5	−0.7	0.4	1.9	−3.3	0.5
56	441.5	0.7	0.2	0.6	22.3	−0.5	0.4	1.3	−3.5	0.3

Reference
1. Libbrecht, K.G. 1989, *ApJ*, **336**, 1092

The dependence of p-mode frequency change Δv of $v(\ell, n)$ on area-weighted average full-disk absolute magnetic field B in Gauss [32] is

$$\Delta v = a(B - 7),$$

with fitted value

$$a = 0.027 \ \mu\text{Hz/G}.$$

The approximate formulas for amplitude $A(v, \ell)$ of p modes [33, 34]

$$A(v, \ell) = 10^{(b+c)/2} \ \text{cm/s},$$

with fitted values

$$
\begin{aligned}
b &= 2.2v - 3.5, & v &< 2.9 \text{ mHz}, \\
&= -0.9v + 5.6, & v &> 2.9 \text{ mHz}, \\
c &= -8.8 \times 10^{-4}\ell, & \ell &< 340, \\
&= -3.1 \times 10^{-3}\ell + 0.75, & \ell &> 340.
\end{aligned}
$$

The observed estimate of absorption fraction α of p-mode power by sunspots is discussed in [35] and listed in Table 14.7.

Table 14.7. *Sunspot absorption.*

k_h (Mm^{-1})	α
0.2	0.10
0.3	0.18
0.4	0.34
≥ 0.5	0.42

Approximate formulas for the full width at half maximum (FWHM) $\Gamma(\nu, \ell)$ of p modes [33, 36, 37] are

$$\Gamma(\nu, \ell) = 1.7 \times 10^{-2}(\ell - 20) + 10^d \ \mu\text{Hz},$$

with fitted values

$$\begin{aligned} d &= \nu - 2.3, & \nu < 2.4 \text{ mHz}, \\ &= 0.1, & 2.4 \text{ mHz} \leq \nu \leq 3.1 \text{ mHz}, \\ &= \nu - 3.0, & 3.1 \text{ mHz} \leq \nu \leq 4.3 \text{ mHz}, \\ &= 0.4\nu - 0.6, & 4.3 \text{ mHz} < \nu. \end{aligned}$$

The dispersion relation for f (fundamental) mode is

$$\omega = \sqrt{gk_h},$$

or equivalently

$$\nu = 99.8569[\ell(\ell + 1)]^{1/4} \ \mu\text{Hz}.$$

The first-order asymptotic approximation for period $P(n, \ell)$ of g (gravity) mode with $n \gg \ell$ [25] is

$$P(n, \ell) = \frac{P_0}{2} \frac{2n + \ell + \phi}{[\ell(\ell + 1)]^{1/2}}.$$

Theoretical estimates of period spacing P_0 and phase ϕ from standard solar models [38] are

$$P_0 = 33.9 \text{ to } 38.0 \text{ min},$$
$$\phi = -0.42 \text{ to } -0.25.$$

Observational estimates [38] are

$$P_0 = 29.9 \text{ to } 42.6 \text{ min},$$
$$\phi = -0.35 \text{ to } +2.$$

Properties of "160-min" oscillation [38] are

$$\text{period} = 160.010 \text{ min},$$
$$\text{amplitude} = 54 \text{ cm/s}.$$

Table 14.8 gives zonal p-mode frequencies for selected n and ℓ values.

Table 14.8. Selected measured zonal p-mode frequencies [1, 2]. All values in μHz.

n	ℓ=0	5	10	25	50	75	100	150	200	300	400	500	600	700	800	900	1000
0										1743.6	1998.0	2227.8	2438	2632	2821	2984	3140
2					1409.99	1655.77	1844.12	2148.50	2395.6	2780.3	3089.6	3359.1	3622	3877	4120	4374	4601
4				1489.44	1833.33	2181.39	2423.47	2811.51	3130.2	3650.1	4092.1	4475.4	4815	5124			
6			1443.40	1847.91	2296.79	2638.73	2928.67	3405.04	3800.2	4458.8	4974.2						
8			1737.84	2181.78	2676.98	3067.43	3397.96	3954.48	4425.0								
10		1823.33	2021.53	2494.35	3045.59	3474.37	3844.40	4475.10									
12	1823.60	2098.45	2302.80	2797.28	3402.83	3868.91	4271.92										
14	2093.50	2371.21	2578.64	3096.17	3753.12	4252.25	4688.09										
16	2362.50	2641.07	2855.70	3390.93	4095.93	4628.75											
18	2629.60	2913.54	3132.88	3684.81	4428.00	4999.33											
20	2899.30	3186.12	3410.10	3977.49	4753.20												
22	3168.60	3459.57	3688.24	4270.76													
24	3439.80	3734.52	3967.99														
26	3711.50	4009.61	4247.68														
28	3984.90																
30	4257.40																
32	4532.30																

References

1. Duvall, Jr., T.L., Harvey, J.W., Libbrecht, K.G., Popp, B.D., & Pomerantz, M.A. 1988, *ApJ*, **324**, 1158
2. Libbrecht, K.G., Woodard, M.F., & Kaufman, J.M. 1990, *ApJS*, **74**, 1129

14.4   PHOTOSPHERIC–CHROMOSPHERIC MODEL
by Eugene Avrett

Table 14.9 gives a model of the average quiet solar atmosphere, from [39]. The height h is the distance above $\tau_{500} = 1$, where τ_{500} is the radial optical depth in the continuum at 500 nm. Hydrostatic equilibrium is assumed so that $m = P_{tot}/g$, where m is the column mass, P_{tot} is the total pressure, and g is the gravitational acceleration at the solar surface. In the photosphere ($-100 < h < 525$ km) and in the chromosphere ($525 < h < 2100$ km) the temperature T has been adjusted empirically so that the computed spectrum is in agreement with the spatially averaged spectrum from quiet areas (away from sunspots and active regions). The temperature distribution in the transition region above $h \approx 2100$ km (up to $T = 10^5$ K) has been determined theoretically by balancing the downflow of energy from the corona (due to thermal conduction and diffusion) with the radiative energy losses. The microvelocity v_t roughly accounts for the Doppler broadening that is observed to exceed the thermal broadening of lines formed at various heights (see [40, 41]). The total pressure P_{tot} is the sum of the gas pressure P_{gas} and the turbulent pressure $\rho v_t^2/2$, where ρ is the gas density.

The table also lists the total hydrogen density n_H and the proton and electron densities n_p and n_e. The number densities and other quantities are determined by solving the coupled radiative transfer and statistical equilibrium equations [without assuming local thermal equilibrium (LTE)], given the T and v_t distributions. The helium to hydrogen abundance ratio is assumed to be 0.1. The abundances of the other contributing elements are from [42].

See [43] and [44] for similar empirical models of the photosphere. Models for faint and bright components of the quiet Sun and for a plage region are given in [39]. See [45] for a theoretical line-blanketed LTE photospheric model, and [46] for theoretical non-LTE line-blanketed chromospheric models. Bifurcated chromospheric models based on a combination of hot and cool components are given in [47] and [48]. Papers in [49] and [50] discuss related studies and include references to earlier work.

Other aspects of the chromosphere, such as infrared and radio data, are referred to in [51–53].

Table 14.9. *Solar atmospheric model.*

h (km)	τ_{500}	m (g cm^{-2})	T (K)	v_t (km s^{-1})	n_H (cm^{-3})	n_p (cm^{-3})	n_e (cm^{-3})	P_{tot} (dyn cm^{-2})	P_{gas}/P_{tot}	ρ (g cm^{-3})
2218.20	0.00×10^{-10}	6.777×10^{-6}	100 000	11.73	5.575×10^{9}	5.575×10^{9}	6.665×10^{9}	1.857×10^{-1}	0.952	1.31×10^{-14}
2216.50	7.70×10^{-10}	6.779×10^{-6}	95 600	11.65	5.838×10^{9}	5.837×10^{9}	6.947×10^{9}	1.857×10^{-1}	0.950	1.37×10^{-14}
2214.89	1.53×10^{-9}	6.781×10^{-6}	90 816	11.56	6.151×10^{9}	6.150×10^{9}	7.284×10^{9}	1.858×10^{-1}	0.948	1.44×10^{-14}
2212.77	2.60×10^{-9}	6.785×10^{-6}	83 891	11.42	6.668×10^{9}	6.667×10^{9}	7.834×10^{9}	1.859×10^{-1}	0.945	1.56×10^{-14}
2210.64	3.75×10^{-9}	6.788×10^{-6}	75 934	11.25	7.381×10^{9}	7.378×10^{9}	8.576×10^{9}	1.860×10^{-1}	0.941	1.73×10^{-14}
2209.57	4.38×10^{-9}	6.790×10^{-6}	71 336	11.14	7.864×10^{9}	7.858×10^{9}	9.076×10^{9}	1.860×10^{-1}	0.938	1.84×10^{-14}
2208.48	5.06×10^{-9}	6.792×10^{-6}	66 145	11.02	8.488×10^{9}	8.476×10^{9}	9.718×10^{9}	1.861×10^{-1}	0.935	1.99×10^{-14}
2207.38	5.81×10^{-9}	6.794×10^{-6}	60 170	10.86	9.334×10^{9}	9.307×10^{9}	1.059×10^{10}	1.862×10^{-1}	0.931	2.19×10^{-14}
2206.27	6.64×10^{-9}	6.797×10^{-6}	53 284	10.67	1.047×10^{10}	1.047×10^{10}	1.182×10^{10}	1.862×10^{-1}	0.925	2.47×10^{-14}
2205.72	7.10×10^{-9}	6.798×10^{-6}	49 385	10.55	1.135×10^{10}	1.125×10^{10}	1.266×10^{10}	1.863×10^{-1}	0.921	2.66×10^{-14}
2205.21	7.55×10^{-9}	6.800×10^{-6}	45 416	10.42	1.233×10^{10}	1.217×10^{10}	1.365×10^{10}	1.863×10^{-1}	0.916	2.89×10^{-14}
2204.69	8.05×10^{-9}	6.801×10^{-6}	41 178	10.27	1.356×10^{10}	1.332×10^{10}	1.491×10^{10}	1.863×10^{-1}	0.910	3.18×10^{-14}
2204.17	8.61×10^{-9}	6.803×10^{-6}	36 594	10.09	1.521×10^{10}	1.483×10^{10}	1.657×10^{10}	1.864×10^{-1}	0.903	3.56×10^{-14}
2203.68	9.19×10^{-9}	6.805×10^{-6}	32 145	9.90	1.724×10^{10}	1.667×10^{10}	1.858×10^{10}	1.864×10^{-1}	0.894	4.04×10^{-14}
2203.21	9.81×10^{-9}	6.807×10^{-6}	27 972	9.70	1.971×10^{10}	1.887×10^{10}	2.098×10^{10}	1.865×10^{-1}	0.883	4.62×10^{-14}
2202.75	1.05×10^{-8}	6.809×10^{-6}	24 056	9.51	2.276×10^{10}	2.154×10^{10}	2.389×10^{10}	1.866×10^{-1}	0.871	5.33×10^{-14}
2202.27	1.13×10^{-8}	6.812×10^{-6}	20 416	9.30	2.658×10^{10}	2.483×10^{10}	2.743×10^{10}	1.866×10^{-1}	0.856	6.23×10^{-14}
2201.87	1.21×10^{-8}	6.815×10^{-6}	17 925	9.13	3.008×10^{10}	2.778×10^{10}	3.049×10^{10}	1.867×10^{-1}	0.843	7.05×10^{-14}
2201.60	1.27×10^{-8}	6.817×10^{-6}	16 500	9.02	3.255×10^{10}	2.979×10^{10}	3.256×10^{10}	1.868×10^{-1}	0.834	7.63×10^{-14}
2201.19	1.36×10^{-8}	6.820×10^{-6}	15 000	8.90	3.570×10^{10}	3.218×10^{10}	3.498×10^{10}	1.869×10^{-1}	0.823	8.36×10^{-14}
2200.85	1.44×10^{-8}	6.823×10^{-6}	14 250	8.83	3.762×10^{10}	3.343×10^{10}	3.619×10^{10}	1.869×10^{-1}	0.816	8.81×10^{-14}
2200.10	1.63×10^{-8}	6.830×10^{-6}	13 500	8.74	4.013×10^{10}	3.441×10^{10}	3.699×10^{10}	1.871×10^{-1}	0.808	9.40×10^{-14}
2199.00	1.90×10^{-8}	6.840×10^{-6}	13 000	8.66	4.244×10^{10}	3.456×10^{10}	3.695×10^{10}	1.874×10^{-1}	0.801	9.94×10^{-14}
2190.00	4.15×10^{-8}	6.936×10^{-6}	12 000	8.48	4.854×10^{10}	3.411×10^{10}	3.663×10^{10}	1.900×10^{-1}	0.785	1.14×10^{-13}
2168.00	9.85×10^{-8}	7.203×10^{-6}	11 150	8.30	5.500×10^{10}	3.619×10^{10}	3.889×10^{10}	1.974×10^{-1}	0.775	1.29×10^{-13}
2140.00	1.76×10^{-7}	7.588×10^{-6}	10 550	8.10	6.252×10^{10}	3.806×10^{10}	4.095×10^{10}	2.079×10^{-1}	0.769	1.46×10^{-13}
2110.00	2.62×10^{-7}	8.063×10^{-6}	9900	7.87	7.314×10^{10}	3.923×10^{10}	4.238×10^{10}	2.209×10^{-1}	0.760	1.71×10^{-13}
2087.00	3.30×10^{-7}	8.483×10^{-6}	9450	7.70	8.287×10^{10}	3.954×10^{10}	4.291×10^{10}	2.324×10^{-1}	0.753	1.94×10^{-13}
2075.00	3.66×10^{-7}	8.724×10^{-6}	9200	7.61	8.882×10^{10}	3.956×10^{10}	4.305×10^{10}	2.390×10^{-1}	0.748	2.08×10^{-13}
2062.00	4.05×10^{-7}	9.005×10^{-6}	8950	7.52	9.569×10^{10}	3.952×10^{10}	4.314×10^{10}	2.467×10^{-1}	0.743	2.24×10^{-13}
2043.00	4.62×10^{-7}	9.453×10^{-6}	8700	7.41	1.055×10^{11}	3.937×10^{10}	4.314×10^{10}	2.590×10^{-1}	0.738	2.47×10^{-13}
2017.00	5.41×10^{-7}	1.014×10^{-5}	8400	7.26	1.203×10^{11}	3.921×10^{10}	4.313×10^{10}	2.778×10^{-1}	0.732	2.82×10^{-13}
1980.00	6.53×10^{-7}	1.128×10^{-5}	8050	7.06	1.446×10^{11}	3.908×10^{10}	4.310×10^{10}	3.092×10^{-1}	0.727	3.39×10^{-13}
1915.00	8.53×10^{-7}	1.387×10^{-5}	7650	6.74	1.971×10^{11}	3.974×10^{10}	4.351×10^{10}	3.800×10^{-1}	0.724	4.62×10^{-13}
1860.00	1.03×10^{-6}	1.676×10^{-5}	7450	6.49	2.547×10^{11}	4.100×10^{10}	4.423×10^{10}	4.593×10^{-1}	0.727	5.97×10^{-13}
1775.00	1.31×10^{-6}	2.298×10^{-5}	7250	6.12	3.788×10^{11}	4.399×10^{10}	4.630×10^{10}	6.297×10^{-1}	0.736	8.87×10^{-13}
1670.00	1.69×10^{-6}	3.510×10^{-5}	7050	5.69	6.292×10^{11}	4.922×10^{10}	5.085×10^{10}	9.616×10^{-1}	0.752	1.47×10^{-12}
1580.00	2.07×10^{-6}	5.186×10^{-5}	6900	5.34	9.900×10^{11}	5.390×10^{10}	5.535×10^{10}	1.421×10^{-10}	0.767	2.32×10^{-12}
1475.00	2.59×10^{-6}	8.435×10^{-5}	6720	4.93	1.726×10^{12}	6.037×10^{10}	6.191×10^{10}	2.311×10^{-10}	0.787	4.05×10^{-12}
1378.00	3.19×10^{-6}	1.363×10^{-4}	6560	4.53	2.970×10^{12}	6.824×10^{10}	7.007×10^{10}	3.735×10^{-10}	0.809	6.96×10^{-12}

Table 14.9. (Continued.)

h (km)	τ_{500}	m (g cm^{-2})	T (K)	v_t (km s^{-1})	n_H (cm^{-3})	n_p (cm^{-3})	n_e (cm^{-3})	P_{tot} (dyn cm^{-2})	P_{gas}/P_{tot}	ρ (g cm^{-3})
1278.00	4.02×10^{-6}	2.312×10^{-4}	6390	4.04	5.393×10^{12}	7.768×10^{10}	7.994×10^{10}	6.335×10^{-10}	0.837	1.26×10^{-11}
1180.00	5.19×10^{-6}	4.022×10^{-4}	6230	3.53	1.002×10^{13}	8.783×10^{10}	9.083×10^{10}	1.102×10^{1}	0.867	2.35×10^{-11}
1065.00	7.43×10^{-6}	8.074×10^{-4}	6040	2.94	2.164×10^{13}	9.992×10^{10}	1.047×10^{11}	2.212×10^{1}	0.901	5.07×10^{-11}
980.00	1.03×10^{-5}	1.396×10^{-3}	5900	2.52	3.931×10^{13}	1.068×10^{11}	1.142×10^{11}	3.824×10^{1}	0.924	9.21×10^{-11}
905.00	1.44×10^{-5}	2.314×10^{-3}	5755	2.19	6.806×10^{13}	1.078×10^{11}	1.192×10^{11}	6.341×10^{1}	0.940	1.59×10^{-10}
855.00	1.85×10^{-5}	3.282×10^{-3}	5650	1.99	9.931×10^{13}	1.051×10^{11}	1.208×10^{11}	8.993×10^{1}	0.949	2.33×10^{-10}
805.00	2.39×10^{-5}	4.710×10^{-3}	5490	1.77	1.481×10^{14}	9.014×10^{10}	1.122×10^{11}	1.290×10^{2}	0.958	3.47×10^{-10}
755.00	3.09×10^{-5}	6.868×10^{-3}	5280	1.54	2.268×10^{14}	6.493×10^{10}	9.690×10^{10}	1.882×10^{2}	0.967	5.31×10^{-10}
705.00	4.00×10^{-5}	1.022×10^{-2}	5030	1.38	3.560×10^{14}	3.637×10^{10}	8.387×10^{10}	2.799×10^{2}	0.972	8.34×10^{-10}
650.00	5.55×10^{-5}	1.624×10^{-2}	4750	1.18	6.033×10^{14}	1.375×10^{10}	9.000×10^{10}	4.451×10^{2}	0.978	1.41×10^{-9}
600.00	8.53×10^{-5}	2.538×10^{-2}	4550	1.00	9.895×10^{14}	5.368×10^{9}	1.255×10^{11}	6.954×10^{2}	0.983	2.32×10^{-9}
560.00	1.40×10^{-4}	3.680×10^{-2}	4430	0.89	1.478×10^{15}	2.825×10^{9}	1.767×10^{11}	1.008×10^{3}	0.986	3.46×10^{-9}
525.00	2.39×10^{-4}	5.125×10^{-2}	4400	0.80	2.078×10^{15}	2.424×10^{9}	2.413×10^{11}	1.404×10^{3}	0.989	4.87×10^{-9}
490.00	4.29×10^{-4}	7.149×10^{-2}	4410	0.72	2.898×10^{15}	2.618×10^{9}	3.300×10^{11}	1.959×10^{3}	0.991	6.79×10^{-9}
450.00	8.51×10^{-4}	1.044×10^{-1}	4460	0.65	4.192×10^{15}	3.600×10^{9}	4.714×10^{11}	2.860×10^{3}	0.993	9.82×10^{-9}
400.00	1.98×10^{-3}	1.664×10^{-1}	4560	0.55	6.549×10^{15}	6.715×10^{9}	7.344×10^{11}	4.558×10^{3}	0.995	1.53×10^{-8}
350.00	4.53×10^{-3}	2.626×10^{-1}	4660	0.52	1.012×10^{16}	1.267×10^{10}	1.134×10^{12}	7.194×10^{3}	0.996	2.37×10^{-8}
300.00	1.01×10^{-2}	4.103×10^{-1}	4770	0.55	1.545×10^{16}	2.604×10^{10}	1.737×10^{12}	1.124×10^{4}	0.995	3.62×10^{-8}
250.00	2.20×10^{-2}	6.344×10^{-1}	4880	0.63	2.331×10^{16}	5.605×10^{10}	2.645×10^{12}	1.738×10^{4}	0.994	5.46×10^{-8}
200.00	4.73×10^{-2}	9.705×10^{-1}	4990	0.79	3.476×10^{16}	1.253×10^{11}	4.004×10^{12}	2.659×10^{4}	0.990	8.14×10^{-8}
175.00	6.87×10^{-2}	1.195	5060	0.90	4.211×10^{16}	2.028×10^{11}	4.945×10^{12}	3.274×10^{4}	0.988	9.87×10^{-8}
150.00	9.92×10^{-2}	1.466	5150	1.00	5.062×10^{16}	3.579×10^{11}	6.153×10^{12}	4.017×10^{4}	0.985	1.19×10^{-7}
125.00	1.42×10^{-1}	1.790	5270	1.10	6.024×10^{16}	7.119×10^{11}	7.770×10^{12}	4.905×10^{4}	0.983	1.41×10^{-7}
100.00	2.02×10^{-1}	2.174	5410	1.20	7.107×10^{16}	1.485×10^{12}	1.003×10^{13}	5.957×10^{4}	0.980	1.67×10^{-7}
75.00	2.87×10^{-1}	2.625	5580	1.30	8.295×10^{16}	3.281×10^{12}	1.353×10^{13}	7.192×10^{4}	0.977	1.94×10^{-7}
50.00	4.13×10^{-1}	3.148	5790	1.40	9.558×10^{16}	7.614×10^{12}	1.980×10^{13}	8.624×10^{4}	0.975	2.24×10^{-7}
35.00	5.22×10^{-1}	3.496	5980	1.46	1.027×10^{17}	1.439×10^{13}	2.779×10^{13}	9.578×10^{4}	0.973	2.40×10^{-7}
20.00	6.75×10^{-1}	3.869	6180	1.52	1.098×10^{17}	2.588×10^{13}	4.064×10^{13}	1.060×10^{5}	0.972	2.57×10^{-7}
10.00	8.14×10^{-1}	4.132	6340	1.55	1.142×10^{17}	3.926×10^{13}	5.501×10^{13}	1.132×10^{5}	0.971	2.68×10^{-7}
0.00	1.00	4.404	6520	1.60	1.182×10^{17}	6.014×10^{13}	7.697×10^{13}	1.207×10^{5}	0.971	2.77×10^{-7}
-10.00	1.25	4.686	6720	1.64	1.219×10^{17}	9.269×10^{13}	1.107×10^{14}	1.284×10^{5}	0.970	2.86×10^{-7}
-20.00	1.61	4.975	6980	1.67	1.246×10^{17}	1.536×10^{14}	1.730×10^{14}	1.363×10^{5}	0.970	2.92×10^{-7}
-30.00	2.14	5.269	7280	1.70	1.264×10^{17}	2.597×10^{14}	2.807×10^{14}	1.444×10^{5}	0.970	2.96×10^{-7}
-40.00	2.95	5.567	7590	1.73	1.280×10^{17}	4.249×10^{14}	4.480×10^{14}	1.525×10^{5}	0.971	3.00×10^{-7}
-50.00	4.13	5.869	7900	1.75	1.295×10^{17}	6.668×10^{14}	6.923×10^{14}	1.608×10^{5}	0.971	3.04×10^{-7}
-60.00	5.86	6.174	8220	1.77	1.307×10^{17}	1.022×10^{15}	1.050×10^{15}	1.691×10^{5}	0.972	3.06×10^{-7}
-70.00	8.36	6.481	8540	1.79	1.317×10^{17}	1.515×10^{15}	1.546×10^{15}	1.776×10^{5}	0.972	3.09×10^{-7}
-80.00	1.20×10^{1}	6.790	8860	1.80	1.325×10^{17}	2.180×10^{15}	2.215×10^{15}	1.860×10^{5}	0.973	3.10×10^{-7}
-90.00	1.70×10^{1}	7.102	9140	1.82	1.337×10^{17}	2.942×10^{15}	2.979×10^{15}	1.946×10^{5}	0.973	3.13×10^{-7}
-100.00	2.36×10^{1}	7.417	9400	1.83	1.351×10^{17}	3.826×10^{15}	3.867×10^{15}	2.032×10^{5}	0.974	3.17×10^{-7}

14.5 SPECTRAL LINES
by William Livingston and Oran R. White

14.5.1 Absorption Features

Selected Fraunhofer absorption features are given in Table 14.10. Equivalent width refers to disk center. Cycle variability, where known, refers to solar irradiance, or Sun as a star [54–64].

Table 14.10. *Absorption features.*

Wavelength (nm)	Name	Species	Equiv. width (nm)	Cycle var. [% (p-to-p)]	Comment
279.54		Mg II	2.2	10	UV emission, high chromosphere
280.23		Mg II			
388.36	(CN band head)	CN	0.03 (index)	3	Photosphere, magnetic field tracer
393.36	K	Ca II	2.0	15	Chromosphere
396.85	H	Ca II	1.5	10	
430.79	G band	CH (Fe I, Ti II)	0.72		Magnetic field tracer
517.27	b_2	Mg I	0.075		Low chromosphere
518.36	b_1	Mg I	0.025		
525.02		Fe I	0.0070	0.3	Photo. magnetic fields ($g = 3$)
537.96		Fe I	0.0079	0.3	Medium photosphere
538.03		C I	0.0025	0.0	Low photosphere
557.61		Fe I			Photo. velocity fields ($g = 0$)
587.56	D_3	He I			Chromo., flares, prominences
589.00	D_2	Na I	0.075		Upper photo., low chrom., prom.
589.59	D_1	Na I	0.056		(same except water blend free)
612.22		Ca I			Photo. magnetic fields ($g \sim 1.5$)
630.25		Fe I	0.0083		Photo. magnetic fields ($g = 2.5$)
656.28	C (Hα)	H I	0.40	6	Chromo., prom., flares
676.78		Ni I			Photo. oscillations
769.89		K I			Photo. oscillations
777.42		O I	0.0066	−1	High photo. (?) (NLTE?)
854.21		Ca I	0.37		Low chromo., prom.
868.86		Fe I	0.014		Photo. magnetic fields ($g = 1.7$)
1006.37		FeH			Umbral (only) mag. fields ($g = 1.22$)
1083.03		He I	0.003	200	High chromosphere
1281.81	H Paschβ	H I	0.19		Chromosphere
1564.85		Fe I	0.0035		Photo. magnetic fields ($g = 3$)
1565.29		Fe I	0.003		Photo. magnetic fields ($g = 1.8$)
2231.06		Ti I			Umbral (only) mag. fields ($g = 2.5$)
4652.55	H Pfundβ	H I			Chromo., electric fields
4666.24		CO			High photo. thermal structure
12318.3		Mg I			High photo., magnetic fields ($g = 1$)

14.5.2 Emission Features

Table 14.11 gives absolute spectral irradiances at the Earth for the UV and EUV with estimates of solar cycle variability where known. Irradiances from both individual lines and integration over bands are given in the table. The irradiance for all entries identified as a "line" in column 3 (bandwidth) is the integral for the line, and is in units of mW m^{-2}. In contrast, irradiances for the "bands" are mean fluxes per nanometer wavelength interval for that band [65, 66].

Table 14.11. *Solar spectral irradiances:* 0.5–300 nm.

Band	Band center (nm)	Bandwidth (nm)	Solar irradiance		Solar cycle variability I_{max}/I_{min}	Species
			Solar max.	Solar min.		
GOES[a]	0.50	0.6	1.9×10^{-2}	0		
4	22.50	5	6.5×10^{-2}	1.6×10^{-2}	4	
5	25.63	line	2.6	7.7×10^{-2}	34	He II, Si X
6	28.42	line	2.9	5.9×10^{-1}	5	Fe XV
7	27.50	5	8.6×10^{-3}	3.9×10^{-3}	2	
8	30.33	line	1.6	1.6×10^{-1}	10	Si XI
9	30.38	line	7.4	3.9	2	He II
10	32.50	5	6.9×10^{-2}	1.1×10^{-2}	6	
11	36.81	line	1.2	1.1×10^{-1}	10	Mg IX
12	37.50	5	1.5×10^{-2}	7.8×10^{-3}	2	
14	46.52	line	7.1×10^{-1}	1.3×10^{-1}	5	Ne VII
15	47.50	5	3.0×10^{-3}	1.5×10^{-3}	2	
17	55.44	line	1.7	5.7×10^{-1}	3	O IV
18	58.43	line	3.5×10^{-1}	1.6×10^{-1}	2	He I
19	57.50	5	1.1×10^{-2}	5.5×10^{-3}	2	
20	60.98	line	9.8×10^{-1}	4.9×10^{-1}	2	Mg X
21	62.97	line	1.5×10^{-1}	5.5×10^{-2}	3	O V
22	62.50	5	9.3×10^{-3}	2.9×10^{-3}	3	
24	70.33	line	1.2×10^{-1}	4.8×10^{-2}	3	O III
25	72.50	5	3.9×10^{-3}	2.1×10^{-3}	2	
27	77.04	line	4.1×10^{-1}	2.0×10^{-1}	2	Ne VIII
28	78.94	line	5.5×10^{-1}	2.2×10^{-1}	3	O IV
29	77.50	5	2.2×10^{-3}	1.0×10^{-3}	2	
33	97.70	line	9.0×10^{-1}	3.6×10^{-1}	2	C III
34	97.50	5	5.4×10^{-2}	2.4×10^{-2}	2	
35	102.57	line	1.8	6.2×10^{-1}	3	H I (Lyβ)
36	103.19	line	1.7	7.0×10^{-1}	2	O VI
37	102.50	5	5.1×10^{-2}	1.7×10^{-2}	3	
	121.50	1	1.0×10^{1}		1.5	H I (Lyα)
	150.00	1	1.0×10^{-1}		1.15	

Note

[a] Geostationary Operational Environmental Satellite.

14.5.3   Line Widths and Heights

See [67] for a detailed description of curve-of-growth analysis techniques. These yield the following results [68–74]:

Atomic thermal velocity	$= (2kT/m_a)^{1/2}$
	$= 1.4\ \mathrm{km\,s^{-1}}$.
Microturbulence (ξ_{mi})	$= 1.1\ \mathrm{km\,s^{-1}}$.
Macroturbulence (ξ_{ma})	$= 1.6\ \mathrm{km\,s^{-1}}$ (vertical)
	$= 2.8\ \mathrm{km\,s^{-1}}$ (horizontal).
Velocity for line breadth	$= (\xi_{th}^2 + \xi_{mi}^2 + \xi_{ma}^2)^{1/2}$
	$= 2.4\ \mathrm{km\,s^{-1}}$ at center of disk
	$= 3.3\ \mathrm{km\,s^{-1}}$ at limb.

Table 14.12 gives heights of formation of spectral lines [75, 76]:

Table 14.12. *Spectral line heights of formation.*

Line (nm)	Optical depth τ (FWHM)	Height (km) (FWHM)
Continuum (388.385)	3.2 to 0.32	−45 to 60
CN 388.33	0.003 to 0.000039	370 to 740
Continuum (500.0)	2.5 to 0.25	−35 to 90
Fe I 537.9	0.35 to 0.0025	60 to 400
C I 538.0	1.6 to 0.16	−20 to 110
H I 656.0		2000 to 3000
Fe I 1564.8		−20 to −30
Fe I 1564.8 (spot)		20 to 80

14.6 SPECTRAL DISTRIBUTION
by Heinz Neckel

F_λ = intensity of the mean solar disk per unit wavelength with spectrum irregularities smoothed (± 50 Å). Thus $F = \int F_\lambda \, d\lambda$.

$\mathcal{F}_\lambda = \pi F_\lambda$ = emittance of the solar surface per unit wavelength range.

$f_\lambda = \mathcal{F}_\lambda (R_\odot/A)^2 = 6.80 \times 10^{-5} F_\lambda$ solar flux outside the Earth's atmosphere per unit area and wavelength range. A = astronomical unit.

F_λ' same as for F_λ but referring to the continuum between the lines. The curve joining the most intense windows between the lines is regarded as the continuum. This may differ appreciably from the continuum in the entire absence of absorption lines. F_λ' does not have any sudden changes (e.g., at the Balmer limit).

$I_\lambda(0)$ = intensity at the center of the Sun's disk with spectral irregularities smoothed (± 50 Å).

$I_\lambda'(0)$ = intensity of the center of the Sun's disk between spectrum lines. This is obtained by interpolation from the most intense windows, as for F_λ'.

$I_\lambda(0)/I_\lambda'(0)$ represents the observed line blanketing for the center of the Sun's disk.

$F_\lambda/I_\lambda(0)$ represents the broadband (100-Å) disk-to-center ratio. It is approximately equal to $F_\lambda'/I_\lambda'(0)$. The solar spectrum is given in Table 14.13.

Table 14.13. *Solar spectral distribution, 0.2–5.0 μm [1–3].*

λ (μm)	F_λ	F_λ'	$I_\lambda(0)$	$I_\lambda'(0)$	f_λ	$I_\lambda(0)/I_\lambda'(0)$	$F_\lambda/I_\lambda(0)$
	(10^3 W m^{-2} sr^{-1} Å^{-1})				(10^{-3} W m^{-2} Å^{-1})		
0.20	0.01	0.014	0.014	0.02	0.65	0.7	0.7
0.22	0.07	0.10	0.13	0.19	4.5	0.7	0.5
0.24	0.08	0.13	0.13	0.21	5.2	0.6	0.6
0.26	0.19	0.27	0.37	0.53	13	0.7	0.5
0.28	0.34	0.68	0.60	1.21	23	0.5	0.56
0.30	0.83	1.48	1.34	2.39	56	0.56	0.62
0.32	1.12	1.97	1.67	2.94	76	0.57	0.67
0.34	1.34	2.39	1.89	3.30	91	0.57	0.71
0.36	1.42	2.56	1.96	3.47	97	0.56	0.72

Table 14.13. *(Continued.)*

λ (μm)	F_λ	F_λ'	$I_\lambda(0)$	$I_\lambda'(0)$	f_λ (10^{-3} W m^{-2} Å^{-1})	$I_\lambda(0)/I_\lambda'(0)$	$F_\lambda/I_\lambda(0)$
	(10^3 W m^{-2} sr^{-1} Å^{-1})						
0.37	1.67	2.67	2.28	3.60	113	0.63	0.73
0.38	1.58	2.99	2.16	4.14	107	0.52	0.73
0.39	1.52	3.21	2.08	4.41	103	0.47	0.73
0.40	2.17	3.35	2.97	4.58	148	0.65	0.73
0.41	2.50	3.42	3.38	4.63	170	0.73	0.74
0.42	2.54	3.47	3.45	4.66	173	0.74	0.74
0.43	2.34	3.50	3.12	4.67	159	0.67	0.75
0.44	2.71	3.49	3.61	4.62	184	0.78	0.75
0.45	2.94	3.47	3.87	4.55	200	0.85	0.76
0.46	3.01	3.41	3.95	4.44	205	0.89	0.76
0.48	2.99	3.28	3.84	4.22	203	0.91	0.78
0.50	2.83	3.20	3.61	4.08	192	0.88	0.78
0.55	2.76	2.93	3.43	3.63	188	0.94	0.80
0.60	2.61	2.67	3.17	3.24	177	0.98	0.82
0.65	2.34	2.41	2.81	2.90	159	0.97	0.83
0.70	2.08	2.13	2.46	2.52	141	0.975	0.85
0.75	1.87	1.92	2.18	2.24	127	0.975	0.86
0.8	1.68	1.71	1.94	1.97	114	0.983	0.87
0.9	1.38	1.39	1.57	1.58	94	0.993	0.88
1.0	1.11	1.12	1.25	1.26	75	0.995	0.89
1.1	0.90	0.90	1.01	1.01	61	1.0	0.89
1.2	0.76	0.76	0.84	0.84	52	1.0	0.90
1.4	0.51		0.56		35	1.0	0.91
1.6	0.37		0.40		25.5	1.0	0.92
1.8	0.25		0.27		16.9	1.0	0.92
2.0	0.17		0.18		11.6	1.0	0.93
2.5	0.076		0.081		5.2	1.0	0.94
3.0	0.039		0.041		2.6	1.0	0.95
4.0	0.0130		0.0135		0.9	1.0	0.96
5.0	0.0055		0.0057		0.4	1.0	0.96

References

1. Allen, C.W., editor, 1973, *Astrophysical Quantities*, 3rd ed. (Athlone Press, London), Secs. 81 & 82
2. Labs, D., Neckel, H., Simon, P.C., & Thuiller, G. 1987, *Solar Phys.*, **90**, 25
3. Neckel, H., & Labs, D. 1984, *Solar Phys.*, **90**, 205

Brightness temperatures for two optical wavelengths are given in Table 14.14, and Table 14.15 gives them for the infrared.

Table 14.14. *Brightness temperatures.*

	4400 Å	5500 Å
F_λ	5850 K	5860 K
F_λ'	6125 K	5940 K
I_λ	6165 K	6155 K
I_λ'	6465 K	6240 K

Mean intensity and brightness temperature in mid- and far-infrared regions with heights from the Vernazza, Avrett, and Loeser (VAL-C) model [77–79]:

Table 14.15. *Infrared brightness temperatures.*

λ (μm)	h (km)	$\log F_\lambda (\simeq I_\lambda \simeq F'_\lambda \simeq I'_\lambda)$ (W m^{-2} sr^{-1} μm)	T_b (K)	
5	70	4.77	5 730	
10	160	3.57	5 140	
20	240	2.36	4 820	
50	340	0.76	4 500	
100	410	−0.45	4 340	
200	450	−1.67	4 200	(temp min.)
1000 = 1 mm		−4.31	5 920	
1 cm			10–23 000	(transition)

14.7 LIMB DARKENING
by Keith Pierce

$I'_\lambda(\theta)$ = intensity of the solar continuum at an angle θ from the center of the disk; θ = angle between the Sun's radius vector and the line of sight.

$I'_\lambda(0)$ = continuum intensity at the center of the disk.

The ratio $I'_\lambda(\theta)/I'_\lambda(0)$, which varies with the wavelength λ, defines *limb darkening*. As far as possible, measurements are made in the continuum between the lines (hence the primes in the notation).

The results may be fitted to the following expressions:

$$I'_\lambda(\theta)/I'_\lambda(0) = 1 - u_2 - v_2 + u_2 \cos\theta + v_2 \cos^2\theta,$$

or

$$I'_\lambda(\theta)/I'_\lambda(0) = A + B\cos\theta + C[1 - \cos\theta \ln(1 + \sec\theta)],$$

where

$$A + B + (1 - \ln 2)C = 1.$$

The ratio of the mean to central intensity is

$$F'_\lambda/I'_\lambda(0) = 1 - \tfrac{1}{3}u_2 - \tfrac{1}{2}v_2,$$

or

$$\mathcal{F}'_\lambda/I'_\lambda(0) = A + C + \tfrac{2}{3}B - 2C(\tfrac{2}{3}\ln 2 - \tfrac{1}{6})$$
$$= A + 0.667B + 0.409C.$$

The ratio of the limb-to-central intensity is

$$I'_\lambda(90°)/I'_\lambda(0) = 1 - u_2 - v_2 \approx 1 - u_1$$
$$= A + C.$$

Table 14.16 presents limb darkening details, and the fit constants are given in Table 14.17.

Table 14.16. $I'_\lambda(\theta)/I'_\lambda(0)$ [1–16].

λ (μm)	$\cos\theta$ $\sin\theta$	1.0 0.000	0.8 0.600	0.6 0.800	0.5 0.866	0.4 0.916	0.3 0.954	0.2 0.980	0.1 0.995	0.05 0.9987	0.02 0.9998
0.20	[7]	1.00	0.85	0.74	0.69	0.65	0.61	0.58			
0.22	[7]	1.00	0.58	0.33	0.26	0.21	0.16	0.12			
0.245	[7]	1.00	0.71	0.49	0.42	0.36	0.31	0.25			
0.265	[7]	1.00	0.68	0.42	0.32	0.24	0.19	0.14			
0.28	[7]	1.00	0.72	0.47	0.38	0.29	0.22	0.16			
0.30	[7]	1.00	0.77	0.57	0.48	0.39	0.30	0.22	0.14		
0.32	[9]	1.00	0.809	0.623	0.532	0.438	0.347	0.262	0.17		
0.35	[9]	1.00	0.837	0.665	0.579	0.487	0.397	0.306	0.21		
0.37	[9]	1.00	0.851	0.687	0.603	0.513	0.421	0.332	0.23	0.19	
0.38	[9]	1.00	0.83	0.66	0.58	0.48	0.39	0.30	0.22	0.18	
0.40	[9]	1.00	0.835	0.663	0.585	0.490	0.403	0.308	0.222	0.18	
0.45	[9]	1.00	0.860	0.714	0.637	0.556	0.468	0.378	0.278	0.21	0.14
0.50	[9]	1.00	0.877	0.744	0.675	0.599	0.513	0.425	0.323	0.26	0.19
0.55	[9]	1.00	0.890	0.769	0.703	0.633	0.556	0.468	0.371	0.31	0.24
0.60	[9]	1.00	0.900	0.788	0.727	0.664	0.587	0.508	0.412	0.35	0.28
0.80	[9]	1.00	0.924	0.843	0.793	0.744	0.681	0.615	0.533	0.47	
1.0	[9]	1.00	0.941	0.870	0.828	0.783	0.731	0.675	0.59	0.54	
1.5	[9]	1.00	0.957	0.902	0.873	0.831	0.789	0.735	0.65	0.58	
2.0	[9]	1.00	0.966	0.922	0.896	0.865	0.826	0.780	0.70	0.61	
3.0	[9]	1.00	0.976	0.944	0.922	0.902	0.873	0.835	0.78	0.67	
5.0	[8]	1.00	0.986	0.963	0.949	0.937	0.916	0.890	0.84	0.76	
10	[8]	1.00	0.992	0.981	0.973	0.964	0.956	0.937	0.90	0.87	
20	[8]	1.00	0.994	0.983	0.975	0.970	0.964	0.957	0.95	0.93	
Total		1.00	0.898	0.787	0.731	0.669	0.602	0.525	0.448	0.39	0.32

References

1. Allen, C.W., editor, 1973, *Astrophysical Quantities*, 3rd ed. (Athlone Press, London), Sec. 81
2. Pierce, A.K., McMath, R.R., Goldberg, L., & Mohler, O.C. 1950, *ApJ*, **112**, 289
3. Pierce, A.K., & Waddell, J.H. 1961, *MNRAS*, **68**, 89
4. Gaustad, J.E., & Rogerson, J.R. 1961, *ApJ*, **134**, 323
5. Mouradian, Z. 1965, *Ann. d'Astrophys.*, **28**, 805
6. Heintz, J.R.W. 1965, *Rech. Astron. Obs. Utrecht*, **17/2**
7. Bonnet, R. 1968, *Ann. d'Astrophys.*, **31**, 597
8. Lena, P. 1970, *A&AS*, **4**, 202
9. Pierce, A.K., & Slaughter, C.D. 1977, *Solar Phys.*, **51**, 25
10. Neckel, H., & Labs, D. 1987, *Solar Phys.*, **110**, 139
11. Neckel, H., & Labs, D. 1994, *Solar Phys.*, **153**, 91
12. Neckel, H. 1996, *Solar Phys.*, **167**, 9
13. Neckel, H. 1997, *Solar Phys.*, **171**, 257
14. Pierce, A.K., Slaughter, C.D., & Weinberger, D. 1977, *Solar Phys.*, **52**, 179
15. Petro, C.D., Foukal, P.V., Rosen, W.A., Kurucz, R.L., & Pierce, A.K. 1984, *ApJ*, **283**, 462
16. Elste, G.H. 1990, *Solar Phys.*, **126**, 37

Table 14.17. *Limb darkening constants.*

λ	u_2	v_2	A	B	C	$\dfrac{\mathcal{F}'_\lambda}{I'_\lambda(0)}$	$\dfrac{I'_\lambda(90°)}{I'_\lambda(0)}$
0.20	+0.12	+0.33	−0.2	0.9	+0.9	0.79	0.54
0.22	−1.3	+1.6	−3.4	2.9	+5	0.51	0.06
0.245	−0.1	+0.85	−1.9	2.0	+3	0.61	0.20
0.265	−0.1	+0.90	−1.9	2.1	+2.7	0.540	0.08
0.28	+0.38	+0.57	−1.3	1.8	+1.8	0.588	0.10
0.30	+0.74	+0.20	−0.4	1.2	+0.5	0.648	0.06

Table **14.17.** *(Continued.)*

λ	u_2	v_2	A	B	C	$\dfrac{\mathcal{F}'_\lambda}{I'_\lambda(0)}$	$\dfrac{I'_\lambda(90°)}{I'_\lambda(0)}$
0.32	+0.88	+0.03	−0.02	0.97	+0.1	0.685	0.08
0.35	+0.98	−0.10	+0.25	0.79	−0.3	0.705	0.11
0.37	+1.03	−0.16	+0.42	0.68	−0.4	0.71	0.13
0.38	+0.92	−0.05	+0.26	0.78	−0.2	0.71	0.13
0.40	+0.91	−0.05	+0.20	0.81	−0.1	0.718	0.13
0.45	+0.99	−0.17	+0.54	0.60	−0.44	0.755	0.11
0.50	+0.97	−0.22	+0.68	0.49	−0.56	0.782	0.16
0.55	+0.93	−0.23	+0.74	0.43	−0.56	0.803	0.20
0.60	+0.88	−0.23	+0.78	0.39	−0.57	0.817	0.24
0.80	+0.73	−0.22	+0.92	0.25	−0.56	0.862	0.39
1.0	+0.64	−0.20	+0.97	0.18	−0.53	0.886	0.48
1.5	+0.57	−0.21	+1.11	0.08	−0.61	0.916	0.56
2.0	+0.48	−0.18	+1.09	0.07	−0.49	0.932	0.60
3.0	+0.35	−0.12	+1.04	0.06	−0.34	0.948	0.72
5.0	+0.22	−0.07	+1.02	0.05	−0.18	0.964	0.81
10.0	+0.15	−0.07	+1.04	0.00	−0.22	0.982	0.87
Total	+0.84	−0.20	+0.72	+0.42	−0.45	0.82	0.32

14.8 CORONA
by Serge Koutchmy

Optical radiation from the corona contains three components:

K = continuous spectrum due to Thomson scattering by electrons of the coronal plasma,
F = Fraunhofer spectrum diffracted and/or scattered by interplanetary dust particles [81],
L = coronal emission of forbidden lines; L is negligible for coronal photometry (about 1%).

The total coronal light beyond $1.03R_\odot$ (for typical lunar disk at eclipse) [82–84] is

$$\text{at sunspot maximum} = 1.5 \times 10^{-6} \text{ solar flux} \simeq 0.66 \text{ full Moon,}$$
$$\text{at sunspot minimum} = 0.6 \times 10^{-6} \text{ solar flux} \simeq 0.26 \text{ full Moon.}$$
$$\text{Total F corona} = 0.3 \times 10^{-6} \text{ solar flux.}$$

Earthshine on Moon at total eclipse [85] = 2.5×10^{-10} mean Sun brightness. The brightness of the sky near the Sun during a total eclipse [82, 84, 86] is

$$6 \times 10^{-10} < S < 10^{-8} \times [\text{mean Sun brightness } (\bar{B}_\odot)].$$

The spectral distribution of K components is similar to the solar spectrum, with $B - V = 0.65$. The F component is slightly redder in the outer corona [87], with $B - V \simeq 0.75$. The base of corona may be taken as the transition region at $r = 1.0025R_\odot$ from the visible limb. Chromospheric extensions are seen up to $r = 1.015R_\odot$.

The coronal ellipticity from isophotes ϵ [83, 88, 89] is

$$\epsilon = (A_3 - P_3)/P_3 \simeq (A_1 - P_1)/A_1,$$

where A_1 and P_1 are equatorial and polar diameters, and for A_3, P_3 the corresponding diameters are averaged with those oriented 22.5° on either side.

ϵ at sunspot max. $\simeq 0.06$,

ϵ at sunspot min. $\simeq 0.26$ near $r = 2R_\odot$ (extrapolated values; the $a + b$ index).

Values are tabulated against $r(R_\odot)$.
The polarization of coronal light $(K + F)$ [82, 90, 91] is

$$p_{tot} = (I_t - I_r)/(I_t + I_r),$$

where I_t and I_r are intensities polarized in the tangential and radial direction.

$$p_{max} = 50\%.$$

Other values tabulated against $r/R_\odot$ are listed in Tables 14.18 and 14.19.

A most relevant parameter to describe the distribution of electron densities in the plasma corona is $p_k = (I_t - I_r)/K$ with $K = (I_t + I_r) - F$; see [90].

Density irregularities in the corona may be specified approximately by an irregularity factor $x = N_e^2/(\bar{N}_e)^2$, where N_e is the electron density. Then rms $N_e = \bar{N}_e x^{1/2}$. In the striated outer corona one might write

$$x \simeq 1/\text{f.f.},$$

where f.f. is the filling factor, which could be very small indeed. Only approximate data exist (see Table 14.18). x varies with $r/R_\odot$.

Temperature of corona:

Loops	$(1.0–3.0) \times 10^6 K$.
Quiet corona T_{max} at $r \simeq 2R_\odot$	1.6×10^6 K.
Coronal condensation	3×10^6 K.
Coronal hole	1×10^6 K.

Table 14.18. *Radial variations of p, ϵ, and x for homogeneous and minimum cycle corona at 0.55 μm [1–3].*

$r/R_\odot$	1.0	1.2	1.5	2	3	5	10	20	25
Polarization in %									
p_{tot} at equator	20	35	41	38	21	10	4	2.6	
p_{tot} at pole	20	25	17	10	3	< 1			
Ellipticity ϵ, minimum corona	0.06	0.10	0.16	0.13	0.11	0.12	0.18	0.25	
Irregularity x				> 2.5	4	8	17	21	25

References
1. Saito, K. 1972, *Ann. Tokyo Astron. Obs. XII*, **53**, 120
2. Koutchmy, S., Picat, J.P., & Dantel, M. 1977, *A&A*, **59**, 349
3. Allen, C.W. 1961, *Solar Corona IAU Symp.*, **16**, 1

Table 14.19. *Smoothed coronal brightness and electron density in average models* [1–5].

$\rho = r/R_\odot$	$\log(\rho - 1)$	Max.	Min. Eq.	Pole	Eq./Pole	Max.	Min. Eq.	Pole
		$10^{-10}B_\odot$					(cm^{-3})	
1.003	−2.5					9.0	9.0	−8.20
1.005	−2.3					8.8	8.8	...
1.01	−2.0	4.9	4.8	4.25	...	8.7	8.7	8.0
1.03	−1.5	4.65	4.6	4.10	...	8.6	8.6	...
1.06	−1.2	4.45	4.35	3.85	...	8.4	8.4	...
1.10	−1.0	4.3	4.20	3.60	3.10	8.25	8.25	7.50
1.2	−0.7	3.9	3.75	3.06	2.90	7.90	7.8	7.10
1.4	−0.4	3.34	3.26	2.5	2.50	7.44	7.35	6.25
1.6	−0.2	2.92	2.88	1.95	2.25	7.05	7.05	5.95
2.0	0.0	2.23	2.25	1.24	1.91/1.82	6.52	6.50	5.0
2.5	+0.2	1.63	1.63	0.7	1.66/1.56	6.00	5.95	4.75
3.0	+0.3	1.23	1.25	0.25	1.48/1.33	5.60	5.50	4.50
4.0	+0.5	0.70	0.61	−0.35	1.23/1.03	5.1	5.05	4.20
5.0	+0.6	0.3	0.2	−0.75	1.0/0.80	4.8	4.75	4.0
10.0	1.0	−0.5	−0.75	...	0.31/0.06	4.10	4.05	
20.0	1.3				−0.33/−0.72	3.2		

The column group headers for this table are: **log (surface brightness)** spanning K (Max. / Min. with sub-columns Eq. and Pole) and F (Eq./Pole); then **$\log N_e$** with Max. and Min. (sub-columns Eq. and Pole).

References

1. Allen, C.W., editor, 1973, *Astrophysical Quantities*, 3rd ed. (Athlone Press, London), Secs. 73, 84, and 85
2. Newkirk, G., Dupree, R.G., & Schmahl, E.J. 1970, *Solar Phys.*, **15**, 15
3. Koutchmy, S., Zirker, J.B., Steinolfson, R.S., & Zhugzda, J.D. 1991, in *Solar Interior and Atmosphere*, edited by A.N. Cox, W.C. Livingston, and M.S. Matthews (University of Arizona Press, Tucson)
4. Blackwell, D.E., & Petford, A.D. 1966, *MNRAS*, **131**, 383
5. Saito, K. 1972, *Ann. Tokyo Astron. Obs.* XII, **53**, 120

14.8.1 Coronal Photometry and Electron Density N_e

Assuming spherical symmetry, the distribution of coronal intensity I_0 as a function of the projected radial distance ρ may be used to determine the distribution of N_e as a function of radial distance r in Table 14.20. The classical Baumbach expressions [92] are

$$10^6 I_0/I_\odot = 0.0532\rho^{-2.5} + 1.425\rho^{-7} + 2.565\rho^{-17},$$

leading to

$$N_e(r) = 10^8(0.036r^{-1.5} + 1.55r^{-6} + 2.99r^{-16})\ cm^{-3}.$$

The temperature in the inner corona is well described by the approximation of hydrostatic equilibrium [89] with $T_{hyd} = 6.08 \times 10^6[d(\log N_e)/d(r^{-1})]^{-1}$ in K, assuming $H/H_e = 10$.

Table 14.20. *Electron densities* ($\log N_e$ (cm^{-3})) *in coronal structures.*

$r/R_\odot$	Coronal streamer	Coronal hole (Void)	Thread	Loop
1.0				
1.1	8.75	7.0	10.0	10.0
1.3	8.25	6.6	9.5	9.0
1.5	7.90	6.2	9.0	
2.0	7.30	5.25	8.25	
2.5	7.0	4.80		
3.0	6.75			
4.0	6.3			
5.0	6.1			
10.0	5.45			

Coronal line spectrum quantities are:

T_m = temperature (K) at which spectrum reaches greatest intensity,

f = energy flux (10^{-6} W cm^{-2}) from the coronal line seen outside the Earth's atmosphere,

W = equivalent width of coronal line in terms of K continuum,

A = transition probability (s^{-1}).

Tables 14.21, 14.22, 14.23, and 14.24 give some permitted, forbidden, and infrared coronal lines.

Table 14.21. *Selected permitted lines*, 1–61 nm [1–4].

λ (nm)	Ion	Transition	f	$\log T_m$
0.92	Mg XI	$1s^2$–$1s2p$	2	6.4
1.21	Ne X, Fe XVII		1	
1.36	Ne IX	$1s^2$–$1s2p$	2	6.20
1.51	Fe XVII	$2p^6$–$2p^53d$	8	6.58
1.69	Fe XVII	$2p^6$–$2p^53s$	9	6.58
1.90	O VIII	$1s$–$2p$	8	6.36
2.16	O VII	$1s^2$–$1s2p$	6	5.9
5.06	Si X	$2p$–$3d$	6	6.14
6.97	Fe XIV	$3p$–$4s$	4	6.27
17.10	Fe IX	$3p^6$–$3p^53d$	85	5.85
17.48	Fe X	$3p^5$–$3p^43d$	90	6.00
17.72	Fe X	$3p^5$–$3p^43d$	33	6.00
18.04	Fe XI	$3p^4$–$3p^33d$	75	6.11
18.83	Fe XI	$3p^4$–$3p^33d$	40	6.11
19.50	Fe XII	$3p^3$–$3p^23d$	60	6.16
20.20	Fe XIII	$3p^2$–$3p3d$	25	6.21
21.13	Fe XIV	$3p$–$3d$	15	6.27
28.41	Fe XV	$3s^2$–$3s3p$	40	6.31
30.34	Si XI	$2s^2$–$2s2p$	30	6.22
33.54	Fe XVI	$3s$–$3p$	20	6.40

Table 14.21. *(Continued.)*

λ (nm)	Ion	Transition	f	$\log T_m$
36.81	Mg IX	$2s^2 - 2s2p$	15	5.97
49.9	Si XII	$2s-2p$	10	6.27
61.0	Mg X	$2s-2p$	12	6.04

References
1. Batstone, R.M., Evans, K., Parkinson, J.H., & Pounds, K.A. 1970, *Solar Phys.*, **13**, 389
2. Walker, A.B.C., & Rugge, R.H. 1970, *A&A*, **5**, 4
3. Jordan, C. 1965, *Commun. Univ. London Obs.*, **68**
4. Freeman, F.F., & Jones, B.B. 1970, *Solar Phys.*, **15**, 288

Table 14.22. *Selected forbidden lines, 100–300 nm [1, 2].*

λ nm	Ion	Transition	$\log T_m$
124.22	Fe XII	$p^3\,{}^4S_{1\frac{1}{2}} - {}^2P_{1\frac{1}{2}}$	6.16
134.96	Fe XII	$p^3\,{}^4S_{1\frac{1}{2}} - {}^2P_{1\frac{1}{2}}$	6.16
144.60	Si VIII	$2p^3\,{}^4S_{1\frac{1}{2}} - {}^2D_{1\frac{1}{2}}$	5.93
146.70	Fe XI	$3p^4\,{}^3P_1 - {}^1S_0$	6.11
212.60	Ni XIII	$3p^4\,{}^3P_2 - {}^1D_2$	6.27
214.95	Si IX	$2p^2\,{}^3P_2 - {}^1D_2$	6.04
216.97	Fe XII	$3p^3p\,{}^4S_{1\frac{1}{2}} - {}^2D_{2\frac{1}{2}}$	6.16

References
1. Jordan, C. 1971, Eclipse of 1970, *COSPAR Symp.*
2. Gabriel, A.H. et al. 1971, *ApJ*, **434**, 807

Table 14.23. *Selected forbidden lines, 300–700 nm [1–3].*

λ (nm)	Ion	Transition	Upper E.P. (eV)	A (s⁻¹)	W (10^{-10} nm × $\bar{B}_\odot$)	$\log T_m$
332.9	Ca XII	$2p^5\,{}^2P_{1\frac{1}{2}} - {}^2P_{\frac{1}{2}}$	3.72	488	0.07	6.19
338.82	Fe XIII	$3p^2\,{}^3P_2 - {}^1D_2$	5.96	87	1.0	6.19
360.09	Ni XVI	$3p\,{}^2P_{\frac{1}{2}} - {}^2P_{1\frac{1}{2}}$	3.44	193	0.13	6.37
423.20	Ni XII	$3p^5\,{}^2P_{1\frac{1}{2}}$	2.93	237	0.11	6.17
530.281	Fe XIV	$3p\,{}^2P_{\frac{1}{2}} - {}^2P_{1\frac{1}{2}}$	2.34	60	2.0	6.27
569.44	Ca XV	$2p^2\,{}^3P_0 - {}^3P_1$	2.18	95	0.03	
637.45	Fe X	$3p^5\,{}^2P_{1\frac{1}{2}} - {}^2P_{\frac{1}{2}}$	1.94	69	0.5	6.00
670.19	Ni XV	$3p^2\,{}^3P_0 - {}^3P_1$	1.85	57	0.12	6.32

References
1. Allen, C.W., editor, 1973, *Astrophysical Quantities*, 3rd ed. (Athlone Press, London), Secs. 73, 84, and 85
2. Livingston, W., & Harvey, J. 1982, *Proc. Ind. Natl. Sci. Acad.*, **48**, Suppl. 3, 18
3. Jefferies, J.T., Orrall, F.Q., & Zirker, J.B. 1971, *Solar Phys.*, **16**, 103

Table 14.24. *Near IR lines* [1, 2].[a]

λ	Ion	f (10^{-2} W m^{-2} sr^{-1})	Transition
789.19	[Fe XI]		$3p^4\,^3P_2-^3P_1$
1074.617	[Fe XIII]		$3p^2\,^3P_0-^3P_1$
1079.783	[Fe XIII]		$3p^2\,^3P_1-^3P_2$
1083.0	He I		$2p\,^3P-2s\,^3S$
1252.0	[S IX]	1.5	$1s^22s^22p^4\,^3P_1-^3P_2$
1283.0	H I	3.00	Paschen (5–3)
1431.0	[Si X]	1.55	$1s^22s^22p^2\,^2P_{\frac{3}{2}}-^2P_{\frac{1}{2}}$
1523.0	[Cr XI]	0.47	$3s^23p^2\,^3P_2-^3P_1$
1856.0	[Cr XI]	0.4	$3s^23p^2\,^3P_1-^3P_0$
1876.0	H I	13.0	Paschen (4–3)
1922.0	[Si XI]	< 0.7	$1s^22s2p\,^3P_2-^3P_1$
2167.0	H I	< 0.5	Brackett (7–4)
2747.0	[Al X]	< 1	$1s^22s2p\,^3P_2-^3P_1$
3019.0	[Mg VIII]	< 1	$1s^22s^22p\,^2P_{\frac{3}{2}}-^2P_{\frac{1}{2}}$

Note

[a]Kuhn [3] points out that many of the IR lines in this table were not observed at the eclipse of 3 Nov. 1994 and questions their reality.

References

1. Olsen, K.H., Anderson, C.R., & Stewart, J.N. 1971, *Solar Phys.*, **21**, 360
2. Penn, M.J., & Kuhn, J.R. 1994, *ApJ*, **434**, 807
3. Kuhn, J. 1995, private communication

14.9 SOLAR ROTATION
by Robert Howard

The inclination of the solar equator to the ecliptic [93–96] is $7°15'$.

The longitude of the ascending node is $75°46' + 84'\,T$, where T is epoch in centuries from 2000.00.

The sidereal differential rotation coefficients from the formulas

$$\omega = A + B\sin^2\phi \text{ deg/day},$$

where ϕ is the latitude, and

$$\omega = A + B\sin^2\phi + C\sin^4\phi \text{ deg/day},$$

are often used for features that extend to higher latitudes. These are given in Table 14.25. See also [97].

Table 14.25. *Empirical rotation coefficients.*

	A	B	C
From tracers			
Individual sunspots [1]	14.522	−2.84	
Sunspot groups [1, 2]	14.39	−2.95	
Plages [3]	14.06	−1.83	
Magnetic field pattern [4]	14.37	−2.30	−1.62
Supergranular pattern [5, 6]	14.71	−2.39	−1.78
(Doppler features)			
Filaments, prominences [7]	14.48	−2.16	
Coronal features [8, 9]	13.46	−2.99	
Small magnetic features [10]	14.42	−2.00	−2.09
From the Doppler effect in solar lines			
Surface plasma [11]	14.11	−1.70	−2.35
Hα line [12]	14.1		

References

1. Howard, R., Gilman, P.A., & Gilman, P.I. 1984, *ApJ*, **283**, 373
2. Balthasar, H., Vazquez, M., & Woehl, H. 1986, A&A, **155**, 87
3. Howard, R.F. 1990, *Solar Phys.*, **126**, 299
4. Snodgrass, H.B. 1983, *ApJ*, **270**, 288
5. Duvall, Jr., T.L. 1980, *Solar Phys.*, **66**, 213
6. Snodgrass, H.B., & Ulrich, R. 1990, *ApJ*, **351**, 309
7. d'Azambuja, M., & d'Azambuja, L. 1948, *Ann. Observ. Paris*, **6**, 1
8. Dupree, A.K., & Henze, Jr., W. 1972, *Solar Phys.*, **27**, 271
9. Henze, Jr., W., & Dupree, A.K. 1973, *Solar Phys.*, **33**, 425
10. Komm, R.W., Howard, R.F., & Harvey, J.W. 1993, *Solar Phys.*, **145**, 1
11. Snodgrass, H.B., Howard, R., & Webster, L. 1984, *Solar Phys.*, **90**, 199
12. Livingston, W.C. 1969a, *Solar Phys.*, **7**, 144; 1969b, **9**, 448

Rotation of solar plasma as a function of depth from oscillation measurements increases from the surface rate by about 0.8 deg/day at a depth from $0.01 R_\odot$ to $0.08 R_\odot$, then decreases slowly with depth [98, 99].

The period of sidereal rotation adopted for heliographic longitudes is 25.38 days. The corresponding synodic period is 27.275 3 days. Conversion factors between different units are given in Table 14.26.

Table 14.26. *Conversion factors.*

To convert from	Multiply by
deg/day to μrad s^{-1}	0.202 01
deg/day to m s^{-1}	140.596 cos ϕ
deg/day to nHz	32.150

Sidereal—synodic rotation = Earth's orbital motion

= 0.985 6 deg/day (averaged over a year).

14.10 GRANULATION
by Richard Muller

The solar surface is covered by a hierarchy of patterns that are convective in origin: granulation, mesogranulation, and supergranulation [98–109]:

Granules

Diameter of granules	$1''4 = 1000$ km
Range about $0''25$ to $3''5$	
Intergranular distance	$1''0$
Number of granules on whole photospheric surface	5×10^6
Corresponding area occupied by a cell	1.5×10^6 km^2
Granule intensity contrast	
Brighter granule/intergranule	1.3
Corresponding temperature difference	300 K
Root-mean-square variations	
Intensity at 550 nm observed	0.09
Corrected	0.15
Temperature	110 K
Mean lifetime of granules	10 min
Upward velocity of brighter granules	1 km s^{-1}

Mesogranulation

Diameter	5000 km
Lifetime	3 h
Vertical velocity	0.06 km s^{-1}
Proper motion	0.4 km s^{-1}

Supergranulation

Diameter	32 000 km
Lifetime	20 h
Horizontal velocity to edge	0.4 km s^{-1}

14.11 SURFACE MAGNETISM AND ITS TRACERS
by Peter Foukal, Sami Solanki, and Jack Zirker

Buoyancy lofts magnetic fields from the solar interior into the photosphere where they emerge as active regions to be dispersed laterally under the influence of convection (various scales) and other large-scale horizontal flows. White light tracers of magnetism are sunspots and faculae. Monochromatic tracers (line weakening) are plage, filigree, the network, internetwork, coronal holes, and prominences. The network and plages are presumed to be composed of aggregates of flux tubes. Prominences are found along magnetic neutral lines or above active regions. Magnetic field details for various surface structures are given in Table 14.27.

Table 14.27. *Magnetic fields.*[a]

	Field strength
Sunspot umbrae	2–4 kG
Sunspot penumbrae	0.8–2 kG
Pores	1.7–2.5 kG
Plage or facular magnetic elements B $(z = 0)$	1.4–1.7 kG
Network magnetic elements B $(z = 0)$	1.3–1.5 kG
Internetwork	≤ 600 G (probably)
	Flux [1]
Ephemeral region	3×10^{19} Mx
Small active region	3×10^{20} Mx
Moderate active region	3×10^{21} Mx
Large active region	$\geq 10^{22}$ Mx
Giant active region	$= (20\text{–}50) \times 10^{22}$ Mx
	Magnetic elements
Diameter [2, 3]	200–300 km
Lifetime [4]	18 min
	Global aspects [1]
Total flux at solar min	$= (15\text{–}20) \times 10^{22}$ Mx
Total flux at solar max	$= (100\text{–}120) \times 10^{22}$ Mx

Note

[a]The field strength is strongly height dependent. See Sec. 14.12 on sunspots for more information on sunspot field gradients. For magnetic elements the field drops from the tabulated values at $z = 0$ (i.e., the quiet Sun continuum forming layer) to roughly 200–500 G (in plage) near the temperature minimum (e.g., [6] and [7]). The magnetic element lifetime [5] is probably only a lower limit, being a lifetime measurement of the brightness structure that probably lives less long than the underlying magnetic structure. There is no permanent dipole field but one develops over solar cycle due to evolution of polar fields; at other times there is a dipole component to lower-latitude extended active-region fields [8, 9]. Mx means maxwell (G cm^2).

References

1. Harvey, K. 1992, in *Proceedings of the Workshop on Solar Electromagnetic Radiation Study for SOLAR CYCLE 22*, edited by R.F. Donnelly (Natl. Info. Tech. Service, Springfield, VA), p. 113
2. Keller, C.U. 1992, *Nature*, **359**, 307
3. Grossmann-Doerth, U., Knölker, M., Schüssler, M., & Solanki, S.K. 1994, *A&A*, **285**, 648
4. Muller, R. 1985, *Solar Phys.*, **100**, 237
5. Deming D., Boyle, R.J., Jennings D.E., & Wiedemann, G. 1988, *ApJ*, **333**, 978
6. Zirin, H., & Popp, B. 1989, *ApJ*, **340**, 571
7. Sheeley, Jr., N.R., & Boris, J.P. 1985, *Solar Phys.*, **98**, 219
8. Wang, Y.M., & Sheeley Jr., N.R. 1989, *Solar Phys.*, **124**, 81

14.11.1 Faculae

Faculae are cospatial with photospheric magnetic fields. They become visible in white light near the limb (i.e., as $\mu = \cos\theta \to 0$). While fragmented and irregular, they do tend to outline the circular boundaries of supergranular cells [112, 113].

The center-to-limb dependence of wide-band facular contrast (integrated over the spectral range 0.35–1.0 μm) can be expressed as

$$C(\mu) - 1 = 0.115(1 - \mu),$$

where

$$C(\mu) = I_{\text{facula}}/I_{\text{photosphere}}$$

[114]. At the highest spatial resolution values of $C(\mu)$ increase by a factor of 3–4 [115].
The wavelength dependence of facular contrast is approximately given by

$$C_\lambda(\mu) - 1 = [C_{5300}(\mu) - 1]0.5\lambda^{-1},$$

where $C_{5300}(\mu)$ is the intensity of the faculae relative to the photosphere at 5300 Å [116].

Life of average faculae	15 days
Life of large faculae (dominating solar variations)	2.7 months

The excess temperature of magnetic elements [117, 118] is given in Table 14.28.

Table 14.28. *Excess temperatures.*

$\log \tau_{5000}$	−5	−4	−3	−2	−1	0
Plage: $T_{\text{Magel}} - T_{\text{phot}}$ (K)	1400	1500	650	500	560	−130
Network: $T_{\text{Magel}} - T_{\text{phot}}$ (K)	1400	1500	700	700	770	460

14.11.2 Plages

Plages or bright flocculi are readily visible in Hα and in the H and K lines of Ca II. The locations agree well with faculae but plages are visible over the whole disk. Measurements of area and eye estimates of intensity (scale 1 → 5) are made regularly [119].

Table 14.29 shows the approximate relation between plage area and sunspot area (both in 10^{-6} hemisphere).

Table 14.29. *Plage and sunspot areas.*

Plage area	500	1000	2000	3000	4000	6000	8000	10 000
Sunspot area	0	30	100	180	280	500	900	2000

Since the duration of the plage is longer than that of the spot, the spot area may be much less than the value given. Normally sunspots are present when the plage intensity is ≥ 3.

The exponential decay time of a plage observed area is 1.6 rotations (43 days). The actual area of a plage expands continuously but the fainter parts are below measurement threshold.

Values for a typical large active region [115] are:

Sunspot area	600×10^{-6} hemisphere.
Plage area	6000×10^{-6} hemisphere.
Plage area at disk center	$12\,000 \times 10^{-6}$ disk.
Plage diameter	3.5 arcmin.

14.11.3   Prominences

Table 14.30 shows the physical conditions in quiescent prominences.

Table 14.30. *Quiescent prominences.*

log [electron density (cm^{-3})]	Temperature (K)
10.48–11.02 [1]	5000–7000
9–10	20 000–600 000 [2]

References
1. Hirayama, T. 1986, *Coronal and Prominence Plasmas*, edited by A.I. Poland (NASA, Washington, DC), p. 2442
2. Orrall, F.Q., & Schmahl, E.J. 1980, *ApJ*, **240**, 908

The temperature varies considerably within a prominence.
The proton-to-hydrogen density ratio is $0.05 < N_p/N_H < 1$ [120].

Sizes

Threads [121]	300–1800 km (diameter).
Height	2000 km (active),
	10 000–50 000 km (quiescent).
Length	50 000–200 000 km.
Thickness	3000–5000 km.
Magnetic field (horizontal)	2–20 G (quiescent) [122],
	10–40 G (active).
Velocity	15–35 km s^{-1} (threads, apparent) [123],
	1–3 km s^{-1} (Doppler, horizontal) [124],
	2–10 km s^{-1} (turbulent).

The angle of the field with the axis of the prominence $\sim 20°$ [122].
Lifetimes are approximately 1 week to 3 months; the average is 2 months.

14.12   SUNSPOTS
by Sami Solanki

The formula for the center-to-limb variation of umbral brightness $(1 \geq \mu > 0.3)$ is

$$i_u(\mu, \lambda) = i_u(\mu = 1, \lambda) - b_u(\lambda)(1 - \cos\theta),$$

$i_u = I_u/I_q$, where I_q is the quiet Sun brightness, and $i_p = I_p/I_q$. Brightness data for sunspots are given in Table 14.31.

Table 14.31. *Center-to-limb variation and λ dependence of umbral and penumbral brightness [1–4].*

λ (μm)	0.387	0.579	0.669	0.876	1.215	1.54	1.67	1.73	2.09	2.35	3.8
i_u $(\mu = 1, \lambda)_{early}$		0.022	0.061	0.191	0.327	0.451	0.507	0.543	0.567	0.565	
i_u $(\mu = 1, \lambda)_{middle}$	0.008	0.066	0.090	0.215	0.345	0.495	0.548	0.577	0.589	0.581	
i_u $(\mu = 1, \lambda)_{late}$		0.110	0.119	0.239	0.358	0.534	0.590	0.612	0.611	0.597	
b_u (λ)	−0.010	0.012	0.009	0.019	0.031	0.087	0.087	0.094	0.090	0.058	
i_p $(\mu = 1, \lambda)$	0.64	0.768	0.794	0.827	0.876		0.914		0.928		0.936

References
1. Albregtsen, F., & Matby, P. 1978, *Mat.*, **274**, 41
2. Albregtsen, F., Jorås, P.B., & Matby, P. 1984, *Solar Phys.*, **90**, 17
3. Maltby, P. 1972, *Solar Phys.*, **26**, 76
4. Matby, P., Avrett, E.H., Carlsson, M., Kjeldseth-Moe, O., Kurucz, R.L., & Loeser, R. 1986, *ApJ*, **306**, 284

A model for the sunspot umbral core is given in Table 14.32.

Table 14.32. *Model of the dark umbral core [1–4].*

log τ	1	0	−1	−2	−3	−4	−5	−6
T (K)	6140	4040	3540	3420	3400	3450	6400	8700
log P_g (cgs)	5.78	5.43	4.91	4.28	3.64	2.95	0.99	−0.61
log P_e (cgs)	2.01	0.52	−0.28	−0.80	−1.28	−1.75	−1.96	−1.04
z (km)	−94	0	95	220	380	600	1115	1850

References
1. Maltby, P., Avrett, E.A., Carlsson, M., Kjeldseth-Moe, O., Kurucz, R.L., & Loeser, R. 1986, *ApJ*, **306**, 284
2. Avrett, E.H. 1981, in *The Physics of Sunspots*, edited by L.E. Cram and J.H. Thomas (Sacramento Peak Obs., Sunspot, NM), p.235
3. Van Ballegooijen 1984, *A&A*, **91**, 195
4. Obridko, V.N., & Staude, J. 1988, *A&A*, **189**, 232

Magnetic field data for sunspots are given in Tables 14.33 and 14.34.

Table 14.33. *Maximum magnetic field B_0 as a function of umbral radius r_u [1, 2].*

r_u (km)	500	1000	2000	4000	6000	8000	10000
B_0 (G)	2000	2000	2000	2300	2700	3100	3500

References
1. Brants, J.J., & Zwaan, C. 1982, *Solar Phys.*, **80**, 251
2. Kopp, G., & Rabin, D. 1992, *Solar Phys.*, **81**, 231

Table 14.34. *Relative magnetic field B/B_0 and its inclination γ' relative to the vertical versus position in spot r for a large symmetric sunspot* [1–5].

r/r_p	0	0.1	0.2	0.3	0.4	0.5	0.6	0.7	0.8	0.9	1.0
B/B_0	1	0.99	0.96	0.92	0.84	0.74	0.62	0.50	0.41	0.35	0.30
γ' (deg)	0	7	15	24	35	48	58	66	73	77	80

References

1. Solanki, S.K., Rüedi, I., & Livingston, W. 1992, *A&A*, **263**, 339
2. McPherson, M.R., Lin, H., & Kuhn, J.R. 1992, *Solar Phys.*, **139**, 255
3. Lites, B.W., & Skumanich, A. 1990, *ApJ*, **348**, 747
4. Kawakami, H. 1983, *PASJ*, **35**, 459
5. Adam, M.G. 1990, *Solar Phys.*, **125**, 37

The azimuthal angle of the field is $\phi \le 20°$ for symmetric sunspots. In the penumbra γ' is an average value, with bright and dark filaments inclined relative to each other by 20°–40° [125–127]. In the outer penumbra the inclination depends on the size of the sunspot, with smaller sunspots having more vertical fields [128].

Table 14.35 gives structure details of the outer parts of sunspots.

Table 14.35. *Superpenumbral canopy: Base height z_c as a function of distance from center of spot r/r_p normalized by the spot radius r_p* [1, 2].

r/r_p	1.0	1.2	1.4	1.6
Base height z_c (km)	0	200	300	350
B/B_0	0.30	0.21	0.15	0.11
γ' (deg)	80	86	89–90	89–90

References

1. Giovanelli, R.G. 1980, *Solar Phys.*, **68**, 49
2. Giovanelli, R.G., & Jones, H.P. 1982, *Solar Phys.*, **79**, 267

Table 14.36 gives the magnetic field gradients in sunspots.

Table 14.36. *Vertical gradient of the field* [1–7].

r/r_p	0.0	0.6	1.0
dB/dz in photosphere (G/km)	2	2	1
dB/dz in photosphere and chromosphere (G/km)	0.5	0.4	0.2

References

1. Bruls, J.H.M.J., Solanki, S.K., Carlsson, M., & Rutten, R.J. 1993, *A&A*, **293**, 225
2. Abdussamator, H.J. 1971, *Solar Phys.*, **16**, 384
3. Henze, N., Jr., Tandberg-Hanssen, E., Hagyard, M.J., Woodgate, B.E., Shine, R.A., Beckers, J.M., Bruner, M., Gurman, J.B., Hyder, L.L., & West, E.A. 1982, *Solar Phys.*, **81**, 231
4. Lee, J.W., Gary, E.E., & Hurford, G.J. 1993, *Solar Phys.*, **144**, 45 and 349
5. Rüedi, I., Solanki, S.K., & Livingston, W. 1994, *A&A*, **293**, 252
6. Whittman, A.D. 1974, *Solar Phys.*, **36**, 29
7. Pahlke, K.-D. 1988, Ph.D. thesis, University of Göttingen, Göttingen, Germany

Wilson depression. The apparent depression of $\tau = 1$ of the umbra seen near the limb [129–132] and derived from MHS equilibrium [133, 134] is

$$z_W = 600 \pm 200 \text{ km}.$$

The relative magnetic flux in umbra and penumbra [135], with Φ_t the total magnetic flux of spot, Φ_u the magnetic flux of the umbra, and Φ_p the magnetic flux of the penumbra, is

$$\Phi_u/\Phi_t = 1/3 - 1/2,$$
$$\Phi_p/\Phi_t = 1/2 - 2/3.$$

The variation of the umbral-to-photosphere intensity ratio ϕ with solar cycle (at $\lambda = 1.67 \ \mu$m) is

$$\phi = 0.44 + 0.15t/t_0,$$

where t is the time elapsed from the starting epoch and t_0 is the length of the solar half-cycle [136].

The average East–West inclination of field lines in spots is

all spots	$-3°4$,
leading	$-2°8$,
following spots	$-3°8$.

The negative angle indicates that the field lines trail the rotation [137].

Sunspot axial tilt angles (individual sunspots) are the angles between the line joining the leading and following spots of a group and the local parallel of latitude. The leading spots on average are closer to the equator than the following spots as a function of latitude with the value of about $2°$ at the equator to about $12°$ at $\pm 35°$ latitude [138–141].

The area distribution of individual sunspots can be described as a two-parameter log-normal distribution [142]:

$$\ln\left(\frac{dN}{dA}\right) = -\frac{(\ln A - \ln\langle A \rangle)^2}{2 \ln \sigma_a} + \ln\left(\frac{dN}{dA}\right)_{\max}$$

in terms of sunspot umbral area A (in units of $10^{-6} 2\pi R_\odot^2$). Values of the three other quantities in the above equation (Table 14.37) depend somewhat on the range of umbral areas used to derive them:

Table **14.37.** *Sunspot area distribution.*

Range	$\langle A \rangle$	σ_a	$(dN/dA)_{\max}$
1.5–141	0.62	3.8	9.2
5.5–116	0.34	4.8	16.4

14.13   SUNSPOT STATISTICS
by Karen Harvey and Robert M. Wilson

The sunspot number is defined as

$$R = k(10g + s),$$

where k is an observatory reduction constant of order unity, g is the number of sunspot groups, and s is the total number of individual spots [143–145]. Prior to January 1981, R was referred to as the *Zurich* sunspot number. From January 1981 on, R has been referred to as the *International* sunspot number.

Monthly values of R are combined to yield the 12-month moving average of R (denoted R_0), which is also known as the *smoothed* sunspot number [146]. For a cycle, the minimum value of R_0 denotes the *sunspot minimum* (R_m), while the maximum value denotes the *sunspot maximum* (R_M). Conventionally, the length of a sunspot cycle is determined from minimum to minimum ($m \leftrightarrow m$) and is comprised of two parts: the ascent interval, the time from minimum to maximum ($m \leftrightarrow M$), and the descent interval, the time from maximum to succeeding cycle minimum ($M \leftrightarrow m$). Occasionally, the time between maxima is also of interest ($M \leftrightarrow M$). Each sunspot cycle is numbered with the most recent sunspot cycle being cycle 22 (R_m occurred in September 1986 and R_M occurred in July 1989).

The sunspot record is of uneven quality [144]. The most reliable sunspot data extend from the present back to about 1850 and 1818 (covering cycles 7–9), while data of poor quality occur for earlier times (cycles before cycle 7). Some evidence exists suggesting that there was an extensive period of time when sunspots were few in number [147]. This interval of time (ca. 1645–1715; cycles −9 to −4) is often referred to as the *Maunder minimum*.

Other information from the sunspot record follows:

Waldmeier effect. The sunspot amplitude (R_M) varies inversely with the ascent duration ($m \leftrightarrow M$).
Hale cycle. The magnetic polarity changes in alternate cycles (even-numbered cycles have leading spots of southern polarity in the northern hemisphere, and vice versa).
Spörer law. The latitude of sunspots progresses equatorward with the phase of the solar cycle (yielding the so-called *butterfly diagram*).
Odd–even effect. The odd-following cycle tends to be of larger amplitude than the even-preceding cycle.
Gleissberg effect. Sunspot cycles vary according to an 8-cycle variation (the so-called 80–100 year variation).

Tables 14.38 and 14.39 list the sunspot number variations over the solar cycle.

Table 14.38. *Variation of the annual sunspot number over the solar cycle (based on the reliable data of cycles [10–21]).[a]*

| Parameter | Elapsed time (yr) from sunspot minimum occurrence year | | | | | | | | | | |
	0	1	2	3	4	5	6	7	8	9	10
Mean	6.2	18.9	60.2	99.4	107.0	98.5	79.1	52.4	36.5	21.2	12.0
Standard deviation	5.9	16.7	38.6	50.0	41.1	36.6	27.6	19.7	19.6	13.4	10.4
High	28.4	89.2	201.3	253.8	202.5	217.4	153.8	108.5	88.4	60.7	55.8
Low	0.0	0.0	10.4	24.5	39.3	17.8	34.4	14.8	0.3	1.6	0.2

Note
[a] Values listed are monthly mean values based on cycles 10–21 only.

Table 14.39. *Variation of the smoothed sunspot number over the solar cycle (based on the reliable data of cycles [10–21]).[a]*

| Parameter | Elapsed time (month) from R_m | | | | | | | | | | |
	0	12	24	36	48	60	72	84	96	108	120	132
Mean	5.1	18.6	61.6	98.0	109.2	99.3	79.9	52.4	34.7	20.4	11.7	11.7
Standard deviation	3.2	6.1	23.9	41.2	41.3	33.9	25.9	14.9	13.9	9.8	8.3	7.4
High	12.2	26.3	118.7	181.0	196.8	169.2	119.6	70.5	60.6	41.3	30.3	15.4
Low	1.5	9.3	35.5	52.5	54.5	56.9	48.0	31.2	13.8	11.5	3.2	2.6

Note
[a] Values listed are smoothed sunspot number values based on cycles 10–21 only.

Characteristics of all the known sunspot cycles are listed in Table 14.40. Mean values are listed in Table 14.41.

Table 14.40. *Characteristics of sunspot cycles* [1].[a]

Data quality	Cycle	Maximum M epoch	Maximum M R_M	Minimum m epoch	Minimum m R_M	Intervals (yr) $m \leftrightarrow m$	Intervals (yr) $m \leftrightarrow M$	Intervals (yr) $M \leftrightarrow m$	Intervals (yr) $M \leftrightarrow M$
P	−12	1615.5		1610.8		8.2	4.7	3.5	
	−11	1626.0		1619.0		15.0	7.0	8.0	10.5
	−10	1639.5		1634.0		11.0	5.5	5.5	13.5
	−9	1649.0		1645.0		10.0	4.0	6.0	9.5
	−8	1660.0		1655.0		11.0	5.0	6.0	11.0
	−7	1675.0		1666.0		13.5	9.0	4.5	15.0
	−6	1685.0		1679.5		10.0	5.5	4.5	10.0
	−5	1693.0		1689.5		8.5	3.5	5.0	8.0
	−4	1705.5		1698.0		14.0	7.5	6.5	12.5
	−3	1718.2		1712.0		11.5	6.2	5.3	12.7
	−2	1727.5		1723.5		10.5	4.0	6.5	9.3
	−1	1738.7		1734.0		11.0	4.7	6.3	11.2
	0	1750.3	92.6	1745.0		10.3	5.3	4.9	11.6
	1	1761.5	86.5	1755.3	8.4	11.2	6.2	5.0	11.1
	2	1769.8	115.8	1766.5	11.2	9.0	3.3	5.7	8.3
	3	1778.4	158.5	1775.5	7.2	9.3	2.9	6.4	8.6
	4	1788.2	141.2	1784.8	9.5	13.6	3.4	10.2	9.8
	5	1805.2	49.2	1798.4	3.2	12.3	6.8	5.5	17.0
	6	1816.3	48.7	1810.7	0.0	12.7	5.6	7.1	11.1
F	7	1829.9	71.7	1823.4	0.1	10.5	6.5	4.0	13.6
	8	1837.3	146.9	1833.9	7.3	9.7	3.4	6.3	7.4
	9	1848.2	131.6	1843.6	10.5	12.4	4.6	7.8	10.9
R	10	1860.2	97.9	1856.0	3.2	11.3	4.2	7.1	12.0
	11	1870.7	140.5	1867.3	5.2	11.7	3.4	8.3	10.5
	12	1840.0	74.6	1879.0	2.2	10.7	5.0	6.3	13.3
	13	1894.1	87.9	1990.3	5.0	11.8	3.8	8.0	10.1
	14	1906.2	64.2	1902.1	2.6	11.6	4.1	7.5	12.1
	15	1917.7	105.4	1913.7	1.5	10.0	4.0	6.0	11.5
	16	1928.3	78.1	1923.7	5.6	10.1	4.6	5.5	10.6
	17	1937.3	119.2	1933.8	3.4	10.4	3.5	6.9	9.0
	18	1947.4	151.8	1944.2	7.7	10.1	3.2	6.9	10.1
	19	1958.3	201.3	1954.8	3.4	10.5	4.0	6.5	10.9
	20	1968.9	110.6	1964.8	9.6	11.7	4.1	7.6	10.6
	21	1980.0	164.5	1976.5	12.2	10.3	3.5	6.8	11.1
	22	1989.6	158.5	1986.8	12.3		2.8		9.6

Note

[a] R denotes a "reliable" data interval, F denotes a "fair" interval, and P denotes a "poor" interval.

Reference

1. Allen, C.W., editor, 1973, *Astrophysical Quantities*, 3rd ed. (Athlone Press, London), Sec. 87

Table 14.41. *Mean values for selected sunspot cycle parameters.*

Parameter	Mean value R	Mean value R + F	Mean value All (R + F + P)
$m \leftrightarrow m$ period (yr)	10.9	10.9	11.0
$M \leftrightarrow M$ period (yr)	10.9	10.8	11.0
$m \leftrightarrow M$ ascent interval (yr)	3.9	4.0	4.7
$M \leftrightarrow m$ descent interval (yr)	7.0	6.8	6.3
R_M	119.6	119.0	112.9
R_m	5.7	5.7	6.0

Table 14.42 shows how certain solar activity characteristics vary throughout the sunspot cycle.

Table 14.42. *Solar activity.*

Year	Minimum					Maximum						
	0	1	2	3	4	5	6	7	8	9	10	11
					Sunspot regions							
R new cycle	68	237	488	547	561	510	360	269	168	99	38	13
R old cycle	9	7	3									
Spot latitude	24	22	19	17	14	13	12	10	9	8	7	6
					Latitude range							
Low	16	7	3	1	0	0	0	0	0	0	1	1
To high	42	39	40	42	38	37	33	28	24	20	16	9

Characteristics of an average size sunspot group:

Sunspot number	$R = 12$.
Number of individual spots	10.
Spot area (umbra + penumbra)	200 millionths of hemisphere, 260 millionths of disk.
Spot radius (if a single spot)	$0.020 R_\odot$.
Ca II plage area	1800 millionths of hemisphere.

14.14 FLARES AND CORONAL MASS EJECTIONS
by Steve Kahler

14.14.1 Flares

Chromospheric (Cool) Component of Flares [148, 149]

The Hα line importance classes are detailed in Table 14.43.

Table 14.43. *Classes of optical importance in the Hα line.*

Hα importance	Area (10^{-6} hemisphere)	Mean duration
S	$A < 200$	few minutes
1	$200 < A < 500$	25 min
2	$500 < A < 1200$	55 min
3	$1200 < A < 2400$	2 hr
4	$A > 2400$	2 hr

Hα brilliance: f = faint, n = normal, b = bright

Temperature $\sim 15\,000$ K
Density $\sim 3 \times 10^{13}$ cm^{-3}

Frequency of observed importance ≥ 1 flares:

Frequency near solar maximum   1000–2000 flares per year.
Frequency near solar minimum   20–60 flares per year.

White-light flares [150, 151]:

Frequency near solar maximum ~ 15 per year.
Luminosity 10^{27}–10^{28} erg s^{-1}.

Coronal (Hot) Component [152]

Classes of soft X-ray (1–8 Å) peak fluxes measured at 1 AU:

$$Bn = n \times 10^{-7} \text{ W m}^{-2}, \qquad Cn = n \times 10^{-6} \text{ W m}^{-2},$$
$$Mn = n \times 10^{-5} \text{ W m}^{-2}, \qquad Xn = n \times 10^{-4} \text{ W m}^{-2}.$$

Frequency of 1–8 Å flares [153]:

Frequency of $\geq$ M1 flares near solar maximum ~ 500 per year.
Frequency of $\geq$ M1 flares near solar minimum ~ 15 per year.

Peak temperatures	$(8$–$22) \times 10^6$ K.
Peak emission measures $(n_e^2 V)$	10^{48}–10^{50} cm^{-3}.
Density	10^{10}–10^{12} cm^{-3}.

Impulsive Component

The duration is from < 1 min to > 30 min; the median duration ~ 100 s. The γ-ray fluence [154] from < 10 to 10^4 γ cm^{-2} at > 300 keV; from < 0.3 to 3×10^2 γ cm^{-2} for 4–8 MeV lines. For hard X-rays [155, 156]:

Peak flux at $E > 20$ keV from $< 10^{-6}$ to $> 10^{-3}$ erg cm^{-2} s^{-1}.
Spectra: $3 < \gamma < 9$, where $N(E) = A E^{-\gamma}$ photons cm^{-2} s^{-1} keV^{-1}.
Thermal fits yield $T \geq 10^8$ K.

For EUV (10–1030 Å) [157]:

Peak fluxes from $< 3 \times 10^{-2}$ to 10 erg cm^{-2} s^{-1}.
Temporal profiles match those of $E > 10$ keV X-rays.

For microwaves (1000–35 000 MHz) [156]:

Peak fluxes from < 10 to $\geq 10^4$ solar flux units (s.f.u.) (10^{-22} W m^{-2} Hz^{-1}).
Temporal profiles match those of $E > 20$ keV X-rays, and flux ($E > 20$ keV) (erg cm^{-2} s^{-1}) $\sim 10^{-7} \times$ flux (3 cm) (s.f.u.).

14.14.2 Coronal Mass Ejections

Most coronal mass ejection (CME) quantities range over about two orders of magnitude. Average values follow [158–161]:

Mass	3×10^{15} g.
Kinetic energy	2×10^{30} erg.

Speeds (of leading edges)

    at solar maximum 450 km s^{-1}.

    at solar minimum 160 km s^{-1}.

  Angular width (plane of sky, subtended to solar disk) $47°$.

  Frequency [162]

    at solar maximum   2–3 CMEs per day.

    at solar minimum   0.1–0.3 CMEs per day.

14.15 SOLAR RADIO EMISSION
by Timothy Bastian

Solar radio emission is expressed quantitatively in terms of the flux density S_ν, usually in solar flux units (s.f.u.), where 1 s.f.u. $= 10^{-22}$ W m^{-2} Hz^{-1}. For observations that spatially resolve the source of radio emission, the intensity of the radiation is often expressed in terms of its brightness temperature T_B, where $S_\nu = 7.22 \times 10^{-51} T_B \nu^2$ W m^{-2} arcsec^{-2} Hz^{-1}. T_C refers to the brightness temperature at the center of the solar disk. The degree of polarization, ρ_C, is defined by the ratio of the Stokes polarization parameters V and I. Expressed in terms of brightness temperature in the orthogonal (right- and left-hand) senses of circular polarization, $\rho_C = (T_{\mathrm{RCP}} - T_{\mathrm{LCP}})/(T_{\mathrm{RCP}} + T_{\mathrm{LCP}})$, where the RCP sense corresponds to a counterclockwise rotation for radiation propagating toward the observer.

14.15.1 Properties of Radio Emission from the Quiet Sun

The brightness temperature of the quiet Sun at disk center may be calculated approximately from the following expressions for millimeter and centimeter wavelengths (T_C in K, $u = \log_{10} \lambda$, λ in cm):

$$\log T_C = 3.9609 + 0.1856u + 0.0523u^2 + 0.13415u^3 + 0.0834u^4,$$

valid between 0.1 and 20 cm;

$$\log T_C = 0.7392 + 4.3185u - 0.9049u^2,$$

valid for $\lambda = 20$–2000 cm. The fits are based on [163–165].

14.15.2 Properties of Radio Emission from Solar Active Regions

Meter and Decameter Wavelengths

Storm continua and type I bursts (see below and [166]) are often associated with solar active regions. Type I storm durations range from hours to days and are distinguished by high values of ρ_C, bandwidths of a few times 10 MHz, and apparent brightness temperatures $< 10^{10}$ K.

Decimeter and Centimeter Wavelengths

Decimetric and microwave emission associated with active regions is characterized by [167, 168] a diffuse morphology for $\lambda \gtrsim 10$ cm and a low to moderate degree of circular polarization $\rho_C \lesssim 15\%$. Its brightness is typical of coronal temperatures [$T_B \sim (1–2) \times 10^6$ K]. For $\lambda \lesssim 10$ cm, the diffuse morphology gives way to one or more compact components associated with sunspot umbrae and penumbrae that possess a degree of polarization that ranges from low ($\rho_C \sim$ few %) to high ($\rho_C \gtrsim 90\%$) values. The brightness of compact components is again near coronal values. Radio emission associated with solar active regions typically possesses a spectral maximum in flux density between 8 and 10 cm [169].

14.15.3 Properties of Solar Radio Bursts (Flares)

Meter Wavelengths

(i) Type I [166, 170]:

Frequency range	150–350 MHz.
Bandwidth	2.5–7 MHz ($\sim 0.025\nu$ MHz; ν in MHz).
Duration	0.2–0.7 s ($\sim 80/\nu$ s).
Brightness	As high as 10^7–10^{10} K.
Polarization	Up to 100% circularly polarized.
Fine structure	Chains, periodic variations.

(ii) Type II [171]:

Frequency range	< 20–150 MHz; harmonic structure in 60%.
Bandwidth	~ 100 MHz.
Frequency drift rate	~ 1 MHz s^{-1}.
Duration	5–15 min.
Brightness	10^7–10^{13} K.
Polarization	Unpolarized or weakly circularly polarized; herringbone structure sometimes displays $\sim 50\%$ circular polarization.
Fine structures	Band splitting, multiple lanes, herringbone structure.

(iii) Type III [172]:

Frequency range	Full range; harmonic structure common, 1–100 MHz.
Frequency drift rate	$-0.01\nu^{1.84}$ MHz s^{-1}.
Duration	$\sim 220\nu^{-1}$ s.
Brightness	10^8–10^{12} K.
Polarization	$\rho_C \lesssim 15\%$ (harmonic); $\rho_C \lesssim 50\%$ (fundamental).
Variants	Type J and type U bursts.

(iv) Type IV [173, 174]:

Frequency range	20–200 MHz.
Bandwidth	Broadband continuum.
Duration	3–45 min.
Brightness	$< 10^8$–10^{10} K.
Polarization	$\rho_C \lesssim 20\%$ (early), often increasing to high values for events with durations longer than 20 min.
Variants	Moving type IV, slow-drift continuum, type II–associated, pulsations.

(v) Type V [172]:

Frequency range	< 10–120 MHz.
Bandwidth	Broadband continuum.
Duration	$\sim 500\nu^{-1/2}$ s.
Brightness	10^7–10^{12} K.
Polarization	$\rho_C \lesssim 10\%$, decreasing from disk center-to-limb, sense of polarization usually opposite to that of preceding type III bursts.

Decimeter Wavelengths [175]

(i) Type III–like or fast-drift bursts:

Bandwidth	Variable.
Duration	0.5–1.0 s.
Drift rate	> 100 MHz s^{-1}.
Variants	Classical type III and type U bursts, dm extensions to type IIIm bursts, narrowband type III bursts (blips), long duration type III bursts.

(ii) Pulsations:

Bandwidth	Few $\times$ 100 MHz.
Periods	Pulses recur periodically or quasiperiodically with separations of 0.1–1.0 s.
Duration	Groups of pulses (10–100 s) last from seconds to minutes.
Variants	Quasiperiodic pulsations (regular, long period), dm pulsations (irregular, short period).

(iii) Diffuse continua or type IV–like bursts:

Bandwidth	Few $\times$100 MHz.
Duration	10 s of seconds to minutes.
Variants	Smooth continua, modulated continua, ridges.

(iv) Spikes:

Bandwidth	Few MHz.
Duration	< 0.1 s individually, with groups (10–10^4) occurring in broadband clusters during some seconds to minutes.
Variants	Type III–associated spikes, type IV–associated spikes.

Centimeter and Millimeter Wavelengths

Solar bursts at centimeter and millimeter wavelengths tend to be broadband continua, moderately polarized, with a brightness of a few $\times\ 10^6$ K to a few $\times\ 10^9$ K. The spectral peak is generally near 8 GHz [176]; roughly 80% of solar radio burst display more than one spectral component [177].

REFERENCES

1. Brown, T.M. & Christensen-Dalsgaard, J. 1998, *ApJL*, **500**, 195
2. Wesselink, A.J. 1969, *MNRAS*, **144**, 297
3. Sofia, S., Heaps, W., & Twigg, L.W. 1994, *ApJ*, **427**, 1048
4. Dicke, R.H., Kuhn, J.R., & Libbrecht, K.G. 1987, *ApJ*, **318**, 451
5. Foukal, P.V. 1990, *Solar Astrophysics* (Wiley-Interscience, New York)
6. Fröhlich, C., Foukal, P.V., Hickey, J.R., Hudson, H.S., & Willson, R.C. 1991, *The Sun in Time*, edited by C.P. Sonett et al. (University of Arizona Press, Tucson), p. 11
7. Hudson, H.S. 1988, *ARA&A*, **26**, 473
8. Stix, M. 1989, *The Sun* (Springer-Verlag, Berlin)
9. Willson, R.C., & Hudson, H.S. 1988, *Nature*, **332**, 810
10. Hardorp, J. 1980a, *A&A*, **88**, 334
11. Hardorp, J. 1980b, *A&A*, **91**, 221
12. Hayes, D.S. 1985, in *Proceedings of the IAU Symposium No. 111*, edited by D.S. Hayes (Reidel, Dordrecht)
13. Neckel, H. 1986, *A&A*, **159**, 175
14. Allen, C.W., editor, 1973, *Astrophysical Quantities*, 3rd ed. (Athlone Press, London), Sec. 75
15. Sackmann, I.-J., Boothroyd, A.I., & Fowler, W.A. 1990, *ApJ*, **360**, 727
16. Guenther, D.B. 1989, *ApJ*, **339**, 1156
17. Scherrer, P.H. 1973, thesis, Stanford University
18. Guenther, D.B., Demarque, P., Kim, Y.-C., & Pinsonneault, M.H. 1992, *ApJ*, **387**, 372
19. Iglesias, C.A., & Rogers, F.J. 1991, *ApJ*, **371**, 408
20. Anders, E., & Grevesse, N. 1989, *Geochim. Cosmochim. Acta*, **53**, 197
21. Bahcall, J.N., & Ulrich, R.K. 1988, *Rev. Mod. Phys.*, **60**, 297
22. Cox, A.N., Guzik, G.A., & Kidman, R.B. 1989, *ApJ*, **342**, 1187
23. Christensen-Dalsgaard, J., Gough, D.O., & Thompson, M.J. 1991, *ApJ*, **378**, 413
24. Guzik, J.A. & Cox, A.N. 1993, *ApJ*, **411**, 394
25. Tassoul, M. 1990, *ApJ*, **358**, 313
26. Pallé, P.L. 1991, in *Solar Interior and Atmosphere*, edited by A.N. Cox, W.C. Livingston, and M.S. Matthews (University of Arizona Press, Tucson), p. 1249
27. Scherrer, P.H., Wilcox, J.M., Christensen-Dalsgaard, J., & Tassoul, M. 1980, *ApJS*, **43**, 469
28. Hathaway, D.H. 1988, *Solar Phys.*, **117**, 1
29. Duvall, Jr., T.L. 1982, *Nature*, **300**, 242
30. Duvall, Jr., T.L., Harvey, J.W., & Pomerantz, M.A. 1986, *Nature*, **321**, 500
31. Libbrecht, K.G. 1989, *ApJ*, **336**, 1092
32. Woodard, M.F., Kuhn, J.R., Murray, N., & Libbrecht, K.G. 1991, *ApJL*, **373**, L81
33. Hill, F. 1991, private communication
34. Rhodes, Jr. E.J., Brown, T.M., Cacciani, S.G., Korzennik, S.G., & Ulrich, R.K. 1991, *Challenges to Theories of the Structure of Moderate-Mass Stars*, edited by D. Gough and J. Toomre (Springer-Verlag, Berlin), p. 277
35. Braun, D.C., Duvall, Jr., T.L., & LaBonte, B.J. 1988, *ApJ*, **335**, 1015
36. Libbrecht, K.G. 1988, *ApJ*, **334**, 510
37. Jefferies, S.M., Duvall, Jr., T.L., Harvey, J.W., Osaki, Y., Korzennik, S.G., & Ulrich, R.K. 1991, in *Challenges to Theories of the Structure of Moderate-Mass Stars*, edited by D. Gough and J. Toomre (Springer-Verlag, Berlin), p. 277
38. Hill, H.A., Fröhlich, C., Gabriel, M., & Kotov, V.A. 1991, in *Solar Interior and Atmosphere*, edited by A.N. Cox, W.C. Livingston, and M.S. Matthews (University of Arizona Press, Tucson), p. 562
39. Fontenla, J.M., Avrett, E.H., & Loeser, R. 1993, *ApJ*, **406**, 319
40. Fontenla, J.M., Avrett, E.H., & Loeser, R. 1991, *ApJ*, **377**, 712
41. Vernazza, J.E., Avrett, E.H., & Loeser, R. 1981, *ApJS*, **45**, 635
42. Anders, E., & Grevesse, N. 1989, *Geochim. Cosmochem. Acta*, **53**, 197
43. Holweger, H., & Müller, E.A. 1974, *Solar Phys.*, **39**, 19
44. Maltby, P., Avrett, E.H., Carlsson, M., Kjeldseth-Moe, O., Kurucz, R., & Loeser, R. 1986, *ApJ*, **306**, 284
45. Kurucz, R.L. 1979, *ApJS*, **40**, 1
46. Anderson, L.S., & Athay, R.G. 1989, *ApJ*, **346**, 1010
47. Ayres, T.R., Testerman, L., & Brault, J.W. 1986, *ApJ*, **304**, 542
48. Solanki, S.K., Steiner, O., & Uitenbroek, H. 1991, *A&A*, **250**, 220
49. Stenflo, J.O., editor, 1990, *Solar Photosphere: Structure, Convection, and Magnetic Fields* (Kluwer Academic, Dordrecht)
50. Cox, A.N., Livingston, W.C., & Matthews, M.S., editors, 1991, *Solar Interior and Atmosphere* (University of Arizona Press, Tucson)
51. Ewell, Jr., M.W., Zirin, H., & Jansen, J.B. 1993, *ApJ*, **403**, 426
52. Solanki, S.K., Livingston, W.C., & Ayres, T. 1994, *Science*, **263**, 64
53. Clark, T.A., Lindsey, C., Rabin, D.M., & Livingston, W.C. 1995, in *Infrared Tools for Solar Astrophysics: What's Next*, edited by J. Kuhn and M.J. Penn (World Scientific, Singapore), p. 133
54. Kurucz, R.L. 1991, in *Solar Interior and Atmosphere*, edited by A.N. Cox et al. (University of Arizona Press, Tucson) p. 663
55. Moore, C.E., Minnaert, M.G.K., & Houtgast, J. 1966, *NBS Monograph* 61 (US Government Printing Office, Washington, DC)
56. Harvey, J.W., & Hall, D. 1975, *BAAS*, **7**, 459
57. Ayres, T.R., & Brault, J.W. 1990, *ApJ*, **363**, 705
58. Solanki, S.K., Biémont, E., & Mürset, U. 1990, *A&AS*, **83**, 307
59. Jefferies, J.T. 1968, *Spectral Line Formation* (Blaisdell, Waltham, MA)
60. Carlsson, M., Rutten, R.J., & Shchukina, N.G. 1992, *A&A*, **253**, 567

61. Livingston, W., & Wallace, L. 1991, An Atlas of the Solar Spectrum in the Infrared from 1850 to 9000 cm^{-1} (1.1 to 5.4 μm), NSO Technical Report No. 91-001 (Tucson)

62. Wallace, L., & Livingston, W. 1992, An Atlas of a Dark Sunspot Umbral Spectrum from 1970 to 8640 cm^{-1} (1.16 to 5.1 μm), NSO Technical Report No. 92-001 (Tucson)

63. Wallace, L., & Livingston, W. 1993, An Atlas of the Photospheric Spectrum from 8900 to 13 600 cm^{-1} (7350 to 11 230 Å), NSO Technical Report No. 93-001 (Tucson)

64. Wallace, L., & Livingston, W. 1994, An Atlas of the Sunspot Spectrum from 470 to 1233 cm^{-1} (8.1 to 21 μm) and the Photospheric Spectrum from 460 to 630 cm^{-1} (16 to 22 μm), NSO Technical Report No. 94-001 (Tucson)

65. Torr, M.R., & Torr, D.G. 1985, *J. Geophys. Res.*, **90**, A7, 6675

66. Tobiska, W.K. 1993, *J. Geophys. Res.*, **98**, No. A11, 18 879

67. Mihalas, D. 1978, *Stellar Atmospheres* (Freeman, San Franciso)

68. Allen, C.W., editor, 1973, *Astrophysical Quantities*, 3rd ed. (Athlone Press, London), Sec. 78

69. Mallia, E.A. 1968, *Solar Phys.*, **5**, 281

70. Elste, G.H.E. 1967, *ApJ*, **148**, 857

71. Aslanov, I.A., Davudov, Yu.D., & Salmanov, I.R. 1968, *Sov. Astron.—AJ*, **12**, 49

72. Parnell, R.T., & Beckers, J.M. 1969, *Solar Phys.*, **9**, 35

73. Gurtovenko, E., & Troyan, V. 1971, *Solar Phys.*, **20**, 264

74. de Jager, C., & Neven, L. 1972, *Solar Phys.*, **22**, 49

75. Steffen, M. 1991, in *Solar Interior and Atmosphere*, edited by A.N. Cox et al. (University of Arizona Press, Tucson)

76. Solanki, S.K. 1995, private communication

77. Allen, C.W., editor, 1973, *Astrophysical Quantities*, 3rd ed. (Athlone Press, London), Secs. 71 and 82

78. Vernazza, J.E., Avrett, E.H., & Loeser, R. 1981, *ApJS*, **45**, 635

79. Kopp, G., Lindsey, C., Roellig, T.L., Werner, W.W., Becklin, E.E., Orrall, F.Q., & Jefferies, J.T. 1992, *ApJ*, **388**, 203

80. Altrock, R.C., & Canfield, R.C. 1972, *Solar Phys.*, **23**, 257

81. Koutchmy, S., & Lamy, P. 1985, *Proc. IAU Coll. No. 85*, 63 (Reidel, Dordrecht)

82. *AQ*. 1973, Secs. 73, 84, and 85

83. van de Hulst, H.C. 1950, *BAN*, **11**, 135; also, 1953, *The Sun*, edited by G. Kuiper (University of Chicago Press, Chicago), p. 25

84. Lebecq, C., Koutchmy, S., & Stellmacher, G. 1985, *A&A*, **152**, 157

85. Koutchmy, S., & Laffineur, M. 1970, *Nature*, **226**, 1141

86. Newkirk, G., Dupree, R.G., & Schmahl, E.J. 1970, *Solar Phys.*, **15**, 15

87. Nambra, O., & Diemel, W.E. 1969, *Solar Phys.*, **7**, 167

88. Ludendorff, H. 1934, *Sitz Preuss. Ak. Wiss.*, **185**, 1928

89. Koutchmy, S., Zirker, J.B., Steinolfson, R.S., & Zhugzda, J.D. 1991, in *Solar Interior and Atmosphere*, edited by A.N. Cox, W. Livingston, and M.S. Matthews (University of Arizona Press, Tucson)

90. Koutchmy, S., Picat, J.P., & Dantel, M. 1977, *A&A*, **59**, 349

91. Blackwell, D.E., Dewhirst, D.W., & Ingham, M.F. 1967, *Adv. Astron. Astrophys.*, **5**, 1

92. Baumbach, S. 1937, *Astron. Nach.*, **263**, 121

93. Woehl, H. 1978, *A&A*, **62**, 165

94. LaBonte, J.B. 1981, *Solar Phys.*, **69**, 177

95. Balthasar, H., Stark, D., & Woehl, H. 1986, *A&A*, **160**, 277

96. Balthasar, H., Stark, D., & Woehl, H. 1986, *A&A*, **174**, 359

97. Woehl, H., & Ye, B. 1990, *A&A*, **240**, 511

98. Hill, F., Gough, D., Toomre, J., & Haber, D.A. 1988, in *Advances in Helio- and Asteroseismology*, edited by J. Christensen-Dalsgaard and S. Frandsen (Reidel, Dordrecht), p. 45

99. Rhodes, Jr., E.J., Cacciani, A., Korzennik, S.G., & Ulrich, R.K. 1991, in *Challenges to Theories of the Structure of Moderate Mass Stars*, Lecture Notes in Physics, edited by D. Gough and J. Toomre (Springer-Verlag, Berlin), p. 285

100. Allen, C.W., editor, 1973, *Astrophysical Quantities*, 3rd ed. (Athlone Press, London), Sec. 90

101. Bahng, J., & Schwarzschild, M. 1961, *ApJ*, **134**, 312

102. Mehltretter, J.P. 1971, *Solar Phys.*, **19**, 32

103. Roudier, T., & Muller, R. 1986, *Solar Phys.*, **107**, 11

104. Macris, C.J., Muller, R., Rosch, J., & Roudier, T. 1984, in *Small-Scale Dynamic Processes in the Quiet Sun*, edited by S. Keil (National Solar Observatory, Sunspot, NM)

105. Bray, R.J., Loughhead, R.E., & Durrant, C.J. 1984, in *The Solar Granulation* (Cambridge University Press, Cambridge)

106. Muller, R. 1988, in *Solar and Stellar Granulation, NATO Conference No. 263*, edited by R.J. Rutten and G. Severino (Reidel, Dordrecht)

107. Muller, R., & Keil, S.L. 1983, *Solar Phys.*, **87**, 243

108. November, L.J., Toomre, J., Gebbie, K.B., & Simon, G.W. 1981, *ApJ*, **243**, L123

109. Simon, G.W., Title, A.M., Tarbell, T.D., Shine, R.A., Ferguson, S.H., & Zirin, H. 1988, *ApJ*, **327**, 964

110. November, L.J. 1989, *ApJ*, **344**, 494

111. Muller, R., Roudier, T., Vigneau, J., Simon, G.W., Frank, Z., Shine, R., & Title, A.M. 1992, *Nature*, **356**, 322

112. Foukal, P. 1990, in *Solar Astrophysics* (Wiley-Interscience, New York), p. 293

113. Rogerson, J.N. 1961, *ApJ*, **134**, 331

114. Foukal, P., Harvey, K., & Hill, F. 1992, *ApJL*, **383**, L89

115. Muller, R. 1985, *Solar Phys.*, **100**, 237

116. Chapman, G., & McGuire, T. 1977, *ApJ*, **217**, 657

117. Solanki, S.K., & Brigljević, V. 1992, *A&A*, **262**, L29

118. Briand C., & Solanki, S.K. 1995, *A&A*, **299**, 596

119. *Solar Geophysical Data*, NOAA/GC2, Boulder, CO 80303

120. Hirayama, T. 1986, *Coronal and Prominence Plas-*

mas, edited by A.I. Poland (NASA, Washington, DC), p. 2442

121. Dunn, R.B. 1960, Ph.D. thesis, Harvard University
122. Leroy, J.L., Bommier, V., & Sahai-Brechot, S. 1984, *A&A*, **131**, 33
123. Engvold, O. 1980, *Solar Phys.*, **61**, 351
124. Malherbe, J.M., Schmieder, B., & Mein, P. 1981, *A&A*, **102**, 124
125. Degenhardt, D., & Wiehr, E. 1991, *A&A*, **252**, 821
126. Schmidt, W., Hofmann, A., Balthasar, H., Tarbell, T.D., & Frank, Z.A. 1992, *A&A*, **264**, L27
127. Title, A.M., Frank, Z.A., Shine, R.A., Tarbell, T.D., Topka, K.P., Scharmer, G., & Schmidt, W. 1993, *ApJ*, **403**, 780
128. Skumanich, A., Lites, B.W., Martínez, & Pillet, V. 1994, in *Solar Surface Magnetism*, edited by R.J. Rutten and C.J. Schrijver (Kluwer Academic, Dordrecht), p. 99
129. Bray, R.J., & Loughhead, R.E. 1964, *Sunspots* (Wiley, New York)
130. Suzuki, Y. 1967, *PASJ*, **19**, 220
131. Wittmann, A., & Schröter, E.H. 1969, *Solar Phys.*, **10**, 357
132. Gokhale, M.H., & Zwaan, C. 1972, *Solar Phys.*, **26**, 52
133. Kopp, G., & Rabin, D. 1992, *Solar Phys.*, **141**, 253
134. Solanki, S.K., Montavon, C.A.P., & Livingston, W. 1994, *A&A*, **283**, 221
135. Solanki, S.K., & Schmidt, H.U. 1992, *A&A*, **267**, 287
136. Albregtsen, F., & Maltby, P. 1981, *Solar Phys.*, **71**, 269
137. Howard, R.F. 1992, *Solar Phys.*, **137**, 205
138. Hale, G.E., Ellerman, F., Nicholson, S.B., & Joy, A.H. 1919, *ApJ*, **49**, 153
139. Brunner, W. 1930, *Astron. Mitt. Zürich No. 124*
140. Wang, Y.-M., & Sheeley, N.R. 1989, *Solar Phys.*, **124**, 81
141. Howard, R.F. 1991, *Solar Phys.*, **136**, 251
142. Bogdan, T.J., Gilman, P.A., Lerche, I., & Howard, R.F. 1988, *ApJ*, **327**, 451
143. Allen, C.W., editor, 1973, *Astrophysical Quantities*, 3rd ed. (Athlone Press, London), Sec. 87
144. Waldmeier, M. 1961, *The Sunspot Activity in the Years 1610–1960* (Schulthess, Zurich)
145. McKinnon, J.A. 1987, Report No. UAG-95, World Data Center, Boulder, CO
146. Howard, R. 1977, in *Illustrated Glossary for Solar and Solar-Terrestrial Physics*, edited by A. Bruzek and C.J. Durrant (Reidel, Dordrecht)
147. Pepin, R.O., Eddy, J.A., & Merrill, R.B. editors, 1980, *The Ancient Sun* (Pergamon, New York)
148. Svestka, Z. 1976, *Solar Flares* (Reidel, Dordrecht)
149. Zirin, H. 1988, *Astrophysics of the Sun* (Cambridge University Press, Cambridge)

150. Neidig, D., & Cliver, E.W. 1983, *Solar Phys.*, **88**, 275
151. Neidig, D. 1989, *Solar Phys.*, **121**, 261
152. Denton, R., & Feldman, U. 1984, *ApJ*, **286**, 359
153. Hirman, J.W., Heckman, G.R., Greer, M.S., & Smith, J.B. 1988, *EOS, Trans. AGU*, **69**, No. 42, 962
154. Rieger, E. 1989, *Solar Phys.*, **121**, 323
155. Dennis, B.R. 1985, *Solar Phys.*, **100**, 465
156. Kane, S.R. 1974, in *Coronal Disturbances*, edited by G. Newkirk (Reidel, Dordrecht), p. 105
157. McClymont, A.N., & Canfield, R.C. 1986, *ApJ*, **305**, 936
158. Hundhausen, A.J. 1993, *J. Geophys. Res.*, **98**, 13177
159. Hundhausen, A.J., Burkapile, J.T., & St.Cyr, O.C. 1994a, *J. Geophys. Res.*, **99**, 6543
160. Hundhausen, A.J., Stanger, A.L., & Serbicki, S.A. 1994b, *Proc. Third SOHO Workshop*, ESA Sp-373, Noordwijk
161. Howard, R.A., Sheeley Jr., N.R., Koomen, M.J., & Michels, D.J. 1985, *J. Geophys. Res.*, **90**, 8173
162. Webb, D.F. 1991, *Adv. Space Res.*, **11** (1), 37
163. Linsky, J.L. 1974, *Solar Phys.*, **28**, 409
164. Sheridan, K.V., & McClean, D.J. 1985, in *Solar Radiophysics*, edited by J.J. McClean and N.R. Labrum (Cambridge University Press, Cambridge), p. 443
165. Zirin, H., Baumert, B.M., & Hurford, G.J. 1991, *ApJ*, **370**, 779
166. Elgarøy, E.O. 1977, in *Solar Noise Storms* (Pergamon, Oxford), p. 366
167. Kundu, M.R. 1985, *Solar Phys.*, **100**, 491
168. Marsh, K.A., & Hurford, G.J. 1982, *ARA&A*, **20**, 497
169. Kundu, M.R. 1965, in *Solar Radio Astronomy* (Wiley, New York), p. 660
170. Kai, K., Melrose, D.B., & Suzuki, S. 1985, in *Solar Radiophysics*, edited by J.J. McClean and N.R. Labrum (Cambridge University Press, Cambridge), p. 415
171. Nelson, G., & Melrose, D.B. 1985, in *Solar Radiophysics*, edited by J.J. McClean and N.R. Labrum (Cambridge University Press, Cambridge), p. 333
172. Suzuki, S., & Dulk, G. 1985, in *Solar Radiophysics*, edited by J.J. McClean and N.R. Labrum (Cambridge University Press, Cambridge), p. 289
173. Robinson, R.D. 1985, in *Solar Radiophysics*, edited by J.J. McClean and N.R. Labrum (Cambridge University Press, Cambridge), p. 385
174. Stewart, R.T. 1985, in *Solar Radiophysics*, edited by J.J. McClean and N.R. Labrum (Cambridge University Press, Cambridge), p. 361
175. Güdel, M., & Benz, A.O. 1988, *A&AS*, **75**, 243
176. Guidice, D.A., & Castelli, J.P. 1975, *Solar Phys.*, **44**, 155
177. Stähli, M., Gary, D.E., & Hurford, G.J. 1989, *Solar Phys.*, **120**, 351

Chapter 15

Normal Stars

John S. Drilling and Arlo U. Landolt

15.1 STELLAR QUANTITIES AND INTERRELATIONS

$\mathcal{M}$ = mass ($\mathcal{M}_\odot$ = Sun's mass).

R = radius ($R_\odot$ = Sun's radius).

L = luminosity ($L_\odot$ = Sun's luminosity) = total outflow of radiation (ergs s^{-1}).

$\bar{\rho}$ = mean density = $\mathcal{M}/(\frac{4}{3}\pi R^3)$.

Sp = spectral classification, which may be combined with a luminosity class.

m = apparent magnitude = -2.5 log apparent brightness. Typical subscripts: V = visual, B = blue, U = ultraviolet, pg = photographic, pv = photovisual, bol = bolometric (total radiation); in general, m_λ = apparent magnitude of spectral region λ.

$U, B, V = m_U, m_B, m_V$ = apparent magnitudes in the UBV system.

M = absolute magnitude = apparent magnitude standardized to 10 pc without interstellar absorption.

$B - V$ = color index; $(B - V)_0$ = intrinsic color index (i.e., no interstellar absorption); or, in general a color index is the difference in the apparent magnitude as measured at two different wavelengths.

BC = bolometric correction = $m_{\text{bol}} - V$ (always negative).

A = space absorption in magnitudes (usually visual).

m_0 = magnitude corrected for absorption = $m - A$.

E_{B-V} = color excess = $(B - V) - (B - V)_0$.

$m - M$ = distance modulus = $5 \log d - 5 + A$.

$m_0 - M$ = distance modulus corrected for absorption = $5 \log d - 5$, where d is distance in parsecs (pc).

F = total radiant flux at stellar surface.

f = radiant flux for a star outside the Earth's atmosphere.

T_{eff} = stellar effective temperature (from $F = \sigma T_{\text{eff}}^4$), where σ is Stefan's constant.

v_{rot} = equatorial rotational velocity.

g = surface gravity (cm s^{-2}).

d = distance, usually in parsecs (pc).

π = parallax in seconds of arc ($''$) = $1/d$, with d in pc.

All the logarithms in this chapter are common logs with a base of 10.

15.1.1 Numerical Relations

$$M = m + 5 + 5 \log \pi - A = m + 5 - 5 \log d - A,$$
$$M_{\text{bol}} = -2.5 \log L/L_\odot + 4.74,$$

where $L_\odot = 3.845 \times 10^{33}$ ergs s^{-1}, and $+4.74$ is the absolute bolometric magnitude of the Sun. The bolometric correction is the difference between the absolute visual and absolute bolometric magnitude:

$$\text{BC} = M_{\text{bol}} - M_V.$$

Bolometric luminosities, radii, and effective temperatures are related by

$$M_{\text{bol}} = 42.36 - 5 \log R/R_\odot - 10 \log T_{\text{eff}},$$

where solar values of $M_{\text{bol}\odot} = 4.74$ and $T_{\text{eff}\odot} = 5777$ K have been adopted.

$$\log L = -3.147 + 2 \log R + 4 \log T_{\text{eff}},$$
$$(m_{\text{bol}} = 0) \text{ star} \equiv 2.48 \times 10^{-5} \text{ erg cm}^{-2} \text{ s}^{-1} \text{ outside the Earth's atmosphere,}$$
$$(M_{\text{bol}} = 0) \text{ star} \equiv 2.97 \times 10^{28} \text{ watts emitted radiation,}$$
$$(M_V = 0) \text{ star} \equiv 2.45 \times 10^{29} \text{ candela.}$$

The zero age main sequence (ZAMS) is represented by [1]

$$\log R/R_\odot = 0.640 \log \mathcal{M}/\mathcal{M}_\odot + 0.011 \quad (0.12 < \log \mathcal{M}/\mathcal{M}_\odot < 1.3),$$
$$\log R/R_\odot = 0.917 \log \mathcal{M}/\mathcal{M}_\odot - 0.020 \quad (-1.0 < \log \mathcal{M}/\mathcal{M}_\odot < 0.12).$$

The mass–luminosity relation may be written [2, 3]:

$$\log \mathcal{M}/\mathcal{M}_\odot = 0.48 - 0.105 M_{\text{bol}} \quad \text{for} \quad -8 \leq M_{\text{bol}} < 10.5,$$

or

$$\log L/L_\odot = 3.8 \log \mathcal{M}/\mathcal{M}_\odot + 0.08 \quad \text{for} \quad \mathcal{M} > 0.2\mathcal{M}_\odot.$$

Another representation is [4]

$$\log \mathcal{M}/\mathcal{M}_\odot = 0.46 - 0.10 M_{\text{bol}}, \qquad M_{\text{bol}} < 7.5,$$
$$\log \mathcal{M}/\mathcal{M}_\odot = 0.75 - 0.145 M_{\text{bol}}, \qquad M_{\text{bol}} > 7.5.$$

The most reliable stellar masses are summarized in [5] and [6]; also, see the discussion in [2].

15.2 SPECTRAL CLASSIFICATION

We define normal stars to be those which can be classified on the MK system (specifically, [7, 8], and more generally [9]), or which are classified as white dwarfs according to the system described in [10]. Table 15.1 gives these classes.

Table 15.1. *MK spectral classes.*

MK spectral class	Class characteristics
O	Hot stars with He II absorption
B	He I absorption; H developing later
A	Very strong H, decreasing later; Ca II increasing
F	Ca II stronger; H weaker; metals developing
G	Ca II strong; Fe and other metals strong; H weaker
K	Strong metallic lines; CH and CN bands developing
M	Very red; TiO bands developing strongly

The spectral classes are further subdivided into decimal subclasses (e.g., B0, B1, B2, etc.), although not all subdivisions are used, and some classes are further subdivided (e.g., O9.5). Table 15.2 lists the MK luminosity classes.

Table 15.2. *MK luminosity classes.*

MK luminosity class	Examples
I supergiants	B0 I
II bright giants	B5 II
III giants	G0 III
IV subgiants	G5 IV
V dwarfs (main sequence)	G0 V

The luminosity classes are further subdivided (e.g., Ia, Iab, Ib, etc.).

The MK classification is based on the appearance of pairs of spectral lines in the blue spectral region at a spectral resolution of approximately 2 Å, as compared to standard stars [7, 8]. The main line pairs are as shown in Table 15.3 and are illustrated in [11], [12], and [13].

Table 15.3. *Line pairs for spectral classes and luminosity.*

Class	Line pairs for class	Class	Line pairs for luminosity
O5 ⇔ O9	4471 He I/4541 He II	O9 ⇔ B3	4116–21 (Si IV, He I)/4144 He I
B0 ⇔ B1	4552 Si III/4089 Si IV	B0 ⇔ B3	3995 N II/4009 He II
B2 ⇔ B8	4128–30 Si II/4121 He I	B1 ⇔ A5	Balmer line wings

Table 15.3. *(Continued.)*

Class	Line pairs for class	Class	Line pairs for luminosity
B8 ⇔ A2	4471 He I/4481 Mg II	A3 ⇔ F0	4416/4481 Mg II
	4026 He I/3934 Ca II		
A2 ⇔ F5	4030−34 Mn I/4128−32	F0 ⇔ F8	4172/4226 Ca I
	4300 CH/4385		
F2 ⇔ K	4300 (G band)/4340 Hγ	F2 ⇔ K5	4045−63 Fe I/4077 Sr II
F5 ⇔ G5	4045 Fe I/4101 Hδ		4226 Ca I/4077 Sr II
	4226 Ca I/4340 Hγ	G5 ⇔ M	Discontinuity near 4215
G5 ⇔ K0	4144 Fe I/4101 Hδ	K3 ⇔ M	4215/4260, Ca I increasing
K0 ⇔ K5	4226 Ca I/4325		
	4290/4300		

Other characteristics sometimes included with MK types:

e = emission lines, e.g., Be;
f = certain O type emission line stars;
p = peculiar spectrum;
n = broad lines;
s = sharp lines;
k = interstellar lines present;
m = metallic line star.

Additional classes [2] are shown in Table 15.4.

Table 15.4. *Additional spectral classes*

Spectral class	Class characteristics
S	Strong bands of ZrO and YO, LaO, TiO
R (or C)	Strong bands of CN and CO instead of TiO in class M
N (or C)	Swan bands of C_2, Na I (D), Ca I 4227, for the rest similar to R

15.2.1 White Dwarf Spectral Classification

The following information on white dwarf spectral classification was provided by J. Liebert and E. Sion ([10] and illustrated examples in [14]). The system consists of: (1) first symbol: an uppercase D for a degenerate star spectrum; (2) second symbol: an uppercase letter designating the primary or dominant ion or type of element in the optical spectrum; (3) third and possible subsequent symbols: (optional) uppercase letters designating any secondary ions or types of elements appearing in the optical spectrum, usually due to species with trace abundances (special secondary symbols are also provided for spectra showing polarized light and magnetic fields, and others with peculiar spectra); and (4) a temperature index defined by $10\theta_{eff}$, which is equal to $50\,400/T_{eff}$. Originally, this index was specified to be a single digit from 0 to 9. This index can be estimated only from at least a rough analysis of spectrophotometric data, using colors, an energy distribution or the strengths of absorption features. In this way, the system differs from traditional, purely spectroscopic methodology. If such information is unavailable or ambiguous, the temperature subtype is omitted.

Definition of Primary Symbols

DA	Hydrogen Balmer lines dominate optical spectrum.
DB	Neutral helium (He I) lines dominate.
DO	Ionized helium lines strongest, He I and/or H may be visible.
DZ	Metal lines dominate, usually with Ca II strongest.
DQ	Carbon features, either molecular or atomic, in any part of the electromagnetic spectrum (often strongest in the ultraviolet).
DX, DXP	Star with unidentified features, presumably due to a strong magnetic field. If light polarized, the secondary symbol "P" is also appropriate.

Secondary Symbols: All of the Above, Plus ...

P	Star showing polarized light.
H	Star known to be magnetic from optical Zeeman features, but not known to be polarized.
V	Star known to be photometrically variable (optional).
PEC	Star with spectral peculiarities.

Examples

DA1	A white dwarf showing only hydrogen lines with $37\,500 < T_{\text{eff}} < 100\,000$ K.
DAO1	Star in same temperature range showing hydrogen and weak He II.
DOZ1	A star showing strong He II, weak He I, H, and N V features at $T_{\text{eff}} = 70\,000$ K.
DBAQ4	A star showing He I, H, and C features in that order of decreasing strengths, near $T_{\text{eff}} = 12\,000$ K.
DXP5	A polarized, magnetic white dwarf with unidentified spectral features, $T_{\text{eff}} \sim 10\,000$ K.
DZA7	A metallic line white dwarf also showing weak hydrogen lines, $T_{\text{eff}} = 8500$ K.
DC9	A featureless, continuous spectrum with an estimated $T_{\text{eff}} = 5500$ K.

15.3 PHOTOMETRIC SYSTEMS

Various photometric systems are used to supplement or replace the spectral classifications referred to in the last section. Optical filters are used to isolate specific spectral features or wavelength ranges, and the fluxes received through these filters are usually expressed in magnitudes,

$$m = -2.5 \log(f/f_0),$$

where f is the measured flux (corrected for atmospheric effects), and f_0 is the corresponding flux for a star with $m = 0$. The system is defined by the magnitudes and color indices (magnitude differences) for a set of standard stars, which have been determined using a particular instrumental setup. The standard stars are used to transform measurements made with other instrumental setups to the standard system. Also important for theoretical studies are the sensitivity functions (response of the original instrumental setup to a source that emits the same flux at all wavelengths) for the various filters as a function of wavelength. The effective wavelengths (peak sensitivity) and widths at half maximum of the sensitivity functions for selected photometric systems in common use at the present time are given in Table 15.5. References containing lists of standard stars, sensitivity functions, and calibrations, are indicated in the last column.

Table 15.5. *Modern photometric systems.*

System	Characteristic wavelength passbands (effective wavelengths and half-widths) (Å)	Designations	References
Strömgren four-color system	3500 (380), 4100 (200), 4700 (200), 5550 (200), plus Hβ (150/30)	$uvby\beta$	[1–6]
Geneva seven-color system	UBV system plus 4020 (170), 4480 (165), 5400 (200), 5810 (210)	$UBVB_1B_2V_1G$	[1, 7–9] [10–13]
Vilnius seven-color system	3450 (400), 3740 (260), 4050 (220), 4660 (260), 5160 (210), 5440 (260), 6550 (200)	*UPXYZVS*	[1, 14–16]
Walraven system	5400 (710), 4300 (540), 3820 (430), 3620 (230), 3250 (140)	*VBLUW*	[1, 7, 17, 18] [19, 20]
Washington system	3910 (1100), 5085 (1050), 6330 (800), 8050 (1500)	CMT_1T_2	[21–26]
DDO five-color system	4886 (186), 4517 (76), 4257 (73), 4166 (83), additional: 3815 (330), 3460 (383)	*C(41–42)* *C(42–45)* *C(45–48)*	[1, 7, 27, 28]
RGU	3593 (530), 4658 (495), 6407 (430)	*RGU*	[1, 7, 9, 29–33]
UBVRI and (RI)$_{KC}$	3600 (700), 4400 (1000), 5500 (900), 7000 (2200), 8800 (2400), 6400 (1750), 7900 (1400)	*UBVRI*	[1, 34–38]

References

1. Schmidt-Kaler, Th. 1982, *Landolt-Börnstein: Numerical Data and Functional Relationships in Science and Technology*, edited by K. Schaifers and H.H. Voigt (Springer-Verlag, Berlin), VI/2b
2. Crawford, D.L. 1975, *AJ*, **80**, 955
3. Crawford, D.L. 1978, *AJ*, **83**, 48
4. Crawford, D.L. 1979, *AJ*, **84**, 1858
5. Olson, E.C. 1974, *PASP*, **86**, 80
6. Strömgren, B. 1966, *ARA&A*, **4**, 433
7. Golay, M. 1974, *Introduction to Astronomical Photometry* (Reidel, Dordrecht)
8. Rufener, F., & Maeder, A. 1971, *A&AS*, **4**, 43
9. Philip, A.G.D., editor, 1979, *Problems of Calibration of Multicolor Photometric Systems* (Davis, Schenectady)
10. Hauck, B. 1985, in *Calibration of Fundamental Stellar Quantities*, edited by D.S. Hayes, L.E. Pasinetti, and A.G.D. Philip (Kluwer Academic), p. 271
11. North, P., & Nicolet, B. 1990, *A&A*, **228**, 78
12. Rufener, F., & Nicolet, B. 1988, *A&A*, **206**, 357
13. Meynet, G., & Hauck, B. 1985, *A&A*, **150**, 163
14. Straizys, V., & Zdanavicius, K. 1970, *Bull. Vilnius Astron. Obs.* No. 29, 15
15. Straizys, V., 1977, *Multicolor Stellar Photometry, Photometric Systems and Methods* (Mokslos, Vilnius)
16. Straizys, V., & Jodinskiene, E. 1981, *Bull. Vilnius Astron. Obs.* No. 56
17. Lub, J., & Pel, J.W. 1977, *A&A*, **54**, 137
18. Pel, J.W. 1976, *A&AS*, **24**, 413
19. De Ruiter, H.R., & Lub, J. 1986, *A&AS*, **63**, 59
20. Brand, J., & Wouterloot, J.G.A. 1988, *A&AS*, **75**, 117
21. Canterna, R. 1976, *AJ*, **81**, 228
22. Canterna, R., & Harris, H.C. 1979, *Dudley Obs. Rep.* No. 14; op. cit. [9], p. 199
23. Harris, H.C., & Canterna, R. 1979, *AJ*, **84**, 1750
24. Geisler, D. 1984, *PASP*, **96**, 723
25. Geisler, D. 1990, *PASP*, **102**, 344
26. Geisler, D., Claria, J.J., & Minniti, D. 1991, *AJ*, **102**, 1836
27. McClure, R.D. 1976, *AJ*, **81**, 182
28. McClure, R.D., & van den Bergh, S. 1968, *AJ*, **73**, 313
29. Steinlin, U.W. 1968, *Z. Astrophys.*, **69**, 276
30. Smith, L.L., & Steinlin, U.W. 1964, *Z. Astrophys.*, **58**, 253
31. Bell, R.A. 1972, *MNRAS*, **159**, 34; 1972, *A&A*, **62**, 411
32. Buser, R. 1978, *A&A*, **62**, 411
33. Buser, R. 1978, *A&A*, **62**, 425
34. Cousins, A.W.J. 1976, *MemRAS*, **81**, 25
35. Landolt, A.U. 1992, *AJ*, **104**, 340
36. Bessell, M.S. 1979, *PASP*, **91**, 589

37. Bessell, M.S. 1976, *PASP*, **88**, 557
38. Menzies, J.W. et al. 1991, *MNRAS*, **248**, 642

Absolute calibration of a star of the spectral type A0 V with the magnitude $V = 0$ [2] on the Johnson system is shown in Table 15.6.

Table 15.6. *Flux calibration for an A0 V star.*

Symbol	Flux (erg cm^{-2} s^{-1} Å^{-1})	λ_0 (μm)
U	4.22×10^{-9}	0.36
B	6.40×10^{-9}	0.44
V	3.75×10^{-9}	0.55
R	1.75×10^{-9}	0.71
I	8.4×10^{-10}	0.97

Useful relations for the UBV system [2]:

$$(U - B)_0 = 0.08 + 3.85(B - V)_0 \text{ unreddened main sequence, } (B - V)_0 < 0 \text{ and } (U - B)_0 < 0,$$

$$Q = (U - B) - 0.72(B - V) \text{ independent of reddening for early-type stars,}$$

$$\frac{E_{U-B}}{E_{B-V}} = \begin{cases} 0.65 - 0.05(U - B)_0 + 0.05E_{B-V}, & (U - B)_0 < 0, \\ 0.64 + 0.26(B - V)_0 + 0.05E_{B-V}, & (B - V)_0 > 0, \end{cases}$$

$$\frac{A_V}{E_{B-V}} = 3.30 + 0.28(B - V)_0 + 0.04E_{B-V},$$

where $E_{U-B} = (U - B) - (U - B)_0$, $A_V = V - V_0$, $E_{B-V} = (B - V) - (B - V)_0$; and V_0, $(B - V)_0$, and $(U - B)_0$ are the magnitude and color indices stars would have if space were transparent.

Useful relations for the $uvby\beta$ system [15–20]:

$$c_1 = (u - v) - (v - b),$$
$$m_1 = (v - b) - (b - y),$$
$$\beta = 2.5 \log(W/N),$$

where W and N are the fluxes measured through interference filters centered on Hβ with half-widths of about 150 and 30 Å, respectively.

$$\left.\begin{array}{l} E(c_1) = 0.20E(b - y), \\ E(m_1) = -0.32E(b - y), \\ E(u - b) = 1.50E(b - y), \end{array}\right\} \text{ color excesses according to standard reddening law,}$$

$$\left.\begin{array}{l} [c_1] = c_1 - 0.20(b - y), \\ [m_1] = m_1 + 0.32(b - y), \\ [u - b] = (u - b) - 1.50(b - y), \end{array}\right\} \text{ reddening independent quantities,}$$

$(b - y)_0 = -0.116 + 0.097c_1$ for an unreddened main-sequence B star,

$(b - y)_0 = 2.946 - 1.0\beta - 0.1\delta c_1$ $(-0.25\delta m_1$ if $m_1 < 0)$ for A stars with
$\quad\quad 2.870 > \beta > 2.720$ and $\delta c_1 < 0.28$,

$(b - y)_0 = 0.222 + 1.11\Delta\beta + 2.7(\Delta\beta)^2 - 0.05\delta c_1 - (0.1 + 3.6\Delta\beta)\delta m_1$ for F stars
$\quad\quad$ with $2.630 < \beta < 2.720$ and $\delta c_1 < 0.28$, or $2.590 < \beta < 2.630$ and
$\quad\quad \delta c_1 < 0.20$,

where $\Delta\beta = 2.720 - \beta$, $\delta c_1 = c_1 - c_{std}$, $\delta m_1 = m_{std} - m_1$; See Section 15.3.2 for c_{std} and m_{std}.

15.3.1 Calibration of MK Spectral Types [2, 21, 22]

Table 15.7 presents the absolute magnitude, color, effective surface temperature, and bolometric correction calibrations for the MK spectral classes. Table 15.8 gives the calibrated physical parameters for stars of the various spectral classes.

Table 15.7. *Calibration of MK spectral types.*

Sp	$M(V)$	$B - V$	$U - B$	$V - R$	$R - I$	T_{eff}	BC
MAIN SEQUENCE, V							
O5	−5.7	−0.33	−1.19	−0.15	−0.32	42 000	−4.40
O9	−4.5	−0.31	−1.12	−0.15	−0.32	34 000	−3.33
B0	−4.0	−0.30	−1.08	−0.13	−0.29	30 000	−3.16
B2	−2.45	−0.24	−0.84	−0.10	−0.22	20 900	−2.35
B5	−1.2	−0.17	−0.58	−0.06	−0.16	15 200	−1.46
B8	−0.25	−0.11	−0.34	−0.02	−0.10	11 400	−0.80
A0	+0.65	−0.02	−0.02	0.02	−0.02	9 790	−0.30
A2	+1.3	+0.05	+0.05	0.08	0.01	9 000	−0.20
A5	+1.95	+0.15	+0.10	0.16	0.06	8 180	−0.15
F0	+2.7	+0.30	+0.03	0.30	0.17	7 300	−0.09
F2	+3.6	+0.35	0.00	0.35	0.20	7 000	−0.11
F5	+3.5	+0.44	−0.02	0.40	0.24	6 650	−0.14
F8	+4.0	+0.52	+0.02	0.47	0.29	6 250	−0.16
G0	+4.4	+0.58	+0.06	0.50	0.31	5 940	−0.18
G2	+4.7	+0.63	+0.12	0.53	0.33	5 790	−0.20
G5	+5.1	+0.68	+0.20	0.54	0.35	5 560	−0.21
G8	+5.5	+0.74	+0.30	0.58	0.38	5 310	−0.40
K0	+5.9	+0.81	+0.45	0.64	0.42	5 150	−0.31
K2	+6.4	+0.91	+0.64	0.74	0.48	4 830	−0.42
K5	+7.35	+1.15	+1.08	0.99	0.63	4 410	−0.72
M0	+8.8	+1.40	+1.22	1.28	0.91	3 840	−1.38
M2	+9.9	+1.49	+1.18	1.50	1.19	3 520	−1.89
M5	+12.3	+1.64	+1.24	1.80	1.67	3 170	−2.73
GIANTS, III							
G5	+0.9	+0.86	+0.56	0.69	0.48	5 050	−0.34
G8	+0.8	+0.94	+0.70	0.70	0.48	4 800	−0.42
K0	+0.7	+1.00	+0.84	0.77	0.53	4 660	−0.50
K2	+0.5	+1.16	+1.16	0.84	0.58	4 390	−0.61
K5	−0.2	+1.50	+1.81	1.20	0.90	4 050	−1.02
M0	−0.4	+1.56	+1.87	1.23	0.94	3 690	−1.25
M2	−0.6	+1.60	+1.89	1.34	1.10	3 540	−1.62
M5	−0.3	+1.63	+1.58	2.18	1.96	3 380	−2.48

Table 15.7. *(Continued.)*

Sp	$M(V)$	$B - V$	$U - B$	$V - R$	$R - I$	T_{eff}	BC
SUPERGIANTS, I							
O9	−6.5	−0.27	−1.13	−0.15	−0.32	32 000	−3.18
B2	−6.4	−0.17	−0.93	−0.05	−0.15	17 600	−1.58
B5	−6.2	−0.10	−0.72	0.02	−0.07	13 600	−0.95
B8	−6.2	−0.03	−0.55	0.02	0.00	11 100	−0.66
A0	−6.3	−0.01	−0.38	0.03	0.05	9 980	−0.41
A2	−6.5	+0.03	−0.25	0.07	0.07	9 380	−0.28
A5	−6.6	+0.09	−0.08	0.12	0.13	8 610	−0.13
F0	−6.6	+0.17	+0.15	0.21	0.20	7 460	−0.01
F2	−6.6	+0.23	+0.18	0.26	0.21	7 030	−0.00
F5	−6.6	+0.32	+0.27	0.35	0.23	6 370	−0.03
F8	−6.5	+0.56	+0.41	0.45	0.27	5 750	−0.09
G0	−6.4	+0.76	+0.52	0.51	0.33	5 370	−0.15
G2	−6.3	+0.87	+0.63	0.58	0.40	5 190	−0.21
G5	−6.2	+1.02	+0.83	0.67	0.44	4 930	−0.33
G8	−6.1	+1.14	+1.07	0.69	0.46	4 700	−0.42
K0	−6.0	+1.25	+1.17	0.76	0.48	4 550	−0.50
K2	−5.9	+1.36	+1.32	0.85	0.55	4 310	−0.61
K5	−5.8	+1.60	+1.80	1.20	0.90	3 990	−1.01
M0	−5.6	+1.67	+1.90	1.23	0.94	3 620	−1.29
M2	−5.6	+1.71	+1.95	1.34	1.10	3 370	−1.62
M5	−5.6	+1.80	+1.60:	2.18	1.96	2 880	−3.47

Table 15.8. *Calibration of MK spectral types.*[a]

Sp	$\mathcal{M}/\mathcal{M}_\odot$	$R/R_\odot$	$\log(g/g_\odot)$	$\log(\bar{\rho}/\bar{\rho}_\odot)$	v_{rot} (km s^{-1})
MAIN SEQUENCE, V					
O3	120	15	−0.3	−1.5	
O5	60	12	−0.4	−1.5	
O6	37	10	−0.45	−1.45	
O8	23	8.5	−0.5	−1.4	200
B0	17.5	7.4	−0.5	−1.4	170
B3	7.6	4.8	−0.5	−1.15	190
B5	5.9	3.9	−0.4	−1.00	240
B8	3.8	3.0	−0.4	−0.85	220
A0	2.9	2.4	−0.3	−0.7	180
A5	2.0	1.7	−0.15	−0.4	170
F0	1.6	1.5	−0.1	−0.3	100
F5	1.4	1.3	−0.1	−0.2	30
G0	1.05	1.1	−0.05	−0.1	10
G5	0.92	0.92	+0.05	−0.1	< 10
K0	0.79	0.85	+0.05	+0.1	< 10
K5	0.67	0.72	+0.1	+0.25	< 10
M0	0.51	0.60	+0.15	+0.35	
M2	0.40	0.50	+0.2	+0.8	
M5	0.21	0.27	+0.5	+1.0	
M8	0.06	0.10	+0.5	+1.2	

Table 15.8. *(Continued.)*

Sp	$\mathcal{M}/\mathcal{M}_\odot$	$R/R_\odot$	$\log(g/g_\odot)$	$\log(\bar{\rho}/\bar{\rho}_\odot)$	$v_{\rm rot}$ (km s^{-1})
GIANTS, III					
B0	20	15	−1.1	−2.2	120
B5	7	8	−0.95	−1.8	130
A0	4	5		−1.5	100
G0	1.0	6	−1.5	−2.4	30
G5	1.1	10	−1.9	−3.0	< 20
K0	1.1	15	−2.3	−3.5	< 20
K5	1.2	25	−2.7	−4.1	< 20
M0	1.2	40	−3.1	−4.7	
SUPERGIANTS, I					
O5	70	30:	−1.1	−2.6	
O6	40	25:	−1.2	−2.6	
O8	28	20	−1.2	−2.5	125
B0	25	30	−1.6	−3.0	102
B5	20	50	−2.0	−3.8	40
A0	16	60	−2.3	−4.1	40
A5	13	60	−2.4	−4.2	38
F0	12	80	−2.7	−4.6	30
F5	10	100	−3.0	−5.0	< 25
G0	10	120	−3.1	−5.2	< 25
G5	12	150	−3.3	−5.3	< 25
K0	13	200	−3.5	−5.8	< 25
K5	13	400	−4.1	−6.7	< 25
M0	13	500	−4.3	−7.0	
M2	19	800	−4.5	−7.4	

Note

[a] A colon indicates an uncertain value.

Also see [23]. An independent absolute magnitude calibration is given in graphical form in [8]. Plots of $(B - V)$ and $(U - V)$ versus M_V for the various white dwarf subclasses are in [24]. Intrinsic colors and absolute magnitudes of the zero-age main sequence (ZAMS) (locus of young stars just starting hydrogen burning) follow [2]. See [25] for an alternative, and plots in Chapter 20. Table 15.9 gives the zero-age main sequence colors and absolute magnitudes.

Table 15.9. *Zero-age main sequence.*

$(B - V)_0$	$(U - B)_0$	M_v	$(B - V)_0$	$(U - B)_0$	M_v
−0.$^{\rm m}$33	−1.$^{\rm m}$20	−5.$^{\rm m}$2	+0.40	−0.01	+ 3.4
−0.305	−1.10	−3.6	+0.50	0.00	+ 4.1
−0.30	−1.08	−3.25	+0.60	+0.08	+ 4.7
−0.28	−1.00	−2.6	+0.70	+0.23	+ 5.2
−0.25	−0.90	−2.1	+0.80	+0.42	+ 5.8
−0.22	−0.80	−1.5	+0.90	+0.63	+ 6.3
−0.20	−0.69	−1.1	+1.00	+0.86	+ 6.7
−0.15	−0.50	−0.2	+1.10	+1.03	+ 7.1
−0.10	−0.30	+0.6	+1.20	+1.13	+ 7.5
−0.05	−0.10	+1.1	+1.30	+1.20	+ 8.0
0.00	+0.01	+1.5	+1.40	+1.22	+ 8.8
+0.05	+0.05	+1.7	+1.50	+1.17	+10.3
+0.10	+0.08	+1.9	+1.60	+1.20	+12.0

Table 15.9. *(Continued.)*

$(B-V)_0$	$(U-B)_0$	M_v	$(B-V)_0$	$(U-B)_0$	M_v
+0.15	+0.09	+2.1	+1.70	+1.32	+13.2
+0.20	+0.10	+2.4	+1.80	+1.43	+14.2
+0.25	+0.07	+2.55	+1.90	+1.53	+15.5
+0.30	+0.03	+2.8	+2.00	+1.64	+16.7
+0.35	0.00	+3.1			

15.3.2 $uvby\beta$ Standard Relations

For the early-type stars, Table 15.10 gives the standard relation between the β index, colors, and the absolute magnitudes.

Table 15.10. $uvby\beta$ *standard relations.*

β	$b-y$	m_1	c_1	M_V	$[m_1]$	$[c_1]$
			B Stars			
2.590	−0.134	0.045	−0.250	−4.65	0.005	−0.223
2.600	−0.126	0.055	−0.128	−4.12	0.017	−0.103
2.620	−0.118	0.075	−0.025	−3.17	0.040	−0.001
2.640	−0.109	0.080	0.065	−2.36	0.047	0.087
2.660	−0.100	0.085	0.150	−1.69	0.055	0.170
2.680	−0.091	0.093	0.235	−1.12	0.066	0.253
2.700	−0.080	0.100	0.321	−0.65	0.076	0.337
2.720	−0.070	0.100	0.404	−0.27	0.079	0.418
2.740	−0.061	0.109	0.491	0.04	0.091	0.503
2.760	−0.050	0.110	0.578	0.30	0.095	0.588
2.780	−0.044	0.116	0.656	0.51	0.103	0.665
2.800	−0.041	0.120	0.724	0.68	0.108	0.732
2.820	−0.039	0.120	0.785	0.83	0.108	0.793
2.840	−0.037	0.123	0.833	0.97	0.112	0.840
2.860	−0.034	0.128	0.878	1.10	0.118	0.885
2.880	−0.029	0.132	0.925	1.24	0.123	0.931
2.900	−0.023	0.138	0.975	1.39	0.131	0.980
2.910	−0.020	0.140	1.000	1.46	0.134	1.004
			A Stars			
2.880	0.066	0.200	0.930	2.30	0.220	0.917
2.870	0.076	0.202	0.910	2.40	0.225	0.895
2.860	0.086	0.205	0.890	2.50	0.231	0.873
2.850	0.096	0.206	0.870	2.57	0.235	0.851
2.840	0.106	0.208	0.850	2.64	0.240	0.829
2.830	0.116	0.207	0.835	2.67	0.242	0.812
2.820	0.126	0.206	0.820	2.70	0.244	0.795
2.810	0.136	0.204	0.800	2.73	0.245	0.773
2.800	0.146	0.203	0.780	2.76	0.247	0.751
2.790	0.156	0.200	0.760	2.79	0.247	0.729
2.780	0.166	0.196	0.740	2.82	0.246	0.707
2.770	0.176	0.192	0.720	2.85	0.245	0.685
2.760	0.186	0.188	0.700	2.88	0.244	0.663
2.750	0.196	0.185	0.680	2.92	0.244	0.641
2.740	0.206	0.182	0.660	2.96	0.244	0.619
2.730	0.216	0.180	0.630	3.03	0.245	0.587
2.720	0.226	0.177	0.600	3.10	0.245	0.555

Table 15.10. *(Continued.)*

β	$b - y$	m_1	c_1	M_V	$[m_1]$	$[c_1]$
			F Stars			
2.720	0.222	0.177	0.580	3.14	0.244	0.536
2.710	0.233	0.174	0.560	3.21	0.244	0.513
2.700	0.245	0.172	0.530	3.29	0.246	0.481
2.690	0.258	0.171	0.495	3.38	0.248	0.443
2.680	0.271	0.170	0.465	3.48	0.251	0.411
2.670	0.284	0.171	0.440	3.60	0.256	0.383
2.660	0.298	0.174	0.415	3.74	0.263	0.355
2.650	0.313	0.178	0.390	3.88	0.272	0.327
2.640	0.328	0.183	0.370	4.04	0.281	0.304
2.630	0.344	0.189	0.350	4.20	0.292	0.281
2.620	0.360	0.196	0.330	4.36	0.304	0.258
2.610	0.377	0.204	0.310	4.52	0.317	0.235
2.600	0.394	0.214	0.290	4.70	0.332	0.211
2.590	0.412	0.226	0.270	4.90	0.350	0.188

See [15–17] and [26]. See also [27] and [28] for grids for determining effective temperatures and surface gravities. Other calibrations may be found in [29–38].

15.3.3 Empirical $UBV(RI)_{KC}$ Calibrations [39]

The colors and spectral classes are given as a function of the surface effective temperature for dwarf and giant stars in Table 15.11.

Table 15.11. *Empirical $UBV(RI)_{KC}$ calibrations.*

T_{eff}	$b - y$	$B - V$	$(V - R)_{KC}$	$(R - I)_{KC}$	$(V - I)_{KC}$	MK
			Dwarfs (V)			
13 000	−0.054	−0.14	−0.050	−0.070	−0.120	B7
12 000	−0.041	−0.10	−0.035	−0.050	−0.085	B8
11 000	−0.027	−0.065	−0.023	−0.032	−0.055	B9
10 000	−0.010	−0.025	−0.008	−0.012	−0.020	A0
9 500	+0.007	+0.005	+0.007	+0.008	+0.015	A1
9 000	+0.035	+0.055	+0.032	+0.040	+0.072	A2
8 500	+0.072	+0.14	+0.071	+0.084	+0.155	A5
8 000	+0.118	+0.22	+0.118	+0.132	+0.250	A7
7 500	+0.165	+0.275	+0.162	+0.168	+0.330	F0
7 000	+0.220	+0.35	+0.208	+0.207	+0.415	F2
6 500	+0.286	+0.45	+0.265	+0.250	+0.515	F5
6 000	+0.360	+0.57	+0.322	+0.303	+0.625	G0
5 500	+0.445	+0.70	+0.396	+0.364	+0.760	G6
5 000	+0.535	+0.88	+0.50	+0.43	+0.93	K2
4 500	+0.60	+1.02	+0.60	+0.51	+1.11	K4
4 000	+0.80	+1.32	+0.79	+0.74	+1.53	K7
3 500	+1.01	+1.53	+1.01	+1.18	+2.19	M2
3 000	+1.22	+1.74	+1.26	+1.77	+3.03	M4.5
2 750	+1.37	+2.0	+1.40	+2.18	+3.58	M6

Table 15.11. (*Continued.*)

T_{eff}	$b - y$	$B - V$	$(V - R)_{KC}$	$(R - I)_{KC}$	$(V - I)_{KC}$	MK
			Giants (III)			
5 000	+0.55	+0.89	+0.497	+0.433	+0.93	G7
4 750	+0.60	+0.98	+0.539	+0.461	+1.00	K0
4 500	+0.68	+1.11	+0.60	+0.510	+1.11	K2
4 250	+0.80	+1.26	+0.68	+0.600	+1.28	K3
4 000	+0.90	+1.43	+0.795	+0.735	+1.53	K5
3 750	+1.00	+1.62	+0.945	+1.025	+1.97	M2
3 500			+1.19	+1.57	+2.76	M4.5
3 250					+3.80	M6

15.4 STELLAR ATMOSPHERES

15.4.1 Model Atmospheres for Normal Stars (Solar Composition) [40]

Table 15.12 lists stellar atmosphere parameters depending on the surface effective temperature and gravity of a star.

Table 15.12. *Model atmospheres for normal stars.*

T_{eff} $\log g$	$\log \tau^a$	$\log x$	T	$\log P$	$\log n_e$	$\log n_a$	$\log \rho$	$\log P_r$	$\dfrac{F_{conv}}{F}$
5 500	−3.0	6.79	4 282	3.23	11.35	15.47	−8.19	0.09	0.00
4	−2.0	7.65	4 487	3.84	11.91	16.05	−7.61	0.10	0.00
	−1.0	7.92	4 846	4.41	12.49	16.59	−7.07	0.17	0.00
	0.0	8.08	6 130	4.92	13.50	16.99	−6.66	0.54	0.01
	1.0	8.14	8 176	5.10	14.94	17.04	−6.62	1.05	0.85
5 500	−3.0	10.65	4 104	1.28	9.53	13.52	−10.13	0.09	0.00
1	−2.0	10.98	4 444	2.09	10.35	14.30	−9.36	0.10	0.00
	−1.0	11.14	4 846	2.73	11.08	14.91	−8.75	0.17	0.00
	0.0	11.22	6 145	3.13	12.55	15.20	−8.46	0.56	0.00
	1.0	11.24	8 431	3.18	14.06	15.07	−8.58	1.10	0.91
6 000	−3.0	7.60	4 667	3.29	11.48	15.49	−8.17	0.24	0.00
4	−2.0	7.90	4 891	3.87	12.04	16.04	−7.61	0.25	0.00
	−1.0	8.08	5 293	4.42	12.62	16.55	−7.10	0.32	0.00
	0.0	8.18	6 789	4.82	13.94	16.85	−6.81	0.70	0.05
	1.0	8.22	8 709	4.95	15.12	16.86	−6.79	1.16	0.88
6 000	−3.0	10.75	4 489	1.26	9.72	13.47	−10.19	0.24	0.00
1	−2.0	11.03	4 869	2.02	10.62	14.19	−9.47	0.25	0.00
	−1.0	11.17	5 318	2.59	11.44	14.72	−8.94	0.33	0.00
	0.0	11.24	6 861	2.89	13.01	14.90	−8.75	0.75	0.00
	1.0	11.25	8 981	2.92	14.11	14.73	−8.93	1.21	0.91
7 000	−3.0	7.63	5 458	3.10	11.87	15.22	−8.44	0.51	0.00
4	−2.0	7.95	5 726	3.67	12.45	15.77	−7.89	0.52	0.00
	−1.0	8.12	6 190	4.17	13.13	16.23	−7.42	0.60	0.00
	0.0	8.20	8 217	4.45	14.63	16.39	−7.26	1.02	0.20
	1.0	8.24	9 911	4.55	15.37	16.37	−7.28	1.38	0.92

Table 15.12. *(Continued.)*

T_{eff} $\log g$	$\log \tau^a$	$\log x$	T	$\log P$	$\log n_e$	$\log n_a$	$\log \rho$	$\log P_r$	$\dfrac{F_{conv}}{F}$
10 000	−3.0	8.34	7 586	1.71	12.84	13.63	−10.03	1.13	0.00
4	−2.0	8.48	8 030	2.36	13.42	14.26	−9.40	1.15	0.00
	−1.0	8.58	8 982	2.86	14.08	14.67	−8.99	1.28	0.00
	0.0	8.65	11 655	3.17	14.62	14.71	−8.95	1.68	0.00
	1.0	8.83	16 287	3.75	15.08	15.12	−8.54	2.25	0.00
20 000	−3.0	8.70	13 060	1.38	12.81	12.84	−10.82	2.34	0.00
4	−2.0	8.90	14 067	2.09	13.49	13.52	−10.14	2.35	0.00
	−1.0	9.02	15 560	2.71	14.07	14.08	−9.57	2.40	0.00
	0.0	9.15	19 521	3.33	14.60	14.60	−9.05	2.63	0.00
	1.0	9.28	27 451	4.03	15.15	15.15	−8.50	3.15	0.00
40 000	−3.0	9.48	28 059	1.19	12.31	12.29	−11.37	3.54	0.00
4	−2.0	9.66	31 336	2.16	13.24	13.21	−10.45	3.55	0.00
	−1.0	9.77	34 855	2.93	13.96	13.93	−9.72	3.62	0.00
	0.0	9.87	40 920	3.55	14.52	14.48	−9.18	3.85	0.00
	1.0	9.97	53 682	4.21	15.06	15.02	−8.64	4.32	0.00

Note

$^a\tau$ = continuum optical depth (5000 Å); x = geometric depth; T = temperature (K); P = pressure; n_e = electron number density; n_a = atom number density; ρ = mass density; P_r = radiation pressure; F_{conv}/F = fraction of flux carried by convection. All units are cgs.

Model atmospheres for metal-deficient stars are given in [40] and [41].

15.4.2 Theoretical Physical Continuum Fluxes [40]

Logarithms of theoretical physical continuum fluxes (ergs cm^{-2} s^{-1} Å^{-1}) for normal stars (solar composition) with $\log g = 4$ [40] are given in Table 15.13.

Table 15.13. *Continuum fluxes for normal stars.*

λ (Å)	T_{eff} (K) = 5500	6000	7000	10 000	20 000	40 000
506	$-\infty$	$-\infty$	$-\infty$	−6.26	4.81	11.19
890	$-\infty$	$-\infty$	$-\infty$	1.11	7.34	10.93
920	−5.80	−4.49	−2.07	3.73	10.28	11.28
1 482	0.05	1.63	3.83	8.28	9.84	10.75
2 012	4.14	5.41	6.88	8.12	9.48	10.35
2 506	5.91	6.55	7.19	7.98	9.22	10.05
3 012	6.80	7.00	7.29	7.89	8.99	9.78
3 636	6.86	7.04	7.27	7.79	8.76	9.49
3 661	6.94	7.16	7.56	8.33	8.93	9.49
4 012	6.94	7.15	7.52	8.21	8.80	9.35
4 512	6.92	7.12	7.46	8.06	8.63	9.17
5 025	6.89	7.08	7.38	7.92	8.46	8.99
5 525	6.86	7.03	7.31	7.79	8.32	8.84
6 025	6.81	6.97	7.24	7.68	8.19	8.70
7 075	6.72	6.86	7.09	7.46	7.94	8.44
8 152	6.62	6.75	6.95	7.26	7.71	8.20
8 252	6.61	6.74	6.95	7.33	7.73	8.19
10 050	6.45	6.56	6.74	7.03	7.41	7.86
14 594	6.13	6.19	6.29	6.47	6.80	7.23

Table 15.13. *(Continued.)*

λ (Å)	T_{eff} (K) = 5500	6000	7000	10 000	20 000	40 000
27 000	5.22	5.27	5.34	5.48	5.77	6.17
50 000	4.20	4.24	4.31	4.45	4.72	5.10
100 000	3.03	3.06	3.13	3.27	3.52	3.89
200 000	1.83	1.87	1.94	2.07	2.32	2.67

15.5 STELLAR STRUCTURE

Age-zero models for $X = 0.70$, $Z = 0.02$, $l/H_p = 1.7$ [1]. l = mixing length; P_c = central pressure (dyn cm^{-2}); T_c = central temperature (K); ρ_c = central density (g cm^{-3}); H_p = pressure scale height; q_{cc} = fraction of stellar mass within convective core; q_{ce} = fraction of stellar mass at bottom of convective envelope are given in Table 15.14.

Table 15.14. *Age-zero models* [1].

$\mathcal{M}/\mathcal{M}_\odot$	$\log L/L_\odot$	$\log T_{eff}$	$\log g$	$R/R_\odot$	$\log P_c$	$\log T_c$	$\log \rho_c$	q_{cc}	q_{ce}
10.0	3.7600	4.4096	4.269	3.8434	16.6331	7.5051	0.974	0.3193	≥ 0.9999
9.0	3.6118	4.3855	4.275	3.6213	16.6688	7.4959	1.024	0.3088	≥ 0.9999
8.0	3.4433	4.3576	4.280	3.3916	16.7116	7.4853	1.081	0.2877	≥ 0.9999
7.0	3.2480	4.3251	4.287	3.1472	16.7617	7.4729	1.148	0.2772	≥ 0.9999
6.0	3.0184	4.2865	4.296	2.8860	16.8243	7.4583	1.229	0.2666	≥ 0.9999
5.0	2.7381	4.2380	4.303	2.6122	16.8995	7.4400	1.326	0.2455	≥ 0.9999
4.0	2.3815	4.1766	4.317	2.2992	16.9982	7.4163	1.450	0.2350	≥ 0.9999
3.0	1.9087	4.0928	4.330	1.9620	17.1280	7.3844	1.614	0.1928	≥ 0.9999
2.5	1.6002	4.0376	4.338	1.7737	17.2085	7.3631	1.716	0.1717	≥ 0.9999
2.0	1.2157	3.9658	4.339	1.5858	17.2943	7.3333	1.834	0.1401	≥ 0.9999
1.8	1.0298	3.9297	4.334	1.5120	17.3260	7.3151	1.883	0.1190	≥ 0.9999
1.6	0.8186	3.8855	4.317	1.4532	17.3476	7.2913	1.930	0.0874	≥ 0.9999
1.4	0.5562	3.8355	4.322	1.3526	17.3274	7.2500	1.948	0.0386	≥ 0.9999
1.2	0.2325	3.7964	4.422	1.1157	17.2645	7.1936	1.941	0.0129	0.9965
1.0	−0.1523	3.7514	4.548	0.8813	17.1851	7.1306	1.924	0.0073	0.9720
0.8	−0.5742	3.7016	4.674	0.6820	17.0651	7.0603	1.882	0.0875	0.8725

Reference
1. Schönberner, D., Blöcker, T., Herwig, F., & Driebe, T. 1996, private communication.

REFERENCES

1. Lacy, C.H. 1977, *ApJS*, **34**, 479
2. Schmidt-Kaler, Th. 1982, *Landolt-Börnstein: Numerical Data and Functional Relationships in Science and Technology*, edited by K. Schaifers and H.H. Voigt (Springer-Verlag, Berlin), VI/2b
3. McCluskey, G.E., Jr., & Kondo, Y. 1972, *A&SS*, **17**, 134
4. Harris, D.L., III, Strand, K.Aa., & Worley, C.E. 1963, in *Stars and Stellar Systems*, III (University of Chicago Press, Chicago), p. 273.
5. Popper, D.M. 1980, *ARA&A*, **18**, 115
6. Andersen, J. 1991, *A&AR*, **3**, 91
7. Morgan, W.W., & Keenan, P.C. 1973, *ARA&A*, **11**, 29
8. Keenan, P.C. 1985, in *Calibration of Fundamental Stellar Quantities*, edited by D.S. Hayes, L.E. Pasinetti and A.G.D. Philip (Kluwer Academic), p. 121
9. Garrison, R.F., editor, 1984, *The MK Process and Stellar Classification* (David Dunlap Observatory, Toronto)
10. Sion, E.M., Greenstein, J.L., Landstreet, J.D., Liebert, J., Shipman, H.L., & Wegner, G.A. 1983, *ApJ*, **269**, 253
11. Morgan, W.W., Abt, H.A., & Tapscott, J.W. 1978, *Revised MK Spectral Atlas for Stars Earlier than the Sun* (Yerkes Observatory)
12. Keenan, P.C., & McNeil, R.C. 1976, *An Atlas of the Spectra of the Cooler Stars: Types G, K, M, S, & C* (Ohio State University Press, Columbus)
13. Yamashita, Y., Nariai, K., & Morimoto, Y. 1977, *An Atlas of Representative Stellar Spectra* (University of Tokyo Press, Tokyo)

14. Wesemael, F., Greenstein, J.L., Liebert, J., Lamontagne, R., Fontaine, G., Bergeron, P., & Glaspey, J.W. 1993, *PASP*, **105**, 761

15. Crawford, D.L. 1975, *AJ*, **80**, 955

16. Crawford, D.L. 1978, *AJ*, **83**, 48

17. Crawford, D.L. 1979, *AJ*, **84**, 1858

18. Strömgren, B. 1966, *ARA&A*, **4**, 433

19. Crawford, D.L., Glaspey, J.W., & Perry, C.L. 1970, *AJ*, **75**, 822

20. Crawford, D.L. 1975, *PASP*, **87**, 481

21. Johnson, H.L. 1966, *ARA&A*, **4**, 193

22. De Jager, C., & Nieuwenhuijzen, H. 1987, *A&A*, **177**, 217

23. Habets, G.M.H.J., & Heintze, J.R.W. 1981, *A&AS*, **46**, 193

24. Greenstein, J.L. 1988, *PASP*, **100**, 82

25. VandenBerg, D.A., & Poll, H.E. 1989, *AJ*, **98**, 1451

26. Perry, C.L., Olsen, E.H., & Crawford, D.L. 1987, *PASP*, **99**, 1184

27. Moon, T.T., & Dworetsky, M.M. 1985, *MNRAS*, **217**, 305

28. Napiwotzki, R., Schönberner, D., & Wenske, V. 1992, in *The Atmospheres of Early-Type Stars*, edited by U. Heber and C.S. Jeffery (Springer-Verlag, Berlin and New York), p. 18

29. Philip, A.G.D., & Egret, D. 1980, *A&AS*, **40**, 199

30. Balona, L.A. 1984, *MNRAS*, **211**, 973

31. Crawford, D.L. 1984, in *The MK Process and Stellar Classification*, edited by R.F. Garrison (David Dunlap Observatory, Toronto), p. 191

32. Balona, L.A., & Shobbrook, R.R. 1984, *MNRAS*, **211**, 375

33. Kilkenny, D., & Whittet, D.C.B. 1985, *MNRAS*, **216**, 127

34. Greenstein, J.L. 1984, *PASP*, **96**, 62

35. Olsen, E.H. 1988, *A&A*, **189**, 173

36. McNamara, D.H., & Powell, J.M. 1985, *PASP*, **97**, 1101

37. Olsen, E.H. 1984, *A&AS*, **57**, 443.

38. Hayes, D.S., Passinetti, L.E., & Philip, A.G.D. 1985, in *Calibration of Fundamental Stellar Quantities*, edited by D.S. Hayes, L.E. Pasinetti, and A.G.D. Philip (Reidel, Dordrecht)

39. Bessell, M.S. 1979, *PASP*, **91**, 589

40. Kurucz, Robert L. 1979, *ApJS*, **40**, 1

41. Bell, R.A., Eriksson, K., Gustafsson, B., & Nordlund, Å. 1976, *A&AS*, **23**, 37

Chapter 16

Stars with Special Characteristics

J. Donald Fernie

16.1   VARIABLE STARS
by Douglas S. Hall

All types of variables are collected in the *General Catalogue of Variable Stars* [1]. Except for the eclipsing variables, all are considered intrinsic variables, with the physical mechanism responsible for the variability being of four main classes. The approximate numbers of the principal types, as of 1990, are given in the following:

Abbreviation	Description	Number
Pulsating (periodic, multiperiodic, quasiperiodic, or nonperiodic)		
DCEP/CEP	Classical Cepheids + those of uncertain type	638
CW	W Virginis stars + BL Herculis stars	172
RR	RR Lyrae stars	6180
DSCT	δ Scuti stars (some called dwarf Cepheids)	100
SXPHE	SX Phe variables, pop. II δ Sct variables	15
BCEP	β Cephei stars $=\beta$ Canis Majoris stars	89
ZZ	ZZ Ceti variables	28
RV	RV Tauri stars	120
SR	Semiregular (sometimes called LPV) variables	3377
L	Slow irregular (sometimes called LPV) variables	2389
M	Mira stars (sometimes called LPV)	5827
RCB	R Coronae Borealis stars	37
Rotating (periodic or quasiperiodic)		
ELL	Ellipsoidal variables	45
ACV	α^2 Canum Venaticorum variables	163
SXARI	SX Ari variables	15
PSR	Optically variable pulsars	2
BY	BY Draconis variables	34
RS	RS Canum Venaticorum binaries	67
FKCOM	FK Comae Berenices stars	4
INT	Orion variables of the T Tauri type	59

		Eruptive (all nonperiodic)	
IN	Orion variables, including T Tauri stars and RW Aurigae stars		898
FU	FU Orionis variables		3
GCAS	γ Cassiopeiae variables		108
RCB	R Coronae Borealis variables		37
UV	UV Ceti variables or flare stars		746
SDOR	S Doradus variables or P Cygni stars		15
	Explosive or Cataclysmic (quasiperiodic or nonperiodic)		
N	Novae		61
NL	Novalike variables		30
NR	Recurrent novae		8
SN	Supernovae		7
UG	U Geminorum variables or dwarf novae		182
ZAND	Symbiotic variables of the Z Andromedae type		46
	Eclipsing (periodic) all types		5074

16.2   CEPHEID AND CEPHEID-LIKE VARIABLES

Descriptions of Cepheid families are given below [2–13]:

IAU designation	Name	Population	Period (days)
DCEP	Classical Cepheids	I	1.5–60
CW	W Vir + BL Her stars	II	1–50
RR	RR Lyrae stars	II	0.4–1
DSCT	δ Scuti	I	0.04–0.2
BCEP	β Cephei stars	I	0.15–0.25

Cepheid mean characteristics as a function of period P are given below. The period is that of the fundamental mode. In the K band $M_K = -2.97 \log P - 1.08$. The Cepheid ratio of radial velocity amplitude to V-light amplitude is $50 \pm 5 \text{ km s}^{-1} \text{ mag.}^{-1}$.

$\log P$	M_V	$B - V$	$\log L/L_\odot$	$\log R/R_\odot$	$\log T_{\rm eff}$	$\log \mathcal{M}/\mathcal{M}_\odot$	$\log g$	Q^a
			Classical Cepheids					
0.4	−2.4	0.49	2.81	1.41	3.76	0.54	2.2	0.036
0.6	−2.9	0.57	3.05	1.55	3.75	0.61	1.9	0.037
0.8	−3.5	0.66	3.30	1.70	3.74	0.68	1.7	0.039
1.0	−4.1	0.75	3.55	1.85	3.72	0.75	1.5	0.040
1.2	−4.7	0.84	3.80	2.00	3.71	0.82	1.3	0.041
1.4	−5.3	0.93	4.06	2.15	3.70	0.89	1.0	0.042
1.6	−5.8	1.01	4.32	2.29	3.70	0.97	0.8	0.044
1.8	−6.4	1.10	4.58	2.44	3.69	1.04	0.6	0.045
			δ Sct/Dwarf Cepheids[b]					
−1.4	+2.7	0.20	0.90	0.07	3.91	0.21	4.5	0.039
−1.0	+1.7	0.25	1.22	0.37	3.88	0.25	3.9	0.037
−0.7	+0.8	0.29	1.58	0.59	3.86	0.31	3.6	0.037
			Population II Cepheids					
0.1	−0.1	0.30	1.95	0.86	3.82	−0.22	2.5	0.049
0.5	−0.9	0.44	2.27	1.08	3.79	−0.22	2.1	0.059
1.0	−1.8	0.60	2.63	1.34	3.75	−0.22	1.5	0.076
1.3	−2.4	0.70	2.87	1.52	3.72	−0.22	1.2	0.080
			RR Lyrae stars[c]					
−0.5	0.7	0.20	1.47	0.54	3.86	−0.26	3.1	0.036
−0.3	0.7	0.25	1.57	0.67	3.82	−0.26	2.8	0.037
−0.1	0.7	0.38	1.59	0.76	3.78	−0.26	2.7	0.043
			β Cephei stars					
−1.0	−2.5	−0.23	3.99	0.84	4.34	1.02	3.8	0.020
−0.8	−3.1	−0.25	4.33	0.93	4.38	1.15	3.7	0.030
−0.6	−3.8	−0.27	4.69	1.03	4.42	1.22	3.6	0.037

Notes

[a] $Q \equiv P(\rho/\rho_\odot)^{0.5}$ is the pulsation constant expressed in days.

[b] Many of these stars show higher-order and/or nonradial modes as well.

[c] M_V depends on metallicity: $M_V = 0.20[{\rm Fe/H}] + 1.0$.

16.3 VARIABLE WHITE DWARF TABLES
by Paul A. Bradley

Tables 16.1, 16.2, and 16.3 give data for the DAV, DBV, and DOV variable white dwarfs, respectively.

Table 16.1. *Names, positions, and magnitudes of the ZZ Ceti (DAV) stars.*

WD No.	Name	Variable star name	α (2000.0)	δ (2000.0)	T_{eff} (10^3 K)	V (mag.)	B (mag.)	Amp. (mag.)	Periods (seconds)
0104 −464	BPM 30551 [1, 2]	AX Phe	01 06 56	−46 10.2	11.3 [46, 47]	15.26	15.43	0.22	823 + others
0133 −116	R 548 [3–5]	ZZ Cet	01 36 10	−11 20.7	12.0 [46–50]	14.16	14.33	0.012	213, 274
0341 −459	BPM 31594 [6–8]	VY Hor	03 43 25	−45 48.7	11.5 [46, 47]	15.03	15.24	0.28	402, 618, C, H[a]
0416 +272	HL Tau 76 [9–11]	V411 Tau	04 18 59	+27 18.4	11.4 [46, 47]	15.20	15.40	0.28	393, 625, 746, C, H
0417 +361	G 38-29 [12]	V468 Per	04 20 18	+36 16.6	11.2 [46, 47]	15.59	15.79	0.22	∼1000, C, H
0455 +553	G 191-16 [13, 14]	BR Cam	04 59 28	+55 25.3	11.4 [46, 47]	15.98	16.1	0.3	510, 600, 710893, C, H
0507 +0435	HS0507+0434B [15]	—	05 10 13	+04 38.6	11.8	15.36	15.6	0.2	355, 445, 560, CH
0517 +307	GD 66 [16, 17]	V361 Aur	05 20 38	+30 48.5	12.0 [46, 49, 50]	15.56	15.78	0.02	197, 272, 302, 819, C, H
0837 +403	KUV 08368+4026 [18]	—	08 40 08	+40 15.1	—	15.55	15.77	0.03	495, 618
0858 +363	GD 99 [19]	VW Lyn	09 01 49	+36 07.2	11.8 [46–49, 51]	14.55	14.74	0.07	350, 481, 592
0921 +354	G117-B15A [20–22]	RY LMi	09 24 17	+35 16.9	11.6 [46–51]	15.50	15.72	0.05	215, 271, 304, C, H
1137 +422	KUV 11370+4222 [18]	—	11 39 41	+42 05.3	—	16.56	16.78	0.006	257 + others
1159 +803	G255-2 [23]	BT Cam	12 01 48	+80 05.0	11.4 [46, 47]	16.04	—	0.04	685, 830 + others
1236 −495	BPM 37093 [24]	V886 Cen	12 38 48	−49 49.0	11.7 [46, 48, 50, 51]	13.96	14.14	0.004	∼600
1307 +354	GD 154 [25, 26]	BG CVn	13 09 58	+35 09.5	11.2 [46, 48–51]	15.33	15.51	0.10	403, 1089, 1186, C, H
1350 +656	G 238-53 [27]	DU Dra	13 52 12	+65 24.9	11.9 [46, 49]	15.51	15.7	0.02	∼206
1401 −1454	EC 14012−1446 [28]	IU Vir	14 03 57	−15 01.1	—	15.67	15.50	0.30	610, 724 + others
1422 +095	GD 165 [29–31]	CX Boo	14 24 40	+09 17.3	12.0 [46, 49]	14.32	14.46	0.10	114, 120 192, 250, C
1425 −811	L 19-2 [32–34]	MY Aps	14 32 18	−81 20.1	12.1 [46, 48–51]	13.75	14.00	0.03	113, 118, 143, 192, 350
1559 +369	R 808 [19]	TY CrB	16 01 21	+36 47.0	11.2 [46, 48]	14.36	14.53	0.15	833, complex, C, H
1647 +591	G 226-29 [35, 36]	DN Dra	16 48 26	+59 03.5	12.5 [46–51]	12.24	12.40	0.006	109
1714 −547	BPM 24754 [37]	—	17 18 54	−54 47.1	12.9	15.60	15.87	0.07	∼1100
1855 +338	G 207-9 [38]	V470 Lyr	18 57 30	+33 57.2	12.0 [46–48, 51]	14.62	14.81	0.06	259, 292, 557, 739, C
1935 +276	G 185-32 [13]	PY Vul	19 37 13	+27 43.3	12.1 [46–48, 50, 51]	12.97	13.15	0.02	141, 215, C, H
1950 +250	GD 385 [7, 39]	PT Vul	19 52 28	+25 09.3	11.7 [46, 47]	15.12	15.32	0.05	256, C
2303 +242	PG 2303+242 [40, 41]	KR Peg	23 06 16	+24 32.0	11.5 [46, 47]	15.50	15.58[b]	0.2	394, 483, 540, 611, 936, C, H
2326 +049	G 29-38 [12, 42–44]	ZZ Psc	23 28 49	+05 15.0	11.8 [46–50]	13.03	13.17	0.27	284, 615, 820, C, H
2349 −244	EC 23487−2424 [45]	HW Aqr	23 51 22	−24 08.2	—	15.33	15.52	0.24	∼800–1000, complex

Notes

[a]C means combination of frequencies and H means harmonics.

[b]Photographic (blue) magnitude.

References

1. Hesser, J.E., Lasker, B.M.., & Neupert, H.E. 1976, *ApJ*, **209**, 853

2. McGraw, J.T. 1977, *ApJ*, **214**, L123

3. Lasker, B.M., & Hesser, J.E. 1971, *ApJ*, **163**, L89

4. Stover, R.J., Hesser, J.E., Lasker, B.M., Nather, R.E., & Robinson, E.L. 1980, *ApJ*, **240**, 865

5. Kepler, S.O. et al. 1995, *Baltic Astr.*, **4**, 238

6. McGraw, J.T. 1976, *ApJ*, **210**, L35
7. O'Donoghue, D. 1986, *MNRAS*, **220**, 19P
8. O'Donoghue, D., Warner, B., & Cropper, M. 1992, *MNRAS*, **258**, 415
9. Page, C.G. 1972, *MNRAS*, **159**, 25P
10. Fitch, W.S. 1973, *ApJ*, **181**, L95
11. Dolez, N., & Kleinman, S.J. 1997 in *White Dwarfs*, edited by J. Isern, M. Hernanz, and E. Garcia-Berro (Kluwer Academic, Dordrecht), p. 437
12. McGraw, J.T., & Robinson, E.L. 1975, *ApJ*, **200**, L89
13. McGraw, J.T., Fontaine, G., Dearborn, D.S.P., Gustafson, J., Lacombe, P., & Starrfield, S.G. 1981, *ApJ*, **250**, 349
14. Vauclair, G. et al. 1991, *A&A*, **215**, L17
15. Jordan, S., Koester, D., Vauclair, G., Dolez, N., Heber, U., Hager, H.J., Reimers, D., Chevreton, M., & Dreizler, S. 1998, *A&A*, **330**, 277
16. Dolez, N., Vauclair, G., & Chevreton, M. 1983, *A&A*, **121**, L23
17. Fontaine, G., Wesemael, F., Bergeron, P., Lacombe, P., Lamontagne, R., & Saumon. D. 1985, *ApJ*, **294**, 339
18. Vauclair, G. et al. 1996, *A&A*, **322**, 155
19. McGraw, J.T., & Robinson, E.L. 1976, *ApJ*, **205**, L155
20. Kepler, S.O., Robinson. E.L., Nather, R.E., & McGraw, J.T. 1982, *ApJ*, **254**, 676
21. Kepler, S.O. et al. 1991, *ApJ*, **378**, L45
22. Kepler, S.O. et al. 1995, *Baltic Astr.*, **4**, 221
23. Vauclair, G., Dolez, N., & Chevreton, M. 1983, *A&A*, **103**, L17
24. Kanaan, A.N. et al. 1992, *ApJ*, **390**, L89
25. Robinson, E.L., Stover, R.J., Nather, R.E., & McGraw, J.T. 1978, *ApJ*, **220**, 614
26. Pfeiffer, B. et al. 1996, *A&A*, **314**, 182
27. Fontaine, G., & Wesemael, F. 1984, *AJ*, **89**, 1728
28. Stobie, R.S. et al. 1995, *MNRAS*, **272**, L21
29. Becklin, E.E., & Zuckerman, B. 1988, *Nature*, **336**, 656
30. Bergeron, P., & McGraw, J.T. 1990, *ApJ*, **352**, L45
31. Bergeron, P. et al. 1993, *AJ*, **106**, 1987
32. O'Donoghue, D., & Warner, B. 1982, *MNRAS*, **200**, 573
33. O'Donoghue, D., & Warner, B. 1987, *MNRAS*, **228**, 949
34. Sullivan, D.J. 1995, *Baltic Astr.*, **4**, 261
35. Kepler, S.O., Robinson, E.R., & Nather, R.E. 1983, *ApJ*, **271**, 744
36. Kepler, S.O. et al. 1995, *ApJ*, **447**, 874
37. Giovannini, O. et al. 1998, *A&A*, **329**, L13
38. Robinson, E.L., & McGraw, J.T. 1976, *ApJ*, **207**, L37
39. Kepler, S.O. 1984, *ApJ*, **278**, 754
40. Vauclair, G., Chevreton, M., & Dolez, N. 1987, *A&A*, **175**, L13
41. Vauclair, G. et al. 1992, *A&A*, **264**, 547
42. Winget, D.E. et al. 1990, *ApJ*, **357**, 630
43. Patterson, J. et al. 1991, *ApJ*, **374**, 330
44. Kleinman, S.J. 1997 in *White Dwarfs*, edited by J. Isern, M. Hernanz, and E. Garcia-Berro (Kluwer Academic, Dordrecht), p. 437
45. Stobie, R.S., Chen, A., O'Donoghue, D., & Kilkenny, D. 1993, *MNRAS*, **263**, L13
46. Bergeron, P. Wesemael, F., Lamontagne, R. Fontaine, G., Saffer, R.A. & Allard, N.F. 1995, *ApJ*, **449**, 258
47. Weidemann, V., & Koester, D. 1984, *A&A*, **132**, 195

48. Kepler. S.O., & Nelan, E.P. 1993, *AJ*, **105**, 608
49. Daou, D., Wesemael, F., Bergeron, P., Fontaine, G., & Holberg, J.B. 1990, *ApJ*, **364**, 242
50. Wesemael, F., Lamontagne, R., & Fontaine, G. 1986, *AJ*, **91**, 1376
51. Koester, D., Allard, N., & Vauclair, G. 1994, *A&A*, **291**, L9

Table 16.2. *Names, positions, and magnitudes of the DBV stars.*

WD No.	Name	Variable star name	α (2000.0)	δ (2000.0)	T_{eff}^a (10^3 K)	V (mag.)	B (mag.)	Amp. (mag.)	Period (seconds)
0513 +261	KUV 0513+2605 [1]	V1063 Tau	05 16 28	+26 08.6	—	16.3	—	0.10	400, complex
0954 +342	CBS 114 [2]	SW LMi	09 57 50	+33 59.7	—		17[b]	0.30	650, complex
1115 +158	PG 1115+158 [3, 4]	DT Leo	11 18 23	+15 33.5	22.5: [13, 14]	—	16.12[b]	0.06	1000, complex
1351 +489	PG 1351+489 [5]	EM UMa	13 53 10	+48 40.4	22.0: [13, 14]	—	16.38[b]	0.05	489+harmonics
1456 +103	PG 1456+103 [6]	CW Boo	14 58 33	+10 08.3	22.5: [13, 14]	—	15.89[b]	0.10	420–860, complex
1645 +325	GD 358 [7–10]	V777 Her	16 47 19	+32 28.5	25.3 ± 0.3 [13, 14]	13.65	13.54	0.10	700, complex
1654 +160	PG 1654+160 [11]	V824 Her	16 56 58	+15 56.4	21.5: [13, 14]		16.15[b]	0.10	150–850, complex
2006 −523	EC 20058−5234 [11]	QU Tel	20 09 40	−52 25.4	—	15.54	—	0.07	134, 195, 204, 257, 281

Notes

[a]Uncertain temperatures are marked by a colon.
[b]Photographic (blue) magnitude.

References

1. Grauer, A.D., Wegner, G., & Liebert, J. 1989, *AJ*, **98**, 2221
2. Winget, D.E., & Claver, C.F. 1989, in *White Dwarfs*, edited by G. Wegner (Springer-Verlag, Berlin), IAU Colloq. 114, p. 290
3. Winget, D.E., Nather, R.E., & Hill, J.A. 1987, *ApJ*, **316**, 305
4. Clemens, J.C., et al. 1993, in *White Dwarfs: Advances in Observation and Theory*, edited by M.A. Barstow (Kluwer Academic, Dordrecht), p. 515
5. Grauer, A.D., Bond, H.E., Green R.F., & Liebert, J. 1988, *AJ*, **95**, 879
6. Winget, D.E., Robinson, E.R., Nather, R.E., & Fontaine, G. 1982, *ApJ*, **262**, L11
7. Winget, D.E. et al. 1994, *ApJ*, **430**, 839
8. Bradley, P.A., & Winget, D.E. 1994, *ApJ*, **430**, 850
9. Provencal, J.L. et al. 1996, *ApJ*, **466**, 1011
10. Winget, D.E., Robinson, E.L., Nather, R.E., & Balachandran, S. 1984, *ApJ*, **279**, L15
11. O'Donoghue D. 1995, *9th European Workshop on White Dwarf Stars*, edited by D. Koester & K. Werner (Springer-Verlag, Berlin), p. 297
12. Liebert, J. et al. 1986, *ApJ*, **309**, 241
13. Thejll, P., Vennes, S., & Shipman, H.L. 1991, *ApJ*, **370**, 355

Table 16.3. Names, positions, and magnitudes of the PG 1159 and PNNV stars.

WD No.	Name	Variable star name	α (2000.0)	δ (2000.0)	T_{eff}^a (10^3 K)	V (mag.)	B (mag.)	Amp. (mag.)	Period (seconds)
Pulsating PG 1159 (DOV) stars									
0122 +200	PG 0122+200 [1–3]	BB Psc	01 25 22	+20 17.8	75	—	16.13	0.10	400–600, complex
1159 −035	PG 1159−035 [4, 5]	GW Vir	12 01 46	−03 45.6	140	14.84	14.21	0.10	~500, complex
1707 +427	PG 1707+427 [6–8]	V817 Her	17 08 48	+42 41.0	100	16.69	16.08	0.10	~450, complex
2131 +066	PG 2131+066 [6, 9]	IR Peg	21 34 08	+06 50.9	80	16.63	16.24	0.10	400–600, complex
2324 +3944	HS 2324+3944 [10–12]	—	23 27 16	+40 01.4	130	14.8	—	0.02	~2100
Nonpulsating PG 1159 stars									
0123 −754	RXJ 0122.9−7521		01 22 55	−75 21.2	180	15.45	15.12		
0130 −196	MCT 0130−1937		01 32 40	−19 21.7	95	15.84	15.55		
0444 +045	HS 0444+0453		04 47 04	+04 58.7	100	—	15.9		
0704 +615	HS 0704+6153		07 09 32	+61 48.3	65	—	16.8		
VV 47	164+31°1		07 57 50	+53 25.0	130	16.83	16.53		
1144 +005	PG 1144+005		11 46 36	+00 12.5	150	—	15.04		
1151 −029	PG 1151−029		11 54 15	−03 12.1	140	—	16.07		
1424 +535	PG 1424+535		14 25 55	+53 15.4	100	16.2	15.86		
1504 +652	H 1504+65		15 02 08	+66 12.5	170	16.24	—		
1517 +740	HS 1517+7403		15 16 46	+73 52.1	95	—	15.9		
1520 +525	PG 1520+525		15 21 47	+52 22.0	150	15.52	15.56		
Pulsating planetary nebula nuclei (PNNVs)									
NGC 246	118−74°1 [13]		00 47 03	−11 52.3	150	11.96	—	~0.002	~1500
NGC 1501 [6]	144+6°1 [13, 14]	CH Cam	04 06 59	+60 55.2	ND	14.20	14.91	~0.15	~1500, complex
NGC 2371-2 [7]	189+19°1 [13, 15]	—	07 25 35	+29 29.4	ND	14.85	15.9	~0.01	~1000, complex
NGC 2867	278−5°1 [13]		09 21 25	−58 18.7	ND	—	16.62[b]	~0.02	~770
Lo-4 [8]	274+9°1 [16, 17]	LV Vel	10 05 45	−44 21.5	120	16.6	16.45	~0.06	1800–2000, complex
NGC 5189	307−3°1 [13]	KN Mus	13 33 33	−65 58.4	ND	—	14.9[b]	~0.003	~690
Sanduleak 3 [15]	—		16 03 08	−35 37.3	130	—	13[b]	~0.01	~1000
K 1-16	94+27°1 [18, 19]	DS Dra	18 21 52	+64 21.9	~140	15.08	14.66	~0.05	1500–1700, complex
NGC 6905	61−9°1 [13, 15]	—	20 22 23	+20 06.3	ND	15.7	—	~0.01	710, 875, + others
2117 +341	RXJ 2117+3412 [20–22]	V2027 Cyg	21 17 07	+34 12.4	170	13.16	—	0.05	~800, complex

Notes

[a]Temperatures for the DOV stars taken from Sion, E.M., & Downes, R.A. 1992, *ApJ*, **396**, L79; Werner, K., 1992, *A&A*, **251**, 147; Werner, K., & Heber, U. 1991, *A&A*, **247**, 476; Werner, K., Heber, U., & Hunger K. 1991, *A&A*, **244**, 437; and estimated by me from comparisons of optical and UV spectra given in Wesemael, F., Green, R.F., & Liebert, J. 1985, *ApJS*, **58**, 379.

[b]Photographic (blue) magnitude.

References

1. Bond, H.E., & Grauer, A.D. 1987, *ApJ*, **321**, L123
2. Vauclair, G. et al. 1995, *A&A*, **299**, 707
3. O'Brien, M.S. et al. 1996, *ApJ*, **467**, 397
4. Winget, D.E. et al. 1991, *ApJ*, **378**, 326
5. Kawaler, S.D., & Bradley, P.A. 1994, *ApJ*, **427**, 415
6. Bond, H.E., Grauer, A.D., Green, R.F., & Liebert, J. 1984, *ApJ*, **279**, 751
7. Fontaine, G. et al. 1991, *ApJ*, **378**, L49
8. Grauer, A.D., Green, R.F., & Liebert, J. 1992, *ApJ*, **399**, 686
9. Kawaler, S.D. et al. 1995, *ApJ*, **450**, 350
10. Silvotti, R. 1996, *A&A*, **309**, L23
11. Dreizler, S. et al. 1996, *A&A*, **309**, 820
12. Handler, G. et al. 1997, *A&A*, **326**, 692
13. Ciardullo, R., & Bond, H.E. 1996, *AJ*, **111**, 2332
14. Bond, H.E. et al. 1996, *AJ*, **112**, 2699
15. Bond, H.E., & Ciardullo, R. 1993, in *The 8th European Workshop on White Dwarf Stars*, edited by M.A. Barstow (Kluwer Academic, Dordrecht), NATO ASI Ser., p. 491
16. Longmore, A.J., 1977, *MNRAS*, **178**, 251
17. Bond, H.E., & Meakes, M.G. 1990, *AJ*, **100**, 788
18. Grauer, A.D., & Bond, H.E. 1984, *ApJ*, **277**, 211
19. Grauer, A.D., Bond, H.E., Green, R.F., & Liebert, J. 1987, in *The Second Conference on Faint Blue Stars*, edited by A.G.D. Philip, D.S. Hayes, and J.W. Liebert (Davis, Schenectady), IAU Colloq. 95, p. 231
20. Watson, T.K. 1992, IAU Circ. 5603
21. Vauclair, G. et al. 1993, *A&A*, **267**, L35
22. Motch, C., Werner, K., & Pakull, M.W. 1993, *A&A*, **268**, 561

16.4 LONG-PERIOD VARIABLES [14–19]

Long-period variables (LPVs) including Mira stars are mostly M-type giant and supergiant stars, usually with emission lines in their spectra. Many carbon stars (C-type) and zirconium (S-type) stars also show this type of variability. OH/IR stars are probably dust-enshrouded Miras having periods $\approx$ 600 to 2000 days and are not visible optically.

16.4.1 Properties of Mira Variables

The visual magnitude variation range is 2.5–10 mag.
   The galactic scale height is 240 pc.
   The mass is $\sim 1\,\mathcal{M}_\odot$

$$\log T_{\text{eff}} \approx 4.255 - 0.35 \log P \quad (100 < P < 500).$$

Luminosity: M_V at maximum light $\approx 0.0040 P - 2.6$ $(200 < P < 500)$.

$$\langle M_K \rangle = -3.47 \log P + 1.0,$$
$$\langle M_{\text{bol}} \rangle = -2.34 \log P + 1.3.$$

Types of variables are given below:

Designation	Type	Pop	P (days)	Sp	M_V	ΔV	⟨gal. lat.⟩	Periodicity
M	Mira, oxygen-rich;	I&II	200–600	M, C, S	in text	in text	in text	regular
SR	Long period,	I&II	100–500	F-M, C, S	−1	≤ 2.5	22°	semiregular
a, b, c, d	semiregular							
L	Slow irregular	I&II	100–500	M, C, S	−1	≤ 2.5	22°	not regular

16.5 OTHER VARIABLES [19–22]

Details of the classes and subclasses are given in [20].
   Types of other kinds of variables are given below:

Designation	Type and features	Pop	P	Sp	M_V	ΔV	Average galactic latitude
RV	RV Tau. Alternating depth of minimum	II	40–150 days	G-K	−2	1.3	23°
RCB	R CrB. Deep fades + pulsation	I?	∼ years 40–70 days	F-G	−4	5 0.2	14°
UU Her	UU Her	I&II	40–70 days	F I	?	0.3	20°

16.6 ROTATING VARIABLES
by Douglas S. Hall

These stars vary in brightness periodically as the star rotates about its axis. Except for the ellipsoidal variables, the star is essentially spherical, and it is a longitudinally asymmetric distribution of surface brightness that causes the variability.

Four basic mechanisms apply here:

1. The ellipticity effect results when one or both stars in a binary is ellipsoidal in shape.
2. The reflection effect in a close binary causes the facing hemisphere of one star to be brighter than its opposite hemisphere.
3. The oblique rotator model explains the ACV, SXARI, and PSR variables, where a strong dipolar magnetic field is not parallel to the rotation axis.
4. Starspots explain the BY, RS, and FKCOM variables and one component of the variability in some INT variables. The spots are magnetic in origin, like sunspots and sunspot groups, but a hundredfold larger in area. The temperature difference, photosphere minus spot, is 1000–2000 K [23]. Variability is periodic with the star's rotation but periods differ by a few percent due to solar-type differential rotation [24]. Physical mechanism for the large spots is strong dynamo action due to rapid rotation and deep convection [24]. Size of spots can be predicted by rotation period, $B - V$, and luminosity class [24]. Full amplitude can be up to 0.5 magnitude [24].

16.6.1 Types of Rotating Variables

ELL. No formal prototype, though b Persei is often considered so. Tidal forces in a close binary make one or both stars ellipsoidal (prolate) in shape. The difference in projected surface area between end-on and broad-side views causes the brightness to vary, with gravity darkening enhancing the effect [25]. There are two maxima and two minima per rotation, but limb-darkening effects can make the two minima unequal in depth [26]. Full amplitude, maximum to deeper minimum, can be up to 0.35 magnitude in *V*.

Reflection variables. These are not defined formally in the GCVS. There is no formal prototype. The sole source of the variability is the reflection effect. Only a few cases are known. Examples are BH Canum Venaticorum [27] and HZ Herculis [28].

ACV. The prototype is α^2 Canum Venaticorum. Signatures of a strong (several kilogauss) magnetic field and an anomalous strengthening of absorption lines of certain elements both vary, along with the brightness, with the star's rotation period. These include the so-called Ap or peculiar A stars. The full amplitude can be up to around 0.1 magnitude in *V* [28].

SXARI. The prototype is SX Arietis. These are high-temperature, spectral type B, analogs of the ACV variables and are sometimes called the helium variables. The full amplitude is up to around 0.1 magnitude in *V* [28].

PSR. There is no formal prototype. These are optically variable pulsars like CM Tauri, the supernova remnant in the Crab Nebula: rapidly rotating neutron stars with very strong (several megagauss) magnetic fields emitting narrow beams of radiation in the radio, optical, and X-ray bands. The rotation periods are between milliseconds and seconds. The amplitude of light pulses can be up to about 1 magnitude in *V* [28]. More details can be found in Section 16.16.

BY. The prototype is BY Draconis. These are defined by [29] and are K Ve or M Ve stars, where e means Hα emission. These can be single or binary. Rapid rotation is a consequence of the star's youth, i.e., recent arrival upon the main sequence, or by tidally enforced synchronism in a binary of short orbital period. There is considerable overlap with the UV variables, i.e., flare stars.

RS. The prototype is RS Canum Venaticorum. These are defined by [24, 30] and are G or K stars, always in binaries, by definition. Rapid rotation is caused by tidally enforced synchronism in a relatively short-period binary orbit, where "short" means days, weeks, or months for luminosity class V, IV, or III [24]. A spotted star typically is post-main-sequence but before Roche lobe overflow.

FKCOM. The prototype is FK Comae Berenices. These stars are defined by [31] and are single, rapidly rotating G or K giants. They may be coalesced W UMa-type binaries [31] or A-type main-sequence stars after evolution into the Hertzsprung gap [32].

INT. Starspots cause one component of the complex variability in some T Tauri stars, i.e., INT variables of spectral type G, K, M. Variability is periodic with the star's rotation [33].

16.7   T TAURI STARS [34–37]
by Gibor Basri

The T Tauri stars are systems containing solar-type pre-main sequence stars. As such, they provide the opportunity to learn something about our own early solar system. Indeed, many of them are now thought to contain circumstellar disks analogous to the solar nebula. Signs of youth include their position in the HR diagram, association with molecular clouds, and undepleted lithium in their spectra. They fall into two basic categories: the so-called "classical" TTS, and the weak-lined TTS. These are roughly distinguished observationally by the strength of their Hα emission. Physically, the distinction is probably between systems containing an active accretion disk or disk extending almost to the stellar surface (the CTTS), and those that have no disk or only an outer disk (the WTTS).

The CTTS show many phenomena associated with accretion disks and (apparently related) strong mass loss, including infrared excesses (from 2 to 100 μm), strong Balmer line (and continuum) emission, other emission lines (including forbidden-line emission in many systems), and an optical and ultraviolet excess thought to be generated by the accretion onto the star. When the accretion is strong, the photospheric absorption lines can be "veiled" by dilution from the accretion-related continuum. In extreme cases the light from the accretion disk can completely mask the stellar light (FU Ori systems).

The WTTS show none of these effects, displaying only the effects of very strong magnetic activity (chromospheres, coronae, starspots). Their properties generally lie along the activity sequence that extends down to old main-sequence stars, but they are among the most active examples known. The underlying stars in the CTTS are probably very similar. The CTTS and WTTS are commingled both spatially and on the HR diagram, except that CTTS much older than 10 Myr are not found. There are generally more WTTS than CTTS in well-surveyed clouds (and the WTTS are more likely to be incompletely sampled).

See Table 16.4 for characteristics of T Tauri stars.

Table 16.4. *T Tauri Star Characteristics.*

Name[a]	V (V range)[b]	EW (Hα)(Å)[c]	$v \sin i$[e]	L_{star}^{f}	
HBC No.; Sp.T.	A_V	Veiling (5500 Å)[d]	P_{rot}^{e}	L_{system}^{g}	Remarks
T Tau	9.9 (9.3–13.5)	50–70	20	12:	Prototype, but somewhat
35; K0 IV,V	1.3	0–0.2	2.8	18.9+4.4	atypical; IR companion
BP Tau	12.1 (10.7–13.6)	30–50	< 10	0.9	A fairly typical case,
32; K7 V	0.5	0.5–1	7.6	1.6	bright spots seen
DF Tau	11.5 (11.5–15)	30–90	22	2.0	Also typical; speckle binary
36; M0,1 V	0.45	1–1.5	8.5	5.1	

Table 16.4. *(Continued.)*

Name[a]	V (V range)[b]	EW (Hα)(Å)[c]	$v \sin i$[e]	L_{star}^{f}	
HBC No.; Sp.T.	A_V	Veiling (5500 Å)[d]	P_{rot}^{e}	L_{system}^{g}	Remarks
DR Tau	11.2 (10.5–16)	40–90	< 10	0.9	Extremely variable,
74; c(M0)	1?	3–8	13?	5.3	many emission lines
RW Aur	10.1 (9.6–13.6)	70–90	15	2.5	Very broad lines, variable veiling
80; c(K5)	1.2	0.5–3	5.4?	4.4	
DG Tau	12.0 (10.5–14.5)	70–100	20	0.9	Optical jet source,
37; c(K7 M0)	1?	3–5	?	10.1	many emission lines
FU Ori	9.6: (9.2–16.5)	27	110:	4?	Prototype of extreme outburst
186; G: I,II	1.85	Very high	?	490	sources; light is declining
					slowly; pure disk spectrum
V410 Tau	10.8 (10.8–12.4)	3	73	2	Large spots, 1 solar mass,
29; K3 V	0.03	0	1.88	2	1 Myr old
V830 Tau	12.2 (12.1–12.4)	3	29	1.2	Typical "weak" or "naked"
405; K7–M0 V	0.4	0	2.75	1.2	T Tauri star

Notes

[a] Herbig and Bell catalog: compendium of 742 pre-main-sequence stars. For spectral types, c indicates a heavily veiled or "continuum" star.

[b] Visual magnitude and range; extinction is not very accurate, somewhat model-dependent (in classical T Tauri stars).

[c] Estimated range of Hα equivalent width (Å) (low-dispersion spectra).

[d] Excess continuum veiling at 5500 Å (in units of photospheric continuum).

[e] Projected rotation velocity in km/s; period (from photometric modulation) in days.

[f] Stellar luminosity (bolometric, in solar luminosities); estimated from flux near 1 μm; not very accurate due to extinction and accretion disk effects, except on weak-lined stars.

[g] Systemic bolometric luminosity (in solar luminosities, from 0.3–100 μm); dominated by IR excess when well above stellar value.

16.8 FLARE STARS [38–42]
by Douglas S. Hall

Flares are observed on K Ve and M Ve (classical UV Cet stars), on spectral classes G and K, with luminosity classes V and IV (RS CVn, W UMa, and Algol binaries, FK Com stars). Flare amplitudes are up to $\approx$ 5 mag. in U. The energy output of any flare in the spectral windows are

$$E_x \approx E_{\text{euv}} \approx E_{\text{opt}} \gg E_{\text{radio}},$$
$$E_{\text{bol}} \approx 6E_{\text{opt}}; \qquad E_{\text{opt}} \approx 4E_U,$$
$$E_U : E_B : E_V = 1.8 : 1.5 : 1.0.$$

The largest flare yet observed is represented by

$$L_B = L_U = 2 \times 10^{33} \text{ erg s}^{-1}, \qquad E_U = 10^{37} \text{ erg}, \qquad E_{\text{bol}} = 2 \times 10^{38} \text{ erg}.$$

The time-averaged bolometric luminosity of flares on a given star is

$$\langle L_{\text{flare}} \rangle = 0.003 L_{\text{star}} \text{ for active stars at saturation level,}$$
$$\langle L_{\text{flare}} \rangle < 0.003 L_{\text{star}} \text{ for less active stars.}$$

The flare duration (rise time and half-life of decay time) in s is

$$\log t_{\text{rise}} = 0.25 \log E_u - 6.0,$$
$$\log t_{1/2} = 0.30 \log E_u - 7.5.$$

Flare colors and temperatures are

$$\langle B - V \rangle = +0.34 \pm 0.44, \qquad \langle U - B \rangle = -0.88 \pm 0.31,$$

$$T = \begin{cases} 3 \times 10^7 \text{ K for dMe stars,} \\ 10^8 \text{ K for RS CVn stars.} \end{cases}$$

The flaring frequency, for a given star, depends on flare energy:

$$\nu \propto E^{-\beta}, \qquad 0.4 < \beta < 1.4.$$

For UV Cet itself, $\nu = 1$ flare/day at $E_B = 10^{31}$ erg.

16.9 WOLF–RAYET AND LUMINOUS BLUE VARIABLE STARS
by Kenneth R. Brownsberger and Peter S. Conti

Wolf–Rayet (WR) stars are highly evolved descendants of massive stars, stars whose initial masses are larger than about $40 \mathcal{M}_\odot$. WR stars are characterized by strong, broad emission lines in their spectra, due to their intense stellar winds. The strengths of the emission lines serve as the basis for the classification of these stars. There are two main WR spectra types, WN and WC, and a third, less numerous type, WO. The WN types have spectra dominated by helium and nitrogen emission lines (see Table 16.5); in WC types, the predominant lines are from helium, carbon, and oxygen. The WO types have very strong oxygen lines (see Table 16.6).

Because the underlying stars are shrouded behind their dense stellar winds, the intrinsic stellar parameters of WR stars are difficult to ascertain. Furthermore, comparisons with non-LTE (local thermodynamic equilibrium) wind models yield a spread of values of the stellar parameters for different stars of the same spectral subtype: Radii of WR stars range from 2 to 20 $R_\odot$, temperatures from 30 000 to 70 000 K, wind velocities from 1000 to 3000 km s^{-1}, and mass loss rates from 10^{-5} to 10^{-4} $\mathcal{M}_\odot$ yr^{-1} [43, 44]. (See Tables 16.7–16.11.)

Luminous blue variables (LBVs) are hot, massive stars that show strong photometric and spectroscopic variations. LBVs include Hubble–Sandage variables, P Cygni stars, and S Doradus stars and can vary by 1 to 2 magnitudes on time scales of a few decades. Occasionally, they can erupt and increase their brightness by more than 3 mag. During their quiescent phases of minimum brightness, LBVs appear to be blue, B-type supergiants. During periods of maximum brightness they resemble late-type (A–F) supergiants [45, 46]. (See Table 16.12.)

Table 16.5. *Classification criteria for WN spectra* [1, 2].

WN subtypes	Nitrogen ions	Other criteria
WN2	N v weak or absent	He II strong
WN2.5	N v present, N IV absent	
WN3	N IV ≪ N v, N II weak or absent	

Table 16.5. *(Continued.)*

WN subtypes	Nitrogen ions	Other criteria
WN4	N IV ≈ N V, N III weak or absent	
WN4.5	N IV > N V, N III weak or absent	
WN5	N III ≈ N IV ≈ N V	
WN6	N III ≈ N IV, N V present but weak	
WN7	N III > N IV, N III λ4640 < He II λ4686	He I weak P Cyg
WN8	N III ≫ N IV, N III λ4640 ≈ He II λ4686	He I strong P Cyg
WN9	N III present, N IV weak or absent	He I, lower Balmer series P Cyg

References

1. van der Hucht, K.A., Conti, P.S., Lundstrom, I., & Stenholm, B. 1981, *SSRv*, **28**, 227
2. Conti, P.S., Massey, P., & Vreux, J.M. 1990, *ApJ*, **354**, 359

Table 16.6. *Classification criteria for WC, WO spectra* [1, 2, 3].

Subtypes	Carbon ions	Other criteria
WC4	C IV strong, C III weak or absent	O V moderate
WC5	C III ≪ C IV	C III < O V
WC6	C III ≪ C IV	C III > O V
WC7	C III < C IV	C III ≫ O V
WC8	C III > C IV	C II absent, O V weak or absent
WC9	C III > C IV	C II present, O V weak or absent

Subtypes	Oxygen ions	Other criteria
WO1	O VI strong	O V ≥ C IV
WO2	O VI strong	O V < C IV

References

1. van der Hucht, K.A., Conti, P.S., Lundstrom, I., & Stenholm, B. 1981, *SSRv*, **28**, 227
2. Conti, P.S., Massey, P., & Vreux, J.M. 1990, *ApJ*, **354**, 359
3. Barlow, J.J., & Hummer, D.G. 1982, in *Wolf–Rayet Stars: Observations, Physics, Evolution* (Reidel, Dordrecht), IAU Symp. 99, p. 387

Table 16.7. *Observed numbers of Wolf–Rayet subtypes* [1–7].

	WNE (WN2–WN5)	WNL (WN6–WN9)	WCE (WC4–WC7)	WCL (WC8–WC9)	WN/WC	WO
Galaxy	30	50	40	27	7	2
LMC	61	25	19		1	1
SMC	5	2				1

References

1. van der Hucht, K.A., Conti, P.S., Lundstrom, I., & Stenholm, B. 1981, *SSRv*, **28**, 227
2. Breysacher, J. 1981, *A&A*, **43**, 203
3. Azzopardi, M., & Breysacher, J. 1979, *A&A*, **75**, 120
4. Lundstrom, I., & Stenholm, B. 1984, *A&A*, **58**, 163
5. Vacca, W.D., & Torres-Dodgen, A.V. 1990, *ApJS*, **73**, 685
6. Conti, P.S., & Vacca, W.D. 1990, *AJ*, **100**(2), 431
7. Morris, P.W., Brownsberger, K.R., Conti, P.S., Massey, P., & Vacca, W.D. 1993, *ApJ*, **412**, 324

Table 16.8. *Observed numbers of single Wolf–Rayet stars and those with companions and/or absorption lines* [1–7].

	WN	WN+abs	WC	WC+abs	WO	WO+abs
Galaxy	64	16	51	15	2	
LMC	64	22	7	12	1	
SMC	1	6				1

References

1. van der Hucht, K.A., Conti, P.S. Lundstrom, I., & Stenholm, B. 1981, *SSRv*, **28**, 227
2. Breysacher, J. 1981, *A&A*, **43**, 203
3. Azzopardi, M., & Breysacher, J. 1979, *A&A*, **75**, 120
4. Lundstrom, I., & Stenholm, B. 1984, *A&A*, **58**, 163
5. Vacca, W.D., & Torres-Dodgen, A.V. 1990, *ApJS*, **73**, 685
6. Conti, P.S., & Vacca, W.D. 1990, *AJ*, **100**(2), 431
7. Morris, P.W., Brownsberger, K.R., Conti, P.S., Massey, P., & Vacca, W.D. 1993, *ApJ*, **412**, 324

Table 16.9. *Average intrinsic colors and absolute magnitudes* [1–3].

	WNE	WNL	WCE	WCL
M_v	−3.8	−5.5	−4.5	−4.8
$(b − v)_0$	−0.2	−0.2	−0.3	−0.3

References

1. Vacca, W.D., & Torres-Dodgen, A.V. 1990, *ApJS*, **73**, 685
2. Conti, P.S., & Vacca, W.D. 1990, *AJ*, **100**(2), 431
3. Morris, P.W., Brownsberger, K.R., Conti, P.S., Massey, P., & Vacca, W.D. 1993, *ApJ*, **412**, 324

Table 16.10. *Selected Wolf–Rayet stars* [1–4].

ID	Star	Other	l^{II}	b^{II}	Sp. Class	v	$E(b − v)$ [5, 6]	$(m − M)$ [7]
AB 5	HD 5980	AzV 229	302.07	−44.95	WN4+O7 I:	11.88	0.05	19.2
Br 08	HD 32257	Sk−69°42	280.81	−35.29	WC4	14.89	0.08	18.5
Br 26	HD 36063	Sk−71°21	282.64	−32.63	WN7	12.68	0.07	18.5
WR 006	HD 50896	EZ CMa	234.76	−10.08	WN5+(cc?)	7.26	0.05	11.3
WR 011	HD 68273	γ^2 Vel	262.80	−7.69	WC8+O9 I	1.74	0.03	8.3
WR 048	HD 113904	θ Mus	304.67	−2.49	WC6+O9.5 I	5.58	0.17	11.9
WR 111	HD 165763	MR 84	9.24	−0.61	WC5	8.25	0.32	11.0
WR 136	HD 192163	MR 102	75.48	+2.43	WN6 (SB1)	7.79	0.45	11.6

References

1. van der Hucht, K.A., Conti, P.S., Lundstrom, I., & Stenholm, B. 1981, *SSRv*, **28**, 227
2. Breysacher, J. 1981, *A&AS*, **43**, 203
3. Azzopardi, M., & Breysacher, J. 1979, *A&A*, **75**, 120
4. Conti, P.S., & Vacca, W.D. 1990, *AJ*, **100**(2), 431
5. Vacca, W.D., & Torres-Dodgen, A.V. 1990, *ApJS*, **73**, 685
6. Morris, P.W., Brownsberger, K.R., Conti, P.S., Massey, P., & Vacca, W.D. 1993, *ApJ*, **412**, 324
7. Lundstrom, I., & Stenholm, B. 1984, *A&AS*, **58**, 163

Table 16.11. *Representative UV-NIR (0.1–1.1 μm) emission lines in Wolf–Rayet stars. WN stars contain He, N, and C; WC stars contain He, C, and O.*

Ion	IP (eV)	λ (Å)	Ion	IP (eV)	λ (Å)
He I	24.6	5876, 6678, 7065, 10830	N III	47.4	1751, 4634, 4641
He II	54.4	1640, 2511, 2734, 3203, 4686, 4860, 5412, 6560, 6683, 6891, 8237, 10124	N IV	77.4	1486, 3480, 4058, 6383, 7115
C II	24.4	1335, 4267, 7236, 9234, 9891	N V	97.9	1240, 1718, 4604, 4620
C III	47.9	1176, 1247, 1909], 2297, 4650, 5696, 6742, 8500, 8665, 9711	O III	54.9	3265, 3708, 3760, 3962
C IV	64.5	1549, 2405, 2530, 4441, 4787, 5017, 5470, 5805, 7061, 7726, 8859	O IV	77.4	3411, 3730
			O V	113.9	2788, 3140, 4933, 5592
			O VI	138.1	3811, 3834

Table 16.12. *Observed properties of luminous blue variables (LBV) [1, 2].*

	LBV Star	T (K)		M_{bol}	Mass loss ($\mathcal{M}_\odot \, yr^{-1}$)
		Minimum brightness	Maximum brightness		
Galaxy	η Car	27 000:		−11.3	10^{-3} to 10^{-1}
	AG Car	25 000:	9000	−10.1	3×10^{-5}
	HR Car [3]	14 000:		−9.4	2×10^{-6}
	WRA 751 [4]	< 30 000:		−9.5	2×10^{-6}
	P Cyg	19 000		−9.9	2×10^{-5}
LMC	S Dor	20 000–25 000	8000	−9.8	5×10^{-5}
	R 71	13 600	9000	−8.8	5×10^{-5}
	R 127	30 000	8500	−10.5	6×10^{-5}
	R 143 [5]	19 000	6500	−10.0	
M33	Var C	20 000–25 000	7500–8000	−9.8	4×10^{-5}
	Var A	35 000	8000	−9.5	2×10^{-4}

References
1. Humphreys, R.M. 1989, in *Physics of Luminous Blue Variables* (Kluwer Academic, Dordrecht), IAU Symp. 157, p. 3
2. Conti, P.S. 1984, in *Observational Tests of Stellar Evolution Theory* (Reidel, Dordrecht), IAU Symp. 157, p. 233
3. Hutsemekers, D., & van Drom, E. 1991, *A&A*, **248**, 141
4. van Genderen, A.M., The, P.S., & de Winter, D. 1992, *A&A*, **258**, 316
5. Parker, J.W., Clayton, G.C., Winge, C., & Conti, P.S. 1993, *ApJ*, **409**, 770

16.10 Be STARS
by Arne Slettebak[†] and Myron Smith

Be stars may be defined as nonsupergiant B-type stars whose spectra have or had at one time Balmer lines in emission; the "Be phenomenon" is the episodic occurrence of rapid mass loss in these stars, resulting in Balmer emission. As a group, Be stars are characterized by rapid rotation, $v \sin i$ up to 400 km/s, but below the critical velocity.

[†]Deceased.

Statistical studies show that Be stars are not exotic objects. They comprise nearly 20% of the B0–B7 stars in a volume-limited sample of field stars, with a maximum incidence at B2 and considerably lower frequencies among late B-types.

Studies of Be stars in clusters show that Be stars may exist anywhere from the main sequence to giant regions in the H-R Diagram, with an average of $\frac{1}{2}$ to 1 magnitudes above the ZAMS. These positions are consistent with core hydrogen-burning rapid rotators.

Balmer (and occasionally He I) emission lines arise from equatorially confined disks of (typically) 5–20 stellar radii, 10 000 K temperature, and 10^{10}–10^{13} cm^{-3} electron density. These disks are expelled from the star on a timescale from days to years.

Be stars are variable on a continuum of timescales from several decades to minutes. Periodic variations in radial velocities and/or flux are often observed. The optical spectra of some Be stars can exhibit continual low-level absorptions and/or emission components which alter the line profile. Disks are formed, sometimes quasi-periodically, on a variety of timescales from days to years and their dispersal from weeks to decades; the fractional amount returning to the star is unknown. It is generally assumed that the disk material is in Keplerian orbit. Departures from axisymmetry and/or inhomogeneous density particle distributions can cause cyclic variations in the Violet and Red emission components in the Balmer lines.

Ultraviolet studies show that another side of the Be phenomenon is the appearance of strong, high-velocity absorption components in the UV resonance lines of C IV, Si IV (and occasionally Al III, N V, O VI) which can be attributed to the acceleration of a radiatively driven wind. The strengthening features are well, but imperfectly, correlated with Balmer emission episodes. Mass loss rates from the UV data are 10^{-11}–10^{-9} $\mathcal{M}_\odot$ yr^{-1}; these are typically 10–50× lower than rates estimated from the slow expansion of the equatorial disks from infrared and radio data.

Two hypotheses are current explanations of the Be phenomenon: surface magnetic activity and nonlinearities in nonradial pulsations. Historically, rapid rotation, stellar winds, nonlinear pulsations, magnetic fields, and binary interactions, singly and in combination, have been invoked to explain this activity. Because there are a few subclasses of Be stars, various combinations of different mechanisms could be responsible in particular cases.

Be-star catalogs (see [47]) list thousands of these objects. We list in Table 16.13 several of the brightest, recently well-studied stars which are not in interacting binaries (significant numbers are also Algol-type).

Table 16.13. *Best-known Be stars* [1–8].

Be star name	HD[a]	Sp. type	$v \sin i$ (km/s)
γ Cas[b]	5 394	B0.5 IVe	230
ψ Per[b]	22 192	B5 IIIe	280
28 Tau[b]	23 862	B8 Ve	320
X Per	24 534	O9.5 IIIe	200
λ Eri	33 328	B2 IIIe	220
ζ Tau[b]	37 202	B1 IVe	220
ν Gem	45 542	B6 IVe	170
β Mon A[b]	45 725	B4 Ve	300
ω CMa	56 139	B2.5 Ve	80
β CMi	58 715	B8 Ve	245
δ Cen	105 435	B2 IVe	220
κ Dra	109 387	B5 IIIe	200
μ Cen	120 324	B2 IV–Ve	155
θ CrB	138 749	B6 IIIe	320
4 Her	142 926	B7 IVe	300

Table 16.13. (*Continued.*)

Be star name	HD[a]	Sp. type	$v \sin i$ (km/s)
48 Lib[b]	142 983	B3 IVe	400
χ Oph	148 184	B1.5 Ve	140
66 Oph	164 284	B2 IV–Ve	240
59 Cyg[b]	200 120	B1 Ve	260
π Aqr	212 571	B1 III–IVe	300
EW Lac[b]	217 050	B3 IVe	300
β Psc	217 891	B5 Ve	100

Notes

[a] Henry Draper Catalogue number.

[b] In addition to the Balmer emission and rotationally broadened lines of neutral helium, these stars' spectra may show hydrogen lines with sharp absorption cores as well as narrow absorption lines of ionized metals during the stars' shell phases.

References

1. Doazan, V. 1982, in *B Stars with and without Emission Lines*, edited by A. Underhill and V. Doazan, Monograph Series on Nonthermal Phenomena in Stellar Atmospheres, NASA-CNRS, NASA SP-456, p. 277
2. *Physics of Be Stars*, 1987, IAU Colloq. 92, edited by A. Slettebak and T.P. Snow (Cambridge University Press, Cambridge)
3. Landolt-Bornstein, Vol. 2, 1982, Part 1, Peculiar Stars; 5.2.1.4,
4. Slettebak, A. 1988, *PASP*, **100**, 770
5. Slettebak, A. 1992, in *The Astronomy and Astrophysics Encyclopedia*, edited by S.P. Maran (Van Nostrand Reinhold, New York), p. 710
6. Balona, L.A., Henrichs, H.F., & Lecontel, J.M. 1994, *Pulsation, Rotation, and Mass Loss in Early-Type Stars*, IAU Symp. No. 162 (Kluwer Academic, Dordrecht)
7. Jaschek, M., & Egret, D. 1981, *A Catalogue of Be Stars*, Centre de Donnees Stellaires (CDS) Microfiche #3067
8. Jaschek, M. & Egret, D. 1982, Catalogue of Special Groups, Part I: The Earlier Groups, CDS Pub. Spec. No. 4

16.11 CHARACTERISTICS OF CARBON-RICH STARS
by Cecilia Barnbaum

Carbon-rich stars (defined as having atmospheric C/O > 1 and C/O = 1 for C and S stars, respectively) make up an odd assortment of peculiar abundance stars. Atmospheric enhancement of carbon is caused either by internal dredge-up of processed material during the late stages of stellar evolution, or by environmental interactions such as mass transfer from an evolved, close companion. Carbon stars (C designation, also classified as spectral types R, N, or J [48, 49]) on the asymptotic giant branch (AGB) have acquired their enriched carbon and s(slow)-process elements through convective dredge-up of the interior processed layers due to thermal pulsing [50]. s-Process elements result from slow neutron capture (and subsequent β decay) due to the low neutron flux in the stellar interior; the s-process produces different elements than the rapid (r-process) neutron capture that takes place in high neutron flux environments, such as in supernova events. An evolutionary sequence M-S-C is possible, but not verified [51]. Some stars with carbon-rich atmospheres have been shown to be the result of mass transfer in a binary system, e.g., the CH and Ba II stars [52, 53], and a number of C and S stars that lack the s-process element ^{99}Tc, a signature of the ABG phase [54]. A few C stars have a great over-abundance of ^{13}C; these are known as the J types. For most carbon stars, ^{12}C/^{13}C is $\sim$ 30–50, whereas for J types it is $\sim$ 3 [55]. Most of these J-type stars are not enriched in s-process elements, although there are a few exceptions (e.g., WX Cyg). Finally, there are the dwarf C stars [56]. These faint carbon-rich stars have a large proper motion, indicating that they are nearby and are therefore under-luminous for the AGB (asymptotic giant branch). They are thought to be main sequence stars that have acquired their carbon-richness by mass transfer from a giant companion that has since evolved into a white dwarf. Table 16.14 gives characteristics of C-rich stars.

Table 16.14. *Characteristics of carbon-rich stars.*

Type	Evol.	Pop.	Chemistry	Variability	Lum.	Special character[a]	References
C (N and late R)	AGB	I, II	$C/O > 1$; CN, C_2, s-pr. enhanced, often Tc	LPV: Lb, M, SR	6×10^3– $7 \times 10^4 \mathcal{L}_\odot$	CSE: CO, dust; $\dot{\mathcal{M}} \sim 10^{-7}$ to $10^{-5} \mathcal{M}_\odot$/yr	[1–3]
C (J and early R)	? pre-AGB	I, II	$C/O > 1$; CN, C_2, ^{13}C isotopic species, not s-pr. enhanced	Lb, M, SR	$< 10^3 \mathcal{L}_\odot$	CSE: CO, dust; $\dot{\mathcal{M}} \sim 10^{-7} \mathcal{M}_\odot$/yr	[4]
S	AGB+?	I, II	$C/O = 1$; ZrO; CN; s-pr. enhanced, often Tc	Lb, M, SR	$10^4 \mathcal{L}_\odot$	CSE: CO, dust; $\dot{\mathcal{M}}$ 6×10^{-8}	[5–7]
Ba II	Giant	I	s-process enhanced, esp. Ba, Sr; no Tc	Var.	$M_V <$ 0 to -3	Binaries, C-rich by mass transfer	[8, 9]
CH	Giant	II	Stronger CN, CH than Ba II stars; s-pr. enhanced but weaker metals than Ba II stars	Var.	M_V 0 to -3	Binaries, C-rich by mass transfer	[9, 10]
sgCH	Subgiant	I, II	CN, CH, s-pr. enhanced esp. Sr and Ba	Var.	Fainter than CH stars	Progenitors of CH stars?	[9, 11]
dC	Main Sequence?	??	CN; some ^{13}C enhanced	?	$M_V \sim 10$	Binaries? Mass transfer?	[3, 12]

Note

[a]CSE = circumstellar envelope.

References

1. Claussen, M.J. et al. 1987, *ApJS*, **65**, 385
2. Dean, C.A. 1976, *AJ*, **81**, 364
3. Kastner, J.H. et al. 1993, *A&A*, **275**, 163
4. Lambert, D.L. et al. 1986, *ApJS*, **62**, 373
5. Jura, M. 1988, *ApJS*, **66**, 33
6. Smith, V.V., & Lambert, D.L. 1988, *ApJ*, **333**, 219
7. Johnson, H.R., Ake, T.B., & Ameen, M.M. 1993, *ApJ*, **402**, 667
8. Jorrison A., & Mayor M. 1988, *A&A*, **198**, 187
9. McClure, R.D. 1989, in *Evolution of Peculiar Red Giant Stars*, edited by H.R. Johnson and B. Zuckerman (Cambridge University Press, Cambridge), p. 196
10. McClure, R.D. 1984, *ApJ*, **280**, L31
11. Luck, R.E., & Bond, H.E. 1982, *ApJ*, **259**, 792
12. Green, P.J., Margon, B., & MacConnell, D.J. 1991, *ApJ*, **380**, L31

16.12 BARIUM, CH, AND SUBGIANT CH STARS
by William Dean Pesnell

Barium stars show absorption at Ba II λ4554, Sr II λ4077 and λ4215, and bands of CH, CN, and C_2. The enrichment of material is due to mass exchange from an evolved companion [57]. The subgiant

CH stars may be the main-sequence progenitors of the barium stars. CH stars have strong bands of CH, CN, and C_2, but less metal enrichment than the Ba stars. See Tables 16.15 and 16.16.

Table 16.15. *The brighter barium stars* [1, 2].

HR	α (2000.0)	δ (2000.0)	m_v	Sp.
774	2 47 47.6	+81 26 55	5.9	G8p
2392	6 32 46.9	−11 09 59	6.3	K0 III
3123	7 59 05.6	−23 18 38	5.1	K2
3842	9 38 01.4	−43 11 27	5.5	G8 II
4474	11 37 52.9	+50 37 04	6.1	K0p
4862	12 49 44.9	−71 59 11	5.5	G8 Ib–II
5058	13 26 07.7	−39 45 19	5.1	K0.5 III
5802	15 36 29.5	+10 00 36	5.3	K0 III
8204	21 26 39.9	−22 24 41	3.7	G4 Ib

References

1. McClure, R.D. 1989, in *Evolution of Peculiar Red Giant Stars*, edited by H.R. Johnson and B. Zuckerman (Cambridge University Press, Cambridge), p. 196
2. MacConnell, D.J., Frye, R.L., & Upgren, A.R. 1972, *AJ*, **77**, 384

Table 16.16. *Subgiant CH stars* [1, 2].

No.	α (2000.0)	δ (2000.0)	m_v	Sp.
HD 89948	10 22 21.9	−29 33 21.1	7.50	G8 III
BD +17°2537	12 47 22.8	+16 49 35.0	8.82	G0
HD 123585	14 09 35.9	−44 22 01.7	9.28	F7 Vwp
HD 127392	14 31 53.5	−31 12 01.1	9.89	Gp
BD −10°4311	16 24 13.2	−11 13 07.5	10.1	G0
CPD −62°6195	21 06 02.8	−61 33 45.3	10.1	G5
HD 207585	21 50 34.8	−24 11 11.4	10.0	Gwp
HD 224621	23 59 17.3	−36 02 37.0	9.59	G0 III/IV

References

1. Hipparcos Input Catalogue (ESA), 1990
2. Luck, R.E., & Bond, H.E. 1991, *ApJS*, **77**, 515

16.13 HYDROGEN-DEFICIENT CARBON STARS
by Warrick Lawson

Hydrogen-deficient carbon stars are luminous, probable born-again post-AGB stars consisting of the cool R Coronae Borealis (RCB) and hydrogen-deficient Carbon (HdC) stars ($T_{eff} \approx 5000$ to 7000 K [58]) and extreme helium (eHe) stars ($T_{eff} \approx 8400$ to 55 000 K [59]). Three hot RCB-like stars may be unrelated [60]. Typically C/H > 10^3 although at least two stars are relatively H-rich [61]. RCB/HdC and cooler eHe stars are unstable to radial pulsations [58–62], whereas higher-temperature eHe stars are nonradial pulsators. RCB stars have pulsation-related declines in light of up to 8 magnitudes due to dust formation [63, 64] and have bright IR excesses [65, 66]. See Table 16.17.

Table 16.17. *Selected hydrogen-deficient carbon stars.*

Star	Type	α (2000.0)	δ (2000.0)	V (at maximum)	B − V	~ T_{eff} (K)	Notes
XX Cam	HdC?	04 08 39	+53 21 39	7.3	0.87	7 000	
HV 5 637	RCB	05 11 32	−67 56 00	14.8	1.23	5 000	LMC
W Men	RCB	05 26 24	−71 11 18	13.9	0.42	7 000	LMC
HV 12 842	RCB	05 45 03	−64 24 24	13.7	0.51	7 000	LMC
SU Tau	RCB	05 49 06	+19 04 00	9.7	1.10	7 000	
BD +37 1 977	eHe	09 24 24	+36 42 54	10.2		55 000	sdO?
UW Cen	RCB	12 43 17	−54 31 41	9.1	0.67	6 800	
DY Cen	RCB?	13 25 34	−54 14 47	12.5	0.35	14 000	
HD 124 448	eHe	14 14 59	−46 17 19	10.0	−0.10	15 500	Nonvariable?
V854 Cen	RCB	14 34 48	−39 33 19	7.1	0.50	7 000	Balmer lines present
R CrB	RCB	15 48 34	+28 09 24	5.8	0.59	7 000	
BD −9 4 395	eHe	16 28 35	−09 19 34	10.5	0.06	28 000	Nonradial pulsator?
HD 148 839	HdC	16 35 46	−67 07 37	8.3	0.93	6 500	C/H ~ 0.05
V2 076 Oph	eHe	17 41 50	−17 54 08	9.8	0.14	31 900	Nonradial pulsator
PV Tel	eHe	18 23 15	−56 37 43	9.3	0.00	12 400	
V348 Sgr	RCB?	18 40 20	−22 54 29	11.8	0.30	20 000	
MV Sgr	RCB?	18 44 32	−20 57 16	12.7	0.26	15 400	
LS IV −14 109	eHe	18 59 39	−14 26 11	11.1	0.33	8 400	
RY Sgr	RCB	19 16 33	−33 31 18	6.2	0.62	7 000	
U Aqr	RCB	22 03 20	−16 37 40	11.2	1.00	5 500	Sr-, Y-rich

16.14   BLUE STRAGGLERS
by Peter J.T. Leonard

Blue stragglers are main-sequence or slightly evolved stars in a stellar system that are apparently much younger than the majority of the stars in the system. Consequently, these stars pose a problem for standard stellar evolutionary theory. Blue stragglers have been discovered everywhere that they could have possibly been discovered, which includes OB associations, open clusters of all ages, globular clusters, the population II field, and dwarf spheroidal galaxies. Theories for these objects include stellar mergers due to physical stellar collisions, stellar mergers due to the slow coalescence of contact binaries, mass transfer in close binary systems, extended main-sequence lifetimes due to internal mixing (which may be induced by rapid rotation, strong magnetic fields, or pulsation), recent star formation, and several others. Table 16.18 provides a sample of blue stragglers found in various stellar systems.

Table 16.18. *Selected blue stragglers.*

Name	Type of the parent stellar system	Characteristics
HD 93843 [1]	Car OB1, OB association	$M_{bol} = -9.5$, O5III(f)var, $v \sin i = 90$ km s^{-1}
HD 152233 [1]	Sco OB1, OB association	$M_{bol} = -9.7$, O6III:(f)p, $v \sin i = 140$ km s^{-1}
HD 60855 [2]	NGC 2422, young open cluster	$M_V = -2.86$, B2IVe, $v \sin i = 320$ km s^{-1}
HD 162586 [2]	NGC 6475, young open cluster	$M_V = -1.08$, B6V, $v \sin i < 40$ km s^{-1}
HD 27962 [3]	Hyades, intermediate-age open cluster	Am(K/H/M=A2/A3:/A5), $v \sin i = 18$ km s^{-1}
HD 73666 [3]	Praesepe, intermediate-age open cluster	A1Vp(Si), $v \sin i = 40$ km s^{-1}
F 81 [4]	M67, old open cluster	$V = 10.04$, B8V
F 184 [4]	M67, old open cluster	$V = 12.22$, F0, $v \sin i = 80$ km s^{-1}
S I, II, 21 [5]	M3, globular cluster	$m_{pv} = 17.39$, $CI = 0.06$

Table 16.18. *(Continued.)*

Name		Type of the parent stellar system	Characteristics
E 39	[6]	ω Cen, globular cluster	0.056-day dwarf cepheid, $V = 17.03$, $B - V = 0.31$
NC 6	[7]	NGC 5053, globular cluster	$g = 18.33$, $g - r = -0.47$
AOL 1	[8]	NGC 6397, globular cluster	$V = 14.42$, $B - V = 0.16$
NH 19	[9]	NGC 5466, globular cluster	0.34-day contact binary, $V = 18.54$, $B - V = 0.15$
HST-1	[10]	47 Tuc, globular cluster	$m_{220} = 16.0$, $m_{140} = 16.7$
BD +25°1981	[11]	Population II field	$V = 9.29$, $B - V = 0.30$, $v \sin i = 9\ \mathrm{km\,s}^{-1}$
BD −12°2669	[11]	Population II field	$V = 10.22$, $B - V = 0.30$, $v \sin i = 31\ \mathrm{km\,s}^{-1}$
MA 308	[12]	Carina dwarf spheroidal galaxy	$V = 20.54$, $B - V = -0.07$
D 227	[13]	Sculptor dwarf spheroidal galaxy	$V = 21.67$, $B - V = 0.14$
S1267	[14]	M67, old open cluster	$P_{\mathrm{orb}} = 846$ days, $e_{\mathrm{orb}} = 0.475 \pm 0.125$
BSS-19	[15]	47 Tuc, globular cluster	$\mathcal{M} = 1.7 \pm 0.4 \mathcal{M}_{\odot}$, $v \sin i = 155 \pm 55\ \mathrm{km\,s}^{-1}$

References

1. Mathys, G. 1987, *A&AS*, **71**, 201
2. Mermilliod, J.-C. 1982, *A&A*, **109**, 37
3. Abt, H.A. 1985, *ApJ*, **294**, L103
4. Mathys, G. 1991, *A&A*, **245**, 467
5. Sandage, A.R. 1953, *AJ*, **58**, 61
6. Jørgensen, H.E., & Hansen, L. 1984, *A&A*, **133**, 165
7. Nemec, J.M., & Cohen, J.G. 1989, *ApJ*, **336**, 780
8. Aurière, M., Ortolani, S., & Lauzeral, C. 1990, *Nature*, **344**, 638
9. Mateo, M., Harris, H.C., Nemec, J., & Olszewski, E.W. 1990, *AJ*, **100**, 469
10. Paresce, F. et al. 1991, *Nature*, **352**, 297
11. Carney, B.W., & Peterson, R.C. 1981, *ApJ*, **251**, 190
12. Mould, J., & Aaronson, M. 1983, *ApJ*, **273**, 530
13. Da Costa, G.S. 1984, *ApJ*, **285**, 483
14. Latham, D.W., & Milone, A.R.R. 1996, *ASP Conf Ser.*, **90**, 385
15. Shara, M.M., Saffer, R.A., & Livio, M. 1997, *ApJ*, **489**, L59

16.15 PECULIAR A AND MAGNETIC STARS [67–70]

The peculiar A stars comprise the following:

1. Ap stars, which extend into B and earlier F types as well. The hotter (but not hottest) ones have unusually strong lines of Mn, Si, and Hg; the cooler ones have similarly strong lines of Si, Cr, Sr, and Eu, and other rare earth elements.

2. Am stars, for which the spectral type varies with the criterion used: the type based on the K-line is earlier than that from the Balmer lines and that, in turn, is earlier than the type from metallic lines. Differences are ≥ 5 subclasses.

Table 16.19 lists some other properties.

Table 16.19. *Other properties.*

	Ap (Mn, Hg)	Ap (Sr, Eu)	Am
Temperature:	10 000–15 000 K	8000–12 000 K	7000–9000 K
Luminosity and mass:	At or near main sequence values		
$v \sin i$ (km/s):	30	30	40
Magnetic field (gauss):	0 or low	10^3–10^4	0 or low
Close binary frequency	Normal	Low	High

Ap stars show spectrum, light, and magnetic variability due to rotational modulation on time scales of days to years. Magnetic Ap stars can also show rapid oscillations (roAp stars) on time-scales of 4 to 15 minutes due to high-overtone, low-degree, nonradial p modes [69].

16.16 PULSARS

by Kaiyou Chen and John Middleditch

Pulsars are believed to be strongly magnetized rotating neutron stars. The radiated spectrum of a pulsar can extend over many decades in wavelength. So far, more than 500 radio pulsars have been discovered. Among these are about 50 so-called *millisecond pulsars*, the majority of which have spin periods less than 10 ms, and all of which are thought to have fields significantly weaker than the so-called *pulsars* with a canonical magnetic field strength near 10^{11}–10^{12} G, typical of the vast majority of known radio pulsars.

A disproportionately large number of millisecond pulsars have been found to belong to the Galactic globular cluster population (about 30 so far). The suggestion that the millisecond pulsar population is the result of recycling of old (radio-dead) neutron stars through accretion from an orbital companion in a low-mass X-ray binary phase has gained a wide acceptance. However, alternative production mechanisms have also been proposed, some of which do not suffer from the vast overpopulation of the millisecond pulsars with respect to that of the low mass X-ray binaries.

There are only about a few dozen accretion-powered X-ray pulsars known to date, with all of these thought to be strongly magnetized. Only two (Her X-1 and 4U1626-67) are known to produce optical pulsations through reprocessing of the pulsed X-ray flux.

There is only one rotation-powered pulsar (Geminga) with strong γ-ray emission, which is not yet known as a radio source. The existence of non-(or slowly)-varying high-energy γ-ray sources discovered by EGRET on the Compton Gamma-Ray Observatory satellite may indicate a larger population of γ-ray pulsars yet undiscovered.

So far there are three neutron star–neutron star binary systems known, two of which belong to the galactic disk population and one in the globular cluster M15. Such binaries provide the best-known tests of general relativity theory in addition to the accurate measurement of the mass of the component neutron stars.

At least one millisecond pulsar (1257+12) is thought to have at least two (few Earth mass) planetary companions in orbits apparently synchronized with a 2:3 period ratio. Further study of such systems may eventually be able to exclude pulsar precession as an alternative explanation to the timing irregularities.

Neutron stars are thought to be the compact remnants of supernova explosions. The pulsar in the Crab nebula is almost certainly such a remnant. The most rapidly spinning pulsar associated with a supernova remnant, the 16 ms pulsar in N157B, may have had an initial spin period of only seven milliseconds. The detection of neutrinos from SN1987A indicated that a neutron star was indeed formed in the original core collapse. However, no evidence for a strong pulsar has yet been detected in this remnant.

See Table 16.20 for characteristics of a few important pulsars and Figure 16.1 for a radio pulsar diagram.

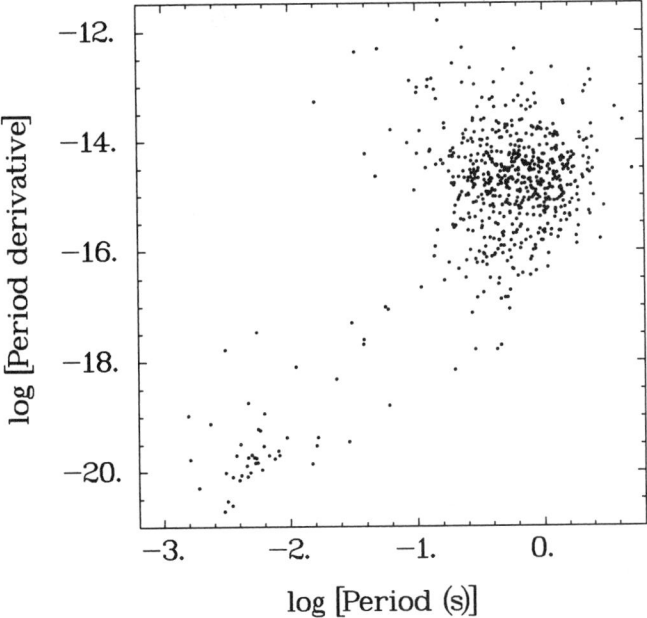

Figure 16.1. The "HR diagram" of 645 pulsars showing the period and the period derivative (seconds per second).

Table 16.20. *Some important pulsars.*

Name	Period (s)	Comments	Reference
0021−72c	0.0058	Nine more pulsars with $P < 6$ ms in the same globular cluster, 47 Tuc	[1]
0531+21 (Crab pulsar)	0.033	Pulsed emission from radio to γ-ray; obvious supernova association	[2]
0538-69 (N157B pulsar)	0.016	In the Large Magellanic Cloud (LMC); fastest known pulsar associated with a supernova remnant	[3, 4]
0540−69 (LMC pulsar)	0.050	Also in the LMC and supernova association; pulsed radio, optical, and X-ray emission	[5]
Geminga (1E0630+17)	0.237	Strong γ-ray pulsar; no radio detection	[6]
0833−45 (Vela pulsar)	0.089	Supernova association; strong γ-ray source	[7]
1257+12	0.0062	Having two companion planets	[8]
1534+12	0.0379	Having a companion neutron star	[9]
Her X-1	1.24	Accretion-powered X-ray pulsar	[10]
1821−24	0.003	First (millisecond) pulsar discovered; in a globular cluster, M28	[11]
1845−19	4.3082	Slowest known pulsar	[12]

Table 16.20. *(Continued.)*

Name	Period (s)	Comments	Reference
1913+16 (Hulse–Taylor pulsar)	0.059	First binary pulsar discovered; evidence of gravitational radiation	[13]
1919+21	1.337	First pulsar discovered;	[14]
1937+21	0.0015	First millisecond pulsar discovered; fastest known pulsar	[15]
1957+20	0.0016	Eclipsed by the evaporating companion every 9.2 hours	[16]

References

1. Manchester, R.N. et al. 1991, *Nature*, **352**, 219
2. Staelin, D.H., & Reifenstein, E.C. 1968, *Science*, **162**, 1481
3. Marshall, F.E. et al. 1998, IAU Circ. No 6810
4. Wang, D.Q., & Gotthelf, E.V. 1998, *ApJ*, **494**, 623
5. Seward, F.D. et al. 1984, *ApJ*, **287**, L19
6. Halpern, J.P., & Holt, S.S. 1992, *Nature*, **357**, 222
7. Large, M.I. et al. 1968, *Nature*, **220**, 340
8. Wolszczan, A., & Frail, D.A. 1992, *Nature*, **355**, 145
9. Wolszczan, A. 1991, *Nature*, **350**, 688
10. Tananbaum, H. et al. 1972, *ApJ*, **174**, L143
11. Lyne, A.G. et al. 1987, *Nature*, **328**, 399
12. Newton, L.M. et al. *MNRAS*, **194**, 841
13. Hulse, R.A., & Taylor, J.H. 1975, *ApJ*, **195**, L51
14. Hewish, A. et al. 1968, *Nature*, **217**, 709
15. Backer, D.C. et al. 1982, *Nature*, **300**, 615
16. Fruchter, A.S. et al. 1988, *Nature*, **333**, 237

16.17   GALACTIC BLACK HOLE CANDIDATE X-RAY BINARIES
by Jonathan E. Grindlay

Black hole candidates in the Galaxy are best defined as X-ray binaries in which the accreting compact object has a probable mass $\mathcal{M}_X \gtrsim 3\mathcal{M}_\odot$, or above the limit for neutron stars, as determined from spectroscopic measurements of the semiamplitude velocity, K, of the companion star with mass $\mathcal{M}_c$. Together with the orbital period P, this defines the mass function of the system: $f(\mathcal{M}_X) = PK^3/2\pi G = \mathcal{M}_X^3 \sin^3 i/(\mathcal{M}_X + \mathcal{M}_c)^2$. The mass function thus gives a firm lower limit for $\mathcal{M}_X$, although with additional constraints on system inclination, $\sin i$, and spectral type and thus mass $\mathcal{M}_c$, the black hole candidate mass $\mathcal{M}_X$ can be measured [71, 72]. The measurement [73] of $\mathcal{M}_X = 7.02 \pm 0.22\mathcal{M}_\odot$ obtained for the galactic "micro-quasar" source, with relativistic jets, GROJ1655-40 = Nova Sco 94 provides the currently (1998) most accurate determination of the mass of a probable black hole in the Galaxy.

Secondary indicators for galactic black hole candidates are their similar hard X-ray spectra, with power law form typically extending out beyond 100 keV and containing a significant fraction of the total luminosity [74] and (in high luminosity states) accompanying ultra-soft X-ray spectral components [75]. The black hole binaries are most often found as transient X-ray sources which are often particularly luminous in their soft X-ray emission at their peak and are thus frequently called soft X-ray transients (despite their nearly universal accompanying hard X-ray emission which dominates the emission during the decay phase). In comparison with transients known to contain neutron stars from their X-ray bursts, the black hole transients show significantly larger increase from their quiescent low states to outburst, consistent with their having an event horizon and advection-dominated accretion flow at the low accretion rates found in quiescence [76].

The most reliable black hole candidates are the eight systems listed in Table 16.21 with lower mass companions for which radial velocities and mass functions were derived when these transient type X-ray sources faded to quiescent optical levels. The 7–8 mag. optical brightening of some of these recurrent ($\sim$ 50 year) transients resemble novae leading to the term X-ray novae for these systems. The three high mass ($\gtrsim 10 \mathcal{M}_\odot$) companion systems included in Table 16.22 are less well determined (with LMC X-1 particularly questionable), although the prototype system Cyg X-1 is most secure [72]. A general summary of the properties of X-ray transients of all types provides constraints on the galactic population of black holes in binary systems [77], and a statistical analysis of the quiescent transients suggests [78] the black hole masses may be clustered near $7\mathcal{M}_\odot$ for all but V404 Cyg.

Table 16.21. *Galactic black hole candidates with low mass companions* [1–5].

Object	Opt. ID	α (2000.0) δ (2000.0)	m_v	Sp. Type	P (hours)	$f(\mathcal{M}_x/\mathcal{M}_\odot)$	$\mathcal{M}_x/\mathcal{M}_\odot$
GROJ0422+32	Nova Per 92	04 21 46.9 32 54 36	22	M0V	5.1	1.21 ± 0.06	3.5–14
A0620-00	Nova Mon 75	06 22 44.5 −00 20 45	18	K5V	7.8	2.91 ± 0.08	2.8–25
GS1124-68	Nova Mus 91	11 26 26.7 −68 40 33	20	K5V	10.4	3.01 ± 0.15	4.5–6.2
4U1543-47	IL Lup	15 47 08.6 −47 40 09	17	A0V	27.0	0.22 ± 0.02	2.7–7.5
GROJ1655-40	Nova Sco 94	16 54 00.2 −39 50 45	21	F4IV	62.9	3.24 ± 0.09	6.5–7.8
H1705-25	Nova Oph 77	17 08 14.2 −25 05 32	21	K3V	12.5	4.86 ± 0.13	4.7–8.0
GS2000+25	Nova Vul 88	20 02 49.6 25 14 11	21	K5V	8.3	4.97 ± 0.10	5.8–18
GS2023+33	V404 Cyg	20 24 03.8 33 52 04	19	G9V	155.3	6.08 ± 0.06	10.3–14

References
1. McClintock, J.E. 1998, in *Accretion Processes in Astrophysical Systems*, AIP Conf. Proc., p. 431.
2. Van Paradijs, J., & McClintock, J.E. 1995, in *X-ray Binaries*, edited by W.H.G. Lewin, J. van Paradijs, and E.P.J. van den Heuval (Cambridge University Press, Cambridge), p. 58.
3. Orosz, J.A., & Bailyn, C.D. 1997, *ApJ*, **477**, 876.
4. Tanaka, Y., & Shibazaki, N. 1996, *ARA&A*, **34**, 607.
5. Bailyn, C.D., Jain, R.K., Coppi, P., & Orosz, J.A. 1998, *ApJ*, **499**, 367.

Table 16.22. *Galactic black hole candidates with high mass companions* [1, 2].

Object	Opt. ID	α (2000.0) δ (2000.0)	m_v	Sp. Type	P (hours)	$f(\mathcal{M}_x/\mathcal{M}_\odot)$	$\mathcal{M}_x/\mathcal{M}_\odot$
LMC X-1	star R148	05 39 38.7 −69 44 36	14	OB	4.2	0.14 ± 0.05	
LMC X-3	star WP	05 38 56.4 −64 05 01	17	B3V	1.7	2.3 ± 0.3	> 7
Cyg X-1	HDE 226868	19 58 21.7 35 12 06	9	O9.7Iab	5.6	0.24 ± 0.01	> 7

References
1. Van Paradijs, J., & McClintock, J.E. 1995, in *X-ray Binaries*, edited by W.H.G. Lewin, J. van Paradijs and E.P.J. van den Heuval, (Cambridge University Press, Cambridge), p. 58.
2. Tanaka, Y., & Shibazaki, N. 1996, *ARA&A*, **34**, 607.

16.18   DOUBLE STARS

Indications are that some 40%–60% of all stars are members of double or multiple systems [79], with some estimates running as high as 85% [80]. As far as selection effects allow, there seems no significant dependence on stellar type. Such effects preclude any reliable determination of the percentage of duplicity as a function of semimajor axis size.

The eccentricity of binary orbits and the orbital period are given below:

$\log P$ (P in days)	0	1	2	3	4	5	6	7
Mean eccentricity	0.03	0.17	0.31	0.42	0.47	0.45	0.64	0.8

For further statistics of binary stars, see [79–82].

16.18.1   Visual binaries

Dawes's rule is the limit of resolution $= 11.6/D$ arcsec (D is objective diameter in centimeters).

The limit of largest refractors under best conditions ~ 0.1 arcsec.

The angular separation beyond which it is unlikely that pairs are physical binaries

$$\log \rho = 2.8 - 0.2V,$$

where ρ is in arcsec and V is the combined magnitude. Reference [83] suggests that a pair is likely not physical if the projected linear separation exceeds 0.01 pc.

A catalogue of orbital elements for ~ 850 visual binaries is given in [84]. See Table 16.23.

Table 16.23. *Selected visual binaries.*

Name	Component	a'' T	ω (deg) e	i (deg) Ω (deg)	P (yr) Equinox	π''	Sp.	M_{bol}	$\mathcal{M}/\mathcal{M}_\odot$
Sirius	A	50.1	7.50	147.3	136.5	0.379	A1 V	0.8	2.28
	B	1894.1	0.59	44.6	1950		DA	11.2	0.98
Procyon	A	40.4	4.50	268.8	31.9	0.290	F5 IV–V	2.6	1.69
	B	1967.9	0.36	284.8	2000		WD	12.6	0.60
α Cen	A	79.9	17.52	231.6	79.2	0.760	G2 V	4.4	1.08
	B	1955.6	0.52	204.9	2000		K0 V	5.6	0.88
70 Oph	A	88.1	4.54	13.2	121.2	0.199	K0 V	5.6	0.90
	B	1984.0	0.50	301.7	2000		K4 V	6.8	0.65
Krüger 60	A	44.6	2.41	217.8	164.5	0.253	dM4	9.6	0.27
	B	1925.6	0.41	161.1	2000		dM6	10.6	0.16

16.18.2   Spectroscopic Binaries

The formula relating the observed radial velocity maximum and period to the eccentricity, semimajor axis, and orbital inclination is

$$a_1 \sin i = 0.013\,75 K_1 P (1 - e^2)^{0.5}$$

for a_1 in units of 10^6 km, K_1 in km s^{-1}, and P in days, similarly for a_2 and K_2.

If only one velocity curve is available, the mass function is

$$f(\mathcal{M}) = (M_2 \sin i)^3/(M_1 + M_2)^2 = 1.036 \times 10^{-7} K_1^3 P(1 - e^2)^{1.5}.$$

If both velocity curves are available, then

$$\mathcal{M}_1 \sin^3 i = 1.036 \times 10^{-7}(K_1 + K_2)^2 K_2 P(1 - e^2)^{1.5},$$

and similarly for M_2 and K_1. The mass is expressed in solar masses.

The catalogue of orbital elements for $\sim$ 1470 spectroscopic binaries is given in [85]. See Table 16.24.

Table 16.24. *Selected spectroscopic binaries.*

Name	P (days)	T (2400000+)	ω (deg)	e	K (km s^{-1})	V_γ (km s^{-1})	$f(\mathcal{M})$ ($\mathcal{M}_\odot$)	$\mathcal{M} \sin^3 i$ ($\mathcal{M}_\odot$)	$a \sin i$ (10^6 km)	Sp.
ζ Phe[a]	1.6698	41 643.689	—	0.0	131.5	17.2	—	3.9	3.02	B6 V
					202.6	11.6	—	2.5	4.65	B8 V
β Aur	3.9600	31 075.759		.0	111.5	—	—	2.1	6.07	A2 IV
					107.5	−17.1	—	2.2	5.85	A2 IV
α Vir	4.0145	40 284.78	142	0.18	120	0	—	7.2	6.52	B1 V
					189	−2	—	4.5	10.3	B3 V
β Lyr	12.9349	42 260.922	—	0.0	184	−17.8	8.4	—	32.7	B8pe
31 Cyg	3784.3	37 169.73	201.1	0.22	14.0	−7.7	—	9.2	711	K4 Ib
					20.8	−12.3	—	6.2	1060	B4 V

16.18.3 Eclipsing binaries

Classification schemes are as follows:

1. By ellipticity:

EA	Algol type	near spherical.	
EB	β Lyr type	$P > 1$ day	ellipsoidal, unequal brightness.
EW	W UMa type	$P < 1$ day	ellipsoidal, equal brightness.

2. By stability within critical equipotential surfaces (Roche lobes). Mass loss occurs when Roche lobes are filled:

D	Detached	Both components are well within Roche lobes.
SD	Semidetached	One component reaches Roche lobes.
C	Contact	Both components reach Roche lobes.
OC	Overcontact	Both components overfill Roche lobes.

The inter-relations

$$EA \rightleftharpoons D, SD, \qquad EB \rightleftharpoons SD(D), \qquad EW \rightleftharpoons C.$$

No comprehensive catalogue of reliable elements for eclipsing binaries is presently available, but see [86] for a selection of 323 eclipsing systems. See Table 16.25.

Table 16.25. *Selected eclipsing binaries (from [1]).*

Name	P (days)	a^a ($R_\odot$)	e	i (deg)	$\langle r \rangle^b$ (unit $= a$)	T (K)	q^c	F^d	Sp.
ER Ori	0.423	2.12	0.00	80.9	0.31	5 800	1.99	1.00	F8 V
					0.43	5 650		1.00	
RY Aqr	1.967	7.61	0.00	82.1	0.17	7 605	0.20	1.00	A3
					0.27	4 520		1.00	
RW Mon	1.906	9.97	0.00	88.0	0.20	10 650	0.37	4.99	B9 V
					0.32	5 055		1.00	
V889 Aql	11.121	34.3	0.37	88.4	0.054	10 200	1.0	2.34	B9
					0.053	10 500		2.34	
AI Phe	24.592	47.75	0.19	88.5	0.037	6 310	1.03	1.49	F7 V
					0.061	5 160		1.49	K0 IV

Notes

[a] a ($= a_1 + a_2$) is the semimajor axis of the relative orbit.

[b] $\langle r \rangle$ is the approximate mean stellar radius.

[c] q is the mass ratio of secondary to primary.

[d] F is the ratio of spin angular speed to mean orbital angular speed.

Reference

1. Terrell, D., Mukherjee, J.D., & Wilson, R.E. 1992, *Binary Stars: A Pictorial Atlas* (Krieger, Florida)

REFERENCES

1. Kholopov, P.N. et al. 1985–1990, *General Catalogue of Variable Stars*, 4th ed. (Moscow State University)
2. Feast, M.W., & Walker, A.R. 1987, *ARA&A*, **25**, 345
3. Fernie, J.D. 1992, *AJ*, **103**, 1647
4. Chiosi, C., Wood, P., Bertelli, G., & Bressan, A. 1992, *ApJ*, **387**, 320
5. Breger, M. 1979, *PASP*, **91**, 5
6. Demers, S., & Harris, W.E. 1974, *AJ*, **79**, 719
7. Harris, H.C. 1981, *AJ*, **86**, 719
8. Rosino, L. 1978, *Vistas Ast.*, **22**, 39
9. Carney, B.W., Storm, J., & Jones, R.V. 1990, *ApJ*, **386**, 645
10. Sandage, A., & Cacciari, C. 1990, *ApJ*, **350**, 645
11. Sterken, C., & Jerzykiewicz, M. 1980, Lecture Notes in Physics, *Nonradial and Nonlinear Stellar Pulsation*, edited by H.A. Hill and W.A. Dziembowski (Springer-Verlag, New York), p. 105
12. Shobbrook, R.R. 1983, *MNRAS*, **205**, 1215
13. Moskalik, P., & Dziembowski, W.A. 1992, *A&A*, **256**, L5
14. Allen, C.W. 1973, *Astrophysical Quantities*, 3rd ed. (Athlone Press, London), p. 218
15. Feast, M.W., Glass, I.S., Whitelock, P.A., & Catchpole, R.M. 1989, *MNRAS*, **241**, 375
16. Jura, M., & Kleinmann, S.G. 1992, *ApJS*, **79**, 105
17. Wood, P.R. 1990, in Astronomical Society of the Pacific Conference Series, *Confrontation Between Stellar Pulsation and Evolution*, edited by C. Cacciari and G. Clementini (ASP, San Francisco), Vol. 11, p. 355
18. Whitelock, P.A. 1990 in Astronomical Society of the Pacific Conference Series, *Confrontation Between Stellar Pulsation and Evolution*, edited by C. Cacciari and G. Clementini (ASP, San Francisco), Vol. 11, p. 365
19. Lloyd Evans, T. 1989, in *Evolution of Peculiar Red Giant Stars*, edited by H.R. Johnson and B. Zuckerman (Cambridge University Press, Cambridge), IAU Colloq. 106, p. 241
20. Kholopov, P.N. et al. 1985–1990, *General Catalogue of Variable Stars*, 4th ed. (Moscow State University) Vol. 1, p. 17
21. Allen, C.W. 1973, *Astrophysical Quantities*, 3rd ed. (Athlone Press, London), p. 219
22. Jura, M. 1986, *ApJ*, **309**, 732
23. Vogt, S.S. 1983, *IAU Colloq.* No. 71, p. 137
24. Hall, D.S. 1991, *IAU Colloq.* No. 130, p. 353
25. Morris, S.L. 1985, *ApJ*, **295**, 143
26. Hall, D.S. 1990, *AJ*, **100**, 554
27. Burke, E.W. et al. 1980, *AJ*, **85**, 744
28. Kholopov, P.N. et al. 1985–1990, *General Catalogue of Variable Stars*, 4th ed., Part I, Introduction (Moscow State University)
29. Bopp, B.W., & Fekel, F.C. 1977, *AJ*, **82**, 490
30. Hall, D.S. 1976, *IAU Colloq.* No. 29, p. 287
31. Bopp, B.W. 1983, *IAU Colloq.* No. 71, p. 343
32. Fekel, F.C., Moffett, T.J., & Henry, G.W. 1986, *ApJS*, **60**, 551
33. Rydgren, A.E., & Vrba, F.J. 1983, *ApJ*, **267**, 191
34. Appenzeller, I., & Mundt, R. 1989, *Rev. Astron. Astrophys.* **1**, 291
35. Bertout, C. 1989, *ARA&A*, **27**, 351
36. Herbig, G.H., & Bell, K.R. 1988, *Lick Obs. Bull.* **1111**, University of California
37. *Protostars and Planets* III, 1993, edited by D.H. Levy and J.I. Lunine (University of Arizona Press, Tucson)
38. Haisch, B.M. 1989, *Sol. Phys.* **121**, 3
39. Pettersen, B.R. 1989, *Sol. Phys.* **121**, 299
40. Shakhovskaya, N.I. 1989, *Sol. Phys.* **121**, 375
41. Holtzman, J.A., & Nations, H.L. 1984, *AJ*, **89**, 391

42. Henry, G.W., & Hall, D.S. 1991, *ApJ*, **373**, L9
43. Hamann, W.R., Koesterke, L., & Wessolowski, U. 1993, *A&A*, **274**, 379
44. Abbott, D.C., & Conti, P.S. 1987, *ARA&A*, **25**, 113
45. Humphreys, R.M. 1989, in *Physics of Luminous Blue Variables* (Kluwer Academic, Dordrecht), IAU Symp. No. 157, p. 3
46. Conti, P.S. 1984, in *Observational Tests of Stellar Evolution Theory* (Reidel, Dordrecht), IAU Symp. No. 105, p. 233
47. Slettebak, A. 1982, *ApJS*, **50**, 55
48. Keenan, P.C. 1993, *PASP*, **105**, 905
49. Barnbaum, C., & Keenan, P.C. 1996, *ApJS*, **105**, 419
50. Iben, I, Jr., & Renzini, A. 1983, *ARA&A*, **21**, 271
51. Lloyd Evans, T. 1984, *MNRAS*, **208**, 447
52. McClure, R.D. 1989, *IAU Colloq.* No. 106, p. 196
53. McClure, R.D. 1984, *ApJ*, **280**, L35
54. Brown, J.A., Smith, V.V., Lambert, D.L., Dutchover Jr, E., Hinkle, K.H., & Johnson, H.R. 1990, *AJ*, **99**, 1930
55. Lambert, D.L, Gustafsson, B., Eriksson, K., & Hinkle, K.H. 1983, *ApJS*, **62**, 373
56. Green, R.F., Margon, B.H., & MacConnell, D.J. 1991, *ApJ*, **380**, L31
57. McClure, R.D. 1989, in *Evolution of Peculiar Red Giant Stars*, edited by H.R. Johnson and B. Zuckerman (Cambridge University Press, Cambridge), p. 196
58. Lawson, W.A., Cottrell, P.L., Kilmartin, P.M., & Gilmore, A.C. 1990, *MNRAS*, **247**, 91
59. Drilling, J.S., Schönberner, D., Heber, U., & Lynas-Gray, A.E. 1984, *ApJ*, **278**, 224
60. Pollacco, D.L., & Hill, P.W., 1991, *MNRAS*, **248**, 572
61. Lambert, D.L., & Rao, N.K., 1994, *JApA*, **15**, 47
62. Lawson, W.A., & Cottrell, P.L. 1997, *MNRAS*, **285**, 266
63. Cottrell, P.L. 1996, in Astronomical Society of the Pacific Conference Series, *Hydrogen Deficient Stars*, edited by C.S. Jeffrey and U. Huber, PASP Conference Series, No. 96, p. 13
64. Clayton, G.C. 1996, *PASP*, **108**, 225
65. Walker, H.J. 1986, in *Hydrogen Deficient Stars and Related Objects*, edited by K. Hunger, D. Schönberner,

and N.K. Rao (Reidel, Dordrecht), IAU Colloq. No. 87, p. 407
66. Feast, M.W., Carter, B.S., Roberts, G., Marang, F., & Catchpole, R.M. 1997, *MNRAS*, **285**, 317
67. *Upper Main Sequence Stars with Anomalous Abundances*, 1985, edited by C.R. Cowley, M.M. Dworetsky, and C. Mégessier (Reidel, Dordrecht), IAU Colloq. No. 90
68. Wolff, S.C. 1983, *The A-Stars: Problems and Perspectives*, NASA SP-463
69. Kurtz, D.W. 1990, *ARA&A*, **28**, 607
70. Preston, G.W. 1974, *ARA&A*, **12**, 257
71. McClintock, J.E. 1998, in *Accretion Processes in Astrophysical Systems*, AIP Conf. Proc., 431
72. Van Paradijs, J., & McClintock, J.E. 1995, in *X-ray Binaries*, edited by W.H.G. Lewin, J. van Paradijs, and E.P.J. van den Heuval (Cambridge University Press, Cambridge), p. 58
73. Orosz, J.A., & Bailyn, C.D. 1997, *ApJ*, **477**, 876
74. Barret, D., McClintock, J.E., & Grindlay, J.E. 1996, *ApJ*, **473**, 963
75. White, N.E., & Marshall, F.E. 1984, *ApJ*, **281**, 354
76. Narayan, R., Garcia, M.R., & McClintock, J.E. 1997, *ApJ*, **478**, 79L
77. Tanaka, Y., & Shibazaki, N. 1996, *ARA&A*, **34**, 607
78. Bailyn, C.D., Jain, R.K., Coppi, P., & Orosz, J.A. 1998, *ApJ*, **499**, 367
79. Herczeg, T. 1982, *Landolt-Börnstein Tables* (Springer-Verlag, Berlin), **2**, p. 381
80. Heintz, W.D. 1969, *JRASC*, **63**, 275
81. Hogeveen, S.J. 1992, *Ap&SS*, **193**, 29
82. Allen, C.W. 1973, *Astrophysical Quantities*, 3rd ed. (Athlone Press, London), p. 227
83. Sinachopoulos, D. *A&AS*, **87**, 453
84. Worley, C.E., & Heintz, W.D. 1983, *Pub. U.S. Naval Obs.* **24**, Part 7
85. Batten, A.H., Fletcher, J.M., & MacCarthy, D.G. 1989, *Pub. Dom. Ap. Obs.* **17**, 1
86. Terrell, D., Mukherjee, J.D., & Wilson, R.E. 1992, *Binary Stars: A Pictorial Atlas* (Krieger, Florida)

Chapter 17

Cataclysmic and Symbiotic Variables

W.M. Sparks, S.G. Starrfield,
E.M. Sion, S.N. Shore,
G. Chanmugam[†], and R.F. Webbink

17.1   TYPES OF CATACLYSMIC VARIABLES

A cataclysmic variable (CV) [1, 2] is a binary star system in which a white dwarf primary accretes hydrogen-rich material usually through an accretion disk from a Roche lobe filling secondary that is on or near the main sequence. The CVs consist of several classes such as classical novae, recurrent novae, nova-likes, dwarf novae, helium CVs, and magnetic CVs. The distributions of their orbital periods are shown in Figure 17.1. Catalogues of CVs are found in [3, 4]. Proceedings of CV conferences [5–9] are bountiful sources of information.

A classical nova [10–12] is a CV that has undergone an outburst (9–15 mag. increase) which ejects a shell of gas at high velocity. Tables 17.1 and 17.2 contain the brightest and best-observed classical novae in our Galaxy. More extensive lists are found in [13] and [3]. Table 17.3 lists the brightest novae from 1991 to 1995. Well-observed novae in the Large Magellanic Cloud are given in Table 17.4. Classical novae are commonly assumed to be caused by a thermonuclear runaway in the accreted material on the white dwarf. The classical novae are also designated as CNO and ONeMg novae according to the composition of the ejecta (see Table 17.5). It is inferred that these novae occur on CO and ONeMg white dwarfs, respectively, and their ejecta include white dwarf material. As their name implies, recurrent novae have been observed to undergo more than one outburst. Although there are currently only nine members listed in this class (see Tables 17.6 and 17.7), it may be necessary to subdivide them according to their type of outburst or their type of secondary when they are better understood. In some systems the outbursts are probably caused by thermonuclear runaways, but in

---

[†]Deceased.

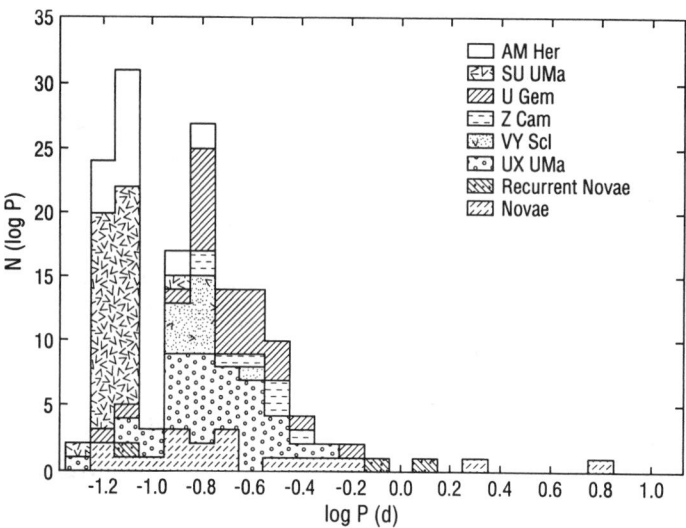

Figure 17.1. The orbital period distributions of the cataclysmic variables.

other systems the outburst may result from an episodic mass transfer accompanied by the release of gravitational energy onto the primary which could be a white dwarf or a main-sequence star [14, 15]. In addition, for some recurrent novae, the secondary is a late-type giant.

The dwarf novae (Table 17.8) [16] undergo a periodic brightening (2–5 mag.) on a time scale of weeks to years with little or no mass ejection aside from the wind outflow during the outburst in most of the systems. Most dwarf novae change from having an emission line spectrum to having an absorption line spectrum during outburst. This phenomenon is normally assumed to be caused by an instability in the accretion disk surrounding the white dwarf. The SU UMa stars are a subclass of dwarf novae that also show semiperiodic outbursts of unusually large amplitude (superoutburst), distinguished by the appearance at outburst maximum of periodic modulations (superhumps) in the light curve with periods a few percent larger than the orbital period. Dwarf novae that show occasional standstills (episodes of intermediate brightness lasting days to years) during decline from maximum are termed Z Cam stars. The remainder of the dwarf novae are called U Gem systems after the original prototype.

The nova-likes [16] are CVs that have the appearance of quiescent classical novae, i.e., they are probably classical novae that have not had a recorded outburst. Table 17.9 contains the best observed nova-likes. Additional listings are found in Ritter [3]. There are two subclasses of nova-likes: UX UMa and VY Scl. The UX UMa systems look like dwarf novae in a permanent outburst state while the VY Scl systems (or antidwarf novae) are normally in a high state but have slow, short excursions to a low state. These variations of luminosity are probably due to changes in the accretion rate. The helium CVs (or AM CVn systems) are transferring helium-rich material instead of hydrogen-rich material. Otherwise they appear to be nova-likes.

The white dwarf in a magnetic CV has a sufficiently strong magnetic field to channel the flow of accreting material at least near the white dwarf's surface [17]. The magnetic CVs may be divided into two subclasses depending on whether the white dwarf is rotating synchronously (Table 17.10) with its binary companion, as in the AM Her binaries or polars, or asynchronously (Table 17.11) as in the DQ Her binaries or intermediate polars. In the AM Her binaries, the magnetic field is sufficiently strong so that the accretion flows via an accretion column and no accretion disk is formed. In the DQ Her binaries, the magnetic field is probably weaker and an accretion disk may form but is disrupted close to the white dwarf's surface.

Being a member of one class of CVs does not prevent a system from being a member of another. For example, GK Per, an old classical nova, also shows dwarf nova outbursts. Nova V1500 Cyg is also an AM Her system. The space density of CVs, ρ_{cv}, is a subject of much controversy. Assuming that the novae, dwarf novae, and nova-likes found in a galactic plane survey [18] represent all the CVs, their space density, ρ_{cv}, is $(5.3–8.2) \times 10^{-7}$ pc^{-3}. However, if novae fade considerably between outbursts, then a higher space density like that of $\rho_{cv} \geq 3 \times 10^{-5}$ pc^{-3} found in a deep but narrow survey [19] may be more realistic.

Many of the following tables make use of the SIMBAD database, operated at CDS (Centre de Donnees Stellaires), Strasbourg, France. Uncertain numbers are followed by a colon.

Table 17.1. *Selected list of classical novae.*

Name (alternate name)	α^a (2000) hr min sec	δ^a (2000) deg min sec	ℓ^b (deg)	b^b (deg)	m^c_{max} m^c_{min}	t^d_3 (days)	Light curve Refs.	Secondary spectral type^e
GK Per (N Per 1901)	03 31 11.82	43 54 16.8	150.55	−10.60	0.2v 13.0v	13	[1]	K2 IV–V [2]
T Aur (N Aur 1891)	05 31 59.06	30 26 45.2	176.79	−2.30	4.1B 14.9B	100	[3]	
RR Pic (N Pic 1925)	06 35 36.05	−62 38 23.4	272.30	−25.71	1.2v 12.3v	150	[4, 5]	
CP Pup (N Pup 1942)	08 11 45.96	−35 21 05.7	252.59	−1.08	0.2v 15.0v	8	[4, 6]	> M6 [7]
GQ Mus (N Mus 1983)	11 52 02.35	−67 12 20.2	296.92	−4.78	7.2v 17.5v	45	[8]	
DQ Her (N Her 1934)	18 07 30.17	45 51 31.9	73.09	26.68	1.3v 14.7v	94	[4, 9]	M3 V [10]
FH Ser (N Ser 1970)	18 30 46.92	02 36 51.5	32.59	6.33	4.4v 16.1v	62	[11]	
V693 CrA (N CrA 1981)	18 41 57.63	−37 31 13.1	357.51	−13.79	6.5v > 19v	12	[12]	
V603 Aql (N Aql 1918)	18 48 54.50	00 35 02.9	32.82	1.37	−1.4v 11.6v	8	[4, 13]	
V1370 Aql (N Aql 1982)	19 23 21.10	02 29 26.1	38.43	−5.43	7.5p 20.0p	13:	[14, 15]	
PW Vul (N Vul 1984 No. 1)	19 26 05.03	27 21 58.3	60.80	5.55	6.4v 17.0v	97	[16]	
HR Del (N Del 1967)	20 42 20.18	19 09 40.3	62.96	−13.64	3.3v 12.1v	230	[17]	K8 [18]
V1500 Cyg (N Cyg 1975)	21 11 36.61	48 09 01.9	89.48	−0.00	2.0B 16.3B	3.6	[19, 20]	
V1668 Cyg (N Cyg 1978)	21 42 35.22	44 01 54.9	90.42	−6.70	6.0v 20:	23	[21, 22]	
OS And (N And 1986)	23 12 05.76	47 28 19.7	105.69	−11.97	6.2v 17.8v	22	[23]	

Notes

^{a} Adapted from Duerbeck [24] and precessed to equinox 2000.

^{b} Galactic coordinates.

^{c} Maximum and minimum magnitudes from Warner [25]. and the light curve references B, v, and p are the blue, visual, and photographic magnitudes.

^{d} The time for the visual light curve to fall three magnitudes after maximum, t_3, was taken from Duerbeck [24].

^{e} The secondary spectral types are from spectroscopic or infrared photometric observations and do not include estimates from mass determinations.

References
1. Sabbadin, F., & Bianchini, A. 1983, *A&AS*, **54**, 393
2. Gallagher, J.S., & Oinas, V. 1974, *PASP*, **86**, 952
3. Leavitt, H.S. 1920, *Harvard Annuals*, **84**, 121
4. Payne-Gaposchkin, C. 1957, *The Galactic Novae* (North-Holland, Amsterdam)
5. Spencer Jones, H. 1931, *Cape Obs. Ann.*, **10**, part 9
6. Pettit, E. 1954, *PASP*, **66**, 142
7. Szkody, P., & Feinswog, L. 1988, *ApJ*, **334**, 422
8. Krautter, J. et al. 1984, *A&A*, **137**, 307
9. Beer, A. 1935, *MNRAS*, **95**, 538
10. Young, P., & Schneider, D.P. 1981, *ApJ*, **247**, 960
11. Rosino, L., Ciatti, F., & Della Valle, M. 1986, *A&A*, **158**, 34
12. Shylaya, B.S. 1984, *A&SS*, **104**, 163
13. Campbell, L. 1919, *Harvard Annuals*, **81**, 113
14. Rosino, L., Iijima, T., & Ortolani, S. 1983, *MNRAS*, **205**, 1069
15. Snijders, M.A.J., Batt, T.J., Roche, P.F., Seaton, M.J., Morton, D.C., Spoelstra, T.A.T., & Blades, J.C. 1987, *MNRAS*, **228**, 329
16. Noskova, R.I., Zaitseva, G.V., & Kolotilov, E.A. 1985, *Pis'ma AZh*, **11**, 613; *Sov. Astron. Lett.*, **11**, 257
17. Drechsel, H., Rahe, J., Duerbeck, H.W., Kohoutek, L., & Seitter, W.C. 1977, *A&AS*, **30**, 323
18. Bruch, A. 1982, *PASP*, **94**, 916
19. Young, P.J., Corwin, H.G., Bryan, J., & de Vaucouleurs, G. 1976, *ApJ*, **209**, 882
20. Tempesti, P. 1979, *ANac.*, **300**, 51
21. Duerbeck, H.W., Rindermann, K., & Seitter, W.C. 1980, *A&A*, **81**, 157
22. Mattei, J.A. 1980, *JRAS Can*, **74**, 185
23. Kikuchi, S., Kondo, M., & Mikami, Y. 1988, *PASJ*, **40**, 491
24. Duerbeck, H.W. 1987, *Sp. Sci. Rev.*, **45**, 1; *A Reference Catalogue and Atlas of Galactic Novae* (Reidel, Dordrecht)
25. Warner, B. 1987, *MNRAS*, **227**, 23

Table 17.2. *Data for selected classical novae.*

Name	$E(B-V)^a$	Nebular shell studies^b	Distance^c (pc)	Max. abs. mag.	Period^d (days)	Rapid optical oscillation period^e (s)	Expansion velocity (km/s)	Outburst spectra.^{f}	Quiescent spectra.^{f}	Description^g
GK Per	0.3 [1]	[2]	490n	−9.2n	1.996 803 [3]	∼ 350 QPO [4]	1200 [5]	[6, 7]	[8]	VF
T Aur	0.6 [9]	[9, 10]	1300n	−8.4n	0.204 378 29 [11]	[12]	655 [13]	[14]	[15]	MF
RR Pic	0.07 [16]	[17]	460n	−7.3n	0.145 025 5 [18]	20–40 QPO [19]	475 [13]	[20, 21]	[22]	S
CP Pup	0.08 [23]	[23]	835n	−9.7n	0.061 43 [24] 0.068 34 [25]		710 [13]	[26, 27]	[24]	VF
GQ Mus	0.45 [28]		4280	−7.4	0.059 4 [29]		800 [28]	[28, 30]		MF CNO no dust
DQ Her	0.11 [31]	[10, 32]	560n	−7.8n	0.193 620 6 [33, 34, 35]	71.074 514 [36, 37]	384 [38]	[39, 40]	[41]	MF CNO dust
FH Ser	0.4 [42]		850n	−6.5n			560 [13]	[43]	[43]	MF C dust
V693 CrA	0.56 [44]		5030	−8.8			2200 [44]	[44, 45]		VF ONeMg
V603 Aql	0.07 [16]	[10, 46]	370n	−9.5n	0.138 15 [47] 0.144 854 [48, 49]		1700 [5]	[50]	[51]	VF
V1370 Aql	0.6 [52]						2800 [53]	[52, 53]		VF ONeMg C, SiC, SiO$_2$ dust
PW Vul	0.45 [54]		2050	−6.6	0.2137 [55]		285 [56]	[57]		MF Solar C dust
HR Del	0.29 [58]	[59, 60]	850n	−7.3n	0.214 165 [61] 0.1775 [62]	[12]	520 [13]	[58, 63]	[64, 65]	VS
V1500 Cyg	0.5 [66]	[67]	1080n	−9.8n	0.139 613 [68, 69]		1180 [13]	[70, 71]	[68]	VF CNO

Table 17.2. *(Continued.)*

Name	E($B-V$)[a]	Nebular shell studies[b]	Distance[c] (pc)	Max. abs. mag.	Period[d] (days)	Rapid optical oscillation period[e] (s)	Expansion velocity (km/s)	Outburst spectra.[f]	Quiescent spectra.[f]	Description[g]
V1668 Cyg	0.4 [72]		3660	−8.1	0.1384 [73]		760 [72]	[74, 75]		F CNO C dust
OS And	0.25 [76]		7200	−8.2			1000 [77]	[78]		F CNO

Notes

[a] The color excess, E($B-V$), is assumed to be related to the visual interstellar extinction, A_v, by $A_v = 3.2\mathrm{E}(B-V)$.

[b] In addition to these nebular shell studies, a short spectroscopic description of the nova remnant is given by Duerbeck and Seitter [79].

[c] The distances and absolute maximum magnitudes that are followed by an "n" have been determined by the nebular expansion parallax method. The angular shell sizes are from Cohen and Rosenthal [13], except for V1500 Cyg [80], DQ Her [38], and FH Ser [42]. The other distances and absolute magnitudes are found from the maximum magnitude $-t_2$ relationship derived by Cohen [80] and assuming $t_2 \sim t_{3/2}$ [81]. The time for the visual light curve to fall after maximum by n magnitudes is denoted by t_n.

[d] The spectroscopic period is the first entry while the photometric period is the second if it is different. Orbital parameters can usually be found in the reference for the spectroscopic period.

[e] If only a reference is given in this column, it means an unsuccessful search.

[f] Only optical references are given. Surveys or catalogues of novae exist in the radio [82, 83], infrared [84–86], visual [87, 88], ultraviolet [89, 90], and X-ray [91–93] spectral regions. References of observations for individual novae can be found in Payne-Gaposchkin [94], Ritter [95–97], Duerbeck [98], and Bode et al. [99]. Finding charts can be found in these last two references and in Williams [87].

[g] The speed class is defined by Payne-Gaposchkin [94] as

Speed class	Rate of decline (mag./day)
Very fast (VF)	> 0.20
Fast (F)	0.18 to 0.08
Moderately fast (MF)	0.07 to 0.025
Slow (S)	0.024 to 0.013
Very slow (VS)	0.013 to 0.008

The type of nova is defined by the strong enhancement above solar values of the chemical composition of the ejecta (CNO or ONeMg) and often implies the composition of the white dwarf. The types of dust formed, if any, is quoted from Gehrz [56].

References

1. Wu, C.-C., Panek, R.J., Holm, A.V., Raymond, J.C., Hartmann, L.W., & Swank, J.H. 1989, *ApJ*, **339**, 443
2. Seaquist, E.R., Bode, M.F., Frail, D.A., Roberts, J.A., Evans, A., & Albinson, J.S. 1989, *ApJ*, **344**, 805
3. Crampton, D., Cowley, A.P., & Fisher, W.A. 1986, *ApJ*, **300**, 788
4. Patterson, J. 1991, *PASP*, **103**, 1149
5. McLaughlin, D.B. 1960, in *Stellar Atmospheres*, edited by J.L. Greenstein (University of Chicago Press, Chicago), p. 585
6. McLaughlin, D.B. 1949, *Mich. Publ.*, **9**, 13
7. Stratton, F.J.M. 1936, *Ann. Solar Phys. Obs. Cambridge*, **4**, part 2
8. Bianchini, A., Sabbadin, F., & Hamzaoglu, E. 1982, *A&A*, **106**, 176
9. Gallagher, J.S., Hege, E.K., Kopriva, D.A., Williams, R.E., & Butcher, H.R. 1980, *ApJ*, **237**, 55
10. Mustel, E.R., & Boyarchuk, A.A. 1970, *Ap&SS*, **6**, 183
11. Beuermann, K., & Pakull, M.W. 1984, *A&A*, **136**, 250
12. Robinson, E., & Nather, R.E. 1977, *PASP*, **89**, 572
13. Cohen, J.G., & Rosenthal, A.J. 1983, *ApJ*, **268**, 689
14. McLaughlin, D.B. 1941, *PASP*, **53**, 102
15. Bianchini, A. 1980, *MNRAS*, **192**, 127
16. Gallagher, J.S., & Holm, A.V. 1974, *ApJ*, **189**, L123
17. Williams, R.E., & Gallagher, J.S. 1979, *ApJ*, **228**, 482
18. Haefner, R., & Metz, K. 1982, *A&A*, **109**, 171
19. Schoembs, R., & Stolz, B. 1981, *Inf. Bull. Var. Stars*, No. 1986

20. Spencer Jones, H. 1931, *Cape Ann.*, **10**, part 9
21. Lunt, J. 1926, *MNRAS*, **86**, 498
22. Wyckoff, S., & Wehinger, P.A. 1977, *Veröff. Remeis-Sternw. Bamberg*, **11**, 201
23. Williams, R.E. 1982, *ApJ*, **261**, 170
24. O'Donoghue, D., Warner, B., Wargau, W., & Grauer, A.D. 1989, *MNRAS*, **240**, 41
25. White, J.C., Honeycutt, R.K., & Horne, K. 1993, *ApJ*, **412**, 278
26. Weaver, H.F. 1944, *ApJ*, **99**, 280
27. Sanford, R.F. 1945, *ApJ*, **102**, 357
28. Krautter, J. et al. 1984, *A&A*, **137**, 307
29. Diaz, M.P., & Steiner, J.E. 1989, *ApJ*, **339**, L41
30. de Freitas Pacheco, J.A., & Codina, S.J. 1985, *MNRAS*, **214**, 481
31. Ferland, G.J., Williams, R.E., Lambert, D.L., Shields, G.A., Slovak, M., Gondhalekar, D.M., & Truran, J.W. 1984, *ApJ*, **281**, 194
32. Williams, R.E., Woolf, N.J., Hege, E.K., Moore, R.L., & Kopriva, D.A. 1978, *ApJ*, **224**, 171
33. Hutchings, J.B., Cowley, A.P., & Crampton, D. 1979, *ApJ*, **232**, 500
34. Smak, J. 1980, *AcA*, **30**, 267
35. Horne, K., Welsh, W.F., & Wade, R.A. 1993, *ApJ*, **410**, 357
36. Walker, M.F. 1958, *ApJ*, **127**, 319
37. Balachandran, S., Robinson, E.L., & Kepler, S.O. 1983, *PASP*, **95**, 653
38. Herbig, G.H., & Smak, J.I. 1992, *AcA*, **42**, 17
39. Stratton, F.J.M., & Manning, W.H. 1939, *Atlas of Spectra of Nova Herculis* (Solar Physics Observatory, Cambridge)
40. McLaughlin, D.B. 1954, *ApJ*, **119**, 124
41. Schneider, D.P., & Greenstein, J.L. 1979, *ApJ*, **233**, 935
42. Duerbeck, H.W. 1992, *AcA*, **42**, 85
43. Rosino, L., Ciatti, F., & Della Valle, M. 1986, *A&A*, **158**, 34
44. Brosch, N. 1982, *A&A*, **107**, 300
45. Shylaya, B.S. 1984, *Ap&SS*, **104**, 163
46. Duerbeck, H.W. 1987, *Ap&SS*, **131**, 461
47. Drechsel, H., Rahe, J., & Wargau, W. 1982, *Mitt. Astron. Ges.*, **57**, 301
48. Haefner, R. 1981, *Inf. Bull. Var. Stars*, No. 2045
49. Bruch, A. 1991, *AcA*, **41**, 101
50. Wyse, A.B. 1940, *Lick Obs. Publ.*, **14**, part 3
51. Drechsel, H., Rahe, J., Holm, A., & Krautter, J. 1981, *A&A*, **99**, 166
52. Snijders, M.A.J., Batt, T.J., Roche, P.F., Seaton, M.J., Morton, D.C., Spoelstra, T.A.T., & Blades, J.C. 1987, *MNRAS*, **228**, 329
53. Rosino, L., Iijima, T., & Ortolani, S. 1983, *MNRAS*, **205**, 1069
54. Duerbeck, H.W., Geffert, M., Nellse, B., Dummler, R., & Nolte, M. 1984, *Inf. Bull. Var. Stars*, No. 2641
55. Hacke, G. 1987, *Inf. Bull. Var. Stars*, No. 2979
56. Gehrz, R.D. 1990, in *Physics of Classical Novae*, edited by A. Cassatella and R. Viotti (Springer-Verlag, Berlin), p. 138
57. Kenyon, S.J., & Wade, R.A. 1986, *PASP*, **98**, 935
58. Drechsel, H., Rahe, J., Duerbeck, H.W., Kohoutek, L., & Seitter, W.C. 1977, *A&AS*, **30**, 323
59. Kohoutek, L. 1981, *MNRAS*, **196**, 87P
60. Solf, J. 1983, *ApJ*, **273**, 647
61. Kürster, M., & Barwig, H. 1988, *A&A*, **199**, 201
62. Kohoutek, L., & Pauls, R. 1980, *A&A*, **92**, 200
63. Andrillat, Y., & Fehrenbach, Ch. 1981, *Ap&SS*, **76**, 149
64. Stephenson, G.B. 1967, *PASP*, **79**, 586
65. Hutchings, J.B. 1980, *PASP*, **92**, 458
66. Ferland, G.J. 1977, *ApJ*, **215**, 873
67. Becker, H.J., & Duerbeck, H.W. 1980, *PASP*, **92**, 792
68. Horne, K., & Schneider, D.P. 1989, *ApJ*, **343**, 888
69. Kaluzny, J., & Semeniuk, I. 1987, *AcA*, **37**, 349
70. Boyarchuk, A.A., Galkina, T.S., Gershberg, R.E., Krasnobabtsev, V.I., Rachkovskaya, T.M., & Shakhovskaya, N.I. 1977, *AZh*, **54**, 458; *Sov. Astron.*, **21**, 257
71. Ferland, G.J., Lambert, D.L., & Woodman, J.H. 1986, *ApJS*, **60**, 375
72. Stickland, D.J., Penn, C.J., Seaton, M.J., Snijders, M.A.J., & Storey, P.L. 1981, *MNRAS*, **197**, 107
73. Kaluzny, J. 1990, *MNRAS*, **245**, 547
74. Klare, G., Wolf, B., & Krautter, J. 1980, *A&A*, **89**, 282
75. Smith, S.E., Noah, P.V., & Cottrell, M.J. 1979, *PASP*, **91**, 775

76. Schwarz, G.J. et al. 1997, *MNRAS*, **284**, 669
77. Andrillat, Y. 1986, *IAU Circ.* No. 4289
78. Austin, S.J., Wagner, R.M., Starrfield, S., Shore, S.N., Sonneborn, G., & Bertram, R. 1996, *AJ*, **111**, 869
79. Duerbeck, H.W., & Seitter, W.C. 1987, *Ap&SS*, **131**, 467
80. Cohen, J.G. 1985, *ApJ*, **292**, 90
81. Warner, B. 1987, *MNRAS*, **227**, 23
82. Seaquist, E.R. 1989, in *Classical Novae*, edited by M.F. Bode and A. Evans (Wiley, New York), p. 143
83. Hjellming, R.M. 1990, in *Physics of Classical Novae*, edited by A. Cassatella and R. Viotti (Springer-Verlag, Berlin), p. 169
84. Bode, M.F., & Evans, A. 1989, in *Classical Novae*, edited by M.F. Bode and A. Evans (Wiley, New York), p. 163
85. Harrison, T.E., & Gehrz, R.D. 1988, *AJ*, **96**, 1001
86. Harrison, T.E., & Gehrz, R.D. 1991, *AJ*, **101**, 587
87. Williams, G. 1983, *ApJS*, **53**, 523
88. Bruch, A. 1984, *A&AS*, **56**, 441
89. Friedjung, M. 1989, in *Classical Novae*, edited by by M.F. Bode and A. Evans (Wiley, New York), p. 187
90. Starrfield, S., & Snijders, M.A.J. 1987, in *Scientific Accomplishments of the IUE Satellite*, edited by Y. Kondo (Reidel, Dordrecht), p. 377
91. Becker, R.H. 1989, in *Classical Novae*, edited by M.F. Bode and A. Evans (Wiley, New York), p. 215
92. Córdova, F.A., & Mason, K.O. 1983, in *Accretion-driven Stellar X-ray Sources*, edited by W.H.G. Lewin and E.P.J. van den Heuvel (Cambridge University Press, Cambridge), p. 147
93. Eracleous, M., Halpern, J., & Patterson, J. 1991, *ApJ*, **382**, 290
94. Payne-Gaposchkin, C. 1957, *The Galactic Novae* (North-Holland, Amsterdam)
95. Ritter, H. 1984, *A&AS*, **57**, 385
96. Ritter, H. 1987, *A&AS*, **70**, 335
97. Ritter, H. 1990, *A&AS*, **85**, 1179
98. Duerbeck, H.W. 1987, *A Reference Catalogue and Atlas of Galactic Novae* (Reidel, Dordrecht); *Space Sci. Rev.*, **45**, Nos. 1–2
99. Bode, M.F., Duerbeck, H.W., & Evans, A. 1989, in *Classical Novae*, edited by M.F. Bode and A. Evans (Wiley, New York), p. 249

Table 17.3. *Recent novae.*

Name (alternate name)	α (2000) hr min sec	δ (2000) deg min sec	Date of discovery	IAU Circ. No.	m_{max}	t_3[a] (days)	$E(B-V)$[b]	FWHM[c] emission vel. (km/s)	Desc.[d]
V351 Pup (N Pup 91)	8 11 38.38	−35 07 30.4	27 Dec 1991	5422	6.4v			3000	ONeMg no dust VF
V4160 Sgr (N Sgr 91)	18 14 13.83	−32 12 28.5	29 July 1991	5313	7v			8000	ONeMg
V838 Her (N Her 91)	18 46 31.48	+12 14 01.8	24 Mar 1991	5222	5.3v	2.8	0.6	6000	ONeMg Dust VF
V1974 Cyg (N Cyg 92)	20 30 31.66	+52 37 50.8	20 Feb 1992	5454	4.9B	43	0.35 ± 0.05	2000	ONeMg No dust MF
V705 Cas (N Cas 93)	23 41 47.25	+57 30 59.7	7 Dec 1993	5902	5.3v				CO Dust
V1425 Aql (N Aql 95)	19 05 26.64	−01 42 03.3	7 Feb 1995	6133	6.2:v	22:	≥ 0.56	1600	ONeMg Dust F

Notes

[a] The time for the visual light curve to fall three magnitudes after maximum.

[b] The color excess.

[c] The full width half maximum velocity of the emission lines in IUE spectra measured by S. Shore.

[d] The description is the same as in Table 17.2 for classical novae.

Table 17.4. *Recent novae in the Large Magellanic Cloud* [1, 2].

Nova	α (2000) hr min sec	δ (2000) deg min sec	IAU Circ. No.	Outburst	V_{max}	Type	Remarks
LMC V3479	5 35 29.33	−70 21 29.4	4569	21 Mar 1988	11.0	Dust, CNO	a
LMC V1161	5 08 01.10	−68 37 37.7	4663	12 Oct 1988	10.4	ONeMg	b
LMC V2361	5 23 21.82	−69 29 48.5	4946	16 Jan 1990	10.6	ONeMg	c
LMC V1341	5 09 58.40	−71 39 51.6	4964	15 Feb 1990	11.9	Recurrent	d
LMC V0850	5 03 44.99	−70 18 13.7	5244	18 Apr 1991	8.9	CNO?	e
LMC 1992	5 19 19.84	−68 54 35.1	5651	11 Nov 1992	10.2	CNO	f
LMC 1995	5 26 50.33	−70 01 23.8	6143	2 Mar 1995	11.3	CNO	

Notes

[a] UV versus optical analysis: Austin, S., Starrfield, S., Saizar, P., Shore, S.N., & Sonneborn, G. 1990, in *Evolution in Astrophysics: IUE in the Era of New Space Missions*, edited by E. Rolfs (ESA SP 310), p. 367. Possible dust-forming nova.

[b] UV description: IAU Circ. No. 4669. First extragalactic ONeMg nova.

[c] t_3(optical) = 5.8 days. Sonneborn, G., Shore, S.N., & Starrfield, S.G. 1990, in *Evolution in Astrophysics: IUE in the Era of New Space Missions*, edited by E. Rolfs (ESA SP 310), p. 439; see also, Starrfield, S., Shore, S.N., Sparks, W.M., Sonneborn, G., Truran, J.W., & Politano, M. 1992, *ApJ*, **391**, L71.

[d] Recurrence of Nova LMC 1968. Dynamics, abundances: Shore, S.N., Starrfield, S., Sonneborn, G., Williams, R.E., Haumy, M., Cassatella, A., & Drechsel, H. 1991, *ApJ*, **370**, 193. First spectroscopically confirmed, extragalactic recurrent nova; U Sco analog (low mass companion, helium rich).

[e] $F_{UV,max} = 1.64 \times 10^{-10}$ erg s^{-1} cm^{-2}; t_3(UV) = 140 days; delay: optical versus UV peak $\approx$ 10 days. This was the intrinsically brightest nova yet observed in the Local Group. Probable CNO nova.

[f] Star is a match to the Galactic nova OS And 1986.

References

1. General reference for LMC novae: van den Bergh, S. 1988, *PASP*, **100**, 1486.
2. General reference for M31 novae: Tomaney, A.B. & Shafter, A.W. 1992, *ApJS*, **81**, 683
3. General reference for extragalactic novae: Artiukhina, N.M. et al. 1995, *General Catalogue of Variable Stars*, Vol. V. *Extragalactic Variable Stars* (Kosmosinform, Moscow)

Table 17.5. *Element abundances in novae (mass fraction).*

Object	Year	X	Y	C	N	O	Ne	Z	Ref.
T Aur	1891	0.47	0.40		0.079	0.051		0.13	[1]
RR Pic	1925	0.53	0.43	0.0039	0.022	0.0058	0.011	0.043	[2]
DQ Her	1934	0.34	0.095	0.045	0.23	0.29		0.57	[3]
DQ Her	1934	0.27	0.16	0.058	0.29	0.22		0.57	[4]
HR Del	1967	0.45	0.48		0.027	0.047	0.0030	0.077	[5]
V1500 Cyg	1975	0.49	0.21	0.070	0.075	0.13	0.023	0.30	[6]
V1500 Cyg	1975	0.57	0.27	0.058	0.041	0.050	0.0099	0.16	[7]
V1668 Cyg	1978	0.45	0.23	0.047	0.14	0.13	0.0068	0.32	[8]
V693 CrA	1981	0.29	0.32	0.046	0.080	0.12	0.17	0.39	[9]
V693 CrA	1981	0.40	0.21	0.0040	0.069	0.067	0.23	0.39	[10]
V1370 Aql	1982	0.053	0.088	0.035	0.14	0.051	0.52	0.86	[11]
GQ Mus	1983	0.37	0.39	0.0080	0.125	0.095	0.0023	0.24	[12]
PW Vul	1984	0.69	0.25	0.0033	0.049	0.014	0.00066	0.067	[13]
PW Vul	1984	0.62	0.25	0.018	0.068	0.044	0.00014	0.13	[14]
QU Vul	1984	0.30	0.60	0.0013	0.018	0.039	0.040	0.10	[15]
QU Vul	1984	0.36	0.19		0.071	0.19	0.18	0.44	[16]
V842 Cen	1986	0.41	0.23	0.12	0.21	0.030	0.00090	0.36	[17]
V827 Her	1987	0.36	0.29	0.087	0.24	0.016	0.00066	0.35	[17]
QV Vul	1987	0.68	0.27		0.010	0.041	0.00099	0.053	[17]
V2214 Oph	1988	0.34	0.26		0.31	0.060	0.017	0.40	[17]
V977 Sco	1989	0.51	0.39		0.042	0.030	0.026	0.10	[17]
V433 Sct	1989	0.49	0.45		0.053	0.0070	0.00014	0.062	[17]
LMC 1990 No. 1	1990	0.53	0.21	0.014	0.069	0.10	0.049	0.26	[18]
V351 Pup	1991	0.37	0.25	0.0056	0.064	0.19	0.11	0.38	[19]

Table 17.5. *(Continued.)*

Object	Year	X	Y	C	N	O	Ne	Z	Ref.
V838 Her	1991	0.80	0.093	0.018	0.019	0.0032	0.068	0.11	[20]
V838 Her	1991	0.60	0.31	0.010	0.012	0.0021	0.056	0.09	[10]
V1974 Cyg	1992	0.30	0.52	0.015	0.023	0.10	0.037	0.18	[21]
V1974 Cyg	1992	0.19	0.32		0.085	0.29	0.11	0.49	[16]
Solar		0.705	0.275	0.003	0.001	0.010	0.002	0.020	[22]

References

1. Gallagher, J.S. et al. 1980, *ApJ*, **237**, 55
2. Williams, R.E., & Gallagher, J.S. 1979, *ApJ*, **228**, 482
3. Williams, R.E. et al. 1978, *ApJ*, **224**, 171
4. Petitjean, P., Boisson, C., & Pequignot, D. 1990, *A&A*, **240**, 433
5. Tylenda, R. 1978, *AcA*, **28**, 333
6. Ferland, G.J., & Shields, G.A. 1978, *ApJ*, **226**, 172
7. Lance, C.M., McCall, M.L., & Uomoto, A.K. 1988, *ApJS*, **66**, 151
8. Stickland, D.J. et al. 1981, *MNRAS*, **197**, 107
9. Williams, R.E., Ney, E.P., Sparks, W.M., Starrfield, S., Wyckoff, S., & Truran, J.W. 1985, *MNRAS*, **212**, 753
10. Vanlandingham, K., Starrfield, S., & Shore, S.N. 1997, *MNRAS*, **290**, 87
11. Snijders, M.A.J. et al. 1987, *MNRAS*, **228**, 329
12. Morisset, C., & Pèquignot, D. 1996, *A&A*, **312**, 135
13. Saizar, P., Starrfield, S., Ferland, G.J., Wagner, R.M., Truran, J.W., Kenyon, S.J., Sparks, W.M., Williams, R.E., & Stryker, L.L. 1991, *ApJ*, **367**, 310
14. Schwarz, G.J., Starrfield, S., Shore, S.N., & Hauschildt, P.H. 1997, *MNRAS*, **290**, 75
15. Saizar, P., Starrfield, S., Ferland, G.J., Wagner, R.M., Truran, J.W., Kenyon, S.J., Sparks, W.M., Williams, R.E., & Stryker, L.L. 1992, *ApJ*, **398**, 651
16. Austin, S.J., Wagner, R.M., Starrfield, S., Shore, S.N., Sonneborn, G., & Bertram, R. 1996, *AJ*, **111**, 869
17. Andreä, J., Drechsel, H., & Starrfield, S. 1994, *A&A*, **291**, 869
18. Vanlandingham, K., Starrfield, S., Shore, S.N., & Sonneborn, G. 1999, *MNRAS*, **308**, 577
19. Saizar, P., Pachoulakis, I., Shore, S.N., Starrfield, S., Williams, R.E., Rotschild, E., & Sonneborn, G. 1996, *MNRAS*, **279**, 280
20. Vanlandingham, K.M., Starrfield, S., Wagner, R.M., Shore, S.N., & Sonneborn, G., 1996, *MNRAS*, **282**, 563
21. Hayward, T.L., Saizar, P., Gehrz, R.D., Benjamin, R.A., Mason, C.G., Houck, J.R., Miles, J.W., Gull, G.E., & Schoenwald, J. 1996, *ApJ*, **469**, 854
22. Anders, E., & Grevesse, N. 1989, *Geochimica et Cosmochimica Acta*, **53**, 197

Table 17.6. *Recurrent novae.*[a]

Name	α^b (2000) hr	min	sec	δ^b (2000) deg	min	sec	l^c (deg)	b^c (deg)	Years of recorded outbursts
LMC 1990 #2	05	09	58.40	−71	39	51.6	283.04	−33.49	1968, 1990
T Pyx	09	04	41.47	−32	22	47.0	256.76	+9.51	1890, 1902, 1920, 1944, 1966
T CrB	15	59	30.09	+25	55	11.4	42.43	+48.66	1866, 1946
U Sco	16	22	30.68	−17	52	42.1	357.29	+22.47	1863, 1906, 1936, 1979, 1987
RS Oph	17	50	13.08	−06	42	28.4	19.48	+10.96	1898, 1933, 1958, 1967, 1985
V745 Sco	17	55	22.13	−33	14	58.3	357.02	−3.40	1937, 1989
V394 CrA	18	00	25.97	−39	00	35.1	352.50	−7.13	1949, 1987
V3890 Sgr	18	30	43.32	−24	01	08.6	8.85	−5.84	1962, 1990
V1017 Sgr	18	32	04.30	−29	23	12.8	4.15	−8.50	1901, 1919, 1973

Notes

[a] Three possible recurrent novae have been found in M31. Two are recorded by Rosino, L. 1973, *A&AS*, **9**, 347, and all three (M31 V0609, M31 V0665, and M31 V0979) by Artiukhina, N.M. et al. 1995, *General Catalogue of Variable Stars*, Vol. V. *Extragalactic Variable Stars* (Kosmosinform, Moscow).

[b] Adapted from Duerbeck, H.W., 1987, *Sp. Sci. Rev.*, **45**, 1, and precessed to equinox 2000.

[c] Galactic coordinates.

Table 17.7. *Recurrent novae data.*

Name	t_3^a (days)	V_{max}	V_{min}	A_v (mag.)	Distance (kpc)	Spectral type secondary	Period[b] (days)	Refs.
LMC 1990 #2	< 7	11.7	> 20	∼ 0.45	55	?		[1, 2]
T Pyx	88	7.0	15.2	∼ 1.0	> 1	?	∼ 0.1	[3–6]
T CrB	6.8	2.0	10.2	∼ 0.35	1	M4.1 ± 0.1 III	227.5	[3, 5,7–9]
U Sco	5	8.9	17.9	0.6	∼ 15:	G3	1.23	[3–5]
RS Oph	9.5	4.6	11.5	2.3	< 1.3	K5.7 ± 0.4 I–III	460	[3, 5,10–13]
V745 Sco	14.9	9.6	19.0	∼ 3	4.6	M4/5 III	?	[5, 14, 15]
V394 CrA	5.0	7.0	18.0	∼ 3	> 10:	K	0.7577	[4, 5,16]
V3890 Sgr	17	8.2	17.0	1.5	∼ 5	M5 III	?	[5, 14, 17]
V1017 Sgr	130	7.0	13.6	1.2	2	G5 III	5.7	[3, 5,18]

Notes

[a] The time for the visual light curve to fall three magnitudes after maximum.

[b] Orbital period.

References

1. Shore, S.N. et al. 1991, *ApJ*, **370**, 193
2. Sekiguchi, K. et al. 1990, *MNRAS*, **245**, 28P
3. Webbink, R.F. et al. 1987, *ApJ*, **314**, 653
4. Schaefer, B.E. 1990, *ApJ*, **355**, L39
5. Duerbeck, H.A. 1987, *A Reference Catalog and Atlas of Galactic Novae* (Reidel, Dordrecht)
6. Schaefer, B.E. et al. 1992, *ApJS*, **81**, 321
7. Kenyon, S.J., & Garcia, M. 1986, *AJ*, **91**, 125
8. Selvelli, P.L., Cassatella, A., & Gilmozzi, R. 1992, *ApJ*, **393**, 289
9. Shore, S.N., & Aufdenberg, J.P. 1993, *ApJ*, **416**, 355
10. Bode, M. 1987, *RS Oph (1985) and the Recurrent Nova Phenomenon* (VNU Science, Utrecht)
11. Garcia, M.R. 1986, *AJ*, **91**, 1400
12. Dobrzycka, D., & Kenyon, S.J. 1994, *AJ*, **108**, 2259
13. Shore, S. et al. 1996, *ApJ*, **456**, 717
14. Harrison, T.E. et al. 1993, *AJ*, **105**, 320
15. Sekiguchi, K. et al. 1990, *MNRAS*, **246**, 78
16. Sekiguchi, K. et al. 1989, *MNRAS*, **236**, 611
17. Gonzalez-Riestra, R. 1992. *A&A*, **265**, 71
18. Sekiguchi, K. 1992, *Nature*, **358**, 563

Table 17.8. *Dwarf novae.*

(1) Name[a,b] (alt. name)	(2) Coord.[c] (2000.0)	(3) DN	(4) P_{orb}	(5) V_{min}	(6) V_{max}	(7) t_{rec}	(8) Incl.	(9) M_{WD}	(10) XRS	(11) EC	(12) QP	(13) WD SP	(14) Wind	(15) Spect. type sec.
WW Cet	0 11 24.77 −11 28 42.7	Z	0.1765	15.0	9.3	31	54 4	0.85 0.11	1.62[d] 1.41[e]	N	N	N	Y	... ...
RX And	1 04 35.55 41 17 58.0	Z	0.209893	12.6	10.9	5–20	51 9	1.14 0.33	3.0[d]	N	Y	Y	Y	... ...
HT Cas	1 10 12.98 60 04 35.9	SU	0.073647	16.4	10.8	30–35 430	81	0.62 0.04	0.7[d] 1.64[e]	Y	Y	Y	N	... ...
FO And	1 15 32.14 37 37 35.5	SU	0.071	17.5	13.5				N	N	N	N	N	... ...
WX Cet (N Cet 1963)	1 17 04.17 −17 56 23.0		0.052:	17.5	9.5	450			N	N	N	N	N	... ...
TY Psc	1 25 39.35 32 23 09.7	SU	0.068:	15.3	12.2	11–35 370			N	N	N	N	N	... ...
AR And	1 45 03.27 37 56 33.3	U	0.19:	16.9	11.0	25			N	N	N	N	N	... ...
WX Hyi	2 09 50.65 −63 18 39.9	SU	0.074813	14.7	12.5	14 140	40 10	0.9 0.3	0.30[d] 0.82[e]	N	N	N	Y	... ...
UV Per	2 10 08.25 57 11 20.6	SU	0.0622:	17.5	11.7	360			N	N	N	N	N	... ...

Table 17.8. *(Continued.)*

(1)	(2)	(3)	(4)	(5)	(6)	(7)	(8)	(9)	(10)	(11)	(12)	(13)	(14)	(15)
Name[a,b] (alt. name)	Coord.[c] (2000.0)	DN	P_{orb}	V_{min}	V_{max}	t_{rec}	Incl.	M_{WD}	XRS	EC	QP	WD SP	Wind	Spect. type sec.
CP Eri	3 10 32.76 −09 45 05.3	U	0.01995	19.7	16.5				N	N	Y	N	N	··· ···
GK Per (N Per 1901)	3 31 11.82 43 54 16.8	DN	1.996803	10.2	0.2		< 73	0.90 0.20	$\sim 18^d$	N	Y	N	N	K0-4 ···
AF Cam	3 32 15.59 58 47 22.1	DN	0.23:	17.0	13.4	75			N	N	N	N	N	··· ···
VW Hyi	4 09 11.34 −71 17 41.1	SU	0.074271	13.4	9.5	27 179	60 10	0.63 0.15	0.20^d 0.71^e	N	Y	Y	N	··· ···
AH Eri	4 22 38.10 −13 21 30.2	U		18.4	13.5				$1.0F^f$	N	Y	N	N	··· ···
TU Men	4 41 40.71 −76 36 46.3	SU	0.1176	> 16	11.6	37 194	65 10	0.6	N	N	N	N	Y	··· ···
AQ Eri	5 06 13.04 −04 08 07.0	SU	0.06094	17.7	12.5	40:			N	N	N	N	N	··· ···
FS Aur	5 47 48.34 28 35 11.1	U	0.059:	16.2	14.4				N	N	N	N	N	··· ···
CN Ori	5 52 07.77 −05 25 00.7	U	0.163199	14.2	11.9	8–22	67 3	0.74 0.10	$< 0.32F^f$	N	Y	N	N	M4-5 ···
SS Aur	6 13 22.44 47 44 25.7	U	0.1828	14.5	10.5	40–75	38 16	1.08 0.40	N	N	N	N	Y	M1-5 ···
CW Mon	6 36 54.53 00 02 16.3	U	0.1762	16.3	11.9	122			N	Y	N	N	N	M3-5 ···
HL CMa (1E0643-1648)	6 45 17.21 −16 51 35.4	Z	0.2145	13.2	11.7	17	45:	1.0:	$\sim 1.20^d$ 1.50^e	N	N	N	Y	··· ···
IR Gem	6 47 34.58 28 06 22.7	SU	0.0684	16.3	11.7	22–48 150			N	N	N	N	N	··· ···
AW Gem	7 22 40.83 28 30 16.1	SU	0.0762	18.8	13.8	98 410			N	N	N	N	N	··· ···
BV Pup	7 49 05.26 −23 34 00.7	U	0.225:	15.6	13.1	19			0.55^e	N	N	N	N	··· ···
U Gem	7 55 05.29 22 00 05.7	U	0.176906	14.0	9.1	118	69.7 0.7	1.12 0.13	$\sim 0.18^d$ 0.40^e	Y	Y	Y	N	M4.5 ···
Z Cha	8 07 28.30 −76 32 01.3	SU	0.074499	15.3	12.4	82 287	81.8 0.1	0.84 0.09	$0.87F^f$ $1.97F^e$ 0.08^e	Y	Y	Y	N	M5.5 ···
YZ Cnc	8 10 56.62 28 08 33.6	SU	0.0868	14.1	11.9	6–16 134	38 3	0.82 0.05	0.24^d 0.95^e	N	Y	N	Y	··· ···
SU UMa	8 12 28.20 62 36 22.6	SU	0.07635	14.2	12.2	5–33 160			$\sim 3.1^d$ 1.13^e	N	N	N	N	··· ···
Z Cam	8 25 13.20 73 06 39.4	Z	0.289840	13.6	10.5	19–28	57 11	0.99 0.15	$2.1F^f$ 18.7^e	N	Y	N	Y	K7 ···
AT Cnc (Ton 323)	8 28 36.92 25 20 02.6	Z	0.238691	15.0B	12.7B	14			N	Y	N	N	N	··· ···
SW UMa	8 36 42.80 53 28 38.2	SU	0.056815	16.5	10.6	459	45 18	0.71 0.22	0.32^e	Y	Y	N	N	··· ···
EI UMa (PG0834+488)	8 38 21.98 48 38 01.7	U	0.26810	14.9B					1.01^e	N	N	N	N	··· ···
BZ UMa	8 53 44.14 57 48 41.1	DN	0.0679	17.8	10.5				N	N	N	N	N	M5.5 ···
CU Vel	8 58 32.87 −41 47 50.8	SU	0.0773	15.5	10.7	113 386			N	N	N	N	N	··· ···
SY Cnc	9 01 03.35 17 53 56.1	Z	0.380	13.5	11.1	22–35	26 6	0.89 0.28	$< 0.51F^f$	N	Y	N	Y	G8-9 ···
AR Cnc	9 22 07.48 31 03 14.6	U?	0.2146	18.7	15.3				N	Y	N	N	N	M4-5 ···
DV UMa (US 943)	9 46 36.67 44 46 45.1	SU?	0.08597	18.6	15.4				N	N	N	N	N	M4.5 ···
X Leo	9 51 01.51 11 52 31.1	U	0.1644	15.8	12.4	8–38			$< 0.21F^f$	N	Y	N	N	M2 ···
OY Car	10 06 22.43 −70 14 04.9	SU	0.063121	15.3	12.4	25–50 300	82.6 0.1	0.90 0.04	0.11^e	Y	N	Y	N	M6 ···
CH UMa (PG10030+678)	10 07 00.57 67 32 46.5	U	0.3448:	15.9	10.7	204	21.0 4.0	1.95 0.30	N	N	N	N	N	··· ···
DO Leo (PG1038+155)	10 40 51.21 15 11 33.7		0.234515	16.0B					N	N	N	N	N	··· ···

Table 17.8. *(Continued.)*

(1)	(2)	(3)	(4)	(5)	(6)	(7)	(8)	(9)	(10)	(11)	(12)	(13)	(14)	(15)
Name[a,b] (alt. name)	Coord.[c] (2000.0)	DN	P_{orb}	V_{min}	V_{max}	t_{rec}	Incl.	M_{WD}	XRS	EC	QP	WD SP	Wind	Spect. type sec.
CY UMa	10 56 57.05 49 41 18.7	SU	0.0583	17.0	11.9	115: 297:			N	N	N	N	N	... ...
V436 Cen	11 14 00.10 −37 40 48.6	SU	0.062501	15.3	12.4	22 335	65: 5	0.7: 0.1	$8.3F^f$	N	Y	N	N	... ...
V442 Cen	11 24 51.92 −35 54 37.7	U	0.46:	16.5	11.9	14–39			N	N	Y	N	N	... ...
RZ Leo	11 37 22.30 01 48 57.8	SU?	0.0708	19.0	11.5				0.41^e	N	N	N	N	... ...
T Leo	11 38 26.96 03 22 08.1	SU	0.058819	15.2	11.0	450	65 19	0.16 0.04	N	N	N	N	N	... ...
DO Dra (PG1140+719)	11 43 38.34 71 41 20.4		0.165	15.6B	10.6B		42 5	0.83 0.18	N	N	Y	N	N	M3-5 ...
TW Vir	11 45 21.13 −04 26 05.9	U	0.18267	15.8	12.1	15–44	43 13	0.91 0.25	0.38^d	N	N	N	Y	M2-4 ...
AL Com	12 32 25.90 14 20 42.5		0.061:	20.8	12.8	225:			N	N	Y	N	N	... ...
BV Cen	13 31 19.55 −54 58 33.6	U	0.610116	12.6	10.5	150	62 5	0.83 0.10	3^d	N	N	N	N	G5-8 ...
LY Hya (1329-294)	13 31 53.84 −29 40 59.1		0.13695	14.4					N	N	N	N	N	... ...
UZ Boo	14 44 01.30 22 00 56.0		0.125:	19.B	11.5	360:			N	N	N	N	N	... ...
TT Boo	14 57 44.74 40 43 42.2	SU	0.077:	< 15.6	12.7	45			N	N	Y	N	N	... ...
EK TrA	15 14 01.47 −65 05 31.3	SU	0.0636	> 17	12.1	231 487			Y	N	N	N	N	... ...
DM Dra	15 34 12.13 59 48 31.9	U	0.087:	20.8	15.5				N	N	N	N	N	... ...
BR Lup	15 35 53.15 −40 34 05.5	SU	0.0793	> 17.5	13.1				N	N	N	N	N	... ...
SS UMi (PG1551+719)	15 51 22.24 71 45 11.9	U	0.088:	16.9	12.6	30–48			N	N	N	N	N	... ...
AH Her	16 44 09.99 25 15 02.1	Z	0.258116	13.9	11.3	7–27	46 3	0.95 0.10	$0.48F^f$ 2.38^e	N	Y	N	Y	K2-M0 ...
V2051 Oph	17 08 19.09 −25 48 30.8	U?	0.062428	15.0	13.0		80.5 2.0	0.44 0.05	N	Y	Y	N	N	... ...
V426 Oph	18 07 51.71 05 51 48.5	Z	0.2853	11.5		17–55	57 11	0.9 0.15	1.07^e	N	N	N	N	K2-4 ...
UZ Ser	18 11 24.90 −14 55 33.9	U	0.1730	15.5	11.9	10–40			N	N	N	N	N	... ...
BD Pav	18 43 11.90 −57 30 44.2	U	0.17930	15.4	12.4		> 55		N	Y	N	N	N	... ...
AY Lyr	18 44 26.73 37 59 51.8	SU	0.07340	18.4B	13.2	8–43 205			$< 0.27F^f$ 30.1^e	N	N	N	N	... ...
EM Cyg	19 38 40.10 30 30 28.0	Z	0.290909	13.3	12.5	13–46	63 10	0.57 0.08	4^d	Y	Y	N	N	K5 ...
AB Dra	19 49 06.50 77 44 23.5	Z	0.15198	14.5	12.3	8–22			1.6^d 3.62^e	N	N	N	Y	... ...
EY Cyg	19 54 36.77 32 21 54.7	U	0.18123:	15.5	11.4	96			N	N	N	N	N	K0 ...
UU Aql	19 57 18.68 −09 19 20.8	U	0.14049:	16.1	11.0	71			N	N	N	N	N	... ...
V4140 Sgr (NSV 12615)	19 58 49.71 −38 56 12.3	SU?	0.061430	17.5	15.5				N	Y	N	N	N	... ...
RZ Sge	20 03 18.49 17 02 52.6	SU	0.0686	16.9	12.8	62–93 266			N	N	N	N	N	... ...
WZ Sge	20 07 36.40 17 42 15.4	SU	0.056688	14.9	7.0:	11876	72 2	0.8: 0.3	$\sim 0.30^d$ 0.21^e	Y	Y	Y	Y	... ...
CM Del	20 24 56.92 17 17 54.3	U	0.162	13.4 15.3			73: 47	0.48 0.15	N	N	N	N	N	... ...
V503 Cyg	20 27 17.44 43 41 23.1	SU	0.07599	17.4	13.4	28			N	N	N	N	N	... ...
VW Vul	20 57 45.08 25 30 26.0	U?	0.0731	13.6		14–29	44 12	0.24 0.06	N	N	N	N	N	... ...

Table 17.8. *(Continued.)*

(1)	(2)	(3)	(4)	(5)	(6)	(7)	(8)	(9)	(10)	(11)	(12)	(13)	(14)	(15)
														Spect.
Name^{a,b}	Coord.^{c}											WD		type
(alt. name)	(2000.0)	DN	P_{orb}	V_{min}	V_{max}	t_{rec}	Incl.	M_{WD}	XRS	EC	QP	SP	Wind	sec.
VY Aqr	21 12 09.20	SU	0.06312	17.1	8.0B				3.96^e	N	N	N	N	...
	−08 49 36.5													...
SS Cyg	21 42 42.66	U	0.275130	11.4	8.2	24–63	37	1.19	86F^f	N	Y	Y	Y	K5
	43 35 09.5						5	0.02	1.12^e					...
RU Peg	22 14 02.58	U	0.3746	12.7	9.0	75–85	33	1.21	18.2^d	N	Y	N	Y	K2-3
	12 42 11.4						5	0.19	7.90^e					...
TY PsA	22 49 39.86	SU	0.08400	16.0	12.0				N	N	Y	N	N	...
(PS 74)	−27 06 54.2													...
GD 552	22 50 39.64		0.07134	16.5			20:	1.4:	N	N	N	N	N	...
	63 28 39.3													...
IP Peg	23 23 08.60	U	0.158206	14.0	12.B	95	68	1.15	N	Y	N	N	N	M4
	18 24 59.4			18.5B				0.10						...

Key definitions of columns

(1) System name. (2) Right ascension, declination (Equinox 2000.0). (3) Dwarf nova sub-type, U Gem, Z Cam (standstills), SU UMa (superoutbursts), DN (undetermined subtype). (4) Orbital period in days (spectroscopic period), colon indicates uncertain value as adapted from [1]. (5) V_{min}: minimum visual brightness in quiescence, B denotes a B magnitude measurement (adapted from [1, 2]). (6) Maximum visual brightness peak at dwarf nova outburst (adapted from [1, 2]). (7) Recurrence time of dwarf nova outbursts in days; the second entry is the approximate recurrence time of super outbursts in the case of SU UMa systems (adapted from [1]). (8) Orbital inclination in degrees; second entry is the ± error estimate in degrees (adapted from [1]). (9) Mass determination for the white dwarf in solar masses; the second entry is ± error estimate (adapted from [3] and [1]). (10) X-ray data. If the system is a detected hard-X-ray source (0.1–4 keV) with the Einstein Observatory (HEAO-B) imaging proportional counter (IPC) [4–7] or has an upper limit detection, then an X-ray luminosity is given in units of 10^{31} ergs/s when a distance estimate is available, otherwise a count rate. If the system is a detected X-ray source with the EXOSAT (2–20 keV) medium energy (ME) experiment [8] or is an upper limit detection, then an X-ray luminosity is given in units of 10^{31} ergs/s. If the system is a detected Einstein IPC source but with no distance estimate, then a count rate is given followed by an F. If the entry is N, then the system has not been observed with either Einstein or EXOSAT, but ROSAT data may exist. (11) Does the system undergo eclipses, yes (Y) or no (N)? (12) Does the system exhibit quasiperiodic oscillations (QPO), yes or no? (13) Is the underlying white dwarf detected spectroscopically during dwarf nova quiescence (i.e., dominates the light in the far UV, EUV (IUE, HST, HUT, EUVE) and/or in the optical), yes or no [9–13] and references therein? (14) Does the system exhibit direct spectroscopic evidence of wind outflow (e.g., P Cygni line structure/shortward-shifted absorption or broad wind emission, during dwarf nova outburst), yes or no [14] and references therein? (15) Spectral type of the cool, normally main sequence, lower mass, secondary star, if known.

Notes

^{a}Finding charts for dwarf nova systems are given in [2]. Other references to finding charts are in [1] and [15].

^{b}References to the key ground-based and space-based spectroscopic studies of dwarf novae are given in [1, 2, 15, 16] and references therein.

^{c}Coordinates for equinox 2000.0 adapted from [1, 2]. Coordinates for 2000.0 measured off the Space Telescope Guide Star plates are given in [2].

^{d} Einstein IPC X-ray luminosity.

^{e} EXOSAT ME data.

^{f} Einstein IPC observed flux.

Informative and stimulating reviews of virtually all aspects of dwarf novae can be found in [13, 14, 17–22]. References to original spectroscopy can be found in [1, 14–18, 21].

References

1. Ritter, H. 1990, *A&AS*, **85**, 1179
2. Downes, R.A., Webbink, R.F., & Shara, M.M. 1997, *PASP*, **109**, 345
3. Webbink, R.E. 1990, in *Accretion-Powered Compact Binaries*, edited by C. Mauche (Cambridge University Press, Cambridge), p. 177
4. Córdova, F.M., & Mason, K.O. 1984, *MNRAS*, **206**, 879
5. Eracelous, M., Halpern, J., & Patterson, J., 1991, *ApJ*, **370**, 330
6. Eracelous, M., Halpern, J., & Patterson, J., 1991, *ApJ*, **382**, 290
7. Patterson, J., & Raymond, J. 1985, *ApJ*, **292**, 535
8. Mukai, K., & Shiokawa, K. 1993, *ApJ*, **418**, 803
9. Panek, R., & Holm, A. 1984, *ApJ*, **277**, 700

10. Sion, E.M. 1987, in *The 2nd Conference on Faint Blue Stars*, IAU Coll. No. 95, edited by A.G.D. Philip, D. Hayes, and J. Liebert (Davis, Schenectady), p. 413

11. Smak, J. 1992, *AcA*, **42**, 323

12. Long, K. et al. 1993, *ApJ*, **405**, 327

13. La Dous, C. 1991, *A&AS*, **252**, 100

14. Patterson, J. 1984, *ApJS*, **54**, 443

15. Williams, G. 1983, *ApJS*, **53**, 523

16. Szkody, P. 1987, *ApJS*, **63**, 685

17. Robinson, E.L. 1980, *ARA&A*, **14**, 119

18. Córdova, F.M. 1995, *X-Ray Binaries*, edited by W.H.G. Lewin, J. van Paradijs, and E.P.J. van den Heuvel (Cambridge University Press, Cambridge)

19. Verbunt, F. 1986, in *The Physics of Accretion onto Compact Objects*, edited by M.G. Watson and N.E. White (Springer-Verlag, Berlin), p. 59

20. Wade, R. 1985, *Interacting Binaries*, edited by P.P. Eggleton and J.E. Pringle (Reidel, Dordrecht)

21. Warner, B. 1995, *Cataclysmic Variable Stars* (Cambridge University Press, Cambridge)

22. Szkody, P. 1985, in *Cataclysmic and Low Mass X-Ray Binaries*, edited by D.Q. Lamb and J. Patterson (Reidel, Dordrecht), p. 385

Table 17.9. *Selected list of nova-likes.*

Name (alt. names)	Coord.[a] (2000)	Galactic coord.	V_{max}[b] V_{min}	$B - V$	$E(B - V)$[c]	Second. spectral class[d]	Period[e] (d)	Rapid oscillation period (s)	Spec.[f] Refs.	Type
TT Ari (BD+14°341)	02 06 53.09 +15 17 43.0	147.69 −44.05	9.5 16.3	−0.04 [1]	0.0 [2]		0.137 551 [3, 4] 0.1329 [5, 6]	1000–1600 QPO [7]	[3, 8]	VY
RW Tri	02 25 36.14 +28 05 51.4	146.34 −30.59	12.6 15.6	0.15 [1]	0.25 [9] 0.1 [2]	K5V [10]	0.231 883 297 [11, 12]		[11]	UX
IX Vel (CPD–48°1577)	08 15 18.90 −49 13 18.3	264.80 −8.09	9.1 10.0	−0.03 [13]	0.03 [13]		0.193 929 [14, 15]	[16–22]	[14, 23]	UX
RW Sex (BD–7°3007)	10 19 56.63 −08 41 56.0	251.32 38.28	10.4 10.8	−0.04 [24]	0.0 [2]		0.245 07 [25]	620, 1280 QPO [26]	[24, 25]	UX
QU Car (HDE 310376) (CD–67°1010)	11 05 42.80 −68 37 58.0	293.35 −7.79	11.1 11.5	0.02 [1]	0.10 [2]		0.454 [27] 0.113 47? [28]		[27]	
UX UMa	13 36 40.97 +51 54 50.3	107.67 63.91	12.7 14.1	0.07 [1]	0.0 [2]	K8V- M6V [29]	0.196 671 26 [30, 31]	28–30 QPO [32]	[33, 34]	UX
MV Lyr (MacRAE+43°1)	19 07 16.30 +44 01 08.4	74.61 16.08	12.1 18.0	−0.13 [1] −0.35 [1]	0.0 [2]	M5V [35]	0.1336 [35] 0.1379 [36]	2800 QPO [36]	[35, 37]	VY
V3885 Sgr (CD–42°14462)	19 47 40.54 −42 00 25.5	357.32 −27.14	9.6 10.3	0.0 [1]	0.0 [2]		0.206–0.259 [38, 39] 0.2163 [40]	29–30 QPO [38, 41]	[39]	UX
V794 Aql	20 17 33.97 −03 39 51.0	39.38 −20.22	13.7 20.2	0.32 [19]	0.0 [2]		0.23? [20]		[21, 42]	VY
VY Scl (PS 141) (SPC Var4)	23 29 00.45 −29 46 46.0	19.84 −71.14	12.9 18.5	−0.10 [16]	0.06 [16]		0.1662 [17]	∼ 500 QPO [18]	[17, 19]	VY

Notes

[a] Adapted from [43].

[b] The range in magnitudes is taken from [44].

[c] The color excess, $E(B - V)$, is assumed to be related to the visual interstellar extinction, A_v, by $A_v = 3.2\, E(B - V)$.

[d] The secondary spectral types are from spectroscopic or photometric observations and do not include estimates from mass determinations.

[e] The spectroscopic period is the first entry while the photometric period is the second if it is different. Orbital parameters can usually be found in the reference for the spectroscopic period.

[f] Only optical references are given. Surveys or catalogues of nova-likes exist in the infrared [45], visual [1, 46, 47], ultraviolet [2, 48, 49], far ultraviolet [50], and X-ray [51, 52] spectral regions. References of observations for individual nova-likes can be found in [43, 53, 54] and finding charts in [43, 46].

References

1. Bruch, A. 1984, *A&AS*, **56**, 441
2. Verbunt, F. 1987, *A&AS*, **71**, 339
3. Cowley, A.P., Crampton, D., Hutchings, J.B., & Marlborough, J.M. 1975, *ApJ*, **195**, 413
4. Thorstensen, J.R., Smak, J., & Hessman, F.V. 1985, *PASP*, **97**, 437
5. Udalski, A. 1988, *AcA*, **38**, 315
6. Volpi, A., Natali, G., & D'Antona, F. 1988, *A&A*, **193**, 87
7. Semeniuk, I., Schwarzenberg-Czerny, A., Duerbeck, H., Hoffmann, M., Smak, J., Stepien, K., & Tremko, J. 1987, *Ap&SS*, **130**, 167
8. Shafter, A.W., Szkody, P., Liebert, J., Penning, W.R., Bond, H.E., & Grauer, A.D. 1985, *ApJ*, **290**, 707
9. Córdova, F.A., & Mason, K.O. 1985, *ApJ*, **290**, 671
10. Longmore, A.J., Lee, T.J., Allen, D.A., & Adams, D.J. 1981, *MNRAS*, **195**, 825
11. Kaitchuck, R.H., Honeycutt, R.K., & Schlegel, E.M. 1983, *ApJ*, **267**, 239
12. Robinson, E.L., Shetrone, M.D., & Africano, J.L. 1991, *AJ*, **102**, 1176
13. Garrison, R.F., Schild, R.E., Hiltner, W.A., & Krzeminsk, W. 1984, *ApJ*, **276**, L13
14. Beuermann, K., & Thomas, H.-C. 1990, *A&A*, **230**, 326
15. Haug, K. 1988, *MNRAS*, **235**, 1385
16. Warner, B. 1976 in *Structure and Evolution of Close Binary Systems*, edited by P. Eggleton, S. Mitton, and J. Whelan (Reidel, Dordrecht), p. 85
17. Hutchings, J.B., & Cowley, A.P. 1984, *PASP*, **96**, 559
18. Burrell, J.F., & Mould, J.R. 1973, *PASP*, **85**, 627
19. Szkody, P. 1987, *ApJS*, **63**, 685
20. Shafter, A.W. 1992, *ApJ*, **394**, 268
21. Szkody, P., Crosa, L., Bothun, G.D., Downes, R.A., & Schommer, R.A. 1981, *ApJ*, **249**, L61
22. Warner, B., O'Donoghue, D., & Allen, S. 1985, *MNRAS*, **212**, 9P
23. Wargau, W., Drechsel, H., & Rahe, J. 1983, *MNRAS*, **204**, 35P
24. Bolick, U., Beuermann, K., Bruch, A., & Lenzen, R. 1987, *Ap&SS*, **130**, 175
25. Beuermann, K., Stasiewski, U., & Schwope, A.D. 1992, *A&A*, **256**, 433
26. Hesser, J.E., Lasker, B.M., & Osmer, P.S. 1972, *ApJ*, **176**, L31
27. Gilliland, R.L., & Phillips, M.M. 1982, *ApJ*, **261**, 617
28. Kern, J.R., & Bookmyer, B.B. 1986, *PASP*, **98**, 1336
29. Frank, J., King, A.R., Sherrington, M.R., Jameson, R.F., & Axon, D.J. 1981, *MNRAS*, **195**, 505
30. Shafter, A.W. 1984, *AJ*, **89**, 1555
31. Rubenstein, E.P., Patterson, J., & Africano, J.L. 1991, *PASP*, **103**, 1258
32. Nather, R.E., & Robinson, E.L. 1974, *ApJ*, **190**, 637
33. Walker, M.F., & Herbig, G.H. 1954, *ApJ*, **120**, 278
34. Schlegel, E.M., Honeycutt, R.K., & Kaitchuck, R.H. 1983, *ApJS*, **53**, 397
35. Schneider, D.P., Young, P., & Shectman, S.A. 1981, *ApJ*, **245**, 644
36. Borisov, G.V. 1992, *A&A*, **261**, 154
37. Voikhanskaya, N.F. 1988, *A&A*, **192**, 128
38. Cowley, A.P., Crampton, D., & Hesser, J.E. 1977, *ApJ*, **214**, 471
39. Haug, K., & Drechsel, H. 1985, *A&A*, **151**, 157
40. Metz, K. 1989, *Inf. Bull. Var. Stars*, 3385
41. Hesser, J.E., Lasker, B.M., & Osmer, P.S. 1974, *ApJ*, **189**, 315
42. Honeycutt R.K., & Schlegel, E.M. 1985, *PASP*, **97**, 1189
43. Downes, R.A., Webbink, R.F., & Shara, M.M. 1997, *PASP*, **109**, 345
44. Ritter, H. 1990, *A&AS*, **85**, 1179
45. Szkody, P. 1977, *ApJ*, **217**,140
46. Williams, G. 1983, *ApJS*, **53**, 523
47. Honeycutt, R.K., Kaitchuck, R.H., & Schlegel, E.M. 1987, *ApJS*, **65**, 451
48. la Dous, C. 1990, *Space Sci. Rev.*, **52**, 203
49. la Dous, C. 1991, *A&A*, **252**, 100
50. Polidan, R.S., Mauche, C.W., & Wade, R.A. 1990, in *Accretion-Powered Compact Binaries*, edited by C.W. Mauche (Cambridge University Press, Cambridge), p. 77
51. Patterson, J., & Raymond, J.C. 1985, *ApJ*, **292**, 535
52. Eracleous, M., Halpern, J., & Patterson, J. 1991, *ApJ*, **382**, 290
53. Ritter, H. 1984, *A&AS*, **57**, 385
54. Ritter, H. 1987, *A&AS*, **70**, 335

Table 17.10. *Synchronously rotating magnetic CVs (AM Her binaries).*[a]

Name (alt. name)	Coord.[b] (2000)	Dist.[c] (pc)	P_{orb}^d (min)	$m_v^{c,d}$	B_1, B_2 (MG)	B_d (MG)	L_{sx} (ergs s^{-1})	Circ. (%)	Lin. (%)	Refs.
BL Hyi (H0139-68)	1 41 00.25 −67 53 27.7	128	113.6	14–18.5	33 P	30 Z	1×10^{31}	17	12	[1, 2]
WW Hor (EXO0234.5-5232)	2 36 11.45 −52 19 13.5	500	114.6	19–21	25: P		4×10^{33}	30	...	[3]
EF Eri (2A0311-227)	3 14 13.03 −22 35 41.4	> 89	81.0	13.5–17.5B		15 Z	$> 1 \times 10^{32}$	20	9	[4, 5]
VY For (EXO032957-2606.9)	3 31 04.58 −25 56 55.5		228	17.5	10–50: P					[6]
UZ For (EXO033319-2554.2)	3 35 28.61 −25 44 22.6	250	126.5	18–20.5	53, 75: C		7×10^{33}	6	3	[7, 8]
BY Cam (H0538+608)	5 42 48.90 60 51 31.8	200	201.9 199.3[h]	14.5->17B	41: C			10	1	[9–12]
VV Pup	8 15 06.73 −19 03 16.8	145	100.4	14.5–18	31.5, 56 C		5×10^{32}	15	15	[13–15]
EK UMa (1E1048+5241)	10 51 35.23 54 04 36.0		114.5	18–20	47: C			20	...	[16, 17]
AN UMa	11 04 25.71 45 03 15.0	> 270	114.8	14.5–19B	36 C		$> 3 \times 10^{32}$	35	...	[18–20]
ST LMi (CW1103+254)	11 05 39.75 25 06 28.9	128	113.9	15.0–17		18 Z	2×10^{32}	20	12	[21–24]
DP Leo (1E1114+182)	11 17 16.00 17 57 41.1	> 380	89.8	17.5–19.5B	30.5, 59 C,Z		$> 1 \times 10^{33}$	35	9	[25, 26]
EU UMa[f] (RE1149+28)	11 49 55.70 28 45 07.5		90: 103:	16.5B						[27]
V834 Cen (1E1405-451)	14 09 07.46 −45 17 17.1	86	101.5	14.0–17	23 Z,C		1×10^{32}	30	10	[28–31]
MR Ser (PG1550+191)	15 52 47.23 18 56 27.6	112	113.6	15–17	24 C	10 Z	5×10^{30}	12	5	[32, 33]
AM Her[g] (3U1809+50)	18 16 13.33 49 52 04.2	75	185.6	12–15.5	14, 28 C	22 Z	9×10^{32}	10	8	[34–37]
EP Dra (H1907+690)	19 07 06.13 69 08 42.4	600:	104.6	18				10	...	[38]
QS Tel (RE 1938-461)	19 38 35.73 −46 12 56.5		140.0	15.5				6	10	[39]
QQ Vul (1E2003+225)	20 05 41.93 22 39 59.1	> 400	222.5	14.5–15.5	10–50: P		$> 4 \times 10^{34}$	8	2	[40]
V1500 Cyg (Nova Cyg 1975)	21 11 36.61 48 09 01.9	1000– 1400	201.0 197.5[i]	17–18	25–50: P			10	...	[41]
CE Gru (Grus V1)	21 37 56.38 −43 42 13.1		108.6	18–21B	20:, 20: P			15	...	[42, 43]

Notes

[a] These binaries contain accreting white dwarfs that are strongly magnetized and rotate essentially synchronously, i.e., rotation period within 2% of the orbital period P_{orb} [44, 45]. They are more commonly known as AM Herculis binaries, or polars, and are characterized by the strong optically polarized radiation they emit.

m_v: Visual magnitude. B indicates blue magnitude. Nova outburst magnitude not given.

B_1, B_2 are the dominant and less dominant accretion poles, respectively.

B_d is the polar field if a dipole structure is assumed to model the Zeeman features.

Z: Zeeman features.

C: Cyclotron features.

P: Magnetic field estimated from polarization.

L_{sx}: Soft X-ray luminosity (see [46, 47] for uncertainties in the estimates).

[b] Adapted from [44].

[c] Adapted from [45].

[d] Adapted from [48].

[e] Adapted from [49].

[f] Shows strong He II λ4686 line and strong EUV emission, characteristics of AM Her binaries.

[g] Radio source.

[h] Rotation period.

[i] White dwarf presumed to have become asynchronous following nova outburst and is now synchronizing.

References

1. Wickramasinghe, D.T., Visvanathan, N., & Tuohy, I.R. 1984, *ApJ*, **286**, 328
2. Piirola, V.V., Reiz, A., & Coyne, G.V. 1987, *A&A*, **185**, 189
3. Bailey, J., Wickramasinghe, D.T., Hough, J.H., & Cropper, M. 1988, *MNRAS*, **234**, 19P
4. Achilleos, N., Wickramasinghe, D.T., & Wu, K. 1992, *MNRAS*, **256**, 80
5. Tapia, S. 1979, IAU Circ. 3327
6. Berriman, G., & Smith, P.S. 1988, *ApJ*, **329**, L97
7. Ferrario, L., Wickramasinghe, D.T., Bailey, J., Tuohy, I.R., & Hough, J.H. 1989, *ApJ*, **337**, 832
8. Schwope, A.D., Beuermann, K., & Thomas, H.-C. 1990, *A&A*, **230**, 120
9. Cropper, M. et al. 1989, *MNRAS*, **236**, 29P
10. Mason, P.A., Liebert, J., & Schmidt, G.D. 1989, *ApJ*, **346**, 941
11. Szkody, P., Downes, R.A., & Mateo, M. 1990, *PASP*, **102**, 1310
12. Silber, A., Bradt, H.V., Ishida, M., Ohashi, T., & Remillard, R.A. 1992, *ApJ*, **389**, 704
13. Barrett, P.E., & Chanmugam, G. 1985, *ApJ*, **298**, 743
14. Wickramasinghe, D.T., Ferrario, L., & Bailey, J. 1989, *ApJ*, **342**, L35
15. Piirola, V., Reiz, A., & Coyne, G. 1990, *A&A*, **235**, 245
16. Cropper M., Mason, K.O., & Mukai, K. 1990, *MNRAS*, **243**, 565
17. Morris, S.L. et al. 1989, *AJ*, **98**, 665
18. Liebert, J., Tapia, S., Bond, H.E., & Grauer, A.D. 1982, *ApJ*, **254**, 232
19. Schmidt, G.D., Stockman, H.S., & Grandi, S.A. 1986, *ApJ*, **300**, 804
20. Schwope, A.D., & Beuermann, K. 1990, *A&A*, **238**, 173
21. Schmidt, G.D., Stockman, H.S., & Grandi, S.A. 1983, *ApJ*, **271**, 735
22. Bailey, J. et al. 1985, *MNRAS*, **215**, 179
23. Cropper, M. 1987, *Ap&SS*, **131**, 651
24. Shore, S.N. et al. 1982, *PASP*, **94**, 682
25. Cropper, M., & Wickramasinghe, D.T. 1993, *MNRAS*, **260**, 696
26. Biermann, P., Schmidt, G.D., Liebert, J., Stockman, H.S., & Tapia, S. 1985, *ApJ*, **293**, 303
27. Mittaz, J.P.D., Rosen, S.R., Mason, K.O., & Howell, S.B. 1992, *MNRAS*, **258**, 277
28. Wickramasinghe, D.T., Tuohy, I.R., & Visvanathan, N. 1987, *ApJ*, **318**, 326
29. Schwope, A.D., & Beuermann, K. 1990, *A&A*, **238**, 173
30. Bailey, J., Axon, D.J., Hough, J.H., Watts, D.J., Giles, A.B., & Greenhill, J. 1983, *MNRAS*, **205**, 1
31. Visvanathan, N., & Tuohy, I.R. 1983, *ApJ*, **275**, 709
32. Wickramasinghe, D.T., Cropper, M., Mason, K.O., & Garlick, M. 1991, *MNRAS*, **250**, 692
33. Liebert J., Angel, J.R.P., Stockman, H.S., Spinrad, H., & Beaver, E.A. 1977, *ApJ*, **214**, 457
34. Chanmugam, G., & Dulk, G.A. 1982, *ApJ*, **255**, L107
35. Bailey, J., Ferrario, L., & Wickramasinghe, D.T. 1991, *MNRAS*, **251**, 37P
36. Wickramasinghe, D.T., Bailey, J., Meggitt, S.M.A., Ferrario, L., Hough, J., & Tuohy, I.R. 1991, *MNRAS*, **251**, 28
37. Tapia, S. 1977, *ApJ*, **212**, L15
38. Remillard, R.A., Stroozas, B.A., Tapia, S., & Silber, A. 1991, *ApJ*, **379**, 715
39. Buckley, D.A.H. et al. 1993, *MNRAS*, **262**, 93
40. Nousek, J.A. et al. 1984, *ApJ*, **277**, 682
41. Schmidt, G.D., & Stockman, H.S. 1991, *ApJ*, **371**, 749
42. Tuohy, I.R., Ferrario, L., Wickramasinghe, D.T., & Hawkins, M.R.S. 1988, *ApJ*, **328**, L59
43. Cropper, M., Bailey, J., Wickramasinghe, D.T., & Ferrario, L. 1990, *MNRAS*, **244**, 34P
44. Downes, R.A., Webbink, R.F., & Shara, M.M. 1997, *PASP*, **109**, 345
45. Cropper, M. 1990, *Space Sci. Rev.*, **54**, 195
46. Watson, M.G. 1986, in *The Physics of Accretion onto Compact Objects*, edited by K.O. Mason, M.G. Watson, and N.E. White (Springer-Verlag, Berlin), p. 97
47. Chanmugam, G., Ray, A., & Singh, K.P. 1991, *ApJ*, **375**, 600
48. Ritter, H. 1990, *A&AS*, **85**, 1179
49. Chanmugam, G. 1992, *ARA&A*, **30**, 143

Table 17.11. *Asynchronously rotating magnetic CVs (DQ Her binaries).*[a]

Name (alt. name)	Coord.[b] (2000)	Dist.[c] (pc)	P_{rot} (min)	P_{orb} (h)	m_v[d]	L_{hx}[c,e] (erg s^{-1})	Refs.
XY Ari[f] (1H0253+193)	2 56 08.1 19 26 34.	200	3.42	6.06	12–13.5K	2×10^{32}	[1, 2]

Table 17.11. (*Continued.*)

Name (alt. name)	Coord.[b] (2000)	Dist.[c] (pc)	$P_{\rm rot}$ (min)	$P_{\rm orb}$ (h)	$m_v{}^d$	$L_{\rm hx}{}^{c,e}$ (erg s^{-1})	Refs.
GK Per[g] (Nova Per 1901)	3 31 11.82 43 54 16.8	525	5.86	47.9	10–14.0	7.4×10^{32}	[3–5]
V471 Tau	3 50 24.79 17 14 47.8	49	9.24	12.51	9–10		[6–8]
TV Col (2A0526-328)	5 29 25.44 −32 49 04.5	> 500	31.9	5.49	13.5–14	$> 6.1 \times 10^{32}$	[9, 10]
TW Pic (H0534-581)	5 34 50.67 −58 01 40.9	650:	126:	6.5:	14–16		[11]
TX Col (1H0542-407)	5 43 20.22 −41 01 55.2	> 500	31.9:	5.72	15.5	$> 2.8 \times 10^{32}$	[12, 13]
BG CMi[h] (3A0729+103)	7 31 28.98 9 56 22.6	700–1000	14.1: 15.2: 28.2:	3.24	14–14.5	$(0.7–1.4) \times 10^{33}$	[14–19]
PQ Gem[i] (RE0751+14)	7 51 17.39 14 44 24.6		13.9:	~ 6:	14.5		[20, 21]
EX Hya	12 52 24.40 −29 14 56.7	76–90	67.0	1.64	10–14	$(0.3–1.8) \times 10^{32}$	[22, 23]
V795 Her[f] (PG 1711+336)	17 12 56.09 33 31 21.4		93.8: 106.4:	2.60	12.5–13B		[24, 25]
DQ Her (Nova Her 1934)	18 07 30.17 45 51 31.9	300–500	1.18	4.65	14–17.5	$< (1.1–3.0) \times 10^{30}$	[26–28]
V533 Her[f] (Nova Her 1963)	18 14 20.34 41 51 21.3	1000:	1.06	5.04	14.5–16	$< 2 \times 10^{31}$	[26, 27, 29]
V1223 Sgr	18 55 02.24 −31 09 48.5	540–660	12.4	3.37	12-> 17	$(0.9–1.3) \times 10^{33}$	[30–33]
AE Aqr[g]	20 40 09.02 −0 52 15.5	28–78	0.55:	9.88	10–11.5	$< (0.5–3.6) \times 10^{30}$	[34–37]
FO Aqr (H2215-086)	22 17 55.43 −8 21 04.6	200–640	20.9	4.85	13–14	$(0.8–8.3) \times 10^{32}$	[38–41]
AO Psc (H2252-035)	22 55 17.97 −3 10 40.4	100–750	13.4	3.59	13.5–15	$(0.02–1.3) \times 10^{33}$	[33, 42–44]

Notes

[a] These binaries are believed to contain accreting magnetized white dwarfs that rotate asynchronously with the rotation period $P_{\rm rot}$ differing by more than 2% from the orbital period $P_{\rm orb}$. They do not in general emit optically polarized radiation and probably have magnetic fields strengths that are weaker than those found in the synchronously rotating magnetic CVs. They are more commonly, but inconsistently, referred to as DQ Herculis binaries and/or intermediate polars. For example, some authors refer to only those binaries with $P_{\rm rot} \ll 0.1 P_{\rm orb}$ as DQ Hers and the others as intermediate polars. $P_{\rm rot}$ is often difficult to identify so that some of the binaries in the table should not actually belong to it. m_v: Visual magnitude. B, K indicate blue, K band magnitudes. Nova outburst magnitude not given. $L_{\rm hx}$: Hard-X-ray luminosity. The main uncertainty is due to that in the distance.

[b] Adapted from [45].

[c] Adapted from [46, 47].

[d] Adapted from [48].

[e] Adapted from [47, 49].

[f] Identification as a magnetic CV uncertain. XY Ari lies behind Lynds dark cloud L 1457 and is not visible optically.

[g] Radio source.

[h] Shows weak but significant optical/IR circular polarization implying a magnetic field of roughly 4 MG.

[i] Shows significant optical/IR linear and circular polarization implying a magnetic field of 8–18 MG.

References

1. Kamata, Y., Tawara, Y., & Koyama, K. 1992, *ApJ*, **379**, L65
2. Zuckerman, B., Becklin, E.E., McLean, I.S., & Patterson, J. 1992, *ApJ*, **400**, 665
3. Seaquist, E.R., Bode, M.F., Frail, D.A., Roberts, J.A., Evans, A., & Albinson, J.S. 1989, *ApJ*, **344**, 805
4. Norton, A.J., Watson, M.G., & King, A.R. 1988, *MNRAS*, **231**, 783
5. Patterson, J. 1991, *PASP*, **103**, 1149

6. Young, A., & Nelson, B. 1972, *ApJ*, **173**, 653

7. Barstow, M.A. et al. 1992, *MNRAS*, **255**, 369

8. Clemens, J.C. et al. 1992, *ApJ*, **391**, 773

9. Barrett, P.E., O'Donoghue, D., & Warner, B. 1988, *MNRAS*, **233**, 759

10. Watts, D.J., Greenhill, J.G., Hill, P.W., & Thomas, R.M. 1982, *MNRAS*, **200**, 1039

11. Mouchet, M., Bonnet-Bidaud, J.M., Buckley, D.A.H., & Tuohy, I.R. 1991, *A&A*, **250**, 99

12. Tuohy, I.R., Buckley, D.A.H., Remillard, R.A., Bradt, H.V., & Schwartz, D.A. 1986, *ApJ*, **311**, 275

13. Buckley, D.A.H., & Tuohy, I.R. 1989, *ApJ*, **344**, 376

14. McHardy, I.M., Pye, J.P., Fairall, A.P., Warner, B., Cropper, M.S., & Allen, S. 1984, *MNRAS*, **210**, 663

15. Penning, W.R., Schmidt, G.D., & Liebert, J. 1986, *ApJ*, **301**, 885

16. Norton, A.J., McHardy, I.M., Lehto, H.J., & Watson, M.G. 1992, *MNRAS*, **258**, 697

17. West, S.C., Berriman, G., & Schmidt, G.D. 1987, *ApJ*, **322**, L35

18. Chanmugam, G., Frank, J., King, A.R., & Lasota, J.-P. 1990, *ApJ*, **350**, L13

19. Patterson, J., & Thomas, G. 1993, *PASP*, **105**, 59

20. Mason, K.O. et al. 1992, *MNRAS*, **258**, 749

21. Piirola, V., Hakala, P., & Coyne, G.V. 1993, *ApJ*, **410**, L107

22. Vogt, N., Krzeminski, W., & Sterken, C. 1980, *A&A*, **85**, 106

23. Rosen, S.R., Mason, K.O., & Córdova, F.A. 1988, *MNRAS*, **231**, 549

24. Shafter, A.W., Robinson, E.L., Crampton, D., Warner, B., & Prestage, R.M. 1990, *ApJ*, **354**, 708

25. Mironov, A.V., Moshkalev, V.G., & Shugarov, S. Yu, 1983, *Inf. Bull. Var. Stars*, No. 2438

26. Córdova, F.A., Mason, K.O., & Nelson, J.E. 1981, *ApJ*, **245**, 609

27. Patterson, J. 1984, *ApJS*, **54**, 443

28. Horne, K., Welsh, W.F., & Wade, R.A. 1990, in *Accretion-Powered Compact Binaries*, edited by C.W. Mauche (Cambridge University Press, Cambridge), p. 383

29. Hutchings, J.B. 1987, *PASP*, **99**, 57; *Binaries*, edited by C.W. Mauche (Cambridge University Press, Cambridge), p. 383

30. Steiner, J.E. et al. 1981, *ApJ*, **249**, L21

31. van Amerongen, S., Augusteijn, T., & van Paradijs, J. 1987, *MNRAS*, **228**, 377

32. Osborne, J.P., Rosen, S.R., Mason, K.O., & Beuermann, K. 1985, *Space Sci. Rev.*, **40**, 143

33. Hellier, C. 1991, *MNRAS*, **251**, 693

34. Bookbinder, J.A., & Lamb, D.Q. 1987, *ApJ*, **323**, L131

35. Bastian, T.S., Dulk, G.A., & Chanmugam, G. 1988, *ApJ*, **324**, 431

36. Patterson, J. 1979, *ApJ*, **234**, 978

37. Robinson, E.L., Shafter, A.W., & Balachandran, S. 1991, *ApJ*, **374**, 298

38. Osborne, J.P., & Mukai, K. 1989, *MNRAS*, **238**, 1233

39. Steiman-Cameron, T.Y., Imamura, J.N., & Steiman-Cameron, D.V. 1989, *ApJ*, **339**, 434

40. Hellier, C., Mason, K.O., & Cropper, M. 1990, *MNRAS*, **242**, 250

41. Norton, A.J., Watson, M.G., King, A.R., Lehto, H.J., & McHardy, I.M. 1992, *MNRAS*, **254**, 705

42. Hassall, B.J.M. et al. 1981, *MNRAS*, **197**, 275

43. van Amerongen, S., Kraakman, H., Damen, E., Tjemkes, S., & van Paradijs, J. 1989, *MNRAS*, **215**, 45P

44. Pietsch, W., Voges, W., Kendziorra, E., & Pakull, M. 1987, *Ap&SS*, **130**, 281

45. Downes, R.A., Webbink, R.F., & Shara, M.M. 1997, *PASP*, **109**, 345

46. Berriman, G. 1987, *A&AS*, **68**, 41

47. Norton, A.J., & Watson, M.G. 1989, *MNRAS*, **237**, 715

48. Ritter, H. 1990, *A&AS*, **85**, 1179

49. Chanmugam, G., Ray, A., & Singh, K.P. 1991, *ApJ*, **375**, 600

17.2 TYPES OF SYMBIOTIC VARIABLES

A symbiotic variable [2, 20–22] is a long-period (> 100 days) binary system with a detached or semidetached cool, red giant or Mira star losing mass via a wind to a hot, luminous white dwarf or main-sequence companion. Table 17.12 lists the positions of the best-known symbiotic stars, Table 17.13 gives their variability, and Table 17.14 supplies their orbital parameters. The symbiotic variables are subdivided into D(ust) types which tend to be associated with Miras and S(tar) types which tend to be associated with red giants. Some symbiotic variables undergo a large outburst which may be due to an accretion event or a shell flash on the hot component. These symbiotic variables are observationally similar to the very slow novae and are called symbiotic novae [23]. Recurrent novae which have a red giant companion may be related to the symbiotic novae [2].

Table 17.12. *Representative symbiotic stars.*[a]

Name	Other name	α hr min sec	δ deg min sec	l_{II}	b_{II}
EG And	HD 4174	00 44 37.18	40 40 45.6	121.5	−22.2
AX Per	MWC 411	01 36 22.73	54 15 02.7	129.5	−8.0
BX Mon	AS 150	07 25 22.72	−03 35 50.8	220.0	5.9
RX Pup	HD 69190	08 14 12.24	−41 42 29.8	258.5	−3.9
SY Mus	HD 100336	11 32 10.16	−65 25 11.2	294.8	−3.8
AG Dra	⋯	16 01 40.98	66 48 10.3	100.3	41.0
RT Ser	MWC 265	17 39 51.94	−11 56 38.8	13.9	10.0
AR Pav	MWC 600	18 20 27.92	−66 04 41.8	328.5	−21.6
BF Cyg	MWC 315	19 23 53.55	29 40 29.3	62.9	6.7
CH Cyg	HD 182917	19 24 33.08	50 14 29.5	81.8	15.6
HM Sge	⋯	19 41 57.07	16 44 39.6	53.6	−3.2
CI Cyg	MWC 415	19 50 11.86	35 41 03.2	70.9	4.7
V1016 Cyg	AS 373	19 57 04.99	39 49 36.5	75.2	5.7
RR Tel	Hen 1811	20 04 18.54	−55 43 32.9	342.2	−32.2
V1329 Cyg	HBV 475	20 51 01.27	35 34 53.3	77.8	−5.5
AG Peg	HD 207757	21 51 01.99	12 37 31.9	69.3	−30.9
Z And	HD 221650	23 33 40.02	48 49 06.1	110.0	−12.1
R Aqr	HD 222800	23 43 49.45	−15 17 04.4	66.5	−70.3

Note

[a]For general systems, see [1–7]. For Miras, see [8] and [9]. For multiwavelength properties, see [10–12]. For outbursts, see [13]. For interpretation of UV spectra, see [14] and [15].

References

1. Payne-Gaposchkin, C. 1957, *The Galactic Novae* (North-Holland, Amsterdam)
2. Allen, D.A. 1984, *Proc. ASA*, **5**, 369
3. Kenyon, S.J. 1986, *The Symbiotic Stars* (Cambridge University Press, Cambridge)
4. Kenyon, S.J. 1992, in *Evolutionary Processes in Interacting Binary Stars*, IAU Symposium No. 151, edited by Y. Kondo et al. (Kluwer Academic, Dordrecht), p. 137
5. Mikolajewska, J. et al. 1988, *The Symbiotic Phenomenon*, IAU Coll. 103 (Kluwer Academic, Dordrecht) (referred to as IAU Coll. 103)
6. Garcia, M.R., & Kenyon, S.J. 1988, in *The Symbiotic Phenomenon*, IAU Coll. 103 (Kluwer Academic, Dordrecht), p. 27
7. Vogel, M. 1990, thesis, ETH, Zürich
8. Whitelock, P. 1987, *PASP*, **99**, 573
9. Whitelock, P. 1988, IAU Coll. 103, p. 47
10. Ivison, R.J., et al. 1991, *MNRAS*, **249**, 374
11. Seaquist, E.R., & Taylor, A.R. 1990, *ApJ*, **349**, 313
12. Van Winckel, H., et al. 1993, *A&AS*, **102**, 401
13. Mürset, U. & Nussbaumer, H. 1994, *A&A*, **282**, 586
14. Shore, S.N., & Aufdenberg, J.P. 1993, *ApJ*, **416**, 355
15. Boyarchuk, A.A. 1993, in *The Realm of Interacting Binary Stars*, edited by J. Sahade, G.E. McCulsky, and Y. Kondo (Kluwer Academic, Dordrecht), p. 189

Table 17.13. *Symbiotic stars: Variability.*

Name	Max	Min	Orbit epoch[a]	Orbit period	Type	Spectrum	Ecl.	Outburst	Refs.
EG And	7.1	7.8	45 380.0	482	S	M2.4 III	E	⋯	[1–3]
AX Per	9.4	13.6	36 667.0	680.8	S	M5 IIIep	E	1980	
BX Mon	9.5	13.4	49 530.0	1401	S	M4.6ep	E?	⋯	
RX Pup[b]	9.0	14.1	⋯	⋯	D	M5-6 IIIpe:	⋯	⋯	
SY Mus	10.2	12.7	36 460.0	621.8	S	M2 III	E	⋯	
AG Dra	8.9	11.8	38 900.0	554.0	S	K1 IIpev	⋯	1980	
RT Ser	10.6	17.0	⋯	⋯	S	M5.5 III	⋯	1909	[4]

Table 17.13. (*Continued.*)

Name	Max	Min	Orbit epoch[a]	Orbit period	Type	Spectrum	Ecl.	Out-burst	Refs.
AR Pav	7.4	13.6	⋯	605	S	≥ M4 III	E	1964	
BF Cyg	9.3	13.4	15 058.0	756.8	S	M5 III	⋯	⋯	[5–7]
CH Cyg[c]	5.6	8.5	47 302.0	756	S	M6 III + WD	⋯	1964	[8]
CH Cyg[c]	5.6	8.5	45 517.0	5294	S	+?	E?	1964	[8]
HM Sge[b]	11.1	18.0	⋯	⋯	D	> M4	⋯	1975	[9]
CI Cyg	9.9	13.1	45 323.8	855.3	S	M5 II	⋯	1975	[10]
V1016 Cyg[b]	10.1	17.5	⋯	⋯	D	> M4 III	⋯	1963	[11]
RR Tel[b]	6.5	16.5	⋯	⋯	D	Pec	⋯	1944	[12, 13]
V1329Cyg	12.1	18.0	24 869.9	964	S	> M4 IIIp	E	1964	[14]
AG Peg	6.0	9.4	42 710.1	816.5	S	M3 III	⋯	1850	[15]
Z And	8.0	12.4	⋯	756.9	S	M3.5 III	E?	1985?	[16]
R Aqr[b]	5.8	12.4	⋯	⋯	S	M7 IIIpev	E?	⋯	[17]

Notes

[a] Date given as JD2400000+.

[b] Mira periods: HM Sge: 540 days; R Aqr: 387 days; RX Pup: 580 days; V1016 Cyg: 450 days; RR Tel: 387 days.

[c] CH Cyg is a possible triple system.

References

1. Skopal, A. et al. 1988, *IAU Coll.* No. 103, p. 289
2. Vogel, M. 1991, *A&A*, **249**, 173
3. Oliversen, N. et al. 1985, *ApJ*, **295**, 620
4. Payne-Gaposchkin, C. 1957, *The Galactic Novae* (North-Holland, Amsterdam)
5. Mikolajewska, J. et al. 1989, *AJ*, **98**, 1427
6. Mikolajewska, J., & Mikolajewski, M. 1988, *IAU Coll.* No. 103, p. 299
7. Slovak, M. et al. 1988, *IAU Coll.* No. 103, p. 265
8. Hinkle, K.H. et al. 1993, *AJ*, **105**, 1074
9. Nussbaumer, H., & Vogel, M. 1990, *A&A*, **101**, 118
10. Kenyon, S.J. et al. 1991, *AJ*, **101**, 637
11. Nussbaumer, H., & Schild, H. 1981, *A&A*, **101**, 118
12. Thackeray, A.D. 1977, *MNRAS*, **83**, 1
13. Pentson, M.V. et al. 1983, *MNRAS*, **202**, 833
14. Nussbaumer, H., & Vogel, M. 1991, *A&A*, **248**, 81
15. Kenyon, S. et al. 1993, *AJ*, **106**, 1573
16. Mattei, J.A. 1978, *JRASC*, **72**, 61
17. Kafatos, M. et al. 1986, *ApJS*, **62**, 853

Table 17.14. *Symbiotic stars: Orbital parameters.*

Name	Period (days)	e	$f(M)$ ($\mathcal{M}_\odot$)	Refs.
EG And	482	0.043 ± 0.021	0.021	[1–5]
AX Per	680.8	0.0:	0.032	[6]
BX Mon[a]	1401 ± 8	0.49 ± 0.1	0.0075	[7]
AG Dra	554.0	0.0:	0.008	[8]
AR Pav	605	0.11	0.14	[9]
BF Cyg[b]	756.8	0.3		[10–13]
CH Cyg	756	0.0:	0.0014	[14]
CH Cyg	5294	0.058 ± 0.035	0.057	[15]
CI Cyg	855.3	0.0:	0.027	[16, 17]
V1329 Cyg	964	0.3:	21:	[18, 19]
AG Peg	816.5	0.0:	0.012	[1, 13, 20]
Z And	756.9	0.20 ± 0.11	0.024	[1]

Notes

[a] BX Mon: mass ratio $q = 6.7 \pm 1.3$, epoch is for cool star in front.

[b] BF Cyg: Mass ratio $q = 1.7 \pm 0.6$.

References

1. Kenyon, S.J. 1992, in *Evolutionary Processes in Interacting Binary Stars*, IAU Symposium No. 151, edited by Y. Kondo et al. (Kluwer Academic, Dordrecht), p. 137
2. Skopal, A. et al. 1988, *IAU Coll.* No. 103, p. 289
3. Munari, U. et al. 1988, *A&A*, **198**, 173
4. Vogel, M. 1991, *A&A*, **249**, 173
5. Oliversen, N. et al. 1985, *ApJ*, **295**, 620
6. Garcia, M.R., & Kenyon, S.J. 1988, in *The Symbiotic Phenomenon*, IAU Coll. No. 103 (Kluwer Academic, Dordrecht), p. 27
7. Dumm, T., Muerset, U., Nussbaumer, H., Schild, H., Schmid, H.M., Schmutz, W., and Shore, S.N. 1998, *A&A*, **336**, 637
8. Mikolajewska, J. et al. 1995, *AJ*, **109**, 1289
9. Kenyon, S.J. 1986, *The Symbiotic Stars* (Cambridge University Press, Cambridge)
10. Boyarchuk, A.A. 1993, in *The Realm of Interacting Binary Stars*, edited by J. Sahade, G.E. McCulsky, and Y. Kondo (Kluwer Academic, Dordrecht), p. 189
11. Mikolajewska, J. et al. 1989, *AJ*, **98**, 1427
12. Mikolajewska, J., & Mikolajewski, M. 1988, *IAU Coll.* No. 103, p. 299
13. Slovak, M. et al. 1988, *IAU Coll.* No. 103, p. 265
14. Hinkle, K.H. et al. 1993, *AJ*, **105**, 1074
15. Mikolajewski, M. 1988, *IAU Coll.* No. 103 (Kluwer Academic, Dordrecht), p. 221
16. Kenyon, S.J. et al. 1991, *AJ*, **101**, 637
17. Mikolajewska, J. 1997, *Physical Processes in Symbiotic Binaries and Related Systems* (Copernicus Foundation for Polish Astronomy, Warsaw), p. 3
18. Vogel, M. 1990, thesis, ETH, Zürich
19. Nussbaumer, H., & Vogel, M. 1991, *A&A*, **248**, 81
20. Kenyon, S. et al. 1993, *AJ*, **106**, 1573

REFERENCES

1. Szkody, P., & Cropper, M. 1988, in *Multiwavelength Astrophysics*, edited by F.A. Córdova (Cambridge University Press, Cambridge), p. 109
2. Shore, S.N., Livio, M., & van den Heuvel, E.P.J. 1994, *22nd Saas Fee Advanced Course: Interacting Binaries*, edited by H. Nussbaumer and A. Orr (Springer-Verlag, Berlin)
3. Ritter, H. 1990, *A&AS*, **85**, 1179
4. Downes, R.A., Webbink, R.F., & Shara, M.M. 1997, *PASP*, **109**, 345
5. Cassatella, A., & Viotti, R. 1990, *Physics of Classical Novae*, IAU Coll. No. 122 (Springer-Verlag, Berlin)
6. Mauche, C.W. 1990, *Accretion-Powered Compact Binaries* (Cambridge University Press, Cambridge)
7. Vogt, N. 1992, *Viña del Mar Workshop on Cataclysmic Variable Stars* (ASP Conference Series, Provo)
8. Regev, O., & Shaviv, G. 1993, *Second Technion Haifa Conference: Cataclysmic Variables and Related Physics* (Institute of Physics, Bristol)
9. Bianchini, A., Della Valle, M., & Orio, M. 1995, *Padova-Abano Conference on Cataclysmic Variables: Inter Class Relations* (Kluwer Academic, Dordrecht)
10. Starrfield, S.G. 1988, in *Multiwavelength Astrophysics*, edited by F.A. Córdova (Cambridge University Press, Cambridge), p. 159
11. Shara, M.M. 1989, *PASP*, **101**, 5
12. Starrfield, S.G. 1993, in *The Realm of Interesting Binary Stars*, edited by J. Sahade, G.E. McClusky, and Y. Kondo (Kluwer Academic, Dordrecht), p. 209
13. Duerbeck, H.W. 1987, *Space Sci. Rev.*, **45**, 1; *A Reference Catalogue and Atlas of Galactic Novae* (Reidel, Dordrecht)
14. Webbink, R.F., Livio, M., Truran, J.W., & Orio, M. 1987, *ApJ*, **314**, 653
15. Selvelli, P.L., Cassatella, A., & Gilmozzi, R. 1992, *ApJ*, **393**, 289
16. Córdova, F.A. 1995, in *X-ray Binaries*, edited by W.G.H. Lewin, J. van Paradijs, and E.P.J. van den Heuvel (Cambridge University Press, Cambridge), p. 331
17. Chanmugam, G. 1992, *ARA&A*, **30**, 143
18. Downes, R.A. 1986, *ApJ*, **307**, 170
19. Shara, M.M., et al. 1990, in *Physics of Classical Novae*, IAU Coll. No. 122 (Springer-Verlag, Berlin), p. 57
20. Kenyon, S.J. 1986, *The Symbiotic Stars* (Cambridge University Press, Cambridge)
21. Mikolajewska, J., Friedjung, M., Kenyon, S., & Viotti, R. 1988, *The Symbiotic Phenomenon*, IAU Coll. No. 103 (Kluwer Academic, Dordrecht)
22. Allen, D.A. 1984, *Proc. ASA*, **5**, 369
23. Nussbaumer, H. 1992, *IAU Symposium* 151 (Kluwer Academic, Dordrecht), p. 429

Chapter 18

Supernovae

J. Craig Wheeler and Stefano Benetti

Supernovae represent the catastrophic endpoint of evolution of stars. Study of them provides clues to the progenitor evolution, to the explosion mechanism, to the origin of heavy elements, and to their use as distance calibrators. To date around 1200 optical supernova outbursts have been discovered, the vast majority in external galaxies. A catalog of supernovae is given by Barbon, Cappellaro, and Turatto [1]. For access to supernova information on the World Wide Web, see http://cssa.stanford.edu/marcos/sne.html.

18.1 SPECTRAL TYPES

Supernovae are primarily classified by their spectral evolution with complementary consideration of their light curve morphology [2]. The traditional categories have been defined as Type II for those

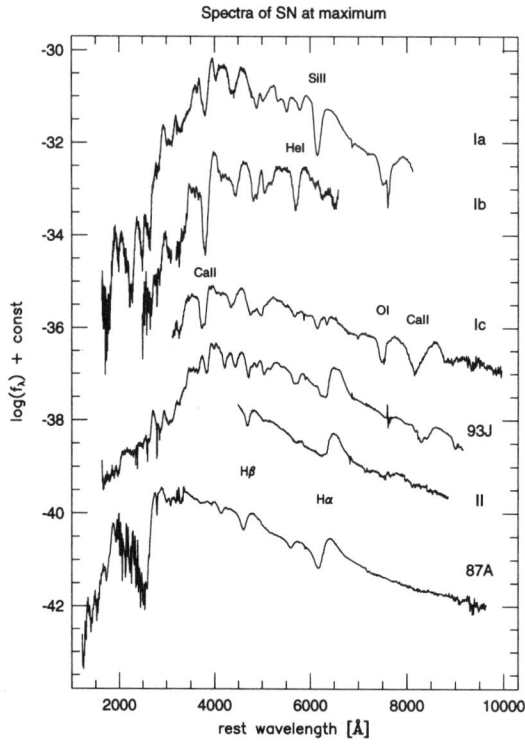

Figure 18.1. Representative spectra near maximum light of Type Ia (SN 1992A [3]), SN 1993J [4], Type II plateau (SN 1992H [5]), SN 1987A [6], Type Ib (SN 1984L [7, 8]), Type Ic (SN 1987M [9]).

that display conspicuous evidence for hydrogen and Type I for those that do not. Here the current nomenclature is maintained that defines three categories of Type I, Types Ia, Ib, and Ic, according to the major aspects of their spectral evolution. An accommodation of the suspected physical characteristics of these categories is made by separating events that are associated with population I environments, Type II, Type Ib and Type Ic, from the Type Ia supernovae that are generally associated with older stellar populations. Figure 18.1 gives a sample of spectra for some recognizable categories near maximum light. At this phase the spectra are predominantly composed of the blended P-Cygni profiles of individual lines. In the later, nebular, phase the features are predominantly due to emission lines.

18.2   OLDER POPULATION, TYPE IA SUPERNOVAE

Type Ia supernovae appear in all morphological types of galaxies. In spiral galaxies they are concentrated in spiral arms only to the extent that the oldest stars are [10, 11]. They show no correlation with giant H II regions [12, 13]. They do not seem to be strongly associated with the halos or bulges of spiral galaxies, and their properties may depend on the distance from the galactic center [14]. Type Ia supernovae show no conspicuous evidence for hydrogen in the spectra at any phase. The majority of Type Ia supernovae that have been observed to date show very similar light curves and spectral evolution. Type Ia supernovae are characterized by elements of intermediate mass, O, Mg, S, Si, and Ca, near maximum, and iron-peak elements, predominantly Fe II, beginning about 20 d after optical maximum. Near maximum light, the blend of lines of Si II λ "6355" make a prominent P-Cygni absorption with a minimum near 6150 Å. This feature is often taken as a defining characteristic of

SNI M_V template light curves

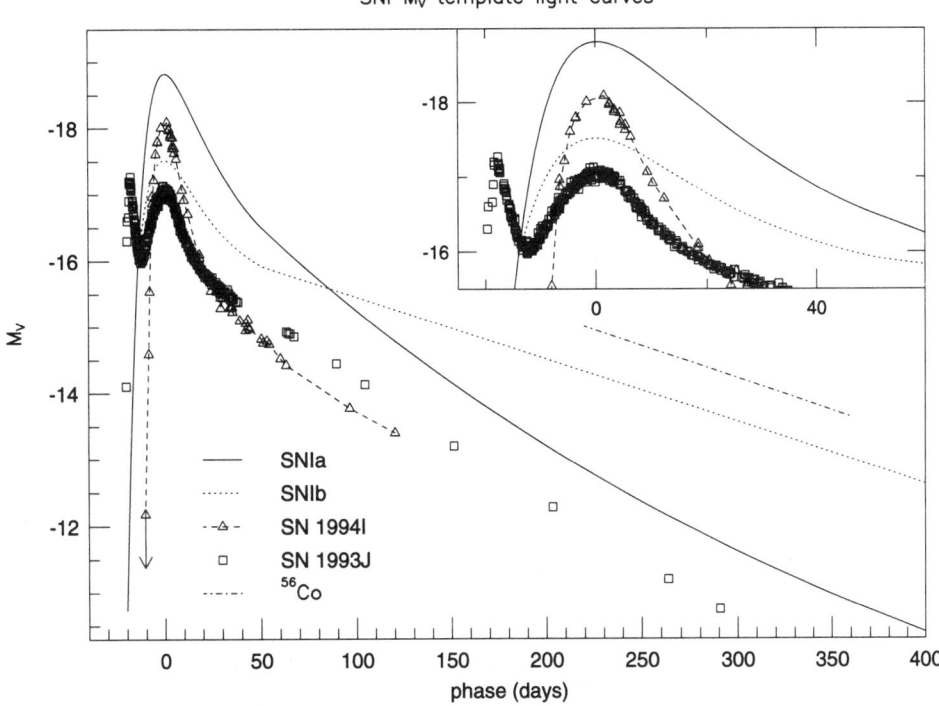

Figure 18.2. Composite V light curves of Type Ia, hydrogen-deficient events like Type Ib SN 1984L [8], hydrogen- and helium-deficient events with steep late declines like Type Ic SN 1994I [28], and the transition event SN 1993J [29]. For the Type Ia, b, c events $H_0 = 75$ km s^{-1} Mpc^{-1} is assumed.

Type Ia supernovae, but care must be exercised not to confuse it with Hα in unreduced spectra. In the later, nebular phase, the spectrum is dominated by [Fe II] and [Fe III], but continues to show Ca II in absorption.

Type Ia supernovae are intrinsically the brightest class of supernovae. The peak blue magnitude of a canonical Type Ia supernova is estimated to be $-18.4 \pm 0.3 + 5 \log_{10} h$ where h is the Hubble constant in units of 100 km s^{-1} Mpc^{-1}. The light curve consists of a peak about 30 d wide followed by an exponential decline with a slope of 0.012–0.015 mag. d^{-1}. Type Ia supernovae are characterized by a secondary maximum or inflection that extends from the V band to at least 2 μm about 20 d after peak light [15]. No radio or X-ray emission has been detected from a Type Ia supernova. Type Ia supernovae are apparently not strongly polarized in the continuum, but may show polarized spectral features (see Table 18.3). A composite V light curve for Type Ia supernovae is given in Figure 18.2. A composite spectral evolution of a typical Type Ia supernova is illustrated in Figure 18.3.

Definite departures from the canonical behavior of Type Ia supernovae have also been observed. One extreme is represented by SN 1991T [16, 17] which was perhaps 0.6 mag. brighter in V at maximum than a typical Type Ia supernova with a broader peak, bluer color, and slower decline from maximum. SN 1991T showed Fe III features at maximum, but not the classical Si II λ6355 blend nor S or Ca. The Si II feature was observed weaker than normal after maximum, as was Ca II. The secondary infrared peak occurred later and dimmer than the average event. The other extreme is represented by SN 1991bg [18, 19], which was dimmer by about 1.4 mag. in V at maximum than a typical Type Ia supernova with a redder, narrower light curve peak and no evidence for a secondary infrared peak. SN 1991bg had a distinct Ti II absorption trough at 4200 Å, a paucity of Fe features, and other spectral

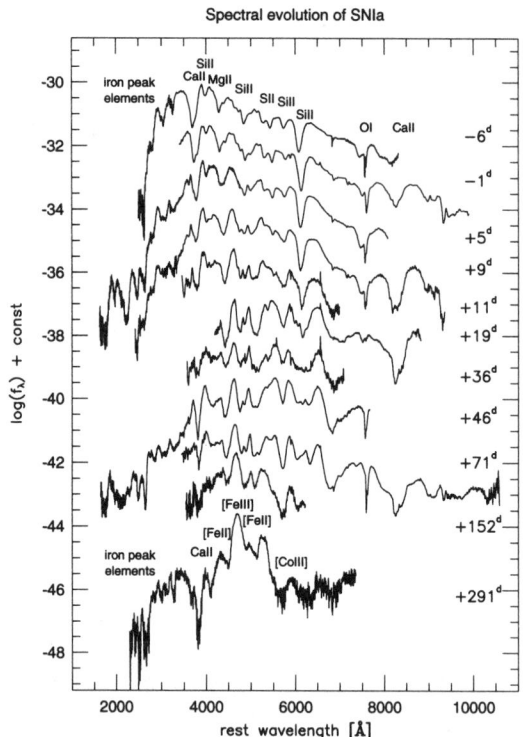

Figure 18.3. Composite spectral evolution of a typical Type Ia supernova. The data are a composite of SN 1992A in the UV and optical [3] and SN 1989B [25] in the optical.

anomolies. It is not known at this time whether or not most Type Ia supernovae are intrinsically of the canonical type with relatively small dispersion in properties and the exceptions are relatively rare, or whether or not the distribution of Type Ia supernova is more uniformly populated between the extremes of SN 1991bg and SN 1991T and the perception of a canonical type is due to selection effects.

The spectra and light curves of Type Ia supernovae are generally thought to be consistent with a thermonuclear explosion of a white dwarf that leaves no compact remnant. The light curve is thought to be powered by radioactive decay of ^{56}Ni and ^{56}Co (Table 18.7) produced in the explosion. The progenitor evolution is thought to involve binary mass transfer onto a white dwarf, but no self-consistent evolutionary scheme has been devised and no direct evidence for duplicity exists.

Prototype events are SN 1937C [20, 21], SN 1972E [22], SN 1981B [23], SN 1989B [24, 25], SN 1992A [3, 26], SN 1991T [16, 17], SN 1991bg [18, 19], and SN 1994D [27].

18.3 YOUNG POPULATION SUPERNOVAE

18.3.1 Type II Supernovae

Type II supernovae are characterized by the evidence for hydrogen in their spectra. The Hα line is prominent. In some cases the very early spectrum is principally a continuum and the Hα line strengthens with time. For canonical Type II supernovae, the later, nebular phase is very different from that of Type Ia supernovae, being dominated by Hα, Na D, [O I] $\lambda\lambda$ 6300, 6364, [Ca II] $\lambda\lambda$7291, 7323, and the Ca II IR triplet. Type II supernovae are associated with spiral arms and H II regions in

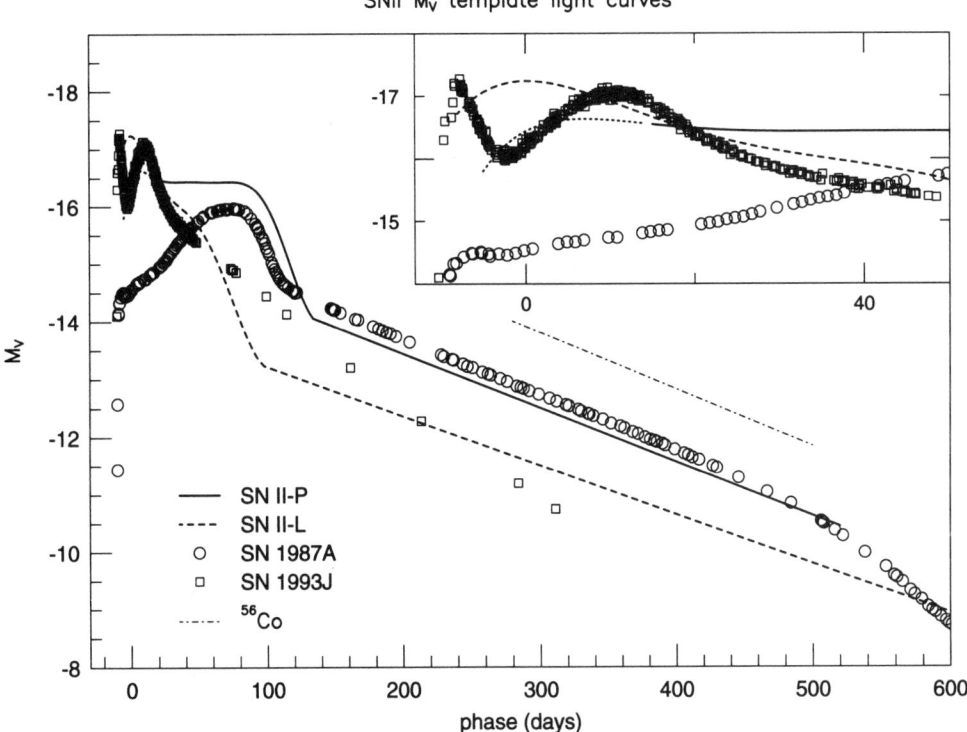

Figure 18.4. Composite V light curves (assuming $H_0 = 75 \, \text{km s}^{-1} \, \text{Mpc}^{-1}$) for Type II plateau and linear supernovae along with SN 1987A [36] and SN 1993J [29].

spiral galaxies and hence with population I stellar environments [12, 13]. None has yet been observed in an elliptical galaxy. A number have been observed in radio (see Table 18.2) and X-rays, the emission of which is attributed to interaction of the ejecta with circumstellar matter. Type II supernovae may typically be polarized at the 1% level (see Table 18.3). Type II supernovae show a large range in peak brightness from nearly as bright as a Type Ia supernova to considerably fainter. Typical Type II supernovae are about 1.5 mag. dimmer than typical Type Ia. Type II supernovae have been subdivided according to light curve and spectral behavior.

Plateau Type II supernovae are characterized by broad Hα and a light curve that shows a prominent plateau. A characteristic light curve of a typical Type II plateau supernova is shown in Figure 18.4. The Hα line generally shows a blue-shifted absorption with net emission in the strong emission component. A representative spectral evolution is given in Figure 18.5. The plateau is reproduced in models in which the explosion occurs in a red supergiant envelope. The plateau represents the phase when a recombination wave moves inward in mass at nearly constant temperature, releasing the energy deposited by the shock. The general interpretation is that these supernovae result from the explosion of a red supergiant of moderately large mass, in excess of about $10\mathcal{M}_\odot$ which retains a substantial portion of its original mass in the hydrogen envelope. Prototype events are SN 1969L [30], SN 1986I [31], SN 1988A [32], SN 1990E [33, 34], and SN 1992H [5].

Type II linear supernovae are characterized by broad hydrogen lines and an exponential decline by about 5 magnitudes in 100 d after maximum with little or no evidence of a plateau. These events may be distinct from the plateau events or the extremum of a continuous distribution of properties. In some linear Type II supernovae, SN 1979C and SN 1980K, the Hα line shows no blue-shifted absorption.

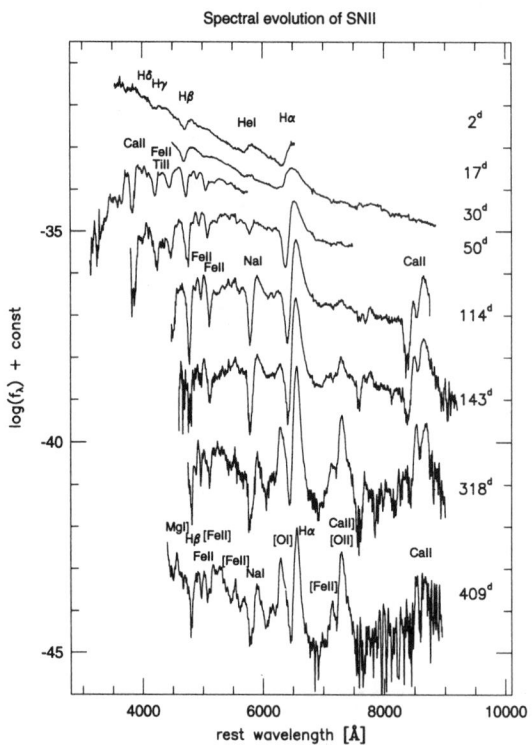

Figure 18.5. Spectral evolution of a typical Type II plateau, SN 1992H [5]. The first spectrum is from SN 1988A (McDonald Observatory, unpublished).

The Hα flux in SN 1980K showed an exponential decline for 2 to 3 years after outburst, but then leveled out. It has been constant for years [35]. There is some evidence that the peak luminosity of most linear events is rather homogeneous with SN 1979C being an especially bright exception [37]. Most of these events are about 1.5 mag. dimmer than a typical Type Ia supernova, but SN 1979C was comparable in brightness. A characteristic light curve for the more common, dimmer Type II linear supernova is given in Figure 18.4. The statistics are poor, especially on the tail, so this composite light curve must be regarded as provisional. The lack of a distinct plateau in the linear events is generally interpreted as evidence for a rather small hydrogen envelope, perhaps a few $\mathcal{M}_\odot$. There is no accepted interpretation for the origin of the implied mass loss. There is some evidence from the modeling of light curves that the outer envelopes of linear events are helium rich. SN 1990K [38] shows evidence for a substantial helium mantle. Prototype events are SN 1979C [39], SN 1980K [40], and SN 1990K [38].

Some Type II supernovae are characterized by narrow emission lines superposed on the normal broad Balmer lines, especially Hα. The narrow component is interpreted as emission from a surrounding circumstellar nebula. An example is SN 1987F [41]. Some events with narrow lines also show a very slow decay. An example is SN 1988Z [42, 43]. Such events are thought to derive a substantial portion of their optical emission from circumstellar interaction.

Despite its historically bright apparent magnitude, SN 1987A (see Section 18.4) established that there is a class of Type II supernovae that are intrinsically subluminous. This is interpreted as requiring a relatively compact structure, e.g., a blue supergiant, so that the initial shock energy is dissipated in adiabatic expansion. The resulting light curve displays a peak that is delayed from the time of

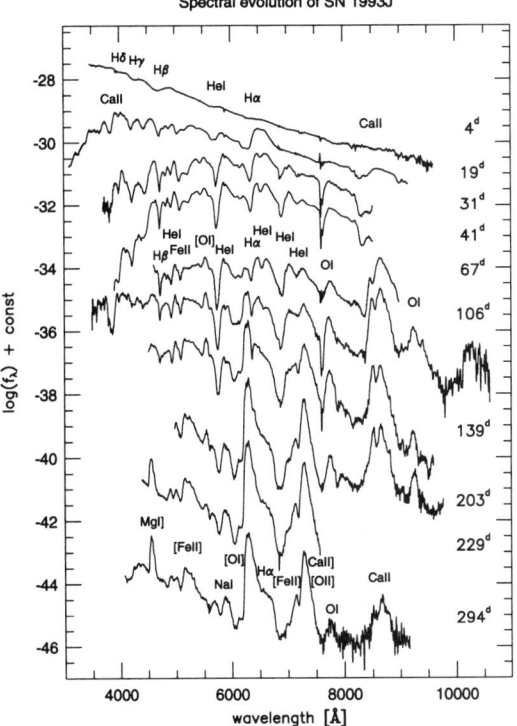

Figure 18.6. Spectral evolution of SN 1993J [4].

explosion by months and that is powered entirely by radioactive decay. In practice, it is somewhat difficult to identify such events because intrinsically low luminosity must be differentiated from galactic extinction.

A number of supernovae, e.g., SN 1987K [44] and SN 1993J [4, 29] (also SN 1996B and SN 1996cb), have been observed to make a transition from an early photospheric phase that shows strong evidence of the Balmer lines of hydrogen but little evidence of helium to a later phase that shows evidence for He I as well. In the nebular phase there is little or no evidence for spectral features of either H or He in SN 1987K, although SN 1993J continued to show a shell of Hα in emission. SN 1954A showed evidence for both Balmer lines and He I lines near maximum and hence might be a candidate for this category [45]. The fading of the high-excitation He I lines may be predominantly an excitation effect, but the early appearance of He I lines and the fading of the Balmer lines in the nebular phase are interpreted as requiring a relatively small hydrogen envelope, perhaps a few tenths of $\mathcal{M}_\odot$, compared to the canonical Type II supernova. The spectral evolution of SN 1993J is shown in Figure 18.6. Both SN 1987K and SN 1993J had light curves with relatively narrow peaks. The V magnitude light curve of SN 1993J is shown in Figures 18.2 and 18.4 in contrast to canonical Type I and Type II supernovae, respectively. Note that SN 1993J showed an early spike due to shock breakout followed by a primary maximum about 30 days later. Spectropolarimetry and line profiles for SN 1993J suggest spatial asymmetries in the explosion. Theoretical models suggest an origin in the evolved primary in a binary system, but there is no firm evidence for duplicity at this writing.

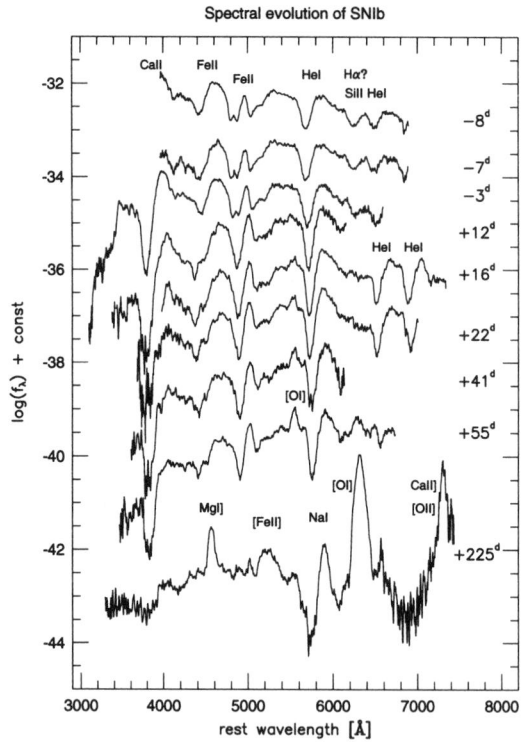

Figure 18.7. Spectral evolution of the prototype Type Ib, SN 1984L [8]. The last spectrum is from SN 1983N [47].

18.3.2 Hydrogen- and Helium-Deficient Supernovae

Some supernovae associated with young population environments are characterized as Type I in terms of the most obvious aspect of their spectral evolution, the absence of conspicuous evidence for hydrogen, but in many ways are generically related to Type II events.

Supernovae classified as Type Ib are characterized by the absence of conspicuous Balmer lines near maximum but the presence of strong absorption lines of He I in the month or so after maximum. Some weak, broad Hα may be present [46]. The late emission line phase is very reminiscent of canonical Type II supernovae, but with no detectable optical evidence for either H or He. The nebular spectrum shows strong lines of [Mg I] λ4571, Na D, [O I] λλ 6300, 6364, [Ca II] λλ 7291, 7323, and the Ca II IR triplet. Only a few cases of Type Ib are known that can be identified by the presence of strong He I lines in the optical spectrum. Among these are SN 1983N and SN 1984L. SN 1985F is often put in this class because of the shape of its optical light curve and the nature of its late-time spectrum, but there were no spectral observations near maximum so an ambiguity remains. SN 1990W is another possible candidate. The spectral evolution of SN 1984L is shown in Figure 18.7. Radio emission was detected from SN 1983N and SN 1984L, suggesting the presence of a rather substantial circumstellar medium (see Table 18.2), but no radio emission was observed from SN 1985F.

The light curves of these helium-rich events tend to be dimmer and redder than Type Ia at maximum and show some dispersion in the width of the peak of the light curve and the slope of the tail. These events do not show the secondary maximum in the IR that characterizes Type Ia supernovae. On the tail, SN 1984L and SN 1985F showed a flat decline of 0.01 mag. d^{-1}, near that expected for the decay

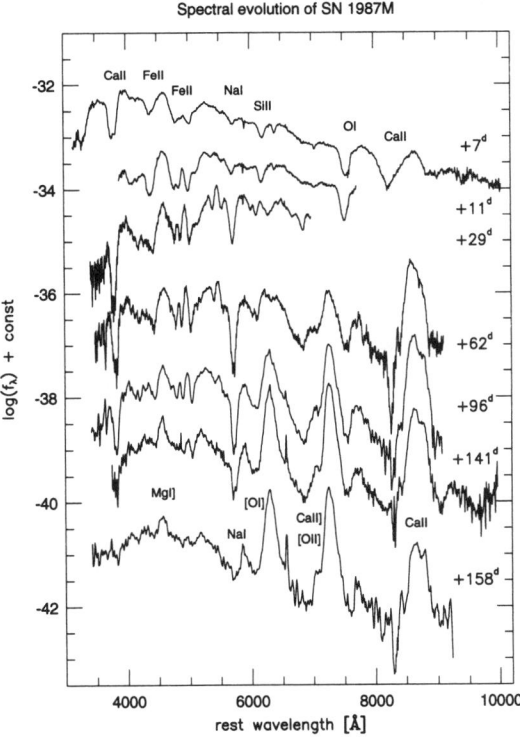

Figure 18.8. Spectral evolution of a well-studied Type Ic supernova, SN 1987M [9].

of ^{56}Co as shown in the composite V light curve given in Figure 18.2. SN 1983N showed a somewhat steeper decline of 0.013 mag. d^{-1} very similar to that of SN 1993J. This steeper decay may represent in part a smaller ejecta mass to trap γ-rays, but other factors may be involved.

The association of Type Ib with spiral arms and H II regions [12, 13] suggest that these helium-rich events arise from relatively massive stars of population I and that they undergo explosion by core collapse. There is no direct evidence for the latter. Association of the helium-rich events with Wolf–Rayet and other helium star progenitors has been discussed, but not proven. Binary mass transfer and stellar winds are thought to play a role in the loss of the hydrogen envelope. There is no direct evidence for duplicity. The weak or absent evidence for H near maximum suggests that these events had substantially less remaining H than even the transition events like SN 1993J.

Prototype events are SN 1983N [7] and SN 1984L [8].

Helium-deficient events spectrally classified as Type Ic are characterized by the absence of evidence for both H and He in optical spectra both near maximum and in the month or two after maximum when H is strong in the transition events like SN 1993J and He I is strong in the helium-rich events like SN 1984L. High velocity He I λ10830 was observed near maximum in SN 1994I, so Type Ic cannot be completely helium deficient [48,49]. Near maximum the spectrum is characterized by a strong absorption of O I λ7774, absorption of Si II λ6355 that is considerably weaker than that for Type Ia, and otherwise by features that are mostly blends of Fe II. In the nebular emission line phase, the spectra resemble those of the helium-rich events. The spectral evolution of Type Ic SN 1987M is shown in Figure 18.8.

The light curves of the helium-deficient events are similar to the helium-rich events in magnitude and color near maximum, but essentially all well-studied events initially decline more rapidly from maximum than either Type Ia or Type Ib supernovae. Some Type Ic supernovae have relatively shallow

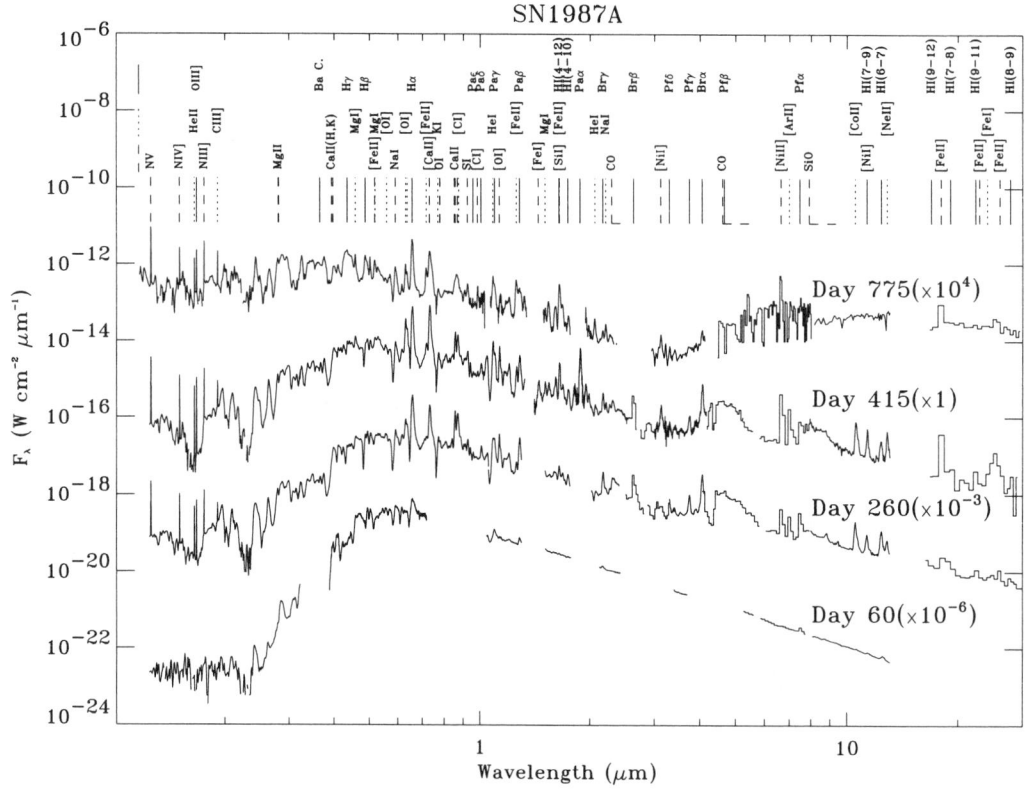

Figure 18.9. Spectral evolution of SN 1987A [53].

late-time decay very similar to SN 1993J and the Type Ib events in this light curve category. Other spectrally identified Type Ic events show late-time light curves with some dispersion but with declines even steeper than Type Ia supernovae. There are perceptible differences in spectral details, but spectral characteristics are as yet inadequate to categorize the light curve behavior. An example of a Type Ic with rapid decline, SN 1994I [28], is given in Figure 18.2.

Circumstantial evidence suggests that Type Ic supernovae explode by core collapse, but again there is no firm evidence. The spectral evolution is qualitatively consistent with a progenitor that has lost both its hydrogen and helium envelopes. Again duplicity may play a role and there may be a connection with Wolf–Rayet or other helium stars, but neither has been confirmed.

Prototype events are SN 1983I [50], SN 1983V [50], SN 1987M [9], and SN 1994I [28, 46, 48, 49].

18.4 SN 1987A

Supernova 1987A in the Large Magellanic Cloud was observed throughout the electromagnetic spectrum and was a detected source of neutrinos. Reviews are given by Arnett et al. [36], Hillebrandt and Höflich [51], Imshennik and Nadyozhin [52], McCray [53], and Phillips and Suntzeff [54].

Figure 18.9 shows the spectral evolution from the UV to the far infrared. Line profiles in the early phase indicate velocities up to $25\,000$ km s^{-1}, and the emission during the nebular phase is characterized by velocities $\lesssim 2500$ km s^{-1}. The optical spectrum was comparable to other Type II supernovae (Figure 18.1). The spectrum gives evidence for mixing of hydrogen and helium at the

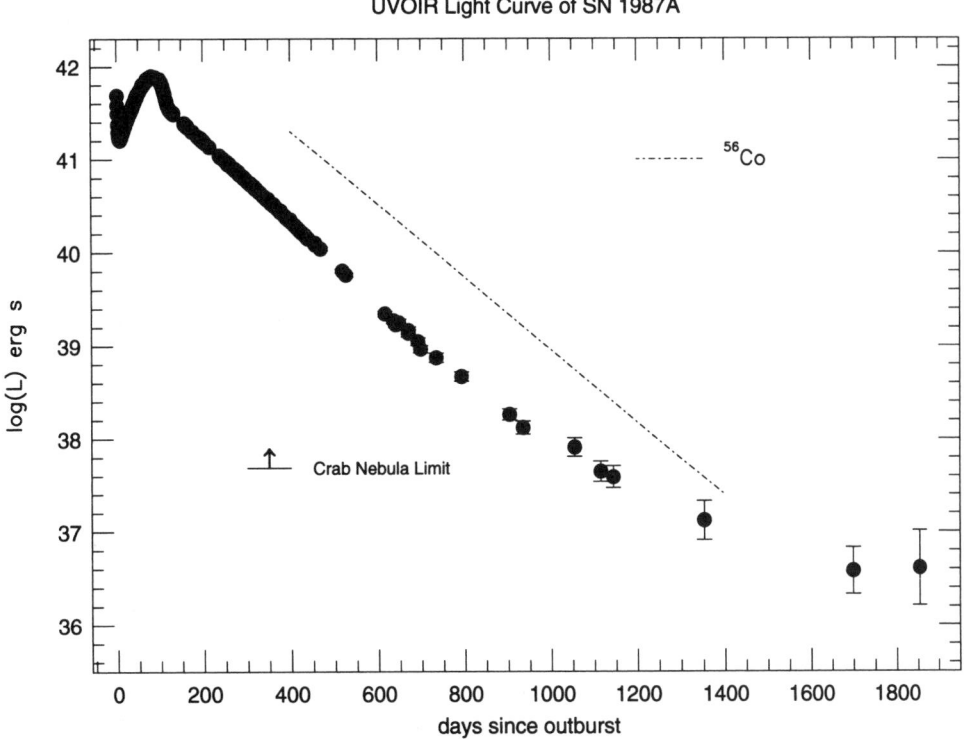

UVOIR Light Curve of SN 1987A

Figure 18.10. Bolometric light curve of SN 1987A. (Courtesy of N. Suntzeff.)

core/envelope interface and for outward mixing of ^{56}Ni. This mixing may be macroscopic rather than microscopically homogeneous. The IR spectra showed evidence for molecules, especially carbon monoxide and silicon monoxide.

There was evidence for dust formation beginning at about 450 days after the explosion. The dust appears to have two components, small grains distributed diffusely that affect short wavelengths and clumps of very optically thick grains the effect of which is wavelength independent. The rate of decline of the light curve decreased after 1000 days, which may be explained by time-dependent recombination. The V light curve is given in Figure 18.4 and the bolometric light curve in Figure 18.10. Figure 18.11 gives the $B - V$ color evolution for SN 1987A in comparison with canonical Type Ia, Type II supernovae, and SN 1993J. The precursor star, Sk–69° 202, was identified but not especially well studied. The fact that the precursor was a blue supergiant accounted for the unorthodox light curve, although the physical reasons for its explosion in this state are still uncertain. The detection of neutrinos confirmed the basic process of core collapse. The information content was not sufficient to usefully constrain the explosion mechanism. At this writing, there is no direct evidence for a compact object. Such a compact remnant, if it exists, must have a bolometric luminosity less than a few times 10^{36} erg s^{-1}, less than about 10% that of the Crab Nebula. The extended neutrino (antielectron neutrino) signal demands that a neutron star remained for about 10 s, but the eventual collapse to form a black hole cannot be precluded.

SN 1987A also provided the first unambiguous test of the theoretically predicted production of ^{56}Ni, and its radioactive decay into ^{56}Co and then ^{56}Fe (Table 18.7). The evidence for this decay was contained in the bolometric light curve, the slope of which closely matched that expected for the Co decay from 125 to 450 d after the explosion, in the detection of the infrared lines of [Co II] at 1.547 μm

B-V curves for different SN types

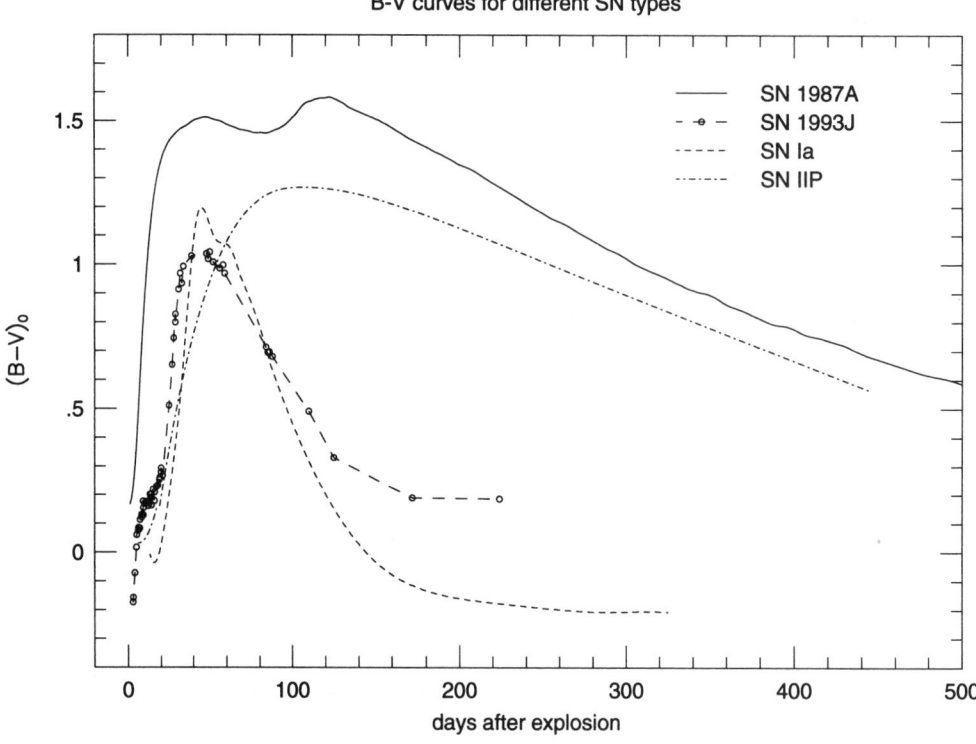

Figure 18.11. The evolution of $B - V$ color for SN 1987A and SN 1993J in comparison with mean curves representing Type Ia and plateau Type II events.

and 10.52 μm and in the direct detection of γ-ray lines of ^{56}Co and ^{57}Co. The ^{56}Co was first detected at about 160 d after outburst and ^{57}Co was detected at 1500 d. The ratio of these lines implies a ratio of the final stable decay products of about 1.5 times the solar value.

Based on the known distance of the Large Magellanic Cloud, the bolometric luminosity of the progenitor is computed to be about $100\,000 L_\odot$. Invoking structural models, this luminosity implies a helium core mass of about $6 \mathcal{M}_\odot$. Standard evolutionary models predict that such a star would have a mass of about $20 \mathcal{M}_\odot$ on the main sequence, but some mixing theories give this core mass for lower initial masses, so the total initial mass is somewhat uncertain. Fits to the light curve and spectral evolution suggest a hydrogen envelope mass of $\sim 10 \mathcal{M}_\odot$ at the time of the explosion. Various arguments suggest that the envelope was helium-enriched with a mass fraction $Y \sim 0.4$. The helium enrichment, enhancement of barium, an s-process element, and the circumstellar rings raise the possibility that the precursor accreted mass from a binary companion, but there is no definite evidence for present or past duplicity.

Early radio emission is interpreted as evidence for interaction of the ejecta with circumstellar matter. Radio flux was first detected on day 2 and reached a peak flux of S_ν (2 GHz) $\approx$ 130 mJy on day 3. Renewed detection occurred 1200 d after the explosion, reached S_ν (4.8 GHz) $\approx$ 11 mJy by day 2050 and continued to rise at about 15% per year a decade after the explosion.

Hard X-ray emission was observed beginning about 130 d after explosion, reaching a peak about day 320. This is interpreted as resulting from Compton down scattering of γ-rays within the ejecta. The maximum flux in lines of ^{56}Co at 0.847 and 1.238 MeV also peaked about the same time. Transient soft X-ray emission was also reported, the origin of which is not well explained, but thought to arise from collision of the ejecta with circumstellar clumps. Soft X-ray emission began to rise again along with the radio about 1200 days after the explosion, reached a luminosity of $L_X \sim 3 \times 10^{34}$ erg s^{-1} a decade after the explosion, and continued to increase.

SN 1987A was surrounded by a complex circumstellar environment. The principal feature is the ring of matter that fluoresces as a result of excitation by the UV emission from the supernova at shock breakout and emits strongly in UV and optical emission lines. Two fainter rings and more complex structures are also present.

There is a significant population of both red and blue supergiants in the Large Magellanic Cloud. The evolutionary state of the blue supergiants is not well established. The nitrogen-rich, narrow fluorescent emission line spectrum from the circumstellar ring around SN 1987A is widely regarded as evidence for the collision of a blue supergiant wind with a previous, slower red supergiant wind and hence evidence that the progenitor was once a red supergiant. The unexplained origin of the ring geometry and the possibility of mixing in early evolutionary phases leave some room for question. Some models propose that the ring is a protostellar remnant or an unbound excretion disk.

Here are summarized some of the properties deduced for SN 1987A:

Progenitor: SK −69° 202 spectral type B3 I.
    $M_V = 12.29 \pm 0.04$; $M_B = 12.32 \pm 0.06$.
    $R = (3 \pm 1) \times 10^{12}$ cm; $T_{eff} = 16\,500 \pm 1500$ K.
    $L = (4.5 \pm 1.5) \times 10^{38}$ erg s^{-1}.
    Main sequence mass $= (15\text{–}20)\mathcal{M}_\odot$.
    Helium core mass $= (6 \pm 1)\ \mathcal{M}_\odot$.
    Hydrogen envelope mass at explosion $\sim 10\mathcal{M}_\odot$.

Explosion:
    Kinetic energy $= (1.3 \pm 0.2) \times 10^{51}$ erg.
    Total neutrino energy $= (2 \pm 1) \times 10^{53}$ erg.
    Neutrino temperature $= (4 \pm 1)$ MeV.
    Mean neutrino energy $= 12.5 \pm 3.0$ MeV.
    Mass of ^{56}Ni $= (0.069 \pm 0.003)\mathcal{M}_\odot$.
    Detected γ-ray lines: ^{56}Co 0.847, 1.238, 2.599, 3.250 MeV; ^{57}Co 122, 136 keV.

Polarization:
    Position angle $\sim 120°$.
    Percent polarization $= 0.6$ (V band, day 40).

Circumstellar ring:
    Semimajor axis $= 0.858 \pm 0.011$ arcsec $= 6.4 \times 10^{17}$ cm at 50 pc.
    Width $= 0.122 \pm 0.022$ arcsec $= (9.0 \pm 1.6) \times 10^{16}$ cm at 50 kpc.
    Tilt angle $= 44 \pm 1°$.
    Position angle major axis $= 89° \pm 3°$.
    Expansion velocity $\simeq 10.3$ km s^{-1}.

18.5 CHARACTERISTIC SPECTRAL LINES

Table 18.1 gives the wavelength in nanometers of characteristic lines observed in the early photospheric and later nebular phases of various types of supernovae.

Table 18.1. *Characteristic spectral features of various supernova types.*

Photospheric phase	Nebular phase
Type Ia	
UV: blends of Fe II, Ni II, Ti II, V II, Co II, Cr II [1]	UV: blends of [Fe II, III, IV], [Co III](?), [Ni II](?) [2]
Mg II 279.8	
Si III 385.9	
Ca II H&K 393.4, 396.8	Ca II H&K 393.4, 396.8
Si II 413.0	
Fe II "422.0"	
Fe II 435.2	
Mg II 448.1	
Si III 456.0	blends of [Fe III] 450–480
Fe I "456.8"	
Fe III "440.4"	
Fe II "455.5"	
Fe II 492.4, 501.8	
Si II "505.1"	
Fe III "512.9"	
Fe II 516.9	
Fe II "521.5"	
Fe II 553.5	
S II "546.8," "561.2," "565.4"	
Na I 589.0, 589.6	Na I 589.0, 589.6(?)
	[Co III] 589.0, 590.8
Si II "597.2"	
Si II "635.5"	
O I 777.4	
Ca II 849.8, 854.2, 866.2	
	O I 926.1, 926.3, 926.6
Type Ib and Ic	
UV: blends of iron peak elements	
Ca II H&K 393.4, 396.8	
Mg II 448.1[a]	
Fe II "440.4"	
Fe II "455.5"	
	[Mg I] 457.1
Fe II 492.4, 501.8, 516.9	Fe II 492.4, 501.8, 516.9
He I 447.1, 501.5, 587.6, 667.8, 706.5, 728.3, 1083[a]	
Na I 589.0, 589.6	Na I 589.0, 589.6
	[O I] 557.7
	[O I] 630.0, 636.4
Si II 635.5	
C II 658.0	
	[Ca II] 729.1, 732.3
	[O II] 732.0, 733.0
O I 777.4	O I 777.4
	O I 844.6
Ca II 849.8, 854.2, 866.2	Ca II 849.8, 854.2, 866.2
	[C I] 873.0(?)
	[O I] 926.1, 926.3, 926.6
	[C I] 982.4, 985.0
	[Fe II], Si I 1.644 μm

Table 18.1. *(Continued.)*

Photospheric phase	Nebular phase
	Type II and SN 1987A
UV: blends of iron peak elements	blends of [Fe II] 240–570 [3]^b
	Mg II 279.5, 280.2
	Mg I 285.2^b
	[O II] 372.6, 372.8^b
Mg II 448.1	
Ca II H&K 393.4, 396.8	
Sr II 407.7, 421.5^b	
Ca I 422.6^b	
Ba II 455.4, 493.4^b	
	Mg I] 457.1
Sc II 467.0^b	
Fe II 492.4, 501.8, 516.9	
Fe I 526.9	
Sc II 552.7, 565.8^b	
Ba I 553.5^b	
	[O I] 557.7
Ba II 585.4^b	
He I 587.6, 1083.0	
Na I 589.0, 589.6	Na I 589.0, 589.6
Ba II 614.2^b	
Sc II 624.5	
	[O I] 630.0, 636.4
Hα, Hβ, Hγ, Hδ, Hε	Hα, Hβ, Hγ
	[Fe II] 700.0–950.0 blends
	[Fe II] 715.5
	Ca II 729.1, 732.3
	[O II] 732.0, 733.0
Ca II 849.8, 854.2, 866.2	Ca II 849.8, 854.2, 866.2
	[C I] 873.0
	[C I] 982.4, 985.0
	Paα, Paβ, Paγ, Paδ, Paε
	Brα, Brβ, Brγ
	Pfγ, Pfδ
Sr II 1.033 μm	
	O I 1.129 μm
	[Fe II] 1.257 μm^b
	[Fe II] 1.533 μm^b
	[Co II] 1.547 μm^b
	[Fe II] Si I 1.644 μm^b
	[Ni] 3.119 μm^b
	[Ni II] 6.634 μm^b
	[Ar II] 6.983 μm^b
	[Co II] 10.52 μm^b
	[Fe II] 17.93 μm^b
	SiO^b
	CO^b
	Dust^b

Notes
^{a}Especially in Type Ib.
^{b}Especially in SN 1987A.

References
1. Kirshner, R.P. et al. 1993, *ApJ*, **415**, 589
2. Ruiz-Lapuente, P. et al. 1995, *ApJ*, **439**, 60
3. Wang, L. et al. 1996, *ApJ*, **466**, 998

18.6 RADIO SUPERNOVAE

Table 18.2 gives supernovae that have been observed in the radio range [55].

Table 18.2. *Radio supernovae.*

Supernova	Type	Galaxy
1923A	II plateau	NGC 5236
1950B	II ?	NGC 5236
1957D	Unknown	NGC 5236 (M83)
1961V	II peculiar	NGC 1058
1968D	II	NGC 6946
1970G	II linear?	NGC 5457 (M101)
1978K	II	NGC 1313
1979C	II linear	NGC 4321 (M100)
1980K	II linear	NGC 6946
1981K	II	NGC 4258
1982aa	II	NGC 6464
1983N	Ib	NGC 5236 (M83)
1984L	Ib	NGC 991
1985L	II linear	NGC 5033
1986E	II linear	NGC 4302
1986J	II	NGC 891
1987A	II peculiar	LMC
1988Z	II narrow line	MCG 03-28-022
1990B	Ic	NGC 4568
1992ad	II	NGC 4411B
1993J	II transition	NGC 3031 (M81)
1994I	Ic	NGC 5194
1995N	II	MCG-2-38-017
1996N	Ib	NGC 1398
1996cb	II transition	NGC 3510
1997X	Ic	NGC 4691

18.7 POLARIZATION

Table 18.3 gives data on supernovae for which polarization information has been obtained.

Table 18.3. *Polarization of supernovae* [1, 2].

Supernova	Type	Polarization (%)
1968L	II	Undetermined
1970G	II	0.5
1972E	Ia	Undetermined
1975N	Ia	< 0.3
1981B	Ia	Undetermined
1983G	Ia	< 0.5
1983N	Ib	Detected
1986G	Ia	< 0.1
1987A	II peculiar	0.6
1992A	Ia	< 0.3
1993J	II transition	1.0 [3]
1994D	Ia	< 0.3
1994Y	II narrow line	1.5
1994ae	Ia	< 0.3
1995D	Ia	< 0.2

Table 18.3. *(Continued.)*

Supernova	Type	Polarization (%)
1995H	II	1.0
1995V	II	1.5
1996W	II	0.7
1996X	Ia	0.3
1997cb	II transition	1.5
1997X	Ic	1.5
1997Y	Ia	< 0.3
1997bp	Ia	0.6
1997bq	Ia	< 0.3
1997br	Ia	< 0.3

References
1. Wang, L., Wheeler, J.C., Li, Z., & Clocchiatti, A. 1996, *ApJ*, **467**, 435
2. Wang, L., Wheeler, J.C., & Höflich, P. 1997, *ApJ*, **476**, L27
3. Tran, H.D. et al. 1997, *PASP*, **109**, 489

18.8 SUPERNOVA RATES

The deduced rate of explosion of supernovae of different types in galaxies of different morphological type must allow for varying visibility due to absolute brightness and light curve width in estimating the probability of detection. Poor statistics make these estimates uncertain. Because of the large variation of supernova rates among galaxies of differing morphological type, some normalizing procedure must be adopted. Table 18.4 gives the estimated rates of supernovae of various spectral types normalized to the blue luminosity of the host galaxy in units of $10^{10} L_\odot$ as a function of the morphological type of the galaxy. Other estimates are given by van den Bergh and Tammann [56] and van den Bergh and McClure [57]. Assuming that the Galaxy is an Sb with a blue luminosity of $2.0 \pm 0.6 \times 10^{10} L_\odot$, the number of supernovae per century are estimated to be: Type Ia—0.3 ± 0.2; Type Ib/Ic—0.2 ± 0.2; Type II—1.7 ± 0.9 [58].

Table 18.4. *Supernova rates per 100 yrs per $10^{10} L_\odot B$ luminosity* [1, 2].[a]

Galaxy Type	Ia	Ib and Ic	II
E	0.11		
S0	0.15		
S0a, Sa	0.30	0.15	0.19
Sab, Sb	0.12	0.12	0.36
Sbc	0.22	0.10	1.19
Sc	0.50	0.54	1.45
Scd, Sc	0.48	0.09	1.87
Sdm-Im	0.20	0.30	0.40
E, S0	0.15 ± 0.06	< 0.3	< 0.3
S0a-Sb	0.20 ± 0.07	0.11 ± 0.06	0.40 ± 0.19
Sbc-Sd	0.24 ± 0.09	0.16 ± 0.08	0.88 ± 0.37

Notes
[a] All rates are proportional to $(H_0/75 \ \mathrm{km\,s^{-1}\,Mpc^{-1}})^2$.

References
1. Cappellaro, E., Turatto, M., Benetti, S., Tsvetkov, D.Yu., Bartunov, O.S., & Makarova, L.N. 1993, *A&A*, **273**, 383
2. Cappellaro, E., Turatto, M., Tsvetkov, D.Yu., Bartunov, O.S., Pollas, C., Evans, R., & Hamuy, M. 1997, *A&A*, **322**, 431

18.9 OLD SUPERNOVAE, HISTORICAL SUPERNOVAE, AND SUPERNOVA REMNANTS

Table 18.5 gives supernovae that have been observed over 5 years after explosion. These may be regarded as supernovae very late in the nebular phase or as very young supernova remnants.

Over 175 high-surface-brightness supernova remnants have been observed in the optical, radio, and X-ray spectra in the Galaxy. A few dozen have been observed in other galaxies, especially the Magellanic Clouds, Andromeda, M33, and M82. A catalog of supernova remnants is given by Green [59]. Table 18.6 gives information on supernovae observed in the historical past.

Table 18.5. *Long-lived supernovae.*

Supernova	Type	Galaxy
1885A	I?	M31
1957D	II?	NGC 5236 (M83)
1968D	II	NGC 6946
1961V	II peculiar	NGC 1058
1970G	II linear(?)	NGC 5457 (M101)
1978K	II	NGC 1313
1979C	II linear	NGC 4321 (M100)
1980K	II linear	NGC 6946
1986J	Unknown	NGC 891
1987A	II peculiar	LMC
1988Z	II narrow line	MCG-03-28-022

Table 18.6. *Historical supernovae.*

Event	Extended remnant	Compact remnant
185		
1006	Shell, Balmer dominated	
1054 (Crab Nebula)	Filled, helium-rich	
1572 (Tycho)	Shell, Balmer dominated	33 msec pulsar
1604 (Kepler)	Shell, Balmer dominated	
~ 1670 (Cas A)	Oxygen rich knots, He, N-rich, quasistationary flocculi	

18.10 RADIOACTIVE DECAY

A principal source of power of the light curve peak and especially the late-time tail of many supernovae is the decay of radioactive ^{56}Ni and its daughter ^{56}Co. Other radioactive species that may contribute late-time power are ^{57}Co and ^{44}Ti. The decay rates for these species are given in Table 18.7.

Table 18.7. *Radioactive decay time scales in supernovae.*

Element	Half-life	e-Fold	Per magnitude
^{56}Ni [1]	6.10 d	8.80 d	8.11 d
^{56}Co [1]	77.12 d	111.3 d	102.3 d
^{57}Co [2]	271.8 d	392.1 d	361.2 d
^{44}Ti [3]	54.2 yr	78.2 yr	72.0 yr

References
1. Huo, J. et al. 1987, *Nuclear Data Sheets*, **51**, 1
2. Burrows, T.W., & Bhat, M.R. 1986, *Nuclear Data Sheets*, **47**, 1
3. Frekers, D. et al. 1983, *Phys. Rev. C*, **28**, 1756

REFERENCES

1. Barbon, R., Cappellaro, E., & Turatto, M. 1989, *A&A*, **81**, 421 (see also: http://www.pd.astro.it/supern/).
2. Filippenko, A.V. 1997, *ARAA*, **35**, 309
3. Kirshner, R.P. et al. 1993, *ApJ*, **415**, 589
4. Lewis, J.R. et al. 1994, *MNRAS*, **266**, L27
5. Clocchiatti, A. et al. 1996, *AJ*, **111**, 1286
6. Phillips, M.M., Heathcote, S.R., Hamuy, M. & Navarrete, M. 1988, *AJ*, **95**, 1087
7. Panagia, N. 1985, *Lect. Notes Phys.*, **224**, 14
8. Harkness, R.P. et al. 1987, *ApJ*, **317**, 355
9. Filippenko, A.V., Porter, A.C., & Sargent, W.L.W. 1990, *AJ*, **100**, 1575
10. Maza, J., & van den Bergh, S. 1976, *ApJ*, **204**, 519
11. McMillan, R.J., & Ciardullo, R. 1996, *ApJ*, **473**, 707
12. Van Dyk, S.D. 1992, *AJ*, **103**, 1788
13. Bartunov, O.S., Tsvetkov, D.Yu., & Filimonova, I.V. 1994, *PASP*, **106**, 1276
14. Wang, L., Höfich, P., & Wheeler, J.C. 1997, *ApJ*, **483**, L29
15. Elias, J.H., Matthews, K., Neugebauer, G., & Persson, S.R. 1985, *ApJ*, **296**, 379
16. Filippenko, A.V. et al. 1992, *ApJ*, **384**, L15
17. Phillips, M.M., Wells, L.A., Suntzeff, N.B., Hamuy, M., Leibundgut, B., Kirshner, R.P., & Foltz, C.B. 1992, *AJ*, **103**, 1632
18. Filippenko, A.V. et al. 1992, *AJ*, **104**, 1543.
19. Leibundgut, B. et al. 1993, *AJ*, **105**, 301
20. Baade, W., & Zwicky, F. 1938, *ApJ*, **88**, 411
21. Minkowski, R. 1939, *ApJ*, **89**, 156
22. Kirshner, R.P., Oke, J.B., Penston, M.V., & Searle, L. 1973, *ApJ*, **185**, 303
23. Branch, D., Lacy, C.H., McCall, M.L., Sutherland, P.G., Uomoto, A., Wheeler, J.C., & Wills, B.J. 1983, *ApJ*, **270**, 123
24. Barbon, R., Benetti, S., Cappellaro, E., Rosino, L., & Turatto, M. 1990, *A&A*, **237**, 79
25. Wells, L. et al. 1994, *AJ*, **108**, 2233
26. Ruiz-Lapuente, P., Kirshner, R.P., Phillips, M.M., Challis, P.M., Schmidt, B.P., Filippenko, A.V., & Wheeler, J.C. 1995, *ApJ*, **439**, 60
27. Cumming, R.J., Lundqvist, P., Smith, L.J., Pettini, M., & King, D.L. 1996 *MNRAS*, **283**, 1355
28. Richmond, M.W. et al. 1996, *AJ*, **111**, 327
29. Wheeler, J.C., & Filippenko, A.V. 1996, in *Supernovae and Supernova Remnants*, edited by R.A. McCray and Z. Wang (Cambridge University Press, Cambridge), p. 241
30. Ciatti, F., Rosino, L., & Bertola, F. 1971, *Mem. Soc. Astron. Italia*, **42**, 163
31. Pennypacker, C.R. et al. 1989, *AJ*, **97**, 186
32. Turatto, M., Cappellaro, E., Benetti, S., & Danziger, I.J. 1994, *MNRAS*, **265**, 483
33. Schmidt, B.P. et al. 1993, *AJ*, **105**, 6
34. Benetti, S., Cappellaro, E., Turatto, M., Della Valle, M., Mazzali, P.A., & Gouiffes, C. 1994, *A&A*, **285**, 147
35. Fesen, R.A., & Becker, R.H. 1990, *ApJ*, **351**, 437
36. Arnett, W.D., Bahcall, J.N., Kirshner, R.P., & Woosley, S.E. 1989, *ARAA*, **27**, 629
37. Young, T.R., & Branch, D. 1989, *ApJ*, **342**, L55
38. Cappellaro, E., Danziger, I.J., Della Valle, M., Gouiffes, C., & Turatto, M. 1995, *A&A*, **293**, 723
39. Branch, D., Falk, S.W., McCall, M.L., Rybski, P., Uomoto, A.K., & Wills, B.J. 1981, *ApJ*, **244**, 780
40. Uomoto, A., & Kirshner, R.P. 1986, *ApJ*, **308**, 685
41. Filippenko, A.V. 1989, *AJ*, **97**, 726
42. Stathakis, R.A., & Sadler, E.M. 1991, *MNRAS*, **250**, 786
43. Turatto, M., Cappellaro, E., Danziger, I.J., Benetti, S., Gouiffes, C., & Della Valle, M. 1993, *MNRAS*, **262**, 128
44. Filippenko, A.V. 1988, *AJ*, **96**, 1941
45. Branch, D. 1972, *A&A*, **16**, L43
46. Wheeler, J.C., Harkness, R.P., Clocchiatti, A., Benetti, S., Depoy, D., & Elias, J. 1994, *ApJ*, **436**, L135
47. Gaskell, C.M., Cappellaro, E., Dinerstein, H.L., Garnett, D.R., Harkness, R.P., & Wheeler, J.C. 1986, *ApJ*, **306**, L77
48. Filippenko, A. et al. 1995, *ApJ*, **450**, L11
49. Clocchiatti, A. et al. 1996, *ApJ*, **462**, 462
50. Wheeler, J.C., Harkness, R.P., Barker, E.S., Cochran, A.L., & Wills, D. 1987, *ApJ*, **313**, L69
51. Hillebrandt, W., & Höflich, P. 1989, *Rep. Prog. Phys.*, **52**, 1421
52. Imshennik, V.S., & Nadyozhin, D.K. 1989, *Sov. Sci. Rev.*, **8**, 1
53. McCray, R.A. 1993, *ARAA*, **31**, 175
54. Phillips, M.M., & Suntzeff, N.B. 1997, *Supernova 1987A: Ten Years After* (Provo, Astronomical Society of the Pacific)
55. Sramek, R.A., & Weiler, K.W. 1990, in *Supernovae*, edited by S.E. Woosley (Springer-Verlag, New York), p. 76 (see also: http://rsd-www.nrl.navy.mil/7214/weiler/sne-home.html)
56. van den Bergh, S., & Tammann, G.A. 1991, *ARAA*, **29**, 363
57. van den Bergh, S., & McClure, R.D. 1994, *ApJ*, **425**, 205
58. Cappellaro, E., Turatto, M., Benetti, S., Tsvetkov, D.Yu., Bartunov, O.S., & Makarova, L.N. 1993, *A&A*, **273**, 383.
59. Green, D.A. 1991, *PASP*, **103**, 207 (see also: http://www.mrao.cam.ac.uk/surveys/snrs/)

Chapter 19

Star Populations and the Solar Neighborhood

Gerard F. Gilmore and Michael Zeilik

19.1 THE NEARBY STARS

Table 19.1 lists the 100 nearest stars to the Earth in order of increasing distance. Positions are calculated from observed positions with corrections for proper motions; they are given in the FK5 system with equinox = J2000.0 and epoch = J2000.0. The first column gives the identifier in the Hipparcos

Catalogue [1]; the last column gives the parallax in milliarcsecs with the associated standard deviation error.

Table 19.1. *Nearby stars.*

HIP	HD number	α (2000.0) δ (2000.0)	Other name	Sp. type	V magnitude	Abs V	B − V	Parallax
70890	N/A	14 29 42.95 −62 40 46.1	α Cen C	M5 Ve	11.01	15.45	+1.81	772.33 ± 2.42
71681	128621	14 39 35.08 −60 50 13.8	α Cen B	K1 V	1.35	5.70	+0.88	742.12 ± 1.40
71683	128620	14 39 36.50 −60 50 02.3	α Cen A	G2 V	−0.01	4.34	+0.71	742.12 ± 1.40
87937	N/A	17 57 48.50 +04 41 36.2	Barnard's star	M5 V	9.54	13.24	+1.57	549.01 ± 1.58
54035	95735	11 03 20.19 +35 58 11.6	Gl 411	M2 Ve	7.49	10.46	+1.50	392.40 ± 0.91
32349	48915	06 45 08.92 −16 41 58.0	α CMa	A1 V	−1.44	1.45	+0.01	379.21 ± 1.58
92403	N/A	18 49 49.36 −23 50 10.4	Gl 729	M4.5 Ve	10.37	13.00	+1.51	336.48 ± 1.82
16537	22049	03 32 55.84 −09 27 29.7	ε Eri	K2 V	3.72	6.18	+0.88	310.75 ± 0.85
114046	217987	23 05 52.04 −35 51 11.1	Gl 887	M2 Ve	7.35	9.76	+1.48	303.90 ± 0.87
57548	N/A	11 47 44.40 +00 48 16.4	Ross 128	M4.5 V	11.12	13.50	+1.75	299.58 ± 2.20
104214	201091	21 06 53.94 +38 44 57.9	61 Cyg A	K5 Ve	5.20	7.49	+1.07	287.13 ± 1.51
37279	61421	07 39 18.12 +05 13 30.0	α CMi	F5 IV–V	0.40	2.68	+0.43	285.93 ± 0.88
104217	201092	21 06 55.26 +38 44 31.4	61 Cyg B	K7 Ve	6.05	8.33	+1.31	285.42 ± 0.72
91772	173740	18 42 46.90 +59 37 36.6	Gl 725B	M5 V	9.70	11.97	+1.56	284.48 ± 5.01
91768	173739	18 42 46.69 +59 37 49.4	Gl 725A	M4 V	8.94	11.18	+1.50	280.28 ± 2.57
1475	1326	00 18 22.89 +44 01 22.6	GX And	M2 V	8.09	10.33	+1.56	280.27 ± 1.05
108870	209100	22 03 21.66 −56 47 09.5	ε Ind	K5 Ve	4.69	6.89	+1.06	275.76 ± 0.69
8102	10700	01 44 04.08 −15 56 14.9	τ Cet	G8 Vp	3.49	5.68	+0.73	274.17 ± 0.80
5643	N/A	01 12 30.64 −16 59 56.3	Gl 54.1	M5.5 Ve	12.10	14.25	+1.85	269.05 ± 7.57
36208	N/A	07 27 24.50 +05 13 32.8	Luyten's star	M3.5	9.84	11.94	+1.57	263.26 ± 1.43
24186	33793	05 11 40.58 −45 01 06.3	Kapteyn's star	M0 V	8.86	10.89	+1.55	255.26 ± 0.86
105090	202560	21 17 15.27 −38 52 02.5	AX Mic	M0 Ve	6.69	8.71	+1.40	253.37 ± 1.13
110893	239960	22 27 59.47 +57 41 45.1	Kruger 60	M2 V	9.59	11.58	+1.61	249.52 ± 3.03
30920	N/A	06 29 23.40 −02 48 50.3	Ross 614	M4.5 Ve	11.12	13.05	+1.69	242.89 ± 2.64
72511	N/A	14 49 34 −26 06 22	N/A	N/A	11.72	13.58	N/A	235.24 ± 22.43
80824	N/A	16 30 18.06 −12 39 45.3	Wolf 1061	M3.5	10.10	11.95	+1.58	234.51 ± 1.82
439	225213	00 05 24.43 −37 21 26.5	Gl 1	M4 V	8.56	10.36	+1.46	229.33 ± 1.08
15689	N/A	03 22 05.50 −13 16 43.8	N/A	N/A	12.16	13.94	N/A	227.45 ± 61.79
3829	N/A	00 49 09.90 +05 23 19.0	Van Maanen 2	DG	12.37	14.15	+0.55	226.95 ± 5.35
72509	N/A	14 49 31.76 −26 06 42.0	Gl 563.2	M3	12.07	13.80	+1.52	221.80 ± 69.07
86162	N/A	17 36 25.90 +68 20 20.9	BD +68 946	M3.5 Vvar	9.15	10.81	+1.50	220.85 ± 0.92

Table 19.1. (Continued.)

HIP	HD number	α (2000.0) δ (2000.0)	Other name	Sp. type	V magnitude	Abs V	B − V	Parallax
85523	N/A	17 28 39.95 −46 53 42.7	Gl 674	M3	9.38	11.10	+1.55	220.43 ± 1.63
114110	N/A	23 06 39 −14 52 19	GJ 293	DC	12.24	13.92	−0.17	216.52 ± 18.28
57367	N/A	11 45 42.92 −64 50 29.5	Gl 440	DQ6	11.50	13.18	+0.20	216.40 ± 2.11
113020	79210	22 53 16.73 −14 15 49.3	Ross 780	M5 V	10.16	11.80	+1.60	212.69 ± 2.10
54211	79211	11 05 28.58 +43 31 36.4	Gl 412A	M2 Ve	8.82	10.40	+1.54	206.94 ± 1.19
49908	88230	10 11 22.14 +49 27 15.3	Gl 380	K2 Ve	6.60	8.16	+1.33	205.22 ± 0.81
82725	N/A	16 54 32.47 −62 24 12.0	N/A	N/A	11.72	13.26	N/A	203.01 ± 29.27
85605	N/A	17 29 36.25 +24 39 14.7	N/A	N/A	11.39	12.92	+1.10	202.69 ± 39.48
106440	204961	21 33 33.98 −49 00 32.4	Gl 832	M1 V	8.66	10.19	+1.52	202.53 ± 1.33
86214	N/A	17 37 03.66 −44 19 09.2	Gl 682	M3.5	10.94	12.43	+1.66	198.32 ± 2.43
19849	26965	04 15 16.32 −07 39 10.3	ω 2 Eri	K1 Ve	4.43	5.92	+0.82	198.24 ± 0.84
112460	N/A	22 46 49.73 +44 20 02.4	EV Lac	M4.5 Ve	10.29	11.77	+1.54	198.07 ± 2.05
88601	165341	18 05 27.29 +02 30 00.4	70 Oph	K0 Ve	4.03	5.50	+0.86	196.62 ± 1.38
97649	187642	19 50 47.00 +08 52 06.0	α Aql	A7 IV–V	0.76	2.20	+0.22	194.44 ± 0.94
1242	N/A	00 15 28.11 −16 08 01.7	L722-22	M4	11.49	12.90	+1.75	191.86 ± 17.24
57544	N/A	11 47 41.38 +78 41 28.2	Gl 445	M4 V	10.80	12.14	+1.57	185.48 ± 1.43
67155	119850	13 45 43.78 +14 53 29.5	Wolf 498	M4 Ve	8.46	9.79	+1.44	184.13 ± 1.27
103039	N/A	20 52 33.02 −16 58 29.1	N/A	N/A	11.41	12.71	+1.65	182.15 ± 3.68
21088	N/A	04 31 11.52 +58 58 37.5	Gl 169.1A	M4	10.82	12.11	+1.65	181.36 ± 3.67
33226	265866	06 54 48.96 +33 16 05.4	Wolf 294	M4 V	9.89	11.18	+1.57	181.32 ± 1.87
53020	128620	10 50 52.06 +06 48 29.3	Wolf 358	M5 V	11.64	12.89	+1.68	177.46 ± 23.00
25878	36395	05 31 27.40 −03 40 38.0	Wolf 1453	M1.5 V	7.97	9.19	+1.47	175.72 ± 1.20
82817	152751	16 55 28.75 −08 20 10.8	Wolf 630A	M3 Ve	9.02	10.23	+1.55	174.23 ± 3.90
96100	185144	19 32 21.59 +69 39 40.2	σ Dra	K0 V	4.67	5.87	+0.79	173.41 ± 0.46
29295	42581	06 10 34.62 −21 51 52.7	Gl 229	M1 Ve	8.15	9.34	+1.49	173.19 ± 1.12
26857	N/A	05 42 09.27 +12 29 21.6	Ross 47	M4	11.56	12.75	+1.62	172.78 ± 3.88
86990	152751	17 46 34.23 −57 19 08.6	Gl 693	M3.5	10.75	11.93	+1.66	172.08 ± 2.22
94761	180617	19 16 55.26 +05 10 08.1	Ross 652	M3.5 Ve	9.12	10.28	+1.46	170.26 ± 1.37
73184	131977	14 57 28.00 −21 24 55.7	Gl 570A	K5 Ve	5.72	6.86	+1.02	169.32 ± 1.67
37766	N/A	07 44 40.17 +03 33 08.8	YZ CMi	M4.5 Ve	11.19	12.32	+1.60	168.59 ± 2.67
76074	155876	15 32 12.93 −41 16 32.1	Gl 588	M3	9.31	10.44	+1.52	168.52 ± 1.42
3821	4614	00 49 06.29 +57 48 54.7	η Cas	G3 V	3.46	4.59	+0.59	167.99 ± 0.62
84478	156026	17 16 13.36 −26 32 46.1	Gl 664	K5 Ve	6.33	7.45	+1.16	167.56 ± 1.06
117473	156026	23 49 12.53 +02 24 04.4	Gl 908	M2 Ve	8.98	10.10	+1.46	167.51 ± 1.49

Table 19.1. (Continued.)

HIP	HD number	α (2000.0) δ (2000.0)	Other name	Sp. type	V magnitude	Abs V	B − V	Parallax
84405	155885	17 15 20.98 −26 36 10.2	36 Oph	K1 Ve	4.33	5.44	+0.86	167.08 ± 1.07
99461	191408	20 11 11.94 −36 06 04.4	Gl 783A	K3 V	5.32	6.41	+0.87	165.24 ± 0.90
15510	20794	03 19 55.65 −43 04 11.2	e Eri	G8 V	4.26	5.35	+0.71	165.02 ± 0.55
99240	190248	20 08 43.61 −66 10 55.4	δ Pav	G8 V	3.55	4.62	+0.75	163.73 ± 0.65
71253	164058	14 34 16.81 −12 31 10.4	Wolf 1481	M3	11.32	12.39	+1.64	163.51 ± 2.77
86961	N/A	17 46 12.63 −32 06 12.8	CD −32 13297	M2 V	10.49	11.53	+1.46	161.77 ± 11.29
86963	N/A	17 46 14.41 −32 06 08.3	CD −32 13298	M2 V	11.39	12.43	+1.44	161.77 ± 11.29
45343	79210	09 14 22.79 +52 41 11.8	BD +53 1320	M0 Ve	7.64	8.68	+1.41	161.59 ± 5.23
99701	191849	20 13 53.40 −45 09 50.5	Gl 784	M0 V	7.97	9.01	+1.11	161.17 ± 1.08
116132	N/A	23 31 52.18 +19 56 14.1	BD +19 5116	M0 Ve	10.05	11.07	+1.19	160.06 ± 2.81
74995	N/A	15 19 26.82 −07 43 20.2	Gl 581	M5 V	10.57	11.58	+1.60	159.52 ± 2.27
120005	79211	09 14 24.70 +52 41 11.0	Gl 338B	M0 Ve	7.70	8.71	+1.42	159.48 ± 6.61
84140	155876	17 12 07.89 +45 39 57.5	Gl 661A	M3	9.31	10.31	+1.49	158.17 ± 3.26
34603		07 10 01.83 +38 31 46.1		M4.5 Ve	11.65	12.63	+1.70	157.24 ± 3.32
54298	N/A	11 06 30.65 −53 16 05.5		N/A	11.69	12.65	N/A	155.28 ± 78.30
82809	N/A	16 55 25.23 −08 19 21.3	Wolfe 629	M4	11.73	12.67	+1.70	153.96 ± 4.04
114622	219134	23 13 16.98 +57 10 06.1	Gl 892	K3 V var	5.57	6.50	+1.00	153.24 ± 0.65
80459	191408	16 25 24.62 +54 18 14.8	Gl 645	M2 V	10.13	11.04	+1.59	151.93 ± 1.11
53767		11 00 04.26 +22 49 58.7	Gl 408	M3	10.03	10.92	+1.52	150.96 ± 1.59
72659	131156	14 51 23.38 +19 06 01.7	ξ Boo	G8 Ve	4.54	5.41	+0.72	149.26 ± 0.76
106106	N/A	21 29 36.81 +17 38 35.8	Gl 829	M4 Ve	10.33	11.19	+1.62	148.29 ± 1.85
114176	N/A	23 07 19 −32 16 05	N/A	N/A	12.28	13.13	N/A	147.95 ± 13.76
113296	216899	22 56 34.81 +16 33 12.4	BD+15 4733	M2 Ve	8.68	9.49	+1.51	145.27 ± 1.22
84709	156384	17 18 57.18 −34 59 23.3	Gl 667A	K3 V	5.91	6.69	+1.08	143.45 ± 17.12
103096	199305	20 53 19.79 +62 09 15.8	BD +61 2068	M2 Ve	8.55	9.31	+1.48	141.95 ± 0.77
12114	16160	02 36 04.89 +06 53 12.7	BD +06 398	K3 V	5.69	6.50	+0.92	138.72 ± 1.04
51317	N/A	10 28 55.55 +00 50 27.6	BD +01 2447	M2.5 V	9.65	10.35	+1.51	138.29 ± 2.13
83945	N/A	17 09 31.54 +43 40 52.9	N/A	M3	11.77	12.47	+1.70	137.84 ± 8.95
3765	4628	00 47 01.46 +11 58 25.9	BD +11 96	G8 II	5.74	6.38	+0.99	134.04 ± 0.86
7981	10476	01 42 29.76 +20 16 06.6	107 Psc	K1 V	5.24	5.87	+0.84	133.91 ± 0.91
2021	2151	00 25 45.07 −77 15 15.3	β Hyi	G2 IV	2.82	3.45	+0.62	133.78 ± 0.51
73182	131976	14 57 26.54 −21 24 41.5	BD −20 4123	M2 V	8.01	8.64	+1.01	133.63 ± 33.56
12781	N/A	02 44 15.51 +25 31 24.1	Gl 109	M3.5 Ve	10.55	11.16	+1.53	132.42 ± 2.48

Table **19.1.** (*Continued.*)

HIP	HD number	α (2000.0) δ (2000.0)	Other name	Sp. type	V magnitude	Abs V	B − V	Parallax
5336	6582	01 08 16.39 +54 55 13.2	μ Cas	G5 Vp	5.17	5.78	+0.70	132.40 ± 0.60
65859	N/A	13 29 59.79 +10 22 37.8	BD +11 2576	M1 V	9.05	9.64	+1.49	131.12 ± 1.29

19.2 THE BRIGHTEST STARS

Table 19.2 lists the 100 brightest stars in the sky in order of right ascension. Positions are given in the FK5 system with equinox = J2000.0 and epoch = J2000.0; they are calculated from observed positions with corrections for proper motions. Parallaxes are from the Hipparcos Main Catalogue [1] and given in milliarcseconds with the associated standard deviation error.

Table **19.2.** *Brightest stars.*

Star name	HD number	α (2000.0) δ (2000.0)	Other name	Sp. type	V magnitude	B − V	V − I	Parallax
Alpheratz	358	00 08 23.26 +29 05 25.6	α And	B9p	2.07	−0.04	−0.10	33.60 ± 0.73
Caph	432	00 09 10.69 +59 08 59.2	β Cas	F2 III–IV	2.28	+0.38	+0.40	59.89 ± 0.56
CD −77 15	2151	00 25 45.01 −77 15 15.3	β Hyi	G2 IV	2.82	+0.62	+0.68	133.78 ± 0.51
Ankaa	2261	00 26 17.05 −42 18 21.5	α Phe	K0 II I	2.40	+1.08	+1.11	42.14 ± 0.78
Schedar	3712	00 40 30.44 +56 32 14.4	α Cas	K0 II–IIIvar	2.24	+1.17	+1.13	14.27 ± 0.57
Diphda	4128	00 43 35.37 −17 59 11.8	β Cet	K0 III	2.04	+1.02	+1.00	34.04 ± 0.82
Cih	5394	00 56 42.53 +60 43 00.3	γ Cas	B0 IV	2.15	−0.05	−0.02	5.32 ± 0.56
Mirach	6860	01 09 43.92 +35 37 14.0	β And	M0 IIIvar	2.07	+1.58	+1.74	16.36 ± 0.76
Achernar	10144	01 37 42.85 −57 14 12.3	α Eri	B3 Vp	0.45	−0.16	−0.17	22.68 ± 0.57
Almach	12533	02 03 53.95 +42 19 47.0	γ And	K3 IIb	2.26	+1.37	+1.37	9.19 ± 0.73
Hamal	12929	02 07 10.41 +23 27 44.7	α Ari	K2 I II	2.01	+1.15	+1.13	49.48 ± 0.99
Polaris	8890	02 31 49.08 +89 15 50.8	α UMi	F7: Ib–IIv	1.97	+0.64	+0.70	7.5 6 ± 0.48
Menkar	18884	03 02 16.77 +04 05 23.0	α Cet	M2 III	2.54	+1.63	+1.97	14.82 ± 0.83
Algol	19356	03 08 10.3 +40 57 20.3	β Per	B8 V	2.09	−0.03	+0.02	35.14 ± 0.90
Mirfak	20902	03 24 19.37 +49 51 40.2	α Per	F5 Ib	1.79	+0.48	+0.63	5.51 ± 0.66
Aldeberan	29139	04 35 55.24 +16 30 33.5	α Tau	K5 III	0.87	+1.54	+1.61	50.09 ± 0.95
Rigel	34085	05 14 32.27 −08 12 05.9	β Ori	B8 Ia	0.18	−0.03	+0.03	4.22 ± 0.81
Capella	34029	05 16 41.36 +45 59 52.8	α Aur	M1 III	0.08	+0.80	+0.83	77.29 ± 0.89
Bellatrix	35468	05 25 07.86 +06 20 58.9	γ Ori	B2 III	+1.64	−0.22	−0.22	13.42 ± 0.98
El Nath	35497	05 26 17.51 +28 36 28.6	β Tau	B7 III	1.65	−0.13	−0.09	24.89 ± 0.88
Mintaka	36486	05 32 00.40 −00 17 56.7	δ Ori	O9.5 II	2.25	−0.18	−0.21	3.56 ± 0.83
Arneb	36673	05 32 43.82 −17 49 20.3	α Lep	F0 Ib	2.58	+0.21	+0.32	2.54 ± 0.72
Alnilam	37128	05 36 12.81 −01 12 06.9	ε Ori	B0 Ia	1.69	−0.18	−0.16	2.43 ± 0.91

Table 19.2. *(Continued.)*

Star name	HD number	α (2000.0) δ (2000.0)	Other name	Sp. type	V magnitude	B − V	V − I	Parallax
Alnitak	37742	05 40 45.53 −01 56 33.3	ξ Ori	O9.5 Ib	1.74	−0.20	−0.18	3.99 ± 0.79
Saiph	38771	05 47 45.39 −09 40 10.6	κ Ori	B0.5 Ivar	2.07	−0.17	−0.14	4.52 ± 0.77
Betelgeuse	39801	05 55 10.31 +07 24 25.4	α Ori	M2 Ib	0.45	+1.50	+2.32	7.63 ± 1.64
Menkalinan	40183	05 59 31.72 +44 56 50 .8	β Aur	A2 V	1.90	+0.08	+0.05	39.72 ± 0.78
Mirzam	44743	06 22 41.99 −17 57 21.3	β CMa	B1 II–III	1.98	−0.24	−0.24	6.53 ± 0.66
Canopus	45348	06 23 57.11 − 52 41 44.4	α Car	F0 Ib	−0.62	+0.16	+0.23	10.43 ± 0.53
Alhena	47105	06 37 42.75 +16 23 57.3	γ Gem	A0 IV	1.93	+0.00	+0.04	31.12 ± 2.33
Sirius A	48915	06 45 08.92 −16 41 58.0	α CMa	A1 V	−1.44	+0.01	−0.02	379.21 ± 1.58
Adhara	52089	06 58 37.55 −28 58 19.5	ε CMa	B2 II	1.50	−0.21	−0.20	7.57 ± 0.57
Wezen	54605	07 08 23.48 −26 23 35.5	δ CMa	F8 Ia	1.83	+0.70	+0.67	1.82 ± 0.56
Aludra	58350	07 24 05.70 −29 18 11.2	η CMa	B5 Ia	2.45	−0.08	+0.01	1.02 ± 0.57
Castor	60178	07 34 35.86 +31 53 17.8	α Gem	A2 V	1.58	+0.03	+0.05	63.27 ± 1.23
Procyon	61421	07 39 18.12 +05 13 30.0	α CMi	F5 IV–V	0.40	+0.43	+0.05	285.93 ± 0.88
Pollux	62509	07 45 18.95 +28 01 34.3	β Gem	K0 IIIvar	1.16	+0.99	+0.97	96.74 ± 0.87
Naos	66811	08 03 35.05 −40 00 11.3	ζ Pup	O5 IAf	2.21	−0.27	−0.22	2.33 ± 0.51
CD −46 3847	68273	08 09 31.95 −47 20 11.7	γ Vel	WC8	1.75	−0.14	−0.14	3.88 ± 0.53
Avoir	71130	08 22 30.84 −59 30 34.1	ε Car	K3 III	1.86	+1.20	+0.16	5.16 ± 0.49
Suhail	78647	09 07 59.76 −43 25 57.4	λ Vel	K4 Ib–II	2.23	+1.67	+1.69	5.69 ± 0.53
Miaplacidus	80007	09 13 11.98 −69 43 01.9	β Car	A2 IV	1.67	+0.07	+0.02	29.34 ± 0.47
Scutulum	80404	09 17 05.41 −59 16 30.8	τ Car	A8 Ib	2.21	+0.19	+0.28	4.71 ± 0.46
CPD −54 2219	81188	09 22 06.82 −55 00 38.4	κ Vel	B2 IV	2.47	−0.14	−0.17	6.05 ± 0.48
Alphard	81797	09 27 35.24 −08 39 31.0	α Hya	K3 III	1.99	+1.44	+1.39	18.40 ± 0.78
Regulus	87901	10 08 22.31 +11 58 01.9	α Leo	B7 V	1.36	−0.09	−0.10	42.09 ± 0.79
Algieba	89484	10 19 58.35 +19 50 29.4	γ Leo	K0 III	2.01	+1.13	+1.17	25.96 ± 0.83
Merak	95418	11 01 50.48 +56 22 56.7	β UMa	A1 V	2.34	+0.03	+0.02	41.07 ± 0.60
Dubhe	95689	11 03 43.67 +61 45 03.7	α UMa	F7 V	1.81	+1.06	+1.03	26.38 ± 0.53
Zosma	97603	11 14 06.50 +20 31 25.4	δ Leo	A4 V	2.56	+0.13	+0.12	56.52 ± 0.83
Denebola	102647	11 49 03.58 +14 34 19.4	β Leo	A3 Vvar	2.14	+0.09	+0.10	90.16 ± 0.89
Phe cda	103287	11 53 49.85 +53 41 41.4	γ UMa	A0 V SB	2.41	+0.04	+0.06	38.99 ± 0.68
Gienah	106625	12 15 48.37 −17 32 30.9	γ Crv	B8 III	2.58	−0.11	−0.10	19.78 ± 0.81
Alpha Crucis	108248	12 26 35.90 −63 05 56.7	α Cru	B0.5 IV	0.77	−0.24	−0.26	10.17 ± 0.67
Gacrux	10890 3	12 31 09.96 −57 06 47.6	γ Cru	M4 III	1.59	+1.60	+2.37	37.09 ± 0.67
Muhlifain	110304	12 41 31.04 −48 57 35.6	γ Cen	A1 IV	2.20	−0.02	−0.01	25.01 ± 1.01
Mimosa	111123	12 47 43.26 −59 41 19.5	β Cru	B0.5 III	1.25	−0.24	−0.27	9.25 ± 0.61

Table 19.2. (*Continued.*)

Star name	HD number	α (2000.0) δ (2000.0)	Other name	Sp. type	V magnitude	B − V	V − I	Parallax
Alioth	112185	12 54 01.75 +55 57 35.4	ε UMa	A0p	1.76	−0.02	−0.04	40.30 ± 0.62
Spica	116658	13 25 11.58 −11 09 40.8	α Vir	B1 V	0.98	−0.24	−0.25	12.44 ± 0.86
Mizar	116656	13 23 55.54 +54 55 31.3	ζ UMa	A2 V	2.23	+0.06	+0.07	41.73 ± 0.61
CPD −52 6655	118716	13 39 53.26 −53 27 59.0	ε Cen	B1 III	2.29	−0.17	−0.23	8.68 ± 0.77
Alcaid	120315	13 47 32.44 +49 18 47.8	η UMa	B3 V	1.85	−0.10	−0.08	32.39 ± 0.74
CD −46 8949	121263	13 55 32.39 −47 17 18.2	ζ Cen	B2.5 IV	2.55	−0.18	−0.18	8.48 ± 0.74
Agena	122451	14 03 49.40 −60 22 22.9	β Cen	B1 III	0.61	−0.23	−0.25	6.21 ± 0.56
Menkent	123139	14 06 40.95 −36 22 11.8	θ Cen	K0 IIIb	2.06	+1.01	+1.01	53.52 ± 0.79
Arcturus	124897	14 15 39.67 +19 10 56.7	α Boo	K2 II Ip	−0.05	+1.24	+1.22	88.85 ± 0.74
CD −41 8917	127972	14 35 30.42 −42 09 26.2	η Cen	B1 Vn	2.33	−0.16	−0.17	10 .57 ± 0.83
GJ 559B	128621	14 39 35.08 −60 50 13.8	α Cen B	K1 V	1.35	+0.90	+0.88	742.12 ± 1.40
Rigil Kent	128620	14 39 36.50 −60 50 02.3	α Cen A	G2 V	−0.01	+0.71	+0.69	742.12 ± 1.40
CD −46 9501	129056	14 41 55.76 −47 23 17.5	α Lup	B1.5 III	2.30	−0.15	−0.21	5.95 ± 0.76
Izar	129989	14 44 59.21 +27 04 27.4	ε Boo	K0 II–III	2.35	+0.97	+0.95	15.55 ± 0.78
Kochab	131873	14 50 42.33 +74 09 19.8	β UMi	K4 IIIvar	2.07	+1.46	+1.46	25.79 ± 0.52
Alphecca	139006	15 34 41.27 +26 42 52.9	α CrB	A0 V	2.22	+0.03	+0.05	43.65 ± 0.79
Dzuba	143275	16 00 20.01 −22 37 18.2	δ Sco	B0.2 IV	2.29	−0.12	−0.09	8.12 ± 0.88
Acrab	144217	16 05 26.27 −19 48 19.6	β Sco	B0.5 V	2.56	−0.06	−0.04	6.15 ± 1.12
Antares	148478	16 29 24.46 −26 25 55.2	α Sco	M1 Ib	1.06	+1.87	+2.90	5.40 ± 1.68
BD −10 4350	149757	16 37 09.54 −10 34 01.5	ζ Oph	O9.5 V	2.54	+0.04	+0.10	7.12 ± 0.71
Atria	150798	16 48 39.89 −69 01 39. 8	α TrA	K2 IIb	1.91	+1.45	+1.45	7.85 ± 0.63
CD −34 11285	151680	16 50 09.81 −34 17 35.6	ε Sco	K2 II Ib	2.29	+1.14	+1.10	49.85 ± 0.81
Sabik	155125	17 10 22.69 −15 43 29.7	η Oph	A2.5 Va	2.43	+0.06	+0.06	38.77 ± 0.86
Shaula	158926	17 33 36.52 −37 06 13.8	λ Sco	B1.5 IV	1.62	−0.23	−0.24	4.64 ± 0.90
Ras-Alhague	159561	17 34 56.07 +12 33 36.1	α Oph	A5 III	2.08	+0.16	+0.17	69.84 ± 0.88
CD −42 12312	159532	17 37 19.13 −42 59 52.2	θ Sco	F1 II	1.86	+0.41	+0.02	11.99 ± 0.84
CD −38 12137	160578	17 42 29.27 −39 01 47.9	κ Sco	B1.5 III	2.39	−0.17	−0.22	7.03 ± 0.73
Eltanin	164058	17 56 36.37 +51 29 20.0	γ Dra	K5 III	2.24	+1.52	+1.54	22.10 ± 0.46
Kaus Australis	169022	18 24 10.32 −34 23 04.6	ε Sgr	B9.5 III	1.79	−0.03	+0.01	22.55 ± 1.02
Vega	172167	18 36 56.34 +38 47 01.3	α Lyr	A0 Vvar	0.03	+0.00	−0.01	128.93 ± 0.55
Nunki	175191	18 55 15.93 −26 17 48.2	σ Sgr	B2.5 V	2.05	−0.13	−0.13	14.54 ± 0.88
Altair	187642	19 50 47.0 +08 52 06.0	α Aql	A7 IV–V	0.76	+0.22	+0.27	194.44 ± 0.94
Sadir	194093	20 22 13.70 +40 15 24.0	γ Cyg	F8 Ib	2.14	+0.67	+0.65	2.14 ± 0.51
Peacock	193924	20 25 38.86 −56 44 06.3	α Pav	B2 IV	1.94	−0.12	−0.10	17.80 ± 0.70

Table 19.2. *(Continued.)*

Star name	HD number	α (2000.0) δ (2000.0)	Other name	Sp. type	V magnitude	B − V	V − I	Parallax
Deneb	197345	20 41 25.91 +45 16 49.2	α Cyg	A2 Ia	1.25	+0.09	+0.16	1.01 ± 0.57
Gienar	197989	20 46 12.68 +33 58 12.9	ε Cyg	K0 III	2.48	+1.02	+1.00	45.26 ± 0.53
Alderamin	203280	21 18 34.77 +62 35 08.1	α Cep	A7 IV–V	2.45	+0.26	+0.26	66.84 ± 0.49
Enif	206778	21 44 11.16 +09 52 30.0	ε Peg	K2 Ibvar	2.38	+1.52	+1.42	4.85 ± 0.84
Al Na'ir	209952	22 08 13.99 −46 57 39.5	α Gru	B7 IV	1.73	−0.07	−0.05	32.16 ± 0.82
CD −47 14308	214952	22 42.40.05 −46 53 04.5	β Gru	M5 III	2.07	+1.61	+2.60	19.17 ± 0.75
Fomalhaut	216956	22 57 39.05 −29 37 20.1	α PsA	A3 V	1.17	+0.14	+0.16	130.08 ± 0.92
Scheat	217906	23 03 46.46 +28 04 58.0	β Peg	M2 II–IIIvar	2.44	+1.66	+2.31	16.37 ± 0.72
Markab	218045	23 04 45.65 +15 12 19.0	α Peg	B9.5 III	2.49	−0.03	+0.00	23.36 ± 0.76

19.3 STELLAR POPULATIONS

The concept of stellar populations arose on observational grounds [2], and was rapidly extended as the underlying astrophysical processes were identified [3–5]. In modern usage the sequence of populations is essentially that of time, with the boundaries between population classes being poorly defined or definable, and often internally inconsistent. The classical scheme and current usage for both the Milky Way and external galaxies are summarized in the first and second parts of Table 19.3. A complementary presentation in astrophysical units is summarized in Table 19.4. It remains uncertain if there is a single sequence of stellar populations in the Milky Way, or two separate sequences, halo–bulge and thick disk–young disk. The age ranges are in units of τ_u, the age of the Universe.

The top part of Table 19.3 presents the classical view of stellar populations in the Milky Way. Each of the three basic population divisions is further subdivided, with defining examples of observed classes of objects listed. The combinations of spatial distributions, spectral types, kinematics, and chemical abundances are all correlated. It is this set of correlations which provides the evidence for the basic physical validity of the population concept. The bottom line of the top part illustrates schematically a classical extension of the populations concept to external galaxies. The bottom part illustrates the current appreciation of stellar populations. The format is similar to that of the top part, as are the essential features. Many more details are shown, together with a finer subdivision. The essential features of the population concept however remain little modified.

In each panel of Table 19.4, the vertical axis represents an observable which is closely related to the physical processes of galaxy evolution. The horizontal axis represents a monotonic, though not necessarily linear, evolutionary sequence related to time. Only in the panel showing the distributions of specific angular momentum is an apparently clear evolutionary connection between stellar populations evident.

Table 19.3. *Classical (top) and current (lower) concepts of stellar populations*

	Population II		Disk	Population I	
Characteristic objects and properties	Halo Pop II subdwarfs globular clusters RR Lyrae $P > 0.^{d}4$	Intermediate stars with $V_z \geq 30$ km s^{-1} LPV's, $P < 250^{d}$	galactic nucleus RR Lyrae $P < 0.^{d}4$ weak-line stars	Old Pop I A stars Me dwarfs strong-line stars	Extreme Pop I gas, spiral structure supergiants Cepheids
Scale height (pc)	2000	500	300	100	60
central concentration	strong	strong	strong	little	little
τ/τ_u	1.0	1.0–0.8	0.8–0.25	0.25–0.05	0.05–0.00
σ_W	75 km s^{-1}	25 km s^{-1}	17 km s^{-1}	10 km s^{-1}	8 km s^{-1}
$Z/Z_\odot$	0.1	0.25	0.5	0.75	1.0
External Galaxies	← Ellipticals →		← Bulges →	← Spiral disks, Irr's →	
Characteristic objects and properties	Extreme Pop II "halo" subdwarfs globular clusters with [Fe/H] < -1 RR Lyrae $\Delta S > 4$ BHB stars	Intermediate Pop II "thick disk" globular clusters with [Fe/H] > -1 RR Lyrae, c-type LPV's, $P \sim 250^{d}$ RHB stars	Bulge/Pop II "bulge" SMR stars = "IR bulge" planetary nebulae = "optical bulge" RR Lyrae $\Delta S < 4$ tri-axial (?)	Pop I "old disk" intermediate age disk stars	Extreme Pop I "young disk" young stars spiral structure Cepheids
$\langle V_{rot} \rangle$	30	170	60	200	220
$\sigma_U : \sigma_V : \sigma_W$	130:100:85	60:45:40	120:120:120	38:25:20	20:10:8
$Z/Z_\odot$	0.03	0.3	0.1–2	0.9	1
τ/τ_u	1.0–0.9	0.9–0.8	1.0–0.5 (?)	0.9–0.1	0.1–0.0
External Galaxies	dE	Sa → ← S0 → gE		← Sbcd, Irr's →	

Table 19.4. *Astrophysical representation of stellar populations.*

[Fe/H]					
−2	Halo				
−1		Thick disk			
0			Old disk →		
		Bulge →		Young disk	
+1					time →

Half mass radius (kpc)					
0					
		Bulge			
2					
	Halo				
4					
6		Thick disk	Old disk	Young disk	
				time →	

Table 19.4. (*Continued.*)

[Fe/H]		
−2	Halo	
−1		Thick disk
0		Young disk Old disk
+1	Bulge	
		log mass →

Vertical velocity disp. (km/s)		
100		Bulge
80	Halo	
40		Thick disk
0		Old disk Young disk
		[Fe/H] →

Angular mom. per mass (km/s kpc)		
0	Halo	Bulge
10^3		Thick disk Old disk
2×10^3		Young disk
		[Fe/H] →

19.4 STAR COUNTS AT HIGH LATITUDES [6–12]

Counts of stars in selected color ranges in the V and photographic B_J magnitude bands are given for the north galactic pole in Tables 19.5 and 19.6. Similar data are given in Table 19.7 for the south galactic pole.

Table 19.5. *The stellar color–magnitude distribution at $b = +90°$ (stars per square degree per mag.).*

V	$B - V \leq 0.2$	0.3	0.5	0.7	0.9	1.1	1.3	1.5	1.7	≥ 1.9	Total
10.5											5
11.5											11
12.5	0.5	0.5	6	8	3	2	1	0	0.5	0.5	20
13.5	0.2	1	8	14	8	2	2	1	0	0.5	36
14.5	0.5	1	9	24	13	6	3	3	0.5	0	60
15.5	1	1	9	35	16	9	8	6	1	0	86
16.5	1	2	21	34	22	13	13	16	3	0	125
17.5	2	10	26	29	21	15	19	27	10	0	159

Table 19.6. B_J *magnitude star counts at the north galactic pole (stars per square degree per mag.).*

B_J	Blue	Red	Total
15.5			60
16.5			100
17.5			125
18.5			160
19.5	50	200	250
20.5	160	240	400
21.5	180	450	630

Table 19.7. V *magnitude star counts at the south galactic pole (stars per square degree per mag.).*

V	$B - V \leq 0.4$	$0.4 \leq B - V \leq 0.8$	$0.8 \leq B - V \leq 1.2$	$1.2 \leq B - V \leq 1.6$	$1.6 \leq 1.6$	Total
16–17	3	55	17	21	0	97
17–18	17	90	62	86	0	255
18–19	31	72	59	79	14	255
19–20	14	93	66	148	48	369
20–21	10	186	79	245	69	590
21–22	24	234	169	341	141	910

There is an excess of stars toward $b = -90°$ over $b = +90°$ which is consistent with a location of the Sun at $z = +40$ pc, a scale error in the data, or differential reddening.

19.5 VERTICAL STELLAR DENSITY PROFILE

The vertical structure of the disk is approximately exponential. Near the plane the scale height is low, as young stars are significant. Far from the plane, the thick disk dominates. Details are in Table 19.8.

Table 19.8. *Stellar density structure.*

Vertical distance range (pc)	Apparent exponential scale height (pc)
$0 \lesssim z \lesssim 50$	$\sim$ constant density
$0 \lesssim z \lesssim 200$	120
$0 \lesssim z \lesssim 300$	175
$0 \lesssim z \lesssim 500$	225
$250 \lesssim z \lesssim 750$	260

For distances between 300 and 4000 pc the density profile is describable by

$$\frac{\rho}{\rho_0} = 0.959 \exp\left(\frac{-z}{250}\right) + 0.041 \exp\left(\frac{-z}{1000}\right).$$

Beyond ~ 4 kpc, the density law is a galactocentric power law of index ~ -3, axis ratio $c/a \sim 0.75$, or a deprojected $r^{1/4}$ law with effective radius $r_e \sim 2.7$ kpc, axis ratio $c/a \sim 0.75$, with c/a probably varying with radius, being smallest in the center. Table 19.9 gives the scale heights as a function of the star luminosity for disk stars.

Table 19.9. *Absolute magnitude–scale height relation.*

Absolute magnitude M_V	Scale height (pc)
$\leq +2$	90
$+3$	150
$+4$	200
$+5$	250

19.5.1 The Standard Model of the Optical Stellar Galaxy

This standard model given in Table 19.10 is in reasonable agreement with data for $|b| \gtrsim 30°$, $|l| \gtrsim 30°$, $10 \lesssim V \lesssim 22$. It predicts too few stars within $\sim 30°$ of the galactic center.

Table 19.10. *Standard model of optical stellar galaxy.*[a,b]

	Old disk	Thick disk	Bulge	Halo
Radial density profile	exponential	exponential	power law	$r^{1/4}$ or power law
Vertical density profile[c]	exponential	exponential	power law	
Radial scale length	3000 pc	3000 pc	500 pc	2700 pc
Vertical scale length	dwarfs = 300 pc giants = 250 pc	1000 pc	300 pc	2000 pc
Stellar luminosity function	local field	metal-rich globular	metal-rich globular	mean globular
Local stellar density	tabulated	$0.02 \times$ disk	(?)	$0.002 \times$ disk
Color–magnitude relation	old open cluster	metal-rich globular	metal-rich globular	metal-poor globular
Mean metallicity	-0.1	-0.6	-0.2	-1.6
Metallicity dispersion	0.3	0.3	0.4	0.5
Radial metallicity gradient	-0.04 dex/kpc	0: dex/kpc	—	-0.02: dex/kpc
Vertical metallicity gradient	-0.3 dex/kpc	0: dex/kpc	—	-0.03: dex/kpc
Velocity dispersions $\sigma_U, \sigma_V, \sigma_W$	38:25:20 km/s	60:45:40 km/s	120:120:120 km/s	130:100:85 km/s
Mean rotation	200 km/s	170 km/s	60 km/s at $R < 1$ kpc	30 km/s

Notes

[a] Solar position $(R, \delta) = (8000, +20)$ pc.

[b] Extinction: In projection, Sandage cosec law, distributed constant density in radius, and vertical exponential scale height $= 100$ pc.

[c] See the scale height versus distance and scale height versus M_V relations given.

Tables 19.11 and 19.12 give typical star counts for the standard galaxy model in the V and I bands. This first table also includes data from the previous *Astrophysical Quantities* edition.

Table 19.11. *Model* $\log N(V)$ *(stars per square degree) averaged over galactic longitude and over the whole sky.*

V	$b = 0$	$b = 5$	$b = 10$	$b = 20$	$b = 30$	$b = 60$	$b = 90$	Sky	AQ3[a]
12.0	1.751	1.752	1.743	1.665	1.544	1.234	1.160	1.556	1.76
13.0	2.275	2.267	2.218	2.085	1.935	1.603	1.524	1.994	2.17
14.0	2.708	2.694	2.618	2.450	2.279	1.925	1.838	2.372	2.56

Table 19.11. *(Continued.)*

V	b = 0	b = 5	b = 10	b = 20	b = 30	b = 60	b = 90	Sky	AQ3[a]
15.0	3.105	3.085	2.985	2.781	2.591	2.209	2.110	2.719	2.94
16.0	3.485	3.460	3.331	3.089	2.875	2.458	2.347	3.047	3.29
17.0	3.842	3.809	3.648	3.363	3.124	2.673	2.553	3.350	3.64
18.0	4.170	4.130	3.939	3.612	3.349	2.869	2.744	3.632	3.95
19.0	4.474	4.428	4.210	3.845	3.559	3.052	2.924	3.897	4.20
20.0	4.776	4.722	4.468	4.057	3.748	3.226	3.092	4.156	4.5
21.0	5.078	5.011	4.705	4.236	3.909	3.389	3.245	4.406	4.7
22.0	5.342	5.262	4.903	4.382	4.046	3.538	3.382	4.626	
23.0	5.536	5.448	5.059	4.510	4.168	3.665	3.500	4.797	
24.0	5.684	5.594	5.195	4.633	4.286	3.772	3.604	4.937	
25.0	5.813	5.722	5.319	4.749	4.393	3.863	3.695	5.061	

Note

[a] From Allen, C.W. 1973, *Astrophysical Quantities* (Athlone Press, London).

Table 19.12. *Model* $\log N(I)$ *(stars per square degree) averaged over galactic longitude and over the whole sky.*

I	b = 0	b = 5	b = 10	b = 20	b = 30	b = 60	b = 90	Sky
12.0	2.466	2.434	2.279	2.011	1.799	1.450	1.374	2.001
13.0	2.922	2.877	2.671	2.355	2.138	1.816	1.728	2.396
14.0	3.316	3.262	3.023	2.678	2.461	2.149	2.045	2.753
15.0	3.650	3.594	3.344	2.991	2.772	2.445	2.326	3.075
16.0	3.951	3.897	3.653	3.301	3.069	2.702	2.575	3.376
17.0	4.243	4.191	3.955	3.594	3.341	2.926	2.799	3.663
18.0	4.548	4.494	4.247	3.861	3.584	3.133	3.005	3.947
19.0	4.868	4.807	4.530	4.103	3.804	3.327	3.191	4.230
20.0	5.191	5.120	4.802	4.326	4.005	3.503	3.353	4.510
21.0	5.479	5.398	5.039	4.515	4.173	3.649	3.488	4.761
22.0	5.710	5.623	5.234	4.673	4.313	3.770	3.604	4.967
23.0	5.909	5.816	5.400	4.806	4.430	3.874	3.706	5.145
24.0	6.092	5.991	5.546	4.916	4.528	3.970	3.797	5.308
25.0	6.258	6.150	5.669	5.002	4.606	4.055	3.875	5.453

19.5.2 Surface Brightness

The surface brightness of the sky, excluding solar system and terrestrial sources, has been measured by Pioneer 10 in two bands (Table 19.13). 3950 Å < λ < 4850 Å ("blue") and 5900 Å < λ < 6900 Å ("red") [13]. Contributions from stars with $V \lesssim 6.5$ have been removed. The flux is quoted in units of the equivalent number of stars of $V = 10$, type G2 V, per square degree, $S_{10}(V)$.

$$S_{10}(V) = 1.16 \times 10^{-9} \text{ erg cm}^{-2} \text{ s}^{-1} \text{ sr}^{-1} \text{ Å}^{-1} \text{ blue band}$$
$$= 1.07 \times 10^{-9} \text{ erg cm}^{-2} \text{ s}^{-1} \text{ sr}^{-1} \text{ Å}^{-1} \text{ red band},$$
$$S_{10}(V) \text{ (blue band)} \equiv 0.265 L_{\odot,B} \text{ pc}^{-2}$$
$$\equiv 28.49 B_{\text{mag.}} \text{ arcsec}^{-2}.$$

Table 19.13. *Pioneer* 10 *sky brightness measurements.*

Region	α (1950)	δ	l^{II}	b^{II}	$S_{10}(V)$ Blue	$S_{10}(V)$ Red	Blue–red (mag.)
NCP	—	$+90°$	$123°$	$27°$	56	77	1.43 ± 0.08
NEP	$18^{\mathrm{h}}\ 0^{\mathrm{m}}$	$+67°$	$96°$	$30°$	66	82	1.32 ± 0.08
NGP	$12^{\mathrm{h}}\ 50^{\mathrm{m}}$	$+27°$	—	$90°$	29	31	1.18 ± 0.15
SCP	—	$-90°$	$303°$	$-27°$	74	94	1.34 ± 0.07
SEP[a]	$6^{\mathrm{h}}\ 0^{\mathrm{m}}$	$-67°$	$277°$	$-30°$	128	125	1.06 ± 0.06
SGP	$0^{\mathrm{h}}\ 50^{\mathrm{m}}$	$-27°$	—	$-90°$	26	36	1.41 ± 0.14

Note
[a] The SEP beam included part of the LMC.

The full data set has been analyzed [13, 14] to derive a surface brightness map of the Milky Way Galaxy with the following properties:

Surface brightness at galactic poles:

$$\mu_B = 24.55 \pm 0.1 \text{ B mag. arcsec}^{-2}$$
$$= 10.1 \pm 1.0 L_{\odot,B} \text{ pc}^{-2},$$
$$\mu_V = 23.71 \pm 0.1 V \text{ mag. arcsec}^{-2}$$
$$= 12.0 \pm 1.5 L_{\odot,V} \text{ pc}^{-2}.$$

Color of poles:

$$B - V = 0.76 \pm 0.15.$$

Face-on disk central surface brightness:

$$\mu_0 = 95 \pm 30 L_{\odot,B} \text{ pc}^{-2}$$
$$= 22.1 \pm 0.3 \text{ B mag. arcsec}^{-2}.$$

Integrated disk color:

$$B - V = 0.84 \pm 0.15.$$

Integrated disk luminosity:

$$L_{\text{tot}} = (1.8 \pm 0.3) \times 10^{10} L_{\odot,B}.$$

Old disk color:

$$B - V = 0.95 \pm 0.15.$$

Old disk luminosity:

$$L_{\text{tot}} = (1.1 \pm 0.2) \times 10^{10} L_{\odot,B}.$$

Integrated halo luminosity:

$$L = 2 \times 10^9 L_{\odot,B}.$$

Halo color:

$$B - V = 0.8 \pm 0.1.$$

19.6 MAIN SEQUENCE FIELD STELLAR LUMINOSITY FUNCTION [15–22]

Table 19.14 gives the main sequence stellar luminosity function.

Table 19.14. $\log \Phi(M_V)$ *stars* pc^{-3} M_V^{-1}.

M_V	Single stars	Unresolved binary/triple systems	Mean globular cluster		
			Principal sequence	BHB	RR Lyrae
−7	−7.98	7.98			
−6	−7.60	−7.60			
−5	−7.27	−7.27			
−4	−6.72	−6.72			
−3	−6.05	−6.05			
−2	−5.43	−5.43	−5.61		
−1	−4.80	−4.80	−4.77		
0	−4.18	−4.18	−4.39		
1	−3.60	−3.60	−4.19	−3.49	−3.49
2	−3.16	−3.16	−3.89		
3	−2.89	−2.89	−3.59		
4	−2.63	−2.63	−2.84		
5	−2.49	−2.49	−2.49		
6	−2.44	−2.44	−2.44		
7	−2.52	−2.52	−2.24		
8	−2.41	−2.42	−2.24		
9	−2.32	−2.48	−2.14		
10	−2.14	−2.26	−1.80		
11	−1.99	−2.01	−1.80		
12	−1.82	−1.95	−1.90		
13	−1.9	−2.25	−2.00		
14	−2.0	−2.53	−2.10		
15	−2.0	−2.69	−2.20		
16	−2.1	−2.67			
17	−2.1	−2.67			
18	−2.2	−2.67			

There is no significant evidence for variation in the field luminosity function from place to place, while systematic changes with metallicity are consistent with those expected from the metallicity dependence of the mass–luminosity relation for a constant initial mass function. The globular cluster values are the mean of published data. The specific frequency of RR Lyraes varies considerably from cluster to cluster, although the mean value is correct for field halo stars.

19.7 WHITE DWARF LUMINOSITY FUNCTION [23]

The space density of white dwarfs is given in Table 19.15.

Table 19.15. *The V magnitude and bolometric white dwarf luminosity function.*

M_V	$\log \Phi$ (stars pc^{-3} M_V^{-1})	M_{bol}	$\log \Phi$ (stars pc^{-3} M_{bol}^{-1})
9.5	−5.91 (+0.18, −0.31)	5.50	−5.91 (+0.18, −0.31)
10.0	−5.00 (+0.14, −0.21)	6.88	−5.00 (+0.14, −0.21)
10.5	−4.67 (+0.13, −0.18)	7.84	−4.67 (+0.13. −0.18)
11.0	−4.02 (+0.12, −0.16)	8.92	−4.02 (+0.12, −0.16)
11.5	−3.92 (+0.11, −0.15)	10.12	−3.92 (+0.11, −0.15)
12.0	−3.82 (+0.11, −0.16)	11.24	−3.82 (+0.11, −0.16)
12.5	−3.54 (+0.11, −0.16)	11.98	−3.54 (+0.11, −0.16)
13.0	−3.22 (+0.20, −0.39)	12.55	−3.22 (+0.20, −0.39)
13.5	−3.06 (+0.18, −0.30)	13.25	−3.18 (+0.20, −0.38)
14.0	−2.93 (+0.17, −0.29)	13.75	−2.95 (+0.18, −0.30)
14.5	−3.03 (+0.26, −0.76)	14.25	−3.00 (+0.25, −0.60)
15.0	−2.98 (+0.18, −0.30)	14.75	−2.78 (+0.13, −0.19)
15.5	−3.09 (+0.15, −0.23)	15.25	−3.35 (+0.19, −0.35)
16.0	−4.14 (+0.25, −0.64)	15.75	−4.47 (+0.30, −∞)
16.5	−4.50 (+0.30, −∞)		
17.0	?		

19.8   LUMINOSITY CLASS DISTRIBUTION FOR NEARBY FIELD STARS [24]

The fraction of the stellar luminosity from the disk and halo versus luminosity is given in Table 19.16.

Table 19.16. *Main sequence fraction.*

M_V	Disk	Thick disk and halo
−6	0.40	0
−5	0.42	0
−4	0.43	0
−3	0.44	0
−2	0.45	0
−1	0.47	0
0	0.51	0
1	0.56	0
2	0.66	0
3	0.82	0
4	1.00	1

19.8.1   Relative Number of Stars by MK Class to $V = 8.5$ in HD Catalogue [24]

The relative fraction of stars in the MK spectral classes is given in Table 19.17.

Table 19.17. *Fraction of stars in MK spectral classes.*

MK type	0	B	A	F	G	K	M
% Stars	1	10	22	19	14	31	3

19.9 MASS DENSITY IN THE SOLAR NEIGHBORHOOD [25–32]

Observed volume mass density

Interstellar matter (ISM)	0.04 ± 0.02	$\mathcal{M}_\odot \text{ pc}^{-3}$
Main Sequence Stars:		
$\quad 0.08 \leq \mathcal{M}/\mathcal{M}_\odot < 1.0$	0.036	$\mathcal{M}_\odot \text{ pc}^{-3}$
$\quad 1.0 \leq \mathcal{M}/\mathcal{M}_\odot < 100$	0.014	$\mathcal{M}_\odot \text{ pc}^{-3}$
$\quad$ Halo stars	0.0001	$\mathcal{M}_\odot \text{ pc}^{-3}$
Evolved stars:		
$\quad$ White dwarfs	0.005	$\mathcal{M}_\odot \text{ pc}^{-3}$
$\quad$ Dark extended halo, local density	0.01	$\mathcal{M}_\odot \text{ pc}^{-3}$
Total	0.10 ± 0.03	$\mathcal{M}_\odot \text{ pc}^{-3}$

Note that $0.01 \mathcal{M}_\odot \text{ pc}^{-3}$ is 0.3 Gev cm^{-3}.

Observed column mass densities, to $|z| = 1.1$ kpc

Neutral ISM	8	$\mathcal{M}_\odot \text{ pc}^{-2}$
Ionized ISM	2	$\mathcal{M}_\odot \text{ pc}^{-2}$
Molecular ISM	3	$\mathcal{M}_\odot \text{ pc}^{-2}$
ISM total	13 ± 3	$\mathcal{M}_\odot \text{ pc}^{-2}$
Stars:		
$\quad$ Disk main sequence	30	$\mathcal{M}_\odot \text{ pc}^{-2}$
$\quad$ Disk white dwarfs	3	$\mathcal{M}_\odot \text{ pc}^{-2}$
$\quad$ Thick disk	2	$\mathcal{M}_\odot \text{ pc}^{-2}$
$\quad$ Halo subdwarfs	< 1	$\mathcal{M}_\odot \text{ pc}^{-2}$
Stellar total	35 ± 5	$\mathcal{M}_\odot \text{ pc}^{-2}$
Observed total	48 ± 8	$\mathcal{M}_\odot \text{ pc}^{-2}$
Extended dark halo		
$\quad |z| < 1.1$ kpc	23	$\mathcal{M}_\odot \text{ pc}^{-2}$
Total	71 ± 6	$\mathcal{M}_\odot \text{ pc}^{-2}$

K dwarfs ($z \lesssim 160$ pc) $\rho_0 = 0.10 \pm 0.03 \mathcal{M}_\odot \text{ pc}^{-3}$.

All determinations are consistent with each other and with zero local unidentified matter at the $\sim 1.5\,\sigma$ level.

Dynamical analysis of the column mass density, $\mathcal{M}_\odot \text{ pc}^{-2}$

K dwarfs ($300 \lesssim z \lesssim 2000$ pc)

$$\sum\nolimits_{\text{tot}}(z \leq 1.1 \text{ kpc}) = 71 \pm 6 \mathcal{M}_\odot \text{ pc}^{-2},$$
$$\sum\nolimits_{\text{disk}} = 48 \pm 9 \mathcal{M}_\odot \text{ pc}^{-2},$$
$$\sum\nolimits_{\text{dark halo}} = 23 \mathcal{M}_\odot \text{ pc}^{-2},$$

Unidentified disk dark matter $= 0 \pm 12 \mathcal{M}_\odot \text{ pc}^{-2}$.

Limit on scale height/local volume density of dark disk matter with scale height H (pc):

$$\rho_{0,\text{dark}} \lesssim 0.017 \left(\frac{H}{300 \text{ pc}} \right)^{-1} \mathcal{M}_\odot \text{ pc}^{-3}$$

Local luminosity density $= 0.10 L_{\odot,V} \text{ pc}^{-3}$.
Mass-to-light ratio for all stars:

$$\mathcal{M}/L|_{\text{stars}} = 0.5 \mathcal{M}/L|_{\odot,V}.$$

Mass-to-light ratio for all local matter:

$$\mathcal{M}/L|_{\text{local}} = 1.0 \mathcal{M}/L|_{\odot,V}.$$

Surface brightness in a column:

$$= 12.0 L_{\odot,V} \text{ pc}^{-2}.$$

Mass density in column, stars, and ISM:

$$= 48 \mathcal{M}_\odot \text{ pc}^{-2}.$$

Mass-to-light ratio of identified mass:

$$= 4.0 \mathcal{M}/L|_{\odot,V}$$
$$= 9.5 \mathcal{M}/L|_{\odot,B}.$$

Extended halo mass, $|z| < 1.1$ kpc:

$$= 23 \mathcal{M}_\odot \text{ pc}^{-2}.$$

Identified matter in a column:

$$= 48 \mathcal{M}_\odot \text{ pc}^{-2}$$
$$= 100 \text{ g/sq. meter (gsm)}$$
$$= 0.010 \text{ g cm}^{-2}$$
$$= 6 \times 10^{21} \text{ H atoms/cm}^2.$$

19.10 STELLAR MASS FUNCTION [28]

The single star and system luminosity functions are consistent with a single stellar initial mass function (IMF):

$$\xi(\mathcal{M}) = \begin{cases} 0.035 \mathcal{M}^{-1.3(\pm 0.6)}, & 0.08 \leq \mathcal{M} \leq 0.50, \\ 0.019 \mathcal{M}^{-2.2}, & 0.50 < \mathcal{M} \leq 1.0, \\ 0.019 \mathcal{M}^{-2.7}, & 1.00 < \mathcal{M} \leq 100, \end{cases}$$

where $\xi(\mathcal{M}) d\mathcal{M}$ is the number of stars in the mass interval $\mathcal{M}$ to $\mathcal{M} + d\mathcal{M}$ in units of $\mathcal{M}_\odot$.

Some properties of the IMF are as follows:

Binary fraction $\sim 50\%$.

Binary primary: secondary ratio—uncorrelated.

Total mass density in IMF at Sun $\simeq 0.05 \pm 0.01 \mathcal{M}_{\odot}$ pc^{-3}.

Mass density in stars with $\mathcal{M} \leq 1 \mathcal{M}_{\odot} = 0.036 \mathcal{M}_{\odot}$ pc^{-3}.

Extrapolation of IMF to zero mass (brown dwarfs) from $0.085 \mathcal{M}_{\odot} = 0.0085 \mathcal{M}_{\odot}$ pc^{-3}.

Fraction of all stars with $0.08 \leq \mathcal{M} \leq 0.5 \mathcal{M}_{\odot}$ ($\mathcal{M}$ dwarfs) $= 77 \pm 10\%$.

Local stellar number density:

$$\text{Total} = 0.087 \text{ stars } \mathcal{M}_{\odot}^{-1} \text{ pc}^{-3},$$

$$0.08 \leq \mathcal{M} \leq 1 \mathcal{M}_{\odot} = 0.13 \text{ stars pc}^{-3},$$

$$1 \leq \mathcal{M} \leq 60 \mathcal{M}_{\odot} = 0.011 \text{ stars pc}^{-3}.$$

19.10.1 Mass–Luminosity Relation [19, 28]

The mass function $\xi(\mathcal{M})$ is related to the luminosity function $\phi(\mathcal{M}_V)$ by $\xi(\mathcal{M}) = (d\mathcal{M}/d\mathcal{M}_V)\Phi(\mathcal{M}_V)$. The mass–luminosity relation for solar main sequence stars is tabulated in Table 19.18 to allow this transformation.

Table 19.18. *Mass luminosity relation.*[a]

M_V	$\mathcal{M}/\mathcal{M}_{\odot}$	M_V	$\mathcal{M}/\mathcal{M}_{\odot}$	M_V	$\mathcal{M}/\mathcal{M}_{\odot}$	M_V	$\mathcal{M}/\mathcal{M}_{\odot}$
18.00	0.0700	8.77	0.595	4.09	1.119	0.0	3.467
16.96	0.0854	8.56	0.610	4.02	1.135	−1.0	5.248
16.13	0.101	8.35	0.626	3.95	1.150	−2.0	7.943
15.45	0.116	8.14	0.641	3.88	1.166	−3.0	12.023
14.87	0.132	7.92	0.656	3.82	1.181	−4.0	18.197
14.36	0.147	7.71	0.672	3.75	1.196	−5.0	26.915
13.93	0.163	7.51	0.687	3.69	1.212	−6.0	41.687
13.54	0.178	7.31	0.703	3.63	1.227	−7.0	63.1
13.21	0.193	7.11	0.718	3.57	1.243		
12.92	0.209	6.93	0.734	3.51	1.258		
12.67	0.224	6.75	0.749	3.46	1.274		
12.47	0.240	6.59	0.764	3.40	1.289		
12.29	0.255	6.43	0.780	3.35	1.304		
12.13	0.271	6.28	0.795	3.30	1.320		
11.99	0.286	6.14	0.811	3.25	1.335		
11.86	0.301	6.01	0.826	3.20	1.351		
11.73	0.317	5.88	0.842	3.15	1.366		
11.60	0.332	5.75	0.857	3.10	1.382		
11.47	0.348	5.63	0.872	3.05	1.397		
11.34	0.363	5.52	0.888	3.01	1.412		
11.21	0.379	5.40	0.903	2.96	1.428		
11.08	0.394	5.29	0.919	2.92	1.443		
10.93	0.409	5.19	0.934	2.87	1.459		
10.79	0.425	5.08	0.950	2.83	1.474		
10.64	0.440	4.98	0.965	2.79	1.490		
10.48	0.456	4.88	0.980	2.75	1.505		
10.31	0.471	4.78	0.996	2.71	1.521		
10.14	0.487	4.68	1.011	2.67	1.536		

Table 19.18. *(Continued.)*

M_V	$\mathcal{M}/\mathcal{M}_\odot$	M_V	$\mathcal{M}/\mathcal{M}_\odot$	M_V	$\mathcal{M}/\mathcal{M}_\odot$	M_V	$\mathcal{M}/\mathcal{M}_\odot$
9.96	0.502	4.58	1.027	2.63	1.551		
9.78	0.517	4.49	1.042	2.59	1.567		
9.59	0.533	4.41	1.058	2.56	1.582		
9.39	0.548	4.32	1.073	2.52	1.598		
9.19	0.564	4.24	1.088	2.00	1.862		
8.99	0.579	4.17	1.104	1.00	2.512		

Note

[a]Three significant figures are provided because the derivative of this relation is important and not because any particular value is that significant.

Table 19.19 presents the mass and the luminosity (in magnitude units, in each of the V, I, and K bands) associated with the stellar mass function. These values are the luminosity functions in V, I, and K bands corresponding to the tabulated mass function.

Table 19.19. *Masses and luminosities of the stellar mass function.*

Mass ($\mathcal{M}_\odot$) center of increment $\overline{\mathcal{M}}/\mathcal{M}_\odot$	Mass increment $\Delta\mathcal{M}$	Mass in that increment mass/n	Luminosity (mag.) $M_V + 2.5\log n$	I-band luminosity (mag.) $M_I + 2.5\log n$	K-band luminosity (mag.) $M_K + 2.5\log n$
0.015	0.01	10.4×10^{-3}	—	—	—
0.055	0.01	9.7×10^{-3}	—	—	—
0.095	0.01	8.2×10^{-3}	19.1	15.0	12.0
0.155	0.01	7.1×10^{-3}	17.50	14.11	11.50
0.205	0.01	6.5×10^{-3}	16.72	13.71	11.30
0.255	0.01	6.1×10^{-3}	16.34	13.52	11.24
0.305	0.01	5.8×10^{-3}	16.13	13.45	11.25
0.355	0.01	5.5×10^{-3}	15.93	13.38	11.25
0.405	0.01	5.3×10^{-3}	15.68	13.26	11.21
0.455	0.01	5.1×10^{-3}	15.35	13.08	11.11
0.505	0.01	4.9×10^{-3}	14.95	12.84	10.98
0.525	0.05	2.4×10^{-2}	13.05	11.02	9.20
0.625	0.05	1.9×10^{-2}	12.13	10.50	8.92
0.725	0.05	1.6×10^{-2}	11.17	9.93	8.58
0.825	0.05	1.4×10^{-2}	10.46	9.53	8.36
0.925	0.05	1.2×10^{-2}	9.97	9.27	8.24
1.05	0.10	2.0×10^{-2}	8.75	8.28	7.40
1.25	0.10	1.5×10^{-2}	8.35	8.16	7.40
1.55	0.10	1.0×10^{-2}	8.07	8.15	7.59
1.75	0.10	8.4×10^{-3}	8.00	8.21	7.72
2.05	0.10	6.4×10^{-3}	7.94	8.30	7.91
2.25	0.10	5.5×10^{-3}	7.90	8.35	8.02
2.55	0.10	4.4×10^{-3}	7.85	8.43	8.17

Table 19.19. *(Continued.)*

Mass ($\mathcal{M}_\odot$) center of increment $\overline{\mathcal{M}}/\mathcal{M}_\odot$	Mass increment $\Delta\mathcal{M}$	Mass in that increment mass/n	Luminosity (mag.) $M_V + 2.5\log n$	I-band luminosity (mag.) $M_I + 2.5\log n$	K-band luminosity (mag.) $M_K + 2.5\log n$
2.75	0.10	3.9×10^{-3}	7.83	8.48	8.26
3.05	0.10	3.3×10^{-3}	7.80	8.55	8.40
3.25	0.10	2.9×10^{-3}	7.79	8.60	8.48
3.55	0.10	2.5×10^{-3}	7.81	8.69	8.61
3.75	0.10	2.3×10^{-3}	7.83	8.76	8.70
4.05	0.10	2.0×10^{-3}	7.87	8.85	8.83
5.05	0.10	1.4×10^{-3}	7.99	9.13	9.20
6.05	0.10	1.0×10^{-3}	8.09	9.36	9.51
7.05	0.10	7.9×10^{-4}	8.17	9.55	9.76
8.05	0.10	6.3×10^{-4}	8.24	9.71	9.98
9.05	0.10	5.1×10^{-4}	8.30	9.86	10.18
9.95	0.10	4.4×10^{-4}	8.35	9.97	10.34
10.5	1.0	4.0×10^{-3}	5.87	7.54	7.92
15.5	1.0	2.1×10^{-3}	6.08	8.03	8.58
20.5	1.0	1.3×10^{-3}	6.21	8.36	9.04
25.5	1.0	8.8×10^{-4}	6.28	8.60	9.38
30.5	1.0	6.5×10^{-4}	6.38	8.83	9.69
35.5	1.0	5.0×10^{-4}	6.48	9.04	9.95
40.5	1.0	4.0×10^{-4}	6.57	9.21	10.18
45.5	1.0	3.3×10^{-4}	6.64	9.36	10.38
50.5	1.0	2.8×10^{-4}	6.70	9.50	10.56
55.5	1.0	2.4×10^{-4}	6.75	9.61	10.72
60.5	1.0	2.0×10^{-4}	6.79	9.71	10.86
70.5	1.0	1.5×10^{-4}	6.9	9.9	11.1
80.5	1.0	1.3×10^{-4}	6.9	10.1	11.3
90.5	1.0	1.0×10^{-4}	7.0	10.2	11.5
99.5	1.0	8.7×10^{-5}	7.0	10.3	11.7

These data in a similar format are plotted in Figure 19.1.

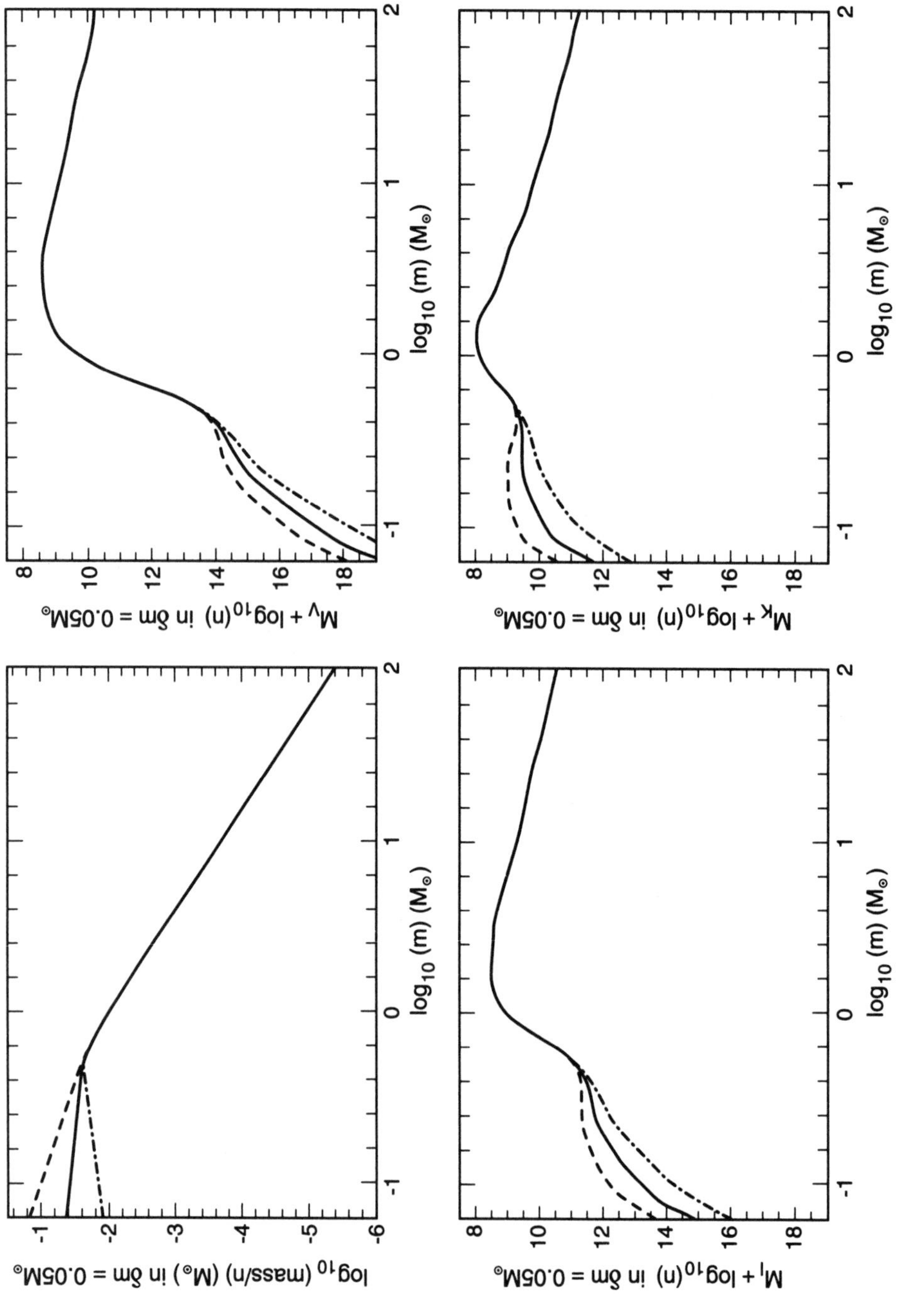

Figure 19.1. Mass–luminosity relations.

19.11 SOLAR MOTION AND KINEMATICS OF NEARBY STARS [33, 34]

The local standard of rest (LSR) is defined as the origin of a velocity system corrected for solar peculiar motion. It is defined empirically, from the mean motion of nearby stars, the kinematic definition, or from the local circular velocity, the dynamical definition. The standard solar motion is an implicit kinematic definition of the LSR from the mean motion of nearby gas and stars. The basic solar motion is an implicit kinematic definition of the LSR from the maximum in the kinematics of nearby stars. The peculiar solar motion is a dynamical definition, derived from extrapolation of the asymmetric drift–velocity dispersion relation to zero dispersion. These motions are given in Table 19.20.

Table 19.20. *Standard, basic, and peculiar solar motion.*

Solar motion	$U_\odot$	$V_\odot$	$W_\odot$	$v_\odot$	Apex of motion α		δ
	(km s^{-1})						
Standard	10.0	5.2	7.2	13	270°	(1900)	+30°
Basic	9	11	6	15.4	267°4	(1950)	+25°
Peculiar	9	12	7	16.6	267°0	(1950)	+28°

The sign convention U is positive toward the galactic center, V is positive in the direction of the galactic rotation, and W is positive toward the North Galactic Pole.

19.11.1 Solar Motion and Velocity Dispersion for Stars of Various Spectral Types [33, 35]

Table 19.21 gives the solar motion and velocity dispersion of stars of various spectral classes.

Table 19.21. *Solar motion relative to stars of various spectral types.*

Spectral type	Solar motion (km s^{-1}) $U_\odot$	$V_\odot$	$W_\odot$	Velocity dispersion (km s^{-1}) σ_U	σ_V	σ_W	σ_u (km s^{-1})	ψ (deg)
			Supergiants					
gO-gB5	+9.0	+13.4	+3.7	12	11	9	19	+36
gF-gM	+7.9	+11.7	+6.5	13	9	7	17	+18
			Giants					
gA	+13.4	+11.6	+10.3	22	13	9	27	+27
gF	+19.7	+18.5	9.5	28	15	9	33	+14
gG	+7.2	+11.1	+6.9	26	18	15	35	+12
gK0	+10.6	+18.6	+6.5	31	21	16	41	+21
gK3	+9.0	+17.6	+6.4	31	21	17	41	+14
gM	+4.5	+18.3	+6.2	31	23	16	42	+7
			Main sequence					
B0	+9.6	+14.5	+6.7	10	9	6	15	−50
dA0	+7.3	+13.7	+7.2	15	9	9	20	+15
dA5	+8.5	+7.8	+7.4	20	9	9	24	+19
dF0	+11.2	+10.8	+7.3	24	13	10	29	+21
dF5	+10.1	+12.3	+6.2	27	17	17	36	+13
dG0	+14.5	+21.1	+6.4	26	18	20	37	+2

Table 19.21. (*Continued.*)

Spectral type	Solar motion (km s^{-1})			Velocity dispersion (km s^{-1})			σ_u (km s^{-1})	ψ (deg)
	$U_\odot$	$V_\odot$	$W_\odot$	σ_U	σ_V	σ_W		
dG5	+8.1	+22.1	+4.3	32	17	15	39	+14
dK0	+10.8	+14.9	+7.4	28	16	11	34	+3
dK5	+9.5	+22.4	+5.8	35	20	16	43	+11
dM0	+6.1	+14.6	+6.9	32	21	19	43	+8
dM5	+9.8	+19.3	+8.6	31	23	16	42	−7

19.11.2 Solar Motion and Velocity Dispersion for Groups of Selected Objects [36–42]

Table 19.22 gives the solar motion for groups of selected objects.

Table 19.22. *Solar motion relative to groups of selected objects.*[a]

Objects	Solar motion (km s^{-1})			Velocity dispersion (km s^{-1})			
	$U_\odot$	$V_\odot$	$W_\odot$	σ_U	σ_V	σ_W	σ_u
Interstellar H I	+12	+15	+9	(	5.7	)[b]	10
Interstellar Ca II	+11	+14	+8	(	6	)[b]	10
Classical Cepheids	+11	+12	+10	8	7	5	12
Carbon stars	+10	+12	+5	30	20	14	39
White dwarfs	+10	+15	+7	42	22	18	50
RR Lyraes							
c-type	0	120	0	100	70	50	
ab, $\Delta s < 5$	0	120	0	130	120	80	
$\Delta s \geq 5$	0	180	0	160	110	90	
all ab's	0	155	0	160	120	90	
Miras, by period P							
$P < 148^d$		33		60	40	—	81
$145 < P < 200^d$		111		90	145	60	180
$200 < P < 250^d$		61		—	70	—	101
$250 < P < 300^d$		33		60	60	—	88
$300 < P < 350^d$		32		45	40	35	69
$350 < P < 400^d$		23		45	25	23	58
$P < 410^d$		15		—	—	—	50

Notes

[a]Missing values have very large errors.

[b]The values for the interstellar gas are line-of-sight velocity dispersions, and cannot be deconvolved into orthogonal components reliably. It is probable that the velocity dispersions are nearly isotropic, so are similar in each component.

19.11.3 Velocity–Age Relation for Disk Stars

The total velocity dispersion σ_V as a function of stellar age τ is adequately described by

$$\sigma_V^3(\tau) = \sigma_{V,\tau=0}^3 + \tfrac{3}{2}\alpha_V\delta_2 T_\delta\left[\exp\left(\frac{\tau}{T_\delta}\right) - 1\right],$$

REFERENCES

1. From the web site: ESA, 1997, The Hipparcos and Tycho Catalogues, ESA SP-1200
2. Baade, W. 1942, *ApJ*, **100**, 137
3. Baade, W. 1942, *ApJ*, **100**, 147
4. Oort, J.H. 1958, *Ric Astron Specola Vaticana*, **5**, 415
5. Blaauw, A. 1965 in *Galactic Structure*, edited by A. Blaauw and M. Schmidt (University of Chicago Press, Chicago), p. 435
6. Sandage, A.R. 1986, *ARAA*, **24**, 421.
7. Bahcall, J.N. 1986, *ARAA*, **24**, 577
8. Gilmore, G., Wyse, R., & Kuijken, K. 1989, *ARAA*, **27**, 555
9. Gilmore, G., & Wyse, R.F.G. 1987, in *The Galaxy*, edited by G. Gilmore and R. Carswell (Reidel, Dordrecht), p. 247
10. Yamagata, T., & Yoshii, Y. 1991, *AJ*, **103**, 117
11. Fenkart, R. 1989, *A&AS*, **81**, 187
12. Majewski, S. 1992, *AJS*, **78**, 87
13. Toller, G., Tanake, H., & Weinberg, J.L., 1987 *A&A*, **188**, 24
14. van der Kruit, P.C. 1986, *A&A*, **157**, 230
15. Dahn, C.C., Liebert, J., & Harrington, R. 1986, *AJ*, **91**, 621
16. Gilmore, G., Reid, N., & Hewett, P. 1985, *MNRAS*, **213**, 257
17. Henry, T., & McCarthy, D. 1990, *AJ*, **350**, 224
18. Kinman, T. 1992, in *Variable Stars and Galaxies*, ASP Conf. Ser. Vol. 30,
19. Scalo, J. 1986, *Fund. Cosmic Phys.*, **11**, 1
20. Stobie, R., Ishida, K., & Peacock, J. 1989, *MNRAS*, **238**, 709
21. von Hippel, T., Gilmore, G., Tanvir, N., Robison, D., & Jones, D. 1996, *AJ*, **112**, 192
22. Wielen, R., Jahreiss, H., & Krüger, R. 1983, in *Nearby Stars and the Stellar Luminosity Function*, edited by A. Davis Philip and A. Upgren (Davis Press, Schenectady, NY), p. 163.
23. Leibert, J., Dahn, C.C., & Monet, D.G. 1988, *AJ*, **332**, 891
24. Allen, C.W., 1963, *Astrophysical Quantities*, 2nd ed. (Athlone Press, London)
25. Bahcall, J. 1984, *ApJ*, **276**, 156
26. Kulkarni, S., & Heiles, C. 1987, in *Interstellar Processes*, edited by H. Thronson and D. Hollenback (Reidel, Dordrecht), p. 87
27. Kuijken, K., & Gilmore, G. 1989, *MNRAS*, **239**, 605
28. Kroupa, I., Tout, C., & Gilmore, G. 1993, *MNRAS*, **262**, 545
29. Bahcall, J., Flynn, C., & Gould, A. 1992, *ApJ*, **389**, 234
30. Kuijken, K. 1991, *ApJ*, **373**, 125
31. Kuijken, K., & Gilmore, G. 1989, *MNRAS*, **239**, 605
32. Kuijken, K., & Gilmore, G. 1991, *ApJ*, **367**, L9
33. Wielen, R. 1982, Landolt-Börnstein Tables, *Astrophys.*, 2C, Sec. 8.4, 202
34. Dehnen, W. & Binney, J.J. 1998, *MNRAS*, **298**, 387
35. Delhaye, J. 1965, in *Galactic Structure*, edited by by A. Blaauw and M. Schmidt, 1982 (University of Chicago Press, Chicago), p. 61
36. Wielen, R., 1982, Landolt-Börnstein Tables, *Astrophys.*, 2C, Sec. 8.4, 29, Table 8
37. Feast, M., Woolley, R., & Yilmaz, N., 1972, *MNRAS*, **158**, 23
38. Hawley, S., Jeffreys, W., Barnes, T., & Wan, Lai 1986, *AJ*, **302**, 626
39. Strugnell, P., Reid, N., & Murray, C. 1986, *MNRAS*, **220**, 413
40. Wielen, R., 1982, Landolt-Börnstein Tables, *Astrophys.*, 2C, 211, Table 2
41. Wielen, R. 1977, *A&A*, **60**, 263
42. Fuchs, B., & Wielen, R. 1987, in *The Galaxy*, edited by G. Gilmore and R. Carswell (Reidel, Dordrecht), p. 375

Chapter 20

Theoretical Stellar Evolution

Arthur N. Cox, Stephen A. Becker, and W. Dean Pesnell

20.1   BASIC EQUATIONS OF STELLAR STRUCTURE
by Stephen A. Becker

20.1.1   List of Symbols

A_a	Atomic mass of particle a
a	Radiation density constant
c	Speed of light in vacuum
d	Distance traveled by overshooting element from the stable convective boundary
G	Gravitational constant
g	Gravitational acceleration
H_P	Pressure scale height
$L(r)$	Energy transferred per second through a sphere of radius r
L_*	Stellar luminosity
$\mathcal{M}(r)$	Mass in a sphere of radius r
$\mathcal{M}_*$	Stellar mass
$\dot{\mathcal{M}}_*$	Stellar mass-loss rate in $\mathcal{M}_\odot$/yr
$\mathcal{M}_\odot$	Solar mass
N_A	Avogadro's number (6.0221×10^{23} mol^{-1})
n_a	Number density of particle a
P	Pressure
Q	Energy generated per nuclear reaction in MeV
R_*	Stellar radius
r	Nuclear reaction rate or radius
S	Entropy per unit mass
T	Temperature
T_{eff}	Effective temperature of the star
t	Time
X_a	Mass fraction of particle a
ϵ	Nuclear energy generation ergs per g s
Γ_2	Second adiabatic exponent
κ	Rosseland mean opacity
ρ	Density
τ_a	Mean lifetime of particle a in a nuclear reaction

All units are in cgs units except where stated otherwise.

For an Eulerian one-dimensional (1D) treatment (r, radius as the independent variable), which assumes hydrostatic equilibrium and neglects rotation and magnetic fields, one has for a nonbinary star the following equations.

For conservation of mass,

$$\frac{d\mathcal{M}(r)}{dr} = 4\pi r^2 \rho(r). \tag{20.1}$$

For conservation of thermal energy,

$$\frac{dL(r)}{dr} = 4\pi r^2 \rho(r) \left[\epsilon(r) - T(r)\frac{dS}{dt} \right]. \tag{20.2}$$

For conservation of momentum (hydrostatic equilibrium),

$$\frac{dP(r)}{dr} = -\rho(r)\frac{G\mathcal{M}(r)}{r^2}. \tag{20.3a}$$

In the relativistic limit such as the case for neutron stars, the Oppenheimer–Volkoff equation of hydrostatic equilibrium applies:

$$\frac{dP(r)}{dr} = -\frac{G\left[\rho(r) + P(r)/c^2\right]\left[\mathcal{M}(r) + 4\pi r^3 P(r)/c^2\right]}{r^2\left[1 - 2G\mathcal{M}(r)/rc^2\right]}. \tag{20.3b}$$

For energy transport,

$$\frac{dT(r)}{dr} = -\frac{3}{4ac}\frac{\kappa(r)\rho(r)}{T^3(r)}\frac{L(r)}{4\pi r^2} \quad \text{(radiative equilibrium)}, \tag{20.4a}$$

$$\frac{dT(r)}{dr} = \left(\frac{dT(r)}{dr}\right)_{\text{ad}} = \frac{\Gamma_2 - 1}{\Gamma_2}\frac{T(r)}{P(r)}\frac{dP(r)}{dr} \quad \text{(adiabatic convection)}. \tag{20.4b}$$

Equation (20.4b) is used whenever

$$-\frac{dT(r)}{dr} > -\left(\frac{dT(r)}{dr}\right)_{\text{ad}} \tag{20.5}$$

(see [1–3]).

Where convection is not adiabatic (such as near the stellar surface), a different expression for Equation (20.4b) should be used [2].

These differential equations also require five specific functions of the local thermodynamic state (at r) obtained from the microphysics:

$$\epsilon = \epsilon(\rho, T, \text{composition}), \tag{20.6a}$$

$$P = P(\rho, T, \text{composition}), \tag{20.6b}$$

$$\kappa = \kappa(\rho, T, \text{composition}), \tag{20.6c}$$

$$S = S(\rho, T, \text{composition}), \tag{20.6d}$$

$$\Gamma_2 = \Gamma_2(\rho, T, \text{composition}). \tag{20.6e}$$

Equation (20.6a), the energy generation rate, includes the effect of neutrino generation as a negative contribution. This relation, as well as Equation (20.6b), the equation of state, and (20.6c), the opacity, are discussed in more detail in the following sections. See [1–3] for more on Equations (20.6d) and (20.6e). Time dependence enters both through Equation (20.2) and the fact that the composition changes with time due to nuclear reactions, which generally allows the stellar model to evolve slowly in a quasistatic manner. When the evolution ceases to be quasistatic, Equations (20.3a) and (20.3b) need to be modified to include an acceleration term.

The basic boundary conditions are

$$\begin{array}{llll} \mathcal{M}(r) \to 0, & L(r) \to 0 & \text{for } r \to 0 & \text{(central conditions)}, \\ \mathcal{M}(r) \to \mathcal{M}_*, & L(r) \to L_* & \text{for } r \to R_* & \text{(surface conditions)}. \end{array}$$

In addition, the variables P and T approach their photospheric values, which are estimated from stellar atmosphere theory [1–4].

The alternative Lagrangian treatment of the stellar structure equations (m, mass as the independent variable) can be found in [4].

A good introduction on how the basic equations of stellar structure are modified for nonspherical rotating stars and on stars with magnetic fields can be obtained in [5] and [6].

20.2   STELLAR NUCLEAR ENERGY GENERATION
by Stephen A. Becker

20.2.1   Two-Body Interactions

For $a + b \rightarrow c + d + e$, where e may or may not be present (for example, $^1\text{H} + {}^2\text{H} \rightarrow {}^3\text{He} + \gamma$, or $^{15}\text{N} + {}^1\text{H} \rightarrow {}^{12}\text{C} + {}^4\text{He}$, or $^1\text{H} + {}^1\text{H} \rightarrow {}^2\text{H} + e^+ + \nu_e$), the reaction rate per g s is

$$r_{ab} = \rho X_a X_b \frac{N_A}{A_a A_b (1 + \delta_{ab})} f(\rho, T) g(T). \tag{20.7a}$$

The mean lifetime in seconds for particle a interacting with b is

$$\tau_a = \left[\rho \frac{X_b}{A_b} f(\rho, T) g(T) \right]^{-1}. \tag{20.8a}$$

The rate of destruction for particle a by b in numbers per cm^3 s is

$$\left(\frac{dn_a}{dt} \right)_b = -(1 + \delta_{ab}) \rho^2 X_a X_b \frac{N_A}{A_a A_b (1 + \delta_{ab})} f(\rho, T) g(T) = -(1 + \delta_{ab}) \rho r_{ab}. \tag{20.9a}$$

The rate of energy generation in erg/(g s) is

$$\epsilon_{ab} = 1.6022 \times 10^{-6} Q_{ab} \rho X_a X_b \frac{N_A}{A_a A_b (1 + \delta_{ab})} f(\rho, T) g(T),$$

$$\epsilon_{ab} = 1.6022 \times 10^{-6} Q_{ab} r_{ab}. \tag{20.10a}$$

In these equations $f(\rho, T)$ is the electron screening correction [7–9], $g(T)$ is the temperature dependence of the nuclear reaction rate as discussed and tabulated in [10, 11]. Q_{ab}, the energy in MeV released or absorbed by a interacting with b, is given in [10].

20.2.2   Three-Body Interactions

Here $a + b + c \rightarrow d + e + f$, where f may or may not be present. For example, $3\,{}^4\text{He} \rightarrow {}^{12}\text{C} + \gamma$, can be considered this way even though this reaction is really a resonant reaction involving a pair of two-body reactions.

The reaction rate per g s is

$$r_{abc} = \rho^2 X_a X_b X_c \frac{N_A}{A_a A_b A_c (1 + \delta_{ab} + \delta_{bc} + \delta_{ac} + 2\delta_{abc})} f(\rho, T) g(T). \tag{20.7b}$$

The mean lifetime for particle a in seconds interacting with particles b and c is

$$\tau_a = \left[\rho \frac{X_b X_c}{A_b A_c} \frac{(1 + \delta_{ab} + \delta_{ac})}{(1 + \delta_{ab} + \delta_{bc} + \delta_{ac} + 2\delta_{abc})} f(\rho, T) g(T) \right]^{-1}. \tag{20.8b}$$

The rate of destruction for particle a by b and c in numbers per cm^3 s is

$$\left(\frac{dn_a}{dt}\right)_{bc} = -(1 + \delta_{ab} + \delta_{ac})\rho^2 X_a X_b X_c \frac{N_A f(\rho, T) g(T)}{A_a A_b A_c (1 + \delta_{ab} + \delta_{bc} + \delta_{ac} + 2\delta_{abc})}$$

$$= -(1 + \delta_{ab} + \delta_{ac})\rho r_{abc}. \tag{20.9b}$$

The rate of energy generation in erg/(g s) is

$$\epsilon_{abc} = 1.6022 \times 10^{-6} \, Q_{abc} \rho^2 X_a X_b X_c \frac{N_A f(\rho, T) g(T)}{A_a A_b A_c (1 + \delta_{ab} + \delta_{bc} + \delta_{ac} + 2\delta_{abc})}$$

$$= 1.6022 \times 10^{-6} \, Q_{abc} r_{abc}. \tag{20.10b}$$

The change in the chemical composition for a given atomic nucleus can be obtained by summing all the production and destruction rates. When nuclear reactions are part of a chain (such as the proton–proton chain) the abundances of different atomic nuclei become interrelated and simplifications become possible when equilibrium is attained.

For a discussion on how nuclear reaction rates are calculated, see [12–14]. For a discussion of the various reactions involved in the hydrogen-, helium-, carbon-, oxygen-, and neon-burning processes, see [15–18].

During post-core helium-burning evolutionary phases, significant nuclear energy can be removed by neutrinos produced by plasma, photoelectron, pair, and bremsstrahlung processes. The rate of energy loss produced by these processes is discussed in [17], [19], and [20].

20.2.3 The Solar Neutrino Problem

Nuclear energy production in the Sun should produce a measurable flux of neutrinos on the Earth primarily due to the following nuclear reactions:

$$^1H + {}^1H \longrightarrow {}^2H + e^+ + \nu_e,$$
$$^7Be + e^- \longrightarrow {}^7Li + \nu_e + \gamma,$$
$$^8B \longrightarrow {}^8Be + e^+ + \nu_e,$$
$$^1H + {}^1H + e^- \longrightarrow {}^2H + \nu_e.$$

Measurement of this neutrino flux would provide confirmation that the nuclear reactions do take place in the Sun and provide important constraints on theoretical models of the Sun, and to this end there are currently four different solar neutrino detector experiments in operation. Unfortunately, the measured neutrino fluxes are less than those predicted by the best theoretical solar models, and this conflict is the basis of the solar neutrino problem. For a detailed discussion of the experiments and the various approaches to modeling the Sun, see [21] and [22], as well as the review papers of Bahcall [23] and [24]. As discussed in [22] and [24], a possible resolution of the solar neutrino problem could arise if neutrinos have mass and they can undergo oscillations from the electron neutrino state into the muon or tau neutrino states, then the measured flux would be less than expected based only on stable electron neutrinos. Future experiments should be able to test this hypothesis.

20.3 EQUATIONS OF STATE
by W. Dean Pesnell

The equation of state (EOS) joins the microscopic world of quantum mechanics and the macroscopic world of astrophysics [25]. Interactions that govern the world of the atom are averaged to form an

EOS. An EOS is usually presented as a table of pressure and internal energy for many values of ρ and T or as a simple formula for easy computation.

The astrophysical EOS has three major parts: P_N (nuclei), P_{rad} (radiation), and P_e (electrons). Other processes, such as Coulomb interactions and statistical correlations, can modify the EOS. See Table 20.1 for components of the EOS.

Table 20.1. *Components of the equation of state.*[a]

Component	Pressure	Internal energy
Black-body radiation	$aT^4/3$	aT^4/ρ
Nondegenerate nuclei[b]	$(\mathcal{R}/\mu)\rho T$	$3(\mathcal{R}/\mu)\rho T/2$
Nondegenerate electrons	$N_e kT = (\mathcal{R}/\mu_e)\rho T$	$\frac{3}{2}N_e kT = \frac{3}{2}(\mathcal{R}/\mu_e)\rho T$
Degenerate electrons[c,d,e,f]	$Af(x)$	$Ag(x)$
Coulomb corrections[g,h] [1–3]	$-P_N h(U_D)$	$-3P_N h(U_D)/\rho$

Notes

[a]Approximate formulas for optically thin radiation are shown in [4]. Nuclei can crystallize in white dwarfs and the crusts of neutron stars [5].
[b]$\mathcal{R} = 8.3145 \times 10^7 \, \text{erg K}^{-1} \, \text{mol}^{-1}$.
[c]$A = 6.0023 \times 10^{22} \, \text{dyn cm}^{-2}$.
[d]$f(x) = x(2x^2 - 3)\sqrt{1 + x^2} + 3\ln(x + \sqrt{1 + x^2})$.
[e]$g(x) = 8x^3(\sqrt{1 + x^2} - 1) - f(x)$.
[f]$Bx^3 = \rho/\mu_e$, $B = 9.739 \times 10^5 \, \text{mol cm}^{-3}$.
[g]$U_D = e^2/(R_D kT)$, $R_D^2 = kT/(4\pi e^2 N_e)$, $R_D = 6.90(T/N_e)^{1/2} \, \text{cm}$.
[h]$h(U_D) = 0.3U_D^{3/2}/(1.03921 + U_D^{1/2})$.

References
1. Koester, D. 1976, *A&A*, **52**, 415
2. Salpeter, E.E., & Zapolsky, H.S. 1967, *Phys. Rev.*, **158**, 876
3. Lai, D., Abrahams, A.M., & Shapiro, S.L. 1991, *ApJ*, **337**, 612
4. Raymond, J.C., Cox, D.P., & Smith, B.W. 1976, *ApJ*, **204**, 290
5. Lamb, D.Q., & Van Horn, H.M. 1975, *ApJ*, **200**, 306

In normal stars the electron contribution to the EOS is the most complicated. Material inside a star is ionized by collisions and radiation. The populations of the electron energy levels are distorted by the close proximity of ionized material. The upper states dissolve and lower energy levels are perturbed from those of an isolated atom. Several ways of approaching this problem are as follows.

Stellingwerf [26, 27] used a simplified atomic model including H, He, and two averaged easily ionized metals. The resulting analytic model is surprisingly accurate and shows that the EOS is insensitive to the upper levels, unlike the opacity.

Eggleton, Faulkner, and Flannery [28] use a fit to the electron degeneracy integrals and force pressure ionization at about the correct density. The EOS uses both tabulated and analytic fits. This EOS is used in Iben's evolution code and in some studies of the Sun.

Mihalas, Däppen, and Hummer [29] use an occupancy probability to smoothly remove the electrons from each state and give a table of values.

Rogers [30] developed a Planck–Larkin partition function that smoothly ionizes the electrons in the usual ionization zone. He also approximates pressure ionization by distorting and removing the electron energy levels as the density of atoms increases.

Both electrons and nuclei can be in a degenerate form. The pressure is greatly enhanced over the nondegenerate value when the repulsive nature of the Fermi–Dirac statistics becomes very large [25]. The electron contribution to the EOS is summarized in Figure 20.1. The degeneracy parameter η has

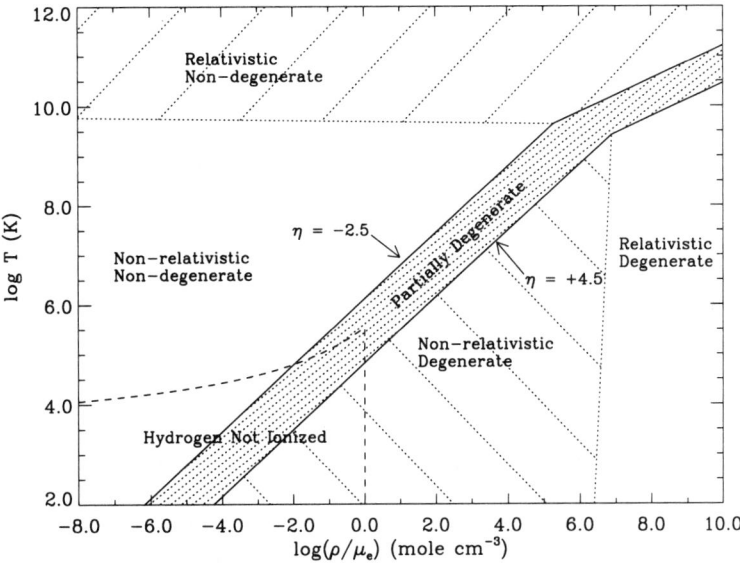

Figure 20.1. The electron contribution to the EOS, defined in the ρ-T plane. Approximate regions of degeneracy, hydrogen ionization, and relativistic effects are shown. The area of partial degeneracy is defined where corrections to P_e in an expansion in η are $< 1\%$. In the nondegenerate, relativistic region ($T > 6 \times 10^9$ K), the effects of $e^- - e^+$ pairs become important. Adapted from [25].

the limits $-\eta \gg 0$ in nondegenerate matter and $\eta \gg 0$ in degenerate matter.

Degenerate nuclei can never be treated as an ideal gas. Nuclear densities ($\rho \geq 10^{14}$ g cm^{-3}) imply average separations on the order of a nucleus. The strong nuclear force becomes important and the EOS must incorporate nonideal gas corrections [31].

For $T < 4000$ K, the effect of molecules becomes important. Statistical equilibrium is used to find the concentration of the molecular and atomic components. At extremely low temperatures and high densities (such as in the interior of gas giant planets), the formation of metallic hydrogen is important [32].

Derivatives of EOS variables are an important part of a table or fitting function. Stellar evolution codes require the compressibilities (χ_ρ and χ_T), adiabatic indices (Γ_1, $\Gamma_3 - 1$, and Γ_2), and specific heats (c_V and c_P). As there are two dependent functions of two independent variables (plus composition, see [33]), four derivatives are required to completely specify the EOS and should be tabulated with the dependent variables to increase the numerical accuracy of their interpolation.

20.4 STELLAR OPACITIES
by Arthur N. Cox

The Los Alamos [34–39] and Livermore [40–45] opacities include the bound–bound, bound–free, free–free, and electron scattering processes. Collective effects of the free-electron scattering have been studied definitively by Boercker [46]. Low-temperature opacities including molecular absorptions have been calculated by Los Alamos workers and by Alexander [47, 48], Alexander and Ferguson [49], and by Kurucz [50]. The composition for elements heavier than hydrogen and helium in the solar mixture is given by Grevesse [51–53]. More opacity information is given in Chapter 5.

20.5 ELECTRON CONDUCTION
by Arthur N. Cox

At high densities in stars at advanced evolution stages, sometimes energy can be significantly transported by the flow of electrons. The effective opacity then is reduced to a total opacity κ_T given by the formula

$$1/\kappa_T = 1/\kappa_R + 1/\kappa_C,$$

with κ_R being the purely radiative opacity and κ_C being the conductive opacity that is related to the conductivity ν_C in the relation

$$F = -\nu_C \, dT/dr.$$

The conductive opacity then is

$$\kappa_C = 4acT^3/3\rho\nu_C.$$

The Cox and Tabor [37] tables included this electron conductivity (and a sharp cutoff for the infinite opacity for photons at frequencies below the plasma frequency). With the improvements from the Mestel [54] treatment to those of Hubbard and Lampe [55] and Canuto [56], more recent opacities no longer allowed for this effect in the opacity tables. A simple fit for the conductive opacity is from Sweigart [57]. Fits to the Hubbard and Lampe (nonrelativistic electrons) and Canuto (relativistic electrons) tables are given by Iben [58]. Recent conductivities for neutron star material are given by Itoh et al. [59].

20.6 ELEMENT DIFFUSION
by Arthur N. Cox

The diffusive settling of all elements, except hydrogen, occurs in stars. In very special cases, the absorption of photons and their outward momentum can result in a net outward acceleration for isotopes that are in specific ionization and excitation states. Usually surface convection zones are slightly drained of helium and the Z composition elements, but that effect can be overcome if the depleted outer layers are simultaneously blown away in a stellar wind. Similarly, enhancement of special elements, as seen mostly in Ap, Bp, and Am stars [60], is reduced by stellar winds. Formulations for calculating diffusive settling are given by Burgers [61] and Michaud, Fontaine, and Charland [62], using diffusion coefficients by Paquette et al. [63, 64]. A good review is by Michaud and Vauclair [65].

20.7 MIXING IN STARS
by Arthur N. Cox

A difficult problem in stellar evolution involves the degree of mixing of isotopes that have been segregated by diffusion or are burned at different rates by nuclear reactions. In deep layers where hydrogen is being converted to helium, helium is being burned to carbon and oxygen, or even further nuclear reactions are occurring for more massive stars, any mixing-in of hydrogen or helium will greatly affect this processing and its time scale. The existence of short-lived radioactive isotopes like ^{99}Tc in some stellar atmospheres provides the strongest evidence for mixing. In surface layers, mixing (dredging up) processed material into a surface convection zone can significantly change the observed stellar composition. Very sensitive spectroscopic determinations of some isotopes can probe aspects of stellar evolution, especially in late evolution stages for the red giants and supergiants with deep surface convection zones. Mixing can produce subtle effects, such as moving carbon into hydrogen- and helium-rich material, increasing its opacity, and then causing a superadiabatic gradient that convects and mixes even more. This gradient of temperature steeper than that given by the adiabatic relations between temperature, density, and pressure cannot exist without convective motions that are very effective in mixing, even in the presence of a magnetic field.

Evidence of transport of angular momentum in the Sun is found in its almost solid body rotation to very deep layers. Its primordial angular momentum is carried somehow to the surface and then magnetically braked. The observation of essentially the primordial beryllium abundance, however, indicates little deep mixing to high-temperature burning layers of solar convection zone material. How angular momentum can be transported without moving matter is a current problem in solar evolution.

Analysis of mixing is complicated. Even a small mass loss can expose compositions that are either depleted or enhanced in some isotopes and confuse conclusions about any diffusive separation or mixing that has occurred.

A good exposition of mixing in general is by Kippenhahn [66].

Some mixing processes are as follows:

- Convection caused by temperature gradients that are superadiabatic [67].

- Overshooting beyond the formal superadiabatic convecting layers [68, 69].

- Semiconvection where mixing occurs just enough to make a subadiabatic gradient [70].

- Merging of isotopes with differing diffusion or radiation levitation rates [71].

- Turbulence resulting from differential rotation shearing layers [72].

- Goldreich–Schubert–Fricke instability in noncylindrically rotating layers [73, 74].

- Meridional (Eddington–Sweet) circulation caused by rotation [5, 75, 76].

- Nonradial pulsations in composition gradient layers (Kato mechanism and Rayleigh–Taylor instability) [77, 78].

- Angular momentum transfer by internal gravity waves [68].

- Diffusion of magnetic fields inside stars [79].

Some effects that impede mixing are as follows:

1. Composition gradients with the normal situation of higher mean molecular weight material being deeper, requiring more superadiabaticity (Ledoux criterion) [25, 80].

2. Magnetic fields that constrain motions across the field lines [81].

20.8 STAR FORMATION
by W. Dean Pesnell

Stars are formed by the collapse of interstellar clouds. Such clouds have $\rho \sim 10^{-20}$ g cm^{-3} and temperatures near 20 K. The gravitational energy released during the collapse either heats the material or is radiated away to space [82]. If the pressure at the center of a spherical cloud of mass $\mathcal{M}$ and radius R exceeds the pressure of an homogeneous sphere,

$$P_c > \frac{G\rho\mathcal{M}}{2R},$$

(20.11)

the cloud cannot collapse. A cloud with $\mathcal{M} = 10\mathcal{M}_\odot$, $R \sim 0.3$ pc, and the density listed above is stable in our Galaxy. Interstellar magnetic fields, rotation, and cloud motion also prevent a cloud from collapsing [83]. In certain cases, magnetic fields can actually carry away angular momentum and promote cloud collapse [84].

Current star formation theories use triggers to overcome the internal pressure and begin the gravitational runaway. Shocks from supernova explosions or density waves can be triggers. Massive stars emit UV radiation, creating ionization shock fronts in the surrounding material, inducing additional collapses.

The biggest hurdle in star formation is rotation. All clouds will rotate, because of the velocity shear in the galaxy if nothing else. As the collapse proceeds, conservation of angular momentum requires the core to spin more quickly. If this is followed to a radius of $1R_\odot$, the star will be rotating faster than the speed of light. The scenario in Table 20.2 is more acceptable.

Table 20.2. *Stages of star formation* [1].

Stage	Feature	Length (yr)
Initial collapse	Disk and central core	10^5
Viscous accretion	Dissipates rotation	10^4–10^5
Cleansing	Strong winds	10^7

Reference
1. Ruden, S.P., & Pollack, J.B. 1991, *ApJ*, **375**, 740

The presence of a disk explains the presence of strong bipolar flows around Herbig–Haro objects [85]. Magnetic fields are examined in [86]. Once contraction has begun, magnetic fields assist the collapse by transporting angular momentum away from the central object.

20.9 PRE-MAIN-SEQUENCE EVOLUTION
by W. Dean Pesnell

Pre-main-sequence evolution follows a protostar to the main sequence. The protostar enters this phase with a high luminosity and large radius. As the star collapses, it moves along the Hayashi track, an almost vertical line in the Hertzsprung–Russell (HR) diagram [87]. When the core becomes radiatively stable, the star turns to the left and, moving along a line of roughly constant radius, contracts in the core until hydrogen burning sets in and the star settles onto the main sequence. Three divisions in mass of the final star are made:

1. Low mass, $\mathcal{M} < 1\mathcal{M}_\odot$ [88–90].
2. Low mass, $1\mathcal{M}_\odot < \mathcal{M} < 8\mathcal{M}_\odot$ [90, 91].
3. High mass, $\mathcal{M} > 8\mathcal{M}_\odot$ [90, 92, 93].

The luminosity in the early stages of evolution comes from an accretion shock at the "surface" of the protostar [94]. Accurate calculations of the evolution require a robust numerical scheme [95] and treatment of deuterium burning [88]. For rotation effects see [96].

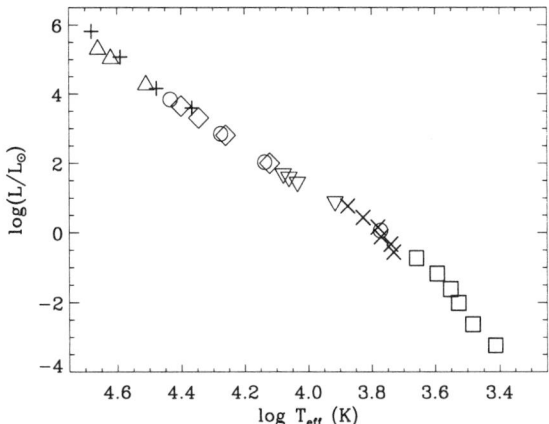

Figure 20.2. A theoretical HR diagram of the zero-age main sequence for Population I stars. The symbols, diamond, triangle, inverted triangle, square, plus sign, circle and × correspond to references [97–103].

20.10 MAIN-SEQUENCE POPULATION I STARS
by W. Dean Pesnell

The main-sequence (MS) is the area of the HR diagram where the greatest density of stars is observed. In the theoretical HR diagram, the MS is the locus of core hydrogen-burning models. Typical surface compositions for Population I stars are $Y = 0.28$ and $Z = 0.02$. Figure 20.2 is an aggregation of theoretical zero-age main-sequence stars (ZAMS) with masses ranging from $0.1\mathcal{M}_\odot$ to $60\mathcal{M}_\odot$. Theoretical luminosity functions are $L \propto \mathcal{M}^{4.5}$ ($\mathcal{M} < 7\mathcal{M}_\odot$) and $L \propto \mathcal{M}^3$ ($M > 7\mathcal{M}_\odot$). The lower limit of the MS is $0.08\mathcal{M}_\odot$, which is not sufficient to begin hydrogen fusion reactions [104]. The upper limit of about $100\mathcal{M}_\odot$ is due to a slowly growing and apparently, not limited in amplitude, pulsational instability driven by the temperature sensitivity of the CNO reaction [105, 106].

Observationally η Car seems to be as massive as $140\mathcal{M}_\odot$ from its place in the Hertzsprung–Russell diagram, but all other stars in our Galaxy and the Magellanic Clouds seem to have less than $100\mathcal{M}_\odot$.

20.11 MAIN-SEQUENCE POPULATION II STARS
by W. Dean Pesnell

Population II stars have a ZAMS slightly below that of Population I stars. Typical surface compositions for Population II stars are $Y = 0.28$ and $Z = 10^{-3}$. Figure 20.3 is an aggregation of theoretical ZAMS models with masses ranging from $0.25\mathcal{M}_\odot$ to $30\mathcal{M}_\odot$. Theoretical luminosity functions are similar to those of Population I stars.

20.12 STELLAR WINDS
by Stephen A. Becker

With the exception of the Sun, the mass-loss rates in Table 20.3 due to stellar winds of main-sequence stars are derived from radio data as described in [109]. For O stars of all luminosity classes, [109] finds the following equation to be accurate to within 50% using common logs:

$$\log(\dot{\mathcal{M}}_*) = 1.69\log(L_*/L_\odot) - 15.41.$$

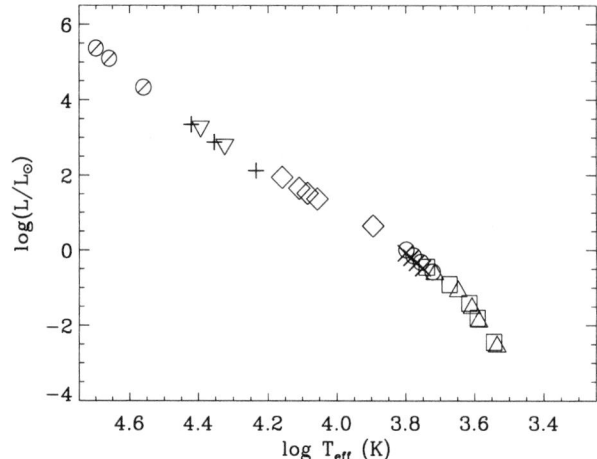

Figure 20.3. A theoretical HR diagram of the zero-age main sequence for Population II stars. The symbols, inverted triangle, circle, diamond, triangle, plus sign, slash in circle, ×, and square correspond to references [98, 107, 108, 100, 97, 98, 107, 108].

Here $\dot{\mathcal{M}}_*$ is in $\mathcal{M}_\odot$/yr. For binaries, this relation may not be appropriate.

Table 20.3. *Observed rates of stellar winds.*

Star name	Spectral class	$\log T_{\text{eff}}$	$\log L_*/L_\odot$	$\mathcal{M}_*/\mathcal{M}_\odot$	$\log_{10} \dot{\mathcal{M}}_*$ ($\mathcal{M}_\odot$/yr)	Reference
CygOB2 #7	O3 If	4.65	6.0	82	< 4.9	[1]
HD15570	O4 If+	4.62	6.2	88	−5.0	[1]
HD190429A	O4 If+	4.62	6.1	77	< 4.7	[1]
HD14947	O5 If+	4.61	5.9	64	< −4.8	[1]
CygOB2 #11	O5 If	4.61	6.0	69	< −5.1	[1]
HD15558	O5 III	4.63	6.0	78	< −4.8	[1]
CygOB2 #9	O6 If+	4.61	6.4		−4.9	[1]
HD210839	O6 I(n)fp	4.62	5.9	55	< −5.5	[1]
CygOB2 #5	O6f+O7f	4.60	6.4		−4.5	[1]
HD57060	O7 Iafpvar	4.56	6.3	46	< −5.3	[1]
HD166734	O7.5f+09 I	4.54	5.9		−4.8	[1]
HD151804	O8 Iaf	4.52	6.1	79	−5.0	[1]
HD152408	O8 Iafpe	4.52	6.0	63	−4.7	[1]
HD149757	O9 V	4.53	4.9	24	< −6.8	[1]
HD36486	O9.5 II	4.49	5.7	23	−6.0	[1]
HD37742	O9.7 Ib	4.48	5.8	60	−5.7	[1]
HD37128	B0 Ia	4.42	5.6	32	−5.6	[1]
HD38771	B0.5 Ia	4.36	5.6	32	−6.0	[1]
HD152236	B1 Ia+	4.30	5.9	29	−5.1	[1]
HD169454	B1 Ia+	4.30	6.0	30	−5.4	[1]
κ Cas	B1 Ia+	4.30	5.4	29	−5.6	[2]
P Cyg	B1 Iap	4.09	5.1	17	−4.8	[2]
ρ Leo	B1 Ib	4.31	4.9		−6.2	[2]
HD190603	B1.5 Ia+	4.28	5.7		−5.5	[1]

Table 20.3. *(Continued.)*

Star name	Spectral class	$\log T_{\rm eff}$	$\log L_*/L_\odot$	$\mathcal{M}_*/\mathcal{M}_\odot$	$\log_{10} \dot{\mathcal{M}}_*$ $(\mathcal{M}_\odot/{\rm yr})$	Reference
χ^2 Ori	B2 Ia	4.24	5.5		−5.7	[2]
σ^2 CMa	B3 Ia	4.17	5.1	26	−5.9	[2]
55 Cyg	B3 Ia	4.15	5.1	26	−6.5	[2]
CygOB2 #12	B8 Ia+	4.06	6.1		−4.5	[1]
β Ori	B8 Ia	4.06	5.5		−6.2	[1]
γ Cyg	A2 Ia	3.96	5.3		< −6.8	[1]
Sun	G2 V	3.76	1.0	1.0	−13.6	[3]

References
1. Howarth, I.D., & Prinja, R.K. 1989, *ApJS*, **69**, 527
2. Underhill, A. 1982, in *Stars With and Without Emission Lines*, edited by A. Underhill and V. Doazan (National Technical Information Service, Springfield, VA), NASA Special Publication SP-456 B, p. 140
3. Noci, G. 1988, in *Mass Outflows from Stars and Galactic Nuclei*, edited by L. Branichi and R. Gilmozzi (Kluwer Academic, Dordrecht), p. 11

Wolf–Rayet stars are a class of very luminous, very hot stars whose spectra have broad emission lines of He and may also show broad emission lines of carbon or nitrogen. WR stars appear to be the hydrogen-exhausted interior of stars which have undergone extreme mass loss. The observed mass-loss rates due to stellar winds for a sample of WR stars which are thermal radio emitters is given in Table 20.4 [110]. Typical mass-loss rates for WR stars are greater than those of O stars.

20.13 STELLAR EVOLUTION TRACKS: MASSIVE AND INTERMEDIATE-MASS STARS
by Stephen A. Becker

The evolutionary track that a stellar model traces in an HR diagram is somewhat code dependent. Many stellar evolution codes are currently being used, and they produce different results *in part* due to uncertainties in the input physics (such as the opacity, the equation of state, and the nuclear reaction rates), uncertainties in the modeling of physical phenomenon (such as the treatment of convective overshoot and semiconvection, the handling of nonadiabatic convection, and how non-quasi-static phases of evolution are approximated), different initial conditions (such as the composition, whether nonsolar abundance ratios are considered in the heavy-element mixture, and mass loss), and modeling techniques. Consequently, stellar evolution tracks in the literature change with time as the input physics improves and modeling approaches change. The state of the art is such that the results of different codes may differ in detail, but the codes do agree on the general qualitative features of the evolutionary tracks. To reflect the diversity of approaches a representative sample of stellar evolution tracks of massive and intermediate-mass stars with the same Population I composition is presented for comparison in Figures 20.4, 20.5, and 20.6. This presentation is by no means all-inclusive, as a search of the literature would reveal.

Table 20.4. *Wolf–Rayet mass loss rates.*

Name (HD)	Spectral class	$\log_{10} \dot{\mathcal{M}}_*$ $(\mathcal{M}_\odot/\mathrm{yr})$
190918	WN4.5+O9.5 IB	−4.52
50896	WN5	−4.12
193077	WN5+OB	−4.73
193576	WN5+O6	−4.62
191765	WN6	−4.13
192163	WN6	−4.02
151932	WN7	−4.31
214419	WN7+O	< −4.61
165763	WC5	−4.50
156385	WC7	<4.46
152270	WC7+O5–8	−3.98
192641	WC7+O5	−4.48
193793	WC7+O4–5	−4.10
192103	WC8	−4.36
68273	WC8+O9 I	−4.09
168206	WC8+O8–9 IV	≤ −4.38
164270	WC9	≤ −4.54

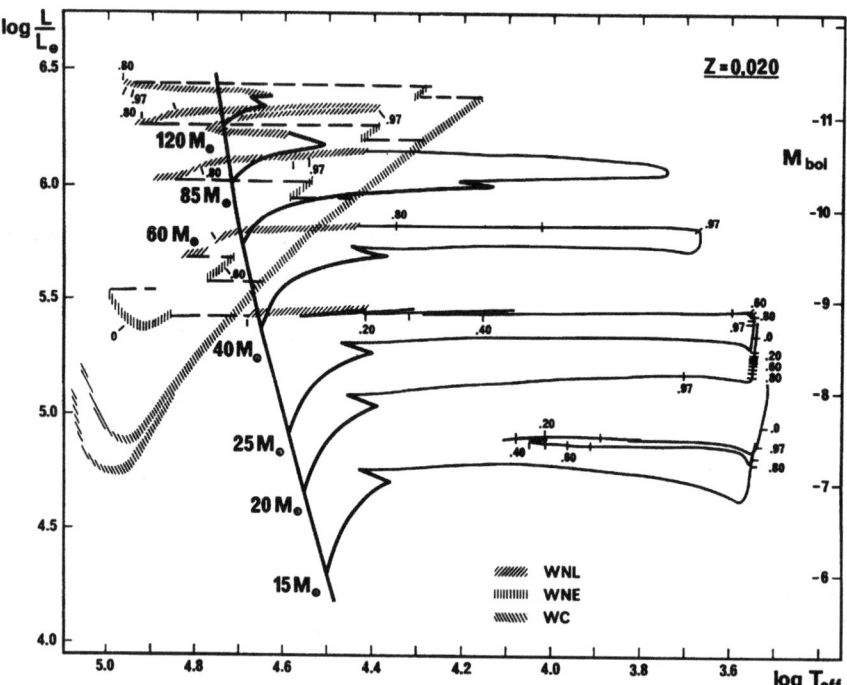

Figure 20.4. Theoretical HR diagram showing evolutionary tracks running from the ZAMS stage to the end of central carbon burning for stellar models of initial mass in the range $(15\text{–}120)\mathcal{M}_\odot$ and initial composition $(X, Y, Z) = (0.70, 0.28, 0.02)$. Evolution is followed with mass loss and includes the effect of moderate overshooting during the different convective core phases [111].

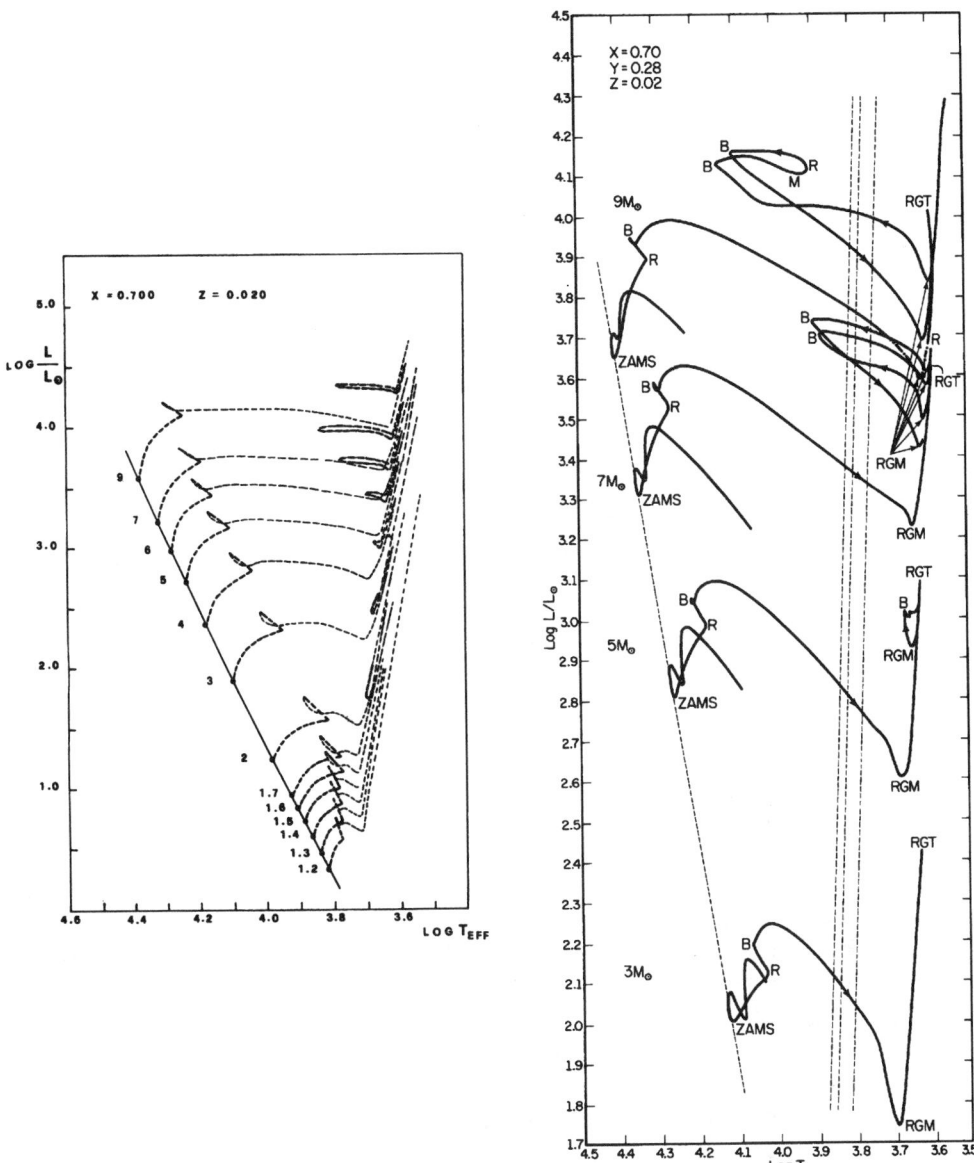

Figure 20.5. (Left) Theoretical HR diagram showing evolutionary tracks running from the ZAMS stage to the start of the thermally pulsing AGB or core carbon ignition for stellar models of initial mass in the range (1.2–9)$\mathcal{M}_\odot$ and initial composition $(X, Y, Z) = (0.70, 0.28, 0.02)$. Evolution is followed without mass loss but with convective overshoot ($d/H_P = 1$) present during the hydrogen and helium burning convective core phases [112].

Figure 20.6. (Right) Theoretical HR diagram showing evolutionary tracks running from the ZAMS stage to the start of the thermally pulsing AGB or core carbon ignition for stellar models of initial mass in the range (3–9)$\mathcal{M}_\odot$ and initial composition $(X, Y, Z) = (0.70, 0.28, 0.02)$. Evolution is followed without mass loss or convective overshoot [97].

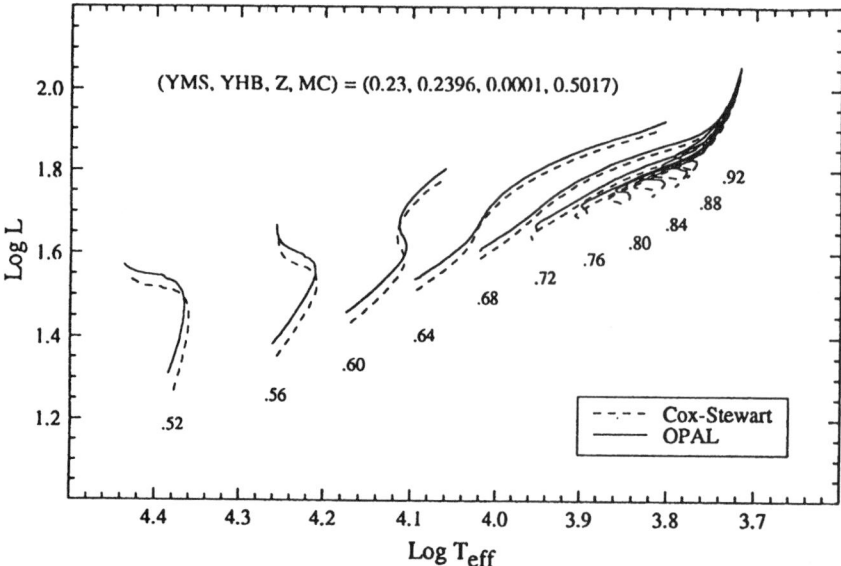

Figure 20.7. Horizontal branch evolution for Population II stars using both the Los Alamos and Livermore opacities. The stellar parameters are the main-sequence helium mass fraction, the increased dredged-up helium content on the zero-age horizontal branch, the heavy-element mass fraction, and the helium core mass (in solar units).

20.14 EVOLUTION TO RED GIANT BRANCH
by W. Dean Pesnell

The evolution of stars with mass less than $3.5\mathcal{M}_\odot$ creates a long-lived evolutionary stage called the asymptotic giant branch (AGB) [113, 114]. The stars either burn helium quiescently or as flashes in a degenerate core. The resulting evolution paths lie very close to the Hayashi track the star followed during its initial collapse. See [99, 108].

20.15 HORIZONTAL BRANCH EVOLUTION
by Arthur N. Cox

Population II stars lose of the order of $(0.1–0.2)\mathcal{M}_\odot$ as red giants in the HR diagram only after igniting helium at their centers and then quickly evolve to the horizontal branch. Details of how this mass loss occurs at the core helium flash and the tracks they follow have not been clearly presented. It is known that the core flash does not occur at the stellar center, and the multidimensional hydrodynamics that occur are complicated. The zero-age horizontal branch extends blueward of the main sequence for the lowest ($\sim 0.5\mathcal{M}_\odot$) masses, and these lowest-mass stars can move directly into the white dwarf region as their helium cores burn to carbon and oxygen. With significant surface hydrogen, the stars evolve redward to the AGB, but for very thin layers the stars are already white dwarfs. Horizontal branch tracks using both the Los Alamos and Livermore opacities are given by Yi, Lee, and Demarque [115] in Figure 20.7. The evolution of the central temperature and density in these models is given in Figure 20.8.

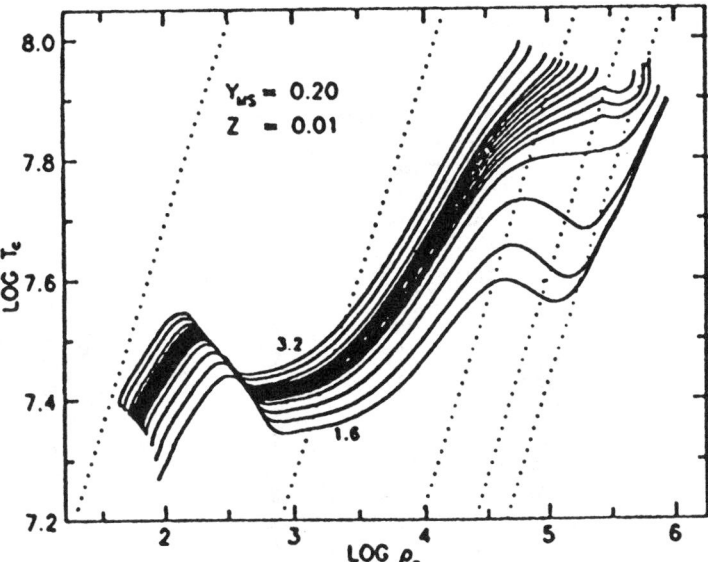

Figure 20.8. Evolutionary dependence of the central temperature (T_c) and density (ρ_c) for the models between 1.6 and $3.2\mathcal{M}_\odot$. The dotted lines are lines of constant degeneracy parameter at the center of the model, $\eta_c = -4$, 0.5, 10, and 15 from left to right. Adapted from [99, 108].

20.16 RED GIANT MASS-LOSS RATES
by Stephen A. Becker

Based on a study of normal red giant and supergiant stars, Reimers [116, 117] has proposed the following semiempirical scaling law for stellar mass loss:

$$\dot{\mathcal{M}}_* = 4 \times 10^{-13} \eta (L_*/L_\odot) \left(\frac{g_*}{g_\odot} \frac{R_*}{R_\odot} \right)^{-1}. \tag{20.12}$$

Here $\dot{\mathcal{M}}_*$ is in units of $\mathcal{M}_\odot$/yr, g is the gravitational acceleration at the stellar surface, and η is a dimensionless factor introduced to take into account the uncertainty in the mass-loss rate determination or behavioral differences between different types of stars. Typically, η is considered to be in the range of 1/3 to 3. Reimers [117] states that his relation should not be applied to stars of distinctly different properties like OH-IR stars [where the rate appears to be an order of magnitude larger than given by Equation (20.12)], F stars, and C stars. The accuracy of this relation can be tested by comparing with the observed mass-loss rates given in Table 20.5. Care should be taken in this comparison because while $\mathcal{M}_*$ and L_* are inferred from observation, the values for R_* and $\mathcal{M}_*$ are generally not, and consequently, the values listed are estimates which can vary considerably from author to author. For $\eta = 1$ the Reimers mass-loss rate agrees with the observed mass loss given in Table 20.5 to within a factor of 3 for most of the stars listed (although there are some cases where it is a factor of 10 off). Examples of mass-loss rates determined for OH-IR stars and C stars are given in Tables 20.6 and 20.7. Other mass-loss studies and formulations have been done (for representative examples, see [118–120]).

Table 20.5. *Measured mass-loss rates for normal giant and supergiant stars and the Sun.*

Star	Spectral type	T_e	$R_*/R_\odot$	$\log L_*/L_\odot$	$\mathcal{M}_*/\mathcal{M}_\odot$	$\dot{\mathcal{M}}_*$ ($\mathcal{M}_\odot$/yr)	Reference
HR 8752	G0 Ia	5000	1000	5.5	30	1×10^{-5}	[1]
α Ori	M2 Iab	3900	860	5.0	10	4×10^{-6}	[2]
α Sco	M1.5ab	3540	625	4.68	~ 18	1×10^{-6}	[3]
χ Her	M6 III	2670	810	4.56	3.1	7.2×10^{-7}	[2]
$\delta^2 2$ Lyr	M4 II	3490	420	4.32	2.8	4.8×10^{-8}	[2]
g Her	MG III	3250	630	4.28	4.0	1.3×10^{-7}	[2]
α HerA	M5 II	3200	460	4.23	7.6	8.2×10^{-8}	[2]
L^2 Pup	M5 IIIe	2825	600	3.92	2.6	4.8×10^{-8}	[2]
31 Cyg	K4 Ib	3800	202	3.91	6.2	4×10^{-8}	[4]
32 Cyg	K5 Iab	3800	188	3.82	8	2.8×10^{-8}	[4]
W Hya	M5 IIIe	2825	510	3.77	1.5	1.1×10^{-7}	[2]
R Dor	M8 Ie	2230	680	3.73	1.5	3.9×10^{-7}	[2]
R Lyr	M5 III	3394	210	3.57	2.2	1.4×10^{-8}	[2]
δ Sge	M2 II	3600	140	3.43	8	2×10^{-8}	[5]
ζ Aur	K4 Ib	3950	140	3.41	8.3	6×10^{-9}	[4]
ρ Per	M4 II–III	3500	150	3.36	5.0	1.2×10^{-8}	[2]
β Peg	M2 II–III	3600	110	3.23	1.7	1.1×10^{-9}	[2]
22 Vul	G3 Ib–II	5200	> 40	2.99	4.3	6×10^{-9}	[6]
α Boo	K1 IIIp	4250	27	2.23	1.1	2×10^{-10}	[2]
Sun	G2 V	5780	1	1	1	2.6×10^{-14}	[7]

References

1. Lambert, D.L., & Luck, R.E. 1978, *MNRAS*, **184**, 405
2. Judge, P.G., & Stencel, R.E. 1991, *ApJ*, **371**, 357
3. Hagen, H-J., Hempe, K., & Reimers, D. 1987, *A&A*, **184**, 256
4. Che, A., Hempe, K., & Reimers, D. 1983, *A&A*, **126**, 225
5. Reimers, D., & Schröder, K-P. 1983, *A&A*, **123**, 241
6. Reimers, D., & Che-Bohenstenzel, A. 1986, *A&A*, **166**, 252
7. Noci, G. 1988, in *Mass Outflows from Stars and Galactic Nuclei*, edited by L. Branichi and R. Gilmozzi (Kluwer Academic, Dordrecht), p. 11

Table 20.6. *Observed mass-loss rates for OH-IR stars.*

Star	$\dot{\mathcal{M}}_*$ ($\mathcal{M}_\odot$/yr)	Reference
PZ Cas	1.1×10^{-5}	[1]
Z Cas	2.3×10^{-7}	[1]
RS Vir	6.8×10^{-7}	[1]
OH 11.5+0.1	3.2×10^{-6}	[1]
OH 12.8−1.9	1.6×10^{-5}	[1]
OH 13.1+5.0	8.7×10^{-5}	[1]
OH 16.1−0.3	1.1×10^{-6}	[1]
OH 18.3+0.4	1.6×10^{-4}	[1]
OH 18.5+1.4	3.3×10^{-6}	[1]
OH 18.8+0.3	1.1×10^{-6}	[1]
OH 20.2−0.1	1.4×10^{-5}	[1]
OH 20.7+0.1	1.3×10^{-4}	[1]
OH 21.3+0.5	2.6×10^{-4}	[1]
OH 25.1−0.3	3.1×10^{-6}	[1]
OH 26.2−0.6	3.8×10^{-7}	[1]
OH 26.5+0.0	1.2×10^{-4}	[1]
OH 30.1−0.7	3.1×10^{-5}	[1]
OH 30.1−0.2	4.0×10^{-6}	[1]

Table 20.6. *(Continued.)*

Star	$\dot{M}_*$ ($M_\odot$/yr)	Reference
OH 30.7−10.4	1.1×10^{-4}	[1]
OH 31.0−0.2	1.6×10^{-5}	[1]
OH 31.0+0.0	6.1×10^{-6}	[1]
OH 32.0−0.5	8.9×10^{-5}	[1]
OH 32.8−0.3	2.4×10^{-4}	[1]
OH 35.6−0.3	3.4×10^{-5}	[1]
OH 36.9+1.3	4.7×10^{-7}	[1]
OH 39.7+1.5	9.7×10^{-7}	[1]
OH 39.9−0.0	7.7×10^{-6}	[1]
OH 44.8−2.3	4.0×10^{-5}	[1]
OH 45.5+0.1	1.8×10^{-5}	[1]
OH 53.6−0.2	2.0×10^{-6}	[1]
OH 75.3−1.8	1.6×10^{-4}	[1]
OH 83.4−0.9	4.6×10^{-6}	[1]
OH 104.9+2.4	3.5×10^{-5}	[1]
OH 127.8−0.0	9.4×10^{-5}	[1]
OH 127.9−0.0	1.4×10^{-4}	[1]
OH 138.0+7.2	2.2×10^{-5}	[1]
OH 138.0+7.3	5.3×10^{-6}	[1]
OH 141.7+3.5	1.1×10^{-5}	[1]
OH 231.8+4.2	1.3×10^{-4}	[2]

References
1. Netzer, N., & Knapp, G.R. 1987, *ApJ*, **323**, 734
2. Knapp, G.R., & Morris, M. 1985, *ApJ*, **292**, 640

Table 20.7. *Mass loss rates for selected carbon stars.*

Star	Type	$\dot{M}_*$ ($M_\odot$/yr)	Reference
CIT 6	C4,3	2.3×10^{-6}	[1]
RY Dra	C4,4	5.1×10^{-6}	[2]
TU Gem	C4,6	9.0×10^{-7}	[2]
UU Aur	C5,3	3.0×10^{-7}	[1]
X Cnc	C5,4	2.5×10^{-7}	[2]
Y Hya	C5,4	3.9×10^{-7}	[2]
Y CVn	C5,4	1.0×10^{-7}	[3]
TX Psc	C6,2	1.6×10^{-7}	[1]
R Scl	C6,II	4.2×10^{-6}	[1]
V Hya	C6,3e	4.0×10^{-6}	[3]
V460Cyg	C6,3	4.0×10^{-7}	[2]
VY UMa	C6,3	1.5×10^{-7}	[2]
U Cam	C6,4e	1.08×10^{-5}	[2]
ST Cam	C6,4	2.2×10^{-7}	[2]
RV Cyg	C6,4	7.4×10^{-7}	[2]
RY Mon	C6,5	8.5×10^{-7}	[2]
T Lyr	C6,5	6.0×10^{-7}	[2]
TW Hor	C7,2	8.0×10^{-8}	[1]
NP Pup	C7,2	1.3×10^{-7}	[1]
U Hya	C7,3	4.9×10^{-7}	[1]
UX Dra	C7,3	7.5×10^{-8}	[2]
T Ind	C7,3	8.0×10^{-8}	[1]
Z Psc	C7,3	9.1×10^{-8}	[2]
IRC+20370	C7,3e	1.0×10^{-5}	[3]

Table 20.7. *(Continued.)*

Star	Type	$\dot{\mathcal{M}}_*$ ($\mathcal{M}_\odot$/yr)	Reference
V Cyg	C7,4e	1.04×10^{-5}	[2]
IRC+40540	C8,3,5	1.4×10^{-5}	[1]
T Dra	C8e	1.3×10^{-6}	[3]
IRC+10216	C9,5	5.5×10^{-5}	[1, 3]
IRC+00499	Ne	4.4×10^{-6}	[3]
IRC+50096	N	6.3×10^{-6}	[3]
R Lep	N6e	2.1×10^{-6}	[3]
CRL 482	C	1.7×10^{-5}	[3]
CRL 865	C	2.3×10^{-5}	[3]
IRC−10236	C	4.7×10^{-6}	[3]
IRC+20326	C	2.3×10^{-5}	[3]
CRL 2155	C	1.7×10^{-5}	[3]
CRL 2199	C	1.3×10^{-5}	[3]
IRC+40485	C	2.6×10^{-6}	[3]
CRL 3068	C	7.3×10^{-6}	[3]
CRL 3099	C	1.5×10^{-6}	[3]
RU Vir	R3ep	9.5×10^{-6}	[3]

References

1. Judge, P.G., & Stencel, R.E., 1991, *ApJ*, **371**, 357
2. Wannier, P.G., Shai, R., Anderson, B.G., & Johnson, H.R. 1990, *ApJ*, **358**, 251
3. Knapp, G.R., & Morris, M. 1985, *ApJ*, **292**, 640

20.17 ASYMPTOTIC GIANT BRANCH EVOLUTION
by Stephen A. Becker

The evolutionary behavior of a low-mass star on the asymptotic giant branch (AGB) in the HR diagram is illustrated in Figure 20.9 from [113]. Such stars are the major source of planetary nebulae and they evolve into white dwarfs. A detailed description of the AGB phase of evolution for stars of low and intermediate mass can also be found in [113]. A more recent review of the evolutionary phase can be found in [121, 122].

20.18 WHITE DWARFS AND NEUTRON STARS
by Arthur N. Cox and Stephen A. Becker

After a star has become a red giant, it can have blue loops in the HR diagram as both Population I and II stars exhibit, it can have no loops, or it can evolve directly to the blue. Thus white dwarfs can be formed either directly from the Population II red giant tip or from the results of heavy mass loss from Population I or II stars after the AGB evolution. After a superwind ejects essentially all surface hydrogen, stars evolve across the top of the HR diagram at about $10^4 L_\odot$ with a rapidly decreasing surface radius. These are the (DO) pre-white dwarfs, the helium (DB) white dwarfs, and the hydrogen (DA) white dwarfs. Neutron stars (or even black holes) are the result of a supernova explosion when the core of the highly evolved star collapses to create an explosion of the outer stellar layers. See Chapter 16 for a list of pulsars that are these neutron stars seen in all photon energy bands as they rotate, exposing photon sources.

A review of the masses and evolutionary states of white dwarfs can be found in [123]. The evolution and cooling of white dwarfs is discussed in [124] and [125]. The equilibrium mass-radius

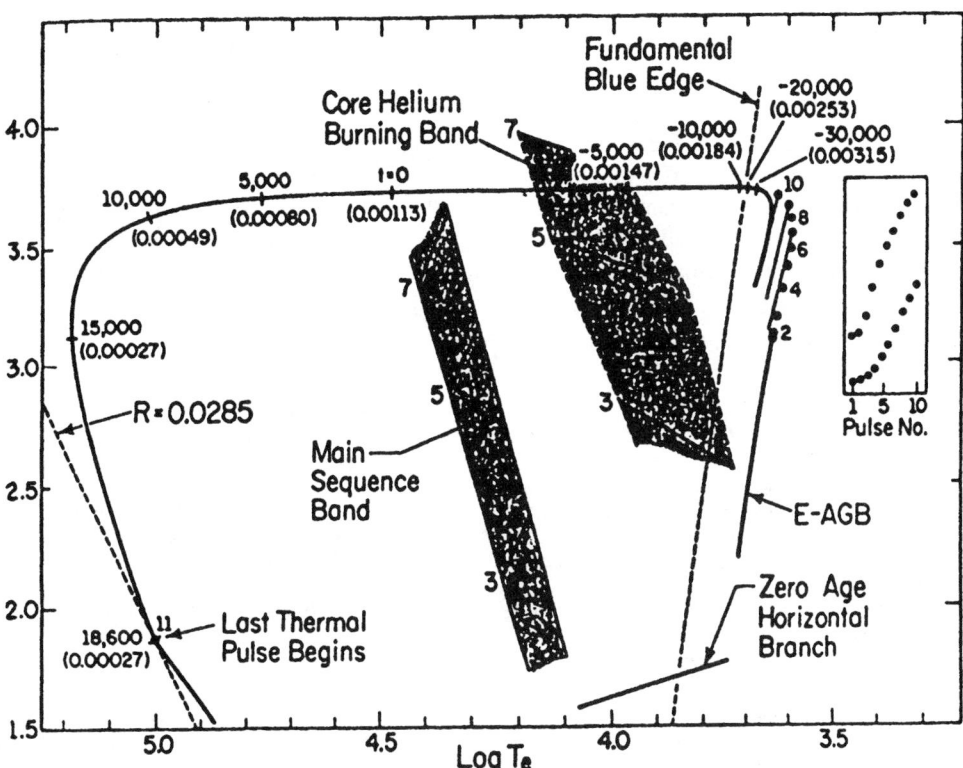

Figure 20.9. Evolutionary track in the HR diagram of an AGB model of mass $0.6\mathcal{M}_\odot$, initial composition $(Y, Z) = (0.25, 0.001)$. After burning helium in its core on the horizontal branch, the model arrives at the early-AGB track to burn helium in a shell; the hydrogen-burning shell is extinguished. The early-AGB phase is terminated when hydrogen reignites and thermal pulsing begins. The location of the model at the start of each pulse is indicated by heavy dots. Excursions in the HR diagram during the extended postflash dip and recovery period are shown for pulses 7, 9, and 10. Dots in the panel in the extreme right-hand portion of the diagram describe the excursion in luminosity during extended dips for all pulses that occur on the AGB track. Evolution time ($t = 0$ when $T_e = 30\,000$ K) and mass in the hydrogen-rich envelope (in parentheses) are shown at various points along the track leaving the AGB phase after the tenth pulse. Time is in years, and $\mathcal{M}_e$ and R are in solar units. A line of constant radius passes through the location of the beginning of the eleventh pulse when the model has become a hot white dwarf. The dashed line is a blue edge for pulsation in the fundamental mode for a model of mass $0.6\mathcal{M}_\odot$ and $(Y, Z) = (0.25, 0.001)$. Shown for orientation purposes are the rough evolutionary tracks, during core hydrogen- and core helium-burning phases for $(Y, Z) = (0.28, 0.001)$ and masses 3, 5, and $7\mathcal{M}_\odot$.

relation for zero-temperature white dwarfs is reviewed in [126]. A general review of the theoretical and observational aspects concerning white dwarfs can be found in [127–129].

20.19 BINARY STAR EVOLUTION
by Stephen A. Becker

The previous sections of this chapter have dealt with the evolutionary behavior of single stars. Considerations of binary star evolution opens an area of investigation which is much more complicated and cannot be adequately summarized in the limited space available here. Reference [130] provides a brief introduction to the subject of binary star evolution, while [122, 131] provide excellent general reviews of this subject.

20.20   THEORY VERSUS OBSERVATION IN THE HR DIAGRAM
by Stephen A. Becker

As noted in Section 20.13, there are various uncertainties in stellar evolution calculations, particularly those associated with convection, and the best approach to constrain these (when better data or better theory is lacking) is through comparison with observational data. The HR diagram provides a useful method of comparison, especially when applied to star clusters and certain types of variable stars. Representative examples of what can be learned from such comparisons are given in [132–136].

REFERENCES

1. Clayton, D.D. 1968, *Principles of Stellar Evolution and Nucleosynthesis* (McGraw-Hill, New York), p. 436
2. Cox, J.P., & Giuli, R.T. 1968, *Principles of Stellar Structure*, Vol. 2, *Application to Stars* (Gordon and Breach, New York), p. 644
3. Collins, G.W. 1989, *The Fundamentals of Stellar Astrophysics* (Freeman, New York), p. 100
4. Kippenhahn, R., & Weigert, A. 1990, *Stellar Structure and Evolution* (Springer-Verlag, Berlin), p. 64
5. Tassoul, J.-L. 1978, *Theory of Rotating Stars* (Princeton University Press, Princeton, NJ)
6. Hansen, C.J., & Kawaler, S.J. 1994, *Stellar Interiors* (Springer-Verlag, New York), p. 330
7. Lang, K.R. 1980, *Astrophysical Formulae* (Springer-Verlag, Berlin), p. 385
8. Cox, J.P., & Giuli, R.T. 1968, *Principles of Stellar Structure*, Vol. 1, *Physical Principles* (Gordon and Breach, New York), p. 467
9. Clayton, D.D. 1968, *Principles of Stellar Evolution and Nucleosynthesis* (McGraw-Hill, New York), p. 357
10. Caughlan, G.R., & Fowler, W.A. 1988, *Atomic Data Nucl. Data Tables*, **40**, 283
11. Fowler, W.A., Caughlan, G.R., & Zimmerman, B.A. 1975, *ARA&A*, **13**, 69
12. Lang, K.R. 1980, *Astrophysical Formulae* (Springer-Verlag, Berlin), p. 375
13. Cox, J.P., & Giuli, R.T. 1968, *Principles of Stellar Structure*, Vol. 1, *Physical Principles* (Gordon and Breach, New York), p. 429
14. Clayton, D.D. 1968, *Principles of Stellar Evolution and Nucleosynthesis* (McGraw-Hill, New York), p. 288
15. Lang, K.R. 1980, *Astrophysical Formulae* (Springer-Verlag, Berlin), p. 419
16. Cox, J.P., & Giuli, R.T. 1968, *Principles of Stellar Structure*, Vol. 1, *Physical Principles* (Gordon and Breach, New York), p. 475
17. Kippenhahn, R., & Weigert A. 1990, Stellar *Structure and Evolution* (Springer-Verlag, Berlin), p. 161
18. Clayton, D.D. 1968, *Principles of Stellar Evolution and Nucleosynthesis* (McGraw-Hill, New York), p. 369
19. Cox, J.P., & Giuli, R.T. 1968, *Principles of Stellar Structure*, Vol. 1, *Physical Principles* (Gordon and Breach, New York), p. 512
20. Meyer-Hofmeister, E. 1982, in *Landolt-Börnstein Numerical Data and Functional Relationships in Science and Technology*, Group VI, Vol. 2b (Springer-Verlag, New York), p. 187
21. Cox, A.N., Livingston, W.C., & Matthews, M.S., editors, 1991, *Solar Interior and Atmosphere* (University of Arizona, Tucson)
22. Balantekin, A.B., & Bahcall, J.N., editors, 1995, *Solar Modeling* (World Scientific, Singapore)
23. Bahcall, J.N. 1989, *Neutrino Astrophysics* (Cambridge University Press, Cambridge)
24. Bahcall, J.N. 1996, *ApJ*, **467**, 475
25. Cox, J.P., & Giuli, R.T. 1968, *Principles of Stellar Structure* (Gordon and Breach, New York), Chapters 9 & 24
26. Stellingwerf, R.E., 1975, *ApJ*, **195**, 441
27. Stellingwerf, R.E., 1975, *ApJ*, **199**, 705
28. Eggleton, P.P., Faulkner, J., & Flannery, B.P. 1973, *AAp*, **23**, 325
29. Mihalas, D., Däppen, W., & Hummer, D.G. 1988, *ApJ*, **331**, 815
30. Rogers, F. 1986, *ApJ*, **310**, 723
31. Shapiro, S.L., & Teukolsky, S.A. 1983, *Black Holes, White Dwarfs, and Neutron Stars: The Physics of Compact Objects* (Wiley-Interscience, New York), Sec. 8
32. Hubbard, W.B. 1990, in *The New Solar System*, edited by J.K. Beatty, C.C. Petersen, and A. Chaikin (Sky Publishing, New York), p. 131
33. Pesnell, W.D. 1986, *ApJ*, **301**, 204
34. Cox, A.N., & Stewart, J.N. 1969, Sci. Info. Astr. Council USSR Acad. Sci. Vol. 15
35. Cox, A.N., & Stewart, J.N. 1970, *ApJS*, **19**, 243
36. Cox, A.N., & Stewart, J.N. 1970, *ApJS*, **19**, 612
37. Cox, A.N., & Tabor, J.E. 1976, *ApJS*, **31**, 271
38. Huebner, W.F., Merts, A.L., Magee, N.H., & Argo, M.F. 1977, LA 6760M, Los Alamos Publication
39. Weiss, A., Keady, J.J., & Magee, N.H. 1990, *Atomic Data Nucl. Data*, **45**, 209
40. Iglesias, C.A. & Rogers, F.J. 1991, *ApJ*, **371**, 408
41. Iglesias, C.A. & Rogers, F.J. 1991, *ApJS*, **371**, L73
42. Iglesias, C.A. & Rogers, F.J. 1992, *ApJS*, **79**, 507
43. Iglesias, C.A., Rogers, F.J., & Wilson, B.G. 1992, *ApJ*, **397**, 717
44. Rogers, F.J. & Iglesias, C.A. 1992, *ApJ*, **401**, 361
45. Iglesias, C.A. & Rogers, F.J. 1996, *ApJ*, **464**, 943
46. Boercker, D.B. 1987, *ApJ*, **316**, L95

47. Alexander, D.R. 1975, *ApJS*, **29**, 363
48. Alexander, D.R., Johnson, H.R., & Rypma, R.L. 1983, *ApJ*, **272**, 773
49. Alexander, D.R. & Ferguson, J.W. 1994, *ApJ*, **437**, 879
50. Kurucz, R.L. *IAU Symposium* 149, 225 (see Chapter 5)
51. Grevesse, N. 1991, *A&A*, **242**, 488
52. Grevesse, N. 1991, *A&A*, **232**, 225
53. Grevesse, N. 1991, private communication
54. Mestel L. 1950, *Proc. Cambridge Philos. Soc.*, **46**, 331
55. Hubbard, W.B., & Lampe, M. 1969, *ApJS*, **18**, 279
56. Canuto, V. 1970, *ApJ*, **159**, 641
57. Sweigart, A.V. 1973, *A&A*, **24**, 459
58. Iben, I. Jr. 1975, *ApJ*, **196**, 525; erratum 1993, *ApJ*, **415**, 767
59. Itoh, N., Kohyama, Y., Mastumoto, N., & Seki, M. 1984, *ApJ*, **285**, 758
60. Wolff, S.C. 1983, *The A Stars; Problems and Perspectives*, NASA Special Publication No. 463
61. Burgers, J.M. 1969, *Flow Equations for Composite Gases* (Academic Press, New York)
62. Michaud, G., Fontaine, G., & Charland, Y. 1984, *ApJ*, **280**, 787
63. Paquette, C., Pelletier, C., Fontaine, G., & Michaud, G. 1986, *ApJS*, **61**, 177
64. Paquette, C., Pelletier, C., Fontaine, G., & Michaud, G. 1986, *ApJS*, **61**, 197
65. Michaud, G., & Vauclair, S. 1991, in *Solar Interior and Atmosphere*, edited by A.N. Cox, W.C. Livingston, and M.S. Matthews (University of Arizona Press, Tucson)
66. Kippenhahn, R. 1974, *Late Stages of Stellar Evolution*, IAU Symposium 66 (Reidel, Dordrecht), p. 20
67. Vitense, E. 1953, *Z. Astrophys.*, **32**, 135
68. Schatzman, E. 1991, in *Solar Interior and Atmosphere*, edited by A.N. Cox, W.C. Livingston, and M.S. Matthews (University of Arizona Press, Tucson)
69. Chan, K.L., Nordlund, A., Steffen, M., & Stein, R.F. 1991, in *Solar Interior and Atmosphere*, edited by A.N. Cox, W.C. Livingston, and M.S. Matthews (University of Arizona Press, Tucson)
70. Schwarzschild, M., & Härm, R. 1958, *ApJ*, **128**, 348
71. Vauclair, S., & Vauclair, G. 1982, *ARA&A*, **20**, 37
72. Zahn, J.P. 1983, in *Astrophysical Processes in Upper Main Sequence Stars*, edited by A.N. Cox, S. Vauclair, and J.P. Zahn (Geneva Observatory, Geneva)
73. Goldreich, P., & Schubert, G. 1967, *ApJ*, **150**, 571
74. Fricke, K.J. 1968, *Z. Astrophys.*, **68**, 316
75. Eddington, A.S. 1925, *Observatory*, **48**, 73
76. Sweet, P.A. 1950, *MNRAS*, **110**, 548
77. Kato, S. 1966, *PASJ*, **18**, 374
78. Dilke, F.W.W., & Gough, D.O. 1972, *Nature*, **240**, 262
79. Parker, E.N. 1984, *ApJ*, **281**, 839
80. Ledoux, P. 1947, *ApJ*, **105**, 305
81. Chapman, S., & Cowling, T.G. 1970, in *The Mathematical Theory of Non-Uniform Gases* (Cambridge University Press, Cambridge)
82. Jeans, J.H. 1902, *Philos. Trans. R. Soc. London* A, **199**, 1
83. Spitzer, L. 1968, *Diffuse Matter in Space* (Inter-Science, New York), Sec. 6
84. Mouschiovios, T.C. 1991, in *Physics of Star Formation*, edited by C.J. Lada and N.D. Kylafis (Kluwer Academic, Dordrecht)
85. Shu, F.H., Adams, F.C., & Lizano, S. 1987, *ARA&A*, **25**, 23
86. Pudritz, R.E., & Carlberg, R.G. 1989, in *Low Mass Star Formation and Pre–Main Sequence Objects* (ESO, Garching), p. 33
87. Hayashi, C. 1966, *ARA&A*, **4**, 171
88. Stahler, S.W. 1988, *PASP*, **100**, 1474
89. Stahler, S.W. 1989, *ApJ*, **347**, 950
90. Iben, I., Jr. 1965, *ApJ*, **141**, 993
91. Palla, F., & Stahler, S.W. 1991, *ApJ*, **375**, 288
92. Lequeux, J. 1985, in *Birth and Infancy of Stars*, edited by R. Luces, A. Omont, and R. Stora (North-Holland, Amsterdam), p. 3
93. Downes, D. 1989, in *Star Forming Regions*, edited by M. Peimbert and J. Jugaku (Reidel, Dordrecht), p. 93
94. Larson, R.B. 1968, *MNRAS*, **145**, 271
95. Winkler, K.-H., & Newman, M.J. 1980, *ApJ*, **236**, 201
96. Pinsonneault, M.H., Kawaler, S.D., Sofia, S., & Demarque, P. 1989, *ApJ*, **338**, 424
97. Becker, S.A. 1981, *ApJS*, **45**, 475
98. Brunish, W.M., & Truran, J.W. 1982, *ApJS*, **49**, 447
99. Sweigart, A.V., Greggio, L., & Renzini, A. 1989, *ApJS*, **69**, 911
100. VandenBerg, D.A., Hartwick, F.D.A., Dawson, P., & Alexander, D.R. 1983, *ApJ*, **266**, 747
101. Maeder, A. 1981, *A&A*, **102**, 401
102. Pesnell, W.D., unpublished
103. Green, E.M., Demarque, P. & King. C.R. 1987, *The Revised Yale Isochrones and Luminosity Functions* (Yale University Observatory: New Haven, CT)
104. D'Atona, F., & Mazzitelli, I. 1985, *ApJ*, **296**, 502
105. Ziebarth, K. 1970, *ApJ*, **162**, 947
106. Stothers, R. 1992, *ApJ*, **392**, 706
107. Iben, I., Jr., & Rood, R.T. 1970, *ApJ*, **159**, 605
108. Sweigart, A.V., Greggio, L., & Renzini, A. 1990, *ApJ*, **364**, 527
109. Howarth, I.D., & Prinja, R.K. 1989, *ApJS*, **69**, 527
110. Prinja, R.K., Barlow, M.J., & Howarth, I.D. 1990, *ApJ*, **361**, 607
111. Maeder, A. 1990, *A&AS*, **84**, 139
112. Bertelli, G., Bressan, A., Chiosi, D., & Angerer, K. 1986, *A&AS*, **66**, 191
113. Iben, I., Jr., & Renzini, A. 1983, *ARA&A*, **21**, 271
114. Sweigart, A.V., & Gross, P.G. 1978, *ApJS*, **36**, 405
115. Yi, S., Lee, Y.-W., & Demarque, P. 1993, *ApJ*, **411**, L25
116. Reimers, D. 1975, in *Problems in Stellar Atmospheres and Envelopes*, edited by B. Baschek, W.H. Kegel, and G. Traving (Springer-Verlag, New York), p. 229
117. Reimers, D. 1988, in *Mass Outflows from Stars and Galactic Nuclei*, edited by L. Bianchi and R. Gilmozzi (Kluwer Academic, Dordrecht), p. 25
118. Garmany, C.D., & Conti, P.S. 1984, *ApJ*, **284**, 705
119. Chiosi, C., & Maeder, A. 1986, *ARA&A*, **24**, 239
120. de Jager, C., Nieuwenhuijzen, H., & van der Hucht, K.A. 1988, *A&AS*, **72**, 259

121. Lattanzio, J.C. 1989, in *Evolution of Peculiar Red Giant Stars*, edited by H.R. Johnson and B. Zuckerman (Cambridge University Press, Cambridge), p. 161

122. Iben, I., Jr. 1991, *ApJS*, **76**, 55

123. Weidemann, Y. 1990, *ARA&A*, **28**, 103

124. Iben, I., Jr., & Tutukov, A.V. 1984, *ApJ*, **282**, 615

125. D'Antona, F., & Mazzitelli, L. 1990, *ARA&A*, **28**, 139

126. Hamada, J., & Salpeter, E. 1961, *ApJ*, **134**, 683

127. Cox, J.P., & Giuli, R.T. 1968, *Principles of Stellar Structure*, Vol. 2, *Application to Stars* (Gordon and Breach, New York), p. 874

128. Kippenhahn, R., & Weigert, A. 1990, *Stellar Structure and Evolution* (Springer-Verlag, New York), p. 366

129. Hansen, C.J., & Kawaler, S.D. 1994, *Stellar Interiors* (Springer-Verlag, New York), p. 338

130. Hansen, C.J., & Kawaler, S.D. 1994, *Stellar Interiors* (Springer-Verlag, New York), p. 72

131. Trimble, V. 1983, *Nature*, **303**, 137

132. Maeder, A., & Renzini, A., editors, 1984, *Observational Tests of the Stellar Evolution Theory*, IAU Symposium 105 (Reidel, Dordrecht)

133. Madore, B.F., editor. 1985, *Cepheids: Theory and Observation*, IAU Colloqium 82 (Cambridge University Press, New York)

134. Renzini, A., & Fusi Pecci, F. 1988, *ARA&A*, **26**, 199

135. Chiosi, C., Bertelli, G., & Bressan, A. 1992, *ARA&A*, **30**, 235

136. Bedding, T.R., Booth, A.J., & Davis, J., editors, 1997, *Fundamental Stellar Properties: The Interations Between Observation and Theory*, IAU Symposium 189 (Kluwer Academic, Dordrecht)

Chapter 21

Circumstellar and Interstellar Material

John S. Mathis

21.1 OVERVIEW OF THE INTERSTELLAR MEDIUM

The bulk of the material presented here was submitted in October 1992, but some updating has been done.

21.1.1 Characteristic Pressures and Energy Densities

$P(\text{gas}) = nkT$; $P(\text{gas})/k \approx 3 \times 10^3$ cm^{-3} K for phases in approximate mechanical equilibrium; $P(\text{magnetic})/k = B^2/8\pi k = 7200(B/5\mu G)^2$ cm^{-3} K; $B \approx 5\mu G$ (diffuse ISM) [1, 2]; energy densities ≈ 0.5 eV cm^{-3} (starlight); 1.5 eV cm^{-3} (cosmic rays); 0.3 eV cm^{-3} (thermal pressure of gas in clouds); ≈ 1 eV cm^{-3} (gas kinetic motions); 0.6 eV cm^{-3} (magnetic field). All are highly variable spatially.

21.1.2 Structures Within the Interstellar Medium (ISM) [3–6]

1. **Molecular clouds:** Defined by hydrogen being in H_2, but observed by many molecules, especially CO; $n(H_2) > 100$ cm^{-3}; $T \approx 10$–20 K. Geometrical structure is very complex and fractal in nature. Contains half of mass of ISM in the Galaxy (the rest in H I, diffuse and in clouds). Scale height in Galaxy: varies with $D_G = $ distance from center of Galaxy: ≈ 40 pc ($D_G \approx 2$ kpc) to 60 pc ($D_G \geq D_{G\odot} = 8.5$ kpc).

2. **Diffuse interstellar clouds:** 10 cm$^{-3} \leq n(H) \leq 1000$ cm^{-3}; $T \approx 80$ K. Defined by H being atomic, and observed by 21 cm line in emission and absorption. Also observed by narrow (≤ 10 km s^{-1}) absorption lines of C II, Si II, Fe II, Mn II, Zn II, Mg II, Ni II, Na I, Ca II, etc., in the spectra of more distant stars. Also seen as dark clouds against background stars because of the dust absorption, or as reflection nebulae when a nearby star illuminates the cloud, or from far-infrared emission (especially in 60 and 100 μm emission as seen by Infrared Astronomical Satellite, IRAS).

3. **Warm intercloud medium:** 0.1 cm$^{-3} < n(H) < 10$ cm^{-3}; $T \approx 8000$ K. Seen by 21 cm emission (but not absorption) and the narrow lines in stellar spectra (see above). Hydrogen is atomic.

4. **Warm ionized medium:** 0.3 cm$^{-3} < n(H^+) < 10$ cm^{-3}; $T \approx 8000$ K. Seen by nebular emission lines, such as Hα, [N II], [S II], etc. The total power radiated is very large, $\approx 10^{41}$ erg s$^{-1} \approx$ total power of all supernovae. The electrons from this medium are responsible for most of pulsar dispersion measures. Extremely faint, covering the sky.

5. **Hot ionized medium:** $n(H^+) < 0.01$ cm^{-3}; $T > 10^{5.5}$ K. Seen by interstellar absorption lines of high stages of ionization (O VI, N V, perhaps Si IV) that could be formed at the interfaces between this gas and the warm ISM. Almost surely seen directly by the soft X-ray background, produced in a cavity ("the local bubble") surrounding the Sun. This medium was thought to be pervasive in the disk of the Galaxy, but is probably confined to local disk regions that have been heated by hot-star stellar winds and supernova shocks. This material has a large scale height (≈ 3 kpc) and is an important part of the galactic "halo."

6. **Supernova remnants:** $n(H^+) > 1$ cm^{-3}; $T \approx 10^4$–10^7 K. Seen by radio synchrotron emission, X-rays, optical spectrum from discrete filaments. Sometimes very peculiar composition, but older ones show ISM swept up by the expanding remnant. Spectra are H, He recombination lines, [O II], [O III], [S II], etc., with relative intensities produced by shocks rather than photoionization. Some show strong far-infrared radiation from dust. Widely varying conditions reflecting strong expansion and cooling throughout evolution.

7. **Large-scale nonthermal structures:** Arcs seen in synchrotron radio emission and X-rays arising from large-scale ordered magnetic fields interacting with cosmic rays. An example is "Loop I," the inner edge of which is the "North Polar Spur," with an angular radius on the sky of about 30° and a physical radius of about 125 pc.

8. **H II regions:** 10 cm$^{-3} < n(H^+) < 10^4$ cm^{-3}; $T \approx 8000$ K. Much denser than warm ionized ISM, so easily detectable. Strongest lines are [O II], [O III], H recombination lines; 52 and 88 μm [O III] fine-structure lines. Large ones powered by clusters of stars.

9. **Ultracompact H II regions:** 10^4 cm$^{-3} < n(H^+) < 3 \times 10^5$ cm^{-3}; $T \approx 8000$ K; excitation dominated by a single type O or early B star within the parent molecular cloud. Radio surface brightnesses much larger than classical H II regions mentioned above. Not seen optically because of dust extinction.

10. **Ring nebulae around Wolf–Rayet (WR) stars:** Ejecta from the rapid mass loss from WRs, but containing some swept-up ISM as well. Sizes $\approx$ few pc; compositions He/H ≈ 0.1–0.2; N/O ≈ 1.

11. **Planetary nebulae:** 100 cm$^{-3} < n(H^+) < 10^4$ cm^{-3}; $T \approx 10^4$ K. Excited by hot central star that has left the Asymptotic Giant Branch on its way to becoming a white dwarf, with $3 \times 10^4 \leq T_* \leq 2 \times 10^5$ K. Spectrum varies with T_*, ranging from very strong [O II], no He II to strong He II and Ne V. Nebula expands at 20–50 km s^{-1}. $\mathcal{M}$(neb) $\approx 0.2 \mathcal{M}_\odot$.

21.1.3 Interstellar Gas and Radiation [7–11]

Table 21.1 gives details of the interstellar gas near the Sun.

Table 21.1. *Interstellar gas densities in solar vicinity* [1].

Component	$\langle n(0)\rangle^a$ (cm^{-3})	Frac. of vol.	T (K)	v (rms) (km s^{-1})	$n(z)^b$	h_0 (pc)	$\sigma_{gas}{}^c$	$N(H)^d$
H$_2$	0.6	0.001	≈ 10	5	Gau	70	3.3	1.6
Cold H I	0.3	0.02	80	6	Gau	135	3.2	1.5
Warm H I (clouds)	0.07	0.05	8000	9	Gau	135	0.7	0.3
Warm H I (intercloud)	0.10	0.3:^{e}	8000	≥ 9	Expt	400	0.9	0.4
Warm H$^+$ (diffuse)	0.025	0.2:	8000	≥ 9	Expt	900	1.5	0.7
Warm H II regions^f	0.015	0.04	8000	9	Expt	70	0.04	0.02
Hot H$^+$	0.002	0.5:	10^6		Expt	3000	0.2	0.1
Total	1.1	1	$\dots$	$\dots$	$\dots$	$\dots$	10	5

Notes

$^a\langle n(0)\rangle$ = midplane density, averaged over the area. Local density at $z = 0$ is $\langle n(0)\rangle$/(Frac. of vol.).
$^b n(z)$ = type of distribution: Gau: $n(z) = n_0 \exp(-z^2/2h_0^2)$; Expt: $n(z) = n_0 \exp(-z/h_0)$.
$^c \sigma_{gas}$ = surface density of gas, including a factor of 1.29 for He, on plane of Galaxy, in $\mathcal{M}_\odot$ pc^{-2}.
$^d N(H)$ = Ave. column density of H (all forms), $z = 0$ to $z = \infty$, in 10^{20} cm^{-2}.
^{e}Colons in this and other tables denote uncertain quantities.
^{e}H$^+$ detected by radio free–free emission, with density ≥ 1 cm^{-3}.

Reference
1. Boulares, A., & Cox, D.P. 1990, *ApJ*, **365**, 544

Table 21.2 gives this hydrogen gas distribution in the Galaxy. All masses include He. z = height above the midplane of the Galaxy. Masses are in $\mathcal{M}_\odot$; surface densities in $\mathcal{M}_\odot$ pc^{-2}.

Table 21.2. *Distribution of hydrogen in the Galaxy.*

$D_G{}^a$ (kpc) =	< 0.4	0.4–3.5	3.5–7	7–8.5	8.5–14	Total
Mass of gas in H$_2$	2×10^8	2.9×10^8	1.4×10^9	3.7×10^8	2×10^8	2.5×10^9
% of total H$_2$	9%	11%	57%	15%	8%	100%
FWHW in z of H$_2$ (pc)	65	75	120	120	300	$\dots$
$n(H_2, z = 0)$ (cm^{-3})	100	2	1.6	0.8	0.03	0.5
Surface dens. in H$_2{}^b$	300	7	9	4	0.4	4
Surface dens. in H I^b	7	1–5	6	6	6	5
Mass of H I ($\mathcal{M}_\odot$)	5×10^6	5×10^7	6×10^8	4×10^8	2×10^9	3×10^9
Surface dens. in H^{+b}, radio H II regions	1.2	1.5	3.2	0.9	0.2	0.9
Surface dens., diffuse H^{+b}	$\dots$	2.7	2.7	1.7	$\dots$	$\dots$

Notes

$^a D_G$ = distance from center of Galaxy. D_G(Sun) is assumed to be 8.5 kpc.
^{b}Units of surface densities are $\mathcal{M}_\odot$/pc^2.

21.1.4   Diffuse Ionized Gas [12]

The average distribution of electrons with height in local Galaxy: $n_e = 0.025$ cm^{-3} exp$(-z/900$ pc$)$; $N(H^+)$ = column density of H^+ or electrons from midplane to infinity = 1×10^{20} cm^{-2}; filling factor of diffuse $H^+ \approx 0.2$.

The intensity of Hα in the galactic plane $\approx$ 10 Rayleighs, where 1 Rayleigh = $10^6/(4\pi)$ photons cm^{-2} s^{-1} sr^{-1}, or 2.4×10^{-7} erg cm^{-2} s^{-1} sr^{-1} at Hα. At the galactic poles, I(Hα) $\approx$ 1 Rayleigh = 2×10^{-7} erg cm^{-2} s^{-1} sr^{-1}.

The total recombination rate producing this Hα $\approx (2-7) \times 10^6$ recombinations cm^{-2} s^{-1} (best fit, 4×10^6), requiring a power of at least 5×10^{-5} erg cm^{-2} s^{-1} and probably twice that. Total power produced by supernovae is $\approx 10^{-4}$ erg cm^{-2} s^{-1}, which is inadequate because of the large radiative losses of SNRs. O, B stars produce about 3×10^7 ionizing photons cm^{-2} s^{-1} (Abbott, D.C. 1982, ApJ, **263**, 723; Mezger, P.G. 1978, A&A, **70**, 565), which is adequate if $\approx$ 15% of the photons can reach the high-latitude gas.

Recombination time of H^+ at $z = 1$ kpc $\approx 1.0 \times 10^6$ yr.

Spectrum of galactic diffuse ionized gas: [N II]6583/Hα $\approx$ 0.3–0.4; [S II]6717/Hα $\approx$ 0.3–0.5; [O III]5007/Hα $\leq$ 0.06; [O I]6300/Hα $\leq$ 0.02.

The spectrum of the local photon radiation near the Sun is given in Table 21.3. The X-ray spectrum is given in Table 21.4.

Table 21.3. *Local interstellar radiation field* [1–3].

λ (μm)	$4\pi J_\lambda{}^a$	λ (μm)	$4\pi J_\lambda{}^a$	λ (μm)	$4\pi J_\lambda{}^a$	λ (μm)	$4\pi J_\lambda{}^a$
0.091	1.07(−2)	0.216	9.17(−3)	1.20	9.26(−3)	25	6.0(−5)
0.10	1.47(−2)	0.23	8.25(−3)	1.8	4.06(−3)	60	4.6(−5)
0.11	2.04(−2)	0.25	7.27(−3)	2.2	2.41(−3)	100	7.3(−5)
0.13	2.05(−2)	0.346	1.30(−2)	2.4	1.89(−3)	200	2.6(−5)
0.143	1.82(−2)	0.435	1.50(−2)	3.4	6.49(−4)	300	5.4(−6)
0.18	1.24(−2)	0.55	1.57(−2)	4	3.79(−4)	400	1.72(−6)
0.20	1.04(−2)	0.70	1.53(−2)	5	1.76(−4)	600	3.22(−6)
0.21	9.61(−3)	0.90	1.32(−2)	12	1.7(−4)	1000	7.89(−6)

Note
^{a}Units of $4\pi J_\lambda$: erg cm^{-2} s^{-1} μm^{-1}.

References
1. Mathis, J.S., Mezger, P.G., & Panagia, N. 1983, A&A, **128**, 212
2. Cox, P., & Mezger, P.G. 1989, A&AR, **1**, 49
3. Wright, E.L. et al. 1991, ApJ, **381**, 200

Table 21.4. *Mean intensity of galactic X-rays* [1].

Energy (keV)	Name of band	λ (μm)	$I(E)^a$	$4\pi J_\lambda{}^a$
0.100	Be	1.24(−2)	118	1.9(−5)
0.163	B	7.61(−3)	73	3.1(−5)
0.232	C	5.34(−3)	52.5	4.5(−5)
0.625	M1	1.98(−3)	28.1	1.8(−4)
0.799	M2	1.55(−3)	20.7	2.1(−4)
1.04	I	1.19(−3)	14.1	2.5(−4)
1.56	J	7.95(−4)	9.78	3.8(−4)
3.3^b	...	3.76(−4)	6.82	1.2(−3)

Notes
[a] $I(E)$ in keV cm^{-2} s^{-1} sr^{-1} keV^{-1}; $4\pi J_\lambda$ in erg cm^{-2} s^{-1} μm^{-1}.
[b] For $E > 3.3$ keV, $I(E) = 11.0E$ (keV)$^{-0.4}$ keV cm^{-2} s^{-1} sr^{-1} keV^{-1}.

Reference
1. McCammon, D., & Sanders, W.T. 1990, *ARA&A*, **28**, 657

Sources of the observed gas and dust in the Galaxy are given in Table 21.5.

Table 21.5. *Sources of gas and dust in the Galaxy* [1, 2].

Stellar type	No. in Galaxy	dM(gas)$/dt$ ($\mathcal{M}_\odot$ yr^{-1})	dM(dust)$/dt$ ($\mathcal{M}_\odot$ yr^{-1})
M stars (Miras)	1.3×10^5	0.01–0.03	$(1–3) \times 10^{-4}$
OH/IR stars	10^4	0.1–0.5	$(1–5) \times 10^{-3}$
C stars	$(3–6) \times 10^4$	0.1–0.5	$(1–5) \times 10^{-3}$
Supernovae	0.02–0.03 yr^{-1}	0.1–0.3	0.001–0.006
M supergiants	5211	0.05–0.5	$(2–50) \times 10^{-4}$
Wolf–Rayets: WN, WC7	2744	0.05	0
WC8, WC9	484	0.01	10^{-4}
Planetary Nebulae	1.5×10^4	0.02–0.2	$(0.7–7) \times 10^{-5}$
Novae	30–50 yr^{-1}	$(0.5–1) \times 10^{-4}$	10^{-5}–10^{-4}
RV Tauri stars	600–1200	0.006–0.01	$(3–5) \times 10^{-6}$
O, B stars	$(2.5–5) \times 10^4$	0.03–0.3	0
Total in Galaxy	⋯	0.3–1.5	0.003–0.015
Star formation rate	⋯	$-(3–10)$	$-(0.03–0.1)$

References
1. Gehrz, R.D. 1989, in *Interstellar Dust*, edited by L.J. Allamandola and A.G.G.M. Tielens (Kluwer Academic, Dordrecht), IAU Symp. 135, p. 445
2. Jura, M., & Kleinmann, S.G. 1992, *ApJS*, **79**, 123

21.2 GALACTIC INTERSTELLAR EXTINCTION

21.2.1 Extinction

If $E(B - V) = A(B) - A(V)$, then $N(\text{H})/E(B - V) = 5.8 \times 10^{21}$ atoms cm^{-2} mag^{-1} [13]. Here $A(\lambda)$ is the extinction, in magnitudes, or $1.086\tau(\lambda)$, where τ is the optical depth in dust. The mean extinction law for interstellar dust can be described [14] as depending upon the optical parameter $R_V = A(V)/[A(B) - A(V)]$. The diffuse ISM has a typical value $R_V = 3.1$; in dense clouds, a typically $R_V = 4$–5. Table 21.6 gives mean values for 3.1 and 5. There is considerable uncertainty in the infrared extinction for $(\lambda > 5 \ \mu$m), perhaps a factor of 2 or more for $\lambda \geq 20 \ \mu$m.

$A(V)/N(\text{H}) \approx 5.3 \times 10^{-22}$ cm^2 mag, where $N(\text{H}) = $ column density of (H$+$H$^+ +$2H$_2$). Johnson filters are indicated in parentheses in Table 21.6.

Table 21.6. $A(\lambda)/A(V)$ *at various wavelengths for* $R_V = 3.1$ *and* 5.[a]

λ (μm)	$R_V = 3.1$	5	λ (μm)	$R_V = 3.1$	5	λ (μm)	$R_V = 3.1$	5
250[b]	4.2(−4)	4.9(−4)	5	0.027	0.031	0.24	2.54	1.68
100	1.2(−3)	1.3(−3)	3.4 (L)	0.051	0.059	0.218	3.18	1.97
60	2.0(−3)	2.3(−3)	2.2 (K)	0.108	0.125	0.20	2.84	1.74

Table 21.6. *(Continued.)*

λ (μm)	$R_V = 3.1$	5	λ (μm)	$R_V = 3.1$	5	λ (μm)	$R_V = 3.1$	5
35	3.7(−3)	4.2(−3)	1.65 (*H*)	0.176	0.204	0.18	2.52	1.52
25	0.014	0.016	1.25 (*J*)	0.282	0.327	0.15	2.66	1.49
20	0.021	0.025	0.9 (*I*)	0.479	0.556	0.13	3.12	1.60
18	0.023	0.027	0.7 (*R*)	0.749	0.794	0.12	3.58	1.74
15	0.015	0.017	0.55 (*V*)	1.00	1.00	0.091[c]	4.85	⋯
12	0.028	0.032	0.44 (*B*)	1.31	1.20	0.073	5.38	⋯
10	0.054	0.063	0.365 (*U*)	1.56	1.33	0.041	2.58	⋯
9.7	0.059	0.068	0.33	1.65	1.35	0.023	2.06	⋯
9.0	0.042	0.051	0.28	1.94	1.42	0.004	0.96	⋯
7.0	0.020	0.023	0.26	2.15	1.50	0.002	0.38	⋯

Notes

[a]Except as noted below, entries are from [1]. Values of $A(\lambda)/A(V)$ for other values of R_V can be determined from that paper.

[b]For $\lambda > 250$ μm, multiply entry for 250 μm by $(250 \ \mu m/\lambda)^2$.

[c]For $\lambda < 0.1$ μm, entries are from [2], increased by 1.15 for continuity at 0.12 μm.

References

1. Cardelli, J.A., Clayton, G.C., & Mathis, J.S. 1989, *ApJ*, **345**, 245
2. Martin, P.G., & Rouleau, F. 1990, in *Extreme Ultraviolet Astronomy*, edited by R.F. Malina and S. Bowyer (Pergamon, Oxford), p. 341

21.2.2 Opacity in X-Ray Region of Spectrum [15]

The cross section per H atom is given by $\sigma = (c_1 + c_2 E + c_3 E^2) E^{-3} \times 10^{-24}$ cm^2 (H atom)$^{-1}$. There are breaks in σ at various energies as detailed in Table 21.7.

Table 21.7. *X-ray opacity of interstellar gas and dust.*

E range (keV)	Edge[a]	c_1	c_2	c_3
0.030–0.100	⋯	17.3	608.1	−2150
0.100–0.284	C	34.6	267.9	−476.1
0.284–0.400	N	78.1	18.8	4.3
0.400–0.532	O	71.4	66.8	−51.4
0.532–0.707	Fe–L	95.5	145.8	−61.1
0.707–0.867	Ne	308.9	−380.6	294.0
0.867–1.303	Mg	120.6	169.3	−47.7
1.303–1.840	Si	141.3	146.8	−31.5
1.840–2.471	S	202.7	104.7	−17.0
2.471–3.210	Ar	342.7	18.7	0.0
3.210–4.038	Ca	352.2	18.7	0.0
4.038–7.111	Fe	433.9	2.4	0.75
7.111–8.331	Ni	629.0	30.9	0.0
8.331–10.00	⋯	701.2	25.2	0.0

Note

[a]The element whose absorption produces a discontinuity at the upper energy of the range.

21.2.3 Diffuse Interstellar Bands (DIBs)

Over 100 interstellar features [16, 17], some broad, are still unidentified. Some profiles are asymmetric to the red; some are symmetric. Observed properties of strongest diffuse interstellar bands. In Table 21.8, W is the equivalent width in nm; Δ = central depth; the FWHM is in nm.

Table 21.8. *Selected diffuse interstellar bands.*

λ (nm)	W^a	Δ^a	FWHM	λ (nm)	W^a	Δ^a	FWHM
443.0	0.34	0.16	2.0	617.7	0.18	0.07	3.0
488.2	0.13	0.06	1.7	628.3	0.20	0.38	0.38
544.9	0.56	0.045	1.4	661.38	0.036	0.24	0.013
577.83	0.095	0.06	1.7	722.40	0.037	0.21	0.016
578.0	0.088	0.37	0.26	792.7	0.036	0.026	1.4
579.7	0.039	0.22	0.13	862.08	0.042	0.084	0.43

Note

$^a W$, depth are for HD 183143, a well-observed star [1].

Reference

1. Herbig, G.H., & Leka, K.D. 1991, *ApJ*, **382**, 193

21.3 ABUNDANCES IN INTERSTELLAR GAS

Table 21.9 gives the elemental abundances for the gas in many objects where observations can be made.

Table 21.9. *Chemical composition of interstellar gas.*

Object	He/H	10^6 O/H	C/O	100 N/O	100 Ne/O	100 S/O	1000 Fe/O	Reference
Orion Neb ^{a}								
($t^2 = 0$)	0.096	384	0.73	13	17	3.0	10^b	[1–4]
($t^2 = 0.035$)	0.100	550	0.62	13	16	2.8	7.8^b	[5]
M 17 ($t^2 = 0$)	0.105	324	1.69	16	19	3.0	1.3^b	[5]
30 Doradus (LMC)	0.89	200	0.15	2.8	22	2.3	$\cdots$	[6–8]
Other LMC H IIs	0.091	230	0.35	5.0	20	3.2	$\cdots$	[6, 7, 9]
SMC H IIs	0.083	135	0.14	2.6	36	5.0	$\cdots$	[6, 7, 9]
Cas A SNR, FMK^c	c	c	< .003	< .007	< 2	6.5	< 1	[10]
Crab Neb	0.7	2000	0.75	7	32	2	$\cdots$	[11]
LMC SNRs	$\cdots$	177	0.26	16	18	2.3	93	[9]
SMC SNRs	$\cdots$	85	$\cdots$	9	9	3	110	[9]
Local B stars	0.100	390	0.42	16	19	3.4	110	[12, 13]
Sun	0.098	850	0.43	12	14	1.91	55	[14]

Notes

$^a t^2$ = temperature fluctuation parameter = $[\langle T^2 \rangle / \langle T \rangle^2 - 1]$ [15].

^{b}Gas-phase Fe only. Most of the Fe is in solid grains.

^{c}FMK = Fast-moving knots. H, He not detected; H/O $\leq$ 0.3, He/O $\leq$ 1.7.

References

1. Baldwin, J.A. et al. 1991, *ApJ*, **374**, 580
2. Rubin, R.H., Simpson, J.P., Haas, M.R., & Erickson, E.F. 1991, *ApJ*, **374**, 564
3. Osterbrock, D.E., Tran, H.D., & Veilleux, S. 1992, *ApJ*, **389**, 305
4. Peimbert, M. 1987, in *Star Forming Regions*, edited by M. Peimbert and J. Jugaku (Reidel, Dordrecht), p. 111

5. Peimbert, M., Torres-Peimbert, S., & Ruiz, M.T. 1992, *Rev. Mex. Astron. Astrof.*, **24**, 155
6. Dufour, R.J., Shields, G.A., & Talbot, R.J., Jr. 1982, *ApJ*, **252**, 461
7. Dufour, R.J. 1984, in *Structure and Evolution of the Magellanic Clouds*, edited by S. van den Bergh and K.S. de Boer (Kluwer Academic, Dordrecht), p. 353
8. Mathis, J.S., Chu, Y.-H., & Peterson, D. 1985, *ApJ*, **292**, 155
9. Russell, S.C., & Dopita, M.A. 1990, *ApJS*, **74**, 93
10. Chevalier, R.A., & Kirshner, R.P. 1979, *ApJ*, **233**, 154
11. Péquinot, D., & Dennefeld, M. 1983, *A&A*, **120**, 249
12. Gies, D.R., & Lambert, D.L. 1992, *ApJ*, **387**, 673
13. Kilian, J. 1992, *A&A*, **262**, 171
14. Anders, E., & Grevesse, N. 1989, *Geochim. Cosmochim. Acta*, **53**, 197
15. Peimbert, M. 1967, *ApJ*, **150**, 825

21.3.1 Isotopic Abundances [18]

The hydrogen, carbon, nitrogen, and oxygen isotopic abundances are given for some places in the Galaxy in Table 21.10.

Table 21.10. *Isotopic abundances in the Galaxy.*

Isotope ratio	Galactic center	Carbon star: · IRC10+216	5 Dusty carbon stars	Solar system	Local ISM	Gradient[a] (dex kpc^{-1})
$10^5\ ^2$H/^{1}H[b]	...	...	...	(1–3)	1.65	...
^{12}C/^{13}C	20	40 ± 8	≥ 30	89	70	6.2 ± 1.6
^{14}N/^{15}N	900	> 515	...	270	380	28 ± 14
^{16}O/^{18}O	250	> 2700	320–1300	490	490	66 ± 24
^{18}O/^{17}O	3.2	< 1	0.6–0.9	5.5	3.2	-0.03 ± 0.05

Notes

[a] $d(\log_{10} X)/dD_G$ in range 4–14 kpc.

[b] ^{2}H/^{1}H from Encrenaz, T., & Combes, M. 1983, *Icarus*, **52**, 54; for local ISM from Linsky, J.L. et al. 1992, *ApJ*, **402**, 694.

21.4 LINE EMISSIONS FROM THE ISM

Relative line intensities, $I(\lambda)/I(\mathrm{H}\beta)$, for various objects corrected for reddening are given in Table 21.11. These objects are: Orion Nebula = NGC 1976, position near θ^1Ori C; 30 Doradus = NGC 2070 = N157, giant H II region in LMC; Crab Nebula = NGC 1952, remnant of SN1054, a He-poor and He-rich filament listed; Cygnus Loop = NGC 6960, NGC 6979, NGC 6992/5, a medium-aged SNR, with high-ionization and low-ionization filaments listed; NGC 7662 = high-excitation PN.

Table 21.11. *Line emission fluxes relative to Hβ.*

λ (nm)	$f(\lambda)^a$	Orion Neb [1–4]	30 Dor [1, 5]	Crab Nebula He-poor [6, 7]	Crab Nebula He-rich [7, 8]	Cygnus Loop High ion [9, 10]	Cygnus Loop Low ion [9]	NGC 7662 [11, 12]
155.5 C IV	−1.17	...	0.012:	6.00	...	5.1	0.76	13.9
164.0 He II	−1.11	...	...	4.80	...	3.1	0.50	6.28
166.3 [O III]	1.10	...	0.15	...	...	4.6	1.5	0.53

Table 21.11. *(Continued.)*

λ (nm)	$f(\lambda)^a$	Orion Neb [1–4]	30 Dor [1, 5]	Crab Nebula He-poor [6, 7]	Crab Nebula He-rich [7, 8]	Cygnus Loop High ion [9, 10]	Cygnus Loop Low ion [9]	NGC 7662 [11, 12]
175.9 [N III]	−1.09	⋯	⋯	< 1.7	⋯	1.4	0.8	0.33
190.8 [C III]	−1.20	0.19	0.15	5.6	⋯	13.4	6.5	4.23
232.6 [C II]	−1.32	0.09	0.07:	< 6.5	⋯	2.91	3.86	⋯
247.0 [O II]	−0.99	0.056	⋯	⋯	⋯	0.69	0.68	⋯
279.8 Mg II	−0.64	< 0.012	< 0.02	< 1.10	⋯	⋯	⋯	⋯
372.7 [O II]	−0.31	0.94	1.38	10.0	12.0	16	12.8	0.10
386.9 [Ne III]	−0.28	0.20	0.44	1.72	0.187	2.0	0.72	0.75
434.0 Hγ	−0.15	0.49	0.46	0.48	0.48	0.48	0.46	0.48
436.3 [O III]	−0.14	0.015	0.041	0.15:	0.26	0.92	0.22	0.161
447.1 He I	−0.11	0.046	0.044	0.23	0.24	⋯	⋯	0.028
468.6 He II	−0.04	⋯	⋯	0.68	0.88	⋯	⋯	0.79
486.1 Hβ	0.00	1.00	1.00	1.00	1.00	1.00	1.00	1.00
495.9 [O III]	0.025		1.70	3.50	4.63	4.21	1.25	3.66
500.7 [O III]	0.037	3.43	4.87	11.0	15.1	13.9	3.38	10.9
587.6 He I	0.20	0.137	0.124	0.74	0.77	⋯	0.07	0.076
630.0 [O I]	0.26	0.007	0.007	0.98	0.45	0.35	0.31	⋯
656.3 Hα	0.29	2.71	2.92	2.98	3.22	3.00	3.00	2.87
658.4 [N II]	0.30	0.44	0.058	3.0	3.79	3.11	2.98	0.051
667.8 He I	0.31	0.036	0.035	0.19	0.12	⋯	⋯	0.022
671.7 [S II]	0.31	0.019	0.050	1.61	1.42	1.26	1.15	0.0039
673.1 [S II]	0.31	0.034	0.042	1.90	1.72	0.85	0.78	0.0060
713.6 [Ar III]	0.37	0.154	0.12	0.26	0.44	0.44	0.17	0.0945
732.5 [O II]	0.39	0.119	0.023	0.77	0.22	0.62	0.44	0.014
906.9 [S III]	0.48	⋯	⋯	1.09	0.94	⋯	⋯	⋯
953.2 [S III]	0.61	1.45	⋯	1.90	1.73	⋯	⋯	⋯
$C(H\beta)^a$	⋯	0.84	0.65	0.74	0.7	0.12	0.12	0.26

Note

^{a}Observed flux ratio $F(\lambda)/F(H\beta) = \text{dex}[C(H\beta)f(\lambda)][I(\lambda)/I(H\beta)]$, where $[I(\lambda)/I(H\beta)]$, the de-reddened flux ratio, is tabulated.

References

1. Dufour, R.J., Shields, G.A., & Talbot, R.J., Jr. 1982, *ApJ*, **252**, 461
2. Baldwin, J.A. et al. 1991, *ApJ*, **374**, 580
3. Torres-Peimbert, S., Peimbert, M., & Daltabuit, E. 1980, *ApJ*, **238**, 133
4. Bohlin, R.C., Harrington, J.P., & Stecher, T.P. 1978, *ApJ*, **219**, 575
5. Mathis, J.S., Chu, Y-H, & Peterson, D.E. 1985, *ApJ*, **292**, 155
6. Davidson, A.F. et al. 1982, *ApJ*, **253**, 696
7. Henry, R.B.C., MacAlpine, G.M., & Kirschner, R.P. 1984, *ApJ*, **278**, 619
8. Fesen, R.A., & Kirshner, R.P. 1982, *ApJ*, **258**, 1
9. Raymond, J.C. et al. 1988, *ApJ*, **324**, 869
10. Blair, W.P. et al. 1991, *ApJ*, **379**, L33
11. Barker, T. 1986, *ApJ*, **308**, 314
12. Aller, L.J., & Czyzak, S.J. 1983, *ApJS*, **51**, 211

21.4.1 Far-Infrared Line Intensities [19]

Luminosities given in Table 21.12 are estimates for the power emitted inside the solar circle only.

Table 21.12. *Far-infrared line luminosities.*

λ (μm)	Source	$10^8 4\pi J^a$	$\log L_G$	λ (μm)	Source	$10^8 4\pi J^a$	$\log L_G$
1302	CO ($J = 2$–1)	11 ± 2	4.9	370.4	[C I]	44 ± 6	5.5
867.2	CO ($J = 3$–2)	15 ± 4	5.1	205.3	[N II]	630 ± 34	6.7
650.4	CO ($J = 4$–3)	16 ± 4	4.1	157.7	[C II]	6550 ± 270	7.7
609.1	[C I]	24 ± 5	5.3	121.9	[N II]	990 ± 140	6.9
519.8	CO ($J = 5$–4)	12 ± 3	5.0	Cont.	Dust	2.2×10^6	10.26

Note

^{a}Units of J (= mean intensity) are erg cm^{-2} s^{-1} sr^{-1}; L_G is in $L_\odot$.

21.4.2 "Unidentified Infrared Bands"

The total power radiated from bright, well-observed objects [20] is > 0.05 of far-infrared. The total power from Galaxy is unknown. These unidentified infrared bands are given in Table 21.13.

Table 21.13. *The unidentified infrared bands.*

λ (μm)	ν (cm^{-1})	FWHM (cm^{-1})	PNe (C-rich)	Refl. Neb.	H II regions	Assignment and comments [2]
3.29	3040	30	0.063	0.057	0.13	Aromatic C–H stretch
5.2	1960–1890	30	0.035	0.054	0.075	C–H out-of-plane and in-plane bend (?)
5.6	1785–1755	40	0.073	0.090	0.058	Overtone of 11.3 μm band; C = O stretch (?)
6.2	1615	30	0.41	0.68	0.58	Aromatic C–C stretch
6.9	1450–1470	30	$\cdots$	0.16	$\cdots$	Aromatic C–C stretch; aliphatic C–H deformation
7.7	1250–1350	70–200	1	1	1	Blending of several aromatic C–C stretches
11.2	890	30	0.29	0.22	0.30	Aromatic out-of-plane bend for nonadjacent, peripheral H atoms

The header spanning "Average strength relative to 7.7 μm [1]" covers the PNe, Refl. Neb., and H II regions columns.

References

1. Cohen, M. et al. 1989, *ApJ*, **341**, 246
2. Allamandola, L.J., Tielens, A.G.G.M., & Barker, J.R. 1989, *ApJS*, **712**, 733

21.5 H$_2$ AND MOLECULAR CLOUDS

The conversion from CO strength to mass: $X = N(\text{H}_2)/\int T(^{12}\text{CO})\,dv \approx 1.8 \times 10^{20}$ cm^{-2} K^{-1} (km/s)$^{-1}$ in solar neighborhood and inner Galaxy [21, 22]; $1/(4 \pm 2)$ as much in outer Galaxy [23]; twice as much for $D_G < 1$ kpc. The value and constancy of X is controversial; some authors use 3.6×10^{20} throughout Galaxy.

Molecular clouds are hierarchical in nature and have very poorly defined edges, structures, and masses. For a large range of sizes they seem to be fractal in nature. The following are statistical relationships that are commonly used, but there is a considerable spread about the values derived from these relations:

Let R = half the average diameter of a cloud; n = number density of H atoms = $2n(\text{H}_2)$; σ = velocity dispersion for a Gaussian distribution (FWHM$^2 = 8 \ln 2\sigma^2$). Then [24]: n(H atoms cm^{-3}) $\approx$

$3.9 \times 10^3 R$ (pc)$^{-1}$; $N(H) = 2nR \approx 3 \times 10^{22}$ cm^{-2}; σ(CO; km s^{-1}) $\approx 0.15(\mathcal{M}/\mathcal{M}_\odot)^{1/4}$; $\mathcal{M}(\mathcal{M}_\odot) \approx 540R$ (pc)^2; σ(CO; km s^{-1}) $\approx 0.71R$ (pc)$^{1/2}$; $\mathcal{M}(\mathcal{M}_\odot) \approx 40L(CO)^{0.8}$.

For the forms given above, clouds approximately obey the virial theorem, but they may be supported magnetically. Other workers have found somewhat different relations; for instance, Larson [25] suggested σ (km s^{-1}) $= 1.4R$ (pc)$^{0.38}$. Mass spectrum of massive molecular clouds [26]: $dN/d\mathcal{M} \propto \mathcal{M}^{-1.5}(7 \times 10^4 \leq \mathcal{M}/\mathcal{M}_\odot \leq 2 \times 10^6)$. This relation implies that most of the mass is in the massive complexes, but it is controversial [27].

Typical parameters for the observed molecular clouds are given in Table 21.14.

Table 21.14. *Typical parameters of molecular clouds* [1].

Type	Size (pc)	Density (cm^{-3})	FWHM[a] (km s^{-1})	T (K)
Giant MC complex	50	100	10	10
Giant MC	4	1000	4	25
Core of MC	1	4000	2	40
Clump within MC	0.5	$> 10^5$	4	100
Dark cloud complex	10	400	3	10
Dark cloud	0.5	10^4	1	12
Dark cloud core	0.2	4×10^4	0.3	10
Circumstellar envelope	0.2	10^2–10^7	20–40	10–100

Note

[a] Full width half maximum in ^{12}CO.

Reference

1. Goldsmith, P.F. 1987, in *Interstellar Processes*, edited by D.J. Hollenbach and H.A. Thronson, Jr. (Reidel, Dordrecht), p. 51

21.5.1 Global Properties, Solar Neighborhood Giant Molecular Cloud Complexes [28, 29]

Mass, $(1–2) \times 10^5 \mathcal{M}_\odot$; volume averaged n(H$_2$) within cloud ≈ 50 cm^{-3}; mean N(H$_2$) $\approx (3–6) \times 10^{21}$ cm^{-2}; mean optical absorption A(V) $\approx 1–3$ mag.; local surface density of complexes ≈ 4 kpc^{-2}; mean separation ≈ 500 pc.

21.5.2 Individual Molecular Clouds (Some of Which Coincide with H II Regions)

Some individual molecular cloud positions (equinox 2000) are given in Table 21.15.

Table 21.15. *Individual molecular clouds.*

Name	α (2000)	δ (2000)	Name	α (2000)	δ (2000)
Barnard 5	3 48	32 54	NGC 2024	05 42	−1 56
L1551	4 31	18 03	ρ Oph	16 26	−24 24
TMC-1	4 42	25 41	M17 SW	18 17	−16 14
OMC-1	5 33	−5 26	Cep A	22 56	62 02
OMC-2	5 36	−5 10	NGC 7538	23 14	61 28

21.5.3 Molecules Found in Clouds and Circumstellar Envelopes [30, 31]

Line frequencies are given in [32].

Inorganic stable molecules: H_2, CO, CS, NO, NS, SiO, SiS, HCN, PN, NaCl*, AlCl*, KCl*, AlF(?)*, H_2O, H_2S, SO_2, OCS, NH_3, SiH_4*.

Organic stable molecules: CH_3OH, C_2H_5OH, H_2CO, CH_3CHO, H_2CCO, $(CH_3)_2CO$ (?), HCN, HCOOH, HNCO, C_2H_2*, C_2H_4*, CH_4, NH_2CHO, NH_2CH, NH_2CN, NH_2CH_3, $HCOOCH_3$, $(CH_3)_2O$, H_2CS, HNCS, CH_3SH, CH_3CN, C_2H_5CN, HC_3N, CH_3C_2H, CH_3C_4H, CH_2NH, CH_2CHCN.

Free radicals: CH, CN, OH, SO, HCO, C_2*, C_2H, C_3*, C_5*, C_3H, C_3N, C_3O, C_4H, C_5H, C_6H, C_2S, CH_2CN, SiC, HNO, HCC_2HO.

Ions: CH^+, HCO^+, N_2H^+, $HOCO^+$, $HCS+$, H_3O^+(?), $HCNH^+$, H_2D^+, SO^+.

Rings: SiC_2*, C_3H_2, C_3H.

Carbon chains and isomers: C_3S, HC_5N, HC^7N, HC^9N, $HC^{11}N$, CH_3C_3N, CH_3C_4N, CH_3C_5N(?), HNC, CH_3NC.

[*Found only in circumstellar material, such as in IRC+10 216 (a carbon star).]

21.5.4 Interstellar and Circumstellar Masers

Molecular masers are observed in both interstellar and circumstellar locations as listed in Table 21.16.

Table 21.16. *Interstellar and circumstellar masers [1, 2].*

Molecule	Freq. (GHz)	Transition	Where found
OH	1.612231	$^3\Pi_{3/2}$, $J = 3/2$, $F = 1\text{--}2$	M stars
	1.665402	$^3\Pi_{3/2}$, $J = 3/2$, $F = 1\text{--}1$	MCs, M stars
	1.667359	$^3\Pi_{3/2}$, $J = 3/2$, $F = 2\text{--}2$	MCs, M stars
	1.720530	$^3\Pi_{3/2}$, $J = 3/2$, $F = 2\text{--}1$	MCs
	4.66042	$^3\Pi_{1/2}$, $J = 1/2$, $F = 0\text{--}1$	MCs
	4.765562	$^3\Pi_{1/2}$, $J = 1/2$, $F = 1\text{--}0$	MCs
H_2O	22.235080	$6_{16}\text{--}5_{23}$	MCs, M stars
H_2CO	4.829	$1_{10}\text{--}1_{11}$	MCs
CH_3OH	12.178	$2_0\text{--}3_{-1}$ E	MCs
SiS	18.155	$1\text{--}0$	C stars
NH_3	23.870	$J, K = 3, 3$	MCs
SiO	43.122	$v = 1$, $J = 1\text{--}0$	MCs, M stars, S stars
	86.243	$v = 1$, $J = 2\text{--}1$	M stars, S stars
HCN	89.087	$v = 2$, $J = 1\text{--}0$	C stars

References
1. Reid, M.J., & Moran, J.M. 1981, *ARA&A*, **19**, 231
2. Genzel, R. et al. 1981, *ApJ*, **247**, 1039

The size of the OH masering region in the W51MAIN molecular cloud [33] $\approx (1\text{--}3) \times 10^{13}$ cm. They are $\approx 10^{14}$ cm in compact H II regions, and 10^{15} cm in circumstellar masers. A catalogue of Herbig–Haro Objects is given in [34].

21.6 NEUTRAL GAS; CLOUDS; DEPLETIONS

Standard cloud parameters for neutral clouds are given in [35]. It is assumed that there are two types of clouds, "standard" and "dense," embedded within a low-density warm intercloud component.

Varying mean densities $\langle n(\mathrm{H}) \rangle$ arise from varying proportions of the three constituents. The suggested properties of the components are: (a) Intercloud component, $n(\mathrm{H}) \approx 0.14$ cm^{-3}, $T \approx 8000$ K. (b) "Diffuse" cold cloud, $N(\mathrm{H}) \approx 3.5 \times 10^{20}$ cm^{-2}; number clouds/kpc ≈ 6.2; reddening in each cloud, $E(B - V) \approx 0.060$ mag.; mean contribution of clouds ≈ 0.70 cm^{-3}. (c) "Large" cold clouds, $N(\mathrm{H}) \approx 1.7 \times 10^{21}$ cm^{-2}, $E(B - V) \approx 0.30$, number/kpc ≈ 0.8; mean contribution ≈ 0.45 cm^{-3}.

21.6.1 Interstellar Depletions from Gas Phase

The depletion, δ, of an element A is defined by

$$\log \delta(\mathrm{A}) \equiv \log(N(\mathrm{A})/N(\mathrm{H}))_{\mathrm{gas}} - \log(N(\mathrm{A})/N(\mathrm{H}))_{\odot},$$

where $N(\mathrm{A})$ and $N(\mathrm{H})$ are column densities of the element and H in all forms (H_2, H^0, and H^+), respectively. In practice, usually various stages of ionization are observed, and theory and velocity information are used to decide whether the stage of ionization is mainly found in the H^0 or H^+ phases. There is no depletion information from deep inside molecular clouds because suitable background stars cannot be observed through the heavy extinction. Depletions are measured by ionic absorption lines. See [36] for an extensive list.

As shown in Table 21.17, depletions vary with $\langle n(\mathrm{H}) \rangle$, the mean gas density along the line of sight to the star: $n(\mathrm{H}) = N(\mathrm{H})/d$, where $N(\mathrm{H})$ is the observed column density and d is the distance to the star whose spectrum shows the interstellar absorption lines of the ion. The $\langle n(\mathrm{H}) \rangle$ is, presumably, only a crude measure of local density. A formula for δ is

$$\log \delta(\mathrm{A}) = d_0 + m[\log \langle n(\mathrm{H}) \rangle + 0.5].$$

Table 21.17. *Parameters of depletion of element A in H I regions* [1].

Elem.	$\log(\mathrm{A}/\mathrm{H})_{\odot}$	d_0	m
C	−3.44	−0.34 ± 0.15	a
N	−4.04	−0.12 ± 0.05	−0.01 ± 0.09
O	−8.08	−0.36 ± 0.07	−0.01 ± 0.11
Fe	−4.50	−1.65 ± 0.02	−0.38 ± 0.05
Ti	−7.26	−1.86 ± 0.04	−0.84 ± 0.06
Cr	−6.30	−1.58 ± 0.09	−0.50 ± 0.11
Zn	−7.40	−0.38 ± 0.07	−0.11 ± 0.09
P	−6.50	−0.43 ± 0.03	−0.32 ± 0.06
Si	−4.45	−1.09 ± 0.11	−0.49 ± 0.15
Mg	−4.40	−0.44 ± 0.02	−0.28 ± 0.05
Mn	−4.50	−1.65 ± 0.02	−1.06 ± 0.02
S	−4.80	≈ 0	≈ 0
Ca	−5.66	−2.4 ± 0.3	−1.2 ± 0.3
Al	−4.57	−1.85 ± 0.2	−0.85

Note

a At time of writing (Sept. 1992) C II, the dominant ion for carbon in the neutral ISM, has only been measured in three stars [2], all of which sample relatively dense gas.

References

1. Jenkins, E.B. 1987, in *Interstellar Processes*, edited by D.J. Hollenbach and H.A. Thronson, Jr. (Reidel, Dordrecht), p. 533
2. Cardelli, J.A., Mathis, J.S., Ebbets, D.C., & Savage, B.D. 1992, *ApJ*, **402**, L17

21.6.2   Reflection Nebulae

Lists, with exciting stars and references, are given in [37–40].

21.6.3   Halo Equivalent Widths

Table 21.18 gives major ions, oscillator strengths (f), and the equivalent widths toward two stars at very large distances above and below the plane: HD18100 ($z = -3.4$ kpc) and HD100340 ($z = 5.2$ kpc), averaged by [41].

Table 21.18. *Equivalent widths of interstellar absorption lines.*

Ion	λ (nm)	f	W (nm)	Ion	λ (nm)	f	W (nm)
H I	121.567	0.41	1.8	Si III	120.651	1.65	0.055
C I	165.693	0.135	0.007	Si IV	140.277	0.26	0.012
	156.031	0.081	0.006		139.376	0.53	0.023
	127.721		0.008	P II	115.28	0.236	
C II	133.4532	0.128	0.070		130.19	0.017	
	232.540	6.0(−8)	...		153.25	7.61(−3)	
C II^a	133.571	0.115	0.012	S II	125.95	0.016	0.035
C IV	154.819	0.194	0.034		125.38	0.011	0.025
	155.076	0.097	0.018		125.06	5.45(−3)	0.014
N I	119.99	0.266	0.030	Ca II	396.847	0.3145	
	115.98	8.51(−6)	...		393.366	0.635	
O I	130.217	0.049	0.060	Cr II	205.62	0.140	0.007
	135.560	1.25(−6)	...		206.22	0.105	
O I^a	130.486	0.049	...	Mn II	257.688	0.351	0.016
Mg I	285.213	1.97	0.030		259.450	0.271	0.014
	202.648	0.11	0.006		260.646	0.193	0.012
Mg II	123.993	2.67(−4)	...	Fe II	160.8451	0.062	0.030
	124.039	1.34(−4)	...		234.421	0.110	0.080
	279.635	0.612	0.135		237.446	0.028	0.040
	280.351	0.305	0.125		238.277	0.301	0.080
Al II	167.079	1.89	0.050		258.665	0.0653	0.075
Al III	186.279	0.27	0.013		260.017	0.224	0.080
	185.472	0.54	0.025		226.078	0.0037	...
Si II	180.801	0.0055	0.013		224.988	0.0025	...
	152.671	0.230	0.045	Ni II	131.722	0.146	
	130.437	1.473	0.047		137.013	0.131	
	126.042	1.007	0.060		174.155	0.103	0.008
	119.329	0.499	0.059	Zn II	202.614	0.515	0.012
	119.042	0.250			206.266	0.253	0.006
Si II^a	119.450	0.623	...				
	126.474	0.903	...				
	153.343	0.229	...				

Note

^{a}Arising from an excited state within a few kelvins excitation from ground; population sensitive to density.

21.7   H II REGIONS, IONIZED GAS, AND THE GALACTIC HALO

21.7.1   Some Galactic H II Regions

W designations are found in [42], and the DR designations are from [43]. Catalogs of H II regions are from [44, 45]. Table 21.19 entries are from [46, 47].

Table 21.19. *Galactic H II regions.*

Name(s)	G desig. (l, b)	Size (') $\alpha \quad \delta$	$2R$ (pc)	$\log(\text{EM})^a$	S_ν 6 cm (Jy)	$\log(n_{\text{ion}})$ (s^{-1})
W 3	133.7+1.2	1.7 1.5	1.3	6.56	80	49.89
M 42, Ori A						
NGC 1976	209.0−19.4	3.8 4.3	0.6	6.86	430	48.84
M 43, NGC 1982	30.8−0.04	2.5 2.5	0.7	5.42	16.7	47.63
NGC 2024, Ori B	206.5−16.4	3.6 3.0	0.6	6.08	65.6	48.10
NGC 2237-8;						
2244, 2246, Rosette	206.4−1.9	80 80	34	3.88	350	49.88
Sgr B2	0.7−0.0	1.8 1.8	9	6.28	21.4	50.40
M 20, NGC 6514						
Trifid	7.0−0.2	5.4 5.8	6.2	4.76	13	50.40
M 8, NGC 6523						
Lagoon	6.1−1.2	8.5 7.6	2.2	5.15	85	48.36
M 17, NGC 6618						
Omega	15.0−0.7	4.1 6.7	5.2	6.65	900	50.31
W49	43.2−0.0	1.5 2.0	7.0	6.23	50	50.96
W 51	49.1−0.4	complex	...	...	400	50.37
DR 21	81.7+0.5	0.3 0.4	0.2	7.70	19	48.62
NGC 7538	111.5+0.8	2.3 1.9	3.0	5.78	26	49.73

Note

[a] EM = Emission measure = $\int n_e^2\, ds$ (cm^{-6} pc); n_{ion} = number of ionizations s^{-1} within nebula.

21.7.2 Galactic Halo and High Stages of Ionization [48–51]

High ionization stages for interstellar elements may be in coronal (collisional) ionization equilibrium. Table 21.20 gives typical column densities, with H I and electrons for comparison.

Table 21.20. *High stages of ionization in the galactic halo.*[a]

Ion	IP$(i-1)$ (eV)	IP(i) (eV)	T_{max}	f_{max}	$n\ (z=0)$ (cm^{-3})	$\log N$ (cm^{-2})	h (kpc)
Si IV	33.5	45.1	0.6×10^5	0.26	4.1×10^{-9}	13.54	2.8
C IV	47.9	64.5	1.0×10^5	0.35	1.1×10^{-8}	14.20	4.7
N V	77.5	97.9	1.8×10^5	0.12	5.1×10^{-9}	13.40	1.6
O VI	113.9	138.1	3.0×10^5	0.22	2×10^{-8}	> 13.27	> 0.3
H I	...	13.6	b	b	0.10	20.39	1.5
e^-	...	...	b	b	0.04	20.06	0.4

Notes

[a] f_{osc} = oscillator strength of transition; IP$(i-1)$ = ionization potential of lower species of ionization, from which this species is produced in coronal (i.e., purely collisional) ionization; IP(i) = ionization potential of this species, above which the species can be destroyed by collisions; T_{max} = temperature at which the maximum amount of species is produced; f_{max} = fraction of element in this stage of ionization when $T = T_{\text{max}}$; $n(z = 0)$ = density in plane; N = total column density, from mid-plane of Galaxy; h = exponential scale height.

[b] Not produced by collisional ionization.

21.7.3 Ultracompact H II Regions

Ultracompact H II regions are ionized nebulae surrounding newly formed stars still embedded within their natal molecular clouds. Optically invisible because of the dust in the cloud, they are observed in the radio and far-infrared spectral regions. About 1700–3000 are seen in the IRAS far-infrared survey of the Galaxy, representing $\approx$ 20% of the total O-star population. Their shapes (from radio maps) are: cometary (an arclike structure), 20%; core-halo (bright central source, the core, surrounded by a more diffuse halo), 16%; shell (brighter at rim than in center, circular shape), 4%; irregular and clumpy, 17%, and spherical or unresolved by VLA, 43%. They probably usually represent bow shocks of hot stars moving $\approx$ 10 km s^{-1} through the molecular gas, as viewed from various directions [52, 53]. Some of these ultracompact H II regions are listed in Table 21.21.

Table 21.21. *Bright ultracompact H II regions.*

Name (l, b)	D (kpc)	τ_ν (15 GHz)	10^{-7} EM^a (pc cm^{-6})	$10^{-4} n_e$ (cm^{-3})	$\log n_{ion}$ (s^{-1})	Δs^b (pc)
G5.89−0.39	2.6	2.67	244.5	23.7	48.65	0.044
G10.62−0.38	6.5	1.27	116.0	15.6	48.96	0.048
G29.96−0.02	9.0	0.23	21.1	4.4	49.34	0.109
G34.26+0.15A	3.7	3.19	291.7	29	48.74	0.035
W51d	7.0	3.71	340	20	49.42	0.084

Notes

^{a}EM = emission measure = $\int n_e^2 \, ds$; n_{ion} = ionizing photons s^{-1}.

$^b\Delta s$ = path length through the emitting region.

21.8 PLANETARY NEBULAE (PNe)

A major uncertainty with PNe is their distances. Many authors refer to the scale of Seaton [54]; others have found distances similar to those of Cudworth [55], but see also [56], which are statistically 1.47 times those of Seaton. Let K = the statistical average of the true distances relative to the Seaton scale. Then $\langle M \rangle \propto K^{5/2}$, n = number density $\propto K^{-3}$, birth rate per volume $\propto K^{-4}$.

21.8.1 General Statistics [57]

The local number density of PNe is 90 kpc^{-3}; scale height above plane of Galaxy = 150 pc; lifetime (as visible PN) $\approx (2–4) \times 10^4$ yr; rate of production, locally, = 2.4×10^{-12} pc^{-3} yr^{-1}; total number in Galaxy (radii $\leq$ 0.6 pc) = 3×10^4; rate of production = 1.3 yr^{-1}; average mass of ionized gas [58]: $0.217 \mathcal{M}_\odot$ [58] for optically thin nebulae.

21.8.2 Main Catalogs of PNe

Perek, L., & Kohoutek, L. 1967, *The Catalogue of Galactic Planetary Nebulae* (Praha, Academia Press, CSSR) (1036 objects); Acker, A. et al. 1982, *Catalogue of the Central Stars of True and Possible Planetary Nebulae*, Publication Spéciale du C.D.S. No. 3, Observatoire de Strasbourg, and Complements I (1982), II (1983), III (1984) (480 stars) [12]. Line intensities and wavelengths of bright PNe: Kaler, J.B. 1976, *ApJS*, 31, 517 [59]. Radio observations of PNe: Higgs, L.A. 1971, *Publications of the Astrophysics Branch, National Research Council of Canada*, Vol. 1, No. 1, NRC 12129.

21.8.3 General Information Regarding Some Bright PNe and Their Central Stars [60]

Table 21.22 gives positions (eqinox 2000) and some data for selected bright planetary nebulae. Other commonly known ones are: NGC 3587 = "Owl"; NGC 6853 = "Dumbbell Nebula"; NGC 7009 = "Saturn Nebula."

Table 21.22. *Bright planetary nebulae.*

Name[a]	Desig.[b]	α (2000) (h m)	δ (2000) (° ′ ″)	Radius (″)	Excit. class[c]	log F(Hβ)[d]	F_ν (6 cm) (mJy)
NGC 40	120 +9°	00 13	72 31 16	18.2	2	−10.64	460
NGC 246	118 −74°	00 47	−11 52 40	118	8p	−10.20	247
IC 418	215 −25°	05 27	−12 41 43	6.2	3	−9.59	1550
NGC 2392	197 +17°	07 29	20 54 48	23.0	8p	−10.39	251
NGC 2440	234 +2°	07 42	−18 12 28	16.2	9	−10.45	410
Abell 30	208 +33°	08 47	17 52 37	64	⋯	−12.19	⋯
NGC 6210	43 +37°	16 44	23 47 56	8.5	5	−10.08	260
NGC 6543	96 +29°	17 58	66 37 59	10.0	5	−9.58	850
NGC 6572	34 +11°	18 13	06 51 15	7.3	6	−9.81	1307
NGC 6720	63 +13°	18 53	33 01 36	35	6p	−10.06	360
BD+30°3639	64 +5°	19 32	30 24 18	2.5	2	−10.01	600
NGC 7027	84 −3°	21 07	42 14 06	4.0	10p	−10.17	6370
NGC 7293	36 −57°	22 29	−20 50 30	360	6	−9.35	1292
NGC 7662	106 −17°	23 26	42 32 05	7.6	⋯	−9.98	600

Notes

[a] Common designations: NGC 6720 = "Ring Nebula"; NGC 7293 = "Helix Neb."

[b] The designation in the Perek–Kohoutek (1967) catalog, or galactic longitude and latitude in degrees, numbered from north to south within each 1° × 1° bin.

[c] Excitation classes range from 1–10, increasing with the excitation as shown in line ratios [O II]/[O III]; [O III]/Hβ; He II/Hβ; Ne V/Ne III. See Aller, L.H., 1956, *Gaseous Nebulae* (Chapman & Hall, London), p. 66. p = peculiar.

[d] The reddened Hβ flux, in erg cm^{-2} s^{-1}.

Other information regarding these nebulae [60] are given in Table 21.23.

Table 21.23. *Further information on bright galactic PNe.*

Name[a]	$E(B-V)$	$10^{-3}n_e$; $10^{-4}T$	Star m_V, type[b]	v_{expan} (km s^{-1})	v_{rad} (km s^{-1})	Dist. (kpc)	R (pc)
NGC 40	0.51, 1.3	0.85	11.65, WC8	29	−13	2.3	0.20
NGC 246	0.0	0.09, ⋯	11.95, OVI	38	−46	0.50	0.29
IC 418	0.20	14.0, 0.85	9.93, Of	12	+43	1.25	0.038
NGC 2392	0.12	3.4, 1.3	10.43, O7f	54	+64	2.0	0.22
NGC 2440	0.3	2.5, 1.4	19	23	+45	2.2	0.17
Abell 30	0.0		14.3, OVI	40		2.2	0.68
NGC 6210	0.04	7.5, 1.0	12.90, O3	21	−18	3.3	0.14
NGC 6543	0.05	4.0, 0.83	11.44, O7+WR	20	−51	1.9	0.092
NGC 6572	0.27	20, 1.05	13.0, Of+WR	16	+10	1.6	0.057
NGC 6720	0.07	0.6, 1.0	15.00, Cont.	30	0.	0.65	0.11
BD+30°3639	0.24	10.0, 0.8	9.95, WC9	26	−13	1.1	0.013

Table 21.23. *(Continued.)*

Name[a]	$E(B-V)$	$10^{-3}n_e$; $10^{-4}T$	Star m_V, type[b]	v_{expan} (km s^{-1})	v_{rad} (km s^{-1})	Dist. (kpc)	R (pc)
NGC 7027	0.93	80, 1.4	19.4	18	+24	0.78	0.015
NGC 7293	0.0		13.43	14	−26	0.30	0.5
NGC 7662	0.11	4.5, 1.25	13.3, Cont.	26	−5	1.95	0.072

Note

[a]WR and WC = Wolf–Rayet spectrum (wide emission lines of H, He, C, N, and O; OV I = a WR spectrum plus strong O^{+5} lines at 381.1, 383.4 nm; Of = spectrum like young Of stars with emission of H, He II, N III, and usually C III; Cont. = continuous, but often taken with a low dispersion spectrogram.

21.9 SUPERNOVA REMNANTS

Large catalogs of SNRs: Green, D.A., *Ap. Space Sci.*, **148**, 3; Saken, J.M., Fesen, R.A., & Shull, J.M. 1992, *ApJS*, **81**, 715.

In general, SNRs have synchrotron radio emission showing linear polarization, with $S_\nu \propto \nu^\alpha$. The spectral index varies with type, as given below. Types of SNRs:

1. **Balmer-dominated:** Steep radio spectrum ($a \leq -0.3$); radial magnetic field when young; thermal optical spectrum of strong Balmer lines, weak [O III], [S II]; full or partial shell arrangement of filaments; thermal X-ray emission with both shell and filled forms possible. Ex.: Kepler's SN (SN 1604); Tycho's SN (1572); SN 1006.

2. **Oxygen-rich:** Steep radio spectrum ($a \leq -0.3$); radial magnetic field when young; optical spectrum of strong Balmer lines, strong [O III], high-velocity dispersion; full or partial shell arrangement of filaments; strong thermal X-ray emission; presumably caused by shock within peculiar internally processed stellar material. Ex: Cas A, Puppis A.

3. **Plerionic-composite:** Extended, filled (centrally concentrated) centers, possibly with surrounding shells ("composite"). Flat radio spectrum ($a \geq -0.3$), possibly steeper ($a \leq -0.3$) in shell, regular in inner part. Thermal optical spectrum plus optical synchrotron emission. Nonthermal X-rays, γ-rays. Inner part probably driven by pulsar. Ex: Crab (SN 1054), 3C 58.

4. **Evolved:** Partial shell form; steep radio spectrum ($F_\nu \propto \nu^\alpha$; $\alpha \leq -0.3$); mixed magnetic field direction; thermal optical line radiation from shocked ISM, with strong [O I], [S II]; thermal X-rays; slow shocks (50–200 km s^{-1}) from shocked cloudlets of ISM. Most galactic SNRs in this class because of long lifetime (final stage of the other types). Ex: Cygnus Loop.

21.9.1 Continuum Spectrum of Crab Nebula (with Pulsar Spectrum Excluded) [61, 62]

The spectrum generally follows a power law between the tabulated points. There are breaks in the exponent of the power law at $\log(\nu) \approx 12.7$, 16, and 19 for the nebula itself, ≈ 8, 14, and 19 for the pulsar. Below $\log(n) \approx 8$, the pulsar flux increases as $\nu^{3.8}$, peaking at $\log(\nu) \approx 8$. Details are given in Table 21.24.

Table 21.24. *Spectrum of the Crab Nebula; ν in Hz, S_ν in Jy.*

$\log \nu$	8	9	10	12	14	14.7	16	19	21	23
$\log S_\nu{}^a$	3.3	3.0	2.7	2.13	1.10	0.88	−0.60	−4.05	−7.0	−10.0
$\log S_\nu{}^b$	1.1	−1.95	< −4.1	$\cdots$	−2.58	−2.50	−3.15	−4.92	−7.24	−9.56

Notes

[a] Flux from the nebula, excluding the pulsed radiation.

[b] Pulsed flux from the system.

21.9.2 Galactic SNRs [63, 64]

For line spectra of the supernova remnants Crab and Cas A, see the section on spectra. Other details of some supernova remnants are given in Table 21.25.

Table 21.25. *Galactic supernova remnants.*

Name	Date or (age in yr)	Type[a] d (kpc)	v_{exp} (km s^{-1})	$2R$ (pc)	$E(B-V)$ (mag)
Cas A	1658	II, O, 2.8	6000	4.1	1.4
Kepler's SNR	1604	I, B 4.4	2000	3.8	1.1
Tycho's SNR	1572	I, B 2.3	1800	5.4	0.7
Crab Nebula	1054	II?, P 2.0	1500	2.7×4.2	0.5
SN 1006	1006	I, B 1.0	> 3000	8.8	0.2
Pup A (W44, 3C392)	(3700)	II?, O 2	400–1600	35	0.12
Cygnus Loop	(15,000)	?, E 0.77	130	50	0.08

Note

[a] First number, type of SN from light curve and H content of ejecta (I = no H; II = H present); second, type of remnant (see above: B = Balmer, O = oxygen-rich, P = plerion, PS = plerion/shell, E = evolved). R (pc) from radio image.

21.10 COSMIC RAYS (Excluding Photons and Neutrinos)
by D.P. Cox and J.F. Ormes

This information was compiled in October 1992.

Cosmic rays observed at Earth are of five types: solar CR, anomalous CR probably arising from the interaction of the solar wind with the interstellar medium, galactic CR with local origins, galactic CR with distant origins, and CR with extragalactic origins.

Galactic CR have several components [65, 66]:

1. Primary nuclei, with source abundances paralleling solar (or local galactic) values, but with a tendency for elements with higher first ionization potential to be under-represented [67].

2. Primary electrons, with a flux about 1% that of protons at the same rigidity (same momentum per unit charge, and gyroradius).

3. Secondary nuclei (easily identifiable by their excess abundances and steeper spectra) generated by spallation of primaries in collisions with interstellar matter.

4. Positrons and an equal number of additional electrons, probably daughters of charged pion secondaries, with fluxes about 10% of the primary electrons at 10 GeV [68, 69].

5. Antiprotons, $< 5 \times 10^{-4}$ as abundant as protons, probably arising as products of high-energy CR collisions on interstellar matter [68–70].

6. Heavier antinuclei, none of which have been found, consistent with the strong limit from γ-rays [71] on antimatter content of the Galaxy.

From measurements of radioactive species produced on Earth and in meteorites by CR impact, the average CR flux has been fairly constant, at a value like that observed at present, over time scales from 10^3 to 10^7 yr, with less certain evidence to about 5×10^8 yr, beyond which it may have been somewhat lower [70, 72]. Except at the shorter time scales, the techniques are insensitive to potential fluctuations from nearby (transient) galactic sources, and none have been unambiguously identified (but see [73]). From the brightness of the γ-ray background, it appears that the CR intensity elsewhere in the Galaxy is not very different from that here, with about a factor of 2 decrease between the inner and outer Galaxy [74].

Given this temporal and spatial uniformity, it is commonly assumed that the CR sample just outside the Solar System is representative of that in the larger neighborhood, i.e., that atypical CRs with origins within a few hundred parsecs are not a major component. See, however, [75–77].

Assuming our sample to be typical of the galactic population, one infers a number of properties, some of which depend on the choice of the model used for propagation and escape (values given here are ranges for leaky box and simple disk diffusion models at 1 GeV):

1. From the ratio of secondary to primary abundances, mean grammage $\bar{X}$ along path [78] $\approx$ $(9 \pm 1)(v/c)$ g cm^{-2}. Grammage decreases with increasing rigidity above ≈ 4.5 GV (GeV/ec).

2. From the ratio of radioactive to stable secondary isotopes (principally ^{10}Be/^{9}Be), the mean trapping (= disk storage) time $t \approx (0.8–2) \times 10^7$ yr [79, 80]; mean path length within the disk = $ct \approx$ 2–6 Mpc.

3. From (1) + (2), the mean density within the trapping volume $\bar{n} \leq 0.3$ cm^{-3}. This value may rise due to the cross sections quoted in [78, 80], but is reduced in diffusion models or when there is a nested disk-halo box.

4. From (3) + the surface density of ISM ($\sigma \approx 2 \times 10^{-3}$ g cm$^{-2} \approx 10 \mathcal{M}_\odot$ pc^{-2} [81]), the vertical extent of trapping region, $\pm h$; $h \geq 0.5$ kpc. (This is for $\bar{n} < 0.3$ cm^{-3}.)

5. From the synchrotron surface brightness perpendicular to the galactic plane, the effective half-thickness of emission region $h \approx 1$–2 kpc [81–83].

6. From the weakness of the radial gradient of γ-ray distribution versus probable cosmic ray sources, the vertical extent of the diffusion region $h > 15$ kpc [74], or sources have weak gradient and $h < 4$ kpc [84]. Models with a radially decreasing diffusion coefficient or an increasing escape time encounter no problem with a shallow radial gradient.

7. From comparison of the trapping volume scale with total path before escape, the scattering mean free path $\lambda \approx 1$ pc $\approx 10^6$ gyro radii at 1 GeV. (With enough convection, the mean free path for diffusion could be smaller. In a true leaky box, diffusion is not the confining mechanism and the mean free path could be much longer.)

8. From the anisotropy (0.1%) of the CR flux, increasing above $10^{15.5}$ eV, the bulk flow speed of CR past the solar system < 65 km/s [85], also providing limits on nearest sources, the diffusion coefficient, and the distance of the Sun from the galactic midplane in various models.

9. From the measured flux and spectrum (corrected for solar modulation), energy density $u \approx$ 2.4×10^{-12} erg cm^{-3} [86]; pressure $\approx 1.0 \times 10^{-12}$ dyn cm^{-2} [81]. Significant contributions to each are made by subrelativistic particles which are subject to large solar modulation.

10. From the energy density, grammage, and ISM surface density, the CR source power per unit area of the galactic disk in the solar neighborhood must equal escaping energy flux: $P/A \approx \sigma uc/\bar{X} \approx$ $(1–2) \times 10^{-5}$ erg cm^{-2} s^{-1}.

11. Summed over the Galaxy, depending on method, the total CR power $P \approx (0.5–2) \times 10^{41}$ erg^{-1}, compared to a total supernova power $\approx (10^{51}$ erg/30 yr) $\approx 10^{42}$ erg s^{-1}.

12. From the abundances of products of ion-driven chemistry in diffuse clouds, and the γ-ray intensities from diffuse and molecular clouds, CR are present at fairly normal intensity throughout the bulk of interstellar matter, where they appear to induce ionization at a rate about 5×10^{-17} s^{-1}

(H atom)$^{-1}$ [74, 87]. It is less clear whether they penetrate easily into the densest cores of molecular clouds (a small fraction of the total mass), although ions present there suggest that exclusion, if any, is not extreme.

Models more physical than the simple leaky box, including diffusion, convective transport, and the interaction between CR pressure and the confining magnetic field tend to imply a larger galactic halo for particle storage. This leads to rather large differences in interpretation of the CR lifetime, mean density in the trapping volume, and scale height [88].

"All particle" spectra exist up to 10^{20} eV, and for individual nuclei and groups to lower energies ($\approx 10^{12}$ eV/nucleon). For energies between $10^{9.5}$ and 10^{12} eV/nucleon, primary nuclei have power law flux spectra: $dJ/dE \propto E^{-\gamma}$, with $\gamma \approx 2.6$. The secondary nuclei have a steeper slope ($\gamma + \delta$) with $\delta \approx 0.6$, interpreted as a decreasing trapping time scale with increasing energy, and implying a primary source spectrum with index ($\gamma - \delta$) ≈ 2 (up to ≈ 2.4) [65]. This spectrum could derive from acceleration by shocks with a compression factor of about 4 (e.g., strong adiabatic shocks in a nonrelativistic monatomic gas), but most models have difficulty reaching sufficiently high energy [89, 90], but see [91, 92].

A downward turn or "knee" appears at $10^{15.5}$ eV in the all-particle spectrum, similar to the energy at which the anisotropy begins to rise. This has been attributed to a breakdown of magnetic trapping or an upper limit on the energy achieved by a major acceleration mechanism [90].

The correlation of the elemental abundances (for primaries, relative to solar) with the first ionization potential, an absence of depletion of those elements common in interstellar dust, an underabundance of nitrogen, and an overabundance of the neutron-rich isotope ^{22}Ne (and perhaps ^{25}Mg and ^{26}Mg) are strong contraints on the injectors of primary CR particles. Stellar flares are a potential source, with a small contribution required from He-burning zones, possibly Wolf–Rayet stars [67], but see [78]. Injection at MeV must then be followed by powerful acceleration, perhaps by a supernova remnant shock, which generates the full source spectrum. Further accelerations in the ISM must be rather uncommon [93], but see [91, 94].

Cosmic rays with energies in excess of 10^{19} eV have high anisotropy, poor trapping in the galactic magnetic field and probably are extragalactic in origin, perhaps from the Virgo supercluster [90].

Scattering of cosmic rays in the ISM is thought to occur via pitch-angle diffusion, possibly on resonant waves created by the cosmic rays themselves [85]. Conditions in the diffuse ISM are too poorly known for the wave damping rate (and thus the scattering rate) to be known with certainty.

REFERENCES

1. Rand, R.J., & Kulkarni, S.R. 1989, *ApJ*, **343**, 760
2. Norris, R.P. 1984, *MNRAS*, **207**, 127
3. Jura, M. 1987, in *Interstellar Processes*, edited by D.J. Hollenbach and H.A. Thronson, Jr. (Reidel, Dordrecht), p. 1
4. Clemens, D.P., Sanders, D.B., & Scoville, N.Z. 1988, *ApJ*, **327**, 139
5. Boulares, A., & Cox, D.P. 1990, *ApJ*, **365**, 544
6. Reynolds, R.J. 1992, *The Astronomy and Astrophysics Encyclopedia* (Van Nostrand, New York), p. 352
7. Scoville, N.Z., & Sanders, D.B. 1987, in *Interstellar Processes*, edited by D.J. Hollenbach and H.A. Thronson, Jr. (Reidel, Dordrecht), p. 21
8. Combes, F. 1991, *ARA&A*, **29**, 195
9. Dickey, J.M., & Lockman, F.J. 1990, *ARA&A*, **28**, 215
10. Mezger, P.G. 1978, *A&A*, **70**, 565
11. Cordes, J.M. et al. 1991, *Nature*, **354**, 121
12. Reynolds, R.J. 1991, *ApJ*, **372**, L17
13. Bohlin, R.C., Savage, B.D., & Drake, J.F. 1979, *ApJ*, **224**, 132
14. Cardelli, J.A., Clayton, G.C., & Mathis, J.S. 1989, *ApJ*, **345**, 245
15. Morrison, R., & McCammon, D. 1983, *ApJ*, **270**, 119
16. Herbig, G.H. 1975, *ApJ*, **196**, 120
17. Herbig, G.H., & Leka, K.D. 1991, *ApJ*, **382**, 193
18. Wilson, T.L., & Matteucci, F. 1992, *A&A Rev.*, **4**, 1
19. Wright, E.L. et al. 1991, *ApJ*, **381**, 200
20. Allamandola, L.J., Tielens, A.G.G.M., & Barker, J.R. 1989, *ApJS*, **712**, 733
21. Lee, Y., Snell, R.L., & Dickman, R.L. 1990, *ApJ*, **355**, 536
22. Maloney, P. 1990, in *The Interstellar Medium in Galaxies*, edited by H.A. Thronson and J.M. Shull (Kluwer Academic, Dordrecht), p. 493
23. Digel, S., Bally, J., & Thaddeus, P. 1990, *ApJ*, **357**, L29

24. Solomon, P., Rivolo, A.R., Barrett, J., & Yahil, A. 1987, *ApJ*, **319**, 730
25. Larson, R.B. 1981, *MNRAS*, **194**, 809
26. Solomon, P., & Rivolo, A.R. 1989, *ApJ*, **339**, 919
27. Lizst, H.S., & Burton, W.B. 1981, *ApJ*, **243**, 778
28. Blitz, L. 1990, in *The Evolution of the Interstellar Medium*, edited by L. Blitz (Astronomical Society of the Pacific, San Francisco), p. 273
29. Blitz, L. 1991, *The Physics of Star Formation and Early Stellar Evolution*, edited by C.J. Lada and N.D. Kylafis (Kluwer Academic, Dordrecht), p. 3
30. Turner, B.E., 1992, in *The Astronomy and Astrophysics Encyclopedia*, edited by S.P. Maran (Van Nostrand Reinhold, New York), p. 378
31. Dalgarno, A., 1992, in *The Astronomy and Astrophysics Encyclopedia*, edited by S.P. Maran (Van Nostrand Reinhold, New York), p. 343
32. Lovas, F.J., Snyder, L.E., & Johnson, D.R. 1979, *ApJS*, **41**, 451
33. Genzel, R. et al. 1981, *ApJ*, **247**, 1039
34. von Hippel, T., Bell Burnell, S.J., & Williams, P.M. 1988, *A&AS*, **74**, 431
35. Spitzer, L., Jr. 1985, *ApJ*, **290**, L21
36. Morton, D.C. 1991, *ApJS*, **77**, 119
37. van den Bergh, S. 1966, *AJ*, **71**, 990
38. van den Bergh, S., & Herbst, W. 1975, *AJ*, **80**, 208
39. Herbst, W. 1975, *AJ*, **80**, 212
40. Racine, R. 1968, *AJ*, **73**, 233
41. Savage, B.D. 1988, in *QSO Absorption Lines*, edited by J.C. Blades and D. Turnshek (Cambridge University Press, Cambridge), p. 195
42. Westerhout, G. 1958, *B.A.N.*, **14**, 215
43. Downes, D., & Rinehart, R. 1966, *ApJ*, **144**, 937
44. Altenhoff, W.J. et al. 1979, *A&AS*, **35**, 23
45. Goss, W.M., & Shaver, P.A. 1970, *Aust. J. Phys. Sup.*, **14**, 1
46. Churchwell, E.B. et al. 1978, *A&A*, **70**, 719
47. Lang, K.R., 1980, *Astrophysical Formulae* (Springer-Verlag, Berlin), p. 121
48. Savage, B.D. 1987, in *Interstellar Processes*, edited by D.J. Hollenbach and H.A. Thronson, Jr. (Reidel, Dordrecht), p. 123
49. Savage, B.D., & Massa, D. 1987, *ApJ*, **314**, 380
50. Reynolds, R.J. 1989, *ApJ*, **339**, L29
51. Sembach, K.R., & Savage, B.D. 1992, *ApJS*, **83**, 147
52. Churchwell, E.B. 1990, *A&AR*, **2**, 79
53. Wood, D.O.S., & Churchwell, E.B. 1989, *ApJS*, **69**, 831
54. Seaton, M.J. 1968, *Ap. Lett.*, **2**, 55
55. Cudworth, K. 1974, *AJ*, **79**, 1384
56. Mallik, D.C.V., & Peimbert, M. 1988, *Rev. Mex. A&A*, **16**, 111
57. Phillips, J.P. 1989, in IAU Symp. 131, *Planetary Nebulae*, edited by S. Torres-Peimbert (Reidel, Dordrecht), p. 425
58. Kingsburgh, R.L., & Barlow, M.J. 1992, *MNRAS*, **257**, 317
59. Mallik, D.C.V., & Peimbert, M. 1988, *Rev. Mex. Astron. Astrof.*, **16**, 111
60. Pottasch, S.R. 1984, *Planetary Nebulae* (Reidel, Dordrecht), p. 322
61. de Jager, O.C., & Harding, A.K. 1992, *ApJ*, **396**, 161
62. Seward, F.D. 1989, *Space Sci. Rev.*, **49**, 389
63. Raymond, J.C. 1984, *ARA&A*, **22**, 75
64. Lang, K.R. 1992, *Astrophysical Data* (Springer-Verlag, Berlin), p. 716
65. Meyer, P. 1992, in *The Astronomy and Astrophysics Encyclopedia* (Van Nostrand, New York), p. 136
66. Lund, N. 1984, in *Cosmic Radiation in Contemporary Astrophysics*, edited by M.M. Shapiro (Reidel, Dordrecht), p. 1
67. Meyer, J.-P. 1985, *ApJS*, **57**, 173
68. Mueller, D., & Tang, J.K. 1987, *ApJ*, **321**, 183
69. Stephens, S.A., & Golden, R.L. 1987, *Space Sci. Rev.*, **46**, 131
70. Reedy, R.C., Arnold, J.R., & Lal, D. 1983, *Annu. Rev. Nucl. Part. Sci.*, **33**, 505
71. Steigman, G. 1976, *ARA&A*, **14**, 379
72. Vogt, S., Herzog, G.F., & Reedy, R.C. 1990, *Rev. Geophys.*, **28**, 253
73. Raisbeck, G.M., Yiou, F., Bourles, D., Lorius, C., Jonzel, J., & Barkov, N.I. 1987, *Nature*, **326**, 273
74. Bloemen, H. 1989, *ARA&A*, **27**, 469
75. Streitmatter, R.E., Balasubrahmanyan, V.K., Protheroe, R.J., & Ormes, J.F. 1985, *A&A*, **143**, 249
76. Clayton, D.D., Cox, D.P., & Michel, F.C. 1986, in *The Galaxy and the Solar System*, edited by R. Smoluchowski, J.N. Bahcall, and M.S. Matthews (University of Arizona, Tucson), p. 129
77. Bhat, C.L., Issa, M.R., Mayer, C.J., & Wolfendale, A.W. 1985, *Nature*, **314**, 515
78. Gupta, M., & Webber, W.R. 1989, *ApJ*, **340**, 1124
79. Simpson, J.A., & Garcia-Muõz, M. 1988, *Space Sci. Rev.*, **46**, 205
80. Shapiro, M.M., Leuw, J.R., Silverberg, R., & Tsao, C.H. 1991, in *The 22nd Internat'l Cosmic Ray Conf.* (Dublin Inst. for Advanced Study, Dublin), pp. 2, 304
81. Boulares, A., & Cox, D.P. 1990, *ApJ*, **365**, 544
82. Bloemen, J.B.G.M. 1987, *ApJ*, **322**, 694
83. Beuermann, K., Kanbach, G., & Berkhuijsen, E.M. 1985, *A&A*, **153**, 17
84. Webber, W.R., Lee, M.A., & Gupta, M. 1992, *ApJ*, **390**, 96
85. Cesarsky, C.J. 1980, *ARA&A*, **18**, 289
86. Webber, W.R. 1987, *A&A*, **179**, 277
87. Black, J.H., van Dishoeck, E.F., Willner, S.P., & Woods, C. 1990, *ApJ*, **358**, 459
88. Berezinskii, V.S., Bulanov, S.V., Dogiel, V.A., Ginzburg, V.L., & Ptuskin, V.S. 1990, *Astrophysics of Cosmic Rays* (North-Holland, Amsterdam)
89. Axford, W.I. 1981, in *Origin of Cosmic Rays*, edited by G. Setti, G. Spada, and A.W. Wolfendale (Reidel, Dordrecht), p. 339
90. Hillas, A.M. 1984, *ARA&A*, **22**, 425
91. Bryant, D.A., Powell, G.I., & Perry, C.H. 1992, *Nature*, **356**, 582
92. Silberberg, R., Tsao, C.H., Shapiro, M.M., & Biermann, P.L. 1990, *ApJ*, **363**, 265
93. Cesarsky, C.J. 1987, *Proc. Intl. CR Conf.* (20th Moscow), **8**, 87
94. Schlickeiser, R. 1984, in *Cosmic Radiation in Contemporary Astrophysics*, edited by M.M. Shapiro (Reidel, Dordrecht), p. 27

Chapter 22

Star Clusters

Hugh C. Harris and William E. Harris

The classification of star clusters as "open" or "globular" was originally based on their appearance in the sky. However, in the Milky Way Galaxy, the common properties of open clusters (location in the galactic disk, disk kinematics, near solar metallicity, and younger age) are fairly distinct from the properties of globular clusters (often in the galactic spheroid, often metal-poor, older age). Here we designate clusters with ages 10 Gyr and younger as "open" and older clusters as "globular." However, cases of ambiguous or inadequate data exist (e.g., for clusters Berkeley 17, Pal 1, BH176, Lyngå 7, see tables below), and there is no single rigorous definition that can be used to classify every star cluster.

22.1 OPEN CLUSTERS

Open star clusters have been cataloged many times, but lists of newly discovered clusters continue to appear. In the Milky Way, the extensive 1987 Lyngå–Lund Observatory catalog [1] containing 1154 clusters is the most recent comprehensive listing. We have started with this catalog, using the cluster's names from this catalog, and updated it with the many recent references through 1996. Beyond the Milky Way, only the Magellanic Clouds are close enough to have relatively complete searches made for open clusters. In this chapter, we concentrate on Milky Way clusters.

Unlike the case for globular clusters, the catalog of open clusters in the Milky Way is far from final. The first deficiency is that almost 200 clusters are noted in the catalog or in other literature as "doubtful." Many (perhaps 20% of the catalog or more) are not true clusters of physically associated and gravitationally bound stars. In Tables 22.2–22.4 clusters noted as "doubtful" have been omitted. The second deficiency of the catalog is incompleteness. At visible wavelengths, deeper sky surveys reveal new clusters, most of which will prove to be distant open clusters. The ESO Uppsala Survey [2] cataloged 129 new clusters south of declination $-17.5°$ and outside of the Magellanic Clouds. Only a few of these clusters have been studied yet, and two (AM2 = 368-SC07 and 092-SC18) have been added to Table 22.4. Inaccurate coordinates for many clusters have added to confusion in cluster identifications that is slowly being resolved (e.g., [3]). Undoubtedly many more new clusters will be found on the POSS II northern sky survey. At present, omitting doubtful clusters, we have 488 clusters with a distance determination (of which 445 also have an age determination) and 498 without a reliable distance.

Interstellar absorption is, of course, the major cause for incomplete discovery of open clusters. Infrared data can help alleviate this problem, and new clusters not already identified at visible wavelengths have been incorporated here (see notes to Table 22.2). For heavily obscured clusters, large uncertainties usually exist in their distances and in their membership (and, hence, in whether they are, or will become, bound clusters rather than unbound groups or associations). From very preliminary studies so far (e.g., [4, 5]), we can guess that several hundred clusters in the second and third quadrants of the Galaxy will be discovered with infrared data, and undoubtedly many more will be found in the first and fourth quadrants. Most of these clusters will have ages less than 10^7 yr.

Age determinations for open clusters can be quite uncertain. An accurate cluster distance is often needed because, for younger clusters, the absolute magnitude of the cluster turnoff is the primary age discriminator. For young clusters, the status adopted for the most bright and hot main-sequence stars (cluster member versus nonmember, and main sequence versus evolved) can have a large effect on the derived cluster age. In at least some young clusters, prolonged star formation complicates assigning a unique age. Here the youngest clusters are all listed with an arbitrary age of 10^6 yr. For clusters with ages $\sim 10^9$ yr and older, the difference in magnitude between the turnoff and red-giant clump can be used [6, 7]. Again, the status adopted for a few stars can affect the derived age. Cluster ages depend ultimately on calibration from stellar evolutionary models. The adopted model parameters (mass loss, convective overshoot, etc.) introduce systematic differences in the derived ages. Generally, recent models have been converging toward reasonably good agreement within about 25% (e.g., [8]). In the tables here, cluster ages have been taken directly from the most recent or most reliable literature study without adjustment.

Few open clusters have been studied to determine their structural and dynamical parameters. Thus, accurate data on radii, velocity dispersions, densities, and masses are lacking for all but a few clusters.

22.1.1   Parameters for Open Clusters

This section summarizes some parameters describing the system of open clusters in the Milky Way. We begin with the surface density projected onto the galactic plane of open clusters of different ages in the solar neighborhood within 1200 pc of the sun (Table 22.2). Note, however, that clusters are distributed nonuniformly. The youngest clusters are more common in regions like the Perseus spiral arm than in the solar neighborhood; the oldest clusters are more common in the outer disk. The last entry, the corrected total (corrected for incompleteness), is based on the distribution of cluster distances in Table 22.2, a distribution that indicates increasing incompleteness beyond 600 pc.

Projected surface density (Table 22.2; see also [9, 10]):

0–19 Myr	(26 clusters)	5.5 kpc^{-2}.
20–49 Myr	(21 clusters)	4.5 kpc^{-2}.
50–99 Myr	(23 clusters)	4.9 kpc^{-2}.
100–199 Myr	(24 clusters)	5.1 kpc^{-2}.
200–499 Myr	(27 clusters)	5.7 kpc^{-2}.
500–1999 Myr	(20 clusters)	4.2 kpc^{-2}.

$\geq$ 2000 Myr	(2 clusters)	0.4 kpc^{-2}.
Unknown Age	(13 clusters)	2.8 kpc^{-2}.
All clusters	(156 clusters)	33.1 kpc^{-2}.
Corrected total		50. kpc^{-2}.

The median age in the solar neighborhood (Table 22.2) is:

$\sim 10^8$ yr.

The cluster formation rate [9–11] is:

~ 0.5 kpc^{-2} Myr^{-1}.

The cluster destruction time scale (e-folding time) in the solar neighborhood [11, 12] is:

$\sim 10^7$ yr for low-mass clusters,
$\sim 10^8$ yr for typical clusters,
$\sim 10^9$ yr for rich (populous) clusters.

There is no significant dependence for metallicity as a function of age [9, 13] nor for metallicity as a function of z distance [9, 13].

The metallicity as a function of galactocentric radius is:

All clusters [9]: -0.07 dex kpc^{-1}.
Old clusters [13, 14]: -0.09 dex kpc^{-1}.

The scale height for clusters is:

Young clusters ($< 3 \times 10^8$ yr), solar neighborhood [9]: 55 pc.
Young clusters ($< 3 \times 10^8$ yr), outer disk [9]: Larger, but not well known.
Old clusters ($> 10^9$ yr) [15]: 375 pc.

Kinematics of young clusters ($< 3 \times 10^7$ yr) are described by [16]:

Oort constant A: 17 km s^{-1} kpc^{-1}.
Second-order term: -2 km s^{-1} kpc^{-2}.
One-dimensional radial velocity dispersion: 11 km s^{-1}.

Kinematics of old clusters (> 1 Gyr) [17] are consistent with the same rotation curve for young clusters, and with

One-dimensional radial velocity dispersion: 28 km s^{-1}.

Dynamical properties of one typical and one rich (populous) open cluster are given in Table 22.1.

Table 22.1. *Dynamical properties of two open clusters.*

Property	Typical (Hyades [18])	Rich (M11 [19])
Core radius (pc):	2.1	0.8
Half-mass radius (pc):	4.3	2.7
Central density ($\mathcal{M}_\odot$ pc^{-3}):	3	500
Central one-dimensional velocity dispersion (km s^{-1}):	0.2	1.2
Half-mass relaxation time (yr):	10^8	10^8
Total mass ($\mathcal{M}_\odot$):	350	7000

Open clusters in the Large Magellanic Cloud (LMC) have:
    Total population [20, 21]: 4200 clusters.
    Peak face-on surface density [20]: 200 kpc^{-2}.
    Median age [22, 23]: $\sim 10^9$ yr.
Open clusters in the Small Magellanic Cloud (SMC) have:
    Total population [24]: 2000 clusters.
    Peak face-on surface density [24]: 200 kpc^{-2}.
    Median age [25]: $\sim 10^9$ yr.

22.1.2 Selected Open Clusters in the Milky Way

This section provides data for selected open clusters in the Milky Way. Data are taken from the literature through the end of 1996, plus distances to 14 clusters measured with the Hipparcos satellite published in 1997 [26, 27]. Table 22.2 lists 156 clusters believed to be within 1200 pc of the Sun. Table 22.3 lists 54 clusters with estimated ages 3×10^6 yr and less. Table 22.4 lists 52 clusters with estimated ages more than 10^9 yr. All tables have the same format.

Column (1) gives the cluster name, usually as given by [1], with other common names. Names changed or added to [1] are indicated in the table notes.

Columns (2) and (3) give the right ascension and declination (J2000). Positions are taken from [1], given to the nearest 1 arcmin in declination, except for clusters south of declination $-17.5°$ measured on the ESO southern sky survey [2], and other clusters measured by [28]. For the latter clusters [2, 28], positions are given to 0.1 arcmin in declination.

Columns (4) and (5) give the galactic latitude and longitude in degrees.

Column (6) gives the distance from the Sun in parsecs, taken first from a recent study in the literature, or second from Janes, Duke, and Lyngå, as given by [1].

Column (7) gives the distance from the galactic center (GC) in kiloparsecs. We adopt $R_0 = 8.0$ kpc [29] for the distance of the Sun from the galactic center.

Column (8) gives the foreground interstellar reddening, $E(B - V)$, taken from the same source as the distance, if available.

Column (9) gives the angular diameter in arc minutes, taken first from [30] measured on sky survey prints, or second from a selected diameter from the literature. These estimates are approximate and can be biased by field-star contamination.

Column (10) gives the approximate number of cluster members estimated by [30].

Column (11) gives the richness class given by [12], defined from 5 (very rich cluster with high star density) to 1 (very sparse cluster).

Column (12) gives the logarithm (base 10) of the cluster age in years.

Table 22.2. *Nearby Galactic open clusters.*

(1)	(2)	(3)	(4)	(5)	(6)	(7)	(8)	(9)	(10)	(11)	(12)
	RA	Dec.									
Cluster	(h m s)	(° ′)	ℓ	b	Dist.	R_{GC}	Red.	Dia.	Mem.	Rich.	Age
Ber 59	0 02 10	67 25.2	118.3	4.9	880	8.45		10	40		6.30
Bla 1(ζ Scl)	0 04 07	−29 49.9	15.6	−79.3	252	7.95	0.02	70	30	1	7.70
NGC189	0 39 36	61 05.7	121.5	−1.8	1080	8.61	0.54	5	15	3	7.30
NGC225	0 43 30	61 47.0	122.0	−1.1	525	8.29	0.25	15	15	2	8.08
NGC381	1 08 18	61 35.0	124.9	−1.2	1060	8.65	0.36	7	50		9.00

Table 22.2. *(Continued.)*

(1)	(2)	(3)	(4)	(5)	(6)	(7)	(8)	(9)	(10)	(11)	(12)
	RA	Dec.									
Cluster	(h m s)	(° ′)	ℓ	b	Dist.	R_{GC}	Red.	Dia.	Mem.	Rich.	Age
NGC559	1 29 30	63 18.0	127.2	0.8	900	8.57	0.90	7	60	4	9.10
Col 463	1 47 36	71 46.0	127.4	9.6	600	8.37	0.19	30	40	2	8.18
NGC752	1 57 48	37 41	137.2	−23.4	440	8.30	0.03	75	60	2	9.26
Sto 2	2 15 00	59 16	133.4	−1.9	316	8.22	0.38	45	50	4	8.23
Mar 6	2 29 40	60 42.4	134.7	0.0	485	8.35	0.52	6	6	1	7.40
Tru 2	2 36 53	55 54.9	137.4	−3.9	600	8.45	0.30	17	20	3	7.89
NGC1039(M34)	2 42 00	42 47	143.6	−15.6	465	8.36	0.10	25	60	3	8.26
NGC1027	2 42 36	61 35.7	135.8	1.5	1000	8.74	0.32	15	40	2	8.54
Mel 20(α Per)	3 24 19	49 51.7	146.9	−7.1	184	8.15	0.09	300	50	4	7.72
NGC1333[a]	3 29 01	31 19	158.3	−20.5	320	8.28	2.00				6.00
NGC1342	3 31 42	37 20	155.0	−15.4	550	8.48	0.27	17	40	2	8.48
IC348	3 44 36	32 18	160.4	−17.7	320	8.29	1.50	8	20	1	6.48
Pleiades(M45)	3 47 29	24 06.3	166.6	−23.5	116	8.10	0.04	120	100	4	8.00
NGC1502	4 07 50	62 19.9	143.6	7.6	875	8.71	0.75	20	45	2	6.85
NGC1528	4 15 19	51 12.7	152.0	0.3	800	8.71	0.29	18	40	3	8.43
NGC1545	4 20 57	50 15.3	153.4	0.2	800	8.72	0.34	12	20	2	8.29
Hyades	4 26 54	15 52	180.1	−22.4	46	8.04	0.00	329		4	8.79
LHA 101[a]	4 30 12	35 15	165.4	−9.0	800	8.77					6.00
NGC1647	4 46 00	19 05	180.4	−16.8	542	8.52	0.33	40	200	3	8.04
NGC1662	4 48 29	10 55.8	187.7	−21.1	400	8.37	0.32	12	35	2	8.48
NGC1664	4 51 06	43 40.6	161.7	−0.4	1200	9.15	0.24	18		4	8.48
NGC1746	5 03 36	23 49	179.0	−10.6	420	8.41		40	20		
NGC1901	5 17 48	−68 26	279.0	−33.6	300	7.96	0.05	40	40	1	8.70
Col 69(λ Ori)	5 35 00	9 56	195.1	−12.0	500	8.47	0.09	70	20		
NGC1981	5 35 12	−4 26.0	208.1	−19.0	400	8.34		28	20		
Trapezium[b]	5 35 16	−5 23.4	209.0	−19.4	435	8.36	0.05	47		1	6.60
Col 70	5 35 36	−1 05	205.0	−17.4	430	8.37		140	100		
NGC2024[a]	5 41 42	−1 54	206.5	−16.4	400	8.35					6.00
NGC2068[a]	5 46 42	0 06	205.3	−14.3	400	8.35					6.00
NGC2071[a]	5 47 12	0 19	205.2	−14.1	400	8.35					6.00
NGC2112	5 53 46	0 24.6	205.9	−12.6	750	8.66	0.60	18	50		9.60
NGC2169	6 08 25	13 57.9	195.6	−2.9	1020	8.99	0.20	6	30	2	7.20
NGC2168(M35)	6 08 48	24 20	186.6	2.2	870	8.86	0.22	25	200	4	8.03
NGC2215	6 20 50	−7 17.0	216.0	−10.1	1000	8.82	0.31	8	40	2	8.55
NGC2232	6 28 02	−4 50.8	214.4	−7.7	320	8.26	0.02	45	20	1	7.30
Col 96	6 30 18	2 52	208.0	−3.4	1100	8.98	0.46	12	15	1	7.40
NGC2264	6 40 59	9 53.7	202.9	2.2	800	8.74	0.04	40	40	2	6.60
NGC2287(M41)	6 46 04	−20 45.5	231.0	−10.4	700	8.45	0.06	40	80	3	8.30
NGC2281	6 48 18	41 04.7	175.0	17.1	460	8.44	0.11	25	30	3	8.48
NGC2301	6 51 46	0 27.6	212.6	0.3	750	8.64	0.04	15	80	1	8.03
NGC2302	6 51 54	−7 05.0	219.3	−3.1	1100	8.88	0.24	5	30	1	7.80
Col 121	6 54 12	−24 24.6	235.2	−10.3	1170	8.71	0.01	90	20	1	6.18
NGC2323(M50)	7 03 12	−8 20	221.7	−1.2	910	8.70	0.26	15	80	3	7.89
NGC2335	7 06 50	−10 01.7	223.6	−1.3	1000	8.75	0.38	7	35	1	8.20
NGC2343	7 08 07	−10 37.0	224.3	−1.2	1000	8.74	0.19	6	20	2	8.00
NGC2353	7 14 31	−10 16.0	224.7	0.4	1200	8.89	0.12	18	30	2	7.90
Col 132	7 15 20	−30 40.8	243.0	−8.8	411	8.19	0.03	80	25	1	7.40
NGC2360	7 17 44	−15 38.5	229.8	−1.4	1070	8.73	0.08	14	80	3	9.00
Col 140	7 23 12	−32 02.2	245.0	−7.9	380	8.17	0.05	30	30	1	7.30
Rup 18	7 24 39	−26 11.9	239.9	−4.9	1028	8.56	0.67	8	40	1	8.10
NGC2395	7 27 13	13 36.5	204.6	14.0	1200	9.07	0.69	15	30	1	7.70
NGC2409[c]	7 31 37	−17 11.4	232.5	0.8	828	8.53	0.19	16	30	1	7.40
NGC2422(M47)	7 36 36	−14 29.0	231.0	3.1	460	8.30	0.08	25	30	1	7.89
NGC2423	7 37 07	−13 52.3	230.5	3.5	870	8.58	0.12	12	40	4	8.55

Table 22.2. (*Continued.*)

(1)	(2)	(3)	(4)	(5)	(6)	(7)	(8)	(9)	(10)	(11)	(12)
	RA	Dec.									
Cluster	(h m s)	(° ')	ℓ	b	Dist.	R_{GC}	Red.	Dia.	Mem.	Rich.	Age
NGC2447(M93)	7 44 31	−23 51.7	240.0	0.1	860	8.46	0.00	10	80	4	8.78
NGC2482	7 55 15	−24 15.7	241.6	2.0	800	8.41	0.03	10	40	2	8.60
NGC2516	7 58 07	−60 45.2	273.8	−15.9	348	7.98	0.13	22	80	4	8.15
Col 173	8 02 49	−46 22.8	261.4	−8.2	330	8.06	0.09	369			
NGC2527	8 04 58	−28 08.8	246.1	1.9	600	8.26	0.09	10	40	2	9.00
NGC2547	8 10 10	−49 12.3	264.5	−8.6	420	8.05	0.05	25	80	2	7.76
NGC2546	8 12 16	−37 35.7	254.9	−2.0	1000	8.32	0.10	70	40	3	7.62
NGC2548(M48)	8 13 44	−5 45.0	227.9	15.4	610	8.41	0.05	30	80	1	8.48
NGC2579	8 22 12	−36 24	254.7	0.3	1000	8.32	0.20	8	20	1	7.10
Pis 4	8 34 36	−44 25.4	262.9	−2.4	600	8.10	0.02	25	45	1	7.40
NGC2632[b]	8 40 24	19 40.0	205.5	32.5	177	8.13	0.04	70	50	4	8.90
IC2391	8 40 32	−53 02.1	270.4	−6.8	148	8.00	0.01	60	30	2	7.48
IC2395	8 42 31	−48 09.0	266.7	−3.6	1050	8.13	0.12	17	40	2	7.20
Col 197	8 44 51	−41 13.4	261.5	1.0	1000	8.21	0.56	25	40	1	6.80
NGC2669	8 46 23	−52 56.8	270.8	−6.1	1000	8.05	0.18	20	40	2	7.80
Tru 10	8 47 54	−42 27.4	262.8	0.7	380	8.06	0.04	30	40	1	7.67
NGC2682(M67)	8 51 24	11 49.0	215.6	31.7	830	8.58	0.03	25	200	5	9.65
NGC2925	9 33 13	−53 24.0	275.9	−1.3	810	7.96	0.05	15	40	2	7.90
NGC2972	9 40 14	−50 19.5	274.7	1.8	1200	7.99	0.33	5	25	1	8.60
NGC3033	9 48 35	−56 25.8	279.6	−2.1	1100	7.89	0.58	12	50	1	8.60
NGC3114	10 02 30	−60 07.8	283.3	−3.9	940	7.84	0.04	35		2	8.18
NGC3228	10 21 22	−51 44.0	280.7	4.5	500	7.92	0.03	5	15	1	7.62
vBH 99	10 37 54	−59 11	286.6	−0.6	425	7.89	0.05	20	40		8.00
IC2602	10 42 58	−64 23.6	289.6	−4.9	147	7.95	0.04	100	60	2	7.48
NGC3532	11 05 48	−58 46.2	289.6	1.3	417	7.87	0.04	50	150	5	8.49
NGC3680	11 25 39	−43 14.1	286.8	16.9	850	7.80	0.08	7	30	2	9.28
Rup 98	11 58 40	−64 35.1	297.3	−2.3	400	7.83	0.16	15	50	1	8.50
Mel 111(Coma)	12 25 06	26 07	221.2	84.0	88	8.01	0.00	120	80	2	8.65
Hog 14	12 28 36	−59 49.1	300.1	2.9	912	7.58	0.28	3		1	8.50
Har 5	12 27 10	−60 45.8	300.0	2.0	1029	7.54	0.18	5		1	7.60
NGC4463	12 29 58	−64 47.5	300.6	−2.0	1180	7.47	0.44	6	30	1	7.40
Rup 108	13 32 11	−58 27.7	308.3	4.0	700	7.59	0.17	10	15	1	8.10
NGC5316	13 53 57	−61 52.1	310.2	0.1	1170	7.30	0.34	15	80	3	8.08
NGC5460	14 07 28	−48 20.5	315.8	12.6	790	7.47	0.14	35	40	2	8.30
Lyn 2	14 24 35	−61 20.0	313.9	−0.4	1100	7.28	0.19	10	30	2	7.50
NGC5662	14 35 38	−56 37.1	316.9	3.4	740	7.48	0.31	30	70	4	7.85
Col 285(UMa)	14 40 54	69 34	109.9	44.7	20	8.00	0.00			1	8.21
Hog 18	14 50 43	−52 16.0	320.8	6.4	1100	7.19	0.50	5	15	1	7.70
NGC5764	14 53 32	−52 40.2	321.0	5.9	1000	7.25	0.48	3	12		
NGC5822	15 04 21	−54 23.8	321.6	3.6	740	7.43	0.15	35	150	4	9.15
NGC6025	16 03 18	−60 25.9	324.5	−5.9	780	7.38	0.16	15	60	3	8.08
NGC6087	16 18 51	−57 56.1	327.7	−5.4	860	7.29	0.20	15	40	3	7.85
NGC6124	16 25 20	−40 39.2	340.7	6.0	563	7.47	0.80	40	100	2	8.00
ρ Oph[a]	16 26 54	−24 35	353.0	16.7	160	7.85					6.00
NGC6134	16 27 46	−49 09.1	334.9	−0.2	760	7.32	0.35	6		4	8.90
NGC6152	16 32 46	−52 38.6	332.9	−3.2	1030	7.10		25	70		
NGC6178	16 35 47	−45 38.6	338.4	1.2	904	7.17	0.24	5	12	1	6.80
NGC6204	16 46 09	−47 01.0	338.6	−1.0	1180	6.92	0.51	6	45	2	8.08
NGC6208	16 49 28	−53 43.7	333.8	−5.8	1000	7.12	0.17	18	60	4	9.00
NGC6242	16 55 33	−39 27.6	345.5	2.5	1120	6.92	0.39	9		3	7.70
NGC6249	16 57 41	−44 48.7	341.5	−1.2	1019	7.04	0.45	6	30	1	7.40
NGC6250	16 57 56	−45 56.2	340.7	−1.9	1020	7.05	0.38	10	60	2	7.15
NGC6268	17 02 10	−39 43.7	346.0	1.3	1100	6.94	0.39	6		3	7.40
NGC6281	17 04 41	−37 59.1	347.7	2.0	502	7.51	0.16	8		3	8.34

Table 22.2. *(Continued.)*

(1)	(2)	(3)	(4)	(5)	(6)	(7)	(8)	(9)	(10)	(11)	(12)
	RA	Dec.									
Cluster	(h m s)	(° ′)	ℓ	b	Dist.	R_{GC}	Red.	Dia.	Mem.	Rich.	Age
NGC6322	17 18 26	−42 56.0	345.3	−3.1	1200	6.85	0.51	5	30	2	7.00
IC4651	17 24 49	−49 56.3	340.1	−7.9	780	7.28	0.14	10	80	4	9.28
NGC6405(M6)	17 40 21	−32 15.3	356.6	−0.8	600	7.40	0.16	20	80	4	7.71
NGC6416	17 44 20	−32 21.7	356.9	−1.5	800	7.20	0.31	15	40	4	8.50
IC4665	17 46 12	5 43	30.6	17.1	350	7.71	0.18	70	30	2	7.78
NGC6425	17 47 02	−31 31.8	357.9	−1.6	800	7.20	0.48	10	35	2	7.80
Bas 5	17 52 27	−30 05.8	359.8	−1.9	850	7.15	0.33	5	30	2	9.10
NGC6475(M7)	17 53 51	−34 47.6	355.9	−4.5	292	7.71	0.06	80	80	2	8.34
NGC6494(M23)	17 57 05	−18 59.1	9.9	2.8	660	7.35	0.36	30	150	3	8.48
Tru 31	17 59 48	−28 09.6	2.3	−2.3	1000	7.00	0.42	5	25		9.10
Boc 14	18 01 59	−23 41.8	6.4	−0.5	1150	6.86	1.62	2	10	1	6.00
NGC6546	18 07 22	−23 17.8	7.3	−1.4	830	7.18	0.72	15	150	2	7.60
NGC6613(M18)	18 19 54	−17 06	14.1	−1.0	1200	6.84	0.42	5	20	2	7.50
NGC6633	18 27 42	6 34	36.1	8.3	292	7.77	0.17	20	30	3	8.80
Ser 1[a]	18 30 00	1 14	31.6	5.3	310	7.74					6.00
IC4725(M25)	18 31 47	−19 06.9	13.7	−4.4	560	7.46	0.48	30	30	4	7.95
IC4756	18 38 54	5 26	36.4	5.3	289	7.77	0.20	40	80	5	8.92
NGC6709	18 51 30	10 20	42.2	4.7	950	7.33	0.32	15	40	3	7.89
NGC6716	18 54 34	−19 54.1	15.4	−9.6	550	7.48	0.17	10	20	2	8.00
Ber 82	19 11 24	13 04	46.8	1.6	980	7.36	1.01	3	20		7.85
NGC6811	19 38 12	46 34	79.4	11.9	900	7.89	0.14	15	70	4	8.73
NGC6866	20 03 42	44 00	79.5	6.9	1200	7.87	0.13	15	80	2	8.36
Ros 5	20 10 00	33 46	71.4	0.3	300	7.91	0.06	50	15		
NGC6885	20 12 00	26 29	65.5	−4.1	590	7.77	0.07	10	30	2	
Dol 6	20 20 48	41 23	78.9	2.7	980	7.87	3.00	6	12		6.00
Ber 87	20 21 36	37 22	75.7	0.3	946	7.82	2.00	10	30		6.60
S 106[a]	20 27 24	37 23	76.4	−0.6	600	7.88					6.30
NGC6997[c]	20 56 30	44 37	85.5	−0.5	500	7.98	0.61	8	40		
Col 428	21 03 12	44 35	86.2	−1.4	480	7.98			10	20		
NGC7031	21 07 18	50 50	91.3	2.3	1000	8.08	0.79	15	50	2	7.75
NGC7039	21 11 12	45 39	88.0	−1.7	675	8.00	0.07	15	50	3	8.10
Bas 14	21 21 12	44 49	88.6	−3.6	1030	8.04	0.62	5	15	3	
NGC7063	21 24 30	36 30	83.1	−9.9	635	7.95	0.08	9	12	3	8.15
NGC7086	21 30 30	51 35	94.4	0.2	1200	8.18	0.66	12	50	3	7.93
NGC7092(M39)	21 32 12	48 27	92.5	−2.3	263	8.02	0.01	30	30	1	8.48
IC1396	21 39 00	57 30	99.3	3.7	800	8.17	0.58	90	50	1	6.00
IC5146	21 53 30	47 16	94.4	−5.5	1000	8.14	0.66	20	20	2	8.36
NGC7160	21 53 48	62 36	104.0	6.4	750	8.21	0.36	5	12	3	6.85
NGC7209	22 05 12	46 30	95.5	−7.3	900	8.13	0.20	15	25	2	8.48
NGC7243	22 15 18	49 53	98.9	−5.6	880	8.18	0.18	30	40	2	8.03
NGC7261	22 20 24	58 05	104.0	0.9	900	8.26	0.96	6	30	3	7.60
NGC7686	23 30 12	49 07	109.5	−11.6	1000	8.38		15	20		
NGC7762	23 49 54	68 01	117.2	5.8	800	8.39	0.88	15	40	1	9.26

Notes

[a]Clusters added to [1] list: NGC1333 (Lada et al. 1996, *AJ*, **111**, 1964); LHA 101 (Barsony et al. 1991, *ApJ*, **379**, 221); NGC2024, NGC2068, and NGC2071 (Lada et al. 1991, *ApJ*, **371**, 171); S 269 (Eiroa et al. 1995, *A&A*, **303**, 87); AM 2 (Ortolani et al. 1995, *A&A*, **300**, 726); 092-SC18 (Kassis et al. 1996, *AJ*, **111**, 820); NGC3576 (Persi et al. 1994, *A&A*, **282**, 474); ρ Oph (Williams et al. 1995, *ApJ*, **454**, 144); GM 24 (Tapia et al. 1991, *A&A*, **242**, 388); Ser 1 (Eiroa et al. 1992, *A&A*, **262**, 468); S 106 (Hodapp et al. 1991, *AJ*, **102**, 1108).

[b]Trapezium = NGC1976 and NGC2632 = Praesepe.

[c]Cluster name changed from that used in [1]. NGC2409 is Boc 4 and NGC6997 is NGC6996.

Table 22.3. *Young galactic open clusters.*

(1)	(2)	(3)	(4)	(5)	(6)	(7)	(8)	(9)	(10)	(11)	(12)
	RA	Dec.									
Cluster	(h m s)	(° ′)	ℓ	b	Dist.	R_{GC}	Red.	Dia.	Mem.	Rich.	Age
Ber 59	0 02 10	67 25.2	118.3	4.9	880	8.45		10	40		6.30
NGC637	1 43 04	64 02.2	128.6	1.7	2880	10.05	0.65	3	20	3	6.30
Ber 7	1 54 12	62 22.2	130.1	0.4	2580	9.86	0.80	4	18		6.30
IC1805	2 32 42	61 27.4	134.7	0.9	2340	9.79	0.87	20	40	3	6.30
IC1848	2 51 11	60 24.1	137.2	0.9	2200	9.73	0.63	18	10	3	6.00
NGC1333[a]	3 29 01	31 19	158.3	−20.5	320	8.28	2.00				6.00
IC348	3 44 36	32 18	160.4	−17.7	320	8.29	1.50	8	20	1	6.48
LHA 101[a]	4 30 12	35 15	165.4	−9.0	800	8.77					6.00
NGC1893	5 22 46	33 25.2	173.6	−1.7	4400	12.38	0.53	25	60	4	6.40
NGC2024[a]	5 41 42	−1 54	206.5	−16.4	400	8.35					6.00
NGC2068[a]	5 46 42	0 06	205.3	−14.3	400	8.35					6.00
NGC2071[a]	5 47 12	0 19	205.2	−14.1	400	8.35					6.00
NGC2175	6 09 39	20 29.3	190.2	0.4	2200	10.17	0.50	22	60	1	6.00
S 269[a]	6 14 36	13 50	196.5	−1.9	4000	11.89					6.30
NGC2244	6 32 19	4 51.4	206.4	−2.0	1880	9.72	0.49	30	100	2	6.30
Col 121	6 54 12	−24 24.6	235.2	−10.3	1170	8.71	0.01	90	20	1	6.18
NGC2367	7 20 09	−21 52.8	235.6	−3.8	2000	9.27	0.33	5	30	1	6.00
NGC2384	7 25 10	−21 01.2	235.4	−2.4	2000	9.28	0.30	5	15	1	6.00
Boc 5	7 31 47	−16 59.8	232.6	0.7	2290	9.57	0.63	5	12	1	6.00
Boc 6	7 32 00	−19 26	234.8	−0.2	4000	10.81	0.70	10	40	1	6.00
Boc 15	7 40 16	−33 32.6	248.0	−5.5	3767	10.03	0.55	3	12	1	6.00
Haf 18	7 52 39	−26 23.0	243.1	0.4	6900	12.71	0.50	2	15	1	6.00
Rup 44	7 58 51	−28 35.0	245.8	0.5	4600	10.74	0.67	10	40	1	6.00
Rup 55	8 12 27	−32 35.1	250.7	0.8	4400	10.33	0.52	6	12	2	6.00
Mar 18	9 00 32	−48 58.9	269.2	−1.8	1600	8.18	0.74	5	30	1	6.00
Wes 2(NGC3247)[b]	10 24 02	−57 45.6	284.3	−0.3	7900	9.76	1.70	2	12	1	6.48
NGC3324	10 37 23	−58 37.4	286.2	−0.2	3300	7.76	0.43	5		2	6.34
Col 228	10 42 04	−59 55.2	287.4	−1.0	2100	7.64	0.64	14		2	6.00
Tru 14	10 43 56	−59 32.8	287.4	−0.6	3240	7.68	0.50	5		3	6.48
Col 232	10 44 59	−59 33.4	287.5	−0.5	3240	7.67	0.50	4			6.48
Tru 16(η Car)	10 44 58	−59 43.0	287.6	−0.7	3240	7.67	0.50	10		2	6.48
Boc 11	10 47 12	−60 05.8	288.0	−0.9	3470	7.67	0.59	22	20	1	6.48
NGC3503(Pis 17)[b]	11 01 17	−59 50.7	289.5	0.1	4200	7.70	0.49	1		1	6.00
Col 240	11 11 40	−60 18.6	290.9	0.2	2559	7.48	0.41	20	30	1	6.00
NGC3576[a]	11 11 33	−61 21.8	291.3	−0.8	2400	7.49	3.00				6.00
NGC3603	11 15 08	−61 15.6	291.6	−0.5	7200	8.57	1.44	4	30	2	6.48
Col 272	13 30 26	−61 19.0	307.6	1.2	2900	6.64	0.87	10	40	3	6.30
ρ Oph[a]	16 26 54	−24 35	353.0	16.7	160	7.85					6.00
NGC6193	16 41 20	−48 45.8	336.7	−1.6	1410	6.73	0.49	14		1	6.48
NGC6200	16 44 07	−47 27.8	338.0	−1.1	2400	5.84	0.60	15	40	1	6.00
Lyn 14	16 55 04	−45 14.4	340.9	−1.1	2300	5.88	1.43	3	15	1	6.00
GM 24[a]	17 17 00	−36 21	350.5	1.0	2000	6.04					6.00
Pis 24	17 25 32	−34 24.5	353.1	0.6	2100	5.92	1.66	5	15	1	6.00
Boc 14	18 01 59	−23 41.8	6.4	−0.5	1150	6.86	1.62	2	10	1	6.00
NGC6530(M8)	18 04 31	−24 21.6	6.1	−1.3	1800	6.21	0.28	14		3	6.48
NGC6611(M16)	18 18 48	−13 45	17.0	0.8	2020	6.10	0.90	6		1	6.30
NGC6618(M17)	18 20 48	−16 09	15.1	−0.7	2000	6.09		25	40		6.00
Ser 1[a]	18 30 00	1 14	31.6	5.3	310	7.74					6.00
Biu 2	20 09 12	35 29	72.8	1.4	1500	7.69	0.41	20	10	1	6.00
Ber 86	20 20 24	38 42	76.7	1.3	1900	7.78	0.80	7	30		6.40
Dol 6	20 20 48	41 23	78.9	2.7	980	7.87	3.00	6	12		6.00
S 106[a]	20 27 24	37 23	76.4	−0.6	600	7.88					6.30
IC1396	21 39 00	57 30	99.3	3.7	800	8.17	0.58	90	50	1	6.00
NGC7380	22 47 00	58 06	107.1	−0.9	3700	9.75	0.64	20	40	3	6.30

Table 22.4. *Old galactic open clusters.*

(1)	(2)	(3)	(4)	(5)	(6)	(7)	(8)	(9)	(10)	(11)	(12)
	RA	Dec.									
Cluster	(h m s)	(° ′)	ℓ	b	Dist.	R_{GC}	Red.	Dia.	Mem.	Rich.	Age
Kin 15	0 33 07	61 51.2	120.8	−0.9	2880	9.79	0.46	3	12		9.48
NGC188	0 47 30	85 14.5	122.8	22.5	1680	8.94	0.12	15	120	5	9.81
Kin 2	0 50 57	58 11.5	122.9	−4.7	5700	12.07	0.31	4	40		9.78
NGC559	1 29 30	63 18.0	127.2	0.8	900	8.57	0.90	7	60	4	9.10
IC166	1 52 22	61 51.3	130.1	−0.2	3300	10.43	0.77	8	120	4	9.18
NGC752	1 57 48	37 41	137.2	−23.4	440	8.30	0.03	75	60	2	9.26
Ber 66	3 04 05	58 44.5	139.4	0.2	5200	12.42	1.23	4	30		9.54
NGC1193	3 05 56	44 23.0	146.8	−12.2	4800	12.20	0.12	3	40		9.69
IC361	4 18 51	58 15.0	147.5	5.7	2559	10.24	0.55	7	60	3	9.10
NGC1817	5 12 27	16 41.0	186.1	−13.1	1750	9.70	0.33	20	60	4	9.08
Ber 17	5 20 30	30 34.7	175.6	−3.7	4400	12.38	0.50	8	100		9.90
Ber 19	5 24 03	29 34.2	176.9	−3.6	4000	11.99	0.40	4	40	2	9.58
Ber 20	5 32 37	0 11.3	203.5	−17.3	8400	15.69	0.12	2	20		9.78
Ber 21	5 51 45	21 48.7	186.8	−2.5	8700	16.66	0.70	6	40	1	9.45
NGC2112	5 53 46	0 24.6	205.9	−12.6	750	8.66	0.60	18	50		9.60
Ber 22	5 58 27	7 45.4	199.8	−8.1	6000	13.74	0.62	2	20		9.48
NGC2141	6 02 56	10 26.8	198.1	−5.8	4200	12.04	0.35	10	100	5	9.40
NGC2158	6 07 26	24 05.8	186.6	1.8	4800	12.78	0.55	5		5	9.08
NGC2204	6 15 35	−18 39.8	226.0	−16.1	3100	10.29	0.08	10	80	5	9.34
NGC2243	6 29 35	−31 17.2	239.5	−18.0	3750	10.28	0.01	5	100	5	9.59
NGC2236	6 29 40	6 49.8	204.4	−1.7	3400	11.18	0.44	8	50	4	9.10
Tru 5	6 36 31	9 28.9	202.9	1.0	2400	10.25	0.67	15	150	5	9.10
Ber 29	6 53 04	16 55.7	198.0	8.0	10500	18.17	0.21	2	20		9.60
Biu 7(Ber 31)	6 57 37	8 17.3	206.3	5.1	5200	12.85	0.13	4	30		9.90
Biu 8(Ber 32)	6 58 07	6 25.9	207.9	4.4	3100	10.83	0.16	5	70		9.78
Tom 2	7 03 05	−20 49.2	232.8	−6.9	6200	12.70	0.30	3	50		9.38
Mel 66	7 26 23	−47 40.3	259.6	−14.3	4475	9.76	0.16	15	200	5	9.65
NGC2420	7 38 24	21 34.4	198.1	19.6	2410	10.18	0.04	6	100	5	9.38
AM 2^a(068-SC07)	7 38 46	−33 50.6	248.1	−5.9	12400	17.02	0.44	3			9.65
Ber 39	7 46 50	−4 40.5	223.5	10.1	4340	11.48	0.10	8	120		9.81
NGC2506	8 00 02	−10 46.2	230.6	9.9	2750	9.94	0.09	12	150	5	9.53
Pis 3	8 31 22	−38 39.3	257.9	0.5	1300	8.37	1.35	6	50		9.30
NGC2682(M67)	8 51 24	11 49.0	215.6	31.7	830	8.58	0.03	25	200	5	9.65
NGC2818	9 16 11	−36 37.7	262.0	8.6	2300	8.62	0.18	8	40	4	9.04
092-SC18^a	10 14 58	−64 36.8	287.1	−6.7	9500	10.42	0.26	5			9.70
NGC3680	11 25 39	−43 14.1	286.8	16.9	850	7.80	0.08	7	30	2	9.28
Har 6(Col 261)	12 37 57	−68 22.4	301.7	−5.5	2300	7.07	0.30	9	100		9.95
NGC5822	15 04 21	−54 23.8	321.6	3.6	740	7.43	0.15	35	150	4	9.15
IC4651	17 24 49	−49 56.3	340.1	−7.9	780	7.28	0.14	10	80	4	9.28
Bas 5	17 52 27	−30 05.8	359.8	−1.9	850	7.15	0.33	5	30	2	9.10
Tru 31	17 59 48	−28 09.6	2.3	−2.3	1000	7.00	0.42	5	25		9.10
NGC6791	19 20 48	37 51	70.0	11.0	3750	7.58	0.17	10	300	5	9.90
NGC6802	19 30 36	20 17	55.3	0.9	1840	7.12	0.80	5	50	3	9.18
NGC6819	19 41 18	40 11	74.0	8.5	2200	7.69	0.45	5		5	9.49
NGC6939	20 31 24	60 38	95.9	12.3	1250	8.22	0.48	10	80	4	9.34
Ber 54	21 03 12	40 28	83.1	−4.1	2300	8.05	0.77	4	30		9.60
IC1369	21 12 12	47 44	89.6	−0.4	1500	8.13	0.50	5	40	4	9.10
NGC7044	21 13 00	42 29	85.9	−4.1	3500	8.49	0.65	7	60		9.18
NGC7142	21 45 54	65 48	105.4	9.4	1900	8.69	0.41	12	100	4	9.70
Kin 11	23 47 48	68 37	117.2	6.5	2190	9.20	1.00	6	50		9.70
NGC7762	23 49 54	68 01	117.2	5.8	800	8.39	0.88	15	40	1	9.26
NGC7789	23 57 00	56 43	115.5	−5.4	2200	9.16	0.35	25	300	5	9.23

22.2   GLOBULAR CLUSTERS IN THE MILKY WAY

Globular star clusters are the old, populous clusters found throughout the halo and spheroid of our Galaxy. According to current stellar models, all of them are older than 10 Gyr, and have lower heavy-element abundances than the Sun by factors anywhere from 2 to 200. Figures 22.1–22.5 show the ranges covered by some important properties. In Table 22.5, basic data are summarized for 146 objects generally accepted to be Milky Way globular clusters, following the catalogs of [31] and [32]. Some globular clusters may yet be discovered, especially near the galactic center at low latitude where the foreground reddening is extremely high, but the possible numbers of these remain quite uncertain. However, recent discussions [31, 33] suggest that the currently known clusters may now represent 80%–90% of the true total in the Milky Way.

Table 22.5 provides basic parameters for known globular clusters in the Milky Way. The literature survey is complete to 1996. The available measurements for individual clusters are constantly improving, and readers should refer to the dynamic catalog of [32] for more comprehensive and recent listings.

Column (1) shows the most often used cluster catalog number (usually NGC), and any other common name. The source list for cluster names and coordinates is [31].

Columns (2) and (3) show the right ascension and declination (J2000). The typical uncertainty of cluster center is 5 arcsec.

Columns (4) and (5) show the galactic latitude and longitude in degrees.

Column (6) lists the apparent V magnitude of the horizontal branch, measured from individual color-magnitude studies of each cluster, or (in a few cases) from the mean magnitude of the RR Lyrae stars or the brightest red giants. Measurement uncertainties in V_{HB} differ widely but are typically ±0.1 mag.; values thought to be more uncertain than about 0.3 mag. are marked with colons.

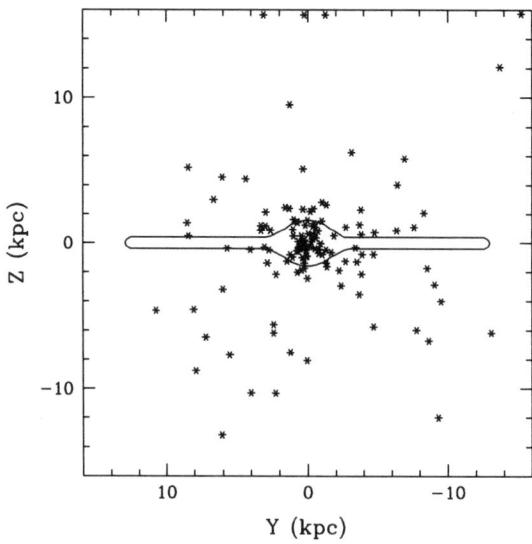

Figure 22.1. Spatial distribution of the globular clusters in the Milky Way Galaxy, projected onto the YZ-plane. (In this graph, we are looking toward the center of the Galaxy along the X-axis, i.e., along the line joining the Sun and the galactic center.) The disk and central bulge of the Milky Way are drawn in schematically. The globular clusters form a roughly spherical distribution through the galactic halo, but with a high central concentration: most clusters are within a few kiloparsecs of the galactic center. Only a few remote clusters lie outside the borders of this plot.

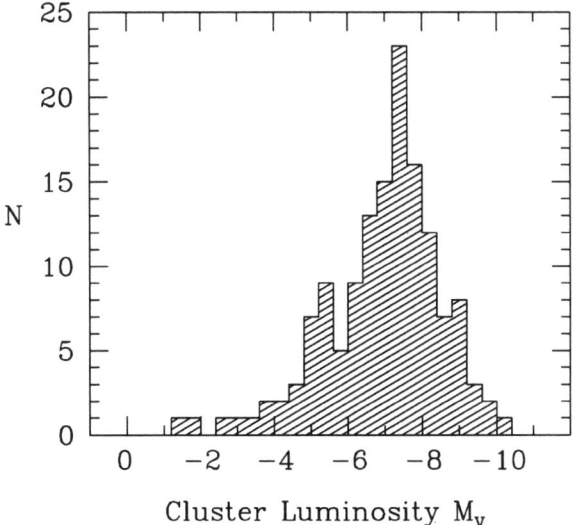

Figure 22.2. Histogram of globular-cluster luminosities M_V. Here N is the number of Milky Way globular clusters per 0.4-magnitude bin. The distribution is strongly peaked at a characteristic absolute magnitude $M_V \simeq -7.2$, with an extended tail to the faint end. These very low-luminosity objects are the sparse Palomar-type clusters found preferentially in the outer halo.

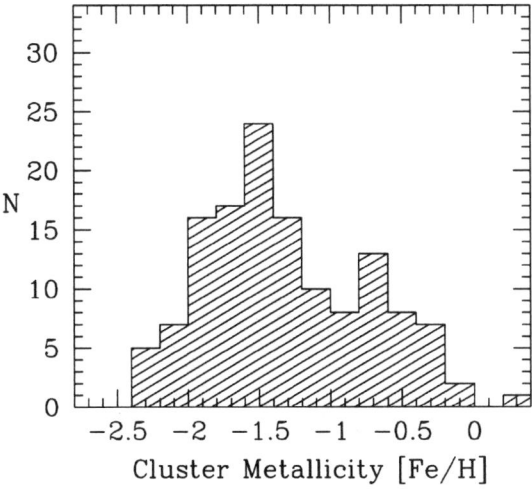

Figure 22.3. Histogram of globular-cluster metallicity [Fe/H] in the Milky Way. This distribution is strongly bimodal, with peaks at [Fe/H] $\simeq -1.5$ and -0.6. Almost all the "metal-rich" clusters ([Fe/H] > -0.8) lie in the central galactic bulge region and form a distinct subpopulation with overall disk-like kinematics; see [34].

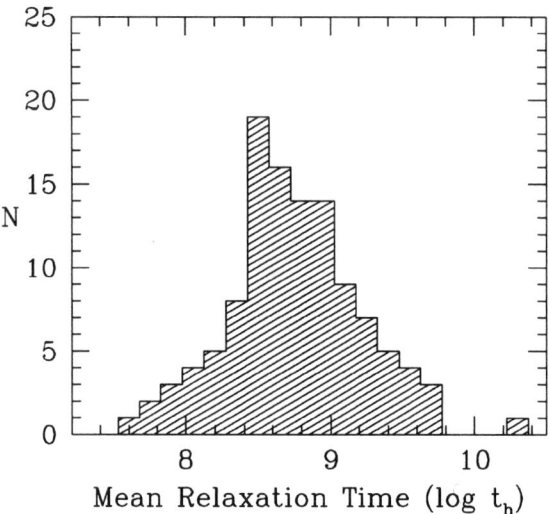

Figure 22.4. Histogram of internal relaxation times for the Milky Way globular clusters. Here t_h (in years) is the dynamical relaxation time due to normal two-body encounters, at the cluster half-mass radius r_h. The majority of clusters have characteristic relaxation times t_h in the range 2×10^8–2×10^9 yr, implying that the inner parts of most globular clusters must be near-Maxwellian relaxed systems.

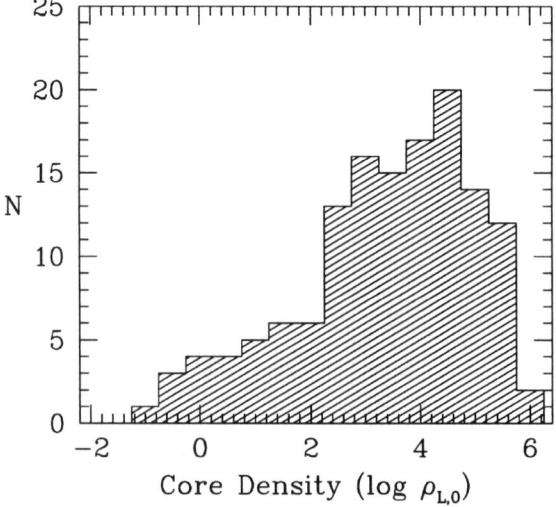

Figure 22.5. Histogram of central luminosity density $\rho_{L,0}$ (in units of solar luminosities per cubic parsec) for the Milky Way globular clusters. The central densities of most clusters fall in the broad range 10^2–$10^6 L_\odot/\text{pc}^3$, with a peak near $10\,000 L_\odot/\text{pc}^3$. See [35] for a general discussion of calculation techniques with references.

Column (7) gives the foreground reddening from [34, 36, 37], and the individual color-magnitude studies. The measurement uncertainty in $E(B - V)$ increases in proportion to the reddening and is typically 15%.

Column (8) gives the heavy-element abundance (metallicity [Fe/H]) of the cluster. Major sources are [34, 38]. Many additional data have been taken from numerous more recent (1986–1993) spectroscopic studies of the giant stars for individual clusters, plus estimates from the individual color magnitude diagram studies. The typical uncertainty in [Fe/H] is ±0.10.

Column (9) gives the integrated spectral type (ST) of the cluster, from [39]. The internal precision is typically one spectral subclass.

Column (10) gives the heliocentric radial velocity in $km\,s^{-1}$. Sources are [40–43], plus numerous other papers published in the interval 1982–1995 for smaller lists of clusters. See [32] for a complete bibliography. Measurement uncertainties in v_r differ widely, from less than $1\ km\,s^{-1}$ up to more than $30\ km\,s^{-1}$ depending on the method and epoch of measurement.

Column (11) gives the cluster number, repeated from column (1).

Column (12) shows the intrinsic distance modulus, calculated from the horizontal-branch magnitude V_{HB}, foreground reddening, and metallicity: $(m - M)_0 = V_{HB} - M_V(HB) - A_V$, where we adopt $A_V = 3.1E(B - V)$ for the ratio of total to selective absorption. The assumed distance scale for the absolute magnitude of the horizontal branch (or RR Lyrae stars) is $M_V(HB) = 0.2[Fe/H] + 1.0$, following both observational and theoretical calibrations [44–46].

Column (13) gives the distance of the cluster from the galactic center, in kiloparsecs. We adopt $R_0 = 8.0$ kpc [29] for the distance of the Sun from the galactic center.

Column (14) gives the integrated V magnitude of the cluster. Sources of integrated photometry include [37, 47, 48]; for a few of the sparser clusters, a variety of individual color-magnitude or luminosity-function studies were used to calculate an integrated magnitude. The typical measurement uncertainty is ±0.15 mag.

Column (15) gives the absolute visual magnitude of the cluster, $M_V = V - (m - M)_V$.

Column (16) shows the integrated apparent color index $(B - V)$ of the cluster (not corrected for reddening) from [49–51]. Uncertainties in the integrated colors are typically ±0.02 mag.

Column (17) gives the half-mass radius r_h, in arc minutes, from [48, 52]. For a few faint or sparse clusters, we have estimated the half-mass radius from the core radius r_c and the correlation between the cluster central concentration c (see below) and the ratio (r_h/r_c). The r_h values have typical uncertainties of 10%.

Column (18) shows the core radius r_c, in arc minutes, For some clusters that have undergone core collapse, r_c is strictly not well defined.

Column (19) gives the central concentration parameter, $c = \log(r_t/r_c)$, where r_t is the cluster tidal radius and r_c the core radius, from [52]. Typical uncertainties in c are ±0.2. Clusters believed to have undergone core collapse are indicated by "c" [52].

Column (20) gives the logarithm (base 10) of the dynamical relaxation time, in years, at the half-mass radius r_h, from [35]. These values are indicative time scales only and may be uncertain by up to a factor of 2 for individual clusters; see the discussion in [35].

Column (21) shows the central surface brightness in V magnitudes per square arc second from [37, 52]. These are uncertain at typically ±0.15 mag. (colons if more than ±0.4 mag.).

Column (22) gives the horizontal-branch population ratio (HBR) $(B - R)/(B + V + R)$, where B is the number of HB stars bluer than the RR Lyrae region, V is the number of RR Lyrae stars, and R is the number of HB stars redder than the RR Lyrae region. Sources are [45, 53–55]. The typical uncertainty is ±0.05.

Column (23) gives the specific frequency of RR Lyrae stars, from [56]. S_{RR} is defined as the number of RR Lyrae variables in the cluster per unit $(M_V = -7.5)$ cluster luminosity.

Table 22.5. *Galactic globular clusters.*[a]

(1)	(2)	(3)	(4)	(5)	(6)	(7)	(8)	(9)	(10)
	RA	Dec.							
Cluster	(h m s)	(° ′ ″)	ℓ	b	V_{HB}	Red.	[Fe/H]	ST	v_r
NGC104(47 Tuc)	00 24 05.2	−72 04 51	305.9	−44.9	14.06	0.04	−0.76	G4	−19
NGC288	00 52 47.5	−26 35 24	152.3	−89.4	15.38	0.03	−1.24		−46
NGC362	01 03 14.3	−70 50 54	301.5	−46.2	15.43	0.05	−1.16	F9	223
NGC1261	03 12 15.3	−55 13 01	270.5	−52.1	16.70	0.02	−1.35	F7	53
Pal 1	03 33 23.0	+79 34 50	130.1	19.0	16.13	.15	−0.80		
AM 1(E1)	03 55 02.7	−49 36 52	258.4	−48.5	20.96	0.00	−1.80		116
Eridanus	04 24 44.5	−21 11 13	218.1	−41.3	20.23	0.02	−1.46		−24
Pal 2	04 46 05.9	+31 22 51	170.5	−9.1	20.70:	1.36			−133:
NGC1851	05 14 06.3	−40 02 50	244.5	−35.0	16.15	0.02	−1.26	F7	321
NGC1904(M79)	05 24 10.6	−24 31 27	227.2	−29.4	16.15	0.01	−1.54	F5	207
NGC2298	06 48 59.2	−36 00 19	245.6	−16.0	16.11	0.13	−1.85	F5	149
NGC2419	07 38 08.5	+38 52 55	180.4	25.2	20.45	0.03	−2.12	F5	−20
Pyxis	09 07 57.8	−37 13 17	261.3	7.0	19.25	0.25	−1.20		
NGC2808	09 12 02.6	−64 51 47	282.2	−11.3	16.19	0.23	−1.37	F7	100
E 3	09 20 59.3	−77 16 57	292.3	−19.0	14.80:	0.30	−0.80		
Pal 3	10 05 31.4	+00 04 17	240.1	41.9	20.45	0.03	−1.66		83
NGC3201	10 17 36.8	−46 24 40	277.2	8.6	14.80	0.21	−1.45	F6	494
Pal 4	11 29 16.8	+28 58 25	202.3	71.8	20.65	0.01	−1.48		75
NGC4147	12 10 06.2	+18 32 31	252.9	77.2	17.01	0.02	−1.83	F2/3	183
NGC4372	12 25 45.4	−72 39 33	301.0	−9.9	15.30	0.45	−2.09	F5	73
Rup 106	12 38 40.2	−51 09 01	300.9	11.7	17.85	0.21	−1.69		−44
NGC4590(M68)	12 39 28.0	−26 44 34	299.6	36.1	15.68	0.04	−2.06	F2/3	−95
NGC4833	12 59 35.0	−70 52 29	303.6	−8.0	15.45	0.33	−1.79	F3	201
NGC5024(M53)	13 12 55.3	+18 10 09	333.0	79.8	16.90	0.01	−2.07	F6	−79
NGC5053	13 16 27.0	+17 41 53	335.7	78.9	16.65	0.03	−2.29		43
NGC5139(ω Cen)	13 26 45.9	−47 28 37	309.1	15.0	14.53	0.12	−1.57	F5	232
NGC5272(M3)	13 42 11.2	+28 22 32	42.2	78.7	15.65	0.01	−1.57	F6	−147
NGC5286	13 46 26.5	−51 22 24	311.6	10.6	16.65	0.24	−1.79	F5	54
AM 4	13 55 50.1	−27 10 22	320.1	33.5	18.00:	0.04	−2.00		
NGC5466	14 05 27.3	+28 32 04	42.2	73.6	16.62	0.00	−2.22		108
NGC5634	14 29 37.3	−05 58 35	342.2	49.3	17.75	0.05	−1.82	F3/4	−45
NGC5694	14 39 36.5	−26 32 18	331.1	30.4	18.50	0.09	−1.86	F4	−146
IC4499	15 00 18.5	−82 12 49	307.4	−20.5	17.65	0.24	−1.62		
NGC5824	15 03 58.5	−33 04 04	332.6	22.1	18.60	0.14	−1.85	F4	−38
Pal 5	15 16 05.3	−00 06 41	0.9	45.9	17.51	0.03	−1.38		−55
NGC5897	15 17 24.5	−21 00 37	342.9	30.3	16.35	0.08	−1.80	F7	102
NGC5904(M5)	15 18 33.8	+02 04 58	3.9	46.8	15.06	0.03	−1.33	F7	52
NGC5927	15 28 00.5	−50 40 22	326.6	4.9	16.60	0.47	−0.37	G2	−100
NGC5946	15 35 28.5	−50 39 34	327.6	4.2	17.80:	0.55	−1.38	F7/8	119
BH 176	15 39 07.3	−50 03 02	328.4	4.3	19.00:	0.77			
NGC5986	15 46 03.5	−37 47 10	337.0	13.3	16.50	0.27	−1.67	F5	92
Lyngå 7	16 11 03.0	−55 18 52	328.8	−2.8	17.25	0.73	−0.62		8
Pal 14(AvdB)	16 11 04.9	+14 57 29	28.8	42.2	20.04	0.04	−1.52		77
NGC6093(M80)	16 17 02.5	−22 58 30	352.7	19.5	15.86	0.18	−1.62	F6	7
NGC6121(M4)	16 23 35.5	−26 31 31	351.0	16.0	13.45	0.36	−1.18	F8	70
NGC6101	16 25 48.6	−72 12 06	317.7	−15.8	16.60	0.04	−1.82	F5:	361
NGC6144	16 27 14.1	−26 01 29	351.9	15.7	16.60	0.32	−1.73	F5/6	189
NGC6139	16 27 40.4	−38 50 56	342.4	6.9	17.80:	0.75	−1.65	F6/7	7
Terzan 3	16 28 40.1	−35 21 13	345.1	9.2	18.80:	0.32			
NGC6171(M107)	16 32 31.9	−13 03 13	3.4	23.0	15.70	0.33	−1.04	G0	−34
1636-283(E452-SC11)	16 39 25.5	−28 23 52	351.9	12.1	16.66	0.50			
NGC6205(M13)	16 41 41.5	+36 27 37	59.0	40.9	14.90	0.02	−1.56	F6	−247
NGC6229	16 46 58.9	+47 31 40	73.6	40.3	18.00	0.01	−1.44	F7	−154
NGC6218(M12)	16 47 14.5	−01 56 52	15.7	26.3	14.90	0.17	−1.48	F8	−43
NGC6235	16 53 25.4	−22 10 38	358.9	13.5	16.70	0.36	−1.40	F9:	87
NGC6254(M10)	16 57 08.9	−04 05 58	15.1	23.1	14.65	0.28	−1.52	F3	75

Table 22.5. (*Continued.*)

(11)	(12)	(13)	(14)	(15)	(16)	(17)	(18)	(19)	(20)	(21)	(22)	(23)
Cluster	DM	R_{GC}	V	M_V	Col.	r_h	r_c	c	$\log t$	μ_V	HBR	S_{RR}
NGC104	13.09	7.3	3.95	−9.26	0.88	2.79	0.37	2.04	9.24	14.43	−0.99	0.4
NGC288	14.53	11.4	8.09	−6.54	0.65	2.22	1.42	0.96	8.99	19.95	0.98	2.4
NGC362	14.51	9.0	6.40	−8.26	0.77	0.81	0.17	1.94c:	8.43	14.88	−0.87	6.4
NGC1261	15.91	17.1	8.29	−7.68	0.72	0.75	0.39	1.27	8.81	17.65	−0.71	16.1
Pal 1	14.82	15.5	13.52	−1.77	0.96	0.48	0.32	1.50	7.53	22.16	−1.00	0.0
AM 1	20.32	117.2	15.72	−4.60	0.72	0.50	0.12	1.23	9.46	23.86	−0.93	43.4
Eridanus	19.46	83.0	14.70	−4.82	0.79	0.40	0.25	1.10		22.81	−1.00	0.0
Pal 2	15.68	21.6	13.04	−6.86	2.08	0.28	0.15	1.91	8.04	19.39		0.0
NGC1851	15.34	16.3	7.14	−8.26	0.76	0.52	0.08	2.24	8.50	14.15	−0.36	10.9
NGC1904	15.43	18.1	7.73	−7.73	0.65	0.80	0.16	1.72	8.66	16.23	0.89	5.7
NGC2298	15.08	15.4	9.29	−6.19	0.75	0.78	0.34	1.28	8.36	18.79	0.93	10.0
NGC2419	19.78	97.7	10.39	−9.48	0.66	0.73	0.35	1.40	10.28	19.83	0.86	6.0
Pyxis	17.72	37.0	12.90	−5.59		1.35	1.38	0.65			−1.00	
NGC2808	14.75	10.7	6.20	−9.26	0.92	0.76	0.26	1.77	8.77	15.17	−0.49	0.4
E 3	13.03	7.6	11.35	−2.61		2.06	1.87	0.75		23.10		0.0
Pal 3	19.69	89.9	14.26	−5.52		0.66	0.48	1.00	9.64	23.08	−0.82	18.5
NGC3201	13.44	8.8	6.75	−7.34	0.96	2.68	1.45	1.31	8.79	18.77	0.08	98.5
Pal 4	19.92	98.8	14.20	−5.75		0.54	0.55	0.78	9.54	23.54	−1.00	0.0
NGC4147	16.31	20.5	10.32	−6.06	0.59	0.43	0.10	1.80	8.37	17.63	0.55	60.5
NGC4372	13.32	6.9	7.24	−7.48	1.10	3.90	1.75	1.30	9.23	20.51	1.00	0.0
Rup 106	16.54	17.7	10.90	−6.29		1.10	1.00	0.70		21.82	−0.82	
NGC4590	14.97	9.9	7.84	−7.25	0.63	1.55	0.69	1.64	8.90	18.67	0.17	51.5
NGC4833	13.78	6.8	6.91	−7.90	0.93	2.41	1.00	1.25	8.77	18.45	0.93	12.5
NGC5024	16.28	18.6	7.61	−8.70	0.64	1.11	0.37	1.78	9.42	17.39	0.76	14.5
NGC5053	16.01	16.5	9.47	−6.64	0.65	3.50	2.25	0.82	9.59	22.19	0.52	22.1
NGC5139	13.47	6.3	3.68	−10.16	0.78	4.18	2.58	1.24	9.72	16.77		13.1
NGC5272	14.93	11.6	6.19	−8.77	0.69	1.12	0.50	1.85	9.02	16.34	0.08	80.4
NGC5286	15.26	8.6	7.34	−8.67	0.88	0.69	0.29	1.46	8.61	16.07	0.86	5.5
AM 4	17.28	24.2	15.90	−1.50		0.42	0.42	0.50		24.75		
NGC5466	16.06	16.6	9.04	−7.02	0.67	2.25	1.96	1.43	9.37	21.28	0.58	43.4
NGC5634	16.96	20.7	9.47	−7.64	0.67	0.54	0.21	1.60	8.98	17.49		6.1
NGC5694	17.59	27.5	10.17	−7.70	0.69	0.33	0.06	1.84	8.79	16.34	1.00	0.0
IC4499	16.23	14.6	9.76	−7.21	0.91	1.50	0.96	1.11	9.45	20.92	0.11	130.1
NGC5824	17.54	26.0	9.09	−8.88	0.75	0.36	0.05	2.45	9.18	15.08	0.82	7.6
Pal 5	16.69	17.2	11.75	−5.04		2.96	2.90	0.74	9.59	24.67	−0.40	48.4
NGC5897	15.46	7.3	8.53	−7.18	0.74	2.11	1.96	0.79	9.31	20.49	0.86	9.4
NGC5904	14.23	6.0	5.65	−8.68	0.72	2.11	0.40	1.87	9.13	16.05	0.38	41.6
NGC5927	14.22	4.5	8.01	−7.66	1.31	1.15	0.42	1.60	8.71	17.45:	−1.00	0.0
NGC5946	15.37	6.7	9.61	−7.47	1.29	0.69	0.08	2.50c	8.84	17.42		3.1
BH 176	15.81	8.8	14.00	−4.20						23.36	−1.00	
NGC5986	15.00	4.5	7.52	−8.31	0.90	1.05	0.63	1.22	8.78	17.56	0.97	4.7
Lyngå 7	14.11	4.2									−1.00	
Pal 14	19.22	64.9	14.74	−4.60		1.15	1.10	0.72	9.66	25.55	−1.00	
NGC6093	14.63	3.0	7.33	−7.85	0.84	0.65	0.15	1.95	8.32	15.19	0.93	4.3
NGC6101	15.84	10.6	9.16	−6.80	0.68	1.71	1.15	0.80	9.22	20.12	0.84	20.9
NGC6121	11.57	6.1	5.63	−7.06	1.03	3.65	0.83	1.59	8.64	17.88	−0.06	76.8
NGC6144	14.95	3.2	9.01	−6.94	0.96	1.62	0.94	1.55	8.94	20.27	1.00	0.0
NGC6139	14.80	3.0	8.99	−8.14	1.40	0.82	0.14	1.80	8.59	17.30		2.8
Terzan 3	17.01	17.7	12.00	−6.00		1.30	1.18	0.70		22.52		
NGC6171	13.89	3.4	7.93	−6.98	1.10	2.70	0.54	1.51	8.75	18.84	−0.73	35.6
1636-283	14.31	2.1	12.00	−3.86						20.75		
NGC6205	14.15	8.2	5.78	−8.43	0.68	1.49	0.88	1.49	8.96	16.80	0.97	2.1
NGC6218	13.67	4.3	6.70	−7.50	0.83	2.16	0.66	1.38	8.86	18.17	0.96	0.0
NGC6229	17.26	27.7	9.39	−7.90	0.70	0.37	0.13	1.61	8.90	16.99	0.25	13.9
NGC6235	14.86	2.5	9.97	−6.01	1.05	0.84	0.36	1.33	8.56	18.98	0.89	11.8
NGC6254	13.09	4.7	6.60	−7.35	0.90	1.81	0.86	1.40	8.56	17.69	0.98	1.1

Table 22.5. (*Continued.*)

(1) Cluster	(2) RA (h m s)	(3) Dec. (° ′ ″)	(4) ℓ	(5) b	(6) V_{HB}	(7) Red.	(8) [Fe/H]	(9) ST	(10) v_r
NGC6256	16 59 32.6	−37 07 17	347.8	3.3	18.20:	0.84	−0.70		
Pal 15	17 00 02.4	−00 32 31	18.9	24.3	20.00	0.40	−1.90		69
NGC6266(M62)	17 01 12.6	−30 06 44	353.6	7.3	16.30	0.47	−1.29	F9:	−68
NGC6273(M19)	17 02 37.7	−26 16 05	356.9	9.4	16.40:	0.37	−1.68	F7	129
NGC6284	17 04 28.8	−24 45 53	358.3	9.9	17.30	0.28	−1.32	F9	30
NGC6287	17 05 09.4	−22 42 29	0.1	11.0	17.00	0.59	−2.05	F5	−208
NGC6293	17 10 10.4	−26 34 54	357.6	7.8	16.50	0.39	−1.92	F3	−99
NGC6304	17 14 32.5	−29 27 44	355.8	5.4	16.25	0.52	−0.59	G3	−105
NGC6316	17 16 37.4	−28 08 24	357.2	5.8	17.78	0.55	−0.55	G2	71
NGC6341(M92)	17 17 07.3	+43 08 11	68.3	34.9	15.10	0.02	−2.33	F2	−120
NGC6325	17 17 59.2	−23 45 57	1.0	8.0	18.30	0.89	−1.17	G0	3
NGC6333(M9)	17 19 11.8	−18 30 59	5.5	10.7	16.30	0.36	−1.78	F5/6	229
NGC6342	17 21 10.2	−19 35 14	4.9	9.7	16.90	0.44	−0.65	G3/4	81
NGC6356	17 23 35.0	−17 48 47	6.7	10.2	17.50	0.29	−0.50	G3	27
NGC6355	17 23 58.6	−26 21 13	359.6	5.4	17.20	0.75	−1.50	G0	−177
NGC6352	17 25 29.2	−48 25 22	341.4	−7.2	15.13	0.21	−0.70	G4	−115
Terzan 2(HP 3)	17 27 33.4	−30 48 08	356.3	2.3	20.10	1.42	−0.25		109
NGC6366	17 27 44.3	−05 04 36	18.4	16.0	15.65	0.69	−0.82		−122
Terzan 4(HP 4)	17 30 38.9	−31 35 44	356.0	1.3			−0.94		−50
HP 1(BH 229)	17 31 05.2	−29 58 54	357.4	2.1	20.10:	1.50	−0.43		53
NGC6362	17 31 54.8	−67 02 53	325.6	−17.6	15.34	0.09	−1.06	G3	−13
Liller 1	17 33 24.5	−33 23 20	354.8	−0.2	25.23:	2.95	+0.22		52
NGC6380(Ton 1)	17 34 28.0	−39 04 09	350.2	−3.4		1.38			
Terzan 1(HP 2)	17 35 47.8	−30 28 11	357.6	1.0	19.95	1.64	−0.35		35
Ton 2(Pismis 26)	17 36 10.5	−38 33 12	350.8	−3.4	18.20:	0.91			
NGC6388	17 36 17.0	−44 44 06	345.6	−6.7	17.25	0.38	−0.60	G2	81
NGC6402(M14)	17 37 36.1	−03 14 45	21.3	14.8	17.20	0.60	−1.39	F4	−111
NGC6401	17 38 36.9	−23 54 32	3.5	4.0	17.70	0.85	−1.12	F9	−65
NGC6397	17 40 41.3	−53 40 25	338.2	−12.0	12.87	0.18	−1.91	F4	19
Pal 6	17 43 42.2	−26 13 21	2.1	1.8	19.70:	1.53	−0.10		201
NGC6426	17 44 54.7	+03 10 13	28.1	16.2	18.00:	0.35	−2.20	G1:	−162
Djorg 1	17 47 28.3	−33 03 56	356.7	−2.5	20.80:	1.70			
Terzan 5(Terzan 11)	17 48 04.9	−24 48 45	3.8	1.7	20.20:	1.87	−0.29		−94
NGC6440	17 48 52.6	−20 21 34	7.7	3.8	18.70:	1.09	−0.34	G4	−79
NGC6441	17 50 12.9	−37 03 04	353.5	−5.0	17.10	0.45	−0.53	G2	18
Terzan 6(HP 5)	17 50 46.4	−31 16 31	358.6	−2.2	21.44	2.04	−0.65		126
NGC6453	17 50 51.8	−34 35 55	355.7	−3.9	17.70:	0.61	−1.53	F8	−84
UKS 1	17 54 27.2	−24 08 43	5.1	0.8	24.14:	2.93	−1.20		
NGC6496	17 59 02.0	−44 15 54	348.0	−10.0	16.47	0.13	−0.64	G4	−98
Terzan 9	18 01 38.8	−26 50 23	3.6	−2.0	20.58:	1.75	−1.00		59
Djorg 2(E456-SC38)	18 01 49.1	−27 49 33	2.8	−2.5	19.50:	1.00			
NGC6517	18 01 50.6	−08 57 32	19.2	6.8	19.10	1.08	−1.37	F8	−40
Terzan 10	18 02 57.4	−26 04 00	4.4	−1.9	23.42:	2.60	−0.70		
NGC6522	18 03 34.1	−30 02 02	1.0	−3.9	16.40	0.50	−1.52	F7/8	−9
NGC6535	18 03 50.7	−00 17 49	27.2	10.4	15.73	0.32	−1.80	G0	−215
NGC6528	18 04 49.6	−30 03 21	1.1	−4.2	17.10	0.62	−0.17	G3	165
NGC6539	18 04 49.8	−07 35 09	20.8	6.8	18.33	1.00	−0.66	G4:	−46
NGC6540(Djorg 3)	18 06 08.6	−27 45 55	3.3	−3.3	15.30:	0.60	−1.00		
NGC6544	18 07 20.6	−24 59 51	5.8	−2.2	14.90:	0.74	−1.56	F9	−17
NGC6541	18 08 02.2	−43 42 20	349.3	−11.2	15.30	0.12	−1.83	F6	−154
NGC6553	18 09 15.6	−25 54 28	5.2	−3.0	16.60	0.84	−0.25	G4	−24
NGC6558	18 10 18.4	−31 45 49	0.2	−6.0	16.70	0.42	−1.44	F7	−144
IC1276(Pal 7)	18 10 44.2	−07 12 27	21.8	5.7	18.40:	0.92			
Terzan 12	18 12 15.8	−22 44 31	8.4	−2.1		1.57			
NGC6569	18 13 38.9	−31 49 35	0.5	−6.7	17.10:	0.56	−0.86	G1:	−28
NGC6584	18 18 37.7	−52 12 54	342.1	−16.4	16.53	0.11	−1.49	F6	223

Table 22.5. *(Continued.)*

(11)	(12)	(13)	(14)	(15)	(16)	(17)	(18)	(19)	(20)	(21)	(22)	(23)
Cluster	DM	R_{GC}	V	M_V	Col.	r_h	r_c	c	log t	μ_V	HBR	S_{RR}
NGC6256	14.74	2.0	11.29	−6.05	1.69	0.85	0.02	2.50c	8.41	17.89		
Pal 15	18.14	35.8	14.00	−5.38		1.21	1.21	0.60		24.87	1.00	0.0
NGC6266	14.10	1.9	6.45	−9.11	1.19	1.23	0.18	1.70c:	8.55	15.35	0.28	19.8
NGC6273	14.59	1.4	6.77	−8.97	1.03	1.25	0.43	1.53	9.28	16.82		0.5
NGC6284	15.70	6.1	8.83	−7.73	0.99	0.78	0.07	2.50c	8.54	16.65		3.2
NGC6287	14.58	1.6	9.35	−7.06	1.20	0.75	0.26	1.60	8.14	18.33	0.98	3.0
NGC6293	14.68	1.3	8.22	−7.66	0.96	0.91	0.05	2.50c	8.52	16.18:	0.90	3.4
NGC6304	13.76	2.5	8.22	−7.15	1.31	1.41	0.21	1.80	8.56	17.34	−1.00	0.0
NGC6316	15.19	3.1	8.43	−8.46	1.39	0.71	0.17	1.55	8.92	17.40	−1.00	
NGC6325	14.77	1.6	10.33	−7.20	1.66	0.94	0.03	2.50c	7.91	17.56		
NGC6341	14.50	9.4	6.44	−8.13	0.63	1.09	0.23	1.81	8.56	15.58	0.91	14.0
NGC6333	14.54	1.7	7.72	−7.94	0.97	0.95	0.58	1.15	8.49	17.40	0.87	6.0
NGC6342	14.67	1.7	9.66	−6.37	1.26	0.88	0.05	2.50c	8.58	17.44:	−1.00	
NGC6356	15.70	6.2	8.25	−8.35	1.13	0.74	0.23	1.54	9.11	17.09	−1.00	0.0
NGC6355	14.18	1.4	9.14	−7.36	1.48	0.87	0.05	2.50c	8.47	18.05:		
NGC6352	13.62	3.5	7.96	−6.31	1.06	2.00	0.83	1.10	8.71	18.42:	−1.00	0.0
Terzan 2	14.75	1.1	14.29	−4.86		1.52	0.03	2.50c	8.55	21.58	−1.00	
NGC6366	12.67	5.1	9.20	−5.61	1.44	2.63	1.83	0.92	8.50	21.24	−0.97	5.7
NGC6362	14.27	5.0	7.73	−6.82	0.85	2.18	1.32	1.10	8.83	19.19	−0.58	61.6
Terzan 4		8.0	16.00									
HP 1	14.54	0.5	11.59	−7.60		3.10	0.03	2.50c	9.49	21.29		
Liller 1	15.04	2.3	16.77	−7.42		0.45	0.06	2.30c:		23.15:	−1.00	0.0
NGC6380			11.31		2.01	0.75	0.34	1.55c:	7.65	19.96		
Terzan 1	13.94	1.9	15.90	−3.12		3.82	0.04	2.50c		25.09		
NGC6388	15.19	3.9	6.72	−9.65	1.17	0.67	0.12	1.70	8.70	14.55		0.7
Ton 2	14.58	1.4	12.24	−5.16		1.08	0.54	1.30		22.16		
NGC6402	14.62	3.7	7.59	−8.89	1.25	1.29	0.83	1.60	9.19	18.41	0.65	23.9
NGC6401	14.29	1.1	9.45	−7.47	1.58	1.91	0.25	1.69	9.13	18.67		22.5
NGC6397	11.69	6.1	5.73	−6.52	0.73	2.33	0.05	2.50c	8.35	15.65	0.98	0.0
Pal 6	13.98	1.8	11.55	−7.17	2.83	1.06	0.66	1.10		21.58	−1.00	
NGC6426	16.36	12.6	11.01	−6.43	1.02	0.96	0.26	1.70	8.86	20.37		32.2
Djorg 1	14.73	3.1	13.60	−6.40		1.26	0.50	1.50		23.10		
Terzan 5	13.46	3.1	13.85	−5.41	2.77	0.83	0.18	1.74		20.33:		
NGC6440	14.39	1.2	9.20	−8.57	1.97	0.58	0.13	1.70	8.15	17.02	−1.00	
NGC6441	14.81	1.7	7.15	−9.06	1.27	0.64	0.11	1.85	8.66	14.99		0.5
NGC6453	15.12	2.7	10.08	−6.93	1.31	0.37	0.07	2.50c	7.97	17.35		0.0
Terzan 6	14.25	1.0	13.85	−6.72		0.44	0.05	2.50c		20.76	−1.00	0.0
UKS 1	14.30	1.0	17.29	−6.09		0.86	0.15	2.10c:		25.52		
NGC6496	15.19	3.9	8.54	−7.06	0.98	1.87	1.05	0.70	8.46	20.10	−1.00	0.0
Terzan 9	14.36	0.8	16.00	−3.78		0.78	0.03	2.50c		23.21		
Djorg 2	15.60	5.2	9.90	−8.80		0.83	0.33	1.50		19.50		
NGC6517	15.03	3.8	10.23	−8.14	1.75	0.62	0.06	1.82	7.89	17.77		
Terzan 10	14.50	0.7	14.90	−7.66						21.28		
NGC6522	14.15	1.3	8.27	−7.43	1.21	1.04	0.05	2.50c	8.25	16.14:	0.71	5.3
NGC6535	14.10	3.9	10.47	−4.62	0.94	0.77	0.42	1.30	8.05	20.22:	1.00	0.0
NGC6528	14.21	1.2	9.60	−6.53	1.53	0.43	0.09	2.29	7.73	16.91	−1.00	0.0
NGC6539	14.36	3.0	9.33	−8.13	1.83	1.67	0.54	1.60	8.38	19.31	−1.00	0.0
NGC6540	12.64	4.6	9.30	−5.20		0.24	0.03	2.50c		16.40		
NGC6544	11.92	5.6	7.77	−6.44	1.46	1.77	0.05	1.63c:	7.82	17.13:	1.00	
NGC6541	14.29	2.2	6.30	−8.37	0.76	1.19	0.30	2.00c:	8.58	15.58	1.00	0.0
NGC6553	13.05	4.0	8.06	−7.59	1.73	1.55	0.55	1.17	8.33	18.15	−1.00	0.9
NGC6558	14.69	1.1	9.26	−6.73	1.11	1.61	0.03	2.50c	9.13	17.08:		6.1
IC1276	14.75	3.4	10.34	−7.26	1.76	2.35	1.08	1.29	9.15	21.66		1.2
Terzan 12			15.63			0.84	0.83	0.57		23.75		
NGC6569	14.54	0.9	8.55	−7.72	1.34	1.33	0.37	1.27	8.57	18.08		8.2
NGC6584	15.49	6.1	8.27	−7.56	0.76	0.80	0.59	1.20	8.53	17.79	−0.15	40.8

Table 22.5. (Continued.)

(1)	(2)	(3)	(4)	(5)	(6)	(7)	(8)	(9)	(10)
	RA	Dec.							
Cluster	(h m s)	(° ′ ″)	ℓ	b	V_{HB}	Red.	[Fe/H]	ST	v_r
NGC6624	18 23 40.5	−30 21 40	2.8	−7.9	16.11	0.27	−0.42	G4/5	54
NGC6626(M28)	18 24 32.9	−24 52 12	7.8	−5.6	15.70	0.41	−1.45	F8	16
NGC6638	18 30 56.2	−25 29 47	7.9	−7.2	16.50	0.40	−0.99	G0	10
NGC6637(M69)	18 31 23.2	−32 20 53	1.7	−10.3	15.85	0.17	−0.71	G2/3	39
NGC6642	18 31 54.3	−23 28 35	9.8	−6.4	16.30:	0.40	−1.35	F8	−57
NGC6652	18 35 45.7	−32 59 25	1.5	−11.4	15.85	0.09	−0.96	G3	−112
NGC6656(M22)	18 36 24.2	−23 54 12	9.9	−7.6	14.15	0.34	−1.64	F5	−149
Pal 8	18 41 29.9	−19 49 33	14.1	−6.8	17.27	0.33	−0.48		−43
NGC6681(M70)	18 43 12.7	−32 17 31	2.9	−12.5	15.60	0.07	−1.51	F5	219
NGC6712	18 53 04.3	−08 42 22	25.4	−4.3	16.25	0.46	−1.01	F9	−108
NGC6715(M54)	18 55 03.3	−30 28 42	5.6	−14.1	18.17	0.15	−1.59	F7/8	142
NGC6717(Pal 9)	18 55 06.2	−22 42 03	12.9	−10.9	15.50	0.21	−1.32	F6	2
NGC6723	18 59 33.2	−36 37 54	0.1	−17.3	15.46	0.04	−1.12	F9	−82
NGC6749	19 05 15.3	+01 54 03	36.2	−2.2	19.20:	1.12			
NGC6752	19 10 51.8	−59 58 55	336.5	−25.6	13.70	0.04	−1.61	F4/5	−27
NGC6760	19 11 12.1	+01 01 50	36.1	−3.9	17.50	0.78	−0.52	G5	−27
NGC6779(M56)	19 16 35.5	+30 11 05	62.7	8.3	16.16	0.20	−1.94	F5	−136
Terzan 7	19 17 43.7	−34 39 27	3.4	−20.1	17.76	0.06	−0.62		166
Pal 10	19 18 02.1	+18 34 18	52.4	2.7	19.50:	1.15			
Arp 2	19 28 44.1	−30 21 14	8.5	−20.8	18.30	0.11	−1.76		115
NGC6809(M55)	19 39 59.4	−30 57 44	8.8	−23.3	14.40	0.07	−1.81	F4	175
Terzan 8	19 41 45.0	−34 00 01	5.8	−24.6	18.03	0.14	−1.87		130
Pal 11	19 45 14.4	−08 00 26	31.8	−15.6	17.35	0.34	−0.39		−68
NGC6838(M71)	19 53 46.1	+18 46 42	56.7	−4.6	14.44	0.25	−0.73	G1	−23
NGC6864(M75)	20 06 04.8	−21 55 17	20.3	−25.7	17.47	0.16	−1.32	F9	−189
NGC6934	20 34 11.6	+07 24 15	52.1	−18.9	16.82	0.11	−1.54	F7/8	−411
NGC6981(M72)	20 53 27.9	−12 32 13	35.2	−32.7	16.90	0.05	−1.54	F7	−289
NGC7006	21 01 29.5	+16 11 15	63.8	−19.4	18.80	0.05	−1.68	F6	−378
NGC7078(M15)	21 29 58.3	+12 10 01	65.0	−27.3	15.83	0.09	−2.22	F3/4	−107
NGC7089(M2)	21 33 29.3	−00 49 23	53.4	−35.8	16.05	0.05	−1.62	F4	−7
NGC7099(M30)	21 40 22.0	−23 10 45	27.2	−46.8	15.10	0.03	−2.12	F3	−184
Pal 12	21 46 38.8	−21 15 03	30.5	−47.7	17.13	0.02	−0.93		28:
Pal 13	23 06 44.4	+12 46 19	87.1	−42.7	17.74	0.05	−1.79		−28
NGC7492	23 08 26.7	−15 36 41	53.4	−63.5	17.63	0.00	−1.51		−188

22.3 GLOBULAR CLUSTERS IN OTHER GALAXIES

Globular clusters can be found in all large galaxies and many small ones, although in quite different numbers. The *specific frequency* S_N [57, 58] is defined as the number of globular clusters per unit galaxy luminosity; if N_{GC} is the total cluster population and M_V^T the visual absolute magnitude of the entire galaxy, then $S_N \equiv N_{GC}10^{0.4(M_V^T+15)}$. S_N is listed in Table 22.6 for 73 galaxies within which old halo globular-cluster populations have been found.

Typically, the "old" globular clusters (that is, objects with ages $\sim$ 10 Gyr or more) make up 0.1%–1% of the total light of a galaxy. They are usually present in much larger numbers within elliptical galaxies than in spiral or irregular galaxies, although even for a given type of galaxy there are large differences in specific frequency that depend roughly—although not exclusively—on environment: E galaxies in rich surroundings like the Virgo or Fornax clusters tend to contain more globular clusters than otherwise-similar galaxies in the field or in small groups. Some very luminous cD-type galaxies, like M87 in Virgo, contain extremely large populations of globular clusters (tens of thousands in one galaxy), extending out into their halos to radial distances of 100 kpc and more.

Table 22.5. *(Continued.)*

(11)	(12)	(13)	(14)	(15)	(16)	(17)	(18)	(19)	(20)	(21)	(22)	(23)
Cluster	DM	R_{GC}	V	M_V	Col.	r_h	r_c	c	log t	μ_V	HBR	S_{RR}
NGC6624	14.36	1.3	7.87	−7.32	1.11	0.82	0.06	2.50c	8.50	15.42	−1.00	1.2
NGC6626	13.72	2.7	6.79	−8.20	1.08	1.56	0.24	1.67	8.78	16.08	0.90	6.8
NGC6638	14.46	1.5	9.02	−6.68	1.15	0.66	0.26	1.40	8.02	17.27	−0.30	42.6
NGC6637	14.47	1.5	7.64	−7.35	1.01	0.83	0.34	1.39	8.69	16.83	−1.00	0.0
NGC6642	14.33	1.7	9.13	−6.44	1.11	0.73	0.10	1.99	8.25	16.68		
NGC6652	14.76	2.0	8.62	−6.42	0.94	0.65	0.07	1.80	8.50	16.31	−1.00	0.0
NGC6656	12.42	5.1	5.10	−8.38	0.98	3.26	1.42	1.31	8.86	17.32	0.91	8.0
Pal 8	15.34	4.5	11.02	−5.35	1.22	0.57	0.40	1.53	9.02	19.83	−1.00	
NGC6681	14.69	2.0	7.87	−7.03	0.72	0.93	0.03	2.50c	8.40	15.28	0.93	3.1
NGC6712	14.03	3.6	8.10	−7.35	1.17	1.37	0.94	0.90	8.64	18.65	−0.64	13.8
NGC6715	17.02	17.8	7.60	−9.89	0.85	0.49	0.11	1.84	9.03	14.82	0.87	8.0
NGC6717	14.11	2.5	9.28	−5.48	1.00	0.68	0.08	2.07c:	8.14	16.48		6.4
NGC6723	14.56	2.4	7.01	−7.67	0.75	1.61	0.94	1.05	8.94	17.92	−0.08	24.7
NGC6749	14.93	5.7	12.44	−5.96	2.14	0.56	0.47	0.75		21.54:		
NGC6752	12.90	5.3	5.40	−7.62	0.66	2.34	0.17	2.50c	8.65	15.20	1.00	0.0
NGC6760	14.19	4.8	8.88	−7.72	1.66	2.18	0.33	1.59	8.80	18.79	−1.00	0.0
Terzan 7	16.70	14.6	12.00	−4.88		0.97	0.61	1.08		20.65	−1.00	0.0
NGC6779	14.93	9.3	8.27	−7.28	0.86	1.16	0.37	1.37	8.70	18.06	0.98	2.5
Pal 10	15.14	8.6	13.22	−5.48		1.52	0.80	1.56		22.03		
Arp 2	17.31	21.8	12.30	−5.35	0.86	1.91	1.59	0.90	9.46	24.21	0.86	0.0
NGC6809	13.54	4.0	6.32	−7.44	0.72	2.89	2.83	0.76	8.89	19.13	0.87	10.5
Terzan 8	16.97	17.9	12.40	−5.00		1.00	1.00	0.60		22.82		
Pal 11	15.37	7.0	9.80	−6.63	1.27	1.49	2.00	0.69	9.06	20.33		0.0
NGC6838	12.81	6.7	8.19	−5.40	1.09	1.65	0.63	1.15	8.10	19.22	−1.00	0.0
NGC6864	16.24	11.7	8.52	−8.21	0.87	0.47	0.10	1.88	8.72	15.55	−0.42	6.2
NGC6934	15.79	11.7	8.83	−7.30	0.77	0.60	0.25	1.53	8.64	17.26	0.28	60.2
NGC6981	16.05	12.2	9.27	−6.94	0.72	0.88	0.54	1.23	8.93	18.90	0.17	63.8
NGC7006	17.98	36.9	10.56	−7.58	0.75	0.38	0.24	1.42	8.97	18.50	−0.28	63.4
NGC7078	14.99	10.2	6.20	−9.07	0.68	1.06	0.07	2.50c	9.01	14.21	0.67	26.3
NGC7089	15.22	10.0	6.47	−8.90	0.66	0.93	0.34	1.80	9.05	15.92	0.96	4.7
NGC7099	14.43	6.9	7.19	−7.33	0.60	1.15	0.06	2.50c	8.55	15.28	0.89	11.7
Pal 12	16.25	14.7	11.99	−4.33	1.07	1.28	0.20	1.94		20.59:	−1.00	0.0
Pal 13	16.94	25.5	13.80	−3.30	0.72	0.46	0.48	0.66		23.36:	−0.20	191.8
NGC7492	16.93	23.5	11.29	−5.64	0.42	1.22	0.83	1.00	9.11	21.33	0.81	16.6

Note

[a] A colon after a number indicates that it is very uncertain.

Measurements of cluster radial velocities, spectroscopic abundance properties, and spatial distributions help to constrain the dynamical and chemical evolution of the early formation stages of large galaxies [57].

Most of the more remote galaxies in this table have no other published information on their globular clusters except for total numbers (S_N values). However, many of the nearby ones have multicolor photometry or even spectroscopic data available, and the sources cited should be consulted for these data. The material in this table is based on the literature survey given in [57], updated to 1996 and recalculated where necessary. The fiducial distance to the Virgo cluster, which provides a fundamental basis of comparison for several other galaxies in the list, is adopted here to be $(m - M)_V = 31.0$. Changes to distances for some individual galaxies with more recent measurements have also been incorporated.

Column (1) gives the NGC number or name of the galaxy. The first 11 galaxies listed are members of the Local Group and are referred to by their most common names. The galaxies outside the Local Group are listed in order of right ascension.

Column (2) gives the group or cluster (if any) of which the galaxy is a member.

Column (3) gives the Hubble type of the galaxy (elliptical E, spiral S, irregular I).

Column (4) shows the radial velocity of the cluster, or (if the galaxy is isolated or very nearby) the individual radial velocity of the galaxy.

Column (5) shows the apparent visual distance modulus ($H_0 = 75$ km s^{-1} Mpc^{-1} for galaxies more distant than Virgo, or individual distance calibrations for nearby ones).

Column (6) shows the absolute visual integrated magnitude of the galaxy.

Column (7) gives the total number of globular clusters estimated to be in the given galaxy (see [57] for method of calculation).

Column (8) gives the specific frequency (number of globular clusters per unit galaxy luminosity) as defined above.

Column (9) gives the reference source for the globular-cluster data.

Table 22.6. *Cluster populations in other galaxies.*[a]

(1) NGC	(2) Cluster	(3) Type	(4) V_0	(5) $(m - M)_V$	(6) M_V^T	(7) N_{GC}	(8) S_N	(9) Reference
Galaxy	Local	Sbc			−21.3	160 ± 20	0.5 ± 0.1	[1]
LMC[b]	Local	Sm[c]	12	18.64	−18.6	15:	0.5:	[1]
SMC[b]	Local	Im[c]	−30	18.92	−16.9	2 ± 1	0.4 ± 0.2	[1]
Fornax	Local	dE0[c]	−51	20.70	−13.7	5 ± 1	17 ± 7	[2]
Sgr	Local	dE[c]	140	17.44	−13:	3 ± 1	19:	[3]
147	Local	dE5[c]	89	24.45	−15.0	4 ± 1	4.0 ± 1.0	[1]
185	Local	dE3[c]	39	24.40	−15.2	8 ± 2	6.5 ± 1.6	[1]
205	Local	dE5[c]	−1	24.49	−16.5	9 ± 1	2.3 ± 0.3	[1]
221	Local	E2	35	24.54	−16.3	< 3	< 0.8	[1]
M31	Local	Sb	−59	24.54	−21.7	350 ± 100	0.7 ± 0.2	[1]
M33	Local	Scd	3	24.64	−19.1	30:	0.6:	[1]
55	Sculptor	Sm[c]	106	25.75	−19.0	2	> 0.1:	[1]
253	Sculptor	Sc	260	26.8	−20.4	22:	0.2 ± 0.1	[1]
524	CfA13	S0	2 600	32.7	−22.1	4430 ± 950	6.4 ± 1.4	[1]
720	T52-9	E5	1 660	31.4	−21.2	660 ± 190	2.2 ± 0.9	[4]
1052	Cetus	E4	1 400	31.3:	−20.8	600 ± 70	3.0 ± 0.4	[1]
1275	Perseus	Ep/cD	5 450	34.9	−23.3	7750 ± 2520	4.3 ± 1.4	[5]
1374	Fornax	E0	1 390	31.0	−19.7	440 ± 200	5.7 ± 2.6	[1]
1379	Fornax	E0	1 390	31.0	−19.9	380 ± 100	4.1 ± 1.1	[1]
1387	Fornax	S0	1 390	31.0	−20.2	385 ± 80	3.3 ± 0.7	[1]
1399	Fornax	E1/cD	1 390	31.0	−21.1	5340 ± 1780	19 ± 6	[6]
1404	Fornax	E1	1 390	31.0	−20.7	950 ± 140	4.3 ± 0.7	[7]
1549	Doradus	E0	1 010	30.7:	−20.8	165 ± 60	0.8 ± 0.3	[1]
1553	Doradus	S0	1 010	30.7:	−21.2	600 ± 134	2.0 ± 0.5	[1]
2403	M81	Scd	260	27.64	−19.5	8:	> 0.1:	[1]
3031	M81	Sab	260	27.8	−21.2	210 ± 30	0.7 ± 0.1	[8]
2683		Sb	370	29.75	−20.8	310 ± 100	1.7 ± 0.5	[1]
3109		Im	130	26.35	−17.3	20:	2.4:	[1]
3115		S0	460	30.2	−21.1	630 ± 150	2.3 ± 0.5	[1]
3115DW1		dE1, N[c]	460	30.2	−17.7	59 ± 23	4.9 ± 1.9	[9]
3226	CfA58	E2	1 210	31.0:	−19.6	480 ± 170	7.0 ± 2.4	[1]
3311	A1060	E0/cD	3 420	33.4	−22.3	$12 400 \pm 5000$	15 ± 6	[10]
3377	Leo	E5	630	30.00	−19.8	210 ± 50	2.6 ± 0.6	[1]
3379	Leo	E1	630	30.00	−20.7	260 ± 140	1.3 ± 0.7	[1]
3384	Leo	S0	630	30.00	−20.0	110 ± 60	1.1 ± 0.5	[1]
3557	T31-10	E3	2 560	33.0	−22.6	400 ± 300	0.4 ± 0.3	[1]
3607	CfA77	S0	1 090	31.0:	−21.0	1000 ± 700	4.0 ± 2.8	[1]
3842	A1367	E3	6 500	34.7	−23.1	$14 000 \pm 2500$	7.7 ± 1.4	[11]

Table 22.6. (*Continued.*)

(1) NGC	(2) Cluster	(3) Type	(4) V_0	(5) $(m-M)_V$	(6) M_V^T	(7) N_{GC}	(8) S_N	(9) Reference
3923	T22-4	E3	1 590	31.9	−22.1	4300 ± 1000	6.4 ± 1.5	[12]
4073	MKW4	E1/cD	6 090	34.6	−23.1	8290 ± 460	4.8 ± 0.3	[13]
4278	Coma I	E1	910	30.0	−19.8	1050 ± 120	12.3 ± 1.4	[1]
4565	Coma I	Sb	1 210	30.0	−21.6	180 ± 45	0.4 ± 0.1	[14]
4216	Virgo	Sb	1 080	31.0	−21.7	620 ± 310	1.3 ± 0.7	[1]
4340	Virgo	SB0	1 080	31.0	−20.0	775 ± 310	8.0 ± 3.2	[1]
4365	Virgo W	E2	1 080	31.4	−21.8	2500 ± 200	5.0 ± 0.4	[15]
4374	Virgo	E1	1 080	31.0	−21.7	3040 ± 400	6.6 ± 0.9	[1]
4406	Virgo	E3	1 080	31.0	−21.8	3350 ± 400	6.3 ± 0.8	[1]
4472	Virgo	E2	1 080	31.0	−22.6	6300 ± 1900	5.6 ± 1.7	[1]
4486	Virgo	E0/cD	1 080	31.0	−22.4	13 000 ± 500	13.9 ± 0.5	[16]
4494	Virgo	E0	1 290	30.8	−21.0	1400 ± 350	5.4 ± 1.3	[14]
4526	Virgo	S0	1 080	31.0	−21.4	2700 ± 400	7.7 ± 1.2	[1]
4552	Virgo	E0	1 080	31.0	−21.2	2400:	8:	[17]
4564	Virgo	E6	1 080	31.0	−20.1	1000 ± 300	10.0 ± 3.0	[1]
4569	Virgo	Sab	1 080	31.0	−21.8	930 ± 300	1.8 ± 0.6	[1]
4621	Virgo	E5	1 080	31.0	−21.2	1900 ± 400	6.3 ± 1.2	[1]
4636	Virgo	E0	1 080	31.0	−21.4	3060 ± 270	8.4 ± 0.8	[18]
4649	Virgo	E2	1 080	31.0	−22.2	5100 ± 160	6.9 ± 0.2	[1]
4697	Virgo	E6	1 080	31.0	−21.7	2500 ± 600	5.0 ± 1.3	[1]
4594	Virgo SE	Sa	840	29.9	−22.3	1900 ± 620	2.3 ± 0.7	[19]
5018	T11-0	E4p	2 800	33.2:	−22.4	800 ± 300	0.9 ± 0.3	[20]
5170	T11-18	Sb	1 350	31.7:	−21.2	390 ± 140	1.2 ± 0.4	[1]
4874	A1656	E0	6 950	34.9	−23.0	22 600 ± 2700	14.3 ± 1.7	[21]
4881	A1656	E0	6 950	34.9	−21.6	390 ± 40	1.0 ± 0.1	[22]
4889	A1656	E4	6 950	34.9	−23.5	17 300 ± 3000	6.9 ± 1.2	[21]
5128	Centaurus	E0p	320	28.25	−22.0	1700 ± 400	2.6 ± 0.6	[1]
5629	AWM3	E/cD	4 600	34.0	−21.7	< 2000	< 5	[13]
5813	CfA150	E1	1 810	32.3	−21.6	3000 ± 750	6.9 ± 1.7	[23]
5846	CfA150	E0	1 810	32.3	−22.1	3120 ± 1850	4.5 ± 2.7	[1]
UGC9799[b]	A2052	E/cD	10 440	35.9	−23.4	48 000 ± 16 000	21 ± 7	[24]
UGC9958[b]	A2107	E/cD	12 630	36.3	−23.4	27 000 ± 13 000	12 ± 5.6	[24]
6166	A2199	E2/cD	9 080	35.6	−23.6	10 000 ± 5000	4 ± 2	[1]
7768	A2666	E2/cD	7 950	35.3	−22.9	4050 ± 2600	2.8 ± 1.8	[24]
7814		Sab	1 110	31.0	−21.0	500 ± 160	2.6 ± 0.8	[25]

Notes

[a] A colon after a number indicates that it is very uncertain.

[b] LMC: Large Magellanic Cloud. SMC: Small Magellanic Cloud. UGC: Uppsala Galaxy Catalog.

[c] Sm, Im: Magellanic-type spiral or irregular. dE: dwarf elliptical.

References

1. Harris, W.E. 1991, *ARA&A*, **29**, 543
2. Zinn, R. 1993, in *The Globular Cluster–Galaxy Connection*, ASP Conference Series, edited by G.H. Smith and J.P. Brodie (ASP, San Francisco), Vol. 48, p. 302
3. Da Costa, G.S., & Armandroff, T.E. 1995, *AJ*, **109**, 2533
4. Kissler-Patig, M., Richtler, T., & Hilker, M. 1996, *A&A*, **308**, 704
5. Kaisler, D., Harris, W.E., Crabtree, D.R., & Richer, H.B. 1996, *AJ*, **111**, 2224
6. Bridges, T.J., Hanes, D.A., & Harris, W.E. 1991, *AJ*, **101**, 469
7. Richtler, T., Grebel, E.K., Domgorgen, H., Hilker, M., & Kissler, M. 1992, *A&A*, **264**, 25
8. Perelmuter, J.-M., & Racine, R. 1995, *AJ*, **109**, 1055
9. Durrell, P.R., McLaughlin, D.E., Harris, W.E., & Hanes, D.A. 1996, *ApJ*, **463**, 543
10. McLaughlin, D.E., Secker, J., Harris, W.E., & Geisler, D. 1995, *AJ*, **109**, 1033
11. Butterworth, S.T., & Harris, W.E. 1992, *AJ*, **103**, 1828
12. Zepf, S.E., Geisler, D., & Ashman, K.M. 1994, *ApJ*, **435**, L117
13. Bridges, T.J., & Hanes, D.A. 1994, *ApJ*, **431**, 625

14. Fleming, D.E.B., Harris, W.E., Pritchet, C.J., & Hanes, D.A. 1995, *AJ*, **109**, 1044
15. Harris, W.E., Allwright, J.W.B., Pritchet, C.J., & van den Bergh, S. 1991, *ApJS*, **76**, 115
16. McLaughlin, D.E., Harris, W.E., & Hanes, D.A. 1994, *ApJ*, **422**, 486
17. Ajhar, E.A., Blakeslee, J.P., & Tonry, J.L. 1994, *AJ*, **108**, 2087
18. Kissler, M., Richtler, T., Held, E.V., Grebel, E.K., Wagner, S., & Capaccioli, M. 1994, *A&A*, **287**, 463
19. Bridges, T.J., & Hanes, D.A. 1992, *AJ*, **103**, 800
20. Hilker, M., & Kissler-Patig, M. 1996, *A&A*, **314**, 357
21. Blakeslee, J.P., & Tonry, J.L. 1995, *ApJ*, **442**, 579
22. Baum, W.A. et al. 1995, *AJ*, **110**, 2537
23. Hopp, U., Wagner, S.J., & Richtler, T. 1995, *A&A*, **296**, 633
24. Harris, W.E., Pritchet, C.J., & McClure, R.D. 1995, *ApJ*, **441**, 120
25. Bothun, G.D., Harris, H.C., & Hesser, J.E. 1992, *PASP*, **104**, 1220

REFERENCES

1. Lyngå, G. 1987, *Catalogue of Open Cluster Data*, 5th ed. (computer-based catalog distributed by NASA Data Center)
2. Lauberts, A. 1982, *The ESO Uppsala Survey of the ESO (B) Atlas* (ESO, Garching)
3. Archinal, B.A. 1993, *The "Non-Existent" Star Clusters of the RNGC*. (Webb Society, Portsmouth)
4. McCaughrean, M. 1993, in *Massive Stars: Their Lives in the Interstellar Medium*, edited by J.P. Cassinelli and E.B. Churchwell (ASP, Provo), p. 80
5. Snell, R., Carpenter, J., Schloerb, F.P., & Skrutskie, M. 1993, in *Massive Stars: Their Lives in the Interstellar Medium*, edited by J.P. Cassinelli and E.B. Churchwell (ASP, Provo), p. 138
6. Carraro, G., & Chiosi, C. 1994, *A&A*, **287**, 761
7. Phelps, R.L., Janes, K.A., & Montgomery, K.A. 1994, *AJ*, **107**, 1079
8. Meynet, G., Mermilliod, J.C., & Maeder, A. 1993, *A&AS*, **98**, 477
9. Janes, K.A., Tilley, C., & Lyngå, G. 1988, *AJ*, **95**, 771
10. van den Bergh, S. 1981, *PASP*, **93**, 712
11. Battinelli, P., & Capuzzo-Dolcetta, R. 1991, *MNRAS*, **249**, 76
12. Janes, K., & Adler, D. 1982, *ApJS*, **49**, 425
13. Friel, E.D., & Janes, K.A. 1993, *A&A*, **267**, 75
14. Thogersen, E.N., Friel, E.D., & Fallon, B.V. 1993, *PASP*, **105**, 1253
15. Janes, K.A., & Phelps, R.L. 1994, *AJ*, **108**, 1773
16. Hron, J. 1987, *A&A*, **176**, 34
17. Scott, J.E., Friel, E.D., & Janes, K.A. 1995, *AJ*, **109**, 1706
18. Gunn, J.E., Griffin, R.F., Griffin, R.E.M., & Zimmerman, B.A. 1988, *AJ*, **96**, 198
19. Mathieu, R.D. 1984, *ApJ*, **284**, 643
20. Hodge, P.W. 1988, *PASP*, **100**, 1051
21. Kontizas, M., Morgan, D.H., Hatzidimitriou, D., & Kontizas, E. 1990, *A&AS*, **84**, 527
22. Elson, R.A.W., & Fall, S.M. 1985, *ApJ*, **299**, 211
23. Mateo, M. 1988, in IAU Symposium 126, *Globular Cluster Systems in Galaxies*, edited by J.E. Grindlay and A.G.D. Philip (Kluwer Academic, Dordrecht), p. 557
24. Hodge, P.W. 1986, *PASP*, **98**, 1113
25. Hodge, P.W. 1987, *PASP*, **99**, 724
26. *In Hipparcos Venice '97*, 1997, edited by B. Battrick (ESA, Noordwijk)
27. Skiff, B.A. 1998, *Sky&Tel*, **95**, 65
28. Skiff, B.A. 1996, unpublished work in progress
29. Reid, M.J. 1993, *ARA&A*, **31**, 345
30. Lyngå, G. 1983, A Reclassification of Open Clusters from Survey Charts, quoted by Lyngå [1]
31. Djorgovski, S., & Meylan, G. 1993, in ASP Conf. Series 50, *Structure and Dynamics of Globular Clusters*, edited by G. Meylan and S. Djorgovski (ASP, San Francisco), p. 325
32. Harris, W.E. 1996, *AJ*, **112**, 1487
33. Racine, R., & Harris, W.E. 1989, *AJ*, **98**, 1609
34. Zinn, R. 1985, *ApJ*, **293**, 424
35. Djorgovski, S. 1993, in ASP Conf. Series 50, *Structure and Dynamics of Globular Clusters*, edited by G. Meylan and S. Djorgovski (ASP, San Francisco), p. 325
36. Reed, B.C., Hesser, J.E., & Shawl, S.J. 1988, *PASP*, **100**, 545
37. Webbink, R.F. 1985, in IAU Symposium 113, *Dynamics of Star Clusters*, edited by J. Goodman and P. Hut (Reidel, Dordrecht), p. 541
38. Armandroff, T.E., & Zinn, R. 1988, *AJ*, **96**, 92
39. Hesser, J.E., & Shawl, S.J. 1985, *PASP*, **97**, 465
40. Hesser, J.E., Shawl, S.J., & Meyer, J.E. 1986, *PASP*, **98**, 403
41. Pryor, C., & Meylan, G. 1993, in ASP Conf. Series 50, *Structure and Dynamics of Globular Clusters*, edited by G. Meylan and S. Djorgovski (ASP, San Francisco), p. 357
42. Webbink, R.F. 1981, *ApJS*, **45**, 259
43. Zinn, R., & West, M.J. 1984, *ApJS*, **55**, 45
44. Carney, B.W., Storm, J., & Jones, R.V. 1992, *ApJ*, **386**, 663
45. Lee, Y.-W. 1990, *ApJ*, **363**, 159
46. Skillen, I., Fernley, J.A., Stobie, R.S., & Jameson, R.F. 1993, *MNRAS*, **265**, 30
47. Peterson, C.J., & Reed, B.C. 1987, *PASP*, **99**, 20
48. van den Bergh, S., Morbey, C., & Pazder, J. 1991, *ApJ*, **375**, 594
49. Peterson, C.J. 1986, *PASP*, **98**, 1258

50. Peterson, C.J. 1993, in ASP Conf. Series 50, *Structure and Dynamics of Globular Clusters*, edited by G. Meylan and S. Djorgovski (ASP, San Francisco), p. 337

51. Reed, B.C. 1985, *PASP*, **97**, 120

52. Trager, S.C., Djorgovski, S., & King, I.R. 1993, in ASP Conf. Series 50, *Structure and Dynamics of Globular Clusters*, edited by G. Meylan and S. Djorgovski (ASP, San Francisco), p. 347. Also see Trager, S.C., King, I.R., & Djorgovski, S. 1995, *AJ*, **109**, 218

53. Fusi Pecci, F., Ferraro, F.R., Bellazini, M., Djorgovski, S., Piotto, G., & Buonanno, R. 1993, *AJ*, **105**, 1145

54. Lee, Y.-W., Demarque, P., & Zinn, R. 1994, *ApJ*, **423**, 248

55. Preston, G.W., Shectman, S.A., & Beers, T.C. 1991, *ApJ*, **375**, 121

56. Suntzeff, N.B., Kinman, T.D., & Kraft, R.P. 1991, *ApJ*, **367**, 528

57. Harris, W.E. 1991, *ARA&A*, **29**, 543

58. Harris, W.E., & van den Bergh, S. 1981, *AJ*, **86**, 1627

Chapter 23

Milky Way and Galaxies

Virginia Trimble

23.1 MILKY WAY GALAXY

23.1.1 Overall Properties

Details of the location of the Sun in the Galaxy are given in Table 23.1.

Table 23.1. *Milky Way parameters.*

	IAU 1964 Standard [1]	IAU 1985 Standard [2]	Other values [3–5]
Solar galactocentric distance (kpc)	10	8.5 ± 1.1	7.0 ± 0.8
LSR rotation speed (km/s)	250	220	190
Oort constant A (km/s kpc)	+15	$+14.4 \pm 1.2$	+13.0
Oort constant B (km/s kpc)	−10	-12.0 ± 2.8	−13.0
$A + B = 0.0$ for a flat rotation curve			

References
1. Allen, C.W. *Astrophysical Quantities*, 3rd ed. (Athlone Press, London), p. 283
2. Kerr, F.J., & Lynden-Bell, D. 1986, *MNRAS*, **221**, 1023
3. Gilmore, G., & Carswell, B., editors, 1987, *The Galaxy* (Reidel, Dordrecht)
4. Allen, C., & Santillan, A. 1991, *Rev. Mexicana Astron. Af.*, **22**, 255
5. Merrifield, M. 1992, *AJ*, **103**, 1442

At the solar circle, the rotation period is 2.4×10^8 yr; the epicyclic period is 1.7×10^8 yr; the vertical period is 6.2×10^7 yr [1].

The total luminosity in the B color band, the integrated color, and the total bolometric luminosity are [1–4]:

$$L_B = (2.3 \pm 0.6) \times 10^{10} L_\odot, \qquad B - V = 0.68 - 0.80,$$

$$L_{bol} = 3.6 \times 10^{10} L_\odot \quad \text{of which 1/3 is in infrared radiation.}$$

The extent of luminous disk, r, is 20–25 kpc [5].
The total mass and $\mathcal{M}/L_B$ are [6]:

$$
\begin{aligned}
\mathcal{M} &= 9.5 \times 10^{10} \mathcal{M}_\odot, & \mathcal{M}/L_B &= 4.2 & \text{to} \quad R_0 &= 8.5 \text{ kpc}, \\
\mathcal{M} &= 4 \times 10^{11} \mathcal{M}_\odot, & \mathcal{M}/L_B &= 18 & \text{to} \quad R &= 35 \text{ kpc}, \\
\mathcal{M} &= 13 \times 10^{11} \mathcal{M}_\odot, & \mathcal{M}/L_B &= 56 & \text{to} \quad R &= 230 \text{ kpc}.
\end{aligned}
$$

The escape velocity [6, 7] at $R_\odot$ is 450–650 km/s, at the edge of the Galaxy is 311 km/s, and at the galactic center is 650–700 km/s.

23.1.2   Galactic Populations

Contents of the components of the Galaxy are given in Table 23.2.

23.1.3   Galactic Backgrounds

Energy densities near the Sun in magnetic fields, galactic cosmic rays, gas cloud turbulence, starlight, and microwave background are each 0.5–2 eV/cm^3.

Cosmic Rays [8]: The flux of particles with energies greater than E (eV) is $10^{20} E^{-1.74}$ particles/m^2 s sr for $E = 10$–10^7 GeV/nucleon. The grammage traversed is 3–6 g/cm^2 (from secondary to primary ratios). The confinement time is 3×10^7 yr (from radioactive nuclides).

In soft X-rays (0.25 kev), the local background is 40–70 keV/cm^2 s sr keV [9].

The ultraviolet (1400–1800 Å) flux [10] is 300–400 photons/(cm^2 s sr Å) at high latitudes and 1200 ± 200 photons/(cm^2 s sr Å) at low latitudes.

The visible light intensity, in units of 1.1×10^{-9} erg/cm^2 s sr Å [11], for sky-averaged starlight in the galactic plane is 200; toward the poles is 50, and in diffuse optical emission is 20. The light absorption in the V band $A_V = 0.11 \pm 0.02$ toward poles [4].

The infrared surface brightness of the Galaxy [4] is $10^{-4.35}$ W/m^2 sr at 100 μm, $10^{-5.35}$ W/m^2 sr at 25 μm, and $10^{-5.1}$ W/m^2 sr at 12 μm.

In the ultraviolet to infrared range, 1000 Å–100 μm, the interstellar radiation field averaged over all directions, $dI/d \ln \nu$, is 10^{-6}–10^{-7} W/m^2 sr and is smaller toward poles by a factor of 10.

Total galactic luminosities [4, 12]: $L(> 100 \text{ MeV}) = (1$–$2) \times 10^{39}$ erg/s, $L(2$–$10 \text{ keV}) = 10^{40}$ erg/s, and $L(4$–$100 \mu\text{m}) = 1.2 \times 10^{10} L_\odot = 4.8 \times 10^{43}$ erg/s.

The B band central surface brightness is 22.1 mag. mag./arcsec^2 [13].

Table 23.2. *Galactic population components.*

Component	Radial scale length (kpc)	Radial extent (kpc)	Vertical scale height or c/a^a	Density near Sun ($\mathcal{M}_\odot\,\mathrm{pc}^{-3}$)	$\mathcal{M}_{tot}$ (units of $\mathcal{M}_\odot$)	L_B^{tot} (units of $L_\odot$)	$\mathcal{M}/L$ within $2R_\odot$	Metallicity [Fe/H]	$V_{rotation}$ at $R_\odot$ (km/s)	Contents
Bulge [1, 2]	0.2	1	0.4	—	10^{10}	2×10^9	5	$+0.3 \pm 0.2$		Old stars (not population II); molecular gas
Spheroid [2–6]	2.9	≥ 40	0.6–1	0.00026	$(2\text{–}10) \times 10^9$	$(1\text{–}2) \times 10^9$	1–5	$-1.5^{+0.5}_{-3.0}$	0–50	Metal-poor globular clusters; old stars (population II)
Disk [2, 4–6]		20–25		0.15	$(3\text{–}6) \times 10^{10}$	1.9×10^{10}	4.6	-0.3 ± 0.3		Metal-rich globular clusters, old stars; hot gas
Thick disk [7–10]	4–5	20–25?	1.3 kpc	0.026	$(2\text{–}4) \times 10^9$	2×10^8	10?	-0.6 ± 0.3	180	
Thin disk [2, 4, 8, 10]	3.5–5	20–25	325 pc	0.124	$(3\text{–}6) \times 10^{10}$	1.7×10^{10}	4.5	-0.1 ± 0.2	220	Population I stars, atomic, molecular gas
Dark halo [2, 4, 5, 6]	2–3	≥ 100	1?	0.009	$(6\text{–}10) \times 10^{11}$		≥ 650		0?	Unknown

Note

^{a}The ratio of the smallest to the largest axis.

References

1. Frogel, J.A. 1988, *ARA&A*, **26**, 51
2. Bahcall, J.N. 1986, *ARA&A*, **24**, 577 and references therein, also private communication
3. Freeman, K.C. 1987, *ARA&A*, **25**, 603
4. van der Kruit, P. 1987, in IAU Symposium 139, *The Galaxy*, edited by G. Gilmore and B. Carswell (Reidel, Dordrecht)
5. Merrifield, M. 1992, *AJ*, **103**, 1442
6. Gilmore, G., & Kuiken, K. 1989, *MNRAS*, **239**, 571
7. Gilmore, G., & Reid, N. 1983, *MNRAS*, **202**, 1025
8. Gilmore, G., & Wyse, R. 1987, in IAU Symposium 139, *The Galaxy*, edited by G. Gilmore and B. Carswell (Reidel, Dordrecht)
9. van der Kruit, P. 1984, *A&A*, **140**, 470
10. Gilmore, G., Wyse, R., & Kuikjen, K. 1989, *ARA&A*, **27**, 555

23.1.4   Galactic Center [14]

The inner parsec contains $\mathcal{M} \approx 3 \times 10^6 \mathcal{M}_\odot$ in some combination of dense star cluster and a supermassive black hole, with compact sources of radio, X-ray, and infrared radiation. Other compact sources, including gamma-ray sources are nearby. The central magnetic field $B \approx 1$ mG, largely perpendicular to the disk.

23.1.5   Spiral Arms [15, 16]

The Milky Way is not a grand design spiral, with two discrete arms extending $360°$ or more. The magnetic field follows the arms with an ordered component of 2–6 μG and a comparable chaotic component. The arms that cut the radius vector through the Sun are Scutum, at 5 kpc from the galactic center; Sagittarius, at 7.5 kpc; Orion, Carina-Cygnus, at 8.8 kpc (the local arm); and Perseus, at 12.3 kpc.

23.1.6   Age [17, 18]

The age of the oldest spheroid (globular cluster) stars and onset of nucleosynthesis can be placed at 13–17 Gyr BP (before present). The oldest disk stars formed 6–13.5 Gyr ago.

23.1.7   Star Formation Rate [19]

The local rate is $(3.5–5)\mathcal{M}_\odot/\text{pc}^3$ Gyr. The global rate is $(0.8–13)\mathcal{M}_\odot/\text{yr}$ (most likely value $3\mathcal{M}_\odot/\text{yr}$).

23.1.8   Gas Content [4, 19–21]

HI: $n(\text{HI}) = 2.9 \times 10^{20} \csc b$ atoms cm^{-2} = $3.25 \mathcal{M}_\odot$ pc^{-2}, or 0.44 atoms cm^{-3} with a scale height of 130 ± 20 pc (at the solar circle).

H$_2$: $\text{H}_2 = 2\mathcal{M}_\odot$ pc^{-2} or $0.013\mathcal{M}_\odot$ pc^{-3} with a scale height of 65 ± 25 pc; the maximum $n(\text{H}_2) = 3 \times 10^{22}$ cm^{-2} at the galactic center and in the 4–6 kpc ring.

The total amount of gas is $(5.7–7.5)\mathcal{M}_\odot$ pc^{-2} = $(2–6) \times 10^9 \mathcal{M}_\odot$; $\mathcal{M}(\text{H}_2)/\mathcal{M}(\text{HI}) \approx 1$.

23.1.9   Galactic Rotation Curve

The basic shape (rapid central rise, extended flat part) has been decomposed into central bulge plus spheroid, disk, and dark halo (corona) in two rather different ways. Table 23.3 shows a version where the disk is the largest contributor between 3 and 11 kpc [22–24]. Figure 23.1 shows a version where the disk is less massive and never dominates [25, 26].

Table 23.3. *Galactic rotation* (km/sec).

R (kpc)	V (bulge + spheroid)	V (disk)	V (halo)
0.25	235.1	22.0	11.4
0.50	247.3	38.5	22.5
0.75	245.3	52.5	32.9
1.0	232.5	64.8	42.5

Table 23.3. *(Continued.)*

R (kpc)	V (bulge + spheroid)	V (disk)	V (halo)
2.0	165.6	101.4	73.3
3.0	134.8	125.1	94.3
4.0	117.2	140.5	109.0
5.0	105.3	149.9	119.9
6.0	96.6	155.2	128.3
7.0	89.8	157.5	135.0
7.7	85.8	157.7	139.0
8.0	84.3	157.5	140.5
8.3	82.8	157.2	142.0
9.0	79.7	156.0	145.2
10.0	75.7	153.5	149.2
12.0	69.4	146.4	155.8
14.0	64.4	138.2	161.0
16.0	60.4	129.9	165.3
18.0	57.1	122.1	169.0
20.0	54.2	115.0	172.1
25.0	48.6	100.6	178.5
30.0	44.5	90.2	183.5
40.0	38.6	76.5	191.1
60.0	31.5	61.7	201.2

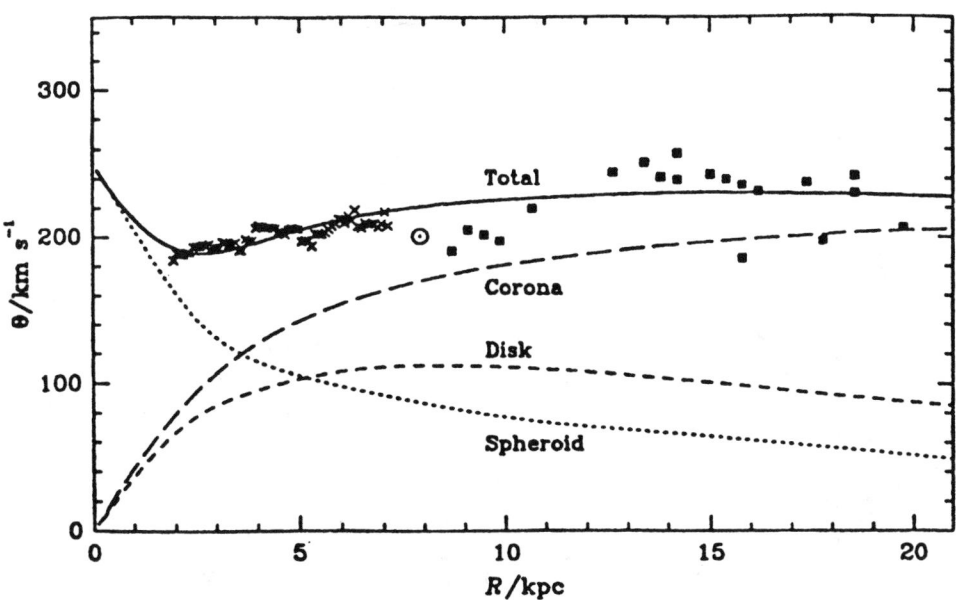

Figure 23.1. A second version of the Galaxy rotation in which the disk is nowhere the dominant component [25, 26].

23.1.10 Acceleration Perpendicular to the Galactic Disk

The high- and low-mass disks of Sec. 23.1.9 correspond to different force laws perpendicular to the disk at the solar position as presented in Figure 23.2. The data are in Table 23.4.

Table 23.4. *Force per unit mass versus height z above galaxy plane.*

z (pc)	K_z (10^{-9} cm/s^2)	K_z [(km/s)^2/pc]
0	0.00	0.00
100	2.46	0.76
200	4.17	1.29
300	5.19	1.60
400	5.81	1.79
500	6.21	1.92
600	6.50	2.01
700	6.72	2.07
800	6.90	2.13
900	7.05	2.18
1000	7.19	2.22
1100	7.31	2.26
1200	7.42	2.29
1300	7.52	2.32
1400	7.62	2.35
1500	7.71	2.38
2000	8.11	2.50
2500	8.42	2.60
3000	8.66	2.67

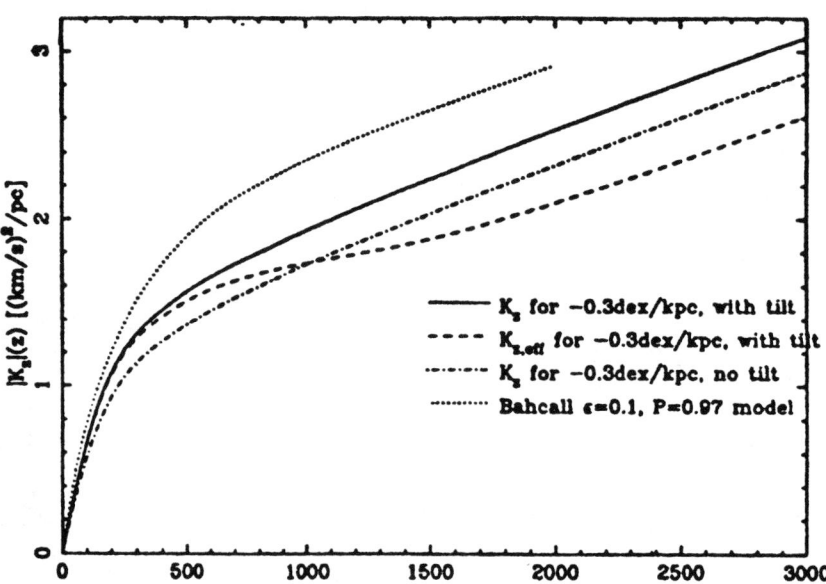

Figure 23.2. Force laws in direction perpendicular to the Galaxy disk versus height above Galaxy disk.

The corresponding mass densities are $(0.10\text{--}0.29)\mathcal{M}_\odot$ pc^{-3} in the solar neighborhood and $(48\text{--}89)\mathcal{M}_\odot$ pc^{-2} perpendicular to the plane.

The mass density and potential anywhere in the Galaxy can be calculated by summing the components that enter into these models [7, 27].

23.1.11 Galactic Coordinates

1. The old system [28] was defined by a North Galactic Pole (NGP) ($b^{\rm I} = 90°$) at

$$\text{RA} = 12^{\rm h}40^{\rm m} = 190°0, \qquad \text{declination} = +28°0 \quad (1900).$$

The ascending node on the equator, at RA $= 280°00 + 1°23T$ (T is the time in centuries after 1900.0), defines $l^{\rm I} = 0°$.

2. The new system [28–30] has been defined for Besselian year 1950 and Julian year 2000 by the coordinates, $l^{\rm II}b^{\rm II}$, of the NGP:

$$
\begin{aligned}
\text{RA} &= 12^{\rm h}49.^{\rm m}0 = 192°15', & \text{declination} &= +27°24'.0 \quad (\text{B1950}), \\
\text{RA} &= 12^{\rm h}51^{\rm m}26^{\rm s}2755 = \alpha_3, & \text{declination} &= +27°7'41''.704 = \delta_3 \\
&= 192°859\,481\,23, & &= +27°128\,251\,20 \quad (\text{J2000}).
\end{aligned}
$$

The corresponding values of $l^{\rm I}$, $b^{\rm I} = 327°41, -1°24'$.

The position of the galactic center defines $l^{\rm II} = 0°$:

$$
\begin{aligned}
\text{RA} &= 17^{\rm h}42.^{\rm m}4 = 265°36', & \text{declination} &= -28°55' \quad (\text{B1950}), \\
\text{RA} &= 17^{\rm h}45^{\rm m}37.^{\rm s}1991 = \alpha_1, & \text{declination} &= -28°56'10''.221 = \delta_1 \quad (\text{J2000}). \\
&= 266°404\,996\,25, & &= -28°936\,172\,42.
\end{aligned}
$$

The galactic longitude of the North Celestial Pole is

$$\theta = 123°00 \quad (1950), \qquad \theta = 122°931\,918\,6 = l_3 \quad (\text{J2000}).$$

The ascending node of the galactic plane on the J2000 equator has the position

$$\text{RA} = \alpha_3 + 6^{\rm h} = 282°86, \qquad l^{\rm II} = \ell_3 - 90° = 32°93.$$

The mutual angle of inclination of the two planes is $90° - \delta_3 = 62°87$.

The conversion between equatorial and new galactic coordinates for (J2000) is given approximately by [31]

$$
\begin{aligned}
\sin b^{\rm II} &= \sin\delta\cos 62°87 - \cos\delta\sin(\alpha - 282°86)\sin 62°87, \\
\cos b^{\rm II}\cos(l^{\rm II} - 32°93) &= \cos\delta\cos(\alpha - 282°86),
\end{aligned}
$$

but see [29] for a more precise formulation.

23.2 NORMAL GALAXIES

23.2.1 Classification

E	Elliptical (spheroidal component dominates), from E0 (round) to E7 (most flattened). $b/a = 1 - n/10$; e.g., $b/a = 0.3$ is E7. gE and dE denote giants and dwarfs, respectively.
S0	Lenticular (spheroid and disk, but no recent star formation).
S	Spiral (spheroid and disk with arms of young stars) from S0a (least dominant arms) through Sa, Sb, Sc, Sd (most dominant arms) to Sm (Magellanic faint spirals). SA = unbarred; SB = barred.
Irr	Irregular (little or no spheroid; young stars are not in coherent arms). The Magellanic type (intrinsically irregular) called IrrI, Im, or IBm (barred). Merger products, disturbed, and active galaxies are labeled Irr II or I0. dIrr denotes dwarf irregulars.
dSph	Dwarf spheroidals (masses and luminosities much less than ellipticals, but much less centrally condensed than globular clusters; many are companions). Probably the commonest kind of galaxy in a volume-limited sample.

Special types:

cD	Supergiant diffuse galaxies at cluster centers,
BCD	Blue compact dwarf galaxies, dIrr with very active star formation.

For additional criteria and refinements, see [32, 33]. Table 23.5 gives more details. Table 23.6 lists the local group members. Table 23.7 gives the local bright galaxies.

23.2.2 Galaxy Counts

Integral Counts

The number of galaxies per square degree brighter than m_B (for apparent magnitude $m_B = 12$–15) in the South Galactic Polar Cap [34]:

$$\log N(m) = 0.62m - 9.7.$$

The number in the North Galactic Polar Cap is larger by $\Delta \log N(m) = 0.4$ at $m_B = 12.0$, tapering to $\Delta \log N(m) = 0.0$ at $m_B = 14$. Here logarithms to the base 10 are used, as elsewhere in this chapter.

Differential Counts

The number per square degree of galaxies with magnitude in the J color band of $m = b_J \pm 0.25$ in the South Galactic Polar Cap [$b_J = B - 0.28(B - V)$] [35] is given in Table 23.8.

Table 23.5. *Representative properties of galaxies by type* [1, 2].[a]

Type	$-M_B$ [3,4]	$-M_B$, range[b] [3,4]	$B-V$[c] [5]	Diameter (kpc) Mean; range [4]	L_x [6]	L_{FIR}/L_B[d] [7]	Gas content[e] phase; amount ($M_\odot$), or fraction [8]	Specific freq. globular clusters[f] [9]	M/L (solar units) Mean; range [4]	M_{tot} ($M_\odot$) [6,10]	M/L Total [6,10]
cD^g [11]	23.0	22.5–25.0	0.93	30–1000	10^{43}		X-ray; (10^{11}–10^{12})	15 ± 3	5–15	10^{12}–10^{13}	180
E	20.0	15.5–22.0	0.91	20.5;	10^{39}–10^{42}	≪ 0.01–0.1	X-ray; (10^{8}–10^{11})		5–15	$10^{11.5}$–10^{13}	10–90
gE		?–18.0		30–60				2–6	5–15		
dE	≤ 14.0	7.6–15.5		1–5				4.8 ± 1			
dSph [12]				0.5–3					10–80		
S0	20.0	16.5–21.5	0.87	17.5; 5.5–71.7		< 0.01–0.1	X-ray; (10^{9}–10^{10}); HI+H$_2$; (10^{7}–$10^{9.5}$)	2–6	4.0; 1–7		
S					10^{38}–10^{41}					10^{10}–10^{12}	10–20
Sa[h]	20.0	16.5–21.5	0.78	16.5; 4.7–48.5		0.03–3.0	HI+H$_2$; 0.04	1.2 ± 0.2	6.5; 1–23		
Sb			0.69	22.4; 5.3–50.9			HI+H$_2$; 0.08		5.8; 1.4–17.3		
Sc	19.5	16.5–21.0	0.54	21.5; 2.8–47.5			HI+H$_2$; 0.15	0.5 ± 0.2	4.2; 1.3–13.9		
Sd	17.5	15.0–20.0	0.47	16.2; 4.2–52.3			HI+H$_2$; 0.25		4.7; 1.0–10.3		
Sm			0.37	13.6; 1.8–32.8							
Irr											
Im	15.5	≤ 14.0–18.0	0.37	8.9; 0.6–8.9	10^{37}–10^{39}	0.1–0.2	HI+H$_2$; 0.25	0.5 ± 0.2	3.1; 0.7–7.0		
dIrr							HI+H$_2$; 0.1–0.5		10–80		
BCD [13]	15.5	14.0–18.0	0.46	1–5		0.1–10.0	HI; 0.2		2–5		

Notes

[a] All quantities are given on the "short" distance scale. None of them is so precisely known that it matters whether this means $H = 85$ or $H = 100$ (km/s)/Mpc. For the "long" scale, distances and masses go up by a factor of 2, luminosities by a factor of 4, and M/L by a factor of 2.

[b] Limits containing galaxies down to 0.1 of the peak space density.

[c] Mean for type, corrected for inclination and internal reddening.

[d] Starburst S galaxies extend up to $2 \times 10^{12} L_\odot$.

[e] Most giant galaxies have $(10^9–10^{10}) M_\odot$ of gas, but it is a cool (HI + H$_2$) disk component in spirals, a hot (keV, X-ray emitting) component in cD and elliptical galaxies, and a mixture of these in S0 galaxies. Young and Scoville give ratios of the mass of disk gas to the total disk mass of the galaxies.

[f] Units are the number of globular clusters per unit of luminosity corresponding to visual absolute magnitude $M_V = -15$.

[g] Luminosity and size of cD galaxies includes the extended halo, much of which lies below the 25^m/sq." contour.

[h] Properties on intermediate lines are averages for Sa + Sb, etc.

References

1. Faber, S.M., editor, 1987, *Nearly Normal Galaxies* (Springer-Verlag, New York)
2. Kron, R.G., editor, 1989, *Evolution of the Universe of Galaxies*, Astronomical Society of the Pacific Conference Series
3. Bingelli, B., Sandage, A., & Tammann, G.A. 1988, *ARA&A*, **26**, 509
4. Tully, R.B. 1987, *Nearby Galaxies Catalog* (Cambridge University Press, Cambridge); generally to 25^m/sq."
5. de Vaucouleurs, G., & Mitra, S. 1991, unpublished
6. Fabbiano, G. 1987, *ARA&A*, **27**, 87
7. Soifer, B.T., Houck, J.R., & Neubebauer, G. 1989, *ARA&A*, **25**, 187

8. Young, J.S., & Scoville, N.S. 1991 ARA&A, **29**, 581; ratios of the mass of disk gas to the total disk mass of the galaxies
9. Harris, W.E. 1991, ARA&A, **29**, 543
10. Trimble, V. 1987, ARA&A, **25**, 425
11. Kormendy, J., & Djorgovski, S. 1989, ARA&A, **27**, 235 and references therein
12. van den Bergh, S. 1992, MNRAS, **255**, 29pp
13. Thuan, T.X., in Ref. [1]

Table 23.6. *The Local Group.[a]*

Galaxy [1]	RA (2000.0) [2]	Decl. [2]	Diam.[b] [2]	b/a[c] [2]	B_T[d] [2]	$(B-V)_0$[e] [2]	V_{rg}[f] [2]	Type [1-3]	Dist. (kpc) [1]	M_V [1]	$V_{rotation}$ (km/s) [4]	L_x[g] (erg/s) [5]	[Fe/H] [6]
Milky Way								Sbc I-II		-20.6	220	3.0×10^{39}	
M31 = NGC224	00h42.7	+41°16'	190.5	0.32	4.36	0.68	-121	Sb I-II	725	-21.1	254	3.6×10^{39}	
M32 = NGC221	00h42.7	+40°52'	8.7	0.74	9.03	0.88	-28	E2	725	-16.4		5.4×10^{37}	
NGC205 = M110	00h40.0	+41°41'	21.9	0.5	8.92	0.82	-60	Sph-E5p	725	-16.3		$< 9 \times 10^{36}$	-0.85
SMC	00h52.7	-72°49'	316.2	0.6	2.70	0.36	+34	Im IV-V	58	-16.2	50:	6.1×10^{37}	-0.75
NGC185	00h38.9	+48°20'	11.7	0.85	10.10	0.73	-64	dSph-dE3p	620	-15.3			-1.20
NGC147	00h33.2	+48°30'	13.2	0.6	10.47	0.78	+28	dSph-dE5	645	-15.1			-1.40
And.I	00h45.5	+38°02'	1.4	1.0	13.5	0.75		dSph-E3p	725	-11.8			-1.80
Sculptor	01h00.2	-33°42'	39.8	0.8	10.50		+115	dSph	78	-10.7			
And.III	00h35.3	+36°31'	0.75	0.4	13.5			dSph	725	-10.3			
IC1613 = DDO8	01h04.8	+02°08'	16.2	0.9	9.88	0.67	-152	IBm/Irr V	715	-14.9			-0.90
Psc = LGS3	01h03.8	+21°51'					-149	dIrr	725				
NGC598 = M33	01h33.9	+30°34'	70.8	0.6	6.27	0.47	-46	Sc,cd II-III	795	-18.9	101	1.1×10^{39}	
And.II	01h16.5	+33°26'	1.6	0.7	13.5			dSph	725	-11.8			
Phoenix	01h51.1	-44°26'	4.9	0.8				dIm/dSph	417	-9.9			
WLM = DDO221	00h02.0	-15°27'	11.5	0.35	11.03	0.31	-61	IBm IV-V	940	-14.1			-2.0
Fornax	02h39.9	-34°27'	63.1	0.7	9.04		-41	dSph	131	-13.7			-1.40
LMC	05h23.7	-69°45'	645.7	0.85	0.91	0.43	+119	SBm-Ir III-IV	55	-18.1	100:	6.6×10^{38}	-0.40
Carina	06h41.6	-50°58'	23.4	0.66			+13	dSph	107	-7.6			-1.52
Leo I	10h08.5	+12°18'	9.8	0.8	11.18	0.63	+60	dSph	229	-11.7			-1.30
Sextans	10h13.0	-01°36'	90.	0.6				dSph	87	-10.0			-1.5?
Leo II	11h13.5	+22°09'	12.0	0.9	12.6	0.59	+36	dSph-E0p	234	-9.4			-1.90
Ursa Minor	15h08.8	+67°12'	30.2	0.6	11.9		-47	dSph	69	-8.9		$< 3 \times 10^{35}$	-2.20
Draco	17h20.1	+57°55'	35.5	0.7	10.8		-87	dSph	76	-8.6			-2.10
NGC6822	19h45.0	-14°48'	15.5	0.9	9.31		+44	IBm IV-V	495	-16.4		1×10^{37}	-0.6
DDO210	20h46.8	-12°50'	2.2	0.5	14.0		-23	IBm V	725	-11.5			
Tucana	22h49.1	-64°25'	5.5	0.55	(14.0R)	0.01		dSph-dE5	890	-9.5			

Notes

[a] Other suggested Local Group members include IC10 and 1927-17 = UGC 594-04 = SagDIG [4] and AndIV [Madore, B.F. in *Observer's Handbook 1992*, edited by B.L. Bishop (Royal Astronomical Society Canada), p. 223].

[b] Arcminutes; isophotal to 25^m/sq.$''$, except And. I, II, and III core radii.

[c] Axial ratio, to 25^m/sq.$''$ isophote.

[d] Entire galaxy, no correction for absorption.

[e] Entire galaxy, corrected for reddening.

[f] Galactocentric radial velocity.

[g] X-ray luminosity, 0.2–3.5 keV.

References

1. van den Bergh, S. 1992, *MNRAS*, **255**, 29pp
2. de Vaucouleurs, G., de Vaucouleurs, A., Corwin, H.G., Buta, R.J., Paturel, G., & Fouque, P. 1990, *Third Reference Catalogue of Bright Galaxies* (University of Texas Press, Austin)
3. Sandage, A., & Tammann, G. 1987, *A Revised Shapley Ames Catalog of Bright Galaxies* (Carnegie Institute of Washington, Washington, DC)
4. Tully, R.B. 1987, *Nearby Galaxies Catalog* (Cambridge University Press, Cambridge)
5. Fabbiano, G. 1989, *ARA&A*, **27**, 87
6. Hodge, P. 1989, *ARA&A*, **27**, 139; Caldwell, N. et al. 1992, *AJ*, **103**, 840

Table 23.7. *Bright Galaxies, $B_T < 10$ (excluding Local Group galaxies).*

Galaxy [1]	RA (2000.0) [2]	Decl. (2000.0) [2]	Diam.[a] [2]	b/a[b] [2]	B_T[c] [2]	$(B-V)_0$[d] [2]	Type [1,2]	V_{rg}[e] [2]	Dist.[f] (Mpc) [3]	M_{BT}^{0i}[g] [3]	V_{rotn}[h] [3]	$\log \mathcal{M}_{tot}$[i] [3]
NGC 55	00^h15^m	$-39°13'$	32.4	0.17	8.42	0.54	SBc/m III	+94	1.3	-18.13	83	9.87
NGC 247	00^h47^m	$-20°46'$	21.4	0.32	9.67	0.54	SABc/d III-IV	+176	2.1	-17.98	92	10.02
NGC 253	00^h48^m	$-25°17'$	27.5	0.25	8.04		SABc II	+251	3.0	-20.02	197	10.87
NGC 300	00^h54^m	$-37°41'$	21.9	0.71	8.72	0.58	Sc/d II-IV	+98	1.2	-16.88	100	9.89
NGC 628 = M74	01^h37^m	$+15°47'$	10.5	0.91	9.95	0.51	SAc II	+753	9.7	-20.32	face on	
NGC 1068 = M77	02^h43^m	$-00°01'$	7.1	0.85	9.61	0.70	SAb II	+1144	14.4	-21.39	face on	
NGC 1291	03^h17^m	$-41°06'$	9.8	0.83	9.39	0.91	SB0/a	+712	8.6	-20.26	face on	
NGC 1313	03^h18^m	$-66°30'$	9.1	0.76	9.20	0.48	SBc/d III-IV	+292	3.7	-18.60	136	10.31
NGC 1316 (Fornax A)	03^h23^m	$-37°12'$	12.0	0.71	9.42	0.87	SAB0/a pec	+1674	16.9	-21.47		
NGC 2403	07^h37^m	$+65°36'$	21.9	0.56	8.93	0.39	Scd III	+226	4.2	-19.68	124	10.67
NGC 2903	09^h32^m	$+21°30'$	12.6	0.48	9.68	0.55	SABbc I-II	+476	6.3	-19.85	196	10.91
NGC 3031 = M81	09^h56^m	$+69°04'$	26.9	0.52	7.89	0.82	SAab I-II	+69	1.4	-18.29	236	10.73
NGC 3034 = M82	09^h56^m	$+69°41'$	11.2	0.38	9.30	0.79	I0/amorphous	+323	5.2	-19.42		
NGC 3115	10^h05^m	$-07°43'$	7.2	0.84	9.87	0.94	S0	+492	6.7	-19.18		
NGC 3521	11^h06^m	$-00°02'$	11.0	0.47	9.83	0.68	SABbc II	+673	7.2	-19.88	295	10.99
NGC 3627 = M66	11^h20^m	$+13°00'$	9.1	0.46	9.65	0.60	SABb II	+643	6.6	-19.66	188	10.74
NGC 4258 = M106	12^h19^m	$+47°18'$	14.8	0.39	9.10	0.55	SABbc II-III	+510	6.8	-20.59	213	11.17
NGC 4449	12^h28^m	$+46°06'$	6.2	0.71	9.99	0.41	IB/Sm IV	+255	3.0	-17.66		
NGC 4472 = M49	12^h30^m	$+08°00'$	10.2	0.81	9.37	0.95	E1-2/S0	+846	16.8	-21.82		

Table 23.7. (Continued.)

Galaxy [1]	RA Decl. (2000.0) [2]		Diam.[a] [2]	b/a[b] [2]	B_T[c] [2]	$(B-V)_0$[d] [2]	Type [1,2]	V_{rg}[e] [2]	Dist.[f] (Mpc) [3]	M_{BT}^{0i}[g] [3]	$V_{\rm rotn}$[h] [3]	$\log \mathcal{M}_{\rm tot}$[i] [3]
NGC 4486 = M87	$12^{\rm h}31^{\rm m}$	$+12°23'$	8.3	0.79	9.59	0.93	cD, E0p	+1229	16.8	−21.64		
NGC 4594 = M104 (Sombrero)	$12^{\rm h}40^{\rm m}$	$-11°37'$	8.7	0.41	8.98	0.45	Sa/ab	+969	20.0	−22.98	369	11.38
NGC 4631	$12^{\rm h}42^{\rm m}$	$+32°32'$	15.5	0.17	9.75	0.55	SBc/d III	+629	6.9	−20.12	142	10.70
NGC 4649 = M60	$12^{\rm h}44^{\rm m}$	$+11°33'$	7.4	0.81	9.81	0.95	S0/E2	+970	16.8	−21.36		
NGC 4736 = M94	$12^{\rm h}51^{\rm m}$	$+41°07'$	11.2	0.81	8.99	0.72	SAab II	+360	4.3	−19.37	186	10.78
NGC 4826 = M64	$12^{\rm h}57^{\rm m}$	$+21°41'$	10.0	0.54	9.36	0.71	Sab II	+403	4.1	−19.15		
NGC 4945	$13^{\rm h}05^{\rm m}$	$-49°28'$	20.0	0.19	9.3		SBcd IV	+383	5.2	−20.65		
NGC 5055 = M63	$13^{\rm h}16^{\rm m}$	$+42°02'$	12.6	0.58	9.31	0.64	SAbc II-III	+571	7.2	−20.42	224	11.15
NGC 5128 (Cen. A)	$13^{\rm h}25^{\rm m}$	$-43°01'$	25.7	0.79	7.84	0.88	S0p	+398	4.9	−20.97		
NGC 5194 = M51	$13^{\rm h}30^{\rm m}$	$+47°12'$	11.2	0.62	8.96	0.53	SAbc I-IIp	+551	7.7	−20.75		
NGC 5236 = M83	$13^{\rm h}37^{\rm m}$	$-29°52'$	12.9	0.89	8.20	0.61	SBc II	+384	4.7	−20.31	face on	
NGC 5457 = M101	$14^{\rm h}03^{\rm m}$	$+54°21'$	28.8	0.93	8.31	0.44	SABcd I	+360	5.4	−20.45	face on	
NGC 6744	$19^{\rm h}10^{\rm m}$	$-63°51'$	20.0	0.65	9.14		Sbc II	+746	10.4	−21.39	185	11.37
NGC 6946	$20^{\rm h}35^{\rm m}$	$+60°09'$	11.5	0.85	9.61	0.40	Scd II	+277	5.5	−20.78	153	10.81
NGC 7793	$23^{\rm h}58^{\rm m}$	$-32°35'$	9.3	0.68	9.63		SAd IV	+228	2.8	−17.69	104	9.95

Notes

[a] Arcminutes; isophotal to $25^{\rm m}$/sq.''

[b] Axial ratio, to $25^{\rm m}$/sq.'' isophote.

[c] Entire galaxy, no correction for absorption.

[d] Entire galaxy, corrected for reddening; inclination corrections revised from RC3 (G. de Vaucouleurs, private communication).

[e] Galactocentric radial velocity, km/s.

[f] Distances larger by factors 1.2–2 and correspondingly brighter absolute magnitudes are given in [1].

[g] Entire galaxy, corrected for absorption and angle of inclination.

[h] 21 cm velocity width, corrected for sin i.

[i] From 21 cm velocity width and angular extent.

References

1. Sandage, A., & Tammann, G.A. 1987, *A Revised Shapley Ames Catalog of Bright Galaxies* (Carnegie Institute of Washington, Washington, DC)

2. de Vaucouleurs, G., de Vaucouleurs, A., Corwin, H.G., Buta, R.J., Paturel, G., & Fouque, P. 1990, *Third Reference Catalogue of Bright Galaxies* (University of Texas Press, Austin) (RC3)

3. Tully, R.B. 1987, *Nearby Galaxies Catalog* (Cambridge University Press, Cambridge)

Table 23.8. *Galaxy counts.*

b_J	N
15.29	0.47
15.77	0.83
16.25	1.51
16.71	2.90
17.29	6.11
17.74	11.7
18.29	24.8
18.72	41.7
19.25	81.3
19.77	140.
20.27	233.
20.77	407.

The fainter counts have been corrected for galaxy spectral shapes (k correction) but generally not for galactic extinction. The numbers at $b \ll 90°$ will be smaller by

$$\Delta \log N = -s A(90°)(\csc b - 1),$$

where $A(90°)$ is the extinction at the pole ($A_B = 0.13$–0.20) and s is the slope of $\log N(m)$ ($= 0.6$ in the Euclidean case).

Most surveys that go fainter than $b_J = 20$ find differential counts per square degree per unit magnitude [35]:

$$\log N(b_J) = 0.45 b_J + C, \quad \text{with normalization } C = 7.7 \text{ to } 8.2.$$

The number of galaxies seen from outside the Milky Way would be larger than the polar cap number $N(m)$ by

$$\Delta \log N = +s A(90°),$$

where $A(90°) = 0$ to 0.20 in this context [32, 36].

23.2.3 Luminosity Function

Away from anomalies due to the Local Supercluster, a Schechter function [37] provides a remarkably good fit [38–41] over the range $M_B = -15$ to -21:

$$\phi(L)\,dL = \phi^*(L/L^*)^\alpha \exp(-L/L^*)\,d(L/L^*),$$

where

$$\phi^* = (0.015 \pm 0.001)h^3 \text{ galaxies Mpc}^{-3},$$

L^* is the luminosity corresponding to $M_B^* = -19.5 \pm 0.2$ ($M_B^\odot = 5.48$), and $\alpha = -1.1 \pm 0.1$ (-1.0 in the field, -1.25 in rich clusters). Such numbers have been corrected for redshift effects (including the k correction) and absorption in the Milky Way, using a variety of prescriptions.

The true luminosity function is a strong function of Hubble morphological type; in particular, that for dwarf galaxies continues to rise faintward of $M_B = -14$, though not so steeply that faint galaxies dominate the cosmic luminosity density [2, 39].

23.2.4 Luminosity Density

The luminosity density of the Universe, for a Schechter luminosity function, is [41]

$$\mathcal{L} = \phi^* \Gamma(\alpha + 2)L^*,$$
$$\mathcal{L} = (1.4\text{–}2.0) \times 10^8 h L_\odot/\text{Mpc}^3 \quad \text{in the } B \text{ band.}$$

Thus the closure density in the Universe corresponds to

$$\mathcal{M}/L_B = (1400\text{–}2000)h(\mathcal{M}/L)_\odot.$$

Notes added in proof.

A number of things have evolved since this chapter was completed in June 1992: the distance to the galactic center [42] and details of galactic structure [43]. New members of the Local Group include a dwarf spheroidal galaxy in Sagittarius [44], a Pegusus dwarf [45], and the nonmembership of Andromeda IV [46]. Galaxy properties as a function of Hubble type are discussed in [47], black holes in the centers of normal galaxies are discussed in [48], and a number of papers on galaxy counts are given in IAU Symposium 161 and 179 [49].

REFERENCES

1. Binney, J., & Tremaine, S.D. 1987, *Galactic Dynamics* (Princeton University Press, Princeton, NJ)
2. van den Bergh, S. 1992, *MNRAS*, **255**, 29pp
3. Fich, M., & Tremaine, S. 1991, *ARA&A*, **29**, 409
4. Beichman, C.A. 1987, *ARA&A*, **25**, 521
5. van der Kruit, P. in *The Galaxy*, edited by G. Gilmore and B. Carswell (Reidel, Dordrecht)
6. Kulessa, A.S., & Lynden-Bell, D. 1992, *MNRAS*, **255**, 105
7. Allen, C., & Santillan, A. 1991, *Rev. Mexicana Astron. Af.*, **22**, 255
8. Zombeck, M.V. 1982, *Handbook of Space Astronomy and Astrophysics* (Cambridge University Press, Cambridge)
9. McCammon, D., & Sanders, W.T. 1990, *ARA&A*, **28**, 657
10. Bowyer, S. 1991, *ARA&A*, **29**, 59
11. Bowyer, S., & Leinert, C., editors, 1990, IAU Symposium 139, *Galactic and Extragalactic Backgrounds* (Kluwer Academic, Dordrecht)
12. Bloemen, H. 1989, *ARA&A*, **27**, 469
13. van der Kruit, P. 1987, in IAU Symposium 139, Ref. [11]
14. Genzel, R., & Townes, C.H. 1987, *ARA&A*, **25**, 377
15. Vallee, J.P., Simard-Normandin, M., & Bignell, R.C. 1988, *ApJ*, **331**, 321
16. Combes, F. 1991, *ARA&A*, **29**, 295
17. Cowan, J.J., Thielmann, F.-K., & Truran, J.W. 1991, *ARA&A*, **29**, 447
18. Wood, M.A. 1992, *ApJ*, **386**, 539
19. Rana, N.C. 1991, *ARA&A*, **29**, 129
20. Dickey, J.M., & Lockman, F.J. 1990, *ARA&A*, **28**, 215
21. Young, J.S., & Scoville, N. 1991, *ARA&A*, **29**, 281
22. Bahcall, J.N., Schmidt, M., & Soneira, R.M. 1983, *ApJ*, **265**, 730
23. Bahcall, J.N., 1984, *ApJ*, **287**, 926
24. Bahcall, J.N., Flynn, C., & Gould, A. 1992, *ApJ*, **389**, 234,
25. Kuiken, K., & Gilmore, G. 1989, *MNRAS*, **239**, 571, 605
26. Merrifield, M. 1992, *AJ*, **103**, 1552
27. Bahcall, J.N. 1986, *ARA&A*, **24**, 577
28. Allen, C.W. *Astrophysical Quantities* (Athlone Press, London), p. 283
29. Murray, C.A. 1988, *A&Ap*, **218**, 325
30. Murray, C.A. 1992, private communication
31. Lang, K.R. 1980, *Astrophysical Formulae* (Springer-Verlag, Berlin), p. 501
32. Sandage, A., & Tammann, G. 1987, *A Revised Shapley Ames Catalog of Bright Galaxies* (Carnegie Institute of Washington, Washington, DC)
33. de Vancouleurs, G., de Vaucouleurs, A., Corwin, H.G., Buta, R.J., Paturel, G., & Fougue, P. 1990, *Third Reference Catalogue of Bright Galaxies* (University of Texas Press, Austin)
34. Sandage, A., Tammann, G.A., & Hardy, E. 1972, *ApJ*, **172**, 261
35. Maddox, S.J. et al. 1990, *MNRAS*, **247**, 1p
36. de Vaucouleurs, G., de Vaucouleurs, A., & Corwin, H.G. 1976, *Second Reference Catalogue of Bright Galaxies* (University of Texas Press, Austin) (RC2)
37. Schechter, P. 1976, *ApJ*, **203**, 297
38. Giovanelli, R., & Haynes, M. 1991, *ARA&A*, **29**, 499 (but with factor 10 error in ϕ^*)
39. Bingelli, B., Sandage, A., & Tammann, G.A. 1988, *ARA&A*, **26**, 509
40. Loveday, J. et al. 1991, *ApJ*, **390**, 338
41. Efstathiou, G., Ellis, R.S., & Peterson, B.A. 1988, *MNRAS*, **232**, 431
42. Reid, M. 1993, *ARA&A*, **31**, 345

43. Majewski, S.R. 1993, *ARA&A*, **31**, 575
44. Ibata, R.A., Gilmore, G., & Irwin, M.J. 1994, *Nature*, **370**, 184
45. Apariacio, A. 1994, *ApJL*, **437**, L27
46. Joseph, J.H. 1993, *AJ*, **105**, 932
47. Robert, M.S., & Haynes, M.P. 1994, *ARA&A*, **32**, 115

48. Kormendy, J., & Richstone, D. 1995, *ARA&A*, **33**, 581
49. 1993, *Astronomy from Wide Field Imaging*, IAU Symposium 161 edited by H.T. MacGillivray (Kluwer Academic, Dordrecht); 1996, *New Horizons from Multiwavelength Surveys*, IAU Symposium 179 edited by B.J. McLean (Kluwer Academic, Dordrecht)

Chapter 24

Quasars and Active Galactic Nuclei

Belinda J. Wilkes

24.1 INTRODUCTION

Quasars were discovered in the 1960s when two strong, compact radio sources in the third Cambridge radio survey, 3C273, 3C48, were identified with blue stellar objects [1–3]. Optical spectra revealed strong, broad, redshifted emission lines, suggesting the objects were at large distances and highly luminous. It was soon realized that the majority ($\sim$ 90%) are radio-quiet [4] and also that they are similar to the lower luminosity *active galactic nuclei* (AGN) discovered $\sim$ 20 years earlier [5]. This chapter includes definitions of many names and terms used for the various types of AGN, parameters used to characterize them, examples of optical spectra, etc., and many references to the literature. The aim is to provide a general reference and starting point for further study in many aspects of quasar research. A recent book [6] covers quasar research in more depth.

Quasars are generally believed to be small but highly luminous sources of radiation, probably powered by a supermassive ($\sim 10^{8-10} \mathcal{M}_\odot$) black hole, embedded in the center of a parent galaxy with the quasar's redshift indicating, via the expansion of the Universe, its distance from our Galaxy. The radio-loud objects (see Table 24.1) also include two major components of radio emission: a compact core and extended lobes. These two components are thought to be linked by a relativistic, beamed jet originating in the core. Lower luminosity active galaxies include a mixture of objects, some of which are weaker versions of quasars and others of which are powered by star formation rather than a central black hole. The relationships between the various types remain a topic of some contention.

24.2   THE TYPES OF ACTIVE GALACTIC NUCLEI

The names and acronyms used for the many types or classes of active galaxies are listed in Table 24.1. Table 24.2 lists the properties of the main classes and Figures 24.1 and 24.2 show their distributions of relative radio–optical–X-ray luminosities and emission line ratios, respectively. Table 24.3 lists a few well-known objects from each of the major classes. Figure 24.3 shows the X-ray–infrared (IR) luminosity correlation for various classes of AGN and also normal galaxies.

Figures 24.4(a)–24.4(f) show optical/ultraviolet (UV) spectra of examples of the five classes: quasar (low and high redshift); BALQSO; BL Lac; Sy1; and NLXG.

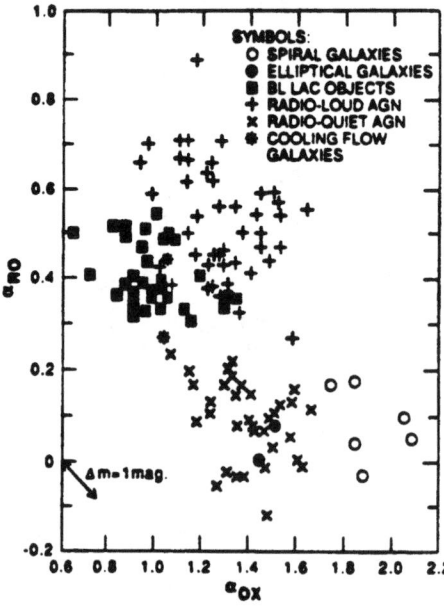

Figure 24.1. Relative radio–optical–X-ray properties of the several types of AGN and galaxies (from [7]). $\alpha_{\rm ro}$, $\alpha_{\rm ox}$ are the effective radio–optical and optical–X-ray slopes; see Sec. 24.4.

Table 24.1. *Common names.*

General	Active galactic nucleus (AGN)	General term for an object containing nonstellar activity in its nucleus and [usually] optical/UV emission lines
	Quasar/QSO[a]	General terms for high luminosity ($M < -23$) AGN with broad emission lines
	QSO	Quasistellar object
	BALQSO	Broad absorption line QSO
Radio	RQQ	Radio-quiet quasar, $R_L{}^b < 1.0$
	RLQ/QSRS	Radio-loud quasar, $R_L{}^b > 1.0$
	CDQ	Core-dominated, radio-loud quasar, $R^b > 1.0$
	LDQ	Lobe-dominated, radio-loud quasar, $R^b < 1.0$
	CSS	Core-dominated, steep spectrum RLQ
	GPS	Gigahertz peaked source, subset of CDQ with narrow (FWHM 1–2 decades) radio continua [1]
	FRI	Edge-darkened radio source, $L(1400\,\text{MHz}) < 10^{25}\,\text{W Hz}^{-1}$, Fanaroff and Riley [2]
	FRII	Edge-brightened radio source, $L(1400\,\text{MHz}) > 10^{25}\,\text{W Hz}^{-1}$
	Superluminal	Radio source containing motion where $v_{\text{app}} > c$
Blazars	Blazar	General term for BL Lacs, OVVs, and HPQs
	BL Lac Object	Active nucleus but no emission lines
	RBL/LBL	Radio-selected/low frequency BL Lacs
	XBL/HBL	X-ray selected/high frequency BL Lacs
	HPQ	Highly polarized quasar, $P \gtrsim 0.03$, generally CDQs
	LPQ	Low polarization quasar, $P \lesssim 0.03$ (normal QSO)
	OVV	Optically violent variable (subset of HPQ)
LLAGN[c]		Low luminosity AGN, $M > -23$ [3, 4]
	Seyfert 1 (Sy1)	LLAGN with broad permitted and narrow forbidden emission lines [5]
	NLS1	Narrow line Sy1 galaxy, permitted lines $< 2000\,\text{km s}^{-1}$ [6]
	Seyfert 2 (Sy2)	LLAGN with narrow permitted and forbidden emission lines [5]
	Seyfert 1.5–1.9	Similar to Sy1 but with progressively weaker broad lines
	BLRG	Broad line radio galaxy ("radio-loud Sy1")
	NLRG	Narrow line radio galaxy ("radio-loud Sy2")
	LINER	Low-ionization nuclear emission line region, present in $\sim 1/3$ of all bright galaxies [7]
	NLXG	Narrow line X-ray galaxy: X-ray strong, $L_x > 10^{41}\,\text{erg}^{-1}$, Sy2 [8]
	Ultraluminous IR galaxy (ULIRG)	Ultraluminous in far-IR, shows starburst and AGN characteristics, $L_{\text{FIR}} > 10^{12} L_\odot$ [9]
	Starburst	Galaxy containing strong starburst activity

Notes

[a]Historically *quasar* was used to indicate radio-loud objects; nowadays both terms are often used interchangeably to describe the whole class of luminous AGN.

[b]For parameter definitions please refer to Section 24.4.

[c]Note that the divisions between the various types of LLAGN (LINERS, starbursts, Sy2, NLXG) are not well defined.

References

1. O'Dea, C.P., Baum, S.A., & Stanghellini, C. 1991, *ApJ*, **380**, 66
2. Fanaroff, B.L., & Riley, J.M. 1974, *MNRAS*, **267**, 31P
3. Filippenko, A.V. 1992, *ASP Conf. Proc.*, **31**, 253
4. Osterbrock, D.E. 1993, *ApJ*, **404**, 551
5. Khachikian, E.Y., & Weedman, D.W. 1974, *ApJ*, **192**, 581
6. Osterbrock, D.E., & Pogge, R.W. 1985, *ApJ*, **297**, 166
7. Heckman, T.M. 1980, *A&A*, **87**, 152
8. Stocke, J.S., Morris, S.L., Gioia, I.M., Maccacaro, T., Schild, R., Wolter, A., Fleming, T.A., & Henry, J.P. 1991, *ApJS*, **76**, 813
9. Sanders, D., Soifer, T.B., Elias, J.H., Madore, B.F., Matthews, K., Neugebauer, G., & Scoville, N.Z. 1988, *ApJ*, **325**, 74

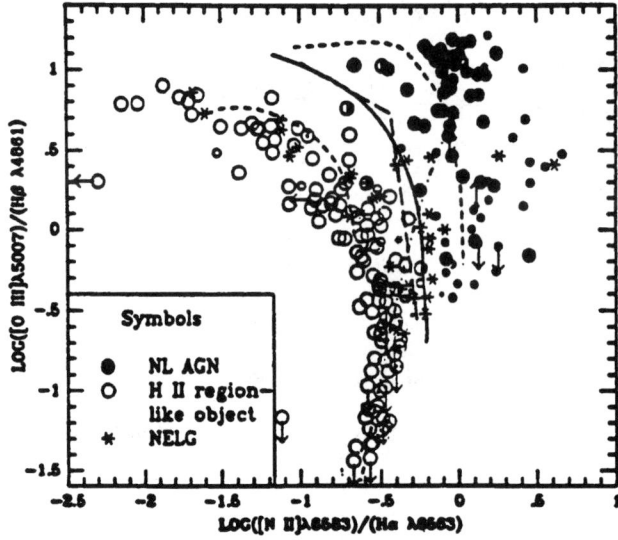

Figure 24.2. An example of a line ratio diagram showing the behavior of various types of LLAGN (from [8, Fig. 1]). Symbols are as shown with open symbols indicating H II and starburst objects. The solid curve divides AGNs from H II region–like objects. Four short-dashed lines are H II region models [9] for $T_* = 56\,000, 45\,000, 38\,500, 37\,000$ K, top to bottom. The long-dashed curve represents H II region models [10].

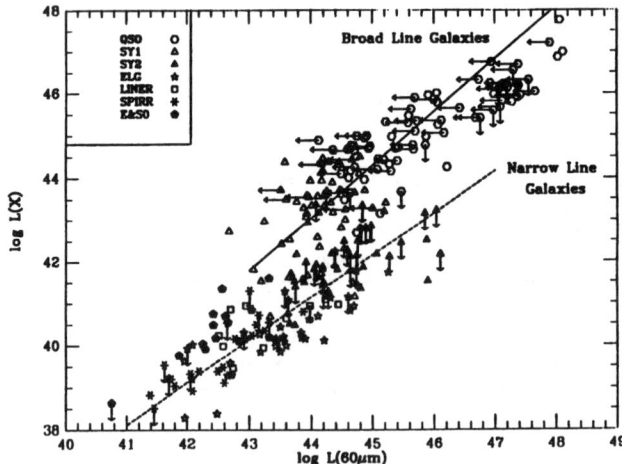

Figure 24.3. The correlation of X-ray and IR luminosities for broad- and narrow-lined objects (from [11]). SPIRR = spiral/irregular galaxy; E&SO = elliptical and SO galaxies; ELG = starburst or extragalactic H II region.

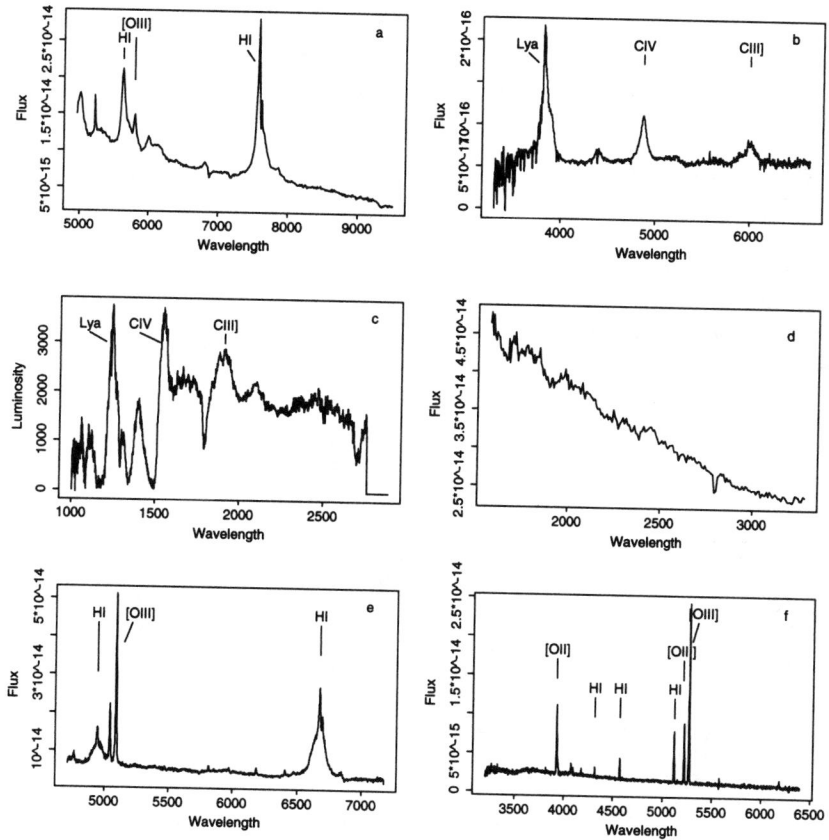

Figure 24.4. Optical/UV spectra of various types of quasar and active galaxy: (a) Optical spectrum of low-redshift quasar 3C273 ($z = 0.158$) observed with the 1.5 m telescope at CTIO, Feb. 1989. (b) Optical spectrum of high-redshift quasar Q1101-264 ($z = 2.144$ [12, 13]). (c) Rest frame ultraviolet spectrum of the BALQSO 1232+1325 ($z = 2.364$) in units of erg s^{-1} Å^{-1}, courtesy of Craig Foltz [14]. (d) HST ultraviolet spectrum of the BL Lac PKS2155-305 in units of erg s^{-1} Å^{-1}, courtesy of Paul Smith [15]. (e) Optical spectrum of the Seyfert 1 galaxy NGC5548 ($z = 0.017$) generated by combining spectra between 13 and 19 April 1993 taken as part of the International AGN Watch monitoring campaign [16] on the SAO 60 in. telescope on Mt. Hopkins, Arizona. (f) Optical spectrum of the starburst galaxy PG0119+229 taken on 9 Jan. 1989 with the SAO Multiple Mirror Telescope (MMT) on Mt. Hopkins, Arizona.

Table 24.2. *Properties of various types of AGN.*

Class	Lines Brd[a]	Lines Nar[b]	RL[c] (%)	P[d] (%)	$\log L_O$[e] (erg s^{-1})	$\log L_x$ (erg s^{-1})	Comments
Quasar	✓	✓	10	< 3	44[f]–48	44–48	LPQ
BALQSO	✓	✓	0	✓[g]	44–48	44–48	
BL Lac Obj.	...	...	100	3–40	44–46	44–46	
HPQ	✓	✓	100	3–40	44–48	44–48	CDQ, γ-ray [1]
GPS	✓	✓	100	< 2	44–48	44–48	CDQ
Seyfert 1/BLRG	✓	✓	10	< 3	42–44[f]	43–45	
Seyfert 2/BLRG	P[h]	✓	10	0–20	41–43	41–43	[O III] > [O II]
LINER	...	✓	...	...	40–44	40–42	[O III] < [O II]
Starburst	...	✓	...	...	...	< 41	[O III] < [O II]
NLXG	...	✓	...	...	40–44	42–43.5	[O III] > [O II]

Notes

[a] Broad emission lines (FWHM $\sim$ 1000–10 000 $km\,s^{-1}$).

[b] Narrow emission lines (FWHM $\sim$ 100–1000 $km\,s^{-1}$).

[c] Rough percentage of radio-loud objects in the class.

[d] Typical percentage polarization in the optical continuum.

[e] Optical continuum luminosity not from stars.

[f] A dividing luminosity ($M_V \sim -23$; $L_O \sim 10^{44}$ erg^{-1}) between quasars and Sy1's is generally used [2].

[g] Frequently have high (> 3%) polarization [3].

[h] Broad lines seen in polarized light for a subset [4].

References

1. Fichtel, C.E. et al. 1994, *ApJS*, **94**, 551
2. Schmidt, M., & Green, R.F. 1983, *ApJ*, **269**, 352
3. Glenn, J., Schmidt, G.D., & Foltz, C.B. 1994, *ApJ*, **434**, L47
4. Miller, J.S. 1994, in *The Physics of Active Galaxies*, edited by G.V. Bicknell, M.A. Dopita, and P.G. Quinn, ASP Conf. Ser. 54, p. 149

Table 24.3. *A selection of well-known objects of various types.*

Name	Class	α (2000)	δ (2000)	Redshift	V	Reference
3C273	RLQ	12 29 06.7	+02 03 09	0.158	13.02	[1]
3C48	RLQ	01 37 41.3	+33 09 35	0.367	16.06	[2]
S5 0014+81	RQQ	00 16 44.4	−04 04 11	3.384	16.5	[3]
NAB0205+024	RQQ	02 07 49.9	+02 42 56	0.155	15.39	[2]
PG1211+143	RQQ	12 14 17.6	+14 03 13	0.085	14.63	[3]
PC1247+3406	RQQ	12 49 42.1	+33 49 52	4.897	20.4	[3]
HS1946+7658	RQQ	19 44 55.0	+77 05 52	3.02	15.8	[3]
PHL5200	BALQSO	22 28 30.3	−05 18 55	1.981	17.7	[2]
1232+1325	BALQSO	12 34 58.3	+13 08 55	2.36	19.5	[4, 3]
BL Lac	BL Lac	22 02 43.3	+42 16 40	0.0686	14.5	[2, 5]
PKS2155−304	BL Lac	21 58 52	−30 13 29	0.116	14	[2, 5]
3C446	OVV	22 25 47.3	−04 57 01	1.404	17.19	[2]
3C345	OVV	16 42 58.8	+39 48 37	0.595	15.96	[2]
3C279	OVV[a]	12 56 11.1	−05 47 22	0.538	17.75	[3]
I Zw 1	Sy1	00 53 34.9	+12 41 36	0.061	14.07	[2]
NGC5548	Sy1	14 17 59.6	+25 08 12	0.017	13.73	[3]
NGC4151	Sy1	12 10 32.5	+39 24 21	0.003	11.85	[3]
NGC4051	Sy1	12 03 09.6	+44 31 53	0.002	12.92	[3]
NGC1068	Sy2	02 42 40.7	−00 00 47	0.003	10.83	[3]
MRK 1	Sy2	01 16 07.2	+33 05 22	0.016	14.96	[3]
3C390.3	BLRG	18 42 08.9	+79 46 17	0.057	15.38	[3]
3C445	BLRG	22 23 49.7	−02 06 13	0.057	15.77	[3]
3C234	NLRG	10 01 49.5	+28 47 09	0.185	17.27	[3]
MRK 348	NLXG	00 48 47.2	+31 57 25	0.014	14.59	[3]
NGC1052	LINER	02 41 04.7	−08 15 21	0.005	12.31	[3]
NGC4258[b]	LINER	12 18 57.5	+47 18 14	0.002	11.65	[3]
PKS2322−12	LINER	23 25 19.7	−12 07 26	0.082	17.2	[3]
IF10214+4724	ULIRG[c]	10 24 34.6	+47 09 10	2.286	20.5	[5]
NGC7714	Starburst	23 36 14.1	+02 09 18	0.009	14.36	[6, 3]
0833+652	Starburst	08 33 57.4	+65 17 46	0.057	13	[7]

Notes

[a] Strong γ-ray source. Hartmann et al. 1992, *ApJ*, **385**, L1.

[b] A central, edge-on ($i = 83°$) Keplerian disk revealed by water maser emission provides the first measured black hole mass in an AGN of $3.6 \times 10^7 \mathcal{M}_\odot$ [8].

[c] Ultraluminous infrared galaxy, gravitational lens [9].

References
1. Hewitt, A., & Burbidge, G. 1993, *ApJS*, **87**, 451
2. Hewitt, A., & Burbidge, G. 1989, *ApJS*, **69**, 1
3. Véron-Cetty, M.-P., & Veron, P. 1993, ESO Scientific Report No. 13
4. Weymann, R.J., Morris, S.L., Foltz, C.B., & Hewett, P.C. 1991, *ApJ*, **373**, 23
5. NASA/IPAC Extragalactic Database (NED); http://nedwww.ipac.caltech.edu
6. Weedman, D.W., Feldman, F.R., Balzano, V.A., Ramsey, L.W., Sramek, R.A., & Wu, C.-C. 1981, *ApJ*, **248**, 105
7. Margon, B., Anderson, S.F., Mateo, M., Fich, M., & Massey, P. 1988, *ApJ*, **334**, 597
8. Miyoshi, M., Moran, J., Herrnstein, J., Greenhill, L., Nakai, N., Diamond, P., & Inoue, M. 1995, *Nature*, **373**, 127
9. Goldrich, R.W., Miller, J.S., Martel, A., Cohen, M.H., Tran, H.D. Ogle, P.M., & Vermeulen, R.C. 1996, *ApJL*, **456**, 9

24.3 CATALOGS AND SURVEYS

In Table 24.4 a list of major optical surveys is given. Note that the number of quasars is a function of magnitude limit and thus numbers are only approximate. The total number of quasars generally reflects the number in the complete part of the survey. In cases where surveys are published in installments, the most recent/complete reference is given.

Table 24.4. *Major optical quasar surveys.[a]*

Name	m_{lim}	z_{max}	Area (deg^2)	n (QSOs)[b]	Reference
BQS("PG")	~ 16	2.2	10^4	114	[1]
Edinburgh UVX	16.5	2.2	330	12	[2]
Hamburg	17.5	3.1	1000	160	[3]
MBQS	17.65	2.2	109	32	[4]
AB	18.25	2.2	35.5	22	[5]
HBQS	18.75	2.2	153	285	[6]
LBQS	18.85	3.4	454	1055	[7]
BF	19.8	2.2	1.72	35	[5]
SA94	19.9	2.2	10.0	94	[8]
CTIO 4m	~ 20	3.4	5.1	66	[9]
WHO	20 (R)	4.5	46	85	[10]
CFHT	20.5	3.4	9.40	268	[11]
SSG2	~ 20.5	4.5	7.84	8	[12]
Durham	21.0	2.2	11.9	397	[13]
SSG1	~ 22	4.5	0.91	10	[14]
(ZM)^{2}B	22.0	2.9	0.5	52	[15]
BJS	22.0	2.9	0.85	62	[16]
KOKR	22.6	3.2	0.29	30	[17]
DMS	23.8	4.36	0.83	55	[18]

Notes
[a] For a conference on Quasar surveys, see [19].
[b] n is the number of QSOs in the survey.

References
1. Schmidt, M., & Green, R.F. 1983, *ApJ*, **269**, 352
2. Goldschmidt, P., Miller, L., LaFranca, F., & Christiani, S. 1992, *MNRAS*, **256**, 65P
3. Reimers, D., Koehler, T. & Wisotzki, L. 1996, *A&AS*, **115**, 235
4. Mitchell, K.J., Warnock, A., III, & Usher, P.D. 1984, *ApJ*, **68**, 449
5. Marshall, H.L., Avni, Y., Bracessi, A., Huchra, J.P., Tananbaum, H., Zamorani, G., & Zitelli, V. 1984, *ApJ*, **283**, 50
6. Cristiani, S. et al. 1995, *A&AS*, **112**, 347

7. Hewett, P.C., Foltz, C.B., & Chaffee, F.H., 1995, *AJ*, **109**, 1498
8. LaFranca, F., Cristiani, S., & Barbieri, C. 1992, *AJ*, **103**, 1062
9. Osmer, P.S. 1980, *ApJS*, **42**, 523
10. Warren, S.J., Hewett, P.C., & Osmer, P.S. 1991, *ApJS*, **76**, 23
11. Crampton, D., Cowley, A.P., & Hartwick, F.D.A. 1989, *ApJ*, **345**, 59
12. Schmidt, M., Schnneider, D.P., & Gunn, J.E. 1986, *ApJ*, **310**, 518
13. Boyle, B.J., Fong, R., Shanks, T., & Peterson, B.A. 1990, *MNRAS*, **241**, 1
14. Schmidt, M., Schnneider, D.P., & Gunn, J.E. 1986, *ApJ*, **306**, 411
15. Zitelli, V., Mignoli, M., Zamorani, G., Marano, B., & Boyle, B.J. 1992, *MNRAS*, **256**, 349
16. Boyle, B.J., Jones, L.R., Shanks, T., Marano, B., Zitelli, V., & Zamorani, G. 1991, in *The Space Distribution of Quasars*, APS Conf. Ser. 21, p. 191
17. Koo, D.C., & Kron, R.G. 1988, *ApJ*, **325**, 92
18. Kennefick, J.D., Osmer, P.S., Hall, P.B., & Green, R.F. 1997, *ApJ*, **114**, 2269
19. Crampton, D. 1991, *The Space Distribution of Quasars*, ASP Conf. Ser. 21

24.3.1   Other Compilations and Surveys

General catalogs	Quasars	Hewitt and Burbidge [17]
	Quasars, BL Lacs, and Sy1s	Veron and Veron [18]
	BL Lacs	Padovani and Giommi [19]
	Emission line galaxies	Hewitt and Burbidge [20]
	Quasar absorption lines	Junkkarinen et al. [21]
Optical	Michigan Tololo QSO survey	MacAlpine, Lewis, and Smith [22]
	Optical spectra of PKS* AGN	Wilkes et al. [23]
	Anderson and Margon	Anderson and Margon [24]
	Parkes ±4 deg, optical	Baldwin, Wampler, and Gaskell [25]
	Liners	Keel [26]
	BALQSOs	Wolfe et al. [27]
	APM* High-Redshift Survey	Irwin, McMahon, and Hazard [28]
	PTGS*	Schneider, Schmidt, and Gunn [29]
Ultraviolet	UV Spectra	Kinney et al. [30]
		Lanzetta et al. [31]
	AGN and galaxies	Kinney et al. [32]
	LLAGN	Ho et al. [33]
Infrared	IRAS PSC* AGN	Low et al. [34]
	IRAS: 12 micron	Rush, Malkan, and Spinoglio [35]
	2MASS	Kleinmann et al. 1994 [36]
SED*	PG (Palomar Green)	Sanders et al. [37]
	Blazars	Impey and Neugebauer [38]
	Einstein quasars	Elvis et al. [39]
X-ray	EMSS (Extended Medium Sensitivity Survey)	Maccacaro et al. [40]
	HEAO 1 A-2 High Latitude	Piccinotti et al. [41]
	HEAO 1 MC-LASS*	Schwartz et al. [42]
	Einstein X-ray Quasar Database	Wilkes et al. [43]
	Einstein Slew Survey	Elvis et al. [44]
	EXOSAT High Galactic Latitude	Giommi et al. [45]

	ROSAT All Sky Survey: BSC*	Voges et al. [46]
	ROSAT/AAT QSO Survey	Boyle et al. [47]
	RASS/LBQS	Green et al. [48]
	CCRSS	Boyle et al. [49]
	RIXOS	Puchnarewicz et al. [50]
	WGACAT	Singh et al. [51]
γ-ray	1st EGRET Catalogue	Fichtel et al. [52]
Radio	PKS* $\pm$ 4-o quasar sample	Masson and Wall [53]
	PG	Kellerman et al. [54]
	1 Jy sample	Kuhr et al. [55]
	NVSS	Condon et al. [56]
	FIRST	Gregg et al. [57]

* PKS: Parkes, APM: Automated Plate Measuring Machine, PTGS: Palomar Transit Grism Survey, PSC: Point Source Catalogue, SED: spectral energy distribution, MC-LASS: Modulation Collimator-Large Area Sky Survey, BSC: Bright Source Catalogue.

24.4 COMMONLY MEASURED PARAMETERS

The following list gives some of the parameters commonly used (or used here) in characterizing quasars and active galaxies and Table 24.5 gives the range of these parameters (where appropriate) for the various types of AGN.

Redshift, $z = (1 + z) = \lambda_{obs}/\lambda_{rest} = \nu_{rest}/\nu_{obs}$ where $\lambda_{obs}, \lambda_{rest}$ = observed, rest wavelengths, and ν_{obs}, ν_{rest} = observed, rest frequency

L_x = Luminosity at 2 keV (rest frame) in erg s^{-1} Hz^{-1}

L_{opt} = Luminosity at 2500 Å (rest frame) in erg s^{-1} Hz^{-1}

α = Spectral index: $F_\nu \propto \nu^\alpha$ (N.B. negative α commonly used in X- and γ-rays)

Γ = Photon index: $-(\alpha + 1)$ (used in X-rays)

α_{ox} = Effective optical to X-ray slope [58]: $\alpha_{ox} = -\dfrac{\log(L_x/L_{opt})}{\log(\nu_x/\nu_{opt})}$, ν_x corresponding to 2 keV; ν_{opt} corresponding to 2500 Å; ratio = 2.605

α_{ro} = Effective radio to optical slope (5 GHz and either 2500 Å [59] or V [60])

ν_c = Critical/turnover frequency, generally in far-IR/radio band

W_λ, EW = Rest frame equivalent width: $W_\lambda(\text{rest}) = W_\lambda(\text{obs})/(1 + z)$

FWHM = Full width at half maximum intensity for a spectral line

Lyman edge = Relative change of the continuum level at the Lyman limit, 912 Å

R_L = Core radio loudness: $\log(F_R(\text{core})/F_B)$ [61]

R_{L_t} = Total radio loudness: $\log(F_R(\text{total})/F_B)$ [62]

R or $R(\theta)$ = Radio core dominance: $\log\left(\dfrac{\text{flux density of beamed component}}{\text{flux density of unbeamed components}}\right)$ (5 GHz, rest frame) [63], where θ is the angle between the line of sight and the direction of approach of the lobes

$R_T = R(\theta = 90°)$ [63]

$C_{UV/IR}$ = Blue bump strength: $L(0.1-0.2 \ \mu\text{m})/L(1-2 \ \mu\text{m})$ [64]

l = Compactness parameter, $\sigma_T L/m_e c^3 R$ [65], where L = luminosity (in region of variability data), m_e = electron mass, R = size of source (from variability), σ_T = Thompson cross section, c = velocity of light

U = Ionization parameter, $\dfrac{\int_{\nu_L}^{\infty} L_\nu \, d\nu / h\nu}{4\pi r^2 c n_e}$ [66] where ν_L = frequency at Lyman limit, n_e = electron density, r = distance from source

$\mathcal{M}_{BH}$ = Derived mass of central black hole

$\dot{\mathcal{M}}$ = Accretion rate

P = Percentage polarization

θ_p = Position angle of polarization

Δm = Amplitude of variation in magnitudes

T_{var} = Variability time scale, e.g., doubling time

γ = Lorentz factor, $= \sqrt{(1 - \beta^2)} = E/mc^2$ where $\beta = v/c$

μ = Angular velocity (mas yr^{-1}, mas = milliarcseconds, used to measure superluminal knots)

β_{app} = v/c where v = apparent linear velocity in the sky plane

$\dfrac{[\text{O\,III}]}{\text{H}\beta} = \dfrac{[\text{O\,III}]\,\lambda 5007 + \lambda 4959}{\text{H}\beta}$

$\dfrac{[\text{N\,II}]}{\text{H}\alpha} = \dfrac{[\text{N\,II}]\,\lambda 6583}{\text{H}\alpha}$

$\dfrac{[\text{O\,I}]}{\text{H}\alpha} = \dfrac{[\text{O\,I}]\,\lambda 6300}{\text{H}\alpha}$

$\dfrac{[\text{S\,II}]}{\text{H}\alpha} = \dfrac{[\text{S\,II}]\,\lambda 6713 + \lambda 6731}{\text{H}\alpha}$

$\dfrac{[\text{O\,II}]}{[\text{O\,III}]} = \dfrac{[\text{O\,II}]\,\lambda 3727}{[\text{O\,III}]\,\lambda 5007}$

$\dfrac{[\text{O\,I}]}{[\text{O\,III}]} = \dfrac{[\text{O\,I}]\,\lambda 6300}{[\text{O\,III}]\,\lambda 5007}$

$\Delta E = 1/3[\Delta E(\lambda 5007/\lambda 4861) + \Delta E(\lambda 6844/\lambda 6563) + \Delta E(\lambda 6300/\lambda 6563)]$ average excess of these ratios above that of an H\,II region [67]

Table 24.5. *Range of values for some common parameters.*

	RQQ	CDQ	LDQ	Reference
z	0.1–4.9	0.1–4.3	0.1–4	
α_{ox}	1–2	1–1.6	1–1.6	[1, 2]
α_X (0.1–3.5 keV)	0.5 to -2	-0 to -1	-0 to -1	[3]
α_X (2–10 keV)	-0.9 to -1.2	-0.5 to -0.9	-0.5 to -0.9	[4][a]
α_γ	0.9 ± 0.05[b]	0.7–1.6	$\cdots$	[5, 6]
α_R	-0.2 to -1.0	< 0.5[c]	> 0.5	[7, 8]
α_O	-1.5 to 1	-1.5 to 1	-1.5 to 1	[9]
$\bar{\alpha}_{IR}$	-1.7	-1.7	-1.7	[10]
α_c	3.75 ± 0.48	1	1	[11, 12]
R_L	< 1	> 1	> 1	[3]
R_{L_t}	< 1	> 1	> 1	[13]
$R(\theta)$	~ 0.6	> 1	< 1	[8]
$C_{UV/IR}$[d]	4.9 ± 2.1	5.6 ± 4.7	5.6 ± 4.7	[14]
l	0 to > 230	0 to > 300	0 to > 200	[15, 16]
P_{opt}	< 3	0–20	0–10	
FWHM (BLR)[e]	$(1–7) \times 10^3$	$(1–5) \times 10^3$	$(1–10) \times 10^3$	[17]
FWHM (NLR)[e]	$(1–10) \times 10^2$	$(1–10) \times 10^2$	$(1–10) \times 10^2$	
W_λ(Fe\,II)[f]	0–120	0–50	0–70	[18]
$\log(\mathcal{M}_{BH}/\mathcal{M}_\odot)$[g]	8–10	8–10	8–10	[19]

Notes

[a] 90% confidence range.

[b] High-energy cutoff generally occurs $\gtrsim$ 100 keV.

[c] Except for steep spectrum, core-dominated sources.

[d] Error is the dispersion in the distribution; CDQs and LDQs not distinguished due to small numbers of objects.

[e] in $km\,s^{-1}$.

[f] $\lambda\lambda$ 4434–4684 Å.

[g] Model dependent.

References

1. Tananbaum, H. et al. 1979, *ApJ*, **234**, L9
2. Wilkes, B.J., Tananbaum, H., Worrall, D.M., Avni, Y., Oey, M.S., & Flanagan, J. 1994, *ApJS*, **92**, 53
3. Wilkes, B.J., & Elvis, M. 1987, *ApJ*, **323**, 243
4. Williams, O.R. et al. 1992, *ApJ*, **360**, 396
5. Fichtel, C.E. et al. 1994, *ApJS*, **94**, 551
6. Gondek, D., Zdziarski, A.A., Johnson, W.N., George, I.M., McNaron-Brown, K., Magdziarz, P., Smith, D. & Gruber, D. 1996, *MNRAS*, **282**, 646
7. Laing, P.A., & Peacock, J.A. 1980, *MNRAS*, **190**, 903
8. Kukula, M.J., Dunlop, J.S., Hughes, D.H., & Rawlings, S. 1998, *MNRAS*, **297**, 366
9. Francis, P.J., Hewett, P.C., Foltz, C.B., Chaffee, F.H., Weymann, R.J., & Morris, S.L. 1991, *ApJ*, **373**, 465
10. Berriman, G. 1990, *ApJ*, **354**, 148
11. Hughes, D.H., Robson, E.I., Dunlop, J.S., & Gear, W.K. 1993, *MNRAS*, **263**, 607
12. Peacock, J.A., & Wall, J.V. 1981, *MNRAS*, **194**, 331
13. Shastri, P., Wilkes, B.J., Elvis, M., & McDowell, J.C. 1993, *ApJ*, **410**, 29
14. McDowell, J.C., Elvis, M., Wilkes, B.J., Willner, S.P., Oey, M.S., Polomski, E., Bechtold, J., & Green, R.F. 1989, *ApJ*, **345**, L13
15. Done, C., & Fabian, A.C. 1989, *MNRAS*, **240**, 81
16. Lightman, A.P., & Zdziarski, A.A. 1987, *ApJ*, **319**, 643
17. Heckman, T.M. 1980, *A&A*, **87**, 152
18. Boroson, T., & Green, R.F. 1992, *ApJS*, **80**, 109
19. Sun, W.-S., & Malkan, M. 1989, *ApJ*, **346**, 68

24.5   EMISSION LINES

The presence of strong emission lines is the most definitive indication of activity in galactic nuclei. A recent conference procedings on this topic is [68] and includes several review articles. An older but very useful review can be found in [66]. Table 24.6 contains a list of multiplet wavelengths for UV–IR observed/candidate emission line features in quasar and Seyfert 1 galaxy spectra. A mean wavelength, assuming optically thin gas, is also given for convenience. For detailed atomic data on the individual lines refer to Chap. 4 and for X-ray features to Chap. 9. Little attempt has been made to include Iron (Fe) features, for comprehensive discussions of optical and ultraviolet Fe emission in quasars and AGN, see [69, 70] and references therein. Figure 24.5 shows a composite UV/optical spectrum with the prominent features labeled.

The study of X-ray emission lines is a new and exciting one which the launch of AXAF in late 1998 promises to open up dramatically. Current studies of broad Fe Kα (6.4 keV) with Ginga and ASCA [71, 72] provide some of the best evidence to date for an accretion disk surrounding a central, supermassive black hole. A comprehensive review of X-ray emission from quasars is provided by [73] and a discussion of X-ray emission lines in AGN in [74].

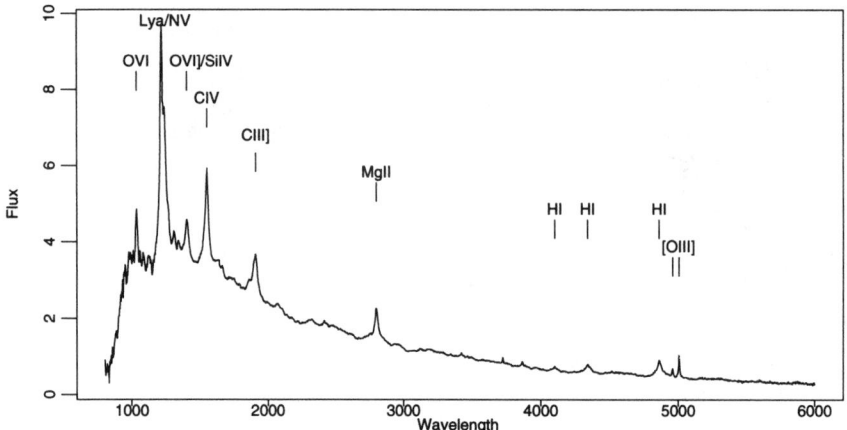

Figure 24.5. Composite optical/UV spectrum of a quasar [75].

Table 24.6. *IR–UV emission line features commonly observed in AGN.*

Element	Mean $\lambda^{a,b}$ (Å)	Component λ's^b (Å)	Reference	Ratio^c to Lyα
S VI	937.06		[1]	
Lyγ	972.54		[1]	
C III	977.02		[1]	
N III	990.98	989.80, 991.51, 991.58	[1]	
Lyβ	1025.72		[1]	
O VI	1033.82	1031.93, 1037.62	[1]	9.3
He II	1084.94		[2]	
Si II	1194.10	1197.39, 1194.50, 1193.29, 1190.42	[1]	
Lα	1215.67		[1]	
O V	1218.34		[1]	100
N V	1240.15	1238.82, 1242.80	[1]	
Si II	1263.31	1260.42, 1264.74, 1265.00	[1]	
O I	1303.49	1302.17, 1304.86, 1306.03	[1]	3.5
Si II	1307.64	1304.37, 1309.28	[1]	
C II	1335.31	1334.53, 1335.66, 1335.71	[1]	2.5
Si IV	1396.75	1393.76, 1402.77	[1]	19
O IV]	1402.34	1397.23, 1399.78, 1401.16, 1404.81, 1407.38	[1]	
N IV]	1486.50		[1]	
Si II	1531.18	1526.71, 1533.43	[1]	
C IV	1549.05	1548.20, 1550.77	[1]	63
He II	1640.42	1640.34, 1640.47	[2]	8
O III]	1664.15	1660.81, 1666.15	[1]	
Al II	1670.79		[1]	
N III]	1750.46	1746.82, 1748.65, 1749.67, 1752.16, 1754.00	[1]	
Si II	1813.98	1808.01, 1816.93, 1817.45	[1]	
Al III	1857.40	1854.72, 1862.79	[1]	
Si III]	1892.03		[1]	
C III]	1908.73		[1]	29
N II]	2141.36	2139.01, 2142.77	[1]	0.34
C II]	2326.58	2324.21, 2325.40, 2326.11, 2327.65, 2328.84	[1]	6.0
[Ne IV]	2418.70	2418.2, 2420.9	[3]	2.2
Mg II	2797.92	2795.53, 2802.71	[1]	34
He II	3203.07	3202.96, 3203.15	[2]	

Table 24.6. *(Continued.)*

Element	Mean $\lambda^{a,b}$ (Å)	Component λ's^b (Å)	Reference	Ratio^c to Lyα
[Ne v]	3345.83		[3]	0.52
[Ne v]	3425.87		[3]	1.0
[O II]	3726.67	3726.03, 3728.82	[3]	0.78
[Ne III]	3868.75		[3]	3.6
He I	3888.65		[4]	
[Ne III]	3967.47		[3]	1.3
Hϵ	3970.07		[4]	
[S II]	4068.60		[3]	
[S II]	4076.35		[3]	2.8
Hδ	4101.73		[4]	
Hγ	4340.46		[4]	13
[O III]	4363.21		[3]	
He II	4685.65	4685.4, 4685.7	[2]	
[Ar IV]	4711.34		[3]	
[Ar IV]	4740.20		[3]	
Hβ	4861.32		[4]	22
[O III]	4958.91		[3]	0.93
[O III]	5006.84		[3]	3.4
[N I]	5199.82^d		[4]	
[Ca v]	5309.18		[5]	
[Fe VII]	5721.11		[6]	
[N II]	5754.57		[3]	
He I	5875.7		[4]	
[Fe VII]	6086.92		[6]	
[O I]	6300.30		[3]	
[O I]	6363.78		[3]	
[Fe X]	6374.53		[3]	
Hα	6562.80		[4]	
[N II]	6548.06		[3]	
[N II]	6583.39		[3]	
[S II]	6716.47		[3]	
[S II]	6730.85		[3]	
O I	8446.5		[4]	
[S III]	9068.9		[3]	
He I	10830.20	10829.09, 10830.25, 10830.34	[4]	
Pα	18751.0		[4]	
Pβ	12818.1		[4]	

Notes

^{a}Mean wavelength assuming optically thin gas, see Chap. 4.

^{b}In vacuo for $\lambda < 2000$ Å, in air $\lambda > 2000$ Å.

^{c}Mean ratio for prominent lines [$\gtrsim 0.5\%$ F(Lyα/NV) blend] based on the composite spectrum of [7].

^{d}Reader & Corliss, [2], give 5197.94 Å.

References

1. Morton, D.C. 1991, *ApJS*, **77**, 119

2. Reader, J., & Corliss, C.H. 1980, *NIST Spectroscopic Properties of Atoms and Atomic Ions Wavelengths*, NSRDS-NBS, Vol. 68, Part I

3. Kaufman, V., & Sugar, J.J. 1986, *Phys. Chem. Ref. Data*, **15**, 321

4. Weise, W.L., Smith, M.W., & Glennon, B.M. 1966, *Atomic Transition Probabilities*, Vol. 1 (National Bureau of Standards, Washington, DC)

5. Weise, W.L., Smith, M.W., & Miles, B.M. 1969, Atomic Transition Probabilities, Vol. 2 (National Bureau of Standards, Washington, DC)

6. Bowen, I.S. 1960, *ApJ*, **132**, 1

7. Francis, P.J., Hewett, P.C., Foltz, C.B., Chaffee, F.H., Weymann, R.J., & Morris, S.L. 1991, *ApJ*, **373**, 465

24.5.1   The Broad Emission Line Region (BLR, BELR)

The broad emission lines that characterize the optical/UV spectra of quasars are thought to originate in gas photoionized by the central continuum source. The smooth profiles and large line widths lead to a popular scenario of large numbers of small clouds moving at high velocity. However the direction of this motion has not been generally determined. For a detailed summary of our knowledge of the BELR, see [76]. Table 24.7 lists the physical parameters of the broad line region and the observational evidence that leads to these numbers. Reverberation mapping studies over the past ~ 10 years have revolutionized these studies by providing a direct measure of the size of the emitting region in a few objects, e.g., NGC5548 [77–79].

Table 24.7. *Parameters of the BLR.*

Parameter	Typical values	Based on
Electron density, n_e	10^{8-12} cm^{-3}	[O III] λ5007, No; [C III] λ1909, Yes[a]
Ionization parameter	$\log U \sim -1$	Line strengths + photoionization models
Temperature	10^4 K	Photoionization models
Size	0.01–0.1 pc	Variability studies in Sy1's [1]
	1 pc	In quasars
Cloud velocities	10^3–10^4 km s^{-1}	Observed linewidths
Covering factor	0.1	Observations of Lyman limit absorption photoionization models
N(H I)	$> 10^{22}$ cm^{-2}	Mg II, Fe II: yes
τ (Lyα)	10^8	Photoionization models

Note

[a]But see [2].

References

1. Netzer, H., & Peterson, B.M. 1997, in *Astronomical Time Series*, edited by D. Maoz, A. Sternberg and E. Leibowitz (Kluwer Academic, Dordrecht), p. 85
2. Mathur, S., Elvis, M.S., & Wilkes, B.J. 1995, *ApJ*, **452**, 230

24.5.2   The Narrow Emission Line Region (NLR, NELR)

Narrow emission lines are present in all varieties of AGN. They also originate mostly in photoionized gas, although collisionally ionized gas often contributes significantly. The gas is further from the continuum source than the BLR and thus has lower ionization and lower velocities. The density is also lower and many forbidden lines are present. Table 24.8 lists the physical parameters of the narrow line region and the observational evidence that leads to these numbers.

Table 24.8. *Parameters of the NLR.*

Parameter	Typical values	Based on
Electron density, n_e	$\sim 10^{3-6}$ cm^{-3}	[O III] λ5007: yes
Ionization parameter	$\log U \sim -2$	Line strengths [1]
Temperature	10^4 K	Emission lines
Size	100 pc–> 1 kpc	Photoionization models
$\log(\mathcal{M}/\mathcal{M}_\odot)$	-6	Emission lines
Cloud velocities	100–1000 km s^{-1}	Observed linewidths
N(H I)	$>$ N(H I) for BLR	Photoionization models

Reference

1. Netzer, H. 1990, in *Active Galactic Nuclei*, edited by T.J.-L. Courvoisier and M. Mayor (Springer-Verlag, Berlin), p. 57

24.5.3 Effects of Lines on Optical Magnitudes

The presence of emission lines within the bandpass of a given filter contribute significantly to the observed magnitudes of an AGN. Since this effect is a strong function of redshift, it is often useful to correct for the presence of the lines thus yielding magnitudes based on the continuum emission of the AGN alone. The correction, based on the line equivalent width, can be expressed (using B magnitude as an example):

$$\Delta B = 2.5 \log_{10}\left(1 + W_\lambda(1+z)\frac{R_B(\lambda)}{\int R_B(\lambda)\,d\lambda}\right), \tag{5}$$

where W_λ is the rest frame equivalent width in Å of the emission line, $\lambda = \lambda_{\text{rest}}(1+z)$ is the observed wavelength of the line at redshift z, and R_B is the response of the B filter in Å^{-1} [43, 80]. Figure 24.6 shows the correction as a function of z for B, V magnitudes assuming equivalent widths for the strongest lines, Lyα, C IV, [C III], Mg II, and Hβ, from Table 24.9 [81]. Differences between the emission line properties of radio-loud and radio-quiet quasars are generally negligible [82, 83], although significantly lower equivalent widths (by $\sim 30\%$) for C IV and Lyα in radio-quiet quasars have been reported [84].

Table 24.9. *Rest frame equivalent widths for emission lines in flat-spectrum, radio-loud quasars* [1].

Line	Wavelength (Å)	Equivalent width [W_λ (Å)]	$\sigma(W_\lambda)$	No.
[O III]	5007	32	28	30
Hβ	4861	47	25	26
[Ne III]	3869	5	2	21
[O II]	3727	8	6	23
[Ne V]	3426	6	6	23
Mg II	2798	27	15	113
C II]	2326	4	3	7
C III]	1909	17	12	96
He II	1640	4	3	7
C IV	1549	32	20	94
O IV]/Si IV	1400	6	3	14
O I	1304	4	3	14
Si II	1264	5	5	21
N V	1240	19	9	34
Lyα	1215	65	34	38
O VI	1034	15	13	12

Reference
1. Wilkes, B.J. 1986, *MNRAS*, **218**, 331

24.5.4 Photoionization Models

The emission lines from both regions are generally believed to arise predominantly in gas photoionized by the central continuum source. Figure 24.7 shows the relative strengths of the prominent emission lines as a function of the ionization parameter from a single zone. Excellent reviews are given in [66, 85], and line strengths for a wide range of cloud conditions are given in [86]. A standard, comprehensive photoionization computer code CLOUDY has been made generally available by anonymous ftp from Gary Ferland at the University of Kentucky, http://www.pa.uky.edu/˜gary/cloudy.

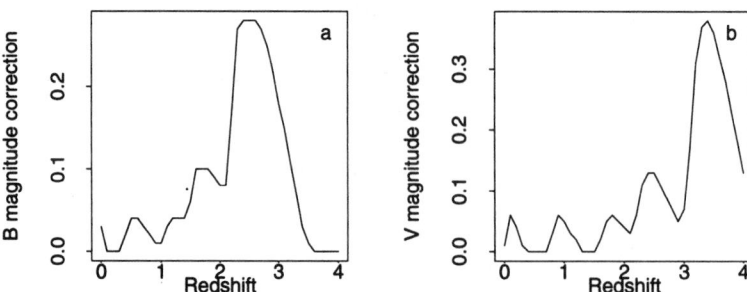

Figure 24.6. Correction to the (a) *B* magnitude and (b) *V* magnitude for the presence of emission lines as a function of redshift *z* using the equivalent widths given in Table 24.9.

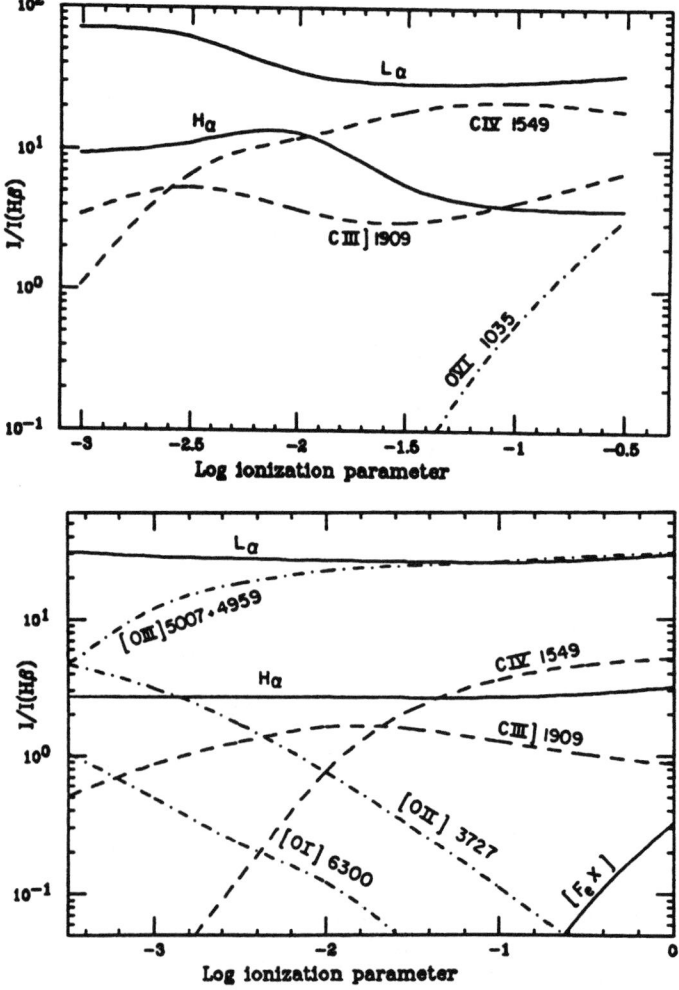

Figure 24.7. The strengths of the broad (upper) and narrow (lower) emission lines relative to Hβ as a function of the ionization parameter of the emitting gas in a single isolated BLR cloud with density 10^{10} cm^{-3} (from [66]).

24.6 ABSORPTION LINES

The spectra of quasars contain a large number of absorption features due to gas clouds along the line of sight between us and the quasar. These features provide our only view of the distribution of matter out to high redshifts apart from the quasars themselves and thus their study is of primary importance to cosmology. Excellent reviews covering all aspects of absorption line research can be found in [87, 88]. Table 24.10 lists the main types of absorption system currently known.

Table 24.10. *Classes of absorption line systems seen in quasar spectra.*

System	$\log N_H$ (cm^{-2})	$\log(N_{C\,IV})$ (cm^{-2})	b^a (km s^{-1})	Ionization	Comments
Lyα forest	13–16^b		20–30	$-3 \gtrsim$ LFN$^c \gtrsim -6$	
Metal–C IV	≥ 15.5	$\geq 13^b$	~ 100	$N_{C\,IV} > N_{C\,II}$	Intervening galaxies
Metal–Mg II	≥ 17.3		~ 100	$N_{C\,IV} > N_{C\,II}$	Intervening galaxies
Lyman limit (LLS)	≥ 17.3		~ 100	$N_{C\,IV} > N_{C\,II}$	$\sim$ Mg II systems
Damped Lyα	19–22		~ 100	$N_{C\,IV} > N_{C\,II}$	
Broad absorption line (BAL)	20–23	15–16	$(5–10) \times 10^3$	$N_{C\,IV} > N_{C\,II}$	Intrinsic?
21 cm	20.5–22		5–50	low	
Associated C IV, Mg II^d	21–22	$\lesssim 100$	$-5 \gtrsim$ LFN$^c \gtrsim -6$	$N_{C\,VI} > N_{C\,IV}$	Intrinsic?

Notes
^{a}Doppler width of H I systems: $b = \sqrt{2}\sigma$.
^{b}The lower limit of $N \sim 10^{13}$ cm^{-2} is a detection limit.
^{c}LFN $= \log(f_{N_{H\,I}})$, log of the fraction of H in the H I state.
^{d}Sometimes including an X-ray warm absorber [1].

Reference
1. Mathur, S., Wilkes, B.J., Elvis, M.S., & Fiore, F. 1994, *ApJ*, **434**, 493

24.6.1 Evolution and Distribution of Absorption Systems

The distribution of H I column densities is parametrized by $d\mathcal{N}/dN \propto N^\beta$, where $\mathcal{N}$ is the number of lines with column density N to $N + dN$ per unit redshift, z, and $\beta = 1.55 \pm 0.05$ for column density range: $12.6 < \log N < 16.0$ and $\langle z \rangle = 3.7$ [89]. A complication to this distribution is the *inverse/proximity* effect whereby the density of lines decreases as the quasar redshift is approached [90, 91]. Table 24.11 gives evolution parameters for the various types of absorption line systems. Evolution with redshift is generally described in terms of a power law: $\mathcal{N}(z) \propto (1 + z)^\gamma$. A value of γ between 0.5 and 1 is consistent with no evolution in the co-moving number density.

Table 24.11. *Evolution of the various absorption line systems.*

System	γ	z range	Reference
Lyα forest	~ 2.5	2.0–3.5	[1, 2]
Lyα forest	0.58 ± 0.50	$z < 1.3$	[3]
C IV	-1.2 ± 0.7	1.3–3.4	[4]
C IV	0.92 ± 0.4	$\bar{z} = 0.3$	[5]
Mg II	0.78 ± 0.42	0.2–2.2	[6]
LLS	1.50 ± 0.39	0.3–4.1	[7]
Damped Lyα	1.3 ± 0.5	2.8–4.4	[8]

References

1. Kim, T.-S., Hu, E.M., Cowie, L.L., & Songaila, A. 1997, *AJ*, **114**, 1
2. Bechtold, J. 1994, *ApJS*, **91**, 1
3. Bahcall, J.N. et al. 1996, *ApJ*, **457**, 19
4. Sargent, W.L.W., Boksenberg, A., & Steidel, C.C. 1989, *ApJS*, **68**, 539
5. Bahcall, J.N., et al. 1993, *ApJS*, **87**, 1
6. Steidel, C.C., & Sargent, W.L.W. 1992, *ApJS*, **80**, 1
7. Stengler-Larrea, E.A. et al. 1995 *ApJ*, **444**, 64
8. Storrie-Lombardi, L.J., Irwin, M.J., & McMahon, R.G. 1996, *MNRAS*, **282**, 1330

24.7   SPECTRAL ENERGY DISTRIBUTIONS (SEDS)

Active galaxies are multiwavelength objects, emitting roughly equal energy in all wave bands throughout the electromagnetic spectrum.   Complete observations can only be made using many different observing techniques, telescopes, satellites, etc., Table 24.12 summarizes flux measurements typically made in each wave band.   Transfer between various flux units is given in Chap. 9 and conversion of magnitude to flux is discussed in Chap. 15.   Figures 24.8 and 24.9 show examples of radio-loud and radio-quiet quasar spectral energy distributions (SED) and the mean SED for low redshift quasars given by [39]. Table 24.13 lists the prominent features of quasar SEDs. Table 24.14 gives the bolometric corrections derived by Elvis et al. [39, Table 21], based upon 43 SEDs in their sample.

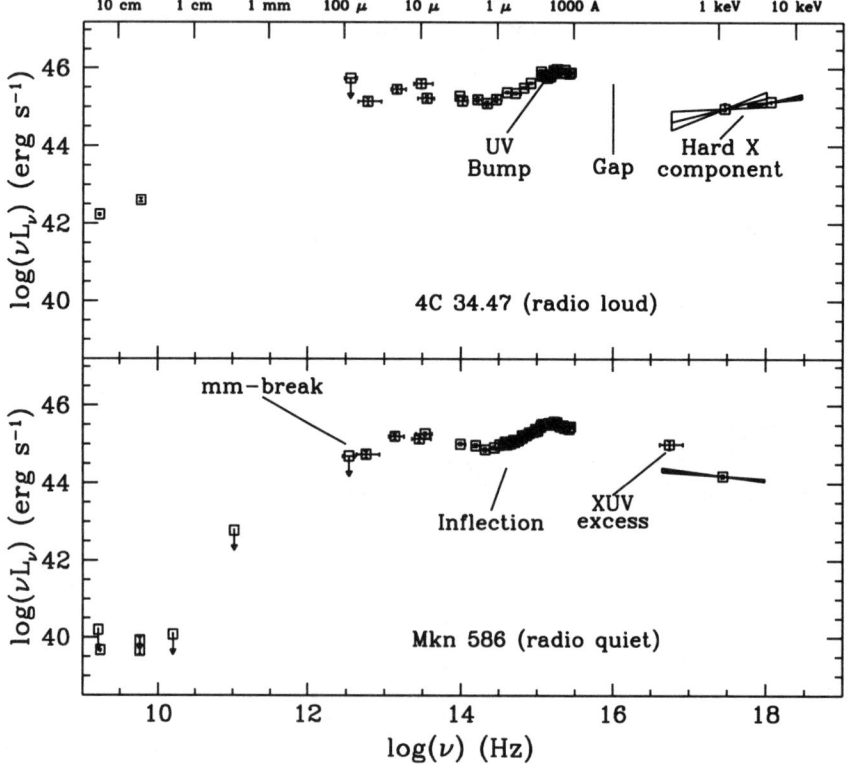

Figure 24.8. The radio–X-ray energy distribution of radio-loud quasar, 4C 34.47, and radio-quiet quasar, Mkn 586 (from [39, Fig. 1]).

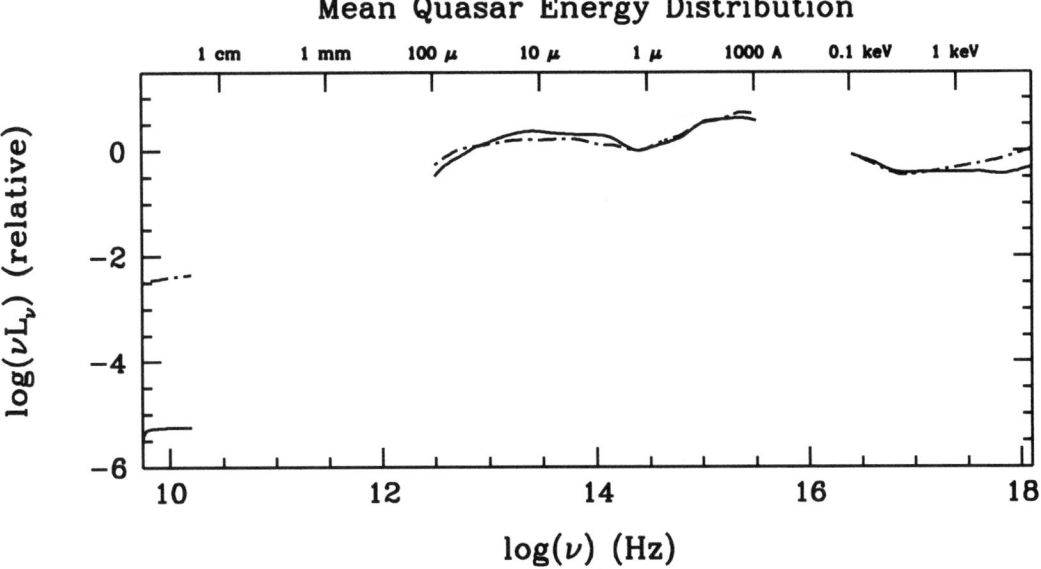

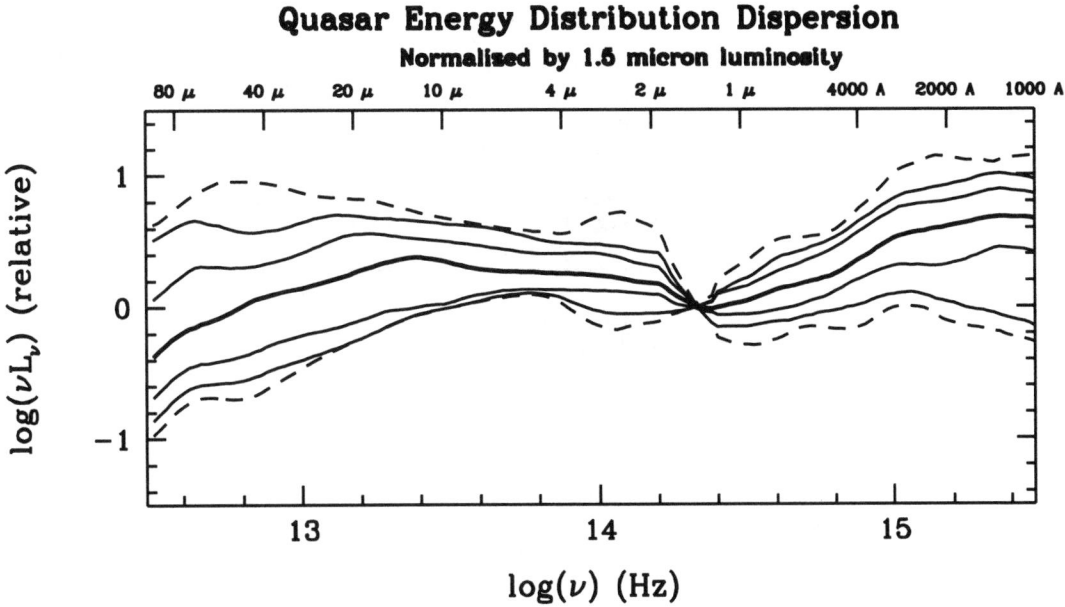

Figure 24.9. (top) The mean radio–X-ray spectral energy distribution of radio-loud (dashed line) and radio-quiet (solid line) quasars, and (bottom) the dispersion around the mean SED in the IR–UV region including 68, 90, and 100 percentile ranges (from [39, Fig. 11]).

Table 24.12. *Typical units for multiwavelength data.*

Wave band	Measurement	Units
Radio	Brightness temperature	Kelvin (K)
	Power, S_ν	Jansky (Jy)
mm	Power, S_ν	Jansky (Jy)
IR	Magnitude	J, H, K, L, N, Q, see Chapter 15
	Spectrum: F_ν/F_λ	$\mathrm{erg\,cm^{-2}\,s^{-1}\,Hz^{-1}/erg\,cm^{-2}\,s^{-1}\,Å^{-1}}$
Optical/UV	Magnitude	U, B, V, R, I
	Spectrum: F_ν/F_λ	$\mathrm{erg\,cm^{-2}\,s^{-1}\,Hz^{-1}/erg\,cm^{-2}\,s^{-1}\,Å^{-1}}$
X-ray	Flux/flux density	$\mathrm{erg\,cm^{-2}\,s^{-1}}$ or $\mathrm{Jy/erg\,cm^{-2}\,s^{-1}\,Hz^{-1}}$
γ-ray	Photon flux	$\mathrm{photons\,cm^{-2}\,s^{-1}}$

Table 24.13. *Features and possible continuum energy generation mechanisms in the spectral energy distributions.*

Component	Mechanism	Reference
All quasars		
OUV blue bump	Thermal: optically thick accretion disk	[1–3]
	optically thin free–free	[4, 5]
1 μm dip	> Dust sublimation temperature	[6]
IR bump	Thermal: cool + warm dust	[6]
	Nonthermal: synchrotron	[7]
X-ray	Seed spectrum, $\alpha_x \sim 1$	[8]
	Compton reflection	[9, 10]
		[11]
	Thermal: accretion disk	[2,3]
Radio-loud quasars		
Radio Core	Synchrotron, flat-spectrum, beamed extends into IR–UV?	
Radio Lobes	Synchrotron, steep-spectrum, isotropic	
X-ray, radio-linked	Nonthermal: synchrotron self-Compton	[12]
	/pair production	[13]
γ-ray	Nonthermal: synchrotron self-Compton	[14, 15]
	Thermal: Comptonization	[16]

References
1. Sun, W.-S., & Malkan, M. 1989, *ApJ*, **346**, 68
2. Czerny, B., & Elvis, M. 1987, *ApJ*, **321**, 305
3. Laor, A. 1990, *MNRAS*, **246**, 369
4. Barvainis, R. 1993, *ApJ*, **413**, 513
5. Ferland, G.F., Korista, K.T., & Peterson, B.M. 1990, *ApJ*, **363**, L21
6. Sanders, D., Phinney, E.S., Neugebauer, G., Soifer, B.T., & Matthews, K. 1989, *ApJ*, **347**, 74
7. Carleton, N.P., Elvis, M., Fabbiano, G., Willner, S.P., Lawrence, A., & Ward, M. 1987, *ApJ*, **318**, 595
8. Haart, F., & Maraschi, L. 1993, *ApJ*, **413**, 507
9. Lightman, A.P., & White, T.R. 1988, *ApJ*, **335**, 57
10. Guilbert, P.W., & Rees, M.J. 1988, *MNRAS*, **233**, 475
11. Pounds, K.A., Nandra, K., Stewart, G.C., George, I.M., & Fabian, A.C. 1990, *Nature*, **344**, 132
12. Zamorani, G. et al. 1981, *ApJ*, **245**, 357
13. Lightman, A.P., & Zdziarski, A.A. 1987, *ApJ*, **319**, 643
14. Fichtel, C.E. et al. 1994, *ApJS*, **84**, 551
15. Bloom, S.D., & Marscher, A.P. 1996, *ApJ*, **461**, 657
16. Zdziarski, A.A., Johnson, W.N., & Magdziarz, P. 1996, *MNRAS*, **283**, 193

Table **24.14**. *Bolometric corrections.*[a]

	Median	Mean, σ	Min.	Max.
$L_{bol}/L_{2500\ \text{Å}}$	5.2	6.2 ± 2.7	2.7	16.8
L_{bol}/L_B	10.4	11.5 ± 4.4	5.1	25.1
L_{bol}/L_V	13.2	13.8 ± 5.3	6.5	29.5
$L_{bol}/L_{1.5\ \mu m}$	24.3	25.4 ± 9.1	8.7	41.8
$L_{UVOIR}{}^{b}/L_{2500\ \text{Å}}$	3.5	4.1 ± 2.2	1.4	12.8
L_{UVOIR}/L_B	7.0	7.5 ± 3.5	4.2	22.7
L_{UVOIR}/L_V	8.2	9.0 ± 4.1	4.7	23.0
$L_{UVOIR}/L_{1.5\ \mu m}$	15.6	16.1 ± 5.6	8.1	29.5
$L_{ion}{}^{c}/L_{bol}$	0.32	0.32 ± 0.13	0.07	0.68
$N_{ion}R^{d}/L_{bol}$	0.11	0.11 ± 0.04	0.02	0.19
$L_{ion}/N_{ion}R$	2.8	3.0 ± 0.8	1.7	5.0

Notes

[a] Bolometric correction factors for UV, visible (V), and IR monochrommatic luminosities [$\nu L(\nu)$ in the rest frame]. Errors in individual energy distributions have been ignored for the purposes of this table [1].

[b] L_{UVOIR}: luminosity in the range 100–0.1 μm.

[c] L_{ion} = ionizing luminosity: 912 Å–10 keV.

[d] $N_{ion}R$ = (number of ionizing photons) $\times$ 1 Ry.

Reference

1. Elvis, M., Wilkes, B.J., McDowell, J.C., Green, R.F., Bechtold, J., Willner, S.P., Cutri, R., Oey, M.S., & Polomski, E. 1994, *ApJS*, **95**, 1

24.8 LUMINOSITY FUNCTIONS AND THE SPACE DISTRIBUTION OF QUASARS

Figure 24.10 shows the optical surface density of quasars from the combined sample discussed in Hartwick and Schade [92], who also provide a comprehensive review of results in this area.

The observational luminosity function $\Phi(L, z)$ (= space density of quasars within a unit luminosity interval and in a limited redshift range) is generally determined using the $1/V_a$ statistic [92, equation (3)].

Pure power law luminosity evolution gives acceptable fits to the data:

$$\Phi(L, z) = \frac{\Phi^*}{[L/L^*(z)]^\alpha + [L/L^*(z)]^\beta}.$$

Evolution is given by

$$L^*(z) = L^*(0)(1 + z)^k \qquad (z < 2)$$

where L^* is the luminosity at the break between the two slopes α ($L \leq L^*$) and β ($L > L^*$). Table 24.15 lists values for these parameters and the relevant references.

Evolution in radio, optical, and X-ray bands slows/stops for $z \gtrsim 2$, although optical results from the LBQS [93] suggest that it continues at a slower rate, $k \sim 1.5$ [94]. Figure 24.11 shows the luminosity function for the LBQS sample based on Table 4 of [95]. Note also that the presence of a range of slopes in the X-ray/optical continua could affect these results [96].

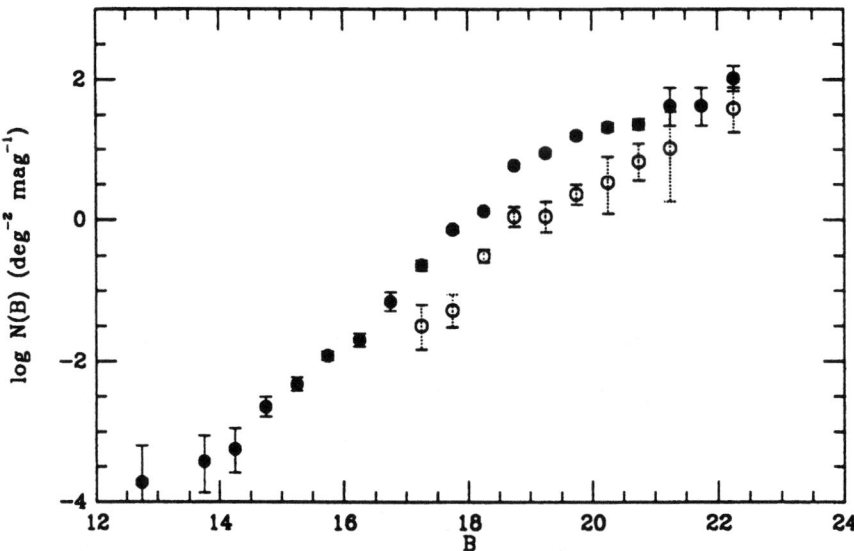

Figure 24.10. The optical surface density of quasars, from [92], solid symbols: $0 < z < 2.2$; open symbols: $2.2 < z < 3.3$.

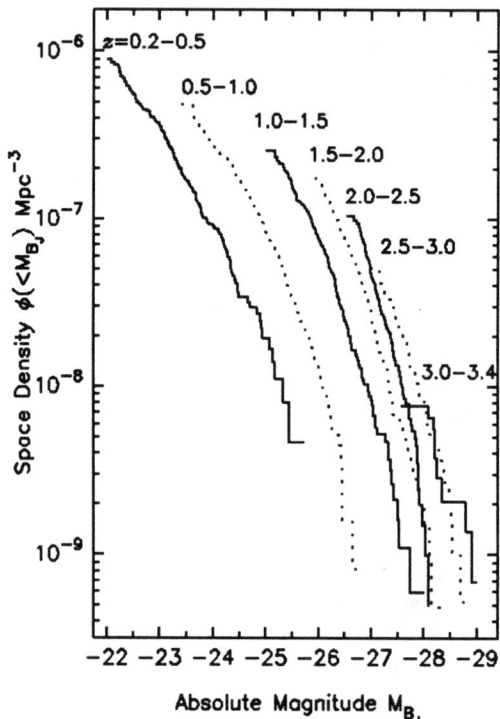

Figure 24.11. The cumulative space density of QSOs derived from the LBQS sample in seven redshift shells as labeled (figure of Table 4 from [95], courtesy of Paul Hewett).

Table 24.15. *Parameters for pure, power law luminosity evolution in various spectral regions, z < 2, for $q_0 = 0.0$.*

Param.	Optical 4400 Å	X-ray 2 keV	Radio (CDQs) 2.7 GHz	Units
α	3.6	3.3 ± 0.2	2.9	
β	1.5	1.6 ± 0.2	1.8	
$L^*(0)$	4.2×10^{29}	0.5×10^{44a}	2.3×10^{33}	$\mathrm{erg\,s^{-1}\,Hz^{-1}}$
Φ^*	5×10^{-7}	1×10^{-6}	2×10^{-9}	$\mathrm{mag.^{-1}\,Mpc^{-3}}$
k	3.50 ± 0.05	$3.0^{+0.27}_{-0.40}$	~ 3	
z_{cut}	1.9	~ 1.6	~ 2	
Reference	[1]	[2]	[3]	

Note

[a] $\mathrm{erg\,s^{-1}}$, 0.3–3.5 keV.

References

1. Boyle, B.J., Jones, L.R., & Shanks, T. 1991, *MNRAS*, **251**, 482
2. Jones, L.R. et al. 1997, *MNRAS*, **285**, 547
3. Dunlop, J.S., & Peacock, J.A. 1990, *MNRAS*, **247**, 19

24.9 BL LACS, HPQS, AND OVVS

The boundaries between these three classes are not always clear. OVVs are the strongly variable subset of HPQs, both of which have strong emission lines in their spectra at some/all epochs. Since in OVVs these lines sometimes disappear, more than one epoch of observations is required to distinguish between a BL Lac and an OVV, for this reason the two classes are often referred to jointly as blazars. However it should be noted that there are observational differences between OVVs and BL Lacs that should discourage such unification [97–100]. BL Lacs, due to their lack of emission lines, are most efficiently found in radio or X-ray surveys resulting in their division into two subcategories: RBL, radio-selected BL Lac (or LBL, low-frequency peaked SED); and XBL, X-ray selected BL Lac (or HBL, high-frequency peaked SED). Discussion continues as to whether or not these are distinct classes.

Table 24.16 lists salient parameters for these objects. A conference proceedings dealing with all aspects of BL Lac Objects and OVVs is [101].

Table 24.16. *General characteristics of highly polarized AGN.*

Property	BL Lac	HPQ	OVV	Reference
Optical morphology	Point source	Point source	Point source	
Continuum	Smooth $\lambda < 10\,\mu$m	Weak BB[a]	Weak BB[a]	
Emission lines	No	Yes	Yes	
Parent population	FRI RGs	FRII RGs	FRII RGs	[1]
P_{opt}	3–40%	3–20%	3–20%	[2, 3]
T_{var}	Days	Days–years	Days	
Δm	1–5	0.5–5	1–5	[2, 4]
μ (angular velocity)	0.1–0.8	0.05–2.7	0.05–2.7	[5]
β_{app}	2–4	1–18.5	1–18.5	[5]

Note

[a] BB = optical/UV blue bump.

References

1. Padovani, P., & Urry, C.M. 1990, *ApJ*, **356**, 75
2. Angel, J.R.P., & Stockman, H.S. 1980, *ARA&A*, **18**, 321
3. Stockman, H.S., Moore, R.L., & Angel, J.R.P. 1984, *ApJ*, **279**, 485
4. Moore, R.L., & Stockman, H.S. 1984, *ApJ*, **279**, 465
5. Zensus, J.A. 1989, in *BL Lac Objects*, Lecture Notes in Physics, 334, edited by L. Maraschi, T. Maccacaro, and M.-H. Ulrich (Springer-Verlag, Berlin), p. 3

24.10   LOW-LUMINOSITY ACTIVE GALACTIC NUCLEI (LLAGN)

These objects have lower luminosity than quasars; $M > -23$ and $z \lesssim 0.1$. A comprehensive review can be found in [102]. Table 24.17 lists values of commonly used parameters for each class of LLAGN. Figure 24.2 shows the line ratio properties in graphical form. Recent results have shown that weak AGN are present in $\sim 20\%$ of all AGN and $\sim 10\%$ of all luminous galaxies [33].

Table 24.17. *Typical values for common parameters.*

	Sy1/BLRG	Sy2/NLRG	LINER	Starburst	Reference
Host galaxy[a]	Sp/E	Sp/E	Sp	Sp	[1]
FWHM[b] $(km\,s^{-1})$	10^3–10^4	10^2–10^3	200–400		
$\log(\mathcal{M}_{BH}/\mathcal{M}_\odot)$	7.5–9.0				[2]
$\log U$	-2^c	-3	-3.5		[3]
$\alpha_x(0.1\text{–}3.5\text{ keV})^d$	0.5–2/0–1[e]	0.5–2/0–1[e]	0.5–2[e]		[4]
$\alpha_x(2\text{–}10\text{ keV})$	0.7–1[e]	~ 0.4			[5, 6]
l	0 to > 200	0 to > 200			[7, 8]
[O III]/Hβ	1.5–3	3–20	0.1–1.5	0.03–8	[3, 9,10]
[N II]/Hα		> 0.5	> 0.5	< 0.5	[3, 9]
[O I]/Hα		> 0.08	> 0.08	< 0.08	[3, 9]
[S II]/Hα		> 0.4	> 0.4	< 0.4	[3, 9]
[O II]/[O III]	≤ 0.5	0.1–1	1–10	> 1	[3]
[O I]/[O III]			> 0.33		[11]

Notes

[a]Most likely morphology of host galaxy: Sp = spiral, E = elliptical.

[b]The FWHM increases with the critical density of the forbidden line [12].

[c]Narrow-line region ($\log U \sim -1$ for the BLR).

[d]90% confidence range.

[e]There is marginal evidence that Sy2, NLRGs, and LINERs have flatter X-ray slopes than Sy1/BLRGs.

References

1. McLeod, K.K., & Rieke, G.H. 1994, *ApJ*, **420**, 58
2. Sun, W.-S., & Malkan, M. 1989, *ApJ*, **346**, 68
3. Netzer, H. 1990, in *Active Galactic Nuclei*, edited by T. Courvoisier and M. Mayer (Springer-Verlag, Berlin), p. 57
4. Kruper, J.S., Urry, C.M., & Canizares, C.R. 1990, *ApJS*, **74**, 347
5. Turner, T.J., & Pounds, K.A. 1989, *MNRAS*, **240**, 833
6. Awaki, H. et al. 1991, *PASJ*, **43**, 195
7. Done, C., & Fabian, A.C. 1990, *MNRAS*, **240**, 81
8. Lightman, A.P., & Zdziarski, A.A. 1987, *ApJ*, **319**, 643
9. Osterbrock, D. 1989, *Astrophysics of Gaseous Nebulae and Active Galactic Nuclei* (University Science Books, Mill Valley)
10. Shuder, J.M., & Osterbrock, D.E. 1981, *ApJ*, **250**, 55
11. Heckman, T.M. 1980, *A&A*, **87**, 152
12. Filippenko, A.V. 1985, *ApJ*, **289**, 475

24.11   AGN ENVIRONMENTS

Low-redshift quasars and AGN are known to be embedded in galaxies. Relatively large samples (with > 20 objects) of nearby quasars have now been studied with ground-based charge-coupled devices (CCDs) [103–108] and infrared arrays [109–112], and with the *Hubble* Space Telescope [113–116].

Many of the closest quasars ($z \lesssim 0.1$), which are in general radio-quiet and low-luminosity objects, live in spiral hosts, although there is a strong bias against edge-on spirals [103, 111]. Elliptical hosts have also been reported for several radio-quiet quasars [113]. Radio-loud quasar hosts are generally thought to be elliptical, though fits to host-galaxy luminosity profiles for these and other high-luminosity quasars are ambiguous. The most luminous quasars are only found in luminous galaxies ($\sim L^*$ [107, 112]). Close companions and/or complex morphology (implying recent interactions) are common [113, 115, 117]. A recent conference proceedings providing a comprehensive review of the subject is [118].

For the few BL Lac objects with data available, the hosts are generally elliptical [119], although a few are spiral [120].

REFERENCES

1. Schmidt, M. 1963, *Nature*, **197**, 1040
2. Hazard, C., Mackay, M.B., & Shimmins A.J. 1963, *Nature*, **197**, 1037
3. Greenstein, J.L., & Schmidt, M. 1964, *ApJ*, **140**, 1
4. Sandage, A.R. 1965, *ApJ*, **141**, 1560
5. Seyfert, C.K. 1943, *ApJ*, **97**, 28
6. Peterson, B.M. 1997, *An Introduction to Active Galactic Nuclei* (Cambridge University Press, Cambridge)
7. Stocke, J.S., Morris, S.L., Gioia, I.M., Maccacaro, T., Schild, R., Wolter, A., Fleming, T.A., & Henry, J.P. 1991, *ApJS*, **76**, 813
8. Veilleux, S., & Osterbrock, D.E. 1987, *ApJS*, **63**, 295
9. Evans, I.N., & Dopita, M.A. 1985, *ApJS*, **58**, 125
10. McCall, M.L., Rybski, P.M., & Shields, G.A. 1985, *ApJS*, **57**, 1
11. Green, P.J., Anderson, S.F., & Ward, M.J. 1992, *MNRAS*, **254**, 30
12. Wilkes, B.J. 1984, *MNRAS*, **207**, 73
13. Kallman, T.R., Wilkes, B.J., Krolik, J.H., & Green, R.F. 1993, *ApJ*, **403**, 45
14. Weymann, R.J., Morris, S.L., Foltz, C.B., & Hewett, P.C. 1991, *ApJ*, **373**, 23
15. Allen, R.G., Smith, P.A., Angel, J.R.P., Miller, B.W., Anderson, S.F., & Margon, B. 1993, *ApJ*, **403**, 610
16. Peterson, B.M. et al. 1992, *ApJ*, **392**, 470
17. Hewitt, A., & Burbidge, G. 1993, *ApJS*, **87**, 451
18. Véron-Cetty, M.-P., & Veron, P. 1993, ESO Scientific Report No. 13
19. Padovani, P. & Giommi, P. 1995, *MNRAS*, **277**, 1477
20. Hewitt, A., & Burbidge, G. 1991, *ApJS*, **75**, 297
21. Junkkarinen, V., Hewitt, A., & Burbidge, G. 1991, *ApJS*, **77**, 203
22. MacAlpine, G.M., Lewis, D.W., & Smith, S.B. 1977, *ApJS*, **35**, 203
23. Wilkes, B.J., Wright, A.E., Peterson, B.A., & Jauncey, D.L. 1983, *Astron. Soc. Aust. Conf. Proc.*, **5**, 2
24. Anderson, S.F., & Margon, B. 1987, *ApJ*, **314**, 111
25. Baldwin, J.A., Wampler, E.J., & Gaskell, C.M. 1989, *ApJ*, **338**, 630
26. Keel, W.C. 1983, *ApJS*, **52**, 229
27. Wolfe, A.M., Turnshek, D.A., Smith, H.E., & Cohen, R.D. 1986, *ApJS*, **61**, 249
28. Irwin, M., McMahon, R.G., & Hazard, C. 1991, ASP Conference Series 21, *The Space Distribution of Quasars*, edited by D. Crampton, p. 117
29. Schneider, D.P., Schmidt, M. & Gunn, J.E. 1994, *AJ*, **107**, 1245
30. Kinney, A.L., Bohlin, R.C., Blades, J.C., & York, D.G. 1991, *ApJS*, **75**, 645
31. Lanzetta, K.M., Turnshek, D.A., & Sandoval, J. 1993, *ApJS*, **84**, 109
32. Kinney, A.L., Bohlin, R.C., Calzetti, D., Panagia, N., & Wyse, R.F.G. 1993, *ApJS*, **86**, 5
33. Ho, L.C., Filippenko, A.V., Sargent, W.L.W., & Peng, C.Y. 1997, *ApJS*, **112**, 391
34. Low, F.J., Cutri, R.M., Huchra, J.P., & Kleinmann, S.G. 1988, *ApJL*, **327**, 41
35. Rush, B., Malkan, M., & Spinoglio, L. 1993, *ApJS*, **89**, 1
36. Kleinmann, S.G. et al. 1994, *Ap&SS*, **217**, 11
37. Sanders, D., Phinney, E.S., Neugebauer, G., Soifer, B.T., & Matthews, K. 1989, *ApJ*, **347**, 29
38. Impey, C.D., & Neugebauer, G. 1988, *AJ*, **95**, 307
39. Elvis, M., Wilkes, B.J., McDowell, J.C., Green, R.F., Bechtold, J., Willner, S.P., Oey, M.S., Polomski, E., & Cutri, R. 1994, *ApJS*, **95**, 1
40. Maccacaro, T., Della Ceca, R., Gioia, I.M., Morris, S.L., Stocke, J.T., & Wolter, A. 1991, *ApJ*, **374**, 117
41. Piccinotti, G., Mushotzky, R.F., Boldt, E.A., Holt, S.S., Marshall, F.E., Serlemitsos, P.J., & Shafer, R.A. 1982, *ApJ*, **253**, 485
42. Schwartz, D.A. et al. 1989, *BL Lac Objects*, in Lecture Notes in Physics, 334, edited by L. Maraschi, T. Maccacaro, and M.-H. Ulrich (Springer-Verlag, Berlin), p. 209
43. Wilkes, B.J., Tananbaum, H., Worrall, D.M., Avni, Y., Oey, M.S., & Flanagan, J. 1994, *ApJS*, **92**, 53
44. Elvis, M., Plummer, D., Schachter, J., & Fabbiano, G. 1992, *ApJ*, **80**, 257
45. Giommi, P. et al. 1989, *BL Lac Objects*, in Lecture Notes in Physics, 334, edited by L. Maraschi, T. Maccacaro, and M-H. Ulrich (Springer-Verlag, Berlin), p. 231
46. Voges, W. et al. 1996, IAUC, 6420
47. Boyle, B.J., Griffiths, R.E., Shanks, T., Stewart, G.C., & Georgantopolous, I. 1993, *MNRAS*, **260**, 49
48. Green, P.J., Schartel, N., Anderson, S.F., Hewett, P.C., Foltz, C.B., Fink, H.H., Brinkmann, W., Trümper, J., & Margon, B. 1995, *ApJ*, **450**, 51
49. Boyle, B.J., Wilkes, B.J., & Elvis, M. 1997, *MNRAS*, **285**, 511
50. Puchnarewicz, E.M., Mason, K.O., Romero-Colmenero, E., Carrera, F.J., Hasinger, G., McMahon,

R., Mittaz, J.P.D., Page, M.F., & Carballo, R. 1996, *MNRAS*, **281**, 1243

51. Singh, K.P., Barrett, P., White, N.E., Giommi, P., & Angelini, L. 1995, *ApJ*, **455**, 456
52. Fichtel, C.E. et al. 1994, *ApJS*, **94**, 551
53. Masson, C.R., & Wall, J.V. 1977, *MNRAS*, **180**, 193
54. Kellerman, K.I., Sramek, R., Schmidt, M., Shaffer, D.B., & Green, R.F. 1989, *AJ*, **98**, 1195
55. Kuhr, H., Pauliny-Toth, I.I.K., Witzel, A., & Schmidt, J. 1981a, *AJ*, **86**, 854
56. Condon, J.J., Cotton, W.D., Greisen, E.W., Yin, Q.F., Perley, R.A., Taylor, G.B., & Broderick, J.J. 1998, *AJ*, **115**, 1693
57. Gregg, M.D., Becker, R.H., White, R.L., Helfand, D.J., McMahon, R.G. & Hook, I.M. 1996, *AJ*, **112**, 407
58. Tananbaum, H. et al. 1979, *ApJ*, **234**, L9
59. Zamorani, G. et al. 1981, *ApJ*, **245**, 357
60. Impey, C.D., & Tapia, S. 1990, *ApJ*, **354**, 124
61. Wilkes, B.J., & Elvis, M. 1987, *ApJ*, **323**, 243
62. Shastri, P., Wilkes, B.J., Elvis, M., & McDowell, J.C. 1993, *ApJ*, **410**, 29
63. Orr, M.J.L., & Browne, I.W.A. 1982, *MNRAS*, **200**, 1067
64. Sziemiginowska, A., Kuhn, O., Elvis, M., Fiore, F., McDowell, J., & Wilkes, B.J. 1995, *ApJ*, **454**, 77
65. Guilbert, P.W., Fabian, A.C., & Rees, M.J. 1983, *MNRAS*, **205**, 593
66. Netzer, H. 1990, in *Active Galactic Nuclei*, edited by T.J.-L. Courvoisier and M. Mayor (Springer-Verlag, Berlin), p. 57
67. Baldwin, J.A., Phillips, M.M., & Terlevich, R. 1981, *PASP*, **93**, 5
68. *Emission Lines in Active Galaxies: New Methods and Techniques*, 1997, edited by B.M. Peterson, F.-Z. Cheng, and A.S. Wilson (Astronomical Society of the Pacific, San Francisco)
69. Boroson, T., & Green, R.F. 1992, *ApJS*, **80**, 109
70. Vestergaard, M. & Wilkes, B.J. 1999, *ApJ*, submitted
71. Nandra, K., George, I.M., Mushotzky, R.F., Turner, T.J., & Yaqoob, T. 1997, *ApJ*, **477**, 602
72. Matt, G., Fabian, A.C. & Ross, R.R. 1996, *MNRAS*, **278**, 1111
73. Mushotzky, R.F., Done, C., & Pounds, K.A. 1993, *ARA&A*, **31**, 717
74. Netzer, H. 1997 in *Emission Lines in Active Galaxies: New Methods and Techniques*, edited by B.M. Peterson, F.-Z. Cheng, and A.S. Wilson (Astronomical Society of the Pacific, San Francisco), p. 20
75. Francis, P.J., Hewett, P.C., Foltz, C.B., Chaffee, F.H., Weymann, R.J., & Morris, S.L. 1991, *ApJ*, **373**, 465
76. Baldwin, J.A. 1997 in *Emission Lines in Active Galaxies: New Methods and Techniques*, edited by B.M. Peterson, F.-Z. Cheng, and A.S. Wilson (Astronomical Society of the Pacific, San Francisco), p. 80
77. Korista, K.T. et al. 1995, *ApJ*, **97**, 285
78. ASP Conference Series, Volume 69, on *Reverberation Mapping of the Broad-Line Region in Active Galactic Nuclei* 1994, edited by P.M. Gondhalekar, K. Horne, and B.M. Peterson (ASP, San Francisco)
79. Ulrich, M-H., Maraschi, L. & Urry, C.M. 1997, *ARA&A*, **35**, 445

80. Marshall, H.M. 1983, Ph.D. thesis, Harvard University
81. Wilkes, B.J. 1986, *MNRAS*, **218**, 331
82. Corbin, M.R. 1992, *ApJ*, **391**, 577
83. Steidel, C.C., & Sargent, W.L.W. 1991, *ApJ*, **382**, 433
84. Francis, P.J., Hooper, E.J., & Impey, C.D. 1993, *AJ*, **106**, 417
85. Ferland, G.F., & Shields, G.A. 1985, in *Astrophysics of Active Galaxies and Quasars*, edited by J. Miller (University Science Books, Mill Valley), p. 157
86. Baldwin, J. et al. 1995, *ApJ*, **455**, L119
87. Blades, C., Turnshek, D., & Norman, C.A., editors, 1988, *QSO Absorption Lines: Probing the Universe* (Cambridge University Press, Cambridge)
88. Meylan, G., editor, 1995, *QSO Absorption Lines: Proceedings of ESO Workshop* (Springer-Verlag, Berlin; New York)
89. Lu, L., Sargent, W.L.W, Woble, D.S., & Takada-Hidai, M. 1996, *ApJ*, **472**, 509
90. Murdoch, H.S., Hunstead, R.W., Pettini, M. & Blades, J.C. 1986, *ApJ*, **309**, 19
91. Weymann, R.J., Carswell, R.F., & Smith, M.G. 1981, *ARA&A*, **19**, 41
92. Hartwick, F.D.A., & Schade, D. 1990, *ARA&A*, **28**, 437
93. Morris, S.L., Weymann, R.J., Anderson, S.F., Hewett, P.C., Foltz, C.B., Chaffee, F.H., Francis, P.J., & MacAlpine, G.M. 1991, *AJ*, **102**, 1627
94. Hewett, P.C., Foltz, C.B., & Chaffee, F.H. 1993, *ApJ*, **406**, L43
95. Hewett, P.C., Foltz, C.B., & Chaffee, F.H. 1995, *AJ*, **109**, 1498
96. Francis, P.J., 1993, *ApJ*, **407**, 519
97. Worrall, D.M., & Wilkes, B.J. 1990, *ApJ*, **360**, 396
98. Smith, P.S., Balonek, T.J., Elston, R., & Heckert, T.A. 1987, *ApJS*, **64**, 459
99. Browne, I.W.A. 1989, *BL Lac Objects*, in Lecture Notes in Physics, 334, edited by L. Maraschi, T. Maccacaro, and M-H. Ulrich (Springer-Verlag, Berlin), p. 401
100. Miller, J.S. 1989, in *BL Lac Objects*, Lecture Notes in Physics, 334, edited by L. Maraschi, T. Maccacaro, and M-H. Ulrich (Springer-Verlag, Berlin), p. 395
101. Maraschi, L., Maccacaro, T., & Ulrich, M-H. 1989, *BL Lac Objects*, Lecture Notes in Physics, 334 (Springer-Verlag, Berlin)
102. Osterbrock, D.E. 1993, *ApJ*, **404**, 551
103. Malkan, M.A., Margon, B., & Chanan, G.A. 1984, *ApJ*, **280**, 66
104. Malkan, M.A. 1984, *ApJ*, **287**, 555
105. Hutchings, J.B. 1987, *ApJ*, **320**, 122
106. Hutchings, J.B., Janson, T., & Neff, S.G. 1989, *ApJ*, **342**, 660
107. Véron-Cetty, M.-P., & Woltjer, L. 1990, *A&A*, **236**, 69
108. Green, R.F., & Yee, H.K.C. 1984, *ApJS*, **54**, 495
109. Dunlop, J.S., Taylor, G.L., Hughes, D.H., & Robson, E.I. 1993, *MNRAS*, **264**, 455
110. Taylor, G.L., Dunlop, J.S., Hughes, D.H., & Robson, E.I. 1996, *MNRAS*, **283**, 930
111. McLeod, K.K., & Rieke, G.H. 1994a, *ApJ*, **420**, 58
112. McLeod, K.K., & Rieke, G.H. 1994b, *ApJ*, **431**, 137
113. Bahcall, J.A., Kirhakos, S., & Schneider, D.P. 1995, *ApJ*, **450**, 486
114. Hutchings, J.B., & Morris, S.C. 1995, *AJ*, **109**, 1541

115. Hutchings, J.B., & Neff, S.G. 1992, *AJ*, **104**, 1

116. Hooper, E.J., Impey, C.D., & Foltz, C.B. 1997, *ApJ*, **480**, L95

117. Boyce, P.J., Disney, M.J., Blades, J.C., Boksenberg, A., Crane, P., Deharveng, J.M., Macchetto, F.D., Mackay, C.D., & Sparks, W.B. 1996, *ApJ*, **473**, 760

118. Clements, D., & Perez-Fournon, I., editors, 1997, *ESO Astrophysics Symposia: Quasar Hosts* (Springer-Verlag, Berlin)

119. Ulrich, M-H. 1989, *BL Lac Objects*, Lecture Notes in Physics, 334, edited by L. Maraschi, T. Maccacaro, and M.-H. Ulrich (Springer-Verlag, Berlin)

120. Abraham, R.G., Crawford, C.S., & McHardy, I.M. 1991, *MNRAS*, **252**, 482

Chapter 25

Clusters and Groups of Galaxies

Neta A. Bahcall

25.1 TYPICAL PROPERTIES OF CLUSTERS AND GROUPS OF GALAXIES

A Hubble constant of $H_0 = 100h$ km s^{-1} Mpc^{-1} is used throughout this chapter.

Table 25.1. *Typical properties of clusters and groups.*

Property[a]	Rich clusters	Groups and poor clusters	Notes
Richness	30–300 galaxies	3–30 galaxies	b
Radius	$(1–2)h^{-1}$ Mpc	$(0.1–1)h^{-1}$ Mpc	c
Radial velocity dispersion	400–1400 km s^{-1}	100–500 km s^{-1}	d
Radial velocity dispersion (median)	~ 750 km s^{-1}	~ 250 km s^{-1}	d
Mass ($r \leq 1.5h^{-1}$ Mpc)	$(10^{14}–2 \times 10^{15})h^{-1}\mathcal{M}_\odot$	$(10^{12.5}–10^{14})h^{-1}\mathcal{M}_\odot$	e
Luminosity (B) ($r \leq 1.5h^{-1}$ Mpc)	$(6 \times 10^{11}–6 \times 10^{12})h^{-2}L_\odot$	$(10^{10.5}–10^{12})h^{-2}L_\odot$	f
$\langle \mathcal{M}/L_B \rangle$	$\sim 300h\,\mathcal{M}_\odot/L_\odot$	$\sim 200h\,\mathcal{M}_\odot/L_\odot$	g
X-ray temperature	2–14 keV	$\lesssim 2$ keV	h
X-ray luminosity	$(10^{42.5}–10^{45})h^{-2}$ erg s^{-1}	$\lesssim 10^{43}h^{-2}$ erg s^{-1}	h
Cluster number density	$(10^{-5}–10^{-6})h^3$ Mpc^{-3}	$(10^{-3}–10^{-5})h^3$ Mpc^{-3}	i
Cluster correlation scale	$(22 \pm 4)h^{-1}$ Mpc ($R \geq 1$)	$(13 \pm 2)h^{-1}$ Mpc	j
Fraction of galaxies in clusters or groups	$\sim 5\%$	$\sim 55\%$	k

Notes

[a] In most entries, the typical range in the listed property or the median value is given. Groups and poor clusters are a natural and continuous extension to lower richness, mass, size, and luminosity from the rich and rare clusters.

[b] Cluster richness (Section 25.4.1): the number of cluster galaxies brighter than $m_3 + 2^m$ (where m_3 is the magnitude of the third brightest cluster galaxy), and located within a $1.5h^{-1}$ Mpc radius of the cluster center [1].

[c] The radius of the main concentration of galaxies (where, typically, the galaxy surface density drops to $\sim 1\%$ of the central density). Many clusters and groups are embedded in larger-scale structures (to tens of Mpc) (Sections 25.4.8, 25.11, and 25.12).

[d] Typical observed range and median value for the radial (line-of-sight) velocity dispersion in groups and clusters (Sections 25.4.10 and 25.12).

[e] Typical dynamical mass range of clusters within $1.5h^{-1}$ Mpc radius sphere (Sections 25.4.11 and 25.12).

[f] Typical blue luminosity range of clusters within $1.5h^{-1}$ Mpc radius sphere (Sections 25.4.11 and 25.12).

[g] Typical mass-to-light ratio of clusters and groups (median value) (Sections 25.4.11 and 25.12).

[h] Typical observed ranges of the X-ray temperature and 2–10 keV X-ray luminosity of the hot intracluster gas (Section 25.9).

[i] The number density of clusters decreases sharply with cluster richness (Sections 25.4.2, 25.8, and 25.12).

[j] The cluster correlation scale for rich ($R \geq 1$, $N_R \geq 50$, $n_c = 0.6 \times 10^{-5}h^3$ Mpc^{-3}) and poor ($N_R \gtrsim 20$, $n_c = 2.4 \times 10^{-5}h^3$ Mpc^{-3}) clusters (Section 25.11).

[k] The fraction of bright galaxies ($\gtrsim L^*$) in clusters and groups within $r \leq 1.5h^{-1}$ Mpc (Sections 25.4.3 and 25.12).

Reference

1. Abell, G.O. 1958, *ApJS*, **3**, 211

25.2 CLUSTER CATALOGS

25.2.1 Abell Catalog of Rich Clusters [1]

1. Identifies the richest, densest clusters to $z \lesssim 0.2$ found on the Palomar Sky Survey red plates. Covers the high-latitude Northern sky and part of the Southern sky ($\delta > -27°$, $|b| \gtrsim 30°$).

2. Sky coverage: 4.26 steradians (2.64 steradians in the north and 1.62 steradians in the south for the statistical sample).

3. Selection criteria: surface density enhancement of galaxies (see Table 25.2).

4. The catalog contains a statistical sample of the richest clusters, and a larger—but incomplete—listing of additional clusters, mostly to a lower richness threshold.

5. Mean number density of richness class $R \geq 1$ clusters: $\sim 6 \times 10^{-6} h^3$ Mpc^{-3}. The density decreases sharply with increasing richness (Section 25.4.2).

6. The distribution of Abell clusters (statistical sample) with distance and richness is presented in Table 25.3. The statistical sample is approximately volume limited.

Table 25.2. *Abell catalog of rich clusters: Selection criteria.*

	Statistical sample	Full catalog	Notes
Number of clusters	1682 clusters	2712 clusters	
Richness N_R	≥ 50 galaxies	≥ 30 galaxies	a
Richness class R	$R \geq 1$	$R \geq 0$	b
Redshift range (estimated, z_{est})	0.02–0.2	0.02 to $\gtrsim 0.2$	c
Sky coverage	$\delta > -27°$, $|b| \gtrsim 30°$	$\delta > -27°$	d

Notes

[a] Richness is the number of member galaxies N_R (above background) that are brighter than $m_3 + 2^m$ (where m_3 is the magnitude of the third brightest galaxy) and located within a projected radius $R_A = 1.7$ arcmin/$z_{est} \sim 1.5 h^{-1}$ Mpc of the cluster center (Section 25.4.1). The richness selection threshold of the catalog is listed here.

[b] Clusters are divided into richness classes R based on their richness count: $R \geq 1$ clusters have $N_R \geq 50$ galaxies; $R \geq 0$ clusters have $N_R \geq 30$ galaxies (see Table 25.3 for a detailed breakdown).

[c] The cluster estimated redshift is obtained from the magnitude of the tenth brightest galaxy. (For a compilation of observed cluster redshifts, see Section 25.3 and references therein.)

[d] For the statistical sample, the exact $|b|$ boundaries are given in Table 1 of [1]. The full catalog contains some clusters at lower latitudes, but most of the galactic plane is excluded.

Reference

1. Abell, G.O. 1958, *ApJS*, **3**, 211

Table 25.3. *Distribution of Abell clusters with distance and richness.*[a]

Distance distribution			Richness distribution		
D	$\langle z_{est} \rangle$	$N_{cl}(R \geq 1)$	R	N_R	N_{cl}
1	0.0283	9	$(0)^b$	(30–49)	$(\sim 10^3)$
2	0.0400	2	1	50–79	1224
3	0.0577	33	2	80–129	383
4	0.0787	60	3	130–199	68
5	0.131	657	4	200–299	6
6	0.198	921	5	≥ 300	1
	Total	1682		Total ($R \geq 1$)	1682

Table 25.3. *(Continued.)*

	Nearby redshift sample[c,d] $D \leq 4$	Distant projected sample[d] $D = 5 + 6$
N_{cl} (total)	104	1547
N_{cl} ($b \geq 30°$)	71	984
N_{cl} ($b \leq -30°$)	33	563
N_{cl} ($R = 1$)	82	1125
N_{cl} ($R \geq 2$)	22	422

Notes

[a] Statistical sample. $|b|$ boundaries as given in Table 1 of [1]. Notation: D = distance group (defined by the estimated redshifts [1]); $\langle z_{est} \rangle$ = average estimated redshift; N_{cl} = number of clusters; R = richness class; N_R = number of galaxies brighter than $m_3 + 2^m$ within $R_A = 1.5h^{-1}$ Mpc (richness count).

[b] $R = 0$ clusters are not part of the statistical sample and are enclosed by parentheses.

[c] Redshifts by Hoessel et al. [2].

[d] This sample is limited to $|b| \geq 30°$ in addition to the $|b|$ boundaries of the statistical sample.

References

1. Abell, G.O. 1958, *ApJS*, **3**, 211
2. Hoessel, J.G., Gunn, J.E., & Thuan, T.X. 1980, *ApJ*, **241**, 486

25.2.2   ACO Catalog of Clusters [2]

1. An extension of the Abell catalog [1] to the Southern hemisphere is given by Abell, Corwin, and Olowin (ACO) [2].

2. Clusters identified on the U.K. Schmidt IIIa–J plates, for $\delta < -17°$.

3. Selection criteria similar to Abell [1] (Section 25.2.1).

4. Number of $R \geq 0$ clusters: $\delta < -17°$ (incomplete sample): 1635,
$\delta < -27°$ (incomplete sample): 1361.
Number of $R \geq 1$ clusters at $\delta < -17°$, $b < -35°$, and $z_{est} \lesssim 0.2$ (a statistical subsample): 622.

5. ACO lists the combined Northern Abell [1] catalog ($\delta > -27°$, 2712 clusters) and the Southern ACO [2] catalog ($\delta < -27°$, 1361 clusters), for a total of 4073 $R \geq 0$ clusters (including three duplicate ACO clusters) over most of the sky (not a statistical sample). (A statistical subsample can be defined, such as $R \geq 1$, $|b| \geq 30°$, $z_{est} \lesssim 0.2$ clusters.)

6. ACO also provides a supplementary list of clusters (Table 5 of [2]; 1174 systems), mostly poor clusters and groups ($N_R < 30$).

25.2.3   Zwicky Catalog of Clusters [3]

1. Identifies clusters on the Palomar Sky Survey plates.

2. Sky coverage: $\delta > -3°$, excluding the galactic plane region (Introduction to Vol. 6 of [3]).

3. Number of clusters: 9700 (includes $\sim$ 6% duplicate clusters in overlap regions).
Cluster selection: $\geq$ 50 galaxies with $m \leq m_1 + 3^m$ and within $r \leq R_c$ (m_1 = magnitude of the brightest cluster galaxy; R_c = cluster contour; see below).
Cluster contour, R_c: isopleth where the projected galaxy density is twice that of the field.
Estimated redshift: $z_{est} \lesssim 0.2$.

4. A subsample of 2230 Zwicky clusters are contained within $\delta > -3°$, $|b| \geq 30°$, $z_{est} \lesssim 0.15$.

5. Zwicky's clusters are poorer, on average, than Abell clusters.

6. Zwicky's cluster selection, unlike Abell's, is distance dependent due to the definition of the cluster contour R_c (which is distance dependent).

25.2.4 Shectman/Lick Catalog of Clusters [4]

1. Automated identification of clusters from the Shane–Wirtanen [5] Lick galaxy survey to 19^m (m refers to magnitude throughout this chapter).
2. Number of clusters: 646; $z_{est} \lesssim 0.1$; sky coverage: $\delta > -22.5°$, $|b| > 40°$.
3. Selection criterion: local surface density maxima of galaxies above a given smoothed threshold.
4. Selection threshold includes poorer clusters than Abell clusters.

25.2.5 Digitized, Automated Cluster Surveys

Large, automated surveys of clusters are currently under construction (e.g., the Sloan Digital Sky Survey of π steradians in the Northern hemisphere will identify complete samples of clusters in both two and three dimensions using accurate CCD imaging, redshifts, and automated selection algorithms). Smaller automated surveys in two dimensions obtained from digitized photographic plates have recently been carried out. These include the following.

25.2.5.1 The Edinburgh–Durham Cluster Catalog [6]

The Edinburgh–Durham Cluster Catalog (EDCC) [6] identifies clusters and groups from digitized U.K. Schmidt IIIa–J survey plates in the Southern hemisphere.

1. Number of clusters: 737 clusters and groups of all richnesses (most are groups and poor clusters, $N_R < 30$). Sky coverage: ~ 1400 deg^2 centered on the South Galactic Pole.
2. Selection criterion: local surface density enhancement of galaxies.
3. A subsample of rich clusters [7] contains 97 clusters with richness count ≥ 22 galaxies (with $m \leq m_3 + 2^m$ and $r \leq 1h^{-1}$ Mpc), $z \lesssim 0.13$, and a space density of clusters of $\sim 1.5 \times 10^{-5} h^3$ Mpc^{-3}. These clusters are comparable to (slightly poorer than) $R \gtrsim 0$ Abell clusters.

25.2.5.2 The Automated Plate Measuring Survey [8, 9]

The automated plate measuring (APM) survey of clusters [8, 9] identifies clusters from digitized U.K. Schmidt IIIa–J plates in the Southern hemisphere.

1. Selection criterion: local surface density enhancement of galaxies. Sky coverage: 4300 deg^2: $z \lesssim 0.1$.
2. Selection threshold includes poorer clusters than Abell clusters.

25.3 CATALOG OF NEARBY RICH CLUSTERS OF GALAXIES

Table 25.4. *Catalog of nearby rich clusters of galaxies.*[a]

Abell[a]	α, δ (2000)[a]	R^b	D^c	$N_R{}^d$	BM[e]	$z_{obs}{}^f$	$\sigma_r{}^g$	$n_v{}^h$	$kT_x{}^i$	L_x (10^{44})[j]
85	00 41.6 −09 20	1	4	59	I	0.0556	749	116	6.2	1.9
88	00 42.9 −26 02	1	3	58	III	0.1096		3		
104	00 49.8 +24 31	1	4	50	II–III:	0.0822		1		
119	00 56.4 −01 15	1	3	69	II–III	0.0440	778	21	5.9	0.61
121	00 57.5 −07 00	1	4	67	III	0.1048		1		
151	01 08.9 −15 25	1	3	72	II:	0.0536	715	22		
154	01 11.0 +17 39	1	3	66	II	0.0638	999	31		0.19*

Table 25.4. *(Continued.)*

Abell[a]	α, δ (2000)[a]	R[b]	D[c]	N_R[d]	BM[e]	z_{obs}[f]	σ_r[g]	n_v[h]	kT_x[i]	L_x (10^{44})[j]
166	01 14.6 −16 16	1	4	76	III:	0.1155		1		
168	01 15.2 −00 14	2	3	89	II–III:	0.0452	581	13	2.6	0.13*
189	10 23.7 +01 38	1	4	50	III	0.0325	259	10		< 0.02*
193	01 25.1 +08 41	1	4	58	II	0.0498	1136	15	4.2	0.35
225	01 38.9 +18 53	1	4	51	II–III	0.0692		1		
246	01 44.7 +05 48	1	4	56	II–III	0.0700		2		
274	01 54.7 −06 16	3	4	140	III	0.1289		1		
277	01 55.8 −07 22	1	3	50	III	0.0947		1		
389	02 51.3 −24 54	2	4	97	II	0.1160		1		0.6*
399	02 57.9 +13 00	1	3	57	I–II	0.0715	1424	29	5.8	2.1
400	02 57.6 +06 01	1	1	58	II–III	0.0238	610	71	2.5	0.06
401	02 58.9 +13 34	2	3	90	I	0.0748	1294	20	7.8	3.8
415	03 06.8 −12 02	1	4	67	II	0.0788		1		
496	04 33.6 −13 14	1	3	50	I:	0.0327	741	148	3.9	0.7
500	04 38.9 −22 06	1	4	53	III	0.0666		1		0.23*
514	04 47.7 −20 25	1	3	78	II–III:	0.0731		2		
787	09 28.6 +74 23	2	4	106	II:	0.1352		3		
957	10 14.0 −00 54	1	4	55	I–II:	0.0450	678	18		0.15*
978	10 20.5 −06 31	1	3	55	II	0.0527		2		
1020	10 27.8 +10 24	1	4	68	II–III:	0.0650		1		
1035	10 32.1 +40 12	2	3	94	II–III:	0.0799		1		
1126	10 54.0 +16 51	1	4	55	I–II:	0.0852		3		0.15*
1185	11 10.8 +28 40	1	2	52	II	0.0321	783	49	3.9	0.05
1187	11 11.7 +39 34	1	3	55	III	0.0791		1		
1213	11 16.5 +29 15	1	2	51	III	0.0468	598	12		0.04*
1216	11 17.7 −04 28	1	4	57	III	0.0524		1		< 0.03*
1228	11 21.5 +34 19	1	1	50	II–III	0.0350	188	8		
1238	11 23.0 +01 05	1	4	63	III	0.0716		1		
1254	11 26.9 +71 04	1	3	58	III	0.1525		1		0.3*
1291	11 32.1 +56 01	1	3	61	III	0.0535	919	7		0.1*
1318	11 36.4 +54 57	1	3	56	II	0.0566	284	6		
1364	11 43.7 −01 45	1	4	74	III	0.1070		1		
1365	11 44.4 +30 54	1	4	51	III	0.0763		1		
1367	11 44.5 +19 50	2	1	117	II–III:	0.0214	822	93	3.5	0.18
1377	11 47.0 +55 44	1	3	59	III	0.0514	488	13		0.12*
1382	11 48.4 +71 26	1	4	57	II:	0.1053		2		0.18*
1383	11 48.2 +54 37	1	4	54	III	0.0603	395	5		
1399	11 51.2 −03 05	2	4	82	III	0.0913		1		
1412	11 55.8 +73 28	2	4	86	III	0.0839		1		
1436	12 00.5 +56 15	1	3	69	III	0.0644		4		
1468	12 05.6 +51 25	1	4	50	I:	0.0844		2		
1474	12 08.0 +14 57	1	4	70	III	0.0791		2		
1496	12 13.4 +59 16	1	4	58	III	0.0941		2		
1541	12 27.4 +08 50	1	4	58	I–II	0.0892		1		
1644	12 57.2 −17 21	1	4	68	II	0.0473	991	92	4.7	0.63
1651	12 59.4 −04 11	1	4	70	I–II	0.0845	965	29	6.6	2.8
1656[k]	12 59.8 +27 58	2	1	106	II	0.0232	880	226	8.1	1.5
1691	13 11.4 +39 12	1	3	64	II	0.0722		1		
1749	13 29.5 +37 37	1	4	55	II:	0.0590		2		
1767	13 36.0 +59 12	1	4	65	II	0.0706	933	16	4.1	0.66
1773	13 42.1 +02 14	1	3	66	III	0.0776		1		
1775	13 41.9 +26 21	2	4	92	I	0.0717	1594	28	4.9	0.56
1793	13 48.3 +32 17	1	4	54	III	0.0849		1		
1795	13 49.0 +26 35	2	4	115	I	0.0622	896	49	5.3	2.3
1809	13 53.3 +05 09	1	4	78	II:	0.0789	249	11		0.43*
1831	13 59.2 +27 59	1	3	67	III	0.0613	316	11		
1837	14 01.8 −11 09	1	4	50	I–II	0.0376		1	2.4	0.1
1904	14 22.1 +48 33	2	3	83	II–III:	0.0708	803	24		0.18*

Table 25.4. *(Continued.)*

Abell[a]	α, δ (2000)[a]	R[b]	D[c]	N_R[d]	BM[e]	z_{obs}[f]	σ_r[g]	n_v[h]	kT_X[i]	L_X (10^{44})[j]
1913	14 26.9 +16 40	1	4	53	III	0.0528	656	16		0.15*
1927	14 31.0 +25 39	1	4	50	I–II:	0.0740		1		
1983	14 52.7 +16 44	1	3	51	III:	0.0449	504	74		0.09*
1991	14 54.5 +18 37	1	3	60	I:	0.0579	510	15	5.4	0.26
1999	14 54.1 +54 18	1	4	68	II–III	0.1032		1		
2005	14 58.7 +27 49	2	4	105	III	0.1257		2		
2022	15 04.3 +28 25	1	3	50	III:	0.0575		3		0.13*
2028	15 09.6 +07 31	1	4	50	II–III	0.0776	434	20		
2029	15 11.0 +05 45	2	4	82	I	0.0768	1411	59	7.8	5.2
2040	15 12.8 +07 25	1	4	52	III	0.0456		1		0.09*
2048	15 15.3 +04 22	1	4	75	III	0.0945		1		
2061	15 21.3 +30 39	1	4	71	III:	0.0782	730	20		
2063	15 23.0 +08 39	1	3	63	II:	0.0355	652	24	4.1	0.32
2065[l]	15 22.7 +27 43	2	3	109	III	0.0722	1082	22	8.4	1.6
2067	15 23.2 +30 54	1	4	58	III	0.0748	761	12		
2079	15 28.1 +28 52	1	3	57	II–III:	0.0656	639	29		0.19*
2089	15 32.7 +28 00	1	4	70	II	0.0733	551	20		< 0.58*
2092	15 33.3 +31 08	1	4	55	II–III	0.0669	504	18		0.10*
2107	15 39.8 +21 46	1	4	51	I	0.0421	536	20	4.2	0.30
2124	15 45.0 +36 03	1	3	50	I	0.0654	852	10		0.20*
2142	15 58.3 +27 13	2	4	89	II	0.0899	1241	15	8.7	6.5
2147[m]	16 02.3 +15 53	1	1	52	III	0.0356	1148	30	4.4	0.43
2151[m]	16 05.2 +17 44	2	1	87	III	0.0368	786	99	3.8	0.11
2152	16 05.4 +16 26	1	1	60	III	0.0374	1244	22		0.05*
2175	16 20.4 +29 54	1	4	61	II	0.0968		2		
2197	16 28.2 +40 54	1	1	73	III	0.0308	593	46		0.02*
2199	16 28.6 +39 31	2	1	88	I	0.0299	794	71	4.5	0.73
2255	17 12.5 +64 05	2	3	102	II–III:	0.0808	1221	35	7.3	1.3
2256	17 03.7 +78 43	2	3	88	II–III	0.0581	1270	90	7.5	2.1
2347	21 29.5 −22 12	1	4	79	III:	0.1196		1		
2382	21 52.0 −15 38	1	4	50	II–III	0.0648		1		
2384	21 52.3 −19 32	1	4	61	II–III	0.0943		1		
2399	21 57.5 −07 47	1	3	52	III	0.0587		1		0.20*
2410	22 02.1 −09 53	1	4	54	III	0.0806		1		0.23*
2457	22 35.8 +01 28	1	4	53	I–II:	0.0597		1		
2657	23 44.8 +09 08	1	3	51	III	0.0414	667	12	3.4	0.31
2670	23 54.2 −10 24	3	4	142	I–II	0.0761	881	220	3.9	0.40
2675	23 55.6 +11 25	1	4	60	II	0.0726		1		
2700	00 03.9 +02 03	1	4	59	II:	0.0978		1		
Perseus[n] (A426)	03 18.6 +41 30	2	0	88	II–III	0.0179	1277	114	6.3	2.8
Virgo[n]	12 30.8 +12 23	(0)	< 0		III	0.0039	757 573	354 (E+Sp) 159 (E)	2.4	0.03

Notes

[a] Abell [1] cluster number and its position [2]. Only the nearest clusters in Abell's statistical sample (Section 25.2) are included [$D = 1$–4 (i.e., $0.02 \leq z \lesssim 0.08$), $R \geq 1$, $|b| \geq 30°$].

[b] R = richness class of the cluster [1, 2].

[c] D = distance class of the cluster [1, 2].

[d] N_R = cluster richness count (see Section 25.4.1) [1, 2].

[e] BM = Bautz–Morgan type (see Section 25.5); the colon indicates uncertain type [1, 2].

[f] z_{obs} = observed redshift of the cluster [3]. See also [4] for additional redshifts.

[g] σ_r = radial (line-of-sight) velocity dispersion of galaxies in the cluster in km s^{-1} [3]. For Virgo [5], two values are listed: one for all galaxies (E + Sp), and one for the elliptical galaxies (E).

[h] n_v = number of galaxy redshifts used to determine the parameters z_{obs} and σ_r [3].

[i] kT_X = X-ray temperature of the cluster, in keV [6]; see also [7, 8] and Section 25.15.

$^{j}L_x(10^{44})$ = X-ray luminosity of the cluster in the 2–10 keV range, in 10^{44} erg s^{-1} [6]; see also [7, 8]. Luminosities marked with an asterisk refer to the 0.5–4.5 keV range, within the cluster central region of $r \leq 0.5h^{-1}$ Mpc [9]. (See also Section 25.15.)

^{k}The Coma cluster.

^{l}The Corona Borealis cluster.

^{m}The Hercules clusters.

^{n}Perseus is not part of the statistical sample (due to its low latitude of $b = -13°$ and $z < 0.02$). Virgo is not a member of the Abell Catalog due to its very low redshift ($z \ll 0.02$).

References

1. Abell, G.O. 1958, *ApJS*, **3**, 211
2. Abell, G.O., Corwin, H., & Olowin, R. 1989, *ApJS*, **70**, 1
3. Stuble, M., & Rood, H. 1991, *ApJS*, **77**, 363
4. Postman, M., Huchra, J., & Geller, M. 1992, *ApJ*, **384**, 404
5. Binggeli, B., Tammann, G.A., & Sandage, A. 1987, *AJ*, **94**, 251
6. David, L.P., Slyz, A., Jones, C., Forman, W., Vrtilek, S., & Arnaud, K. 1993, *ApJ*, **412**, 479
7. Henry, J.P., & Arnaud, K.A. 1991, *ApJ*, **372**, 410
8. Edge, A., Stewart, G.C., Fabian, A.C., & Arnaud, K.A. 1990, *MNRAS*, **245**, 559
9. Jones, C., & Forman, W. 1999, *ApJ*, **511**, 65

25.4   CLUSTER PROPERTIES

25.4.1   Richness

1. Standard usage follows Abell's [1] definition: the richness count, N_R, represents the number of member galaxies in a cluster, above background, brighter than $m_3 + 2^m$ (where m_3 is the magnitude of the third brightest cluster galaxy), and located within a projected radius $R_A = 1.5h^{-1}$ Mpc of the cluster center.

2. This richness count (N_R) is an intrinsic cluster property, independent of cluster distance.

3. A richness class R is assigned to clusters according to their galaxy count N_R as specified in Table 25.3. R and N_R are listed for rich clusters in [2]. They are also listed for a nearby sample of rich clusters in Section 25.3.

4. *Rich* clusters are clusters with $N_R \gtrsim 30$ ($R \geq 0$) [or, frequently, $N_R \gtrsim 50$ ($R \geq 1$)]. Poor clusters and groups: $N_R < 30$.

5. Richer clusters are, on average, more luminous and more massive than poorer clusters.

6. The number density of clusters (number of clusters per unit volume) decreases sharply with increasing richness [10]; Section 25.4.2.

25.4.2   Number Density of Clusters

The number density of clusters is a strong function of cluster richness. Integrated cluster densities, n_c ($> N_R$), represent the number density of clusters above a given richness threshold. These cluster densities and the associated mean cluster separation, d ($\equiv n_c^{-1/3}$), are listed in Table 25.5 [10].

Table 25.5. *Number density of clusters* [1].

R	N_R	n_c $(> N_R)h^3$ (Mpc^{-3})^a	d $(> N_R)h^{-1}$ (Mpc)
≥ 0	≥ 30	13.5×10^{-6}	42
≥ 1	≥ 50	6.0×10^{-6}	55
≥ 2	≥ 80	1.2×10^{-6}	94
≥ 3	≥ 130	1.5×10^{-7}	188

Note

^{a}Approximate uncertainties are $10^{\pm 0.2}$ for the $R \geq 0, 1, 2$ densities and $10^{\pm 0.3}$ for $R \geq 3$.

Reference

1. Bahcall, N.A. & Cen, R. 1993, *ApJL*, **407**, L49

25.4.3   Fraction of Galaxies in Clusters

Fraction of galaxies in $R \gtrsim 0$ clusters: $\sim 5\%$ (within $R_A = 1.5h^{-1}$ Mpc). The fraction of galaxies that belong in clusters increases with increasing radius R_A and with decreasing cluster richness threshold.

25.4.4   Average Number of Galaxies per Cluster

1. For $R \geq 0$ clusters:
$$\langle N_R \rangle_{\text{median}} \simeq 50, \qquad \langle N_R \rangle_{\text{mean}} \simeq 56.$$

For $R \geq 1$ clusters:
$$\langle N_R \rangle_{\text{median}} \simeq 60, \qquad \langle N_R \rangle_{\text{mean}} \simeq 75,$$

(within $R_A = 1.5h^{-1}$ Mpc and $m \leq m_3 + 2^m$).

2. The number of galaxies increases to fainter luminosities following the Schechter [11] luminosity function (Section 25.7).

25.4.5   Galaxy Overdensity in Rich Clusters

1. Average number density of bright $(\gtrsim L^*)$ galaxies in $R \gtrsim 0$ clusters (within $R_A = 1.5h^{-1}$ Mpc):
$$n_g(\text{cluster}) \sim 3h^3 \text{ galaxies Mpc}^{-3}.$$

2. Average overall (field) number density of bright $(\gtrsim L^*)$ galaxies [12, 13]:
$$n_g(\text{field}) \sim 1.5 \times 10^{-2}h^3 \text{ galaxies Mpc}^{-3}.$$

3. Average galaxy overdensity in rich $(R \geq 0)$ clusters:
$$n_g(\text{cluster})/n_g(\text{field}) \sim 200.$$

4. Typical threshold galaxy overdensity in
$$R \geq 0 \text{ clusters}: \quad n_g(\text{cluster})/n_g(\text{field}) \gtrsim 100,$$
$$R \geq 1 \text{ clusters}: \quad n_g(\text{cluster})/n_g(\text{field}) \gtrsim 200.$$

5. Galaxy overdensity in the cores of typical compact rich clusters:
$$n_g(\text{cluster core})/n_g(\text{field}) \sim 10^4 - 10^5.$$

25.4.6 Density Profile

1. The radial density distribution of galaxies in a rich cluster can be approximated by a bounded Emden isothermal profile [14, 15], or by its King approximation [16] in the *central* regions.

2. In the central regions, the King approximation for the galaxy distribution is

$$n_g(r) = n_g^0(1 + r^2/R_c^2)^{-3/2}, \quad \text{spatial profile,}$$
$$S_g(r) = S_g^0(1 + r^2/R_c^2)^{-1}, \quad \text{projected profile.}$$

$n_g(r)$ and $S_g(r)$ are, respectively, the space and projected profiles (of the number density of galaxies), n_g^0 and S_g^0 are the respective central densities, and R_c is the cluster core radius [where $S(R_c) = S^0/2$]. Typical central densities and core radii of clusters are listed in the following subsections.

3. The projected and space central densities relate as

$$S_g^0 = 2R_c n_g^0.$$

4. A bounded Emden isothermal profile of galaxies in clusters [14, 15] yields a profile slope that varies approximately as [15]

$$S_g(r \lesssim R_c/3) \sim \text{constant,}$$
$$S_g(R_c \lesssim r \lesssim 10R_c) \propto r^{-1.6};$$

therefore

$$n_g(R_c \lesssim r \lesssim 10R_c) \propto r^{-2.6}.$$

5. The galaxy–cluster cross-correlation function [17, 18] also represents the average radial density distribution of galaxies around clusters. For $R \geq 1$ clusters, and r in h^{-1} Mpc:

$$\xi_{gc}(r) \simeq 130r^{-2.5} + 70r^{-1.7} \quad [17]$$

or

$$\xi_{gc}(r) \simeq 120r^{-2.2} \quad [18].$$

The average density profile in clusters thus follows, approximately (see also Section 25.15)

$$n_g(r) \propto r^{-2.4 \pm 0.2} \quad \text{(spatial),} \qquad r > R_c$$
$$S_g(r) \propto r^{-1.4 \pm 0.2} \quad \text{(projected),} \qquad r > R_c.$$

6. Some substructure (subclumping) in the distribution of galaxies exists in a significant fraction of rich clusters ($\sim 40\%$) [19].

25.4.7 Central Density

1. Central number density of galaxies in rich compact clusters [15, 20, 21] for galaxies in the brightest 3 magnitude range:

$$n_g^0(\Delta m \simeq 3^m) \sim 10^3 h^3 \text{ galaxies Mpc}^{-3}.$$

The central density reaches $\sim 10^4 h^3$ galaxies Mpc^{-3} for the richest compact clusters.

2. Typical central *mass* density in rich compact clusters, determined from cluster dynamics:

$$\rho_0(\text{mass}) \simeq 9\sigma_{r,c}^2/4\pi G R_c^2$$
$$\sim 4 \times 10^{15} \mathcal{M}_\odot \, \text{Mpc}^{-3}[(\sigma_{r,c}/10^3 \, \text{km s}^{-1})/(R_c/0.2 \, \text{Mpc})]^2 h^2,$$

where $\sigma_{r,c}$ is the radial (line-of-sight) central cluster velocity dispersion (in km s^{-1}) and R_c is the cluster core radius (in Mpc).

3. Typical central density of the hot intracluster gas in rich clusters (Section 25.9):

$$n_e \sim 10^{-3} \, \text{electrons cm}^{-3}.$$

25.4.8 Size

1. Core radii of typical compact rich clusters, determined from their galaxy distribution [20–22]:

$$R_c \simeq (0.1-0.25)h^{-1} \, \text{Mpc}.$$

2. Core radii of the X-ray emitting intracluster gas of rich clusters [22, 23]:

$$R_c(\text{X-ray}) \simeq (0.1-0.3)h^{-1} \, \text{Mpc}.$$

3. Typical radius of the main concentration of the cluster galaxies (where the surface density of galaxies typically drops to $\sim 1\%$ of the central density):

$$R \sim 1.5h^{-1} \, \text{Mpc}.$$

4. Gravitational radius of a cluster ($R_G \equiv 2GM/3\sigma_r^2$, where $\mathcal{M}$ is the cluster mass and σ_r is the line-of-sight velocity dispersion of the cluster):

$$R_G \simeq 1.5h^{-1} \, \text{Mpc}[(\mathcal{M}/5 \times 10^{14}\mathcal{M}_\odot)/(\sigma_r/10^3 \, \text{km s}^{-1})^2].$$

5. Cluster outskirts, or the aspherical large-scale structure in which many clusters are embedded, can extend to tens of Mpc [24, 25].

25.4.9 Galactic Content

1. The fraction of elliptical, S0, and spiral galaxies in rich clusters differs from that in the field, and depends on the classification type, or density, of the cluster (see Section 25.5) [15, 26–29]. See Table 25.6.

2. The fraction of elliptical (E) and S0 galaxies increases and the fraction of spirals decreases toward the central cores of rich compact clusters. The fraction of spiral galaxies in the dense cores of some rich clusters (e.g., the Coma cluster) may be close to zero [28].

3. The galactic content of clusters as represented in Table 25.6 is part of the general density–morphology relation of galaxies [29, 30]; as the local density of galaxies increases, the fraction of E and S0 galaxies increases and the fraction of spirals decreases. For local galaxy densities $n_g \lesssim 5$ galaxies Mpc^{-3}, the fractions remain approximately constant at the average "Field" fractions listed in Table 25.6.

Table 25.6. *Typical galactic content of clusters ($r \lesssim 1.5h^{-1}$ Mpc).*

Cluster type	E	S0	Sp	(E+S0)/Sp
Regular clusters (cD)	35%	45%	20%	4.0
Intermediate clusters (spiral-poor)	20%	50%	30%	2.3
Irregular clusters (spiral-rich)	15%	35%	50%	1.0
Field	10%	20%	70%	0.5

25.4.10 Velocity Dispersion

1. Typical radial (line-of-sight) velocity dispersion of galaxies in rich clusters (median):

$$\sigma_r \sim 750 \text{ km s}^{-1}.$$

2. Typical range of radial velocity dispersion in rich clusters [31, 32]:

$$\sigma_r \sim 400\text{--}1400 \text{ km s}^{-1}.$$

3. A weak correlation between σ_r and richness exists; richer clusters exhibit, on average, larger velocity dispersion [32].

4. Measured velocity dispersions for a nearby sample of rich clusters are listed in Section 25.3; for additional clusters, see [31, 33].

5. The observed velocity dispersion of galaxies in rich clusters is generally consistent with the velocity implied by the X-ray temperature of the hot intracluster gas (Section 25.9), as well as with the cluster velocity dispersion implied from observations of gravitational lensing by clusters (Section 25.14).

6. Velocity dispersion and temperature profiles as a function of distance from the cluster center have been measured only for a small number of clusters so far. The profiles are typically isothermal $[\sigma_r^2(r) \sim T_x(r) \sim \text{constant}]$ for $r \lesssim 0.5\text{--}1h^{-1}$ Mpc, and drop at larger distances (Section 25.15).

25.4.11 Mass, Luminosity, and Mass-to-Luminosity Ratio

1. Typical dynamical mass of rich clusters within $1.5h^{-1}$ Mpc radius sphere (determined from the virial theorem for an isothermal distribution):

$$\mathcal{M}_{\text{cl}}(\leq 1.5) \simeq \frac{2\sigma_r^2 \times (1.5h^{-1} \text{ Mpc})}{G} \simeq 0.7 \times 10^{15} \left(\frac{\sigma_r}{1000}\right)^2$$

$$\simeq 0.4 \times 10^{15} h^{-1} \mathcal{M}_\odot \qquad (\text{for } \sigma_r \sim 750 \text{ km s}^{-1}).$$

2. Approximate range of masses for $R \gtrsim 0$ clusters (within $1.5h^{-1}$ Mpc):

$$\mathcal{M}_{\text{cl}} (\leq 1.5) \sim (0.1\text{--}2) \times 10^{15} h^{-1} \mathcal{M}_\odot.$$

3. Comparable cluster masses are obtained using the X-ray temperature and gas distribution of the hot intracluster gas as tracers of the cluster potential [10, 34].

4. Typical (median) blue luminosity of rich clusters (within $1.5h^{-1}$ Mpc):

$$L_{\text{cl}} (\leq 1.5) \sim 10^{12} h^{-2} L_\odot.$$

5. Approximate range of rich cluster blue luminosities:

$$L_{cl} (\leq 1.5) \sim (0.6\text{–}6) \times 10^{12} h^{-2} L_{\odot}.$$

6. Typical mass-to-luminosity ratio of rich clusters (for total corrected blue luminosity):

$$(M/L_B)_{cl} \sim 300 h (\mathcal{M}_{\odot}/L_{\odot}).$$

7. Inferred mass density in the Universe based on cluster dynamics:

$$\Omega_{dyn} \sim 0.2$$

(if mass follows light, $M \propto L$, on scales $\gtrsim 1 h^{-1}$ Mpc). $\Omega = 1$ corresponds to the critical mass density needed for a closed universe and $\mathcal{M}/L_B$ ($\Omega = 1$) $\simeq 1500 h$.

25.4.12 Characteristic Times

1. Cluster crossing time, t_{cr}:

$$t_{cr} = R/\sigma \simeq 6 \times 10^8 \text{ yr} \, [(R/\text{Mpc})/(\sigma_r/10^3 \text{ km s}^{-1})],$$
$$\simeq 10^9 \text{ yr} \quad (\text{for } R \simeq 1.5 h^{-1} \text{ Mpc}, \ \sigma_r \sim 10^3 \text{ km s}^{-1}),$$

where R is the crossing radius, and σ and σ_r are, respectively, the galaxy velocity and radial velocity in the cluster ($\sigma^2 \simeq 3\sigma_r^2$).

2. Two-body relaxation time for galaxies in clusters [35, 36]:

$$t_r = \sigma^3/(4\pi G^2 m_g^2 n_g \ln \Lambda)$$
$$\simeq 2 \times 10^{10} \text{ yr} \quad [(\sigma_r/10^3 \text{ km s}^{-1})^3/(m_g/10^{12} \mathcal{M}_{\odot})^2 (n_g/10^3 \text{ galaxies Mpc}^{-3}) \ln \Lambda],$$

where σ and m_g are the velocity and mass of the galaxy, n_g is the number density of galaxies in the cluster, and $\ln \Lambda$ is the natural logarithm of the ratio of maximum to minimum impact parameters. ($n_g \sim 10^3$ galaxies Mpc^{-3} is a typical galaxy density in cluster cores; see Section 25.4.7.)

3. Collision time between galaxies in clusters:

$$t_{coll} = (2^{1/2} \sigma n_g \pi R_g^2)^{-1} \sim 10^9 \text{ yr} \, [(\sigma_r/10^3 \text{ km s}^{-1})(n_g/10^3 \text{ galaxies Mpc}^{-3})(R_g/10 \text{ kpc})^2]^{-1},$$

where R_g is the galaxy radius in kpc.

4. Cooling time of intracluster gas by bremsstrahlung emission:

$$t_{br} = 9 \times 10^7 \text{ yr} \, (T_8^{1/2} n_e^{-1}),$$

where T_8 is the gas temperature in 10^8 K, and n_e is the electron density in particles cm^{-3}. For a typical rich cluster with $T_8 \sim 0.4$ and $n_e \sim 10^{-3}$, $t_{br} \sim 5 \times 10^{10}$ yr. At the center of some clusters, the cooling time is shorter than the Hubble time, and cooling flows are observed (Section 25.9).

25.5 CLUSTER CLASSIFICATION

Rich clusters are classified in a sequence ranging from early- to late-type clusters, or equivalently, from regular to irregular clusters. Many cluster properties (shape, concentration, dominance of brightest galaxy, galactic content, density profile, and radio and X-ray emission) are correlated with position in this sequence. A summary of the sequence and its related properties is given in Table 25.7. Some specific classification systems are described first.

25.5.1   Bautz–Morgan (BM) Classification [37]

1. Classifies clusters based on the relative contrast (dominance in extent and brightness) of the brightest galaxy to the other galaxies in the cluster.

BMI: Cluster is dominated by a single, centrally located, cD galaxy (Section 25.6) (e.g., A401, A2199).

BMII: Brightest members are intermediate in appearance between cD galaxies (which have extended envelopes) and normal giant ellipticals galaxies (e.g., the Coma cluster).

BMIII: Cluster contains no dominant galaxies (e.g., Virgo, Hercules).

Two intermediate types, **BMI–II** and **BMII–III**, are also used.

2. The BM types of nearby rich clusters are listed in Section 25.3 [2, 38].

3. $\sim 40\%$ of rich clusters are BMI, I–II, and II; $\sim 60\%$ of rich clusters are BMII–III and III.

4. The absolute metric magnitude of first brightest cluster galaxy dims toward later BM type (Section 25.7):

$$\langle M_1(\text{BMIII})\rangle - \langle M_1(\text{BMI})\rangle \simeq 0.4^m.$$

Table 25.7. *Classification schemes of clusters and related characteristics.*

Property	Regular (early) type clusters	Intermediate clusters	Irregular (late) type clusters
Zwicky type	Compact	Medium-compact	Open
BM type[a]	I, I–II, II	(II), II–III	(II–III), III
RS type[a]	cD, B, (L, C)	(L), (F), (C)	(F), I
Shape symmetry	Symmetrical	Intermediate	Irregular shape
Central concentration	High	Moderate	Low
Galactic content	Elliptical-rich	Spiral-poor	Spiral-rich
E fraction	35%	20%	15%
S0 fraction	45%	50%	35%
Sp-fraction	20%	30%	50%
E:S0:Sp	3:4:2	2:5:3	1:2:3
Radio emission	$\sim 50\%$ detection rate	$\sim 50\%$ detection rate	$\sim 25\%$ detection rate
X-ray luminosity	High	Intermediate	Low
Fraction of clusters	$\sim 1/3$	$\sim 1/3$	$\sim 1/3$
Examples	A401, Coma	A194	Virgo, A1228

Note
[a]Parentheses indicate less certain designations.

25.5.2   Rood–Sastry (RS) Classification [39]

1. Classifies clusters based on the distribution of the ten brightest members.

2. Can be represented by a "tuning-fork" diagram:

$$cD - B \begin{array}{c} \diagup\ L - F \\ \diagdown\ C - I \end{array}$$

cD (= cD galaxy): cluster is dominated by a cD galaxy (Section 25.6) (e.g., A401, A2199).

B (= binary): cluster is dominated by a bright "binary" system (e.g., the Coma cluster).

L (= line): several of the brightest members are arranged in a line (e.g., Perseus).

C (= core): At least four of the brightest members are located with comparable separations in the cluster core (e.g., A2065).

F (= flat): several of the bright galaxies are distributed in a flattened configuration (e.g., A397).

I (= irregular): irregular distribution of galaxies, with no well-defined center (e.g., Virgo).

3. Rood and Sastry [39] find the following frequency distribution of cluster classification:

cD: 21%;   B: 9%;   L: 9%;   C: 14%;   F: 18%;   I: 29%.

25.5.3   Zwicky Classification [3]

This method classifies clusters based on their compactness:

Compact: Single outstanding concentration among the bright member galaxies. Ten or more bright galaxies appear in actual contact. Many of these clusters display a high degree of spherical symmetry.

Medium compact: Single concentration where the ten brightest galaxies are not in contact but separated by several diameters, *or* several distinct concentrations, some of which may be compact.

Open: No obvious condensations, but in various locations the galaxy surface density is at least five times as great as in the surrounding field.

25.6   cD GALAXIES

1. cD galaxy: A galaxy with the nucleus of a giant elliptical surrounded by an extended, slowly decreasing low surface brightness envelope (designated as D in Morgan's classification scheme, with c indicating location in a cluster) [40, 41].

2. Mostly found at the centers of rich clusters [40, 27]. The cD dominates the other cluster galaxies in size and brightness.

3. Surface brightness profile [42, 15]:

$$S_{cD}(r) \propto r^{-1.6}.$$

4. Size: radius to which envelope is traced (to $S_v \sim 28$ mag./arcsec^2) ranges from ~ 100 kpc to $\sim 1h^{-1}$ Mpc [42, 43].

5. Luminosity: mean absolute metric magnitude within $22h^{-1}$ kpc radius, corrected for aperture effect, K dimming, and galactic reddening [44]:

$$\langle M_v \rangle_{cD} = -22.2 \pm 0.1 + 5\log_{10} h.$$

Absolute magnitude of cD galaxies including their halos [42, 45] reaches $M_v(cD) \simeq -24$.

6. cD galaxies frequently contain multiple nuclei [40, 45–47].

7. The total luminosity of the cD galaxy, L_{cD}, and the luminosity of its envelope, L_{env}, are correlated with the cluster luminosity, L_{cl} [42, 43]:

$$L_{cD} \propto L_{cl}^{1.25}, \qquad L_{env} \propto L_{cl}^{2.2}.$$

25.7   LUMINOSITY FUNCTION OF GALAXIES IN CLUSTERS

1. The luminosity function of galaxies in clusters is approximated by a Schechter [11] function:

$$\phi(L)\,dL = \phi^*(L/L^*)^\alpha \exp(-L/L^*)d(L/L^*).$$

This represents the number density of cluster galaxies in the luminosity interval L to $L + dL$.

2. Best-fit parameters for rich clusters are [11, 48] (see also Section 25.15):

$$\alpha \simeq -1.25,$$
$$L_B^* \simeq 10^{10} h^{-2} L_\odot \qquad [M_{J(24.1)}^* \simeq -19.9 \pm 0.1 + 5 \log h; \; M_B^* \simeq -19.5 + 5 \log h].$$

3. The amplitude ϕ^* is proportional to the cluster luminosity and richness.

4. cD galaxies (Section 25.6) are brighter than given by the bright end of this luminosity function [11, 48].

5. Mean absolute metric magnitude of the brightest cluster galaxy (within a $9.6h^{-1}$ kpc radius) in visual intrinsic (Vi) magnitudes (magnitude at 5456 Å in the galaxy rest frame) corrected for galactic obscuration [47]:

$$\langle M_{\text{Vi}} \rangle_1 = -21.57 \pm 0.03 + 5 \log h, \qquad \text{dispersion } \sigma = \pm 0.35^m \quad [15, 44, 47].$$

6. The average magnitude of the brightest cluster galaxy (Table 25.8) depends slightly on cluster richness and type [44, 45, 47].

Table 25.8. *Brightest galaxy magnitudes.*

R	$\langle M_{\text{Vi}} \rangle_1$	BM	$\langle M_{\text{Vi}} \rangle_1$
0	-21.55 ± 0.10	I	-21.82 ± 0.08
1	-21.51 ± 0.04	I–II	-21.63 ± 0.10
2	-21.75 ± 0.07	II	-21.70 ± 0.06
3	-21.88	II–III	-21.53 ± 0.07
		III	-21.41 ± 0.06

7. Number of cluster galaxies brighter than L:

$$N_g (> L) = \int_L^\infty \phi(L') \, dL' = \phi^* \Gamma(\alpha + 1, L/L^*),$$

where Γ is the incomplete gamma function.

8. Total luminosity of a cluster:

$$\begin{aligned} L_{\text{cl}} &= \int_0^\infty \phi(L) L \, dL \\ &= \phi^* L^* \Gamma(\alpha + 2) \\ &= 1.225 \phi^* L^* \qquad \text{for } \alpha = -1.25 \\ &\simeq 10^{12} h^{-2} L_\odot \qquad \text{for } R \simeq 1 \text{ clusters within } R_A = 1.5h^{-1} \text{ Mpc}. \end{aligned}$$

9. Some differences in the luminosity function among different cluster types are suggested [28, 29, 48].

10. The galaxy luminosity function in groups is comparable to that of rich clusters, with a slightly shallower faint-end slope of $\alpha \simeq -1$ (Section 25.12.2).

25.8 MASS FUNCTION OF CLUSTERS

The integrated mass function of clusters and groups of galaxies (Table 25.9) represents the number density of clusters with mass larger than $\mathcal{M}$. Bahcall and Cen [10] determined the cluster mass

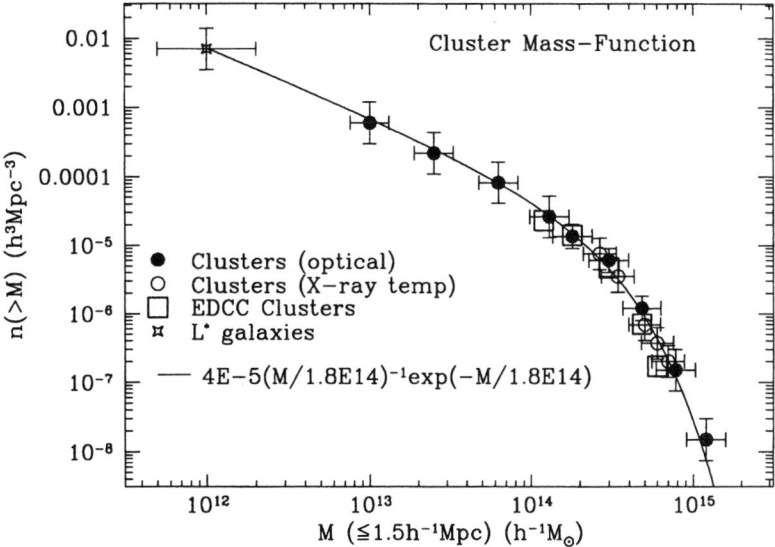

Figure 25.1. The mass function of clusters and groups of galaxies (Section 25.8) as determined from optical and X-ray temperature observations [10]. The best-fit function is shown by the solid curve.

function, and represent it approximately (for mass within $1.5h^{-1}$ Mpc radius) as

$$n_c\,(> \mathcal{M}) \simeq 4 \times 10^{-5}(\mathcal{M}/\mathcal{M}^*)^{-1}\exp(-\mathcal{M}/\mathcal{M}^*)h^3\;\mathrm{Mpc}^{-3},$$

with $\mathcal{M}^* \simeq 1.8 \times 10^{14}h^{-1}\mathcal{M}_\odot$, for $10^{13} \lesssim \mathcal{M}(\leq 1.5) \lesssim 2 \times 10^{15}h^{-1}\mathcal{M}_\odot$ (see Figure 25.1).

Table 25.9. *Cluster mass function* [1].

R	$\mathcal{M}\,(10^{15}h^{-1}\mathcal{M}_\odot)$	$n_c\,(> \mathcal{M})\,(h^3\;\mathrm{Mpc}^{-3})$
	Rich clusters[a]	
≥ 0	0.18	13.5×10^{-6}
	0.26^b	7.5×10^{-6}
≥ 1	0.30	6.0×10^{-6}
	0.34^b	3.5×10^{-6}
≥ 2	0.48	1.2×10^{-6}
	0.50^b	6.9×10^{-7}
	0.60^b	3.7×10^{-7}
	0.70^b	2.0×10^{-7}
≥ 3	0.78	1.5×10^{-7}
≥ 4	1.20	1.5×10^{-8}
	Groups[a]	
	0.010	6.0×10^{-4}
	0.025	2.2×10^{-4}
	0.063	8.2×10^{-5}
	0.13	2.6×10^{-5}

Notes

[a] R is the cluster richness and $\mathcal{M}$ is the corresponding cluster mass threshold (within a $1.5h^{-1}$ Mpc radius sphere). Estimated mean uncertainties: $10^{\pm 0.2}$ for the $R = 0, 1, 2$ densities; $10^{\pm 0.30}$ for the $R = 3, 4$ and the groups densities; and $10^{\pm 0.23}$ for the X-ray determined densities.

[b] Values for cluster mass and densities are from the X-ray temperature function [1, 2]. All other values are optical determinations [1].

References

1. Bahcall, N.A., & Cen, R. 1993, *ApJL*, **407**, L49
2. Henry, J.P., & Arnaud, K.A. 1991, *ApJ*, **372**, 410

25.9 X-RAY EMISSION FROM CLUSTERS

All rich clusters of galaxies produce extended X-ray emission due to thermal bremsstrahlung radiation from a hot intracluster gas [22, 23, 49–59]. The cluster X-ray luminosity emitted in the photon energy band E_1 to E_2 by thermal bremsstrahlung from a hot (T_x degrees) intracluster gas of uniform electron density n_e and a radius R_x is

$$L_x \propto n_e^2 R_x^3 T_x^{0.5} g (e^{-E_1/kT_x} - e^{-E_2/kT_x}).$$

The Gaunt factor correction g (of order unity) is a slowly varying function of temperature and energy [22]. The bolometric thermal bremsstrahlung luminosity of a cluster core can be approximated by

$$L_x(\text{core}) \simeq 1.4 \times 10^{42} [n_e(\text{cm}^{-3})]^2 [R_c(\text{kpc})]^3 [kT_x(\text{keV})]^{0.5} h^{-2} \text{ erg s}^{-1}.$$

Some of the main properties of the hot intracluster gas emission are summarized in Table 25.10.

25.9.1 X-Ray Properties of Clusters

Table 25.10. *X-ray properties of rich clusters.*

Property	Typical value or range	Notes
L_x (2–10 keV)	$\sim (10^{42.5} - 10^{45}) h^{-2}$ erg s^{-1}	a
$I_x(r)$	$I_x(r) \propto [1 + (r/R_c)^2]^{-3\beta + 1/2}$	b
$\langle \beta \rangle$	~ 0.7	c
$\rho_{\text{gas}}(r)$	$\rho_{\text{gas}}(r) \propto [1 + (r/R_c)^2]^{-3\beta/2} \propto [1 + (r/R_c)^2]^{-1}$	d
kT_x	~ 2–14 keV	e
T_x	$\sim 2 \times 10^7$–10^8 K	e
$\beta_{\text{spect}} = \sigma_r^2 / (kT_x / \mu m_p)$	~ 1	f
$R_c(x)$	$\sim (0.1$–$0.3) h^{-1}$ Mpc	g
n_e	$\sim 3 \times 10^{-3} h^{1/2}$ cm^{-3}	h
$\mathcal{M}_{\text{gas}}$ ($\lesssim 1.5 h^{-1}$ Mpc)	$\sim 10^{13.5} \mathcal{M}_\odot$ [range: $(10^{13}$–$10^{14}) h^{-2.5} \mathcal{M}_\odot$]	i
$\mathcal{M}_{\text{gas}} / \mathcal{M}_{\text{cl}}$ ($\lesssim 1.5 h^{-1}$ Mpc)	~ 0.07 (range: 0.03–$0.15 h^{-1.5}$)	i
Iron abundance	~ 0.3 solar (range: 0.2–0.5)	j

Notes

[a] The X-ray luminosity of clusters (2–10 keV band). $\langle L_x \rangle$ increases with cluster richness and with cluster type (toward compact, elliptical-rich clusters) [1–9].

[b] X-ray surface brightness distribution, $I_x(r)$; R_c is the cluster core radius [3, 4].

[c] Mean $\langle \beta \rangle$ from observations of X-ray brightness profiles [3, 4].

[d] Implied spatial density profile of the hot gas in the cluster [from $I_x(r)$; isothermal].

[e] Range of observed X-ray gas temperature in rich clusters [1, 4, 10–12].

$^f \beta_{\text{spect}}$ is the ratio of galaxy to gas velocity dispersion: μ is mean molecular weight in amu ($\mu \simeq 0.6$), m_p is mass of the proton, σ_r is radial velocity dispersion of galaxies in the cluster, and T_x is the X-ray temperature of the gas [3, 4, 8, 13].

^{g} Cluster core radius determined from the X-ray distribution in the cluster [9].

^{h} Typical intracluster gas density in rich cluster cores [3, 4, 9].

^{i} Typical mass (and range of masses) of hot gas in rich clusters and its fraction of the total (virial) cluster mass ($\mathcal{M}_{\text{gas}}/\mathcal{M}_{\text{cl}}$) within $r \lesssim 1.5 h^{-1}$ Mpc of the cluster center [8, 9, 14].

^{j} Typical iron abundance (and range) of the intracluster gas (in solar units) [8, 9].

References

1. David, L.P., Slyz, A., Jones, C., Forman, W., Vrtilek, S., & Arnaud, K. 1993, *ApJ*, **412**, 479
2. Burg, R., Giacconi, R., Forman, W., & Jones, C. 1994, *ApJ*, **422**, 37
3. Sarazin, C.L. 1986, *Rev. Mod. Phys.*, **56**, 1; 1988, *X-Ray Emission from Clusters of Galaxies* (Cambridge University Press, Cambridge)
4. Jones, C., & Forman, W. 1984, *ApJ*, **276**, 38
5. Giacconi, R., & Burg, R. 1990, in *Clusters of Galaxies*, STScI Symposium No. 4, edited by W.R. Oegerle et al. (Cambridge University Press, Cambridge), p. 377
6. Bahcall, N.A. 1977, *ApJ*, **217**, L77
7. Bahcall, N.A. 1977, *ApJ*, **218**, L93
8. Edge, A., & Stewart, G.C. 1991, *MNRAS*, **252**, 428
9. Jones, C., & Forman, W. 1992, in *Clusters and Superclusters of Galaxies*, NATO ASI Ser. No. 366, edited by A.C. Fabian (Kluwer Academic, Dordrecht), p. 49
10. Henry, J.P., & Arnaud, K.A. 1991, *ApJ*, **372**, 410
11. Edge, A., Stewart, G.C., Fabian, A.C., & Arnaud, K.A. 1990, *MNRAS*, **245**, 559
12. Arnaud, M., Hughes, J.P., Forman, W., Jones, C., Lachieze-Rey, M., Yamashita, K., & Hatusukade, I. 1992, *ApJ*, **390**, 345
13. Lubin, L., & Bahcall, N.A. 1993, *ApJL*, **415**, L17
14. White, D., & Fabian, A. 1995, *MNRAS*, **273**, 72

25.9.2 X-Ray–Optical Correlations of Cluster Properties

Some observed correlations between X-ray and optical properties are listed in Table 25.11 [49, 53, 54, 56–59].

Table 25.11. *Correlations between X-ray and optical properties.*^{a}

Properties	Correlation
σ_r-T	$\sigma_r \; (\text{km s}^{-1}) \simeq (332 \pm 52)[kT \; (\text{keV})]^{0.6 \pm 0.1}$
T-$N_{0.5}$	$kT \; (\text{keV}) \simeq 0.3 N_{0.5}^{0.95 \pm 0.18}$
L_x-$N_{0.5}$	$L_x(\text{bol}) \simeq 1.4 \times 10^{40} N_{0.5}^{3.16 \pm 0.15} h^{-2}$
L_x-f_{sp}	$L_x(\text{bol}) \simeq 0.6 \times 10^{43} f_{\text{sp}}^{-2.16 \pm 0.11} h^{-2}$
f_{sp}-T	$f_{\text{sp}} \simeq 1.2[kT \; (\text{keV})]^{-0.94 \pm 0.38}$
T-L_x	$kT \; (\text{keV}) \simeq 0.3[L_x(\text{bol}) h^2 / 10^{40}]^{0.297 \pm 0.004}$

Note

$^a \sigma_r$ is the galaxy line-of-sight velocity dispersion in the cluster (km s^{-1}). T is the temperature of the intracluster gas [kT (keV)]. $N_{0.5}$ is the central galaxy density in the cluster (number of galaxies brighter than $m_3 + 2^m$, within $r \leq 0.5 h_{50}^{-1} = 0.25 h^{-1}$ of the cluster center [1]). $L_x(\text{bol})$ is the bolometric X-ray luminosity of the cluster (erg s^{-1}). f_{sp} is the fraction of spiral galaxies in the cluster ($\lesssim 1.5 h^{-1}$ Mpc) [2,3]. Typical uncertainties of the coefficients are $\sim 50\%$ (see references).

References

1. Bahcall, N.A. 1977, *ApJ*, **217**, L77
2. Bahcall, N.A. 1977, *ApJ*, **218**, L93
3. Edge, A., & Stewart, G.C. 1991, *MNRAS*, **252**, 428

25.9.3 The X-Ray Luminosity Function of Clusters

1. The observed X-ray luminosity function of clusters (the number density of X-ray clusters with X-ray luminosity L_x to $L_x + dL_x$) is approximately [51]

$$\Phi_x(L_x)\,dL_x \simeq 2.7 \times 10^{-7}(L_x/10^{44})^{-1.65}\exp(-L_x/8.1 \times 10^{44})(dL_x/10^{44})\ \mathrm{Mpc}^{-3}\quad (h=0.5),$$

where L_x is the 2–10 keV X-ray luminosity in units of $\mathrm{erg\,s}^{-1}$ (for $h = 0.5$).

2. The luminosity function can also be approximated as a power law [51]:

$$\Phi_x(L_x)\,dL_x \simeq 2.2 \times 10^{-7}(L_x/10^{44})^{-2.17}(dL_x/10^{44})\ \mathrm{Mpc}^{-3}\quad (h=0.5).$$

3. The number of X-ray clusters with X-ray luminosity brighter than L_x is approximately

$$n_c\,(> L_x) \simeq 2 \times 10^{-7}(L_x/10^{44})^{-1.17}\ \mathrm{Mpc}^{-3}\quad (h=0.5).$$

4. The observed evolution of the X-ray cluster luminosity function suggests somewhat fewer high-luminosity clusters in the past ($z \gtrsim 0.5$) [51, 52] (see also Section 25.15).

25.9.4 Cooling Flows in Clusters [22, 60]

1. Cooling flows are common at the dense cores of rich clusters; X-ray images and spectra of $\sim 50\%$ of clusters suggest that the gas is cooling rapidly at their centers.

2. Typical inferred cooling rates: $\sim 100\mathcal{M}_\odot$/yr.

3. The gas cools within $r \lesssim 100h^{-1}$ kpc of the cluster center (generally centered on the brightest galaxy).

4. The cooling flows often show evidence for optical line emission, blue stars, and in some cases evidence for colder material in HI or CO emission, or X-ray absorption.

25.10 THE SUNYAEV–ZELDOVICH EFFECT IN CLUSTERS

The Sunyaev–Zeldovich effect [61] is a perturbation to the spectrum of the cosmic microwave background radiation as it passes through the hot dense intracluster gas. It is caused by inverse Compton scattering of the radiation by the electrons in the cluster gas.

At the long-wavelength side of the background radiation spectrum, the hot gas lowers the brightness temperature seen through the cluster center by the fractional decrement

$$\frac{\delta T}{T} = -2\tau_0\frac{kT_x}{m_e c^2},$$

where $T = 2.73$ K is the microwave radiation temperature, τ_0 is the Thomson scattering optical depth through the cluster ($\tau_0 = \sigma_T \int n_e\,dl$, where σ_T is the Thomson scattering cross section and dl is the distance along the line of sight), T_x is the intracluster gas temperature, and m_e is the electron mass.

For typical observed rich cluster parameters of $L_x \sim 10^{44}h^{-2}\ \mathrm{erg\,s}^{-1}$, $R_c \sim 0.2h^{-1}$ Mpc, and $kT_x \simeq 4$ keV, the bremsstrahlung relation ($L_x \propto n_e^2 R_c^3 T_x^{0.5}$, Section 25.9) implies a central gas density of $n_e \simeq 3 \times 10^{-3}h^{1/2}$ electrons cm^{-3}, thus yielding $\tau_0 \simeq 3 \times 10^{-3}h^{-1/2}$ [$\tau_0 = 0.0064 n_e(\mathrm{cm}^{-3})R_c(\mathrm{kpc})$]. Therefore

$$\frac{\delta T}{T} \sim -6 \times 10^{-5}h^{-1/2}.$$

This temperature decrement remains constant over the cluster core diameter

$$\theta_c \simeq \frac{2H_0 R_c}{cz} \simeq \frac{0.5}{z} \text{ arcmin}$$

and decreases at larger separations.

The effect has been detected in observations of rich, X-ray luminous clusters (e.g., Coma, A665, A2163, A2218, Cl 0016+16) [62–66]. See also Section 25.15.

25.11 CLUSTERS AND LARGE-SCALE STRUCTURE

Rich clusters are efficient tracers of the large-scale structure of the Universe [24, 12].

25.11.1 The Cluster–Cluster Correlation Function

1. The two-point spatial correlation function of clusters, $\xi_{cc}(r)$, is defined by

$$dP_c(r) = n_c[1 + \xi_{cc}(r)]\,dV,$$

where $dP_c(r)$ is the probability of finding a cluster in a volume element dV at a separation r from another cluster in the sample; the average space density of clusters in the sample is n_c.

2. The two-point correlation function for a sample of objects i (galaxies or clusters) is generally expressed as

$$\xi_{ii}(r) = A_{ii} r^{-\gamma} = [r/r_0(i)]^{-\gamma},$$

where A_{ii} is the correlation amplitude, $r_0(i) = A_{ii}^{1/\gamma}$ is the correlation scale of the sample i [$\xi(r_0) = 1$], and $\gamma \simeq 1.8$ is the observed slope.

3. The two-point cluster correlation function for $R \geq 1$ clusters [67–69, 33] is

$$\xi_{cc}(r)(R \geq 1) \simeq 250[r(\text{Mpc})]^{-1.8}, \qquad r(\text{Mpc}) \lesssim 50h^{-1} \text{ Mpc}.$$

This can be compared with the Galaxy correlation function [70, 12]

$$\xi_{gg}(r) \simeq 20[r(\text{Mpc})]^{-1.8}, \qquad r(\text{Mpc}) \lesssim 20h^{-1} \text{ Mpc}.$$

4. The rich-cluster correlation scale, r_0, for $R \geq 1$ clusters [67–69, 33] is

$$r_0(R \geq 1) \simeq (22 \pm 4)h^{-1} \text{ Mpc}.$$

This can be compared with the Galaxy correlation scale [70, 12]

$$r_0(g) \simeq (5.4 \pm 1)h^{-1} \text{ Mpc}.$$

(See also Table 25.12 and Section 25.15.)

5. The cluster correlation amplitude A_{cc} [where $\xi_{cc}(r) = A_{cc} r^{-1.8}$] increases with cluster richness [67, 24, 69]:

$$A_{cc} \simeq 4\langle N_R \rangle_{\text{median}},$$

where $\langle N_R \rangle_{\text{median}}$ is the median richness of the cluster sample (Sections 25.4.1 and 25.4.4).

6. The richness-dependent cluster correlation function [24, 69] is thus

$$\xi_{cc}(r)(> N_R) \simeq 4\langle N_R \rangle_{\text{median}}[r(\text{Mpc})]^{-1.8}.$$

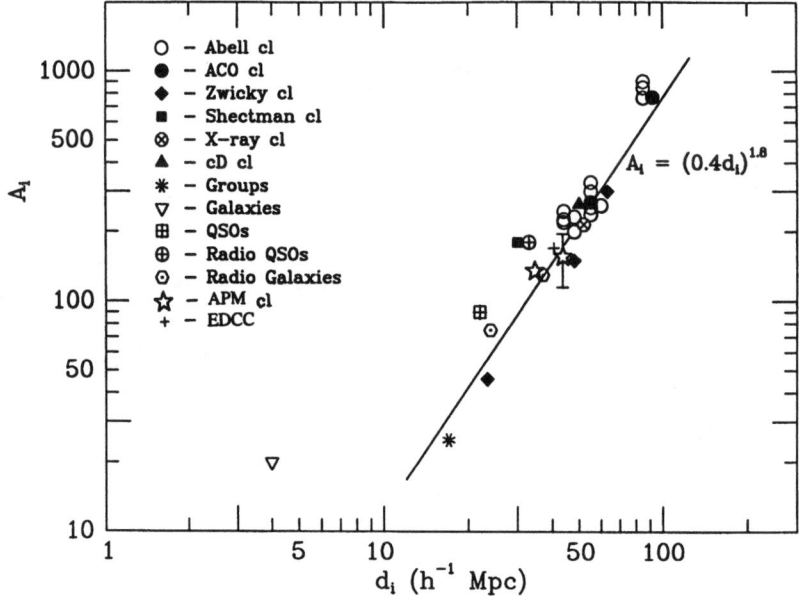

Figure 25.2. The universal dimensionless cluster correlation function: the dependence of correlation amplitude (A_i) on mean cluster separation (d_i) (Section 25.11.1) [69]. The data points represent different samples and catalogs of clusters and groups [including quasars and radio galaxies as represented by their parent groups (Section 25.13)]. Uncertainties in A_i are typically in the range of $\pm 20\%$–40%.

7. The correlation amplitude increases with cluster mean separation d ($d \equiv n_c^{-1/3}$, Section 25.4.2) [69, 71, 72],

$$A_{cc} \simeq (0.4d)^{1.8}, \qquad \text{i.e., } r_0 \simeq 0.4d,$$

implying a universal dimensionless cluster correlation function (Figure 25.2), to $d \lesssim 90h^{-1}$ Mpc:

$$\xi_{cc}(r/d) \simeq 0.2(r/d)^{-1.8}.$$

Table 25.12. *Cluster correlation scales (for* -1.8 *slope)* [1, 2].

Clusters	N_R (galaxies)	n_c (h^3 Mpc^{-3})	d (h^{-1} Mpc)	$0.4d$ (h^{-1} Mpc)	r_0 (observed) (h^{-1} Mpc)	Reference
Abell $R \geq 2$	≥ 80	1.2×10^{-6}	94	38	42 ± 10	[3–6]
Abell $R \geq 1$	≥ 50	6.0×10^{-6}	55	22	22 ± 4	[3–5, 7]
EDCC	$\gtrsim 30$	1.5×10^{-5}	41	16	16 ± 4	[8]
APM	$\gtrsim 20$	2.4×10^{-5}	35	14	13 ± 2	[9]

References
1. Bahcall, N.A., & Cen, R. 1992, *ApJL*, **398**, L81
2. Bahcall, N.A., & West, M. 1992, *ApJL*, **392**, 419
3. Bahcall, N.A. 1988, *ARA&A*, **26**, 631
4. Bahcall, N.A., & Soneira, R.M. 1983, *ApJ*, **270**, 20
5. Peacock, J., & West, M. 1992, *MNRAS*, **259**, 494
6. Postman, M., Geller, M., & Huchra, J. 1986, *AJ*, **91**, 1267
7. Postman, M., Huchra, J., & Geller, M. 1992, *ApJ*, **384**, 404
8. Nichol, R., Collins, C.A., Guzzo, L., & Lumsden, S.L. 1992, *MNRAS*, **255**, 21pp
9. Dalton, G.B., Efstathiou, G., Maddox, S.J., & Sutherland, W. 1992, *ApJL*, **390**, L1

25.11.2 Superclusters

Superclusters (SCs), defined as clusters of rich clusters of galaxies at a given spatial density enhancement f, are observed to scales of $\sim 150h^{-1}$ Mpc [73, 33] [$f \equiv n_c(SC)/n_c$, where $n_c(SC)$ is the number density of clusters in a supercluster and n_c is the average number density of clusters]. See Tables 25.13 and 25.14.

Table 25.13. *Global properties of Bahcall–Soneira superclusters* [1].

Property	$f = 20$ superclusters
Number density of SCs	$\sim 10^{-6}h^3$ Mpc^{-3}
Mean separation of SCs	$\sim 100h^{-1}$ Mpc
Number of clusters per SC	2–15 clusters
Fraction of clusters in SCs	54%
Size of largest SC	$\sim 150h^{-1}$ Mpc
SC shape	Flattened
Volume of space occupied by SCs	$\sim 3\%$

Reference
1. Bahcall, N.A., & Soneira, R.M. 1984, *ApJ*, **277**, 27

Table 25.14. *Bahcall–Soneira supercluster catalog ($f = 20$)* [1].[a]

BS	α, δ (1950)	z	Abell cluster members ($R \geq 1$) $f = 20$	Common name
1	00 53 −12 38	0.0541	85, 151	
2	01 12 +02 04	0.0433	119, 168, 189, 193	
3	01 22 +17 59	0.0652	154, 225	
4	02 56 +13 07	0.0738	399, 401	
5	04 42 +21 22	0.0682	500, 514	
6	10 49 +40 12	0.0795	1035, 1187	
7	11 13 +31 51	0.0347	1185, 1228	
8	11 42 +55 47	0.0581	1291, 1318, 1377, 1383, 1436	Ursa Major
9	11 45 −02 10	0.0992	1364, 1399	
10	11 20 +24 15	0.0218	1367, 1656	Coma SC (Great Wall)
11	13 45 +03 56	0.0782	1773, 1809	
12	15 27 +30 40	0.0710	1775, 1793, 1795, 1831, 1927, 2022, 2061, 2065, 2067, 2079, 2089, 2092, 2124, 2142, 2175	Corona Borealis SC
13	14 49 +15 06	0.0509	1913, 1983, 1991, 2040	
14	15 09 +06 05	0.0831	2028, 2029, 2048	
15	15 49 +16 15	0.0388	2063, 2107, 2147, 2151, 2152	Hercules SC
16	16 27 +40 21	0.0308	2197, 2199	

Note
[a]Catalog is complete to $z \lesssim 0.08$, $\delta > -27°$, $|b| \gtrsim 30°$, Abell $R \geq 1$ clusters.

Reference
1. Bahcall, N.A., & Soneira, R.M. 1984, *ApJ*, **277**, 27

A map of the superclusters is presented in Figure 25.3.

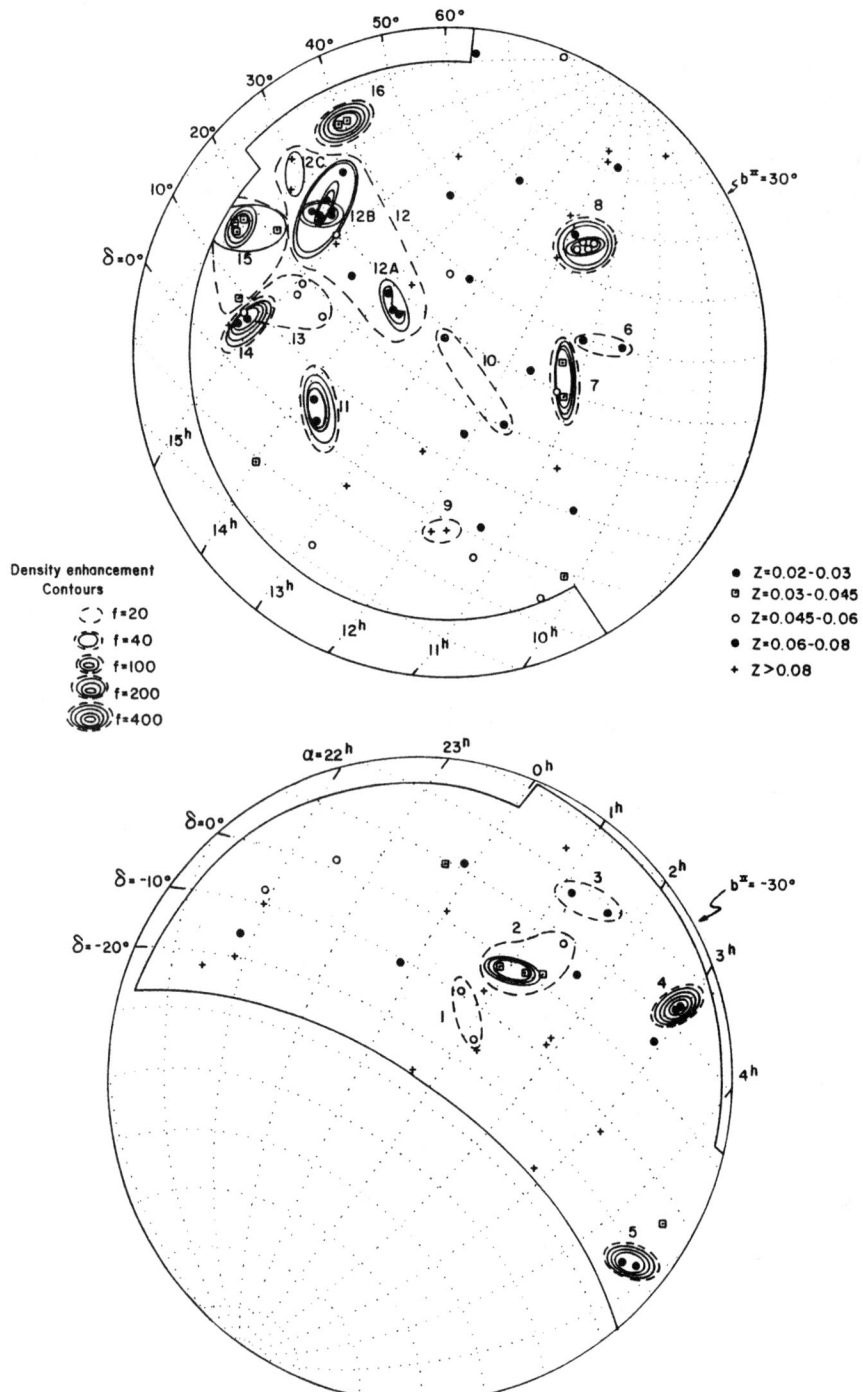

Figure 25.3. Projected map of the Bahcall–Soneira [73] superclusters (Section 25.11.2) for the $R \geq 1$, $D \leq 4$ cluster sample. Each cluster in the sample is presented by a symbol according to its redshift. The spatial density enhancement $f = n_c(SC)/n_c$ of each supercluster is shown. The North and South galactic hemispheres are presented, respectively, in the top and bottom parts of the figure.

25.12 GROUPS OF GALAXIES

25.12.1 Group Catalogs

Several catalogs of groups of galaxies are listed below (and in references therein).

25.12.1.1 Morgan Groups [74, 75]

1. Twenty-three groups and poor clusters; some contain a cD galaxy.
2. Identified on the Palomar Sky Survey.

25.12.1.2 Turner–Gott Groups [76–78]

1. One hundred and three nearby groups of galaxies (statistical sample).
2. Groups identified from the Zwicky galaxy catalog [3] at $\delta > 0°$, $b \geq 0°$, for galaxies with $m_p \leq 14^m$.
3. Selection criterion: surface number density enhancement, $f = 10^{2/3}$.

25.12.1.3 Center for Astrophysics (CfA) Groups [79, 80]

1. One hundred and seventy-six nearby groups with ≥ 3 galaxies (statistical sample).
2. Identified from the CfA redshift survey of Zwicky galaxies ($m_p \leq 14.5^m$, $\delta > 0°$, $b \geq 40°$; and $\delta > -2.5°$, $b \leq -30°$).
3. Selection criterion: spatial number density enhancement of galaxies, $f = 20$.
4. A deeper survey ($\leq 15.5^m$) over a smaller part of the sky ($\alpha = 8^h$–17^h, $\delta = 26.5°$–$38.5°$) yields 128 groups with $N_{gal} \geq 3$ [80]. Of these, 92 groups with $N_{gal} \geq 3$ and 36 groups with $N_{gal} \geq 5$ comprise statistical subsamples.

25.12.1.4 Southern Sky Redshift Survey Groups [81]

1. Southern Sky Redshift Survey (SSRS) [82] group catalog [81]. Eighty-seven groups with $N_{gal} \geq 3$.
2. Covers 1.75 steradians at the South Galactic Cap.
3. Selection criterion: spatial number density enhancement of galaxies, $f = 20$.

25.12.1.5 Hickson's Compact Groups [83]

1. One hundred very compact nearby groups of galaxies.
2. Identified on the red Palomar Sky Survey using a compactness and isolation selection criteria.
3. Small groups; very high galaxy density:

$$N_{gal} \sim 4, \qquad R_{group} \sim 50\text{–}100 \text{ kpc}, \qquad n_{gal} \sim 10^3\text{–}10^4 \text{ galaxies Mpc}^{-3}.$$

25.12.1.6 ACO—Supplementary Catalog [2]

See Section 25.2.2.

25.12.1.7   EDCC [6]

See Section 25.2.5.

25.12.2   Main Properties of Groups of Galaxies

Table 25.15. *Main properties of galaxy groups.*

Property	Typical range or median	Notes
Richness	3–30 galaxies	a
Radius	$(0.1–1)h^{-1}$ Mpc	b
Radial velocity dispersion	250 km s^{-1}	c
Compact groups:		
Richness	~ 4 galaxies	d
Radius	$\sim 0.1h^{-1}$ Mpc	
Velocity dispersion	$\sim 200 \text{ km s}^{-1}$	
Galaxy density	$\sim 10^3$ galaxies Mpc^{-3}	
Number density of groups	$(10^{-3}–10^{-5})h^3 \text{ Mpc}^{-3}$	e
Mass	$(10^{12.5}–10^{14})h^{-1}\mathcal{M}_\odot$	f
Optical blue luminosity	$(10^{10.5}–10^{12})h^{-2}L_\odot$	f
$\langle \mathcal{M}/L_B \rangle$	$\sim 200h\mathcal{M}_\odot/L_\odot$	g
Luminosity function	$\alpha \simeq -1$	
	$L^* \simeq 10^{10}h^{-2}L_\odot$	h
X-ray emission:		i
Luminosity	$\lesssim 10^{43}h^{-2} \text{ erg s}^{-1}$	
Gas temperature	$\lesssim 2 \text{ keV} (\sim 1 \text{ keV})$	
Fraction of galaxies in:		j
Rich clusters	$\sim 5\%$	
Groups	$\sim 55\%$	
Binaries	$\sim 10\%$	
"Singles"	$\sim 30\%$	

Notes

[a] Number of galaxies ($\gtrsim L^*$) within the main concentration of the group (generally $r \lesssim 1h^{-1}$ Mpc).

[b] Radius of the main concentration of galaxies (spatial galaxy density enhancement $\gtrsim 20$). The groups are generally embedded within larger-scale structures.

[c] The median radial velocity dispersion of galaxies in groups.

[d] Typical parameters of compact groups [1].

[e] The number density of groups decreases with increasing richness (Section 25.8).

[f] Typical range of the dynamical mass and luminosity of groups.

[g] Typical (median) $\mathcal{M}/L_B$ of galaxy groups; large scatter exists [2–4].

[h] The luminosity function of galaxies in groups approximates a Schechter function (Section 25.7) with the listed parameters of α and L^* [5].

[i] The X-ray luminosity and temperature of the hot intracluster medium [6–8].

[j] The fraction of galaxies in groups, rich clusters, binaries, and "singles" [3, 9, 10]. (It is possible that all galaxies belong to some groups within a broader definition.)

References

1. Hickson, P. 1982, *ApJ*, **255**, 382
2. Gott, J.R., & Turner, E.L. 1977, *ApJ*, **213**, 309
3. Geller, M., & Huchra, J. 1983, *ApJS*, **52**, 61; **54**, 422(E)
4. Ramella, M., Geller, M.J., & Huchra, J. 1989, *ApJ*, **344**, 57
5. Turner, E.L., & Gott, J.R. 1976, *ApJ*, **209**, 6
6. Jones, C., & Forman, W. 1984, *ApJ*, **276**, 38
7. Jones, C., & Forman, W. 1992, in *Clusters and Superclusters of Galaxies*, NATO ASI Ser. No. 366, edited by A.C. Fabian (Kluwer Academic, Dordrecht), p. 49

8. Mulchaey, J. 1995, in AIP Conf. Proc. 366, *Dark Matter*, edited by S.S. Holt and C.L. Bennett (AIP, New York), p. 243

9. Turner, E.L., & Gott, J.R. 1976, *ApJS*, **32**, 409

10. de Vaucouleurs, G. 1975, in *Stars and Stellar Systems*, edited by A. Sandage et al. (University of Chicago Press, Chicago)

25.13 QUASAR–CLUSTER ASSOCIATION

Imaging and spectroscopic data [84–86] indicate that quasars are found in environments significantly richer than those of average galaxies. The data show a positive association of quasars with neighboring galaxies.

1. Optically selected quasars to $z \lesssim 0.7$ exhibit a quasar–galaxy cross-correlation function amplitude A_{qg} that is approximately 2.3 times stronger than the galaxy–galaxy correlation amplitude (to separations, $r \lesssim 0.25h^{-1}$ Mpc):

$$\langle A_{qg} \rangle \simeq 2.3 \langle A_{gg} \rangle \simeq 46.$$

This excess correlation suggests that the quasars are typically located in groups of galaxies with a *mean* richness

$$\langle N_R \rangle = n_g \int_0^{1.5} A_{qg} r^{-1.8} 4\pi r^2 \, dr \simeq 12 \text{ galaxies}$$

(where $n_g \simeq 0.015$ Mpc^{-3} is the mean density of galaxies). The range of individual group richnesses is, however, wide.

2. Radio-loud quasars at $z \lesssim 0.5$ are found in similar environments to those of the optical quasars above. At $0.5 \lesssim z \lesssim 0.7$, the radio quasars appear to be located in richer environments, with

$$\langle A_{qg} \rangle \simeq 8 \langle A_{gg} \rangle \simeq 160 \qquad \text{(radio quasars, } 0.5 \lesssim z \lesssim 0.7\text{).}$$

This cross-correlation amplitude corresponds to a *mean* environment of rich clusters ($R \sim 0$, $N_R \sim 40$). Radio quasars at these redshifts are thus typically found in rich clusters.

3. The average galaxy velocity dispersion of the parent clusters associated with the quasars is [85, 86]

$$\sigma_r \sim 500 \text{ km s}^{-1}.$$

4. The observed autocorrelation function of optically selected quasars is approximately [87]

$$\xi_{qq}(r, z \sim 0) \simeq 10^{2\pm0.2} [r(\text{Mpc})]^{-1.8}.$$

The quasar correlation strength is intermediate between the correlation strength of individual galaxies and the correlation of rich clusters. This correlation strength is consistent with the quasar location in groups of the above mean richness, as would be suggested by the richness-dependent cluster correlation function (Section 25.11). The quasars may thus trace the correlation function of their parent clusters [88].

5. Similar results are observed for the association of radio galaxies with groups and clusters. This association explains the observed increase in the strength of the radio galaxy correlation function over the general galaxy correlations [88].

25.14   CLUSTERS AS GRAVITATIONAL LENSES

Rich clusters of galaxies can act as gravitational lenses; the mass in clusters at moderate redshifts ($z \simeq 0.2$–1) distorts images of background galaxies that lie near the line of sight to the cluster to form large arcs (near the cluster center), as well as numerous small-distortion elongated images in the weak-lensing regime [89–95].

1. The arcs are long and narrow optical images lying in the cores of rich clusters, stretched along circles centered on the cluster. The arc is an image of a distant galaxy, highly distorted by the strong gravitational potential of the cluster center. The lensed arc observations provide direct estimates of the projected mass density in the cluster core.

2. It is expected that all rich clusters above a critical projected core mass density will exhibit arcs. Examples of some clusters that are known to contain such arcs include [92] A370, A963, A1689, A2218, A2390, C0024+17, C0302+17, C0500−24, and C2244−22. The observations of cluster arcs are rapidly increasing. See Section 25.15.

3. Arclets and weak-lensing distortions are elongated objects, tangentially oriented with respect to the cluster mass center. The elongated images are often faint and blue, representing gravitationally distorted images of the faint background population of galaxies [91, 93–95]. See also Section 25.15

4. Analysis of the distorted images and their distribution serves as a direct measure of the projected mass density and mass distribution in the cluster [91, 93–95]. On average, cluster masses determined from lensing are consistent with the dynamical and the X-ray cluster mass estimates (Sections 25.4.11 and 25.8). The central core region ($r \lesssim 100$ kpc) of some clusters may suggest a more compact mass distribution (smaller core radius) than suggested by optical or X-ray data.

25.15   RECENT RESULTS

Some recent results that became available after the completion of this chapter in 1994 are added in proof below. These developments refer to the period 1994–1997, and supplement the previous sections.

1. Catalog of X-Ray Bright ACO Clusters (XBAC): optical ACO clusters (Section 25.2.2) detected in X-rays with the ROSAT All Sky Survey [96].
Palomar Deep Cluster Survey (PDCS): candidate distant optical clusters obtained from deep imaging of five square degree area [97].

2. Dynamical properties of clusters: velocity dispersions, velocity dispersion profiles, density profiles, masses, mass-to-light ratios (supplements Sections 25.4.6–25.4.11) [98–103].

3. Luminosity function of galaxies in clusters (supplements Section 25.7); the faint-end slope of the luminosity function and the L^* luminosity are determined for various samples and luminosity limits. A faint-end slope with typical values in the range of −1.1 to −1.6 is suggested [104–110].

4. Evolution of the cluster abundance, or mass-function, with redshift; only a mild "negative" evolution is suggested [111–116].

5. X-ray properties of clusters (supplements Section 25.9): temperatures, X-ray luminosities, core radii [96, 116–120].

6. X-ray luminosity function of clusters and the evolution of the luminosity function with redshift (supplements Section 25.9.3). Little or no evolution is observed for low to intermediate X-ray luminosity clusters; mild "negative" evolution may be suggested for highest luminosity clusters (to $z \sim 1$) [118–120].

7. Temperature function of X-ray clusters and its evolution with redshift to $z \sim 0.3$ [116]. Only mild evolution is suggested.

8. Sunyaev–Zeldovich effect in clusters (supplements Section 25.10): the S–Z effect has been accurately mapped and studied in a large number of rich clusters using interferometric observations [121–123]. The derived S–Z gas-mass fraction in clusters is $\mathcal{M}(gas)/\mathcal{M} = 0.06h^{-1}$, consistent with the gas-mass fraction determined from the cluster X-ray emission (Section 25.9).

9. Clusters and large-scale structure (Section 25.11): the power spectrum and correlation function of clusters [124–127].

10. Gravitational lensing by clusters: many rich clusters are detected by gravitational lensing, both in the strong and weak lensing regimes (Section 25.14) [128–132]. Cluster masses determined by gravitational lensing are consistent with those determined independently from the galaxy motions in clusters and from the hot intracluster gas.

REFERENCES

1. Abell, G.O. 1958, *ApJS*, **3**, 211
2. Abell, G.O., Corwin, H., & Olowin, R. 1989, *ApJS*, **70**, 1
3. Zwicky, F., Herzog, E., Wild, P., Karpowicz, M., & Kowal, C.T. 1961–1968, *Catalog of Galaxies and Clusters of Galaxies* (California Institute of Technology, Pasadena), Vols. 1–6
4. Shectman, S. 1985, *ApJS*, **57**, 77
5. Shane, C.D., & Wirtanen, C.A. 1967, *Publ. Lick Obs.* **22**, 1
6. Lumsden, S.L., Nichol, R.C., Collins, C.A., & Guzzo, L. 1992, *MNRAS*, **258**, 1
7. Nichol, R., Collins, C.A., Guzzo, L., & Lumsden, S.L. 1992, *MNRAS*, **255**, 21p
8. Dalton, G.B., Efstathiou, G., Maddox, S.J., & Sutherland, W. 1994, *MNRAS*, **269**, 151
9. Dalton, G.B., Maddox, S.J., Sutherland, W., & Efstathiou, G. 1997, *MNRAS*, **289**, 263
10. Bahcall, N.A., & Cen, R. 1993, *ApJL*, **407**, L49
11. Schechter, P.L. 1976, *ApJ*, **203**, 297
12. Peebles, P.J.E. 1993, *Principles of Physical Cosmology* (Princeton University Press, Princeton, NJ)
13. Efstathiou, G., Ellis, R., & Peterson, B. 1988, *MNRAS*, **232**, 431
14. Zwicky, F. 1957, *Morphological Astronomy* (Springer-Verlag, Berlin)
15. Bahcall, N.A. 1977, *ARA&A*, **15**, 505
16. King, I. 1966, *AJ*, **71**, 64; 1972, *ApJL*, **174**, L123
17. Peebles, P.J.E. 1980, *The Large Scale Structure of the Universe* (Princeton University Press, Princeton, NJ)
18. Lilje, P.B., & Efstathiou, G. 1988, *MNRAS*, **231**, 635
19. Geller, M.J. 1990, in *Clusters of Galaxies*, STScI Symposium No. 4, edited by W.R. Oegerle et al. (Cambridge University Press, Cambridge), p. 25
20. Bahcall, N.A. 1975, *ApJ*, **198**, 249
21. Dressler, A. 1978, *ApJ*, **226**, 55
22. Sarazin, C.L. 1986, *Rev. Mod. Phys.*, **58**, 1; 1988, *X-Ray Emission from Clusters of Galaxies* (Cambridge University Press, Cambridge)
23. Jones, C., & Forman, W. 1984, *ApJ*, **276**, 38
24. Bahcall, N.A. 1988, *ARA&A*, **26**, 631
25. Geller, M., & Huchra, J. 1989, *Science*, **246**, 897
26. Zwicky, F. 1938, *PASP*, **50**, 218
27. van den Bergh, S. 1962, *Z. Astrophys.*, **55**, 21; 1975, *ApJL*, **198**, Ll
28. Oemler, A. 1974, *ApJ*, **194**, 1
29. Dressler, A. 1980, *ApJ*, **236**, 351; 1978, *ApJ*, **223**, 765
30. Postman, M., & Geller, M. 1984, *ApJ*, **281**, 95
31. Struble, M., & Rood, H. 1991, *ApJS*, **77**, 363
32. Bahcall, N.A. 1981, *ApJ*, **247**, 787
33. Postman, M., Huchra, J., & Geller, M. 1992, *ApJ*, **384**, 404
34. Hughes, J.P. 1989, *ApJ*, **337**, 21
35. Chandrasekhar, S. 1942, *Principles of Stellar Dynamics* (Dover, New York), p. 313
36. Spitzer, L., & Hart, M. 1971, *ApJ*, **171**, 399
37. Bautz, L.P., & Morgan, W.W. 1970, *ApJL*, **162**, L149
38. Leir, A.A., & van den Bergh, S. 1977, *ApJS*, **34**, 381
39. Rood, H.J., & Sastry, G.N. 1971, *PASP*, **83**, 313
40. Matthews, T.A., Morgan, W.W., & Schmidt, M. 1964, *ApJ*, **140**, 35
41. Morgan, W.W., & Lesh, J.R. 1965, *ApJL*, **142**, 1364
42. Oemler, A. 1976, *ApJ*, **209**, 693
43. Kormandy, J. 1989, *ARA&A*, **27**, 265
44. Sandage, A. 1976, *ApJ*, **205**, 6
45. Hoessel, J.G., & Schneider, D.P. 1985, *AJ*, **90**, 1648; also Hoessel, J.G. 1980, *ApJ*, **241**, 493
46. Lauer, T.D. 1988, *ApJ*, **325**, 49
47. Hoessel, J.G., Gunn, J.E., & Thuan, T.X. 1980, *ApJ*, **241**, 486
48. Binggeli, B., Sandage, A., & Tamman, G. 1988, *ARA&A*, **26**, 509
49. David, L.P., Slyz, A., Jones, C., Forman, W., Vrtilek, S., & Arnaud, K. 1993, *ApJ*, **412**, 479
50. Henry, J.P., & Arnaud, K.A. 1991, *ApJ*, **372**, 410
51. Edge, A., Stewart, G.C., Fabian, A.C., & Arnaud, K.A. 1990, *MNRAS*, **245**, 559
52. Henry, J.P., Gioia, I.M., Maccacaro, T., Morris, S.L., Stocke, J.T., & Walter, A. 1992, *ApJ*, **386**, 408
53. Edge, A., & Stewart, G.C. 1991, *MNRAS*, **252**, 428
54. Jones, C., & Forman, W. 1992, in *Clusters and Superclusters of Galaxies*, NATO ASI Series, No. 366, edited by A.C. Fabian (Kluwer Academic, Dordrecht), p. 49
55. Böringer, H. 1994, in *Cosmological Aspects of X-Ray Clusters of Galaxies*, NATO ASI Series, edited by W.C. Seilter (Kluwer Academic, Dordrecht)
56. Bahcall, N.A. 1977, *ApJ*, **217**, L77
57. Bahcall, N.A. 1977, *ApJ*, **218**, L93
58. Lubin, L., & Bahcall, N.A. 1993, *ApJL*, **415**, L17
59. Mushotzky, R. 1984, *Phys. Scr.*, **T7**, L157
60. Fabian, A.C. 1992, in *Clusters and Superclusters of Galaxies*, NATO ASI Series No. 366, edited by A.C. Fabian (Kluwer Academic, Dordrecht), p. 151
61. Sunyaev, R.A., & Zeldovich, Ya. B. 1972, *Comments Astrophys. Space Phys.*, **4**, 173
62. Birkinshaw, M., Gull, S.F., & Hardebeck, H.E. 1984, *Nature*, **309**, 34
63. Herbig, T., Lawrence, C.R., Readhead, A.C.S., & Gulkis, S. 1995, *ApJL*, **449**, L1

64. Wilbanks, T.M., Ade, P.A.R., Fischer, M.L., Holzapfel, W.L., & Lange, A. 1994, *ApJL*, **427**, L75

65. Jones, M., Saunders, R., Alexander, P., Birkinshaw, M., & Dillon, N. 1993, *Nature*, **365**, 320

66. Moffet, A.T., & Birkinshaw, M. 1989, *AJ*, **98**, 1148

67. Bahcall, N.A., & Soneira, R.M. 1983, *ApJ*, **270**, 20

68. Peacock, J., & West, M. 1992, *MNRAS*, **259**, 494

69. Bahcall, N.A., & West, M. 1992, *ApJL*, **392**, 419

70. Groth, E., & Peebles, P.J.E. 1977, *ApJ*, **217**, 385

71. Bahcall, N.A., & Burgett, W. 1986, *ApJL*, **300**, L35

72. Szalay, A., & Schramm, D.N. 1985, *Nature*, **314**, 718

73. Bahcall, N.A. & Soneira, R.M. 1984, *ApJ*, **277**, 27

74. Morgan, W.W., Kayser, S., & White, R.A. 1975, *ApJ*, **199**, 545

75. Albert, C.E., White, R.A., & Morgan, W.W. 1977, *ApJ*, **211**, 309

76. Turner, E.L., & Gott, J.R. 1976, *ApJS*, **32**, 409

77. Turner, E.L., & Gott, J.R. 1976, *ApJ*, **209**, 6

78. Gott, J.R., & Turner, E.L. 1977, *ApJ*, **213**, 309

79. Geller, M., & Huchra, J. 1983, *ApJS*, **52**, 61; **54**, 422

80. Ramella, M., Geller, M.J., & Huchra, J. 1989, *ApJ*, **344**, 57

81. Maia, M.A.G., da Costa, L.N., & Latham, D.W. 1989, *ApJS*, **69**, 809

82. da Costa, L.N., Pellegrini, P.S., Sargent, W.L., Tonry J., Davis, M., Meiksin, M, Latham, D.W., Menzies, J.W., & Coulson, I.A. 1988, *ApJ*, **327**, 544

83. Hickson, P. 1982, *ApJ*, **255**, 382

84. Yee, H.K.C., & Green, R.F. 1987, *ApJ*, **319**, 28

85. Ellingson, E., Yee, H.K.C., & Green, R.F. 1991, *ApJ*, **371**, 45

86. Yee, H.K.C. 1992, in *Clusters and Superclusters of Galaxies*, NATO ASI Series No. 366, edited by A.C. Fabian (Kluwer Academic, Dordrecht), p. 293

87. Shaver, P. 1988, in *Large-Scale Structure of the Universe*, IAU Symposium No. 130, edited by J. Audouze et al. (Reidel, Dordrecht), p. 359

88. Bahcall, N.A., & Chokshi, A. 1991, *ApJL*, **380**, L9; 1992, *ApJL*, **385**, L33

89. Lynds, R., & Petrosian, V. 1989, *ApJ*, **336**, 1; 1986, *BAAS*, **18**, 1014

90. Soucail, G., Fort, B., Mellier, Y., & Picat, J.P. 1987, *A&A*, **172**, L14

91. Tyson, J.A., Valdes, F., & Wenk, R.A. 1990, *ApJL*, **349**, Ll

92. Soucail, G. 1992, in *Clusters and Superclusters of Galaxies*, NATO ASI Series No. 366, edited by A.C. Fabian (Kluwer Academic, Dordrecht), p. 199

93. Kaiser, N., & Squires, G. 1993, *ApJ*, **404**, 441

94. Smail, J., Ellis, R., Fitchett, M., & Edge, A. 1995, *MNRAS*, **273**, 277

95. Tyson, J.A., & Fischer, P. 1995, *ApJL*, **446**, L55

96. Ebeling, H., Voges, W., Bohringer, H., Edge, A.C., Huchra, J.P., & Briel, U.G. 1996, *MNRAS*, **281**, 799

97. Postman, M., Lubin, L., Gunn, J.E., Oke, J.G., Schneider, D.P., & Christensen, J.A. 1996, *AJ*, **111**, 615

98. Carlberg, R.G., Yee, H.K.C., & Ellingson, E. 1996, *ApJ*, **462**, 32

99. Carlberg, R.G., Yee, H.K.C., & Ellingson, E. 1997, *ApJ*, **478**, 462

100. Carlberg, R.G., Yee, H.K.C., & Ellingson, E. 1997, *ApJL*, **485**, L13

101. Mazure, A. et al. 1996, *A&A*, **310**, 31

102. Fadda, A.D., Girardi, M., Giuricin, G., Mardirosian, F., & Messetti, M., 1996, *ApJ*, **473**, 670

103. Den Hartog, R., & Katgert, P., 1996, *MNRAS*, **279**, 349

104. Lumsden, S.L., Collins, C.A., Nichol, R.C., Eke, V.R., & Guzzo, L., 1997, *MNRAS*, **290**, 119

105. Wilson, G., Smail, I., Ellis, R., & Couch, W., 1997, *MNRAS*, **284**, 915

106. Gaidos, E.J., 1997, *AJ*, **113**, 117

107. Berenstein, G.M., Nichol, R.C., Tyson, J.A., Ulmer, M., & Wittman, D., 1995, *AJ*, **110**, 1507

108. Valloto, C.A., Nicotra, M.A., Muriel, H., & Lambas, D.G., 1997, *ApJ*, **479**, 90

109. Trentham, N., 1997, *MNRAS*, **286**, 133

110. Trentham, N., 1998, *MNRAS*, **294**, 193

111. Carlberg, R.G., Morris, S.L., Yee, H.K.C., & Ellingson, E., 1997, *ApJL*, **479**, L19

112. Bahcall, N.A., Fan, X., & Cen, R., 1997, *ApJL*, **485**, L53

113. Fan, X., Bahcall, N.A., & Cen, R., 1997, *ApJL*, **490**, L123

114. Bahcall, N.A., & Fan, X. 1998, *ApJ*, **504**, 1; 1998, *Proc. Nat. Academy of Sciences*, **95**, 5956

115. Eke, V.R., Cole, S., & Frenk, C.S., 1996, *MNRAS*, **282**, 263

116. Henry, J.P., 1997, *ApJL*, **489**, L1

117. Mushotzky, R.F., & Scharf, C.A., 1997, *ApJL*, **482**, L13

118. Ebeling, H., Edge, A.C., Fabian, A.C., Allen, S.W., & Crawford, C.S., 1997, *ApJL*, **479**, L101

119. Vikhlinin, A., McNamara, B.R., Forman, W, Jones, C., Quintana, H., & Hornstrup, A., 1998, *ApJ*, **502**, 558; 1998, *ApJL*, **498**, L21

120. Rosati, P., Della Ceca, R., Norman, C., & Giacconi, R., 1998, *ApJL*, **492**, L21

121. Carlstrom, J. et al. 1996, *AAS*, **189**, 115.05

122. Carlstrom, J. et al. 1997, *AAS*, **191**, 116.02

123. Myers, S.T., Baker, J.E., Readhead, A.C.S., Leitch, E.M., & Herbig, T, 1997, *ApJ*, **485**, 1

124. Einasto, J., Einasto, M., Gottlober, S., Muller, V., Saar, V., Starobinsky, A.A., Tago, E., Tucker, D., Andernach, H., & Frisch, P., 1997, *Nature*, **385**, 139

125. Retzlaff, J., Borgani, S., Gottlober, S., & Muller, V., 1999, *New Astronomy*, in press (astro-ph 970944)

126. Tadros, H., Efstathiou, G., & Dalton, G., 1998, *MNRAS*, **296**, 995

127. Croft, R.A.C., Dalton, G., Efstathiou, G., & Sutherland, W., 1997, *MNRAS*, **291**, 305

128. Smail, I., Ellis, R.S., Dressler, A., Couch, W.J., Oemler, A., Sharples, R., & Butcher, H., 1997, *ApJ*, **479**, 70

129. Squires, G., Newman, D.M., Kaiser, N., Arnaud, M., Babul, A., Bohringer, H., Fahlman, G., & Woods, D., 1997, *ApJ*, **482**, 648

130. Fischer, P., & Tyson, J.A., 1997, *AJ*, **114**, 14

131. Colley, W.N., Tyson, J.A., & Turner, E.L., 1996, *ApJL*, **461**, L83

132. Luppino, G., & Kaiser, N., 1997, *ApJ*, **475**, 20

Chapter 26

Cosmology

Douglas Scott, Joseph Silk, Edward W. Kolb, and Michael S. Turner

26.1   FRIEDMANN–ROBERTSON–WALKER METRIC AND DISTANCE MEASURES

The standard metric for homogeneous and isotropic spaces can be written in many forms [1–5], including this version for spherical curvature:

$$ds^2 = c^2\,dt^2 - R^2(t)\left\{ dr^2 + \mathcal{R}^2 \sin^2(r/\mathcal{R})(d\theta^2 + \sin^2\theta\,d\phi^2) \right\}.$$

Here t is cosmic time, $R(t)$ is the scale factor, r is the comoving radial distance coordinate, and $\mathcal{R}$ is the radius of curvature at the present epoch [for more realistic spaces of negative curvature (open models), we can write $\mathcal{R} \to i\mathcal{R}$ and $\sin \to \sinh$]. We can then define an *effective distance* [6] (also known as transverse comoving or proper motion distance)

$$D = \mathcal{R}\sin(r/\mathcal{R}).$$

The Hubble and deceleration parameters can be defined at any epoch as

$$H = \frac{\dot{R}}{R} \quad \text{and} \quad q = -\frac{\ddot{R}R}{\dot{R}^2}.$$

The subscript 0 is used to denote the present value of a quantity, so that, for example, H_0 is the Hubble constant today.

The Einstein field equations lead to the Friedmann equation:

$$\dot{R}^2 = \tfrac{8}{3}\pi G\rho R^2 - KR_0^2 c^2 + \tfrac{1}{3}\Lambda R^2,$$

where $\rho(t)$ is the density, K is the spatial curvature, and Λ is the cosmological constant.

For a universe dominated by pressureless matter and with $\Lambda = 0$, the Friedmann equation gives

$$\frac{dz}{dt} = -H(z)(1+z) = -H_0(1+z)^2(1+\Omega_0 z)^{1/2},$$

where the definition of redshift is

$$1 + z \equiv \lambda_{\text{obs}}/\lambda_{\text{em}} = R_0/R,$$

curvature can be written

$$K \equiv (R_0\mathcal{R})^{-2} = H_0^2(\Omega_0 - 1)/c^2,$$

and the density parameter, $\Omega = 8\pi G\rho/3H^2$, scales as

$$\Omega(z) = \Omega_0(1+z)/(1+\Omega_0 z).$$

Deceleration is related to density and the cosmological constant through

$$q = \Omega/2 - \Lambda/3H^2,$$

and a spatially flat ($K = 0$) model means

$$\Omega + \Lambda/3H^2 = 1.$$

Because of this, the dimensionless cosmological constant is often written $\lambda = \Lambda/3H^2$.

Using the Friedmann equation and the definition of D we can derive (for $\Lambda = 0$)

$$D = \frac{2c}{H_0 \Omega_0^2 (1+z)} \left\{ \Omega_0 z + (\Omega_0 - 2) \left[(1 + \Omega_0 z)^{1/2} - 1 \right] \right\},$$

where

$$D = \frac{2c}{H_0} \left\{ 1 - (1+z)^{-1/2} \right\} \qquad \text{for} \quad \Omega_0 = 1$$

and

$$D \to \frac{2c}{H_0 \Omega_0} \qquad \text{for} \quad (\Omega_0 z)^{1/2} \gg 1.$$

Conventional distance measures can then be defined in terms of D; e.g., angular diameter distance $d_A = D/(1+z)$ and luminosity distance $d_L = D(1+z)$. More useful, however, are specific examples of how observed quantities depend on redshift and cosmological parameters. The angular size subtended by a physical scale ℓ is

$$\theta = \frac{\ell (1+z)}{D},$$

for small angles. Bolometric flux is related to bolometric luminosity through

$$S_{\text{bol}} = \frac{L_{\text{bol}}}{4\pi D^2 (1+z)^2},$$

whereas flux density or monochromatic flux

$$S_\nu(\nu_{\text{obs}}) = \frac{L_\nu(\nu_{\text{em}})}{4\pi D^2 (1+z)},$$

with $\nu_{\text{obs}} = \nu_{\text{em}}/(1+z)$, and if $L_\nu \propto \nu^{-\alpha}$, then

$$S_\nu(\nu_{\text{obs}}) = \frac{L_\nu(\nu_{\text{obs}})}{4\pi D^2 (1+z)^{1+\alpha}}.$$

The distance modulus for this quantity is

$$m - M = 25 + 5 \log[3000(1+z)H_0 D] - 5 \log h + K,$$

where $h = H_0/100 \,\text{km s}^{-1}\text{Mpc}^{-1}$ and the K-correction [7] is

$$K = -2.5 \log[(1+z)L_\nu(\nu_{\text{em}})/L_\nu(\nu_{\text{obs}})].$$

The solid angle ω subtended by the proper area A is

$$\omega = \frac{A(1+z)^2}{D^2},$$

so that the bolometric surface brightness

$$I_{\text{bol}}(\text{observed}) = \frac{S_{\text{bol}}}{\omega} = \frac{L_{\text{bol}}}{4\pi A(1+z)^4} = \frac{I_{\text{bol}}(\text{emitted})}{(1+z)^4}.$$

And for monochromatic surface brightness

$$I_\nu(\nu_{\text{obs}}) = \frac{I_\nu(\nu_{\text{em}})}{(1+z)^3}.$$

Comoving volume can be expressed as

$$dV = 4\pi D^2 dr = 4\pi D^2 \frac{c\, dz}{H(z)}.$$

So number counts per steradian, for a comoving number density n_0, are given by $dN/dz = n_0 D^2 c/H(z)$. Note that neither D, nor d_A, d_L, etc., are additive quantities, although r (the radial distance if it could be measured at fixed time today) is.

26.2   THE AGE OF THE UNIVERSE

For a matter-dominated model, with $\Lambda = 0$ and negligible pressure, the age is

$$t(z) = \frac{1}{H_0} \left[\frac{(1 + \Omega_0 z)^{1/2}}{(1 - \Omega_0)(1 + z)} - \frac{\Omega_0}{2(1 - \Omega_0)^{3/2}} \cosh^{-1} \left\{ \frac{2(1 - \Omega_0)}{\Omega_0(1 + z)} + 1 \right\} \right] \quad \text{for} \quad \Omega_0 < 1$$

$$= \frac{2}{3H_0} (1 + z)^{-3/2} \quad \text{for} \quad \Omega_0 = 1$$

The look-back time at a number of different redshifts is given in Table 26.1 in units of $10^9 h^{-1}$ years (note the dependence on the Hubble constant) for different values of Ω_0.

Table 26.1. *Look-back time in h^{-1} Gyr.*

Redshift	Ω_0						
	2.0	1.0	0.5	0.3	0.2	0.1	0.0
1	3.81	4.21	4.50	4.64	4.71	4.80	4.89
2	4.65	5.26	5.75	6.00	6.15	6.32	6.52
3	4.98	5.70	6.29	6.62	6.82	7.05	7.33
4	5.16	5.94	6.59	6.96	7.20	7.47	7.82
5	5.26	6.07	6.77	7.18	7.43	7.75	8.15
10	5.45	6.34	7.12	7.60	7.91	8.31	8.89
∞	5.58	6.52	7.37	7.91	8.28	8.78	9.78

The age of the Universe today is given in Table 26.2 in units of 10^9 years.

Table 26.2. *Age of the Universe in Gyr (if $\Lambda = 0$)*

H_0 (km s^{-1} Mpc^{-1})	Ω_0						
	2.0	1.0	0.5	0.3	0.2	0.1	0.0
100	5.58	6.52	7.37	7.91	8.28	8.78	9.78
90	6.20	7.24	8.19	8.79	9.19	9.75	10.86
80	6.98	8.15	9.21	9.88	10.34	10.97	12.22
70	7.97	9.31	10.52	11.29	11.82	12.54	13.97
60	9.30	10.86	12.28	13.17	13.79	14.62	16.30
50	11.16	13.03	14.73	15.81	16.54	17.54	19.56
40	13.95	16.29	18.41	19.75	20.67	21.90	24.44

Including a cosmological constant, the age of a matter-dominated Universe is given by

$$t(z) = \frac{1}{H_0} \int_z^\infty \frac{dz'}{(1+z')\left\{(\Omega_0 z' + 1 - \lambda_0)(1+z')^2 + \lambda_0\right\}^{1/2}}.$$

For a flat Universe with nonzero λ_0 (i.e., $\lambda_0 = 1 - \Omega_0$) the age today is

$$t = \frac{2}{3H_0} \frac{1}{\sqrt{1 - \Omega_0}} \ln\left\{\frac{1 + \sqrt{1 - \Omega_0}}{\sqrt{\Omega_0}}\right\},$$

and is given in Table 26.3 in units of 10^9 years.

Table 26.3. *Age of the Universe in Gyr (if $\Omega_0 + \lambda_0 = 1$)*

H_0 (km s^{-1} Mpc^{-1})	Ω_0 (or $1 - \lambda_0$)				
	1.0	0.5	0.3	0.2	0.1
100	6.52	8.13	9.43	10.52	12.49
90	7.24	9.03	10.47	11.69	13.87
80	8.15	10.16	11.78	13.15	15.60
70	9.31	11.61	13.46	15.02	17.83
60	10.86	13.54	15.70	17.52	20.79
50	13.03	16.24	18.84	21.02	24.93
40	16.29	20.31	23.54	26.26	31.13

26.3 CONVERSION FACTORS FOR THE EARLY UNIVERSE

Numerical factors in this and subsequent sections are given to five significant figures, which will almost always be more than sufficient for cosmological calculations. Anyone interested in the uncertainties associated with such values will be able to trace them through the earlier chapter on constants and units, or through, e.g., [8]. Less well-determined quantities will be quoted using an appropriate number of significant figures. Any quantity involving the gravitational constant G will be known with a little less accuracy than implied by the five figures. For definiteness we have used $G = 6.672\,59 \times 10^{-8}$ cm^3 g^{-1} s^{-2} $\equiv m_{\mathrm{Pl}}^{-2}$. On the other hand, quantities involving $G\mathcal{M}_\odot$ (e.g., the gravitational length scale of the Sun, or the critical density in $\mathcal{M}_\odot$ Mpc^{-3}) will be known to many more than five figures. The following conversions are useful for early Universe work, where "natural" units are used, so that $\hbar = c = k = 1$, and there is one fundamental dimension, energy, with the conventional unit of GeV [3]. Note that going from Tesla to Gauss is not a conversion in the normal sense, since they come from systems of units which differ by factors of 4π, etc. [9].

Energy:	$1\,\mathrm{GeV} = 1.6022 \times 10^{-3}$ erg
Temperature:	$1\,\mathrm{GeV} = 1.1604 \times 10^{13}$ K
Mass:	$1\,\mathrm{GeV} = 1.7827 \times 10^{-24}$ g
Length:	$1\,\mathrm{GeV}^{-1} = 1.9733 \times 10^{-14}$ cm
Time:	$1\,\mathrm{GeV}^{-1} = 6.5821 \times 10^{-25}$ s
Power:	$1\,\mathrm{GeV}^2 = 2.4341 \times 10^{21}$ erg s^{-1}
Number density:	$1\,\mathrm{GeV}^3 = 1.3015 \times 10^{41}$ cm^{-3}
Mass density:	$1\,\mathrm{GeV}^4 = 2.3201 \times 10^{17}$ g cm^{-3}

Energy density:	$1\,\mathrm{GeV}^4 = 2.0852 \times 10^{38}\,\mathrm{erg\,cm}^{-3}$
Volume emissivity:	$1\,\mathrm{GeV}^5 = 3.1680 \times 10^{62}\,\mathrm{erg\,cm}^{-3}\mathrm{s}^{-1}$
Cross section:	$1\,\mathrm{barn} = 10^3\,\mathrm{mb} = 10^{-24}\,\mathrm{cm}^2$
	$1\,\mathrm{mb} = 2.5681\,\mathrm{GeV}^{-2}$
Magnetic field:	$1\,\mathrm{Tesla} = 10^4\,\mathrm{Gauss}$
Field energy:	$1\,(\mathrm{Gauss})^2/8\pi = 1.9081 \times 10^{-40}\,\mathrm{GeV}^4$
	$1\,(\mathrm{Tesla})^2/2 = 1.9081 \times 10^{-32}\,\mathrm{GeV}^4$

26.4 OTHER USEFUL CONVERSION FACTORS

This is an assortment of conversion factors which are sometimes useful in various branches of cosmology.

Wavelength/energy:	$\lambda = 12\,398\,\text{Å}/E\,(\mathrm{eV}) = 1239.8\,\mathrm{nm}/E\,(\mathrm{eV})$
Energy of 1 μm photon:	$h\nu = 1.9864 \times 10^{-19}\,\mathrm{W\,s}$
Astronomical unit:	$1\,\mathrm{AU} = 1.4960 \times 10^{13}\,\mathrm{cm}$
	$= 214.94\,\mathcal{R}_\odot = 7.5812 \times 10^{26}\,\mathrm{GeV}^{-1}$
Parsec:	$1\,\mathrm{pc} \equiv 648\,000/\pi\,\mathrm{AU}$
	$= 3.2615\,(\text{sidereal})\,\text{light-yr}$
	$= 3.0857 \times 10^{18}\,\mathrm{cm}$
Megaparsec:	$1\,\mathrm{Mpc} = 10^6\,\mathrm{pc} = 3.0857 \times 10^{24}\,\mathrm{cm}$
	$= 1.5637 \times 10^{38}\,\mathrm{GeV}^{-1}$
Grav. scale of $1\mathcal{M}_\odot$:	$G\mathcal{M}_\odot/c^2 = 1.4766\,\mathrm{km}$
Energy in $1\mathcal{M}_\odot$:	$\mathcal{M}_\odot c^2 = 1.1157 \times 10^{57}\,\mathrm{GeV}$
Proton. equiv. of $1\mathcal{M}_\odot$:	$\mathcal{M}_\odot = 1.1891 \times 10^{57}\,\mathrm{protons}$
Surface density:	$1\,\mathcal{M}_\odot\,\mathrm{pc}^{-2} = 1.1718 \times 10^{20}\,\mathrm{GeV\,cm}^{-2}$
Mass density:	$1\,\mathcal{M}_\odot\,\mathrm{pc}^{-3} = 37.975\,\mathrm{GeV\,cm}^{-3}$
Sidereal day:	$1\,\mathrm{day}\,(\text{sidereal}) = 86\,164\,\mathrm{s}$
	$= 1.3091 \times 10^{29}\,\mathrm{GeV}^{-1}$
Sidereal year:	$1\,\mathrm{yr}\,(\text{sidereal}) = 3.1558 \times 10^7\,\mathrm{s}$
	$= 4.7945 \times 10^{31}\,\mathrm{GeV}^{-1} = 10^{-9}\,\mathrm{Gyr}$
Speed:	$1\,\mathrm{km\,s}^{-1} = 1.0227\,\mathrm{kpc\,Gyr}^{-1}$

There are many conventions for measurements of brightness, intensity, etc., in different wavebands. We list some useful conversions below, where m is an apparent magnitude, $d\omega$ is here an element of solid angle, $1\,\mathrm{Jy} = 10^{-23}\,\mathrm{erg\,cm}^{-2}\,\mathrm{s}^{-1}\,\mathrm{Hz}^{-1}$, and one S_{10} unit corresponds to one magnitude 10 star per square degree or 27.78 magnitudes per square arcsecond [10].

Solid angle:	$4\pi\,\text{steradians} \equiv 360^2/\pi\,\text{degree}^2$
	$= 5.3464 \times 10^{11}\,\text{arcsec}^2$
Specific intensity:	$I_\lambda = S_\lambda/d\omega = (\nu/\lambda)S_\nu/d\omega$
Solar magnitudes:	$m_{\mathrm{bol},\odot} = -26.83;\quad M_{\mathrm{bol},\odot} = 4.74$
	$m_{B,\odot} = -26.09;\quad M_{B,\odot} = 5.48$
Bolometric flux:	$S_{\mathrm{bol}} = 2.52 \times 10^{-5 - 0.4 m_{\mathrm{bol}}}\,\mathrm{erg\,cm}^{-2}\,\mathrm{s}^{-1}$

Flux density (Blue):
$$S_\lambda(\lambda \simeq 4400\,\text{Å}) = 6.76 \times 10^{-9-0.4m_B}\ \text{erg cm}^{-2}\,\text{Å}^{-1}\,\text{s}^{-1}$$
$$S_\nu(\lambda \simeq 4400\,\text{Å}) = 4.37 \times 10^{3-0.4m_B}\ \text{Jy}$$

Bolometric brightness:
$$m_{\text{bol}}\ \text{mag. arcsec}^{-2} = m_{\text{bol}}\,\mu$$
$$= 2.52 \times 10^{-5-0.4m_{\text{bol}}}\ \text{erg cm}^{-2}\,\text{arcsec}^{-2}\,\text{s}^{-1}$$
$$= 3.35 \times 10^{16-0.4\mu}\,\mathcal{L}_\odot\,\text{kpc}^{-2}$$

Blue brightness:
$$m_B\ \text{mag. arcsec}^{-2} = m_B\,\mu_B$$
$$= 6.76 \times 10^{-9-0.4m_B}\ \text{erg cm}^{-2}\,\text{arcsec}^{-2}\,\text{Å}^{-1}\,\text{s}^{-1}$$
$$= 6.62 \times 10^{16-0.4\mu_B}\,\mathcal{L}_\odot\,\text{kpc}^{-2}$$

Night sky:
$$22\mu_B \simeq 1.1 \times 10^{-17}\ \text{erg cm}^{-2}\,\text{arcsec}^{-2}\,\text{Å}^{-1}\,\text{s}^{-1} = 205 S_{10}$$

26.5 COSMOLOGICAL PARAMETERS

Many parameters depend on the precise value of the cosmic microwave background (CMB) temperature, so we scale by the quantity $T_{2.73} = T_{\gamma 0}/2.73$ K. Current observations indicate that $T_{2.73} = 0.9993 \pm 0.0007$ [11, 12].

Hubble constant:
$$H_0 = 100h\ \text{km s}^{-1}\,\text{Mpc}^{-1}$$
$$= 2.1331 \times 10^{-42}h\ \text{GeV}\quad(0.5 \lesssim h \lesssim 0.85)\quad[4]$$

Hubble time:
$$H_0^{-1} = 3.0857 \times 10^{17}h^{-1}\ \text{s} = 9.7778h^{-1}\ \text{Gyr}$$

Hubble distance:
$$cH_0^{-1} = 2997.9h^{-1}\ \text{Mpc} = 9.2506 \times 10^{27}h^{-1}\ \text{cm}$$

Critical density:
$$\rho_c \equiv 3H_0^2/8\pi G = 1.8788 \times 10^{-29}h^2\ \text{g cm}^{-3}$$
$$= 8.0980 \times 10^{-47}h^2\ \text{GeV}^4 = 1.0539 \times 10^4 h^2\ \text{eV cm}^{-3}$$
$$= 1.1233 \times 10^{-5}h^2\ \text{protons cm}^{-3}$$
$$= 2.7754 \times 10^{11}h^2\mathcal{M}_\odot\,\text{Mpc}^{-3}$$

Photons:
$$T_{\gamma 0} \equiv 2.73\,T_{2.73}\ \text{K} = 2.3525 \times 10^{-13}T_{2.73}\ \text{GeV}$$
$$n_{\gamma 0} = 412.77\,T_{2.73}^3\ \text{cm}^{-3}$$
$$\rho_{\gamma 0} = 2.0154 \times 10^{-51}T_{2.73}^4\ \text{GeV}^4$$
$$= 4.6760 \times 10^{-34}T_{2.73}^4\ \text{g cm}^{-3}$$
$$\Omega_{\gamma 0} = 2.4888 \times 10^{-5}h^{-2}T_{2.73}^4$$

Neutrinos:
$$T_{\nu 0} = (4/11)^{1/3}T_{\gamma 0} = 1.9486\,T_{2.73}\ \text{K}$$
$$= 1.6792 \times 10^{-13}T_{2.73}\ \text{GeV}$$
$$n_{\nu 0} = \tfrac{3}{11}n_{\gamma 0} = 112.57\,T_{2.73}^3\ \text{cm}^{-3}\quad\text{(per species)}$$
$$\rho_{\nu 0} = \tfrac{21}{8}(4/11)^{4/3}\rho_{\gamma 0} = 0.681\,32\rho_{\gamma 0}\quad\text{(3 species)}$$
$$\Omega_{\nu 0} = 1.6957 \times 10^{-5}h^{-2}T_{2.73}^4\quad\text{(3 massless species)}$$
$$\Omega_{\nu 0} = (m_\nu/93.625\,\text{eV})h^{-2}T_{2.73}^3\quad\text{(1 massive species)}$$

Entropy:
$$s_0 = [43\pi^4/(11 \times 45 \times \zeta(3))]n_{\gamma 0} = 7.0394\,n_{\gamma 0}$$
$$= 2905.7\,T_{2.73}^3\ \text{cm}^{-3}$$

Cosmological constant:
$$\Lambda \lesssim 3H_0^2 = 9.1556 \times 10^{-122}h^2 t_{\text{Pl}}^{-2}$$

Age of Universe:
$$t_0 \simeq \tfrac{2}{3}H_0^{-1} = 3.8161 \times 10^{60}h^{-1}t_{\text{Pl}}$$

26.6   FRIEDMANN–LEMAÎTRE MODEL

For the standard, homogeneous, isotropic, expanding cosmologies the following relationships may be useful (expressed below in early Universe units). Here the temperature of some species X which decoupled at temperature T_D is denoted by T_X. The subscript "eq" will refer to the equality epoch when $\rho(\text{radiation}) = \rho(\text{matter})$. The baryon-to-photon number density ratio is denoted by η. Note that an accurate conversion between Ω_B and n_B or η will depend on the fraction of helium, electron mass, binding energy, etc. Here we quote numbers using simply the proton mass.

We first define the factors g_* and g_{*s} to be the total number of effectively massless degrees of freedom which contribute to the radiation density and entropy density, respectively. At high energies $g_{*s} \to g_*$. Again subscript 0 refers to the present-day value.

Relativistic degrees of freedom (ρ_R):	$g_* = \sum_{i=\text{boson}}(T_i/T)^4 g_i + \frac{7}{8}\sum_{i=\text{fermion}}(T_i/T)^4 g_i$
	$g_{*0} = 2 + \frac{42}{8}(4/11)^{4/3} = 3.3626$ (3 ν species)
Relativistic degrees of freedom (s):	$g_{*s} = \sum_{i=\text{boson}}(T_i/T)^3 g_i + \frac{7}{8}\sum_{i=\text{fermion}}(T_i/T)^3 g_i$
	$g_{*s0} = 2 + \frac{42}{8}(4/11) = 3.9091$ (3 ν species)
Radiation density:	$\rho_R = (\pi^2/30)g_* T^4$
Entropy density:	$s = (2\pi^2/45)g_{*s} T^3$
Decoupled species:	$T_{X0} = [3.9091/g_{*s}(T_D)]^{1/3}T_{\gamma 0}$
Scale factor:	$R/R_0 = 3.7059 \times 10^{-13} T_{2.73} g_{*s}^{-1/3}(\text{GeV}/T)$
Ω (relativistic):	$\Omega_{\gamma\nu\bar\nu}h^2 = 4.1844 \times 10^{-5}T_{2.73}^4$
Ω (baryons):	$\Omega_B h^2 = 3.67 \times 10^7 T_{2.73}^3 \eta$
	$= 2.59 \times 10^8 T_{2.73}^3 n_B/s$
Entropy in the horizon:	$S_{\text{hor}} \equiv (4\pi/3)\, t^3 s$
	$= 5.0196 \times 10^{-2} g_{*s} g_*^{-3/2}(m_{\text{Pl}}/T)^3$ $(t \ll t_{\text{eq}})$
	$= 2.8548 \times 10^{87}(\Omega_0 h^2)^{-3/2}(1+z)^{-3/2}$ $(t \gg t_{\text{eq}})$
Baryons in the horizon:	$N_{B-\text{hor}} \equiv (n_B/s)S_{\text{hor}}$
	$= 1.94 \times 10^{-10}(\Omega_B h^2)g_{*s}g_*^{-3/2}(m_{\text{Pl}}/T)^3$ $(t \ll t_{\text{eq}})$
	$= 1.10 \times 10^{79}(\Omega_B h^2)(\Omega_0 h^2)^{-3/2}(1+z)^{-3/2}$ $(t \gg t_{\text{eq}})$

26.7   EPOCHS OF INTEREST

We can define several quantities at the Planck epoch. The Planck redshift is computed assuming there has been no inflationary period.

Planck mass:	$m_{\text{Pl}} \equiv (\hbar c/G)^{1/2} = 2.1767 \times 10^{-5}$ g
Planck energy:	$m_{\text{Pl}}c^2 \equiv (\hbar c^5/G)^{1/2} = 1.2210 \times 10^{19}$ GeV
Planck time:	$t_{\text{Pl}} \equiv (\hbar G/c^5)^{1/2} = 5.3906 \times 10^{-44}$ s
Planck length:	$l_{\text{Pl}} \equiv (\hbar G/c^3)^{1/2} = 1.6160 \times 10^{-33}$ cm
Planck density:	$\rho_{\text{Pl}} \equiv c^5/\hbar G^2 = 5.1575 \times 10^{93}$ g cm^{-3}
Planck redshift:	$T(z_{\text{Pl}}) \equiv m_{\text{Pl}}$ at $1 + z_{\text{Pl}} = 3.2948 \times 10^{31} g_{*s}^{1/3} T_{2.73}^{-1}$
cH_0^{-1} at Planck epoch:	$cH_0^{-1}z_{\text{Pl}}^{-1} = 2.8076 \times 10^{-4}$ cm

The age of the Universe can be followed analytically through the equality epoch (at least to $\mathcal{O}(R_{eq}/R_0)$). Under the assumption that we live in a flat, matter-dominated Universe today, with both matter and radiation contributing in the past, we can rewrite the Friedmann equation as $\dot{a} = H_0(a + a_{eq})^{1/2}/a$, where, for conciseness, we have used $a \equiv R/R_0$. From this can be derived

$$H_0 t = \tfrac{2}{3}\left\{(a + a_{eq})^{1/2}(a - 2a_{eq}) + 2a_{eq}^{3/2}\right\},$$

which is an exact expression for $t(R)$, although it cannot be easily inverted to give $R(t)$.

For conformal time (defined by $d\eta \equiv dt/a$, not to be confused with the baryon-to-photon ratio) we also have

$$H_0 \eta = 2\left\{(a + a_{eq})^{1/2} - a_{eq}^{1/2}\right\},$$

which can, in fact, be inverted to give $a = \tfrac{1}{4}H_0^2\eta^2 + \sqrt{a_{eq}}H_0\eta$. The conformal time can be thought of as the comoving size of the horizon.

Similarly, for the expansion time, $t_{exp} \equiv R/\dot{R} \equiv H^{-1}$, we have

$$H_0 t_{exp} = a^2(a + a_{eq})^{-1/2}.$$

At the epoch of equality we have

$$H_0 t_{eq} = \tfrac{2}{3}(2 - \sqrt{2})a_{eq}^{3/2}, \qquad H_0 \eta_{eq} = 2(\sqrt{2} - 1)a_{eq}^{1/2}, \qquad \text{and} \qquad H_0 t_{exp,eq} = a_{eq}^{3/2}/\sqrt{2},$$

again with $a \equiv R/R_0 \equiv (1+z)^{-1}$. The approximate scaling $t \propto \Omega_0^{-1/2}$ can also be used for $\Omega_0 z \gtrsim 1$. The time scales of the Universe can then be written in many useful ways in either the radiation- or matter-dominated limits:

Age of Universe: $t = 0.301\,18g_*^{-1/2}m_{Pl}/T^2$

$(T \gg T_{eq})$:
$$= 2.4206 \times 10^{-6}g_*^{-1/2}(\text{GeV}/T)^2 \text{ s}$$
$$= 2.4206 g_*^{-1/2}(\text{MeV}/T)^2 \text{ s}$$
$$= 1.7625 \times 10^{19}g_{*s}^{2/3}g_*^{-1/2}(1+z)^{-2} \text{ s}$$

$(T \ll T_{eq})$: $t = 0.401\,57g_*^{-1/2}m_{Pl}/(T^{3/2}T_{eq}^{1/2})$
$$= 7.4228 \times 10^{11}(\Omega_0 h^2)^{-1/2}T_{2.73}^{3/2}(\text{eV}/T)^{3/2} \text{ s}$$
$$= 2.0571 \times 10^{17}(\Omega_0 h^2)^{-1/2}(1+z)^{-3/2} \text{ s}$$

Conformal time: $\eta = 1.9847 \times 10^{31}T_{2.73}^{-1}g_{*s}^{1/3}g_*^{-1/2}m_{Pl}/T$

$(T \gg T_{eq})$:
$$= 1.2984 \times 10^7 T_{2.73}^{-1}g_{*s}^{1/3}g_*^{-1/2}(\text{GeV}/T) \text{ s}$$
$$= 3.4247 \times 10^5 T_{2.73}^{-2}g_{*s}^{2/3}g_*^{-1/2}(1+z)^{-1} \text{ Mpc}$$

$(T \ll T_{eq})$: $\eta = 3.9693 \times 10^{31}T_{2.73}^{-1}g_{*s}^{1/3}g_*^{-1/2}m_{Pl}/(T^{1/2}T_{eq}^{1/2})$
$$= 6.0088 \times 10^{15}(\Omega_0 h^2)^{-1/2}T_{2.73}^{1/2}g_{*s}^{1/3}(\text{eV}/T)^{1/2} \text{ s}$$
$$= 5995.8(\Omega_0 h^2)^{-1/2}(1+z)^{-1/2} \text{ Mpc}$$

Expansion time: $t_{exp} = (1 + z_{eq})^{1/2}(2 + z + z_{eq})^{-1/2}(1+z)^{-3/2}H_0^{-1}\Omega_0^{-1/2}$
$$= 4.7702 \times 10^{19}(2 + z + z_{eq})^{-1/2}(1+z)^{-3/2}T_{2.73}^{-2} \text{ s}$$

Several epochs relevant for structure formation and the microwave background spectrum can also be useful. Note that the definition of z_{eq} depends on the number of neutrino species included.

Recombination through the visibility function (or $e^{-\tau}d\tau$ for Thomson scattering), and is centered around $z \simeq 1100$ largely independent of cosmological parameters. We avoid using the term "decoupling" to refer to the last scattering epoch, since it might more physically be applied to the epoch when the matter temperature can depart from the radiation, i.e., when the Compton cooling time is short. Other epochs are defined in terms of the redshift when the rate (generally $|\dot{\epsilon}|/\epsilon$) is equal to the expansion rate, although definitions differing by constant factors could also have been used. Compton cooling and Compton drag depend on the evolution of the ionized fraction, $x_e(z)$. At redshifts much above z_C, Compton scattering will relax spectral distortions to a Bose–Einstein form, whereas much above the smaller of z_{DC} or z_{Br} there will be complete thermalization. Reviews of these topics have been presented in, e.g., [16–18]. Here the factor $(1 - Y_p/2)$ accounts for the helium contribution, assuming it is fully ionized.

End of neutrino free-streaming:	$3T_\nu(z_{nr}) \equiv m_\nu$ at $$1 + z_{nr} = 59\,553(m_\nu/30\,\text{eV})T_{2.73}^{-1}$$
Radiation-matter equality:	$R_{eq} = 4.1845 \times 10^{-5}(\Omega_0 h^2)^{-1}T_{2.73}^4 R_0$ $$1 + z_{eq} = 23\,898(\Omega_0 h^2)T_{2.73}^{-4} \quad \text{(for } \gamma\text{'s} + 3\nu\text{'s)}$$ $$1 + z_{eq} = 40\,180(\Omega_0 h^2)T_{2.73}^{-4} \quad \text{(for } \gamma\text{'s only)}$$ $$T_{eq} = 5.6222(\Omega_0 h^2)T_{2.73}^{-3}\,\text{eV}$$ $$t_{eq} = 3.2618 \times 10^{10}(\Omega_0 h^2)^{-2}\,T_{2.73}^6\,\text{s}$$ $$\eta_{eq} = 16.066(\Omega_0 h^2)^{-1}T_{2.73}^2\,\text{Mpc}$$
Recombination:	$1 + z_{rec} \equiv 1100;\ R_{rec} \equiv R_0/1100$ $T_{rec} = 0.259\,T_{2.73}\text{eV}$ $t_{rec} = 5.64 \times 10^{12}(\Omega_0 h^2)^{-1/2}\text{s}$ (valid if $\Omega_0 h^2 \gtrsim 0.04\,T_{2.73}^4$)
Compton cooling:	$t_{cool}(z) \equiv m_e c n_B/4\sigma_T a T^4 n_e = t_{exp}(z)$ at $1 + z_{cool} \simeq 500(\Omega_B h^2)^{2/5}$ (for standard rec.) $1 + z_{cool} \simeq 5.7(\Omega_0 h^2)^{1/5}x_e^{-2/5}(1 - Y_p/2)^{-2/5}$ (for $x_e \equiv n_e/n_B = $ const.)
Radiation drag:	$t_{drag}(z) \equiv (m_p/m_e)t_{cool} = t_{exp}(z)$ at $1 + z_{drag} \simeq 120(\Omega_0 h^2)^{1/5}x_e^{-2/5}(1 - Y_p/2)^{-2/5}$ (for $x_e = $ const.) [19]
Double-Compton thermalization:	$1 + z_{DC} \simeq 2.8 \times 10^5(1 - Y_p/2)^{-2/5}(\Omega_B h^2)^{-2/5}T_{2.73}^{1/5}$ (when $t_{DC} = t_{exp}$, for $z \gg z_{eq}$)
Bremsstrahlung thermalization:	$1 + z_{Br} \simeq 4.7 \times 10^4(1 - Y_p/2)^{-4/5}(\Omega_B h^2)^{-6/5}T_{2.73}^{13/5}$ (when $t_{Br} = t_{exp}$, for $z \gg z_{eq}$)
Compton thermalization:	$1 + z_C \simeq 7.1 \times 10^3(1 - Y_p/2)^{-1/2}(\Omega_B h^2)^{-1/2}T_{2.73}^{1/2}$ (valid for $\Omega_B h^2 \lesssim 0.09(1 - Y_p/2)^{-1/2}(\Omega_0 h^2)^2 T_{2.73}$)

26.8 AGE LIMITS

In this and subsequent sections, observationally determined quantities are quoted with the 1σ, 2σ, or 95% Confidence Limit error bars taken from the references. Quantities without error bars should be considered as more generic or approximate. For nuclear chronology and for the ages of the oldest globular clusters we give two representative estimates of the range, to indicate that different authors can be more or less conservative.

Oldest Earth rocks:	$t_\oplus = 3.962 \pm 0.003\,\mathrm{Gyr}\,(1\sigma)$ [20]
Oldest meteorites:	$t_{\mathrm{meteor}} = 4.53 \pm 0.02\,\mathrm{Gyr}\,(1\sigma)$ [21]
Oldest lunar rock:	$t_{\mathrm{moon}} = 4.6 \pm 0.1\,\mathrm{Gyr}\,(1\sigma) \simeq t_\odot$ [22]
Nuclear chronology:	$t_{\mathrm{Galaxy}} = 5.4 \pm 1.5\,\mathrm{Gyr}\,(1\sigma) + t_\odot$ [23, 24]
	$10\,\mathrm{Gyr} < t_{\mathrm{Galaxy}} < 20\,\mathrm{Gyr}$ [25]
Oldest globular clusters:	$t_{\mathrm{glob}} = 11.5 \pm 1.3\,\mathrm{Gyr}\,(1\sigma)$ [26, 27]
	$12\,\mathrm{Gyr} < t_{\mathrm{glob}} < 20\,\mathrm{Gyr}\,(2\sigma)$ [28]
White dwarfs:	$t_{\mathrm{disk}} = 9.5 \pm 1\,\mathrm{Gyr}\,(1\sigma)$ [29, 30]

26.9 COSMOLOGICAL TESTS: H_0

This list is not intended to be the ultimate authority on the subject, but should be taken as an indication of values derived by different methods, based largely upon some recent reviews [31–35]. Note that most of these methods require more than one step, e.g., many use the calibration of Cepheid variables. Two distinct estimates are given for the SNe Ia standard candle method. The lensing estimate from quasar 0957+561 [36] has uncertainty dominated by the lens model. The Sunyaev–Zel'dovich (S–Z) value includes some estimate of possible systematic errors. Estimates from, e.g., [31] use the derived distances to Virgo, an assumed ratio of the Coma to Virgo distance of 5.6 ± 0.5 and a Coma velocity of $7160 \pm 200\,\mathrm{km\,s^{-1}}$ after correction for Virgocentric infall. All of the methods are actually attempts to measure the distance to some distant object (e.g., Virgo or Coma), and so would change if a different velocity or distance ratio was adopted.

Type Ia supernovae:	$H_0 = 58 \pm8\,\mathrm{km\,s^{-1}\,Mpc^{-1}}\,(1\sigma)$ [37, 38]
	$H_0 = 64 \pm6\,\mathrm{km\,s^{-1}\,Mpc^{-1}}\,(1\sigma)$ [39, 40]
Type II supernovae:	$H_0 = 73 \pm9\,\mathrm{km\,s^{-1}\,Mpc^{-1}}\,(1\sigma)$ [41]
Tully–Fisher:	$H_0 = 81 \pm 11\,\mathrm{km\,s^{-1}\,Mpc^{-1}}\,(1\sigma)$ [31]
Planetary nebula lum. fn.:	$H_0 = 83 \pm 10\,\mathrm{km\,s^{-1}\,Mpc^{-1}}\,(1\sigma)$ [31]
Globular cluster lum. fn.:	$H_0 = 68 \pm 15\,\mathrm{km\,s^{-1}\,Mpc^{-1}}\,(1\sigma)$ [31]
Novae:	$H_0 = 61 \pm 13\,\mathrm{km\,s^{-1}\,Mpc^{-1}}\,(1\sigma)$ [31]
Surface brightness fluctns.:	$H_0 = 80 \pm8\,\mathrm{km\,s^{-1}\,Mpc^{-1}}\,(1\sigma)$ [31]
$D_n - \sigma$:	$H_0 = 76 \pm 13\,\mathrm{km\,s^{-1}\,Mpc^{-1}}\,(1\sigma)$ [31]
Gravitational lensing:	$H_0 = 64 \pm 13\,\mathrm{km\,s^{-1}\,Mpc^{-1}}\,(95\%\,\mathrm{CL})$ [36]
Sunyaev–Zel'dovich:	$H_0 = 55 \pm 17\,\mathrm{km\,s^{-1}\,Mpc^{-1}}\,(1\sigma)$ [42–45]

26.10 COSMOLOGICAL TESTS: q_0

"Classical" cosmology has dealt with methods of determining the deceleration parameter q_0 as well as H_0. Here it is still conventional to use q_0 in place of $\Omega_0/2$, although occasionally $\Omega_0/2 - \lambda_0$ is meant. The variation of two observable quantities is used to determine the best fit q_0, specifically apparent magnitude versus redshift (m, z), number versus redshift (N, z), number versus apparent magnitude (N, m), angular diameter versus redshift (θ, z), and other geometrical methods of distance determination. However, all of these tests are subject to large evolutionary corrections and other redshift-dependent effects. They therefore require much interpretation in order to estimate q_0.

There may be some evidence from (N, m) that $q_0 \lesssim 0.15$ [46], although other values can also fit [47], while (θ, z) may indicate $q_0 \simeq 0.5$ [48], and (m, z) for distant supernovae prefers q_0 (actually $(\Omega_0 - \lambda)/2) \simeq 0$ [49, 50]. More concretely, age constraints give, for example, $q_0 \leq 0.08$ if $t_0 \geq 13$ Gyr and $h \geq 0.65$ (and assuming $\Lambda = 0$).

26.11 OTHER COSMOLOGICAL PARAMETERS

The best limit on Λ comes from considerations of gravitational lensing, with distant supernovae now competing. $\dot{G}$ limits come from laser and radar ranging experiments in the solar system, from the constancy of neutron star masses, and from Big Bang Nucleosynthesis. Brans–Dicke theories are also constrained by lunar ranging data. The global rotation ω_{rot} and global shear σ of the Universe are constrained by CMB anisotropy measurements.

Cosmological constant:	$\lambda_0 < 0.66$ (95% CL, assuming $\Omega_0 + \lambda_0 = 1$) [51, 52]
Variable G:	$\lvert\dot{G}\rvert/G \leq 4 \times 10^{-12}\,\text{yr}^{-1}\,(1\sigma)$ or $\lvert\dot{G}\rvert/(GH_0) \leq 0.04h^{-1}$
	(from ranging measurement) [53]
	$\dot{G}/G = (-0.6 \pm 4.2) \times 10^{-12}\,\text{yr}^{-1}$ or $\lvert\dot{G}\rvert/(GH_0) \leq 0.05h^{-1}$
	(95% CL, from neutron star masses) [54]
	$\lvert\dot{G}\rvert/G \leq 9 \times 10^{-13}\,\text{yr}^{-1}\,(2\sigma)$ or $\lvert\dot{G}\rvert/(GH_0) \leq 0.009h^{-1}$
	(assuming $G \propto t^{-x}$ from BBN) [55]
Brans–Dicke	$\omega_{\text{BD}} > 600\,(1\sigma)$ [53]
coupling constant:	($\omega_{\text{BD}} \to \infty$ gives General Relativity)
Global rotation:	$(\omega_{\text{rot}}/H)_0 < 6 \times 10^{-8}$ (95% CL, for $0.1 \leq \Omega_0 \leq 1$) [56, 57]
Global shear:	$(\sigma/H)_0 < 10^{-9}$ (95% CL) [56]

26.12 PRIMORDIAL NUCLEOSYNTHESIS AND NEUTRINOS

The following quantities relate to the formation of the light elements during Big Bang Nucleosynthesis (BBN), and are a combination of observational limits and the results of computer simulations.

Nucleon-to-photon ratio:	$2.0 \times 10^{-10} \leq \eta \leq 6.5 \times 10^{-10}$ (95% CL) [58–60]
Baryon density:	$0.007 \leq \Omega_B h^2 \leq 0.024$ (95% CL) [59]
Primordial ^{4}He:	$0.221 \leq Y_p \leq 0.243$ (95% CL) [58]
Primordial D + ^{3}He:	$(\text{D} + {}^3\text{He})/\text{H} \leq 1.1 \times 10^{-4}$ (95% CL) [58]
Primordial ^{7}Li:	$0.7 \times 10^{-10} \leq {}^7\text{Li/H} \leq 3.5 \times 10^{-10}$ (95% CL) [58]

The number and masses of neutrinos can potentially be important for cosmology: $\Omega_{\nu 0} = (\sum m_\nu/93.625\,\text{eV})h^{-2}T_{2.73}^3$. Limits on the number of light neutrino species come from both particle accelerators (e.g., LEP at CERN), where "light" means $\lesssim m(Z^0)/2 \simeq 46\,\text{GeV}$, and from BBN where "light" means $\lesssim 1\,\text{MeV}$.

N_ν from BBN:	$N_\nu < 4$ (conservative limit) [59, 60]
N_ν from LEP:	$N_\nu = 2.993 \pm 0.011$ (1σ) [8]
Electron neutrino:	$m_{\nu_e} < 15\,\text{eV}$ (conservative limit) [8]
Muon neutrino:	$m_{\nu_\mu} < 170\,\text{keV}$ (90% CL) [8, 61]
Tau neutrino:	$m_{\nu_\tau} < 24\,\text{MeV}$ (95% CL) [62, 8]

26.13 POWER SPECTRUM OF DENSITY FLUCTUATIONS

We begin with some definitions and conventions (note that these can vary significantly between authors).

Definition:	$P(k) = \langle |\delta_{\mathbf{k}}|^2 \rangle$
Fourier transform convention:	$\delta_{\mathbf{k}}(t) = (1/V) \int \delta_{\mathbf{x}} e^{-i\mathbf{k}\cdot\mathbf{x}}\, d^3x$
Harrison–Zel'dovich:	$P(k) = Ak^n$ with $n = 1$
Two-point correlation fn.:	$\xi(x) = (V/2\pi^2) \int_0^\infty P(k) j_0(kx) k^2\, dk$
First-moment:	$J_2(x) = (V/2\pi^2) \int_0^\infty P(k)(1 - \cos(kx))\, dk$
Second-moment:	$J_3(x) = (V/2\pi^2) \int_0^\infty P(k)(kx)^2 j_1(kx)\, dk/k$
Density variance:	$\sigma_\rho^2(x) = (V/2\pi^2) \int_0^\infty P(k)(3 j_1(kx)/kx)^2 k^2\, dk$

Here V is the (large) volume used to conveniently apply periodic boundary conditions for the Fourier transform; it should be considered as merely a bookkeeping device.

A useful parametrization of Cold Dark Matter (CDM) power spectra is given by

$$P(k) = \frac{Ak}{\left\{ 1 + \left[ak + (bk)^{3/2} + (ck)^2 \right]^\nu \right\}^{2/\nu}},$$

where $a = (6.4/\Gamma)h^{-1}\,\text{Mpc}$, $b = (3.0/\Gamma)h^{-1}\,\text{Mpc}$, $c = (1.7/\Gamma)h^{-1}\,\text{Mpc}$, $\nu = 1.13$, and the fit is specifically for $\Omega_B = 0.03$ [63]. There are other parametrizations of the transfer function which differ slightly at large k [64–66]. The normalization given by *COBE* is [67]

$$A = (5.9 \pm 1.1) \times 10^5 \frac{(h^{-1}\,\text{Mpc})^4}{V}.$$

For the standard, scale-invariant, adiabatic CDM model $\Omega_0 = 1$ and $\Gamma \simeq h$ above, with a favored value around 0.5. Flat CDM models with nonzero Λ are well fitted using $\Gamma \simeq \Omega h$. Generally Γ can be used as a shape parameter and $\Gamma \simeq 0.2$ provides the best fit to large-scale structure observations [68]. Explicit parametrizations of other theoretical power spectra can be found in, e.g., [64, 69], and can be obtained using codes such as `cmbfast` [70].

Figure 26.1 indicates the approximate status and range of scales spanned by current data relating to the matter power spectrum, and also highlights the relative differences between "reasonable" models. The solid points are from the compilation of [68], which represents an average of several surveys (dominated by the APM galaxy survey [71]), and is explicitly presented here for $\Omega_0 = 1$ models with IRAS galaxies assumed to be unbiased. It is worth stressing that the normalization and the redshift-space corrections would change in open or Λ-dominated cosmologies, and that there are normalization uncertainties due to different bias factors between, e.g., IRAS and optically selected galaxies. The

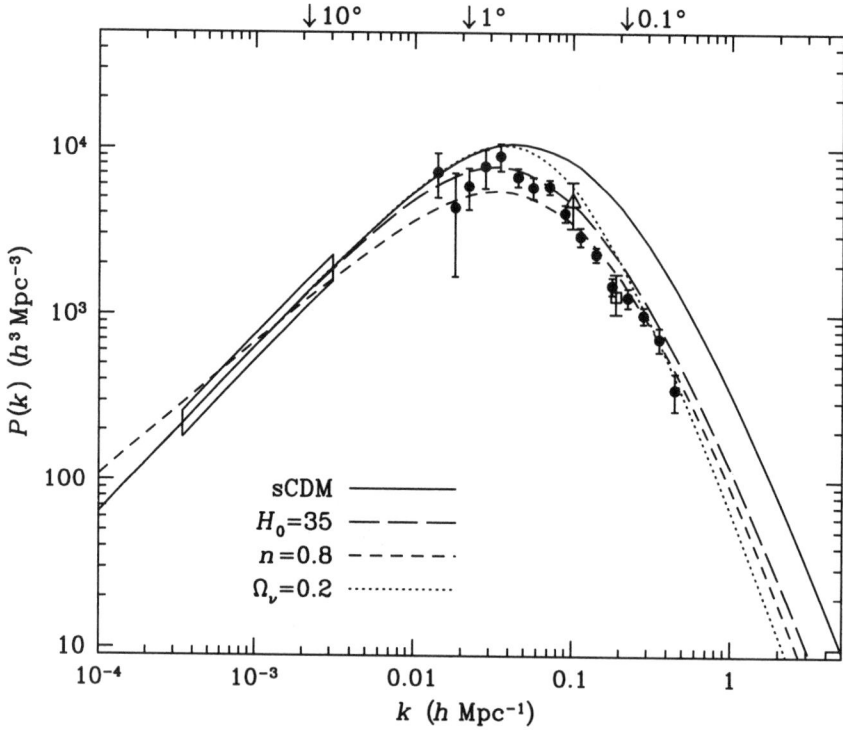

Figure 26.1. Measurements of the power spectrum of density fluctuations.

triangle is derived from peculiar velocities [72], and the square from cluster abundance [73], with error bars indicating 1σ ranges. Both these points are independent of bias, but would move somewhat for different values of Ω_0. An approximation to the COBE error box is plotted at low k. The arrows along the top indicate the angles probed by CMB experiments with given FWHM beams. The curves plotted are the standard CDM (solid) mass power spectrum, together with three representative examples of changes brought about by varying a single cosmological parameter: lowering H_0 (long dashes); tilting the initial conditions (short dashes); and introducing some massive neutrinos (dots). These models are in no sense fits to the data, but demonstrate that variations in *several* parameters (including Ω_0 and Λ, which would require separate figures) can lead to acceptable fits.

26.14 STRUCTURE FORMATION SCALES

There are many physical scales which can be important for the formation of structure in different cosmological scenarios, and which are collected below. Here $T_{\text{hor}}(\lambda)$ and $t_{\text{hor}}(\lambda)$ refer to the temperature and age of the Universe when the comoving scale λ had a physical size equal to c times that of the age of the Universe ($R\lambda/R_0 = ct$). The comoving scale which crosses the horizon at matter-radiation equality is $\lambda_{\text{eq}} = 7.5733\, T_{2.73}^2\, (\Omega_0 h^2)^{-1}$ Mpc, as computed using the exact expression for t_{eq}. Using either the matter- or radiation-dominated expressions would give significantly different values [3].

It is worth pointing out that there are various definitions for what is meant by "horizon size", differing by constant factors. For a matter-dominated Universe the radius of the particle horizon is given by $3ct$ or $2/H$. So the angle subtended by this scale is twice that subtended by the Hubble radius, etc. Typically the physically relevant scale is when $k\eta \simeq 1$, and so there can also be ambiguous factors of 2π when calculating a scale rather than a wave number.

Baryon mass in the horizon:
$$\mathcal{M}_{\text{B-hor}} \equiv m_N N_{\text{B-hor}}$$
$$= 2.97 \times 10^{-10}(\Omega_B h^2) g_{*s} g_*^{-3/2}(\text{GeV}/T)^3 \mathcal{M}_\odot \quad (t \ll t_{\text{eq}})$$
$$= 9.25 \times 10^{21}(\Omega_B/\Omega_0^{3/2} h)(1+z)^{-3/2} \mathcal{M}_\odot \quad (t \gg t_{\text{eq}})$$

Mass within comoving scale λ:
$$\mathcal{M}(\lambda) \equiv \pi\lambda^3 \rho_{\text{nr}}/6$$
$$= 1.4532 \times 10^{11}(\Omega_0 h^2)\lambda_{\text{Mpc}}^3 \mathcal{M}_\odot$$

Physical size of comoving scale λ:
$$R\lambda/R_0 = 1.1435 \times 10^{12}\lambda_{\text{Mpc}} T_{2.73} g_{*s}^{-1/3}(\text{GeV}/T) \text{ cm}$$
$$= 5.7951 \times 10^{25}\lambda_{\text{Mpc}} T_{2.73} g_{*s}^{-1/3}(\text{GeV}/T) \text{ GeV}^{-1}$$

Horizon crossing:
$$T_{\text{hor}}(\lambda) = 63.459 \, T_{2.73}^{-1} g_{*s}^{1/3} g_*^{-1/2}\lambda_{\text{Mpc}}^{-1} \text{ eV} \quad (\lambda \ll \lambda_{\text{eq}})$$
$$t_{\text{hor}}(\lambda) = 6.0108 \times 10^8 \, T_{2.73}^2 g_{*s}^{-2/3} g_*^{1/2}\lambda_{\text{Mpc}}^2 \text{ s}$$
$$T_{\text{hor}}(\lambda) = 939.71(\Omega_0 h^2)^{-1} T_{2.73}\lambda_{\text{Mpc}}^{-2} \text{ eV} \quad (\lambda \gg \lambda_{\text{eq}})$$
$$t_{\text{hor}}(\lambda) = 2.5768 \times 10^7 (\Omega_0 h^2)\lambda_{\text{Mpc}}^3 \text{ s}$$

Curvature scale:
$$L_{\text{curv}} \equiv \mathcal{R}R_0 = 2997.9(1-\Omega_0)^{-1/2} h^{-1} \text{ Mpc}$$

The angle subtended on the sky by the Hubble distance H^{-1}, by a comoving scale λ, and by the curvature scale L_{curv}, can all be expressed in simple form for redshifts $z \gg 1$.

Angle subtended by Hubble radius:
$$\Delta\theta(H^{-1}, z \gg 1) = \Omega_0^{1/2} z^{-1/2}/2$$
$$= 0°\!.87\Omega_0^{1/2}(z/1100)^{-1/2}$$

Angle subtended by comoving scale λ:
$$\Delta\theta(\lambda, z \gg 1) = (R_0\lambda/H_0^{-1})\Omega_0/2$$
$$= 34''\!.4(\Omega_0 h)\lambda_{\text{Mpc}}$$
$$= 65''\!.4(\Omega_0^{2/3} h^{1/3})(M/10^{12}\mathcal{M}_\odot)^{1/3}$$

Angle subtended by curvature scale L_{curv}:
$$\Delta\theta(L_{\text{curv}}, z \gg 1) = \Omega_0/2(1-\Omega_0)^{1/2}$$
$$= 28°\!.6\Omega_0(1-\Omega_0)^{-1/2}$$

The following scales are relevant for various "damping" processes. Below the Jeans scale, baryon fluctuations oscillate, rather than grow, and below the photon damping scale they are dissipated. Neutrino free-streaming erases fluctuations below the neutrino damping length. Microwave background fluctuations tend to be damped on scales below that of the last scattering surface thickness.

Sound speed:
$$c_s = (c/\sqrt{3})\left\{(3\rho_m/4\rho_\gamma) + 1\right\}^{-1/2} \quad (z \gtrsim z_{\text{rec}})$$
$$c_s = (5kT/3m_p)^{1/2} \quad (z \lesssim z_{\text{rec}})$$

Jeans length:
$$\lambda_J \equiv c_s(\pi/G\rho_m)^{1/2}$$
$$\simeq 51(\Omega_0 h^2)^{-1} \text{ Mpc} \quad \text{(comoving) just before } z_{\text{rec}}$$
$$\simeq 10(\Omega_0 h^2)^{-1/2} \text{ kpc} \quad \text{(comoving) just after } z_{\text{rec}}$$

Jeans mass:	$\mathcal{M}_J \equiv \pi \lambda_J^3 \rho_m / 6$
	$\simeq 2.0 \times 10^{16} (\Omega_0 h^2)^{-2} \mathcal{M}_\odot$ just before z_{rec}
	$\simeq 1.5 \times 10^5 (\Omega_0 h^2)^{-1/2} \mathcal{M}_\odot$ just after z_{rec}
Photon damping length:	$L_D \simeq 3.2 (\Omega_B h^2)^{-1/2} (\Omega_0 h^2)^{-1/4}$ Mpc [74]
Photon damping mass:	$\mathcal{M}_D \simeq 4.7 \times 10^{12} (\Omega_B h^2)^{-3/2} (\Omega_0 h^2)^{1/4} \mathcal{M}_\odot$
Neutrino rms velocity:	$\langle v_\nu^2 \rangle^{1/2} \simeq 5.0 (30 \,\text{eV}/m_\nu)(1+z) T_{2.73} \text{ km s}^{-1}$
Neutrino Jeans length:	$L_{J\nu} \simeq 41 \text{ Mpc}(30 \,\text{eV}/m_\nu)$ [75, 76]
or damping length:	$\lambda_\nu u \simeq 13 (\Omega_0 h^2)^{-1}$ Mpc
Neutrino Jeans mass:	$\mathcal{M}_{J\nu} (\sim m_{\text{Pl}}^3/m_\nu^2) \simeq 3 \times 10^{15} \mathcal{M}_\odot (30 \,\text{eV}/m_\nu)^2$
Horizon at t_{eq}:	$\lambda_{eq} \equiv c t_{eq} = 7.5733 (\Omega_0 h^2)^{-1} T_{2.73}^2$ Mpc (comoving)
Horizon at t_{rec}:	$\lambda_{rec} \equiv c t_{rec} \simeq 57 (\Omega_0 h^2)^{-1/2}$ Mpc (comoving)
Thickness of last	$\Delta z_{lss} \equiv 80$ (approx. Gaussian σ) [15]
scattering surface:	$\Delta L_{lss} = 6.6 \Omega_0^{-1/2} h^{-1}$ Mpc
	$\Delta \theta_{lss} = 3\rlap{.}'8 \, \Omega_0^{1/2}$

26.15   COSMIC MICROWAVE BACKGROUND ANISOTROPIES

The temperature fluctuations can be expanded in spherical harmonics

$$\frac{\Delta T}{T} = \sum_{\ell, m} a_{\ell m} Y_{\ell m}(\theta, \phi).$$

Theory predicts the rms values of the multipole moments $C_\ell = \langle |a_{\ell m}|^2 \rangle$. On large angular scales

$$C_\ell = \frac{1}{2\pi} \left(\frac{H_0}{c} \right)^4 \int_0^\infty \frac{P(k)}{k^2} j_\ell^2 (2ck/H_0) \, dk,$$

if the fluctuations are adiabatic, and assuming $\Omega_0 = 1$. For an $n = 1$ spectrum $P(k) = Ak$,

$$C_\ell = \frac{1}{4\pi} \left(\frac{H_0}{c} \right)^4 \frac{A}{\ell(\ell+1)}.$$

The dipole is considered to be due mainly to local velocities, i.e., the motion of the Local Group. The quadrupole coefficient is related to C_2 through $Q_{rms}^2/T_{\gamma 0}^2 = 5C_2/4\pi$. Data from the *COBE* satellite's DMR experiment [77] are summarized below.

Dipole:	$D_{obs} = 3.372 \pm 0.007 \text{ mK}$ (95% CL)
	toward $(\alpha, \delta) = (11^h \, 12^m \pm 0\rlap{.}^m 4, -7\rlap{.}°22 \pm 0\rlap{.}°08)$ (1σ) [11, 12, 78–80]
Quadrupole:	$6 \,\mu\text{K} < Q_{obs} < 17 \,\mu\text{K}$ (68% CL) [81, 82]
rms amplitude:	$\sigma_{obs}(10°) = 29 \pm 1 \,\mu\text{K}; \quad \sigma_{obs}(7°) = 35 \pm 2 \,\mu\text{K}$ (1σ) [83]
Best-fit amplitude:	$\langle Q \rangle = 18 \pm 1.6 \,\mu\text{K}$ $(1\sigma$, assuming $n = 1)$ [82, 84]
Best-fit slope:	$n = 1.2 \pm 0.3$ (1σ) [82, 84]

Detections and upper limits on smaller angular scales are still in a state of flux, with many experimental results at the level of about 1–3×10^{-5} over a range of scales roughly $1'$–$10°$, and have been summarized in [85–87].

Experimental results can be quoted in many ways; one approach is to derive the amplitude of a *flat* (i.e., $\ell(\ell+1)C_\ell = $ constant) power spectrum through the experimental window function, and quote the equivalent quadrupole. On the largest angular scales COBE fixes $Q_{\text{flat}} \simeq 20\,\mu\text{K}$. For $\ell \sim 100$ several experiments indicate $Q_{\text{flat}} \simeq 40\,\mu\text{K}$. Upper limits at $\ell \sim 1000$ are $\simeq 30\,\mu\text{K}$ [88]. Detailed C_ℓ determinations hold great promise for cosmological parameter estimation.

It is worth repeating here that for $z \gg 1$ the Hubble radius $(c/H(z))$ subtends an angle of $0°87\Omega_0^{1/2}(z/1100)^{-1/2}$; if there has been a period of reionization leading to optical depth unity at some z, then this will give approximately the scale up to which fluctuations could be erased. Without a period of reionization the main "acoustic peak" occurs at $\ell \simeq 220\,\Omega_0^{-1/2}$.

26.16 LARGE-SCALE STRUCTURE

We begin with the definition of the two-point correlation function and its moments, and give some measured quantities for the two- and three-dimensional variants. The three-point function has also been measured, but estimates for the higher point functions are still crude [4]. Here R is the Abell richness class [89] of a cluster.

Two-point correlation fn.:	$\xi(\mathbf{x}) \equiv \langle \delta\rho(\mathbf{x}'+\mathbf{x})\delta\rho(\mathbf{x}')/\bar{\rho}^2 \rangle$
enhanced probability:	$\delta P \equiv \bar{n}\,[1 + \xi(r)]\,\delta V$
Nth moment:	$J_N(r) \equiv \int_0^r \xi(\mathbf{x})x^{(N-1)}\,dx$
J_2 normalization:	$J_2(r \to \infty) = 164e^{\pm 0.15}h^{-2}\,\text{Mpc}^2$ [90]
J_3 normalization:	$J_3(10h^{-1}\,\text{Mpc}) \simeq 270h^{-3}\,\text{Mpc}^3$ [91]
	$4\pi J_3(20h^{-1}\,\text{Mpc}) = 10\,000h^{-3}\,\text{Mpc}^3$ [92, 93]
	$J_3(r \to \infty) = 596e^{\pm 0.21}h^{-3}\,\text{Mpc}^3$ [90]
Power-law fit:	$\xi(r) = (r_0/r)^\gamma$
galaxy–galaxy (ξ_{gg}):	$r_0 = 5.1 \pm 0.2(1\sigma)h^{-1}\,\text{Mpc};$ $\gamma = 1.71 \pm 0.05(1\sigma)$
optical:	(over $0.2\,\text{Mpc} \lesssim hr \lesssim 20\,\text{Mpc}$) [91–94]
IRAS:	$r_0 = 3.76 \pm 0.22(1\sigma)h^{-1}\,\text{Mpc};$ $\gamma = 1.66 \pm 0.11(1\sigma)$
	(for $hr \lesssim 20\,\text{Mpc}$) [95, 96]
galaxy–cluster (ξ_{gc}):	$r_0 = 8.8 \pm 0.4(1\sigma)h^{-1}\,\text{Mpc};$ $\gamma = 2.21 \pm 0.04(1\sigma)$
	(for $R \geq 1$, over $0.2\,\text{Mpc} < hr < 10\,\text{Mpc}$) [97]
cluster–cluster (ξ_{cc}):	$r_0 \simeq 25 \pm 6(1\sigma)h^{-1}\,\text{Mpc};$ $\gamma \simeq 1.8 \pm 1.8(1\sigma)$
	(for $R \geq 1$, over $5\,\text{Mpc} \lesssim hr \lesssim 150\,\text{Mpc}$) [98]
	$r_0 \simeq 13.2e^{\pm 0.3}(95\%\,\text{CL})h^{-1}\,\text{Mpc};$ $\gamma \simeq 1.9 \pm 0.3(95\%\,\text{CL})$
	(for $R \gtrsim 0$, over $2\,\text{Mpc} \lesssim hr \lesssim 100\,\text{Mpc}$) [99, 100]
Angular two-point fn.:	$w(\theta) = A\theta^{1-\gamma};$
	$A = 0.0684 \pm 0.0057(1\sigma);$ $\gamma = 1.741 \pm 0.035(1\sigma)$
	(for $m_B \lesssim 18.5$, $\theta < 2°$, steeper for $\theta \gtrsim 2°5$) [94, 101, 102]
faint galaxies:	$\log \omega(1°) = (3.3 \pm 0.2) - (0.27 \pm 0.01)R(1\sigma)$
	(for $18 \lesssim R \lesssim 25$) [103, 104]

Three-point corrln. fn.: $\zeta = Q \left[\xi(r_{12})\xi(r_{23}) + \xi(r_{23})\xi(r_{31}) + \xi(r_{31})\xi(r_{12}) \right]$
with $Q = 0.88 \pm 0.11 (1\sigma, \text{weighted average})$
(for $100\,\text{kpc} \lesssim hr \lesssim 20\,\text{Mpc}$)   [105–108]

Angular three-point fn.: $z(\theta_1, \theta_2, \theta_3) = P \left[w(\theta_1)w(\theta_2) + w(\theta_2)w(\theta_3) + w(\theta_3)w(\theta_1) \right]$
with $P = 1.56 \pm 0.22 (1\sigma)$ for $\theta \lesssim 3°$ [94]

Galaxies may be clustered more or less strongly than the mass, the difference being described by the "bias" parameter b, the ratio of rms galaxy fluctuations to rms mass fluctuations. The simplest mathematical model is linear bias, $(\delta\rho/\rho)_g \equiv b\,(\delta\rho/\rho)_\rho$, i.e., $\xi_{gg} = b^2 \xi_{\rho\rho}$ with $b = $ constant. However, in principle the relation between galaxy and mass fluctuations may be nonlinear, and the bias parameter may depend on scale. The correlation function of galaxies implies rms fluctuations of about unity in spheres of radius $8h^{-1}$ Mpc, so

$$\sigma_{8,g} \simeq 1.0 = b\sigma_{8,\rho}.$$

This is commonly used to describe the normalization of the power spectrum. One can also normalize fluctuations using the J_2 or J_3 observationally determined values listed above. For a standard ($\Omega = 1$, $h = 0.5$) CDM power spectrum, COBE fluctuations imply [67]

$$\sigma_{8,\rho} = 1.22 \pm 0.11 (1\sigma)$$

(scaling approximately as h), whereas for optically selected galaxies [92, 109]

$$\sigma_{8,g} = 0.94 \pm 0.03 (1\sigma),$$

or for IRAS-selected galaxies [95]

$$\sigma_{8,\text{IRAS}} = 0.69 \pm 0.04 (1\sigma),$$

and from cluster masses and abundances [73]

$$\sigma_{8,c} \simeq (0.6 \pm 0.1)\Omega_0^{-\alpha} (1\sigma),$$

with $\alpha \simeq 0.4$ for open CDM and $\alpha \simeq 0.45$ with a cosmological constant.

26.17 DENSITIES

This collection of measurements and estimates of various densities may be useful for quick calculations. Here L_* ($\simeq 8.5 \times 10^9 h^{-2} \mathcal{L}_\odot$, or $M_{b_J} \simeq -19.5$) is the characteristic scale of the galaxy luminosity function, which is well fitted by the Schechter form.

Cluster density: $n_{\text{cluster}} \simeq 6 \times 10^{-6} h^3 \text{ Mpc}^{-3}$ ($R \geq 1$)   [98]

Galaxy density: $n_{\text{gal}} \simeq 0.01 h^3 \text{ Mpc}^{-3}$ (for $\sim L_*$ galaxies)   [4]
$n_{\text{gal}} \simeq 4.70 \times 10^{-2} h^3 \text{ Mpc}^{-3} (\pm \sim 5\%)$
(for $-15 > M_{b_J} > -22$)   [92, 93]

Schechter lum. fn.: $\phi(L)\,dL = \phi^*(L/L^*)^\alpha \exp(-L/L^*)\,d(L/L^*)$
with $M_{b_J}^* = -19.50 \pm 0.13 + 5\log h$; $\alpha = -0.97 \pm 0.15$;
and $\phi^* = 0.0140 \pm 0.0017 (h^{-1}\,\text{Mpc})^{-3} (1\sigma)$ [92, 93]

Galaxy surface density: $dN/dm = 10^{-5.70 \pm 0.10 + 0.6m}$ mag.$^{-1}$ sr^{-1}
(for $14 \lesssim m_B \lesssim 18$)   [110]

Faint galaxy density:	$N_{gal} \simeq 1\,500\,000\,(degree)^{-2}\,(V_{AB} \lesssim 29)$ [111]
Radio gal. surf. density:	$N_{rg} \simeq 100\,(degree)^{-2}\,(> 1\,mJy\ at\ 20\,cm)$ [112]
Quasar surface density:	$N_{QSO} \simeq 95e^{\pm 0.2}\,(degree)^{-2}\,(1\sigma)\,(B \lesssim 22)$ [113, 114]
Quasar density:	$n_{QSO} \simeq 2.7(\pm 0.5) \times 10^{-6}(h^{-1}\,Mpc)^{-3}\,(1\sigma)$
	$(M_B \lesssim -26)$ for $z \simeq 2$ [113, 115]
AGN density:	$n_{AGN} \geq 5\% n_{gal}\,(B < 13.4)$ [116, 117]
Radio galaxy density:	$n_{rg} \simeq 10^{-6}(h^{-1}\,Mpc)^{-3}$
	(for $P_{2.7\,GHz} \gtrsim 10^{25}\,W\,Hz^{-1}\,sr^{-1}$ with $\Omega_0 = 1$) [118]
Gamma-ray bursts:	rate $\sim 10^{-6}\,yr^{-1}$ per L_* galaxy [119]
Optical (B band) light:	$j_B = \phi^* \Gamma(\alpha + 2)L_B^* = (1.76 \pm 0.21) \times 10^8 h\mathcal{L}_\odot\,Mpc^{-3}$ [92, 93, 120]
Mass-to-light ratio:	$(\mathcal{M}/L)_{crit} = 1600h(\pm 10\%)\mathcal{M}_\odot/\mathcal{L}_\odot$
Stars in galaxies:	$\Omega_* \simeq 0.0050h^{-1}e^{\pm 0.5}$ [121, 122]
Gas fraction in clusters:	$\Omega_{gas}/\Omega_{tot} = (0.056 \pm 0.014)h^{-3/2}$ [123]
Baryonic matter:	$\Omega_B = (0.007\text{–}0.024)h^{-2}$ (95% CL) [58–60]
Dynamics, $\lesssim 10h^{-1}$ Mpc:	$\Omega_d = 0.24 \pm 0.10(1\sigma)$ [4, 120, 121]
Dynamics, $\gtrsim 30h^{-1}$ Mpc:	$\Omega_\ell \sim 0.2$ to 1 [4, 121]

26.18 VELOCITIES

Various dipole and other velocity measurements can only be understood as a vector sum of different relative velocities between the Sun ($\odot$), the Galactic Center (GC), the Local Standard of Rest (LSR), the Local Group (LG), and the Cosmic Microwave Background (CMB) [80]. The error bars here are all 1σ.

$\odot$ – CMB:	$370.6 \pm 0.4\,km\,s^{-1}$	
	toward $(\ell, b) = (264°31 \pm 0°17, 48°05 \pm 0°10)$ [11, 12, 79, 80]	
LSR – GC:	$222.0 \pm 5.0\,km\,s^{-1}$ toward $(\ell, b) = (91°1 \pm 0°4, 0°)$ [124]	
$\odot$ – LSR:	$20.0 \pm 1.4\,km\,s^{-1}$ toward $(\ell, b) = (57° \pm 4°, 23° \pm 4°)$ [125]	
$\odot$ – LG:	$308 \pm 23\,km\,s^{-1}$ toward $(\ell, b) = (105° \pm 5°, -7° \pm 4°)$ [126]	
LG – CMB:	$627 \pm 22\,km\,s^{-1}$ toward $(\ell, b) = (276° \pm 3°, 30° \pm 3°)$ [78, 80]	

Converting from a heliocentric velocity to a galactocentric velocity conventionally involves

$$V_G = V_H + 9\cos\ell\cos b + 232\sin\ell\cos b + 7\sin b$$

in $km\,s^{-1}$ [127].

Data on the relative peculiar velocities of galaxies at projected separation r_p indicate

Rel. pec. vel. of galaxies: $540 \pm 180(hr_p/1\,Mpc)^{(0.13\pm 0.04)}\,km\,s^{-1}(1\sigma)$

(for $10\,kpc \lesssim hr_p \lesssim 1\,Mpc$) [91, 128]

Pairwise velocities of galaxies on small scales are apparently only consistent with standard Cold Dark Matter models if the bias parameter is $1.5 \lesssim b \lesssim 2.5$ [64, 129], but issues relating to biasing are unresolved [130].

Comparisons of large-scale velocity and density fields tend to measure the parameter $\beta \equiv \Omega^{0.6}/b$. Current constraints, particularly using IRAS-selected galaxies [131–134], imply $\beta = 0.5\text{–}1.2$.

26.19 INTERGALACTIC MEDIUM

The Gunn–Peterson [135] test uses a limit on the integrated $Ly\,\alpha$ optical depth, shortward of $Ly\,\alpha$ emission in quasar spectra, to limit the proper number density:

$$n_{\mathrm{HI}} = 2.42 \times 10^{-11} \tau_{\mathrm{GP}}(z) h (1 + z)(1 + \Omega_0 z)^{1/2} \ \mathrm{cm}^{-3}.$$

We quote values for different redshifts using $\Omega_0 = 1$ below:

Gunn–Peterson tests:			
Ω (at $z \lesssim 4$)	$\Omega_{\mathrm{HI}} \lesssim 2 \times 10^{-7} h^{-1}(1+z)^{-2}(1+\Omega_0 z)^{1/2}$		[136, 137]
Number density	$n_{\mathrm{HI}}(\langle z \rangle = 2.6) < 1.7 \times 10^{-13} h \ \mathrm{cm}^{-3} \ (1\sigma)$		[136, 138]
(comoving):	$n_{\mathrm{HI}}(\langle z \rangle = 3.8) < 1.2 \times 10^{-13} h \ \mathrm{cm}^{-3} \ (1\sigma)$		[137]
	$n_{\mathrm{HI}}(\langle z \rangle = 0.1) < 2.1 \times 10^{-12} h \ \mathrm{cm}^{-3} \ (1\sigma)$		[139]
Helium:	$\tau_{\mathrm{He\,II}}(\langle z \rangle = 3.2) > 1.7 \ (90\% \mathrm{CL})$		[140]
	$\tau_{\mathrm{He\,II}}(\langle z \rangle = 2.4) = 1.0 \pm 0.1 \ (90\% \mathrm{CL})$		[141]

The ionizing background estimates usually derive from the "proximity effect" [142, 143] and are quoted as the intensity at the Lyman limit

$$\text{Ionizing flux:} \quad \log(J_\nu) \quad (\text{for } 1.7 < z < 4.1)$$
$$= -21.5 \pm 0.5 \ \mathrm{erg\,cm}^{-2}\,\mathrm{s}^{-1}\,\mathrm{Hz}^{-1}\,\mathrm{sr}^{-1} \ (1\sigma) \quad [143\text{–}145]$$

For the $Ly\,\alpha$ absorption systems there are several separate classes of systems, with various parameters determined for each. They tend to be classified by column density into "forest" systems ($10^{12} \ \mathrm{cm}^{-2} \lesssim N_{\mathrm{HI}} \lesssim 10^{15} \ \mathrm{cm}^{-2}$), "intermediate" systems ($10^{15} \lesssim N_{\mathrm{HI}} \lesssim 10^{20}$) and "damped" systems ($N_{\mathrm{HI}} \gtrsim 10^{20}$). We have attempted to concentrate on some of the more general and robust quantities, while the specifics have been reviewed elsewhere Note that some quantities may be extremely sensitive to selection criteria, e.g., the rest equivalent width threshold. Also note that there is no distinction between a Gunn–Peterson effect and absorption by low N_{HI} clouds.

$Ly\,\alpha$ cloud distribution:	$d(\mathrm{Prob}) = B N_{\mathrm{HI}}^{-\beta}$: $\quad \log B = 8\text{–}8.5; \quad \beta = 1.5\text{–}1.8$ (for $10^{13} \lesssim N_{\mathrm{HI}} \lesssim 10^{22} \ \mathrm{cm}^{-2}$) [150–154]
$Ly\,\alpha$ forest Doppler parameter:	$b \equiv \sqrt{2}\sigma = (2kT/m)^{1/2}$ $\langle b \rangle \simeq 30 \ \mathrm{km\,s}^{-1}$ [151, 155]
$Ly\,\alpha$ forest distribution:	$d\mathcal{N}/dz = A(1+z)^\gamma$: $\quad A = 2\text{–}3; \quad \gamma = 2\text{–}3$ (for $10^{14} \ \mathrm{cm}^{-2} \lesssim N_{\mathrm{HI}} \lesssim 10^{16} \ \mathrm{cm}^{-2}$ and $z \gtrsim 1.6$) [153, 156–158]
$Ly\,\alpha$ forest transverse size:	$61 h^{-1} \ \mathrm{kpc} < r < 533 h^{-1} \ \mathrm{kpc} \ (99\% \ \mathrm{CL})$ (assuming spherical clouds) [159]
Ly-limit systems:	$d\mathcal{N}/dz \simeq 3.3$ at $z = 4$ (for $N_{\mathrm{HI}} \geq 1.6 \times 10^{17} \ \mathrm{cm}^{-2}$) [160]
Damped $Ly\,\alpha$ clouds:	$\Omega_{\mathrm{damped}}(z \simeq 2.5) \simeq (1.45 \pm 0.25) \times 10^{-3} h^{-1} \ (1\sigma)$ [161, 162]

26.20   EXTRAGALACTIC DIFFUSE BACKGROUNDS

Figure 26.2 (adapted from the 1990 overview by [163]) shows a compilation of much of the current data relevant to the extragalactic background. This should be taken as illustrative of the status at various wavelengths, and is not meant to be comprehensive. The extreme radio and γ-ray measurements should be considered as upper limits to a possible extragalactic component. The solid lines on the plot indicate spectral fits to the microwave, far-infrared and X-ray data. Some convenient estimates for specific wavebands are listed below. Note that galactic foreground contamination can be considerable, so that many infrared, optical, and ultraviolet measurements should be treated as upper limits. Another useful unit here is $\text{nW m}^{-2}\,\text{sr}^{-1} = 10^{-6}\,\text{erg cm}^{-2}\,\text{s}^{-1}\,\text{sr}^{-1}$.

Radio (30 cm): $\nu I_\nu \lesssim 6 \times 10^{-11}\,\text{erg cm}^{-2}\,\text{s}^{-1}\,\text{sr}^{-1}$ [164]

Sub-mm (400–1000 μm): $\nu I_\nu \simeq 3.4 \times 10^{-6}(\lambda/400\,\mu\text{m})^{-3}\,\text{erg cm}^{-2}\,\text{s}^{-1}\,\text{sr}^{-1}$ [165–167]

Far Infrared (240 μm): $\nu I_\nu = (1.7 \pm 0.4) \times 10^{-5}\,\text{erg cm}^{-2}\,\text{s}^{-1}\,\text{sr}^{-1}$ (95% CL)   [10, 168]

Near Infrared (3.5 μm): $\nu I_\nu < 5.6 \times 10^{-5}\,\text{erg cm}^{-2}\,\text{s}^{-1}\,\text{sr}^{-1}$ (1σ)   [10, 166]

Optical (4400 Å): $\nu I_\nu < 3.2 \times 10^{-5}\,\text{erg cm}^{-2}\,\text{s}^{-1}\,\text{sr}^{-1}$ (1σ)   [10, 169]

Ultraviolet (1600 Å): $\nu I_\nu = 0\text{–}2.2 \times 10^{-6}\,\text{erg cm}^{-2}\,\text{s}^{-1}\,\text{sr}^{-1}$ [10, 170]

X-ray (3 keV): $\nu I_\nu \simeq 2.5 \times 10^{-8}\,\text{erg cm}^{-2}\,\text{s}^{-1}\,\text{sr}^{-1}$ [171]

γ-ray (50 MeV): $\nu I_\nu = (4.5 \pm 1.0) \times 10^{-9}\,\text{erg cm}^{-2}\,\text{s}^{-1}\,\text{sr}^{-1}$ (1σ)   [172]

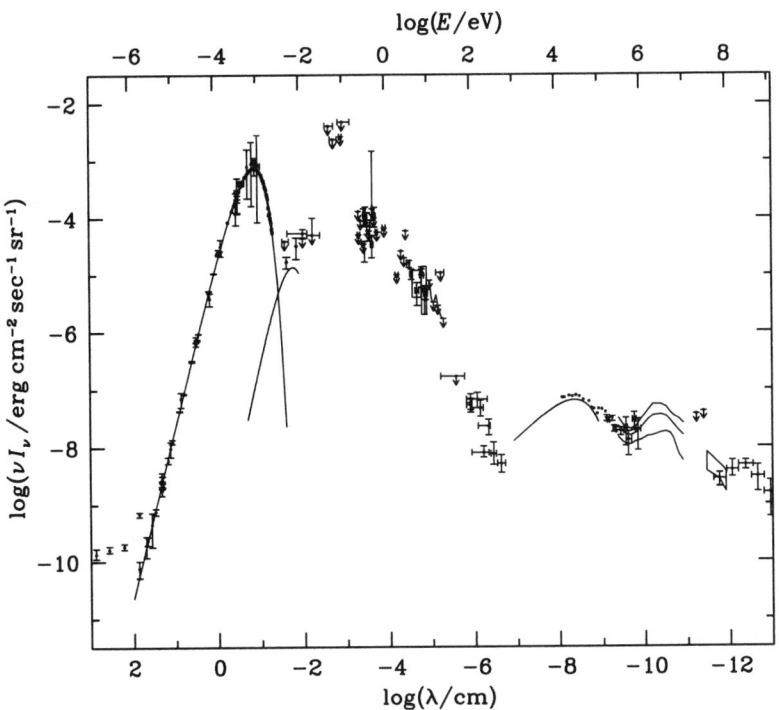

Figure 26.2. The diffuse extragalactic background.

The data specifically for the microwave background are listed below. The parameters y, Y_{ff}, and μ describe Compton scattering, free–free emission and Bose–Einstein spectral distortions, respectively.

Microwave background:
$$T_{\gamma 0} = 2.728 \pm 0.004 \,\text{K} \ (95\% \,\text{CL}) \quad [11, 12]$$
$$u_\gamma = 0.26224 \, T_{2.73}^4 (1 + z)^4 \,\text{eV cm}^{-3}$$
$$B_\gamma(\text{rms}) = 3.2496 \times 10^{-6} T_{2.73}^2 (1 + z)^2 \,\text{Gauss}$$
$$\nu_{\text{peak}} = 160.50 T_{2.73} \,\text{GHz}$$
$$I_\nu(\nu_{\text{peak}}) = 384.94 \,\text{MJy sr}^{-1} \quad (\text{for } T_{2.73} = 1)$$

Comptonization param.:
$$|y| < 1.5 \times 10^{-5} \ (95\% \,\text{CL}) \quad [11, 12]$$
Free–free parameter:
$$Y_{ff} < 1.9 \times 10^{-5} \ (95\% \,\text{CL}) \quad [173]$$
Chemical potential:
$$|\mu| < 9 \times 10^{-5} \ (95\% \,\text{CL}) \quad [11, 12]$$

REFERENCES

1. Weinberg, S. 1972, *Gravitation & Cosmology* (Wiley, New York)
2. Zel'dovich, Ya.B., & Novikov, I.D. 1983, *Relativistic Astrophysics*, Vol. 2: *The Structure and Evolution of the Universe* (University of Chicago Press, Chicago)
3. Kolb, E.W., & Turner, M.S. 1990, *The Early Universe* (Addison Wesley, Redwood City, CA)
4. Peebles, P.J.E. 1993, *Principles of Physical Cosmology* (Princeton University Press, Princeton, NJ)
5. Padmanabhan, T. 1993, *Structure Formation in the Universe* (Cambridge University Press, Cambridge)
6. Longair, M.S. 1984, *Theoretical Concepts in Physics* (Cambridge University Press, Cambridge), Chap. 15
7. King, C.R., & Ellis, R.S. 1985, *ApJ*, **288**, 456
8. Particle Data Group 1996, *Phys. Rev.*, **D54**, 1
9. Jackson, J.D. 1975, *Classical Electrodynamics*, 2nd ed. (Wiley, New York), Appendix
10. Leinert, C. et al. 1998, *A&AS*, **271**, 1
11. Mather, J.C. et al. 1994, *ApJ*, **420**, 439
12. Fixsen, D.J. et al. 1996, *ApJ*, **473**, 576
13. Peebles, P.J.E. 1968, *ApJ*, **153**, 1
14. Zel'dovich, Ya.B., Kurt, V.G., & Sunyaev, R.A. 1969, *Sov. Phys. JETP*, **28**, 146
15. Jones, B.J.T., & Wyse, R.F.G. 1985, *A&A*, **149**, 144
16. Danese, L., & De Zotti, G. 1977, *Riv. Nuovo Cimento*, **7**, 277
17. Bond, J.R. 1988, in *The Early Universe*, edited by W.G. Unruh and G.W. Semenoff (Reidel, Dordrecht), p. 283
18. Hu, W., & Silk, J. 1993, *Phys. Rev.*, **D48**, 485
19. Rees, M.J. 1977, in *The Evolution of Galaxies and Stellar Populations*, edited by B.M. Tinsley and R.B. Larson (Yale University Obs. Publ., New Haven)
20. Dalrymple, G.B. 1991, *The Age of the Earth* (Stanford University Press, Stanford, CA)
21. Wasserburg, G.J., Papanastassiou, D.A., Tera, F., & Huneke, J.C. 1977, *Philos. Trans.*, **A285**, 7
22. Wasson, J.T. 1985, *Meteorites: Their Record of Early Solar-System History* (Freeman, New York)
23. Fowler, W.A. 1987, *QJRAS*, **28**, 87
24. Fowler, W.A. 1989 in *14th Texas Symposium on Relativistic Astrophysics*, edited by E.J. Fenyves, *Ann. N.Y. Acad. Sci.*, **571**, 68
25. Cowan, J.J., Thielemann, F.-K., & Truran, J.W. 1991, *ARA&A*, **29**, 447
26. Chaboyer, B., Kernan, P.J., Krauss, L.M., & Demarque, P. 1996, *Science*, **271**, 957
27. Chaboyer, B., Demarque, P., Kernan, P.J., & Krauss, L.M. 1998, *ApJ*, **494**, 96
28. VandenBerg, D.A., Bolte, M., & Stetson, P.B. 1996, *ARA&A*, **34**, 461
29. Oswalt, T.D., Smith, J.A., Wood, M.A., & Hintzen, P. 1996, *Nature*, **382**, 692
30. D'Antona, F., & Mazzitelli, I. 1990, *ARA&A*, **28**, 139
31. Jacoby, G.H. et al. 1992, *PASP*, **104**, 599
32. van den Bergh, S. 1996, *PASP*, **108**, 1091
33. Tammann, G.A. 1996, *PASP*, **108**, 1083
34. Fukugita, M., Hogan, C.J., & Peebles, P.J.E. 1993, *Nature*, **366**, 309
35. Hogan, C.J. 1996, *Phys. Rev.*, **D54**, 112
36. Kundić, T. et al. 1997, *ApJ*, **487**, 75
37. Saha, A. et al. 1996, *ApJS*, **107**, 693
38. Saha, A., et al. 1997, *ApJ*, **486**, 1
39. Riess, A.G., Press, W.H., & Kirshner, R.P. 1996, *ApJ*, **473**, 88
40. Hamuy, M. et al. 1996, *AJ*, **112**, 2398
41. Schmidt, B.P. et al. 1994, *ApJ*, **432**, 42
42. Rephaeli, Y. 1995, *ARA&A*, **33**, 541
43. Birkinshaw, M., & Hughes, J.P. 1994, *ApJ*, **420**, 33
44. Hughes, J.P., & Birkinshaw, M. 1998, *ApJ*, **501**, 1
45. Myers, S.T. et al. 1997, *ApJ*, **485**, 1
46. Guiderdoni, B., & Rocca-Volmerange, B. 1990, *A&A*, **227**, 362
47. Shanks, T. 1990, in *The Galactic and Extragalactic Background Radiation*, IAU Symp. No. 139, edited by S. Bowyer and G. Leinert (Kluwer Academic, Dordrecht), p. 269
48. Kellermann, K.I. 1993, *Nature*, **134**, 361
49. Perlmutter, S. et al. 1997, *ApJ*, **483**, 565
50. Garnavich, P.M. et al. 1998, *ApJ*, **493**, L53

51. Carroll, S.M., Press, W.H., & Turner, E.L. 1992, *ARA&A*, **30**, 499
52. Kochanek, C. 1996, *ApJ*, **473**, 595
53. Will, C.M. 1993, *Theory and Experiment in Gravitational Physics*, rev. ed. (Cambridge University Press, Cambridge)
54. Thorsett, S.E. 1996, *Phys. Rev. Lett.*, **77**, 1432
55. Accetta, F.S., Krauss, L.M., & Romanelli, P. 1990, *Phys. Lett.*, **248**, 146
56. Kogut, A., Hinshaw, G., & Banday, A.J. 1997, *Phys. Rev.*, **D55**, 1901
57. Bunn, E.F., Ferreira, P., & Silk, J. 1996, *Phys. Rev. Lett.*, **77**, 2883
58. Copi, C.J., Schramm, D.N., & Turner, M.S. 1995, *Science*, **267**, 192
59. Copi, C.J., Schramm, D.N., & Turner, M.S. 1995, *Phys. Rev. Lett.*, **75**, 3981
60. Fields, B.D., Kainulainen, K., Olive, K.A., & Thomas, D. 1996, *New Astron.*, **1**, 77
61. Assamagan, K. et al. 1996, *Phys. Rev.*, **D53**, 6065
62. Buskulic, D. et al. 1995, *Phys. Lett.*, **B349**, 585
63. Efstathiou, G., Bond, J.R., & White, S.D.M. 1992, *MNRAS*, **258**, 1p
64. Bardeen, J.M., Bond, J.R., Kaiser, N., & Szalay, A.S. 1986, *ApJ*, **304**, 15
65. Peacock, J.A. 1991, *MNRAS*, **253**, 1P
66. Liddle, A.R., & Lyth, D.H. 1993, *Phys. Rep.*, **231**, 1
67. Bunn, E.F., & White, M. 1997, *ApJ*, **480**, 6
68. Peacock, J.A., & Dodds, S.J. 1994, *MNRAS*, **267**, 1020
69. Holtzman, J.A. 1989, *ApJS*, **71**, 1
70. Seljak, U., & Zaldarriaga, M. 1996, *ApJ*, **469**, 437
71. Baugh, C.M., & Efstathiou, G. 1993, *MNRAS*, **265**, 145
72. Zaroubi, S. et al. 1997, *ApJ*, **486**, 21
73. Viana, P.T.P., & Liddle, A.R. 1996, *MNRAS*, **281**, 323
74. Silk, J. 1968, *ApJ*, **151**, 459
75. Bond, J.R., Efstathiou, G., & Silk, J. 1980, *Phys. Rev. Lett.*, **45**, 1980
76. Bond, J.R., & Szalay, A.S. 1983, *ApJ*, **274**, 443
77. Smoot, G.F. et al. 1992, *ApJ*, **396**, L1
78. Smoot, G.F. et al. 1991, *ApJ*, **371**, L1
79. Lineweaver, C. et al. 1996, *ApJ*, **470**, L38
80. Kogut, A. et al. 1993, *ApJ*, **419**, 1
81. Kogut, A. et al. 1996, *ApJ*, **470**, 653
82. Bennett, C.L. et al. 1996, *ApJ*, **464**, L1
83. Banday, A.J. et al. 1997, *ApJ*, **475**, 393
84. Górski, K.M. et al. 1996, *ApJ*, **464**, L11
85. Readhead, A.C.S., & Lawrence, C.R. 1992, *ARA&A*, **30**, 653
86. Bond, J.R. 1995, in *Cosmology and Large-Scale Structure*, Les Houches Session LX, edited by R. Schaeffer (Elsevier Science Press, Netherlands)
87. White, M., Scott, D., & Silk, J. 1994, *ARA&A*, **32**, 319
88. Smoot, G.F., & Scott, D. 1996, *Phys. Rev.*, **D54**, 118
89. Abell, G.O. 1958, *ApJS*, **3**, 211
90. Clutton-Brock, M., & Peebles, P.J.E. 1981, *AJ*, **86**, 1115
91. Davis, M., & Peebles, P.J.E. 1983, *ApJ*, **267**, 465
92. Loveday, J., Peterson, B.A., Efstathiou, G., & Maddox, S.J. 1992, *ApJ*, **390**, 338
93. Loveday, J., Maddox, S.J., Efstathiou, G., & Peterson, B.A. 1995, *ApJ*, **442**, 457
94. Groth, E.J., & Peebles, P.J.E. 1977, *ApJ*, **217**, 385
95. Fisher, K.B. et al. 1994, *MNRAS*, **266**, 50
96. Saunders, W., Rowan-Robinson, M., & Lawrence, A. 1994, *MNRAS*, **258**, 134
97. Lilje, P.B., & Efstathiou, G. 1988, *MNRAS*, **231**, 635
98. Bahcall, N.A. 1988, *ARA&A*, **26**, 631
99. Dalton, G.B., Efstathiou, G., Maddox, S.J., & Sutherland, W.J. 1992, *ApJ*, **390**, L1
100. Nichol, R.C., Collins, C.A., Guzzo, L., & Lumsden, S.L. 1992, *MNRAS*, **255**, 21P
101. Maddox, S.J., Efstathiou, G., Sutherland, W.J., & Loveday, J. 1990, *MNRAS*, **242**, 43P
102. Collins, C.A., Nichol, R.C., & Lumsden, S.L. 1992, *MNRAS*, **254**, 295
103. Brainerd, T.G., Smail, I., & Mould, J.R. 1995, *MNRAS*, **275**, 781
104. Roche, N., Shanks, T., Metcalfe, N., & Fong, R. 1993, *MNRAS*, **263**, 360
105. Fry, J.N., Melott, A.L., & Shandarin, S.F. 1993, *ApJ*, **412**, 504
106. Gaztañaga, E. 1992, *ApJ*, **398**, L17
107. Meiksin, A., Szapudi, I., & Szalay, A. 1992, *ApJ*, **394**, 87
108. Fry, J.N. 1984, *ApJ*, **277**, L5
109. Loveday, J., Efstathiou, G., Peterson, B.A., & Maddox, S.J. 1992, *ApJ*, **400**, L43
110. Shanks, T., Stevenson, P.R.F., Fong, R., & MacGillivray, H.T. 1984, *MNRAS*, **206**, 767
111. Williams, R.E. et al. 1996, *AJ*, **112**, 1335
112. Becker, R.H., White, R.L., & Helfand, D.J. 1995, *ApJ*, **450**, 559
113. Hartwick, F.D.A., & Schade, D. 1990, *ARA&A*, **28**, 437
114. Zitelli, V. et al. 1992, *MNRAS*, **256**, 349
115. Hewett, P.C., Foltz, C.B., & Chaffee, F.H. 1993, *ApJ*, **406**, L43
116. Huchra, J.P., & Burg, R. 1992, *ApJ*, **393**, 90
117. Maiolino, R., & Rieke, G.H. 1995, *ApJ*, **454**, 95
118. Dunlop, J., & Peacock, J.A. 1990, *MNRAS*, **247**, 19
119. Narayan, R., Paczyński, B., & Piran, T. 1992, *ApJ*, **395**, L83
120. Carlberg, R. et al. 1996, *ApJ*, **462**, 32
121. Trimble, V. 1987, *ARA&A*, **25**, 425
122. Fukugita, M., Hogan, C.J., & Peebles, P.J.E. 1998, *ApJ*, **503**, 518
123. White, D. & Fabian, A.C. 1995, *MNRAS*, **273**, 72
124. Fich, M., Blitz, L., & Stark, A. 1989, *ApJ*, **342**, 272
125. Kerr, F.J., & Lynden-Bell, D. 1986, *MNRAS*, **221**, 1023
126. Yahil, A., Tammann, G.A., & Sandage, A. 1977, *ApJ*, **217**, 903
127. de Vaucouleurs, G., et al. 1991, *Third Reference Catalogue of Bright Galaxies* (Springer-Verlag, New York)
128. Marzke, R.O., Geller, M.J., da Costa, L.N., & Huchra, J.P. 1995, *AJ*, **110**, 477
129. Frenk, C.S., White, S.D.M., Efstathiou, G., & Davis, M. 1990, *ApJ*, **351**, 10
130. Carlberg, R.G. 1991, *ApJ*, **367**, 385
131. Strauss, M.A. et al. 1992, *ApJ*, **397**, 395

132. Kaiser, N. et al. 1991, *MNRAS*, **252**, 1
133. Dekel, A. 1994, *ARA&A*, **32**, 371
134. Strauss, M.A., & Willick, J.A. 1995, *Phys. Rep.*, **261**, 271
135. Gunn, J.E., & Peterson, B.A. 1965, *ApJ*, **142**, 1633
136. Steidel, C.C., & Sargent, W.L.W. 1987, *ApJ*, **318**, L11
137. Webb, J.K., Barcons, X., Carswell, R.F., & Parnell, H.C. 1992, *MNRAS*, **255**, 319
138. Giallongo, E., Cristiani, S., & Travese, D. 1992, *ApJ*, **398**, L9
139. Bahcall, J.N. et al. 1991, *ApJ*, **377**, L5
140. Jakobsen, P. 1994, *Nature*, **370**, 35
141. Davidsen, A.F., Kriss, G.A., & Zheng, W. 1996, *Nature*, **380**, 47
142. Carswell, R.F. et al. 1982, *MNRAS*, **198**, 91
143. Bajtlik, S., Duncan, R.C., & Ostriker, J.P. 1988, *ApJ*, **327**, 570
144. Giallongo, E. et al. 1996, *ApJ*, **466**, 46
145. Lu, L., Sargent, W.W., Womble, D.S., & Takada-Hidai, M. 1996, *ApJ*, **472**, 509
146. Blades, J.C., Turnshek, D.A., & Norman, C.A., editors, 1988, *QSO Absorption Lines: Probing the Universe* (Cambridge University Press, Cambridge)
147. Weymann, R.J. 1993, in *The Evolution of Galaxies and Their Environment*, the 3rd Teton Summer School, edited by M. Shull and H. Thronson (Kluwer Academic, Dordrecht)
148. Steidel, C.C. 1992, *PASP*, **104**, 843
149. Wolfe, A.M. 1993, in *16th Texas Symposium on Relativistic Astrophysics*, edited by C.W. Akerlof and M.A. Srednicki, *Ann. N.Y. Acad. Sci.*, **688**, 281
150. Tytler, D. 1987, *ApJ*, **321**, 49
151. Hu, E. et al. 1995, *AJ*, **110**, 1526
152. Petitjean, P. et al. 1993, *MNRAS*, **262**, 499
153. Kirkman, D., & Tytler, D. 1997, *ApJ*, **484**, 672
154. Meiksin, A., & Madau, P. 1993, *ApJ*, **412**, 34
155. Rauch, M. et al. 1992, *ApJ*, **387**, 404
156. Murdoch, H.S., Hunstead, R.W., Pettini, M., & Blades, J.C. 1986, *ApJ*, **309**, 19
157. Bechtold, J. 1994, *ApJS*, **91**, 1
158. Bahcall, J.N. et al. 1993, *ApJS*, **87**, 1
159. Fang, Y., Duncan, R.C., Crotts, A.P.S., & Bechtold, J. 1996, *ApJ*, **462**, 77
160. Storrie-Lombardi, L.J., McMahon, R.G., Irwin, M.J., & Hazard, C. 1994, *ApJ*, **427**, L13
161. Lanzetta, K.M. et al. 1991, *ApJS*, **77**, 1
162. Storrie-Lombardi, L.J., McMahon, R.G., & Irwin, M.J. 1996, *MNRAS*, **283**, L79
163. Ressell, M.T., & Turner, M.S. 1990, *Comments on Astrophys.*, **14**, 323
164. Wall, J.V. 1990, in *The Galactic and Extragalactic Background Radiation*, IAU Symposium No. 139, edited by S. Bowyer and G. Leinert (Kluwer Academic, Dordrecht), p. 327
165. Puget, J.-L., 1996, *A&A*, **308**, L5
166. Hauser, M.G. et al. 1998, *ApJ*, **508**, 25
167. Fixsen, D.J. et al. 1998, *ApJ*, **508**, 123
168. Schlegel, D.J., Finkbeiner, D.P., & Davis, M. 1998, *ApJ*, **500**, 525
169. Toller, G.N. 1983, *ApJ*, **266**, L79
170. Martin, C., Hurwitz, M., & Bowyer, S. 1991, *ApJ*, **379**, 549
171. Boldt, E. 1987, *Phys. Rep.*, **146**, 215
172. Fichtel, C.E., Simpson, G.A., & Thompson, D.J. 1978, *ApJ*, **222**, 833
173. Bersanelli, M. et al. 1994, *ApJ*, **424**, 517

Chapter 27

Incidental Tables

Alan D. Fiala, William F. van Altena,
Stephen T. Ridgway, and Roger W. Sinnott

27.1 THE JULIAN DATE
by A.D. Fiala

The Julian Day Number (JD) is a sequential count that begins at Noon 1 Jan. 4713 B.C. Julian Calendar.

27.1.1 Julian Dates of Specific Years

Noon 1 Jan. 4713 B.C. = JD 0.0
Noon 1 Jan. 1 B.C. = Noon 1 Jan. 0 A.D. = JD 172 1058.0
Noon 1 Jan. 1 A.D. = JD 172 1424.0
A Modified Julian Day (MJD) is defined as JD   240 0000.5.
Table 27.1 gives the Julian Day of some centennial and decennial dates in the Gregorian Calendar.

Table 27.1. *Julian date of selected years in the Gregorian calendar* [1, 2].

Jan. 0.5	JD	Jan. 0.5	JD	Jan. 0.5	JD	Jan. 0.5	JD
1500	226 8923	1910	241 8672	1960	243 6934	2010	245 5197
1600	230 5447	1920	242 2324	1970	244 0587	2020	245 8849
1700	234 1972	1930	242 5977	1980	244 4239	2030	246 2502
1800	237 8496	1940	242 9629	1990	244 7892	2040	246 6154
1900	241 5020	1950	243 3282	2000	245 1544	2050	246 9807

Julian day at noon (UT) on 0 January, Gregorian calendar

Century years evenly divisible by 400 (e.g., 1600, 2000) are leap years. Others are not.

References

1. *Explanatory Supplement to the Astronomical Almanac.* 1992, edited by P.K. Seidelmann (University Science Books, Mill Valley, CA), pp. 55, 56, 580, 581, 600–604

2. *Explanatory Supplement to the Astronomical Ephemeris and the American Ephemeris and Nautical Almanac.* 1961, (Her Majesty's Stationery Office, London), pp. 434–439

27.1.2   Conversion Algorithms

Several algorithms for converting among Julian Calendar Date, Gregorian Calendar Date, Islamic Tabular Calendar Date, Indian Civil Calendar, and Julian Day Number, and computing day of the week, are given in [1]. Probably the most useful of these is the conversion from Gregorian Calendar Date to Julian Day Number, as follows [2]:

Julian Day Numbers run from noon to noon. Define the following integer variables:

$$JD = \text{Julian Day Number},$$
$$Y = \text{calendar year},$$
$$M = \text{month},$$
$$D = \text{day of month}.$$

Given Y, M, D, compute JD:

$$JD = (1461 \times (Y + 4800 + (M - 14)/12))/4 + (367 \times (M - 2 - 12 \times ((M - 14)/12)))/12$$
$$- (3 \times ((Y + 4900 + (M - 14)/12)/100))/4 + D - 32075.$$

$JD \geq 0$, that is, the date is after -4713 November 23.

27.2   STANDARD EPOCHS
by A.D. Fiala

The beginning of the Besselian (fictitious) solar year is the instant when the right ascension of the fictitious mean sun, affected by aberration and measured from the mean equinox, is 18^h40^m.

The Julian year begins at noon on January 0.

Table 27.2 gives the Julian Date of several years.

Table 27.2. *Julian dates of Julian and Besselian years* [1, 2].

Julian year		Besselian year	
JY	JD	BY	JD
		B1850.0	239 6758.203
J1900.0	241 5020.0	B1900.0	241 5020.313
J1950.0	243 3282.5	B1950.0	243 3282.423
		B1975.0	244 2413.478
J2000.0	245 1545.0	B2000.0	245 1544.533
		B2025.0	246 0675.588
J2050.0	246 9807.5	B2050.0	246 9806.643
J2100.0	248 8070.0	B2100.0	248 8068.753

References

1. *Explanatory Supplement to the Astronomical Ephemeris and the American Ephemeris and Nautical Almanac.* 1961, (Her Majesty's Stationery Office, London), pp. 434–439
2. *Explanatory Supplement to the Astronomical Almanac.* 1992, edited by P.K. Seidelmann (University Science Books, Mill Valley, CA), p. 8

27.3 REDUCTION FOR PRECESSION
by A.D. Fiala

Approximate formulas for the reduction of coordinates and orbital elements referred to the mean equinox and equator or ecliptic of date (t) are as follows, as given in the *Astronomical Almanac* [3], page B19, in all years since 1984:

For reduction to J2000.0	For reduction from J2000.0
$\alpha_0 = \alpha - M - N \sin\alpha_m \tan\delta_m$	$\alpha = \alpha_0 + M + N \sin\alpha_m \tan b_m$
$\delta_0 = \delta - N \cos\alpha_m$	$\delta = \delta_0 + N \cos\alpha_m$
$\lambda_0 = \lambda - a + b \cos(\lambda + c') \tan\beta_0$	$\lambda = \lambda_0 + a - b \cos(\lambda_0 + c) \tan\beta$
$\beta_0 = \beta - b \sin(\lambda + c')$	$\beta = \beta_0 + b \sin(\lambda_0 + c)$
$\Omega_0 = \Omega - a + b \cos(\Omega + c') \cot i_0$	$\Omega = \Omega_0 + a - b \sin(\Omega_0 + c) \cot i$
$i_0 = i - b \cos(\Omega + c')$	$i = i_0 + b \cos(\Omega_0 + c)$
$\omega_0 = \omega - b \sin(\Omega + c') \csc i_0$	$\omega = \omega_0 + b \sin(\Omega_0 + c) \csc i$

where α and δ are the right ascension and declination; λ and β are the ecliptic longitude and latitude; and Ω, i, and ω are the orbital elements (referred to the ecliptic) longitude of the node, inclination, and argument of perihelion; the subscript zero refers to epoch J2000.0; and α_m and δ_m refer to the mean epoch.

With sufficient accuracy:

$$\alpha_m = \alpha - \tfrac{1}{2}(M + N \sin\alpha \tan\delta),$$
$$\delta_m = \delta - \tfrac{1}{2}N \cos\alpha_m,$$
$$\alpha_m = \alpha_0 + \tfrac{1}{2}(M + N \sin\alpha_0 \tan\delta_0),$$
$$\delta_m = \delta_0 + \tfrac{1}{2}N \cos\alpha_m.$$

The precessional constants M, N, etc., are given by

$$M = 1°281\,2323T + 0°000\,3879T^2 + 0°000\,0101T^3$$
$$N = 0°556\,7530T - 0°000\,1185T^2 - 0°000\,0116T^3$$
$$a = 1°396\,971T + 0°000\,3086T^2$$
$$b = 0°013\,056T - 0°000\,0092T^2$$
$$c = 5°123\,62 + 0°241\,614T + 0°000\,1122T^2$$
$$c' = 5°123\,62 - 1°155\,358T - 0°000\,1964T^2$$

where $T = (t - 2000.0)/100 = (\text{JD} - 245\,1545.0)/36\,525$.

27.4   SOLAR COORDINATES AND RELATED QUANTITIES
by A.D. Fiala

27.4.1   The Sun's Coordinates and the Equation of Time: Low-Precision Formulas

The following formulas from any recent *Astronomical Almanac* [3], page C24, give the apparent coordinates of the Sun to a precision of $0°01$ and the equation of time to a precision of 0^m1 between 1950 and 2050; on this page the time argument n is the number of days from J2000.0.

$$n = \text{JD} - 245\,1545.0$$
$$= -2557.5 + \text{day of year (B2-B3)} + \text{fraction of day from } 0^h \text{ UT}.$$

The mean longitude of the Sun, corrected for aberration, is

$$L = 280°460 + 0°985\,647\,4n.$$

The mean anomaly is $g = 357°528 + 0°985\,600\,3n$.
Put L and g in the range $0°$ to $360°$ by adding multiples of $360°$.

The ecliptic longitude is $\lambda = L + 1°915 \sin g + 0°20 \sin 2g$.
The ecliptic latitude is $\beta = 0°$.
The obliquity of ecliptic is $\epsilon = 23°439 - 0°000\,000\,4n$.
The right ascension (in the same quadrant as λ) is $\alpha = \tan^{-1}(\cos \epsilon \tan \lambda)$.
Alternatively, α may be calculated directly from

$$\alpha = \lambda - ft \sin 2\lambda + (f/2)t^2 \sin 4\lambda,$$

where

$$f = 180/\pi \quad \text{and} \quad t = \tan^2(\epsilon/2).$$

The declination is $\delta = \sin^{-1}(\sin \epsilon \sin \lambda)$.
The distance of the Sun from the Earth, in AU, is

$$R = 1.000\,14 - 0.016\,71 \cos g - 0.000\,14 \cos 2g.$$

The equatorial rectangular coordinates of the Sun, in AU, is

$$x = R \cos \lambda, \qquad y = R \cos \epsilon \sin \lambda, \qquad z = R \sin \epsilon \sin \lambda.$$

The equation of time (apparent time minus mean time) is

$$E(\text{in minutes of time}) = 4(L - \alpha), \text{ where } L - \alpha \text{ is in degrees.}$$

The horizontal parallax is $0°\!.0024$.
The semidiameter is $0°\!.2666/R$.
The light time is $0^{\text{d}}\!.0058$.
Similar formulas for the Moon are given in any recent *Astronomical Almanac* [3], page D46.

27.4.2 The Sun's Coordinates and the Equation of Time: Tables

In Table 27.3 the Sun's position and the equation of time are evaluated to low precision to represent a typical year. In addition, the UT of the transit of Aries is given and also the right ascension of Aries at 0^{h} UT; i.e., the Universal Time of the beginning of the sidereal day, and the sidereal time of the beginning of the civil day.

The quantities are calculated using the reference in the table. They are the approximate mean of values over a four-year cycle, with range on the order of two units in the last significant digit.

Table 27.3. *Sun's coordinates and equation of time* [1].

Date	Sun's geocentric apparent α (h m)	δ (deg)	Long. (deg)	Dist. (AU)	Equation of time = app − mean $E - 12^{\text{h}}$ (m)	Transit of Aries (h m)	RA of midnight meridian $\cong R$ (h m)
Jan. 1	18 44	−23.0	280.2	0.9833	−3.2	17 16	6 41
Jan. 16	19 50	−21.0	295.5	0.9837	−9.5	16 17	7 40
Feb. 1	20 57	−17.3	311.8	0.9853	−13.5	15 14	8 43
Feb. 16	21 57	−12.5	327.0	0.9878	−14.1	14 15	9 43
Mar. 1	22 47	−7.7	340.3	0.9908	−12.5	13 23	10 35
Mar. 16	23 43	−1.9	355.3	0.9947	−8.8	12 24	11 34
Apr. 1	0 41	+4.4	11.2	0.9992	−4.0	11 21	12 37
Apr. 16	1 36	+10.0	25.9	1.0035	+0.1	10 22	13 36
May 1	2 32	+15.0	40.5	1.0075	+2.9	9 23	14 35
May 16	3 31	+19.0	55.0	1.0110	+3.7	8 24	15 34
Jun. 1	4 35	+22.0	70.4	1.0140	+2.3	7 22	16 37
Jun. 16	5 37	+23.3	84.8	1.0158	−0.5	6 23	17 37
Jul. 1	6 39	+23.1	99.1	1.0167	−3.7	5 24	18 36
Jul. 16	7 41	+21.4	113.4	1.0164	−6.0	4 25	19 35
Aug. 1	8 44	+18.1	128.6	1.0150	−6.3	3 22	20 38
Aug. 16	9 41	+13.8	143.0	1.0126	−4.4	2 23	21 37
Sep. 1	10 40	+8.4	158.5	1.0093	−0.1	1 20	22 40
Sep. 16	11 34	+2.8	173.0	1.0054	+5.0	0 21	23 39
Oct. 1	12 28	−3.0	187.7	1.0012	+10.2	23 18	0 38
Oct. 16	13 23	−8.8	202.5	0.9969	+14.3	22 19	1 38
Nov. 1	14 24	−14.3	218.5	0.9926	+16.4	21 16	2 41
Nov. 16	15 25	−18.7	233.5	0.9889	+15.3	20 17	3 40
Dec. 1	16 28	−21.7	248.7	0.9861	+11.1	19 18	4 39
Dec. 16	17 33	−23.3	263.9	0.9842	+4.6	18 19	5 38

Reference
1. *MICA (Multiyear Interactive Computer Almanac)*, 1990–2005. 1998, U.S. Naval Observatory (Willmann-Bell, Richmond, VA).

With quantities from the table, for a given longitude λ_{east}, the local hour angle $\text{HA}_\odot$ of the Sun or HA_* of a star that has right ascension RA_* is found as follows:

$$\text{HA}_* = \text{UT} + R - \text{RA}_* + \lambda_{\text{east}},$$
$$\text{HA}_\odot = \text{UT} + E + \lambda_{\text{east}}.$$

27.4.3   The Sun's Disk

Table 27.4 gives positions on the solar disk for some times during the year. These mean values are averaged over a four-year cycle 1991–1994.

P = position of North point of Sun's axis measured eastward from North point of disk.

B_0 = heliographic latitude of Earth or central point of disk.

Table 27.4. *Sun's disk* [1].

Date	P (degrees)	B_0 (degrees)
Jan. 6	−0.2	−3.6
Feb. 5	−13.6	−6.3
Mar. 7	−22.9	−7.25
Apr. 6	−26.3	−6.3
May 6	−23.3	−3.7
Jun. 5	−14.1	−0.2
Jul. 7	0.0	3.5
Aug. 6	12.7	6.1
Sep. 5	22.0	7.2
Oct. 5	26.2	6.5
Nov. 4	24.0	4.1
Dec. 4	15.0	0.5

Reference

1. *MICA (Multiyear Interactive Computer Almanac)*, 1990–2005.   1998, U.S. Naval Observatory (Willmann-Bell, Richmond, VA).

27.5   CONSTELLATIONS
by A.D. Fiala

Table 27.5 lists data for the constellations in the sky.

Table 27.5. *Constellation names, genitive endings, English meaning, three-letter contractions, approximate positions, and areas on the sky* [1–4].

Constellation	Genitive ending	Meaning	Contraction	α (h)	δ (deg)	Area (deg^2)
Andromeda	-dae	Chained maiden	And	1	40 N	722
Antlia	-liae	Air pump	Ant	10	35 S	239
Apus	-podis	Bird of paradise	Aps	16	75 S	206
Aquarius	-rii	Water bearer	Aqr	23	15 S	980

Table 27.5. *(Continued.)*

Constellation	Genitive ending	Meaning	Contraction	α (h)	δ (deg)	Area (deg^2)
Aquila	-lae	Eagle	Aql	20	5 N	652
Ara	-rae	Altar	Ara	17	55 S	237
Aries	-ietis	Ram	Ari	3	20 N	441
Auriga	-gae	Charioteer	Aur	6	40 N	657
Boötes	-tis	Herdsman	Boo	15	30 N	907
Caelum	-aeli	Chisel	Cae	5	40 S	125
Camelopardus	-di	Giraffe	Cam	6	70 N	757
Cancer	-cri	Crab	Cnc	9	20 N	506
Canes Venatici	-num -corum	Hunting dogs	CVn	13	40 N	465
Canis Major	-is -ris	Great dog	CMa	7	20 S	380
Canis Minor	-is -ris	Small dog	CMi	8	5 N	183
Capricornus	-ni	Sea goat	Cap	21	20 S	414
Carina	-nae	Keel	Car	9	60 S	494
Cassiopeia	-peiae	Lady in chair	Cas	1	60 N	598
Centaurus	-ri	Centaur	Cen	13	50 S	1060
Cepheus	-phei	King	Cep	22	70 N	588
Cetus	-ti	Whale	Cet	2	10 S	1231
Chamaeleon	-ntis	Chamaeleon	Cha	11	80 S	132
Circinus	-ni	Compasses	Cir	15	60 S	93
Columba	-bae	Dove	Col	6	35 S	270
Coma Berenices	-mae -cis	Berenice's hair	Com	13	20 N	386
Corona Australis	-nae -lis	S crown	CrA	19	40 S	128
Corona Borealis	-nae -lis	N crown	CrB	16	30 N	179
Corvus	-vi	Crow	Crv	12	20 S	184
Crater	-eris	Cup	Crt	11	15 S	282
Crux	-ucis	S cross	Cru	12	60 S	68
Cygnus	-gni	Swan	Cyg	21	40 N	804
Delphinus	-ni	Dolphin	Del	21	10 N	189
Dorado	-dus	Dorado fish	Dor	5	65 S	179
Draco	-onis	Dragon	Dra	17	65 N	1083
Equuleus	-lei	Small horse	Equ	21	10 N	72
Eridanus	-ni	River Eridanus	Eri	3	20 S	1138
Fornax	-acis	Furnace	For	3	30 S	398
Gemini	-norum	Heavenly twins	Gem	7	20 N	514
Grus	-ruis	Crane	Gru	22	45 S	366
Hercules	-lis	Kneeling giant	Her	17	30 N	1225
Horologium	-gii	Clock	Hor	3	60 S	249
Hydra	-drae	Water monster	Hya	10	20 S	1303
Hydrus	-dri	Sea serpent	Hyi	2	75 S	243
Indus	-di	Indian	Ind	21	55 S	294
Lacerta	-tae	Lizard	Lac	22	45 N	201
Leo	-onis	Lion	Leo	11	15 N	947
Leo Minor	-onis-ris	Small lion	LMi	10	35 N	232
Lepus	-poris	Hare	Lep	6	20 S	290
Libra	-rae	Scales	Lib	15	15 S	538
Lupus	-pi	Wolf	Lup	15	45 S	334
Lynx	-ncis	Lynx	Lyn	8	45 N	545
Lyra	-rae	Lyre	Lyr	19	40 N	286
Mensa	-sae	Table (mountain)	Men	5	80 S	153
Microscopium	-pii	Microscope	Mic	21	35 S	210
Monoceros	-rotis	Unicorn	Mon	7	5 S	482
Musca	-cae	Fly	Mus	12	70 S	138
Norma	-mae	Square (level)	Nor	16	50 S	165
Octans	-ntis	Octant	Oct	22	85 S	291
Ophiuchus	-chi	Serpent bearer	Oph	17	0	948
Orion	-nis	Hunter	Ori	5	5 N	594
Pavo	-vonis	Peacock	Pav	20	65 S	378

Table 27.5. *(Continued.)*

Constellation	Genitive ending	Meaning	Contraction	α (h)	δ (deg)	Area (deg^2)
Pegasus	-si	Winged horse	Peg	22	20 N	1121
Perseus	-sei	Champion	Per	3	45 N	615
Phoenix	-nisis	Phoenix	Phe	1	50 S	469
Pictor	-ris	Painter's easel	Pic	6	55 S	247
Pisces	-cium	Fishes	Psc	1	15 N	889
Piscis Austrinus	-is -ni	S fish	PsA	22	30 S	245
Puppis	-ppis	Poop (stern)	Pup	8	40 S	673
Pyxis (=Malus)	-xidis	Compass	Pyx	9	30 S	221
Reticulum	-li	Net	Ret	4	60 S	114
Sagitta	-tae	Arrow	Sge	20	10 N	80
Sagittarius	-rii	Archer	Sgr	19	25 S	867
Scorpius	-pii	Scorpion	Sco	17	40 S	497
Sculptor	-ris	Sculptor	Scl	0	30 S	475
Scutum	-ti	Shield	Sct	19	10 S	109
Serpens (Caput and	-ntis	Serpent, Head	Ser	16	10 N	429
Cauda		Tail		18	5 S	+208
Sextans	-ntis	Sextant	Sex	10	0	314
Taurus	-ri	Bull	Tau	4	15 N	797
Telescopium	-pii	Telescope	Tel	19	50 S	252
Triangulum	-li	Triangle	Tri	2	30 N	132
Triangulum Australe	-li -lis	S Triangle	TrA	16	65 S	110
Tucana	-nae	Toucan	Tuc	0	65 S	295
Ursa Major	-sae -ris	Great Bear	UMa	11	50 N	1280
Ursa Minor	-sae -ris	Small Bear	UMi	15	70 N	256
Vela	-lorum	Sails	Vel	9	50 S	500
Virgo	-ginis	Virgin	Vir	13	0	1294
Volans	-ntis	Flying fish	Vol	8	70 S	141
Vulpecula	-lae	Small fox	Vul	20	25 N	268

References

1. *Transactions I.A.U.*, **1**, 158 (for names and coordinates)
2. Davis, G.A., Barton, S.G., & McHugh, D.J. 1943, *Pop. Astron.*, **50**, 356 (for meanings and pronunciations)
3. Oravec, E.G. 1958, *Sky & Tel.*, **17**, 220 (for meanings and pronunciations)
4. Levin, A.E. 1935, in *Handbook of the British Astronomical Association*, 34 (for constellation area)

27.6   THE MESSIER OBJECTS
by A.D. Fiala

Table 27.6 gives information for the Messier objects.

Table 27.6. *The Messier objects* [1].

Messier M	NGC IC	Type^a	Con.	α 2000.0 (h m)	δ 2000.0 (° ′)	m_V	Name, etc.
1	1952	Crab	Tau	5 34.5	+22 01	8.4	Crab nebula
2	7089	Glob	Aqr	21 33.5	− 0 49	6.5	
3	5272	Glob	CVn	13 42.2	+28 23	6.4	
4	6121	Glob	Sco	16 23.6	−26 32	5.9	
5	5904	Glob	Ser	15 18.6	+ 2 05	5.8	
6	6405	Op Cl	Sco	17 40.1	−32 13	4.2	Butterfly

Table 27.6. (*Continued.*)

Messier M	NGC IC	Type[a]	Con.	α 2000.0 (h m)	δ 2000.0 (° ′)	m_V	Name, etc.
7	6475	Op Cl	Sco	17 53.9	−34 49	3.3	
8	6523	Neb	Sgr	18 03.8	−24 23	5.8	Lagoon nebula
9	6333	Glob	Oph	17 19.2	−18 31	7.9	
10	6254	Glob	Oph	16 57.1	− 4 06	6.6	
11	6705	Op Cl	Sct	18 51.1	− 6 16	5.8	Wild duck cluster
12	6218	Glob	Oph	16 47.2	− 1 57	6.6	
13	6205	Glob	Her	16 41.7	+36 28	5.9	Hercules
14	6402	Glob	Oph	17 37.6	− 3 15	7.6	
15	7078	Glob	Peg	21 30.0	+12 10	6.4	
16	6611	Op Cl	Ser	18 18.8	−13 47	6.0	Eagle nebula
17	6618	Neb	Sgr	18 20.8	−16 11	7.5	Swan or Omega nebula
18	6613	Op Cl	Sgr	18 19.9	−17 08	6.9	
19	6273	Glob	Oph	17 02.6	−26 16	7.2	
20	6514	Neb	Sgr	18 02.6	−23 02	8.5	Trifid nebula
21	6531	Op Cl	Sgr	18 04.6	−22 30	5.9	
22	6656	Glob	Sgr	18 36.4	−23 54	5.1	
23	6494	Op Cl	Sgr	17 56.8	−19 01	5.5	
24	6603	Op Cl	Sgr	18 16.9	−18 29	4.5	
25	I 4725	Op Cl	Sgr	18 31.6	−19 15	4.6	
26	6694	Op Cl	Sct	18 45.2	− 9 24	8.0	
27	6853	Plan	Vul	19 59.6	+22 43	8.1	Dumbbell nebula
28	6626	Glob	Sgr	18 24.5	−24 52	6.9	
29	6913	Op Cl	Cyg	20 23.9	+38 32	6.6	
30	7099	Glob	Cap	21 40.4	−23 11	7.5	
31	224	Gal Sb	And	0 42.7	+41 16	3.4	Andromeda nebula
32	221	Gal E	And	0 42.7	+40 52	8.2	
33	598	Gal Sc	Tri	1 33.9	+30 39	5.7	
34	1039	Op Cl	Per	2 42.0	+42 47	5.2	
35	2168	Op Cl	Gem	6 08.9	+24 20	5.1	
36	1960	Op Cl	Aur	5 36.1	+34 08	6.0	
37	2099	Op Cl	Aur	5 52.4	+32 33	5.6	
38	1912	Op Cl	Aur	5 28.7	+35 50	6.4	
39	7092	Op Cl	Cyg	21 32.2	+48 26	4.6	
40		2 stars	UMa	12 22.4	+58 05	8.0	Winnecke 4
41	2287	Op Cl	CMa	6 46.1	−20 46	4.5	
42	1976	Neb	Ori	5 35.4	− 5 27	4.0	Orion nebula
43	1982	Neb	Ori	5 35.6	− 5 16	9.0	Orion nebula
44	2632	Op Cl	Cnc	8 40.2	+19 43	3.1	Praesepe (Beehive)
45		Op Cl	Tau	3 47.0	+24 07	1.2	Pleiades
46	2437	Op Cl	Pup	7 41.8	−14 49	6.1	
47	2422	Op Cl	Pup	7 36.6	−14 30	4.4	
48	2548	Op Cl	Hya	8 13.8	− 5 48	5.8	
49	4472	Gal E	Vir	12 29.8	+ 8 00	8.4	
50	2323	Op Cl	Mon	7 02.8	− 8 23	5.9	
51	5194	Gal Sc	CVn	13 29.9	+47 12	8.1	Whirlpool
52	7654	Op Cl	Cas	23 24.2	+61 35	6.9	
53	5024	Glob	Com	13 12.9	+18 10	7.7	
54	6715	Glob	Sgr	18 55.1	−30 29	7.7	
55	6809	Glob	Sgr	19 40.0	−30 58	7.0	
56	6779	Glob	Lyr	19 16.6	+30 11	8.2	
57	6720	Plan	Lyr	18 53.6	+33 02	9.0	Ring nebula
58	4579	Gal SBb	Lyr	12 37.7	+11 49	9.8	
59	4621	Gal E	Vir	12 42.0	+11 39	9.8	
60	4649	Gal E	Vir	12 43.7	+11 33	8.8	
61	4303	Gal Sc	Vir	12 21.9	+ 4 28	9.7	
62	6266	Glob	Oph	17 01.2	−30 07	6.6	

Table 27.6. (Continued.)

Messier M	NGC IC	Type[a]	Con.	α 2000.0 (h m)	δ 2000.0 (° ')	m_V	Name, etc.
63	5055	Gal Sb	CVn	13 15.8	+42 02	8.6	Sunflower
64	4826	Gal Sb	Com	12 56.7	+21 41	8.5	Black Eye
65	3623	Gal Sa	Leo	11 18.9	+13 05	9.3	
66	3627	Gal Sb	Leo	11 20.2	+12 59	9.0	
67	2682	Op Cl	Cnc	8 51.4	+11 49	6.9	
68	4590	Glob	Hya	12 39.5	−26 45	8.2	
69	6637	Glob	Sgr	18 31.4	−32 21	7.7	
70	6681	Glob	Sgr	18 43.2	−32 18	8.1	
71	6838	Glob	Sge	19 53.8	+18 47	8.3	
72	6981	Glob	Aqr	20 53.5	−12 32	9.4	
73	6994	Op Cl	Aqr	20 58.9	−12 38		4 stars
74	628	Gal Sc	Psc	1 36.7	+15 47	9.2	
75	6864	Glob	Sgr	20 06.1	−21 55	8.6	
76	650	Plan	Per	1 42.4	+51 34	11.5	Little Dumbbell
77	1068	Gal Sb	Cet	2 42.7	− 0 01	8.8	
78	2068	Neb	Ori	5 46.7	+ 0 03	8.0	
79	1904	Glob	Lep	5 24.5	−24 33	8.0	
80	6093	Glob	Sco	16 17.0	−22 59	7.2	
81	3031	Gal Sb	UMa	9 55.6	+69 04	6.8	
82	3034	Gal Irr	UMa	9 55.8	+69 41	8.4	
83	5236	Gal Sc	Hya	13 37.0	−29 52	10.1	
84	4374	Gal E	Vir	12 25.1	+12 53	9.3	
85	4382	Gal So	Com	12 25.4	+18 11	9.3	
86	4406	Gal E	Vir	12 26.2	+12 57	9.2	
87	4486	Gal Ep	Vir	12 30.8	+12 24	8.6	Radio gal
88	4501	Gal Ep	Com	12 32.0	+14 25	9.5	
89	4552	Gal E	Vir	12 35.7	+12 33	9.8	
90	4569	Gal Sb	Vir	12 36.8	+13 10	9.5	
91	4548	Gal S	Com	12 35.4	+14 30	10.2	
92	6341	Glob	Her	17 17.1	+43 08	6.5	
93	2447	Op Cl	Pup	7 44.6	−23 52	6.2	
94	4736	Gal Sb	CVn	12 50.9	+41 07	8.1	
95	3351	Gal SBb	Leo	10 44.0	+11 42	9.7	
96	3368	Gal Sa	Leo	10 46.8	+11 49	9.2	
97	3587	Plan	UMa	11 14.8	+55 01	11.2	Owl nebula
98	4192	Gal Sb	Com	12 13.8	+14 54	10.1	
99	4254	Gal Sc	Com	12 18.8	+14 25	9.8	
100	4321	Gal Sc	Com	12 22.9	+15 49	9.4	
101	5457	Gal Sc	UMa	14 03.2	+54 21	7.7	Pinwheel
102	5866	Gal Sa	Dra	15 06.5	+55 46	10.0	
103	581	Op Cl	Cas	1 33.2	+60 42	7.4	
104	4594	Gal Sa	Vir	12 40.0	−11 37	8.3	Sombrero
105	3379	Gal E	Leo	10 47.8	+12 35	9.3	
106	4258	Gal Sb	CVn	12 19.0	+47 18	8.3	
107	6171	Glob	Oph	16 32.5	−13 03	8.1	
108	3556	Gal Sb	UMa	11 11.5	+55 40	10.0	
109	3992	Gal SBc	UMa	11 57.6	+53 23	9.8	

Note

[a] Op Cl = open cluster; Glob = globular cluster; Plan = planetary nebula; Neb = diffuse nebula; Gal = galaxy (with classification).

Reference

1. The Observer's Handbook—1991, (Royal Astronomical Society of Canada), p. 209

27.7 ASTROMETRY
by W.F. van Altena

27.7.1 Astrometric Catalogues

Summary statistics and bibliographies for astrometric catalogs are given in Table 27.7. Newly planned catalogs are described in Table 27.8. Table 27.9 lists the major Schmidt telescope astrometric surveys. Table 27.10 summarizes the data available to the authors on the status of the major Schmidt telescope surveys and the process of their digitization as of May 1993. See also the paper and table by Morgan [4].

Tables 27.11–27.13 summarize data on catalogs providing surveys of relative proper motions, catalogs compiling stars with proper motions larger than some limit, and catalogs that compile varieties of data for a specific type of star, e.g., all bright stars, stars nearer than some distance, etc. The columns in each table are similar and list, respectively, a reference to the source where the catalog may be found, an acronym for the catalog, number of stars, declination coverage, date of publication (T_{pub}), magnitude limit, parallax limit (π), proper motion or some other kind of limit (μ limit), and a brief description of the catalog. More detailed compilations may be found in [5–10].

Most of the acronyms used in the tables are: FK3 = Third Catalogue of the Astronomisches Gesellshaft; FK5 = Fifth Fundamental Catalogue; IRS = International Reference Stars; KSZ = Faint Star Catalogue; ACRS = Astrographic Catalogue Reference Stars; PPM = Positions and Proper Motions; CPC = Cape Photographic Catalogue; BKAT = Bright Star Catalogue; NPM = Northern Proper Motion survey; SPM = Southern Proper Motion survey; HIC = Hipparcos Input Catalogue; FHST = Fixed Head Star Tracker catalogue; SAOSC = Smithsonian Astrophysical Observatory Star Catalogue; FASTT = Flagstaff Astrometric Scanning Transit Telescope; POSS = Palomar Observatory Sky Survey; AAO-SES = Anglo Australia Observatory Second Epoch Survey; SERC-I/SR = Science and Engineering Research Council Infrared and Short Red survey; GSC = Hubble Space Telescope Guide Star Catalogue; GAC = HST CGS second epoch positions; IFRS = Intermediate Fundamental Reference Stars; FON = Four-Fold Coverage of Northern Sky; EKAT = Equatorial Catalogue; DENIS = Deep Near Infrared Southern Sky Survey; ESO-B = European Southern Observatory Blue Survey; ESO-R; ESO Red Survey; SERC-J/R = Science and Engineering Research Council J and R Survey; STSCIdig = Space Telescope Science Institute digitization; ROEdig = Royal Observatory Edinburgh digitization; CAMdig = Cambridge digitization; APS = University of Minnesota Automated Plate Scanner; USN PMM = U.S. Naval Observatory Precision Plate Measuring Machine; BPM = Bruce Proper Motion Survey; Lowell N = Lowell Proper Motion Survey Northern Hemisphere Catalogue; Lowell S = Lowell Proper Motion Survey Southern Hemisphere Catalogue; LP = Luyten Palomar Proper Motion Survey; Calan = Cerro Calan Observatory, Chile; NLTT = New Luyten Catalogue of stars with proper motions greater than Two Tenths of an arc second; LHS = Luyten Half Second Catalogue; BSC V = Yale Bright Star Catalogue; BSC Supp. = Yale Bright Star supplement; CNS 3 = Catalogue of Nearby Stars, Heidelberg; CNSG = Catalogue of Nearby Stars with Ground-based parallaxes; CNSH = Catalogue of Nearby Stars with Hipparcos-based parallaxes; YPC = Yale Parallax Catalogue; WDS = Washington Double Stars; CHARA = Center for High Angular Resolution. Georgia State; Orbits = Visual double Star Orbit Catalogue; ADC CDL = Astronomical Data Center CD-ROM.

Machine-readable versions of all catalogues may be obtained at the Astronomical Data Center at Goddard Space Flight Center [11], at Data Centers in other countries, and normally at World Wide Web sites at each institution. Recent reviews are also given by van Altena [12] and [13, 14]. Astrometric data and references for open clusters, globular clusters, and radio astrometry may be found in [15], pages 215 and 250.

Table 27.7. Catalogues of positions and proper motions.[a]

Reference	Catalogue	Number stars	Coverage	T_{pub}	m	σ_p (T_{pub})	σ_p (2000)	σ_μ	Description
[1]	Hipparcos	118 218	All sky	1991	12.4	0.8	6.2	0.7	On the ICRS/ICRF[b,c] system
[1]	Tycho	1 058 332	All sky	1991	11.5	25	...	...	On the Hipparcos system
[2]	AC 2000	4 621 836	All sky	1907	11.5	220	...	2.7	Re-reduced Astrographic Catalogue on Hipparcos system
[2]	ACT	988 758	All sky	1991	11.5	25	36	2.7	Proper motions from [1, 2] on Hipparcos system
[3]	FK5	1 535	All sky	1950	7.0	20	40	0.7	The fundamental reference frame
[4]	FK5 Ext.	3 117	All sky	1944	9.5	55	156	2.6	Faint extension to the FK5
[4]	IRS I	29 163	All sky	1950	9.5	80	225	4.2	International Reference Stars
[5]	IRS II	7 064	$\delta \geq -30$	1950	9.5	120	346	6.5	Lower precision extension of IRS I
[6]	KSZ	21 817	All sky	1960	9	150	...	...	Included in the IRS Catalogue [4, 5]
[7]	ACRS I	250 052	All sky	1950	10.5	88	253	4.7	Positions and proper motions
[8]	ACRS II	70 159	All sky	1950	10.5	140[d]	404	7.5[d]	Lower precision extension of ACRS I
[9]	PPM N	181 731	North	1931	10.5	100	313	4.3	Positions and proper motions, north
[10]	PPM S	197 179	South	1962	10.5	73	135	3.0	Positions and proper motions, south
[11]	CPC2	276 131	South	1967	10.5	54	...	...	Positional catalogue
[12]	CPC	454 875	South	Var.	10	Var.	Var.	Var.	The Cape Photographic Catalogues for 1950
[13]	Yale Zone	221 338	All sky	Var.	10	Var.	Var.	Var.	The Yale Zone Catalogues, integrated version available
[14]	AGK3	183 173	North	1944	10.5	210	589	10	Positions; motions from AGK3–AGK2
[15]	AGK3U	170 464	North	1951	10.5	118	308	5.8	Positions; motions from AGK3, AGK2, and the revised [25]
[16]	BKAT	4 949	All sky	1969	5.9	80	96	1.7	Bright star catalogue of positions and proper motions
[17]	NPM1	148 940	+90 to −23	1968	18	150	200	5	Absolute proper motions with respect to galaxies, outside Milky Way, position error for 1950
[18]	SPM-1.0	50 000	−22 to −45	1980	18	39	82	4	Absolute proper motions with respect to galaxies The South Galactic Pole
[19]	HIC	118 209	All sky	Var.	12.5	...	300	4	Hipparcos Input Catalogue (positional error for 1990)
[20]	FHST	219 859	All sky	Var.	9.0	Var.	Var.	Var.	HST star tracker catalogue, complete to 8.5
[21]	SAOSC	258 997	All sky	1930	9	200	1070	15	Accuracy at publication date, not mean epoch

Table 27.7. (*Continued.*)

Reference	Catalogue	Number stars	Coverage	T_{pub}	m	σ_p (T_{pub})	σ_p (2000)	σ_μ	Description
[22]	SKY2000	300 000	All sky	1991	9	0.8–25	6.2–36	0.7–2.7	NASA operations catalogue; data from [1, 2]
[23]	SKYMAP	248 558	All sky	1930	9	200	1070	15	NASA operations catalogue
[24]	MSX IR	177 860	All sky	Var.	8	Var.	Var.	Var.	Reference catalogue for infrared, mag. limit ≥ 8 in K
[25]	HST GSC	2×10^7	All sky	1980	16	250	...	...	HST Guide Star Catalogue, Table 27.9 [3, 9, 10]
[26]	USNO-A1.0	4.9×10^8	All sky	Var.	20–21	250	...	...	USNO PMM Star Catalogue, North from [30, 31] South from [32, 33]
[27]	USNO-A2.0	4.7×10^8	$\delta \geq -18$	...	...	150	...	...	USNO A1.0 on the ICRS system[b]
[28]	USNO-TAC N	705 679		1981	12	90	...	...	USNO Twin Astrograph Catalogue
[29]	USNO-FASTT	661 591	Equatorial	1995	17.8	45	...	...	CCD transit telescope catalogue, $|\delta| < 1.6$ degrees

Notes

[a] T_c is the central epoch, m the magnitude limit, $\sigma_p(T_c)$ the error in milli-arcsec (mas) of a position at the central epoch, $\sigma_p(2000)$ the error of a position in mas at 2000, σ_μ the error of a proper motion in mas/yr.

[b] The extragalactic reference system of the International Earth Rotation Service (ICRS), Arias, E.F. et al. 1995, *A&A*, **303**, 604. The ICRS defines the axes of the system using 212 of the best-observed extragalactic sources. These axes are to be maintained by the International Earth Rotation Service.

[c] *Definition and Realization of the International Celestial Reference System by VLBI, Astrometry of Extragalactic Objects*, edited by Ma, C. and Feissel M., IERS Technical Note 23, 1997. The ICRF is a larger set of sources (about 610) that provide the fiducial points for the system.

[d] Estimated value.

References

1. ESA-SP 1200, 1997; *Proceedings of the Hipparcos '97 Venice Conference*, ESA SP-402 1997
2. Urban, S.E., Corbin, T.E., & Wycoff, G.L., 1997, U.S. Naval Observatory, Astrographic Catalogue positions rereduced using the ACRS and then placed on the Hipparcos system [1]. Proper motions from [1, 2].
3. Fricke, W. et al. 1988, *Fifth Fundamental Catalogue (FK5)*, Part I: *The Basic Fundamental Stars* (Veröff. Astron. Rechen-Instituts, Heidelberg), **32**
4. Fricke, W. et al. 1991, *The FK5 Extension: New Fundamental Stars* (Veröff. Astron. Rechen-Instituts, Heidelberg), **33**
5. Corbin, T. 1991, U.S. Naval Observatory. IRS lower precision data
6. This incomplete catalogue was incorporated into the IRS [5]
7. Corbin, T., & Urban, S.E. 1991, U.S. Naval Observatory. This catalogue and [8] were used for the rereduction of the Astrographic Catalogue [2], so they contain no AC positions and emphasize the inclusion of early epoch positions; both hemispheres are averaged together to form the average errors
8. Corbin, T., & Urban, S.E. 1991, U.S. Naval Observatory.
9. Röser, S., & Bastian, U. 1991, *PPM Star Catalogue*, Vols. I and II, (Astron. Rechen-Instituts, Heidelberg). This catalogue and [10] are intended to include all available data and hence provide the most precise positions and proper motions; all data are averaged to give the average errors rather than high- and low-precision data separately, as for [7] and [8].
10. Röser, S., & Bastian, U. 1993 *PPM Star Catalogue (The Southern Hemisphere)* (Astron. Rechen-Instituts, Heidelberg)

11. de Vegt, C. et al. 1992, *A&AS*, **97**, 985; Zacharias, N. 1995, *AJ*, **109**, 1880

12. Publications of the Royal Observatory, Cape of Good Hope

13. Yale Photographic Catalogues. 1926–83, *Trans. Astron. Obs. Yale University*, Vols. 3–32; and Yale Zone Catalogues—Integrated Version

14. Dieckvoss, W. et al. 1975, *The AGK3 Catalogue of Positions and Proper Motions North of −2.5 Declination*, Vols. 1–8, (Hamburg Bergedorf).

15. Bucciarelli, B. et al. 1992, *AJ*, **103**, 1689, AGK3U = Update to [14] using a rereduction of [25] for third epoch positions

16. Chrutskaya, E.V. 1985, *Sov. Astron. J.*, **62**, 605.

17. Hanson, R.B., 1988, *Proper Motions and Galactic Astronomy*, edited by Roberta M. Humphreys, ASP Conf. Series 127: Klemola, A.R., Jones, B.F., & Hanson, R.B. 1987, *AJ*, **94**, 501. See also [34], p. 235; Cambridge: Galactic and Solar System Optical Astrometry, 1995 (held at Robinson College, Cambridge University in June 1993), edited by L.V. Morrison and G. Gilmore, (Kluwer Academic, Dordrecht), p. 20; and [35], p. 45

18. Girard, T.M. et al. 1998, *AJ*, **115**, 855; and 1997, *BAAS*, **29**, 1383; Platais, I. et al. 1998, *A&A*, **331**, 1119; van Altena, W.F. et al. 1990, *The Yale–San Juan Southern Proper Motion Program* (SPM), in IAU Symposium 141, 1991, *Inertial Coordinate System on the Sky*, edited by J.H. Lieske and V.K. Abalakin, (Kluwer Academic, Dordrecht), p. 419: van Altena, W.F. et al. in Cambridge: Galactic and Solar System Optical Astrometry, 1995 (held at Robinson College, Cambridge University in June 1993), edited by L.V. Morrison and G. Gilmore (Kluwer Academic, Dordrecht), p. 26

19. Turon, C. et al. 1992, and in [35], p. 77, *The Hipparcos Input Catalogue*, ESA SP-1136 (ESTEC, Noordwijk)

20. Yang, T.-g. et al. 1992. FHST, A New Fixed Head Star Tracker Catalogue for the HST, in [34], p. 235

21. Smithsonian Institution 1966.

22. NASA Goddard Space Flight Center, Greenbelt. Major revision and expansion of SKYMAP [23] operations catalogue for guidance and tracking NASA spacecraft. Astrometric data derived from Hipparcos and Tycho catalogues.

23. NASA Goddard Space Flight Center, Greenbelt. An operations catalogue for guidance and tracking NASA spacecraft

24. Egan, M.P., & Price, S.D., 1996, *AJ*, **112**, 2862. Astrometric positions for infrared objects have been collected into a catalogue designed to aid in the determination of positions of objects found in infrared surveys; 61,242 of the objects have been identified in infrared catalogues.

25. Lasker, B.M. et al. 1990, *AJ*, **99**, 2019, and CD-ROM as GSC 1.0; an updated version GSC 1.1 was issued on CD-ROM on 1 August 1992. Positional error is the average local precision

26. Monet, D.G. 1996, US Naval Observatory, Washington, DC: USNO (10 CD ROMs). Derived from scans of the indicated surveys with the USNO's Precision Measuring Machine (PMM) at Flagstaff. Reduced into the system of the HST GSC

27. Monet, D.G., the USNO-A1.0 catalogue reduced into the system of the ACT

28. Zacharias, N. et al. 1996, *AJ*, **112**, 2336. Observed photographically in the blue and visual bandpasses. Higher accuracy anticipated once the catalogue has been rereduced into the Hipparcos system.

29. Stone, R.C., 1997, *AJ*, **114**, 2811; Sixteen astrometric calibration equatorial regions with dimensions of 7.6 × 3.2 degrees established with the US Naval Observatory's Flagstaff Astrometric Scanning Transit Telescope (FASTT) in the magnitude range 9.5 < R < 17.8.

30. Minkowski, R.L., & Abell, G.O. 1963, in *Basic Astronomical Data*, edited by K.Aa. Strand; Vol. III of *Stars and Stellar Systems*, edited by G. Kuiper and B. Middlehurst, p. 481

31. Whiteoke, J. 1969, *Whiteoke Fields at Declination Zones −36 and −42 Degrees* (Calif. Instit. Tech., Pasadena)

32. Lasker, B.M., & Cannon, R.D. 1989, in *Digitized Optical Sky Surveys*, edited by C. Jaschek, in Bull. d'Info. du CDS, No. 37 (Obs. de Strasbourg, Strasbourg), p. 13; and Morgan, D.H. et al. 1992, in *Digitized Optical Sky Surveys*, edited by H.T. MacGillivray and E.B. Thomson (Kluwer Academic, Dordrecht), p. 11.

33. Hartley, M., & Dawe, J.A. 1981, *Proc. Astron. Soc. Australia*, **4**, 251.

34. IAU XXIB: *Transactions of the IAU XXIB* 1991, edited by J. Bergeron (Kluwer Academic, Dordrecht)

35. IAU 166: IAU Symposium 166, *Astronomical and Astrophysical Objectives of Sub-Milliarcsecond Optical Astrometry*, edited by E. Hoeg and P.K. Seidelmann (Kluwer Academic, Dordrecht)

Table 27.8. *Future catalogues and surveys in progress.*[a]

Reference	Catalogue	Number stars	Coverage	T_{pub}	m	σ_p (T_{pub})	σ_p (2000)	σ_μ	Description
[1]	Tycho-Ext.	3 000 000	All sky	1991	12	100	...	...	Extension of the Tycho Catalogue from Hipparcos
[2]	GAC	4 000 000	All sky	1954	12.5	200	350	7	[11] plus [3]
[3]	IFRS	3 000	All sky	...	13	...	...	...	Intermediate Fund. Stars ($9.5 \leq V \leq 13$) (finding list)
[4]	NPM2	150 000	+90 to −23	1968	18	120	200	5	Absolute proper motions with respect to galaxies, inside Milky Way
[5]	SPM-2.0	250 000	−22 to −48	1980	18	39	82	4	Absolute proper motions Extension of SPM 1.0
[6]	SPM-3.0	10^7	−17 to −90	1985	18	30	55	3	Absolute proper motions Completion of SPM
[7]	FON	36 000 000	+90 to −15	1990?	16	150	...	...	Possibly completed by year 2000
[8]	EKAT	1 000 000	+20 to −20	1993?	12	150	...	...	Equatorial catalogue, observations 60% completed
[9]	AC2	4 000 000	All sky	?	13	150	?	?	First epoch is [3] + new epoch; project on hold
[10]	HST GSC II	10^9	North	1988	20–21	47	82	3	μ from [18–22]
			South	1973	21–22	47	92	6	μ from [23, 24]
[11]	USNO-B1.0	10^9	North	1988	20–21	47	82	3	μ from [18–22]
			South	1973	21–22	47	92	6	μ from [23, 24] US Naval Observatory PMM Catalogue
[12]	UCAC-S	40×10^6	−2 to −90	1999	16.5	70	...	...	USNO Southern CCD survey
[13]	DENIS	10^9	−2 to −88	1998	18.5	1000	...	...	Near-Infrared Sky Survey see refs. for detailed limits
[14]	2MASS	??	All sky	2000	14.3	200	...	...	Infrared All Sky Survey
[15]	Sloan DSS	10^9	N. Polar Cap	2002	22.5	50	...	...	Sloan Digital Sky Survey
[16]	Quest-I:	8×10^6	16	1999	20	100	...	...	Quasar Equatorial Survey Team
[17]	Quest-II:	10^8	122	2002	21	100	...	...	Extension of Quest-I with new camera

Note

[a]See Table 27.7 for a description of the column headings. *Sources for new positions*: See Triennial Reports of IAU Commissions 8 and 24, World Wide Web sites and the publications of the respective observatories. *Meridian circles*: Carlsberg Automatic Meridian Circle (−45 to +90, La Palma, Islas Canarias, published annually by Copenhagen Univ. Obs., Royal Greenwich Obs., and Real Inst. y Obs. de la Armada, San Fernando); U.S. Naval Observatory (all sky, Washington and Black Birch, N.Z.); additional northern meridian circles in Belgrade, Bordeaux, Pulkova, San Fernando, Spain (transferred to El Leoncito, Argentina in 1996), Shanghai, Yunnan (China) and southern meridian circles in San Juan (Argentina) and Santiago (Chile).

Astrolabes: Northern astrolabes exist in Beijing (currently in San Juan, Argentina), Belgrade, Bucharest, CERGA (France), Paris, Pulkova, San Fernando (Spain), Shanghai, Washington, and Yunnan (China), while southern astrolabes exist in Rio Grande and San Juan (Argentina), Santiago (Chile), and Sao Paulo (Brazil).

References

1. Hog, E., 1997, ESA SP-402, *Hipparcos Venice '97* (ESTEC, Noordwijk), p. 25. Expected positional error at $V = 11$ is 50 mas.

2. Urban, S.E. 1993, in [25], p. 145. [26] second epoch positions with Astrographic Catalogue first epoch positions AC 2000 in [27]

3. Corbin, T.E., & Urban, S.E. 1990, Faint Reference Stars, in [28], p. 433. See also Report of the Working Group on Star Lists, in [29] p. 142. Note that this is a finding list for observers and not a catalogue. IFRS = Intermediate Fundamental Reference Stars

4. Hanson, R.B., 1997, in [30], p. 23. See NPM1 for other references. This is the extension of the NPM1 into the galactic plane.

5. See SPM-1.0 for references. This is an extension of the SPM-1.0 outside the galactic plane for the −35 and −40 degree in addition to fields with second-epoch plates.

6. See SPM-1.0 and -2.0 for references. This is the planned extension to the south celestial pole using CCDs as detectors to obtain the second-epoch exposures.

7. Kolchinsky, I.G., Garilov, I.V., & Onegina, A.B. 1978, in *IAU Colloq. 48, Modern Astrometry*, edited by F.V. Prochazka and R.H. Tucker (Univ. Obs., Vienna), p. 479

8. Polojentsev, D.D., Potter, H.I., Yagudin, L.I., & Zalles, R.F. 1991, in *Proceedings of the First Spain–USSR Workshop on Positional Astronomy and Celestial Mechanics*, edited by A. López Garcia, R.F. López Machi, and A.G. Sokolsky (Univ. Valencia Obs. Astron., Valencia), p. 63.

9. Proposed by H.I. Potter. This catalogue will use the Astrographic Catalogue as the first epoch and new observations with similar astrographs for the second epoch. The project has been suspended for the time being

10. Lasker, B.M. in 1991, [31], p. 87

11. Monet, D.G. in [30], see also [32, 33]

12. Gauss, F.G. et al. 1996, *BAAS*, **28**, 1282; Observations for stars brighter than 14 have an average accuracy of 20 mas.

13. Epchtein, N. et al. 1997, The Deep Near-Infrared Southern Sky Survey (DENIS), *The ESO Messenger*, No. 87, 27; Epchtein, N., 1997, *Proc. workshop on "The Impact of Large Scale Near-IR Sky Surveys"*, held at Puerto de la Cruz (Tenerife), Apr. 6. Kluwer ASSL Ser. 210, p. 15. Limiting magnitude in Ks-band (18.5), I (16.3), and J (14). Detections of stars in I, 10^9, and 10^8 in Ks; of galaxies in I, 5×10^6, J, 2.5×10^6 and Ks, 10^5. Positional accuracy may be improved in final processing

14. Two-Micron All Sky Survey (2MASS). Limiting magnitude is in the K band

15. Hindsley, R., [30], p. 33. The limiting magnitude is in the r' band, and the survey will cover about 1 steradian centered on the North Galactic Pole; spectra will be obtained for about 1 million galaxies and 50 000 stars, in addition to photometry in five passbands for all of the stars.

16. Snyder, J. et al. 1998, in *Proceedings of SPIE*, Vol. 3355 (Astronomical Telescopes and Instrumentation, Kona, HI, 20–28 March 1998), paper 36. Phase I limiting mag. in V for a signal-to-noise of 10; survey covers about 4000 square degrees at high galactic latitudes within about 6 degrees of the celestial equator.

17. See reference for Quest-I. Phase II is planned with improved detectors covering a larger area.

18. Minkowski, R.L., & Abell, G.O. 1963, in *Basic Astronomical Data*, edited by K.Aa. Strand; Vol. III of *Stars and Stellar Systems*, edited by G. Kuiper and B. Middlehurst, p. 481. POSS = Palomar Observatory Sky Survey

19. Whiteoke, J. 1969, *Whiteoke Fields at Declination Zones −36 and −42 Degrees* (Calif. Instit. Tech., Pasadena)

20. Reid, I.N. 1991, *PASP*, **103**, 661

21. Monet, D.G., & Westerhout, G. 1989, in *Digitized Optical Sky Surveys*, edited by C. Jaschek, in Bull. d'Info. du CDS, No. 37, (Obs. de Strasbourg, Strasbourg), p. 75. The USNO QJ is a two-minute exposure on unhypersensitized IIIa-J emulsions to be used for the determination of positions of bright stars.

22. West, R.M., & Schuster, H.-E. 1982, AAS, **49**, 577

23. Lasker, B.M., & Cannon, R.D. 1989, in *Digitized Optical Sky Surveys*, edited by C. Jaschek, in Bull. d'Info. du CDS, No. 37, (Obs. de Strasbourg, Strasbourg), p. 13; and Morgan, D.H. et al. 1992, in *Digitized Optical Sky Surveys*, edited by H.T. MacGillivray and E.B. Thomson, (Kluwer Academic, Dordrecht), p. 11

24. Hartley, M., & Dawe, J.A. 1981, *Proc. Astron. Soc. Australia*, **4**, 251

25. *Proper Motions and Galactic Astronomy*, edited by Roberta M. Humphreys, ASP Conf. Series 127

26. Lasker, B.M., Sturch, C.R., McLean, B., Russell, J.L., Jenkner, H., & Shara, M. 1990, *AJ*, **99**, 2019

27. Urban, S.E., Corbin, T.E., & Wycoff, G.L., 1997, U.S. Naval Observatory, Astrographic Catalogue positions rereduced using the ACRS and then placed on the Hipparcos system

28. IAU Symposium 141, 1991, *Inertial Coordinate System on the Sky*, edited by J.H. Lieske and V.K. Abalakin, (Kluwer Academic, Dordrecht)

29. *Transactions of the IAU*, XXIB 1991, edited by J. Bergeron (Kluwer Academic, Dordrecht)

30. Swarthmore: *Workshop on Databases for Galactic Structure*, 1993, edited by A.G. Davis Philip, B. Hauck, and A.R. Upgren (L. Davis Press, Schenectady)

31. *Digitized Optical Sky Surveys*, edited by H.T. MacGillivray and E.B. Thomson (Kluwer Academic, Dordrecht)

32. Monet, D.G. 1996, US Naval Observatory, Washington, DC: USNO (10 CD ROMs). Derived from scans of the indicated surveys with the USNO's Precision Measuring Machine (PMM) at Flagstaff. Reduced into the system of the HST GSC

33. Monet, D.G., the USNO-A1.0 catalogue reduced into the system of the ACT

Table 27.9. *Major Schmidt surveys (prepared in collaboration with Barry M. Lasker).[a]*

Survey	Name	Color	Emulsion	Filter	δ Centers	Corr.	Centers	N	Epoch	Archive	Dist.	Mag.	Ref.
1	POSS-I	Blue	103a-O	None	+90, −30	Old	6	936	1950–58	CIT	CT GrArt	21	[1]
1	POSS-I	Red	103a-E	Red	+90, −30	Old	6	936	1950–58	CIT	CT GrArt	20	[1]
2	Whiteoke	Red	103a-E	Red	−36, −42	Old	6	100	1963–64	CIT	CT GrArt	19	[2]
3	Luyten	Red	103a-E	Red	+90, −30	Old	6	936	1962–70	Minn	None	21	[3]
4	Hoessel	IR	IV-N	WR88a	$|b| \leq 10$	Old	6	80	1975–79	CIT	None	19	[4]
5	Quick V	Visual	IIa-D	WR12[b]	+90, +6	Old[c]	6	613[d]	1982	ST ScI	None	19	[5]
6	POSS-II	Green	IIIa-J	GG385	+90, 0	New	5	894	1987–	CIT	ESO	22	[6]
7	POSS-II	Red	IIIa-F	RG610	+90, 0	New	5	894	1987–	CIT	ESO	21	[6]
8	POSS-II	IR	IV-N	RG9	+90, 0	New	5	894	1987–	CIT	ESO	19	[6]
9	USNO QJ	Green	IIIa-J	GG385	+90, 0	New	5	894	1987–	CIT	None	13	[6,7]
10	ESO	B	IIa-O	GG385	−90, −20		5	606	1973–78	ESO	ESO	20	[8]
11	ESO	R	IIIa-F	RG630	−90, −20	Old[c]	5	606	1978–90	ESO	ESO	21	[9]
12	SERC-J	Green	IIIa-J	GG395	−90, −20	New[c]	5	606	1974–87	ROE	ESO	22	[10]
13	SERC-EJ	Green	IIIa-J	GG395	−15, 0	New[c]	5	288	1979–	ROE	ROE	22	[10]
14	SERC-ER	Red	IIIa-F	OG590	−15, 0	New[c]	5	288	1984–	ROE	ROE	21	[10]
15	AAO SES-R	Red	IIIa-F	OG590	−90, −20	New[c]	5	606	1990–	ROE	None	21	[11]
16	SERC-I/SR	IR	IV-N	RG715	$|b| \leq 10$	New[c]	5	163	1978–85	ROE	ROE	19	[12]
17	SERC-IR	IR	IV-N	RG715	$|b| > 10$	New[c]	5	731	1980–	ROE	Future	19	[11]

Notes

[a]*Explanation of the columns:* The columns in this table list sequentially an identification number for the survey that is used in Table 27.10, an acronym ("Name") for the survey, color passband, emulsion and filter used, inclusive declination centers for the survey, the Schmidt corrector used (note that this changes for some of the surveys), the distance in degrees between the declination zones, number of plates, epoch of the plates, location of the plate archive, source of distribution for copies, if any, the magnitude limit, and the reference number listed below where more information may be found. The field size is 6.5° for Palomar and U.K. Schmidt and 5.5° for ESO Schmidt.

[b]Most plates were taken with the Wratten 12 and a few with a GG495 filter.

[c]Corrector changed during observational program.

[d]There are 583 plates in the main survey and 30 in the +3 degree zone, to cover the tilt between the 1855 and 1950 grids.

References

1. Minkowski, R.L., & Abell, G.O. 1963, in *Basic Astronomical Data*, edited by K.Aa. Strand; Vol. III of *Stars and Stellar Systems*, edited by G. Kuiper and B. Middlehurst, p. 481. POSS = Palomar Observatory Sky Survey

2. Whiteoke, J. 1969, *Whiteoke Fields at Declination Zones −36 and −42 Degrees* (Calif. Instit. Tech., Pasadena)

3. Luyten: Luyten, W.J. 1963–87, Proper Motion Survey with the Forty-Eight Inch Schmidt Telescope (University of Minnesota, Minneapolis); Luyten: Luyten, W.J. 1987, My First 72 Years of Astronomical Research, Minneapolis, pp. 26–28

4. Hoessel, J.G. et al. 1979, *PASP*, **91**, 41. Palomar Infrared survey

5. Lasker, B.M. et al. 1990, *AJ*, **99**, 2019

6. Reid, I.N. et al. 1991, *PASP*, **103**, 661

7. Monet, D.G., & Westerhout, G. 1989, in *Digitized Optical Sky Surveys*, edited by C. Jaschek, in Bull. d'Info. du CDS, No. 37, (Obs. de Strasbourg, Strasbourg), p. 75. The USNO QJ is a two-minute exposure on unhypersensitized IIIa-J emulsions to be used for the determination of positions of bright stars.

8. West, R.M., & Schuster, H.-E. 1982, *AAS*, **49**, 577

9. West, R.M. 1984, in *IAU Colloq.* No. 78, p. 13.

10. Cannon, R.D. 1984, in *IAU Colloq.* No. 78, *Astronomy with Schmidt Telescopes*, edited by M. Capaccioli (Reidel, Dordrecht), p. 25

11. Lasker, B.M., & Cannon, R.D. 1989, in *Digitized Optical Sky Surveys*, edited by C. Jaschek, in Bull. d'Info. du CDS, No. 37 (Obs. de Strasbourg, Strasbourg), p. 13; and Morgan, D.H. et al. 1992, in *Digitized Optical Sky Surveys*, edited by H.T. MacGillivray and E.B. Thomson (Kluwer Academic, Dordrecht), p. 11.

12. Hartley, M., & Dawe, J.A. 1981, *Proc. Astron. Soc. Australia*, **4**, 251

Table 27.10. *Digitization of Schmidt plates (prepared in collaboration with Barry M. Lasker).*[a]

Site	Survey	Pixel	Status	Reference	Distribution	Net access
STScI	2, 5, 12, 13	1.7	Done	[1]	CD-ROM	Yes
	6–8, 15	1.0	Scanning	[1]	CD-ROM	Yes
ROE	12, 13, 16	1.1	Done	[2, 3]	CD-ROM	Yes
	1, 6–8, 14, 15	0.5	Scanning	[1]	CD-ROM	Yes
Cam	1, 2, 12, 13	0.5	Done	[4]	Via collaboration	No
	6–8, 15	0.5	Scanning	[4]	Via collaboration	No
Minn	1, 2, 3	0.5	Done	[5, 6]	ADS	Yes
	6–8	0.5	Scanning	[6]	ADS	Yes
USNO	1, 2, 11, 12	0.9	Done	[7]	CD-ROM	Yes
	6–9, 15	0.9	Scanning	[7]	CD-ROM	Yes

Note

[a] *Explanation of the columns*: This table summarizes the status of Schmidt plate digitization known to the authors as of April 1998. The columns list information for each digitization center, giving an acronym for the center ("Site"); the surveys being digitized (the number refers to the survey number given in Table 27.9) the digitization pixel size is in arcseconds; the status of the project (either completed or in the process of being scanned); references to the project, listed below; plans for distribution of the data; and whether or not the data will be available over electronic networks of some kind ("Net access"). For more details, see [8, 9, 10, 11].

References

1. STScI dig.: Lasker, B.M. 1992 in *Digitized Optical Sky Surveys*, edited by H.T. MacGillivray and E.B. Thomson, (Kluwer Academic, Dordrecht), p. 87; see also Lasker, B.M. 1993, in *Workshop on Databases for Galactic Structure*, edited by A.G. Davis, B. Hauck, and A.R. Upgren (L. Davis Press, Schenectady), p. 77; and Sturch, C.R. et al. in *Workshop on Databases for Galactic Structure*, edited by A.G. Davis, B. Hauck, and A.R. Upgren (L. Davis Press, Schenectady), p. 201.
2. ROE dig.: Yentis, D.J. et al. 1992, in *Digitized Optical Sky Surveys*, edited by H.T. MacGillivray and E.B. Thomson (Kluwer Academic, Dordrecht), p. 67
3. ROE dig.: Working Group on Wide-Field Imaging. 1992, *Newsletter* **2**, p. 10
4. Cam. dig.: Irwin, M.J., & McMahon, R. 1992, Working Group on Wide-Field Imaging, *Newsletter* **2**, p. 31. Plans to scan POSS II and AAO SES indicated in private communication.
5. Minn APS: Pennington, R.L. et al. 1993, *PASP*, **105**, 521; see also Pennington, R.L. et al. 1992, in *Digitized Optical Sky Surveys*, edited by H.T. MacGillivray and E.B. Thomson (Kluwer Academic, Dordrecht), p. 77. Minn.
6. Minn APS: Humphreys, R.M. 1993, in *Workshop on Databases for Galactic Structure*, edited by A.G. Davis, B. Hauck, and A.R. Upgren (L. Davis Press, Schenectady), p. 87, and Humphreys, R.M. et al. in *Workshop on Databases for Galactic Structure*, edited by A.G. Davis, B. Hauck, and A.R. Upgren (L. Davis Press, Schenectady), p. 197; see also Odewahn, S.C. et al. 1992, *AJ*, **103**, 318; and Annual Reports for the University of Minnesota 1993, *BAAS*, **25**, 320
7. USN PMM: Monet, D.G., & Westerhout, G. 1989, in *Digitized Optical Sky Surveys*, edited by C. Jaschek, in Bull. d'Info. du CDS, No. 37 (Obs. de Strasbourg, Strasbourg), p. 75. The USNO QJ is a two-minute exposure on unhypersensitized IIIa-J emulsions to be used for the determination of positions of bright stars.
8. Morgan, D.H., & Tritton, S.B. 1988, in *Mapping the Sky*, IAU Colloq. 133, edited by S. Debarbat, J.A. Eddy, H.K. Eichhorn, and A.R. Upgren (Kluwer Academic, Dordrecht), p. 349
9. *Digitized Optical Sky Surveys Newsletter* and *IAU Working Group on Wide-Field Imaging Newsletters*, both edited by H.T. MacGillivray (ROE)
10. Lasker, B.M. 1995, *PASP*, **107**, 763; and 1995, in *Future Utilization of Schmidt Telescopes*, edited by R.D. Cannon et al. (ASP, San Francisco)
11. *Workshop on Databases for Galactic Structure*, 1993, edited by A.G. Davis, B. Hauck, and A.R. Upgren (L. Davis Press, Schenectady), pp. 215 and 250

Table 27.11. *Modern relative proper motion surveys.*

Reference	Catalogue	No. stars	Coverage	T_{pub}	m	μ limit	Description
[1]	BPM	94 263	South +	1963	15	0.10	Bruce Proper Motion Survey, 1928–63
[2]	Lowell	8 991	North	1971	16	0.26	Lowell Proper Motion Survey, 1958–70
[3]	Lowell	2 758	South	1978	16	0.20	Lowell Southern section, 1959–78

Table 27.11. *(Continued.)*

Reference	Catalogue	No. stars	Coverage	T_{pub}	m	μ limit	Description
[4]	LP	450 000	North −	1979	21	0.08	Luyten Palomar 48-Inch Schmidt Survey, 1963–77
[5]	Calan	830	South	1998	21	0.15	Selected regions, Maksutov at Cerro El Roble, Chile

References

1. Luyten, W.J. 1963, University of Minnesota, Minneapolis. Mostly Southern hemisphere, but includes some in the north
2. Giclas, H.L., Burnham Jr., R., & Thomas, N.G. 1971, Lowell Proper Motion Survey, Northern Hemisphere Catalogue, Lowell Obs., Flagstaff
3. Giclas, H.L., Burnham Jr., R., & Thomas, N.G. 1978, *Lowell Obs. Bull.* No. 164, Vol. VIII, p. 89
4. Luyten, W.J. 1963–87, Proper Motion Survey with the Forty-Eight inch Schmidt Telescope, University of Minnesota, Minneapolis; and 1987, My First 72 Years of Astronomical Research, Minneapolis, pp. 26–28. Proper Motion Survey. Mostly Northern Hemisphere, but includes some in the south
5. Wroblewski, H., & Torres, C. 1997, *A&AS*, **122**, 447; 1998, *A&AS*, **128**, 457

Table 27.12. *Relative proper motion summary catalogues.*

Reference	Catalogue	No. stars	Coverage	T_{pub}	m	μ limit	Description
[1]	NLTT	58 855	All sky	1979	All	0.18	Stars with $\mu \geq 0.18$ arcsec/yr
[2]	LHS Cat.	4 447	All sky	1979	All	0.50	Stars with $\mu \geq 0.50$ arcsec/yr, 2nd ed.
[3]	LHS Atlas	3 040	All sky	1979	All	0.50	Stars with $\mu \geq 0.50$ arcsec/yr, Finding Charts for LHS Catalogue

References

1. Luyten, W.J. 1979, 80, *New Luyten Catalogue of Stars with Proper Motions Larger than Two Tenths of an Arcsecond* (University of Minnesota, Minneapolis)
2. Luyten, W.J. 1979, *Luyten Half-Second Catalogue* (University of Minnesota, Minneapolis)
3. Luyten, W.J., & Albers, H. 1979, *Luyten Half-Second Atlas* (University of Minnesota, Minneapolis). Finding charts for stars without published charts

Table 27.13. *Stellar compilation catalogues.*

Reference	Catalogue	No. stars	Coverage	T_{pub}	m	Limits	Description
[1]	BSC V[a]	9 110	All sky	1998	6.5	$V \leq 6.5$	The Bright Star Catalogue, 5th ed.; Stars with $V \leq 6.5$
[2]	BSC Supp.	2 603	All sky	1983	7.1	$V \leq 7.1$	The BSC Supplement; Stars with $V \leq 7.1$ photoelectric
[3]	CNS 3	3 803	All sky	1996	All	$\pi \geq 0.040$	Catalogue of Nearby Stars, 3rd prelim. ed.; stars with $\pi \geq 0.040$
[4]	CNSG	2 542	All sky	1997	All	$\pi \geq 0.040$	Catalogue of Nearby Stars, Ground-based parallaxes
[5]	CNSH	2 678	All sky	1997	9	$\pi \geq 0.040$	Catalogue of Nearby Stars, Hipparcos-based parallaxes
[6]	YPC	8 112	All sky	1995	All	$\cdots$	Gen. Cat. Trig. Parallaxes; all stars with measured π
[7]	WDS	78 100	All sky	1984/96	All	$\cdots$	Washington Double Star Catalogue; all measures
[8]	CHARA	7 598	All sky	1998	10	$\cdots$	Interferometric measures of double stars; 3rd catalogue
[9]	Orbits	847	All Sky	1984	All	$\cdots$	Visual Double Star Orbit Catalogue

Note

^{a}An electronic version of the Bright Star Catalogue, often referred to as the BSC V, has circulated, but the final BSC V will not be released until late 1998 or 1999.

References

1. Hoffleit, E.D., & Warren Jr., W.H. 1998, *The Bright Star Catalogue*, 5th rev. ed. (NASA Astronomical Data Center, Greenbelt); see also Hoffleit, E.D. (with the collaboration of Jaschek, C.) 1982, *The Bright Star Catalogue*, 4th rev. ed. (Yale Univ. Obs., New Haven), and [10]

2. Hoffleit, E.D., Saladyga, M., & Wlasuk, P. 1983, *A Supplement to the Bright Star Catalogue* (Yale Univ. Obs., New Haven), and [10]

3. Jahreiß, H., & Gliese, W. 1996, *The Catalogue of Nearby Stars*, 3rd ed. (Veröff. Astron. Rechen-Instituts, Heidelberg); see also [10] for the 1991 Preliminary Edition

4. Jahreiß, H., & Wielen, R. 1997, *Hipparcos-Venice '97*, ESA SP-402 (ESTEC, Noordwijk), p. 675. This version of the CNS is derived from ground-based parallaxes and is probably more complete.

5. Jahreiß, H., & Wielen, R. 1997, *Hipparcos-Venice '97*, ESA SP-402 (ESTEC, Noordwijk), p. 675. This version of the CNS is derived from Hipparcos-based parallaxes and is biased by the bright limiting magnitude of Hipparcos, but it is more accurate in identifying which stars are included within the 25 pc limit.

6. van Altena, W.F., Lee, J.T., & Hoffleit, E.D. 1995, *General Catalogue of Trigonometric Parallaxes*, 4th ed. (Yale Univ. Obs., New Haven); see also [10] for the 1991 Preliminary Edition, and van Altena et al. for a discussion of the system of the YPC in [11], p. 65 and [12], p. 50.

7. Worley, C.E., & Douglass, G.G. 1984, *Washington Catalog of Visual Double Stars* (U.S. Naval Obs., Washington, DC); see Web site for up-to-date version.

8. McAlister, H.A., Hartkopf, W.I., & Mason, B. 1997, *Third Catalog of Interferometric Measurements of Binary Stars*; see Web site for up-to-date version.

9. Worley, C.E., & Heintz, W.D. 1984, *Fourth Catalog of Orbits of Visual Binary Stars*, Publ. U.S. Naval Obs. (2) 24, Part VII; see Web site.

10. Brotzman, L.E. et al. 1991, *Astronomical Data Center CD-ROM Selected Astronomical Catalogs*, Vol. I, (NASA Goddard Space Flight Center, Greenbelt). See also subsequent volumes in this series.

11. *Swarthmore: Workshop on Databases for Galactic Structure*, edited by A.G. Davis, B. Hauck, and A.R. Upgren (L. Davis Press, Schenectady), p. 1993

12. *Cambridge: Galactic and Solar System Optical Astrometry* (held at Robinson College, Cambridge University in June 1993); 1995, edited by L.V. Morrison and G. Gilmore (Kluwer Academic, Dordrecht)

27.8 OPTICAL AND INFRARED INTERFEROMETRY
by S.T. Ridgway

Optical interferometry is qualitatively similar to radio interferometry, though differing in many quantitative details related to atmospheric turbulence and detection. Optical interferometry has now reached a state of development similar to radio interferometry of about 30 years ago.

Independent telescopes functioning as an interferometer, with a telescope separation of B, operating at wavelength λ, will enable angular resolution up to spatial frequency B/λ and the capability of resolving structures with angular extent of order λ/B. This will allow model-dependent size measures of sources smaller than λ/B by a factor approximately equal to the signal-to-noise ratio.

Interferometers measure the mutual coherence of light from the source by detecting the interference fringes formed in the combined light beams from two or more apertures. Pairs of telescopes can be employed, as in the classical Michelson experiment, to determine the modulus of the fringe visibility. Three or more telescopes can additionally measure (in part) the complex visibility and subject to certain approximations, images can be obtained by numerical manipulation of the mutual coherence data, using most of the techniques developed for image restoration at radio wavelengths [17].

The implementation of interferometry is constrained by the requirement to detect the fringes in a time short enough to "freeze" the fringes in the presence of optical path drifts. In practice, the sensitivity of an interferometer will be limited by the number of source photons in the coherence volume, $\Phi^2 cT\delta_\lambda$, depending on the maximum collecting aperture Φ and time T over which photons in bandwidth δ_λ may be combined coherently for a measurement [18].

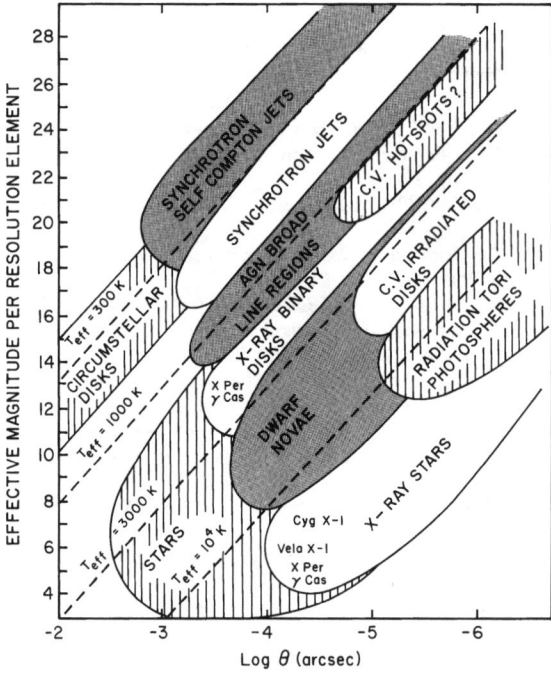

Figure 27.1. The effective magnitude per resolution element versus $\log \theta$ diagram shows how the surface brightness and size are related through the radiometric temperature for various sources [22]

The atmosphere limits Φ to approximately r_0, the Fried length, which is

$$r_0 \approx 0.1 \left(\frac{\lambda}{0.5 \, \mu\text{m}} \right)^{1.2} \text{m}. \tag{1}$$

The atmosphere limits T to times of order r_0/v, where v is a weighted mean wind speed in the turbulent atmosphere. Values of a few milliseconds are common in the visible range. Furthermore, δ_λ is constrained by the residual optical path differences to the source and the source and observing geometry.

There does not appear to be any fundamental limit to the use of optical interferometry from the ground, though the practical difficulties are severe. Interferometry with r_0 size apertures can be carried out on bright sources ($M \approx 10$). In order to reach faint limits several techniques are available. Φ can be increased to the aperture size of the available telescope by use of adaptive optics (phasing a telescope). T can be increased by actively stabilizing the optical path differences (cophasing telescopes). Through coherencing [19] and/or absolute metrology [20] it should be possible to maintain a condition of approximate cophasing for any point on the sky. With laser reference beacons, it may be possible to achieve adaptive phasing and cophasing with faint reference sources. The practical performance limits for optical interferometry from the ground are still a matter of some speculation, as these are intimately tied to the effective implementation of adaptive telescopes with large aperture telescopes, and this effort is itself in an early stage of development. However, extension of interferometry to imaging of complex sources and sources much fainter than $M = 18$ may be exceedingly difficult.

In space, the problems of phasing and cophasing should be ameliorated. Instrument changes due to thermal or tidal effects should be relatively benign and can be monitored with internal metrology. The wavefront quality and optical path difference to the source can be monitored with bright reference

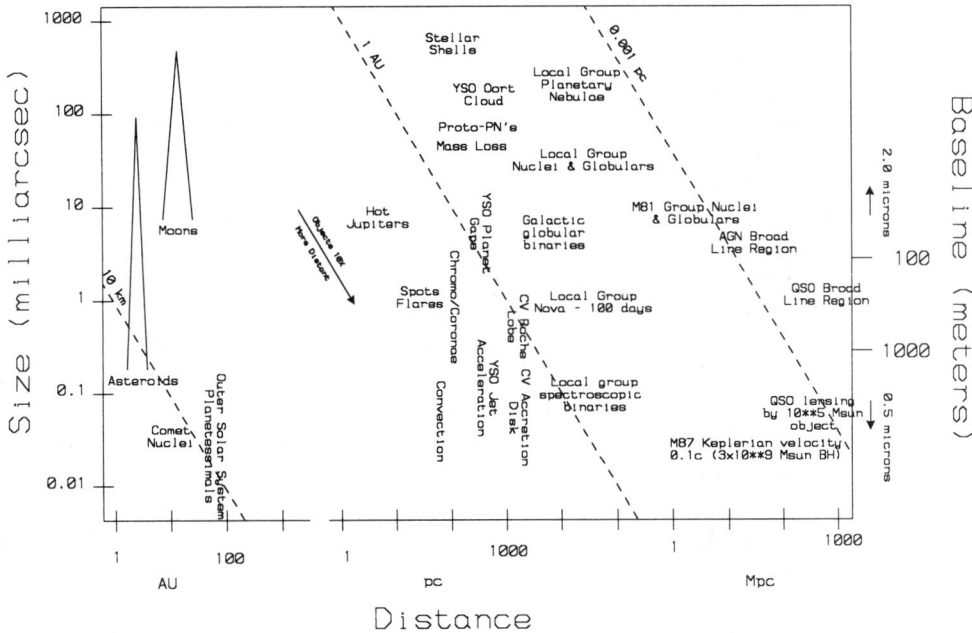

Figure 27.2. Overview of science topics for optical interferometry showing the typical apparent sizes of various types of sources. The angular resolution at 1 μm wavelength, for several telescope separations, is shown on the right-hand side of the figure (for a different wavelength, shift the baseline scale as shown).

stars at large angle from the source of interest. Potentially, T can be greatly increased. However, this gain must be traded against the loss in collecting aperture compared to possible ground-based interferometers [21].

The amount of image information collected in an interferometer is related to the number of independent baselines in the array, which can be as large as $N(N-1)/2$ for N telescopes, providing most radio interferometers a quadratic improvement with number of telescopes. However, in the absence of noiseless optical amplifiers, optical interferometry imposes a signal-to-noise penalty for the combination of multiple beams. For more than about $N = 5$, the gain in observing efficiency will be approximately linear with N.

The most important parameters in an interferometric measurement are the source brightness per spatial resolution element, and the required telescope separation. Figures 27.1 and 27.2 show a schematic representation of this information.

Ground-based interferometry has been used to achieve wide-angle astrometry with errors of a few milliarcseconds, and may be extended to achieve errors as small as tens of microarcseconds for very-narrow-angle astrometry [24]. Interferometry from space is expected to achieve absolute astrometric precision of a few microarcseconds [25] for wide- and narrow-angle measurements. This advantage of space over ground appears to be fundamental. See also [26] and [27].

27.9   THE WORLD'S LARGEST OPTICAL TELESCOPES
by Roger W. Sinnott and Kari Parker

Table 27.14 lists the largest telescopes designed to observe the sky at visible and infrared wavelengths. The instruments are grouped into three main categories and arranged in descending order of aperture. Included are the biggest reflecting telescopes, Schmidt cameras, and refracting telescopes currently in use or in an advanced stage of construction or planning. *Important note*: Many additional reflectors are in use with apertures between 1.3 and 2.3 m; they have been omitted from the table, even though they exceed the existing Schmidt and refracting telescopes in light-gathering power.

Large reflectors stand at the forefront of astronomical research today. They collect starlight using either a single concave monolithic mirror, a segmented-mirror mosaic, or several separate mirrors arranged to send their beams to a combined focus. Generally their angular coverage is small (only a few arcminutes), but such a field is well suited to the needs of spectroscopy and the small physical size of modern electronic detectors, such as the charge-coupled devices (CCDs) that have largely superseded photographic plates on this class of instrument.

Schmidt telescopes are photographic cameras providing exquisite star images on a curved focal surface centered within the tube. Because light first enters through a corrector lens, the diameter of this lens determines the overall light grasp. The primary mirror is always made substantially larger than the corrector so the instrument can photograph an unvignetted field spanning 5° or more.

Refractors have a two-element objective lens instead of a primary mirror to collect light. The difficulty of manufacturing glass with the necessary purity and homogeneity in large sizes has set a natural limit to refractor size—a limit that, in fact, was attained around the turn of the twentieth century. Yet many of these venerable instruments remain in service; they have been of particular value in high-accuracy astrometry.

The first column of Table 27.14 gives the official telescope name, the observatory to which it belongs, and the geographical site, along with the instrument's common name (if any). Sometimes the common name quotes a mirror size that differs slightly from the clear aperture.

The second column lists the clear aperture in meters, the optical design, and the main person(s) or firm(s) involved in making the optics, along with the available focal ratios (that is, effective focal length divided by aperture). In the case of reflectors, the smallest focal ratio usually refers to the primary mirror itself; some of the instruments have wide-field correctors for prime-focus work, and these correctors may alter the focal ratio slightly. Additional focal ratios refer to Cassegrain, Nasmyth, or coudé foci, where IR means a focus used for infrared observations.

Next comes the style of the telescope's mounting and the firm(s) associated with its manufacture. Here, as in the second column, no attempt has been made to recognize the full hierarchy of contractors by which these complex instruments have come into being. Among the abbreviations used, LOMO refers to Leningrad Optical Equipment Works, Russia; SNACRP is the Société Nouvelle des Ateliers et Chantiers de La Rochelle-la-Pallice, France; REOSC is Recherches et Etudes d'Optique et des Sciences Connexes, France; KPNO is Kitt Peak National Observatory, Arizona, US; NOAO is the National Optical Astronomy Observatories, Arizona, US.

The listed latitude and longitude refer to the telescope, as opposed to the observatory headquarters. These values have been rounded off to the nearest arcminute in all cases, and elevations above sea level are expressed in meters. In a few cases accurate values were not available; a colon (:) signifies that the value is uncertain.

The year refers either to an official commissioning ceremony, "first light" through the instrument, or the start of regular observations. For telescopes still under construction at the time this table was compiled (mid-1998), the projected year of completion is given in parentheses. For such telescopes, not only the year but also the instrument's specifications may change as the project unfolds.

The final column gives additional comments, such as the glass type used for the primary mirror and other notes about the instrument.

We are grateful to the staffs at dozens of observatories for returning the questionnaires we sent out in mid-1992. These responses were our primary source of information. Also very helpful was an earlier compilation [28]. Whenever possible, telescope coordinates have been taken from the extensive listings in the *Astronomical Almanac* [3] for 1981–84 (US Naval Observatory and Royal Greenwich Observatory). Telephone calls, electronic mail, and published articles have filled in additional details, especially for instruments still under construction.

Table 27.14. *The World's Largest Optical Telescopes.*

Telescope / Observatory / Place / (Common Name)	Clear aperture / Optical design / Maker / Focal ratios	Mounting / Manufacturer	Latitude / Longitude / Elevation	Year	Notes
Very Large Telescope / European Southern Observatory / Cerro Paranal, Chile / (VLT)	16-m equiv. aperture / Ritchey–Chrétien / REOSC / f/13.5, 15 (each mirror)	Four separate altazimuth mounts and domes	24° 51'.S / 70° 27'.W / 2640 m	(2001)	Four separate 8.2-m reflectors to be used together; first unit became operational in 1998; mirrors are Schott Zerodur
Large Binocular Telescope / Columbus Project / Mount Graham, Arizona, US / (LBT)	11.8-m equiv. aperture / Cassegrain / R. Angel, B. Martin / f/1.14, 5.4, 15 (each)	Altazimuth "two-shooter" mount, with telescopes 14.4 m apart (center to center)	32° 42'.N / 109° 51'.W / 3170 m	(2004)	Twin 8.4-m reflectors; a project of Univ. of Arizona, Arcetri Astrophysical Obs., and Research Corp. (Tucson)
Gran Telescopio Canarias / Obs. del Roque de los Muchachos / La Palma, Canary Islands / (GTC)	10-m segmented mirror / Ritchey–Chrétien / ... / f/1.75. 15, 25	Altazimuth	28° 45' N / 17° 54' W / 2400: m	(2003)	Design similar to that of the Keck instruments; funded by Spain with international partners
Keck I Telescope / W.M. Keck Observatory / Mauna Kea, Hawaii, US / (Keck I)	9.82-m segmented mirror / Ritchey–Chrétien / Itek, Tinsley / f/1.75, 15, 25 IR	Altazimuth TIW Systems	19° 49' N / 155° 28' W / 4150 m	1991	Uses 36 hexagonal mirror segments (Zerodur); operated by Univ. of California and Calif. Institute of Technology
Keck II Telescope / W.M. Keck Observatory / Mauna Kea, Hawaii, US / (Keck II)	9.82-m segmented mirror / Ritchey–Chrétien / Itek, Tinsley / f/1.75	Altazimuth TIW Systems	19° 49' N / 155° 28' W / 4150 m	1996	Located 85 m from Keck I; plans include using them as an optical interferometer mirror
Hobby–Eberly Telescope / University of Texas / Mount Fowlkes, Texas, US / (Spectroscopic Survey Tel.)	9.2-m equiv. aperture / Spherical figure / Univ. of Texas, Penn. State / (f/1.4) f/4.7	Fixed-altitude mount rotates in azimuth	30° 41' N / 104° 01' W / 2002 m	1997	A project of five universities in US and Germany; 91 spherical mirrors with combined focus for spectroscopy
Southern African Large Telescope / South African Astronomical Obs. / Sutherland, South Africa / (SALT)	9.1-m equiv. aperture / Spherical figure / ... / (f/1.4) f/4.7	Fixed-altitude mount rotates in azimuth	32° 23' S / 20° 49'E / 1798 m	(2004)	A twin of the Hobby–Eberly Telescope
Subaru Telescope / National Astronomy Obs. (Japan) / Mauna Kea, Hawaii, US / (Subaru)	8.3-m mirror / Ritchey–Chrétien / ... / f/1.8, 12.5, 35	Altazimuth Mitsubishi	19° 49' N / 155° 28' W / 4215 m	(1999)	Thin primary (Corning ULE) under active control

Table 27.14. (*Continued.*)

Telescope / Observatory / Place / (Common Name)	Clear aperture / Optical design / Maker / Focal ratios	Mounting / Manufacturer	Latitude / Longitude / Elevation	Year	Notes
Gemini Telescope (north) / Joint Astronomy Center / Mauna Kea, Hawaii, US / ...	8.1-m mirror / Ritchey–Chrétien / ... / f/1.8, 16 IR, 19.6	Altazimuth	19° 49' N / 155° 28' W / 4100: m	(1999)	Corning ULE meniscus primary mirror; this northern instrument is to be optimized for IR work
Gemini Telescope (south) / Cerro Tololo Inter-American Obs. / Cerro Pachon, Chile / ...	8.1-m mirror / Ritchey–Chrétien / ... / f/1.8, 6	Altazimuth	30° 21': S / 70° 49': W / 2725 m	(2000)	Partners in the Gemini project are the US, UK, Canada, Chile, Brazil, and Argentina
MMT Observatory Telescope / MMT Observatory / Mount Hopkins, Arizona, US / ...	6.5-m mirror / ... / R. Angel, B. Martin / f/1.25, 5.4, 9, 15	Altazimuth / de Bartolomeis	31° 41' N / 110° 53' W / 2608 m	(1999)	Spin-cast borosilicate honey-comb primary mirror (Ohara E6); joint facility of Smithsonian Institution and Univ. of Arizona
Magellan Project / Las Campanas Observatory / Las Campanas, Chile / (Magellan I)	6.5-m mirror / Cass. and Gregorian / R. Angel, B. Martin / f/1.25, 11, 15	Altazimuth / L & F Industries	29° 00' S / 70° 42' W / 2300 m	(1999)	Will be operated by Carnegie Institution of Washington; a twin, Magellan II is to be built in 2001
Bol'shoi Teleskop Azimutal'nyi / Special Astrophysical Obs. / Mount Pastukhov, Russia / (6 m)	6.00-m mirror / Ritchey–Chrétien / LOMO / f/4, 30	Altazimuth / LOMO	43° 39' N / 41° 26' E / 2100 m	1975	Primary mirror replaced in about 1984, and a third made of Sitall glass was figured in 1992
George Ellery Hale Telescope / Palomar Observatory / Palomar Mountain, Calif., US / (200 inch)	5.08-m mirror / Cassegrain / J.A. Anderson / f/3.3, 16, 30	Horseshoe yoke mount / Westinghouse	33° 21' N / 116° 52' W / 1706 m	1948	California Institute of Technology; ribbed primary of Corning Pyrex
William Herschel Telescope / Obs. del Roque de los Muchachos / La Palma, Canary Islands / ...	4.2-m mirror / — / Grubb–Parsons / f/2.5, 11	Altazimuth / Grubb–Parsons	28° 46' N / 17° 53' W / 2332 m	1987	Operated by the Royal Greenwich Observatory; primary mirror is Owens-Illinois Cer-Vit mirror
Victor M. Blanco Telescope / Cerro Tololo Inter-American Obs. / Cerro Tololo, Chile / (CTIO 4 meter)	4.001-m mirror / Ritchey–Chrétien / KPNO Optical Shop / f/2.8, 8.0	Horseshoe equatorial / Western Gear Corp.	30° 10' S / 70° 49' W / 2215 m	1976	Cer-Vit primary mirror
Anglo-Australian Telescope / Anglo-Australian Observatory / Siding Spring Mtn., Australia / (AAT)	3.893-m mirror / ... / Grubb–Parsons / f/3.3, 8, 15, 36	Horseshoe equatorial / Mitsubishi	31° 17' S / 149° 04' E / 1149 m	1975	Cer-Vit primary mirror

Table 27.14. (*Continued.*)

Telescope Observatory Place (Common Name)	Clear aperture Optical design Maker Focal ratios	Mounting Manufacturer	Latitude Longitude Elevation	Year	Notes
Nicholas U. Mayall Reflector Kitt Peak National Obs. Kitt Peak, Arizona, US (Kitt Peak 4 m)	3.81-m mirror Ritchey–Chrétien KPNO Optical Shop f/2.7, 8, 15.7 IR, 190	Horseshoe equatorial Western Gear Corp.	31° 58′ N 111° 36′ W 2120 m	1973	Fused-quartz primary mirror
United Kingdom Infrared Telescope Joint Astronomy Centre Mauna Kea, Hawaii, US (UKIRT)	3.802-m mirror Cassegrain Grubb–Parsons f/2.5, 36 IR	English-yoke equatorial Hadfields Ltd. (Sheffield)	19° 50′ N 155° 28′ W 4194 m	1978	Cer-Vit primary mirror; used for infrared work only
Canada–France–Hawaii Telescope Canada–France–Hawaii Telescope Corp. Mauna Kea, Hawaii, US (CFHT)	3.58-m mirror ... Dominion Astrophys. Obs. f/3.8, 8, 20, 35	Horseshoe-yoke equatorial SNACRP	19° 49′ N 155° 28′ W 4200 m	1979	Cer-Vit primary mirror
Telescopio Nazionale Galileo Obs. del Roque de los Muchachos La Palma, Canary Islands (Galileo)	3.58-m mirror Ritchey–Chrétien Zeiss f/2.5, 6, 11	Altazimuth	28° 45′ N 17° 54′ W 2370 m	(1998)	Zerodur mirror
ESO 3.6-meter Telescope European Southern Observatory La Silla, Chile (ESO 3.6 m)	3.57-m mirror ... REOSC f/3.0, 8.1, 32	Horseshoe-fork equatorial Creusot–Loire	29° 16′ S 70° 44′ W 2387 m	1977	Fused-silica primary mirror
New Technology Telescope European Southern Observatory La Silla, Chile (NTT)	3.50-m mirror Ritchey–Chrétien Zeiss f/2.2, 11	Altazimuth	29° 16′ S 70° 44′ W 2353 m	1989	Thin Zerodur primary mirror; figure controlled by 78 active supports
3.5-meter Telescope Calar Alto Observatory Calar Alto, Spain ...	3.50-m mirror Ritchey–Chrétien Zeiss (Ober.) f/3.5, 3.9, 10, 35	Horseshoe yoke Voith, Heidenheim; Zeiss (Ober.)	37° 13′ N 2° 32′ W 2168 m	1984	Operated by German–Spanish Astronomical Center; Zerodur primary mirror
Astrophysical Research Consortium Telescope Apache Point, New Mexico, US (ARC 3.5 m)	3.5-m mirror ... R. Angel, B. Martin f/1.75	Altazimuth	32° 47′: N 105° 49′: W 2800 m	1994	Spin-cast borosilicate honey-comb primary mirror (Ohara E6)

Table 27.14. (Continued.)

Telescope Observatory Place (Common Name)	Clear aperture Optical design Maker Focal ratios	Mounting Manufacturer	Latitude Longitude Elevation	Year	Notes
Wisconsin–Indiana–Yale–NOAO Tel. WIYN Observatory Kitt Peak, Arizona, US (WIYN 3.5 m)	3.5-m mirror ... Charles Harmer/NOAO (f/1.75) f/6.3	Altazimuth L & F Industries	31° 57′ N 111° 36′ W 2089 m	1994	Spin-cast borosilicate primary mirror by Steward Observatory Mirror Lab (R. Angel)
C. Donald Shane Telescope Lick Observatory Mount Hamilton, Calif., US (120 inch)	3.05-m mirror Cassegrain Don O. Hendrix f/5, 17, 36	Equatorial fork Judson Pacific–Murphy Corp.	37° 21′ N 121° 38′ W 1290 m	1959	Pyrex primary mirror
NASA Infrared Telescope Facility Mauna Kea Observatory Mauna Kea, Hawaii, US (NASA IRTF)	3.00-m mirror Cassegrain KPNO Optical Shop f/2.5, 35, 120 IR	English yoke equatorial de Bartolomeis	19° 50′ N 155° 28′ W 4208 m	1979	Infrared telescope; Cer-Vit primary mirror
Harlan J. Smith Telescope McDonald Observatory Mount Locke, Texas, US (107 inch)	2.72-m mirror Ritchey–Chrétien Davidson Optronics f/3.9, 8.8, 18	Cross-axis equatorial Westinghouse	30° 40′ N 104° 01′ W 2075 m	1969	Fused-silica primary mirror
UBC–Laval Telescope Univ. of B.C. and Laval Univ. Vancouver, BC, Canada (LMT)	2.7-m mirror (liquid) Paraboloid, field corr. P. Hickson f/1.887	Fixed vertical mount Univ. of British Col.	49° 07′ N 122° 35′: W 50 m	1992	Mercury primary mirror (rotating); views a 21′ field at local zenith (centered on declination +49°1)
Shajn 2.6-m Reflector Crimean Astrophysical Obs. Nauchny, Ukraine (Crimean 102 inch)	2.64-m mirror f/3.8, 15.7, 16.4, 40	Fork equatorial	44° 44′ N 34° 00′ E ...	1960	...
Byurakan 2.6-meter Reflector Byurakan Observatory Mount Aragatz, Armenia (Byurakan 102 inch)	2.64-m mirror f/3.6, 16, 40	Fork equatorial LOMO	40° 20′ N 44° 18′ E 1500 m	1976	...
Nordic Optical Telescope Obs. del Roque de los Muchachos La Palma, Canary Islands (NOT)	2.56-m mirror Cassegrain Optics Labs (Tartu) (f/2.0), 11.0	Altazimuth mount in rotating building	28° 45′ N 17° 53′ W 2382 m	1989	Zerodur primary mirror

Table 27.14. *(Continued.)*

Telescope Observatory Place (Common Name)	Clear aperture Optical design Maker Focal ratios	Mounting Manufacturer	Latitude Longitude Elevation	Year	Notes
Irénée du Pont Telescope Las Campanas Observatory Las Campanas, Chile (du Pont 100 inch)	2.54-m mirror Ritchey–Chrétien Donald A. Loomis f/3.0	Fork equatorial Bruce H. Rule	29° 00′ S 70° 42′ W 2282 m	1976	Owned by Carnegie Inst. of Washington; fused-silica primary mirror
Hooker Telescope Mount Wilson Observatory Mount Wilson, Calif., US (100 inch)	2.5-m mirror Cassegrain G.W. Ritchey f/5, 16, 30	English mount F.G. Pease and Fore River Shipyards	34° 13′ N 118° 03′ W 1742 m	1917	Plate-glass primary mirror from Saint-Gobain (Paris); telescope out of service 1985–92
Isaac Newton Telescope Obs. del Roque de los Muchachos La Palma, Canary Islands (Isaac Newton 98 inch)	2.5-m mirror … Grubb–Parsons f/3, 15	Polar-disk equatorial Grubb–Parsons	28° 46′ N 17° 53′ W 2336 m	1984	Operated by Royal Greenwich Observatory; originally set up in England in 1967
Sloan Digital Sky Survey Tel. Astrophys. Res. Consortium Obs. Apache Point, New Mexico, US (Sloan 2.5 m)	2.5-m mirror Ritchey–Chrétien	Altazimuth L & F Industries and University of Washington	32° 47′ N 105° 49′ W 2800 m	(1998)	To take CCD imagery of a quarter of sky in 5 colors (u, g, r, i, z) and measure redshifts of 1 million galaxies
Hubble Space Telescope Space Telescope Science Inst. Baltimore, Maryland, US (HST)	2.4-m mirror Ritchey–Chrétien Perkin Elmer f/12.9, 30, 48, 96	3-axis-stabilized spacecraft Lockheed	(Earth orbit)	1990	Primary is of Corning ULE glass
Hiltner Telescope Michigan–Dartmouth–MIT Obs. Kitt Peak, Arizona, US (Hiltner 2.3 m)	2.34-m mirror Ritchey–Chrétien Contraves (USA) f/2.07, 13.5	Equatorial fork, friction-disk drives DFM Engineering, L & F Industries	31° 57′ N 111° 37′ W 1938 m	1986	Mirrors repolished 1991; Cer-Vit primary mirror
2.3-meter Telescope Vainu Bappu Observatory Kavalur, Tamil Nadu, India (Vainu Bappu 2.3 m)	2.33-m mirror … Indian Inst. Astrophys. f/3.25, 13, 43	Horseshoe equatorial Walchandnagar Industries	12° 35′ N 78° 50′ E 725 m	1985	Zerodur primary mirror
Mount Stromlo 2.3-meter Mt. Stromlo and Siding Spring Obs. Siding Spring Mtn., Australia …	2.3-m mirror … Norman Cole f/2.09, 18	Altazimuth Australian National Univ., Newcastle Dockyard	31° 16′ S 149° 03′ E 1149 m	1984	Cer-Vit primary mirror

Table 27.14. (*Continued.*)

SCHMIDT TELESCOPES

Telescope / Observatory / Place / (Common Name)	Clear aperture / Optical design / Maker / Focal ratios	Mounting / Manufacturer	Latitude / Longitude / Elevation	Year	Notes
2-meter Telescope / Karl Schwarzschild Observatorium / Tautenberg, Germany / (Tautenberg Schmidt)	1.34-m corrector / Schmidt / Zeiss (Jena) / f/3.00	Equatorial fork Zeiss (Jena)	50° 59' N / 11° 43' E / 331 m	1960	2-m primary mirror of Schott ZK-7 glass; can also be used in Cassegrain and coudé modes
Oschin 48-inch Telescope / Palomar Observatory / Palomar Mountain, Calif., US / (Oschin Schmidt)	1.24-m corrector / Schmidt / Don O. Hendrix / f/2.47	Equatorial fork made at California Institute of Technology	33° 21' N / 116° 51' W / 1706 m	1948	1.83-m primary mirror; a new achromatic corrector plate (Grubb–Parsons) was installed in about 1984
United Kingdom Schmidt Tel. Unit / Royal Observatory, Edinburgh / Siding Spring Mtn., Australia / (U.K. Schmidt)	1.24-m corrector / Schmidt / Grubb–Parsons / f/2.5	... Grubb–Parsons	31° 16' S / 149° 04' E / 1145 m	1973	Cer-Vit 1.83-m primary mirror
Kiso Schmidt Telescope / Kiso Observatory / Kiso, Japan / ...	1.05-m corrector / Schmidt / Nikon / f/3.1	Equatorial fork Nikon	35° 48' N / 137° 38' E / 1130 m	1975	Operated by Univ. of Tokyo; Cer-Vit 1.5-m primary mirror; alternate secondary mirror offers an f/22.6 Cass. focus
3TA-10 Schmidt Telescope / Byurakan Astrophysical Observatory / Mount Aragatz, Armenia / (Byurakan Schmidt)	1.00-m corrector / Schmidt / LOMO / f/2.13	... LOMO	40° 20' N / 44° 30' E / 1450 m	1961	1.5-m primary mirror; this telescope has three 1-m objective prisms
Kvistaberg Schmidt Telescope / Uppsala University Observatory / Kvistaberg, Sweden / (Uppsala Schmidt)	1.00-m corrector / Schmidt / Uppsala Univ. Obs. / f/3.00	... Various Swedish factories	59° 30' N / 17° 36' E / 33 m	1963	1.35-m primary mirror
ESO 1-meter Schmidt Telescope / European Southern Observatory / La Silla, Chile / (ESO Schmidt)	1.00-m corrector / Schmidt / Zeiss (Ober.) / f/3.06	... Heidenreich & Harbeck	29° 15' S / 70° 44' W / 2318 m	1972	1.6-m primary mirror of Schott Duran 50
Venezuela 1-meter Schmidt / Centro "F.J. Duarte" / Llano del Hato, Merida, Venezuela / ...	1.00-m corrector / Concentric Schmidt / Askania / f/3.0	Bent-yoke equatorial Askania	8° 47' N / 70° 52' W / 3610 m	1978	1.52-m primary mirror; has a 1-m objective prism

Table 27.14. (Continued.)

Telescope / Observatory / Place / (Common Name)	Clear aperture / Optical design / Maker / Focal ratios	Mounting / Manufacturer	Latitude / Longitude / Elevation	Year	Notes
Télescope de Schmidt / Observatoire de Calern / Calern, France / (Calern Schmidt)	0.90-m corrector / Schmidt / Jean Texereau / f/3.5	... / C.M.G—Paris	43° 45' N / 6° 56' W / 1270 m	1981	Cer-Vit 1.52-m primary
Télescope Combiné de Schmidt / Observatoire Royal de Belgique / Uccle, Bruxelles, Belgium / ...	0.84-m corrector / Schmidt / Cox, Hargreaves, Thomson / f/2.5	Zeiss (Jena)	50° 48' N / 4° 21' E / 105 m	1958	1.2-m borosilicate primary mirror
Schmidt Telescope / Radioastrophysical Observatory / Riga, Latvia / ...	0.80-m corrector / Schmidt / Zeiss (Jena) / f/3.0	... / Zeiss (Jena)	56° 47' N / 24° 24' E / 75 m	1968	Operated by the Latvian Academy of Sciences; 1.2-m primary mirror
Calar-Alto-Schmidtspiegel / Calar Alto Observatory / Calar Alto, Spain / (Calar Alto Schmidt)	0.80-m corrector / Schmidt / Zeiss (Jena) / f/3.0	Equatorial fork / Grubb–Parsons	37° 13' N / 2° 32' W / 2168 m	1980	Tube and optics moved from Hamburg, Germany, where the instrument had been in use since 1955; 1.2-m primary mirror
REFRACTORS					
Yerkes 40-inch Refractor / Yerkes Observatory / Williams Bay, Wisconsin, US / (40 inch)	1.016-m doublet / Visual refractor / Alvan Clark & Sons / f/19.04	German equatorial / Warner and Swasey	42° 34' N / 88° 33' W / 334 m	1897	University of Chicago; a focal reducer also provides an f/3 focus
36-inch Refractor / Lick Observatory / Mount Hamilton, Calif., US / (36 inch)	0.895-m doublet / Visual refractor / Alvan Clark & Sons / f/19.7	German equatorial / Warner and Swasey	37° 20' N / 121° 39' W / 1290 m	1888	Front surface of crown element refigured in 1987
Meudon Refractor / Observatoire de Paris / Meudon, France / (33 inch)	0.83-m doublet / Visual refractor / Henry brothers / f/19.5	... / P. Gautier	48° 48' N / 2° 14' E / 162 m	1889	
Potsdam Refractor / Zentralinstitut für Astrophysik / Telegrafenberg, Potsdam, Germany / ...	0.80-m doublet / Visual refractor / C.A. Steinheil / f/15.0	... / Repsold	52° 23' N / 13° 04' E / 107 m	1899	

Table 27.14. (*Continued.*)

Telescope / Observatory / Place / (Common Name)	Clear aperture / Optical design / Maker / Focal ratios	Mounting / Manufacturer	Latitude / Longitude / Elevation	Year	Notes
The Thaw Refractor / Allegheny Observatory / Pittsburgh, PA, US / (30 inch)	0.76-m doublet / Red-light refractor / R.E. Sumner / f/18.6	German equatorial / Warner and Swasey	40° 29′ N / 80° 01′ W / 380 m	1985	Originally had a Brashear visual objective (1914); present one is corrected for red light
Lunette Bischoffscheim / Observatoire de Nice / Mont Gros, France	0.74-m doublet / Visual refractor / Henry Brothers / f/24.2	German equatorial / P. Gautier	43° 43′ N / 7° 18′ E / 372 m	1886	
28-inch Visual Refractor / Old Royal Observatory / Greenwich, London, England / (Greenwich refractor)	0.711-m doublet / Visual refractor / Grubb / f/11.9	English equatorial / Ransomes and Sims	51° 29′ N / 00° 00′ / 47 m	1894	Dome under repair in 1992
Grosser Refraktor / Archenhold-Sternwarte / Alt Treptow, Berlin, Germany	0.68-m doublet / Visual refractor / C.A. Steinheil / f/30.9	German equatorial / Hoppe-Berlin	52° 29′ N / 13° 29′ E / 41 m	1896	
Grosser Refraktor / Astronomisches Inst., Univ. Obs. / Vienna, Austria	0.67-m doublet / Visual refractor / Grubb / f/15.7	German equatorial / Grubb	48° 14′ N / 16° 20′ E / 241 m	1880	
McCormick Refractor / Leander McCormick Observatory / Charlottesville, VA, US / (26 inch)	0.667-m doublet / Visual refractor / Alvan Clark & Sons / f/14.9	German equatorial / Warner and Swasey	38° 02′ N / 78° 31′ W / 264 m	1883	
26-inch Equatorial / US Naval Observatory / Washington, DC, US / (26 inch)	0.66-m doublet / Visual refractor / Alvan Clark & Sons / f/15.0	German equatorial / Warner and Swasey	38° 55′ N / 77° 04′ W / 92 m	1873	
Thompson Refractor / Royal Greenwich Observatory / Herstmonceux, Sussex, England	0.66-m doublet / Visual refractor / Grubb / f/10.4	German equatorial / Grubb	50° 52′ N / 0° 20′ E / 34 m	1897	

REFERENCES

1. *Explanatory Supplement to the Astronomical Almanac*, 1992, edited by P.K. Seidelmann (University Science Books, Mill Valley, CA)
2. Fliegel, H.F., & Van Flandern, T.C. 1968, *Commun. ACM* **11**, 657
3. *Astronomical Almanac*, post-1984 (U.S. Government Printing Office, Washington, DC)
4. Morgan et al. 1992, *DOSS*, p. 11.
5. Collins, M. 1977, *Astronomical Catalogues* 1951–1975, INSPEC Series No. 2 (INSPEC, London)
6. Eichhorn, H. 1974, *Astronomy of Star Positions* (Frederick Ungar, New York)
7. Hawkins, G.S., & Rosenthal, S.K. 1967, 5,000 *and* 10,000-*Year Star Catalogs, Smithsonian Contrib. to Astrophysics*, **10**, 2
8. Jaschek, C. 1985, *Standard Values and Information in Data Banks*, IAU Symposium No. 111, 331
9. Sevarlic, B.M., Teleki, G., & Szadeczky-Kardoss, G. 1978, Bibliography of the Catalogues of Star Positions, *Publ. Dep. Astr. Belgrade*, No. 7, 69
10. Sevarlic, B.M., Teleki, G., & Knezevic, Z. 1982, Bibliography of the Photographic Catalogues of Star Positions, *Publ. Obs. Belgrade*, **29**, 71
11. Brotzman, L.E. et al. 1991, ACD CD = Astronomical Data Center (CD-ROM) Selected Astronomical Catalogues, **1** (Goddard Space Flight Center, Greenbelt)
12. *Swarthmore: Workshop on Databases for Galactic Structure*, 1993, edited by A.G. Davis Philip, B. Hauck, and A.R. Upgren (L. Davis Press, Schenectady)
13. IAU 166: IAU Symposium 166, *Astronomical and Astrophysical Objectives of Sub-Milliarcsecond Optical Astrometry*, edited by E. Hoeg and P.K. Seidelmann (Kluwer Academic, Dordrecht)
14. IAU XXIB: *Transactions of the IAU XXIB* 1991, edited by J. Bergeron (Kluwer Academic, Dordrecht)
15. PMGA: *Proper Motions and Galactic Astronomy*, edited by Roberta M. Humphreys, ASP Conf. Series 127.
16. Haniff, C.A. 1989, in *Diffraction Limited Imaging with Very Large Telescopes*, edited by D.M. Alloin and J.-M. Mariotti (Kluwer Academic, Dordrecht), p. 171
17. Cornwell, T.J. 1989, in *Diffraction Limited Imaging with Very Large Telescopes*, edited by D.M. Alloin and J.-M. Mariotti (Kluwer Academic, Dordrecht), p. 273
18. Mariotti, J.-M. 1994, *Adaptive Optics for Astronomy*, NATO ASI Series C, 423, edited by D.M. Alloin and J.-M. Mariotti (Kluwer Academic, Dordrecht), pp. 309–320
19. Beckers, J.M. 1991, *Expt. Astron.*, **2**, 57–71
20. Shao, M., & Colavita, M.M. 1992, *ARA&A*, **30**, 457–498
21. Ridgway, S.T. 1990, *Astrophysics from the Moon*, AIP Conf. Proc. No. 207, 495–499
22. Begelman, M. 1991, Working Papers—Astronomy and Astrophysics Panel Reports, National Research Council, V-6
23. Ridgway, S.T. 1992, see Mariotti 1993, pp. 19–24
24. Shao, M., & Colavita, M.M., 1993, *A&A*, **262**, 353
25. Reasenberg, R.D., Babcock, R.W., Chandler, J.F., Gorenstein, M.V., & Huchra, J.P. 1988, *AJ*, **96**, 1731–1745
26. Roddier, F. 1981, *Prog. Opt.*, **19**, 281–393
27. Tango, W.J., & Twiss, R.Q. 1980, *Prog. Opt.*, **17**, 239–277
28. Classen, J., & Sperling, N. 1981, Telescopes for the Record, *S&T*, **61**, 307

Index

701

Allen's Astrophysical Quantities

Fourth Edition

Also available electronically on-line and on CD-ROM

An electronic version of Allen's Astrophysical Data is available on CD-ROM for use with Windows 95/98/NT operating systems. It is also available on a subscription basis via password to anyone with a browser (such as Netscape or Internet Explorer) that accesses HTML files. Both the CD-ROM and website* offer fully searchable text and accompanying interactive tables and graphs. The latter can be found on Springer-Verlag New York's WWW home page — (URL: http://www.springer-ny.com/)

For more information, to order a CD-ROM, or to arrange for a subscription, please contact Springer-Verlag customer service:

Tel: 1-800-SPRINGE(R)
Fax: 201-348-4505
E-Mail: custserv@springer-ny.com

Springer-Verlag New York, Inc.
Customer Service
PO Box 2485
Secaucus, NY 07096-2485

*The Website will be updated on an ongoing basis, and will thus carry more recent information than the books or CD-ROM.